ELECTRONIC DEVICES AND CIRCUITS

ELECTRONIC DEVICES AND CIRCUITS

Prof. Dr. S.L. Kakani, M.Sc. (Phys.), Ph.D.

Executive Director, Institute of Technology and Management
(Affiliated to Rajasthan Technical University, Kota), Bhilwara, Rajasthan

K.C. Bhandari

Associate Professor of Physics (Retired)
M.L. Sukhadia University, Udaipur, Rajasthan

New Delhi | Mumbai | Chennai | Kolkata | Bengaluru | Hyderabad | Kochi | Guwāhati

Publisher's note

First Published 2012

Viva Books Private Limited

- 4737/23, Ansari Road, Daryaganj, New Delhi 110 002
  Tel. 42242200, 23258325, 23283121, Email: vivadelhi@vivagroupindia.net
- 76, Service Industries, Shirvane, Sector 1, Nerul, Navi Mumbai 400 706
  Tel. 27721273, 27721274, Email: vivamumbai@vivagroupindia.net
- Jamals Fazal Chambers, 26 Greams Road, Chennai 600 006
  Tel. 28294241, 28290304, Email: vivachennai@vivagroupindia.net
- B-103, Jindal Towers, 21/1A/3 Darga Road, Kolkata 700 017
  Tel. 22816713, Email: vivakolkata@vivagroupindia.net
- 7, GF, Sovereign Park Aptts., 56-58, K. R. Road, Basavanagudi, Bengaluru 560 004
  Tel. 26607409, Email: vivabangalore@vivagroupindia.net
- 101-102, Moghal Marc Apartments, 3-4-637 to 641, Narayanguda, Hyderabad 500 029
  Tel. 27564481, Email: vivahyderabad@vivagroupindia.net
- First Floor, Beevi Towers, SRM Road, Kaloor, Kochi 682 018
  Tel. 0484-2403055, 2403056, Email: vivakochi@vivagroupindia.net
- 232, GNB Road, Beside UCO Bank, Silpukhuri, Guwahati 781 003
  Tel. 0361-2666386, Email: vivaguwahati@vivagroupindia.net

www.vivagroupindia.com

ISBN: 978-81-309-1772-6

Published by Vinod Vasishtha for Viva Books Private Limited, 4737/23 Ansari Road, Daryaganj, New Delhi 110 002.

Printed and bound by Raj Press, R-3, Inderpuri, New Delhi 110 012.

E0000/E0900

Contents

Preface

In the past few decades, explosive development and exciting progress has happened in the field of electronics. The need of students and of teachers is a single textbook covering the subject matter of electronic devices and circuits in accordance with the latest syllabi of engineering and sciences for the major universities.

This book will, thus, serve as first-hand text for students of B.Tech., Diploma, AMIE and Graduate IETE courses and as useful reference for those preparing for GATE, NET, SLET and other competitive examinations. The book will also be useful for students of B.Sc. (Hons in Electronics) and M.Sc. (Electronics).

Organized in 21 chapters, this book starts with a discussion on electronic materials and components. Chapters 2 and 3 discuss the basic properties of semiconductors and their conduction process and semiconductor diodes, respectively. Chapters 4, 5 and 6 are devoted to diode circuits, transistors and thermal stability. Chapters 7 to 11 are devoted to amplifiers. Chapter 13 discusses ionosphere. Chapters 14 to 16 are devoted to optical fibres, communication systems, television and radar. Chapters 17 and 18 are devoted to electronic instruments and optoelectronic devices. Chapters 19 and 20 are devoted to digital electronics and microprocessors. The book concludes with a discussion on integrated circuits in Chapter 21.

Concepts and explanations to theories have been well supported by diagrams carrying necessary technical details. Equal emphasis has been given for essential mathematical derivations as well as their physical interpretations. A good number of worked-out problems are provided for each chapter. Summary at the end of each chapter helps quick revision. Review questions, exercises, short answer questions and objective questions with answers will be quite helpful to the students for self-evaluation.

We are grateful to Viva Books Private Limited, New Delhi, for the confidence they have in our authorship.

We welcome suggestions and corrections for further improvement of the book.

S.L. Kakani
K.C. Bhandari

Chapter 1

Electronic Materials and Components

1.1 Introduction

Electronics is the science and technology of motion of electrons (or other such charges) in conductors, semiconductors and other similar materials, including in gas and vacuum. Materials commonly used in electronic industry include:

(a) Conducting materials
(b) Insulating materials
(c) Semiconducting materials
(d) Magnetic materials
(e) Superconducting materials
(f) Nanomaterials
(g) Smart materials, etc.

Some materials and their applications in electronics are summarized in Table 1.1.

Electronics play a vital role in communication and entertainment, control and instrumentation, defence and military, industry, medical sciences, automobiles, etc.

Table 1.1 Important materials and their application in electronics

Type of material	Application
Conducting materials	
Copper, aluminium, gold, silver	For making wires, fuse wires, lead and other components
Iron, steel	For making cores of inductors and transformers
Nichrome	For making wire-wound resistors, heating coils, vacuum tubes, etc.
Carbon	For making thermistors and wire-wound resistors
Tungsten	For making filaments of lamps and cathodes of vacuum tubes
Copper	For making anode of vacuum tubes

Table 1.1 (Continued)

Type of material	Application
Insulating materials	
Mica, paper, ceramic, plastics, wax	For making capacitors and core of inductors and resistors
Polymers, e.g., epoxy resins	For making printed circuit board (PCB) laminates
Synthetic resins	Radio, TV, computer cabinet, etc.
Semiconductor materials	
Germanium, silicon	For making diodes, transistors, silicon controlled rectifier, IC and other components
Magnetic and other materials	
Ferrites	For making memory chips (data storage) in computers
Silicon, steel, lead-tin alloy	For making cores of inductors and transformers and as fuse and soldering materials

The comparison of resistivity of conducting, semiconducting and insulating materials is provided in Table 1.2.

Table 1.2 Resistivity of conducting, semiconducting and insulating materials

Type of material	Resistivity	No. of electrons in the last orbit	Type of bond	Forbidden band gap
Conducting materials	$10^{-8} - 10^{-2}$	Less than 4	Metallic	0
Semiconducting materials	$10^{0} - 10^{6}$	Exactly 4	Covalent	1 eV
Insulating materials	$10^{12} - 20$	More than 4	–	15 eV

1.2 Components of an Electronic Circuit

An electronic circuit may appear complex, but some components in it are active and some are passive. PN-junction diodes, transistors, SCRs, FETs, etc. are active components because they are capable of changing the shape of electrical signals. Resistors, inductors and capacitors are passive components as they do not change the shape of the signal. Active components can perform the function of the circuit independently with the help of passive components. Figure 1.1 depicts the active and passive components.

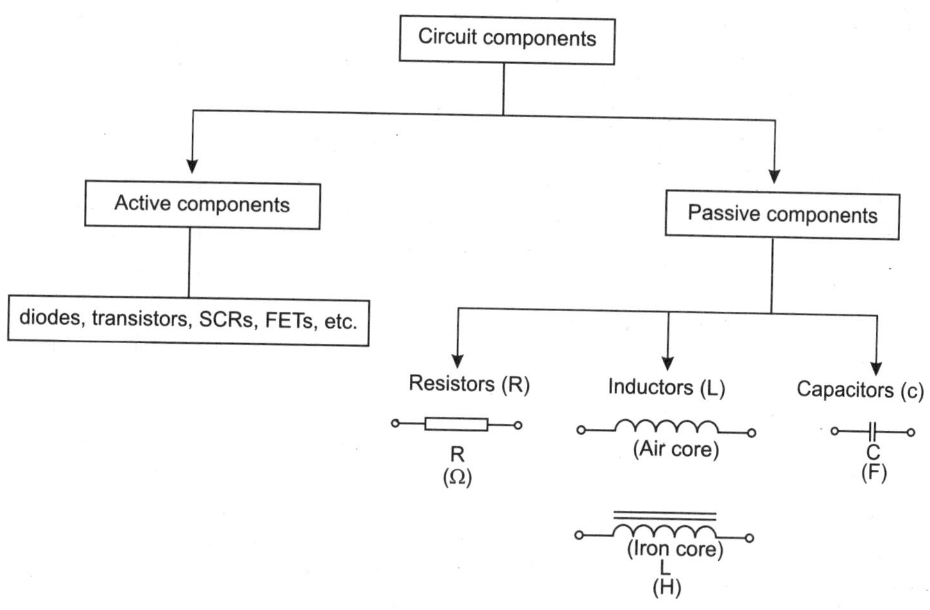

Figure 1.1 Components of an electronic circuit

1.2.1 Active Components

These are used in earlier electronic circuits were of tube-type, e.g. vacuum diode, vacuum triode, vacuum pentode, gas diode, thyraton, etc. Later these bulky tube devices have been replaced by semiconductor devices like diodes, transistors or ICs, etc. Figure 1.2 give an brief account of commonly used semiconductor devices.

1.2.2 Passive Components

1.2.2.1 Resistors

These are components that resist the flow of current, i.e. limit the amount of current flowing through a circuit. Resistance (R) is measured in ohms (Ω). Resistor in a circuit can be represented as shown in Figure 1.3.

Resistors may be pure at low frequencies but have inductive or capacitive impedance at higher frequencies. The frequency upto which resistors are pure, i.e. have only resistivity, is called their frequency range. Figure 1.4 shows the circuit of a resistor at higher frequencies.

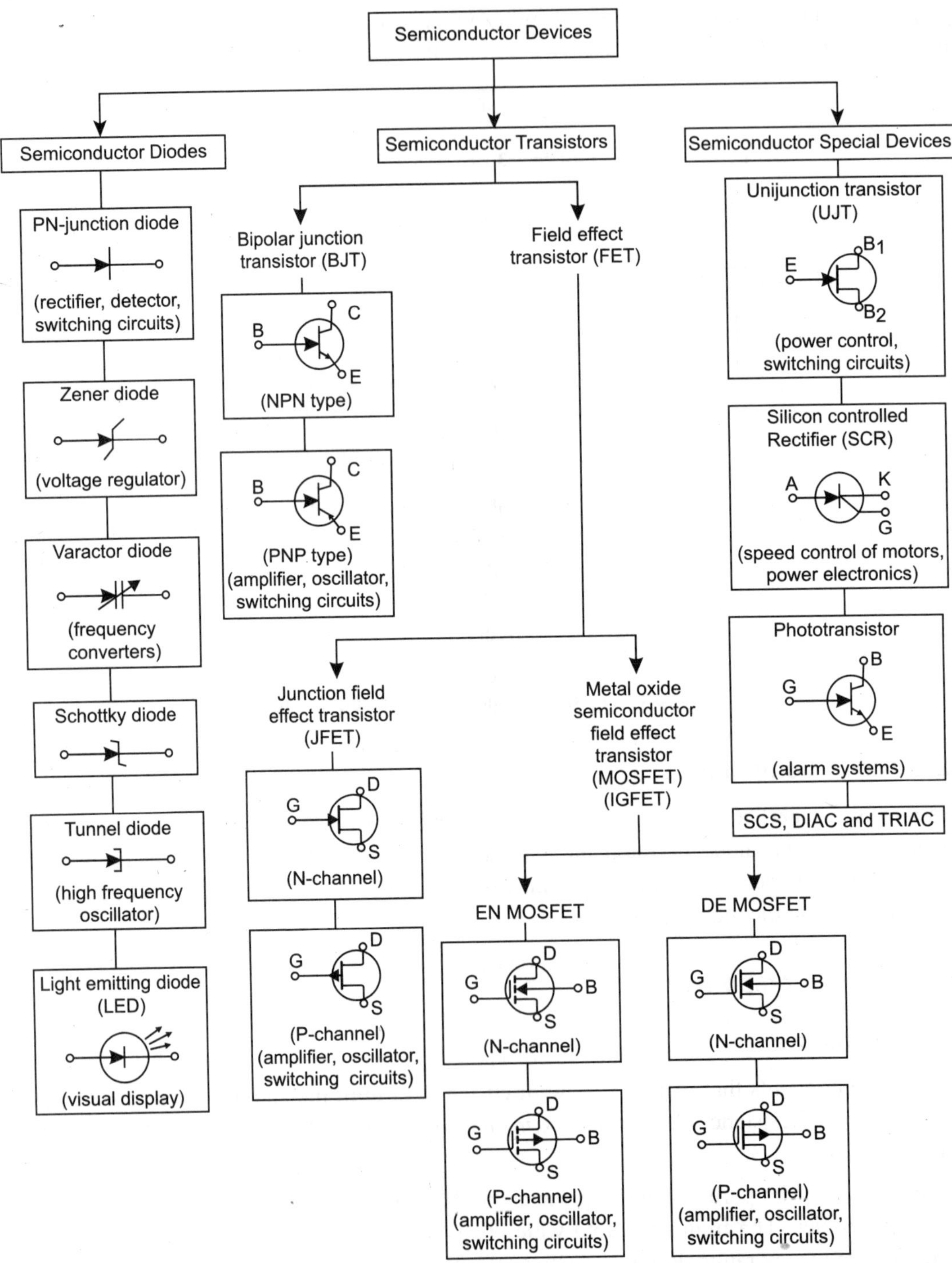

Figure 1.2 Common semiconductor devices along with their symbols

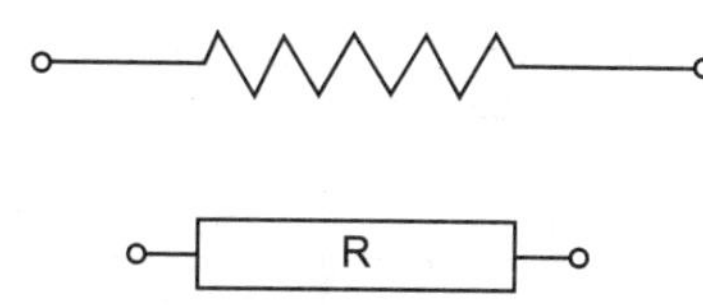

Figure 1.3 Resistor

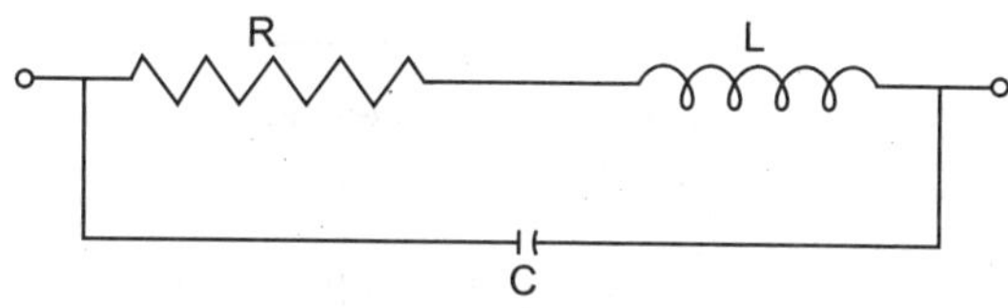

Figure 1.4 Resistor circuit at higher frequencies

When at work in a circuit, resistors often produce noise, which depends upon its resistance as well as temperature.

Resistors have a power rating. Power rating stands for maximum power that can be dissipated without raising the temperature too much. For example, a 1 W resistor having resistance 100 Ω can pass a maximum current of 100 mA. On the other hand, a 1/4 W resistor with same resistance can handle a current of only 50 mA. However, if current flowing through it exceeds this limit for any length of time, the resistor would get overheated and might even burn out.

The standard power ratings used in electronic circuits are 1/4 W, 1/2 W, 1 W and 2 W. Thus, if you require a resistor to dissipate 0.6 W, you should select a 1 W resistor since the operation value could not exceed the power rating.

The size of a resistor increases with its wattage. This should help the resistor withstand higher dissipation losses.

If current passing through the resistor is proportional to the potential difference (pd) across its terminals, then the resistor is said to be linear; otherwise it is nonlinear. It may be noted that resistors made from semiconductor materials are usually non-linear.

Resistors may be of two types: (a) fixed resistors and (b) variable resistors.

(a) Fixed resistors

The resistance value of fixed resistors is fixed. These include carbon resistors and wire-wound resistors.

(a) Carbon resistors The resistive material for carbon resistors is of carbon-clay composition (ground carbon, resin binder and filler) and leads of this type are available with values ranging from a few ohms to about 22 MΩ, with tolerance range of 5–20% and wattage of 1/4 W, 1/2 W and 1 W. The value of resistance is determined using the carbon-filler ratio. The entire arrangement is covered by plastic or lacquer coating to provide electrical insulation and protection from moisture (Figure 1.5).

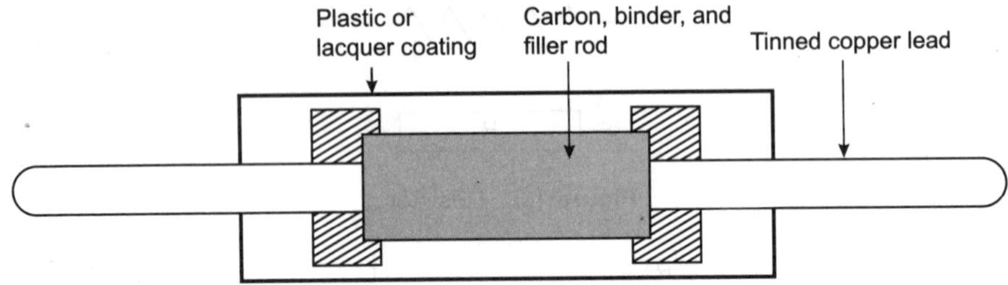

Figure 1.5 Carbon resistors

To indicate the value of their resistance, carbon resistors are provided with coloured rings. One can read the value by looking at the coloured rings. Each colour has a numerical value. Table 1.3 gives the codes coding of carbon resistors.

Table 1.3 Colour codes of carbon resistors

Colour	Numerical value	Multiplier
Black	0	$10^0 = 1$
Brown	1	$10^1 = 10$
Red	2	$10^2 = 100$
Orange	3	$10^3 = 1000$
Yellow	4	$10^4 = 10000$
Green	5	$10^5 = 100000$
Blue	6	$10^6 = 1000000$
Violet	7	$10^7 = 10000000$
Grey	8	$10^8 = 100000000$
White	9	$10^9 = 1000000000$

Tolerance is shown by gold and silver rings. See Table 1.4 for tolerance of carbon resistors.

Table 1.4 Tolerance of carbon resistors

Colour band	Percentage tolerance
Gold	±5%
Silver	±10%
No ring	±20%

Sometimes other colours are also used to indicate tolerance as given in Table 1.5.

Table 1.5 Tolerance of carbon resistors

Colour band	Tolerance
Black	20%
Red	1%
Yellow	2%

It may be noted that resistance value and tolerance could sometimes be printed on the resistor instead of colour codes. A letter may be used to represent tolerance as given below:

F = ±1%

G = ±2%

J = ±5%

K = ±10%

M = ±20%

For example, 8 M 8M is a 8.8 MΩ ± 20% resistor and 20 KK is a 20 KΩ ± 10% resistor (Figure 1.6).

Few other examples are:

(a) A 0.56 resistor is marked as R56.
(b) A 1 Ω resistor is marked as 1RO.
(c) A 56 Ω resistor is marked as 56R.
(d) A 1200 Ω resistor is marked as 1K2.
(e) A 0.5 MΩ resistor is marked as M5.

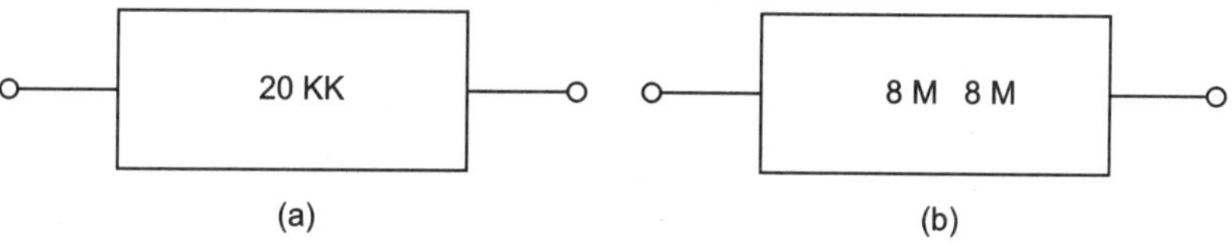

Figure 1.6

(i) Carbon film resistors This resistor (Figure 1.7) is manufactured by depositing a carbon film on a ceramic substrate (base). In this process, one is able to achieve only approximate value of resistance by cutting grooves along its length. After this, caps are fixed at both ends and tinned copper of leads should be welded to these caps. These resistors are available from 10 Ω to 10 MΩ with power rating upto 2 W. Contrary to carbon resistors, carbon film resistors are less noisy as well as cheaper.

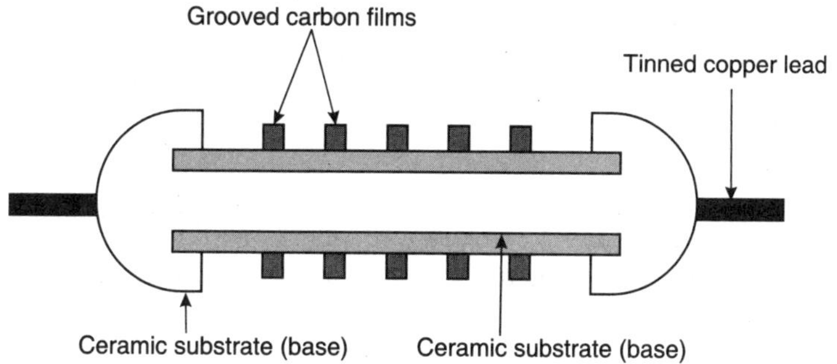

Figure 1.7 Carbon film resistor

(ii) Metal film resistors The development of a these resistors is similar to carbon film resistors except that the film is made of tin–antimony alloy. These resistors may be 'thin film' or 'thick film' type.

(b) Wire-wound resistors These resistors are available as low power resistors of upto 2 W, and high power resistors of more than 5 W.

The value of these resistances may be in the range from 1 Ω to a few kilo-ohms. These resistors are manufactured by winding a suitable length of wire made of a higher resistivity material, (e.g. nichrome) on an insulating frame say (e.g. clay, ceramic or porcelain) and then end caps are provided to which leads have to be connected. To protect from moisture, the entire resistor is covered with an insulating material like lacquer. The values of these resistors and power may be printed on the resistor itself. Figure 1.8 shows wire-wound resistors.

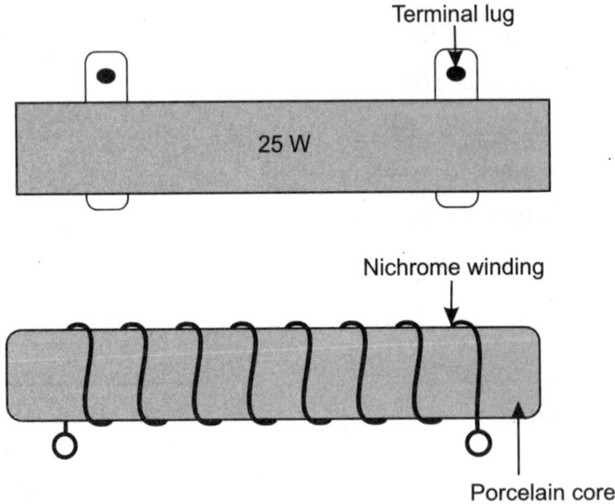

Figure 1.8 Wire-wound resistors

The distribution of capacitance in carbon and wire-woundresistors is as shown in Figure 1.9.

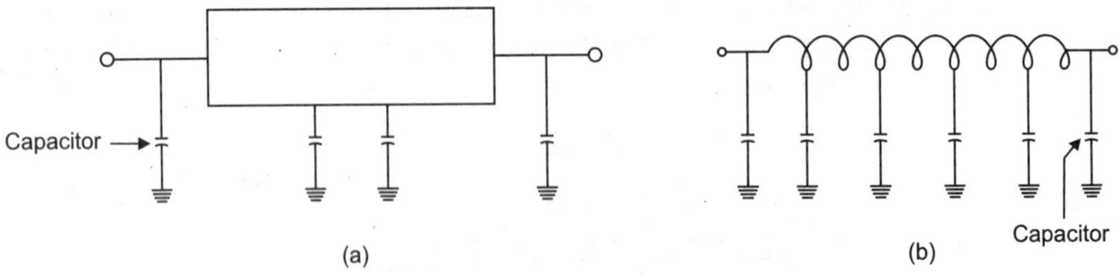

Figure 1.9 Distribution of capacitance in (a) carbon and (b) wire-woundresistors

Specification of a fixed resistor The main specifications of a fixed resistor are as follows:

(a) Value of resistance
(b) Power rating

(c) Voltage rating
(d) Temperature coefficient
(e) Voltage coefficient
(f) Noise coefficient
(g) Frequency range
(h) Tolerance
(i) Stability

One can find temperature coefficient of resistance from the following relation

$$\alpha_T = \frac{R_T - R_{TO}}{R_{TO}(T - T_O)}$$

and voltage coefficient of resistance from the following relation:

$$\alpha_V = \frac{R_V - R_{VO}}{R_{VO}(V - V_O)}$$

where the symbols have usual meanings. We may note that the value of voltage coefficient is usually small. However, the value of voltage coefficient may be significant in resistors made of composite materials.

One can obtain the rms value of noise voltage of a resistor from

$$V_n = (4kTRB)^{1/2}$$

where T is temperature in Kelvin, R is resistance in ohms, B is frequency in Hz and k is Boltzmann's constant.

When a rating of more than 1W is required, one should prefer wire-wound resistors.

We may note that in most electronic circuits, it is not necessary to use resistances of exact values.

Even if resistances differ from marked values by as much as ±20%, the circuit works satisfactorily. Obviously, one does not have the need to manufacture resistors for all values. Standard values of commercially available resistors are given in Table 1.6.

Table 1.6 Standard values of commercially available resistors (tolerance ≈ 10%).

Ohm (Ω)			**Kilo ohm (kΩ)**			**Mega ohm (MΩ)**	
1.0	10	100	1.0	10	100	1.0	10
1.2	12	120	1.2	12	120	1.2	12
1.5	15	150	1.5	15	150	1.5	15
1.8	18	180	1.8	18	180	1.8	18
2.2	22	220	2.2	22	220	2.2	22
2.7	27	270	2.7	27	270	2.7	
3.3	33	330	3.3	33	330	3.3	

Table 1.6 (Continued)

Ohm (Ω)			Kilo ohm (kΩ)			Mega ohm (MΩ)
3.9	39	390	3.9	39	390	3.9
4.7	47	470	4.7	47	470	4.7
5.6	56	560	5.6	56	560	5.6
6.8	68	680	6.8	68	680	6.8
8.2	82	820	8.2	82	820	8.2

(b) Variable resistors

Variable resistors are the resistors for which the value of resistance varies with the applied 'stimulus'. Big size variable resistors are usually called rheostats, where as small size variable resistors are called potentiometers (usually abbreviated 'pots'). From the equation

$$R = \rho\frac{L}{A},$$

one finds that the stimulus has to change one or more of these quantities to vary the value of resistance. Figure 1.10(a) shows the basic construction of a pot. A resistance wire is wound over a dough-shaped core of bakelite or ceramic. There is a rotating shaft at the centre of the core. The shaft moves end to end of the resistance element. The outer two terminals are end points of the resistance element and the middle leads to the rotating contact. Figure 1.10(b) shows its symbol, where the arrow indicates the movable contact.

We may note that the variation of resistance in a potentiometer ('pot') can be either linear or nonlinear (Figure 1.11). The linear type has a core (or former) of uniform height, whereas the core of the non-linear type has a tapered trip. In sound equipment, pots used for volume control are generally of nonlinear type (with logarithmic variation).

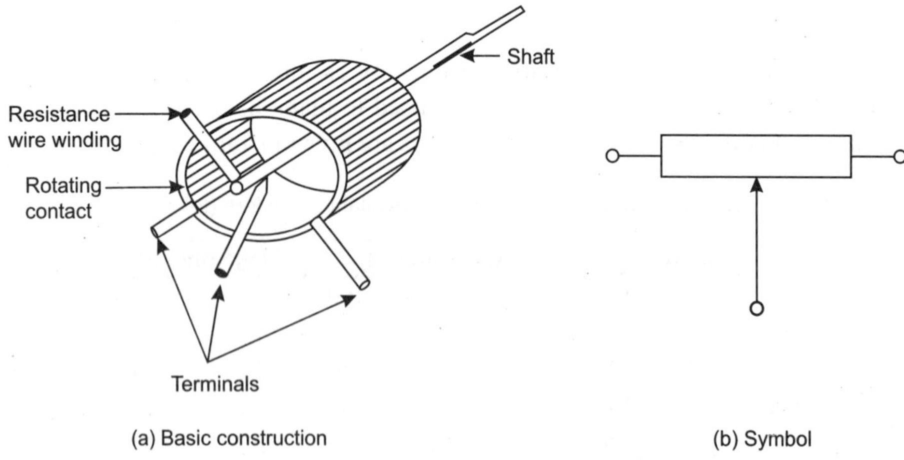

Figure 1.10 Variable resistor 'pot' (potentiometer)

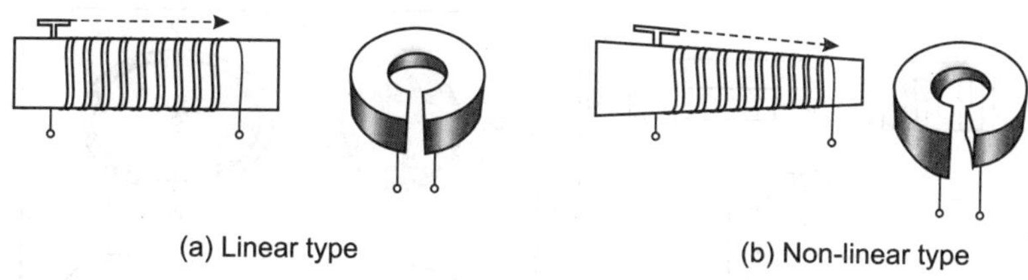

Figure 1.11 Wire-wound potentiometer

Special variable resistors There are some resistors which have special characteristics and are used in specific applications. In terms of stimuli, besides the mechanically variable resistors such as rheostat and potentiometer, we find three types of special variable resistors:

(i) Thermally variable resistors (thermistors) These resistors are usually semiconductors whose resistance decreases with increasing temperature, i.e. they have negative temperature coefficient of resistance, whereas conventional wire-woundmetallic resistors have positive temperature coefficient of resistance, i.e. their resistance increases with increase in temperature. Thermistors are highly resistive to temperature and, therefore, are used in precision measurements and thermal relays. Commercial thermistors are made of sintered mixtures of Mn_2O_3, NiO_2 and Co_2O_3. Thermistors may be:

(a) PTC thermistors These exhibit a high positive temperature coefficient (PTC) of resistance. Usually conductors like iron wire evaporated in hydrogen atmosphere are used in PTC thermistors.

(b) NTC thermistors These exhibit negative temperature coefficient (NTC) of resistance. Usually semiconductor materials like Mn_2O_3, NiO_2 and Co_2O_3 are used. The expression for temperature coefficient for these resistors is as follows:

$$R_T = R_o \left[B\left(\frac{1}{T_O} - \frac{1}{T} \right) \right]$$

where B is bandwidth and other symbols have usual meanings.

Specification of a typical thermistor is as follows:

Resistance at 25°C	: 60 Ω
Resistance at 125°C	: 15 K
Maximum temperature coefficient +5%	: 1°C
Maximum voltage	: 22 V
Tolerance	: 32%
Maximum power dissipation	: 0.5 W

Figure 1.12 shows different types of thermistor structures.

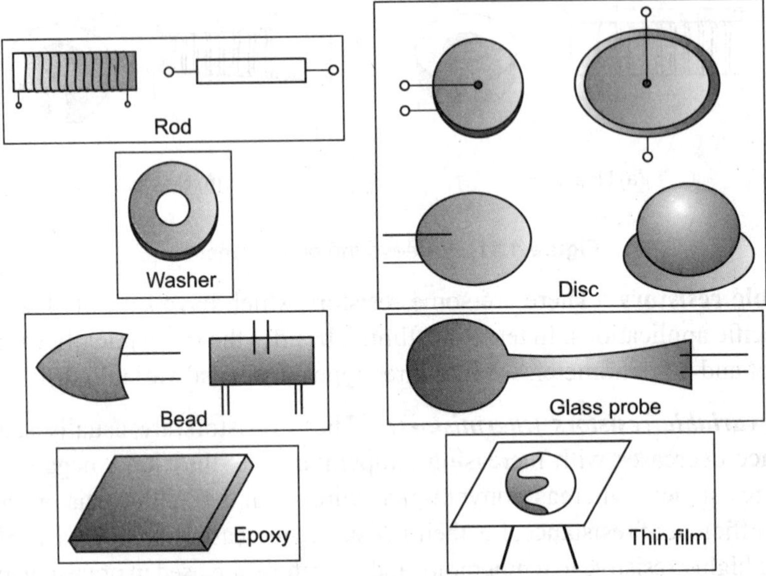

Figure 1.12 Thermistor structures

In electronic circuits where change of resistance with temperature cannot be tolerated, thermistors are used to compensate for change in the resistance of ordinary components.

(ii) Voltage-dependent resistors (varistors or VDR) In varistors, the '–' value of resistance changes with applied voltage. This can be achieved by choosing a material in which charge carrier transportation depends upon field electron emission. Varistors are usually made from conductor, insulator or semiconductor composite materials, e.g. composite of silicon carbide, graphite and sodium silicate, in which resistivity decreases with increase in applied voltage. This happen due to increased transport of charge carriers across the thin insulating layer of sodium silicate surrounding silicon carbide and graphite. Resistance varies with applied voltage in accordance with the following expression:

$$R_0 = RV_0\ [B_V(V - V_0)]$$

where B_V is voltage coefficient of resistance (negative) and other symbols have usual meanings. Specification of a typical varistor are as follows:

Voltage range	: 8-300 V
Current	: 1-120 mA
Power dissipation	: 0.8-2.8 W
Voltage coefficient resistance (B_V)	: 0.5

Varistors are also called **voltage sensitive resistors** (VSR). Varistors may be symmetrical or asymmetrical. Symmetrical varistors have the same resistance for both positive and negative voltages.

Symmetrical varistors are made of SiC. Like thermistors, symmetrical varistors are also available in the form of rod, disc and bead with volt rating from 50 V to 10,000 V. Their power rating is 40 mW/cm^2. Symmetrical varistors are quite useful for spark suppression in inductive loads and for increasing meter sensitivity.

Asymmetrical varistors do not have the same resistance for negative and positive applied voltages. They are like rectifiers having low forward and high reverse resistance, e.g. copper oxide rectifiers (Figure 1.13(b)). A copper oxide rectifier is made by coating copper oxide on a copper disc and is used as a low voltage rectifier (10 V per unit). A selenium oxide rectifier is made by coating selenium oxide on a nickel plated steel and is used as a high voltage rectifier (40 V per unit). By connecting varistors in series, these can be used even at higher voltages.

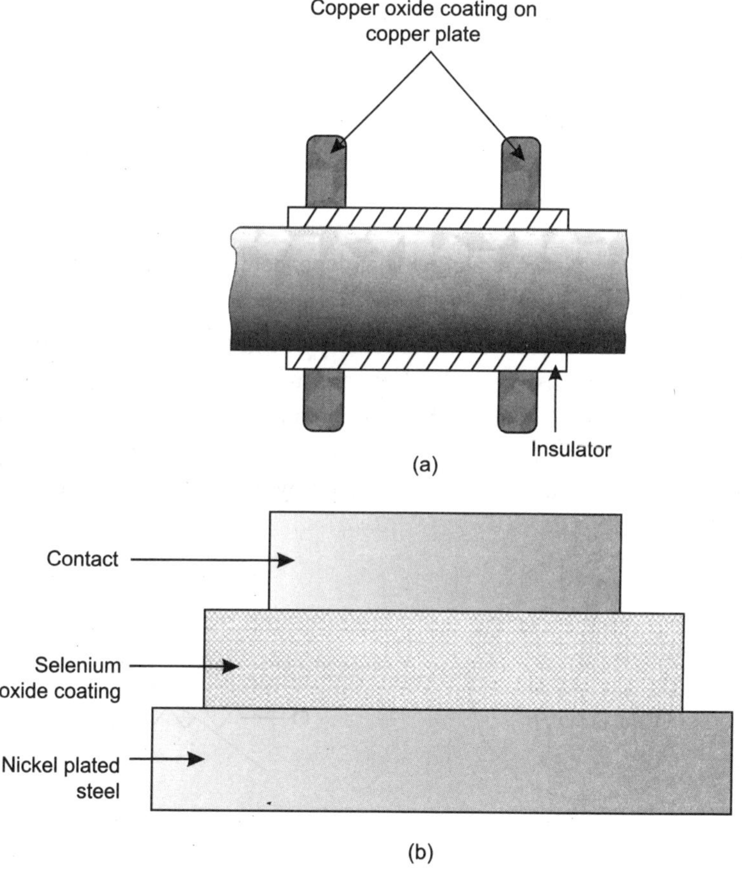

Figure 1.13 (a) Rectifiers, and (b) selenium oxide rectifiers

One can have a device in which resistance between its two terminals is dependent on the voltage at its third terminal. The junction field effect transistor (JFET) has three terminals: drain (D), source (S) and gate (G). Normally the source terminal is grounded. For a given value of gate voltage (V_{GS}), drain current (I_D) varies linearly with drain voltage (V_{DS}), till pinch-off occurs. This region is called **ohmic** or **triode region**. It we change V_{GS}, current I_D still remains linearly dependent on V_{DS} (Figure 1.14). However, the slope of the straight line changes. One finds that resistance between drain and source depends upon the gate voltage in this region.

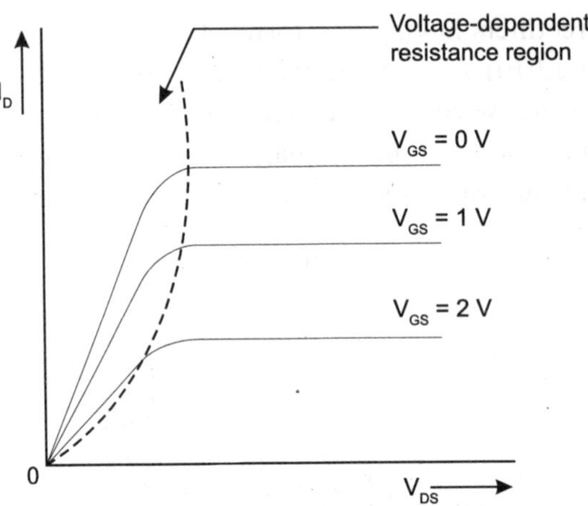

Figure 1.14 Drain characteristics of a JFET light-dependent, optically variable resistors (LDR or photoresistors)

The resistance of LDRs depends on the intensity of light incident on them. An LDR is also called photoresistor or photoresistive cell. LDRs are made of a ceramic substrate over which a photoresistive material (usually Cds or CdSe) is deposited as zigzag strips to increase its length. The structure and symbol of an LDR is shown in Figure 1.15. The ends of the strips are attached to the external pins. The entire assembly is protected by a transparent plastic or glass cover.

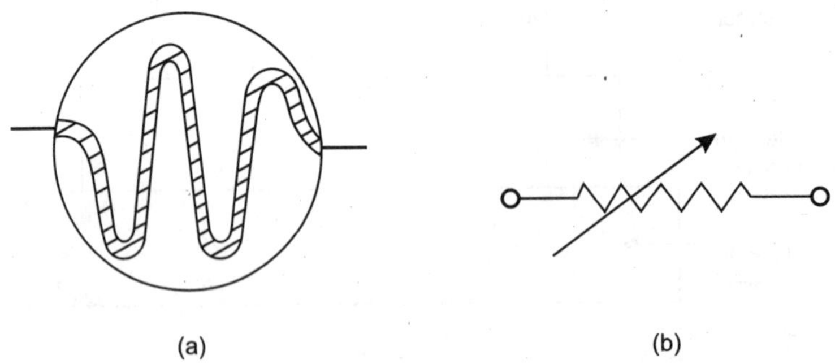

Figure 1.15 (a) A light-dependent resistor (LDR) and its (b) symbol of LDR

For a typical LDR, the variation of resistance versus light intensity is shown in Figure 1.16. This variation of resistance with intensity is governed by the following relation:

$$R = AL^{-\alpha}$$

where L is intensity of light, A is a constant and α is another constant, whose value lies between 0.7 and 0.9. The resistance of an LDR can vary over a wide range (from 100 kΩ to a few hundred ohms). The high resistance of an LDR when not illuminated is called **dark resistance**.

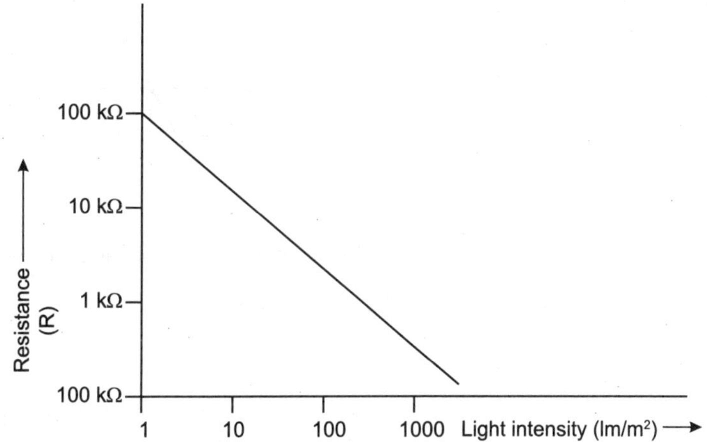

Figure 1.16 Variation of resistance with light intensity for a typical LDR

LDRs are quite useful use in light meters, automatic door openers, safe lockers, etc. Figure 1.17 shows an LDR.

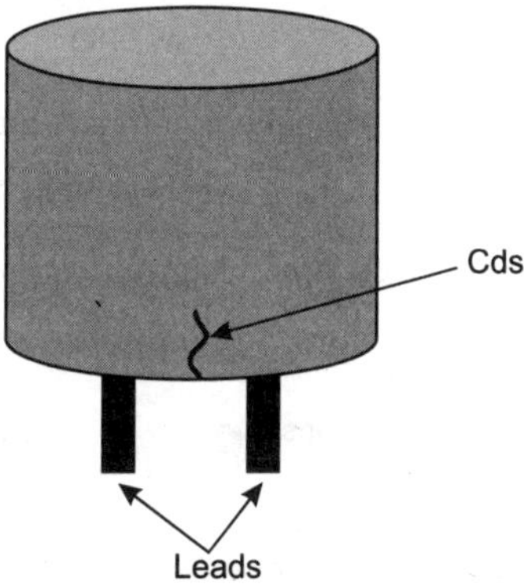

Figure 1.17 LDR

1.2.2.2 Inductors

Inductors are made by winding a conducting wire (of very low resistance) to form a coil. It may have a air core or iron core. Current flowing through the core of an inductor produces a magnetic field, which then opposes any change in current by developing an induced emf. The unit of inductance (L) is henry (H). The inductance of a circuit is 1 H when an emf of 1 V is induced by the current changing at a rate of 1 A/s. Obviously, an inductor is a component which oppose any change in the magnitude or direction of current. In DC circuits, inductance is absent.

An inductor is popularly known as a choke. Figure 1.18 shows its symbol for fixed and variable inductance.

Sometimes, we have two or more windings on the same core. It is then called a transformer, or a set of (magnetically) coupled coils.

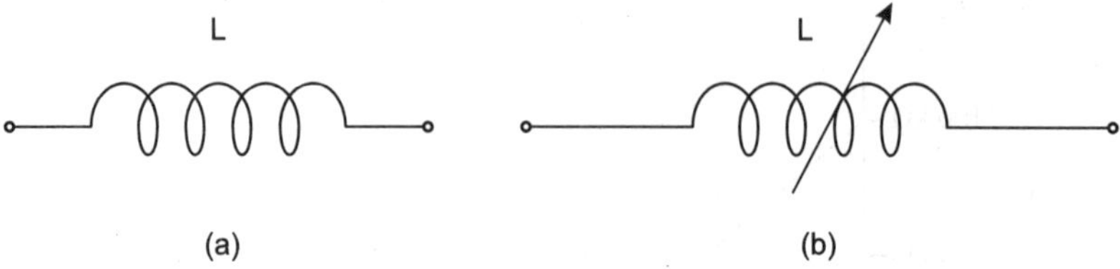

Figure 1.18 Symbols for inductance: (a) fixed, (b) variable

Inductors based on the type of core used are as follows:

Air core inductors

These inductors use air as the core and are mostly used in radio frequency applications. Their inductance is usually in micro-henry. These inductors are light but causes much loss (Figure 1.19).

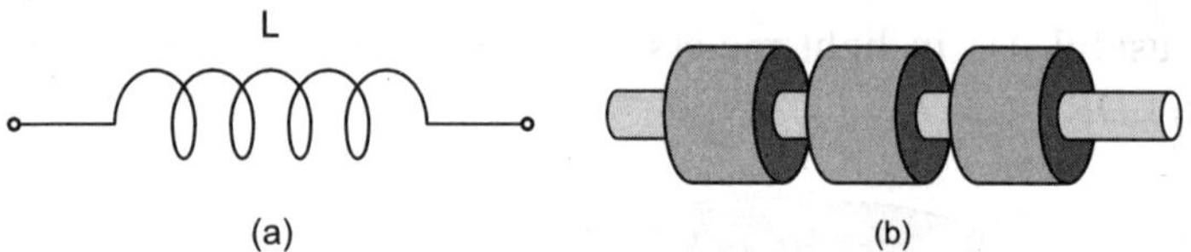

Figure 1.19 (a) Air core inductor, (b) Inductor with air core or ferromagnetic core

Iron core inductors

Silicon steel laminations are generally used for these inductors (Figure 1.20). These inductors find application in filter circuits.

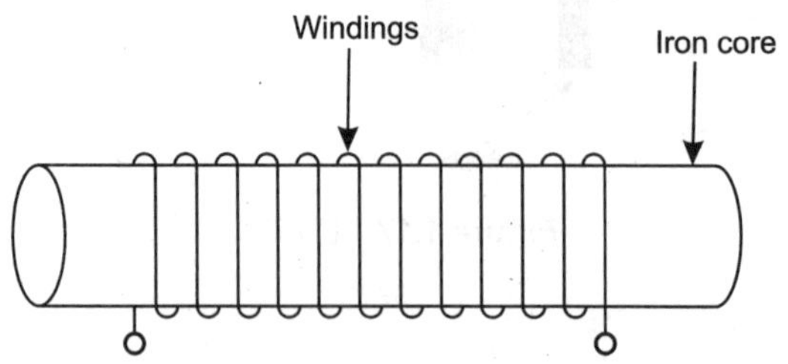

Figure 1.20 Iron core inductor

Ferrite inductors

In these inductors, ferrite cores are used. Ferrite cores provide stable inductance (Figure 1.21(a)) and complete screening. Moreover, the loss can be reduced. The circuit symbol for an inductor with adjustable ferromagnetic core is shown in Figure 1.21(b).

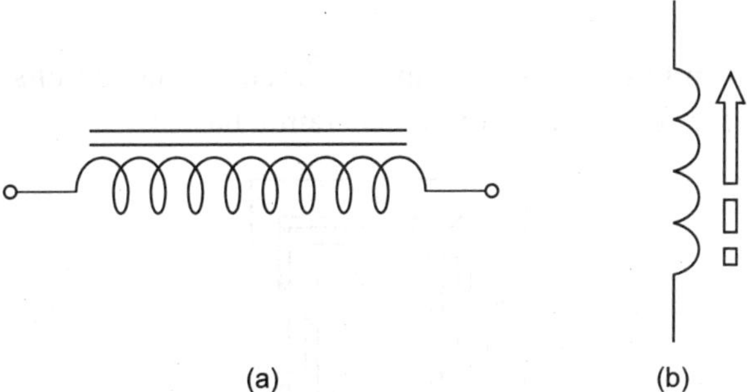

Figure 1.21 (a) Ferrite inductors and (b) circuit symbol for an inductor with adjustable ferromagnetic core

Power core inductors

These inductors can be used at much higher frequencies and losses are low in comparison to other inductors.

Molded inductors

Figure 1.22 shows a molded inductor. Typical values are in the range from 1.2 μH to 10 mH, and maximum current is 70 mA. Usually inductors are marked with their value on the body but some inductors are also coded. Table 1.7 shows the colour codes for inductors in μH.

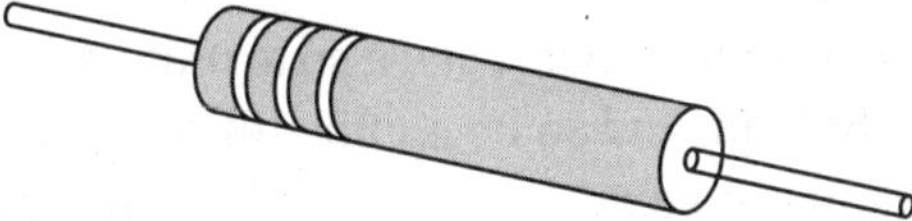

Figure 1.22 Molded inductor

Table 1.7 Colour codes for inductors (μH)

Colour	Significant digits		Multiplier	Tolerance
	1	2		
Black	0	0	1	–
Brown	1	1	10	–
Red	2	2	10^2	–
Orange	3	3	10^3	–
Yellow	4	4	10^4	–
Green	5	5	–	–
Blue	6	6	–	–
Violet	7	7	–	–
Grey	8	8	–	–
White	9	9	–	–
Gold	–	–	–	+5%
Silver	–	–	–	±10%
No band	–	–	–	±20%

Thin-film inductor

Figure 1.23 shows a thin-film inductor used in certain electronic circuits. This is simply a thin metal film deposited in the form of a spiral on a ceramic base.

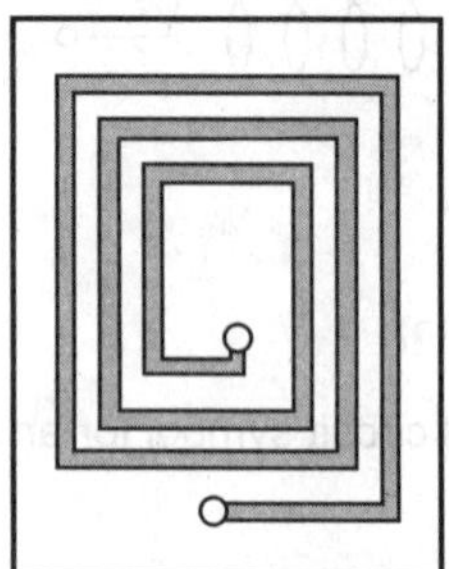

Figure 1.23 Thin-film inductor

Laboratory inductors

In a laboratory inductor, precision inductors are switched into or out of a circuit by means of rotary switches. Alternatively, one can employ two coupled coils as a variable inductor. The coils may be connected in series or in parallel, and total inductance is controlled by adjusting the position of one coil relative to the other.

To protect components against flux leakage and also to protect the coil from external magnetic fields, the coil is wound on a bobbin, and so its number of turns is easily modified. Figure 1.24 shows a ferrite pot core containing some low-current, high frequency inductors wound on a bobbin. The ferrite core increases winding inductance and screens the inductor.

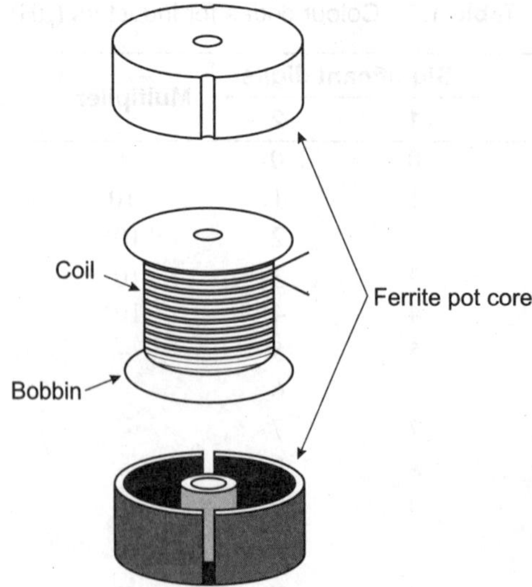

Figure 1.24 Some low-current, high-frequency inductors are wound on a bobbin

When mutual inductance between two adjacent coils is not known, one can determine it by measuring total inductance of the coils in series-adding and series-opposing connections. We have,

$$L_a = L_1 + L_2 + 2m \text{ (For series-adding)}$$

$$L_b = L_1 + L_2 - 2\,m \text{ (For series-opposing)}$$

Subtracting, one obtains

$$L_a - L_b = 4\,M$$

$$\therefore \; M = \frac{L_a - L_b}{4} \qquad (1)$$

$$M = k\sqrt{L_1 L_2} \qquad (2)$$

One can determine the coefficient of coupling of two coils from eqs. (1) or (2).

Stray inductance

Inductance is change in flux linkages/change in current. Obviously, every current-carrying conductor has some self-inductance, and every pair of conductors has inductance. Stray inductance is normally unimportant in DC applications; however in radio frequency AC circuits, stray inductance can cause considerable nuisance. One can minimize stray inductance by keeping connecting wires as short as possible.

Inductors (chokes) may be of different types based on the applications they are put into use:

(i) Filter chokes Filter chokes are used in filter circuits. Thin laminations of silicon steel are used as core (Figure 1.25). Its values are in the range 5–25 H with current capacity 0.25 A.

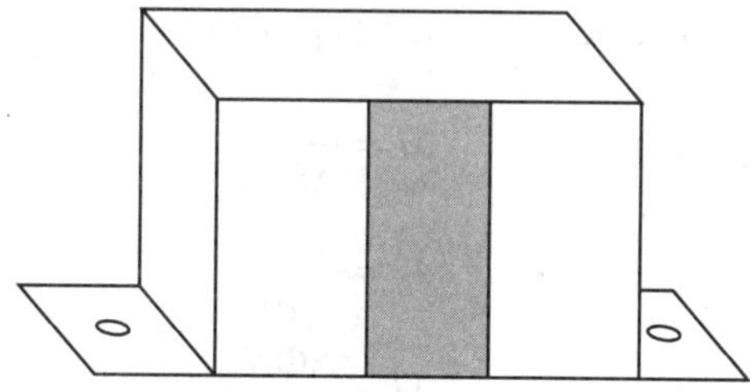

Figure 1.25 Filter choke

(ii) Radio frequency choke (RFC) RFC is a variable inductor having a moving shaft that moves within the coil to vary the inductance (Figure 1.26). The size of RFCs is small as compared to filter chokes. Their values lie in between the range 0.25 mH and 250 mH. They are used above 1000 kHz (radio frequencies).

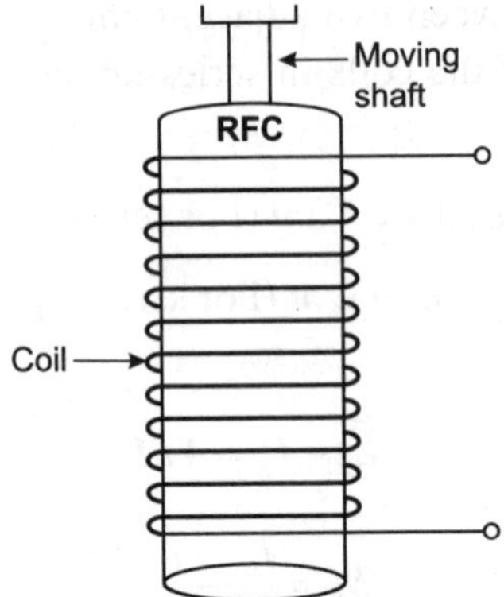

Figure 1.26 Radio frequency choke

(iii) Audio frequency choke (AFC) AFC is used to provide high impedance in 20 Hz to 20 kHz range.

Important relations between inductance and induced emf are summarized in Table 1.8.

Table 1.8 Important relations between inductance and induced emf

Induced emf	$e_L = \dfrac{\Delta\Phi}{\Delta t}$
Inductance	$L = \dfrac{e_L}{\Delta i / \Delta t}$
Inductance	$L = \dfrac{\Delta\Phi\, N}{\Delta i}$
Flux change	$\Delta\Phi = \mu, \mu_0 \Delta i\, N \dfrac{A}{t}$
Self-inductance	$L = \mu, \mu_0\, \Delta i N^2 \dfrac{A}{t}$
Mutual inductance	$M = \dfrac{e_L}{\Delta i / \Delta t}$
Induced emf	$e_L = \dfrac{\Delta\Phi\, N_s}{\Delta t}$
Mutual inductance	$M = \dfrac{\Delta\Phi N_s}{\Delta i}$
Mutual inductance	$M = k \dfrac{\Delta\Phi N_s}{\Delta i}$
Mutual inductance	$M = k\sqrt{L_1 L_2}$
Energy stored	$W = \dfrac{1}{2} L I^2$

Table 1.8 (Continued)

Energy stored	$W = \dfrac{B^2 Al}{2\mu_0}$
Inductances in series	$L_s = L_1 + L_2 + L_3 + \cdots$
Inductances in parallel	$\dfrac{l}{L_P} = \dfrac{1}{L_1} + \dfrac{1}{L_2} + \dfrac{1}{L_3} + \cdots$
Total inductance (series-adding)	$L = L_1 + L_2 + 2M$
Total inductance (series-opposing)	$L = L_1 + L_2 - 2M$
Mutual inductance	$M = \dfrac{L_o - L_b}{4}$

Transformers

It is basically a device in which two inductors (coils) are placed on the same core. One of the inductors is known as 'primary winding' and the other inductor is called secondary winding.

A transformer transfers electrical power from one circuit to another. Obviously in an ideal case input power fed at the primary is equal the power obtained at the secondary (Figure 1.27). Moreover, supply frequency remains the same. If V_1, I_1 are voltage and current, respectively, in the primary and V_2, I_2 on the secondary side, then for an ideal transformer, we have

$$V_1 I_1 = V_2 I_2$$

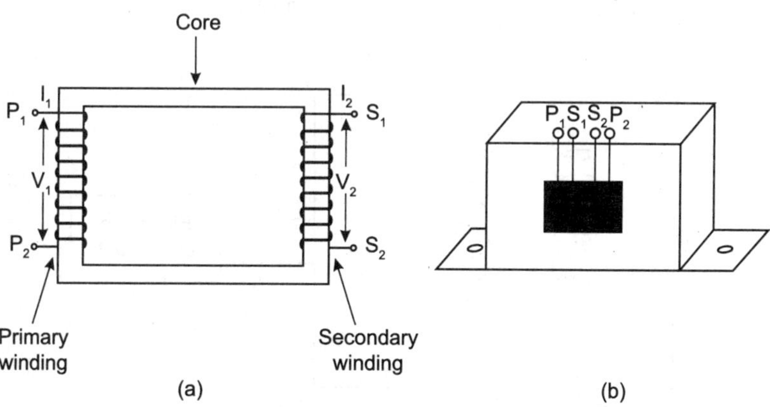

Figure 1.27 (a) Ideal transformer $V_1 I_1 = V_2 I_2$ and (b) small assembled transformer

When we need more output voltage, current will have to reduce proportionately and the transformer is called **step-up transformer**. On the other hand, if we need more current on the output side, voltage will be reduced and the transformer is said to **step-down transformer**.

In a step-up transformer, the number of turns in the output side will be more, and in step-down transformer the number of turns will be less on the output side. Figure 1.27(b) shows a small assembled transformer with P_1 and P_2 its primary terminals and S_1 and S_2 its secondary terminals.

Thus, we have two types of transformers:

(i) Step-up $N_2 > N_1$

(ii) Step-down $N_2 < N_1$

and the voltages of secondary and primary are related as

$$\frac{V_2}{V_1} = \frac{N_2}{N_1}$$

Types of transformers

Generally, the following types of transformers are used in electronic circuits:

(i) **Power transformers** These transformers are used to step-up or step-down supply voltage. These transformers find application in vacuum tubes.

(ii) **Output impedance transformers** These transformers are used in radio receivers at the output stage for impedance matching.

(iii) **Intermediate frequency transformers (IFT)** These transformers handle intermediate frequencies. Similar to IFT, we have radio frequency transformers (RFT).

(iv) **Isolation transformers** These transformers are used to provide isolation between two circuits without changing any electrical parameters. The number of turns in the primary and secondary is equal in these transformers (Figure 1.28).

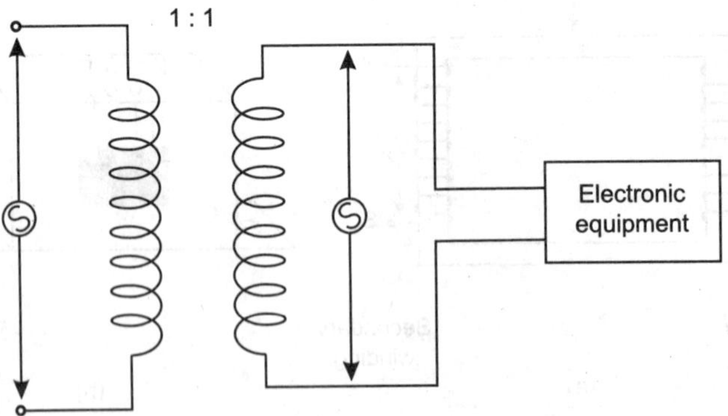

Figure 1.28 Isolation transformer

(v) **Instrument transformers** These transformers are called potential transformers (PT) and current transformers (CT) (Figure 1.29) and are used to measure high values of voltage and current.

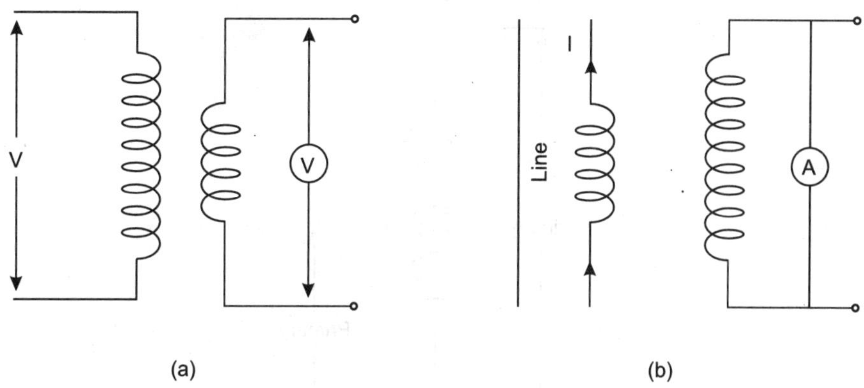

Figure 1.29 (a) Potential transformer (PT) and (b) current transformer (CT)

(vi) **Trigger transformers** These transformers are used to trigger thyristors and other electronic devices.

(vii) **Audio transformers** These transformers are used to transfer signals in the audio frequency range from one circuit to another and are connected at the final stage of an audio amplifier. A loudspeaker (LS) as a load is connected to the audio transformer (Figure 1.30).

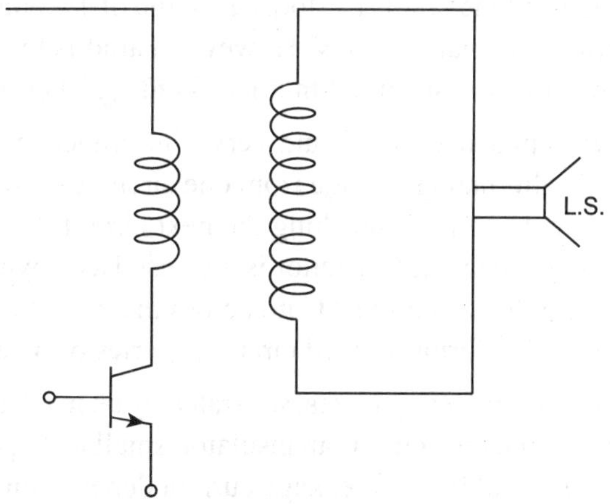

Figure 1.30 Audio transformer

(viii) **Video transformers** These transformers are used to transfer signals in the video range from one circuit to another.

(ix) **Pulse transformers** These are wide band transformers and transfer pulses from one circuit to another.

(x) **Auto transformers** These transformers have only one winding. One can get variable voltages using these transformers.

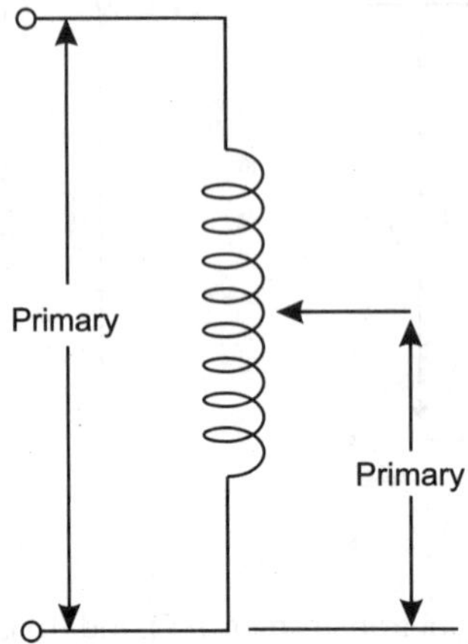

Figure 1.31 Auto transformer

1.2.2.3 Capacitor

This is a device that can store energy in its electronic field and release it whenever desired. A capacitor opposes any change in potential difference (pd) applied across its terminals. The farad (F) is the SI unit of capacitance. One farad (F) stores a charge of 1 Coulomb when the potential difference between the terminals (plates) is 1 V. However, Farad is too large a unit, and hence for practical purposes, capacitors are specified in microfarads (μF) or picofarads (pF).

A capacitor offers low impedance to AC and very high impedance to DC, and therefore capacitors are used to couple alternating voltage from one circuit to another circuit, while at the same time blocking the DC voltage from reaching the next circuit. In such uses, the capacitor is called coupling or blocking capacitor. Capacitor is also used as a bypass. A bypass capacitor bypass the AC through it without letting the AC go across the circuit to which it is connected. With an inductor, a capacitor also forms a tuned circuit in series or parallel.

A capacitor consists of two conducting plates, separated by an insulating material known as a dielectric. Despite the fact that the dielectric is an insulator, small leakage current flows between the plates of a capacitor. The actual level of leakage current depends on insulation resistance of the dielectric. A plastic film capacitor may have insulation resistance higher than 100,000 mW, whereas an electrolytic capacitor may have a micro amp (or more) of leakage current, with only 10 V applied to its terminals.

Electrolytic capacitors normally have one terminal identified as positive terminal. Obviously, they are said to be polarized. Polarization usually limits their application to situations where polarity of applied voltage will not change.

We may note that the capacity of a capacitor depends upon the dielectric material used (Table 1.9).

Table 1.9 Dielectric properties of some materials

Material	Dielectric constant	Dielectric strength in volts/mm
Air / vacuum	1	20
Plastic	2–3.5	1000–5000
Oil	2–5	500–1000
Paper	2.5–6	500–1000
Mica	3–8	600–1500
Glass	4–7	500
Electrolytic	7–11	2000
Ceramics	80–12000	100–300

An ideal capacitor has a dielectric material that has infinite resistance and plates that have zero resistance. However, an ideal capacitor cannot be achieved in practice, as all dielectrics have some leakage current and all capacitor plates have some resistance. Figure 1.32(a) depicts the complete equivalent circuit for a capacitor. The circuit consists of an ideal capacitor C in series with a resistance R_D (resistance of plates), and in parallel with resistance R_L (leakage resistance of the dielectric). Usually, plate resistance can be completely neglected, and the equivalent circuit can be as shown in Figure 1.32(b). With capacitors that have a very high leakage resistance, e.g. mica and plastic film capacitors, the parallel resistor is usually omitted in the equivalent circuit, and the capacitor is then treated as an ideal capacitor. However, this cannot normally be done for electrolytic capacitors, which have relatively low leakage resistances. The parallel $R - C$ circuit is shown in Figure 1.32(b) which is equivalent to series R_S circuit as shown in Figure 1.32(c).

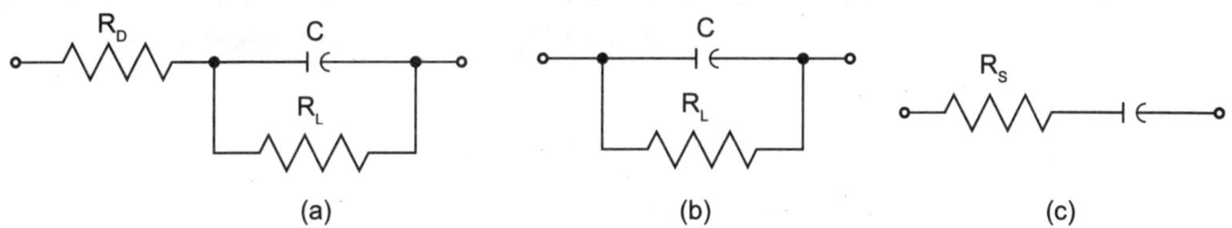

Figure 1.32 (a) Complete equivalent circuit, (b) parallel equivalent circuit and (c) series equivalent circuit

A capacitor equivalent circuit consists of capacitance C, leakage resistance R_L in parallel with C, and plastic resistance R_D in series as with C and R_L.

Capacitors, like resistors, can either be fixed or variable. Fixed capacitors are commonly made from mica, ceramic, paper and electrolytic. Variable capacitors are mostly air-gang capacitors. A variable air capacitor is made up of fixed plates separated by air.

Capacitors are generally identified by the type of dielectric used.

Mica capacitors

These capacitors consists of layers of mica sheets alternated with plates of metal foil (aluminium) for capacitor leads, and the entire assembly is encapsulated in a molded plastic jacket (Figure 1.33). The plates are connected to two electrodes as shown in Figure 1.33.

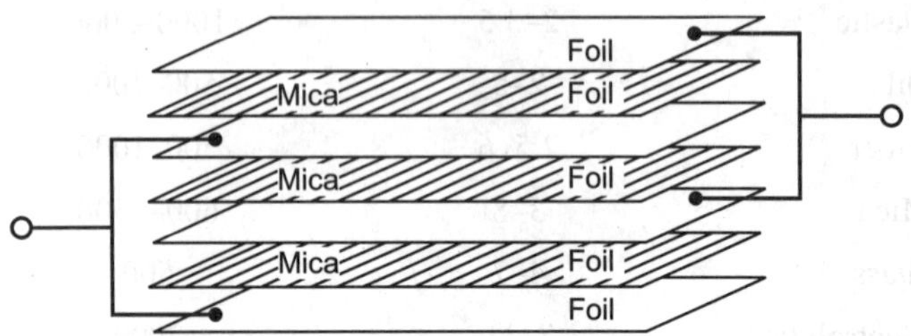

Figure 1.33 Construction of a mica capacitor

Mica capacitors exhibit excellent characteristics under temperature variations and high voltage applications. Typical capacitance values range from 1 pF to 0.1 μF and voltage ratings as high as 35,000 V are possible. Its leakage current is very small ($R_{leakage} \approx 1000\ M\Omega$). Precise capacitance values and wide operating temperatures can be obtained with mica capacitors. In a variation of the method, silvered mica capacitors use films of silver deposited on mica layers instead of metal foil.

Ceramic capacitors

Ceramic capacitors are made in different shapes and sizes. Figure 1.34 shows the construction of a typical ceramic capacitor. A ceramic disc is coated on two sides with a metal, e.g. copper or silver, or films of metal are deposited on each side of a thin ceramic disc, and copper wise terminals are connected to the metal. After attaching tinned-wire leads, the entire unit is encapsulated in a protective coating of plastic. Colour coding is similar to that used for resistances.

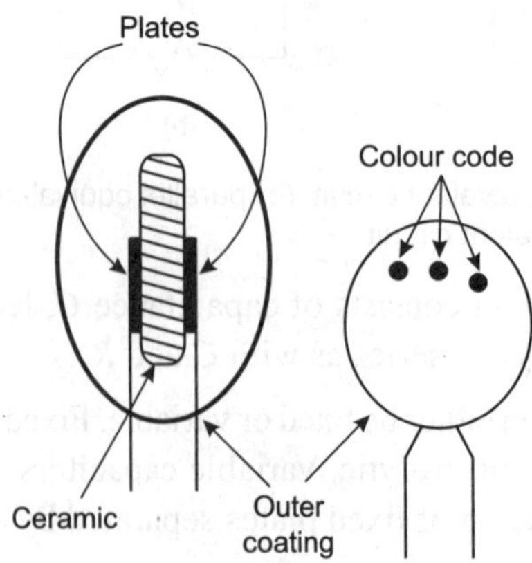

Figure 1.34 Basic construction of a ceramic capacitor

A ceramic disc silvered on each side makes a ceramic capacitor.

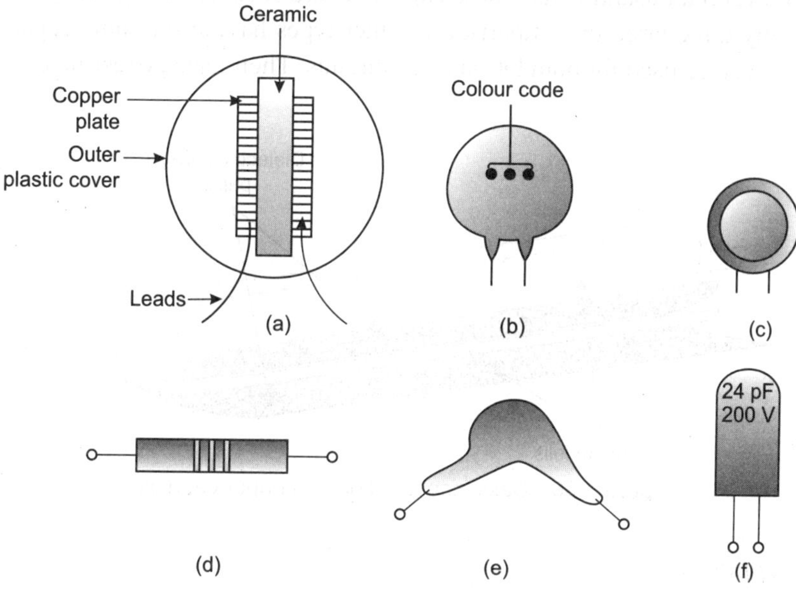

Figure 1.35 Different shapes of ceramic capacitors

Two different ceramic materials are used, one with extremely high relative permittivity. These capacitors are much smaller than paper or mica capacitors but have the same value of capacitance. One major disadvantage of ceramic dielectric is that its leakage resistance is not as high in comparison to other types of capacitors. Another type of ceramic gives leakage resistances of the order of 7500 MΩ. Because of its lower permittivity, this ceramic produces capacitors that are relatively large for a given value of capacitance. Ceramic capacitors are very versatile due to their working voltage range extending from 3 V (for use with transistors) upto 6000 V. The range of capacitance available for these capacitors is typically 1 pF to 0.1 μF with DC working voltages upto 1000 V. Ceramic capacitors have a very low leakage current ($R_{\text{leakage}} \simeq$ 1000 MΩ) and can be used in both DC and AC circuits. Different shapes of ceramic capacitors are shown in Figure 1.35(a–f).

Paper capacitors

A paper capacitor in its simplest form consists of a layer of paper between two layers of metal (aluminium) foil. This paper is impregnated with a dielect material such as wax, plastic or oil. The metal foil and paper are rolled up (Figure 1.36) and hence it is possible to concentrate a large plate area in a small volume. External connections are dipped in wax or plastic. A variation of this is the metalized paper, in which the foil is replaced by thin films of metal deposited on the surface of the paper. This does not mean that the device is polarized but simply identifies the terminal that connects to the outside metal film, and thus it can be grounded to avoid pickup of unwanted signals.

Paper capacitors have usually capacitances ranging from 500 pF to 50 μF and power rating from 100 V to several thousand volts. These capacitors are cheaper for a given capacitance value but are physically quite large in comparison to other types having the same capacitance value. Paper capacitors can be used for both DC and AC circuits. Their leakage resistance is of the order of 100 MΩ.

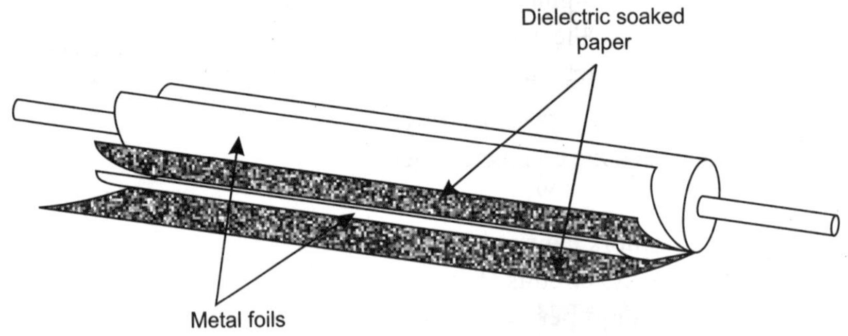

Figure 1.36 Basic construction of a paper capacitor

Electrolytic capacitors

These capacitors can have a very large capacitance in a physically small container, e.g., a capacitance of 5000 μF can be obtained in a cylindrical package approximately 5 cm long and having diameter 2 cm. In this case, the DC working voltage is only 10 V. Similarly, a 1 F capacitor is available in a 22 cm-by-7.5 cm cylinder, with a working voltage of only 3 V. Typical values for these capacitors range from 1 μF through 100,000 μF.

An electrolytic capacitor consists of an aluminium-foil electrode with an aluminium-oxide film covering on one side (Figure 1.37(a)). The aluminium plate serves as the positive plate and the oxide film as the dielectric. When assembled, a direct voltage is applied to the capacitor terminals, and this causes a thin layer of aluminium oxide to form on the surface of the positive plate next to the electrolytic (Figure 1.37(b)). Aluminium oxide is the dielectric, and the electrolyte and the positive sheet of foil are capacitor plates. The externally thin oxide dielectric gives a very large value of capacitance.

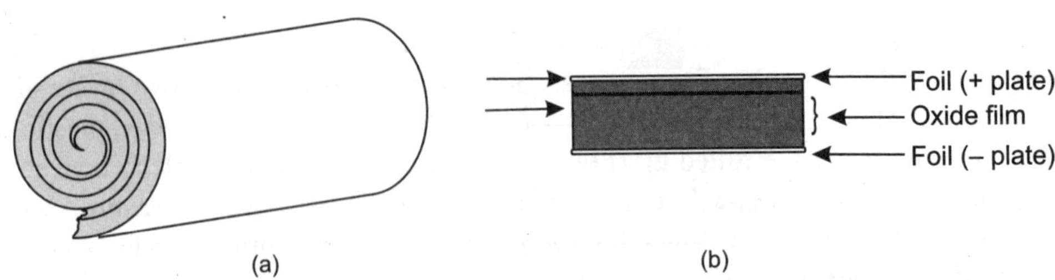

Figure 1.37 (a) An electrolyte capacitor made of rolled-up foil sheets separated by electrolyte soaked gauze, (b) the dielectric has a layer of aluminium oxide at the positive plate.

The aluminium oxide layer is very thin and, therefore, the electrolytic capacitor has larger capacitance in a small volume. It has a high capacitance-to-size ratio. It is important that electrolytic capacitors be connected with correct polarity. It is primarily designed for use in circuits where only DC voltages are applied across the capacitor. The terminals are marked +ve and –ve. When it is incorrectly connected, gas forms within the electrolyte and the capacitor may explode. The terminal designated as +ve should be connected to the most positive of the two points in the circuit where the capacitor is to be installed. Some circuit situations where the capacitor must be correctly connected are shown in Figure 1.38.

One can also obtain non-polarized electrolytic capacitors. Such capacitors consist essentially of two capacitors in one package connected back to back, so that one of the oxide films is always correctly biased.

Normally these capacitors cannot be used with AC circuits. However, some electrolytic capacitors can be used in AC circuits (for starting motors) and in cases where polarity of the DC voltage reverses for a short period of time.

The capacitance value of these capacitors may range from 1 μF to several thousand microfarads. The voltage rating may range from 1 V to 500 V or more. These capacitors are available with DC working voltages greater than 400 V, but in this case capacitance values do not exceed 100 μF. In addition to their low working voltage and polarized operation, another disadvantage of these capacitors is their relatively high leakage current.

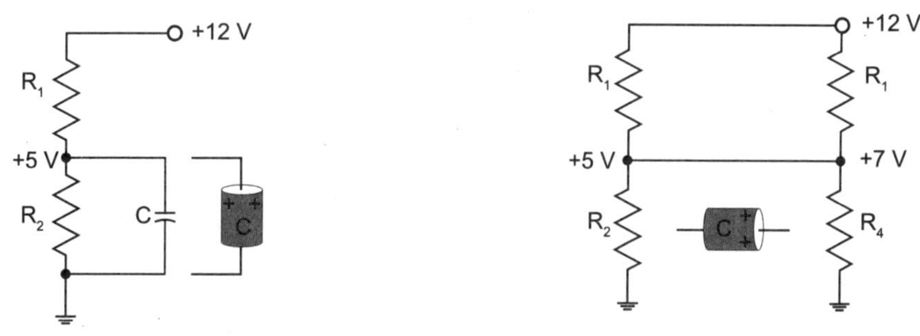

(a) Capacitor connected between +5 V and ground (b) Connected between +7 V and +5 V

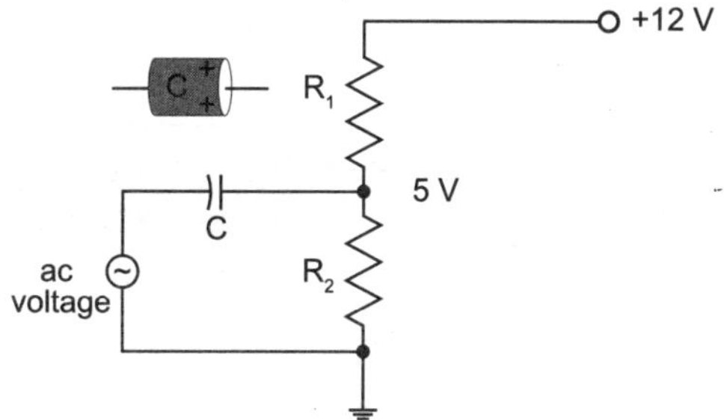

(c) Connected between +5.7 V and a grounded AC voltage source

Figure 1.38 Circuits exhibiting where capacitor is correctly connected

Plastic film capacitors

A plastic film capacitor is similar to that of paper capacitor except that paper is replaced by a thin film that is typically polystyrene or mylar. This type of dielectric gives insulation resistance greater than 10^5 MΩ. Working voltages of plastic film capacitors are as high as 600 V, with a capacitor surviving 1500 V surges for a brief period. Capacitance tolerance of ±2.5% is typical as are temperature coefficients of 60 to 150 ppm/°C.

These capacitors are physically smaller but more expensive than paper capacitors. These are available in values ranging from 5 pF to 0.47 μF. The characteristics of different types of these capacitors are summarized in Table 1.10.

Table 1.10 Characteristics of plastic capacitors

Characteristic	Polystyrene	Polyester	Polycarbonate
Permittivity	2.5	3.3	2.9
Dissipation factor	0.0002	0.005	0.001
Dielectric strength (V/μm)	200	300	186
Insulation resistance (Ω)	10^6	3×10^4	6×10^4
Dielectric absorption (%)	0.01	0.2	0.05
Maximum operating temperature (°C)	80	125	120
Film molecular structure	Non-polar	Polar	Polar

Tantalum capacitors

These capacitors are another type of electrolytic capacitor. Powdered tantalum is sintered (or baked) into a cylindrical shape. The resulting solid is quite porous, so that when immersed in a container of electrolyte, the electrolyte is absorbed into tantalum. Now tantalum has a large surface area in contact with the electrolyte as shown in Figure 1.37.When we apply a DC forming voltage, a thin oxide film is formed throughout the electrolyte–tantalum contact area. This results in a large capacitance value in a small volume.

A typical tantalum capacitor in a cylindrical shape 2 cm × 1 cm, might have a capacitance of 100 μF and a DC working voltage of 20 V. Other types of capacitors are available with a working voltage upto 630 V, but with capacitance values in the order of 3.5 μF. Similar to aluminium foil electrolytic capacitors, tantalum capacitors must be connected with correct polarity of the inductor, so that maximum current can be anything from 50 mA to 1 A. In such an inductor, the core may be made adjustable, so that it can be screwed into or partially out of the coil. Obviously the coil inductance is variable. The graphic symbol for an inductor with an adjustable core is shown in Figure 1.38(b).

Important properties of different capacitors are given in Table 1.11.

Table 1.11 Types of capacitors

Type of capacitor	Range (μF)	AC voltage	DC voltage	Resonant frequency (f_r)
Paper capacitor	10 nF – 10 μF	250/500	600	100 kHz
Mica capacitor	5 pf – 10 nf	–	upto 600	10 μHz
Ceramic capacitors	5 nF – 47 μF	250	upto 400	upto 100 μHz
Aluminium electrolytic capacitor	1 – 105 μF	–	upto 100	50 kHz
Tentalumn electrolytic capacitor	1 – 2000 μF	–	upto 50	0.1 μHz
Plastic capacitor	0.005 – 0.5 μF	–	500 V	–

Variable capacitors

Air-gang capacitor is the most common variable capacitor (Figure 1.39). Air acts as dielectric for this capacitor. By rotating the shaft at one end, one can change the common area between the movable and fixed set of plates. The greater the common area, the larger the capacitance.

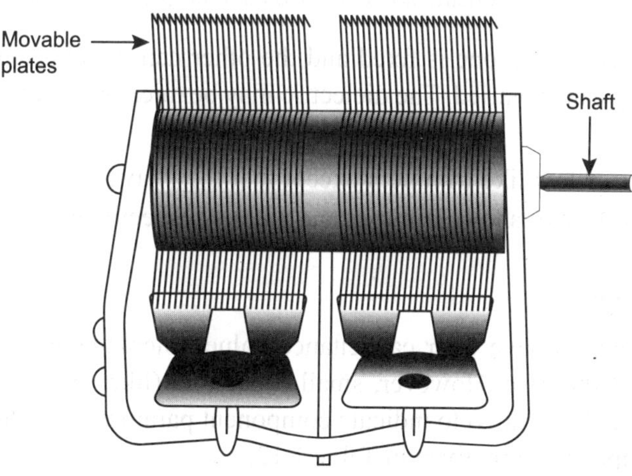

Figure 1.39 Air capacitor

In some applications the variation in capacitance is not required frequently. For all normal operations, only one setting may be sufficient. In such situations, one uses a variable capacitor called a trimmer (sometimes called padder). For trimmer capacitors, both mica and ceramic are used as dielectric (Figure 1.40(a)).

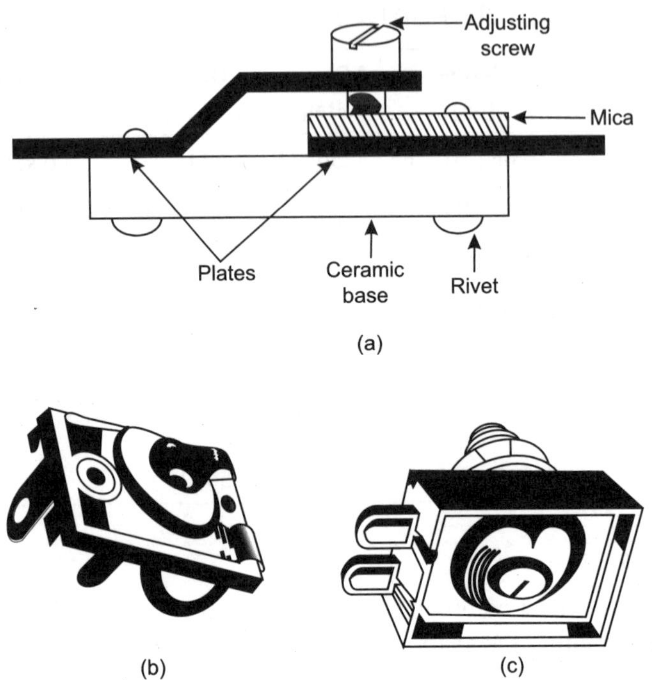

Figure 1.40 (a) Basic construction of a mica trimmer; (b), (c) trimmer and padder

Trimmers have two plates, one is fixed and the other can be adjusted with the help of a screw, i.e. the capacity can be varied. The dielectric may be mica or ceramic. They are available upto 50 pF.

Padders are more or less like trimmers except that they have large capacity and available upto 600 pF, Figure 1.40(b,c) shows trimmer and padder respectively.

Color codes for capacitors

Normally, large capacitors have their capacitance value, tolerance and DC working voltage printed on the side of the case. However, small capacitors (like small resistors) use colored bands (or sometimes colored dots) to indicate component parameters. Colour coding for mica, ceramic and paper capacitors are given in Table 1.12.

Different color codes are in use and the most common one is as shown in Figure 1.41.

Table 1.12 Colour coding chart for mica, ceramic and paper capacitors

Colour band	Capacitance in pF		Multiplier	Tolerance (±%)
	1st digit	2nd digit		
Black	0	0	1	20
Brown	1	1	10	1
Red	2	2	10^2	2
Yellow	4	4	10^4	–
Green	5	5	–	5
Blue	6	6	–	–
Violet	7	7	–	–
Grey	8	8	0.01	–
White	9	9	0.1	10
Gold	–	–	0.1	1
Silver	–	–	0.01	10

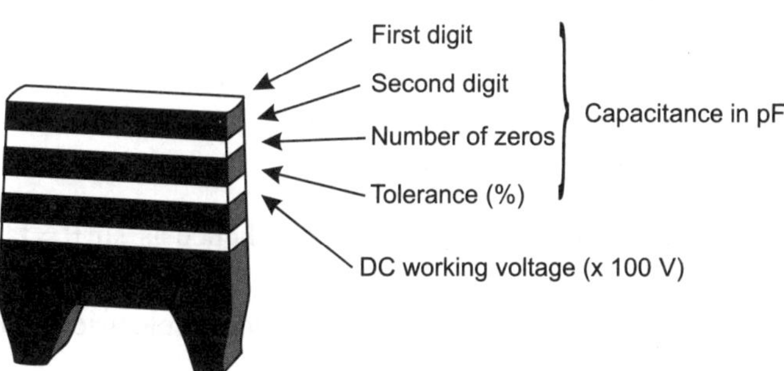

Figure 1.41 Colour coding for capacitors

Losses in the capacitor

In a pure capacitor, on applying an AC voltage, we see that current leads voltage by 90° and power loss is zero, since cos ϕ = cos 90 = 0. However, a practical capacitor possesses resistance and inductance also, and hence angle ϕ is not exactly 90°, but equal to $(90 - \theta)$. Figure 1.42(a) shows an equivalent circuit, and Figure 1.42(b) shows the phasor diagram of the capacitor.

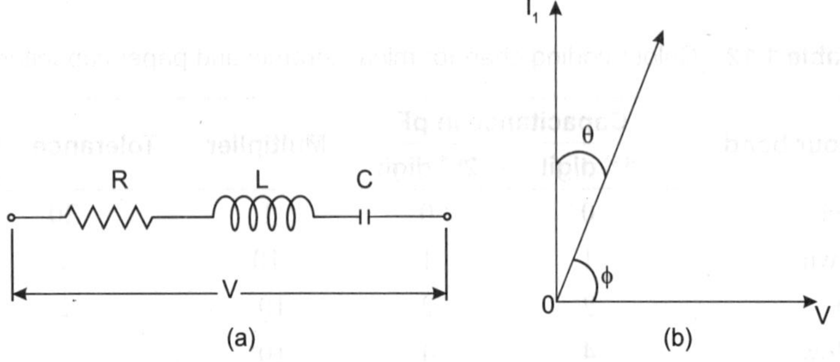

Figure 1.42 (a) Equivalent circuit and (b) phasor diagram for a capacitor

Clearly, a practical capacitor causes a loss as current does not lead voltage exactly by 90°, i.e. this angle will be less than 90°. The power loss, also called dielectric loss, will depend upon the angle of lead ϕ. The complimentary angle ($\theta = 90 - \phi$) is named as **loss angle**.

Power loss in a practical capacitor can be derived as:

$$P = VI\cos\phi$$

$$= V\frac{V}{X_C}\cos\phi \quad (\because I = \frac{V}{X_C}, \text{where } X_C \text{ is capacitive reactance})$$

$$= \frac{V^2}{X_C}2\pi fC\cos\phi \quad (X_C = \frac{1}{2\pi fC})$$

$$= V^2 2\pi fC\sin\theta[\cos\phi - \cos(90-\theta)]$$

$$= 2V^2\pi fC\tan\theta \quad (\because \theta \text{ being very small and hence as } \sin\theta \simeq \tan\theta)$$

Obviously, dielectric loss is proportional to the tangent of dielectric angle, i.e. tan θ, called **power factor** of the capacitor. We must remember that:

(i) Below resonance (when $X_C = X_L$), impedance of a capacitor is capacitive. At resonance, it is resistive and above resonance, it becomes inductive. Clearly one will have to keep inductance low to increase the working frequency range of the capacitor.

(ii) Resistance present in the capacitor is a measure of heat dissipation, and inductance is a measure of application of the capacitor at high frequency.

Capacity of capacitors

1. Parallel plate capacitor with air as dielectric (Figure 1.43(a)):

   $C = \frac{\varepsilon_0 A}{d}$ where ε_0 is permittivity of air, A is area of plates and d is distance between plates

2. Parallel plate capacitors with a dielectric of permittivity ε_r (Figure 1.43(b)):

   $$C = \frac{\varepsilon_0 \varepsilon_r A}{d}$$

3. Parallel plate capacitor with composite dielectric (Figure 1.43(c)):

   $$C = \frac{\varepsilon_0 A}{(d_1/\varepsilon_{r_1}) + (d_2/\varepsilon_{r_2})}$$

4. Parallel plate capacitor with multiple plates (Figure 1.43(d)):

   $$C = \frac{4\varepsilon_0 \varepsilon_r A}{d}$$

5. Two wires separated by a dielectric of ε_r (Figure 1.43(e)):

   $$C = \frac{\pi \varepsilon_0 \varepsilon_r A}{2.3 \log d/r} \text{ per unit length}$$

   where ε_0 is absolute permittivity of air

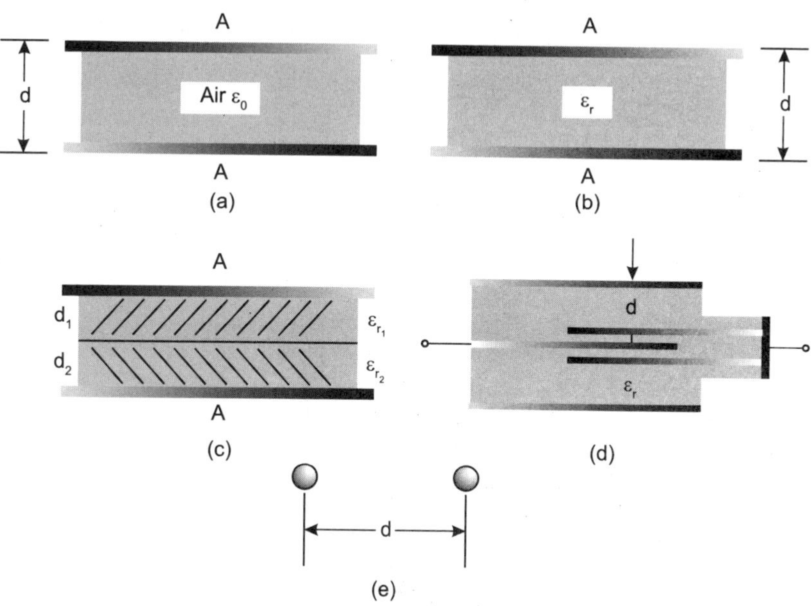

Figure 1.43 Capacitors

1.3 Voltage and Current Dividers

In analyzing networks, the concepts of voltage divider and current divider are quite useful.

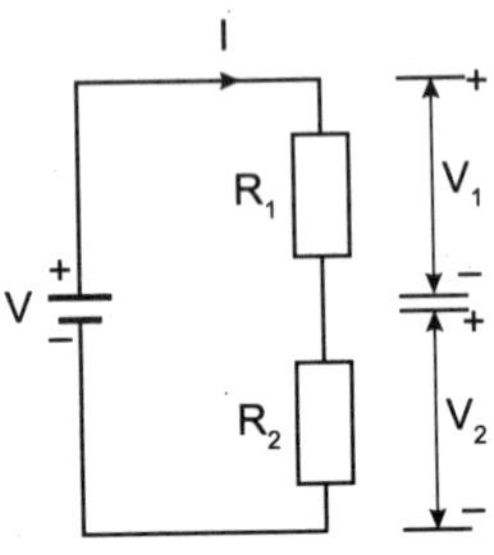

Figure 1.44 Voltage divider

Figure 1.44 shows a simple voltage divider circuit. Voltage across series resistors is obtained as

$$V_1 = R_1 I = R_1 \frac{V}{R_1 + R_2}$$

or

$$V_1 = \frac{R_1}{R_1 + R_2} V \quad (1)$$

Similarly,

$$V_2 = \frac{R_2}{R_1 + R_2} V \quad (2)$$

Current in parallel resistors shown in Figure 1.45 is obtained as

$$I_1 = \frac{R_2}{R_1 + R_2} I \quad (3)$$

$$I_2 = \frac{R_1}{R_1 + R_2} I \quad (4)$$

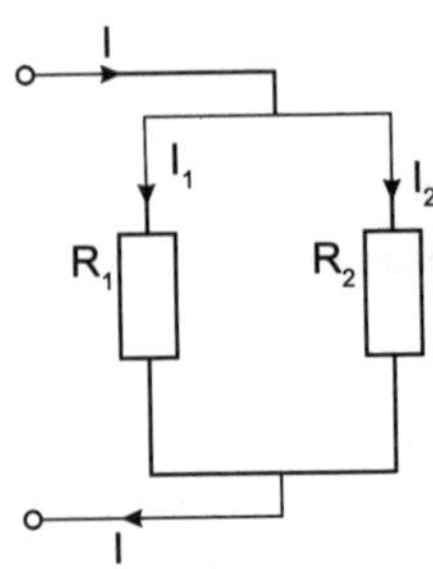

Figure 1.45 Current divider

In Eq. (1) for V_1, we have R_1 in the numerator, but in Eq. (3) for I_1 we have R_2 in numerator.

1.4 Electrical Power Sources

An electrical power source supplies power to the load. These are of two types: DC and AC. Battery, DC generator and rectifier type DC supply are examples of DC power sources. Examples of AC power sources are alternators, oscillators and signal generators.

A power source has an emf (electromotive force). Emf represents the energy expended during the passing of a unit charge through the source.

We may note that emf and potential difference are similar quantities and both are measured in volts. However, an emf is always active, i.e. it tends to produce an electric current in the circuit. On the other hand, the potential difference (pd) may be either passive or active. A pd is said to be passive when it has no tendency to create current in the circuit.

The current flow leaves the source at positive terminal and obviously is in the same direction as the emf as indicated by an arrow in Figure 1.46. Current enters the load (passive elements) at positive terminal, and in a load, both current and pd are in opposite directions.

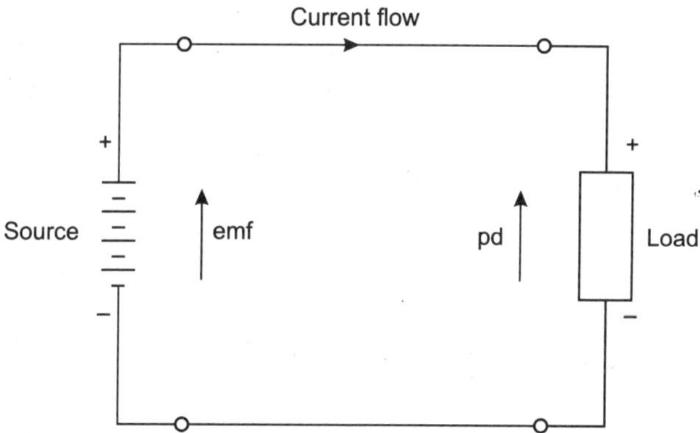

Figure 1.46 Transfer of energy from source to load

1.4.1 Ideal Voltage Source

An ideal voltage source maintains a prescribed voltage independent of its output current. Even if current drawn from such a source varies from zero value (open-circuit condition) to infinity (short-circuit condition), its terminal voltage remains unchanged. Obviously, an ideal voltage source is capable of supplying unlimited amount of power. However, there is no ideal voltage source. Symbols for ideal voltage sources are shown in Figure 1.47.

1.4.2 Practical Voltage Source

When we draw some current from a voltage source, its terminal voltage is found to drop by some amount. The larger the current drawn, the larger is the voltage drop. Figure 1.48 show a practical voltage source.

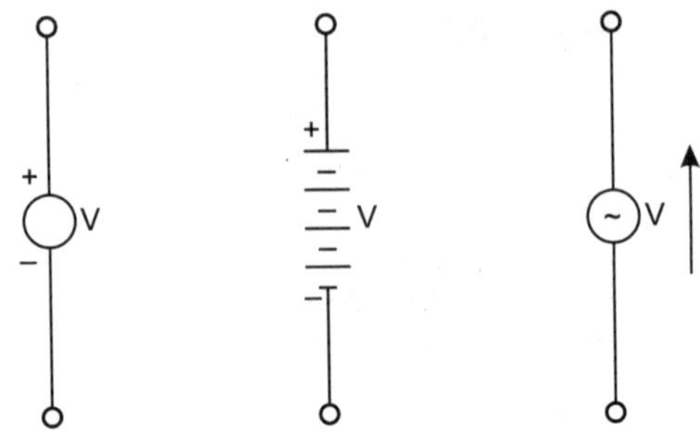

Figure 1.47 Symbols for ideal voltage sources

To understand lowering of terminal voltage, a practical voltage source is represented as an ideal voltage source in series with resistance (or impedance, in case of an AC source). This resistance is termed as internal resistance of the source. R_{SV} (Figure 1.48) represents internal resistance of a practical voltage source.

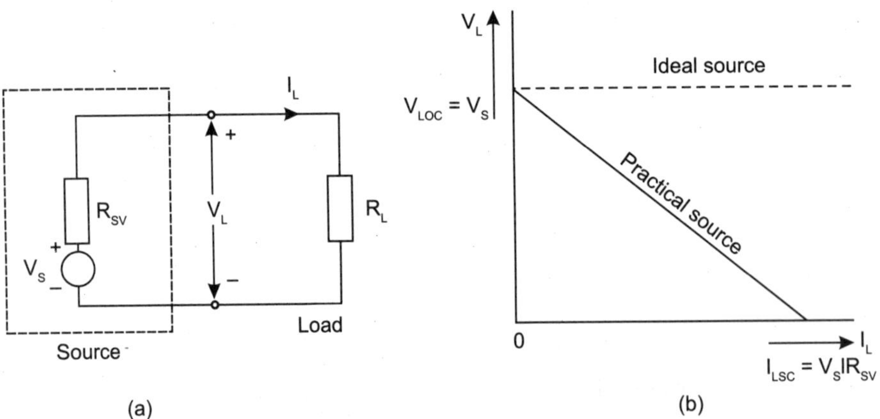

Figure 1.48 A practical voltage source. R_{SV} represents internal resistance

One can express the linear relationship between load voltage (V_L) and load current (L_L) as

$$V_L = V_S - R_{SV} I_L \tag{5}$$

Open-circuit voltage (V_{LOC}) and short-circuit current (I_{LSC}) are given by

$$V_{LOC} = V_S \tag{6}$$

$$I_{LSC} = \frac{V_S}{R_{SV}} \tag{7}$$

1.4.3 Ideal Current Source

An ideal current source is one that produces its prescribed current, independent of its output voltage. Obviously, the output current of an ideal current source remains unchanged from zero load (open-circuit condition) to infinite load (short-circuit condition). However, this could be possible only if the source is capable of supplying unlimited amount of power. Figure 1.49 shows the symbols for ideal current sources.

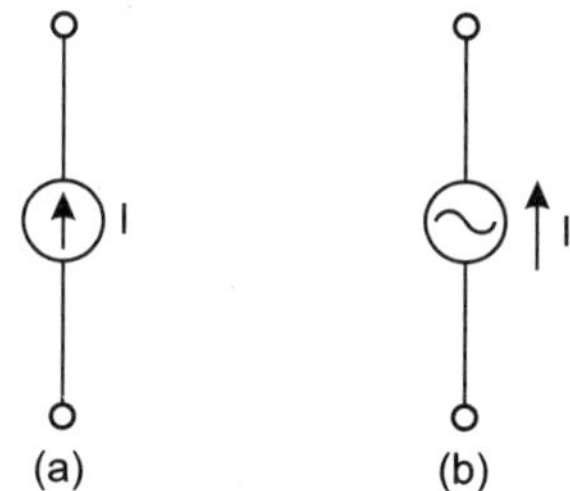

Figure 1.49 Symbols for ideal current source: (a) DC and (b) AC

1.4.4 Practical Current Source

An ideal current source, like an ideal voltage source, is also non-existent in real sense. However, certain transistor circuits (Chapter 5, Common Base) can deliver constant current to a limited range of load. Figure 1.50 shows a practical current source modeled as an ideal current source in parallel to an internal resistance (R_{SI}).

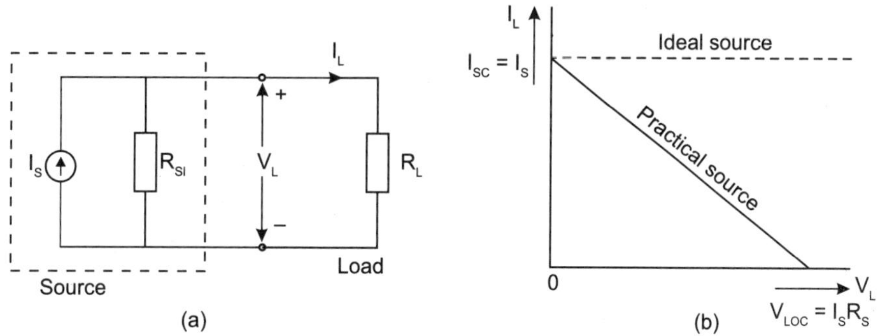

Figure 1.50 A practical current source. R_{SI} is internal resistance

One obtains load current as

$$I_L = I_S - \frac{V_L}{R_{SI}} \tag{8}$$

Open-circuit voltage (V_{LOC}) and short-circuit current (I_{LSC}) are given by

$$V_{LOC} = I_S R_{SI} \tag{9}$$

and

$$I_{LSC} = I_S \tag{10}$$

1.4.5 Voltage Source vs. Current Source

We discussed that an ideal voltage source as well as an ideal current source are non-existent in reality. One can treat a voltage source or a current source as a practical source, depending upon which type suits better for the analysis of the network.

Voltage source and current source would be equivalent provided they produce identical values of I_L and V_L when they are connected to the same load R_L, whatever be its value. However, the two equivalent sources should also provide the same open-circuit voltage and short-circuit current. One can find the conditions of equivalence. Equating open-circuit voltage (V_{LOC}) in both the cases (from Eqs. 6 and 9), one obtains

$$V_S = I_S R_{SI} \tag{11}$$

Now, equating short-circuit currents in the two cases (from eqs. 7 and 10), one obtains.

$$\frac{V_S}{R_{SV}} = I_S \tag{12}$$

From Eqs. (11) and (12), one finds

$$R_{SV} = R_{SI} = R_S \text{ (say)}$$

$$V_S = R_S I_S \tag{13}$$

Here R_S represents internal resistance of either of the sources.

Two practical sources will be equivalent with respect to what happens at load terminals (Figure 1.51). However, the two sources are not equivalent internally.

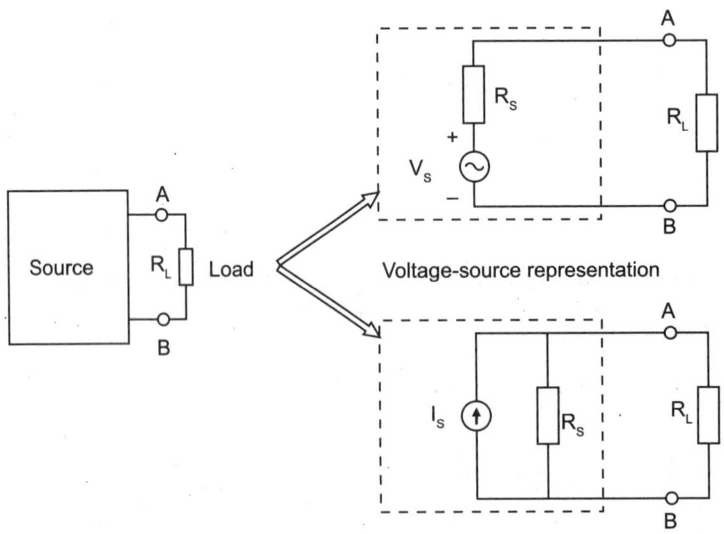

Figure 1.51 Voltage source and current source connected to a load

1.5 Active Components

Electronic devices that perform the basic functions of a circuit (with the help of passive components) are termed as active components, e.g. vacuum tube, semiconductor diodes, etc. A list of important active components along with their symbols and use is given in Table 1.13.

Table 1.13 Important active components, their use and symbol

Name and use	Symbol
(1) Tube devices	
(a) Vacuum tubes	
Vacuum tube diode It is used as a rectifier and detector in electronic circuits. Its terminals are plate (*P*) and cathode (*C*).	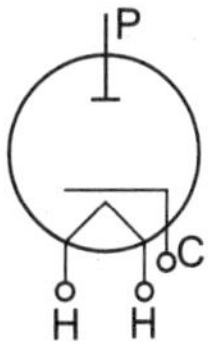
Vacuum tube triode It is used as an amplifier. Terminals are plate (*P*), cathode (*C*) and grid (*g*).	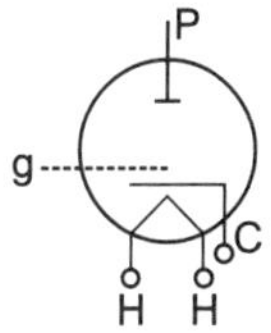
Vacuum tube tetrode It is also used for amplification. It has a plate (*P*), cathode (*C*) and two grids (g_1, g_2).	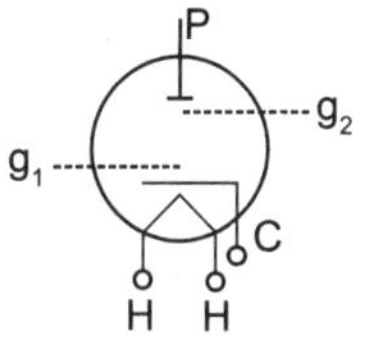
Vacuum tube pentode It is also used for amplification. Its amplification factor is more than that a triode or tetrode. It has plate, cathode and 3 grids (g_1, g_2, g_3).	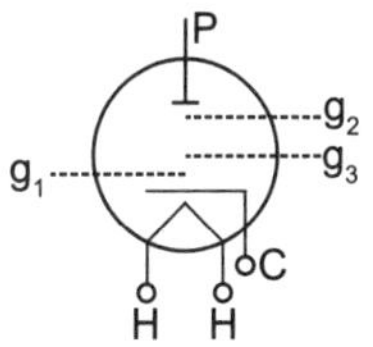
(b) Gas tubes	
Gas diode It is used as a voltage regulator.	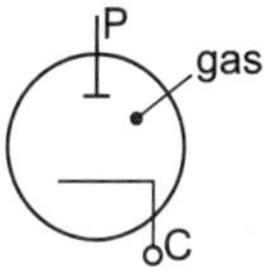
Gas triode or thyratron It is used as controlled rectifier.	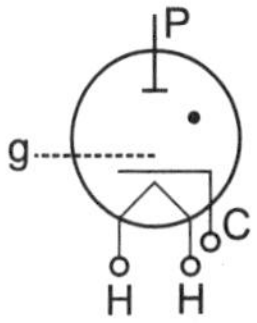

Table 1.13 (Continued)

Name and use	Symbol
(2) Semiconductor devices	
Junction diode It has anode (a) and cathode (c). It is used as rectifier and switch.	 a c
Zener diode It is used as voltage regulator (stabilizer).	 a c
Bipolar junction transistor It is used as amplifier. It may be PNP or NPN type. Its terminals are emitter (E), collector (C) and base (B).	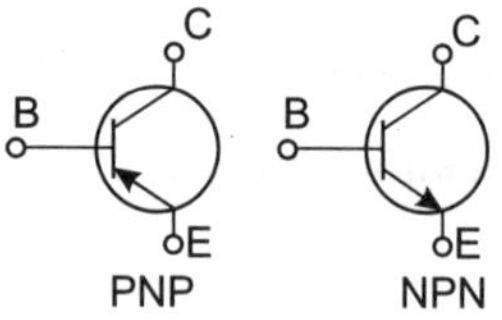 C B E PNP; C B E NPN
Field effect transistor (FET) It is used as buffer amplifier. It can be N-FET and P-FET. Its terminals are source (S), drain (D) and gate (G).	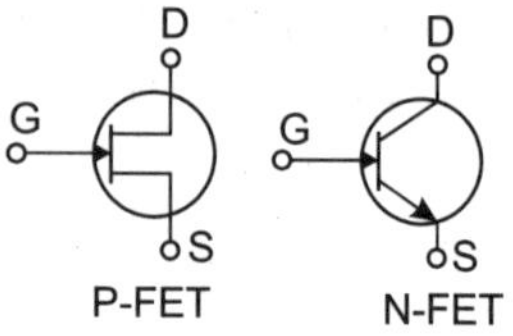 D G S P-FET; D G S N-FET
Unijunction transistor It is used in power controls. It has three terminals, emitter (E), base 1(B_1) and base 2 (B_2).	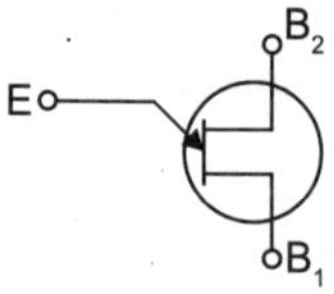 B_2 E B_1
Silicon controlled rectifier It is used for controlled rectification and speed control of DC machines. It has anode, cathode and gate.	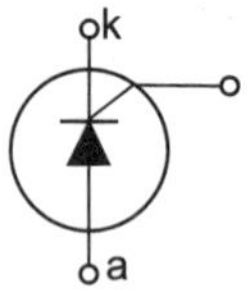 k a
Triac It is used as an AC switch and to obtain regulated AC supply. It has grid and main terminals 1 and 2.	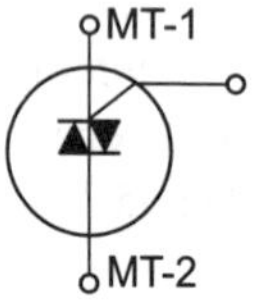 MT-1 MT-2
Diac It is used to fire (start) a triac.	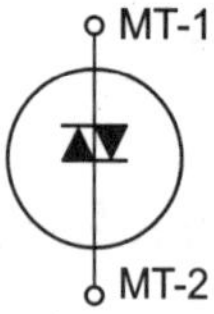 MT-1 MT-2

Example 1

(a) A carbon resistor has the following band sequences—red, black, blue and gold. Show that the value of the resistor is in between 19 to 21 MΩ. [Diploma, B.Tech.]

Solution

Hold the resistor (Figure 1.52) such that the tolerance ring is on your extreme right. Now the value of the resistor is equal to:

Red	Black	Blue	Gold
2	0	6	±5%

i.e., $20 \times 10^6 = 20000000\ \Omega = 20\ \text{M}\Omega \pm 5\%$

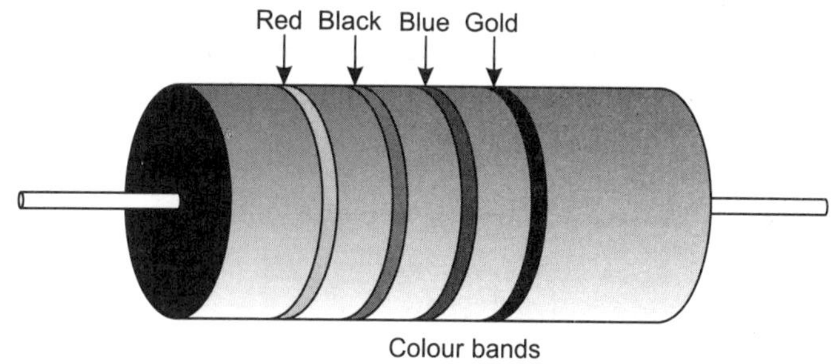

Figure 1.52

The value of the resistor is in between (20 – 1) and (20 + 1), i.e., 19 to 21 MΩ.

Example 2 A resistor has the following colour band sequence: yellow, violet, orange and gold. Find the range in which its value must lie so as to satisfy the manufacturer's tolerance.

[B.Tech]

Solution

With the help of the colour coding table, one finds

1st band	2nd band	3rd band	4th band
Yellow	Violet	Orange	Gold
4	7	10^3	±5%

i.e., $47\ \text{k}\Omega \pm 5\%$

$$\text{Now, } 5\% \text{ of } 47\,\text{k}\Omega = \frac{47 \times 10^3 \times 5}{100}\,\Omega = 2.35\ \text{k}\Omega$$

Thus, resistance should be within the range $47\ \text{k}\Omega \pm 2.35\ \text{k}\Omega$, or between 44.65 kΩ and 49.35 kΩ.

Example 3 A resistor has the following colour band sequence: gray, blue, gold and silver. What is the range in which its value must lie so as to satisfy the manufacturer's tolerance?

[B.Tech]

Solution The specification of the resistor from the colour coding table is as follows:

1st band	2nd band	3rd band	4th band
Gray	Blue	Gold	Silver
8	6	10^{-1}	±10%

i.e., $86 \times 0.1\ \Omega \pm 10\% = 8.6\ \Omega \pm 10\%$

10% of $8.6\ \Omega = 0.86\ \Omega$

Obviously, resistance should lie between the values $(8.6 - 0.86) = 7.74\ \Omega$ and $(8.6 + 0.86) = 9.46\ \Omega$.

Example 4 Find the equivalent current source representation of the DC voltage source given in Figure 1.53. Take load resistance of 1 Ω and show that the two equivalent source representations give the same load current and load voltage. [Diploma, B.Tech.]

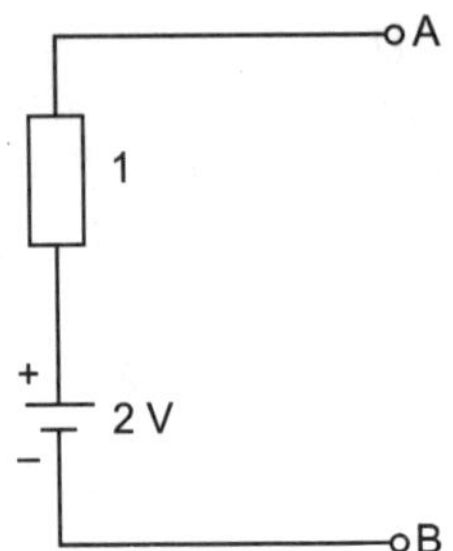

Figure 1.53 A voltage source

Solution Current I_S of the equivalent current source is

$$I_S = \frac{V_S}{R_S} = \frac{2\text{ V}}{1\Omega} = 2\text{ A}$$

Internal resistance remains the same but it is now connected in parallel with the current source *IS*, as shown in Figure 1.54(a).

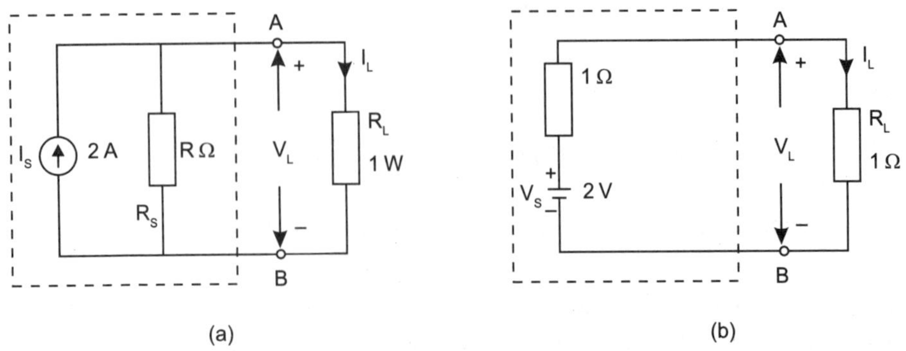

Figure 1.54 Equivalent representations

Now, we connect a load resistance $R_L = 1\ \Omega$ across the terminals of two representations, and find I_L and V_L. From Figure 1.54(b) and using the current–divider concept, one obtains

$$I_L = I_S \frac{R_S}{R_S + R_L} = 2 \times \frac{1}{1+1} = 1\,\text{A}$$

$$\therefore V_L = I_L R_L = 1 \times 1 = 1\,\text{V}$$

From Eq. 53(b), using the voltage-divider concept, one obtains

$$V_L = V_S \frac{R_L}{R_S + R_L} = 2 \times \frac{1}{1+1} = 1\,\text{V}$$

$$I_L = \frac{V_L}{R_L} = \frac{1}{1} = 1\,\text{A}$$

Thus, one finds that the two representations give the same values of I_L and V_L.

Example 5 Consider a practical AC voltage source having an open-circuit voltage of 10 V and internal resistance of 100 Ω. Find the percentage variation in load current and load voltage, if the load connected across its terminals varies (a) from 1 Ω to 10 Ω (b) from 1 kΩ to 10 kΩ.

[Diploma]

Solution

(a) R_L varies from 1 Ω to 10 Ω (Figure 1.55(a)).
Currents for the two extreme values of R_L are

$$I_{L_1} = \frac{V_S}{R_{L_1} + R_S} = \frac{10}{1+100} = 0.0990\,\text{A}$$

$$I_{L_2} = \frac{V_S}{R_{L_2} + R_S} = \frac{10}{10+100} = 0.0909\,\text{A}$$

$$\therefore \text{Percentage variation in current} = \frac{0.0990 - 0.0909}{0.0990} \times 100 = 8.1\%$$

Now, load voltage for the two extreme values of R_L are

$$V_{L_1} = I_{L_1} R_{L_1} = 0.0990 \times 1 = 0.099\ \text{V}$$

$$V_{L_2} = I_{L_2} R_{L_2} = 0.0909 \times 10 = 0.909\ \text{V}$$

$$\therefore \text{Percentage variation in voltage} = \frac{0.909 - 0.099}{0.909} \times 100 = 89.1\%$$

We find that percentage variation in load current is much less than that in load voltage. Thus, the behaviour of the source is nearer to the ideal current source under this condition.

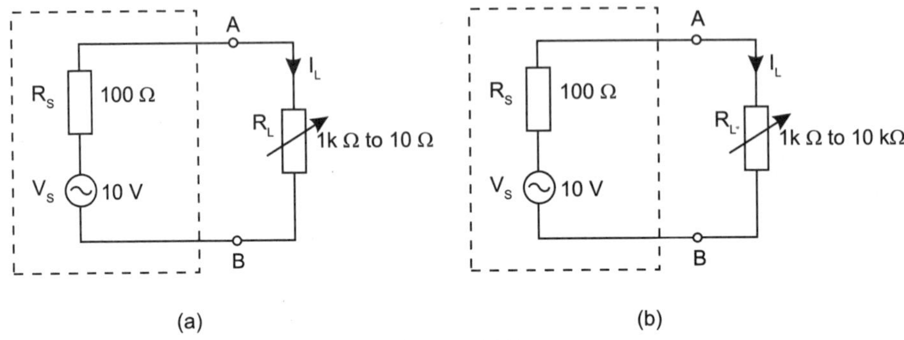

Figure 1.55

(b) R_L varies from 1 Ω to 10 kΩ (Figure 1.55(b)).

Currents for the two extreme values or R_L are

$$I_{L_1} = \frac{V_S}{R_{L_1} + R_S} = \frac{10}{1000+100} = 0.00909 \text{ A} = 9.09 \text{ mA}$$

$$I_{L_2} = \frac{V_S}{R_{L_2} + R_S} = \frac{10}{10000+100} = 0.000999 \text{ A} = 0.999 \text{ mA}$$

$$\therefore \text{Percentage variation in current} = \frac{9.09 - 0.999}{9.09} \times 100 = 89\%$$

Review Questions

1. Mention the names of few important electronic materials. Classify them and give a brief description of each.
2. What do you understand by passive and active components? Explain clearly.
3. Discuss variable resistors and mention where they are used?
4. What do you understand by linear and non-linear resistors?
5. What are polar and non-polar electrolytic capacitors?
6. Explain the construction and working of air-gang capacitors.
7. Write a short note on carbon composition resistor, carbon film resistor and metal film resistor?
8. What do you understand by dielectrics? Explain dielectric loss in capacitors.
9. Write short notes on: (a) power transformer; (b) paper, mica and ceramic capacitors; (c) resistors; (d) inductors; (e) capacitors; (f) logarithmic resistor; (g) transformers.

10. Compare power loss in an ideal capacitor and an inductor.
11. Explain why the use of inductors and transformers is avoided in electronic circuits.
12. Explain why carbon resistors are popular as compared to wire-woundresistors?

Fill in the Blanks.

1. The rating of resistors is in ______________. [**Ans:** ohms]
2. Farad is the unit of ______________. [**Ans:** capacitance]
3. Components perform the basic function in an electronic circuit are called ______________ components. [**Ans:** active]
4. If a polar electrolytic capacitor is connected without considering polarity, it will be ______________. [**Ans:** damaged]
5. Non-polar capacitors do not need any consideration of ______________. [**Ans:** polarity]
6. A FET is an ______________ component. [**Ans:** active]
7. A transformer is a ______________ components. [**Ans:** passive]
8. IC stands for ______________. [**Ans:** integrated circuit]
9. IFT stands for ______________. [**Ans:** intermediate frequency transformer]
10. A triode functions as a ______________. [**Ans:** amplifier]
11. When there is no tolerance band, tolerance of the resistor is ______________. [**Ans:** ±20%]
12. If tolerance band is silver, tolerance of the resistor is ______________. [**Ans:** ±10%]
13. A 100 μF capacitor is ______________. [**Ans:** electrolytic]

Chapter

2 Semiconductor Physics

2.1 Introduction

Solid state devices, e.g. diodes, transistors and integrated circuits (ICs), have had a great impact on electronics engineering and also on our society. These devices are fabricated using semiconductor materials and are, therefore, called semiconductor devices. Today semiconductor devices are a part of all modern electronic circuits. Therefore, to understand electronic circuits and their operations, one must have a thorough knowledge in basic semiconductor physics.

Solid materials are classified into three groups mainly based on their current carrying capabilities: **conductors, semiconductors** and **insulators**. Conductors have abundant free electrons that act as charge carries, which means they have high conductivity. An insulator, on the other hand, has hardly any free electrons and offers very low level of conductivity. A semiconductor is a material that shares the characteristics of a good conductor as well as an insulator. The term resistivity (inverse of conductivity) is often used when comparing the resistance levels of different materials. Typical resistivity values for three broad categories of materials are given in Table 2.1.

Table 2.1 Typical resistivity values of select few materials at 20°C (Ω – m)

Conductor	Semiconductor	Insulator
Copper ~ 1.7×10^{-8}	Silicon ~ 50×10^{1}	Mica ~ 7×10^{10}
Silver ~ 1.6×10^{-8}	Germanium ~ 50×10^{-2}	Glass ~ $10^{6} - 10^{7}$
Aluminium ~ 2.8×10^{-8}	Indium - Antimonide ~ 200×10^{-2}	Diamond ~ 10^{10}
Iron ~ 10×10^{-8}		
Nichrome ~ 100×10^{-6}		

At 20°C, resistivity of semiconductors lies between that of metals and insulators. However, at a very low temperature (~ absolute zero), a semiconductor behaves as an insulator. We may note that semiconductors in very pure form (intrinsic) have negative temperature coefficient of resistance, i.e. their resistivity decreases (or conductivity increases) rapidly with increase of temperature. This means that the number of charge carriers in pure semiconductors increases rapidly with temperature. On the contrary, conductors have positive temperature coefficient of resistance. We may further note that semiconductors have two types of charge carriers: **electrons** and **holes**. "Holes" are absence of electrons in covalent bonds and act equivalently as positive charge carriers. Another unique property of semiconductors is that their conductivity increases by several orders of magnitude even by minute amount of doping with certain impurities. Such semiconducting materials are called **extrinsic semiconductors**. This property of semiconductors has been exploited to develop electronic devices. Silicon (Si), Germanium (Ge) and Gallium Arsenide (GaAs) semiconductors have received the greatest attention in the development of semiconductor devices. Indium Phosphide (InP), Indium Arsenide (In As), Lead Sulphide (PbS), Cadmium Sulphide (CdS), Lead Telluride (PbTe), etc. are other semiconductor materials of interest. Ga As is used in transistors, lasers, and microwave and millimeter wave devices. PbS and PbTe find application in infrared detectors, CdS in light meters, CdTe in the detection of nuclear radiation, and so on. Alloys of compound semiconductors also have important applications. For example, gallium arsenide phosphide ($GaAs_xP_{1-x}$) is used in light-emitting diodes (LED); indium gallium arsenide (In_{1-x}, Ga_xAs) in microwave and opto-electronic devices; cadmium mercury telluride ($Cd_xHg_{2-x}Te$) in infrared detectors, etc. Moreover, the characteristics of semiconductors can be significantly altered by the application of light or heat. This discovery led to the application of semiconducting materials in light- and heat-sensitive devices. The resistivity of semiconductors varies within wide limits, i.e., 10^{-4} to 10^4 Ω-m, and can be reduced greatly with an increase in temperature (according to the exponential law).

Elemental semiconductors, e.g. Ge and Si etc. are covalent and belong to Group IV of periodic table. They have diamond like structure. As regard applications, the **compound** semiconductors appear to be the most promising. Amongst them are the intermetallic compounds of Group III and V elements such as InSb, GaAs, InP. The energy gap of these compounds ranges from a few tenths of an eV to several electron volts, thereby widening their possibility of applications as electronic devices at wider ranges of temperature than Si or Ge. The other group of compound semiconductors are formed between Group IV and VI such as PbS, PbTe and PbSe and between Group II and VI, e.g. CdS, ZnS and CdTe. The conductivity of these compounds may originate from the presence of defects in crystalline structure or deviation in from stoichiometry.

In Mendeleev's periodic table, semiconductors form the group IV of elements. The most typical and extensively studied semiconductors are Germanium (Ge), Silicon (Si) and Tellurium (Te). The study of their electrical properties reveals that

(i) Semiconductors have negative temperature coefficient of resistance, i.e., resistance decreases with increase in temperature and vice versa. For example, germanium is an insulator at low temperatures and a good conductor at high temperatures. This is contrary to the positive temperature coefficient of resistance exhibited by any metal.

(ii) The resistivity of semiconductors lies within the range 10^{-4} to 10^4 Ω–m.

(iii) The electrical conductivity of a semiconductor can be affected when a suitable metallic impurity, e.g., Arsenic, Gallium, Indium, etc, is added to it. This property of semiconductors is the most important from the point of view of applications.

In addition, semiconductors have the following characteristics:

(i) Upon irradiation a semiconductor exhibits a photovoltage or a change of resistance.

(ii) Semiconductors have high thermo-electric power of signs both positive and negative with respect to a metal.

(iii) The junction formed between a *p*-type and an *n*-type semiconductor possesses rectification properties.

We know that semiconductors form the backbone of electronics. Transistors, integrated circuits, lasers and detectors, sensors and other semiconductor devices are indispensable in the digital era.

A transistor is a three terminal device having properties similar to those of a vacuum tube (triode), but it requires no cathode power and at the same time offers long life. The complete electronic circuit, containing many transistors, resistors, diodes, capacitors, etc, is possible to be fabricated on a single chip called an integrated circuit (IC). Recently, a new type of transistor called metal oxide semiconductor (MOS) transistor has been very useful in microelectronic circuits. The MOS transistor can be also used as a capacitor. Taking advantage of silicon integrated circuit technology, a new technology has emerged, which attempts to combine the sensor, aculator and control circuit into an integrated unit, to be called microsystem technology (MST) or micro-electromechanical system (MEMS). It is now possible to fabricate miniature mechanical devices such as gears, motors, springs, etc. This technology is doing to mechanical systems what the invention of transistors and ICs did to electronic systems. MEMS devices are small, light weight, low cost and reliable. A large number of materials are used in their fabrication. Silicon forms the backbone of these devices due to its excellent mechanical properties as well as it supports micro-fabrication techniques such as lithography, etching and bonding. As compared to electronic valves, semiconductor devices offer the following advantages: (i) low weight and small size; (ii) no power needed for the filament; (iii) long service life (thousands of hours); (iv) mechanical ruggedness; (v) low power losses, and (vi) low supply voltage. However, semiconductor devices suffer from a number of disadvantages: (i) deterioration in performance with time (ageing); (ii) higher noise level than in electronic valves; (iii) unsuitability at frequencies over tens of megahertz; (iv) low input resistance as compared with vacuum triodes; (v) inability to handle high power; (vi) deterioration in performance on exposure to radioactive emissions.

Efforts are in progress however, to eliminate or minimize many of the demerits of semiconductor devices. Replacement of germanium with silicon makes crystal diodes and transistors suitable for operation at temperatures upto 125°C. There are transistors that work at hundreds of megahertz, and also microwave devices such as gunn diode, tunnel diode, etc. The field of semiconductors is rapidly changing. We expect organic semiconductors to gather research attention during this decade. Diamond as semiconductor will also be important. Optoelectronic devices will provide three-dimensional integration of circuits, and optical computing.

Semiconductor devices impose a very small drain on power sources and make it possible to miniaturize or even micro-miniaturize components and whole circuits. The minimum power for an electronic valve is 0.1 W, while for a transistor it may be one microwatt or one hundred thousandth part of the former.

Of course, semiconductor devices will not replace electronic valves in all applications. Simply, crystal diodes and transistors may be advantageous in some uses and electronic valves in others.

Rapid progress in the fabrication of semiconductor structures has resulted in the reduction of three-dimensional bulk systems to two-dimensional, one-dimensional and finally zero-dimensional systems. These reduced dimensional systems are expected to improve semiconductor lasers and microelectronics.

Quantum Dots (QDs) represent ultimate reduction in the dimensionality of semiconductor devices. These are three-dimensional semiconductor structures, only nanometers in size, confining electrons and holes. QDs operate at the level of a single electron, which is certainly the ultimate limit for an electronic device. QDs are used in quantum dot lasers, QD memory devices, QD photodetectors and even quantum cryptography. Emission wavelength of a quantum dot is a function of its size. Obviously, by making quantum dots of different sizes, we can create light of different colours.

Semiconductors work based on the number of charges and their energy. Electronic devices such as transistors work due to flow of charge. An electron can be assumed as a tiny rotating bar magnet with two possible orientations: spin-up or spin-down. An applied magnetic field can flip electrons from one state to another. In this way, spin can be measured and manipulated to represent the 0's and 1's of digital programming. The study of electron spin in materials is called **spintronics**. Spintronics is based on the direction of spin and spin coupling.

A revolutionary new class of semiconductor electronics based on the spin degree of freedom could be created. The performance of conventional devices is limited by speed and dissipation whereas spintronic devices are capable of much higher speed at very low power.

When electron spins are aligned in one direction, these create a net magnetic moment as observed in magnetic materials like, iron and cobalt. Magnetism is an intrinsic physical property associated with the spins of electrons in a material. Electron spin may exist not only in the up or down state but also in infinite intermediate states because of its quantum nature depending on the energy of the system. This property may be useful in highly parallel computation which could make a **quantum computer** work much faster for certain types of calculations. In quantum mechanics, an electron can be in both spin-up and spin-down states, at the same time. The mixed state could form the base of a computer built around not binary bits but quantum bits or qubit. It is any combination of a 1 or a 0. A simple device using spin-dependent effect is a sandwich of two magnetic layers surrounding a non-magnetic metal or insulator. If the two magnetic layers are different, then magnetization direction of one can be rotated with respect to the other. This lends to the utilization of these structures as sensor and memory elements. Scientists and engineers are now trying to use the property of electron-like spin to develop a new generation of microelectronic devices that are more versatile and robust than silicon chips and circuit elements. Spins appear to be remarkably robust and move easily between semiconductors. In case of electron transport from one material to another, spins do not lose its orientation or scatter from impurities or structural effects.

Spintronic devices work on a simple principle that information is stored into the spins at a particular spin orientation (up or down); the spins attached to mobile electrons carry information along the wire and information is read at the terminal. Two recent discoveries include:

(i) Optically induced, long-lived coherent spin-state in semiconductors, and

(ii) Ferromagnetic semiconducting material, GaMnAs, is expected to makes revolutionary advances in **photonics** and electronics, including very fast, very dense memory and logic at extremely low power; spin quantum devices like spin-LEDs and spin-RTDs; and quantum computing in conventional semiconductors at room temperature.

To understand the working of semiconductor devices, it is necessary to study the fundamentals of semiconductor materials in greater detail.

2.2 Atomic Structure and Energy Level

We know that matter is composed of compounds and elements. Elements are the basic materials found in nature. When elements are combined to form a new material, we have a compound. The smallest particle that an element can be reduced to and still retain its properties is called an atom.

Although atoms of different elements have different properties, they all contain the same subatomic particles. A number of different subatomic particles have been discovered, but only three of these are of interest in basic electronics—the proton, the neutron and the electron.

Protons and neutrons are contained in the nucleus of the atom, and electrons revolve around the nucleus along specific orbits. Electrons and protons are particles that have electrical properties. Neutrons have no electrical charge. Usually atoms have same number of electrons and protons, and thus they are electrically neutral. If an atom has more electrons, it is called a negative ion. If it has more protons, it is called a positive ion.

Figure 2.1 shows the representation of atomic structures of H, B, Si, P and Ge. Figure 2.1(a) represents the structure of a boron atom. Its nucleus contains 5 protons and 5 neutrons. There are 5 electrons revolving around the nucleus in different orbits: 2 electrons in the first orbit, 2 electrons in the second orbit and 1 electron in the outermost or valence orbit.

Figure 2.1(b) represents the structure of a silicon atom. It contains 14 protons and 14 neutrons in the nucleus. There are 14 electrons revolving around the nucleus in different orbits: 2 electrons in the first orbit, 8 electrons in the second orbit and 4 electrons in the valence orbit.

Figures 2.1(c) and 2(d) represent the structure of phosphorus and germanium atoms, respectively. Phosphorus contains 5 electrons in the outermost orbit whereas Ge atom contains 4 electrons in the outermost orbit.

Electrons in the inner orbits of an atom do not normally leave the atom. However, electrons in the outermost orbit do not always remain confined to the same atom. Some of these valence electrons in metals move in a random manner and may travel from one atom to another in a crystal lattice. These electrons are called free electrons. It is due to the presence of these free electrons that, conduction is possible. Electrons in the inner orbits of an atom remain bound to the nucleus and are, therefore, called bound electrons.

The tendency of an atom to give up its valence electrons depends on its chemical stability. When an atom is stable, it resists giving up electrons, and when it is unstable, it tends to give up electrons. The level of stability is determined by the number of valence electrons, because the atom strives to have its outermost or valence shell completely filled.

If an atom's valence orbit is more than half filled, then the atom tends to fill its orbit. Since 8 is the maximum number of electrons that can be held in the valence orbit, elements with 5 or more valence electrons make good insulators since they tend to accept rather than give up electrons. On the other band, atoms with less than 4 valence electrons tend to give up electrons, making the valence shell empty, which allows the next shell to behave as the outermost shell. These atoms are the best electrical conductors. Si and Ges have 4 valence electrons, and are neither good conductors nor good insulators. These are called **semiconductors**.

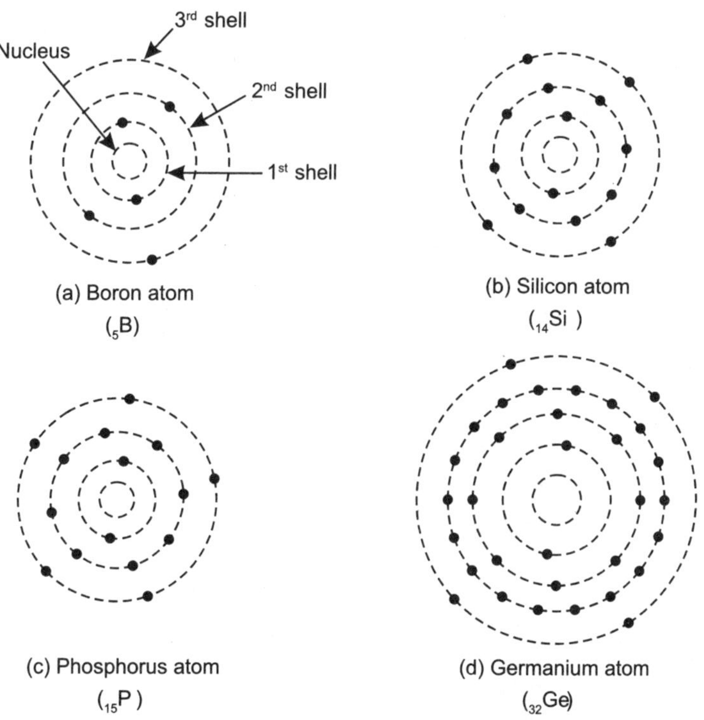

Figure 2.1 Atomic structure of few atoms

Most characteristics of a semiconductor can be studied by means of an energy level diagram. We know that each isolated atom has only a certain number of orbits available. These orbits represent energy levels for electrons in the atom. According to Bohr's theory of atomic structure, only discrete values of electron energies are possible. Electron energy is usually expressed in electron volt (1 eV = 1.6×10^{-19} J = 1.6×10^{-12} erg). An electron can have only certain permissible values. The energy level diagram for hydrogen atom is shown in Figure 2.4. The permissible energy levels for hydrogen atom are numbered $n = 1, 2, 3\ldots$ in increasing order of energy.

An electron orbiting very close to the nucleus in the first orbit is tightly bound to the nucleus and possesses only a small amount of energy. The greater the distance of an electron from the nucleus, the greater is its total energy. The total energy of an electron includes kinetic and potential energies. Obviously, an electron orbiting far from the nucleus would have greater energy, and hence it can be easily knocked out of its orbit. This makes it clear that valence electrons having the maximum energy take part in chemical reactions and in bonding the atoms to form solids.

When radiations impinge on an atom, the energy of electrons increases. As a result, electrons are excited to higher energy levels. The excited state does not last long and the electron after emitting out energy in the form of heat/light fall back to the original energy level.

2.3 Energy Bands

The crystalline structure is common among most metals and semiconductors. We know that a crystal is made up of a space array of molecules in regular repetition in three dimensions of some fundamental building block. For a gaseous element, electronic energy levels are the same as for a single free atom because individual atoms in a gas are well apart and have negligible influence on each other. However, in a crystal individual atoms are so closely packed that the resulting energy levels are modified due to interactions between atoms. When atoms form crystals, energy levels of inner-shell electrons are not appreciably affected, however, the levels of outer-shell electrons are considerably altered as these electrons are shared by the adjacent atoms in the crystal. As a result of the interaction between outer-shell electrons, energy levels spread up to form a band of energy as shown in Figure 2.2.

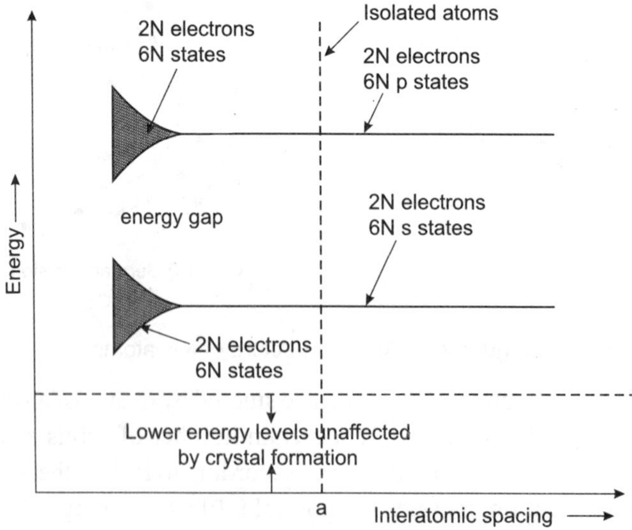

Figure 2.2 Formation of energy bands in a crystal

In order to understand the formation of energy bands in a crystal, let us consider a silicon crystal made up of N atoms. We shall assume that the interatomic spacing can be varied without affecting the fundamental crystal structure. For very large interatomic spacing, say '*a*' in Figure 2.2, the interaction between adjacent atoms is negligible, and energy levels are same as that of isolated atoms. In silicon, the outermost subshells, namely 3s and 3p, contain 2 electrons each. In a silicon crystal consisting of N atoms, the outermost subshells 3s and 3p consist of 2N electrons each. Thus, the 3s sub shell has 2N electrons completely occupying the available 2N states, and the 3p subshell has only 2N electrons partially occupying the available 6N states, all at the same energy level.

If interatomic spacing is gradually decreased, i.e. moving from right to left in Figure 2.2, there will be a gradual increase in the interaction between neighbouring atoms. Due to this interaction, atomic wave functions overlap, and the crystal becomes an electronic system that obeys Pauli's exclusion principle. Hence the 2N *s* states spread out to form a band of energy. Actually, the separation between levels is small but since N is very large, of the order of 10^{23} cm^{-3}, the total spread between minimum and maximum energy levels becomes large. This spread will have several electron volts of energy and is referred to as energy band as indicated by the lower shaded region in Figure 2.2. The 2N states in this band are completely filled with 2N electrons.

Similarly, at the same value of interatomic spacing at *p*-level, 6N *p* states spread up to form a band. This band is shown as the upper shaded region in Figure 2.2. Although a total of 6N states are available in this band, only 2N states are occupied and 4N states remain unoccupied.

Energy gap exists between the two energy bands. This energy gap is called **forbidden energy gap,** as no electrons can occupy any state in this gap. This forbidden energy gap decreases as atomic spacing is decreased and becomes zero with further reduction in interatomic spacing, say at '*b*', as shown in Figure 2.3. That is, the two energy bands will overlap when interatomic spacing is small enough. Under such circumstances, the 6N *p* states in the upper band merge with the 2N *s* states in the lower band, giving a total of 8N states. Half of these 8N states are occupied by 4N available electrons. These 4N electrons no longer belong to either *p* subshell or s subshell but belong to the crystal as a whole. Thus, at this interatomic spacing, each atom in the crystal can contribute 4 electrons to the crystal. The band occupied by these contributed electrons is called valence band.

If interatomic spacing is reduced further, say '*c*' as shown in Figure 2.3, the interaction between atoms becomes extremely large, and the energy band structure assumes the shape shown in Figure 2.4. The exact energy band structure depends upon (i) the orientation of atoms relative to one another in space and (ii) the atomic number of the atom, and may be obtained from the solution of Shrodinger's wave number. For spacing '*c*', 4 *n* states in the valence band are completely filled by 4 *n* electrons, and this valence band is separated from the upper unfilled or empty band by a forbidden energy gap E_g. The forbidden energy gap contains no allowed states. The upper band has 4N unfilled states and is referred to as the conduction band.

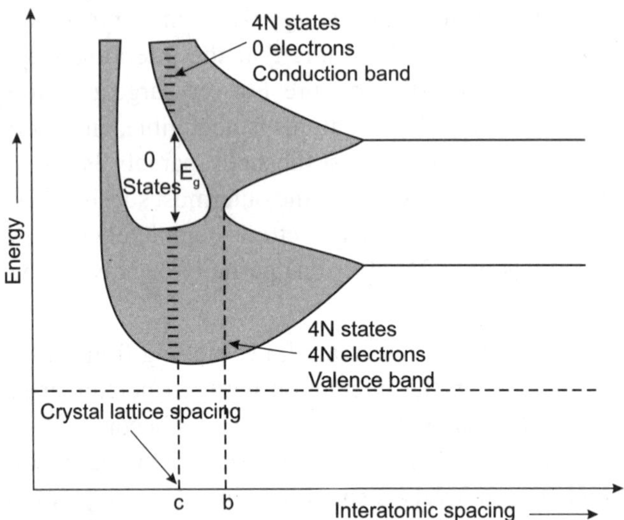

Figure 2.3 Energy band in a crystal as a function of interatomic spacing

The simple energy level diagram of hydrogen (Figure 2.4) for electron energies is no longer applicable for solids. A solid is formed when atoms bond together. In a solid, the orbit of an electron is influenced not only by the charges in its own atom but by electrons and nuclei of every atom in the solid. Each electron in a solid occupies a different position inside the solid and hence no two electrons can have exactly the same pattern of surrounding charges. Obviously, the orbits of electrons in a solid are different.

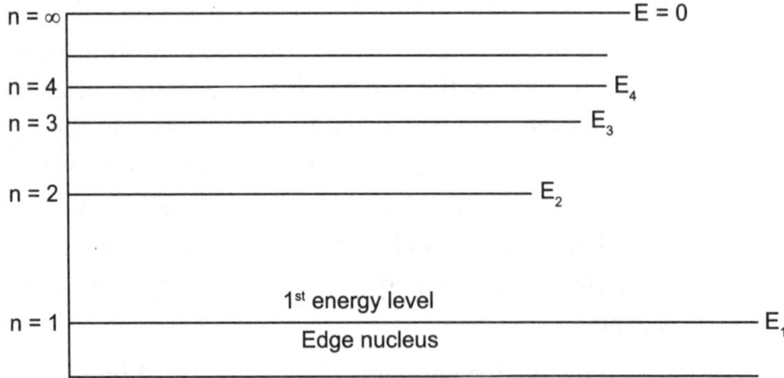

Figure 2.4 Energy levels of an isolated hydrogen atom

When a solid in bulk is considered, the simple energy level diagram in Figure 2.4 modifies to that shown in Figure 2.5. All electrons belonging to the first orbits have slightly different energy levels because no two electrons can have the same charge environment. Since there are billions of first-orbit electrons, the slightly different energy levels form a group or band. Similarly, the billions of second-orbit electrons, all with slightly different energy levels, form the second energy band. Similarly, all third-orbit electrons form the third band.

Silicon is a commonly used semiconductor with atomic number 14. Obviously, it has 4 electrons in its outermost orbit. Clearly, the third band becomes the valence band. In Figure 2.5, all the three bands are shown completely filled. Although the third shell of an isolated atom of silicon is not completely filled as it has only 4 electrons whereas it could accommodate a maximum of 8 electrons, the third energy band or valence band of a bulk silicon material behaves as if completely filled. It is so, because in solid silicon each atom positions itself between four other silicon atoms, and each of these neighbours share an electron with the central atom. In this manner, each atom now has 8 electrons associated with it thereby filling the valence band completely, i.e. all permissible energy levels in the band are occupied by electrons. Obviously, no electron in a filled band can move and hence an electron in a completely filled band cannot contribute to electric current. At absolute zero temperature, electrons cannot move through the solid silicon material and hence it acts like a perfect insulator. Beyond the valence band there is a conduction band. At absolute zero temperature, the conduction band is empty, i.e., no electron has enough energy to go into a conduction band. Definite amount of energy is needed to shift an electron from the valence band to the conduction band. This amount differs from one substance to another and helps to classify them as conductors, insulators and semiconductors.

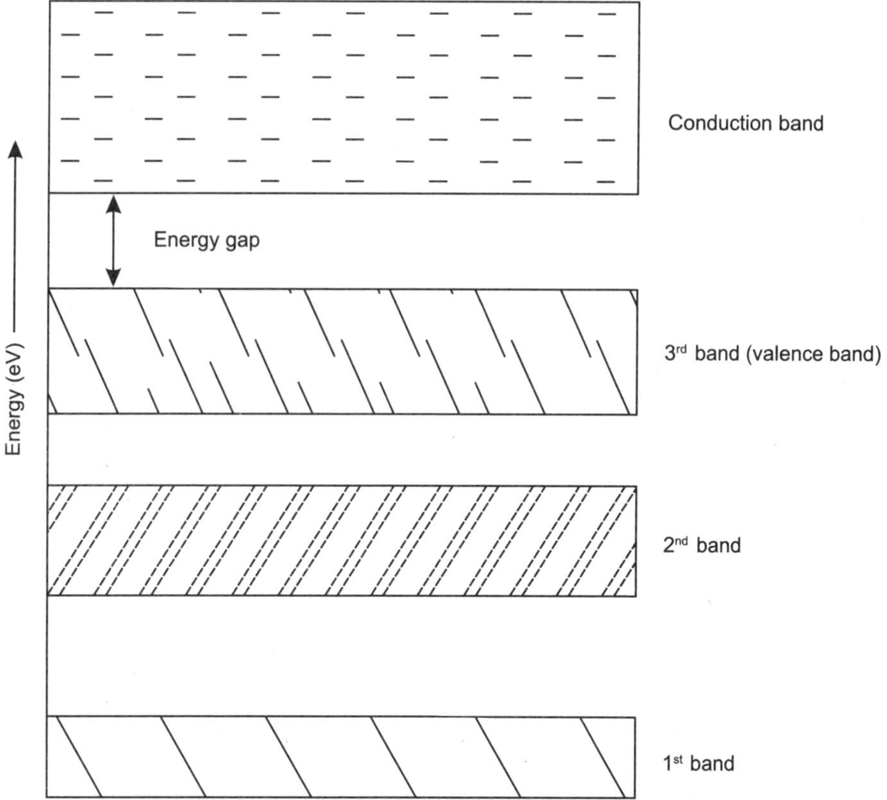

Figure 2.5 Energy bands of silicon at absolute zero

2.4 Conductors, Semiconductors and Insulators

The band structure determines whether a solid is an insulator, a conductor or a semiconductor. The bands are filled upto a certain level by electrons within each atom. The highest band in which electrons are still predominantly attached to their atoms is called the valence band. This is the band in which outermost electrons from each atom will be located. These electrons are possible carriers of electricity. However, in order for an electron to conduct, it must get up to a slightly higher energy so that it is free from the grip of its atom.

Insulator Assume the valence band is full, i.e. there are no more energy levels available. Then, a valence electron must jump (increase its energy) into the next higher band in order to be free. If energy gap between the valence band and the conduction band is too large, then the electron will not be able to make that jump. Such a material will not be a good conductor of electricity, and is called an insulator. Forbidden energy gap in an insulator is about 5 eV or more (Figure 2.6(a)). The band theory of solids tells us that an insulator is a material in which the valence bands are filled and forbidden energy gap between the valence band and the conduction band is too great since the valence electrons cannot jump at normal temperatures from VB to CB. An insulator does not conduct at room temperature because there are no conduction electrons in it. However, an insulator may conduct if its temperature is very high or if high voltage is applied across it. This is known as breakdown of the insulator.

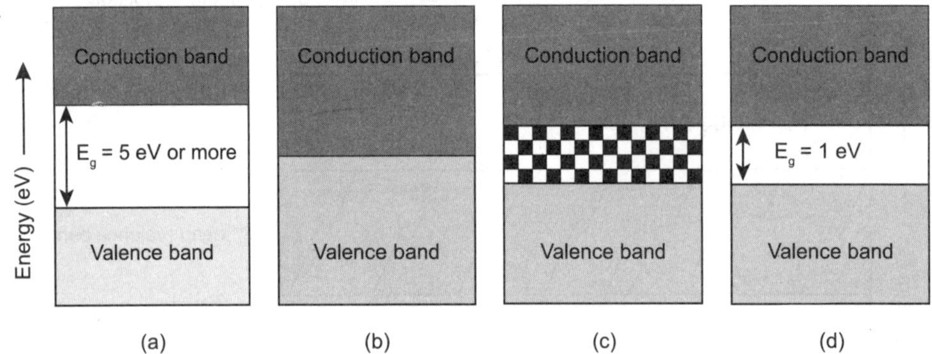

Figure 2.6 Energy band diagram for (a) insulators, (b) and (c) conductors, and (d) semiconductors

Conductor There are two possibilities for a substance to behave as a conductor. One is that its valence band is not completely filled. Then an electron in the valence band can get free of its atom by simply jumping to a higher energy level within the same band. This jump requires a small amount of energy, and many electrons can make that jump (Figure 2.6(b)). Another situation in which a conductor can result arises from the fact that the size of energy gaps between the valence band and the conduction band is very small and different for different materials. It can even disappear when the valence band and the conduction band overlap (Figure 2.6(c)). Therefore, the band theory tells us that we have a conductor when

(i) The valence band is not filled, and so electrons can move to higher states in the valence band and be free, or

(ii) When there is no energy gap between the valence band and the conduction band, and so electrons can easily make the transition from valence to conduction band, e.g., metals.

Semiconductor A material with intermediate properties is called a semiconductor. In a semiconductor,

(i) The valence band is filled, and

(ii) Although there is energy gap between the valence band and the conduction band, yet the energy gap is not very large.

The energy band diagram for a semiconductor is shown in Figure 2.6(d). In this case, the forbidden energy gap is of the order of 1 eV (for silicon, E_g = 1.12 eV and for germanium, E_g = 0.72 eV). Silicon and germanium are very good semiconductors. The gaps between their valence and conduction bands are sufficiently small that normal thermal energy of the solid at room temperature is enough to knock a few electrons into the conduction band. These electrons conduct current, but not as much as a real conductor. The conductivity of these substances has a number of peculiarities:

(i) The dependence of conductivity on temperature is opposite to that of metals. The conductivity of semiconductors, in contrast to that of conductors, may decrease rapidly with temperature. At low temperatures, a semiconductor may become an insulator. Hence the distinction between semiconductors and insulators is purely quantitative and, to a large extent, conventional. The resistance of most semiconductors is considerably more sensitive to changes in temperature than for metals. Compact temperature meters may be constructed using semiconducting thermal resistors (thermistors).

(ii) In a number of cases, semiconductors may possess positive as well as negative temperature coefficients.

2.5 Fermi Level

As far as solid-state theory is concerned, only the upper energy bands (valence bands) are of interest, since electrons at lower levels practically do not take part in interactions among atoms. Since we are dealing with a very large number of electrons, it is natural to use statistical physics methods and consider an aggregate of such electrons as a kind of gas, usually called electron gas.

The state of each electron in such a gas may be represented by a point (p_x, p_y, p_z) in momentum space. The direction of motion of an electron is parallel to its radius vector P and the energy of an electron depends on its momentum. Let us consider a crude approximation that electrons in a solid behave like free particles, i.e. we neglect the potential energy of the field in which electrons move and the interaction between electrons.

If electrons are free, the relationship between their energy and momentum is given by $E = \left(\frac{1}{2}m\right)P^2$. This means that in momentum space, a surface of total energy is a sphere. Such a sphere is usually called a Fermi sphere. One may call the Fermi surface of maximum energy as the states of an electron gas are contained in a sphere of radius

$$P_{max} = \sqrt{2mE_{max}}$$

It is important to determine how electrons distribute themselves in a band among energies from zero up to E_{max}. The number of electrons per unit volume that can be accommodated up to energy E is given by

$$n = \frac{16\pi(2m_e^3)^{1/2}}{3h^3}E^{3/2} \tag{1}$$

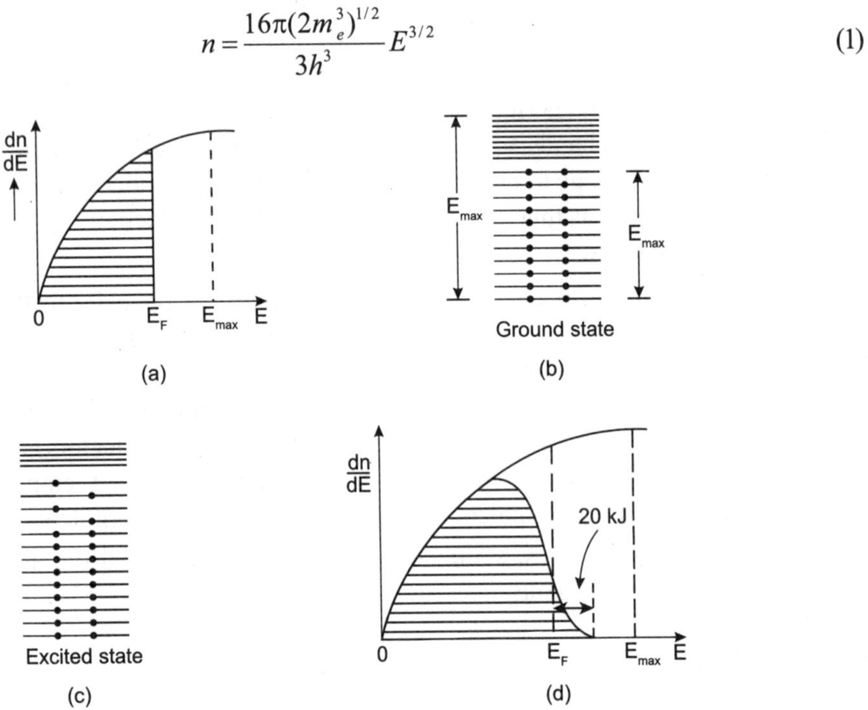

Figure 2.7 (a) Density of energy states of free electrons in a solid; (b) and (c) Distribution of free electrons among energy states in the conduction band; (d) Occupation of energy states at a temperature different from absolute zero

If a metal is in its ground state, which occurs at absolute zero, all electrons occupy the lowest possible energy levels according to Pauli exclusion (i.e., each can accommodate two electrons: one with spin-up and one with spin-down) as indicated in Figure 2.7(b). If the total number of electrons per unit volume, n_0, is less than the total number of energy levels available in the band, the electrons will then occupy all energy states up to a maximum energy, designated by E_F, called Fermi energy. Thus, Fermi energy E_F is defined as energy of the topmost filled level in the ground state of n electrons. If we set $E = E_F$ in Eq. (1), we must have $n = n_0$. Therefore, for Fermi energy, we obtain the value

$$E_F = \frac{h^2}{8m_e}\left(\frac{3n_0}{\pi}\right)^{\frac{2}{3}} \tag{1(a)}$$

Energy distribution of electrons in a metal in ground state corresponds to the shaded area in Figure 2.7(a). When Fermi energy is equal to energy bandwidth, the band is fully occupied.

When the band is not completely filled, a small amount of energy is enough to excite the uppermost electrons to nearby energy levels as indicated in Figure 2.7(c). However, only the uppermost electrons can be thermally about 0.025 eV, which is very small compared with E_F, and the exclusion principle makes it impossible for low-energy electrons to be excited into nearby occupied states. The distribution of electrons among the energy level in a thermally excited state of the lattice corresponds to the shaded area in Figure 2.7(d). Electrons that have been thermally excited are those with energy greater than E_F. The states occupied by excited electrons fall in an energy region of the order of 20 kT about E_F.

We can now conclude that the concept of Fermi level serves the reference energy level from which all other energies are conveniently measured. The probability $F(E)$ of a state corresponding to energy E being occupied by an electron at temperature T K is given by

$$F(E) = \frac{1}{1 + \exp\left(\frac{E - E_F}{kT}\right)} \tag{1(b)}$$

Here k is Boltzmann constant. Three cases of interest are:

At $T = 0$ K, if $E > E_F$ then $F(E) = 0$, i.e., energy state is empty.

At $T = 0$ K, if $E < E_F$ then $F(E) = 1$, i.e., energy state is occupied by an electron.

At $T \neq 0$ K, and $E = E_F$ then $F(E) = 1/2$, i.e., energy state is 50% occupied.

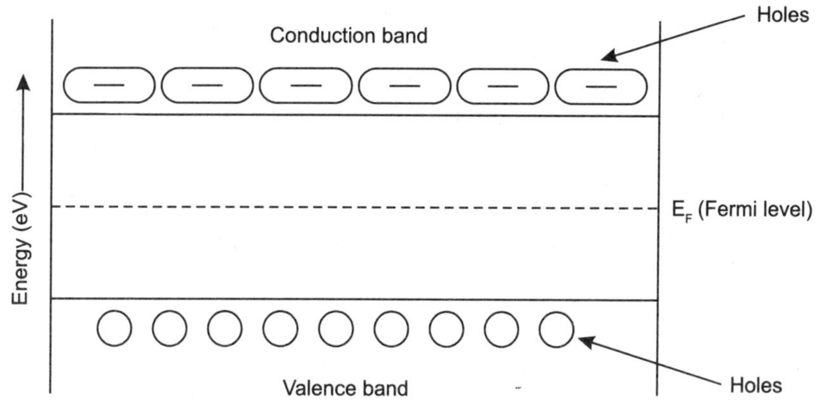

Figure 2.8 Fermi level in an intrinsic semiconductor at 0 K

Obviously, one finds that at $T = 0$ K, all energy states above E_F are empty, whereas all those below E_F are filled with electrons. With increase in temperature, states can no more remain empty. They are occupied by electrons to some extent. Figure 2.8 shows the position of Fermi level in energy band diagram in the case of pure or intrinsic semiconductors. It depicts that the Fermi level for an intrinsic semiconductor is situated in the middle of forbidden gap (See Eq. (9a)), i.e. between conduction and valence band, and the position of Fermi level is independent of temperature. When temperature is raised, there is a greater possibility of electrons being found above the Fermi level with an equal possibility of finding a 'hole' below the Fermi level. We will discuss it in later sections.

2.6 Intrinsic Semiconductors

Semiconductor devices are made from a single crystal of semiconductor material, e.g. germanium or silicon. A semiconductor in its purest form is called an intrinsic semiconductor. A semiconductor is not truly intrinsic unless its impurity content is less than 1 part in 100 million parts of the semiconductor. At 0 K, Fermi level (E_F) in intrinsic semiconductors lies nearly midway between the valence and conduction band.

The conductivity of an intrinsic semiconductor is almost zero at 0 K and it behaves like an insulator.

As temperature is increased some electrons near the top of the valence band are thermally excited into the lowest levels of the conduction band. The conduction band now contains a few electrons, and the valence band has a few missing electrons or holes*. Thermal ionization, therefore, is said to create hole-electron pairs, since there is one hole for each electron. As both electrons and holes transport charge through the semiconductor, they are collectively termed charged carriers. In applied electric and magnetic fields, a hole acts as if it is a positive electron. To understand the phenomenon of conduction of current in a semiconductor, it is essential to study its crystal structure.

2.6.1 Crystal Structure of Semiconductor

It is the nature of some solid materials to form crystals which have characteristic geometric shapes. These are crystalline substances. On the other hand, solids which form into shapeless masses are said to be plastic, noncrystalline, or amorphous. Quartz is a familiar example of a crystalline substance whereas a block of rubber is amorphous. Elements and compounds can be crystalline, so are metals and nonmetals. Virtually all minerals are crystalline.

Externally, a crystal has several flat faces arranged symmetrically w.r.t. each other. Internally, it has a orderly arrangement of atoms in a repeating system called a lattice. Both externally and

---

* When an electron moves from the valence band, it leaves behind a vacant energy state with a positive charge. Moreover, the electron moving from the valence band has a negative effective mass and its absence in the valence band is associated with a positive mass. The vacant energy state in the valence band, therefore, has a positive charge and a positive mass. The vacant energy state is called a 'hole'.

internally, a single crystal of a given true crystalline material looks like all other crystals of that material. A single crystal may be very large or it may be so small to be visible only with the aid of a magnifier.

Inside the crystal lattice, certain loosely bound electrons (called valence electrons) in the outer orbit of one atom align themselves with similar electrons in adjacent atoms to form covalent bonds which hold the atoms together in an orderly structure in the lattice. Thus, in any covalent bond, one can find shared electrons.

Germanium and silicon (Si) are typical semiconductors. In germanium an atom has 32 electrons distributed in four orbits whereas in silicon there are 14 electrons distributed in three orbits. The outermost orbit called the valence orbit contains 4 electrons in each case (Figure 2.1). The crystals are tetrahedral in structure and each atom shares one of its electron with its neighbour. Such a sharing of electrons between two neighbouring atoms is called a covalent bond. A simplified representation of the crystalline structure of a Ge semiconductor at absolute zero is shown in Figure 2.9.

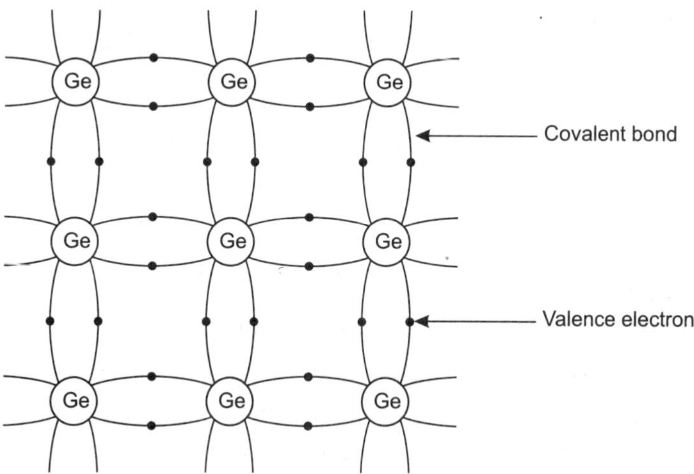

Figure 2.9 A simplified representation of the crystalline structure of a semiconductor (Ge) at absolute zero

At near absolute zero, all the electrons in the atoms are tied up strongly by these bonds, but with increase in temperature, a few electrons break away from some of the covalent bonds and get themselves freed, creating vacant spaces deficient of electrons known as holes. **A hole is equivalent to a net positive charge equal to that of electron**. Whenever a free electron is generated, a hole is created simultaneously, i.e. free electrons and holes are always generated in pairs. Obviously, the concentration of free electrons and holes will always be equal in an intrinsic semiconductor. This type of generation of free electron–hole pairs in semiconductors is called thermal generation.

Owing to thermal vibrations the free electrons in a semiconductor crystal jump from one bound position to another. This is equivalent to the motion of a hole relatively in opposite direction and thus gives an electric current in the direction of motion of the hole. Conductivity in a semiconductor is, therefore, due to the motion of holes and of electrons (Figure 2.10(b)). Such intrinsic semiconductors produce very weak electric current.

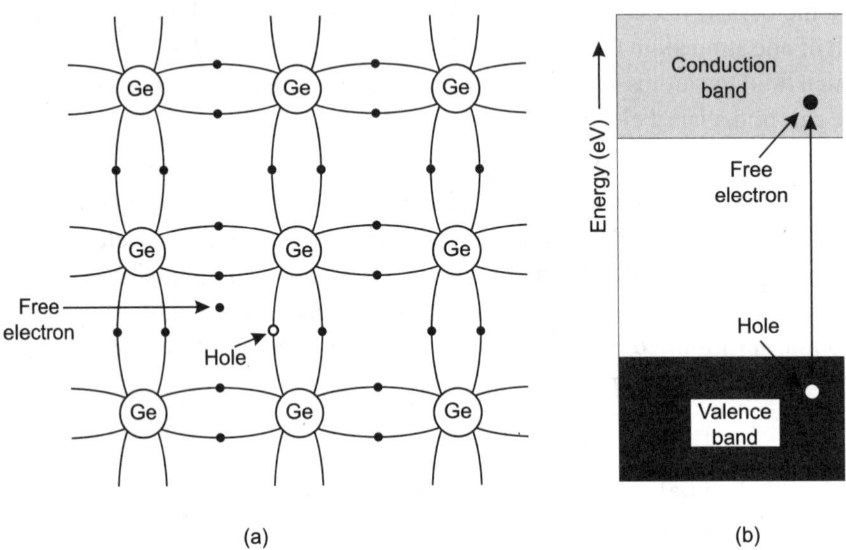

Figure 2.10 (a) Crystalline structure of Ge at room temperature; (b) generation of electron-hole pair in an intrinsic semiconductor (Ge)

2.6.2 Carrier (or Electron and Hole) Concentrations in Intrinsic Semiconductors

In order to calculate intrinsic carrier concentration, we first calculate the number of electrons excited into the conduction band at any temperature (T K) and which in turn are free to migrate in the crystal lsttice. In carrying out these calculations, it will be assumed that electrons in the conduction band behave as if they are free with effective mass m_e^*, and energy will be measured from the top of the valence band. We further take electron density of states in the conduction band equal to that for free electrons, i.e.

$$N(E) = \frac{1}{(2\pi)^2}\left(\frac{2m_e^*}{h^2}\right)^{\frac{1}{2}}(E - E_g)^{\frac{1}{2}} \tag{2}$$

In addition to the density of states function, we also require the probability function for calculating electron density in the conduction band. The appropriate probability function is the Fermi function $F(E)$ given by Eq. (3). As stated earlier, Fermi level E_F is in the middle of the band gap for an intrinsic semiconductor. If $(E - E_F) < 4\,kT$, i.e., the lower edge of the conduction band is about $4k_BT$ above E_F, one can neglect 1 in the denominator of Eq. (1), and thus Fermi function $F(E)$ reduces to

$$F(E) = \frac{\exp(E_F - E)}{kT} \tag{3}$$

The number of electrons per unit volume having energy in the range dE in the conduction band can be obtained from

$$n_e(E)\,dE = N_e(E)F(E)dE \tag{4}$$

$$\text{or } n_e = \int_{E_g}^{\infty} n_e(E)dE$$

This leads to

$$n_e = N_e \exp\left(\frac{E_F - E_g}{kT}\right) \tag{5}$$

Here

$$N_e = 2\left(\frac{2\pi m_e^* kT}{h^2}\right)^{\frac{3}{2}}$$

is called effective density of states in the conduction band. Equation (5) gives us electron concentration in the conduction band.

In order to calculate hole density n_h where h refers to holes, we assume that holes near the top of the valence band behave as if they are free particles with effective mass m_h^*. One can express the density of hole states in the valence band as

$$N_h(E)dE = \frac{1}{2\pi^2}\left(\frac{2m_h^*}{h^2}\right)^{\frac{3}{2}}(-E)^{1/2}\,dE \text{ per unit volume}$$

Here, one must remember that energy is measured positive upwards from the top of the valence band. If $f_h(E)$ represents the probability of occupation of holes in the valence band, then it must be equal to the probability of electron states being unoccupied in the valence band, i.e.,

$$f_h(E) = 1 - f_e(E) = 1 - \frac{1}{\exp[(E - E_F)kT] + 1}$$

$$= \frac{\exp[(E - E_F)/kT]}{\exp[(E - E_F)/kT] + 1} \tag{6}$$

Since $E < E_F$, being in the valence band, so exp $[(E - E_F)/kT] << 1$ and hence we can always neglect exp $[(E - E_F)/kT]$ in comparison to 1 in the denominator and one obtains

$$F_h(E) = \exp\left(\frac{E - E_F}{kT}\right) \tag{7}$$

Using the relation $n_h(E)dE = N_h(E)F_h(E)dE$, one obtains hole concentration in the valence band as

$$n_h = N_v \exp\left(\frac{-E_F}{kT}\right) \tag{8}$$

where

$$N_v = 2\left[\frac{2\pi m_h^* kT}{h^2}\right]$$

is called effective density of states in the valence band.

For intrinsic semiconductors, we have $n_e = n_h$ and, therefore, from Eqs. (5) and (8), one obtains

$$m_e^{*3/2} \exp\left(\frac{(E_F - E_g)}{kT}\right) = m_h^{3/2} \exp\left(\frac{-E_F}{kT}\right)$$

or

$$E_F = \frac{E_g}{2} + \frac{3}{4} kT \log\left(\frac{m_h^*}{m_e^*}\right) \tag{9}$$

At $T = 0$ K, Eq. (9) reduces to

$$E_F = \frac{E_g}{2} \tag{9a}$$

2.6.3 Fermi Level in Intrinsic Semiconductors

From Eq. (9) it follows that at $T = 0$ K, $m_e^* = m_h^*$, and Fermi level in the intrinsic semiconductor lies exactly mid-way in the forbidden gap, i.e.

$$E_F = \frac{E_g}{2}$$

Obviously, Fermi level E_F, which can be the highest occupied energy level at $T = 0$ K, lies near the middle of the band gap for an intrinsic semiconductor (Figure 2.10(c)). The Fermi-Dirac (FD) distribution function $f(E)$ at 0 K is plotted in Figure 2.10(d).

In general, $m_h^* > m_e^*$, i.e. m_h^* is slightly greater than m_e^* and so the Fermi level is raised slightly as temperature increases. However, for all practical purposes, the Fermi level in intrinsic semiconductors can be assumed to be constant over a wide range of temperatures. It is worthwhile to mention that Fermi level in semiconductors is not static, but a dynamic one, since it changes appreciably with change in temperature and presence of impurities. Variation of Fermi level with impurity concentration enables the operation of various semiconductor devices.

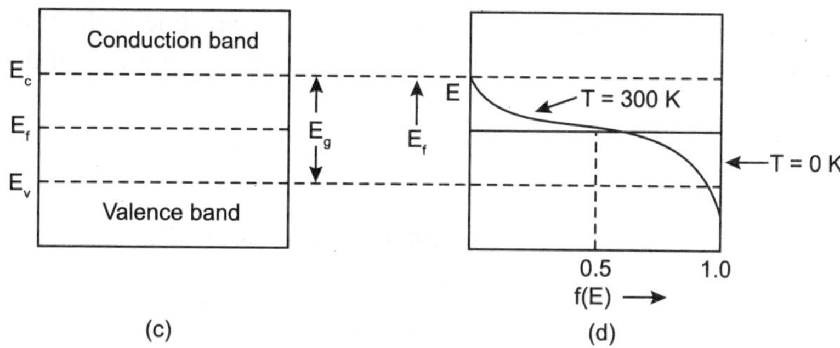

Figure 2.10 (c) Energy band diagram (d) FD distribution function *f*(*E*) for an intrinsic semiconductor

2.6.4 Law of Mass Action

Multiplying Eqs. (5) and (8), one obtains.

$$np = n_e n_h = 4\left(\frac{2\pi kT}{h^2}\right)^3 (m_e^* m_h^*)^{\frac{3}{2}} \exp\left(-\frac{E_g}{kT}\right)$$

where $n_e = n$ and $n_h = p$ and E_g is band gap energy. Since E_g is constant and hence the product np is also constant, i.e.

$$np = n_i^2 = \text{const}\left[T^3 \exp\left(\frac{-E_g}{kt}\right)\right] \tag{10}$$

Here n_i is called intrinsic density of either carriers. From Eq. (10), it is evident that n_i^2 or np is a constant depending on the temperature and width of the forbidden gap. It does not depend on the impurities introduced as long as the impurities do not change the width of forbidden energy gap. Equation (10) is called the law of mass action. The result also holds good in the presence of impurities.

Because np is a constant independent of impurity concentration at a given temperature, the introduction of a small proportion of a suitable impurity to increase n must decrease p. This result is important in practice as one can reduce total carrier concentration $(n + p)$ in an impurity crystal, sometimes enormously, by controlled introduction of suitable impurities. Such a reduction is called compensation of one impurity type by adding another.

2.6.5 Electrical Conductivity

Electrical conductivity of intrinsic semiconductors at very low temperature (~0 K) is due to intrinsic charge carriers, i.e. due to electrons and holes. Such conductivity is sometimes termed as intrinsic conductivity.

Since there are two types of carriers in an intrinsic semiconductor (i.e., electrons and holes), its specific conductance is the sum of conductivity $\sigma_e = |e| n\mu_e$ due to free electrons, with the concentration n and mobility μ_e, and of conductivity $\sigma_h = |e|\, p\mu_h$ due to the presence of holes, with concentration p and mobility μ_h. Mobility is defined as the magnitude of drift velocity per unit electric field

$$\mu = \frac{|v|}{E} \tag{11}$$

Mobility is defined to be positive for both electrons and holes, although their drift velocities are opposite.

Electrical conductivity of an intrinsic semiconductor is given by

$$\sigma = |e|(n\mu_e + p\mu_h) \tag{12}$$

where n is density of charge carriers and μ is mobility of charge carriers. Obviously the conductivity of semiconductors depends on (i) the density of charge carriers and (ii) the mobility of charge carriers. It is well known that if density of charge carriers in an intrinsic semiconductors is proportional to $\exp\left(\frac{-E_g}{k_B T}\right)$, then mobility of charge carriers

$$n = n_o \exp\left(\frac{-E_g}{k_B T}\right)$$

where k_B is Boltzmann constant and E_g is forbidden energy gap.

Since for an intrinsic semiconductor $n = p$, we have

$$\sigma_i = |e|\, n(\mu_e + \mu_h) \tag{13}$$

Here σ_h denotes intrinsic conductivity. Substituting the value of n, one obtains

$$\sigma_i = 2|e|\left(\frac{2\pi kT}{h^2}\right)^{\frac{3}{2}} (m_e^* m_h^*)^{\frac{3}{4}} \exp[-E_g / 2kT](\mu_e + \mu_h) \tag{14}$$

It is worthwhile to mention that the value of E_g is more in the case of silicon ($E_g = 1.12$ eV) than in the case of germanium ($E_g = 0.72$ eV). Obviously, less number of electron-hole pairs will be generated in silicon than in germanium at room temperature. This means that conductivity of silicon will be less than that of germanium at room temperature.

When a battery is connected across a semiconductor (Figure 2.11), electrons experience a force towards the positive terminal of the battery, and holes towards the negative terminal. The random motion of electrons and holes gets modified. Over and above the random motion, there also occurs a net movement called drift. The random motion does not contribute to any electric current. Free electrons drift towards the positive terminal of the battery, while holes towards the negative terminal. Electric current flows through the semiconductor in the same direction as the holes are moving. Since electrons are negatively charged, the direction of conventional electric current is opposite to the direction of their motion as shown in Figure 2.11. Although the two types of charge carriers move in opposite directions, the two currents are in the same direction.

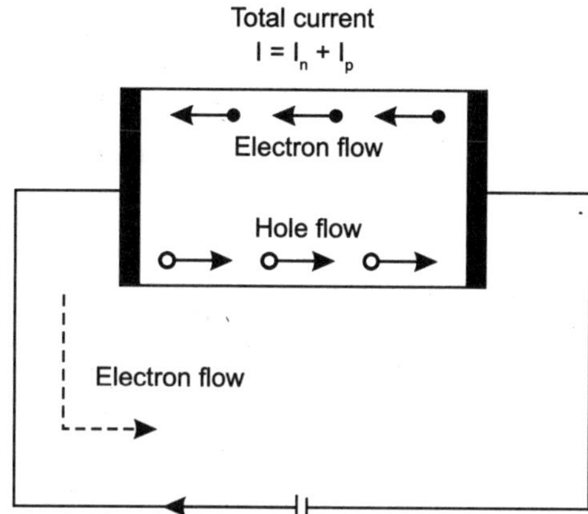

Figure 2.11 Electric current in an intrinsic semiconductor

When the flow of charge carriers in a semiconductor is due to an applied voltage, the resultant current is called drift current. A second type of current called diffusion current also exists in a semiconductor. Diffusion current flows as a result of a gradient of carrier concentration, i.e. the difference of carrier concentration from one region to another in a semiconsuctor. Diffusion current is also due to the motion of both electrons and holes.

Let us now investigate the effect of temperature on intrinsic semiconductors. From Eq. (14), it is evident that the exponential term $\exp\left(\frac{-E_g}{2kT}\right)$ dominates all other temperature dependence. Writing Eq. (14) as

$$\log\sigma_i = -\frac{E_g}{2kT} + \log\left[2|e|\left(\frac{2\pi kT}{h^2}\right)^{\frac{3}{2}}(m_e^* m_h^*)^{\frac{3}{4}}(\mu_e + \mu_h)\right]$$

or $$\log\rho = -\log\sigma_i$$

$$= -\frac{E_g}{2kT} - \log\left[2|e|\left(\frac{2\pi kT}{h^2}\right)^{\frac{3}{2}}(m_e^* m_h^*)^{\frac{3}{4}}(\mu_e + \mu_h)\right] \quad (15)$$

where ρ is resistivity.

An intrinsic semiconductor usually has a small conductivity. In a sample of germanium at room temperature, intrinsic carrier concentration is $2.5 \times 10^{19}/\text{m}^3$. With increase in temperature, conductivity increases, i.e. resistivity decreases. Figure 2.12 shows a plot of log ρ vs. *I*/*T* for some intrinsic semiconductors.

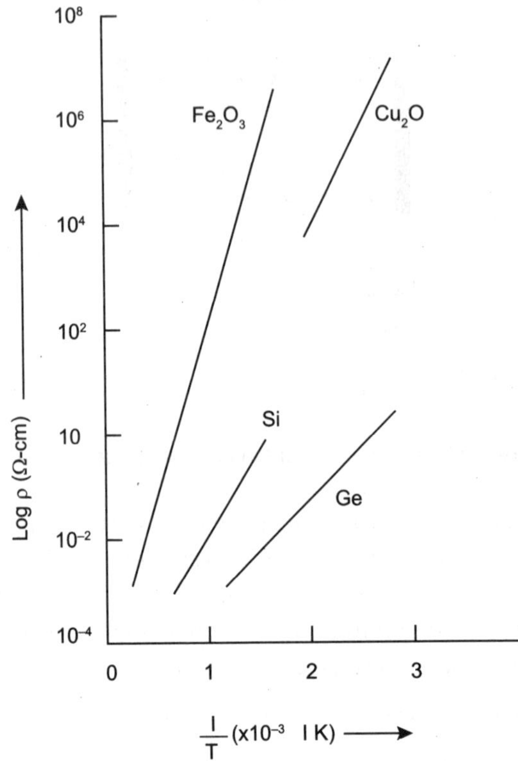

Figure 2.12 Plot of log ρ vs. 1/T for some intrinsic semiconductors

2.7 Extrinsic Semiconductors

We have seen that the conductivity of intrinsic semiconductors is very small and hence it may not be suitable for any useful work except as heat- or light-sensitive resistances. Conductivity can, however, be enormously increased by the addition of suitable impurity in a very small proportion, i.e., nearly 1 in 10^8 parts of the semiconductor. In general, only one impurity atom is added to about 10^8 atoms of the intrinsic semiconductor. This impurity rate is so low that it does not significantly affect the crystal structure of the pure semiconductor. This process is called doping. Doping is done after the semiconductor material has been refined to a high degree of purity. A doped or impure semiconductor is known as an extrinsic semiconductor.

2.7.1 N-type Semiconductors

Germanium and silicon are tetravalent atoms. However, impurity atoms may be either pentavalent or trivalent, i.e. from group V and III of the periodic table. If a small quantity of a pentavalent impurity (having 5-electrons in the outermost orbit) like Arsenic, Antimony or Phosphorus is introduced in Germanium, it replaces equal number of Germanium atoms without changing the physical state of the crystal. Four out of five valency electrons of the impurity, say of Arsenic, enter into a covalent bond with Germanium, while the fifth valency electron is set free to move from one atom to the other as shown in Figure 2.13. The impurity is called donor impurity as it donates electrons and the crystal is called N-type semiconductor. A small amount of Arsenic (impurity) injects billions of free electrons into Germanium, thus increasing its conductivity enormously. In N-type semiconductors, the majority carriers of charge are electrons, and holes are minority carriers. This is because, when donor atoms are added to a semiconductor, the extra free electrons give the semiconductor a greater number of free electrons than it would normally have. Donor electrons do not produce holes. As a result, current carriers in a semiconductor doped with pentavalent impurities are primarily negative electrons. The impurity atom has five valence electrons. After donating one electron, it is left with +1 excess charge. It then becomes a positively charged immobile ion. It is immobile because it is held tightly in the crystal by four covalent bonds as shown in Figure 2.13.

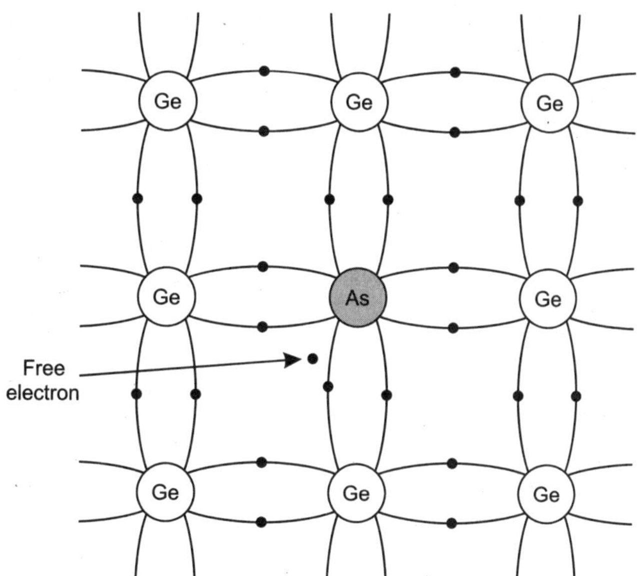

Figure 2.13 N-type semiconductor

In an N-type semiconductor, although electrons (negative charges) are majority carriers, the semiconductor doped with impurity remains electrically neutral. Free electrons and holes are generated in pairs due to thermal energy, and negative charge of electrons donated by impurity atoms is exactly balanced by positive charge of immobile ions. Representation of an N-type semiconductor in shown in Figure 2.14. One should assume it as a continuous structure over the whole background. The fixed or immobile ions are regularly distributed in the crystal structure. The electrons and holes, being free to move, are shown randomly distributed at any moment.

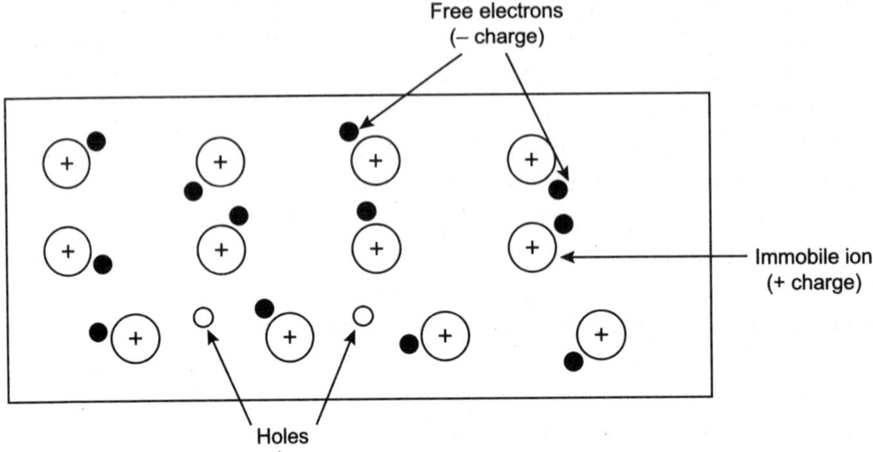

Figure 2.14 Representation of an N-type semiconductor

Since N-type semiconductors have extra free electrons, and pure semiconductors do not, the energy band diagram for a doped semiconductor is slightly different from that of a pure semiconductor. In effect, another energy level exists, a level for the donor electron E_{FN}, which is closer to the conduction band. The forbidden band for the donor electron is much narrower than the forbidden band for the valence electron, and thus one can conclude that it is much easier to cause electron flow in an N-type semiconductor (Figure 2.15).

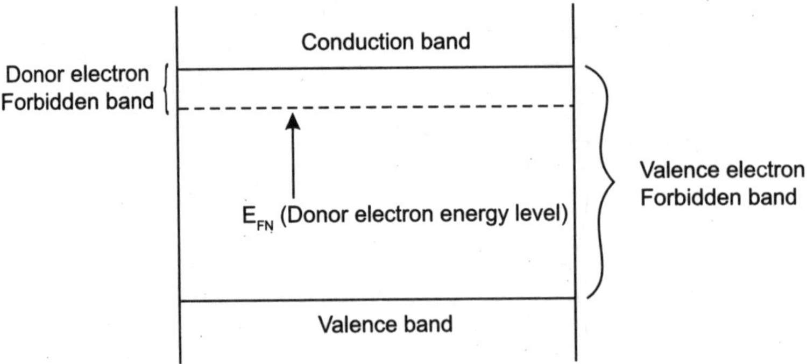

Figure 2.15 Excess free electrons in an N-type semiconductors produce a donor energy level E_{FN} close to the conduction band

2.7.2 P-type Semiconductors

When a trivalent impurity (having 3 electrons in the outermost orbit) like Indium, Boron or Gallium is added to a Germanium intrinsic semiconductor, the impurity atoms will displace some of the Germanium atoms in the crystal during its formation as shown in Figure 2.16. In this case, only three out of four possible covalent bonds are filled while the fourth bond is vacant and the vacancy acts as a hole. Hence a hole moves relative to the electron in a direction opposite to the direction of the electron when an electron moves from one bond to the other. This trivalent impurity known as acceptor or P-type impurity injects into the crystal billions of holes and majority carriers of the charges are holes responsible for the conductivity of the crystal. For this reason, such crystals are called P-type semiconductors or P-type crystals. A P-type semiconductor can be represented as shown in Figure 2.17

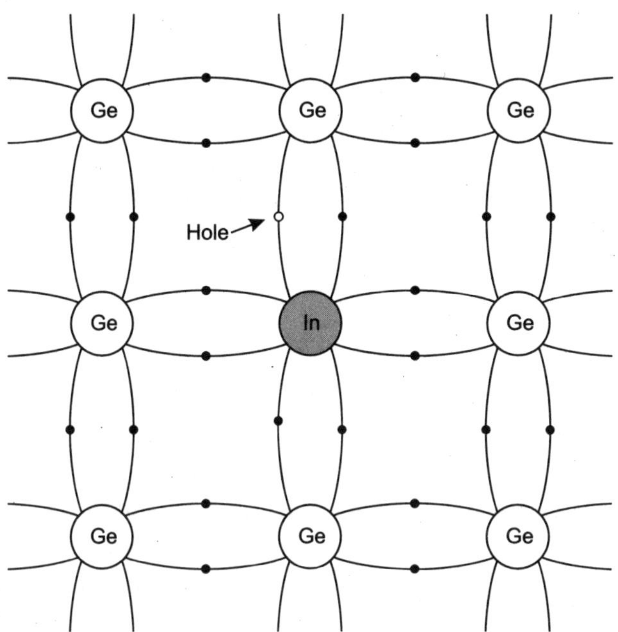

Figure 2.16 P-type semiconductor

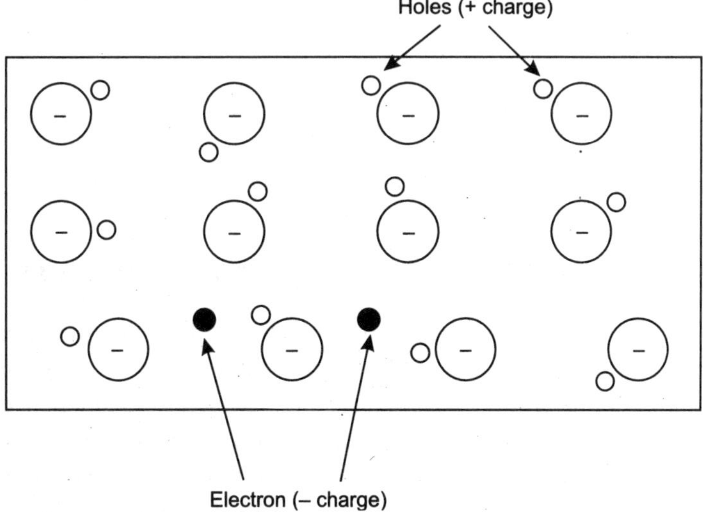

Figure 2.17 Representation of a P-type semiconductor

The energy band diagram of a P-type semiconductor also differs from that of a pure superconductor. Since there is an extra number of holes which tend to attract electrons, they aid in starting current flow. As a result, the acceptor energy level is also somewhat higher than that of the valence band. However, it is not as high as the donor level (Figure 2.18(a)). P-type semiconductors will conduct easily than pure semiconductors, but not quite as easy as N-type semiconductors.

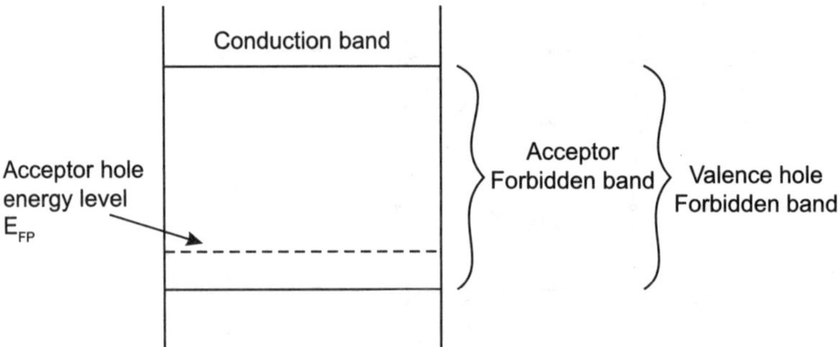

Figure 2.18 (a) Excess holes in a P-type semiconductors introduce an acceptor energy level lose to valence level

The conductivity of a P-type semiconductor is $\sigma = n_a e\mu_h$

where n_a is the density of acceptor atoms or holes and μ_h is the mobility of holes. Energy difference between acceptor energy level E_{FP} and valence band (VB) is extremely small (0.01–0.05 eV) so that with little thermal excitation, electrons from VB can easily be promoted to acceptor energy levels. This leaves behind 'holes' in the VB. These 'holes' can move in the applied field and exhibit a P-type semiconductor in the material.

Since $n_h > n_e$, temperature dependence of 'hole' charge carriers of a P-type extrinsic semiconductor is

$$n_a = n_o \exp\left(\frac{-E_a}{2k_B T}\right)$$

where E_a is energy difference between acceptor VB energy levels.

2.8 Important Properties of Extrinsic Semiconductors

We know that the introduction of impurities in pure semiconductors will increase the density of one type or another type of charge carriers. The product of holes and electrons in a semiconductor is constant depending on the width of energy gap and temperature, and hence the introduction of impurities will result in an increase in the density of one type of carriers and a reduction in the density of other type of carriers. In an extrinsic semiconductor, carriers introduced by the impurities are called majority carriers, and others are called minority carriers. It is important to note that the low value for minority carrier density is due to added recombination.

Let N_d be donor impurity density, N_a acceptor impurity density, p density of holes and n density of electrons in an extrinsic semiconducting material. From the condition of charge neutrality, we have

$$N_d + p = N_a + n$$

or
$$n = (N_d - N_a) + p = (N_d - N_a) + \frac{n_i^2}{n} \quad (\because np = n_i^2)$$

or
$$n^2 - (N_d - N_a)n - n_i^2 = 0 \tag{16}$$

Solving the quadratic equation in n, one obtains

$$n = \frac{(N_d - N_a) \pm \sqrt{(N_d - N_a)^2 + 4n_i^2}}{2} \tag{17}$$

Considering (+) sign

$$n = (N_d - N_a) \text{ when } (N_d - N_a) \gg n_i$$

Obviously, electron density (n) in an N-type semiconductor equals the difference in donor and acceptor impurity densities when they are large compared to intrinsic density n_i. Similarly, hole density (p) in a P-type semiconductor is given by

$$p = (N_a - N_d) \text{ when } (N_a - N_d) \gg n_i$$

2.8.1 Donor and Acceptor States

When an impurity atom from group V of the periodic table, say phosphorous, is added to a pure Ge or Si crystal as a pentavalent impurity, these impurity atoms enter the lattice by substitution for normal atoms, and not in interstitial positions. These impurity atoms contribute five electrons per atom to the valence band, i.e. we have an extra electron per impurity atom. These additional electrons (which cannot be accommodated in the valence band of the original lattice) occupy some discrete energy levels just below the conduction band, the separation may be a few tenths of an electron volt. These excess electrons are released by the impurity atoms and excited into the conduction band. The excited electrons then contribute to the electrical conductivity of the semiconductor. Conversely, the impurity may consist of atoms having fewer electrons than that of a semiconductor (Figure 2.18(b)). For cases in which Si and Ge are host substances, the impurity atoms could be boron or aluminium, each of which contributes only three electrons. In this situation, the impurity introduces vacant discrete energy levels very close to the top of the valence band. Therefore, it is easy to excite some of the more energetic electrons in the valance band into the impurity levels. This process produces vacant states or holes, in the valence band. These holes then act as positive electrons (Figure 2.18(c)). In order to produce significant changes in the conductivity of a semiconductor, it is sufficient to have about one impurity atom per million semiconductor atoms.

We have already seen that a crystal as a whole remains neutral because the electron remains in the crystal lattice. We have further seen that the band diagram of an N-type or P-type semiconductor differs from that of pure a semiconductor. The band diagram for a N-type or P-type semiconductor explains clearly why conduction becomes possible in impure or extrinsic semiconductors at comparatively low temperatures.

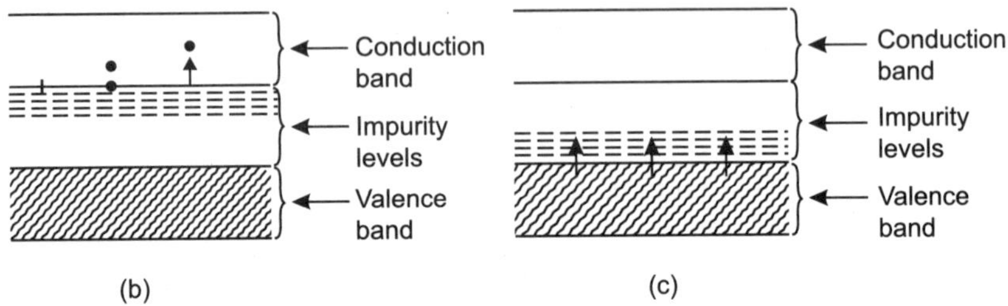

Figure 2.18 Impurities in a semiconductor: (b) donors or N-type, (c) acceptors or P-type

2.8.2 Fermi Level in Extrinsic Semiconductor

We know that Fermi level is situated in the middle of the band gap in an intrinsic semiconductor since the electron and hole densities are equal. When an intrinsic semiconductor is doped by an impurity, carrier densities change, and consequently the position of the Fermi level also changes. The shift in the position of the Fermi level can easily be related to majority carrier density in an extrinsic semiconductor, if it is assumed that the addition of impurities does not affect the densities of energy states in conduction and valence bands.

Let N_c and N_v denote the density of states in the conduction band and valence band, respectively. We have for an intrinsic semiconductor

$$n = N_e \exp\left[\frac{E_{fi} - E_c}{kT}\right]; \; p = N_v \exp\left[\frac{E_c - E_{\beta}}{kT}\right] \tag{18}$$

Here E_{β} is energy associated with the Fermi level in an intrinsic semiconductor. For an intrinsic semiconductor, we have $n = p$ and therefore from Eq. (18), we have

$$\frac{N_c}{N_v} = \exp\left[\frac{E_c - E_v - 2E_{fi}}{kT}\right]$$

Let E_{fn} be energy associated with the Fermi level in an N-type semiconductor having electron density n. We have

$$n = N_c \exp\left[\frac{E_{fn} - E_c}{kT}\right]$$

and

$$p = N_c \exp\left[\frac{E_v - E_{fn}}{kT}\right]$$

$$\therefore \frac{n}{p}=\frac{n^2}{n_i^2}=\frac{N_c}{N_v}=\exp\left[\frac{2E_{fn}-E_c-E_v}{kT}\right]=\exp\left[\frac{2(E_{fn}-E_{fi})}{kT}\right]$$

or
$$n=n_i \exp\left[\frac{E_{fn}-E_{fi}}{kT}\right] \tag{19}$$

Similarly, one obtains for a N-type semiconductor,

$$p=n_i \exp\left[\frac{E_{fi}-E_{fp}}{kT}\right] \tag{20}$$

Thus, the shift in the Fermi level in N- and P-type of semiconductors can be expressed as

$$\left.\begin{aligned} E_{fn}-E_{fi} &= kT \text{ In}\frac{n}{n_i} \\ E_{fi}-E_{fp} &= kT \text{ In}\frac{p}{p_i} \end{aligned}\right\} \tag{21}$$

Figure 2.19 represents the shift in the Fermi level in N- and P-type of semiconductors.

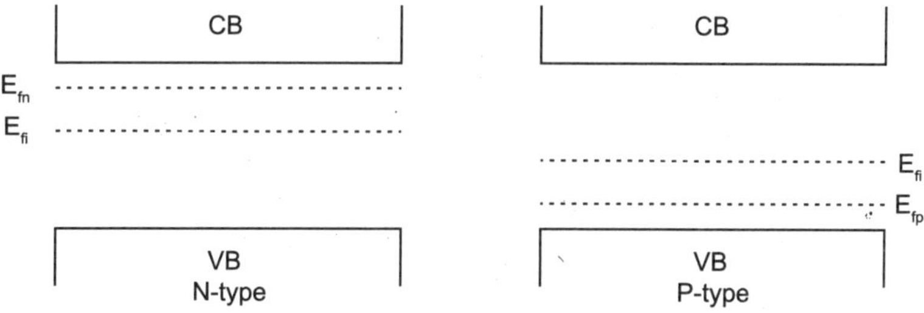

Figure 2.19 Shift in the Fermi level in N- and P-type of semiconductors

2.8.3 Thermal Ionization of Extrinsic Semiconductors

When the temperature of an extrinsic semiconductor is raised to above 0 K, the impurity atoms get ionized. Due to ionization, donor impurity atoms give rise to electrons in the conduction band, and acceptor impurity atoms give rise to holes in the valence band. These electrons or holes along with those generated by the intrinsic action then serve as current carriers at a temperature. Two cases of interest are:

(1) Conduction electron concentration is equal to

$$n=n_0 \exp\left[\frac{(E_F-E_g)}{kT}\right]=N_d \tag{22}$$

This shows that the concentration of conduction electrons is approximately equal to the concentration of donors under the present conditions. This means that all donors are ionized.

Equation (22) suggests that at room temperature, impurity concentration for Si and Ge can go upto 10^{14} to 10^{16} donors per cc, i.e. if we dope an intrinsic semiconductor crystal with this concentration of donors, one will have ~ 10^{16} electrons/cc.

At higher temperatures, carrier concentration is proportional to $\sqrt{N_d}$.

2.8.4 Charge Densities in Extrinsic Semiconductors

The density of impurity atoms in N- and P-materials is so low as compared to the density of semiconductor atoms, and that the rate of thermal pair generation is not affected appreciably by the presence of impurity atoms. In case of intrinsic semiconductors, we have seen that the concentration product was constant at a given temperature.

$$np = n_i^2 \tag{22(a)}$$

Assuming all impurity atoms in an extrinsic semiconductor to be ionized at usual operating temperature, free charge densities in the impurity material can be based upon N_D, the donor atom density in N-material, or N_A the acceptor atom density in P-material.

Electrical neutrality of the material demands.

$$p + N_D = n + N_A \tag{22(b)}$$

The L.H.S. of Eq. 22(b) gives the total positive charge as the sum of holes in the valence band and the positive charge associated with donor atoms that have given up electrons and become positive ions. The R.H.S. of Eq. 22(b) sums the negative charge of free electron density n and the negative charge due to electrons held by ionized acceptor atoms. Using Eq. 22(a), one can write

$$n = \frac{n_i^2}{p} \text{ and } p = \frac{n_i^2}{n} \tag{22(c)}$$

Only donor impurities are introduced in the N-type material, and so $N_A = 0$. Donor density will be made much larger than the density of intrinsic holes, or

$$N_D \gg p$$

and electron density in the N-type material is written from Eq. 22(b) as

$$n \cong N_D \tag{22(d)}$$

With similar reasoning for the P-type material, we have $N_D = 0$ and $N_A \gg n$. One obtains a relation in the P-type material as

$$n = \frac{n_i^2}{N_A} \tag{22(e)}$$

From the above results, we can conclude that the density of majority carriers is approximated to impurity atom density at usual ambient temperatures, and the density of minority carriers is reduced to below the intrinsic level. This means that increased electron density in the N-type material raises the probability that an electron will meet and recombine with a hole, and so hole density is decreased to maintain n^2 constant.

2.8.5 Effect of Temperature on Extrinsic Semiconductors

Semiconductivity of an extrinsic semiconductor is given by $\sigma = ne\mu$ where n and μ are found to be temperature dependent.

Mobility μ_e of electrons and holes is influenced by scattering. The chief source of scattering in a semiconductor are 'phonons' and ionized impurity atoms (donors or acceptors).

At low temperatures, it is found that the mobility of charge carriers in a semiconductor varies as $T^{3/2}$. At high temperatures, the converse holds. This reveals that 'ionic' scattering (due to dopants) tends to dominate in extrinsic semiconductors at low temperatures, and phonon at high temperatures. The variation of conductivity, σ, of these P-type elemental semiconductors with temperature is due to thermally activated mobility, μ, of localized electrons or holes in the host lattice.

Carrier density, n, of a semiconductor is sensitive to temperature. Since conduction electrons and holes are provided by thermal excitation, n is proportional to $\exp\left(-\frac{E}{2k_B T}\right)$ and conductivity varies as $\sigma = ne\mu$.

2.9 Semiconductor Devices

A semiconductor device can be defined as a unit which consists partly or wholly, of semiconducting materials and which can perform useful functions in electronic apparatus and solid state research. Examples of semiconductor devices include semiconductor diodes (P-N junction), transistors, integrated circuits, etc. Si, Ge and GaAs are the most commonly used materials in the fabrication of semiconductor devices. The properties of some important semiconductors are highlighted in Table 2.2.

Table 2.2 Properties of Si, Ge, GaAs semiconductors

Property	Symbol	Unit	Value		GaAs
			Germanium	Silicon	
Atomic numbers			32	14	
Atomic weight			72.6	28.08	144.63
Density		kg/m^3	5.32×10^3	2.33×10^3	5.32×10^3
Atom concentration		atoms/m^3	4.4×10^{28}	5×10^{28}	2.21×10^{28}
Relative dielectric constant	ε_r		16	11.8	10.9
Band gap at 0 K	E_{g0}	eV	0.785	1.21	1.43

Table 2.2 (Continued)

Property	Symbol	Unit	Value		GaAs
			Germanium	Silicon	
Band gap at 300 K	E_g	eV	0.72	1.1	1.32
Intrinsic carrier (generation) at 300 K	n_i	m^3	2.5×10^{19}	1.5×10^{16}	9.0×10^{12}
Crystal structure			Diamond	Diamond	Zincblende
Lattice constant	a	Å	5.65748	5.43086	5.6534
Melting point		°C	936	1420	1250
Minority carrier life time		second	$\sim 10^{-3}$	$\sim 2.5 \times 10^{-3}$	$\sim 10^{-8}$
Breakdown field		V/m	$\sim 10^{-7}$	$\sim 3 \times 10^{7}$	$\sim 4 \times 10^{5}$
Diffusion constant	D_n (electrons) D_n (holes)	m^2/s	0.009842 0.004662	0.003367 0.001295	0.001036 0.000906
Effective density of states in the conduction band	N_c	m^{-3}	1.04×10^{25}	2.8×10^{25}	4.7×10^{22}
Effective density of states in the valence band	N_v	m^{-3}	6.4×10^{24}	1.02×10^{25}	7.0×10^{22}
Intrinsic conductivity	σ_i	Sm^{-1}	2.2428	0.4325×10^{-3}	1.2832×10^{-6}
Mobility (drift) at 300 K	μ_n (electron)	$\frac{m^2}{V-s}$	0.38	0.13	0.85
	μ_p (holes)	$\frac{m^2}{V-s}$	0.18	0.05	0.04
Work function	W	V	4.4	4.8	4.7
Raman phonon energy		eV	0.037	0.063	0.035

Almost all semiconductor devices consist of a single crystal semiconductor incorporating two or more semiconducting regions of different impurity density. The difference in electric fields and carrier densities associated with differently doped regions, called junctions, permits a wide range of nonlinear conductivity effects in devices incorporating two, three or more distinct regions. Most semiconductor devices can be understood through the simplest of such junctions, called the P-N junction, which is a system of two semiconductors in physical contact, one with excess of electrons (N-type) and other with excess of holes (P-type).

2.10 Advantages of Semiconductor Devices

Semiconductor devices have the following advantages:

(i) These devices do not need vacuum for operation.

(ii) These devices have no filaments and, therefore, do not require a heating source.

(iii) The performance of these devices is quite reliable.

(iv) Life expectancy of these devices is high.

(v) A low voltage power supply is required for these devices; the power consumption of such devices is not large.

(vi) These devices are mechanically rugged, light and small.

(vii) These devices have a fast response after the supply voltages are switched on because these devices require very little warm-up time.

2.11 Hall Effect

It is an effect whereby a conductor carrying electric current perpendicular to the strong magnetic field develops a voltage gradient which is transverse to both current and the magnetic field. This phenomenon was discovered by E.H. Hall in 1879. One can study the nature of conduction in semiconductors and metals through this effect. Whether a given sample of semiconductor material is N- or P-type can be determined by observing the Hall effect. If electric current is caused to flow through a sample of semiconductor material and magnetic field is applied in a direction perpendicular to the current, charge carriers are crowded to one side of the sample giving rise to an electric field perpendicular to both current and the magnetic field. This development of converse electric field is known as Hall effect. The effect is also used in the Hall probe for the measurement of magnetic fields and in magnetically operated switching devices.

The strength of electric field E_H produced is given by the relationship $E_H = R_H jB$, where j is current density, B is magnetic flux density, and R_H is a constant called Hall coefficient. The value of R_H can be shown to be I/ne, where n is the number of charge carriers per unit volume and e is the electronic charge.

2.11.1 Hall Effect in Semiconductors

Hall effect in semiconductors originates from the motion of both electrons and holes. This provides the best experimental evidence of the motion of holes, i.e. the production of a transverse electric field by the motion of charge carriers in a magnetic field. With an applied electrical field and a transverse component of magnetic field, both electrons and holes are deflected transversely toward the same side of the sample as shown in Figure 2.20.

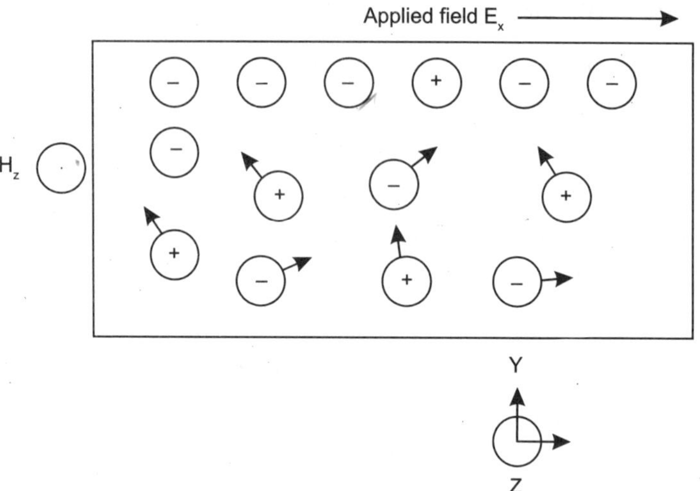

Figure 2.20 Hall effect in semiconductors

Drifting carriers experience Lorentz forces in the magnetic field H_z, and thereby suffer deflections. Both electrons and holes are deflected towards the same side of the sample as shown in Figure 2.20. The situation in all sections is similar. The two types of carriers in semiconductors under this arrangement, called Hall geometry, tend to combine and cancel each other at the front surface. However, this cancellation is incomplete, and thus there is a net charge which accumulates on the front surface of the sample. Because of these surface charges on the front and rear surfaces of the sample, an electric field is produced in Y direction, known as Hall effect in semiconductors. ε_y may be calculated as follows.

Electron and hole current densities along X axis can be expressed as

$$\left.\begin{aligned} J_{xe} &= n_e\mu_e\varepsilon_x \\ J_{xh} &= p_e\mu_h\varepsilon_x \end{aligned}\right\} \tag{1}$$

Total current density J_x is

$$J_x = J_{xe} + J_{xh} = e(n\mu_e = p\mu_h)\varepsilon_x \tag{2}$$

(Small changes in mobility due to magnetic resistance are ignored.)

Electron current density (J_{ye}) along Y direction acting on an electron is $-e(v_e \times B) = eV_eB_z$, where v_e is one drift velocity of the electron. This Lorentz force is equivalent to an electric field $-v_eB_z$ since $v_e = -\mu_e\varepsilon_x$, and therefore this field may be expressed as $B_z\,\mu_e\varepsilon_\chi$. Using (1), one obtains

$$J_{ye} = ne\mu_e(B_z\mu_e\varepsilon_x)$$

Replacing B_z by H_z, we have

$$J_{ye} = ne\mu_e^2H_z\varepsilon_\varepsilon$$

and hole current density is given by

$$J_{vh} = pe\mu_n^2 H_z \varepsilon_x$$

Total transverse current density is

$$J_y = J_{ye} + J_{yh} = (e\mu_e^2 - p\mu_h^2) H_z \varepsilon_x \tag{3}$$

A Hall field ε_y is required to make the net Y current zero. The ratio $\frac{\varepsilon_y}{\varepsilon_x}$ is thus the same as $\frac{J_y}{J_x}$.

$$e(n\mu_e + p\mu_h)\varepsilon_y = e(n\mu_e^2 - p\mu_h^2) H_z \varepsilon_x \tag{4}$$

$$\therefore \varepsilon_y = \frac{p\mu_h^2 - n\mu_e^2}{n\mu_e + p\mu_h} H_z E_x \tag{4a}$$

Hall coefficient R_H is given by

$$R_H = \frac{\varepsilon_y}{H_z J_x} = \frac{p\mu_h^2 - n\mu_e^2}{e(n\mu_e + p\mu_h)^2} \tag{5}$$

Equation (5) is similar as for a metal if the semiconductor is extrinsic ($n \gg p$ or $p \gg n$). Measurements of a and R_H for strongly P-type and N-type extrinsic semiconductors permit the determination of Hall mobilities from Eq. (5). Since μ_n, μ_p, n and p all are temperature dependent, Hall effect in a semiconductor exhibits wide temperature variations, corresponding to the increase of intrinsic carriers at higher temperatures.

From Eq. (5), the magnitude of Hall coefficient is inversely proportional to carrier concentration. The coefficient of proportionality involves a factor which depends on the energy distribution of carriers and the way in which the carriers are scattered in their motion. However, the value of this factor normally does not differ from unity by more than the factor of two. The situation is more complicated when more than one type of carrier is important for conduction. The Hall coefficient then depends on the concentration of various types of carriers and their relative mobilities.

The product of Hall coefficient and conductivity is proportional to the mobility of carriers when one type of carrier is dominant. The proportionality involves the same factor which is contained in the relationship between Hall coefficient and carrier concentration. The value obtained by taking this factor to be unity is referred to as Hall mobility.

2.11.2 Experimental Determination of Hall Coefficient

To measure carrier concentration directly, the most common method is to use the Hall effect. Figure 2.21 shows the basic setup, where a rectangular specimen of the semiconductor of width b and thickness t is placed between the pole pieces of an electromagnet such that a magnetic field B is applied along the Z-direction. With the help of a battery, current I is allowed to pass through the sample in X-direction (Figure 2.21). One can measure Hall voltage V_H with the help of two probes placed at the centres of the top and bottom surfaces of the sample. We have

$$V_H = E_H t \tag{6}$$

But $E_H = R_H J B$ and hence Eq. (6) takes the form

$$V_H = R_H Jt \tag{7}$$

We have current density

$$J = \frac{I}{bt} \tag{8}$$

where *bt* is the cross-sectional area of the sample. Making use of Eq. (8), Eq. (7) becomes

$$V_H = \frac{R_H IB}{b}$$

or

$$R_H = \frac{V_H b}{IB} \tag{9}$$

One can measure the quantities appearing on the R.H.S. of Eq. (9) and hence R_H can be readily determined. In S.I. system, R_H is obtained in m^3/Coulomb. We must note that polarity of V_H is opposite for N-type and P-type of materials and hence R_H has opposite signs for the two types of semiconductors.

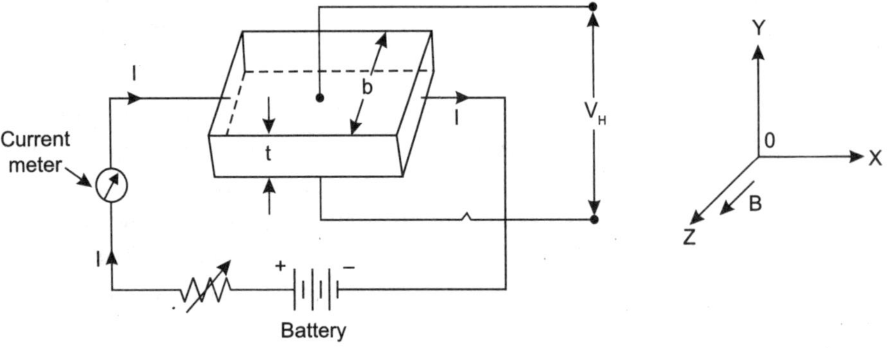

Figure 2.21 Basic setup for the determination of Hall coefficient

2.11.3 Uses of Hall Effect

Determination of semiconductor type Hall voltage V_H is measured by two probes at the centres of the top and bottom face of the sample. If magnetic flux density is $B \frac{Wb}{m^2}$, then we have $n = \frac{1}{R_H e}$, where $R_H = \frac{1}{ne}$.

Since R_H is positive for P-type and negative for N-type semiconductor, the sign of R_H is used to determine the type of semiconductor specimen.

Determination of carrier concentration We have $n = \frac{1}{|eR_H|}$, where $R_H = \frac{1}{ne}$. Obviously, by measuring R_H, one can find electron concentration n and hole concentration p.

Determination of mobility of carriers If conduction is due to one type of carriers, e.g. electrons, we have

$$\sigma = ne\mu_n$$

$$\mu_n = \frac{\sigma}{ne} = \sigma R_H$$

i.e.

$$\mu_n = \sigma\left(\frac{V_H b}{IB}\right) \qquad \text{[using Eq. 9]}$$

Obviously, by knowing σ, one can determine μ_n. This mobility, given by $\sigma\,|\,R_H\,|$, is referred to as Hall mobility.

Measurement of magnetic flux density We have seen that Hall voltage V_H is proportional to magnetic flux density B for a given current I passing through the sample. Thus, knowing the sample dimensions and R_H, one can determine the magnetic field by measuring I and V_H. One can also use it to design a magnetic flux density meter.

Measurement of power in an electromagnetic wave We know that in an electromagnetic wave in free space, magnetic field H and electric field E are at right angles to each other. Thus, if a semiconductor sample is placed parallel to E, it will derive current I in the semiconductor. The semiconductor sample is subjected simultaneously to a transverse magnetic field H producing a Hall voltage across the semiconductor sample. Hall voltage will be proportional to the product E and H across the sample, i.e. to the magnitude of the pointing vector of the electromagnetic wave. Obviously, Hall Effect can be used to determine power flow in an electromagnetic wave.

Hall effect multiple If magnetic field B is produced by passing current I through an air-core coil, B will be proportional to current I. Hall voltage V_H is thus proportional to the product, II. This forms the basis for the design of a multiplier.

2.11.4 Quantum Hall Effect

Klitzing *et.al.* (1980), observed that on application of a large magnetic field on a silicon MOSFET at liquid helium temperature, Hall voltage developed (V_H) is a function of gate voltage (V_G), and there appear plateaus around those values for which potential drop V_{pp} along the sample parallel to the current vanishes (Figure 2.23). Figure 2.23 shows a plot of V_{pp} against V_G. It shows uniformly spaced zeros. We may note that the difference between Hall voltages for two successive plateau is exactly 1/25813 Ω and serves as the definition of ohm in terms of Planck's constant (h) and electron charge (e).

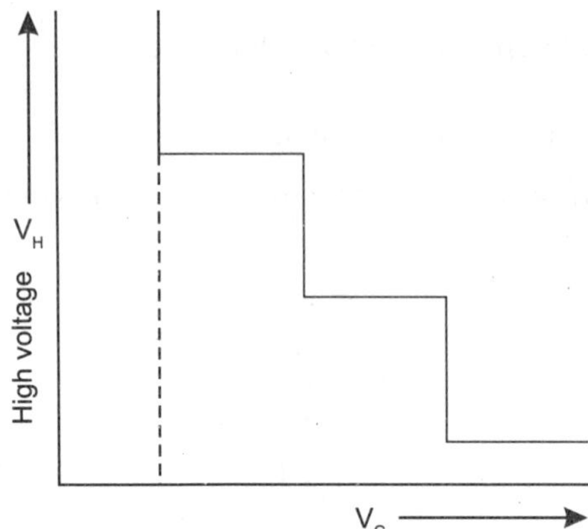

Figure 2.22 Hall voltage (V_H) versus gate voltage (V_G) in a silicon MOSFET

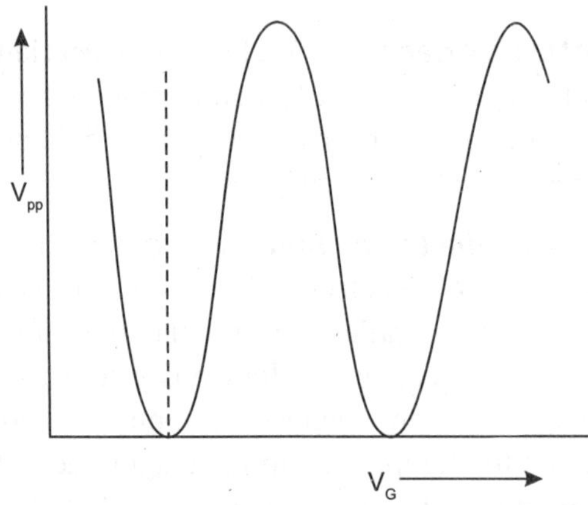

Figure 2.23 In a silicon MOSFET, potential drop along the current is parallel to current vs gate voltage (V_G)

In the channel of MOSFET, electrons form a two-dimensional electron gas and in a strong magnetic field. Say along Z-axis, electron energies in the two-dimensional electron gas are quantized into Landau levels of energy $\left(n+\frac{1}{2}\right)\hbar\omega_c$, where ω_c is cyclotron frequency and n being an integer. When a Landau level of energy coincides with the Fermi level, potential drop V_{pp} tends to zero, because electrons cannot be scattered to the already occupied states of energy. We may note that gate voltage V_G controls the number of electrons per unit area of the channel and, therefore, the Fermi energy. This is why minima in V_{pp} and plateau in ϕ_H occur at interval $\hbar\omega_c$. With a sufficiently large magnetic field, it is observed that only a fraction of the lowest Landau level is occupied. This phenomenon is called fractional quantum Hall effect.

Example 1 Find conductivity and resistivity of a pure silicon crystal at 300 K. The density of electron-hole pair per cc at 300 K for a pure silicon crystal is 1.072×10^{10} and the mobility of electrons $\mu_n = 1350 \text{ cm}^2/\text{Volt-sec}$, and mobility $\mu_h = 480 \text{cm}^2/\text{Volt-sec}$.

Solution Conductivity of pure silicon crystal is given by

$$\sigma = n_i e(\mu_e + \mu_h) \; n_i = 1.072 \times 10^{10}$$

$$\sigma_i = 1.072 \times 10^{10} \times 1.6 \times 10^{-19}(1350 + 480) e = 1.6 \times 10^{-19} \text{C}$$

$$\mu_h = 480 \text{cm}^2/\text{Volt-sec}$$

Resistivity of silicon crystal is given by

$$\rho_i = \frac{1}{\sigma_i} = \frac{1}{3.14 \times 10^{-6}} = 3.18 \times 10^5 \,\Omega\text{-cm}$$

Example 2 A silicon wafer is doped with phosphorus of concentration 10^{13} atoms/cm^3. If all the donor atoms are active, what is its resistivity at room temperature? Electron mobility is 1200 cm^2/Volt-sec. Charge on the electron is 1.6×10^{-19} Coulomb.

Solution

$$\sigma = \mu e n$$

$$\mu = 1200 \text{cm/Volt-sec}$$

$$= 1200 \times 1.6 \times 10^{-10} \times 10^{13} = 19.2 \times 10^{-4} \,\Omega^{-1}/\text{cm}$$

$$n = 10^{-13} = N_p$$

$$e = 1.6 \times 10^{19} \text{C}$$

$$\text{Resistivity, } \rho = \frac{1}{\sigma} = \frac{1}{19.2 \times 10^{-4}} = 5.2 \times 10^2 \,\Omega\text{-cm}$$

Example 3 Find the resistance of an intrinsic germanium rod 1 cm long, 1 mm wide and 1 mm thick at temperature of 300 K. For germanium $n_i = 2.5 \times 10^{13}$, $\mu_n = 3900$ cm^2/Volt-sec at 300 K.

Solution

$$\sigma = n_i e(\mu_e + \mu_h)$$

$$= 25 \times 10^{13} \times 1.6 \times 10^{-19}(0.39 + 0.19)$$

$$= 2.32 \,\Omega/\text{m}$$

$$\rho = \frac{1}{232} \,\Omega\text{m}$$

$$\text{Now resistance, } R = p_i \left[\frac{\text{Length}}{\text{Area of cross section}}\right] = 2.32 \times \frac{10^{-6}}{10^{-2}}$$

$$R = \frac{1}{2.32 \times 10^{-4}} \Omega = 4.31\ k\Omega$$

Example 4 A sample of Germanium in made of *P* material by adding acceptor atoms at a rate of one atom per 4 × 10[8] Germanium atoms. Acceptor density is assumed to be zero and $n_i = 2.5 \times 10^{19}$ per m[3] at 300 K. There are 4.4 × 10[28] Germanium atoms/m[3]. Acceptor density is found to be 1.1 × 10[20] atoms/m[3].

Solution

$$n_i^2 = np = n_p N_a$$

$$n_p = \frac{n_i^2}{N_a} = \frac{6.25 \times 10^{38}}{1.1 \times 10^{20}} = 5.6 \times 10^{18}$$

$$\frac{n_p}{n_i} = \frac{5.6 \times 10^{18}}{2.5 \times 10^{19}} = 0.22$$

Example 5 Determine the number density of donor atoms which have to be added to an intrinsic germanium semiconductor to produce an N-type semiconductor of conductivity 5 mho/cm. The mobility of electrons in the N-type semiconductor is 3850 cm[2]/Volt-sec.

Solution

$$\sigma_n = nq\mu_n = N_D q\mu_n = 5 \text{ mho/cm}$$

$$q = 1.6 \times 10^{-19}\text{C}$$

$$\therefore n = N_D = \sigma_n / q\mu_n = \frac{5}{1.6 \times 10^{-19} \times 3850} = 0.8 \times 10^{18} \text{ per cm}^3$$

$$\mu = 3850 \text{cm}^2\text{/Volt-sec}$$

Example 6 Will Silicon and Germanium semiconductors at 1000 K ambient temperature still be semiconductors?

Solution We know forbidden energy gap between conduction and valence bands for a semiconductor is nearly 1 eV. For Ge and Si, energy gap is 0.785 eV and 1.21 eV respectively at 0 K. Energy gap decreases with increase in temperature which is represented by the expression. Obviously, Si and Ge will remain semiconductors at 1000 K ambient temperature.

$$E_g(T) = 1.21 - 3.60 \times 10^{-4}\ T \text{ (for Si)}$$

$$E_g(1000) = 1.21 - 3.60 \times 10^{-4}(1000) = 0.85\text{eV}$$

Example 7 Electrons in the conduction band in silicon have an effective mass of 0.259 m_0 and mobility of 0.1350 m[2]/V. Holes in one of the valance bands have an effective mass of 0.547 m_0 and mobility of 0.0480 m[2]/V. Here m_0 is free electron mass. Find the average relaxation time for (a) electrons and (b) holes.

Solution Relaxation time in terms of mobility is given by

$$\tau = \frac{m\mu}{e}$$

Taking effective mass of electron and holes in consideration, relaxation time is given by

$$\tau = \frac{m^*\mu}{e}$$

(a) For electrons, $m^* = 0.259\ m_0$.

$$\therefore\ \tau_e = \frac{0.359 m_0 \mu}{e} = \frac{0.259 \times 9.1 \times 10^{-31} \times 0.1350}{1.6 \times 10^{-19}} = 1.99 \times 10^{-13}\ \text{s}$$

(b) For holes in the valance bands, $m = 0.537\ m_0$.

$$\tau = \frac{0.537 \times 9.1 \times 10^{-31} \times 0.048}{1.6 \times 10^{-19}} = 1.47 \times 10^{-13}\,\text{s}$$

Example 8 A sample of Ge is doped to the extent of 10^{14} donor atoms/cm^3 and 5×10^{13} acceptor atoms/cm^3. At 300 K, the resistivity of intrinsic Ge is 60 Ω/cm. If applied electric field is 2V/cm, find the total conduction current density. Assume $\mu_p / \mu_n = 1/2$ and $n = 2.5 \times 10^{13}$/cm^3 at 300 K. [B.Tech, 2009]

Solution Conductivity of an intrinsic material is given by

$$\sigma = en_i\mu_n\left(1 + \frac{\mu_p}{\mu_n}\right)$$

Given

$$\sigma_i = \frac{100}{60} = \frac{5}{3}\text{S/m}$$

$$n_i = 2.5 \times 10^{19}\ /\ \text{m}^3 \text{ and } \frac{\mu_p}{\mu_n} = \frac{1}{2}$$

$$\mu_n = \frac{\sigma_i}{en_i\left(1 + \dfrac{\mu_p}{\mu_n}\right)}$$

$$= \frac{5}{3 \times 1.6 \times 10^{-19} \times 2.5 \times 10^{19} \times \dfrac{3}{2}}$$

$$= 0.2778\,\text{m}^2\ /(V.S)$$

and

$$\mu_p = \frac{\mu_n}{2} = 0.1389\,\text{m}^2\ /(V.S)$$

Let n be electron concentration and p be hole concentration for the doped sample. Since the sample is electrically neutral, we have

$$N_d + p = N_a + n$$

where N_d is doner concentration and N_a is acceptor concentration, assumed to be fully ionized. From mass action law, we have $np = n_i^2 S_o$

$$N_d + \frac{n_i^2}{n} = N_a + n$$

$$\text{or } n^2 + (N_a - N_d)n - n_1^2 = 0$$

$$\text{or } n = \frac{1}{2}\left\{[N_d - N] \pm \sqrt{(N_d - N_a)^2 + 4n_i^2}\right\}$$

n is positive and hence we drop the negative sign before the radical.

Thus $$n = \frac{1}{2}\left\{[N_d - N] \pm \sqrt{(N_d - N_a)^2 + 4n_i^2}\right\}$$

Here $$N_d = 10^{20}\,/\text{m}^3,\ N_a = 5\times10^{19}\,\text{m}^3$$

and $$n_i = 2.5\times10^{19}\,\text{m}^3$$

$$\therefore n = \frac{10^{19}}{2}\left[5 + \sqrt{25 + 4\times6.25}\right]$$

$$= 6.036\times10^{19}\,/\text{m}^3$$

and $$p = \frac{n_1^2}{n} = \frac{6.25}{6.036}\times10^{19}$$

$$= 1.035\times10^{19}\,\text{m}^3$$

Conductivity of doped sample is

$$\sigma = e(n\mu_n + P\mu_p)$$

$$= 1.6(6.036\times0.2778 + 1.035\times0.1389)$$

$$= 2.95\ \text{S/m}$$

We have assumed that carrier mobilities in the doped sample are the same as those in the intrinsic material. This assumption is justified by the low doping level, and Coulomb scattering by the ionized impurities is weak at 300 K.

Applied electric field

$$F = \frac{2V}{cm} = 200\frac{V}{m}$$

Thus, total conduction current density

$$J = \sigma F = 2.9 \times 200 = 580\,\text{A/m}^2$$
$$= 0.58\,\text{A/cm}^2$$

Example 9 Light doping of a P-type semiconductor material is defined as the case for which $p \gg n$ is not valid ($n \gg p$ is not valid for an N-type semiconductor). Derive a procedure to determine the number of mobile carriers for the case of light doping. Consider N_D and N_A are immobile donor and acceptor ions, respectively.

Solution For P-type doping, $N_D = 0$. By charge neutrality and mass action law,

$$p + N_D = p = n + N_A = \frac{n_i^2}{n}$$

or
$$n^2 + N_A n - n_i^2 = 0$$

Solving the quadratic equation for n and discharging the extraneous negative root, one obtains

$$n = \frac{1}{2}(-N_A + \sqrt{N_A^2 + 4n_i^2})$$

Knowing n, one obtains from mass action law $p = \dfrac{n_i^2}{n}$

For N-type doping, $N_A = 0$. By analogous procedure,

$$p = \frac{1}{2}(-N_D + \sqrt{N_D^2 + 4N_i^2})\ ; \quad n = \frac{n_i^2}{p}$$

Example 10 Samples of silicon are doped with 10^{17} phosphorus atoms/cm^3. What would you expect to measure for its resistivity? What Hall voltage would you expect in a sample 100 μm thick if $I_x = 1$ mA and $B_z = 1$ kg ($= 10^{-5}$ b/cm^2)?

Solution Mobility for silicon is 700 cm^2/V^{-s}. Obviously, conductivity is

$$\sigma = e\mu_n n = 1.6 \times 10^{-19} \times 700 \times 10^{17} = 11.2(\Omega - \text{cm})^{-1}$$

Resistivity,
$$p = \sigma^{-1} = 0.0893\ \Omega - \text{cm}$$

Hall coefficient,
$$R_H = -(\text{en})^{-1} = -62.5\,\text{cm}^3/\text{C}$$

Hall voltage,
$$V_H = \frac{I_x B_z}{x} R_H = \frac{10^{-3} \times 10^{-5}}{10^{-2}} \times (-62.5) = -62.5\ \mu\text{V}$$

Example 11 A Hall effect experiment is performed on a heavily doped N-type semiconductor sample of length 2.65 cm, width 1.70 cm and thickness 0.052 cm in a magnetic induction field of 0.5 T. Current in the sample along its length is 200μA. Potential difference along the length of the sample in 195 mV and potential difference across its width is 21.4 mV.

(a) What is the concentration of charge carriers?

(b) What is their mobility?

Solution (a) Hall co-efficient is

$$R_H = \frac{E_H}{J_x B}$$

Here $E_H = \frac{V_H}{b}$, where V_H is Hall voltage = 21.4 m; $V = 21.4 \times 10^{-3}$ V; b is breadth = 1.7 cm = 1.7×10^{-2} m.

$$\text{Current density, } J_x = \frac{1}{bt}$$

$$\begin{aligned} \text{where } I \text{ is current} &= 200\,\mu\text{A} = 200 \times 10^{-6}\text{ amp} \\ t \text{ is thickness} &= 0.052\,\text{cm} = 0.052 \times 10^{-2}\,\text{m} \\ B \text{ is magnetic field} &= 0.5\text{ T} \end{aligned}$$

$$\therefore R_H = \frac{21.4 \times 10^{-3}}{1.7 \times 10^{-2}} \times \frac{1.7 \times 10^{-2} \times 0.052 \times 10^{-2}}{200 \times 10^{-6} \times 0.5} = 0.11128\,\text{m}^3/\text{C}$$

(b) Electrons per unit volume,

$$n = \frac{1}{eR_H} = \frac{1}{1.6 \times 10^{-19} \times 0.11128} = 5.6 \times 10^{19}\text{ electrons/m}^3$$

Resistance of the sample,

$$R = \frac{V}{I} = \frac{195 \times 10^{-3}}{200 \times 10^{6}} = 975\,\Omega$$

Since $R = \frac{\ell}{A\sigma} = \frac{\ell}{Aen\mu}$

where σ is conductivity and μ is mobility

we have

$$\mu = \frac{\ell}{AenR}$$

$$= \frac{2.65 \times 10^{-2}}{1.7 \times 10^{-2} \times 0.052 \times 10^{-2} \times 1.6 \times 10^{-19} \times 5.6 \times 10^{19} \times 975}$$

$$= 0.343\,\text{m}^3/\text{Vs}$$

Example 12 The band gap of a specimen of gallium arsenide phosphate is 1.98 eV. Determine the wavelength of electromagnetic radiation that is emitted upon direct recombination of electrons and holes in this sample. Show that the colour of emitted radiation is red. [B.Tech 2008]

Solution $\lambda = \frac{C_h}{E_g}$

We have $C = 3\times10^8$ m/s, $h = 6.6\times10^{-34}$ Js

Band gap, $E_g = 1.98\,\text{eV} = 1.98\times1.6\times10^{-19}\,\text{J}$

$$\therefore \lambda = \frac{3\times10^8\times6.6\times10^{-34}}{1.98\times1.6\times10^{-19}}\,\text{m}$$
$$= 625\,\text{nm}$$

Since λ is in the red region of visible light and hence the colour of emitted radiation is **red**.

Summary

1. Semiconductors constitute a large class of substances characterized by negative temperature coefficient of resistance.
2. Semiconductors have half-filled outer shells, and are neither good insulators nor good conductors.
3. Semiconductors have a forbidden band that represents the amount of energy needed to move a valence electron from the valence band to the conduction band to cause conduction.
4. The crystal structure of semiconductors, e.g. germanium and silicon, reveals that a pair of electrons is shared between atoms to complete the valence shells of individual atoms. Such a bonding between atoms is termed as covalent bonding.
5. In a pure semiconductor, the number of holes and electrons are equal. Such a semiconductor is termed intrinsic.
6. Conduction in a pure semiconductor consists of electron flow in the conduction band and hole flow in the valence band. Hole conductivity is the result of motion of bound electrons along the bonds.
7. Impurities are added to pure semiconductors, in a process called doping, to increase the number of current carriers. Trivalent or acceptor impurities like Indium have only 3 valence electrons and form holes in the bands to accept electrons from the semiconductor material, making it positive or P-type. Pentavalent or donor impurities like Arsenic have 5 valence electrons and produce excess electrons that make the semiconductor material a negative or N-type.
8. P-type and N-type refer to the majority current carriers produced by doping; the overall semiconductor remains electrically neutral. Minority current carriers, produced by thermal energy, are opposite to majority carriers.
9. Majority current carriers in an N-type material are electrons, while minority carriers are holes. The reverse is true for a P-type material. The doped semiconductor as a whole remains electrically neutral.

10. A doped semiconductor acts like a resistor. The resistance of doped semiconductor is called bulk resistance. Greater the doping, lesser the bulk resistance.
11. A host of non-elemental materials exhibit semiconductivity above 0 K. These materials are known as non-elemental or compound semiconductors, e.g. Ga As, In Sb, PbTe, GaP, etc. A host of oxide materials such as ZnO, NiO, FeO, $V_2 O_5$, etc. are semiconductors.
12. Ga As and Si are termed as **direct** and **indirect** band gap (B.G.) semiconductors respectively. In a direct semiconductor, an electron in the conduction band can fall to an empty state in the valence band giving off energy difference (E_g) as a single photon. On the other hand, in an indirect semiconductor, an electron in the conduction band minimum cannot fall directly to the valence band maximum, but must undergo a change in momentum as well as change in energy, i.e. it involves a defect state of phonons. Thus for indirect transion

$$h\upsilon = E_g + \hbar\omega$$

Where $h\upsilon$ is energy of photon and $\hbar\omega$ is the energy of phonons. The energy difference is given up as heat to the lattice. This concept is important for deciding which semiconductors can be used for devices requiring light output, e.g. LED's and lasers.

Review Questions

1. What are semiconductors? How do they differ from conductors and insulators? Why an increase in temperature should increase conductivity in a semiconductor?
2. Explain with suitable diagrams the conduction band, valence band and forbidden band and hence explain the behaviour of conductor, semiconductor and insulator. Explain the contribution of electrons and holes to electrical conduction.
3. Derive the expression for conductivity of an intrinsic semiconductor.
4. What are intrinsic and extrinsic semiconductors? Discuss the locations of Fermi levels under suitable limiting conditions and give the necessary theory.
5. What is impurity conduction in semiconductors? Explain how the presence of a little impurity in a semiconductor modifies its conduction properties.
6. Explain the movement of electrons and holes in a semiconductor. In what respect N-type semiconductors differ from each other.
7. What is drift current and diffusion current in a semiconductor? Write the expression for total electrons current density in a semiconductor.
8. Explain the mechanism of electrical conduction in a typical semiconductor like Germanium or Silicon. How the conductivity of a pure semiconductor is affected by adding traces of trivalent and pentavalent impurities?
9. What is meant by doping a semiconductor? What basic properties must these doping elements possess to make a pure semiconductor a P-type or N-type? Explain the physical principles involved.

10. What is Hall effect? Briefly discuss the origin of Hall effect. Mention the uses of Hall effect.
11. What is Hall coefficient? Show that for a P-type semiconductor Hall coefficient R_H is given by, $R_H = 1/eP$. Describe an experimental setup for the measurement of Hall voltage, V_H.
12. For a semiconductor, show that Hall coefficient is given by

$$R_H = \frac{1 p\mu p^2 - n\mu_n^2}{e(p\mu_p + n\mu_n)^2}$$

where μ_p and μ_n are mobilities of holes and electrons respectively. Show that for a intrinsic semiconductor the above expression reduces to

$$R_H = \frac{1}{n,e}\left[\frac{\mu_n - \mu_p}{\mu_n + \mu_p}\right]$$

[**Hint:** $N = p = n_i$ for an intrinsic semiconductor]

13. Mobilities of electrons and holes in a sample of intrinsic germanium at room temperature are 3600 cm²/volt-sec and 1700 cm²/volt-sec, respectively. If electron and hole densities are each equal to 2.5×10^{13} per cc, calculate its conductivity. [**Ans:** 2.12 mho/m]
14. The concentration of acceptor atoms in a P-type germanium crystal is 4×10^{15} per cc. Find the conductivity of the crystal at 300 K. Hole mobility in germanium at 300 K is 1900 cm² per volt-sec. All acceptor atoms are ionized at this temperature. [**Ans:** 121.6 mho/m]
15. Calculate the current produced in a small germanium plate of area 1 cm² and of thickness 0.3 mm. A potential difference of 2 V is applied across the faces. Given concentration of free electrons in germanium is 2×10^{19} m³ and mobilities of electrons and holes are 0.36 m²/volt-sec and 0.17 m²/volt-sec, respectively. [**Ans:** 1.13 amp]
16. A specimen of pure germanium at 300 K has a density of charge carriers of 2.5×10^{19} per m³. It is doped with donor impurity atoms at the rate of one impurity atom for every 10^6 atoms of germanium. All impurity atoms may be supposed to be ionized. The density of germanium atom is 4.2×10^{28} atoms/m³. Find the resistivity of doped germanium if electron mobility is 0.36 m²/volt-sec. [**Ans:** 0.41×10^{-3} Ω-m]
17. What concentration of donor atoms per cc is required in a germanium crystal so that the conductivity of the crystal is 2 mho/cm at 300 K. For germanium $\mu_n = 3900$.

    [**Ans:** 320.5×10^{19} m³]
18. Find the conductivity of germanium which is doped with donor atoms of concentration $N_d = 10^{17}$ per cm³ and compare it to that of intrinsic case. Given $n_i = 2.5 \times 10^{13}$/cm³, $\mu_n = 3800$ cm²/volt-second. [**Ans:** 6.25×10^{15}m³]
19. Compare the number of electron-hole pairs per cm³ in a pure silicon crystal at the temperature of (i) –27°C and (ii) 57°C. For silicon $E_g = -1.13$. eV. [**Ans:** 8.9]
20. Determine donor concentration in N-germanium having a resistivity of 0.015 Ω-m, $\mu_p = 0.36$ m²/volt-sec. Repeat for P-germanium of equal resistivity, $\mu_n = 0.1$m²/volt-sec.

21. (a) Show that the conductivity of a semiconductor is minimum when it is lightly doped with P-type impurity such that

$$P = n_i\sqrt{\frac{\mu_n}{\mu_p}}$$

(b) Show that minimum conductivity is $2n_i\sqrt{\mu_n\mu_p}q$

(c) With the help of the above result, determine minimum conductivity of silicon

[Ans: (c) $\sigma_{mn} = 3.87 \times 10^{-6}$ $(\Omega\text{-cm})^2$]

Short Answer Questions

1. What is the order of energy for forbidden band in (i) diamond (ii) silicon (iii) germanium, and (iv) aluminum? **[Ans:** (i) 9 eV (ii) 1.2 eV (iii) 0.7 eV (iv) zero]
2. What happens to the resistance and conductance of a semiconductor on heating?
   [**Ans:** Resistance decreases but conductance increases.]
3. What is the order of resistivity of a metal, insulator and semiconductor?
   [**Ans:** Resistively of (i) metal ≈ 10^{-6} Ω-cm, (ii) insulator ≈10^{13} Ω-cm and (iii) semiconductor lies between 10^{13} Ω and 10^{-6} Ω-cm]
4. What are the important characteristics of a semiconductor?
   [**Ans:** (i) it has covalent bonding; (ii) it is crystalline; (iii) it has a negative temperature coefficient of resistance; (iv) its conductivity increases with the addition of impurities.]
5. What is mobility? Write its unit?
   [**Ans:** Drift velocity per unit electric field is called mobility (μ). Its unit is $N^{-1}S^{-1}$ m C or N^{-1} mA.]
6. Why metallic bodies are always opaque?
   [**Ans:** We know that metallic solids have partially filled conduction bands. The energy of photons in the visible light region varies between *I* eV and 3 eV. As light is incident to metallic solids, free electrons of the conduction band absorb the energy of the incident photons. Obviously, no photons are allowed to pass through, the metallic solids behave as though opaque.]
7. A small portion of indium (In) is incorporated in germanium (Ge). Is the crystal N-type or P-type?
   [**Ans:** Obviously, the crystal is P-type as indium is trivalent and one of the covalent bonds will remain without electrons giving excess holes]
8. What are the charge carriers in N-type and P-type semiconductors?
   [**Ans:** The charge carriers in N-type semiconductors are the electrons while in P-type the charge carriers are holes.]

9. Why a semiconductor is damaged by a strong current?

   [**Ans:** The safety limits of temperature for germanium and silicon are about 80°C and 200°C, respectively. A strong current passing through a semiconductor heats it up beyond these temperature. At these temperatures a large number of charge carriers are available for conduction and specific resistance of the crystal becomes very low and thus they are damaged, i.e. lose the property of semiconductors.]

10. What is the difference between hole-current and electron flow?

    [**Ans:** Hole current is the direction of conventional current which is due to the flow of positive charges, while electron flow is opposite to the conventional current as it is due to the flow of negative charges.]

Objective Questions

1. N-type germanium is obtained on doping intrinsic germanium by
   (a) Phosphorous (b) Aluminium
   (c) Boron (d) Gold [**Ans:** (a)]
2. Depletion region is a zone which contains
   (a) holes only (b) electrons only
   (c) both electrons and holes (d) neither electrons nor holes [**Ans:** (d)]
3. In a semiconductor diode, arrow represents
   (a) N-type material (b) P-type material
   (c) both P- and N-type materials (d) none of the above [**Ans:** (b)]
4. When holes leave the P materials to fill electrons in the N material, the process is called
   (a) diffusion (b) depletion
   (c) avalanche breakdown (d) backward diode [**Ans:** (c)]
5. Which atoms can act as donors in Si
   (a) Ga (b) As (c) P (d) Al [**Ans:** (b)]
6. The conductivity of a semiconductor lies in the range
   (a) 10^5 to 10^9 S/m (b) 10^5 to 10^{-4} S/m
   (c) 10^{-4} to 10^{-8} S/m (d) 10^5 to 10^{-9} S/m [**Ans:** (b)]
7. The band gap of a semiconductor lies in the range
   (a) 0.01 to 0.1 eV (b) 5 to 10 eV
   (c) 0.2 to 2.5 eV (d) 1 to 5 eV [**Ans:** (c)]
8. The Fermi level of an N-type semiconductor lies
   (a) at the middle of the forbidden gap (b) near the conduction band edge
   (c) near the valence band edge (d) none of the above [**Ans:** (b)]
9. Intrinsic concentration of a semiconductor at 0 K is
   (a) 10^{19} m^{-3} (b) zero (c) infinite (d) none of the above [**Ans:** (b)]

Mark the Following Statements True/False.

1. Solid state materials can be classified into three major groups on the basis of their electrical conductivity: metals or conductors, semiconductors and insulators. [True]
2. Semiconductors posses very high electrical conductivity and low resistivity. [False]
3. Semiconducting materials are neither good conductor nor good insulator. [True]
4. The conductivity of semiconducting material is generally sensitive to temperature. [True]
5. At very low temperature (T = O K) a semiconductor behaves as an conductor. [False]
6. The conductivity of the semiconductor at a given temperature (at room temperature) can be enhanced by adding some impurities (elements). [True]
7. Semiconductors are useful in electronics because their electrical properties can be greatly altered in a controlled manner by adding small amount of impurities. [True]
8. A semiconductor material in which impurities are added, are called intrinsic semiconductors. [False]
9. In p-type semiconductor holes in valence band are responsible for current in the semiconductor. Those impurities that cause holes in valence band are called donor impurities. [False]
10. A pure silicon semiconductor material doped with Vth group elements is called n-type semiconductor. [True]

Fill in the Blanks.

1. Semiconductors are primarily of two types _______________ and _______________.

   [**Ans:** intrinsic, extrinsic]

2. The most useful feature of semiconductors is that their electrical conductivity generally _______________ with increasing purification in contrat to _______________.

   [**Ans:** decreases, metals]

3. In a semiconductor, the gap separating the filled and empty bands is very ____________.

   [**Ans:** small]

4. Silicon and germanium are prototypical covalently bonded ______________.

   [**Ans:** semiconductors]

5. A semiconductor material that has been subjected to the doping process is called an ______________ semiconductor.

   [**Ans:** extrinsic]

6. There are two types of extrinsic semiconductors ___________ type and ___________ type, which are of immense importance to semiconductor device fabrication.

   [**Ans:** p, n]

7. The p-type extrinsic semiconducting material is formed by doping Ge or Si crystal with impurity atoms having ___________ valence ___________.

   [**Ans:** three, electrons]

8. In n-type semiconductor, there is a high concentration of electrons in the conduction band compared with the ___________ concentration in the ___________ band.

   [**Ans:** hole, valence]

9. In an intrinsic semiconductor, Fermi level lies at the ___________ of the band gap.

   [**Ans:** middle]

10. n-type semiconductor is created by introducing impurity elements that have ___________ valence electrons.

    [**Ans:** five]

Chapter 3

Semiconductor Diodes

3.1 Introduction

We know that extrinsic semiconductors, i.e. P-type and N-type, are used in fabricating electronic devices such as diodes, transistors and integrated circuits (ICs). We also know that a single piece of P-type or N-type semiconductor conducts electricity in much the same way as an ordinary conductor. However, when a piece of P-type semiconductor is joined to a piece of N-type semiconductor, some remarkable properties are observed at this P-N junction, in that current flows easily in one direction through the P-N junction and hardly at all in the opposite or other direction. When P and N extrinsic semiconductors are joined, the device formed is called a P-N junction diode, or semiconductor diode, or simply diode. Most semiconductor devices have at least one such junction. To understand the operations of electronic devices, it is essential to study P-N junction diodes. A diode is a two-terminal (or two-lead) device that has a P-N junction and acts as a one-way conductor. When forward-biased, the diode conducts. When reverse-biased, conduction is so small that it is usually considered zero. There are other types of diodes, e.g. Zener, light-emitting. In the present chapter, we will discuss the basic concepts of diode, different types of diodes and their working.

3.2 P-N Junction

When a P-type semiconductor is brought into contact with an N-type semiconductor the resulting combination is called a P-N junction. This junction has important properties and forms the basis of modern semiconductor theory and practice. Most semiconductor devices contain one or more P-N junctions. The most important characteristic of a P-N junction is its ability to conduct current in one direction only. In the reverse direction, it offers very high resistance.

Formation of a P-N junction Figure 3.1 shows three types of such junctions. P and N regions have been grown into a block of germanium by mixing acceptor and donor impurities, respectively, into a single crystal during its formation. This is known as grown junction. It is worthwhile to mention that the grown type of P-N junction is not a sandwich made by attaching a P block to an N block, but it actually consists of P and N layers in a single piece of germanium.

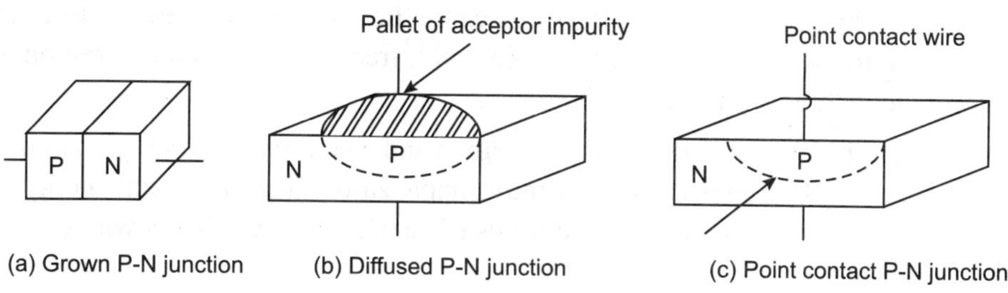

Figure 3.1 Types of P-N junction

The diffused junction, as shown in Figure 3.1(b), is made by placing a pellet of acceptor impurity, such as indium, on one face of a wafer of N-type germanium and then heating the combination to melt the impurity. Under proper conditions of temperature and time, a portion of the impurity metal will diffuse a short distance into the wafer, thereby creating a region of P-type germanium in intimate association with **Z-type** bulk. This is called an alloyed junction or fusion-alloy junction from the fact that a small amount of pellet material alloys with germanium.

Figure 3.1(c) shows a point contact type P-N junction. Here a fine, pointed wire (catwhisker) makes pressure contact with the face of an N-type germanium wafer. After assembly, the device is electroformed by passing a high-current surge momentarily across the junction of wafer and whisker. Heat generated during the short interval drives a few electrons from the atoms in the region of s point contact, leaving holes and thus converting sa small volume of germanium immediately into P-type under and around the point. Silicon P-N junction can also be produced in a similar manner. In most instances, silocon has been processed to make it P-type. To create the junction, an N-type material is either inserted at a proper point in the crystal process or in the diffused, junction process. Like germanium diodes, silicon diodes also are produced both in junction and point contact types.

P-N junction with no external voltage (unbiased P-N junction) Figure 3.2 shows a P-N junction just immediately after it is formed. There is no external voltage connected to the P-N junction. Since an N-type material has a high concentration of free electrons while a P-type material has a high concentration of holes, the following processes are initiated.

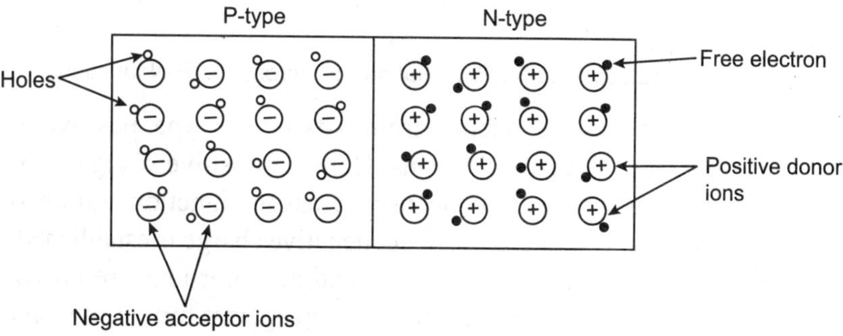

Figure 3.2 P-N junction when just formed

At the junction, holes from the P region diffuse into the N region and free electrons from the N region diffuse into the P region. Holes combine with free electrons in the N region whereas electrons combine with holes in the P region.

The diffusion of holes from P region into N region and electrons from N region into P region across the junction takes place because they move haphazardly due to thermal energy and also because there is difference in their concentrations (P region has more holes whereas N region has more free electrons).

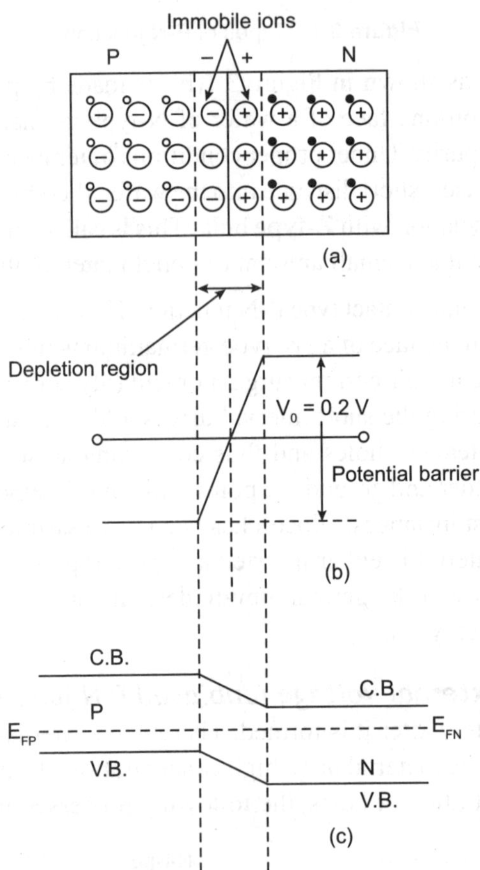

Figure 3.3 (a), (b) and (c) Potential barrier across the P-N junction

As free electrons move across the junction from N-type to P-type, positive donor ions are uncovered, i.e. they are robbed off free electrons. Hence a positive charge is built on the N side of the junction. At the same time, free electrons cross the junction and uncover negative acceptor ions by filling in the holes. Therefore, a net negative charge is established on the P side of the junction. When a sufficient number of donor and acceptor ions are uncovered, further diffusion is stopped. It is because a barrier has been set up against further movement of charge carriers. This is called potential barrier or junction barrier. Potential barrier is of the order of 0.1

to 0.3 volt. Figure 3.3(b) shows electrostatic potential difference across the P-N junction. How this potential barrier has developed? When a sufficient number of donor and acceptor ions are uncovered, further diffusion is prevented. It is because positive charges (ions) on N side repels holes to cross from P-type to N-type and negative charges (ions) on N side repels free electrons to enter from N-type to P-type. Because of this, difference in potential exists between the two sections, which inhibits further electron-hole combinations at the junction, and the Fermi level E_{FN} of two sides is the same as shown in Figure 3.3(c).

The region across the P-N junction in which the potential changes from positive to negative is called **depletion region**. The width of this region is of the order of 6×10^{-8}m. Since this region consists of immobile (fixed) ions which are electrically charged, it is also called space-charge or transition region. Outside this region on each side of the junction, the material is still neutral. The formation of depletion region is illustrated in Figure 3.3(d).

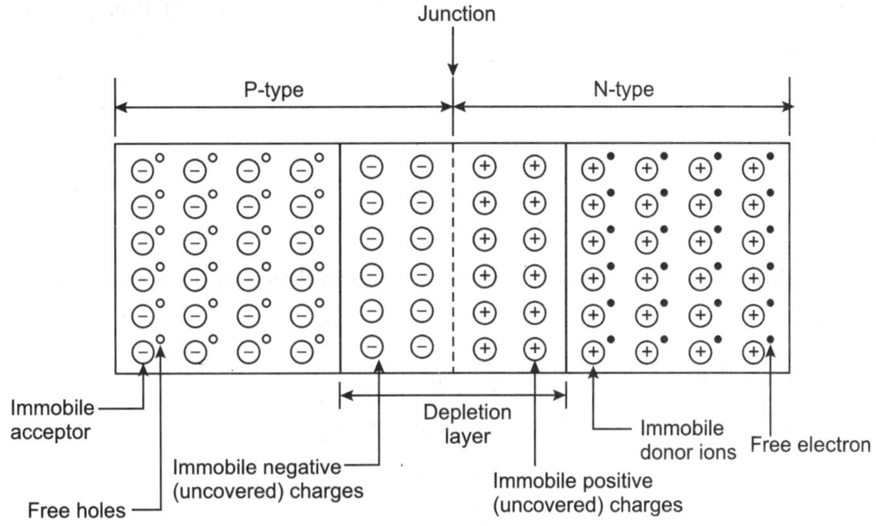

Figure 3.3 (d) Formation of depletion layer in the vicinity of an unbiased P-N junction

Potential barrier for a silicon P-N junction is about 0.7 V, whereas for a germanium P-N junction it is approximately 0.3 V.

A potential barrier discourages the diffusion of majority carriers across the junction. However, it helps minority carriers (few free electrons in the P region and few holes in the N region) to drift across the junction. Minority carriers are constantly generated due to thermal energy. However, electric current cannot flow since no circuit has been connected to the P-N junction.

3.2.1 Width of Depletion Layer and Height of Potential Barrier

The width of depletion layer in the P-N junction depends on the level of impurity doping in N and P regions. Higher the level of impurity doping, the thinner the depletion layer. In a higher doping level, more atoms of impurity are added to the intrinsic semiconductor and this increases the concentration of majority carriers in N and P regions. Obviously, under higher doping levels, when a majority carrier crosses the P-N junction, it finds a very large number of free charges with which it can recombine, e.g. when a hole crosses the junction from the P region to the N region, it gets a large number of free electrons, and recombines with one of them. Immobile uncovered charges are quickly produced and are close to the P-N junction. This reduces the width of the depletion region. However, when the level of doping is low, there will be a small number of mobile charges in the two regions. Due to this, mobile charges should move deeper into the region, after crossing the P-N junction, to get a free mobile charge for recombination. This increases the width of the depletion region. Let x_1 and x_2 be the width of depletion region in P and N sides, respectively, of a P-N diode junction, then the total width of depletion region is

$$x = x_1 + x_2$$

Similarly, if V_1 and V_2 are potential barriers in P and N sides of P-N junction, then net junction potential barrier (Figure 3.3(b)) is

$$V_0 = V_1 + V_2$$

Poisson's equation in one dimension is

$$\frac{d^2V}{dx^2} = \frac{\rho_1}{\varepsilon} \tag{i}$$

where ε is permittivity of the medium. Charge density in P side of depletion layer is given by

$$\rho = -eN_a \tag{ii}$$

where N_a represents the density of acceptor ions.

Equation (i) for P side of depletion layer is given by

$$\frac{d^2V}{dx^2} = \frac{\rho N_a}{\varepsilon} \tag{iii}$$

On integration, one obtains

$$\frac{dV}{dx} = \frac{eN_a}{\varepsilon}x + C_1 \tag{iv}$$

where C_1 is constant of integration using boundary condition.

i.e. at $$x = -x_1, \frac{dV}{dx} = 0$$

(since there is no variation of potential in the region excluding depletion layer). Equation (iv) gives

$$C_1 = \frac{eN_a}{\varepsilon}x_1$$

$$\therefore \frac{dV}{dx} = \frac{eN_a}{\varepsilon} x_1 + \frac{eN_a}{\varepsilon} x_1 = \frac{eN_a}{\varepsilon}(x + x_1) \tag{v}$$

On integration, one obtains

$$V = \frac{eN_a}{\varepsilon}\left(\frac{x^2}{2} + x_1 x\right) + C_2$$

where C_2 is another constant of integration. Now at $x = 0, V = 0$ (at junction)

$$\therefore C_2 = 0$$

Hence $V = \frac{eN_a}{\varepsilon}\left(\frac{x^2}{2} + x_1 x\right)$

At $x = -x_1, V = V_1$

$$\therefore V_1 = \frac{eN_a}{\varepsilon}\left[\frac{x_1^2}{2} - x_1^2\right] = \frac{eN_a x_1^2}{2\varepsilon}$$

$$\therefore |V_1| = \frac{eN_a}{2\varepsilon} x_1^2 \tag{vi}$$

If N_d is density of donor ions in N side of depletion layer, then Poission's equation for this becomes

$$\frac{d^2V}{dx^2} = -\frac{\rho}{\varepsilon} = -\frac{eN_d}{\varepsilon} \quad (\because \rho = eN_d) \tag{vii}$$

Proceeding as above and applying boundary conditions that at $x = 0$, $V = 0$ and at $x = x_2$, $\frac{dV}{dx} = 0, V = V_2$, one obtains

$$V_2 = \frac{eN_d x_2^2}{2\varepsilon} \tag{viii}$$

We have seen that a depletion layer is formed around the junction in a P-N diode when the diode is unbiased. This is due to the presence of immobile oppositely charged ions on the two sides of the P-N junction. Due to this, an electric potential develops across the P-N junction. This electric potential is called the **potential barrier** or **junction potential**. A practical significance of potential barrier is that a mobile charge may only cross the P-N Junction. Then it acquires enough energy to overcome the repelling force exerted on it by the electric field around the P-N Junction. For germanium and silicon, the heights of potential barriers are 0.3 V and 0.7 V at room temperature, respectively. In a P-N diode, this is also sometimes termed as knee voltage or cut-in voltage of the P-N diode.

The height of potential barrier across the junction is obtained as

$$V_a = |V_1| + |V_2| = e\left[\frac{N_a x_1^2 + N_a x_2^2}{2\varepsilon}\right] \tag{ix}$$

Since the crystal as a whole is electrically neutral, the number of charge carriers on both sides must be equal, i.e.

$$N_a x_1 A = N_d x_2 A$$

where A is the area of crystal.

$$\therefore \frac{x_1}{x_2} = \frac{N_d}{N_a} \qquad \text{(x)}$$

or

$$x_2 = \frac{N_a}{N_d} x_1$$

Substituting the value of x_2 in (ix), one obtains

$$V_o = \frac{\varepsilon N_a}{2\varepsilon} x_1^2 \left(1 + \frac{N_a}{N_d}\right)$$

This gives

$$x_1 \left[\frac{2\varepsilon V_o}{\varepsilon N_a \{1 + (N_a / N_d)\}}\right]^{\frac{1}{2}} = \left[\frac{2\varepsilon V_o}{e} \frac{(N_a / N_d)}{(N_a + N_d)}\right]^{\frac{1}{2}}$$

and

$$x_2 \left[\frac{2\varepsilon V_o}{e N_d \{1 + (N_d / N_a)\}}\right]^{\frac{1}{2}} = \left[\frac{2\varepsilon V_o}{e} \frac{(N_a / N_d)}{(N_a + N_d)}\right]^{\frac{1}{2}}$$

or

$$x = x_1 + x_2 = \left[\frac{2\varepsilon V_o}{e}\right]^{\frac{1}{2}} = \left[\left\{\frac{(N_d / N_a)}{(N_a + N_d)}\right\}^{\frac{1}{2}} + \left\{\frac{(N_a / N_d)}{(N_a + N_d)}\right\}^{\frac{1}{2}}\right]$$

or

$$x = \left(\frac{2\varepsilon V_0}{e}\right)^{\frac{1}{2}} \left[\left(\frac{N_a + N_d}{N_a N_d}\right)^{\frac{1}{2}}\right] \qquad \text{(xi)}$$

Eq. (xi) gives the width of depletion layer.

The width of depletion region changes with the square root of voltage. In order to exemplify the order of thickness commonly met with, let us consider the example of germanium P-N diode in which $N_d = 10^{21}/\text{m}^3$ and $N_a = 10^{23}/\text{m}^3$. For these concentrations across the junction $V_0 = 0.31$ volt, and since for germanium $\varepsilon_r = 16$ (here $\varepsilon = \varepsilon_r \varepsilon_0$), one finds that the width of the unbiased junction is $x = 0.75 \times 10^{-6}$ m or 0.75 micron. In this numerical example, $N_a \gg N_d$. Obviously, Eq. (xi) reduces to

$$x = \left(\frac{2\varepsilon V_0}{e N_d}\right)^{\frac{1}{2}} \qquad \text{(xii)}$$

This shows that heavier doping on the side of the junction extends the depletion layer to the other side. Moreover, the depth of penetration or width of depletion region varies inversely as the square root of doping density on the opposite side.

Because of the presence of space charges of opposite kinds across the junction, there is a voltage gradient confined within the depletion region. Obviously, the transition region thus acts like a parallel plate capacitor in which the plates are separated by distance x.

3.2.2 Unidirectional Property of a P-N Junction Diode

Unidirectional conduction property of a P-N junction diode is illustrated in Figure 3.3(e) and 3.3(f). Figure 3.3(e) shows that the positive terminal of battery E is connected to the P side and the negative terminal (through a bulb) to the N side of the P-N junction. The P-N junction allows the flow of current and the bulb glows. The P-N junction diode works like a closed circuit and we can say that the P-N junction is forward-biased.

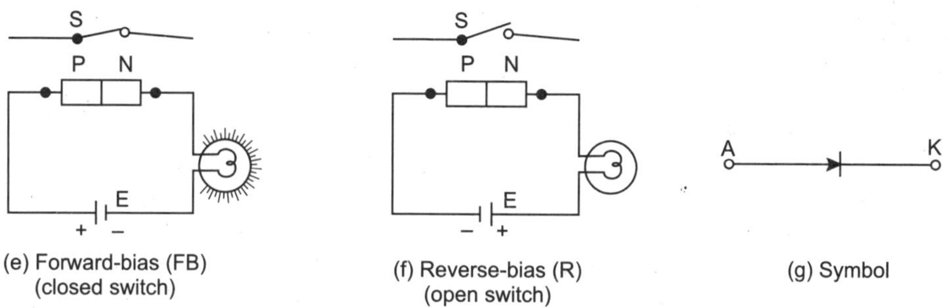

(e) Forward-bias (FB) (closed switch)

(f) Reverse-bias (R) (open switch)

(g) Symbol

Figure 3.3 (e) and (f) Unidirectional conduction property of a P-N junction diode

In Figure 3.3(f), the positive terminal of the battery is connected to the N side and the negative terminal to the P side. Obviously, the P-N diode is reverse-biased and no current can flow through the P-N junction diode and the bulb does not glow. Clearly, the P-N junction diode has a unidirectional conduction property.

3.2.3 Junction Capacitance of a P-N Junction Diode

We have seen that a depletion region is develops across the junction of a P-N diode as soon as the P-N junction is fabricated. When the P-N junction is unbiased, the depletion region offers a very high resistance to the mobile charges of in both sides of the junction, i.e. the depletion region behaves like an insulator because it is non-conducting and does not allow any charges to move across it. On the other hand, the P and N regions on either side of the depletion region have mobile charges, and hence offer low resistance to the flow of charges. One can compare this construction with a parallel plate capacitor, which has two metallic, conducting parallel plates and a dielectric non-conducting medium in between the two plates. The two regions N and P can be thought of as two parallel plates, and the depletion region between them may be thought of as a **dielectric**. Clearly, N and P regions along with the depletion region form a capacitor around the junction of a P-N diode as shown in Figure 3.3(h).

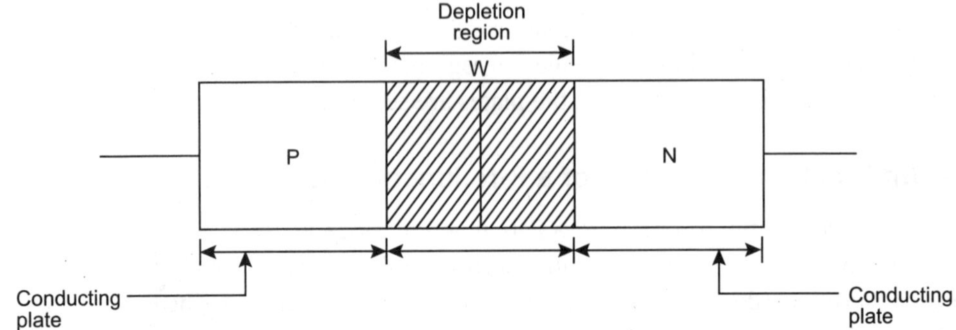

Figure 3.3 (h) P-N junction as a parallel plate capacitor

Junction capacitance is

$$C_j = \frac{dQ}{dV} = \frac{\varepsilon}{x} \text{ A Farad} \quad \text{(xiii)}$$

where A is area of the junction. Substituting (x) from (xi), one obtains

$$\frac{C_J}{A} = \left[\frac{\varepsilon e}{2V_0} \frac{N_a N_d}{N_a + N_d} \right]^{\frac{1}{2}} \frac{\text{Farad}}{\text{m}^2} \quad \text{(xiv)}$$

For a special case, $N_a \gg N_d$. Eq.(xiv) reduces to

$$\frac{C_J}{A} = \left[\frac{\varepsilon e}{2V_0} N_d \right]^{\frac{1}{2}} \quad \text{(xv)}$$

By substituting Eq. (xii) in Eq. (xiii), one obtains junction capacitance as

$$\frac{C_J}{A} = \left[\frac{\varepsilon e}{2V_0} N_d \right]^{\frac{1}{2}} \quad \text{(xvi)}$$

3.2.4 Effect of Temperature on Barrier Voltage

Barrier voltage depends on doping density and temperature. For a given junction, doping density remains constant; therefore, barrier voltage depends on temperature. With an increase in temperature, more minority charge carriers are produced, leading to a drift across the junction. Consequently, equilibrium occurs at a slightly lower barrier potential. Obviously, barrier potential decreases with increase of temperature. It has been found that for both germanium and silicon semiconductors,

$$\Delta V_0 \simeq -2t \text{ mV}$$

Here t°C denotes change in temperature, i.e. barrier potential decreases by about 2 mV per °C.

3.2.5 Mechanism of Current Flow in a P-N Junction

Electric current can flow in a P-N junction diode by following two distinct mechanisms:

(i) Drift current In a semiconductor (say N-type), free electrons continuously move in a random manner, and the direction of movement is changed after each collision with heavy immobile ions. Let us imagine a plane that cuts through the semiconductor as shown in Figure 3.3(i). During a given time, the number of free electrons crossing this plane from left to right is the same as that crossing from right to left. Obviously, the net number of electrons crossing the plane is zero, i.e. the average current is zero.

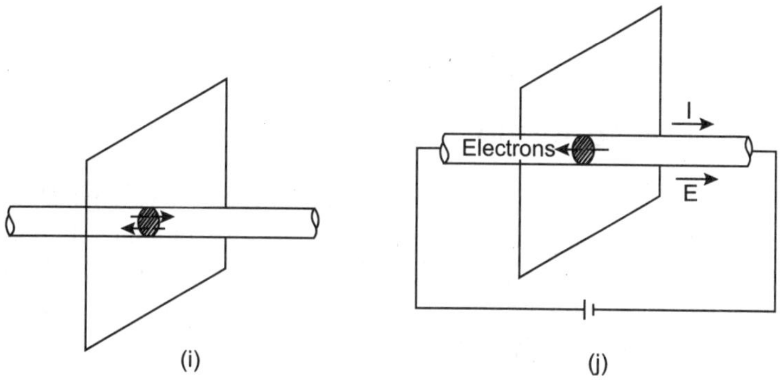

Figure 3.3 (i) and (j) Drift current in an N- type semiconductor

Now, let us suppose that we connect a battery across the semiconductor (Figure 3.3(j)). Clearly, a constant electric force is created, and due to this force free electrons now tend to move towards the positive terminal of the battery. In addition to random motion, these electrons drift towards the positive terminal of the battery (or against the direction of electric field E inside the semiconductor). This results in a net flow of electrons. Current I_{dr} that results from the drift of electrons (or holes in a P-type semiconductor) due to the electric field is termed as **drift current**. We may note that conduction of current takes place in metals due to this phenomenon.

(ii) Diffusion current In addition to drift current, the transport of charge-carriers may also occur due to diffusion. The phenomenon of diffusion is not ordinarily seen in metals. We have seen in a P-N junction that the concentration of charge carriers is not the same. The concentration of holes on the left side of the P-N junction is much higher than that on the right side. Due to thermal energy, holes are in random motion. One may expect that, in a given duration of time, more holes cross the P-N junction from left to right than from right to left. This net transport of holes across the P-N junction results in diffusion current. Clearly, diffusion current is the result of nonuniform concentration of charge carriers in the semiconductor.

Similarly, one can see that there is a net transport of electrons across the P-N junction from right to left. We may note that holes and electrons diffuse in opposite directions across the P-N junction. However, diffusion currents due to both is in the same direction (from left to right).

If there was no formation of potential barrier, the diffusion process would have continued till the concentration of holes (and of free electrons) becomes same on two sides of the P-N junction. Potential barrier opposes the diffusion of majority carriers across the P-N junction. However, the same barrier helps minority carriers to drift across the P-N junction. Minority carriers are constantly generated due to thermal energy. Obviously, there should be constant drift current due to these minority carriers crossing the P-N junction. Actually, this drift current (I_{dr}) is counterbalanced by diffusion current (I_{dif}) due to crossing of some majority carriers having high kinetic energy. We know that in a semiconductor (or in a metal), all charges do not possess the same kinetic energy. Some have very high energy, whereas some have quite low energy. The average energy depends upon the temperature of the sample. Net current across the P-N junction is zero, as it should be

$$I_{dr} + I_{dif} = 0$$

3.3 Biasing the P-N junction

One can change the behaviour of the P-N junction by applying a bias voltage. Bias voltage changes the width of the P-N junction's depletion region and, therefore, changes its resistance. The P-N junction conducts when it is forward-biased and does not conduct when it is reverse-biased.

3.3.1 Forward- and Reverse-Biasing

(i) Forward-biasing We have seen that the natural tendency of majority carriers (free electrons in the N-section and holes in the P-section) was to combine at the junction. This is how depletion region and potential barrier were formed. Actually, the combination of electrons and holes at the junction allows electrons to move in the same direction in both P- and N-sections. In the N-section, free electrons move toward the junction, while in the P-section, for holes to move towards the junction, valence electrons move away from the junction. Therefore, electron flow in both sections is in the same direction. This, of course, would be the basis of current flow.

With P-N junction alone, the action stops because there is no external circuit as well as due to potential barrier that has built up. For current to flow, a battery can be connected to the diode to overcome potential barrier. Polarity of the battery should be such that majority carriers in both sections are driven toward the junction. When the battery is connected in this way, it provides forward-bias, causing high current to flow, because it allows majority carriers to provide current flow.

To apply forward-bias, the positive terminal of the battery is connected to P-type and negative terminal to N-type as shown in Figure 3.4. The applied forward potential establishes an electric field which acts against the potential barrier field. Obviously, the resultant field is weakened, and barrier height is reduced at the P-N junction as shown in Figure 3.4(b) and (c).

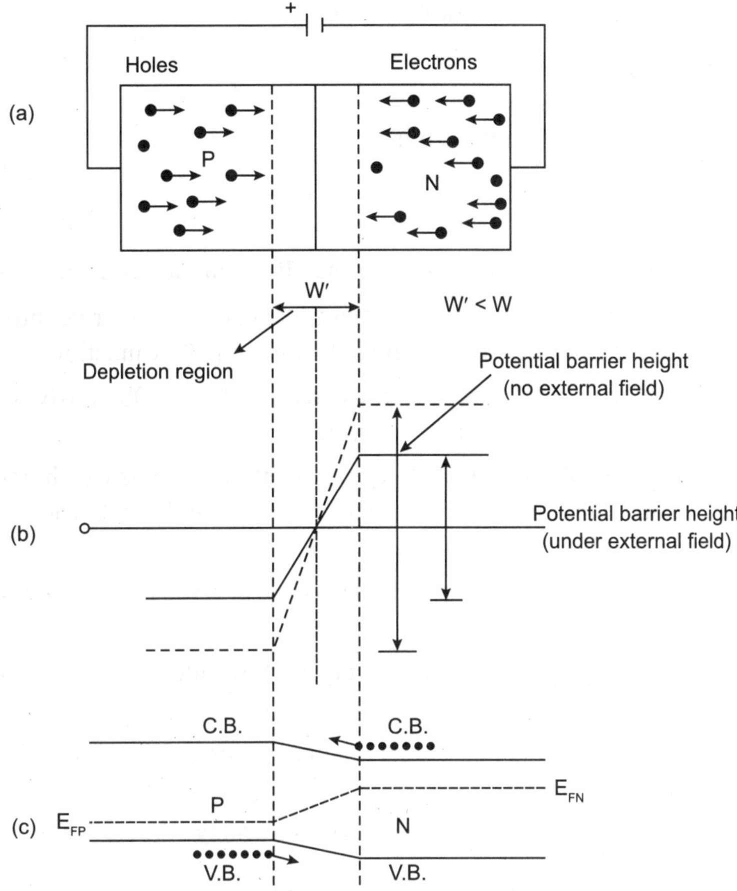

Figure 3.4 P-N junction under forward-bias

Since potential barrier height is very small (~0.2 V), a small forward voltage is sufficient to completely eliminate the barrier. Obviously, at some forward voltage, potential barrier at the P-N junction can be eliminated altogether. Then junction resistance will become almost zero and a low resistance path is established for the entire circuit. Thus, a large current is generated in the circuit even when a small potential difference is applied. Such a circuit is called forward-bias circuit, and the current is called forward current. The salient features of forward-bias circuit are summarized below:

At some forward voltage, potential barrier is eliminated altogether.

The P-N junction offers low forward resistance (r_f) to the current flow.

The magnitude of current in the circuit due to the establishment of low resistance path depends upon the applied forward voltage, and it reduces as voltage is increased there-by forward current increases (Figure 3.5). It is given by

$$I = I_0(e^{V/\eta V_T} - 1) \tag{i}$$

where V is applied voltage,

$V_T = \dfrac{T}{11600}$, here T is temperature in Kelvin,

$\eta = 1$ for Ge and 2 for Si.

The mechanism of current flow in a forward-biased P-N junction is as follows:

(i) Free electrons from the negative battery terminal continue to arrive into the N region while free electrons in the N region move towards the P-N junction.

(ii) Electrons travel through the N region as free electrons. Obviously, current in the N region of P-N junction is by free electrons.

(iii) When these free electrons reach the P-N junction, they combine with holes and become valence electrons. Since a hole is in the covalent bond and hence when a free electron combines with a hole, it becomes a valence electron.

(iv) Current in the P region is by holes. Electrons travel through the P region as valence electrons.

(v) These valence electrons after leaving the crystal flow into the positive terminal of the battery.

(vi) The typical amount of external voltage required to forward-bias the junction is 0.7 V (Si) and 0.3 V (Ge).

(vii) Power semiconductors may have greater forward voltage drops, depending on current magnitude.

(viii) When the junction is forward-biased, the depletion layer decreases and the P-N junction conducts.

As a consequence of forward-biasing, the following development takes place:

(i) Applied voltage(V_f) opposes the contact electric potential (V_0).

(ii) The net barrier potential (V_B) is reduced and the depletion layer becomes thin.

(iii) Under forward-bias, majority carriers which have less kinetic energy are now able to cross the reduced potential barrier. Obviously, diffusion current increases.

(iv) Drift current (due to crossing of minority carriers) also reduces slightly.

(v) Forward-biasing current is the difference of diffusion current and drift current, i.e.

$$I_f = I_{dif} - I_{dr}$$

(vi) With increase in external applied voltage V_f, forward current increases sharply.

Current flow in a forward-biased P-N junction is illustrated in Figure 3.5.

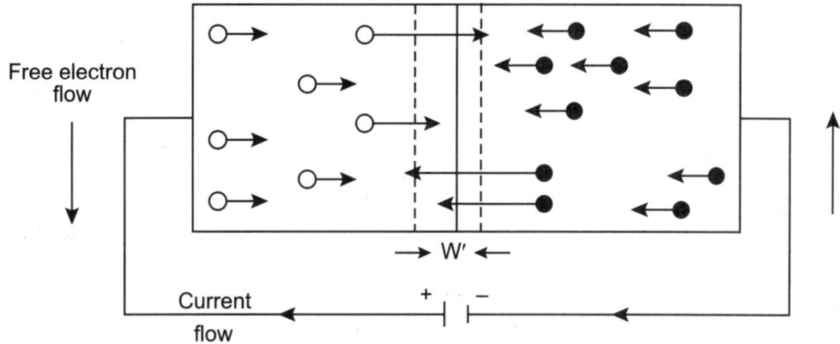

Figure 3.5 Current flow under forward-bias

(ii) Reverse-biasing For forward current flow, the battery must be connected to drive majority carriers towards the junction where they combine to allow electrons to enter and leave the P-N junction. If battery connections are reversed, the potential at the N side will draw free electrons away from the junction, and negative potential at the P side will attract holes away from the junction. With this battery connection, majority carriers cannot combine at the junction, and majority current cannot flow. For this reason, when a voltage is applied in this way, it is called reverse-bias.

However, reverse-bias can cause reverse current to flow, because minority carriers are present in the semiconductor sections. Although the P-section is doped to have excess holes, yet some electrons are freed because of thermal agitation. Also, although the N-section is doped to have excess free electrons, some electrons are freed to produce holes in the N-section. Free electrons in the P-section and holes in the N-section are minority carriers. Now, with reverse-bias, one can see that battery potentials repel minority carriers toward the junction. As a result, these minority carriers cross the P-N junction in exactly the same way that majority carriers did with forward-bias. However, since there are much fewer minority carriers than are majority carriers, this minority current or reverse current is much less and is of the order of μA, with the same voltage than majority or forward current would be. Reverse-biases P-N junction is shown in Figure 3.6. The salient features of reverse-biased P-N junction are as follows:

The height of potential barrier is increased and the width of depletion region also increases (Figure 3.6 (b) and (c)).

The reverse-biased P-N junction offers very high resistance to current flow. This resistance is called **reverse resistance** (R_r).

Due to the establishment of high resistance path, very small current flows in the circuit. This current is usually called leakage current or reverse saturation current.

Current in the above situation is given by

$$I = I_0(e^{-V/\eta V_T} - 1) \tag{ii}$$

Since applied voltage V is negative and as it increases, the first term in R.H.S. reduces very fast and current from Eq. (ii) reduces to $I = (-)I_0$, which is quite low and constant. Here I_0 is reverse saturation current.

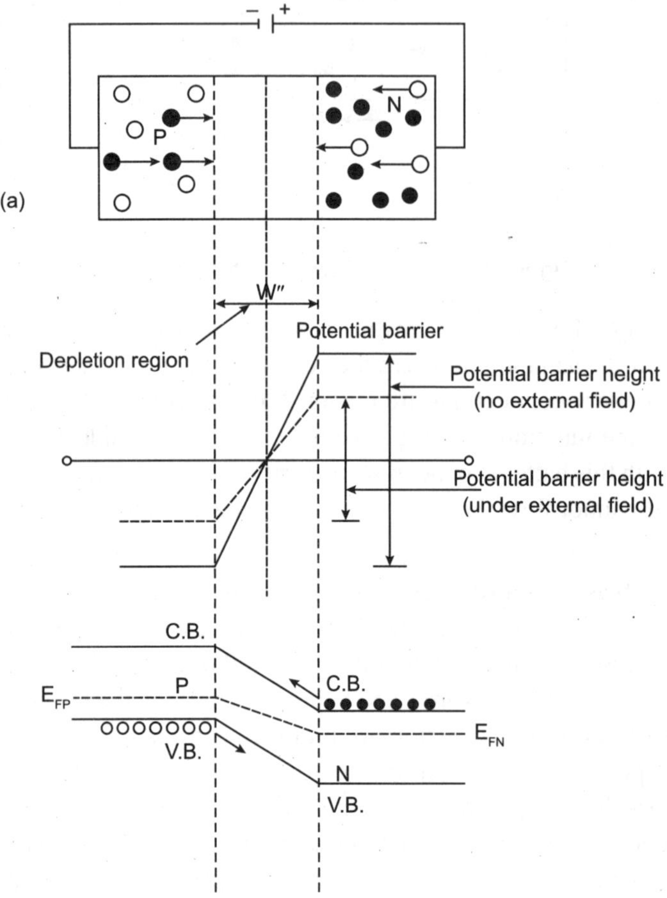

Figure 3.6 (a) P-N junction under reverse-bias

Due to reverse-biasing, the following things happen:

(i) Applied voltage (V_r) strengthens contact potential (V_0).

(ii) Net barrier potential (V_B) is increased, and the depletion layer becomes thick.

(iii) Diffusion current due to flow of majority carriers reduces to almost zero.

(iv) There is slight increase in drift current due to the flow of minority carriers.

(v) With increase in external applied voltage (V_r), current (I_r) almost remains constant until breakdown occurs

$$I_r = I_{dr} - I_{dif} = I_{dr}$$

As soon as a minority carrier is generated due to thermal energy, it is swept (or drifted) across the P-N junction due to high potential barrier. Current due to the flow of minority carriers does not change with applied voltage V_r. This is why this current is termed as **reverse saturation current** I_s. I_s is very small.

3.3.2 Reverse Saturation Current (I_s or I_0)

In a reverse-biased diode, there is practically no current due to majority carriers. However, there is small current due to the flow of minority carriers across the P-N junction. Minority carriers (holes in N-type region and electrons in P-type region) are produced due to thermal agitation. The applied voltage drives these minority carriers across the P-N junction (whether forward-biased or reverse-biased). Obviously, these minority carriers constitute a small current called reverse current or reverse saturation current. This reverse saturation current **(I_s or I_0)** is extremely **temperature dependent**. We must remember that forward current is also temperature dependent but to a much less degree because minority current forms a very small percentage of majority current. It is found that for every 10°C rise in temperature for G_e, I_0 becomes double, whereas for Si, I_0 becomes double for every 6°C.

3.3.3 Summary of Biasing Conditions

Three biasing (n bias, forward-bias and reverse-bias) conditions for a semiconductor diode are depicted in Figure 3.6 (b to d). With no bias ($V_D = 0$ V) diode current $I_D = 0$. With forward-bias ($V_D > 0$ V), the current I_D flows in the same direction as the arrow of symbol indicates. With reverse-bias ($V_D < 0$ V), direction of I_s is against the arrow of the symbol.

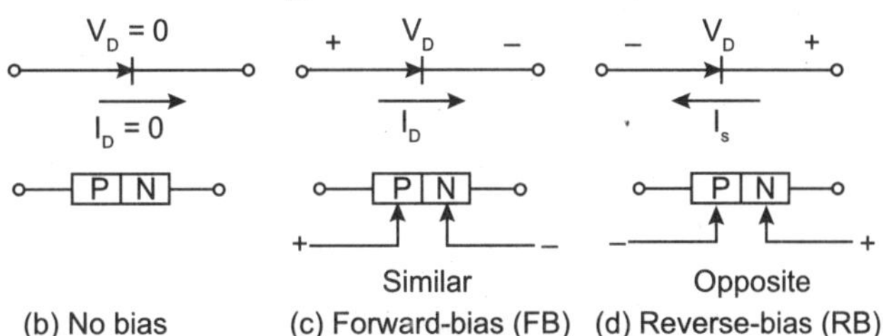

Figure 3.6 Biasing conditions for a semiconductor P-N junction diode

3.3.4 Reverse Breakdown Voltage

Excessive reverse-bias will breakdown and possibly destroy the P-N junction. When the voltage is large enough, the P-N junction's resistance drops rapidly, a very high current develops, and the P-N junction is destroyed. The voltage at which this happens is called reverse breakdown voltage, peak inverse voltage or avalanche voltage.

There are certain types of semiconductor devices designed to be operated in the reverse breakdown area. Zener diode is the most common of these devices.

3.3.5 Rating Characteristics

Diodes have several characteristics that must be considered when using them in circuits. The most common ones are outlined in Table 3.3.

Table 3.3 Diode characteristics (ratings)

Characteristic (rating)	Description
Peak inverse voltage (PIV)	Maximum recurrent (or repetitive) reverse voltage the diode is rated to withstand. Exceeding this voltage will destroy the diode. Sometimes called peak inverse voltage or **avalanche voltage**.
Average forward current (I_0)	Maximum allowed DC forward current. Exceeding this current will destroy the diode.
Forward power dissipation [$P_{D(\max)}$]	Maximum power that can be dissipated by the diode when it is forward-biased. Exceeding this power limit will destroy the diode.

3.4 Volt-Ampere Characteristic of P-N Junction

The P-N junction can be represented as shown in Figure 3.7. The arrow head represents the P-section of the crystal and shows the direction of flow of holes or conventional current. Since the P-N junction diode's resistance changes according to the direction of current flow, it is called a nonlinear device. Basically, its nonlinearity is dependent on the polarity of applied voltage. For current in forward direction, it has a resistance of only a few hundred ohms. In reverse direction, its resistance is often close to 100,000 Ω.

Figure 3.8 shows the circuit arrangement for drawing the V-I characteristics of a P-N junction diode.

Figure 3.7 Symbol of P-N junction

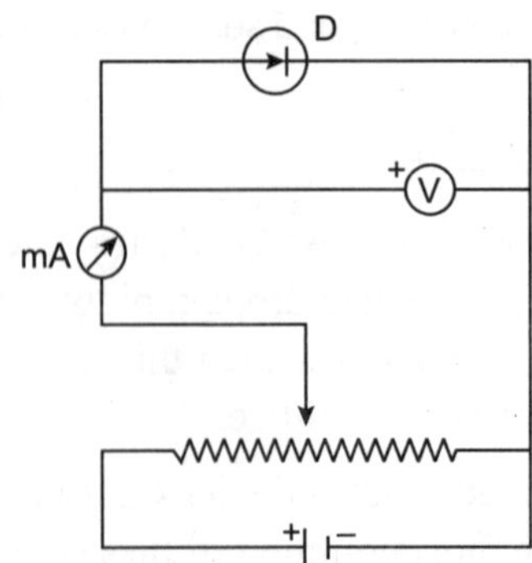

Figure 3.8 Circuit arrangement for drawing V-I characteristics of a P-N junction

A graph between potential difference across the P-N junction and current through the junction is called the V-I characteristic of the P-N junction diode, and is shown in Figure 3.9.

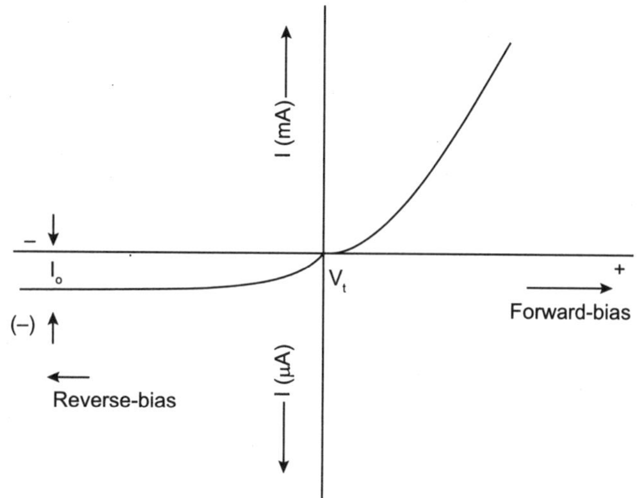

Figure 3.9 V-I characteristic of a P-N junction diode

In forward direction, more current flows and current for the most part increases linearly as bias voltage is increased. In forward direction, the P-N junction can be considered a linear device over a large portion of its operating curve. The small portion of the curve that is just above zero bias is nonlinear and current does not flow till the applied forward-bias voltage attains a minimum value V_t known as cut-in voltage. This results because both majority and minority current actually comprise the overall current. Since minority carriers are low energy carriers, majority current starts first, and then as voltage is increased in reverse direction, minority current joins in, causing a nonlinear rise in current. But as voltage is increased further (in negative direction), minority current becomes saturated and attains a value I_o since there are only few minority carriers.

Because of non-linearity of the curve, if a very small signal voltage is applied to the diode so that it only operates around the knee, the signal will be distorted. The signals must be large enough so that they operate mostly over the linear part of the curve.

When reverse-bias is applied, a slight reverse current flows. This reverse current increases only negligibly as bias voltage is increased.

3.5 Diode Equation (V-I Equation)

Figure 3.10 shows an open circuit P-N junction with switch S open. Let p_p and n_p represent holes and electron densities, respectively in P region, and n_n and p_n represent electrons and hole densities in N region, respectively. The densities of holes in P region and densities of holes in N region are related by Boltmann relation as

$$p_p = p_n \exp(V_B / V_T) \tag{1}$$

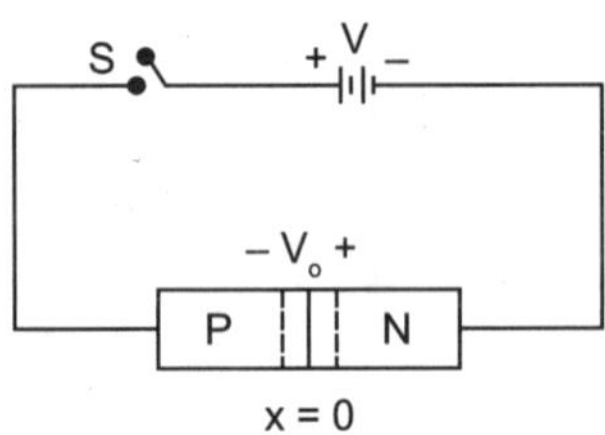

Figure 3.10 (a) Open circuit P-N junction with switch *S* open

where V_B is barrier potential across the depletion layer and V_T is volt equivalent of temperature defined as

$$V_T = \frac{\bar{k}T}{e} = \frac{T}{11,600}$$

where $\bar{k}$ is Boltzmann constant in joule per Kelvin and k is Boltmann constant in electron volt per Kelvin ($\bar{k} = 1.60 \times 10^{-19}$ K). Obviously, $\bar{k} = 1.381 \times 10^{-23}$ J/K and $k = 8.620 \times 10^{-5}$ eV/K. For an open-circuit P-N junction, we have $V_B = V_0$. Thus,

$$p_p = p_n \exp(V_0 / V_T) \tag{2}$$

Let us consider that the junction is biased in forward direction by applying voltage V, i.e. by closing switch S. Now barrier voltage V_B is decreased from its equilibrium value V_0 by an amount V, i.e. $V_B = V_0 - V$. With forward-bias, hole density in the P region remains constant up to the depletion region. While in N region at the junction, it increases from p_n to $p_n + \Delta p_n$ due to the diffusion of holes across the P-N junction. As holes diffuse further in N region, they combine with electrons, and hence their density decreases with increase in distance from the P-N junction. Ultimately, at a large distance, it becomes same as p_n. Now, one can express hole density in the N region by Boltzmann relation as

$$p_p = (p_n + \Delta p_n) \exp \frac{(V_0 - V)}{V_T}$$

$$= (p_n + \Delta p_n) \exp(V_0 - V) \exp\left(\frac{V}{V_T}\right) \tag{3}$$

Now, substituting the value of p_p from (1) in (3), one obtains

$$p_n \exp\left(\frac{V_0}{V_T}\right) = (p_n + \Delta p_n) \exp\left(\frac{V_0}{V_T}\right) \exp\left(\frac{V}{V_T}\right)$$

or

$$p_n = (p_n + \Delta p_n) \exp\left(\frac{-V}{V_T}\right)$$

or $$p_n \exp\left(\frac{V}{V_T}\right) = (p_n + \Delta p_n)$$

or $$\Delta p_n = p_n \left[\exp\left(\frac{V}{V_T}\right) - 1\right] \tag{4}$$

From eq. (1), we have

$$p_n = p_n \exp\left(\frac{-V_0}{V_T}\right) \tag{5}$$

Using (5) in (4), one obtains

$$\Delta p_n = p_p \exp\left(\frac{-V_0}{V_T}\right)\left[\exp\left(\frac{V}{V_T}\right) - 1\right] \tag{6}$$

We know that the diffusion of hole constitutes hole current and hole current is proportional to Δp_n. Thus,

$$I_p \alpha \ \Delta p_n$$

or $$I_p \, \alpha \, \exp\left(\frac{-V_0}{V_T}\right) \exp\left[\frac{V}{V_T} - 1\right]$$

or $$I_p = I_{sp} \exp\left[\frac{V}{V_T} - 1\right] \tag{7}$$

where I_{sp} denotes the constant of proportionality.

Similarly, one can obtain an expression for electron current due to diffusion of electrons from N region to P region as

$$I_n = I_{sn} \exp\left[\frac{V}{V_T} - 1\right] \tag{8}$$

Total current I becomes

$$I = I_p + I_n$$

$$= I_{sp} \exp\left[\frac{V}{V_T} - 1\right] + I_{sn} \exp\left[\frac{V}{V_T} - 1\right]$$

$$= I_0 \exp\left[\frac{V}{V_T} - 1\right] \tag{9}$$

where I_0 is called saturation current. Eq. (9) is the diode current equation and generally expressed as

$$I = I_0 \exp\left[\left(\frac{V}{\eta V_T}\right) - 1\right] \tag{10}$$

where η is constant, which depends upon the material property and has a value 1 for Ge and 2 for Si.

3.5.1 For Forward-Biased Junction

The value of V will be positive when the junction is forward-biased and we have

$$I_f = I_0 \exp\left(\frac{V}{\eta V_T}\right) \tag{11}$$

(assuming V to be large)

Eq. (11) shows that for a given temperature, forward current increases exponentially with voltage V except for a small value of V.

3.5.2 For Reverse-Biased Junction

When the junction is reverse-biased, we have

$$I_r = I_0 \exp\left[\left(-\frac{V}{\eta V_T}\right) - I\right]$$

For reverse-bias whose magnitude is large compared with V_T (~26 mV) at room temperature, one finds

$$\exp\left(\frac{-V}{\eta V_T}\right) \to 0$$

or

$$I_r \to -I_0 \tag{12}$$

Here (–) sign shows the direction of current.

Hence I_0 is called **reverse saturation current**. We may note that this is constant independent of the applied reverse voltage as shown in Figure 3.9.

3.6 Static and Dynamic Resistance of a Diode

One of the most important properties of the diode is its resistance in forward- and reverse-biasing. An ideal diode must offer zero resistance in forward-bias and infinitely large resistance in reverse-bias. In reality no diode can act as an ideal diode, i.e. a diode does not behave as a perfect conductor when forward-biased and as a perfect insulator when reverse-biased. We consider two resistances of the diode: (i) DC or static resistance, (ii) AC or dynamic or small signal resistance.

(i) DC or static resistance When a P-N junction diode is forward-biased, it offers a definite resistance in the circuit. This resistance is called DC or static resistance (R_F) of the diode. DC resistance of a diode is simply the ratio of DC voltage across the diode to DC current flowing though it at a particular instant.

$$R_F = \frac{V}{I}$$

where R_F is resistance of the resistor and V is voltage drop across the resistor. R_F is called static or DC resistance of the diode. It is resistance offered by a diode to the DC current.

(ii) AC or dynamic resistance AC or dynamic resistance of a diode, at a particular DC voltage, is equal to the reciprocal of the slope of the characteristic at that point, i.e.

$$r_{ac} = \frac{\text{Change in voltage}}{\text{Resulting change in current}} = \frac{\Delta V}{\Delta I} = \frac{1}{(dI/dV)}$$

AC resistance (r_{ac}) of an ideal diode at operating current may be determined with the help of Figure 3.10(b).

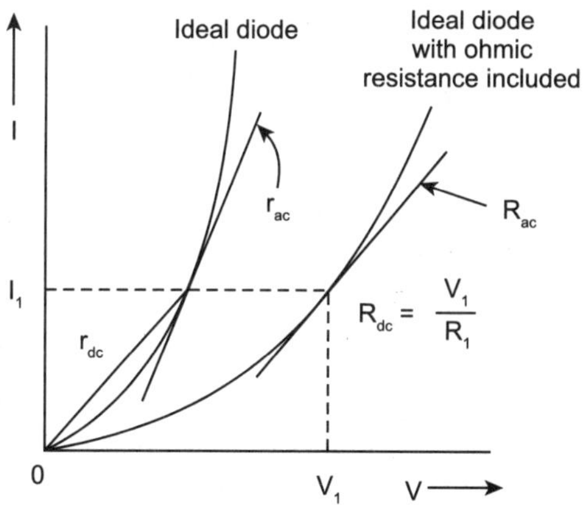

Figure 3.10 (b) Resistance of diode

It is not a constant quantity but depends on the operating voltage. For forward-bias greater than internal junction voltage, dynamic resistance varies inversely with current, but for reverse-bias its value is much larger.

Under forward-bias for a large applied voltage, diode current at junction temperature T is given as

$$I = I_0\left[\exp\left(\frac{eV}{\eta kT}\right) - 1\right]$$

where $\eta = 1$ for Ge and 2 for Si. Differentiating the above w.r.t. V, one obtains

$$\frac{dI}{dV} = I_0\left(\frac{e}{\eta kT}\right)e^{eV/\eta kT} = I_0\left(\exp\frac{eV}{\eta kT}\right)$$

$$\therefore r_f = \frac{dV}{dI} = \frac{\eta kT}{Ie} = \frac{\eta kT}{Ie} = \eta\left(\frac{k}{e}\right)\frac{T}{I}$$

$$= \eta\left(\frac{1.38\times10^{-23}}{1.6\times10^{-19}}\right)\frac{T}{I}$$

or
$$r_f = 8.625\times10^{-5}\,\eta\left(\frac{T}{I}\right)$$

AC resistance of a real diode is then

$$R_{ac} = r_{ac} + R_P + R_N$$

At room temperature, i.e. at 300K, the value of r_{ac} is $\frac{0.026}{I}\,\Omega$. For $I = 1$ mA, $r_{ac} = 26\ \Omega$.

3.7 Space Charge (or Depletion Region or Transition Region) Capacitance

When no external voltage is applied, width (d) of the depletion region of a P-N junction diode comes out to be about 5×10^{-7} m and is mostly in the N-type material. The movement of majority carriers, i.e. holes and electrons. across the junction causes opposite charge to be stored at distance '*d*' apart. This is effectively a parallel plate capacitor whose capacitance C_T (often called space-charge capacitance) is approximately 20 pF with no external bias.

As forward-bias is applied, the depletion region decreases and capacitance (C_T) increases. Under reverse-bias conditions, the depletion region increases and C_T decreases. This property of voltage variable capacitance is made use of in varactors, varicaps or voltacaps (refer section 3.11.10).

3.7.1 Space Charge or Transition Capacitance

The thickness of depletion layer changes with voltage applied across the junction. Reverse-biasing of a P-N junction, causes majority carriers to move away from the junction, thereby uncovering more immobile charges. This increases the thickness of depletion region. Similarly, forward-biasing of a P-N junction decreases the amount of uncovered charge and hence the thickness of depletion layer. This change in uncovered charge with applied voltage may be considered a capacity effect. It gives rise to transition capacitance or junction capacitance C_T given by

$$C_T = \frac{dQ}{dV} = \frac{\varepsilon A}{x}$$

Here x is thickness of the depletion region and A is junction cross-sectional area. One can show that width x of the depletion region is given by

$$x = \left[\frac{2\varepsilon V_0}{e}\frac{N_a + N_d}{N_a N_d}\right]^{\frac{1}{2}}$$

$$\therefore\ C_T = \left[\frac{\varepsilon e}{2V_0}\frac{N_a N_d}{N_a + N_d}\right]^{\frac{1}{2}} A$$

or $$C_T \propto V_0^{-1/2}$$

Obviously, transition or junction capacitance is not a constant but depends on applied voltage and hence it is also known as voltage-variance capacitance. One can also calculate C_T by the following relation:

$$C_T = \frac{K}{(V_B - V)^n}$$

where K is constant whose value depends on the type of semiconducting material (Si or Ge). V_B is barrier potential whose value is 0.3 V for Ge and 0.7 for Si. V is applied reverse-biased voltage and n is a constant whose value depends on the mechanism of fabrication of the P-N junction. A typical value of C_T under reverse-biased conditions is 8 pF and it is about 20 pF under unbiased conditions.

3.7.2 Diffusion or Storage Capacitance (C_D)

When forward-biased, capacitance around is called diffusion or storage capacitance (C_D). This arises due to injected charge stored near the junction outside the transition region and is defined as the rate of change of injected charge with applied voltage, i.e.

$$C_D = \frac{dQ}{dV} = \frac{\lambda I}{\eta V_T}$$

C_D is found to be proportional to current (I) and is much larger, and $C_T C_D$ has a larger value in forward direction and smaller value in reverse direction. The value of C_D in forward direction varies from 500 to 200 pF. Obviously, C_D predominates under forward-bias and arises due to minority carriers, while C_T predominates under reverse-bias and arises due to majority carriers. Here λ is the mean life of majority carriers in the depletion region, I is forward current across the junction at the time of abrupt change in biasing conditions, η is a constant having value 1 and 2 for Ge and Si, respectively, and V_T is volt equivalent of temperature. $V_T = 0.026$ V at room temperature (300 K).

3.8 Effect of Temperature on P-N Junction Diodes

We know that an increase in temperature will boost the generation of electron-hole pairs in semiconductors and increase their conductivity. Considering the energies of carriers crossing the depletion region in a P-N junction diode, an involved calculation yields the following relation between diode current, voltage and temperature.

$$I = I_0\left[\exp\left(\frac{V}{\eta V_T}\right) - 1\right]$$

Here

I = diode current (forward if positive, reverse if negative)

I_0 = diode reverse current, called reverse saturation current, at temperature T

V = diode voltage, positive for forward-bias, negative for reverse-bias; in volts

η = 1 for germanium and 2 for silicon

V_T = T/11600, a quantity in volts, dependent upon temperature and is known as volt-temperature equivalent

T = temperature of the diode junction (K).

On increasing the temperature, the forward characteristic (Figure 3.9) shifts to the left, showing an increase in current for the same voltage, and it shifts to the right when temperature is decreased, showing biased current. Similarly, the reverse characteristic shifts downwards, i.e. for the same voltage current increases.

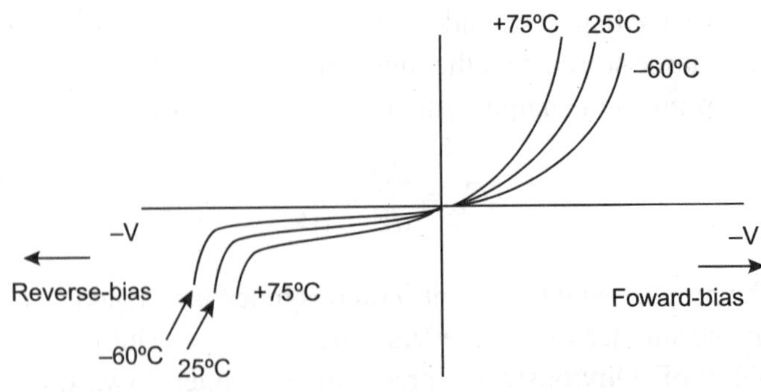

Figure 3.10 (c) Variation of diode characteristic with temperature

3.9 Breakdown of a P-N Junction Diode (Junction Breakdown)

In reverse-biased condition, minority electrons of the P region in a P-N junction drift towards the junction, due to negative potential of the DC source. Similarly, minority holes of the N region drift towards the junction under the influence of positive potential of the DC source. Minority carriers that cross the P-N Junction constitute a small current across the P-N junction. This current is called **reverse saturation current**. Now, if reverse voltage is increased, reverse saturation current also increases due to the crossing of P-N junction by more minority carriers. However, if reverse-bias voltage is further increased, a situation develops when reverse-saturation current abruptly shoots up to a very high value. This happens due to the breakdown of lattice structure of the P-N junction diode as a result of the development of very high electric field across the P-N junction. This situation is termed as breakdown of the P-N diode. The diode will continue to regulate the voltage until diode current falls to knee current I_{Zk} (Figure 3.9(c)). Depending upon the nature of the semiconductor and its doping, breakdown voltage in the diode ranges from about 3 volt to several hundred volts. The breakdown phenomenon is reversible and harmless so long as safe operating temperature is maintained. The mechanism of diode breakdown at reverse voltage is explained hereunder. The possible ways for diode breakdown are (i) avalanche breakdown and (ii) zener breakdown.

(i) Avalanche breakdown In this mechanism, minority charge carriers (electrons in P-type and holes in N-type) acquire sufficient energy from the applied reverse voltage to produce new charge carriers by removing valence electrons from covalent bonds. The new charge carriers in turn produce additional charge carriers and the process multiplies to give a large reverse current. The diode is then said to be in the region of avalanche breakdown; usually a junction with a wide depletion region (low field intensity) breaks down by this mechanism. With an increase in temperature, vibrations of atoms in the crystal increase the possibility of collisions of charge carriers with lattice atoms and reduce the possibility for carriers to gain sufficient energy to start the avalanche process. Thus, at higher voltages, the avalanche process is prominent to cause diode breakdown. Operating voltages in such diodes range from several volts to a few hundred volts.

(ii) Zener breakdown In this mechanism, the breakdown is initiated through direct rupture of covalent bonds rather than s due to the existence of a strong electric field across the s depletion layer. A junction having a narrow depletion region (and hence high field intensity $E = V_r/d$) will break down by this mechanism.

An increase in temperature will increases the energy of valence electrons and makes it easier for these electrons to escape from covalent bonds. Smaller applied voltage is, therefore, required to pull these electrons from the crystal lattice. The zener effect plays an important role only in diodes with breakdown voltage below 6 volts. The zener diode is always used in reverse-biased condition (Figure 3.29b). The range of voltages above the breakdown voltage in which a zener diode conducts in the reverse direction is called **tolerance voltage**.

3.9.1 Silicon vs. Germanium Diode

The V-I characteristics of typical silicon and germanium diodes are shown in Figure 3.11(a). We may note that there are different scales of voltage and current in forward-bias and reverse-bias regions.

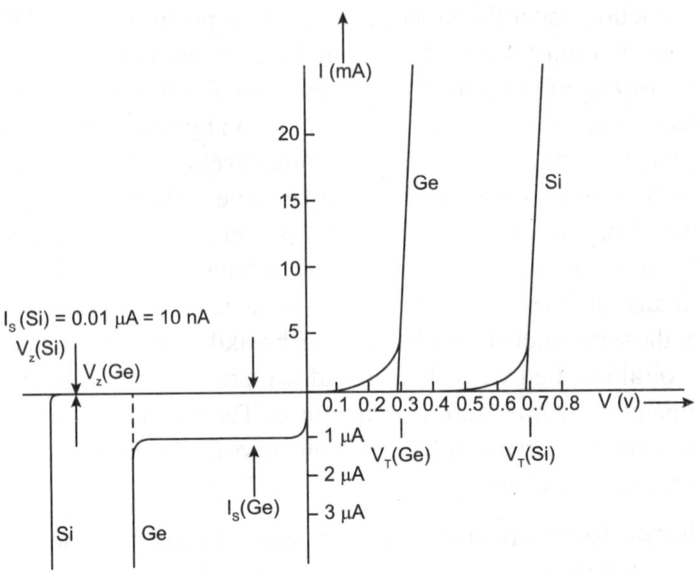

Figure 3.11 (a) Silicon vs germanium diodes: a comparison

In the figure, the vertical scale in forward-bias is in mA, whereas in reverse-bias it is in μA. It is thus easy to see the details of current variation in reverse-bias region. We note that the horizontal scale in forward-bias region has a maximum of 1 volt. Clearly, the voltage across a forward-biased diode never exceeds 1V. We may also note how current rises quickly beyond the knee of the curve. The shape of the curve in this region is just like a knee and this is why voltage V_T is also called **knee voltage**. (V_T = 0.7 V for Si diodes, V_T = 0.3 V for Ge diodes).

The only disadvantage in an Si diode as compared to a Ge diode is the higher value of V_T (= 0.7 V). The closer the forward-bias curve to the vertical axis (i.e, less the value of V_T), the more "ideal" is the diode.

An ideal diode should not conduct at all in reverse-bias condition. However, for a silicon diode, reverse saturation current I_s is not zero, although it is much lower than that for a germanium diode. I_s for Si diode is of the order of 10 or 20 μA, where as for Ge diode I_s is of the order of 1–10 μA.

A Si diode has much higher breakdown voltage as compared to Ge diode. This is why Si diodes find much use in applications where temperatures may rise to over 200°C, whereas Ge diode has a lower maximum temperature rating (85°C). One can easily understand why reverse saturation is lower for Si diodes. Forbidden energy gap for Si (E_g = 1.12 eV) is more than that for Ge (0.72 eV). Obviously, at a given temperature, less number of bonds will be broken in Si. Reverse saturation current is due to the drift of minority carriers and so its value is comparatively less in a Si diode than in a Ge diode. For most commercial applications, Si diodes are considered a better choice.

3.10 Diode Circuit Analysis

Diode circuit analysis may be more difficult than it might appear at first. Remember that many of the circuits analysis tools you have previously used expect the circuit to be bilateral; that is, all the elements must work the same both ways. Clearly, diodes do not behave in this way. Therefore, tools such as superposition and Kirchhoff's laws must be used very carefully.

Example 1 The following illustrates the effect of 0.7 V barrier voltage in a simple series circuit. Referring to Figure 3.11(b), calculate voltage drop for R_1 and D_1 (silicon) and total circuit current. [B.Tech]

Solution

Using Ohm's law,

$$V_T = V_{D1} + V_{R1}$$

where

$$V_{D1} = 0.7 \text{ volts (silicon)}$$

$$V_{R1} = V_T - V_{D1} = 12 - 0.7 = 11.3\text{V}$$

Ohm's law:

$$I_T = \frac{V_{R1}}{R_1} = \frac{11.3\text{V}}{1.200\ \Omega} = 9.42 \text{ mA}$$

Note: One can also calculate voltage drop for a germanium diode with barrier voltage of 0.3 V.

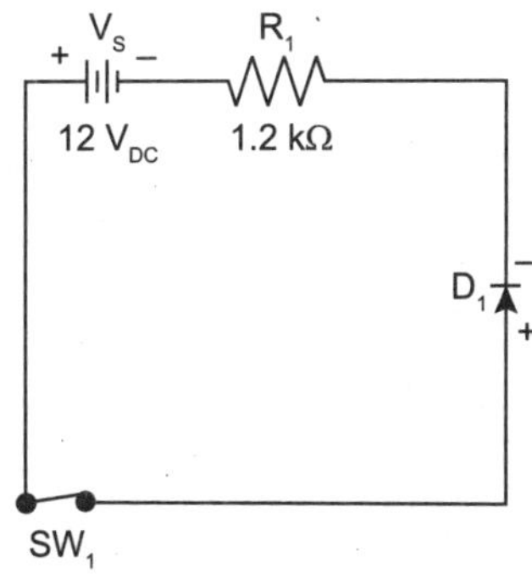

Figure 3.11 (b) A simple DC diode circuit

Diode and AC circuit analysis We know that a diode is polarity sensitive. When forward-biased, the cathode is negative and the anode is positive. Likewise, when reverse-biased, the cathode is positive and the anode is negative. This polarity can be used in an AC circuit for rectification of the output signal of the circuit. This happens because a diode placed in a circuit with AC power supply will conduct (when forward-biased) only half the time. Figure 3.11(c) shows a diode connected to an AC power source in a series circuit and the resulting output waveform.

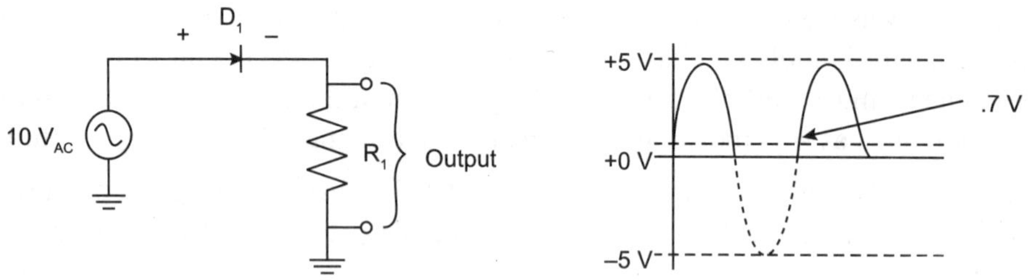

Figure 3.11 (c) Diode used to rectify AC

3.10.1 P-N Diode Switching Times

A P-N junction diode may be used as a switch in electrical circuits. The circuit can be made ON or OFF by forward-biasing or reverse-biasing the P-N junction diode. In both the cases, diode response is accompanied by a transient. The interval of time before the diode recovers its steady state is called **recovery time**. When the diode is switched from reverse-bias condition to forward-bias condition, the time interval is called forward recovery time (t_{rr}). This is defined as the time interval between 10% diode voltage to the point this voltage attains 90% of the final voltage. It has been found that t_{rr} usually does not possess a serious practical problem and hence one can consider the more important situation of reverse recovery time (t_{rr}).

Figure 3.11(d) shows the variation of steady-state minority carrier density with distance x in a forward-biased P-N junction.

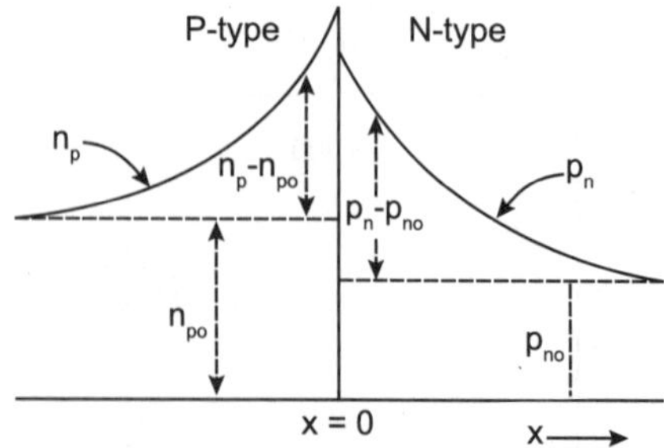

Figure 3.11 (d) Variation of steady-state minority carrier density with distance x in a forward-biased P-N junction

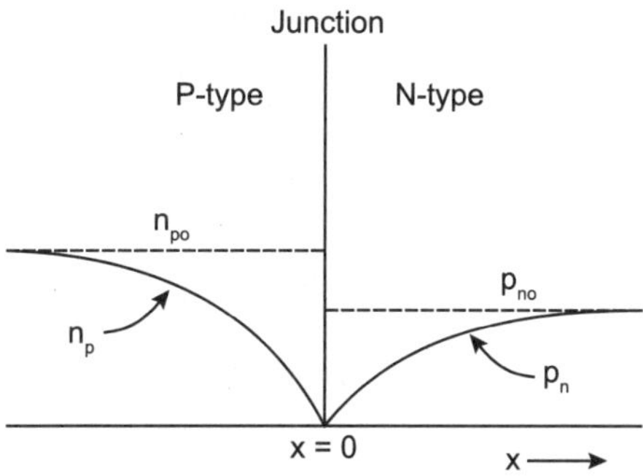

Figure 3.11 (e) Reverse-biased junction

Holes in the N region are supplied from the P region while minority carrier electrons in the P region are supplied from the N region. The concentration p_n of holes in the N region is increased above thermal equilibrium value, P_{n0}. Similarly, the concentration n_p of electrons in the P region is increased above its thermal equilibrium value, n_{p0}. The excess of holes (i.e., $p_n - p_{n0}$) and excess of electrons (i.e., $n_p - n_{p0}$) are shown in figure.

When the P-N junction is reverse-biased, the situation is shown in Figure 3.11(e). As minority carriers approach the junction, they are rapidly swept across, and the density of minority carriers diminishes to zero at the junction. Reverse saturation current I_0 is small because the density of thermally generated minority carriers is very small.

Let us consider a case when the diode carrying current in forward direction is suddenly reverse-biased. By doing so, diode current will not immediately fall to its steady state because the distribution of minority carriers has to change to steady state situation as shown in Figure 3.11(e) from that shown in Figure 3.11(d). The diode will continue to conduct for a time, known as diode reverse recovery time t_{rr}, until excess minority carrier density ($p_n - p_{n0}$) is achieved or until ($n_p - n_{p0}$) has dropped normally to zero. In this situation, current is determined by external resistance in the circuit, storage and transition times. Figure 3.12 shows the various events occurring in sequence on reverse-biasing a conducting diode.

Let us consider that, at a time t_1, input voltage V_i applied to a diode-resistance circuit of Figure 3.12(a) is reversed abruptly. Upto a time t_1, the diode is conducting in forward direction and voltage V_i is equal to forward voltage V_F. For a large value of load resistor R_L, voltage drop across R_L is large in comparison to the drop across the diode, and hence current flowing through R_L is given by $i = (V_F / R_L) = I_F$. At time $t = t_1$ voltage V_i reverses abruptly to a value V_R. Now current does not drop to zero, but instead reverses and remains at a value $i = (-V_R / R_L) = I_R$ until time $t = t_2$. At $t = t_2$, the injected minority carrier density at $x = 0$ has reached its equilibrium state as shown in Figure 3.12(c). Now consider diode voltage V. At time t_1 diode voltage falls slightly by $R_d(I_F + I_R)$ but does not reverse. Here R_d is diode ohmic resistance. At time $t = t_2$, the process of sweeping the excess minority carriers in the vicinity of the junction has completed. So diode voltage begins to reverse and diode current magnitude begins to reduce. During the time interval t_1 to t_2, the stored minority charge remains stored and hence this interval is called storage time, denoted by t_s. This is shown in Figure 3.12(e).

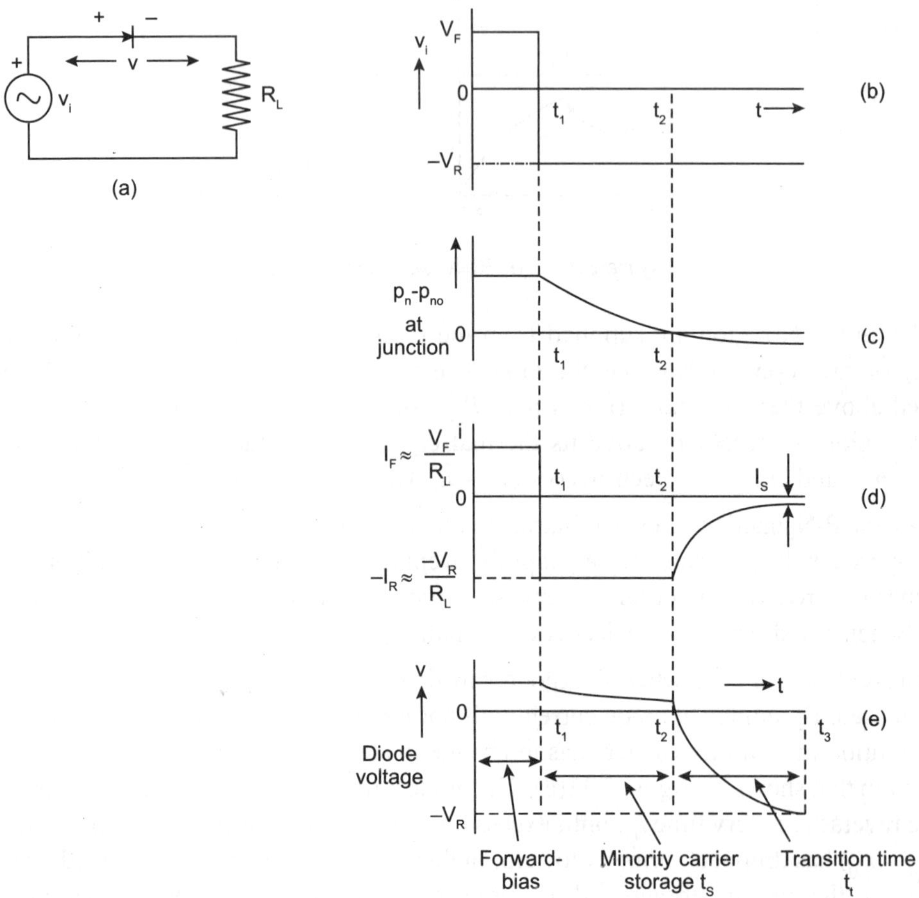

Figure 3.12 Waveform in a diode on reversing the bias: (a) circuit, (b) applied voltage to the diode circuit, (c) excess carrier density, (d) diode current and (e) diode voltage

After all excess minority carriers have been swept out or recombined in the neighborhood of the junction, diode voltage begins to reverse. Now the magnitude of diode current begins to decrease exponentially to the normal steady-state reverse saturation current value. The time interval between t_2 and t_3 when the diode has recovered nominally is called transition time t_t. This recovery time is required for minority carriers, which are at some distant from the junction, to diffuse to the junction and cross it. In addition, junction transition capacitance C_T across the reverse-biased junction has changed through the external resistor R_L to voltage $-V_R$.

Under typical operating conditions, manufacturers normally specify the reverse recovery time t_{rr} of a diode. It is the interval from current reversal at $t = t_1$ until the diode has recovered to a specified extent either in terms of diode current or in terms of diode resistance. For commercial switching diodes, reverse recovery time t_{rr} lies in the range from a low value of less than a nanosecond to a value exceeding 1 μs.

3.10.2 Current Components in a P-N Diode

Figure 3.13(a) shows a P-N junction in forward-bias by external voltage V. Due to the applied voltage, there exists a potential gradient in P and N materials. Now holes from the P region and electrons from the N region drift towards the junction. Holes that drift from the P region towards the junction enter the N region where they represent minority carriers. Similarly, electrons from the N region that enter the P region represent minority carriers. Minority carriers diffuse away from the junction (Figure 3.13(b)) exponentially with distance. Their concentration reduces steadily because of recombination with electrons and holes, respectively. We know that diffusion current of minority carriers is proportional to concentration gradient and hence this must also vary exponentially with distance. Minority current components are shown in Figure 3.13(b).

Let us assume that

(i) Space charge region is negligibly small.

(ii) Doping of the two regions is unequal. Acceptors concentration is far greater than donor concentration. Hence hole diffusion current in the N region far exceeds electron diffusion current in the P region.

(iii) Minority carrier drift current is negligible.

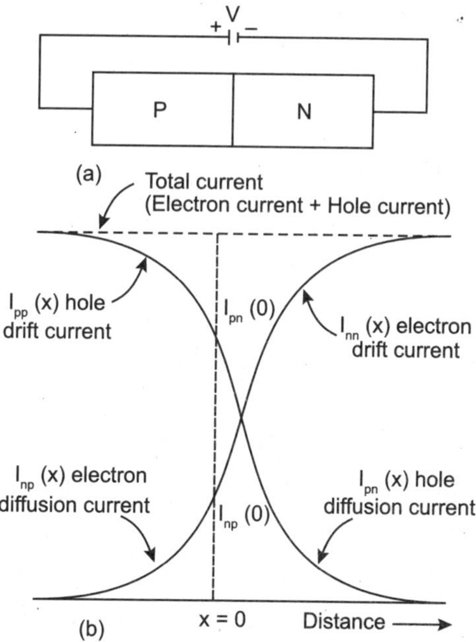

Figure 3.13 (a) Forward-biased P-N junction, (b) current components in forward-biased unsymmetrical junction

The symbols used are:

$I_{pn}(x)$ = hole current in N material

$I_{pn}(0)$ = hole current at junction (x-0)

$I_{np}(x)$ = electron current in P material

$I_{np}(0)$ = electron current at junction (x-0)

(Here the first letter refers to the type of carrier and the second to the type of material.)

At junction $x = 0$, electrons crossing from right to left constitute a current in the same direction as holes crossing from left to right. Thus, total current I at junction is given by

$$I = I_{pn}(0) + I_{np}$$

Since current is the same throughout, I is independent of x. This is indicated by a horizontal line in Figure 3.13(b).

Majority (electron) current I_{nn} is given by

$$I_{nn}(x) = I - I_{pn}(x) \quad (2)$$

Majority (hole) current I_{pp} is given by

$$I_{pp}(x) = I - I_{np}(x) \quad (3)$$

Far away from the P-N junction, current is drift current I_{pp} of holes as caused by a small electric field in the semiconductor. As holes approach the P-N junction, some of them recombine with electrons which have been injected into P region from N region. So current I_{pp} deceases towards the P-N junction. The rate of reduction is such that total current remains constant independent of the distance. The reduced hole current at P-N junction now enters the N region and becomes hole diffusion current I_{pn}. Similar remarks apply to I_{nn}. Thus, for a forward-biased P-N diode, current enters the P side as hole current and leaves the N side as electron current of the same magnitude.

It is clear from the earlier discussion that current in a P-N junction diode is bipolar in nature because current is constituted by both positive and negative charge carriers and total current is constant throughout the device, but the proportions of current carried by holes and electrons vary with distance as shown in Figure 3.13(b).

3.10.3 Quantitative Theory of P-N Junction Diode Currents

We now derive the expression for total current as a function of applied voltage (*the volt-ampere characteristic*). When a P-N diode is forward-biased, holes are injected from P side into the N material. Figure 3.14 shows several components of hole concentration in the N side of a forward-biased P-N diode. It is obvious from the figure that hole concentration decreases exponentially with distance.

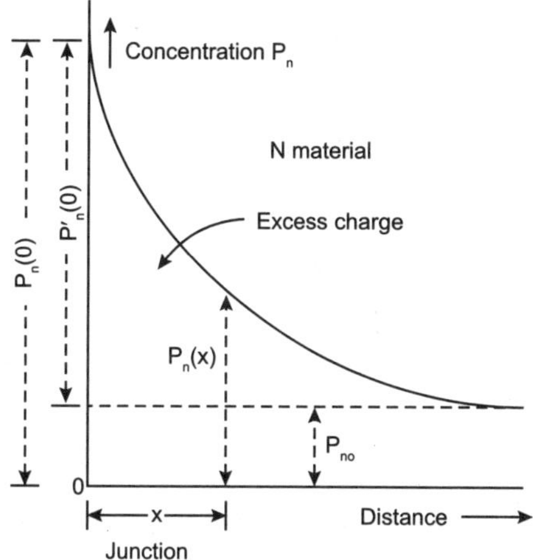

Figure 3.14

The following points should be noted:

(i) Hole concentration in N material is given by

$$p_n(x) = P_{no} + p_n(0)e^{-x/L_p} \tag{1}$$

where P_{no} = thermal equilibrium concentration

L_p = diffusion length of holes in N material

x = distance from the junction where concentration is considered

From Figure 3.14, the excess of hole concentration at $x = 0$, i.e., $p_n'(0)$ is given by

$$p_n'(0) = p_n(0) - p_{no}$$

(ii) We know that diffusion hole current in N side is given by

$$I_{pn}(x) = \frac{AqD_p p'_n(0)}{L_p} e^{-x/L_p} \tag{2}$$

At P-N junction, i.e., $x = 0$

$$I_{pn}(0) = \frac{AqD_p p'_n(0)}{L_p} \tag{3}$$

where, A = area of material

q = magnitude of charge

L_p = diffusion length of holes in N-material

$p_n'(0)$ = excess of hole concentration at P-N junction

Substituting the value of $p_n'(0)$ from Eq. (2) in Eq. (3), one obtains

$$I_{pn}(0) = \frac{AqD_p}{L_p}[p_n(0) - P_{no}] \tag{4}$$

(iii) Using Boltzmann relationship of kinetic theory of gases, one finds

$$p_n(0) = p_{n0}e^{V/V_T} \tag{5}$$

This is known as the law of P-N junction. Here

V = applied voltage, and

V_T = volt equivalent of temperature $= \frac{kT}{q} = \frac{T}{11,600}$

where k is Boltzmann constant.

Total diode current Total diode current I at $x = 0$ is given by

$$I = I_{pn}(0) + I_{np}(0) \tag{6}$$

where $I_{pn}(0)$ = current caused by holes entering N region

$I_{np}(0)$ = current caused by electrons entering P region

Total value of $I_{pn}(0)$ can be calculated using Eqs. (4) and (5). Thus, one obtains

$$I_{pn}(0) = \frac{AqD_p}{L_p}\left[p_{n0}e^{V/V_T} - p_{n0}\right]$$

or

$$I_{pn}(0) = \frac{AqD_p p_{n0}}{L_p}\left[e^{V/V_T} - 1\right] \tag{7}$$

Similarly,

$$I_{np}(0) = \frac{AqD_n n_{p0}}{L_n}\left[e^{V/V_T} - 1\right] \tag{8}$$

Substituting the values of $I_{pn}(0)$ and $I_{np}(0)$ from Eqs. (7) and (8) in Eq. (6), one obtains

$$I = \left[\frac{AqD_p p_{n0}}{L_p} + \frac{AqD_n p_{p0}}{L_n}\right](e^{V/V_T} - 1)$$

or

$$I = I_0(e^{V/V_T} - 1) \tag{9}$$

where

$$I_0 = \left[\frac{AqD_p p_{n0}}{L_p} + \frac{AqD_n n_{p0}}{L_n}\right] \tag{10}$$

Let W_p and W_n be the widths of P and N materials, respectively. In the above derivation, it was assumed that $W_p \gg L_p$ and $W_n \gg L_n$. In case the widths are much smaller than diffusion lengths, then L_p and L_n in Eq. (10) are replaced by W_p and W_n, respectively. Of course, the expression for I_0 remains valid.

Reverse saturation current Equation (9) is equally valid for negative values of V, i.e. for reverse-bias voltage. For reverse-bias of magnitude greater than V_T (≈26 mV) at room temperature, current $I \to -I_0$. Accordingly, I_0 is called **reverse saturation current**.

Let p_p = holes or majority carriers in P region

n_p = electrons or minority carriers in P region

n_n = electrons or majority carriers in N region

p_n = holes or minority carriers in N region

Then, $p_p \times n_p = n_i^2 = p_i^2 = p_n \times n_n$

Under biased conditions, hole and electron densities in P region are p_p and n_p respectively.

So in an N-type semiconductor, we have

$$p_{n0} = \frac{n_i^2}{n_{n0}} = \frac{n_i^2}{N_D} \tag{11}$$

Similarly, in a P-type semiconductor, we have

$$n_{p0} = \frac{n_i^2}{p_{p0}} = \frac{n_i^2}{N_A} \tag{12}$$

where N_D = concentration of donor atoms

N_A = concentration of acceptor atoms

Substituting the values of p_{n0} and n_{p0} from Eqs. (11) and (12) in Eq. (10), one obtains

$$I_0 = Aq\left[\frac{D_p}{L_p N_D} + \frac{D_n}{L_n N_A}\right] n_i^2 \tag{13}$$

The value of n_i^2 is given by

$$n_i^2 = A_0 T^3 \exp\left[\frac{-E_{G0}}{kT}\right] = A_0 T^3 \exp\left[\frac{-V_{G0}}{V_T}\right] \tag{14}$$

where V_{G0} = voltage, which is numerically equal to forbidden gap energy E_{G0} (in eV)

V_T = voltage equivalent of temperature = $T/11{,}600$

For germanium, diffusion constant D_p and D_n is approximately inversely proportional to temperature. Hence

$$I_0 = K_1 T^2 \exp\left[-\frac{V_{G0}}{V_T}\right] \tag{15}$$

where K_1 is a constant independent of temperature.

Note: In the above discussion, the effects of carrier generation and recombination have been ignored. However, this is valid for germanium devices but not for silicon devices. One can show for a silicon device that,

$$I = I_0\left[\exp\left(\frac{V}{\eta V_T}\right) - 1\right] \tag{16}$$

where $\eta = 1$ for large currents and $\eta = 2$ for small (rated) currents, and

$$I_0 = K_2 T^{1.5} \exp\left[V_{G0}/2V_T\right] \tag{17}$$

where K_2 is a constant independent of temperature.

3.11 Types of Diodes

Diodes are widely used in electronic circuits for various applications. Depending on the application, its characteristics and construction are modified from those of a basic diode. One may use diodes in power supplies, communication systems, digital systems and several other devices. Almost every field of electronics makes use of diodes. There are may variants of a diode. Some of frequently used doides in various electronic circuits are: (i) power diodes, (ii) zener diodes, (iii) varactor diodes, (iv) point contact diodes, etc.

3.11.1 Power Diodes or Rectifier Diodes

Power diodes are extensively used in rectifiers. A rectifier is a circuit (refer Chapter 4) which converts AC voltage into DC voltage. Rectifiers are used to construct DC power supply for operating electronic devices and circuits. Obviously, rectifier diodes are a mandatory for any electronic system.

Diodes used in the rectification of AC voltages into DC voltages are termed as power diodes. These diodes are called so because they are used to build power supplies. Examples of power diodes include:

101 KL: This power diode can handle forward current up to 100 A

251 UL: This power diode can handle forward current up to 250 A

401 PDL: This power diode can handle forward current up to 400 A

Power diodes are used at low powers and are also called P-N junction diodes. These diodes can handle small currents and voltage, and can only produce low-voltage DC power supplies. However, there are some electronic circuits that require high-voltage DC power supplies up to a few hundred ampere. In such cases, diodes are so designed to handle currents up to a few hundred amperes. Usually, such diodes are termed as **power diodes**.

The basic construction of a power diode is similar to that of P-N junction diode. However, in power diodes, extra precaution is taken so that high current that flows through the diodes may not burn them out. A power diode is encapsulated in a metallic body so that heat generated due to high current flow in the diode is absorbed.

Silicon material is used in the construction of power diode because silicon can withstand high temperature and high current. The cut-in voltage of a silicon diode is 0.7 V, whereas for germanium it is 0.3 V.

3.11.2 Zener Diode

Zener diode is one of the more prevalent and useful forms of diodes made today. These diodes are unique in that they normally operate in reverse-bias mode. When forward-biased, zener diodes operate as normal semiconductor (rectifier) diodes. We know that in the breakdown region, large changes in diode current produce only small changes in diode voltage. Obviously, a semiconductor P-N diode designed to operate in the breakdown region may be employed as a constant voltage device. Diodes used in such a manner are called zener diodes. It operates based on operation depends on Zener breakdown pheneomenon. Normally, zener diodes are fabricated using silicon, which has a higher temperature and current rating than germanium. In reverse-biased condition, the breakdown of P-N diode occurs at the point when its crystal structure is broken. This happens at breakdown or zener voltage. Due to the breakdown, a very large current flows through the P-N diode such that a very small change occurs in the voltage. This characteristic of a zener diode makes it useful in regulating the DC voltage of a power supply so that the output remains steady even if there is change in input voltage or load current.

Construction and working of zener diode It is possible to fabricate a zener diode by controlling the doping level during the construction. Generally the level of doping in zener diodes is high. With increase in the levels of doping, zener or breakdown voltage gets reduced. Commercially, zener diodes are available with zener voltage of 2.4 to 200 volts. The characteristics of zener diode are the same as that of a simple P-N junction diode.

The range of voltage above the breakdown voltage in which zener diode conducts in reverse direction is called **tolerance**. The tolerance of zener diodes is typically 5%.

During manufacturing, it is very difficult to achieve the exact doping levels. Obviously, the breakdown voltages of zener diodes of the same type also differ slightly. This difference in breakdown voltages is referred to as **tolerance**. For example, a particular type of zeners marked 9V can have 10% tolerance. These zener diodes are expected to have a breakdown voltage between 8.1 V (9.0 – 0.9) and 9.9 V (9 + 0.9) instead of 9V.

A zener diode is represented as shown in Figure 3.15(a). The equivalent circuit of an ideal zener diode when reverse-biased is shown in Figure 3.15(b).

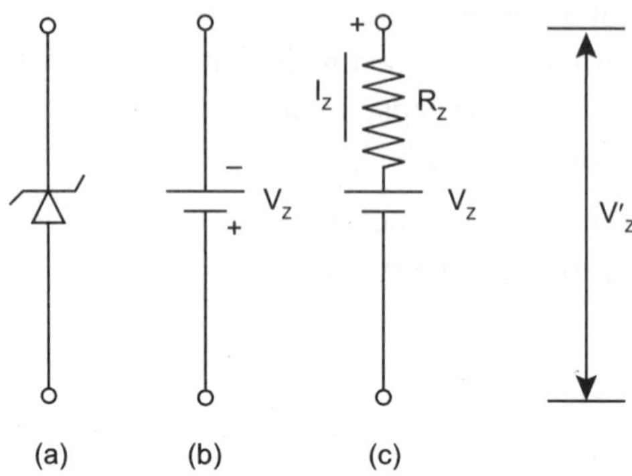

Figure 3.15 Symbols of zener diode and its equivalent circuit

An ideal zener diode is equivalent to a DC battery whose rating, V_z, is equal to the zener voltage of the diode. Zener voltage may range from 1.8 V at 1/4 W to 200 V at 50 W. The equivalent circuit of a practical diode is shown in Figure 3.12(c). A practical zener diode is equivalent to a DC battery connected in series with resistance. The voltage of the DC battery is known as zener voltage, V_z, and the value of series resistance, R_z, is equal to zener resistance under breakdown conditions. Generally, zener resistance is very small, of the order of a few ohms. When zener current, I_z, flows through the zener diode, there is some voltage drop across R_z. Then, terminal voltage of the zener, is given by

$$V_z' = I_z R_z + V_z \tag{1}$$

When R_z is very small, terminal voltage can be taken as zener voltage of the diode, because the drop across the zener diode is negligible. Then, one finds

$$V_z' = V_z$$

Breakdown voltage (V_{BR}) depends upon the operating temperature of zener diode. It is found to decrease with increase in junction temperature. This is due to increased reverse current (i.e., increase in minority carrier) that flows with increasing temperature. The decrease in breakdown voltage is about 2 mV per degree centigrade rise in temperature. V_{BR} is a manufacturing characteristic of the zener diode. It is also called **knee voltage**. As stated above, zener voltage is usually indicated as V_Z. The maximum power which a zener diode can dissipate (or handle) without damage is referred to as its power rating and denoted by P_{zM}. Zener diodes available commercially have a power rating between 1/4 W and 50 W. Wattage rating refers to the maximum power dissipation allowed for the zener before it will be damaged. Zener diodes rated at greater than 1 W usually come in stud mount packages as shown in Figure 3.15(e). Heat sinks can be used to dissipate heat and increase power capability of the diode. The most common zener values are in the range 3.3 V to 75 V.

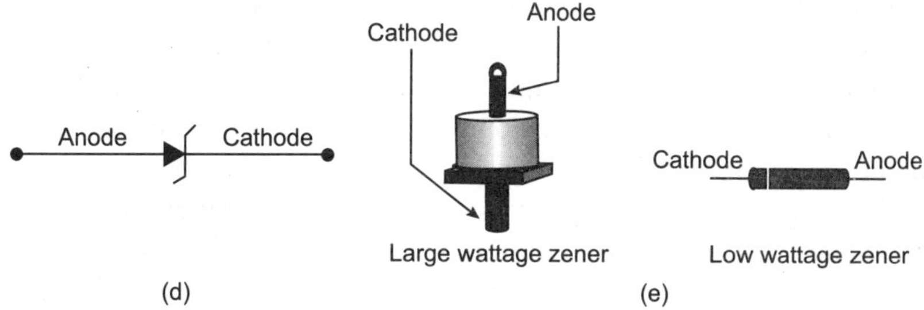

Figure 3.15 (d) Symbol and (e) typical examples of zener diode

Power rating (P_{zM}) of a zener diode is the product of maximum current I_{zT} a zener diode can handle and rated or operating voltage of a zener diode (V_z), i.e.

$$P_{zM} = I_{zT} V_z \tag{2}$$

A data sheet supplied by the manufacturer sometimes includes the maximum current rating of a zener diode. The maximum value of current a zener diode can handle at its rated voltage without damage is referred to as its **maximum current rating** (I_{zM}).

When a zener diode is operated in the breakdown region, an increase in current produces a slight increase in voltage. This means that a zener diode has a small AC resistance. This AC resistance is called **zener resistance** (*Rz*) (often referred to as **zener impedence** Z_z).

For any zener circuit, zener current must stay between two limits:

(i) The minimum zener current must be sufficient to bias the zener into its zener region.
(ii) The maximum zener current must be low enough that zener power does not exceed its maximum.

A zener diode may be taken out of its zener range by either of the following two conditions:

(i) Insufficient current
(ii) Insufficient supply voltage

In fact, both these conditions give the same result. The circuit conditions cause the available voltage and current to be below the zener level. In this condition, the zener behaves like any reverse-biased diode.

Application of zener diodes Few important applications of zener diodes include: (i) voltage regulator, (ii) fixing reference voltage in electronic circuits, such as power supplies and transistor biasing, (iii) clippers in wave-shaping circuits, (iv) square wave generators.

Zener diode as a voltage regulator A voltage regulator is a circuit that maintains the output voltage of a DC power supply constant against variations in: input AC voltage and load current.

By using rectifiers, the AC voltage is converted into DC voltage in a DC power supply. This output voltage is filtered to remove AC ripples. The output of the rectifier varies if the input AC voltage varies. After filtration, the DC output will also vary accordingly. If there is a variation in load resistor of the DC power supply, the output DC voltage will also vary. This is due to the fact that load current changes with variation in load resistor. The change in load current will vary the output DC voltage developed across the load resistor. Obviously, it is required that the output DC voltage is regulated at a constant value. To supply a constant DC voltage to the load resistor, a zener diode can be used as a voltage regulator. Figure 3.16(a) shows the circuit diagram of a zener regulator. One can explain the operation of a zener regulator by separately considering the variations in input supply and load current. These two situations are explained as follows:

Varying input voltage Figure 3.16(b) shows the circuit diagram of a zener regulator for varying input voltages. In this circuit, input voltage V_s is shown as a variable DC supply. This input voltage is unregulated as it varies due to variations in the input AC supply of the rectifier. In addition, it is also assumed that the load resistor is fixed under this condition.

The zener diode is reverse-biased. The essential condition for the operation of the circuit is that the zener diode should always remain in the breakdown region, throughout the operation. Resistor, R_s in Figure 3.16 is called the **current limiting resistor**. Total current, I_s, supplied by input supply, V_s, passes through the resistor, R_s. I_s supply is divided into zener current, I_z, and load current I_L, i.e.

$$I_s = I_z + I_L \tag{3}$$

The basic idea behind the operation of this circuit is that zener current (I_z) varies under breakdown condition without an appreciable change in voltage drop across the zener diode. The voltage across the zener diode practically remains constant at V_z and it varies only when different zener diodes are used. The output DC voltage also remains constant because the zener diode is connected in parallel with the load resistor.

The input DC voltage (V_s) may increase or decrease. In case if DC voltage increases, supply current (I_s) also increases. Load current will increase under this condition if the zener diode is not connected in the circuit. However, in the presence of the zener diode, the additional current passes through the zener diode such that load current (I_L) remains constant.

One can write the output DC voltage of Figure 3.16(b) as

$$V_0 = V_s - I_s R_s \tag{4}$$

When current I_s increases, voltage drop across the resistance, R_s, also increases.

Varying load current Figure 3.16(c) shows the circuit conditions for a zener regulator at varying load current. As shown in Figure 3.16(b), load current, I_L, will vary if load resistor, R_L is variable. Load current increases when load resistance decreases and vice versa. If the load increases, the input DC voltage supplies more supply current, I_s, than under normal conditions.

The zener diode takes this additional current, and therefore, load current, I_L, is reduced to a constant value. Moreover, due to the increase in supply current, I_s, voltage drop across the resistor R_s increases. This action reduces the output voltage, V_0, according to Eq. (4). Obviously, the output DC voltage is maintained at a constant value.

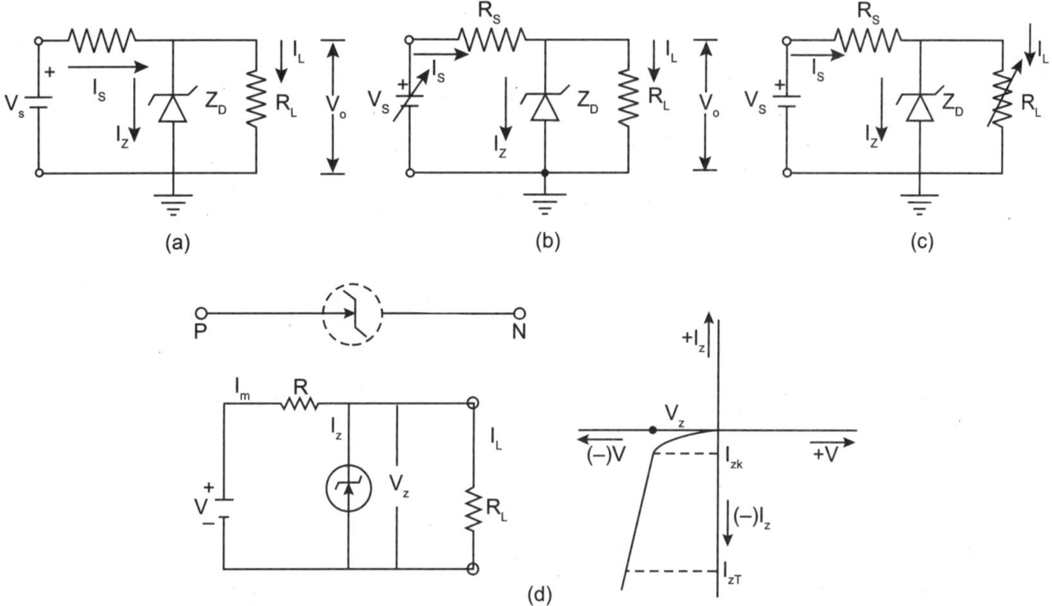

Figure 3.16 (a) Circuit diagram of a zener regulator, (b) zener regulator with varying input voltage V_S, (c) zener regulator with varying load resistor R_L, (d) zener diode as a voltage regulator.

If load current decreases, the output DC voltage will also reduce. Under this condition, the output DC supply supplies less current I_s. The zener diode will accommodate its current such that load current, I_L, is increased by reducing zener current, I_L. This will increase load current, I_L, such that voltage drop across R_L is maintained to a constant output DC voltage (V_Z) (for a detailed discussion about voltage regulation, refer Chapter 4).

3.11.3 Point Contact Diodes

When a P-N junction is made using a semiconducting material and a thin metallic point, the diode thus formed is called a **point contact diode**. Figure 3.17 shows a point contact diode. An N-type silicon or germanium semiconductor wafer is used as the base, termed as a substrate of the diode. A very thin spring, made from phosphor-bronze alloy, is pressed at a point in the N-type substrate. This phosphor-bronze spring is creatively called the cat whisker. It has bends to allow expansion when a high current produces heat, without the whisker getting broken or destroyed. To impart sufficient mechanical strength to the diode, the assembly is encapsulated in a glass or ceramic envelope.

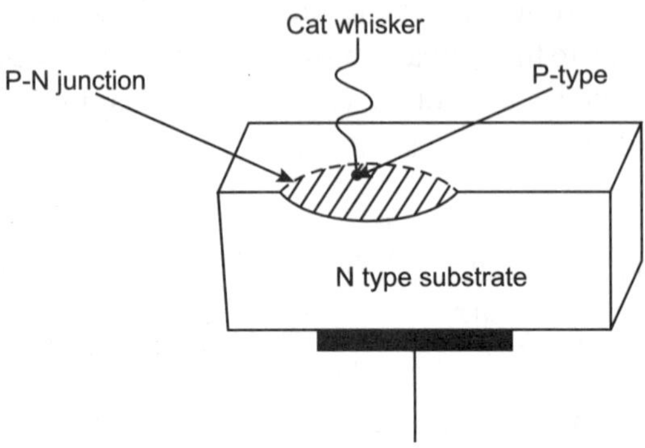

Figure 3.17 Point contact diode

The value of current that is allowed to pass through the assembly for the point contact diode for about 1 to 100 milliseconds may be about 200 mA. Due to this current, some atoms in the material of the cat whisker are injected into the N-type substrate so that a P-N junction is formed at the contact point. The P-N junction so formed has a very small area and junction capacitance. The value of junction capacitance may be 1 pF or even less than this. The current rating of this small P-N junction is very low.

The point contact diode finds use in the following applications:

(i) Microwave circuits

(ii) AM diode detection in AM receivers.

3.11.4 Constant Current Diode

These are diodes that work exactly opposite to a zener diode. These diodes keep the current constantly flowing through them when voltage changes, i.e. instead of holding the voltage constant, these diodes hold the current constant. The range of voltage over which a diode can keep the current constant is known as voltage compliance. These diodes are optimized for a particular voltage compliance.

3.11.5 Tunnel Diode

It is a device just like a P-N junction that offers negative resistance under certain bias conditions. Negative resistance of the diode is due to tunneling and it is called a **tunnel diode** or **Esaki diode**. It is made very much like an ordinary P-N junction diode, except that both the P and N regions are heavily doped, about thousand times more than an ordinary diode. It is used as an active device in electronic circuits in the frequency range of few megahertz. Semiconductors with very high impurity concentration are referred to as **degenerate semiconductors**. The symbol for tunnel diode is shown in Figure 3.18(a) and a typical tunnel diode characteristic is shown in Figure 3.18(b).

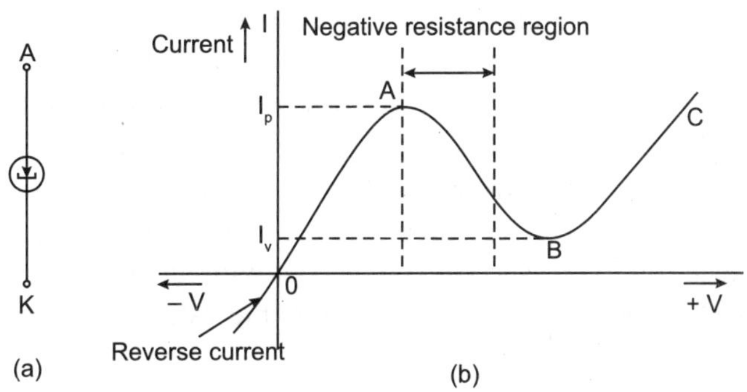

Figure 3.18 (a) Symbol, (b) V-I characteristic of tunnel diode

As a result of high impurity levels, contact potential is high, depletion region is very narrow and Fermi levels lie in the conduction band for N side and in the valence band for P side (Figure 3.15(b)). Under these circumstances, a very abrupt transition from P to N material is achieved within the crystal. Since the depletion region is very narrow, this gives rise to extremely large electric fields.

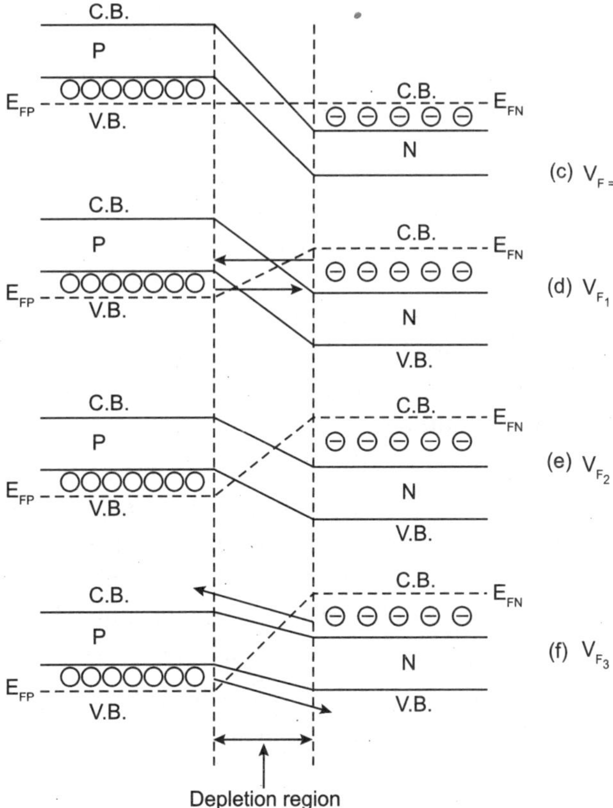

Figure 3.18 (c), (d), (e), (f) Tunnel diode energy levels under forward-bias [$V_{F_3} > V_{F_2} > V_{F_1}$]

Tunnel effect When a P-N semiconductor is heavily doped, it has many majority carriers and ions, and when forward-biased the hole and valence electrons' random drift is heavy. As a result, it is not uncommon for a large number of electrons to fill the holes and release energy to only a few other valence electrons. These few valance electrons then have their energy levels raised considerably so that they can cross from N to P-section (Figure 3.18(d)) and current increases as shown by OA (Figure 3.18(b)) even with little or no applied voltage. This action which seems to allow a valence electron to cross a potential barrier without enough applied external energy is called the tunnel effect because it seems as though the valence electron tunnels through the forbidden band.

On further increasing the voltage, the barrier height decreases (Figure 3.18(e)) and fermi level E_{FN} is much raised so that current decreases as shown by AB (Figure 3.18(b)). This region is called **negative resistance region**. On further increasing the voltage, the conduction band of N-type is in level with CB of P-type, (Figure 3.18(f)) and free electrons of N-type easily cross to P-type thereby current BC again increases like in an ordinary diode as shown by BC (Figure 3.18(b)).

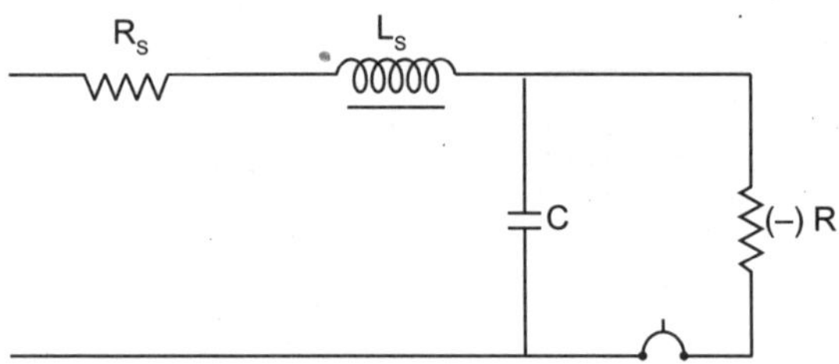

Figure 3.19 (a) Equivalent circuit of a tunnel diode

The negative resistance region AB (Figure 3.18(b)) allows the diode to be used as an oscillator. It can also be used as an electronic switch since it has a good response in negative resistance region. By its nature, the tunnel diode has a rather high reverse current, but is not generally operated under this condition.

The equivalent circuit of a tunnel diode is shown in Figure 3.19(a). Series resistance, R_s, is ohmic resistance. Series inductance, L_s, depends upon lead length and geometry of the diode package. Junction capacitance, *C*, depends upon the bias and is usually measured at the valley point. Negative resistance [$-R_n$] has a minimum at the point of inflection between peak current I_P and valley current I_V. Typical values for these parameters for a tunnel diode having peak current value I_p= 10 mA are R_n = (–)30 Ω, R_s = 1 Ω, L_s = 5 nH and C = 20 pF.

It is a voltage controllable device (like the tetrode). Since between I_p and I_v (Figure 3.18 (b)) we have three values of voltage for a given current and on account of its multivalued feature, it is used in pulse and digital circuiting.

Our main interest in the tunnel diode is its application as a very high-speed switch. Since tunneling takes place at the speed of light, transient response is limited only by total shunt capacitance (junction plus stray wiring capacitance) and peak driving current. Switching times of the order of a nanosecond are reasonable and time as low as 50 p have been obtained.

Figure 3.19(b) shows the use of tunnel diode as an oscillator. This application of tunnel diode is commonly found in communication receivers for which low-power, high-frequency signals are required. Tunnel diodes are normally operated in the **negative resistance** region. They can also operate at **higher temperatures** than do silicon or germanium diodes. Another use of tunnel diodes is in high-frequency oscillators.

The advantages of a tunnel diode are (i) low cost, (ii) low noise, (iii) high speed, (iv) environmental immunity and (v) low power.

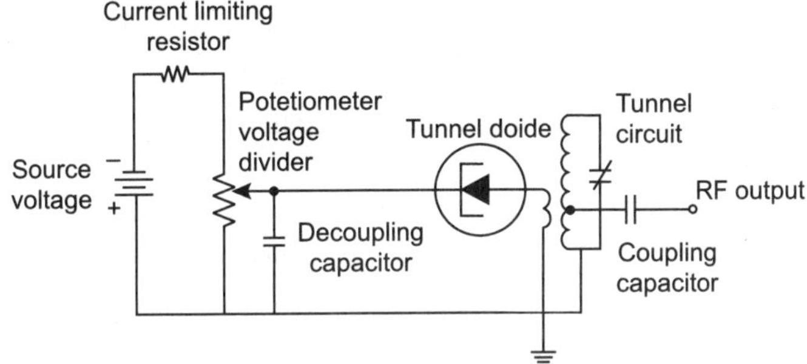

Figure 3.19 (b) Tunnel diode oscillator circuit

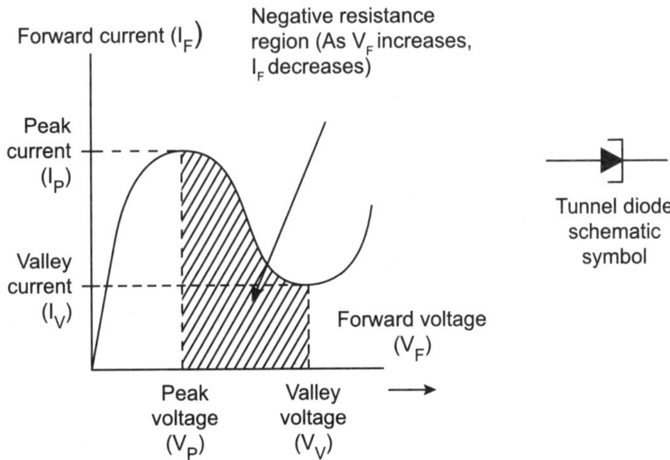

Figure 3.19 (c) Tunnel diode characteristic curve

The disadvantages of the diode are its low output voltage swing and the fact that it is a two-terminal device unlike ordinary diodes in which current flows only when forward-biased. Because of the latter, there is no isolation between the input and output and it leads to serious circuit design problems.

However, we may have a special type of tunnel diode whose peak value current (I_B) is of the order of valley current (I_V). Such a diode is called **backward diode**. The V-I characteristic of a backward diode is shown in Figure 3.20. We may use it under reverse-biased condition an ordinary diode with the advantage that the diode current is available just beyond zero volt, whereas in an ordinary diode, current is not available until the applied voltage is more than (break-over voltage) 0.7 volt in case of silicon, and so in the case a back diode, it is conveniently used as a rectifier for very small (mV) input signals.

A comparison between zener and tunnel diodes is provided in the following table:

Feature	Zener	Tunnel
Construction	Made up of silicon with two layers (P-N)	Made up of germanium or gallium arsenide with two layers (P-N)
Doping	Heavily doped	Heavily doped (1:10^4)
Operation	Acting as a breakdown device at reverse-bias	Exhibits negative resistance at forward-bias
Applications	Constant voltage source, voltage regulator	Microwave oscillator, ultra high speed switching device

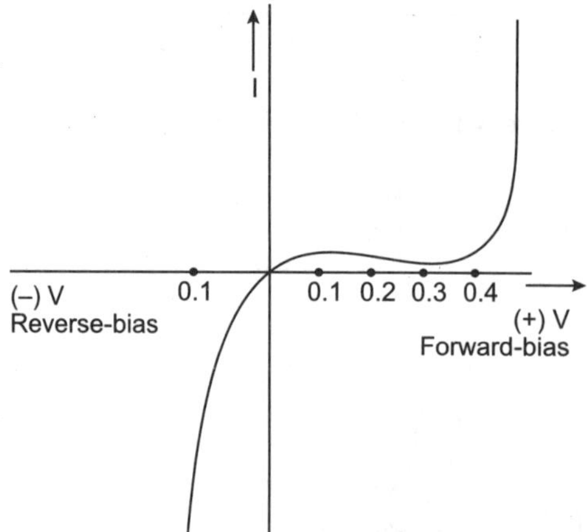

Figure 3.20 V-I characteristic of backward diode

3.11.6 Photodiode

A photo-diode is a P-N junction diode packed into a transparent plastic packet working under reverse-biased condition.

Principle A very small current flows through a P-N junction diode when it is reverse-biased. It is because minority charge carriers take part in conduction. The number of minority carriers depends upon the working temperature, and under no illumination condition current flowing is called **dark current** (Figure 3.22). However, when photons (light) of suitable frequency is incident on the junction, the number of charge carriers increases. But this happens only during light radiations at the junction, which is not the case in ordinary diodes.

Working In a photodiode, visible light is focused on the junction through a lens (Figure 3.21). Light on being incident produces free electrons and holes. Thus the number of charge carriers increases and current increases. Generally, a photodiode is optimized for its sensitivity to light. As light intensity increases more and more, charge carriers are generated and reverse-bias current increases (Figure 3.21).

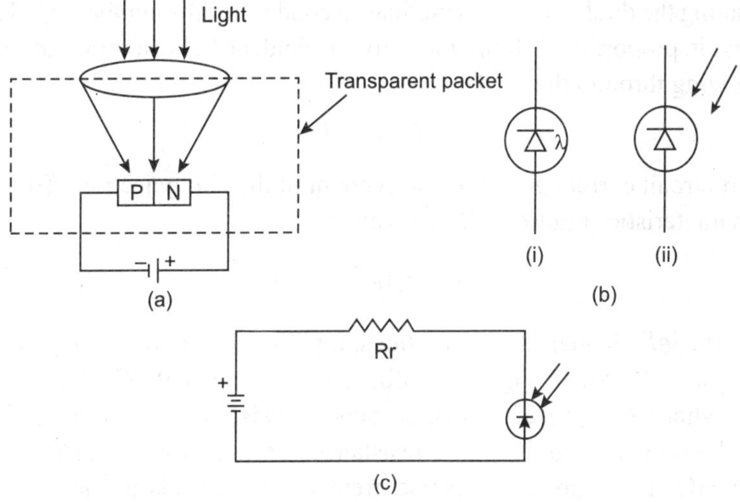

Figure 3.21 Equivalent circuit of a photodiode

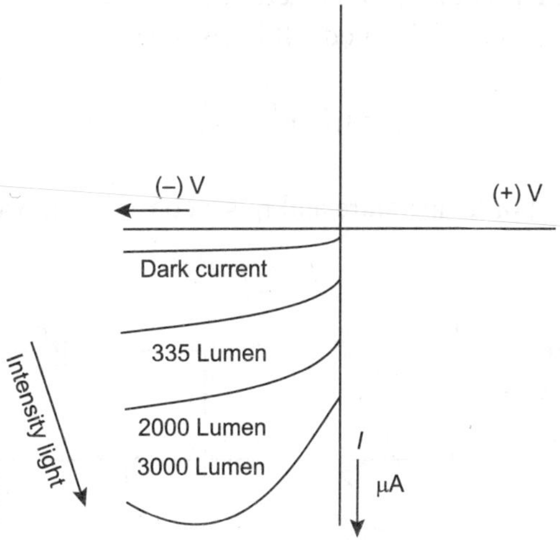

Figure 3.22 V-I characteristic of a photodiode

In this respect, a photodiode acts as a photodetector, a device which converts incoming light signal into electrical signal (see Chapter 4).

Volt-Ampere characteristic Under normal reverse-bias, current in a P-N junction diode is given by

$$I = I_0(e^{-V/\eta kT} - 1)$$

since applied voltage V is negative. So at higher values of V (i.e. more negative V), the first term in the bracket goes to zero.

and $$I \approx I_0$$

On illuminating the diode under reverse-biased conditions, the number of hole-electron pairs increases, which is proportional to the intensity of incident light (number of photons). Hence total current flowing through the diode

$$I = I_0 + I_s$$

where I_s is short circuit current generated on account of the photoelectric effect. In general, the volt-ampere characteristic (Figure 3.22) is given by

$$I = I_s + I_0(e^{-V/\eta kT} - 1) \tag{5}$$

Small signal model Under ideal conditions, a photodiode under reverse-biased condition is shown in Figure 3.23(a). In Figure 3.23(b), parasitic elements C, R and r are taken into consideration, where R represents reverse-biased resistance, C represents the transition capacitance and r represents bulk ohmic resistance. Barrier capacitance C is of the order of 10 pF, R = 50 MΩ. In Figure 3.22, dark current I_0 corresponds to the reverse current due to thermally generated electron-hole pairs. From Eq. (5), we see that if diode terminals are open circuited, then total current I has to be zero. Under such conditions, a voltage known as photovoltaic V_{oc} will appear across the diode. It is given by

$$V_{oc} = \eta V_T \ln\left(1 + \frac{I_s}{I_0}\right) \tag{6}$$

where V_T is volt equivalent of temperature and η is a constant equivalent to 1 for Ge and 2 for Si.

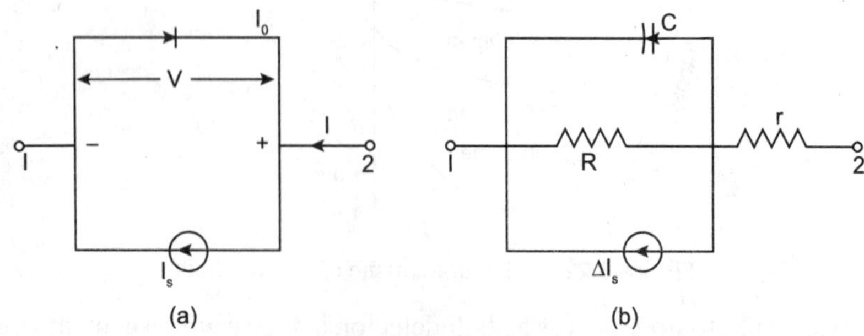

Figure 3.23 Photodiode under (a) reverse-biased condition, (b) parasitic elements

Thus, V_{oc} increases logarithmically with I_s or with intensity of incident light. Now, if load (R_L) is connected across the diode, voltage drop across it will correspond to incident light intensity. Output voltage generated is of the order of few hundred milivolts across R_L (few kiloohms). Large-area photovoltaic P-N junction banks have been used for the conversion of solar energy into electrical energy, which is the basis of solar cells.

Solar cells Solar cells, which convert the energy from sunlight directly into electricity, at present are the most important source of power supply for satellites and space vehicles. These cells have also been successfully employed in small-scale terrestrial applications. Solar cells can covert sun's energy directly into electricity with high conversion efficiency, and can provide nearly permanent power at low operating cost and is virtually free of pollution. Recently, flat panel solar cells, thin film devices, concentrator systems, and many innovative concepts have come into reality. We can expect that in near future solar power plants will be economically feasible for large-scale use of solar energy.

The first solar cell was developed by Chapin et al. in 1954 using a diffused silicon P-N junction. Subsequently, the cadmium-sulphide solar cell was developed by Raynolds et al. in 1954. To date, solar cells have been made from many other semiconductors, employing single-crystal, polycrystal and amorphous thin film structures.

Photo voltaic effect and solar cells The photo voltaic effect can be observed in nature in a variety of materials, but materials that have shown the best performance are semiconductors. The principal advantages associated with solar cells are that they have no moving parts, require little maintenance, and work quite satisfactorily with beam or diffuse radiation.

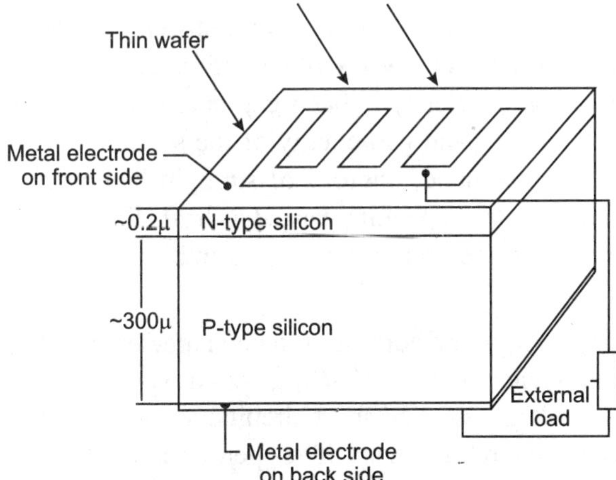

Figure 3.24 (a) Schematic representation of a silicon solar cell

A typical schematic representative of a conventional silicon solar cell is shown in Figure 3.24(a). Silicon cells are thin wafers about 300 μm in thickness and 3 to 4 cm in diameter sliced from a single crystal of N-type or P-type doped silicon. A shallow P-N junction is formed at one end by diffusion. On the front and back side of the cell are attached metal electrodes made from a Ti-Ag solder. On the front side, the electrode is in the from of a metal grid with fingers which permit sunlight to go through, while on the backside the electrode completely covers the surface. An antireflection film of SiO having thickness of about 0.1 μm is also put on the top surface of the cell.

When a monochromatic radiation of wavelength λ is incident on the front surface of a cell, it is absorbed and electron-hole pairs are created. The generation rate of electron-hole pairs at distance x from the semiconductor is given by

$$G(\lambda, x) = \alpha(\lambda)F(\lambda)[1 - R(\lambda)]\exp[-\alpha(\lambda)x] \tag{6}$$

where $\alpha(\lambda)$ is the absorption coefficient, $F(\lambda)$ the number of incident photons/cm^2/s per unit bandwidth, and $R(\lambda)$ the fraction of these photons reflected from the surface. The positive and negative charges are separated because of the P-N junction. Direct current thereby produced is collected by the metal electrodes and flows to the external load. The total photocurrent at a given wavelength is given by

$$J(\lambda) = J_p(\lambda) + N_n(\lambda) + J_{dr}(\lambda) \tag{7}$$

where J_p and J_n are photocurrent densities due to holes and electrons, respectively, and J_{dr} is the photocurrent per unit bandwidth, representing photocurrent generation taking place within the depletion region. Conversion efficiency that can be derived from the device is given by

$$\eta = \frac{I_m V_m}{P_{in}} \tag{8}$$

where V_m and I_m are voltage and current at maximum power point and P_{in} is power density of sunlight. For silicon cells, n ranges between 10 and 15 percent. An important property of the material affecting conversion efficiency is band gap energy (E_g). Photons of solar radiations having energy (E) less than the band gap energy of the semiconductor material used are not usually absorbed. On the other hand, photons of sunlight having $E > E_g$ have their energy partially utilized with quantity ($E–E_g$) being wasted as heat. The value of E_g for silicon is 1.20 eV. This value is quite close to the optimum value desirable from the stand point of obtaining maximum efficiency.

Apart from band gap energy, the current-voltage characteristic (Figure 3.24(b)) of a solar cell also influences its efficiency. Figure 3.24(b) shows a typical characteristic for a solar cell. Current obtained at zero voltage is called short circuit current, while voltage V_{oC} obtained with an open circuit is called open circuit voltage. The open-circuit voltage V_{oC} and the short-circuit current I_{sc} are determined for a given light level by the cell properties. The product of these quantities is the ideal power of the cell. The maximum useful power is the area of the largest rectangle that can be formed under the I-V curve. Calling these values of voltage and current, V_m and I_m, we see that the maximum delivered power illustrated by the rectangle in Figure 3.23(b) is less than $V_{oc}I_{sc}$. The ratio $\frac{I_m V_m}{I_{sc} V_{oc}}$ is called the fill factor, and is a figure of merit for solar cell design. Typical values of the above quantities for a silicon cell are as follows: V_{oc} = – 450 to 600 mV, I_{sc} = – 30 to 50 mA/cm^2 and fill factor = 0.65 to 0.80.

A well-designed Si cell can have about 10 percent efficiency for solar energy conversion, providing approximately 100 W/m^2 of electric power under full illumination. This is the modest amount of power per unit solar cell area, considering the effort in fabricating a large area of Si cells. One approach to obtain more power per cell is to focus considerable light onto the cell using mirrors. Although Si cells lose efficiency at the resulting high temperatures, GaAs and related compounds can be used at 100°C or higher. In a solar concentrator system, more effort and investment can be put into solar cell fabrication since fewer cells are required. For example, a GaAs-AlGaAs hetrofunction cell provides good conversion efficiency and operates at elevated temperatures common in solar concentrator systems.

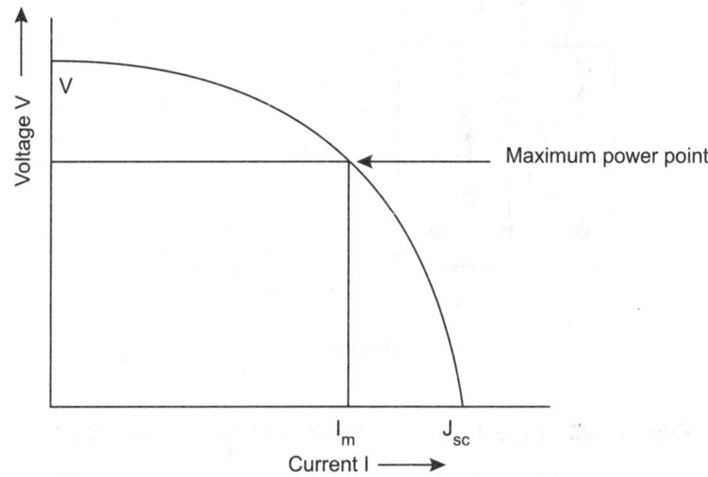

Figure 3.24 (b) A typical current-voltage characteristic for an ideal solar cell

Applications Some of the important applications of photodiodes are (i) photodetection (both visible and invisible), (ii) demodulation, (iii) logic circuits, (iv) switching, (v) optical communication system.

3.11.7 Light-Emitting Diode (LED)

LED is a solid-state (P-N junction diode) light source which has replaced incandascent lamps in many day-to-day applications.

LED is just not an ordinary P-N junction diode where silicon is used. Here we use elements like gallium, arsenic and phosphorus which are semi-transparent, unlike silicon which is opaque (gallium-arsenide gives infrared radiations and gallium-arsenide-phosphide gives visible light either red or yellow).

Principle Whenever a P-N junction is forward-biased, free electrons and holes recombine at the junction. Here we may understand the situation by assuming that electrons fall into a hole.

While jumping, the electron releases (radiates) energy since electron jumps from a higher energy level (N-type) to a hole (P-type) in lower energy level. The energy radiated is in the form of heat and light (Figure 3.25(a)) but none of the light is visible since silicon is opaque and so no light can come out. Instead of silicon, if we use a semitransparent material (like gallium, arsenic compounds etc.) then light emitted due to the recombination of free electrons and holes at the junction will escape to the surrounding. The colour of radiations coming out will depend upon the material used. Mostly we find red, green or amber when the material used is gallium arsenide phosphide. We get infrared radiations also when the compound used is gallium arsenide.

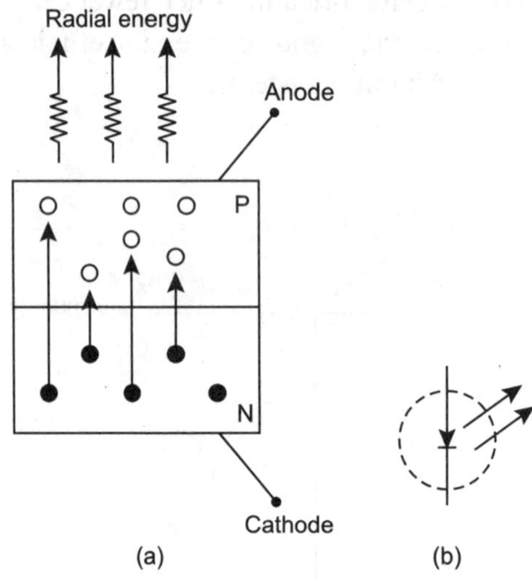

Figure 3.25 (a) Schematic construction of LED, (b) symbol

Working An LED is forward-biased with a small resistance in series (Figure 3.26(a)). The construction of an LED is such that it will have a typical voltage drop of 1.5 to 2.5V for current between 10 and 50 mA when forward-biased.

The exact voltage drop depends on the colour, tolerance and other factors of construction.

Incidentally LED's have low reverse voltage ratings. A typical value is 3 V for TIL 221 (red LED). So, if an LED is reversed-biased, then voltage greater than 3 V may destroy it. One way to protect, is to connect another ordinary diode in parallel with the given LED (Figure 3.26(b)).

Applications LEDs that produce visible radiations are used in instrument displays, calculators, digital clocks, etc.

Here we have LED arrays. An LED array is a group of LEDs that display numbers, letters or other symbols. The most common LED array is a seven-segment display shown in Figure 3.24(a). The display contains seven rectangular LEDs (A to G). Each LED is called a segment because it forms a part of the character being displayed. Figure 3.27(b) shows the schematic arrangement of seven LEDs constituting a display. Their anodes have been joined to a common terminal (we can also have a common terminal for connecting all the cathodes and keeping anodes seperately). So a positive voltage when applied to it drives all the anodes. By grounding one or more cathodes, the diodes are forward-biased, which in turn glow and form any digit from 0 to 9. For instance, by grounding the cathodes of A, B and C, the number '7' is displayed as shown in Figure 3.27(a), or by grounding the cathodes of A, B, C, D and G we display the number '3'.

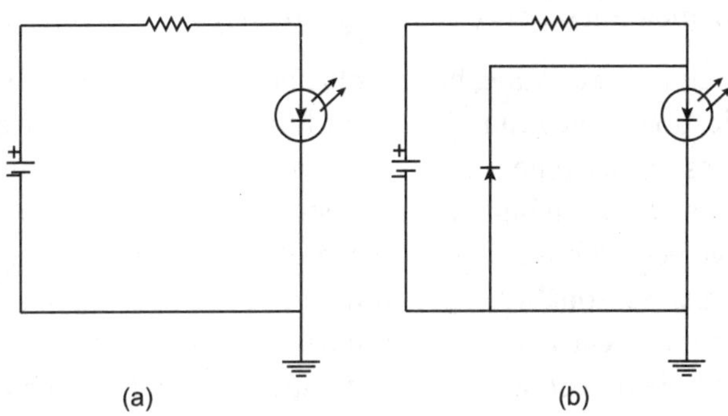

Figure 3.26 LED circuit

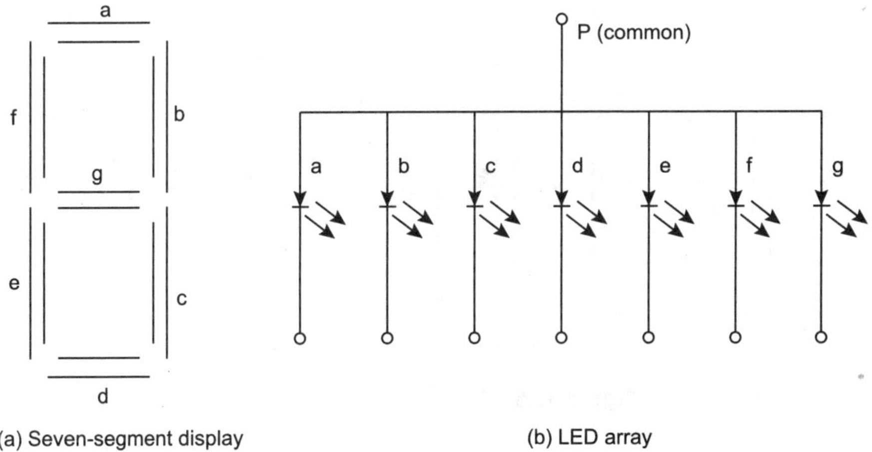

Figure 3.27 LED Array

A seven-segment array can also display capital letters 'A', 'E', 'F' plus small letters 'b' and 'd', which are commonly used in microwave ovens, stereo tuners and microprocessor trainer kits. A typical display TIL 312 shown in Figure 3.28(b) represents the left and right decimal points also. Figure 3.28(a) shows the pin out. The advantages of an LED are:

(i) Low voltage of operation

(ii) Long life (more than 20 years)

(iii) Fast ON-OFF switching (10^{-9} sec)

The other uses of an LED are:

(i) For indicating power ON/OFF (power level indicators)
(ii) Optical switching applications
(iii) Solid state video displays
(iv) Optical communication (energy coupling circuits).

LEDs in red and green are available, but the blue one is uncommon. Blue light emitters have several potential applications. They can be used for full colour displays in a large area. They can be used in traffic lights as a replacement for ordinary bulbs, resulting in a huge power saving. The use of blue lasers can also result in higher density storage of information in optical CD-ROMs. Blue light emitters have been fabricated using several materials including ZnO, ZnSe and SiC. These attempts have been reasonably successful. However, maximum success in the last few years has been achieved with GaN. This material has been proved to be very useful as it is also capable of operating at high power density, high temperature and unfriendly environment. GaN having a band gap of 3.4 eV can give a continuously varying band gap by combining with AIN to get upto 6.2 eV. The high thermal conductivity and superior stability of this material makes it ideal for several applications over other competing materials.

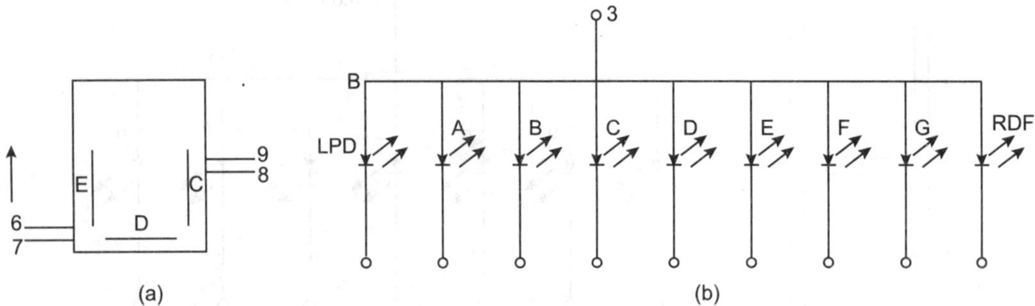

Figure 3.28 A typical display TIL 312

3.11.8 Heterojunction High-Intensity LEDs

A P-N junction between two differently doped semiconductors that are of the same material, i.e. the same band gap E_g, is called a **homo junction**. A junction between two different bandgap semiconductors is called a **heterojunction**. A semiconductor device that has junctions between different bandgap materials is called a heterostructure device.

LED constructions for increasing the intensity of output light make use of the double heterostructure. Figure 3.29(a) shows a double-heterostructure (DH) device based on two junctions between different semiconductor materials with different bandgaps. In this case, semiconductors are AlGaAs with $E_g \approx 2$ eV and GaAs with $E_g \approx 1.4$ eV. The double heterostructure in Figure 3.29(a) has an N^+P heterojunction between N^+-AlGaAs and P-GaAs. There is another heterojunction between P-GaAs and P-AlGaAs. The P-GaAs region is a thin layer, typically a fraction of a micron, and it is lightly doped.

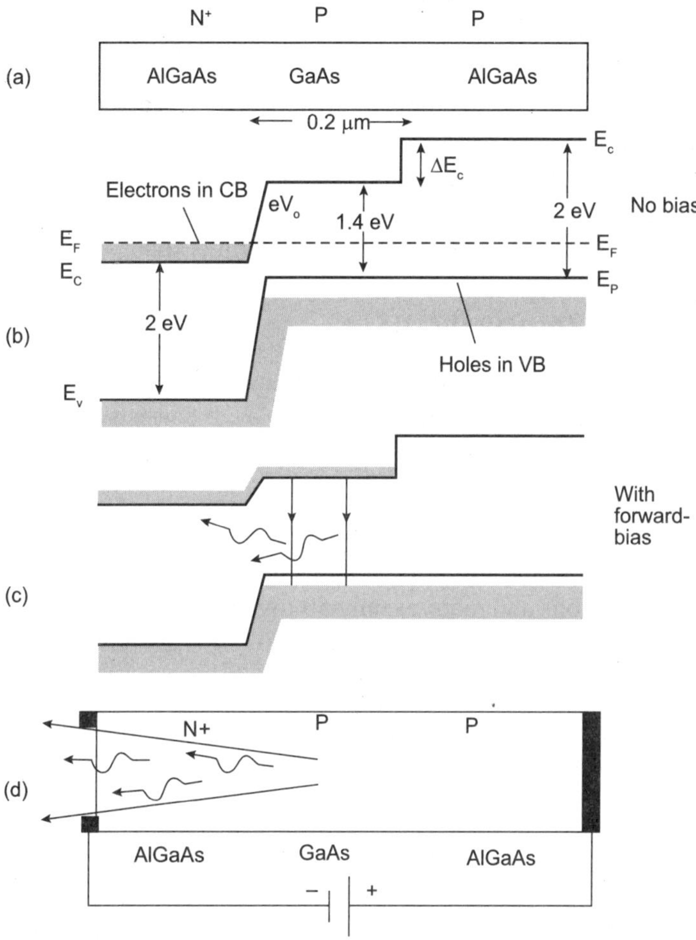

Figure 3.29 (a) A double heterostructure diode has two junctions which are between two different bandgap semiconductors (GaAs and AlGaAs), (b) a simplified energy band diagram with exaggerated features. EF must be uniform, (c) forward-biased simplified energy band diagram, (d) forward-biased LED, schematic illustration of photons escaping reabsorption in the AlGaAs layer and being emitted from the device.

Figure 3.29(b) shows a simplified energy band diagram for the whole device in the absence of an applied voltage. The Fermi level, E_F, is continuous throughout the whole structure. There is a potential energy barrier eV_o for electrons in the CB of N^+-AlGaAs against diffusion into P-GaAs. There is a bandgap change at the junction between P-GaAs and P-AlGaAs which results in a step change. ΔE_C is effectively a **potential energy barrier** that prevents any electrons in the CB in P-GaAs from passing to the CB of AlGaAs. We may note that there is also a step change ΔE_V but this is small and therefore we have not shown.

When a forward-bias is applied, most of this voltage drops between N^+-AlGaAs and P-GaAs and reduces the potential energy barrier eV_0, just as in a normal P-N junction. This allows electrons in the CB of N^+-AlGaAs to be injected into P-GaAs (Figure 3.29(c)). These electrons, however, are confined to the CB of P-GaAs since there is a barrier ΔE_C between P-GaAs and P-AlGaAs. Wide bandgap AlGaAs layers, therefore, act as confining layers that restricts injected electrons to the P-GaAs layer. The recombination of injected electrons and holes already present in this P-GaAs layer results in spontaneous photon emission. Since the bandgap E_g of AlGaAs is greater than that of GaAs, the emitted photons do not get reabsorbed as they escape the active region and reach the surface of the device as depicted in Figure 3.29(d). Since light is also not absorbed in P-AlGaAs, it can be reflected to increase light output.

3.11.9 Liquid Crystal Display (LCD)

Just like LED, we have another type of display that uses seven segments and is called liquid crystal display, also known as electroluminescent display. It consists of a thin layer of normally transparent liquid crystal material such as cholesteryl nonanoate and p-azoxyanisole between two electrodes.

When an electric field is applied, the liquid crystal material between the two electrodes becomes turbulent, reflecting and scattering ambient light. It provides excellent brightness under high ambient light conditions and requires only 50 μw of power per segment, thereby the total power for one complete display of 7 segments is 350 μw. This power is much less than that of an LED display, but life expectancy is not as high as that for LED, which is 10,000 hours minimum.

These displays are used in watches, pocket calculators, pocket televisions and portable instrument displays.

3.11.10 Varactor Diode

An ordinary P-N junction diode under reverse-biased condition causes majority charge carriers to move away from the junction, thereby uncovering more immobile charge, i.e. ions. Hence the thickness of the space charge region at the junction increases with reverse voltage. This increase in uncovered charge with applied voltage may be considered a capacitive effect. We may define incremental capacitance C_T by as

$$C_T = \frac{dQ}{dV} \tag{9}$$

where dQ is the increase in covered charge caused by an increase (dV) in reverse voltage. The quantity C_T is generally referred to as **transition region** or **space charge** or **barrier capacitance** or **transition capacitance**.

Transition capacitance can be calculated as

$$C_T = \frac{K}{(V_B - V)^n} \tag{10}$$

where K is a constant, depending on the semiconducting material, V is applied basing voltage, V_B is cut-in voltage of the diode and n is a constant, depending on the manufacturing process of the diode.

This quantity is not constant but depends upon the magnitude of reverse voltage. A variation of C_T with applied reverse voltage is shown in Figure 3.30.

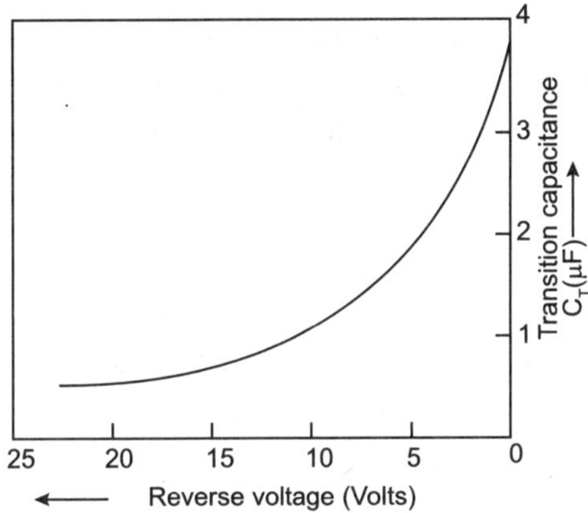

Figure 3.30 Variation of C_T with reverse voltage for a typical silicon diode at 25°C

The larger the reverse voltage, the larger is the space charge width and smaller the capacitance C_T. Similarly, for an increase in voltage in the forward-biased condition, the space charge width will decrease and C_T will increase.

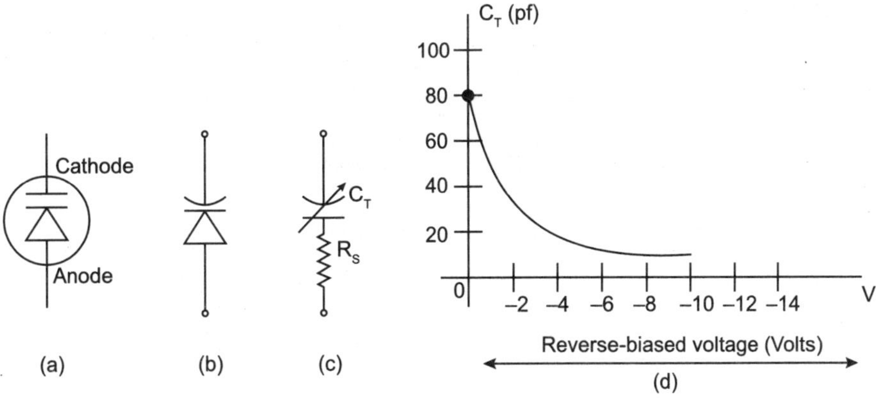

Figure 3.31 (a, b) Varactor diode symbol, (c) equivalent circuit of varactor diode, (d) typical varactor diode voltage-capacitance characteristics (drawn in 1st quadrant)

Diodes that are based on voltage variable capacitance are called **varactors**. A varactor is also called varicap, voltacap, epicap or turning diode. A schematic symbol of a varactor is shown in Figure 3.31(a).

Resistance R_s represents the body (ohmic) series resistance of the diode. A typical value of this is 8.5 Ω at 4 V. Reverse diode resistance R_r = 1 MΩ and that is why it may be treated as an open circuit across C_T.

Capacitance of a varactor is given by

$$C = \frac{\varepsilon A}{W} \tag{11}$$

where ε is the permittivity of depleted region of the semiconductor diode, A is the area of junction and W is the width of depletion region. Since the width of space charge (or depleted region) is approximately proportional to the square root of reverse-bias voltage V, while area A and permittivity ε being constant, we have

$$C = \frac{K}{\sqrt{V}}, \text{ where } K = \varepsilon A$$

Equivalent circuit of a varactor diode is shown in Figure 3.31(b) and a typical varactor diode voltage capacitance characteristics is shown in Figure 3.31(c).

Applications Such diodes are useful in a number of circuits:

(i) Voltage tuning of LC resonant circuit (electronic tuning circuit in radio, television, etc.)
(ii) Self balancing bridge circuits
(iii) Parametric amplifiers
(iv) Automatic frequency control device
(v) Adjustable band pass filter
(vi) FM modulators

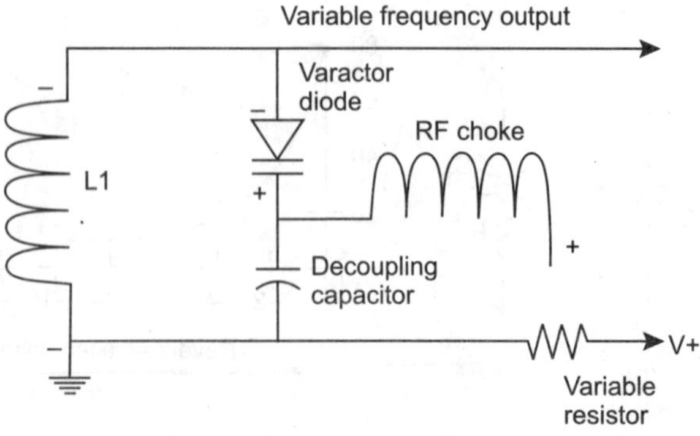

Figure 3.32 Varactor diode utilized as a tank circuit capacitor

Varactor diodes are used primarily as voltage-controlled capacitors in high-frequency tuned circuits. Figure 3.32 shows a typical varactor tuning circuit. In voltage tuning circuits, on changing the applied reverse voltage, the capacitance of the circuit (including varactor diode) will change, i.e. increasing voltage shall reduce the capacitance. Hence resonant frequency of the L-C circuit will increase. This frequency can be measured by using a variable oscillator. Thus, there is linear relationship between applied reverse voltage and resonant frequency upto a certain limit.

We may note that the DC path from the varactor anode is to be grounded. The positive bias at the varactor cathode is furnished from the potentiometer armatured by way of the RF choke. The remaining capacitor decouples the tuned circuit from the variable DC tuning voltage. The varactor remains reverse-biased throughout the range of potentiometer travel.

One advantage of this arrangement is that the tuning potentiometer may be located quite distant from the tuned circuit. Should forward-bias occur, the varactor will not function as a capacitor but as a low-resistance device. Varactor front-to-back resistance may be tested in the same manner as a conventional diode. Be aware that all P-N junctions exhibit the varactor.

3.11.11 Schottky or Shockley Diode

When discussing rectification, we have seen that an ordinary diode is turned off when the bias changes from forward to reverse (Chapter 4). It happens only in case of lower frequency signals. However, as frequency increases, a point reaches where an ordinary diode could not turn off fast enough to prevent noticeable current during a part of the reverse half cycle. One finds that this happens due to the storage of charges around the junction during forward-biasing.

When a diode is forward-biased, the charges are stored in different energy bands near the junction because of the lifetime of minority carriers. This effect is called charge storage. When we suddenly reverse-bias or forward-bias a diode, a large reverse current exists for a while because of this storage charge. Obviously, the charge storage prolongs the reverse recovery time (t_{rr}). Reverse recovery time is the time taken to turn off a forward-biased time. More precisely, it is time taken by a diode to drop the reverse current to 10% of forward current.

For ordinary diodes, t_{rr} is approximately 4 ns. At low frequencies, i.e. below 10 MHz, its effect is not even noticed. However, at higher frequencies, i.e. more than 10 MHz, its effect is noticeable, since the output signal begins to deviate from its normal shape. This problem is eliminated in Schottky diode. This high-frequency response is made possible by one of this diode's unique characteristics. That is it uses metallic flakes (sort of a doping) on top of the silicon semiconductor-type material to produce its junction.

When a semiconductor is brought into contact with a metal, a barrier layer is formed in the semiconductor from which charge carriers are severely depleted. The barrier layer is also called a depletion layer or exhaustion layer. Schottky diode is based on this effect. Schottky diodes are also known as **hot-carrier** or **surface-barrier** diodes. For the construction of a Schottky diode, a metal like gold silver or platinum is used on one side of the junction and doped silicon

(N-type) is used on the other side. The Schottky diode can also have two P-N junctions (PNPN), but only two electrodes (anode and cathode). Figure 3.33(a) shows the symbology of typical diode packaging. When Schottky diode is unbiased, free electrons on N side are in smaller orbit than free electrons on the metal side. This difference in the size of orbit creates a barrier potential (0.25 V only). This barrier potential is called Schottky barrier. The characteristic curve of a schottky diode is shown in Figure 3.33(b).

Advantages over P-N junction diode This diode has the following advantages over an ordinary P-N junction diode.

(i) It is a unipolar device because there are no minority carriers in the reverse direction (i.e. no significant current from the metal to semiconductor when the diode is reverse-biased). The delay due to hole-electrons recombination present in the junction diode is thus absent. Here it should be remembered that an ordinary junction diode is a bipolar device because it has both electrons and holes as majority carriers.

(ii) The junction contact area between the semiconductor and metal is larger than in point contact diode and hence forward resistance is lower (i.e., noise is comparatively lower).

(iii) Since no holes are available in metals, there is no depletion layer or stored charges to worry about. So the Schottky diode can switch OFF faster than a bipolar diode.

When a Schottky diode is forward-biased, free electrons on N side gain sufficient energy to travel in larger orbits. Due to this, free electrons just cross the P-N junction and enter the metal, producing a larger forward current. We know that a metal has no holes and hence there is no charge storage around the junction and reverse recovery time (t_{rr}) is zero. This means the Schottky diode is switched off at once when it is reverse-biased. This is why Schottky diodes can easily rectify signals of frequency above 30 MHz without distortion. The output of a half wave-rectifier using Schottky diode is shown in Figure 3.33(d) when the input signal frequency is 100 MHz.

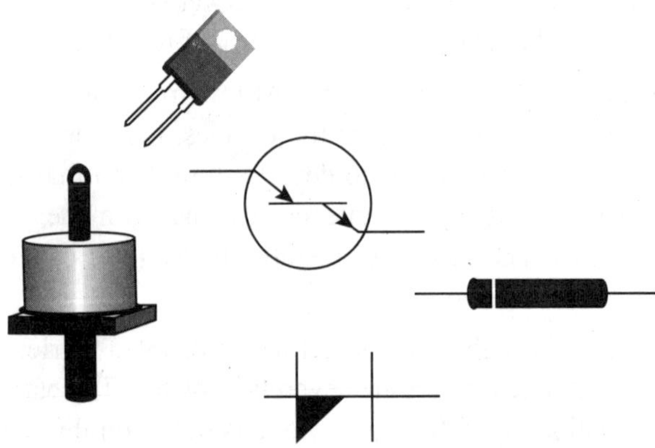

Figure 3.33 (a) Schottky symbols and typical examples

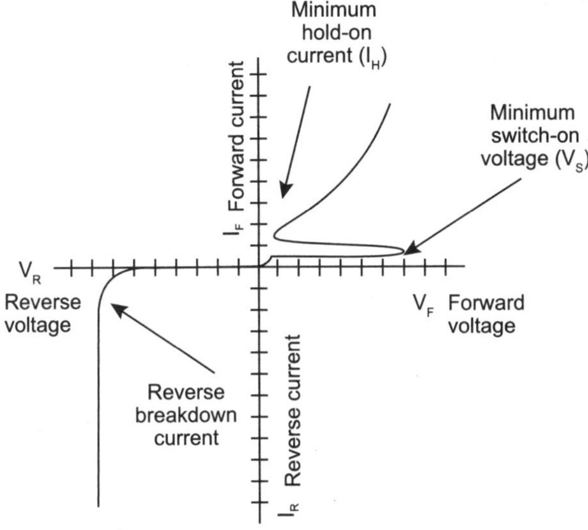

Figure 3.33 (b) Characteristic curve of a schottky diode

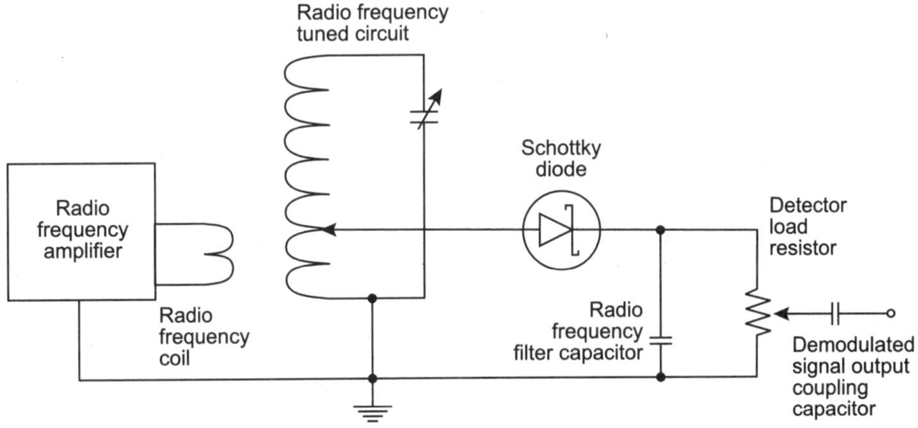

Figure 3.33 (c) AM radio detector circuit

To maximize power output and efficiency, IMPATT diodes are usually designed. IMPATT stands for "impact ionization avalanche transit time". IMPATT diodes employ impact-ionization and transit-time properties of semiconductor structures to produce negative resistance at microwave frequencies. Negative resistance arises from delays which cause current to lag behind voltage. One is "avalanche delay" caused by finite building time of the avalanche current, and the other is "transit-time delay" due to finite time for the carriers to cross the drift region. When these two delays add up to half-cycle time, diode resistance is negative at the corresponding frequency.

Schottky diodes are used in switching circuits and microwaves. We mention below few of them.

(i) We know that the speed of digital computers depends upon the switching time of their diodes and transistors. Obviously, Schottky diodes find their best application in digital computers due to their fast switching operation.

(ii) Schottky diodes are best suited for low voltage rectification due to their small barrier potential (0.25 V).

(iii) Schottky can also have two P-N junctions (PN-PN) but only two electrodes (anode and cathode). The usual forward-biased resistance for the Schottky is small, just a few ohms. This is considerably less than the forward-biased resistance of normal rectifier diodes. The Schottky is rated by its threshold (trigger voltage). This is the voltage required to forward-bias or turn on the diode. A Schottky can have a **third electrode**. If a third electrode, called a gate, is used to "fire" or turn on the diode, the Schottky is called an SCR. Figure 3.33(c) shows an AM radio detector circuit. Schottky diodes are especially suited to these circuits because of the low threshold voltage, which allows the detection of weaker signals that would not be possible with conventional PN diodes.

3.11.12 Step Recovery Diodes or Charge Storage Diode or Snapback Diode

In contrast to fast-recovery diodes, a charge storage diode is designed to store charge while conducting in the forward direction and upon switching to conduct for a short period in the reverse direction. A particularly interesting charge storage diode is the step-recovery diode or snapback diode which conducts in the reverse direction for a short period and then abruptly cuts off the current as stored charges have been dispelled. This cutoff occurs in the range of picoseconds and results in a fast-rising wave rich in harmonics.

The symbol of a step recovery diode is shown in Figure 3.34(a). A step recovery diode has unusual doping profile in which the density of carriers decrease near the junction. This unusual distribution of carriers causes a phenomenon called reverse snap-off, and that is why it is also called snapback diode.

When a high-frequency signal (say 20 MHz) is applied in the circuit (Figure 3.34(b)) during positive half cycle, the diode conducts like an ordinary silicon diode. However, during the negative half cycle, reverse current exists for a while because of stored charges, and then suddenly reverse current drops to zero (Figure 3.34(c)). It is as if the diode is snapped off and hence it is also known as **snapback diode**.

We must note that the output of this diode is non-sinusoidal, periodic wave which contains a sharp spike in the negative half cycle. From the Fourier theorem, we know that any non-sinusoidal periodic wave is equivalent to the superposition of sinusoidal components called harmonics. These harmonics have the frequencies f, $2f$, $3f$,... nf, where f is the fundamental frequency of the waveform.

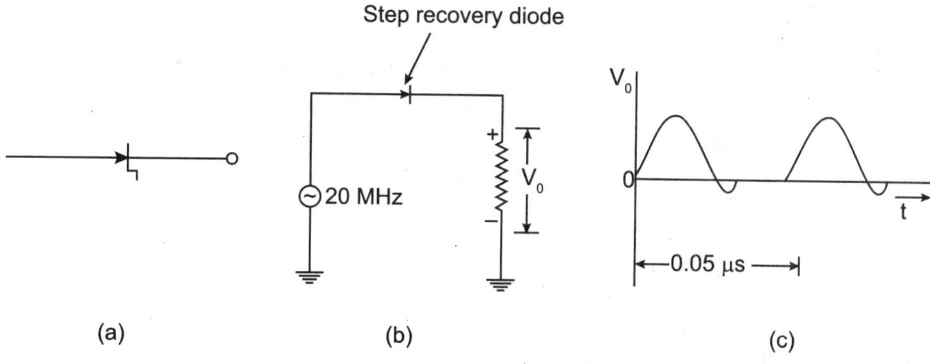

Figure 3.34 Step recovery diode

The output of this diode (Figure 3.34(c)) is rich in harmonics and can be filtered to produce a sine wave of required higher frequency. Because of these characteristics, these diodes are used as harmonic generators and pulse formers. Most step recovery diodes are made from Si with relatively long minority carrier life times ranging from 0.5 to 0.5 μs. We must note that the lifetimes are about 1000 times longer than for fast recovery diodes.

3.11.13 Backward Diode

It is just a zener diode in which the doping level is increased to such an extent that its zener effect, i.e. reverse breakdown, occurs near to zero.

Usually, zener diodes have breakdown voltages more than 2 V. Backward diodes are prepared to have reverse breakdown voltage near to zero (–0.1 V). Obviously, in backward diodes, forward conduction still occurs around +0.7 V but reverse conduction starts approximately at –0.1 V. This means that it conducts better in the reverse direction than in the forward direction.

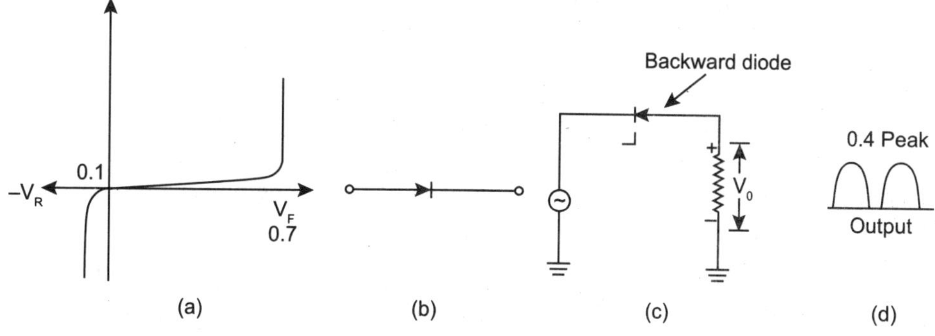

Figure 3.35 Backward diode

Figure 3.35(a) shows the V-I characteristics of backward diodes. Its symbol is similar to zener diode and is shown in Figure 3.35(b).

A backward diode can be used for a rectification of small signals, microwave detection, and mixing. Similar to the tunnel diode, a backward diode has good frequency response because there is no minority carrier storage effect. In addition, the current-voltage characteristic is insensitive to temperature and radiation effect, and a backward diode has very low I/*f* noise. Figure 3.35(c) shows the backward diode circuit and Figure 3.35(d) shows its output.

3.11.14 Thermistors and Barretters

We know that the electrical conductivity of a semiconductor changes significantly with temperature and has a negative temperature coefficient of resistivity. This property is utilized in a device called thermistor, whose resistance is temperature-sensitive, usually decreasing with temperature. Conventional wire-wound metallic resistors have positive temperature coefficients of resistivity. Commercial thermistors are usually made of sintered mixtures of Mn_2O_3, NiO_2 and CO_2O_3. A thermistor consists of a semiconductor bead of approximately 0.04 cm in diameter. Two thin wires are attached to the bead to provide for two terminals. The diameter of the wire is approximately 0.25 μm.

Thermistors find use in control systems operated by temperature changes, in the measurement of microwave power, in thermometry, and as a thermal relay. In electronic circuits, thermistors have been used to compensate for change in resistance with temperature of ordinary component where variation of component values cannot be tolerated.

The I-V characteristic of a thermistor has a negative slope. Devices that exhibit, in some regions, negative slope in their I-V characteristics are useful for making oscillators, amplifiers and switching circuits. However, thermistors are not suitable in these applications because their response characteristics are too slow. There are certain bulk semiconducting compounds which have negative resistance characteristics over a limited range of operating parameters utilizing a mechanism unrelated to temperature sensitivity of resistivity. These materials have been used to obtain devices based on **Gunn effect**.

A heavily doped semiconductor shows metallic properties. It has a positive temperature coefficient of resistivity owing to the decrease of carrier mobility with increasing temperature. Such a device is called **sensistor**. Thermistors also find extensive use as sensing elements in microwave power measuring equipment and as temperature sensors in electronic thermometers. Thermistors are capable of yielding power information over the power range of 10^{-5} mW to 20 mW with a typical burn-out level 400 mW.

A barretter has a positive temperature coefficient of resistance, and consists of a mounted piece of "Wollaston" platinum wire having diameter of approximately 1.25 μm. Barretters are capable of yielding power information over the range of 10^{-5} mW to 20 mW. A typical burn-out level is 20 mW. For low-level *rf* power application below 10^3 MHz, 0.001 A, **Littlefuse** may be used as a barretter.

3.11.15 Photoconductor

It consists of a slab of semiconductor (in bulk or thin film) with ohmic contacts affixed to opposite ends as shown in Figure 3.36. When light is incident on the semiconductor, electrons are excited from the valance bands into the conduction band. Obviously, electron-hole pairs in excess of those generated thermally are produced. The conductivity of the material increases owing to the increased number of carriers. Under the influence of light on the surface of a photoconductor, carriers are generated either by band-to-band transitions (intrinsic) or by transitions involving forbidden gap energy levels (extrinsic), resulting in an increase in conductivity.

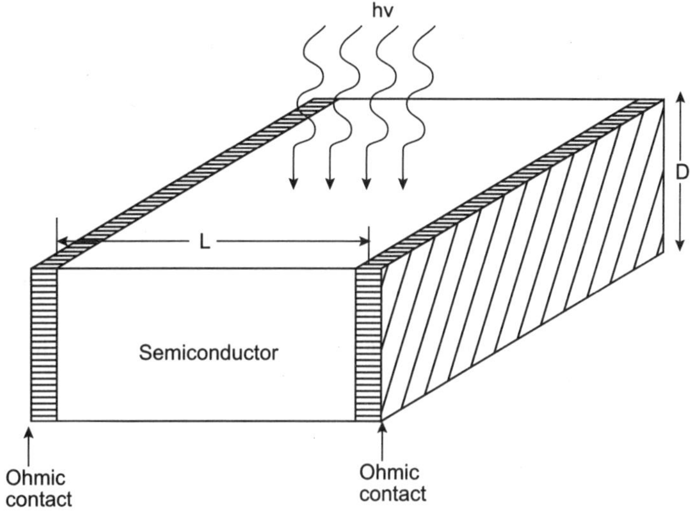

Figure 3.36 Photoconductor

For an intrinsic photoconductor, conductivity is given by $\sigma = e(\mu_n + \mu_p p)$, and the increase of conductivity under illumination in mainly due to the increase in the number of carriers. The corresponding long wavelength cut-off of the photon of frequency υ_c in this case is given by

$$\lambda_c = \frac{hc}{E_g} = \frac{1.24}{E_g(\text{eV})}(\mu\text{m}) \tag{12}$$

where λ_c is wavelength corresponding to the semiconductor band gap E_g, c is the speed of light. For wavelength (λ) shorter than λ_c, the incident radiation is absorbed by the semiconductor, and hole-electron pairs are generated. For an extrinsic case, photoexcitation may occur between a band edge and an energy level in the energy gap. Photoconductivity can take place by the absorption of photons of energy equal to or greater than the energy separation of the band gap levels and the conduction or valence band. In this case, the long wavelength cut off is determined by the depth of forbidden gap energy level.

Figure 3.37 shows the photo response of an intrinsic as well as extrinsic semiconductor e.g., materials that find application as sensitive photoconductors are CdS and CdTe.

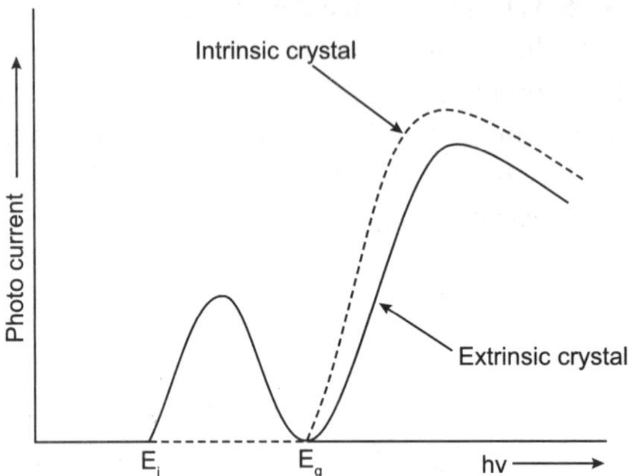

Figure 3.37 Photoresponse of a photoconductor

The performance of a photoconductor is measured in terms of (i) photo conductivity gain and (ii) response time of the detector. Typical gain and response time for a photoconductor are 10^5 and 10^{-3} seconds, respectively. Photoconductivity devices are used as relays in digital or control circuits, as light meters, and to record modulating light intensity such as in a sound track.

3.11.16 Gunn Effect and Gunn Diode

Ridley-Watkins and Hilsum independently predicted that semiconductor materials under certain conditions offer differential negative resistance. This differential negative resistance is a bulk effect and is due to the transfer of electrons from one valley to another in the conduction band. Gunn, while experimenting on a sample of N-type GaAs and some other III-V compounds, found that current through the sample increased linearly with voltage till a certain threshold voltage. Beyond the threshold voltage, a number of current pulses appeared with a time interval proportional to the length of the sample. This threshold field is high (~400 V/mm). The oscillations lie in the microwave range and set in the negative differential conductance (NDC) region where current decreases with increase in electric field E (Figure 3.38). This behaviour is a consequence of the band structure of these materials. In the region of velocity field, where v decreases with increasing E, differential mobility (dv/dE) of the electrons becomes negative. The reason is as follows.

n-GaAs has a direct energy band gap. The band gap is 1.4 eV. Free electrons in n-GaAs normally occupy the lowest energy states in the conduction band. The effective mass of electrons in this situation and their mobility μ_1 is high (≈ 0.02 m^2/v-s). As a result of transfer of the electrons, current begins to decrease with increase in field because $T = n_1 e\mu_1 E + n_2 e\mu_2 E$, where n_1 and n_2 are the concentration of electrons having mobility μ_1 and μ_2 respectively. The average drift velocity of an electron is, $(n_1 + n_2)$ being the total electron concentration, is

$$v = \left(\frac{n_1\mu_1 + n_2\mu_2}{n_1 + n_2} \right) E \tag{13}$$

If the transition of electrons from the high-mobility state to the low-mobility state occurs rapidly over a range of field E, one observes that v diminish with field E beyond a certain threshold field E_{th}, shown by region AB in Figure 3.38. When all the electrons move to the low-mobility state, drift velocity is $v = \mu_2 E$. One finds that v continues to increase slowly with E. This builds up successive oscillations.

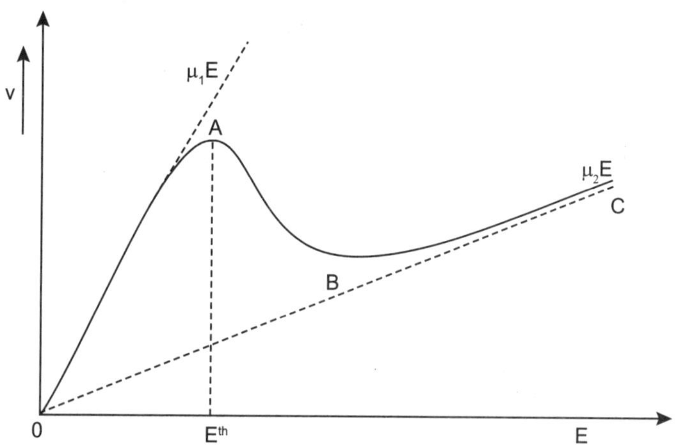

Figure 3.38 A possible characteristic of electron drift velocity vs. field for a semiconductor exhibiting transferred electron mechanism

Owing to the occurrence of NDM in the velocity-field curve, Gunn diodes are used as sources of microwaves especially where high power is not a requirement. Gunn diodes are also used as local oscillators for mixers in microwave receivers over the frequency range of 1 to 100 GHz. A Gunn device can carry just before the threshold voltage one of two possible currents, depending on the presence or absence of a domain. This can, therefore, be used as a high-speed binary logic.

Gunn diode is a versatile semiconductor device. These are commercially available for pulsed operations yielding a power output of 5 W in the frequency range 5.0 to 12.0 GHz.

3.11.17 IMPATT, TRAPATT and QWITT Diodes

IMPATT (impact ionization avalanche transit time) diodes employ impact-ionization and transit time properties of the semiconductor structures to produce negative resistance at microwave frequencies. A TRAPATT (trapped plasma avalanche triggered) diode is an IMPATT-related device. These are quite new devices and used as oscillators in the microwave region. These devices work well in the breakdown region. Although IMPATT operation can be obtained in simpler structures, the read diode is best suited for the illustration of basic principles.

A read diode consists essentially of two regions: (i) the n^+-p region at which avalanche multiplication occurs and (ii) the i (essentially intrinsic) region through which the generated holes must drift in moving to the p^+ contact. One can build similar devices in the p^+-n-i-n^+ configuration, in which electrons resulting from avalanche multiplication drift through the i region, taking advantage of the higher mobility of electrons compared to holes.

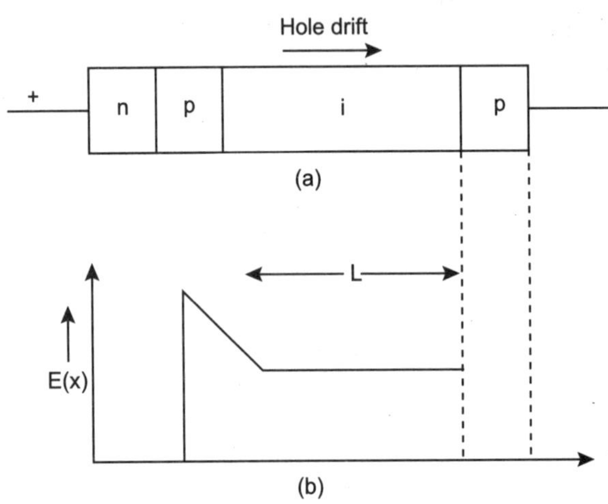

Figure 3.39 Read diode: (a) basic device configuration, (b) electric field distribution in the device under reverse-bias

The device operates in negative conductance mode, when the AC component of current is negative over a portion of cycle during which the AC voltage is positive, and vice versa. Negative conductance occurs because of two processes causing current to lag behind voltage in time: (i) a delay due to avalanche process and (ii) a further delay due to the transit time of carriers across the drift region. When these two delay times combine to produce a net phase-shift between 90° and 270°, diode resistance is negative at the corresponding frequency. Consequently, negative conductance occurs and the device can be used for oscillation and amplification.

Structures that have been successfully employed for IMPATT devices are PN, PIN, P^+ NN^+, P^+ NIN^+ and N^+ PIP^+ (called read diodes). These devices are usually constructed from silicon and gallium arsenide. At present, IMPATT diode can generate the highest CW (continuous wave) power output at millimeter wave frequencies, i.e. microwave power above 30 GHz. A major drawback of the IMPATT diode for very high frequency operation is the fact that the avalanche process, which depends on random impact ionization events, is inherently noisy.

In TRAPATT diode, the TRAPATT mode of operation, i.e. operating frequency, is substantially lower than transit time frequency, and efficiency is considerably higher. Under large-signal conditions, periodic avalanching of the diode begins at the high-field side and sweeps rapidly across the diode, leaving it substantially filled by a highly conducting plasma of holes and electrons whose space-charge depresses the voltage to very low values. Since the plasma cannot rapidly escape, this mode is called TRAPATT mode. This diode has been tried experimentally in the pulsed transmitter of phased array radar systems.

A variety of approaches have been investigated to find alternative methods for injecting carriers into the drift region without relying on the avalanche mechanism. A particularly interesting device is QWITT (quantum well infection transit time) diode, which employs resonating tunneling through a quantum well to inject electrons into the drift region. The device consists of a single GaAs quantum well between two $Al_xGa_{1-x}as$ barriers in series with a drift region of undoped GaAs. This structure is then placed between two n^+-GaAs regions to form two contacts (Figure 3.40). In this device, one can achieve maximum resonant tunneling of electrons through the well if the DC bias is properly adjusted.

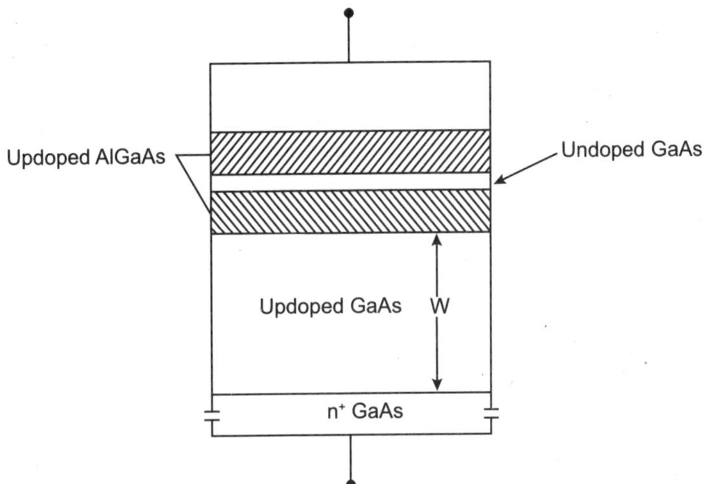

Figure 3.40 Structure of a QWITT diode

The QWITT diode is a low-noise injection mechanism with high-frequency characteristics. The length of transit time as well as the shape of current pulse can be optimized to obtain the best power-frequency performance from the QWITT diode. This diode should extend the normal frequency limit associated with transit-time devices, while providing higher output power than simple quantum-well RTD oscillations.

3.11.18 PIN Diodes

This is a P-N junction with greatly improved switching times. Obviously, this is a P-N junction with a doping profile tailored so that an intrinsic layer the '*i* region' is sandwiched between a P-layer and an N-layer as shown in Figure 3.41. A PIN diode differs from a P-N junction diode because it does not have a P-N junction. Instead, it consists of three materials: P-type, N-type and a pure (intrinsic) silicon slice inbetween. Since the intrinsic silicon slice is between the P-type and N-type materials, there is no P-N junction.

Intrinsic silicon has very high resistance. The thicker the intrinsic material layer, the higher the resistance and greater the breakdown voltage.

Another distinct characteristic of the PIN diode is that it does not have an actual knee voltage. Unlike the P-N junction diode, there is no abrupt "turn-on" point but rather a gradual current.

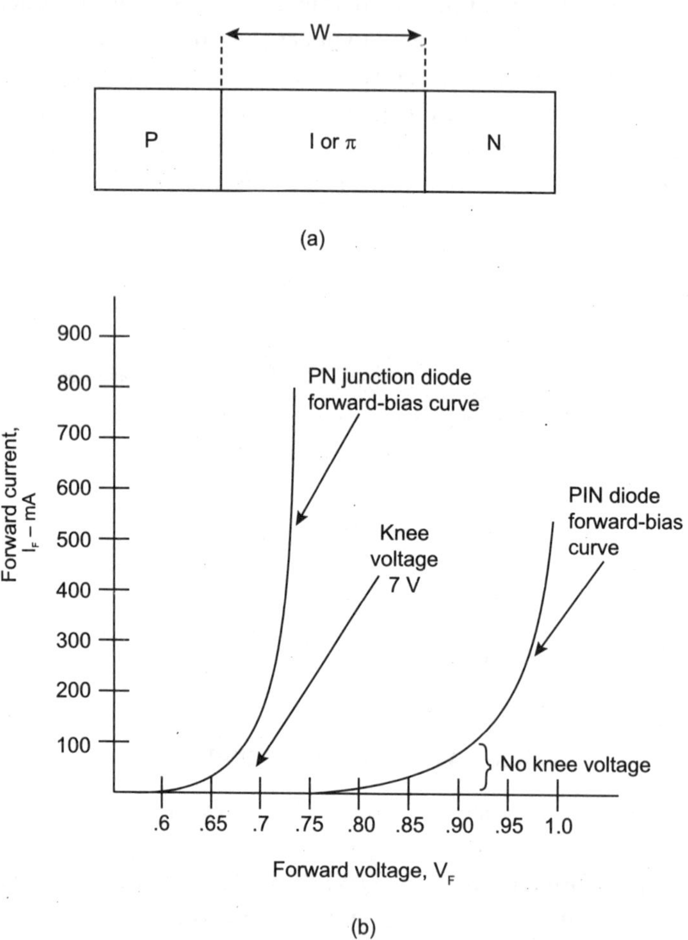

Figure 3.41 (a) PIN diode, (b) PIN diode characteristic

In practice, however, the idealized *I* (intrinsic) region is approximated by either a high resistivity *P* layer (referred to as π-layer) or a high resistivity N-layer (referred to as *v* layer). Resistivity of the *I* layer is typically 10^3 Ω-m.

When no external voltage is applied between the terminals of a PIN diode, concentration gradient across the junction causes diffusion of carriers. The width of the depletion region in the *I*-layer is relatively large. Now, if a reverse-bias is applied and increased gradually, the depletion region becomes thicker. At **swept-out voltage** (a particular value of the reverse-bias), all the free carriers are swept out of the intrinsic layer. With further increase of reverse-bias, the depletion region widens in the highly doped semiconductor region. The PIN diode offers very high resistance under reverse-bias. We must note that the breakdown voltage of the diode under reverse-bias is also very large.

The most common uses of the PIN diode is in radio frequency (RF), microwave, modulator circuits and as a switch.

The PIN diode when used as a switch operates between the ON and OFF states. In ON state, the diode is forward-biased when minority carriers are infected into the high resistivity region between the highly doped regions. The increase, in the densities of carriers reduces the resistance, and diode impedance is low. This causes a current flow. Forward resistance of the diode varies with forward-bias. In the OFF state the high resistivity region is completely swept out and diode impedance is very high as capacitance is very low.

The PIN diode has found wide applications in microwave circuits. It can be used as a microwave switch with essentially constant depletion layer capacitance and high power handling capability. When PIN diodes are used as microwave switches and when they are biased in the OFF condition, the bias is usually beyond the swept-out voltage (usually 2 V). PIN diodes can also be used as a **variolosser** (variable attenuator) by controlling the device resistance which varies approximately linearly with forward current. PIN diodes can also modulate signals upto the GHz range. We must note that the forward characteristics of a thyrister in its ON state closely resemble those of the PIN diode.

Applications of PIN diodes Few important applications of PIN diodes are:

1. It can be used as an alternator.
2. It is used as a constant impedance device.
3. It can be used in the construction of phase modulator and amplitude modulator.
4. It can be used as a phase shifter.
5. It can be used as T-R switch in radar applications.
6. It can be used as a photodetector in optical fibres.

3.11.19 Silicon Controlled Rectifier (SCR)

An SCR is a thyristor having four alternate layers of P-type and N-type silicon. Figure 3.42 (a) shows the structure of an SCR and its symbolic representation is shown in Figure 3.42 (b). Figure 3.42(a) shows that an SCR consists of three junctions J_1, J_2 and J_3. Junctions J, and J_3 operate in the forward direction, while J_2 operates in the reverse direction, and the three terminals are called anode (A), cathode (K) and gate (G). The function of gate (G) is to control the firing of SCR. Since it conducts only in one direction, i.e. from anode to cathode, this constitutes a unidirectional device.

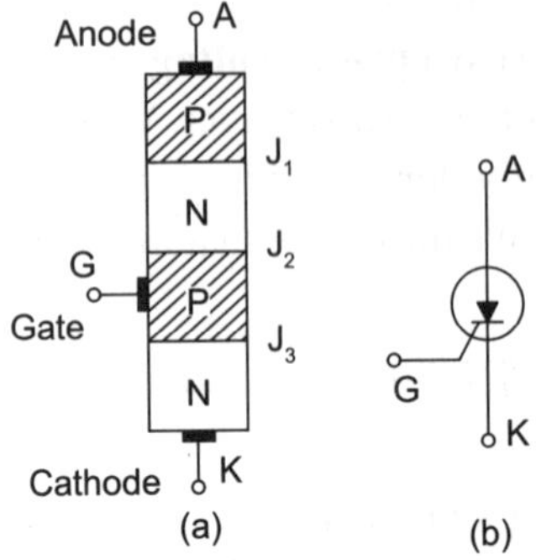

Figure 3.42 (a) Structure and (b) symbolic representation of a silicon controlled rectifier

Operation In this device, a load is connected in series with the anode, which is kept at positive potential. One can study the operation of this device when gate (*G*) is open and when gate (*G*) is positive with respect to the cathode. Figure 3.43 shows these situations.

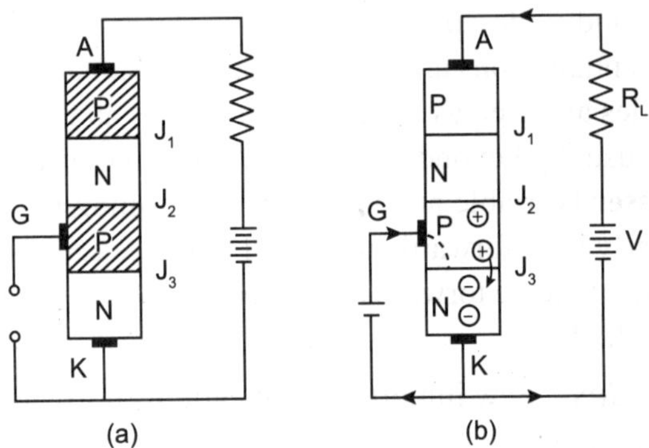

Figure 3.43 Operation of a silicon controlled rectifier (SCR): (a) gate is open, (b) gate is positive w.r.t. cathode

When gate (G) of SCR is open, no voltage is applied at the gate. Under this condition, junctions J_1, and J_2 of SCR are forward-biased while junction J_2 is reverse-biased. Due to the reverse-biasing of junction J_2, no current flows through R_L and therefore SCR is cut-off. However, when we increase the anode voltage, a certain critical value (breakdown voltage) is reached when junction J_2 breaks down. In this situation, the SCR conducts heavily and it is said to be in ON state. Here SCR offers extremely low forward resistance (0.1 – 1.0 Ω) so that voltage across it drops to a low value (~ 1 V). The current is now limited only by the power supply and load resistance (R_L). We may note that current keeps flowing indefinitely untill the circuit is open. We should note the following points.

(i) The anode (A) is made negative w.r.t. the cathode. SCR is blocked due to two reverse-biased junctions J_1 and J_2. On increasing the voltage, a stage is reached when **zener breakdown** occurs and the SCR is destroyed. Obviously, SCR is an **unidirectional electronic device**.

(ii) SCR behaves like a switch, i.e. it either does not conduct or conducts heavily.

(iii) In order to make the SCR non-conducting, the supply voltage is reduced to zero.

(iv) Normally a small positive voltage is applied at the gate to close the SCR due to the fact that breakdown voltage is usually much greater than supply voltage.

(v) One can turn on the SCR either by keeping the gate open and making the supply voltage equal to the breakover voltage or by making the supply voltage less than the breakover voltage and applying a small voltage (~ 1.5 V) to the gate.

3.11.20 Two-Transistor Analogy of SCR

One can explain the operation of SCR by dividing it into two transistors: a PNP transistor (T_1) and another NPN transistor (T_2). Figure 3.44 (a, b) shows its electrical equivalent circuit. It is obvious from the figure that the collector of each transistor is coupled to the base of the other. This makes a positive feedback loop.

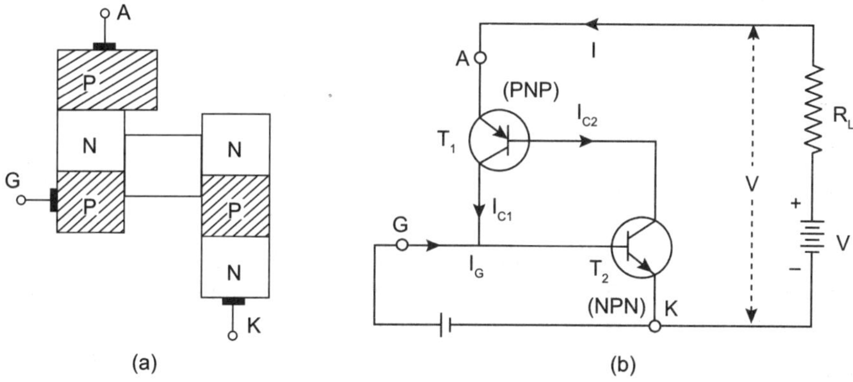

Figure 3.44 (a) Transistor, (b) Equivalent circuit of SCR

Let us suppose that supply voltage V is less than breakover voltage. With the gate open, the base current of T_2 is zero. Therefore, no current flows in the collector of T_2 and hence that of T_1. Under this condition, the SCR is open. With the gate closed, a small gate current flows through the base of T_2. This increases its collector current. As the collector current of T_2 is the base current of T_1, the collector current of T_1 increases. Further, the collector current of T_1 is also the base current of T_2. Therefore, an increase in current of one transistor causes an increase of current in the other transitor. This process is accumulative and both transistors are driven to saturation. Now, a heavy current flows through the load and SCR is in ON condition.

3.11.21 V-I Characteristics of SCR

The volt-ampere characteristic of the SCR for $I_G = 0$. When the anode is positive w.r.t. the cathode, the characteristic is known as forward characteristic. The forward characteristic is shown by the curve *OABC*. When the supply voltage is increased from zero, a point *a* is reached when the SCR starts conducting. The voltage corresponding to *A* is called forward breakover voltage.

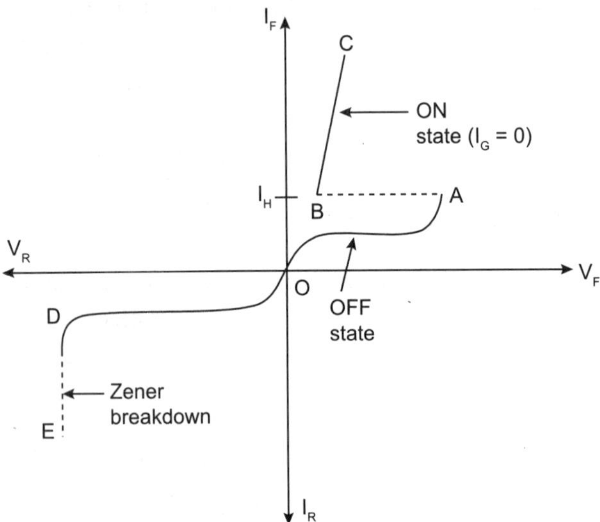

Figure 3.45 Volt-ampere characteristic of SCR

At this voltage, junction J_2 breaks down and SCR switches suddenly to a high conducting state. When the gate is open, *the breakover voltage is defined as the minimum forward voltage at which SCR starts conducting heavily*. Under this condition, the voltage across SCR suddenly drops as shown by dotted curve *AB*. Now most of the supply voltage appears across load resistance. The current corresponding to point *B* is denoted by I_H and is known as holding current. *The Holding current is the maximum anode current, gate being open, at which SCR is turned OFF from ON condition*. If the holding current of an SCR is 5 mA, it means that if anode current is made less than 5 mA, SCR will be turned OFF. Above the holding current, SCR is in ON state and the current keeps flowing indefinitely until the current is opened briefly. *There is a maximum value of anode current which an SCR is capable of passing without destruction*. This is known as forward current rating and is expressed by I_F. Any attempt to exceed this value will result in the destruction of SCR. Commercially available SCRs have forward current ratings from about 30 A to 100 A.

When the anode is negative w.r.t to the cathode, the characteristic is called reverse characteristic. When the anode voltage is gradually increased, at first the anode current remains small (i.e., leakage current) but at a particular value, avalanche breakdown occurs and SCR starts conducting in the reverse direction. This is shown by curve *DE*. The voltage corresponding to point *D* is known as reverse breakdown voltage.

So far, we have the effect of gate current I_G not taken into account. Figure 3.46 shows the SCR forward characteristic, taking into account the gate current. It is obvious from the figure that when gate current is supplied, the SCR fires even with much lower anode voltage. Higher the value of gate current, lower the value of anode voltage at which breakover occurs. Once the SCR has fired, the gate loses its controlling action and the SCR cannot be turned off by gate signals. The only way to bring SCR back to normal OFF condition is to reduce anode voltage sufficiently for anode current to drop below holding current I_H.

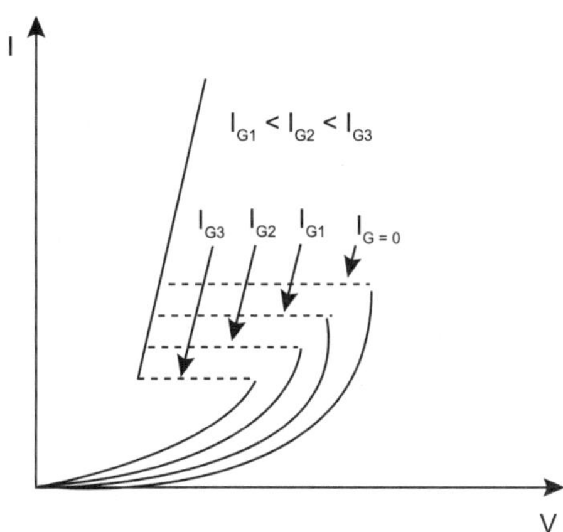

Figure 3.46 SCR forward characteristic with gate current

3.12 SCR Switching

SCR has only two states, i.e., ON state and OFF state. When a proper gate current is passed, it starts conducting heavily. The SCR remains in this position indefinitely even if gate voltage is removed. This is known as ON state. On the other hand, when the anode current is reduced to the holding current, it is turned OFF. So SCR works as a switch. Here, we shall discuss the methods used to turn on and turn off an SCR.

SCR turning off There is a minimum value of gate voltage at which the SCR is turned ON. This is called gate triggering voltage and is denoted by V_{GT}. For most SCRs, the value of V_{GT} = +2 to +10 V. So, when this value of V_{GT} is applied at the gate, the SCR is turned ON. We shall discuss the following three methods only:

(i) **DC gate triggering** Figure 3.47 shows the circuit of DC gate triggering. In this process of triggering, a DC voltage of proper magnitude and polarity is applied between gate and cathode in such a way that the gate becomes positive with respect to the cathode. When this voltage is sufficient to t produce the required gate current (V_{GT}), the device starts conducting. The resistance is used to limit the gate current.

(ii) **Voltage triggering** With increasing applied voltage between anode and cathode, the depletion width of reverse-biased junction starts decreasing. At a particular voltage (break-over voltage), the depletion layer vanishes. This causes the reverse-biased junction to collapse. So, the number of charge carriers increases and thus the device starts conducting.

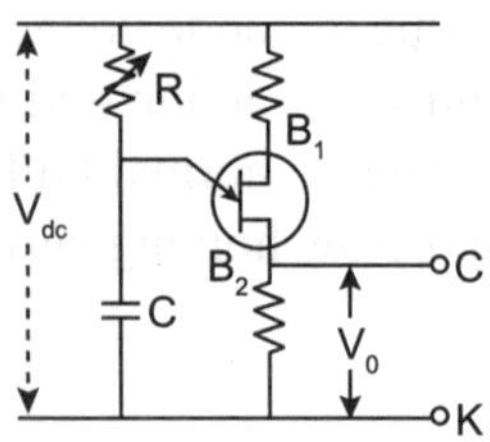

Figure 3.47

(iii) **Pulse gate triggering** In this process, a pulse voltage is applied at the gate for triggering the device as shown in Figure 3.48. The most frequently used method of producing pulses for gate triggering is by means of a unijunction transistor (UJT) in the form of relaxation oscillator. The advantage of this method is that there is no need of applying continuous signals, thus reducing gate power losses.

SCR turning OFF Once the SCR is fired, it remains in the ON state even if the triggering pulse is removed. So, the SCR cannot be turned OFF simply by removing the gate pulse. A number of techniques are employed for turning OFF an SCR. Some of them are discussed here:

(i) **Anode current interruption** In the characteristics of SCR, we have seen that when the anode current is reduced below a minimum value called holding current, I_H, the SCR is turned off. So, a simple way to turn off the SCR is to open switch as shown in Figure 3.49.

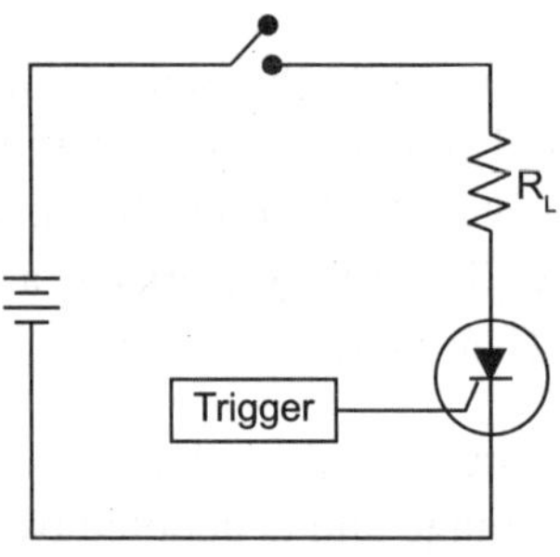

Figure 3.48

(ii) **Commutation of SCR** The method of switching OFF an SCR is known as its commutation. There are two commutation methods:

(a) *Natural commutation process:* When current passing through a device is alternating is nature, it passes through natural zero at regular intervals. At this stage, current through the device goes to zero and simultaneouly a reverse polarity voltage is developed across the device (say a thyristor). This immediately turns OFF the device. The process is called natural commutation. This is only possible in case of AC applications.

(b) *Forced commutation:* For switching OFF of SCR in case of DC circuits, the forward current should be forced to become zero by means of some external circuit, and hence the name forced commutation. The external circuit required for this purpose is known as commutation circuit. The components (viz. L, C, etc.) used in a commutation circuit are called commutation components. The principle of forced commutation is that a reverse voltage is developed across the SCR by means of a commutation circuit which immediately brings the forward current to zero, thus turning it OFF. Figure 3.49 shows forced commutation of SCR where the capacitance performs commutation. Assume SCR-1 is conducting and SCR-2 is in OFF state. Capacitor *C* is charged up to a supply voltage through resistance *R*. The right plate of capacitor *C* is positively charged while the left plate is negatively charged. When SCR-2 is trigged on, *C* is effectively paralleled across SCR-1. The charge on *C* is then opposite to SCR-1 forward voltage. Thus, it is turned OFF. Now the current is transferred to SCR-2 path.

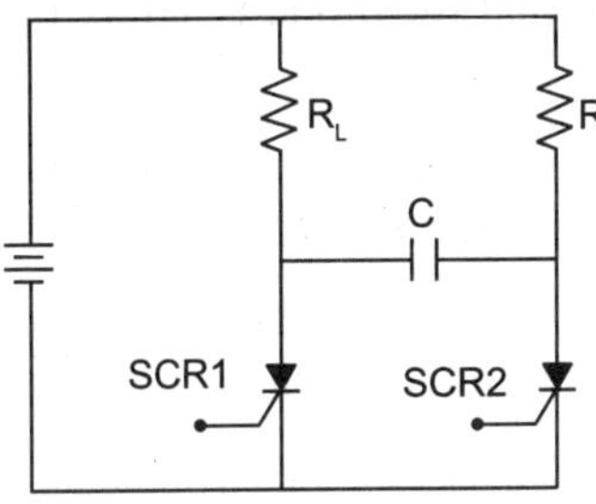

Figure 3.49

3.12.1 Applications of SCR

The main applications of SCR are as under:

1. Converter (conversion of AC voltage into DC voltage)
2. Inverter (conversion of DC voltage into AC voltage)
3. Chopper (fixed DC voltage is converted into variable DC voltage)
4. Cycloconverter (AC voltage at one frequency f_1 is converted into AC at other frequency f_2 where $f_1 > f_2$).

Other common areas of its application include:

1. Speed control of AC motors
2. Speed control of DC motors
3. Voltage regulators
4. Heater control
5. Phase control
6. Regulated power supplies
7. AC circuit breaker
8. DC circuit breaker

3.13 Working with Diode Circuits

When working with circuits in which diodes or diode circuits are used, one should be familiar with testing and troubleshooting techniques. Besides, one can select a diode for a particular application with the help of diode datasheet.

3.14 Datasheet of Diode

The datasheet of a diode contains values of different parameters of diode for a particular application. It also provides the maximum ratings of various parameters. When designing diode circuits, this information is used so that the diode is safely operated within the specified rating of its parameters.

The datasheet contains the following two categories for ratings of the parameters: (i) maximum rating and (ii) electrical characteristics

Tables 3.1 and 3.2 shows a part of the datasheet of a general purpose, low-power rectifier diode. This data sheet is for diode 1N 4001 manufactured by Motorola.

Table 3.1 Maximum ratings of diode 1N 4001

Rating	Symbol	Value for 1N 4001	Unit
Peak repetitive reverse voltage	V_{RWM}	50	Volt
Working peak reverse voltage	V_{RWM}		
DC blocking voltage	V_R		
Non-repetitive peak reverse voltage	V_{RSM}	60	Volt
RMS reverse voltage	$V_{R(RMS)}$	35	Volt
Average rectified forward current	I_0	1.0	Amp
Non-repetitive peak surge current	I_{FSM}	30 (For one cycle)	Amp
Operating and storage junction temperature range	$T_J T_{Stg}$	–65 to + 175	°C

Table 3.2 Electrical characteristics of 1N 4001 diode

Rating	Symbol	Type	Max	Unit
Maximum instantaneous forward voltage drop	V_F			Volts
Maximum full-cycle average forward voltage drop	$V_{F(AV)}$			Volts
Maximum reverse current (rated DC voltage) $T_J = 25°C$	I_R	0.05		1/4 A
$T_J = 100°C$		1.0	50	
Maximum full-cycle average reverse current	$I_{R(AV)}$	-	30	1/4 A

We now present a discussion on the various parameters of the datasheet.

Peak repetitive reverse voltage (V_{RRM}) This rating of diode signifies the peak-inverse voltage (PVI). PVI denotes the maximum repetitive voltage that can be applied to a diode in the reverse- biased condition. In the data sheet, the manufacture represents PVI as peak repetitive reverse voltage (V_{RRM}), and for 1N 4001 diodes, it is 50 V.

Working peak reverse voltage (V_{RWM}) V_{RWM} also signifies the maximum reverse voltage that can be applied to the diode under working conditions, with a value of 50 V for 1N 4001 diodes.

DC blocking voltage (V_R) This is the maximum value of the DC reverse voltage which can be applied to the diode. We may note that in this case the voltage is DC, not necessarily repetitive voltage. The value of V_R for 1N 4001 diodes is 50 V.

Nonrepetitive peak reverse voltage (V_{RSM}) This is the maximum amplitude that can be applied across the diode in one cycle of a repetitive voltage, and this single cycle of repetitive voltage is termed as nonrepetitive voltage. It is designated as V_{RSM} with a value of 60 V for 1N 4001 diodes.

RMS reverse voltage ($V_{R(RMS)}$) This is the rms value of V_{RRM} rating or PVI. For 1N 4001, the value of V_{RRM} is 50 V and obviously its rms value is 35 V.

Average rectified forward current (I_o) When the diode is forward-biased, I_0 is the average value of rectified current. For 1N 4001 diode, the value I_0 is 1 ampere.

Non-repetitive peak surge (I_{FSM}) This rating for diode is used to refer to the maximum or peak value of one cycle that is applied to the diode in forward-biased conditions. For 1N 4001 diodes, the value of I_{FSM} is 30 amperes.

Maximum instantaneous forward-voltage drop (V_F) Under forward-biased condition, V_F is instantaneous drop across the diode, when the forward current is 1 ampere at a junction temperature of 25°C. For 1N 4001 diodes, F_F has a maximum value of 1.1 V.

Maximum full-cycle average forward voltage drop ($V_{F(AV)}$) When the diode is under forward-biased condition and the average is taken over one full cycle, $V_{F(AV)}$ is the maximum voltage drop across the diode. The value of $V_{F(AV)}$ is 0.8 V for 1N 4001 diodes.

Maximum reverse current (rated DC voltage) (I_R) When the diode is reverse-biased with the rated DC voltage, the maximum current across the diode is defined as the maximum reverse current (I_R). For 1N 4001 diodes, I_R has a maximum value of 10 1/4 A at a junction temperature of 25°C.

Maximum full-cycle average reverse current ($I_{R(AV)}$) When the average is taken over one full cycle of the AC voltage, $I_{R(AV)}$ is the maximum current flowing through the diode under reverse-biased conditions. $I_{R(AV)}$ has a value of 10 1/4 A for 1N 4001 diode.

3.15 Testing and Troubleshooting of Diodes

Diodes find wider applications in various electronic circuits, e.g. rectifiers, regulated power supplies, detection of modulated signals in communication systems, etc. When one attempts to troubleshoot an electronic circuit, the probability of finding a faulty diode is very high. Obviously, one must know to test a diode and to troubleshoot a circuit with diodes.

3.15.1 Testing of Diodes

When silicon and germanium diodes fail, they usually fail in one of two ways (or modes).

Open failure mode When forward-bias is exceeded, current no longer flows. This often happens when I_{output} is greatly exceeded. The diode may appear to be burned or cracked.

Short failure mode When reverse breakdown voltage is found to have decreased dramatically, the same physical symptoms as for open failure mode may be present. The characteristics of silicon and germanium diodes may be determined with a semiconductor curve tracer. However, most modern digital multimeters (DMM) will detect these two failure modes with their diode testing function.

One can easily test a diode by using an analog multimeter. This is due to the fact that the movement of the pointer of an analog multimeter is easily visualized as compared to the digital display of a digital multimeter. While testing a diode, the analog multimeter is set to ohm meter, so that the internal battery becomes a part of the circuit. This battery is used to bias the diode under testing to verify whether the diode is good or bad. The following two steps are required to complete the test:

(i) Test for a forward-biased diode The diode is forward-biased by connecting the positive terminal (red lead) of the multimeter to the anode (A) of the diode, and the negative terminal (black lead) of the multimeter to the cathode (K) of the diode (Figure 3.50(a)). Under this condition, the diode is said to be forward-biased by the internal battery and the resistance of the diode becomes very small. The pointer of the analog multimeter moves toward the left, showing a very small resistance on the scale of the multimeter. This test confirms that the diode is properly working under forward-biased situation.

(ii) Test for a reverse-biased diode For testing the reverse-biased diode, the polarities of the multimeter are inverted (Figure 3.50(b)). The red lead is connected to the cathode and the black lead of the multimeter is connected to the anode of the diode (Figure 3.50(b)). Thus, the diode is reverse-biased through this circuit arrangement. The diode offers a very high resistance under reverse-biased conditions and the pointer of the multimeter moves slightly toward the right side from its rest or zero position. This clearly indicates that the diode is good and properly working under reverse-biased conditions. Diodes become faulty due to following reasons: (i) diode is short, (ii) diode is open.

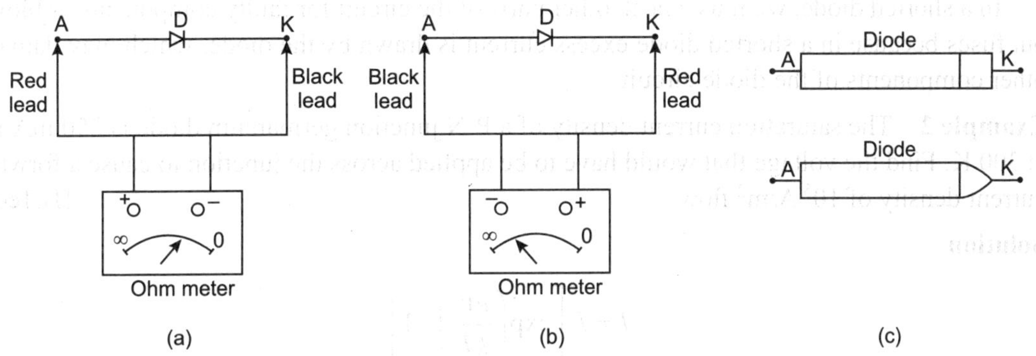

Figure 3.50 Testing of a given diode: (a) test for a diode while forward-biased, (b) test for a diode when reverse-biased, (c) identification of terminals, i.e. anode (A) and cathode (K) of a diode

If a given diode is faulty, then the pointer of the multimeter shows a small resistance in both the above tests. If the diode is open, then the pointer will exhibit a high resistance in both the tests.

Terminal identification In order to identify the anode and cathode of a diode, the manufacturer either prints a ring on the circular body of the diode near the cathode, or the circular body has a pointed shape toward the cathode. Figure 3.50(c) shows these two identifications. There are other packages for different diodes. The manufacturer usually marks the cathode of the diode using various identification styles, e.g. a dot near the cathode.

3.15.2 Troubleshooting Diodes

One can test the operation of diodes in a circuit by taking voltage measurement. Current flows through the diode only when the diode is forward-biased by an external voltage source that is at least equal to its cut-in voltage. As mentioned earlier, cut-in voltage in a germanium diode is 0.3 V and in a silicon diode is 0.7 V. In order to check the working of a diode in a given circuit, one will have to measure the anode and cathode voltages of the diode with respect to the circuit ground. The difference of the anode and cathode voltages must be at least equal to its cut-in voltage. According to the circuit design, voltage difference between anode and cathode may be more than the cut-in voltage, but it should not be less than the cut-in voltage, and then voltages at the other connected points of the circuit are checked.

The voltage at the anode and cathode of the diode also indicates a shorted or open diode. However, if anode voltage is exactly equal to cathode voltage, then it is an indication of a shorted diode. On the other hand, if voltage at the anode is available and voltage at the cathode is zero, then it is most likely that the diode is open. In this case, we must make sure that the cathode of the diode is not connected to the circuit ground. In both the cases, we have to replace the diode with a new one.

In a shorted diode, we must check other parts of the circuit for faulty components or blown out fuses because in a shorted diode excess current is drawn by the diode, which may damage other components of the diode circuit.

Example 2 The saturation current density of a P-N junction germanium diode is 250 mA/m^2 at 300 K. Find the voltage that would have to be applied across the junction to cause a forward current density of 10^5 A/m^2 flow. [B.Tech]

Solution

$$I = I_s\left[\exp\left(\frac{eV}{kT}\right) - 1\right]$$

$$J = J_s\left[\exp\left(\frac{eV}{kT}\right) - 1\right]$$

$$\therefore \exp\left(\frac{eV}{kT}\right) - 1 = J/J_s$$

$$\therefore \exp\left(\frac{eV}{kT}\right) - 1 = \frac{10^5}{250\times 10^{-3}} = 4\times 10^5 \quad J_s = 250 \text{ mA/m}^2$$

$$J = 10^5 \text{ A/m}^2$$

or

$$\frac{eV}{kT} = \log_e(4\times 10^5) = 129$$

$$V = \frac{12.9\times 1.38\times 10^{-23}\times 300}{1.6\times 10^{-19}} = 0.33 \text{ V}$$

Example 3 Find the value of applied forward voltage for a P-N junction diode if $I_s = 50$ micro amp/cm, $I = 2$ amp/cm^2 and $e/kT = 40$ V.

Solution

$$I_s = V(e^{eV/kT} - 1)$$

$$e^{40V}\ \frac{J}{J_s} - 1 \simeq \frac{2\times 10^6}{30}$$

$$V = \frac{2303}{40}\log_{10}\frac{2\times 10^5}{3} = 0.277 \text{ V}$$

Example 4 The approximate value of P-N junction current under forward-bias is given by $I = I_s e^{eV/kT}$. Show that the incremental resistance R_e defined by $\Delta V/\Delta I$ is equal to 1/40 I at the room temperature ($e/kT = 40/V$).

Solution

$$I = I_s \exp\left[\frac{eV}{kT}\right] = I_s e^{40V}$$

$$R_e = \frac{\Delta V}{\Delta I} = \frac{1}{40I}$$

Dividing throughout by volume, one obtains the equation in the form of current density as

$$J = J_s\left[\exp\left(\frac{eV}{kT}\right) - 1\right]$$

$$\therefore\ \exp\left(\frac{eV}{kT}\right) - 1 = J/J_s$$

$$\therefore\ \exp\left(\frac{eV}{kT}\right) - 1 = \frac{10^5}{250\times10^{-3}} = 4\times10^5\ J_s = 250\ \text{mA/m}^2$$

$$J = 10^5\ \text{A/m}^2$$

$$\frac{eV}{kT} = \log_e(4\times10^5) = 129$$

or

$$V = \frac{12.9\times1.38\times10^{-23}\times300}{1.6\times10^{-19}} = 0.33\ \text{V}$$

Example 5 A P-N junction in series with a resistance of 5 × 10^3 Ω is connected across a 50 and 1 mΩ respectively. Show that the forward- and reverse-bias currents in the circuit are 9.9 mA and 4.975 ×10^{-2} mA respectively.

Solution (i) Forward-bias:

$$\text{Current} = \frac{\text{Applied voltage}}{\text{Junction resistance} + \text{external resistance}} = \frac{50}{5000+50} = 9.9\ \text{mA}$$

(ii) Reverse-bias:

$$\text{Current} = \frac{50}{5000\times10^6} = 4.975\times10^{-2}\,\text{mA}$$

Example 6 The reverse-bias saturation current for a P-N junction diode is 1 μA at 300 K. Determine its AC resistance at 150 mV forward-bias.

Solution

We know that

$$r_{ac} = \frac{dV}{dI} - \frac{1}{(dI/dV)} = \frac{1}{(I_o/KT)\exp(V/KT)}$$

At temperature 300 K, $kT = 8.62 \times 10^{-5} \times 300$

$$= 25.86 \times 10^{-3}\text{V}$$

$$\therefore r_{ac} = \frac{1}{(1\mu A/25.86\times10^{-3}V)\exp(0.15/25.86\times10^{-3}V)}$$

$$= \frac{25.86\times10^{-3}}{10^{-6}\times\exp(0.15/25.86\times10^{-3})}$$

$$= \frac{25.86\times10^{-3}}{332.66} = 77.65\Omega$$

Example 7 The leakage current in a certain diode is 25 μA at 25°C. Show that the change in temperature required to have leakage current of 40 μA is 6.77°C. [B.Tech]

Solution

We know that $(I_0)T_2 = (I_0)T_1\,(2)^{(T_2-T_1)/10}$

Substituting the given values, we have

$(40 \times 10^{-6}) = (25 \times 10^{-6})(2)^x$ where $(T_2-T_1)/10 = x$

$\therefore\ (2)^x = 1.6$

Taking log on both sides, one obtains

$x \log(2) = \log(1.6)$ or $x = \log(1.6)/\log(2)$

$$\text{or } x = \frac{0.204}{0.301} = 0.677$$

$$\text{Now } x = \frac{T_2 - T_1}{10} \text{ or } 0.677 = \frac{T_2 - 25}{10}$$

$\therefore\ T_2 = 6.77 + 25 = 31.77°\text{C}$

So, change in temperature required = 31.77 – 25 = 6.77°C

Example 8 Find the forward current for germanium diode at room temperature 22°C when voltage across it is 0.3 volt, and compare this current with current when temperature rises to 72°C. Express answers in terms of I_0. [B.Tech]

Solution

Forward current I is given by

$$I = (I_0)\left[\exp\left(\frac{V}{\eta V_T}\right) - 1\right]$$

$$= (I_0)_{22}\,(e^{0.3/1.0.025} - 1)$$

$$\because \eta = 1 \text{ for Ge and } V_T = \frac{T}{11600} = \frac{22+273}{11600} = 0.025$$

$$= (I_0)_{22}(e^{12} - 1)$$

$$= (I_0)_{22}(162754.79 - 1)$$

$$= (I_0)_{22}(162753.79)$$

When temperature rises to 72°, then

$$V_T = \frac{T}{11600} = \frac{72+273}{11600} = 0.0298$$

Temperature rise = $T_2 - T_1 = 72 - 22 = 50°C$

$\therefore (I_0)_{72} = (I_0)_{22}\,(2)^{50/10} = (I_0)_{22} \times 32$

or $(I_0)_{72} = 32(I_0)_{22}$

New current (I')is given by

$$I' = (I_0)_{72}(e^{V/\eta VT} - 1)$$

$$= 32(I_0)_{22}(e^{0.3/0.029} - 1)$$

$$= 32(I_0)_{22}(23555 - 48 - 1)$$

$$= 753743.36(I_0)_{22}$$

Thus, at 72°C, forward current rises by

$$\frac{I'}{I} = \frac{753743.36}{162753.79} = 4.63$$

Example 9 Find the static and dynamic resistance of a P-N junction germanium diode if temperature is 27°C and saturation current is 1 μA for an applied forward voltage of 0.2 volt.

Solution

Here $\quad \eta = 1, V_T = \frac{T}{11600} = \frac{27+273}{11600} = 0.02586$

$$\therefore \frac{V}{\eta V_T} = \frac{0.2}{1 \times 0.02586} = 7.7339$$

Now $I = I_0\,(e^{7.7339} - 1) = 1 \times 10^{-6} \times (2283 - 49)$ Amp = 2.283 mA

$$\text{Static resistance} = \frac{V}{I} = \frac{02}{2.283\,\text{mA}}$$

$$= 0.0876\ \text{k}\Omega = 87.6\Omega$$

$$\text{Dynamic resistance} = \frac{V_T}{(I + I_0)}$$

$$= \frac{1 \times 0.02586}{2.283\,\text{mA} + 1\,\mu A} = \frac{1 \times 0.02586}{2.283\,\text{mA} + 0.001\,\text{mA}}$$

$$= 0.01132\,\text{k}\Omega$$

$$= 11.32\ \Omega$$

Example 10 The saturation current density of a P-N junction Ge-diode is 250 mA/m^2 at 300 K. Find the voltage that would have to be applied across the junction to cause forward current density of 10^5 amp/m^2. [B.Tech]

Solution

We know that $$I = I_0 \left[\exp\left(\frac{V}{\eta V_T} \right) - 1 \right]$$

Dividing on both sides by area A, one obtains

$$\frac{I}{A} = \frac{I_o}{A} \left[\exp\left(\frac{V}{\eta V_T} \right) - 1 \right]$$

or $$J = J_0 \left[\exp\left(\frac{V}{\eta V_T} \right) - 1 \right]$$

Here $\mu = 1$ and $$V_T = \frac{300}{11600} = 0.02586$$

Now $10^5 = (250 \times 10^{-13})(e^{V/0.02586}\_1)$

$4 \times 10^5 - e^{38\text{-}66V}\_1$

or $e^{38-66V} = 4 \times 10^5 + 1$

Taking log of both sides, one obtains

$38.66\ V = 128992 + 0$

$\therefore V = 0.3336$ Volt

Example 11 Assume that a silicon diode in the following circuit requires a minimum current of 1 mA to be above the knee point (0.7 V) of its I-V characteristics. Also assume that voltage across the diodes is independent of current above the knee point.

(i) If $V_B = 5$ V, what should be maximum value of R so that the voltage is above the knee point?

(ii) If $V_B = 5$ V, what should be the value of R to establish a current of 5 mA in the circuit?

(iii) What is the power dissipated in the resistance R and in the diode when a current of mA flows in the circuit at $V_B = 6$ V.

(iv) If $R = 1$ kΩ, what is the minimum voltage V_B required to keep the diode above the knee point? [B.Tech]

Solution

The minimum voltage across the diode is 0.7 V. Obviously, this is above the knee point of the characteristic curve. Thus, voltage across resistance $R = v' = (5 - 0.7)$ V $= 4.3$ V. The minimum current $I = 1$ mA $= 10^{-3}$ A.

$$R = \frac{V^1}{i} = \frac{4.3}{10^{-3}}\Omega = 4.3\text{k}\Omega$$

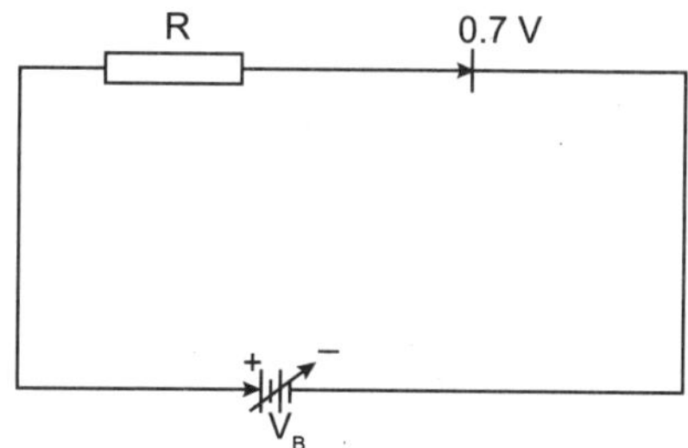

Current through resistance $R = 5$ mA $= 5 \times 10^{-3}$ A and also voltage drop across $R = (5 - 0.7)$ $V = 4.3$ V.

$$\therefore\ R = \frac{4.3}{5 \times 10^{-3}} = 0.86 \times 10^{-3}\Omega = 860\Omega$$

We have $V_B = 6$ V, and hence voltage V' across the resistance is $V' = 6 - 0.7 = 5.3$ V and current $i = 5$ mA $= 5 \times 10^{-3}$ A.

∴ Power (P) dissipated through resistance R is

$$P = iV' = 5 \times 10^{-3} \times 5.3 \text{ W}$$
$$= 26.5 \times 10^{-3} W = 26.5 \text{ W}$$

Power P' dissipated in the diode is

$$p' = i \times 0.7\,\text{W} = 5 \times 10^{-3} \times 7 \times 10^{-1}\,\text{W}$$
$$= 3.5 \times 10^{-3}\,\text{W} = 3.5\,\text{mW}$$

To keep the diode above the knee point, the minimum current required is 1 mA.

$$V_R = \text{voltage drop across } R(=1\text{ k}\Omega)$$
$$= 1\times10^3 \times 10^{-3} = IV$$

We have minimum voltage drop across the diode = 0.7 V

$\therefore\ V_B = (1 + 0.7)\text{V} = 1.7\text{ V}$

Example 12 Determine the DC resistance of a silicon diode whose V-I characteristics is shown in Figure 3.44, at (a) $I_D = 2$ mA, (b) $I_D = 20$ mA, and (c) $V_D = -10$ V. [B.Tech]

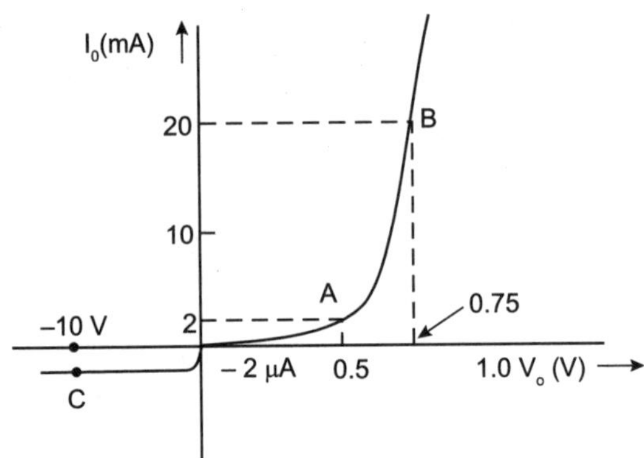

Figure 3.51

Solution

At $I_D = 2$ mA, $V_D = 0.5$ V

$$\therefore\ R_F = \frac{V_D}{I_D} = \frac{0.5\text{V}}{2\text{mA}} = 250\Omega$$

At $I_D = 250$ mA, $V_D = 0.75$ V (from the given curve in Figure 3.44)

$$\therefore\ R_F = \frac{V_D}{I_D} = \frac{0.75\text{V}}{2\text{mA}} = 37.5\Omega$$

At $V_D = 10$ V, $I_D = 2$ µA (from the given curve in reverse-bias region)

$$R_R = \frac{V_D}{I_D} = \frac{-10\text{ V}}{2\ \mu\text{A}} = 5\text{ M}\Omega$$

Example 13 For the circuit shown in Figure 3.45, find the current through diode D assuming the diode to be (a) ideal, (b) real with piecewise linear circuit model having threshold voltage $V_T = 0.3$ V, and average resistance $r_f = 25\ \Omega$. [B.Tech]

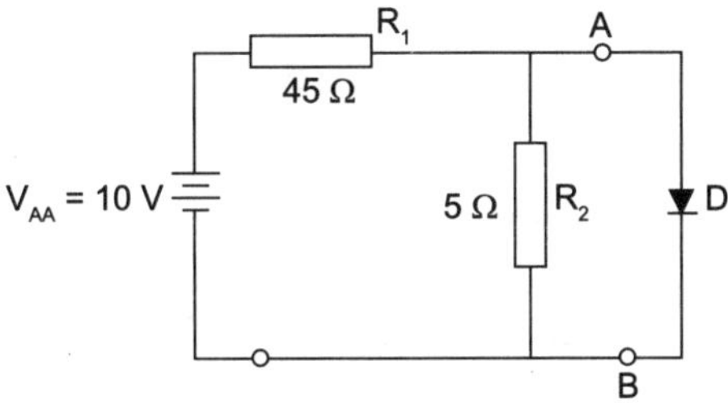

Figure 3.52

Solution

(a) **Assuming diode *D* to be ideal:** Ignoring diode *D*, voltage across R_2 is given as (by applying potential divider concept),

$$V_{AB} = V_{AA}\frac{R_2}{R_1 + R_2} = 10 \times \frac{5}{45+4} = 1.0\,\text{V}$$

Thus, diode *D* is forward-biased. It conducts, offering zero resistance. Hence no current would flow through the parallel branch R_2. The circuit is equivalent to that shown in Figure 3.46(a). Therefore, diode current is

$$ID = \frac{V_{AA}}{R_1} = \frac{10}{45} = 222\,\text{mA}$$

(b) **Assuming the diode to be real:** Voltage V_{AB} is much larger than V_T. Hence, the diode conducts. It is replaced by its equivalent as shown in Figure 3.46(b). To determine current I_D through the diode, we first find the Thevenin's equivalent of the circuit on the left of *AB*.

V_{TH} = open circuit voltage across *AB*

$$= V_{AA} \times \frac{R_2}{R_1 + R_2} = 10 \times \frac{5}{45+5} = 1.0\,\text{V}$$

$$R_{TH} = \frac{R_1 R_2}{R_1 + R_2} = \frac{45 \times 5}{45+5} = 4.5\,\Omega$$

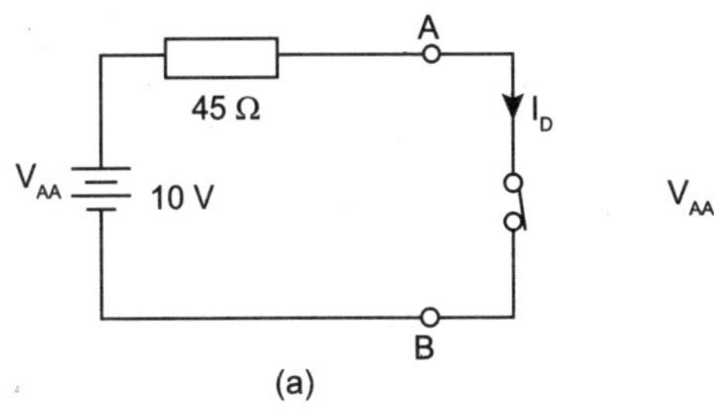

(a)

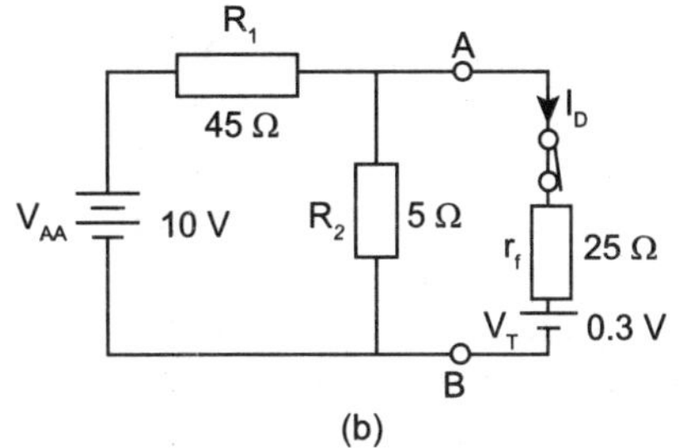

(b)

(c)

Figure 3.53

Thus, the equivalent circuit becomes as shown in Figure 3.46(c). Thus, diode current is

$$I_D = \frac{V_{TH} - V_T}{R_{TH} + T_f} = \frac{1.0 - 0.3}{4.5 + 25} = \frac{0.7}{29.5} = 23.7\,\text{mA}$$

Example 14 Determine the current through the 90 Ω resistor in the circuit of Figure 3.47. The diodes are silicon diodes, and forward resistance of each diode is 5 Ω. [B.Tech]

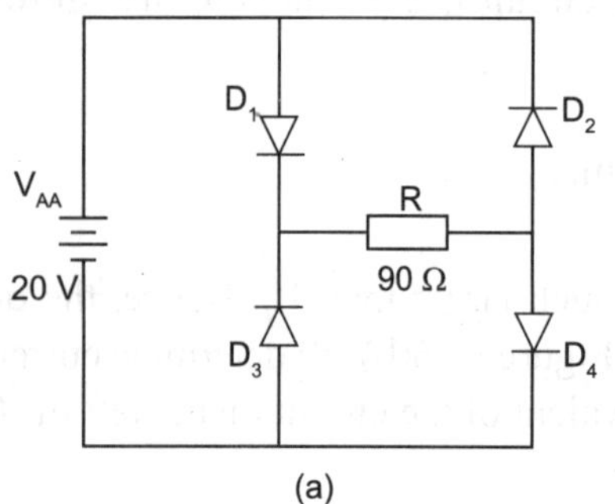

(a)

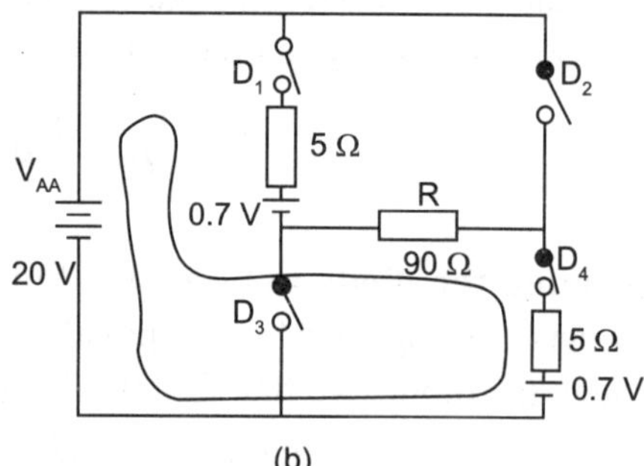

(b)

Figure 3.54

Solution Diodes D_2 and D_3 are reverse-biased. Therefore, these are like open-switches. Diodes D_1 and D_4 are forward-biased. These are replaced by their equivalent circuits, as shown in Figure 3.47. Since the diodes are silicon diodes, $V_T = 0.7$ V. There exists only one series circuit, having net emf $= V_{AA} - V_T - V_T = 20 - 0.7 - 0.7 = 18.6$ V and total resistance $= 5 + 90 + 5 = 100$ Ω.

Therefore, current through 90 Ω resistor is

$$I = \frac{18.6}{100} = 0.186 \text{ A} = 186 \text{ mA}$$

Example 15 Calculate the current drawn from a 10 V battery in the circuit of Figure 3.48(a), if the diode connected is (a) forward-biased, (b) reverse-biased. Assume the diode to be ideal. [B.Tech.]

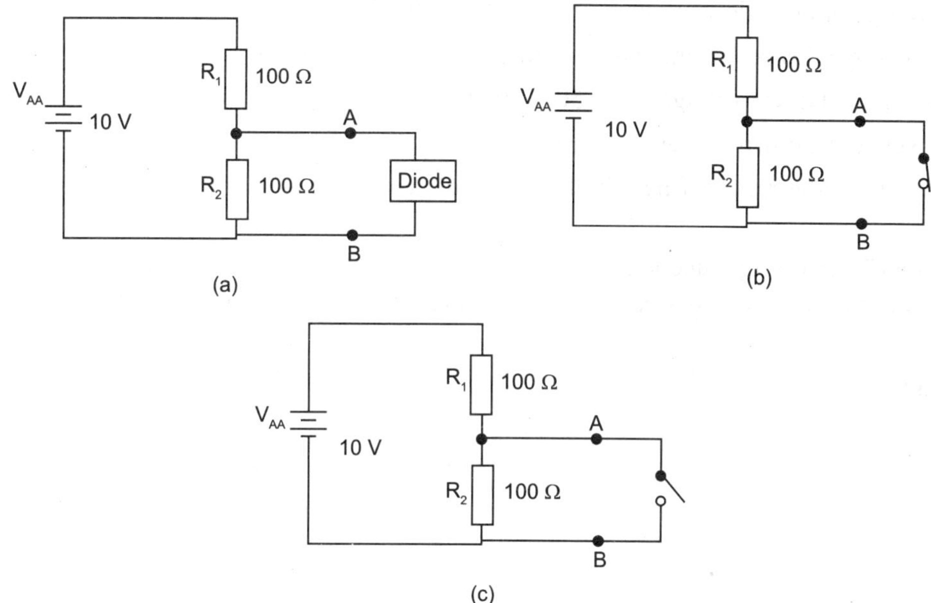

Figure 3.55

Solution

(a) When the diode is forward-biased (Figure 3.48(b), it offers zero resistance. It is like a shorted switch. This shorted switch across AB also short-circuits the resistance R_2. Obviously, a parallel combination of the diode and R_2 is equivalent to a resistance of zero ohms. Thus, net resistance, $R = R_1 = 100\ \Omega$.

$$\therefore \text{Current drawn from the battery} = \frac{V_{AA}}{R} = \frac{10}{100} = 0.1 \text{ A}$$

(b) When the diode is reverse-biased (Figure 3.48(c)), the diode is like an open-switch. Obviously, it then does not make any difference whether the diode is connected or not. Thus, net resistance, $R = R_1 + R_2 = 100 + 100 = 200\ \Omega$.

$$\therefore \text{Current drawn from the battery} = \frac{V_{AA}}{R} = \frac{10}{100} = 0.05 \text{ A}$$

Example 16 A zener diode (BC-147) has a zener voltage of 9 V with 10% tolerance at 25°C. Find the range of voltage ratings at which it will break down? If the temperature is raised to 75°C, what will be the new breakdown voltage when decrease in zener voltage is about 2 mV per degree centigrade? [B.Tech]

Solution Given average breakdown voltage $V_z = 9$ V

$$\text{Tolerance} = 10\% = \frac{10}{100} \times 9 = 0.9 \text{ V}$$

∴ Range of breakdown voltage = (9 – 0.9) to (9 + 0.9) V = 8.2 to 9.9 V.

In the second case, the increase in temperature = 75 – 25 = 50°C

Fall in breakdown voltage = 2 × 50 = 100 mV = 0.1 V

∴ New breakdown voltage = 9 – 0.1 = 8.9 V

Obviously, the range of new breakdown voltage = (8.1 – 0.1) to (9.9 – 0.1) V = 8.0 to 9.8 V.

Example 17 An abrupt junction diode has transition temperature of 20 pF when reverse-biased at 5 V. If the voltage is increased by 1 V, show that the capacitance will decrease by 1.7 pF.

[B.Sc. (Hons.)]

Solution

$$C = \frac{K}{\sqrt{V}} \; (K = eA)$$

$$20 = \frac{K}{\sqrt{5}}$$

$$\therefore K = 20\sqrt{5} = 44.8$$

$$\therefore \text{ Capacitance for 1V increase} = \frac{44.8}{\sqrt{5+1}} = 18.3\,\text{pF}$$

Example 18 Two germanium diodes D_1 and D_2 are connected in series across 5 V battery as shown in Figure 3.56. (a) Find the voltage across each diode assuming that breakdown voltage of diodes is greater than 5 V. What is the effect of temperature? (b) Find current if $V_Z = 4.9$ V and $I_0 = 5\mu$A.

[B.Tech, B.Sc. (Hons.)]

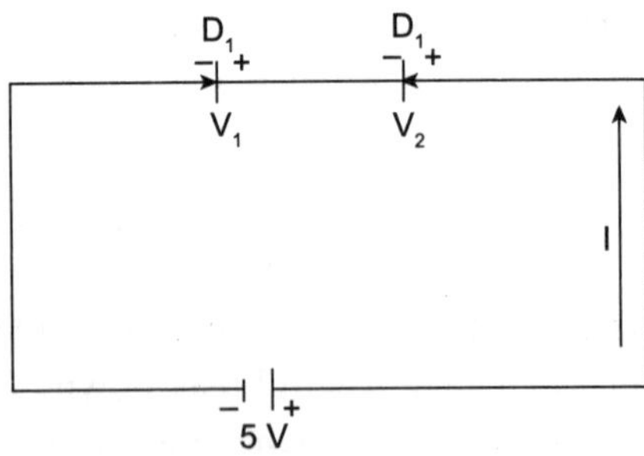

Figure 3.56

Solution (a) The two diodes are connected in series and hence the same current I flows in D_1 and D_2. Obviously, it is in forward direction through D_2 and in reverse direction through D_1. Since D_2 diode is forward-biased, V_2 will be very small and hence V_1 (= 5 – V_2) will be very much large than $V_T(= 0.026\text{ V})$. This means the current will be equal to reverse saturation current I_o. Now, we consider diode D_2. We have

$$I = I_0\left[\exp\left(\frac{V}{V_T}\right) - 1\right]$$

Putting $I = I_0$ and $V = V_2$, we have

$$I_0 = I_0\left[\exp\left(\frac{V_2}{V_T}\right) - 1\right]$$

or
$$\exp\left(\frac{V_2}{V_T}\right) - 1 = 1$$

or
$$V_2 = V_T \ln 2 = 0.026 \times 0.693 = 0.018\text{ V}$$

$$\therefore\ V_1 = 5 - 0.018 = 4.892\text{ V.}$$

Effect of temperature: $V_2 = V_T \ln 2 = kT \ln 2$.

So V_2 will increase with temperature.

(b) If V_2 is 4.9 V then D_1 will break down. This means $V_1 = 4.9$ V.

$\therefore\ V_2 = 5 - 4.9 = 0.1$ V.

Now using $I_o = 5\ \mu\text{A}$, $V_2 = 0.1$ V, one obtains

$$I = 5\left[\exp\left(\frac{0.1}{0.026}\right) - 1\right] = 229\ \mu\text{A}$$

Example 19 An LED has a greater forward voltage drop than does a common signal diode. A typical LED can be modeled with a constant forward voltage drop $V_D = 1.6$ V. Its luminous intensity I_V varies directly with forward current and is described by $I_V = 40\ i_D$ millicandela (mcd). A series circuit consists of such an LED, a current-limiting resistance R, and a 5 V DC source V_s. Find the value of LED current i_D such that the luminous intensity is I mcd.

Solution We have $I_V = 40\ i_D$

$$\therefore\ i_D = \frac{I_V}{40} = \frac{1}{40} = 25\text{mA}$$

Example 20 Estimate the value of AC resistance that would be offered by a semiconductor diode at (a) $I_D = 10$ mA, and (b) $I_D = 20$ mA. [B.Tech]

Solution (a) At $I_D = 10$ mA,

$$r_{ac} = \frac{25(\text{mV})}{I_D(\text{mA})} = \frac{25\,\text{mV}}{20\,\text{mA}} = 1.25\,\Omega$$

(b) At $I_D = 20$ mA,

$$r_{ac} = \frac{25(\text{mV})}{I_D(\text{mA})} = \frac{25\,\text{mV}}{20\,\text{mA}} = 1.25\,\Omega$$

Example 21 The reverse-bias saturation current for a P-N junction diode is 1 μA at 300 K. Show that its AC resistance at 150 mV forward-bias is 77.65Ω. [B.Tech.]

Solution

$$r_{ac} = \frac{dV}{dI} = \frac{1}{(dI/dV)} = \frac{1}{(I_s/kT)\exp(V/kT)}$$

Given $T = 300$ K $\therefore kT = 8.62 \times 10^{-5} \times 300 = 25.86 \times 10^{-3}$ V

$$\therefore r_{ac} = \frac{1}{(1\mu A/25.86\times10^{-3}V)\exp[0.15/25.86\times10^{-3}V]}$$

$$= \frac{25.86\times10^{-3}}{332.66} = 77.65\,\Omega$$

Summary

1. A semiconductor diode is fabricated using a single crystal of semiconductor (Si or Ge) wafer. One half of the semiconductor wafer is doped with a trivalent atom like In to act as a P-type and the other half is doped with a pentavalent atom like to be converted into an N-type material. Finally we have a semiconductor P-N junction diode.
2. An ideal diode is a two-terminal device which has the following properties: (i) conducts with zero resistance when forward-biased, and (ii) offers infinite resistance when reverse-biased.
3. A junction is formed in a process which may be of the grown, alloyed or diffused.
4. A junction has a built-in potential hill or barrier that exists across the depletion region and may be increased or decreased by the application of an external voltage.
5. A forward-biased P-N junction has the positive terminal applied to the P side and the negative terminal to the N side. The reverse is true for a reverse-biased junction.
6. A diode can be used either in forward-biased or reverse-biased conditions.
7. Current only flows in forward-based condition. No current or very small current flows through the diode in the reverse-biased condition, unless breakdown of the diode occurs to allow a large current to flow. A breakdown in diodes occurs either due to zener or avalanche breakdowns.

8. Forward voltage at which current through the P-N junction starts increasing rapidly is called **cut-in voltage**.
9. The maximum value of voltage that the diode can bear in forward-biasing is called **maximum forward voltage** (V_f). Its value is about 1 volt.
10. The maximum value of current in forward direction a diode can conduct without damaging its junction is called its maximum forward current (I_f). It is in mA.
11. The reverse voltage at which a diode breaks down and its junction is destroyed permanently is called breakdown voltage. This is also known as **avalanche voltage**.
12. **Drift current** is caused due to an electric field across the specimen. **Diffusion current** results from the nonuniform concentration of charge carriers.
13. The voltage in reverse direction a diode can bear without damaging its junction is called **peak inverse voltage** (PIV). This is very important when diodes are used as rectifiers. The value of PIV is less than that of breakdown voltage.
14. If the reverse-bias is too large over a P-N junction, **zener breakdown**, or **avalanche breakdown**, or both may occur.
15. The sum of resistance of P region and N region is called bulk resistance of the diode. Its value depends upon the doping of the regions. Its value is from 1 to 25 ohms.

$$r_b = r_p + r_n$$

16. Avalanche breakdown is the runaway increase in minority current carriers during reverse current flow. It occurs when many valence electrons are freed with sufficient force to dislodge other valence electrons. It is used in switching diodes and controlled rectifiers.
17. Zener breakdown is the runaway increase in minority carriers during reverse current flow. It is also caused by the release of high-energy valence electrons due to the breaking of covalent bonding. It is used in voltage regulating diodes.
18. A capacitance called space-charge capacitance is formed around the junction of a diode under forward-biased conditions. When biasing across the diode is abruptly changed from forward- to reverse-biasing, the capacitance formed around the junction is called diffusion capacitance.
19. A zener diode, when used in the forward-biased condition, behaves like a simple P-N junction diode, and its characteristics are similar to that of a simple diode.
20. A diode offers static resistance to a DC voltage and offers dynamic resistance to an AC voltage. A diode offers a very high resistance in reverse-biased condition.
21. The **tunnel effect** is the movement of valence electrons from the valence energy band to the conduction band of other region with little applied forward voltage. Valence electrons seem to tunnel through the forbidden energy band. The tunnel effect provides a negative resistance region in the tunnel diode; increasing voltage results in decreasing current. Tunnel diodes are used as switch in oscillatory circuits.

22. A germanium device is seldom used at a temperature higher than 75°C, while a silicon device can be used upto 175°C.
23. A P-N junction exhibiting a transition capacitive effect of several picrofarad in reverse-bias may have many microfarads of diffusion capacitance in the forward-biased state.
24. A photodiode is a two-element semiconductor, light-sensitive device. It converts light into electrical energy directly.
25. An LED is a semiconductor diode which emits visible light from its P-N junction when it is forward-biased. They are used as readouts in alphanumeric displays. They are also used in the pilot lamps in most electronic and electrical devices.
26. Schottky diodes are also known as hot-carrier or surface-barrier diodes, and they are used in switching circuits and microwaves. These diode uses only majority carriers and are intended for high-frequency operations.
27. In some compounds (III-IV), e.g. GaAs and InP, after an electric field in the material reaches a threshold level, the mobility of electrons decreases as the electric field is increased, thereby producing negative resistance. This diode is called Gunn diode.
28. A type of diode that emits light when there is forward current is called light emitting diode (LED).
29. A diode in which the reverse current varies directly with the amount of light is called photodiode.
30. A solar cell is a photodiode operated in its photovoltaic mode. It is used into convert light energy to electrical energy and thereby serves as a DC voltage source.
31. The PNPN diode is a four-layer solid state device having only two electrodes (anode and cathode). Current starts flowing through it only when the applied forward-biased voltage is more than the breakdown voltage, and once current starts flowing, the voltage across the diode drops and current flow will be very high.
32. An SCR is a three-element, four-layer (NPNP) solid state device which permits current through it in one direction only (i.e. when forward-biased) and the bias is equal to or greater than forward breakover voltage. The third element called gate controls the breakover voltage. Once the SCR is ON, the gate loses control. It continues to conduct as long as it is forward-biased and as long as current through it is equal to greater than the holding current.
33. A varactor diode is a special type of solid state diode whose capacitance varies with inverse applied voltage.

Review Questions

1. Draw the characteristics of a semiconductor diode and label it?
2. How an abrupt junction is formed?
3. What happens to the depletion region in a P-N diode under forward-bias and reverse-bias condition? Explain.

4. Explain the working of a P-N junction. Discuss forward- and reverse-biasing of the P-N junction diode.
5. Explain with diagrams forward- and reverse-biasing of a P-N junction. What is meant by avalanche breakdown?
6. Explain with reference to zener diode characteristic curve the following:

   (i) I_{zK} (ii) I_{zT} (iii) Z_Z whose symbols have usual meaning.
7. What do you understand by zener breakdown? Draw a typical characteristic of a zener diode. Identify the breakdown region and explain it.
8. A zener diode is a P-N junction, yet it is different from an ordinary P-N junction. Explain.
9. Describe an experiment for drawing the characteristic of a zener diode.
10. Draw the characteristics of a tunnel diode; label it and discuss its applications.
11. Write short notes on the following:

    (i) avalanche breakdown, (ii) zener breakdown, (iii) tunnel diode,

    (iv) varactor diode, (v) light emitting diode, (vi) solar cell,

    (vii) schotty diode, (viii) backward diode
12. Give the structure of a light emitting diode. Explain its working and give some of its uses.
13. Explain why zener breakdown voltage has a negative temperature coefficient, whereas the avalanche breakdown voltage has a positive temperature coefficient? Why dynamic resistance of an ideal zener diode is zero but its DC-resistance is not zero.
14. Draw the volt-ampere characteristic of a tunnel diode. Explain the occurrence of negative differential resistance in the characteristic of a tunnel diode. Write some of the uses of a tunnel diode.

Problems

1. Find the value of the P-N junction diode current in the forward-bias condition when the applied voltage across it is 0.05 volt and $I_s = 50\ \mu V$. Given $e/kT = 40$/volt.

   [**Ans:** 0.319 mA]
2. A silicon diode operates at a fixed bias of 0.4 V. Determine the factor by which the current will get multiplied when its temperature is raised from 25°C to 150°C. [**Ans:** 638]
3. When a diode is reverse-biased with 8 V, it has junction capacitance of 15 pF. When the reverse-bias is increased to 12 V, capacitance drops to 13.05 pF. Find whether it is abrupt or graded junction? [**Ans:** graded junction type]
4. (a) For what voltage will the reverse current in a P-N junction Ge diode reach 90% of its saturation value at room temperature?

   (b) If reverse saturation current is 10 μA, calculate the forward currents for a voltage of 0.2 and 0.3 V, respectively. [**Ans:** (a) –0.017 V, (b) 1.01A]

5. Find the value of DC resistance and AC resistance of Ge P-N junction diode, if the temperature is 25°C and $I_0 = 20\ \mu A$ with an applied voltage of 0.1 V.
[Ans: r_{dc}=109.2Ω, r_{ac}=26.3Ω**]**

6. Find the width of the depletion layer in a germanium P-N junction diode which has the following specification: Area $A = 0.001\ cm^2$, $\sigma_n = 1$ mhos/cm, $\sigma_p = 100$ mhos/cm, $\mu_n = 3800$ cm²/sec and μ_p =1800 cm²/sec **[Ans:** 0 .083 × 10^{-4} cm**]**

7. Calculate the dynamic forward and reverse resistance of a P-N junction diode when the applied voltage is 0.25 V for Ge diode. Given $I_0 = 1\ \mu A$ and T =300 K
[Ans: r_f= 1.734 Ω, r_r = 389.7M Ω**]**

8. The transition capacitance of an abrupt P-N junction diode is 20 pF at 5 V. Find the value of decrease in capacitance for 1 volt increase in bias. **[Ans:** 1.75 pF**]**

9. What is the ratio of the current for a forward-bias of 0.05 V to the current for the same magnitude of reverse-bias for a Ge diode? **[Ans:** –6.83**]**

10. A resistor R is placed parallel to a Ge tunnel diode. The tunnel diode has

$$\left(\frac{d_{id}}{dV}\right)_{max} = 0.1\Omega$$

Show that the value of R should be at least 10 Ω so that the combination does not exhibit negative resistance region in the volt-ampere characteristic.

11. Over what range of input voltage will a zener regulator circuit maintain 30 V across 2 kΩ resistor, assuming $R_s = 200\ \Omega$ and maximum zener current is 25 mA. **[Ans:** 38 V**]**

Short Answer Questions

1. What is the zener voltage of germanium diode?

   Ans: When a reverse-bias of about 25 volts is applied to the germanium diode, the excessively high temperature destroys the covalent structure of germanium and the reverse current rises sharply. This breakdown voltage is called zener voltage (V_z).

2. What do you understand by 'tolerance' of a zener diode?

   Ans: The range of voltages above the breakdown voltage in which a zener diode conducts in reverse direction is called tolerance.

3. What do you understand by power rating of a zener diode?

   Ans: The maximum power which zener diode can handle without damage is known as its power rating.

4. What do you understand by maximum current rating of a zener diode?

   Ans: The maximum value of current which a zener diode can handle at its rated voltage without damage is known as its maximum current rating.

5. Which type of diode is used for the rectification of very high frequency (MHz)?
   Ans: Schottky diode.
6. Which diode is best suited for the rectification of very small peak voltage (~0.5 V)?
   Ans: Backward diode.
7. Which diode has negative resistance?
   Ans: Tunnel diode.

Objective Questions

1. Zener diode is used for
   (a) rectification (b) amplification
   (c) stabilization (d) none of the above [**Ans:** (c)]
2. For a tunnel diode, a decrease in current causes
   (a) voltage constancy (b) decrease in voltage
   (c) increase in voltage (d) none of the above [**Ans:** (c)]
3. A P-N junction is formed when a P-type semiconductor and N-type semiconductor are joined
   (a) together
   (b) physically
   (c) to get homogeneous material chemically
   (d) in such a manner that electrons and holes diffuse to give depletion layer [**Ans:** (d)]
4. The depletion region of a junction diode is formed
   (a) when forward-bias is applied to it
   (b) when the temperature of the junction is reduced
   (c) under reverse-bias
   (d) during the manufacturing process [**Ans:** (d)]
5. The width of the depletion layer of a junction
   (a) is independent of applied voltage
   (b) is increased under reverse-bias
   (c) decreases with light doping
   (d) increases with heavy doping [**Ans:** (b)]
6. The light emitting diode
   (a) is made from one of the two basic semiconducting materials, silicon or germanium
   (b) is made from the semiconducting compound gallium arsenide phosphide
   (c) emits light when forward-biased
   (d) emits light when reverse-biased [**Ans:** (b), (c)]

7. The P side of a junction diode is earthed and the N side is given a potential of 2V. The diode will
   (a) not conduct (b) conduct partially
   (c) breakdown (d) conduct fully [**Ans:** (d)]
8. For detecting light intensity, we use a/an
   (a) photodiode in reverse-bias (b) photodiode in forward-bias
   (c) LED in reverse-bias (d) ED in forward-bias [**Ans:** (a)]
9. When a p-n junction diode is forward-biased, the flow of current across the junction is mainly due to
   (a) diffusion of charges
   (b) drift of charges
   (c) depends on the nature of the material
   (d) both drift and diffusion of charges [**Ans:** (a)]
10. A p-n junction diode cannot be used
   (a) as a rectifier
   (b) for increasing the amplitude of an AC signal
   (c) for getting light radiation
   (d) for converting light energy into electrical energy [**Ans:** (b)]
11. A strong electric field across a P-N junction that causes covalent bonds to break apart is called
   (a) reverse breakdown (b) avalanche breakdown
   (c) lever breakdown (d) low voltage breakdown [**Ans:** (b)]
12. A light emitting diode produces light when
   (a) forward-biased (b) reverse-biased
   (c) unbiased (d) none of the above [**Ans:** (a)]
13. A solar cell is an example of
   (a) photoemissive cell (b) photoradiation cell
   (c) photovoltaic cell (d) photoconductive cell [**Ans:** (c)]
14. A diode which has zero breakdown voltage is known as
   (a) tunnel diode (b) zener diode
   (c) Schottky diode (d) backward diode [**Ans:** (a)]
15. A zener diode is used as
   (a) a coupler (b) a rectifier
   (c) an amplifier (d) a voltage regular [**Ans:** (d)]

16. A zener diode
    (a) is useful as an amplifier
    (b) has negative resistance
    (c) has high forward-voltage rating
    (d) has sharp breakdown at low reverse voltage [**Ans:** (d)]

17. The turn-on voltage of a silicon diode is about
    (a) 1.0 V (b) 0.7 V
    (c) 0.3 V (d) 0.1 V [**Ans:** (b)]

18. When used in a circuit, a zener diode is always
    (a) reverse-biased (b) forward-biased
    (c) connected in series (d) none of the above [**Ans:** (a)]

19. Leakage current in a junction diode
    (a) increases with rise in temperature
    (b) decreases with fall in temperature
    (c) depends on the method of fabrication
    (d) flows due to majority carriers [**Ans:** (c)]

20. The capacitance of a varactor diode can be varied by
    (a) varying its temperature
    (b) varying its forward-bias
    (c) varying its reverse-bias
    (d) varying its doping level [**Ans:** (c)]

21. When reverse-bias is applied to a junction diode, it
    (a) raises the potential barrier
    (b) lowers the potential barrier
    (c) increases minority carrier current to a great extent
    (d) increases majority carrier current to great extent [**Ans:** (a)]

22. The reverse saturation current in a junction diode is the current that flows when
    (a) there is no potential barrier
    (b) the junction is unbiased
    (c) only minority carriers are crossing the junction
    (d) only majority carriers are crossing the junction [**Ans:** (c)]

23. If the junction current in a P-N junction diode is zero, then this means that
    (a) there are no carriers crossing the P-N junction
    (b) there is no potential barrier

(c) the number of holes diffusing from the P region equals the number of electrons diffusing from the N region

(d) the number of majority carriers crossing the junction equals the number of minority carriers crossing the P-N junction **[Ans:** (d)]

24. Reverse saturation current I_0 of a Si diode varies as

(a) T (b) T^2

(c) T^3 (d) $T^{1.5}$

(T → Temperature) **[Ans:** (d)]

25. The cut-in voltage for Si diode is about

(a) 1.1 V (b) 0.6 V

(c) 0.2 V (d) 0 V **[Ans:** (b)]

Mark the Following Statements True/False.

1. Under reverse-bias current in a P-N junction, the diode reaches saturation. **[Ans:** True]
2. Both LED and zener diode are forward-biased for usual operation. **[Ans:** False]
3. Zener diode has a breakdown voltage but a P-N does not have a breakdown voltage. **[Ans:** False]
4. Both zener diode and photo diode are reverse-biased for normal work. **[Ans:** True]
5. The width of the depletion region in a P-N diode increases on reverse-biasing. **[Ans:** True]
6. The capacitance of a varactor diode increases on reverse-biasing. **[Ans:** False]
7. The value of breakdown voltage in a tunnel diode is zero. **[Ans:** True]
8. The power consumption in a LCD is more than that in an LED. **[Ans:** False]
9. A P-N junction diode can be used as a rectifier for all frequencies of the input signal. **[Ans:** False]

Fill in the Blanks.

1. A potential barrier in a P-N junction diode ____________ with forward-bias and ____________ with reverse-bias. **[Ans:** decrease, increase]
2. Current in a P-N junction diode mostly constitutes of ____________ under forward-bias and of ____________ under reverse-bias. **[Ans:** majority carriers, minority carriers]
3. Under ____________, LED emits light when ____________ combine with ____________ in lower energy level. **[Ans:** forward-bias, electron, holes]
4. For high-frequency (in MHz range) input signals, ____________ diode is used as ____________. **[Ans:** Schottky, rectifier]

5. ____________ in a *p-n* diode consists of ____________ ions.

   [**Ans:** depletion region, immobile]

6. Tunnel diode ____________ characteristic shows ____________ resistance.

   [**Ans:** V-I, negative]

7. On increasing the temperature of a *p-n* diode, ____________ current and ____________ current both ____________.

   [**Ans:** forward, reverse, increase]

8. The semiconductor material employed in LED is ____________ and it ____________ emits ____________ light when ____________ biased.

   [**Ans:** gallium, arsenic phosphide, red, forward]

Chapter 4

Diode Circuits

4.1 Introduction

In the previous chapter, we saw that a *p-n* junction conducts easily when forward-biased and practically does not conduct (very small current flows) when reverse-biased. We can put this property of or *p-n* junction as having low resistance to current in one direction and relatively high resistance to current in the reverse direction.

4.2 Ideal Diode

The V-I characteristics of an ideal *p-n* junction diode may be looked upon as made up of two straight lines (Figure 4.1(a)). Such a volt-ampere characteristics is said to be piece-wise linear since it consists of line segments. Thus it acts like a switch in ON position (Figure 4.1(b)) when it is conducting and in OFF position when it does not conduct.

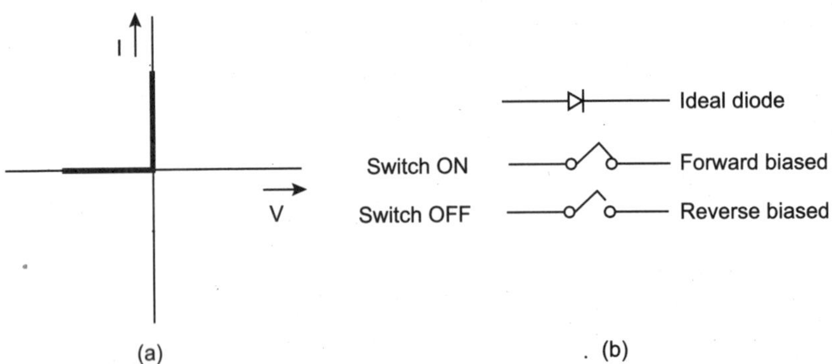

Figure 4.1 (a) V–I characteristic of an ideal diode, (b) ideal diode acting like a switch

Actually the V-I characteristics of *p-n* junction diode is different and is as shown in Figure (4.2(a)). However, for practical or mathematical purposes it may be viewed as shown in Figure (4.2(b)).

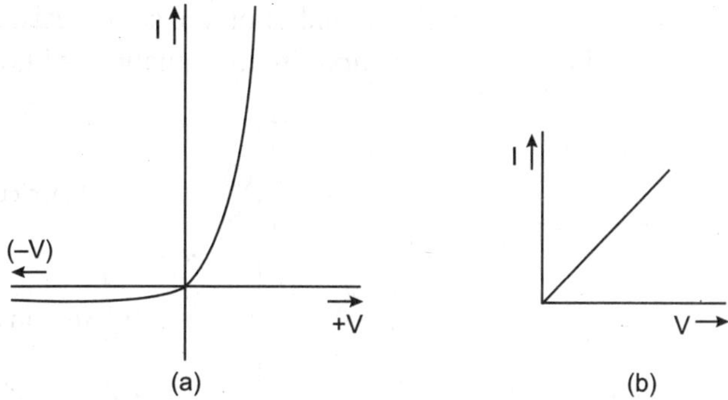

Figure 4.2 V-I characteristic of a *P-N* junction diode

4.3 The Second Approximation

The ideal approximation serves for mathematical purposes where simplification is desired. But many a times we require more accurate current and voltage in a circuit. This is where second approximation comes into play. In reality no current flows through a *p-n* junction diode till a voltage drop of 0.7 V (silicon diode) appears across it and then on increasing the applied voltage the current flows through the circuit. It is to be noted that voltage drop across the diode remains 0.7 V for any value of current through it.

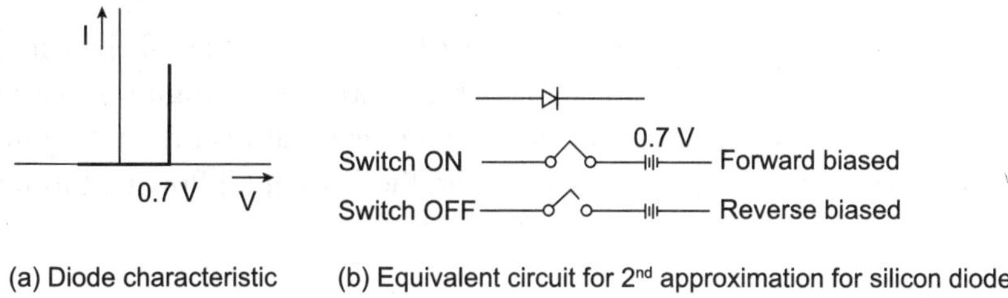

(a) Diode characteristic (b) Equivalent circuit for 2nd approximation for silicon diode

Figure 4.3

4.4 Load Line Analysis of a Diode Circuit

Load line is a tool to find the exact value of diode current and voltage. It is very useful in transistors (Chapter 5).

4.4.1 Load Line and Operating Point Q

DC Applied Voltage A DC voltage V is applied across diode and load R_L. Consider the circuit (Figure 4.4(a)).

The current flowing through the circuit will depend upon the static resistance of the diode and external resistance R_L. The V-I characteristic curve in this situation is known as static curve.

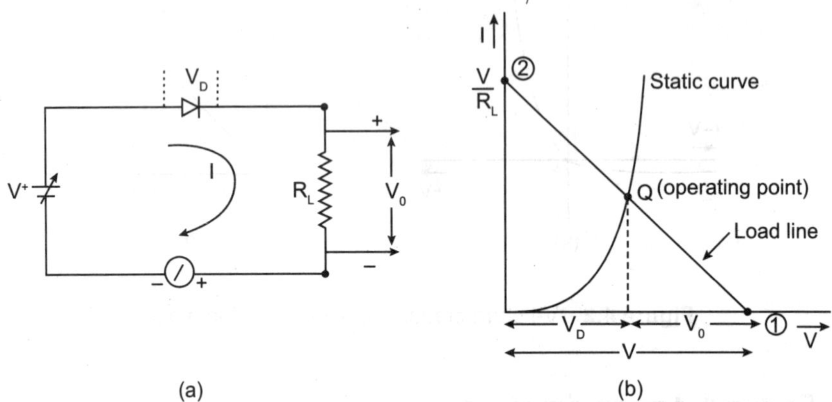

Figure 4.4 (a) Diode circuit (b) V, I characteristic curve with load line

Applying Kirchoff's Law to the circuit we have

$$V = V_D + V_O = V_D + IR_L \tag{1}$$

or

$$I = (-)\frac{1}{R_L}V_D + \frac{V}{R_L} \tag{2}$$

Equation (2) represents a straight line $y = mx + c$ having slope $(-)\frac{1}{R_L}$ called **load line.**

Q point The load line as represented by equation (2) is drawn on the V-I characteristic of diode. The point of intersection of the load line and characteristic curve (Figure 4.4(b)) is Q which is also called operating point or **Quiescent Point** which means at rest. By noting the coordinates of the point, Q we know the diode voltage (V_D) and the current (I) flowing through it.

4.4.2 AC Applied Voltage

Consider the circuit of Figure 4.5(a) where an alternating voltage $V = V_m \sin \omega t$ is applied: The applied voltage across the diode and resistor combination is varying so equation (2) cannot be applied for drawing load line. In such a case the process of drawing load line is repeated for instantaneous values of V. However the load R_L is constant so all the respective load lines (Figure 4.5(b)) are parallel to each other. The intersection of these load lines with static curve will give current flowing in the circuit corresponding to each instantaneous value of applied voltage. Instead of using these varying values of current to obtain the output current wave form, it is more convenient to construct a dynamic curve (Figure 4.5(b)).

$$I = (-)\frac{1}{R_L}V_D + \frac{V_m}{R_L}$$

Consider an applied voltage v at its maximum value V_m. The intersection of the load line given by $I = (-)\frac{1}{R_L}V_D + \frac{V_m}{R_L}$ corresponding to this condition with static curve is shown to occur at A giving a peak current I_m in the circuit. Drawing a horizontal line through A will intersect a vertical line drawn from V_m at A'.

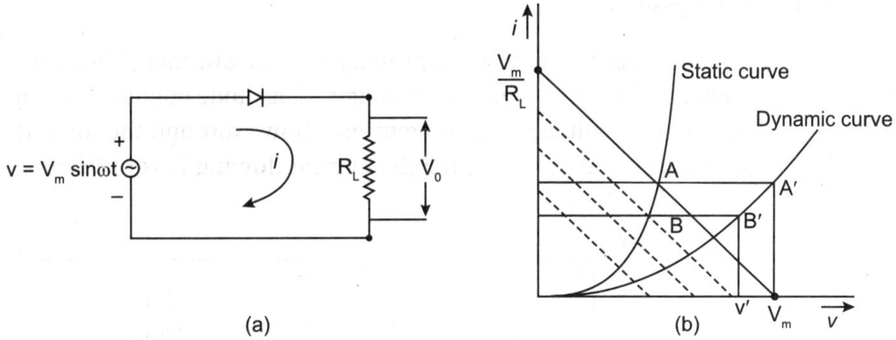

Figure 4.5

Now take any other value of v, say v', and draw a load line parallel to first. The intersection with static curve at B gives the current i' corresponding to v'. Drawing a horizontal line through B and a vertical line from v' gives the new intersecting point B'. Continuing this process for arbitrary values of v will yield sufficient points to construct the dynamic curve (Figure 4.5b).

It must be emphasized that this dynamic curve applies only to a particular circuit having a particular load.

The manner in which output current wave form is obtained using the dynamic curve for a given load R_L is indicated in Figure 4.6, which is self-explanatory.

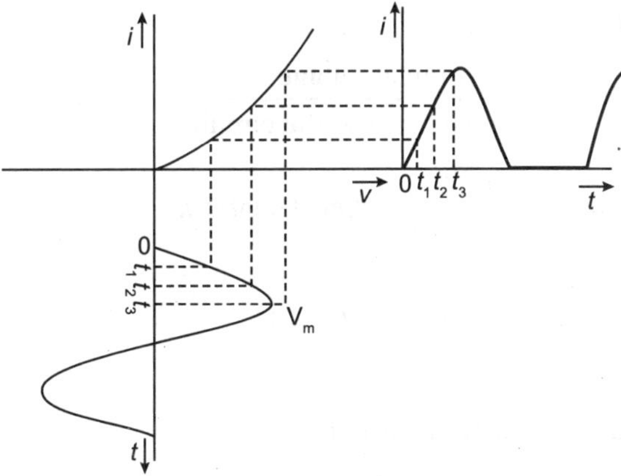

Figure 4.6

4.5 Rectifier

A rectifier may be defined as a device which converts AC voltage/current into DC voltage/current. It is of two types: (i) half wave rectifier (ii) full wave rectifier.

4.5.1 Half Wave Rectifier

An AC signal $V = V_m \sin \omega t$ is applied across the primary of a transformer (Figure 4.7(a)). The diode and load R_L is connected in series with the secondary. The diode conducts during positive half cycle $(0 - \pi)$ of the applied voltage, thereby current i flows through the load. It does not conduct through negative half cycle $(\pi - 2\pi)$, thereby current flowing is zero (Figure 4.7(b)).

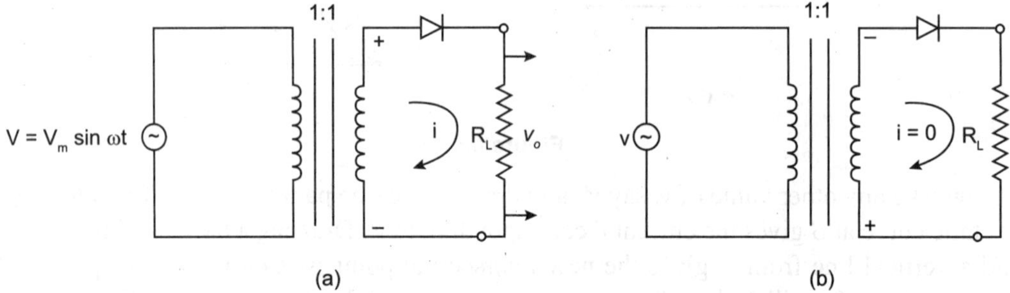

Figure 4.7 Half wave rectifier

Thus, with such a circuit, half wave is rectified. The shape of current wave through load R_L consists of a half sinusoidal wave as shown in Figure 4.8(a) so also is the shape of output voltage v_o.

Current during positive half cycle

$$i = \frac{\text{Applied voltage}}{\text{Total resistance of the circuit}}$$

$$= \frac{V_m \sin \omega t}{R_S + r_f + R_L} \quad \text{when } 0 < \omega t < \pi \tag{1}$$

and $$i = 0 \text{ when } \pi < \omega t < 2\pi$$

Here,

R_S – Resistance of secondary coil of transformer

r_f – Forward resistance of diode

R_L – Load

If R_s and r_f are very small in comparison with load R_L, then Eqn. (1) reduces to

$$i = \frac{V_m}{R_L}\sin\omega t \quad \text{when } 0 < \omega t < \pi \tag{1a}$$

$$= I_m \sin\omega t \tag{1b}$$

where $I_m = \dfrac{V_m}{R_L}$ is called peak value of current.

From eqn. (1a) it is clear that the shape of current wave is half sinusoidal wave (Figure 4.8(a)).

Output voltage across load R_L is similar to it.

$$v_o = i\ R_L = (I_m R_L)\sin\omega t$$

$$= V_0 \sin\omega t \ \text{ when } 0 < \omega t < \pi \tag{1c}$$

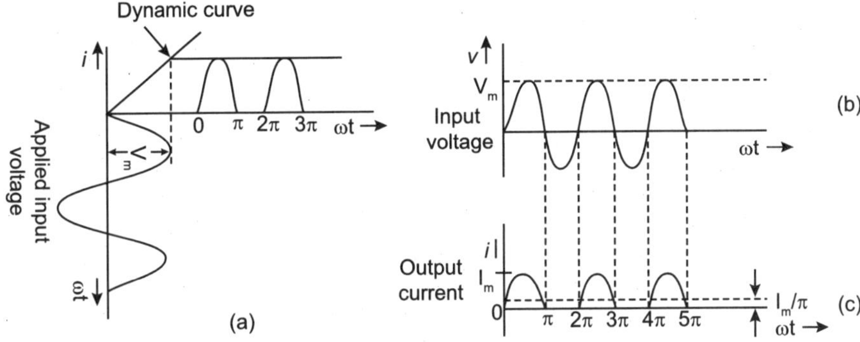

Figure 4.8 Half wave rectification

A Fourier analysis of half sinusoidal voltage pulse (Eqn. 1(c)) appearing across the load yields

$$v_o = \frac{V_0}{\pi} + \frac{V_0}{2}\sin\omega t - \frac{2V_0}{3\pi}\cos 2\omega t - \frac{2V_0 \cos 4\omega t}{15\pi} \tag{2}$$

Here the first term $[V_0/\pi]$ of R.H.S. is free from ω (frequency); it is zero frequency signal or DC component of output voltage. Its value is $\dfrac{V_0}{\pi} = 0.318V_0$.

Eqn. (2) suggests that the diode is a frequency converter where the fundamental frequency is changed to a large number of output frequencies.

If terms R_S and r_s in Eqn. (1) cannot be neglected, then DC component of output voltage

$$V_{dc} = \frac{V_0}{\pi} - I_{dc}[R_S + r_f] \tag{2a}$$

This analysis indicates that the DC output voltage depends not only upon load but also on diode resistance and transformer secondary coil resistance too.

DC or average value of current is given by

$$I_{dc} = \frac{1}{2\pi}\left[\int_{o}^{\pi} i\, d(\omega t) + \int_{\pi}^{2\pi} i\, d(\omega t)\right]$$

The second term in the bracket is zero. Since the diode does not conduct during this period,

$$I_{dc} = \frac{1}{2\pi}\int_{o}^{\pi} i\ d(\omega t)$$

$$= \frac{1}{2\pi}\int_{o}^{\pi} \frac{V_0}{R_L} \sin \omega t\ d(\omega t)$$

$$= (-)\frac{V_0}{2\pi R_L}[\cos \omega t]_{o}^{\pi}$$

$$= \frac{V_0}{\pi R_L} = \frac{I_m}{\pi} = 0.318\ I_m \tag{3}$$

where $I_m = \dfrac{V_0}{R_L}$

This current is shown as dashed straight line (Figure 4.8(c)).

The effective value of current in the output is the RMS value of current and given by

$$I_{rms} = \sqrt{\frac{1}{2\pi}\int_{o}^{2\pi} i^2\ d(\omega t)}$$

$$= \sqrt{\frac{1}{2\pi}\int_{o}^{\pi} i^2\ d(\omega t) + \int_{\pi}^{2\pi} i^2\ d(\omega t)}$$

Substituting for i and noting that the diode does not conduct during $(\pi - 2\pi)$, we have

$$I_{rms} = \sqrt{\frac{1}{2\pi}\int_{o}^{\pi}\left(\frac{V_0}{R_L}\sin \omega t\right)^2 d(\omega t)}$$

$$= \frac{1}{2\pi}\sqrt{\int_{o}^{\pi} I_m^2 \sin^2 \omega t\ \ d(\omega t)}$$

$$= \frac{I_m}{2} \tag{4}$$

Efficiency of rectifier Efficiency of a rectifier is defined as the ratio of DC output power to AC input power.

$$P_{dc} = I_{dc}^2\ R_L \tag{5}$$

AC input power is what a wattmeter would indicate if placed in the circuit with its voltage terminals connected to the secondary.

$$P_{ac} = (I_{rms})^2[r_f + R_L]$$

Substituting for I_{rms}, we have

$$P_{ac} = \left[\frac{I_m}{2}\right]^2 (r_f + R_L)$$

Hence efficiency of rectification

$$\eta = \frac{P_{dc}}{P_{ac}} = \frac{[I_m/\pi]^2 R_L}{\left(\frac{I_m}{2}\right)^2 (r_f + R_L)}$$

$$= \frac{4}{\pi^2}\left[\frac{1}{\frac{r_f}{R_L}+1}\right] \tag{6}$$

If $\frac{r_f}{R_L} \ll 1$, then

$$\eta = \frac{4}{\pi^2} = 0.406 \tag{6a}$$

Thus, the maximum theoretical value of efficiency of rectification of a half wave rectifier is 40.6% under no diode loss.

Ripple factor The aim of a rectifier is to convert AC into DC. A measure of purity of DC output is the ripple factor γ (Greek symbol gamma). It is defined as the ratio of the two current (or voltage) components

$$\gamma = \frac{\text{Effective (or RMS) value of AC component}}{\text{Respective DC component}}$$

or

$$r = \frac{(I_r)_{rms}}{I_{dc}} = \frac{(V_r)_{rms}}{V_{dc}}$$

Now

$$I_{rms}^2 = I_{dc}^2 + (I_r)_{rms}^2$$

or

$$(I_r)_{rms} = I_{rms}^2 - I_{dc}^2$$

$$\therefore\ r = \frac{(I_r)_{rms}}{I_{dc}} = \frac{\sqrt{I_{rms}^2 - I_{dc}^2}}{I_{dc}}$$

$$= \sqrt{\left(\frac{I_{rms}}{I_{dc}}\right)^2 - 1}$$

Substituting for I_{rms} and I_{dc}, we have ripple factor

$$\gamma = \sqrt{\left(\frac{I_m/2}{I_m/\pi}\right)^2 - 1}$$

$$= \sqrt{\frac{\pi^2}{4} - 1}$$

$$= 1.21 \tag{7}$$

Expressed as a percentage this leads to γ = 121%. Thus, an half wave rectifier is not an efficient device for converting AC into DC.

Output frequency (Ripple frequency) In case of a half wave rectifier, the output frequency is the same as the input frequency, i.e. frequency of applied input AC voltage. From Figure 4.8(b, c) it is clear that each cycle of input voltage produces one cycle of output voltage, so $f_{out} = f_{in}$.

Voltage regulation (VR) It is another term often used in rectification. It is defined as the degree to which a power supply (rectifier) varies in output voltage under conditions of load variations. It is usually expressed as a percentage:

$$\%\,\text{Voltage regulation} = \frac{V_{\text{No load}} - V_{\text{Full load}}}{V_{\text{Full load}}} \times 100 \tag{8}$$

A Thevenin equivalent circuit for the halfwave rectifier is shown in Figure 4.9 where R_{INT} is defined as internal resistance of power supply (rectifier) and it is equal to $R_s + r_f$.

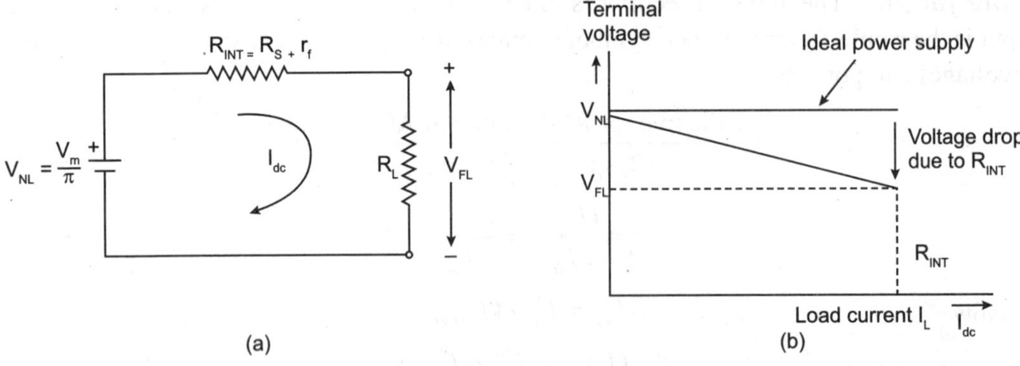

Figure 4.9 (a) Thevenin equivalent of half wave rectifier (b) graph showing variation of terminal voltage with current (I_{dc}) drawn

It is to be noted that a power supply having good voltage regulation has low % regulation.

Peak inverse voltage (PIV) It is the term often used in rectifier (power supply). It is defined as voltage across the diode during its nonconducting period in the rectifier, i.e. during negative half cycle of the AC input voltage. Thus, in a half wave rectifier its value is (–) V_m during π to 2π interval when applied input voltage is $V = V_m \sin \omega t$.

Transformer utilization factor (TUF) The actual voltage coming out of a powerpoint may vary from place to place and many times it depends on timing also. So it is not safe to use the supply directly in electronic equipment. Generally, we use a transformer of desired turn ratio, i.e. either step up or step down.

In a power supply design, it is necessary to determine the rating of a transformer. This is done with the knowledge of DC power to be delivered to the load and the type of rectifying circuit to be used.

$$\text{TUF} = \frac{\text{DC power delivered to the load}}{\text{AC rating of transformer secondary}}$$

$$= \frac{P_{dc}}{P_{ac}\text{ rated}} = \frac{I_{dc}^2 R_L}{\frac{V_m}{\sqrt{2}} \times I_{rms}} = \frac{\frac{I_m^2}{\pi^2} \times R_L}{\frac{V_m}{\sqrt{2}} \times \frac{\text{I}_\text{m}}{2}}$$

but $V_m = I_m (r_f + R_L)$ negating R_s

$$\therefore \text{ TUF} = \frac{(I_m^2/\pi^2)R_L}{\frac{I_m(r_f + R_L)}{\sqrt{2}} \times \frac{I_m}{2}} = \frac{2\sqrt{2}}{\pi^2} \times \frac{R_L}{r_f + R_L}$$

if $$R_L \gg r_f$$

$$\text{TUF} = \frac{2\sqrt{2}}{\pi^2} = 0.287 \tag{9}$$

Example 1 In a half wave rectifier, input signal voltage is 110 V, having frequency of 50 Hz. Calculate (i) peak value of source voltage (ii) peak output voltage and DC load voltage assuming (a) ideal diode (b) second approximation.

Solution

Peak source voltage $V_m = \frac{V_{rms}}{0.707} = \frac{110}{0.707} = 155\,\text{V}$

(a) With an ideal diode

(i) $V_{pout} = V_{pin} = V_m = 155\,\text{V}$

(ii) DC load voltage

$$V_{dc} = \frac{V_m}{\pi} = \frac{155}{3.14} = 49.29\,\text{V}$$

(b) With second approximation

(i) $V_{pout} = V_{pin} - 0.7$

$= 155 - 0.7 = 154.3\,\text{V}$

(ii) $V_{dc} = \frac{V_{pout}}{\pi} = \frac{154.3}{\pi} = 49\text{ V}$

Example 2 In a power supply, full load voltage is 20 V. Calculate its no load voltage if voltage regulation is (a) 0% (b) 100%.

Solution

$$VR = \frac{V_{\text{No load}} - V_{\text{Full load}}}{V_{\text{Full load}}} \times 100\%$$

(a) VR = 0% $\qquad V_{\text{No load}} = V_{\text{Full load}} = 20\text{ V}$

(b) VR = 100% $\qquad 100 = \frac{V_{NL} - 20}{20} \times 100$

$$\therefore V_{NL} = 20 + 20 = 40\text{ V}$$

Example 3 What is the necessary AC input power from the transformer secondary used in a half wave rectifier to deliver 500 W of DC power to load.

Solution

Ratio of rectification or efficiency of halfwave rectifier,

$$\eta = 0.406 = \frac{\text{DC power delivered to load}}{\text{AC input power from transformer secondary}}$$

$\therefore$ AC input power from transformer secondary

$$= \frac{\text{dc power delivered to load}}{0.406}$$

$$= \frac{500}{0.406} = 1232\text{ W}$$

4.5.2 Full Wave Rectifier

The circuit for a full wave rectifier is shown in Figure (4.10). Here the signal $V = V_m \sin \omega t$ is applied across the primary of the transformer. The secondary which is centre tapped, is connected to two diodes D_1 and D_2. The load R_L is connected between the centre tapping of secondary and common point of the diodes. The voltage across the secondary is $2V_m$. During positive half cycle $(0 - \pi)$ of input signal, one half voltage i.e. V_m appears across diode D_1 and load R_L which are series.

Diode D_1 being forward-biased (Figure 4.10(a)) conducts and current flows along the path A, D_1, P, Q, C, thereby the upper end of load R_L is positive. Diode D_2 does not conduct, it being reverse-biased.

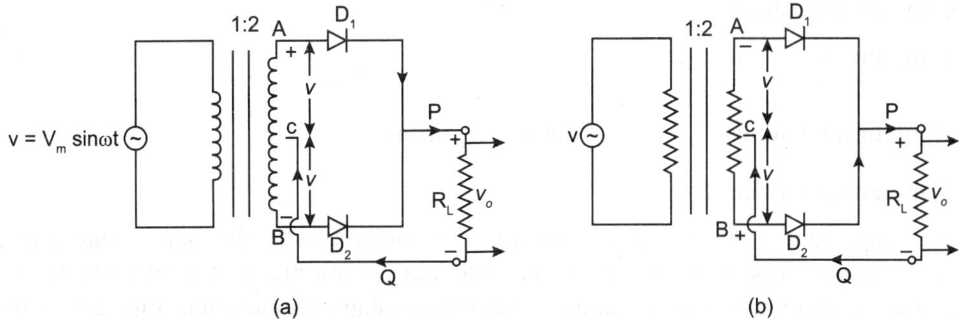

Figure 4.10 Full wave rectifier

The shape of current wave through load R_L consists of a half sinusoidal wave (Figure 4.11(b)) like that in a half wave rectifier (Figure 4.8).

During negative half cycle ($\pi - 2\pi$), the lower end B of secondary is positive (Figure 4.10(b)) thereby diode D_2 is forward-biased, and current flows along the path B, D_2, P, Q, C. Again it is to be noted that the upper-end of load R_L is positive. Diode D_1 does not conduct, it being reverse-biased. The shape of current wave through load R_L consists of a half sinusoidal wave (Figure 4.11(b)).

Thus, output voltage v_o is also unidirectional and pulsating (Figure 4.11(c)).

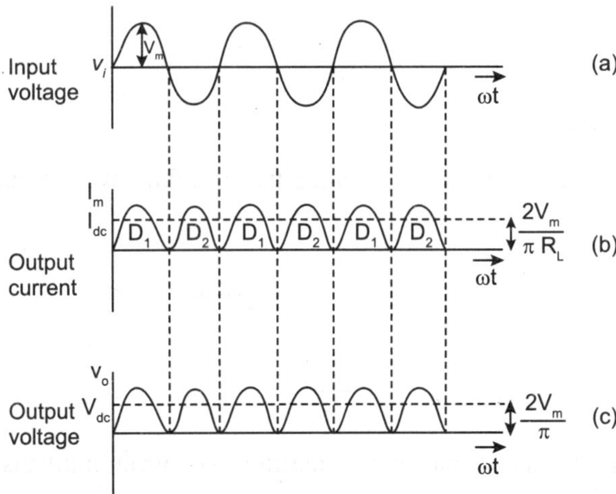

Figure 4.11

Assuming ideal diodes, current flow is given by eqn. (1a).

During interval $\pi < \omega t < 2\pi$:

$$\text{current through diode } D_1 = i_1 = \frac{V_m}{R_L}\sin\omega t = I_m \sin\omega t$$

current through diode $D_2 = 0$ (10)

During interval $\pi < \omega t < 2\pi$

current through diode $D_2 = i_2 = \frac{V_m}{R_L}\sin\omega t = I_m \sin\omega t$

current through diode $D_1 = 0$. (11)

Thus, each diode operates independently and under exactly the same conditions as in half wave rectifier. However, current through the load is present for the entire cycle of input voltage and it is unidirectional. So output voltage wave shape which is pulsating across the load (Figure 4.11(c)) is similar to that of current wave. Further it is to be noted that the frequency of full wave output signal is double the input signal frequency.

DC or average value of current is given by

$$I_{dc} = \frac{1}{2\pi}\int_{o}^{2\pi} i\, d(\omega t)$$

$$= \frac{1}{2\pi}\left[\int_{0}^{\pi}\frac{V_m}{R_L}\sin\omega t\, d(\omega t) + \int_{\pi}^{2\pi}\frac{V_m}{R_L}\sin\omega t\, d(\omega t)\right]$$

Integrating, we obtain

$$I_{dc} = \frac{2V_m}{\pi R_L} = \frac{2I_m}{\pi} \tag{12}$$

as shown in Figure 4.11(b).

It may be noted that the value of I_{dc} is twice that of a half wave rectifier (Eqn. 3). The DC output voltage

$$V_{dc} = I_{dc}\, R_L = \frac{2I_m}{\pi} R_L$$

$$= \frac{2V_m}{\pi} \tag{12a}$$

which is double than in the case of half wave rectifier. A Fourier analysis of voltage wave form (Figure 4.11(c)) gives

$$v_o = \frac{2V_m}{\pi} - \frac{4V_m}{3\pi}\cos 2\omega t - \frac{4V_m}{15\pi}\cos 4\omega t \tag{13}$$

Here the first term of R.H.S. represents the DC component since it is free from ωt term.

Its value is $\frac{2V_m}{\pi} = 0.636V_m$ and shown as a dash dash horizontal line in Figure 4.11(c).

The second term consists of 2 ωt term. It suggests that output consists of 2nd harmonic and even higher harmonics.

Eqn. (13) suggests that the rectifier is a frequency converter where a fundamental frequency is converted into a large number of output frequencies where 2nd harmonic is most prominent.

DC output voltage Following a similar discussion to that used in case of half wave rectifier, DC output voltage across the load is

$$V_{dc} = \frac{2V_m}{\pi} - I_{dc}(R_s + r_f)$$

and so at no load, output voltage is $V_{dc} = \frac{2V_m}{\pi}$ as derived in eqn. (12a).

Peak inverse voltage (PIV) It is the maximum reverse voltage that occurs across a diode (Figure 4.10(a) or (b)) when it is not conducting, i.e. when diode D_1 is conducting, D_2 is not conducting and when D_2 is conducting, D_1 is not conducting. The peak voltage that occurs across load R_L is V_m and at the same time peak voltage from B to C is also V_m. So maximum reverse voltage across diode D_2 is $2V_m$. So is the case across diode D_1 when it is not conducting.

It is amply clear that PIV in case of full wave rectifier is double than in case of half wave rectifier.

Ripple factor As defined earlier, ripple factor

$$r = \sqrt{\left(\frac{I_{rms}}{I_{dc}}\right)^2 - 1}$$

Here $I_{rms} = \frac{I_m}{\sqrt{2}}$ and $I_{dc} = \frac{2I_m}{\pi}$, where I_m is the peak value of current pulse.

Substituting these values, we have

$$r = \sqrt{\left[\frac{I_m/\sqrt{2}}{2I_m/\pi}\right]^2 - 1}$$

$$= 0.482 \tag{14}$$

which suggests that there is a drastic improvement from 1.21 in case of half wave rectifier. Expressed as percentage this leads to $r = 48.2\%$.

Efficiency of rectification

$$\text{Efficiency of rectification} = \frac{P_{dc}}{P_{ac}} = \frac{(I_{dc})^2 R_L}{I_{rms}^2 (r_f + R_L)}$$

$$\eta = \frac{(2I_m/\pi)^2 R_L}{\frac{I_m^2}{2}(r_f + R_L)}$$

Assuming $r_f \ll R_L$,

$$\eta = 0.812 \tag{15}$$

It is just double that of a half wave rectifier.

Transformer utilization factor (TUF) In a Full wave rectifier, the average of TUF is found by considering the primary and secondary windings separately. This yields to a value of 0.693 as compared with 0.287 for a half wave rectifier.

Frequency components of output current/voltage It is to be noted that output current/voltage is pulsating but unidirectional and has the same shape from 0 to π and π to 2π when the two diodes are conducting turn by turn.

Current during the interval π to 2π has a phase difference of π from that of I_m sin ωt (which flows from 0 to π) since the negative half cycle is inverted. Making use of Fourier analysis for the half sinusoidal wave, we have

during 0 to π,

$$i_1 = I_m\left[\frac{1}{\pi} + \frac{1}{2}\sin\omega t - \frac{2}{3\pi}\cos 2\omega t - \frac{2}{15}\cos 4\omega t + \ldots\right]$$

during π to 2π,

$$i_2 = I_m\left[\frac{1}{\pi} + \frac{1}{2}\sin(\omega t + \pi) - \frac{2}{3\pi}\cos(2\omega t + \pi) - \frac{2}{15}\cos(4\omega t + \pi) + \ldots\right]$$

$$= I_m\left[\frac{1}{\pi} - \frac{1}{2}\sin\omega t - \frac{2}{3\pi}\cos 2\omega t - \frac{2}{15}\cos 4\omega t + \ldots\right]$$

Hence, load current

$$i = i_1 + i_2 = I_m\left[\frac{2}{\pi} - \frac{4}{3\pi}\cos 2\omega t - \frac{4}{15}\cos 4\omega t + \ldots\right] \tag{16}$$

Thus, load current wave form is composed of continuous positive half cycles.

From Eqn. (16), it is to be noted that

(i) DC component $I_{dc} = \dfrac{2I_m}{\pi}$

(ii) First harmonic frequency component is absent

(iii) Rest of the terms consist of even harmonics.

These terms can be filtered easily and thereby output current/voltage is DC only.

4.5.3 Bridge Rectifier

A full wave rectifier discussed earlier has two demerits:

(i) It has a centretapped transfer which is quite bulky, thereby it occupies a large space in the equipment.

(ii) The peak inverse voltage requirement for the two diodes is twice that of the peak of input voltage when transformer turn ratio is 1 : 2. So the capacitor used for filtering should have a rating for $2V_m$.

The above demerits have been removed in a bridge rectifier shown in Figure 4.11(d, e).

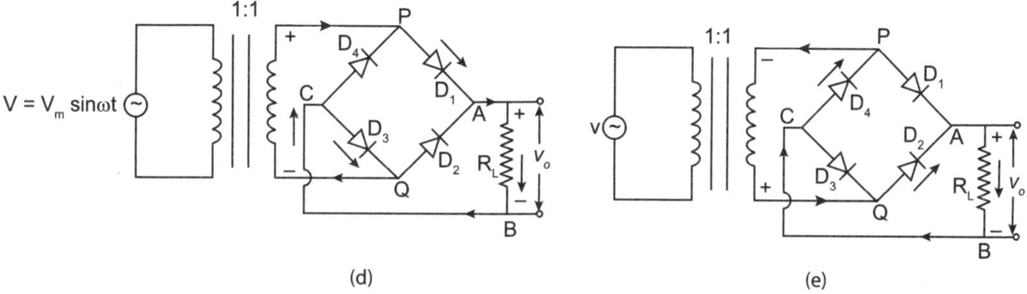

Figure 4.11 Bridge rectifier

Here we use four identical *p-n* junction diodes D_1 D_2 D_3 and D_4 arranged in the form of a wheatstone's bridge. The terminals of secondary of transformer are connected at *P* and *Q* respectively and load R_L is connected between *A* and *C*.

During positive half cycle of the input voltage, diodes D_1 and D_3 conduct being forward-biased while D_2 and D_4 do not conduct (being reverse-biased) and current flows along path *PABCQ* (Figure 4.11(d)). So the upper terminal *A* of load R_L is positive.

During negative half cycle of the input voltage, diodes D_2 and D_4 conduct (being forward-biased) while D_1 and D_3 do not conduct (being reverse-biased) and current flows along path *QABCP* (Figure 4.11e). Again the upper terminal *A* of load is positive.

We note that output current flows through the load in one direction during the entire cycle of input voltage keeping its upper terminal positive. So we find the output voltage unidirectional and pulsating (Figure 4.11(c)) like that in a full wave rectifier.

Merits and Demerits

(a) Merits

(i) A transformer without a centre tap is used.

(ii) Current flowing in primary as well as in secondary is sinusoidal, and thus we require a smaller transformer than that required in a full wave rectifier for the same output voltage.

(iii) The peak inverse voltage rating of diode is half that of the diode used in full wave rectifier for the same output voltage. This circuit is useful for high voltage requirement.

(iv) Crystal diodes are usually used and the assembly of four diodes is available in a package. Hence the circuit is more compact.

(b) Demerits

(i) Here we require four diodes whereas in a Full wave we require only two diodes.

(ii) Since current during one half cycle flows in two diodes in series with load, large power is dissipated than in a full wave rectifier.

(iii) Vacuum diodes cannot be used here. Bridge circuits generally employ selenium and copper oxide rectifier.

The above points have been summarized in Table 4.1.

Table 4.1

Particular	Half wave rectifier	Full wave rectifier	Bridge rectifier
No. of diodes	1	2	4
Rectifier input at secondary (peak value)	V_m	(i) V_m (transformer turn ratio 1 : 2) (ii) 0.5 V_m (transformer turn ratio 1 : 1)	V_m
Peak output voltage (ideal)	$V_0(p) = V_m$	(i) $V_0(0) = V_m$ (ii) $\frac{V_0(p)}{2} = \frac{V_m}{2}$	$V_m = V_0(\text{p})$
Peak output voltage (practical)	$[V_{op} - 0.7]V$	(i) $V_0(p) - 0.7$ V (ii) $\frac{V_0(p)}{2} - 0.7\text{V}$	$V_0(p) = 1.4$ V
DC output	$\frac{V_0(p)}{\pi}$	(i) $\frac{2V_0(p)}{\pi}$ (ii) $\frac{V_0(p)}{\pi}$	$2V_0(p)/\pi$
Output/ripple frequency	ω	2ω	2ω
PIV	$(-)V_m$	$(-)2\ V_m$	$(-)V_m$
Ripple factor	121%	48.2%	48.2%
Efficiency of rectification	40.6%	81.2%	81.2%

Example 4 A half wave rectifier has a load of 3.5 kΩ. If diode resistance and secondary coil resistance together have a resistance of 800 Ω and peak value of input voltage is 240 V. Calculate:

(i) Peak, average and *rms* value of current flowing

(ii) DC output power

(iii) AC input power

(iv) Efficiency of rectifier

Solution

(i) (a) Peak value of current flowing

$$I_m = \frac{V_m}{r_f + R_L} = \frac{240}{800+3500} = 55.81 \text{ mA}$$

(b) Average or DC current flowing,

$$I_{dc} = \frac{I_m}{\pi} = \frac{55.81}{3.14} = 17.78 \text{ mA}$$

(c) R.M.S value of current flowing

$$I_{rms} = \frac{I_m}{2} = \frac{55.81}{2} = 27.90 \text{ mA}$$

(ii) DC output power,

$$P_{dc} = (I_{dc})^2 R_L = (17.78\times10^{-3})^2 \times 3500$$
$$= 1.1 \text{ Watt}$$

(iii) AC input power,

$$P_{ac} = (I_{rms})^2 (r_f + R_L)$$
$$= (27.90\times10^{-3})(4300) = 3.35 \text{ Watt}$$

(iv) Efficiency of rectifier,

$$\eta = \frac{P_{dc}}{P_{ac}} = \frac{1.1}{3.35}\times100 = 32.9\%$$

Example 5 Four diodes are used in a bridge rectifier, each having $V_r = 0.7$ V. Load resistor is 820 Ω and peak value of input voltage is 40 V. Calculate (i) peak output voltage, (ii) average output current and (iii) diode dissipation.

Solution

(i) Peak output voltage: Current flows through load only when two diodes conduct. While conducting, there is voltage drop across the diode.

Voltage drop across two diodes = 2 × 0.7 = 1.4

∴ Peak voltage drop across the load

V_m = Peak input voltage – Voltage drop across two diodes

= 40 – 1.4 = 38.6 V

(ii) Average output current $= \dfrac{\left[\dfrac{2V_m}{\pi}\right]}{R_L}$

$$I_{dc} = \frac{2 \times 38.6}{3.14 \times 820} = 30\,\text{mA}$$

(iii) Diode dissipation $= I_{dc} \times V_r$

$= 30 \times 10^{-3} \times 0.7$

$= 21$ mW

Example 6 In a half wave rectifier, transformer turn ratio is 5 : 1. Input voltage is 120 V, 60 Hz. Calculate (i) peak secondary voltage, (ii) peak inverse voltage of diode, (iii) peak load voltage, (iv) DC load voltage

Also determine (i) peak load voltage, (ii) DC load voltage assuming second approximation.

Solution

RMS value of secondary voltage $= \dfrac{V_i}{5} = \dfrac{120}{5} = 24\,\text{V}$

Peak secondary voltage, $V_m = 24\sqrt{2} = 33.6 \approx 34\,\text{V}$

Peak inverse voltage of diode = (–) V_m = (–) 34 V

Peak load voltage = 34 V

DC load voltage $= \dfrac{V_m}{\pi} = \dfrac{34}{3.14} = 10.8$ V

Assuming second approximation

$V'_m = V_m - V_r$

Peak load voltage, $V'_m = 34 - 0.7 = 33.3$ V

DC load voltage $= \dfrac{V'_m}{\pi} = \dfrac{33.3}{3.14} = 10.6\,\text{V}$

Example 7 In a full wave rectifier having shunt capacitor filter with a load of 5 kΩ the transformer turn ratio is 5 : 1. Calculate (i) peak value of voltage across the load, (assume input voltage 120 V, 60 Hz), (ii) DC load voltage, (iii) DC load current.

Solution

Secondary RMS voltage $= \dfrac{120}{5} = 24\,\text{V}$

Secondary peak voltage $= 24\sqrt{2} = 34\,\text{V}$

Half the secondary voltage is input to the half section.

So input to half section $= \frac{34}{2} = 17\,\text{V}$

Peak voltage across load = 17 V

DC voltage across load = 17 V. Since the capacitor gets changed up to peak value,

DC load current, $I_{dc} = \frac{V_{dc}}{R_L} = \frac{17}{5000} = 3.4\,\text{mA}$

4.6 Filters

All electronic equipment require DC having minimum ripple. We usually separate ripple from the unregulated output current of a rectifier using a filter. This process of separating ripple from the unregulated (pulsating) output current is called smoothing and the circuit employed for smoothing is called a filter.

A filtering circuit generally consists of either inductor or capacitor or both. They are of many types. The main circuits among them are (i) shunt capacitor filter, (ii) series inductor filter, (iii) L-section filter, (iv) π filter.

Shunt Capacitor Filter (R-C Filter) The circuit of *R–C* filter or shunt capacitor filter along with a full wave rectifier is shown in Figure (4.12). The *AC* signal $V = V_m \sin \omega t$ to be rectified is connected across the primary of the transformer and the filter is connected in the secondary across the centre tapping of secondary tapping and the common point of the two diodes D_1 and D_2. The output V_o is observed across *PQ*.

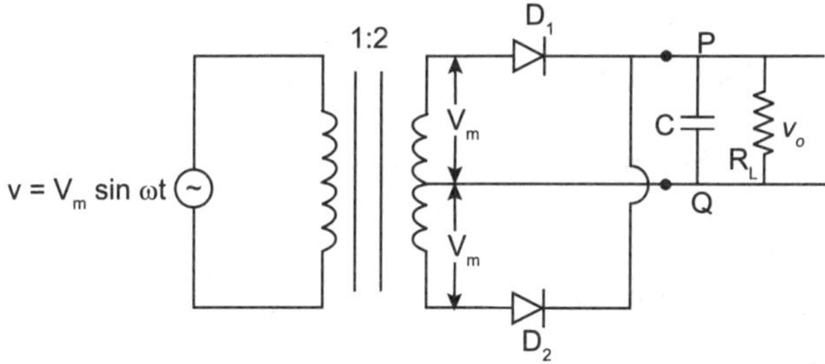

Figure 4.12 Full wave rectifier with C filter

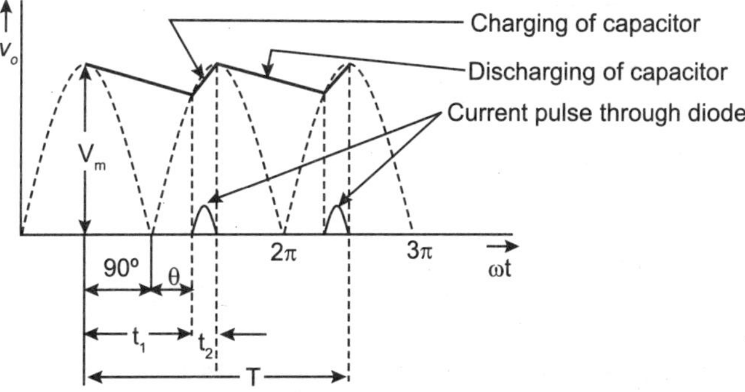

Figure 4.13 Charging and discharging of capacitor

Here the value of reactance, X_C, of capacitor is so selected that its value $\frac{1}{\omega c}$ is much less than the value of load R_L. In this situation, the ripple component (*ac*) of output current is bypassed through *C*.

During this process, the capacitor *C* gets charged until voltage attains a maximum value V_m and stores energy (Figure 4.13).

When the voltage pulse decreases, then the capacitor gets discharged through load R_L and thereby voltage across R_L is kept nearly constant.

The time constant of discharging circuit is R_LC, i.e. it depends upon the value of load. The value of voltage across the capacitor at any instant is

$$v = V_m \exp\left[-\frac{t}{R_L c}\right] \tag{17}$$

If the value of time constant R_LC is very large, then Eqn. (17) can be expressed as

$$v = V_m\left[1 - \frac{t}{R_L C}\right] = V_m - \frac{V_m t}{R_L C} \tag{18}$$

Suppose the time interval between two consecutive peaks of output voltage is T (Figure 4.13), we can assume that the capacitor again begins to recharge after this interval. Charging of capacitor is through the diode which conducts when forward-biased. The forward resistance of diode is very small so time (t_2) for charging the capacitor is also very small. The recharging process of capacitor begins when its voltage is $\left[V_m - \frac{V_m t}{R_L C}\right]$ across load R_L.

In this way there is fluctuation of voltage across load R_L between the peak value V_m of voltage and $\left[V_m - \frac{V_m t}{R_L C}\right]$ as shown in Figure 4.14.

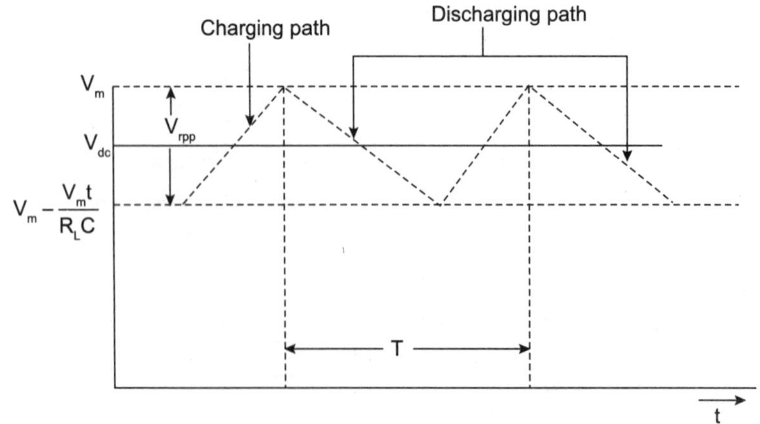

Figure 4.14 Voltage fluctuation across capacitor

4.6.1 Ripple Factor

To determine ripple factor in case of a shunt capacitor filter we make use of the approximation that the value of C is such that at output frequency $\omega CR_L \gg 1$.

Under the above condition, the capacitor discharges very slowly and the conduction time t_2 for diode is very small (Figure 4.13). So we can assume that discharging period t_1 of capacitor is nearly equal to $T/2$ where T is the time period of applied sinusoidal voltage.

Charge lost by the capacitor in the nonconducting period of diode is

$$q = \text{Average current} \times \text{time}$$
$$= I_{dc} \times t_1$$
$$= I_{dc} \times \frac{T}{2}$$

due to the charge lost, the voltage of capacitor falls.

The fall of voltage $\Delta V = \dfrac{\text{Charge lost}}{\text{Capacitance}} = \dfrac{I_{dc}T}{2C}$

or $$\Delta V = V_{rpp} = \frac{I_{dc}}{2fC} \quad \text{where } f = \frac{1}{T} \tag{18a}$$

Average voltage across the capacitor from Figure 4.14 is

$$V_{dc} = V_m = \frac{\Delta V}{2} = V_m - \frac{V_{rpp}}{2}$$
$$= V_m - \frac{I_{dc}}{4fC} \tag{18b}$$

Eqn. (18a) suggests that average voltage depends upon current drawn by the load; so one can assume $\dfrac{1}{4fC}$ as internal resistance of voltage source. No load value of voltage across capacitor is V_m.

Hence percent voltage regulation $\dfrac{V_{NL} - V_{\text{Full load}}}{V_{\text{Full load}}} \times 100$

Percentage voltage regulation $= \dfrac{V_m - I_{dc} R_L}{I_{dc} R_L} \times 100$

$$= \frac{I_{dc} R_o}{I_{dc} R_L} \times 100 \qquad \text{where } R_o \text{ is internal resistance} = \frac{1}{4fC}$$

$$= \frac{100}{4fCR_L}\%$$

From Figure 4.14, it is clear that variation of voltage V_{rpp} is a measure of ripple voltage. which can be reduced by increasing frequency f or capacitance C. Also the trace of variation of voltage can be taken as sawtooth voltage.

The rms value of such a voltage is given by $\frac{\Delta V}{2\sqrt{3}}$.

Substituting for ΔV from Eqn. 18(a),

$$[V_r]_{rms} = \frac{I_{dc}}{4\sqrt{3}fC}$$

$$\text{Ripple factor } r \text{ is defined as} = \frac{[V_r]_{rms}}{V_{dc}} = \frac{I_{dc}/4\sqrt{3}\,fC}{I_{dc}\,R_L}$$

$$= \frac{1}{4\sqrt{3}\,fC\,R_L} \tag{19}$$

For (i) f = 60 Hz, C in μF and R_L in ohms,

$$r = \frac{2410}{CR_L} \tag{19a}$$

(ii) f = 50 Hz

$$r = \frac{2890}{CR_L} \tag{19b}$$

Equation (19) suggests that ripple factor in case of capacitor filter is inversely proportional to (i) frequency of applied voltage, (ii) load R_L and (iii) capacitor value C.

If the value of load R_L is infinity, i.e. terminals are open, then the value of r reaches zero and the value of DC voltage across the capacitor reaches the peak-value of input voltage V_m.

Assuming $T = \frac{1}{2f}$, the value of V_{dc} (Eqn. 18a) reduces to

$$V_{dc} = V_m - \frac{V_m}{4R_L\,f\,C}$$

This filter is preferred for high value R_L (light load). In such a situation, V_{dc} approaches V_m.

Alternative To determine ripple factor in case of, R-C filter, we shall use an approximation. The value of DC voltage across the load can be determined by knowing the mean value of variation $\Delta V_0 = V_{rpp}$ (Figure 4.14).

$$\text{Mean value of } V'_{rpp} = \frac{\left[V_m - \left(V_m - \frac{V_m t}{R_L C}\right)\right]}{2}$$

$$\therefore V_{dc} = V_m - \frac{V_m t}{2R_L C} \tag{19c}$$

The trace of fluctuating voltage (Figure 4.14) can be approximately taken as sawtooth voltage.

The RMS value of such a AC voltage is given by $[V_{ac}]_{\text{RMS}} = \dfrac{V_m T}{2\sqrt{3}\, R_L\, C}$

$$\text{Ripple factor is defined as } r = \frac{\text{RMS value of ripple (AC) voltage}}{\text{DC voltage}}$$

$$= \frac{[V_{ac}]_{\text{RMS}}}{V_{dc}}$$

$$r = \frac{V_m T / 2\sqrt{3}\, R_L C}{V_m - \dfrac{V_m T}{2R_L\, C}} = \frac{V_m T / 2\sqrt{3}\, R_L\, C}{V_m \left[1 - \dfrac{T}{2R_L C}\right]}$$

Example 8 A full wave rectifier having diodes of negligible forward resistance uses a load of 2 kΩ. A capacitor of capacitance 100 μF is in parallel with the load. The transformer secondary voltage is 40 V (rms) with respect to centretapping. Assuming supply frequency to be 50 Hz calculate (a) ripple factor, (b) DC output voltage, (c) percentage regulation.

Solution

(a) Ripple factor for a full wave rectifier

$$r = \frac{1}{4\sqrt{3}\, f\, C\, R_L} \qquad \text{Here, } f = 50 \text{ Hz},\ C = 100\ \mu\text{F},\ R_L = 2\ \text{k}\Omega$$

$$= \frac{1}{4\sqrt{3}\, 50 \times 100 \times 10^{-6} \times 2000}$$

$$= 0.014$$

(b) DC output voltage

$$V_{dc} = V_m - \frac{I_{dc}}{4 fc}$$

$$= V_m - \frac{V_{dc}}{4 fR_L C}$$

$$V_{dc}\left[1 + \frac{1}{4 f\, R_L C}\right] = V_m$$

$$V_{dc} = \frac{V_m}{1 + \dfrac{1}{4 fR_L C}} = \frac{56.4}{1 + \dfrac{1}{4 \times 50 \times 2000 \times 100 \times 10^{-6}}}$$

$$= 55.2\ \text{V}$$

(c) Percentage voltage regulation $= \dfrac{100}{4 \times f \times R_L C} = \dfrac{100}{4 \times 50 \times 2000 \times 100 \times 10^{-6}}$

$$= 2.5\%$$

4.6.2 Series Inductor Filter

A high inductance coil L is used in series with load R_L connected to the rectifier (Figure 4.15).

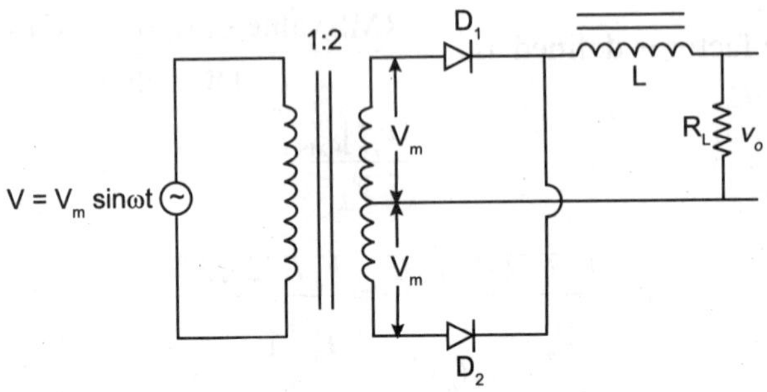

Figure 4.15 Full wave rectifier with inductor filter

When output current of the rectifier increases by more than the average or DC current, then energy due to increase in current is stored as magnetic energy in the inductor. So when output current from the rectifier decreases, then an additional induced e.m.f. is generated due to the self-inductance of the coil and it opposes the decrease in current.

Fluctuations in the output current are minimized by using an inductor filter. Fourier analysis gives the output voltage of the rectifier as

$$V_o = \frac{2V_m}{\pi} - \frac{4V_m}{3\pi}\cos 2\omega t - \frac{4V_m}{15\pi}\cos 4\omega t$$

The inductor offers negligible resistance to the DC component of current. So DC voltage output across load R_L is

$$V_{dc} = \frac{2V_m}{\pi}$$

DC current flowing through the load is

$$I_{dc} = \frac{V_{dc}}{R_L} = \frac{2V_m}{\pi R_L} = \frac{2I_m}{\pi}$$

Here resistance offered by the secondary coil and diode is neglected.

The AC component (ripple) of the output voltage consists of 2 ωt and 4 ωt terms. The ratio of 2nd harmonic to 4th harmonic is 1 : 1/5 and the ratio of reactance offered by the inductor to these components is 1 : 2 nearly (it is assumed that $\omega L \gg R_L$). So the ratio of 2nd harmonic current to 4th harmonic current is 1 : 1/10. Hence the 2nd harmonic is important and rest of the harmonics can be neglected.

The effective value of 2nd harmonic component of AC voltage is

$$V_{ac} = \frac{4V_m}{3\pi\sqrt{2}} \tag{20}$$

This voltage appears across the combination of inductor (L) and load (R_L) since they are in series. So current flowing through them is

$$I_{ac} = \frac{V_{ac}}{|Z|} = \frac{4V_m}{3\pi\sqrt{2}} \frac{1}{\sqrt{R_L^2 + (2\omega L)^2}} = \frac{4V_m}{3\pi\sqrt{2}\sqrt{R_L^2 + 4\omega^2 L^2}}$$

$$= \frac{4V_m}{3\pi\sqrt{2}(2\omega L)} \text{ since } 2\omega L \gg R_L \tag{21}$$

Here reactance of inductor is determined for 2ω.

$$\text{Ripple factor } r = \frac{[I_{ac}]_{rms}}{I_{dc}} = \frac{4V_m}{3\pi\sqrt{2}(2\omega L)} \times \frac{\pi R_L}{2V_m}$$

or
$$r = \frac{R_L}{3\sqrt{2}(\omega L)}$$

$$\therefore r = \frac{0.236}{(\omega L / R_L)} \tag{22}$$

$$\text{for (i) } f = 60 \text{ Hz}; \; r = \frac{R_L}{1600L} \tag{22a}$$

$$\text{(ii) } f = 50 \text{ Hz}; \; r = \frac{R_L}{1300L} \tag{22b}$$

Eqn. (22) suggests that with increasing load R_L the value of ripple factor increases, and on decreasing load R_L the value of ripple factor decreases. This behaviour of inductor filter is opposite to that of capacitor filter.

With shunt capacitor filter we get higher voltage for lower ripple factor, but with series inductor filter we get higher current with lower ripple factor.

The variation of output current with inductor filter is shown in Figure 4.16.

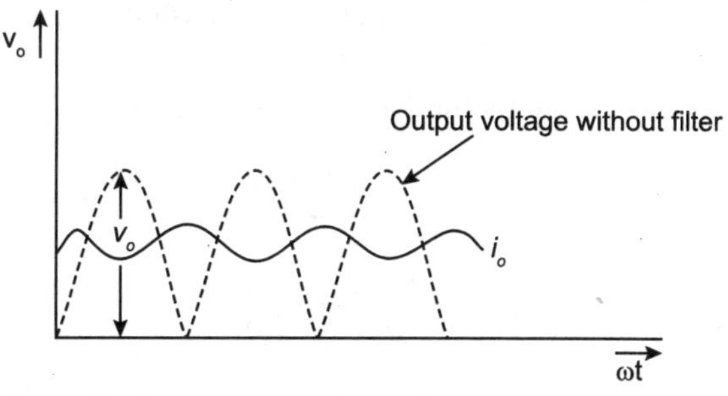

Figure 4.16

4.6.3 L-Section Filter or Choke Input *L–C* Filter

As discussed in the previous section, a capacitor filter is preferred for high load (light load) and an inductor filter is preferred for small load (heavy load) and each cannot lower the value of ripple factor.

To keep minimum ripple factor for a given load, we use *L–C* filter (Figure 4.17). Here inductance '*L*' is connected in series with load R_L and capacitor *C* is connected across the load.

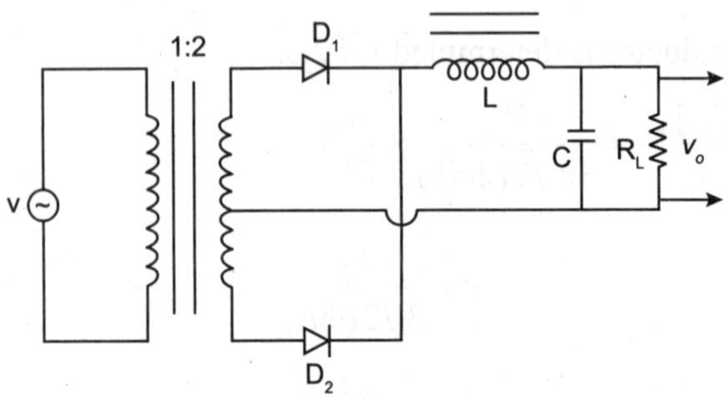

Figure 4.17 Full wave rectifier with *L–C* filter

When an unidirectional pulsating output current from the rectifier flows through the inductor *L*, then the DC component passes easily, since the inductor offers negligible resistance to it, but the AC component does not pass easily since the inductor offers a high reactance to it. However, if a fraction of the AC component passes through the inductor, it finds a bypass through the capacitor C connected before the load, since the reactance of capacitor offered to the AC component is quite low in comparison with resistance of the load.

At the end we find that the AC component is filtered out and only the DC component flows through the load, thereby we get a nearly constant voltage across the load.

From the above discussion, it is clear that, inductor should offer high reactance to the AC component and a capacitor should offer a low reactance to it.

Output voltage of the rectifier is

$$v_o = \frac{2V_m}{\pi} - \frac{4V_m}{3\pi}\cos 2\omega t - \frac{4V_m}{15\pi}\cos 4\omega t + \ldots \tag{23}$$

The first term of R.H.S. represents the DC component; so

$$V_{dc} = \frac{2V_m}{\pi}$$

and

$$I_{dc} = \frac{2V_m / \pi}{R_L} = \frac{2V_m}{\pi R_L}$$

The second and third term in Eqn. (23) represent the 2nd harmonic and 4th harmonic of AC voltage, respectively, which are to be filtered out; here also only the 2nd harmonic is effective since its amplitude is quite large in comparison with other higher harmonics amplitude, and so they are negligible.

$$v_{ac} = \frac{4V_m}{3\pi}\cos 2\omega t \tag{24}$$

Reactance of capacitor $X_C = \frac{1}{2\omega c} \ll R_L$ current flowing through the combination of L and C is

$$i_{ac} = \frac{v_{ac}}{X_L - X_C} = \frac{v_{ac}}{[2\omega L - 1/2\omega C]}$$

Substituting for v_{ac}, we have

$$i_{ac} = \frac{4V_m \cos 2\omega t}{3\pi\left[2\omega L - \frac{1}{2\omega c}\right]}$$

$$= \frac{8V_m\,\omega c}{3\pi[4\omega^2 Lc - 1]}\cos 2\omega t \tag{25}$$

The effective value or R.M.S value of current is

$$I_{rms} = \frac{8V_m\omega c}{3\pi(4\omega^2 Lc - 1)}\sqrt{\frac{1}{T}\int_0^T \cos^2 2\omega t}$$

$$= \frac{8V_m\,\omega c}{3\pi(4\omega^2 Lc - 1)} \times \frac{1}{\sqrt{2}} \tag{26}$$

Voltage across the capacitor is

$$V_{ac} = I_{rms}\left[\frac{1}{2\omega c}\right]$$

$$= \frac{8V_m\,\omega c}{3\pi(4\omega^2 Lc - 1)} \times \frac{1}{2\sqrt{2}\omega c}$$

$$= \frac{4V_m}{3\pi\sqrt{2}(\omega^2 Lc - 1)} \tag{27}$$

$$\text{Ripple factor, } \gamma = \frac{\text{Effective value of ripple voltage}}{\text{DC voltage}}$$

$$= \frac{v_{ac}}{v_{dc}} = \frac{4V_m}{3\pi\sqrt{2}(\omega^2 Lc - 1)} \times \frac{\pi}{2V_m}$$

$$\therefore r = \frac{\sqrt{2}}{3}\left[\frac{1}{4\omega^2 Lc - 1}\right] = \frac{0.47}{(4\omega^2 Lc - 1)} \tag{28}$$

The above relation (eqn. 28) suggests that the value of ripple factor does not depend upon the value of load R_L whereas it is dependent on load in case of other filters.

Its value can be further reduced if $4\omega^2 LC \gg 1$;then

$$r = \frac{0.47}{4\omega^2 LC} = \frac{0.118}{\omega^2 LC} \tag{29}$$

for (i) $f = 60$ Hz, C in μF, $r = \frac{0.83}{LC}$

and (ii) $f = 50$ Hz, C in μF, $r = \left(\frac{1.2}{LC}\right)$

Hence in case of an L–C filter ripple factor is inversely proportional to the value of inductor as well as capacitor. So to reduce the value of ripple factor, their values be kept large.

Critical value of inductor In an L–C filter, the capacitor is more effective for a high value of load R_L or a small value of current. The capacitor gets charged upto a maximum (peak) value of applied voltage but it does not discharge completely through load, rather the voltage across capacitor is slightly less than the peak value. So average or DC voltage across the load, is nearly constant. On increasing the current through load i.e. when load is small, the capacitor discharges through it quite fast, thereby the DC voltage reduces a lot. When current increases, the effect of reactance of the inductor increases, and after a certain value of current I_K (Figure 4.18) the filter acts like a series inductor filter where the voltage remains nearly constant.

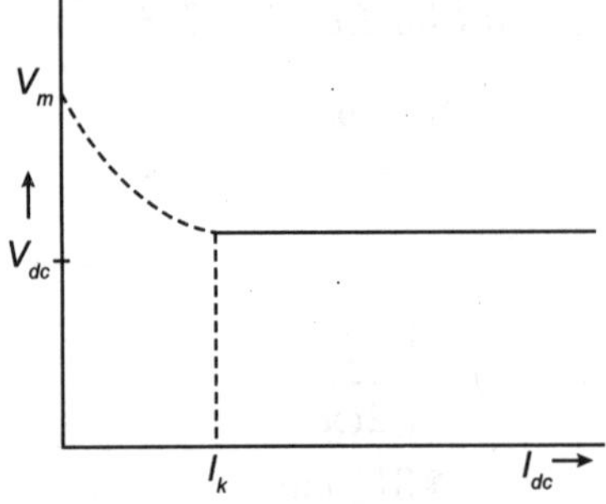

Figure 4.18

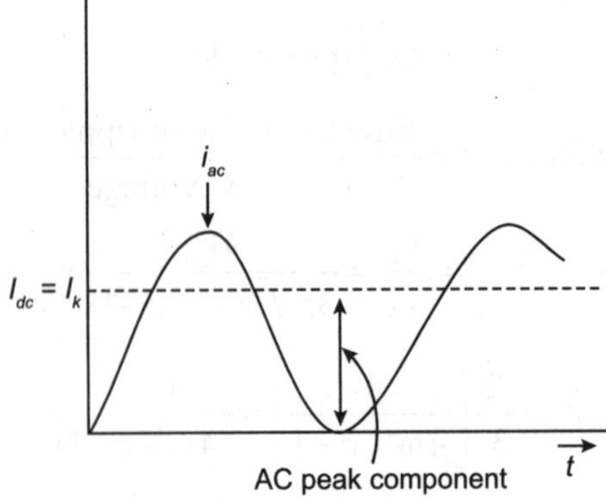

Figure 4.19

We must ensure that the value of I_{dc} (Figure 4.19) exceeds the negative peak of 2nd harmonic component of ripple (AC) current.

For 2nd harmonic, the peak value of ripple current = $\dfrac{4V_M}{3\pi|Z_2|}$ (30)

where Z_2 represents the impedance offered by the filter circuit to the 2nd harmonic component of current.

Hence $I_{dc} \geq I_{ac}$ peak

or
$$\frac{2V_m}{\pi R_L} \geq \frac{4V_m}{3\pi|Z_2|}$$

or
$$R_L = 1.5|Z_2| \tag{31}$$

for 2nd harmonic component.

$$|Z_2| = \left[2\omega L - \frac{1}{2\omega C}\right]$$

Normally the value of $X_L \gg X_C$, so value of $|Z_2| = 2\omega L_k$

Substituting this value of $|Z_2|$ in Eqn. (31), the value of load R_L under critical condition is

$R_L \leq 1.5(\omega L_k)$ or critical value of inductor $L_k \geq \dfrac{R_L}{3\omega}$. (31a)

If we assume frequency of applied input voltage to be 50 Hz, then $\omega = 100\,\pi$.

So, from Eqn. (31a),
$$L_k \geq \frac{R_L}{942}$$
$$\geq \frac{R_L}{1000} \tag{32}$$

If $f = 60$ Hz, then $L_k \geq \dfrac{R_L}{1100}$ (32a)

DC current flowing is $I_{dc} = I_k = \dfrac{2V_m}{3\pi\omega L_k}$ (33)

To keep I_{dc} low, we have to use a high value of inductor.

Bleeder resistance When applied load R_L is high, then current flow is quite small and may be less than I_k (Figure 4.18). In this situation, output DC voltage increases. To minimize this increase in voltage, we generally connect a high resistance called **bleeder resistance** R_B (Figure 4.20) across the given load. So, the effective value of load connected across the rectifier decreases and current flow increases, i.e. it is nearly equal to I_k thereby output voltage remains practically constant V_{dc} (Figure 4.18).

If the value of choke (inductor) is greater than L_k, bleeder resistance R_B is a must and it is kept across the capacitor (Figure 4.20) so a minimum current I_k will always be drawn even in the absence of load, and output voltage $V_o = \frac{2V_m}{\pi}$ will not jump to V_m, for this $R_B < 900\ L_{max}$.

4.6.4 π Filter or Capacitor Input *L–C* Filter

To keep output voltage E_{dc} nearly constant in case of a *L–C* input filter, the current drawn must be larger than critical current I_k. From Figure 4.18, it is to be noted that the value of V_{dc}, is quite less than the peak value of applied input voltage. To increase the value of V_{dc}, we employ another capacitor C_1 connected across the *L–C* combination (Figure 4.20). In this situation, the form of filter employed is like π and that is why it is called π filter.

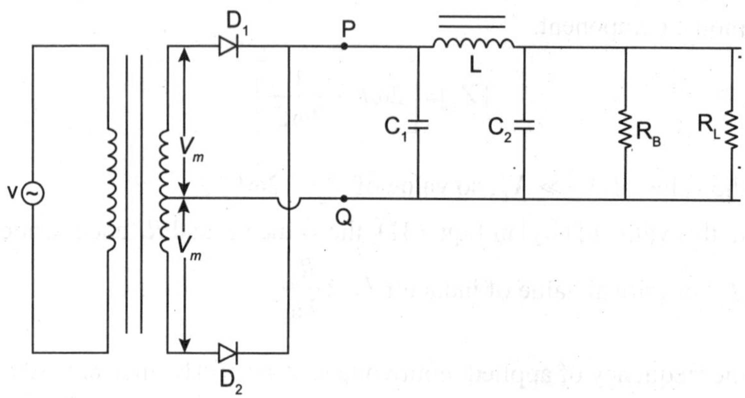

Figure 4.20 Full-wave rectifier with π filter

The above filter may be considered as made up of two sections:

(i) Shunt capacitor filter – made up of capacitor C_1

(ii) *L*-section filter – made up of inductor (choke) *L* and capacitor C_2

Whenever one of the diodes connected in the rectifier, say D_1, conducts then the voltage at *P* has a peak value of V_m; due to this capacitor C_1 also charges up to a maximum voltage of V_m.

When the value of rectified voltage pulse decreases (Figure 4.21), capacitor C_1 begins to discharge through the combination of *L* and R_L. This process of discharge of capacitor C_1 continues till capacitor C_1 again begins to get charged by the rising pulse of voltage due to the conduction of diode D_2 (Figure 4.21).

Finding the actual value of current due to discharge of C_1 is very difficult, but we can find its approximate value by assuming the shape of ripple voltage of capacitor charging and discharging as sawtooth.

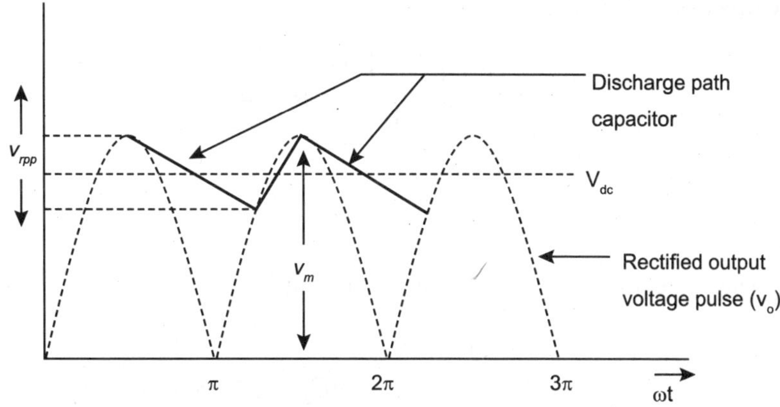

Figure 4.21

Employing Fourier analysis, sawtooth voltage pulse (due to charge and discharge of capacitor C_1) may be expressed as

$$v_r = \frac{V_{Rpp}}{\pi}\sin 2\omega t - \frac{V_{Rpp}}{2\pi}\sin 4\omega t \tag{34}$$

Since current related to higher harmonics flowing through the inductor is negligible, voltage related to 2nd harmonic is important.

$$\therefore\ v_r = \frac{V_R}{\pi}\sin 2\,\omega t \tag{35}$$

Reactance $\left[X_c = \dfrac{1}{2\omega C_2}\right]$ of capacitor C_2 at 2ω frequency is negligible in comparison with load R_L, so the entire ripple flows through capacitor C_2 only.

$$\therefore \text{ Ripple current, } i_{ac} = \frac{v_r}{\text{Reactance of } L \text{ and } C_2 \text{ combination}}$$

$$\text{or } i_{ac} = \frac{v_r}{2\omega L - \dfrac{1}{2\omega C_2}}$$

Substituting for v_r, we have

$$i_{ac} = \frac{\left(\dfrac{V_r}{\pi}\right)\sin 2\,\omega t}{2\omega L - \dfrac{1}{2\omega C_2}} = \frac{2V_r\,\omega C_2}{\pi[4\omega^2 L C_2 - 1]}\sin 2\omega t$$

The rms value of ripple current is

$$I'_{rms} = \frac{2V_r\,\omega C_2}{\pi[4\omega^2 LC_2 - 1]}\left\{\sqrt{\frac{1}{T}\int_0^T \sin^2 2\omega t\,dt}\right\}$$

$$= \frac{2V_r\omega C_2}{\sqrt{2}\,\pi[4\omega^2 LC_2 - 1]} \tag{36}$$

AC voltage across capacitor C_2 due to ripple current is

$$V_{ac} = I'_{rms}X_c = \frac{2V_r\omega C_2}{\sqrt{2}\pi[4\omega^2 LC_2 - 1}\times\frac{1}{2\omega C_2} = \frac{V_r}{\sqrt{2}\pi[4\omega^2 LC_2 - 1]} \tag{37}$$

This ripple voltage would also appear across load R_L since it is across capacitor C_2.

Peak value of ripple voltage $= \dfrac{V_m T}{2R_L C_1}$

$$= \frac{V_m\pi}{2\omega R_L C_1}$$

since $T = \dfrac{\pi}{\omega}$ for full wave.

DC voltage,

$$V_{dc} = V_m - \frac{V_{rpp}}{2} = V_m - \frac{V_m\pi}{2\omega R_L C_1}$$

$$= V_m\left[1 - \frac{\pi}{2\omega R_L C_1}\right] \text{ neglecting second term}$$

$$\approx V_m$$

$$\therefore \text{Ripple factor, } r = \frac{V_{ac}}{V_{dc}}$$

$$= \frac{V_r / \sqrt{2}\,\pi[4\omega^2 LC_2 - 1]}{V_m}$$

$$= \frac{\dfrac{V_m\pi}{2\omega R_L C_1[\sqrt{2}\,\pi(4\omega^2 LC_2 - 1)]}}{V_m}$$

$$= \frac{1}{\sqrt{2}\,\omega R_L C_1(4\omega^2 LC_2 - 1)} \tag{38}$$

The above discussion of the circuit of a fullwave rectifier equally applies to a bridge rectifier.

4.7 Power Supply

A rectifier together with a filter makes a power supply. Generally, a DC power supply should produce a DC output voltage which must remain constant under all circumstances. However, DC voltages derived from rectifier and filter circuits have some ripple voltage content and the output level is affected by (i) changes in input voltage (ii) load current drawn.

To minimize variation in output voltage due to the above reasons, we make use of regulating circuits. In a way, a good power supply consists of (i) rectifying circuit (ii) filter circuit (iii) regulating circuit.

Line regulation: It defines how DC output voltage varies when AC supply voltage (input voltage or line voltage) changes by a given amount.

Usually a change of ±10% in supply is employed.

$$\text{Line regulation} = \frac{\Delta V_0 \text{ for } 10\% \text{ chage in } V}{V_0} \times 100\%$$

Load regulation: It defines how power supply output voltage changes when load current changes by a given amount ΔI_L. Load current changes from zero to $I_{L(\max)}$.

$$\text{Load regulation} = \frac{(\Delta V_0 \text{ for } \Delta I_L = I_{\max}) \times 100\%}{V_0}$$

Often a power supply with a regulating circuit is called regulated power supply.

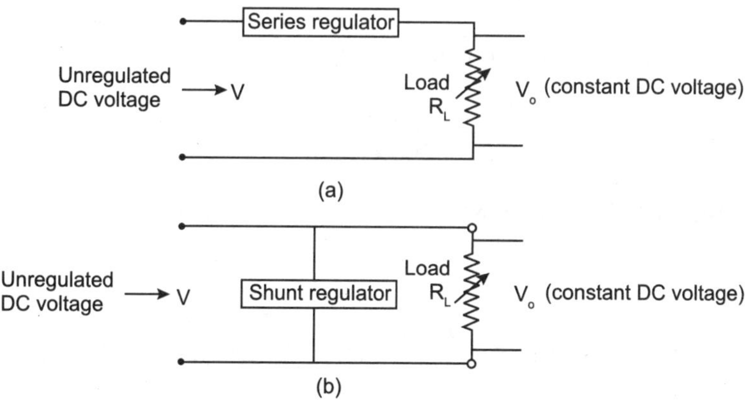

Figure 4.22

Regulating circuits are classified on the basis of their connection with the power supply. They are connected in two ways, (i) series (ii) parallel. Thus, we have (i) series regulator (ii) shunt regulator (Figure 4.22).

Example 9 In a fullwave rectifier, 0.2 A current flows through a load at 300 V, 60 Hz. Calculate ripple factor in case of shunt capacitor filter having 10 μF capacitor.

Solution

$$\text{Connected load} = R_L = \frac{V_{dc}}{I_{dc}} = \frac{\frac{2V_m}{\pi}}{0.2} = \frac{2\times 300\times\sqrt{2}}{3.14\times 0.2} = (955.6)\sqrt{2}\ \text{ohms}$$

$$\text{Ripple factor in case of shunt capacitor filter} = \frac{2410}{C\,R_L}$$

$$= \frac{2410}{10\times 955.6\sqrt{2}} = \frac{2410}{10\times 1337.8} = 0.18$$

Example 10 A 12.6 V centretapped transformer is used in a full wave rectifier with 100 μF capacitor to provide filtering for a 1 kΩ load. Assume $f = 60$ Hz. Determine: (a) percent ripple in the output, (b) DC output voltage.

Solution

Since transformer is centretapped, the R.M.S. value of voltage across half the secondary coil $= \dfrac{12.6}{2} = 6.3$ V.

Peak value of voltage, $V_m = 6.3\sqrt{2} = 8.82$ V

$$V_{dc} = \frac{V_m}{1+\frac{1}{4fR_LC}} = \frac{8.82}{1+\frac{1}{4\times 60\times 1000\times 100\times 10^{-6}}} = \frac{8.82}{1+\frac{1}{24}} = \frac{8.82}{1.04} = 8.55\,\text{V}$$

$$\text{Ripple with a capacitor filter} = \frac{2410}{C\,R_L} = \frac{2410}{100\times 1000} = 0.024 = 2.4\%$$

Example 11 Design a full rectifier with an L–C filter to provide 9 V DC at 100 mA with maximum ripple of 2%.

Solution

$$\text{Ripple factor with an } L\text{–}C \text{ filter, } r = \frac{0.83}{LC} \quad \text{where } L \to \text{Henry},\ C \to \mu\text{F}$$

$$\therefore LC = \frac{0.83}{\gamma} = \frac{0.83}{0.02} = 41.5 \approx 42$$

$$\text{Load connected to the filter, } R_L = \frac{V_{dc}}{I_{dc}} = \frac{9}{100\times 10^{-3}} = 90\ \Omega$$

$$\text{Critical value of inductor, } L_K \geq \frac{R_L}{900} = \frac{90}{900} = 0.1$$

$$\therefore C = \frac{LC}{L_K} = \frac{42}{0.1} = 420\,\mu F$$

4.7.1 Voltage Stabilization Using Zener Diode

All voltage regulating circuits use zener diode for voltage stabilization. As such we shall first discuss the function of zener diode.

When reverse-bias applied across the diode is equal to the breaking voltage or zener voltage V_Z (Figure 4.23), the current flowing through it suddenly increases but voltage across the diode remains practically constant, equal to V_Z.

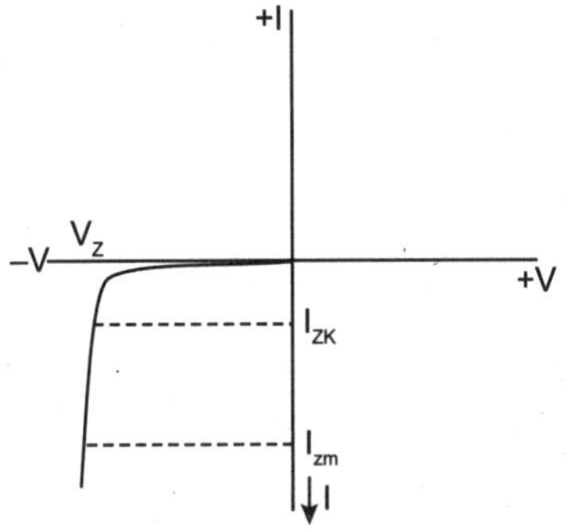

Figure 4.23

From diode characteristic, it is clear that if we keep the current flowing through diode between I_{ZK} and I_{ZM}, then voltage across the diode is nearly constant and equal to V_Z.

A voltage regulating circuit using zener diode is shown in Figure 4.24. Resistance R_S is connected in series.

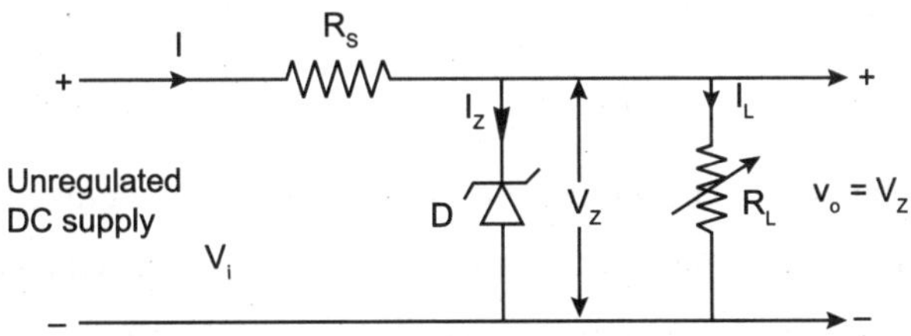

Figure 4.24 Voltage regulating circuit

Load R_L is connected across the diode. Zener Diode D is connected in the reverse bias condition, i.e. its N terminal (cathode) is connected to the positive terminal of unregulated DC supply through series resistance R_S.

If current flowing through the circuit is I and applied DC voltage is V_i, then

$$V_i = I\,R_S + V_Z \tag{39}$$

or $$V_Z = V_i - I R_S$$

and $$I = I_Z + I_L \tag{40}$$

We know that in an unregulated power supply, output voltage V_S changes with current I_L drawn by the load R_L connected across it. But in case of a voltage regulating circuit (Figure 4.24) using a Zener diode, voltage across load R_L remains $V_0 = V_Z$ although I_L may change. Any change in I_L is compensated by I_Z so that Eqn. (40) is satisfied.

If input voltage V_i (i.e. output from unregulated DC supply) changes by ΔV_i, then current in the circuit changes by ΔI. This change in current causes a change in Zener current, and load current I_L remains constant. Thereby voltage across R_L remains constant. Let the input voltage decrease by ΔV_i so current in the circuit also decrease by ΔI, and Eqn. (39) can be written as

$$(V_i - \Delta V_i) = (I_S - \Delta I_S)R_S + I_L \tag{41}$$

and $$(I_S - \Delta I_S) = (I_Z - \Delta I_Z) + I_L \tag{42}$$

since $$I_S = I_Z + I_L$$

So $$\Delta I_S = \Delta I_Z \tag{43}$$

When change in Zener current exceeds its limit by I_{ZM}, then voltage across Zener does not remain constant and it does not function as a voltage regulator.

Series resistor R_S is used to limit circuit current I, which finally governs the Zener current I_Z.

Let current flowing through Zener $I_Z = 0.2\, I_{ZM}$. Substituting this value in Eqn. (40),

$I = I_L + 0.2\, I_{ZM}$ where $I_{ZM} \rightarrow$ Maximum Zener current

$$R_S = \frac{V_i - V_Z}{I}$$

Substituting for I in Eqn. (39), we have

$$R_S = \left[\frac{V_i - V_Z}{I_L + 0.2\, I_{ZM}}\right] \tag{44}$$

The efficiency of the above circuit decreases with higher load current.

Also there is variation in voltage across the diode due to its resistance.

On account of these two reasons, such circuits are not in much use at present. To make voltage regulation more accurate, we may use two Zener diodes (Figure 4.25) instead of one.

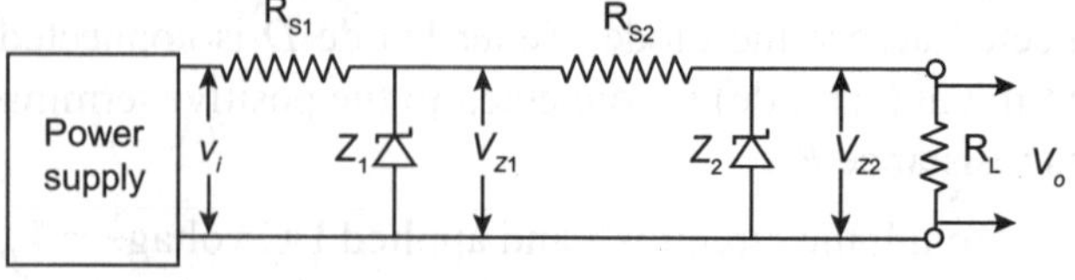

Figure 4.25

From the circuit, it is clear that there is very little change in V_0 since V_{Z2} is practically constant.

Here, the first Zener diode Z_1 acts as a preregulator circuit driving a second Zener diode Z_2 which acts as a regulator. Voltage V_{Z1} of first Zener diode is kept greater than the second diode voltage V_{Z2} and acts as input for the second diode Z_2. Thus input is much more regulated than V_S, so the final output voltage V_0 is extremely well-regulated.

4.7.2 Series Voltage Regulator Circuit

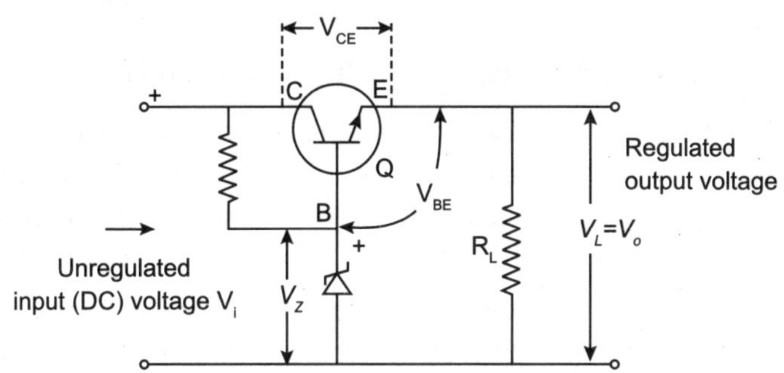

Figure 4.26 Series voltage regulator circuit

Consider the circuit shown in Figure 4.26, which is a very simple circuit to explain the function of voltage Regulation.

Here, transistor Q and load R_L are connected in series. The above circuit is that of a emitter follower. Transistor Q acts as a variable resistor, whose resistance depends upon base current.

From Figure (4.26), load voltage $V_0 = V_L$ is the difference of Zener voltage V_Z and V_{BE}, so

$$V_L = V_0 = V_Z - V_{BE}$$

or
$$V_{BE} = V_Z - V_L \tag{45}$$

Suppose applied load R_L increases thereby output load voltage increases. As load voltage V_L increases, the value of voltage drop V_{BE} decreases, thereby forward-bias across the transistor decreases and its conduction decreases. This leads to an increase in collector to emitter voltage V_{CE}. A decrease in voltage V_{CE} leads to a decrease in load current and it will nullify the increase in load voltage. In this way, voltage across the load remains constant and it is nearly equal to V_Z since the value of V_{BE} in Eqn. (45) is very small.

4.7.3 Shunt Voltage Regulator Circuit

Here the regulating transistor is in shunt with the output terminals of unregulated power supply.

Consider the working of circuit shown in Figure 4.27(a).

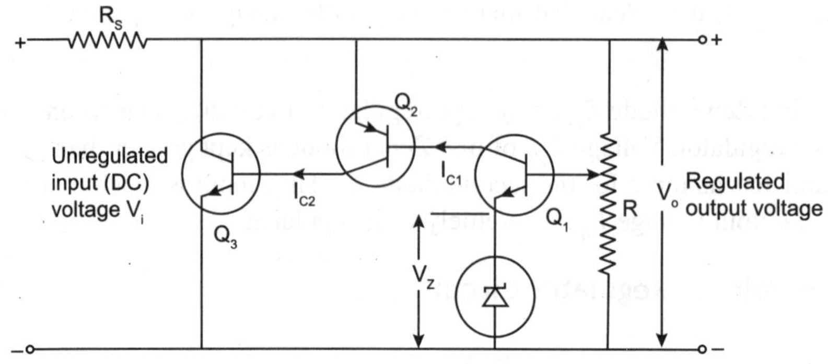

Figure 4.27 **(a)** Shunt voltage regulator circuit

Suppose output voltage V_0 increases, the base-emitter junction voltage of transistor Q_1 increases and it is more forward-biased, thereby its collector current I_{C1} increases. This collector current is also the base current I_B of transistor Q_2. An increase in base current Q_2 increases its collector current I_{C2} which is the amplified version of base current I_B or collector current I_{C1}.

Collector current I_{C2} of Q_2 is also the base current of Q_3. Any enhancement of base current of Q_3 increases its collector current. Thus, any increase in output voltage V_0 causes an increase in collector current of transistor Q_3, this ultimately increases line current and voltage drop across R_S, reducing the output voltage V_0 which is thus stabilized. Transistor Q_3 has to handle a large collector current, so it should be power transistor.

Another version of shunt voltage regulated circuit is shown in Figure 4.27(b).

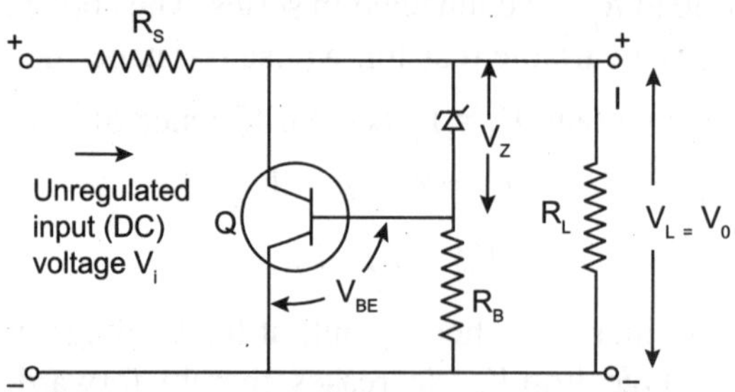

Figure 4.27 **(b)** Shunt voltage regulator circuit

From the circuit, it is clear that output voltage $V_0 = V_L$ is the sum of Zener voltage (V_Z) and base emitter voltage V_{BE} of transistor Q.

$$V_0 = V_L = V_Z + V_{BE}$$

or

$$V_{BE} = V_L - V_Z \quad \text{(i)}$$

Since voltage V_Z across a Zener diode is practically constant, any increase or decrease in output voltage or load voltage will have to be compensated by base to emitter voltage V_{BE}.

Suppose that unregulated input DC voltage V_i increases and because of this output voltage or voltage across the load is affected, consequently V_{BE} increases. Any increase in V_{BE} causes the base current I_B to increase and as a result collector current I_C increases, which means input circuit current I_S increases thereby voltage drop across the series resistor R_S increases. The increase in voltage drop V_{RS} across R_S causes load voltage V_L to decrease since the sum of two voltages is equal to input voltage V_i at all times.

$$V_i = V_{RS} + V_L$$

or

$$V_L = V_i - V_{RS}$$

The above discussion may be summarized as given below:

$$V_i \uparrow \rightarrow V_L \uparrow \rightarrow V_{BE} \uparrow \rightarrow I_C \rightarrow I_S \rightarrow V_{RS} \uparrow \rightarrow V_L \downarrow$$

Here, arrow up means increase and arrow down means decrease. So we find that the increase or decrease in load voltage is nullified.

This type of circuit is suitable for the load through which current varies from a minimum value to a finite value.

Presently monolithic voltage regular circuits are available which are fabricated on a single chip. Such chips are small in size, low cost, and have high performance. Hence they are easy to be used along with power supply. Moreover, they are available for a wider voltage range. The setup of such a chip with power supply is shown in Figure 4.27(c).

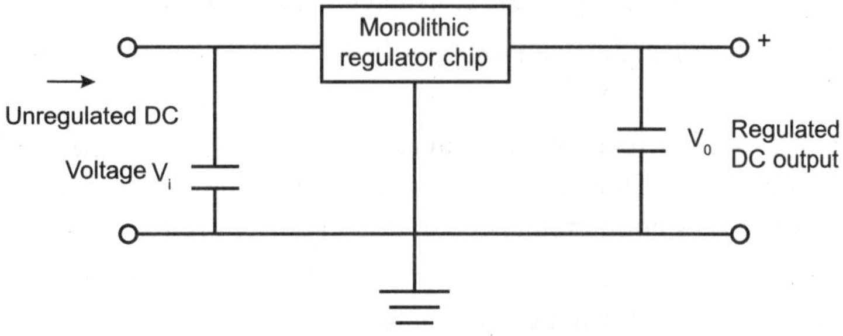

Figure 4.27(c)

4.7.4 Current Regulator

So far, we have discussed voltage regulating circuits which provide a constant DC voltage. In some applications, we require a constant current, and the circuit which gives a constant current irrespective of load resistance is referred to as current regulator.

A simple circuit of a current regulator is shown in Figure 4.28.

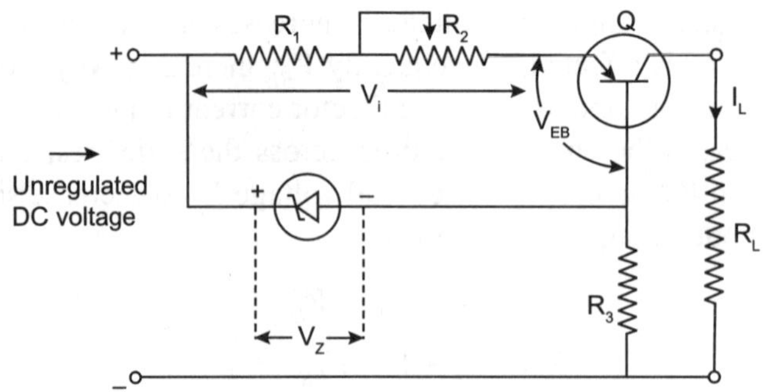

Figure 4.28 Series current regulator circuit

Load current I_L is carried by transistor Q, therefore the circuit is a series current regulator. If load current I_L changes due to negative feedback in the circuit, the internal resistance of transistor is adjusted to counter the change in load current.

Applying KVL to the circuit, we have

Zener voltage, $$V_Z = V_1 + V_{EB} \tag{ii}$$

or $$V_{EB} = V_Z - V_1$$

where V_{EB} is emitter base voltage of transistor Q, V_1 is voltage drop across resistance R_1 and R_2 combination.

Suppose load current I_L increases due to change in R_L. I_L is the collector current of transistor Q; so an increase in I_L will increase the emitter current of Q. An increase in emitter current causes a increase in voltage drop V_1 across resistance R_1, R_2 combination.

As per Eqn. (ii), V_Z is constant so an increase in V_1 must be counterbalanced by a decrease in V_{EB}.

A decrease in V_{EB} will cause a reduction in collector current of transistor Q. This collector current is also load current I_L. Thus, a decrease in V_{EB} causes a reduction in I_L. In this way, the increase in load current I_L is counterbalanced and load current remains constant, which is the aim of the circuit.

Resistance R_2 is made variable to keep the output current at a desired level. The circuit then maintains current at this level as load resistance R_L changes. Resistance R_3 connected to the base of transistor Q gives the current for Zener diode operation. A sufficient current must be carried by R_3 for the Zener diode to remain in operation as R_L changes. Load current remains fairly constant till load resistance R_L becomes so large that voltage drop across it equals voltage drop across R_3.

4.7.5 Power Supply

Thus to get a constant DC voltage we have to employ a full wave rectifier with a filter section and a voltage regulating circuit within. The composite equipment consisting of the above sections is called a power supply.

Example 12 An unloaded Zener regulator has a source voltage of 20 V, a series resistance of 330 Ω and a Zener diode of Zener voltage 12 V. What is its Zener current?

Solution

Voltage across series resistor is 20 – 12 = 8 V

$$\text{Current through series resistor} = \frac{\text{Voltage across resistor}}{\text{Resistance}}$$

$$= \frac{8}{330} = 24.2 \text{ mA}$$

Since Zener diode is in series with resistor, current through it is equal to current flowing through resistor, i.e. 24.2 mA.

Example 13

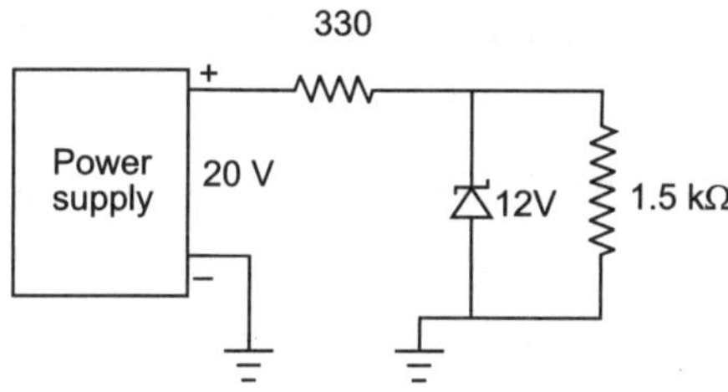

Figure 4.29

Calculate (i) series current through resistor 330 Ω (ii) zener current (iii) load current and (iv) respective wattage of elements used.

Solution

Voltage across series resistor = 20 – 12 = 8 V

(i) Current through series resistor $I_S = \dfrac{\text{Voltage across resistor}}{\text{Resistance}}$

$$I_S = \frac{8}{330} = 24.2 \text{mA}$$

(ii) Current through series load $I_L = \dfrac{\text{Voltage across load}}{\text{Resistance}} = \dfrac{12}{1500}$

$$I_L = 8 \text{mA}$$

(iii) Current through zener diode, $I_Z = I_S - I_L$

$$= (24.2 - 8) = 16.2 \text{ mA}$$

(iv) Respective wattage of elements used:

(a) Series resistor $\Rightarrow W = V_S \times I_S$

$$= 8 \times 24.2 \times 10^{-3} = 193.6 \text{ mW}$$

(b) Zener diode $\Rightarrow W = V_Z \times I_Z$

$$= 12 \times 16.2 \times 10^{-3} = 194.4 \text{ mW}$$

(c) Load resistor $\Rightarrow W = V_L \times I_L$

$$= 12 \times 8 \times 10^{-3} = 96 \text{ mW}$$

Example 14

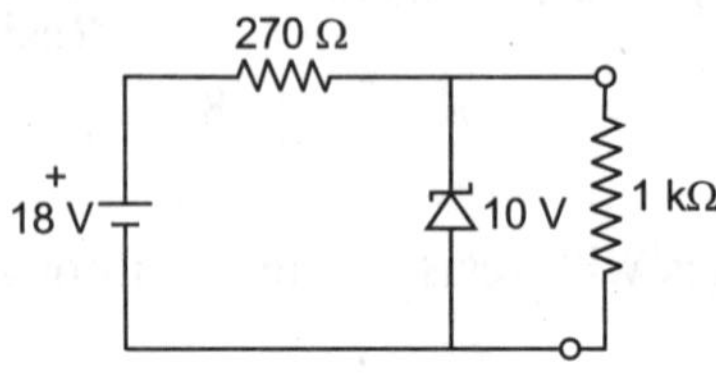

Figure 4.30

Is the Zener diode in the circuit operating in the breakdown region?

Solution

Applying Thevenin's theorem, Thevenin voltage across the Zener diode,

$$V_{Th} = \frac{R_L}{R_L + R_S} V_S$$

$$= \frac{1000}{1000 + 270} 18 = 14.2 \text{ V}$$

Thus, V_{Th} is greater than V_Z (Zener voltage), i.e. 14.2 > 10. So the Zener diode is operating in the breakdown voltage.

Example 15 Design a voltage regulator circuit using a Zener diode to have output voltage of 15 V when load current varies from 10 to 20 mA. Assume input voltage to be 20 V, and Zener diode having Zener voltage 15 V and $I_{ZT} = 6$ mA.

Solution

$$\text{Average load current, } I_L = \frac{10 + 20}{2} = 15 \text{mA}$$

$$\text{Total current entering the circuit} = I_S = I_L + I_{ZT}$$

$$= 15 + 6 = 21 \text{ mA}$$

$$\text{Series resistor, } R_S = \frac{V_{in} - V_Z}{I_{in}} = \frac{20 - 15}{21 \times 10^{-3}} = 238 \ \Omega$$

$$\text{Power rating of resistor} = \frac{V_S^2}{R_S} = \frac{(5)^2}{238} = 0.1 \text{ W}$$

Fuse current Every electrical/electronic device must be provided with a fuse for safety of the device.

To, determine the current carrying capacity of the fuse, in a power supply we need to calculate the primary current since it determines the adequacy of fuse.

For determining (fuse current) primary current, we assume that input power equals output power. The following example will make things clear.

Let the specification of power supply be 15 V, 1.2 A. So, output power

$$P_{out} = (15V)[1.2A] = 18W$$

Ignoring power loss in the rectifier and transformer

$$P_{in} = P_{out}$$
$$= 18 \text{ W}$$

Let input supply voltage or line voltage be 220 V.

$$\therefore \text{ Primary current} = \frac{P_{in}}{\text{Input voltage}} = \frac{18}{220}$$
$$= 0.082 \text{ A}$$
$$= 82 \text{ mA}$$

The actual primary current will be higher than this and it is nearly 5 – 20% more than it.

The required primary current is 86 mA (assuming 20% higher).

Hence fuse of nearly 100 mA or 0.1 A is adequate.

If reverse current is to be avoided or minimized, the diode must switch off relatively slowly.

Thus, if input frequency is such that

$$\frac{1}{f} = T = 2t_{rr}$$

Then the diode is conducting almost as much in reverse as it is in forward direction. In such a situation, it is no longer behaving as one-way device (switch). To minimize the effect of reverse current, the time period of input frequency should be

$$\frac{1}{f} = T = 10t_{rr}; \; t_{rr} - \text{reverse recovery time of diode}$$

Example (a) For 1N917 diode t_{rr} = 10 ns.

So, input frequency f must not be greater than

$$= \frac{1}{10(t_{rr})}$$

$$= \frac{1}{10 \times 10 \times 10^{-9}} = 10 \text{ MHz}$$

(b) For 1N917 diode, $t_{rr} = 3$ ns.

So, input frequency f must not be greater than

$$= \frac{1}{10(t_{rr})} = \frac{1}{10 \times 3 \times 10^{-9}}$$

$$= 33\,\text{MHz}$$

4.7.6 Diode Equivalent Circuit and Frequency Response

In the previous discussion of diode as rectifying device, it has been assumed that the diode conducts when it is forward-biased and does not conduct when reverse-biased i.e. it behaves like a switch. But in actual practice, it is not the case as there is a limit for input signal frequency (time period).

4.7.7 Diode Equivalent Circuit

The depletion layer of *p-n* junction (Chapter 3) is a region depleted of charge carriers. So, we can assume a dielectric medium situated between two blocks *p* and *n* of low resistance and it behaves like a capacitor. Depletion layer capacitance is designated as C_{pn} (Figure 4.31(a)) and its value may be calculated from the usual equation for parallel plate capacitor. Its typical value is 40 pF. Depletion layer capacitance is normally considered when the junction is reverse-biased.

A reverse-biased *p-n* junction can be simply represented by a reverse resistance R_r having very high value (but varying) in parallel with depletion layer capacitance C_{pn}.

Consider a forward-biased diode carrying current I_f. If applied voltage V_F is suddenly reversed, current I_f ceases immediately, leaving some majority carriers in the depletion region. These charge carriers must flow back out of the depletion region, which is widened since it is reverse-biased. The result is that when a forward-biased diode is suddenly reversed biased, a reverse current flows which is large initially and slowly decreases to a very small value equal to reverse saturation current I_r of the order of microampere. This effect may be likened to the discharge of a capacitor, so it is represented by a capacitor of known diffusion capacitance C_d. Its typical value is 0.02 μF.

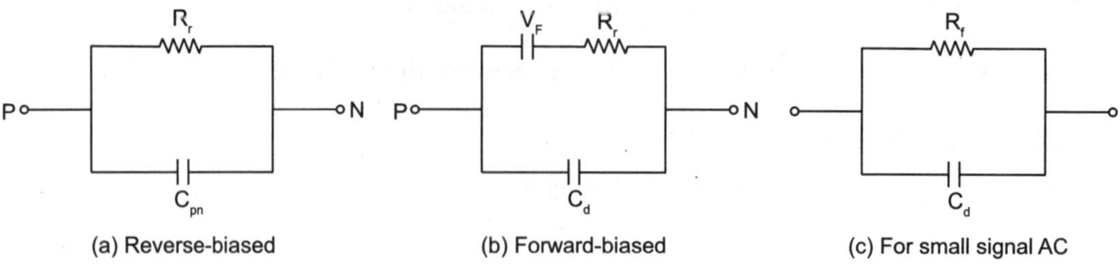

Figure 4.31 Equivalent circuit for diode

The equivalent circuit for a forward-biased diode is represented by dynamic resistance R_f along with battery V_F and in parallel with diffusion capacitance C_d (Figure 4.31 (b)).

The above circuit of a diode may be modified for a small signal AC equivalent circuit (Figure 4.31(c)).

There are many circuit applications for diode other than rectification. Some of these require the diode to switch very rapidly from forward to reverse bias and vice versa. Most diodes will switch rapidly into forward-biased condition however there is always a longer switch off time due to diffusion capacitance.

The switch off time, designated as reverse recovery time t_{rr}, limits the maximum frequency at which the device can be operated. This reverse recovery time is the time required for forward current to decrease to reverse saturation current. A typical value of t_{rr} is 4–50 ns.

4.7.8 Limitation of *P–N* Junction Diode

As the frequency of input signal in a rectifier increases, the action of small-signal rectifier diodes begins to detoriate. They are no longer able to switch off fast enough to produce a well-defined half wave signal. Figure 4.32 illustrates the energy bands of a forward-biased *p-n* junction diode.

The free electrons lying in the conduction band of *n*-type region diffuse across the junction and enter into the *p* region before combining with holes (path A). Similarly, holes of *p* region have crossed the junction and diffused into *n* region before recombination (path B).

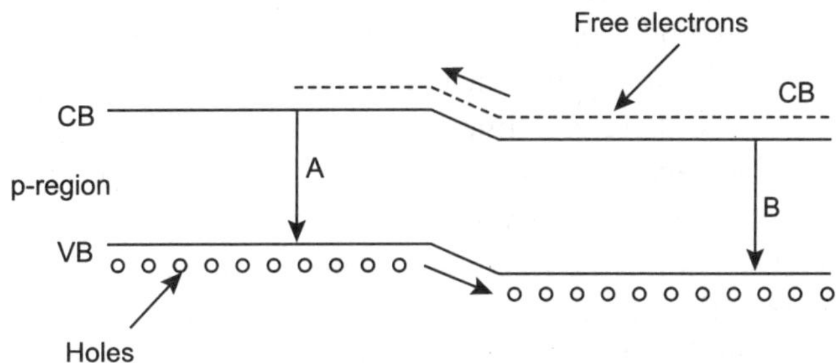

Figure 4.32

The greater the life time, the farther the charge carriers enter into the respective region, before recombination; for instance electrons after entering into *p* region remain temporarily stored in higher energy band (CB of *p*-type region). Similarly, holes penetrate deeply into *n* region and are stored temporarily in lower energy band (VB of *n*-type region). This temporary storage of free electrons in upper energy band and holes in lower energy band is known as charge storage. On applying an AC voltage across the junction we try to switch a diode from ON (forward-bias) to OFF (reverse-bias); these stored charge carriers create a problem, because if we suddenly reverse bias a diode, the stored charges flow in the reverse direction for a while. The greater the life time, the longer these charges can contribute to reverse current. It can exist for a while as shown in Figure 4.33.

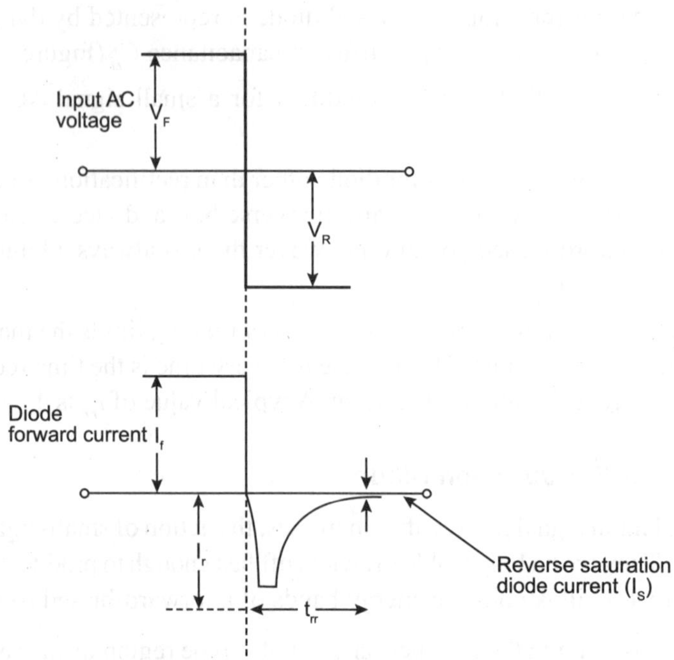

Figure 4.33

We find that on applying AC voltage across a diode, when the pulse goes negative, instead of switching off sharply, the diode conducts in reverse direction as shown in Figure 4.33. When reverse current I_r exists and initially it is nearly equal to I_F but gradually falls off to reverse saturation current level I_S. So there is a time delay before the reverse-biased diode attains the value I_S. This time is called reverse recovery time 't_{rr}' and related to diffusion capacitance C_d of diode junction. It is of the order of nanoseconds and typical values range from 4 to 50 ns. Switching time obliviously limits the maximum operating frequency of the diode. If reverse current is to be avoided or minimized, the diode must be switched off slowly. If input frequency be such that its time period $T = 2\ t_{rr}$, then the diode is conducting almost as much in reverse direction as in forward direction. In this case it is no longer a one-way device.

To minimize the effect of reverse current, the time period of operating (applied) signal should be ten times the t_{rr} value.

$$T > 10t_{rr}$$

We find that we do not notice reverse recovery time because it is so short in case of ordinary diode that we do not notice it at applied signal frequencies below 10 MHz. It is only noticeable when applied signal frequency is greater than 10 MHz. The rectified output current or voltage in case of half wave rectifier is as shown in Figure 4.34, where we observe a tail during negative input (half) cycle and non-zero voltage or current as shown in Figure 4.34.

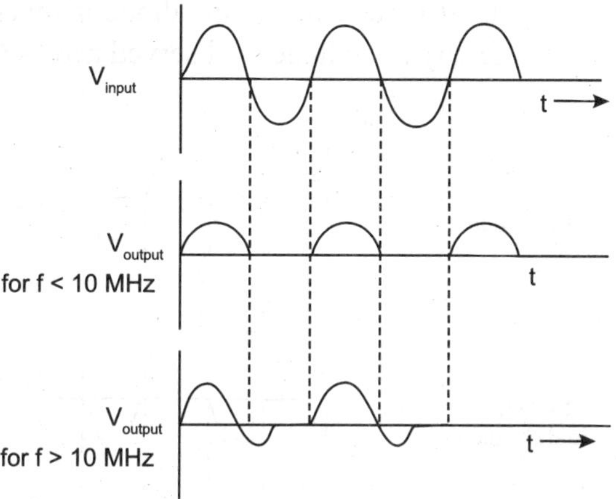

Figure 4.34

The problem of tail current in a rectifier at applied high frequencies is solved by special-purpose diode called Schottky diode.

4.8 Clipping and Clamping Circuits

Clipping circuit Such circuits are used to clip either the positive half or negative half or both of input voltage at certain level. They are of four types:

(i) Positive clipper
(ii) Negative clipper
(iii) Biased clipper
(iv) Both positive and negative clipper

4.8.1 Positive Clipper

It may be divided into two groups: (a) series (b) shunt. The circuit for series position clipper is shown in (Figure 4.35).

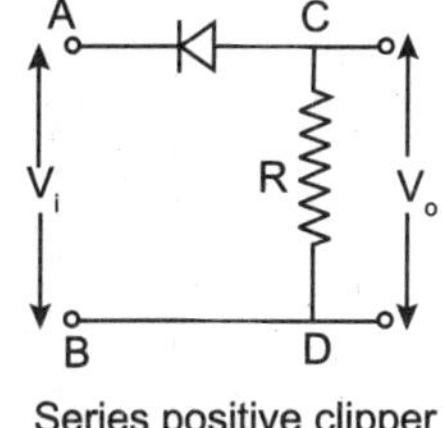

Series positive clipper

Figure 4.35

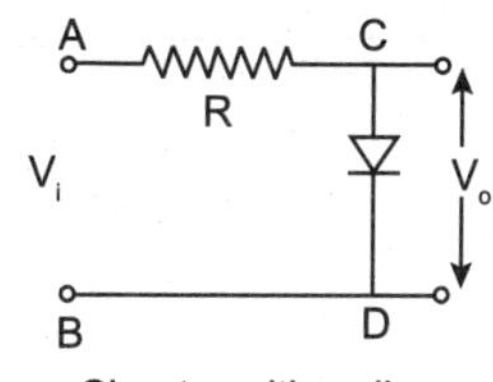

Shunt positive clipper

Figure 4.36

Suppose the input voltage V_i is sinusoidal and applied across the terminals *A*, *B* and output voltage V_0 is studied across terminals *C*, *D*.

During the positive half cycle of input voltage, the diode is reverse-biased so it does not conduct under ideal condition, thereby no voltage is observed across *C*, *D* (Figure 4.37(b)).

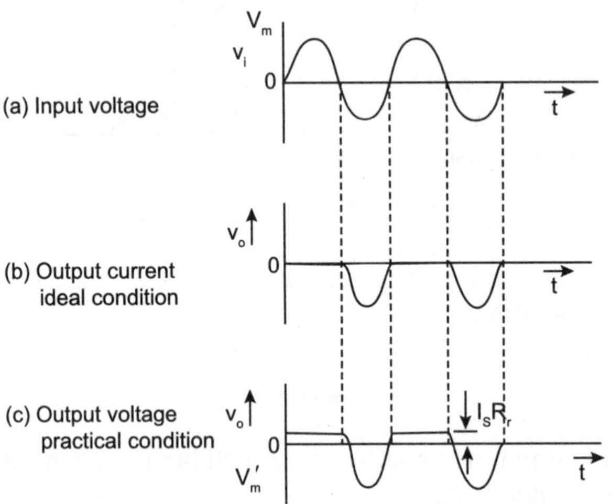

Figure 4.37 Wave shape positive clipper

During the negative half cycle of input voltage, the diode is forward-biased so current flows through it and there is current through *R*. Hence a negative voltage wave appears across resistor *R* (Figure 4.37(b)).

In actual practice, during positive half cycle when a diode is reverse-biased some current I_S (saturated value) flows through it thereby there is voltage drop $V_0 = I_S R$ across resistor *R* as shown in Figure 4.37(c). Thus, we find the positive half cycle is clipped by such a circuit.

Similar is the situation with a shunt clipper (Figure 4.36).

4.8.2 Negative Clipper

It may be divided into two groups: (a) series (b) shunt. The circuit for series negative clipper is shown in (Figure 4.38(a)).

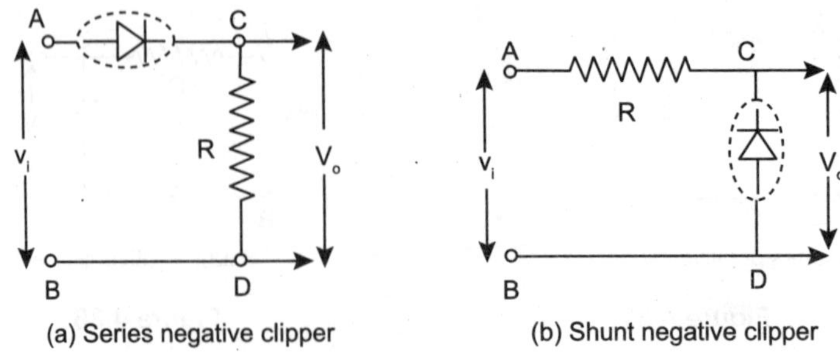

Figure 4.38

Suppose input voltage v_i is sinusoidal and it is applied across terminals *A* and *B* (Figure 4.38) and the output is studied across terminals *C* and *D*.

During the positive half cycle of input voltage, the diode is forward-biased so it conducts and current flows through resistor *R*, thereby there is voltage drop across it and we observe a positive voltage across *R*. (Figure 4.39(b)).

During the negative half cycle, the diode is reverse-biased and it does not conduct under ideal condition thereby no current flows through resistor *R* so voltage drop across it is zero (Figure 4.39(b)).

In actual practice, when a diode is reverse-biased some current I_S (saturated value) flows through it and thereby there is voltage drop I_SR across resistor *R* as shown in Figure 4.39(c).

Similar is the situation with a shunt clipper (Figure 4.38(b)).

We find that either the positive cycle or negative cycle is completely clipped. However, if we desire to clip them at a particular level, we use another circuit called biased clipping circuit.

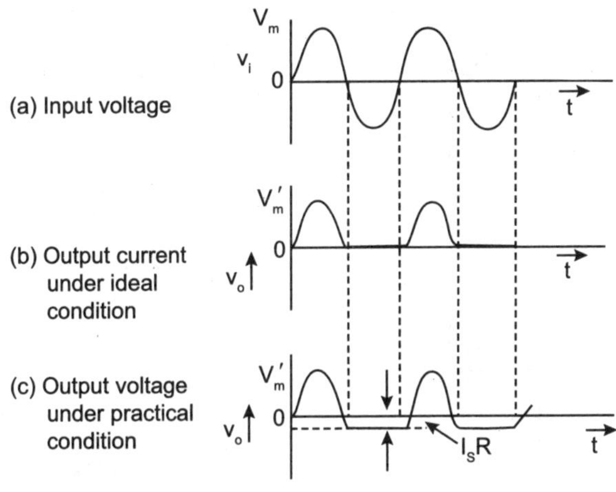

Figure 4.39

4.8.3 Biased Clipping Circuit

(a) Positive clipper

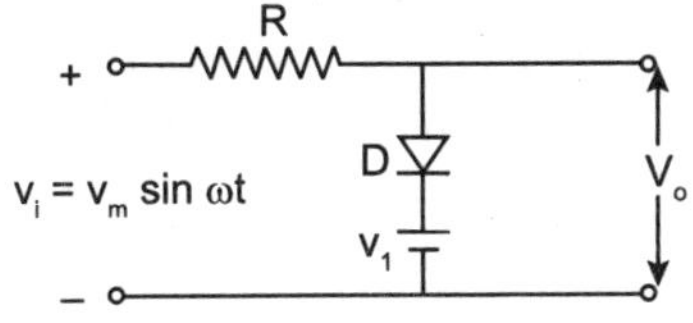

Figure 4.40 Biased positive clipper

The circuit for Biased Positive clipper is shown in (Figure (4.40)).

Here we have a positive bias (battery) voltage $V_1 < V_m$ (peak value of input sinusoidal voltage) connected in series with the diode so that the diode is reverse-biased.

The diode D will not conduct during the positive half of input signal till its instantaneous value $v_i < V_1 + V_r$ where V_r is the cutin voltage of diode D. If $V_r \ll V_1$ (as is the usual case), then we can neglect it in comparison with V_1. Output voltage v_o is equal to input voltage V_1 till the diode does not conduct so till this stage the transfer characteristic of diode is a straight line AOB passing through the origin as shown in Figure 4.41. The slope of this line called transfer characteristic is unity.

When v_i is greater than V_1 (biasing voltage), the diode conducts and behaves like a battery of voltage V_r having internal resistance R_f (forward-biased resistance of diode).

Let δv_i be the increment in input voltage v_i. The corresponding increment in output voltage v_o is given by

$$\delta v_o = R_f \frac{\delta v_i}{R_f + R}$$

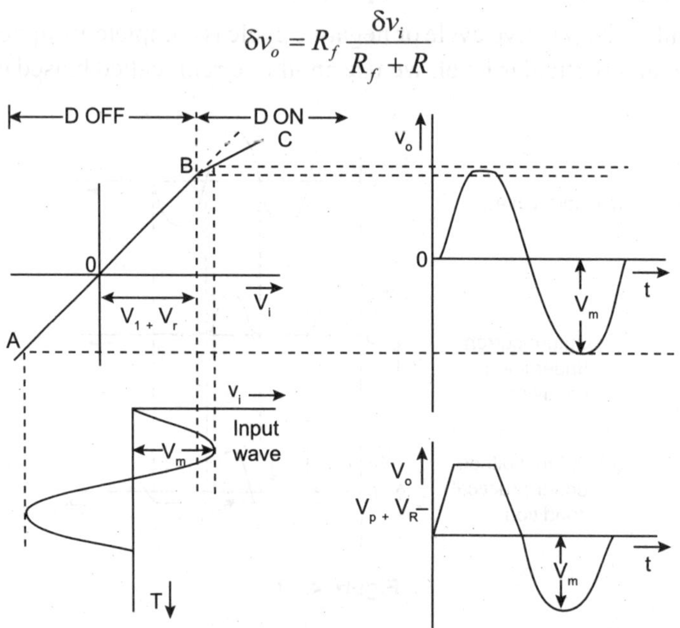

Figure 4.41

As a result, the transfer characteristic slope is no longer unity and given by ($R_f/R_f + R$).

It is clear from Figure 4.41 that the positive half of cycle output is no longer sinusoidal but rather its peak is suppressed. If $R \gg R_f$, then positive peak is sharply suppressed or clipped at voltage $V_R + V_r$.

(b) Negative clipper

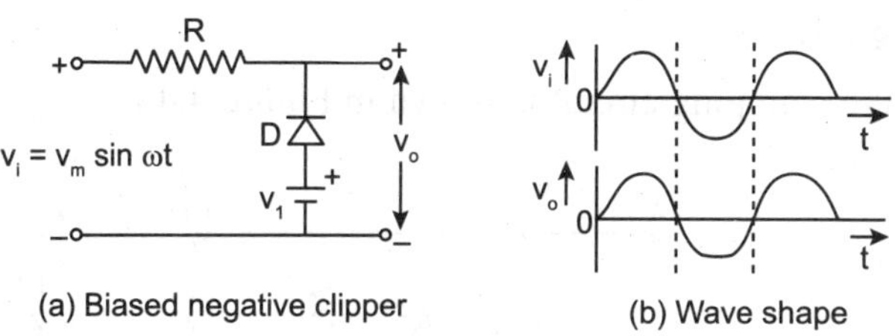

(a) Biased negative clipper (b) Wave shape

Figure 4.42

Figure 4.42 shows a biased negative clipper. Here diode D is forward-biased due to battery V_1, whereas it was reversed-biased in positive clipper. When $v_i > V_1 - V_r$, the diode is OFF but it is ON when $v_i < V_1 - V_r$.

When the diode is off, the input cycle appears across output terminals, i.e. positive half cycle appears, but when the diode is ON the output voltage is just constant and equal to $V_1 - V_r$ as shown in Figure 4.42.

(c) Combination biased clipper

In such a circuit, both positive and negative half cycle of input voltage are clipped at a given voltage.

The circuit for a combination biased clipper is shown in Figure 4.43.

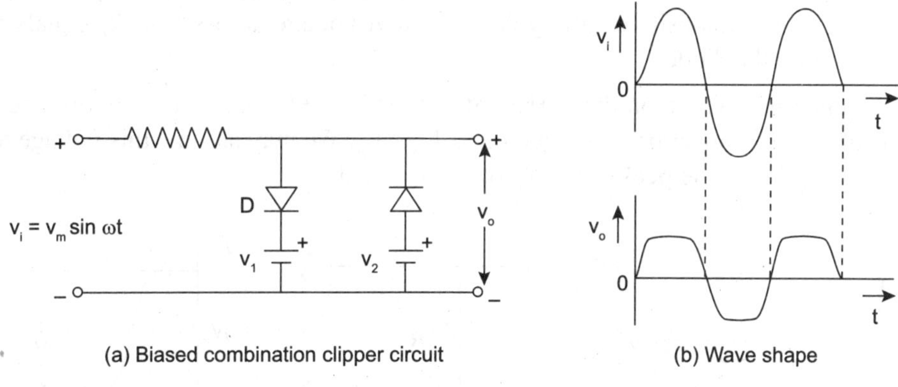

(a) Biased combination clipper circuit (b) Wave shape

Figure 4.43

4.8.4 Clamping Circuits

Unlike clipper circuits these circuits, do not change the shape of an input signal (wave) but they change its DC level. They are also known as DC restorers.

These circuits are of two types:

(a) Positive clamping circuit

The circuit of a positive clamping circuit is shown in Figure 4.44.

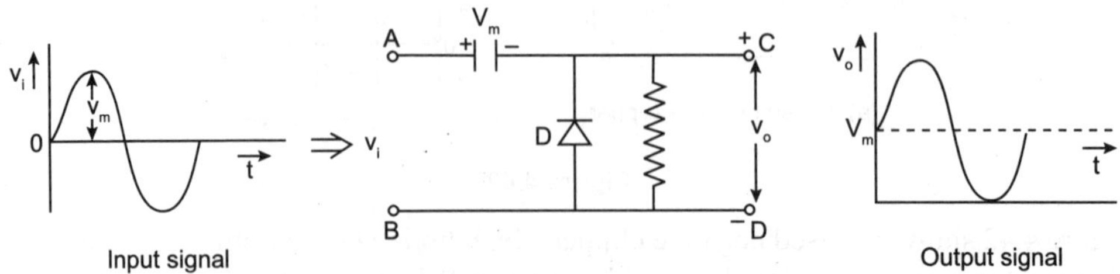

Figure 4.44 Positive clamping circuit

The input signal ($v_i = V_m$ sin ωt) is applied across terminals *A* and *B*.

During its positive cycle, no current flows through the diode since it is reverse-biased but capacitor gets charged up to the peak value V_m of input signal with polarity as shown.

(b) Negative clamping circuit

The circuit of a negative clamping circuit is shown in Figure 4.45.

The input signal ($V_i = V_m$ sin ωt) is applied across terminals *A* and *B*.

During its positive cycle, the diode is forward-biased and capacitor gets charged with polarity as shown; so during the positive half cycle the output voltage across load R_L equals forward voltage $+V_F$ across the diode.

At this time, the voltage on the R.H.S. of capacitor is $+V_F$, while on the left side is $+V_m$. Strictly speaking, the capacitor is charged up to $V_m - V_F$. We may assume this voltage to be V_m if V_F is quite small than the peak value V_m of input signal.

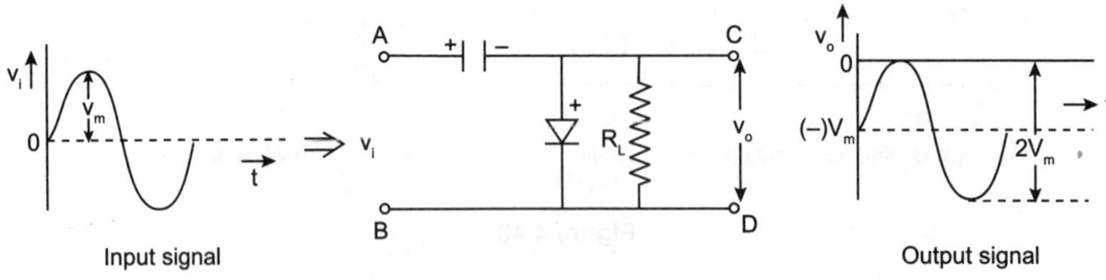

Figure 4.45 Negative clamping circuit

When the input cycle goes negative, the diode is reverse-biased and has no further effect on capacitor voltage.

Also R_L is assumed to be high so the capacitor cannot discharge very much through it during the negative (or positive) cycle of input signal.

While the input signal is negative, output voltage v_o is the sum of V_m peak values of input voltage, and capacitor voltage is assumed to be V_m, i.e. $v_o = -V_m - V_m = (-)2V_m$.

Since the polarity of capacitor voltage is the same as the (negative) input, the output is a negative voltage which is larger than the peak input level as shown in figure.

From the figure, it is clear that peak-to-peak voltage levels:

$$\text{p-to-p output} = V_F - [-(2V_m - V_F)]$$
$$= 2V_m$$

It is seen that the amplitude of output wave form from negative voltage clamper circuits is exactly the same as that of the input signal.

The function of R_L is to gradually discharge the capacitor over several cycles of input signal.

The difference between clamping and clipping circuit is now apparent. While the clipper clips off unwanted portion of the input (signal) waveform, the clamper simply clamps the maximum positive or negative peak of the input to desired DC level.

4.8.5 Voltage Multipliers

It is an electrical device which, making use of diodes and capacitor raises (multiplies) the low AC voltage to a high DC voltage. Like rectifier circuits, these form two groups: (i) half wave multiplier (ii) full wave multiplier.

Under each group, we can have three types of multipliers:

(i) Doublers

(ii) Triplers

(iii) Quadruplers

The above classification is according to the ratio of output to input voltage. The function of a multiplier circuit is similar to that of a battery connected in series.

(a) Half wave voltage doubler

Half wave voltage doubling circuit is shown in Figure (4.46) where two diodes D_1 and D_2 along with capacitor C_1 and C_2 have been used.

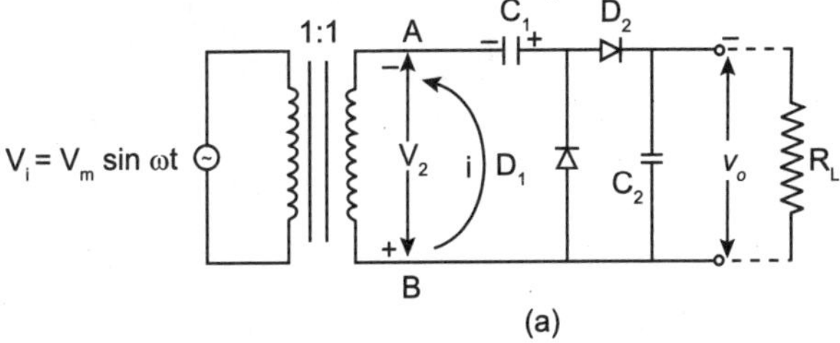

(a)

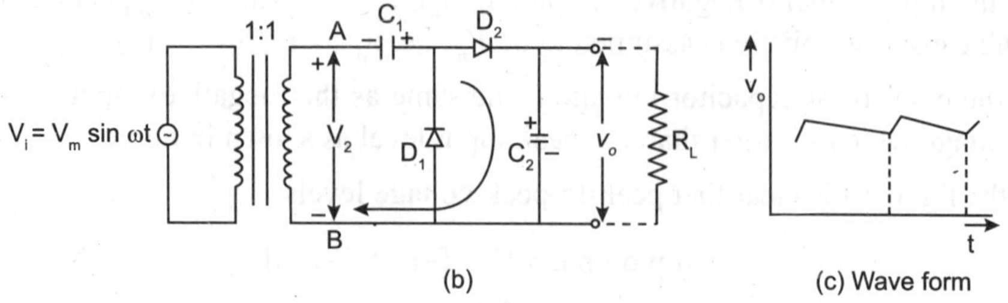

(b)

(c) Wave form

Figure 4.46 Half wave voltage doubler

Working Let the input voltage be $V_i = V_m$ sin ωt which is connected across the primary of the transformer. The peak voltage across the secondary is $V_2 = V_m$. Suppose the peak value voltage is 100 V.

During the first half positive cycle of input sinusoidal voltage the upper terminal of secondary of transformer is negative there by diode D_1 is forward-biased so it conducts as shown in Figure 4.46(a). As current flows capacitor C_1 gets charged upto peak value V_2 (= V_m = 100 V). The polarity of voltage across C_1 is as shown.

During second half negative cycle of input voltage the polarity of voltage across secondary of transformer is as shown in Figure 4.46(b). The diode D_1 is reverse-biased but diode D_2 is forward-biased and it conducts charging capacitor C_2 upto the peak value of voltage. After one full cycle of input voltage the peak value of output voltage V_0 is [voltage across C_1 + voltage across secondary] $(V_2 + V_2) = 2V_2$ (= 200 V) and the circuit behaves like two batteries connected in series, each having voltage V_m (100 V). Total ouput voltage is 2 V_m which is double of the input voltage (peak value). Thus the circuit behaves like a voltage doubler.

(b) Half wave voltage tripler

The circuit of half wave voltage tripler is shown in Figure 4.47 where three diodes D_1, D_2 and D_3 along with three capacitors C_1, C_2 and C_3 have been used.

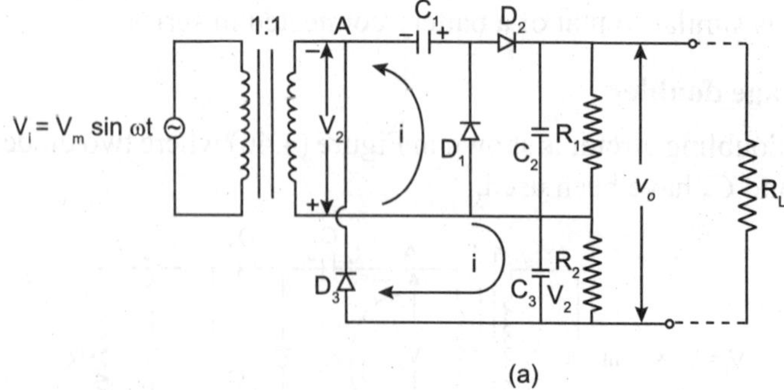

(a)

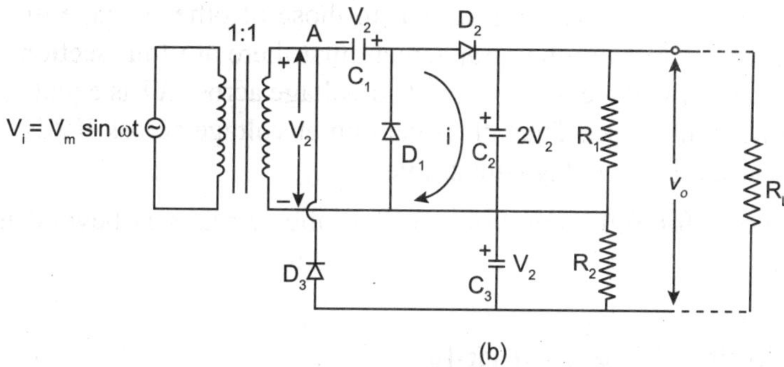

(b)

Figure 4.47 Half wave voltage tripler

Working Let the input voltage be $V_i = V_m \sin \omega t$, which is connected across the primary of transformer. The peak value of voltage across secondary is $V_2 = V_m$ (= 100 V). During the positive half cycle of input voltage, the upper terminal of secondary is negative. The diode D_1 is forward biased, so also is the case for diode D_3. Current flows through them. Due to the conduction of diode D_1 and D_3, capacitor C_1 and C_3 each gets charged up to peak value of voltage across secondary i.e. $V_2 = V_m$. The polarity of voltages across C_1 and C_3 is as shown in (Figure 4.47(a)).

During negative half cycle of input voltage the upper terminal of secondary is positive (Figure 4.47(b)). In this situation diode D_2 is forward biased (but diodes D_1 and D_3 are reverse biased) so it conducts and current flows along the path $AC_1D_2C_2B$ as shown, there by capacitor C_2 gets charged up to the peak value V_2 (= V_m). Voltage across secondary and capacitor C_1 are in series.

Thus, after one full cycle of input signal, the peak value of voltage across the two capacitors C_2 and C_3 is $2V_2$ and V_2 respectively. Thus, voltage across the combination of capacitors is $(2V_2 + V_2) = 3V_2$. This is also the value of output voltage V_0. Thus the circuit behaves like a voltage tripler. The ratio of two resistors R_1 and R_2 connected across the capacitors is kept in the ratio of voltage across capacitor C_2 and C_3 respectively i.e. it must be 2 : 1.

(c) Half wave voltage quadrupler

The circuit for half wave quadrupler is as shown in Figure (4.48) where four diodes D_1, D_2, D_3 and D_4 along with four capacitors C_1, C_2, C_3 and C_4 have been used.

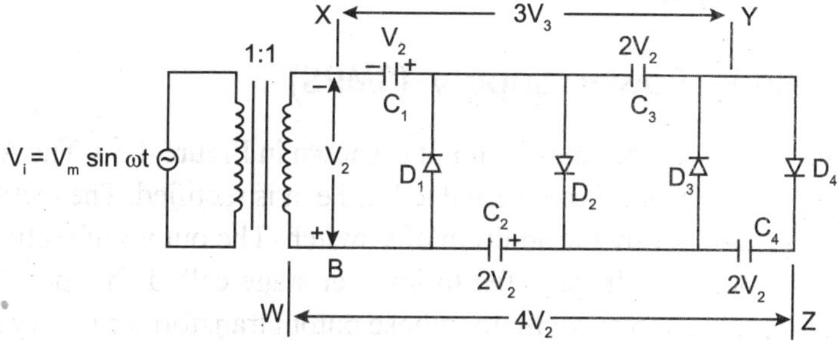

Figure 4.48 Half wave voltage quadrupler

The working of the circuit is quite similar to those of other voltage multiplying circuits described above. It is clear from the arrangement that there are four sections in cascade. The first three sections act as voltage tripler circuit so voltage across XY is equal to $3V_2$. The fourth section adds another voltage V_2. Thus, the total output voltage across WZ is $(3V_2 + V_2) = 4V_2$ and the circuit behaves like a voltage quadrupler.

Note: The discussion for full wave voltage multiplier circuits is beyond the scope at this stage.

4.9 Dual Polarity Power Supply

The technique employed for voltage multipliers can also be employed to give split or dual polarity supply yielding DC positive and negative voltages of equal magnitude with respect to ground.

One such dual polarity power supply circuit is shown in Figure 4.49.

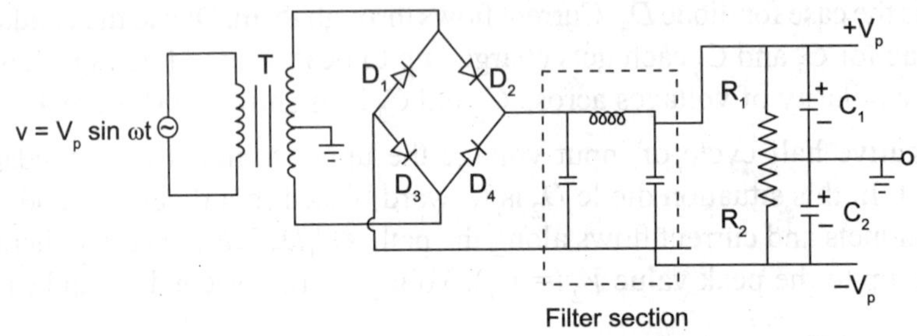

Figure 4.49 Dual polarity power supply

As shown in Figure 4.49, we have a full wave rectifier comprising of transformer T and bridge rectifier employing the four diodes D_1, D_2, D_3 and D_4 along with filter section and load having two equal capacitors C_1 and C_2 whose junction is connected to the center of resistor R_1 and R_2, which is grounded.

Just like a full wave rectifier, both halves of the input supply are used here in each winding of the transformer, thereby the two capacitors C_1 and C_2 get charged up to peak value V_P of input supply. The polarity of C_1 is (+) and that of C_2 is (–) with respect of ground. Thus we get $(+)V_p$ and $(-)V_p$ in the output.

4.10 Switch Mode Power Supply (SMPS)

The block diagram of a SMPS having AC input is shown in Figure 4.50. The AC mains is fed to a rectifier with filter. Here the AC is converted to DC i.e. it is rectified. The rectifier circuit can be configured as voltage doubler by the addition of a switch. The output of rectifier is unregulated DC voltage. The DC output voltage is fed to inverter stage called chopper. It converts DC to AC by running it through a power oscillator whose output transformer is very small having few windings at tens or hundreds of K Hz. The frequency is usually chosen to be 20 K Hz. The output voltage is optically coupled to input. The switching is implemented as a multistage MOSFET amplifier to achieve high gain. The inverted AC is used to drive the primary winding of a high frequency transformer called output transformer. This converts the input voltage to either high or low AC voltage to the required output level. The AC output voltage from the secondary of transformer is fed to output rectifier with filter to get required DC voltage. This rectifier uses silicon diodes for tens volts or so but for low DC voltage. Schottky diodes are used since they have faster recovery times. Thus we get a DC output after proper filtration. A feedback circuit monitors the output voltage.

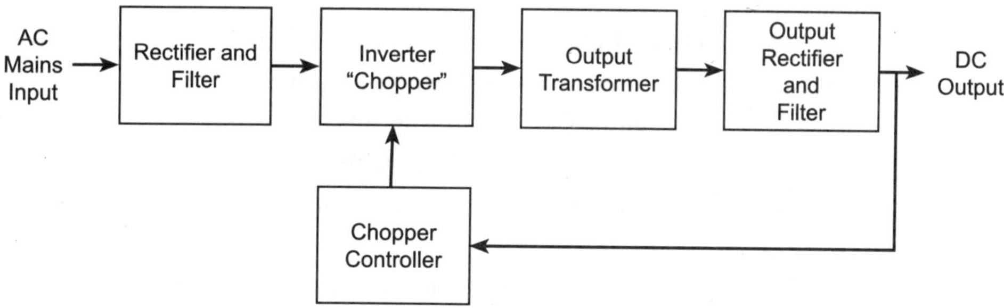

Figure 4.50 Block diagram of a SMPS (AC input)

As we have studied above a linear voltage regulator provides the desired output voltage by dissipating excess power in Ohmic losses (e.g. in a resistor or in the collector-to-emitter region of the pass transistor in the active mode). Its maximum efficiency is always less than 100% since efficiency is given as

$$\eta = \text{output voltage/input voltage}$$

In case of SMPS (Switch Mode Power Supply) it regulates either output voltage or current by switching ideal storage elements like inductor or capacitor, into and out of different electrical configurations. Similarly ideal switching elements like transistors operating outside of their active mode have no resistance when "closed" and they carry no current when "open" so such devices operate with 100% efficiency since no power is lost as heat like that of a passive resistor. It is to be noted that input to SMPS can be with AC or DC voltage.

Summary

1. A rectifier is a device that converts alternating current into direct current.
2. A half wave rectifier uses a single diode having load in series with secondary of transformer, whose primary is connected across the alternating voltage signal.
3. A full wave rectifier uses two diodes having load in series with the secondary centretapped transformer so that only one diode conducts during half cycle. The output consists of unidirectional pulsating current. Its primary, like that of half wave rectifier, is connected across the alternating voltage signal.

4. A bridge rectifier uses four diodes connected across the secondary in the form of a bridge such that during the 1st half cycle only two diodes conduct and remaining two diodes conduct during the 2nd half cycle of the signal. Here too the output consists of unidirectional pulsating current.
5. Efficiency of a rectifier is defined as the ratio of DC output power to AC input power. It is 40.6% in case of half wave rectifier and 81.2% in case of full wave rectifier.
6. The output of a rectifier consists of DC component and AC component called ripple. The percentage of ripple is quite large, i.e. 121% in half wave rectifier and 48% in case of full wave rectifier.
7. The peak inverse voltage of the diode used in a half wave rectifier is (–) V_m and its value is (–) $2V_m$ is case of full wave rectifier. But its value is (–) V_m in case of bridge rectifier, where V_m is peak value across the secondary of transformer used.
8. The ripple or AC component present in the output of a rectifier can be eliminated or minimized by making use of filters along with the rectifier circuit. The most common filter circuits are (i) *R–C* (ii) *R–L* (iii) *L–C* (iv) π.
9. The ripple from a capacitor filter decreases as load increases and it provides effective filtering for loads which draw 50 mA or less. The ripple factor in such a filter is given by $\frac{2410}{CR_L}$ for input frequency of 60 Hz.
10. An inductor filter has ripple factor that increases with load. So it is used in case of small loads. The ripple factor in such a filter is given by $\frac{R_L}{1600L}$ for input frequency of 60 Hz.
11. In an $L-C$ filter the ripple factor is independent of load value. However, the critical inductor used in this filter depends upon the load and given by $L_K \geq \frac{R_L}{900}$ for input frequency of 60 Hz.
12. A π or *C–L–C* filter reduces the ripple factor a lot.
13. All filters must be accompanied by a bleeder resistance.
14. Rectifiers with an appropriate filter do give DC output voltage but the output voltage varies with varying load and input voltage. So to keep the output voltage constant, we make use of a voltage regulating circuit.
15. All voltage regulating circuits make use of zener diode. The choice of diode depends upon the desired output voltage.
16. The output wave shape (normally a sinusoidal wave) can be clipped to a desired level either in positive half or negative half or both halves by using a clipping circuit.
17. A voltage multipler circuit raises the low AC voltage to a high DC voltage, by making use of diode and capacitor. With such circuits we can get DC voltage of the order of kilovolt.
18. Clamping circuits clamp the entire input wave (signal) at a particular DC level.

19. The clamper circuit shifts a signal positively or negatively by adding a DC voltage to the signal.
20. Often we require high voltage to run certain equipment. For this purpose we use a voltage multiplier circuit where we can get double or triple or quadruple times the peak value of rectified voltage of main voltage as required.

Review Questions

1. Sketch the characteristics for an ideal *p-n* junction diode and a practical diode. Briefly explain their shape.
2. Sketch a half wave rectifier circuit. Explain its operation and show the output waveforms that would result for a sinusoidal alternating input voltage.
3. Sketch a full wave rectifier circuit using two diodes. Explain its operation and show the output waveforms that would result for a sinusoidal alternating input voltage.
4. Sketch the circuit of a bridge rectifier. Explain its operation. Show the output waveforms that would result for a sinusoidal alternating input voltage.
5. Explain the working of a two-diode full wave rectifier having resistive load. Calculate ripple factor and efficiency of the rectifier. What is PIV across each diode?
6. Explain the working of a bridge rectifier having resistive load. Plot the wave shape of current and output voltage when input voltage is an alternating sinusoidal voltage.
7. Explain the working of a full wave rectifier having capacitor input filter. Derive necessary relation for ripple factor. When such filter is useful? Plot the output waveform when input voltage is an alternating sinusoidal voltage.
8. Explain the working of a full wave rectifier having inductor filter. Derive necessary relation for ripple factor. When such a filter is useful.
9. Explain the working of a full wave rectifier having L–C filter. Derive necessary relation for ripple factor. When such a filter is useful. Why a critical value of inductor is necessary in such a filter?
10. Explain the working of full wave rectifier having π filter. Derive an expression for the ripple factor. When such a filter is used. Draw the output waveform resulting from a sinusoidal alternating input voltage.
11. Define the terms: (i) voltage regulation (ii) efficiency of rectification (iii) ripple factor (iv) peak inverse voltage as related to a rectifier.
12. Why we prefer a bridge rectifier in comparison with a two-diode full wave rectifier? Explain. What is PIV of each diode in both cases.
13. Sketch the circuit of a simple Zener diode voltage regulator. Explain the working of the circuit. Why a series resistor is necessary in such a circuit? How its value is determined? What are the limitations of such a circuit?
14. Sketch the circuit of an emitter follower series voltage regulator. Explain its working and give necessary theory. What are its limitations?

15. Draw the circuit of a transistor shunt voltage regulator. Describe its working. When such circuit is useful?
16. How does ripple voltage vary with load in the following circuits:

    (a) *R–L* filter (b) *R–C* filter (c) *L–C* filter
17. Why all rectifying circuits use a bleeder resistance along with a filter? Explain.
18. Can a half wave rectifier circuit be considered as a negative clipper? If so, explain its working.
19. What is a clipping circuit? Explain the working of a series positive clipper. How it differs from that of a shunt positive clipper? Where such circuits are used?
20. What is a clipping circuit? Draw the circuit for (a) positive clipper (b) negative clipper. Explain the working of the clipping circuit and draw necessary output waveform resulting from a sinusoidal input signal.
21. What is a combination or biased clipper? Draw necessary circuit and explain the output wave shape resulting from a sinusoidal input voltage.
22. What is a clamper circuit? Draw the circuit of a positive clamping circuit and explain its working. Draw the final output wave.

Short Answer Questions

1. What are the advantages of using a bridge rectifier in a power supply?

   Ans: Following are the advantages of bridge rectifier in comparison to other rectifiers:

   We do not require a centretapped transformer which is quite bulky and difficult to have an electrical centretap. Here all the secondary voltage is used as input to the rectifier. Given the same transformer, we get twice as much peak voltage and twice as much DC voltage as with a full wave rectifier. The PIV of diode is half that of the diode used in full wave rectifier with same secondary voltage. Its efficiency is more than that of other rectifiers.

2. Why should we use a voltage multiplier instead of a transformer with higher turn ratio and an ordinary rectifier?

   Ans: A voltage multiplier is suitable wherever we require very high DC voltage and low current like CRO tube, etc. Since very high voltage can be obtained by bulky transformers using a rectifier along with it they are not preferred.

3. Mention where we use a *R–C* filter and an *L–C* filter?

   Ans: A *R–C* filter is preferred where we have a high load so that ripple is less. But we cannot reduce the ripple to a very low value since in that case the capacitor would be very large. In such situation, we generally use more than one section of *R–C* filter.

   An *L–C* filter is preferred where we require a large load current. The ripple factor in these filters is independent of load current.

4. Why a bleeder resistance is used in a power supply?

   Ans: (i) A bleeder resistance is used in power supply so that minimum current is always drawn from the supply; its value $R_B = 900\, L_{\text{max}}$ and output voltage remains nearly constant specially in case of *L–C* filter where current drawn is more than critical value I_K. Thus it maintains good voltage regulation.

   (ii) Often before we switch off the power supply from load, the capacitor gets charged and after dislocating the applied load it remains charged thereby it may give a shock to the user if there is no bleeder resistance.

5. For a bridge rectifier, list three possible faults.

   Ans: If the load is shorted, the fuse will be blown, and possibly the transformer or diode may be damaged. If the fuse is open, there will be no voltage anywhere in the circuit. If one of the diodes is open, then the DC load voltage will be less (low) because there will be only half wave rectification.

6. Mention the limit of frequency of applied input voltage in a rectifier.

   Ans: The frequency of applied input voltage depends upon the diode used. It should not be greater than frequency *f* given by the relation $f = \dfrac{1}{10[t_{rr}]}$ where t_{rr} is the reverse recovery time of the diode used; e.g. for 1N917 diode t_{rr} is 3 ns. So the applied voltage frequency should not be greater than 33 MHz.

7. Mention ripple frequency in case of (i) half wave rectifier (ii) full wave rectifier (iii) bridge rectifier.

   Ans: Ripple frequency in different rectifiers is as given below:

   (i) for half wave its value is f_{in}

   (ii) for full wave its value is $2f_{\text{in}}$

   (iii) for bridge rectifier its value is $2f_{\text{in}}$

8. If we assume some finite resistance of Zener diode in a Zener diode voltage regulator circuit. What is its effect on output voltage?

   Ans: Ideally load voltage is equal to Zener breakdown voltage. But when we assume Zener resistance R_Z in series with Zener diode then there is a voltage drop across this resistance when Zener current flows. This additional voltage drop across R_Z will slightly increase the voltage across load, i.e. $V_L = V_Z + I_Z R_Z$.

   Hence, increase in load voltage, $\Delta V_L = I_Z R_Z$.

Problems

1. A half wave rectifier consists of a diode having dynamic resistance of 1 Ω at its operating current and transformer whose open circuit secondary voltage is 12.6 V, 60 Hz. It has a secondary resistance of 3 Ω.
   (i) What is the no-load DC output voltage?
   (ii) What is the DC output voltage when load draws full DC current of 100 mA?
   (iii) What is percentage voltage regulation of the rectifier?
   (iv) What is internal resistance of the rectifier?

   [**Ans:** 5.66 V, 5.26 V, 7.6 %, 4 Ω]

2. Determine the rating of a transformer to deliver 100 W of DC power to a load under the following conditions:
   (a) half wave rectifier
   (b) full wave rectifier
   (c) bridge rectifier

   [**Ans:** 348 VA, 144 VA, 123 VA]

3. A 120 V, 60 Hz is applied to the primary of a 5 : 1 transformer whose secondary is centretapped, allowing a load of 1 kΩ to be connected to the rectifier using two diodes. Neglect voltage drop across the diodes. Calculate:
   (i) DC voltage across load
   (ii) DC current through load
   (iii) DC power to the load
   (iv) VA rating of transformer secondary
   (v) AC input power to the transformer assuming an 80% efficient transformer and ratio of rectification of the circuit to be 0.812
   (vi) Ripple voltage across the load and its frequency
   (vii) PIV across each diode

   [**Ans:** 270 V, 270 mA, 73 W, 105 VA, 113 W, 130 V, 120 Hz, 848 V]

4. (a) Determine the R.M.S. value of voltage on the secondary of a transformer to provide no load DC voltage of 9 V when connected to a bridge rectifier.
   (b) The secondary has a resistance of 3 Ω and each diode has dynamic resistance of 1 ohm. What will be the DC output voltage when 90 Ω load is connected to the rectifier.

   [**Ans:** 10V, 8.525V]

5. A full wave rectifier having a transformer of turn ratio 5 : 1 is connected to mains 120 V, 60 Hz. It has a *R*–*C* filter. Assume $R = 5$ kΩ, $C = 100$ μF. Calculate output DC voltage and ripple.

   [**Ans:** 17 V, 0.28 V PP]

6. A bridge rectifier has a transformer of turn ratio 15:1 and an *R*–*C* filter. Assume $R = 500\ \Omega$ and $C = 4700\ \mu F$. Calculate DC load voltage and ripple using second approximation of the diode. The transformer is connected to mains 120 V, 60 Hz.

   [**Ans:** 9.9 V, 35 mV PP]

7. A bridge rectifier having transformer turn ratio 8 : 1 is connected to mains 120 V, 60 Hz. (a) What is the peak output voltage if the diodes are assumed to be ideal? Calculate the average output voltage and DC output voltage and output frequency if load of 470 Ω is connected across it. (b) Repeat the preceeding problem using second approximation of the diodes.

   [**Ans:** (a) 21.4 V, 13.6 V, 13.6 V, 120 Hz
   (b) 19.81, 12.6 V, 12.6 V 120 Hz]

8. In a half wave rectifier the peak value of secondary voltage is 20 V. The rectifier uses an *L*–*C* filter. $X_L = 5\ k\Omega$ and $X_C = 25\ \Omega$. What is peak to peak ripple across the capacitor?

   [**Ans:** 0.1 V]

9. A half wave rectifier connected to mains 120 V, 60 Hz has a transformer of turn ratio 5 : 1 and a capacitor input filter. Assume $C = 100\ \mu F$ and $R = 5\ k\Omega$. Calculate DC load voltage and ripple voltage.

   [**Ans:** 34 V, 1.1 V PP]

10. A full wave rectifier connected to mains 120 V, 60 Hz has a transformer of turn ratio 5 : 1 and a capacitor input filter where $C = 100\ \mu F$ and $R = 5\ k\Omega$. Calculate DC load voltage and ripple voltage.

    [**Ans:** 17 V, 0.228 V PP]

11. What should be the AC input power in case of, half wave rectifier to deliver 500 W DC power to load. What will be its value in case of full wave rectifier having the same load.

    [**Ans:** (a) 1232 W, (b) 616 W]

12. A diode negative series clipper circuit has an input voltage of ± 25 V. The minimum forward current for the device is 1 mA. Determine suitable value of load R_L.

    [**Ans:** 24.3 kΩ]

13. The input voltage of a positive shunt clipper is ±15 V. The output current is 5 mA and output voltage is (–14) V. Determine the value of series resistor to be connected.

    [**Ans:** 200 Ω]

15. Calculate the maximum operating frequency for diodes having reverse recovery times of (a) 8 ns (b) 5 ns.

    [**Ans:** (a) 12.5 mH; (b) 20 MHz]

16. A diode is to operate at a maximum frequency of 5 MHz. Determine the maximum suitable reverse recovery time.

    [**Ans:** 20 ns]

17. A transistor series regulator circuit connected to an unregulated supply of 15 V to 20 V has to supply a load current of 0 to 12 mA at 12 V. Assume β of transistor to be 100 and

$V_{BE} = 0.5$ V, Zener voltage 12.5 V and $I_{ZK} = 1$mA. Find the maximum value of resistor in series of Zener diode and its power rating.

[Ans: $R_{max} = 119\ \Omega$, $P_Z = 0.79$ W, $P_R = 0.47$ W]

18. Design a voltage regulator circuit to provide 12 V across a load whose current varies from 5 mA to 35 mA. Assume an unregulated 18 V DC source and a Zener diode of Zener voltage 12 V when $I_{ZT} = 20$ mA. Calculate the value of series resistor and rating of diode.

[Ans: 150 Ω, 700 mW]

Objective Questions

1. The maximum voltage across a diode during its non conductingperiod in a bridge rectifier having $v = V_m \sin \omega t$ as input signal is

   (a) $\frac{V_m}{2}$ (b) $\frac{V_m}{\sqrt{2}}$ (c) V_m (d) $\sqrt{2}\ V_m$ **[Ans: (c)]**

2. In an $L-C$ filter applied across a full wave rectifier, the value of inductor, for I_{dc} to be greater than peak value of AC component, is selected to be equal to

   (a) $\frac{R_L}{10000}$ (b) $\frac{R_L}{900}$ (c) $\frac{R_L}{2}$ (d) R_L **[Ans: (b)]**

3. Bleeder resistance in a full wave rectifier with L (inductor) filter is always provided so that

   (a) minimum current always flows and output voltage remains constant

   (b) minimum current always flows but output voltage may vary

   (c) maximum current flows and output voltage remains constant

   (d) maximum current flows but output voltage may vary **[Ans: (a)]**

4. The PIV across each diode in a bridge rectifier with secondary voltage 24 V rms is

   (a) 24 V (b) 33.8 V (c) 12 V (d) 48 V **[Ans: (b)]**

5. The line frequency is 60 Hz. The output frequency of bridge rectifier is

   (a) 30 Hz (b) 60 Hz (c) 120 Hz (d) 240 Hz **[Ans: (c)]**

6. With the same secondary voltage and input frequency, the maximum ripple is found in

   (a) half wave rectifier

   (b) full wave rectifier

   (c) bridge rectifier

   (d) cannot be predicated **[Ans: (a)]**

7. The line frequency in half wave rectifier is 60 Hz, its output frequency is

   (a) 30 Hz (b) 60 Hz (c) 120 Hz (d) 240 Hz **[Ans: (c)]**

8. On increasing the capacitance in a shunt filter applied to a rectifier, the ripple will
   (a) remain constant (b) decrease
   (c) increase (d) none of the above [**Ans:** (b)]
9. In a Zener voltage regulator circuit, if we assume resistance of zener diode, then output voltage will
   (a) remain constant
   (b) decrease
   (c) increase
   (d) cannot be predicted [**Ans:** (b)]
10. The output frequency in a bridge rectifier is
    (a) same as that of half wave rectifier
    (b) double as that of half wave rectifier
    (c) half as that of half wave rectifier
    (d) four times as that of half wave rectifier [**Ans:** (b)]
11. Current in each diode in a full wave rectifier flows
    (a) for one complete cycle of input signal
    (b) for half cycle of input signal
    (c) for more than half cycle but less than one complete cycle of input
    (d) for positive half cycle of the input signal [**Ans:** (b)]
12. If no load voltage and full load voltage are equal in a rectifier then voltage regulation is
    (a) zero (b) 100% (c) 50% (d) 10% [**Ans:** (a)]
13. In a full wave rectifier having *R–L* filter the value of inductor is 2 H and load is 1 kΩ. If input frequency is 60 Hz, ripple factor is
    (a) 50% (b) 100% (c) 33% (d) 31% [**Ans:** (c)]
14. In a clipping circuit the diode conducts only through positive half cycle of the input signal. It can be
    (a) positive clipping circuit only
    (b) negative clipping circuit only
    (c) both positive and negative clipping circuit
    (d) it is not a clipping circuit [**Ans:** (b)]
15. In a combination clipper we have
    (a) two diodes both are in parallel but with opposite polarity
    (b) two diodes both are in parallel but with same polarity
    (c) two diodes both are in series with same polarity
    (d) two diodes both are in series with opposite polarity [**Ans:** (a)]

16. In a voltage multiplier circuit
    (a) output voltage is an integral multiple of the input voltage, with current being sub multiple
    (b) output voltage is a multiple of input voltage and current also a multiple
    (c) output voltage is a multiple of input voltage and low current having no relation
    (d) output voltage is multiple of input voltage and high current having no relation

[**Ans:** (a)]

Mark the Following Statements True/False.

1. The DC output voltage of a half wave rectifier is half that of a full wave rectifier. [**Ans:** True]
2. The maximum value of rectification efficiency of a full wave rectifier is 100%. [**Ans:** False]
3. A Zener diode functions under reverse-biased condition in a voltage regulator circuit. [**Ans:** True]
4. The PIV of diode in a bridge rectifier is half that of a diode in two-diode full wave rectifier. [**Ans:** True]
5. In a rectifier circuit the voltage regulation factor can be zero. [**Ans:** True]
6. In a *L–C* filter the critical value of inductor used has a relation with load present. [**Ans:** True]
7. We cannot have a positive as well as negative clipper circuit simultaneously. [**Ans:** False]
8. Half wave rectifier circuit is a negative clipper. [**Ans:** True]
9. Voltage multiplier circuit has a transformer having turn ratio 1 : 1. [**Ans:** False]
10. In a bridge rectifier circuit having a filter and voltage regulating circuit the output does not consist of a ripple voltage. [**Ans:** False]
11. Voltage multiplier circuit can give both high voltage and high current. [**Ans:** False]
12. A clamping circuit can be either positive or negative. [**Ans:** True]
13. The output frequency in a full wave rectifier is equal to input frequency. [**Ans:** False]

Fill in the Blanks.

1. PIV of a diode in half wave rectifier is ____________ and in full wave rectifier is ____________. [**Ans:** V_m, $2V_m$]
2. Efficiency of rectification in a full wave rectifier is ____________%. [**Ans:** 81.2%]
3. Ripple factor in a half wave rectifier is ____________. [**Ans:** 121%]
4. To have regulated voltage, we must use ____________ along with a rectifier circuit. [**Ans:** Zener diode]
5. In an *L–C* filter the value of critical inductor is approximately ____________ times the ____________. [**Ans:** $\frac{1}{900}$, load]
6. The load current wave form in a full wave rectifier is ____________ when input voltage is sinusoidal. [**Ans:** positive half sinusoidal]
7. The reactance of reactive elements in a filter connected to a full wave rectifier is calculated at ____________ of the input signal frequency. [**Ans:** twice]
8. The DC output voltage of a half wave rectifier is ____________ that of full wave rectifier having ____________ input voltage. [**Ans:** half, same]
9. An *L–C* filter is always accompanied by a ____________ resistance with a rectifier. [**Ans:** bleeder]
10. A voltage regulating circuit connected to a power supply can have a ____________ load. [**Ans:** variable]

Chapter 5

Transistors

5.1 Introduction

The invention of transistors (BJT) was a landmark in the field of electronics. Although it was invented by Walter H. Brattian and John Bardeev in 1948 at Bell Laboratory in USA, its application as amplifier was explained by Shockey later on. For their contributions, they were jointly awarded the Nobel Prize for Physics in the year 1956.

The term transistor originates from the phrase "transfer" + "resistor" "transfer of resistance". It has revolutionized the field of electronics by replacing vacuum tubes which were in predominant use previously. Transistors are more useful and easy to use due to their small size, long life and minimum requirement for heating.

5.1.1 Bipolar Junction Transistor

The term bipolar transistor is on account of the fact that there are two types of charge carries, namely electrons (negative) and holes (positive), which take part in conduction.

A junction transistor has three doped regions as shown in (Figure 5.1). Among the three doped regions two external regions are of same type either *p*-type or *n*-type and the middle region is of different type. The middle region (say *n*-type Figure 5.1(a)) called Base (*B*) is sandwiched between the external *p*-regions called Emitter (*E*) and Collector (*C*) respectively. Figure 5.1(b) shows a *n*-*p*-*n* type transistor in which Base is *p*-type. For simplicity we have shown block diagram actually the construction is very different.

The emitter region is heavily doped and the base region is lightly doped wheareas collector is doped intermediate.

We shall confine our study to *p*-*n*-*p* type transistor since the direction of conventional current matches with that of holes. The opposite is the situation for *n*-*p*-*n* type device where electron flow direction is opposite to that of conventional current.

Circuit symbols for *p*-*n*-*p* and *n*-*p*-*n* transistors are shown in Figure 5.2. The arrow head on each symbol always identifies the emitter terminal for the transistor. In each case, its direction indicates the conventional direction of current flow.

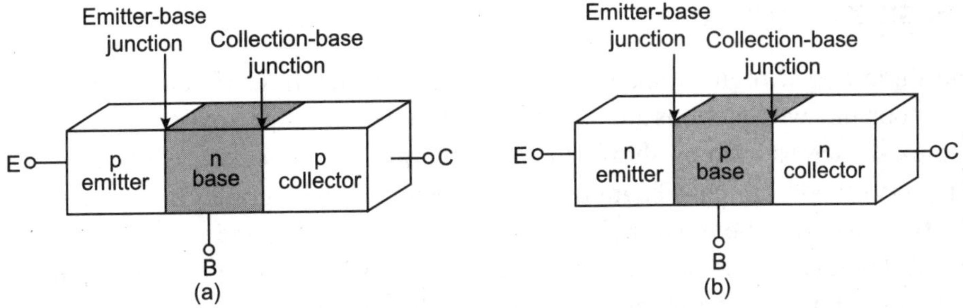

Figure 5.1 Block diagram of transistor: (a) *p-n-p* transistor, (b) *n-p-n* transistor

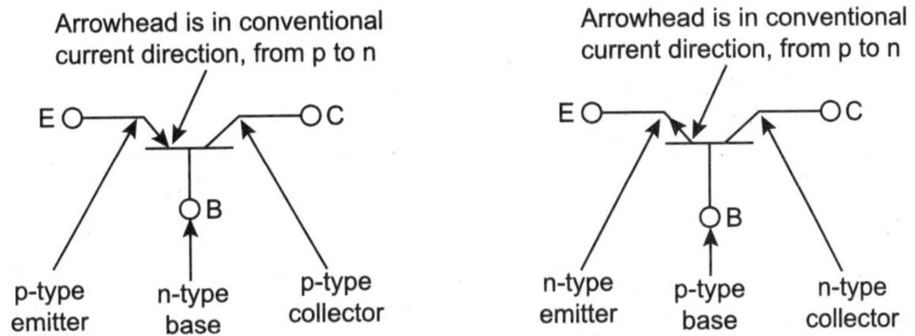

Figure 5.2 Symbols of transistors, (a) *p-n-p,* (b) *n-p-n*

We know that a depletion region (layer) is formed at a *p-n* junction. Here, in case of a transistor, there are two junctions, J_1 and J_2, so a depletion region is formed at each junction (Figure 5.3). It is to be noted that the depletion region penetrates deeply into the base since doping in either emitter or collector region is much more than at base region which is very lightly doped. Due to the penetration of depletion region in the base, the distance between emitter base depletion region and collector base depletion region is minimized. Figure 5.3 also illustrates barrier potential, and it is to be noted that barrier potentials are positive on the base and negative on the emitter and collector for a *p-n-p* transistor (Figure 5.3(a)) and negative on the base and positive on the emitter and collector for a *n-p-n* transistor (Figure 5.3(b)).

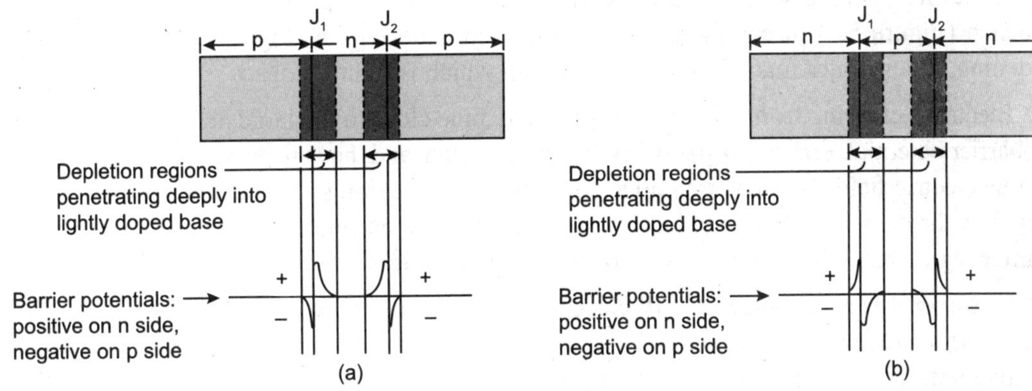

Figure 5.3 Formation of depletion region and barrier potential in a transistor under zero bias: (a) *p-n-p* transistor, (b) *n-p-n* transistor

5.2 Transistor Function

When the three doped regions are joined together, two junctions J_1 and J_2 (Figure 5.3) are formed. As soon the two junctions are formed depletion region (layer) exists there and a potential difference exists across each as shown in Figure 5.4(a). This situation is under zero bias (no external battery) condition. This diagram is called potential hill diagram. On applying proper bias across the two junctions, i.e. forward biasing emitter junction (J_1) and reverse biasing collector junction (J_2), the level of the potential hill change as shown by dashed line (Figure 5.4(b)). As is clear from the figure that forward biasing of emitter junction (J_1) lowers the emitter base hill height by V_{EB} whereas reverse biasing of collector junction (J_2) increases the potential hill height by (V_{CB}).

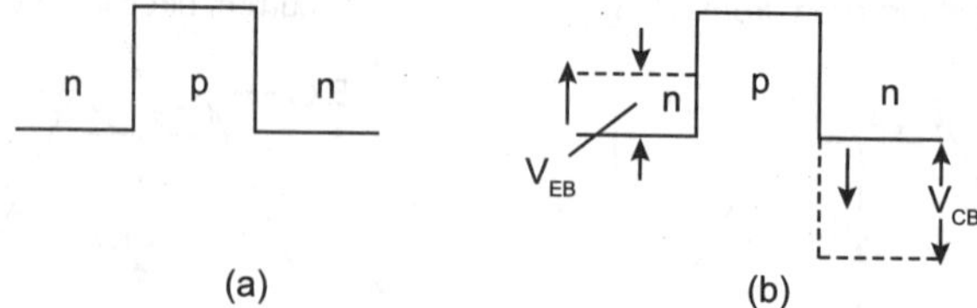

Figure 5.4 Potential hill in a transistor: (a) zero bias across the two junctions, (b) proper bias across the two junctions

(a) N-P-N type Transistor

We shall first discuss the action of N-P-N type transistor when it is properly biased.

Forward-bias across the emitter base junction (J_1) causes electrons (majority carries) to flow from *n*-type emitter to *p*-type base since the potential hill across the junction has reduced so they cross the hill easily. Here, electrons have been emitted (injected) into the base, so the name of the region is "emitter." But the base region is very thin and much more lightly doped so chances of recombination of electrons and holes in the base region are minimum and almost all electrons injected into the base reach the collector base junction (J_2).

Junction J_2 is reverse-biased and the reverse-bias causes the depletion region much more widened (Figure 5.5) in comparison to normal depletion region (Figure 5.3(b)). Due to the widening, it penetrates much into the base region which is lightly doped.

Electrons crossing from emitter to base arrive quite close to the large negative electric field (or barrier potential, Figure 5.4(b)) at CB depletion region, and they are drawn across the junction by this electric field, i.e. they are collected by this outer region so it is called "collector", and they flow through the collector to battery forming collector current; thus electrons coming from emitter region reach back to it via bias batteries and circuit is completed.

However it is to be noted that all electrons emitted from emitter do not reach collector. A few electrons recombine with holes in base region and a very few electrons flow towards the positive terminal of battery V_{EE}, forming base current.

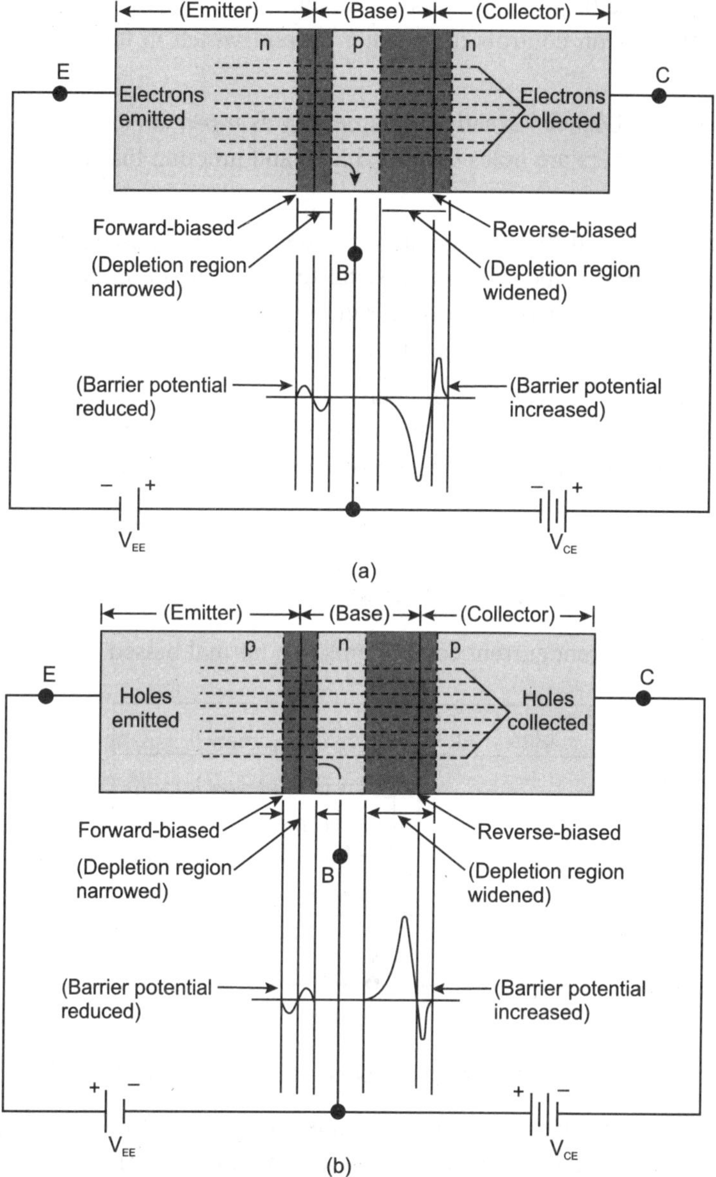

Figure 5.5 Depletion region and barrier potential: (a) normal biasing of *n-p-n* transistor, (b) normal biasing of *p-n-p* transistor

It is to be noted that EB junction (J_1) is forward-biased, so it has a characteristic of forward-biased diode, and sustained current will not flow until Forward-bias is about 0.7 V for silicon device and 0.3 V for germanium device. Reducing the level of EB bias voltage in effect reduces *p-n* junction forward-bias and thus reduces the current that flows from emitter through base to collector. Reducing the bias to zero cuts off the current. Thus, variation of small Forward-bias voltage on EB junction controls the emitter current which in turn controls the collector current.

The *p-n-p* transistor behaves exactly the same way as *n-p-n* device with the exception that in this case majority carriers are holes (Figure 5.5(b) and junction biasing is opposite to *n-p-n* transistor.

Further, it is to be noted that the dynamic resistance of EB junction is very low since it is forward-biased. However, the dynamic resistance of CB junction is quite high since it is reverse-biased. Hence in a transistor, current from low resistance input circuit is transferred to high resistance output circuit with almost unchanged magnitude.

As explained previously, two types of charge carriers are involved in current flow through any type of junction transistor; consequently such devices are called bipolar junction transistors. This is to distinguish them from field effect transistors where only one type of charge carrier is involved.

5.3 Current Components in a Transistor

Figure 5.6 shows the different current components in a normal biased *p-n-p* transistor.

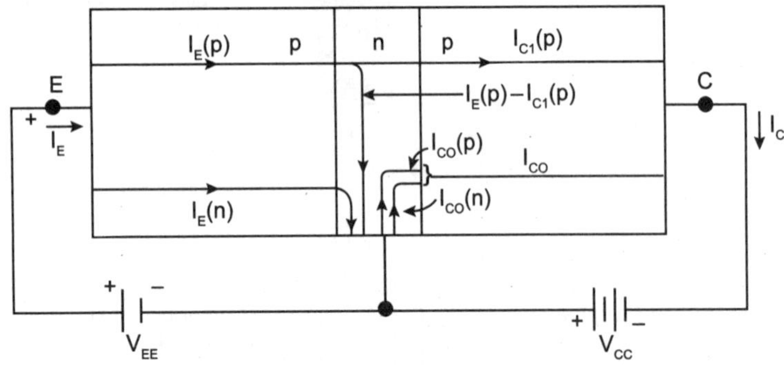

Figure 5.6 Current components in a *p-n-p* transistor

Emitter current I_E is made up of two components. The component $I_E(p)$ due to holes injected from the emitter region to the base region and the component $I_E(n)$ due to electrons crossing from base to emitter region. So,

$$I_E = I_E(p) + I_E(n)$$

The direction of component $I_E(n)$ has been shown similar to that of $I_E(p)$ because flow of electrons is opposite to that of holes. Total current flow is taken be in the direction of flow of holes.

In the above relation,

$$I_E(p) \gg I_E(n)$$

Since doping of emitter region is much higher than that of base region, emitter current is almost due to injected holes and we can have

$$I_E \approx I_E(p)$$

As explained previously, most of the injected holes crossing emitter base junction (J_1) reach the collector base junction (J_2) producing a hole component $I_{C1}(p)$ of collector current. A few holes recombine with electrons in the base region giving recombination hole current $I_E(p) - I_{C1}(p)$ leaving the base.

When emitter base junction is open circuit, i.e. $V_{EE} = 0$, and collector base junction (J_2) is reverse-biased due to V_{CC} the emitter current $I_E = 0$ and collector current is reverse saturation current I_{CO} of reverse-biased diode at J_2. This reverse saturation current has two components: (i) $I_{CO}(n)$ consisting of electrons moving from p side (collector) to n side (base) and (ii) $I_{CO}(p)$ due to holes travelling from n side (base) to p side (collector). With the chosen direction of current, we can write

$$(-)\, I_{CO} = I_{CO(n)} + I_{CO(p)}$$

Total collector current when emitter base junction is forward-biased is

$$I_C = I_{CO} - I_{C1(p)}$$

Note: Here direction of I_C is taken positive when it enters the collector.

Usually I_{CO} is of the order of μA and I_C is of the order of mA (since it is due to emitter current).

We can have

$$I_{CO} \ll I_{C1(p)}$$

Hence $I_C \approx (-)I_{C1(p)}$

Now, we can say that emitter current flows into the transistor which both I_C and I_B flow out of the transistor as shown in Figure 5.6; therefore

$$I_E = I_C + I_B \tag{1}$$

Almost all emitter current reaches the collector and only a fraction of it flows out of the base terminal.

The portion of I_E which flows across collector base junction is designated as $\alpha_{dc} I_E$, where α_{dc} represents the fraction and its typical value is 0.96 to 0.99. Thus, I_C is equal to $\alpha_{dc} I_E$ and reverse saturation current I_{CBO} is also termed as collector to base leakage current which is made up of minority carriers (i.e. holes in case of n type base). Hence we can write

Total collector current $$I_C = \alpha_{dc} I_E + I_{CBO} \tag{2}$$

or $$\alpha_{dc} = \frac{I_C - I_{CBO}}{I_E}$$

Since I_{CBO} is very small in comparison to I_C,

$$\therefore \qquad \alpha_{dc} = \frac{I_C}{I_E} \qquad (3)$$

Hence α_{dc} is approximately equal to the ratio of collector current to emitter current. It is referred as common base current gain factor and its value is nearly 0.96 to 0.99. When we use '*h*' parameters, it is designated at h_{FB}. Using (1) to substitute for I_E in Eqn. (2), we have

$$I_C = \alpha_{dc}[I_C + I_B] + I_{CBO}$$

$$= \alpha_{dc} I_C + \alpha_{de} I_B + I_{CBO}$$

or

$$I_C(1-\alpha_{dc}) = \alpha_{dc} I_B + I_{CBO}$$

$$\therefore I_C = \frac{\alpha_{dc} I_B}{1-\alpha_{dc}} + \frac{I_{CBO}}{1-\alpha_{dc}} \qquad (4)$$

All the above equations hold good for *n-p-n* transistor also.

Eqn. (4) can be written as

$$I_C = \beta_{dc} I_B + (\beta_{dc}+1) I_{CBO} \qquad (5)$$

where

$$\beta_{dc} = \frac{\alpha_{dc}}{1-\alpha_{dc}} = \frac{\alpha}{1-\alpha} \qquad (6)$$

Since I_{CBO} is much less than I_B, Eqn. (5) can be written as

$$\beta_{dc} = \frac{I_C}{I_B} \qquad (7)$$

Hence β_{dc} is approximately equal to the ratio of collector current to base current. It is also referred as DC common emitter current gain factor. When we use *h* parameters, it is designated as h_{FE}.

Example 1 Calculate collector current I_C and emitter current I_E for a transistor with $\alpha_{dc} = 0.97$ and $I_{CBO} = 10\ \mu A$. Base current is measured as 50 μA.

Solution

$$\text{Collector current, } I_C = \frac{\alpha_{dc} I_B}{1-\alpha_{dc}} + \frac{I_{CBO}}{1-\alpha_{dc}}$$

$$= \frac{0.97 \times 50\mu A}{1-0.97} + \frac{10\mu A}{1-0.27}$$

$$= 1.95 \text{ mA}$$

$$\text{Emitter current, } I_E = I_C + I_B$$

$$= 1.95\,\text{mA} + 0.05\,\text{mA}$$

$$= 2.0\,\text{mA}$$

Example 2 For a certain transistor, $I_C = 5.255$ mA, $I_B = 100$ μA and $I_{CBO} = 5$ μA. Calculate: α_{dc}, β_{dc} and I_E. Determine the new level of I_B so that $I_C = 15$ mA.

Solution

$$\text{Collector current, } I_C = \frac{\alpha_{dc} I_B}{1-\alpha_{dc}} + \frac{I_{CBO}}{1-\alpha_{dc}}$$

or

$$I_C(1-\alpha_{dc}) = \alpha_{dc} I_B + I_{CBO}$$

or

$$\alpha_{dc}[I_B + I_C] = I_C - I_{CBO}$$

$$\alpha_{dc} = \frac{I_C - I_{CBO}}{I_B + I_C} = \frac{5.255\,\text{mA} - 0.005\,\text{mA}}{5.255\,\text{mA} + 0.01\,\text{mA}}$$

$$= \frac{5.26}{5.265} = 0.98$$

$$\beta = \frac{\alpha_{dc}}{1-\alpha_{dc}} = \frac{0.98}{1-0.98} = \frac{0.98}{0.02} = 49$$

$$\text{Emitter current, } I_E = I_C + I_B$$

$$= 5.255 + 0.01$$

$$= 5.265\,\text{mA}$$

Example 3 A transistor has the following characteristics in a circuit. $I_C = 12.427$ mA, $I_B = 200$ μA, $I_{CBO} = 7$ μA. Determine the values of I_E, β_{dc} and α_{dc}. Determine the new level of I_C which will result when $I_B = 150$ μA flows in the circuit.

Solution

$$I_C = \beta_{dc} I_B + (\beta_{dc} + 1) I_{CBO}$$

$$12.427\,\text{mA} = \beta_{dc} \times 200\,\mu\text{A} + (\beta_{dc} + 1)7\,\mu\text{A}$$

$$= \beta_{dc} \times [200 + 7]\,\mu\text{A} + 7\,\mu\text{A}$$

$$\therefore \beta_{dc} = \frac{[12.427 - 0.007]\,\text{mA}}{0.207\,\text{mA}} = \frac{12.434}{.207} = 61$$

$$I_E = I_C + I_B$$

$$\therefore I_E = 12.42\,\text{mA} + .2\,\text{mA} = 12.6\text{ mA}$$

$$I_C = \alpha_{dc} I_E + I_{CBO}$$

$$12.42 = \alpha_{dc} \times 12.6\,\text{mA} + 0.007\,\text{mA}$$

$$\therefore \alpha_{dc} = \frac{12.42 - 0.007}{12.6} = \frac{12.413}{12.6} = 0.98$$

$$I_C = \beta_{dc} I_B + (\beta + 1) I_{CBO}$$
$$= 61 \times 150\,\mu\text{A} + (61+1)7\,\mu\text{A}$$
$$= 61 \times .15\,\text{mA} + 62 \times .007\,\text{mA}$$
$$= 9.584\,\text{mA}$$

5.4 Action of Transistor as Amplifier

The principle of operation of a transistor can be understood by considering Figure 5.7.

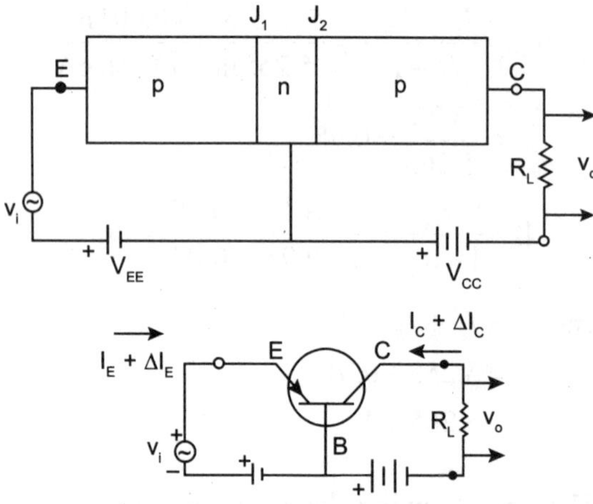

Figure 5.7 Transistor amplifier

Consider circuit of Figure 5.7 where a *p-n-p* transistor is properly biased and an AC voltage signal v_i has been connected along with emitter bias V_{EE}. A load R_L is connected in the collector circuit. On applying voltage signal v_i an additional emitter current ΔI_E flows through the emitter junction J_1, this additional emitter current gives rise to extra collector current ΔI_C in the collector circuit.

Voltage across load R_L due to the extra collector current.

$$v_o = \Delta I_C R_L$$
$$= [\alpha_{dc}\, \Delta I_E] R_L$$

since by definition $\Delta I_C = \alpha_{dc}\, \Delta I_E$

Signal voltage, $v_i = r_i\, \Delta I_E$

$$\text{Voltage gain, } A_v = \frac{\text{Output voltage}}{\text{Input voltage}} = \frac{\Delta I_C R_L}{\Delta I_E r_i}$$

$$= \frac{v_o}{v_i} = \frac{[\alpha_{dc}\, \Delta I_E] R_L}{r_i\, \Delta I_E}$$

$$\text{or} \quad \frac{v_o}{v_i} = \frac{\alpha_{dc} R_L}{r_i} \tag{8}$$

Generally, $R_L \gg r_i$

So output voltage $v_o \gg v_i$ although the value of α_{dc} is less than 1.

Thus eqn. (8) represents a voltage gain (amplification) and in this respect the device behaves like a voltage amplifier.

The ratio of (–) $\Delta I_C/\Delta I_E$ is called current amplification, A_i, of the transistor. The direction of collector current is opposite in a transistor, so A_i is positive. This leads A_v (voltage gain) to be positive.

Power gain A_p of a transistor is defined as the ratio of output power to input power.

$$A_p = \frac{\text{Output power}}{\text{Input power}} = \frac{(\Delta I_C)^2 R_L}{(\Delta I_E)^2 r_i} = \left(\frac{\Delta I_C}{\Delta I_E}\right)\left[\frac{\Delta I_C R_L}{\Delta I_E r_i}\right] = [A_i][A_v]$$

5.4.1 Transistor Configuration

While discussing the action of transistor as amplifier, we have taken the base as common terminal for input as well as output (Figure 5.7).

However, it is not always necessary that it works as a amplifier when the above situation is satisfied. We can have any terminal as common terminal for input as well as output for it to operate as a amplifier.

As such, we have three configurations:

(i) Common base (Figure 5.8(a)) or CB configuration

(ii) Common emitter (Figure 5.8(b)) or CE configuration

(iii) Common collector (Figure 5.8(c)) or CC configuration

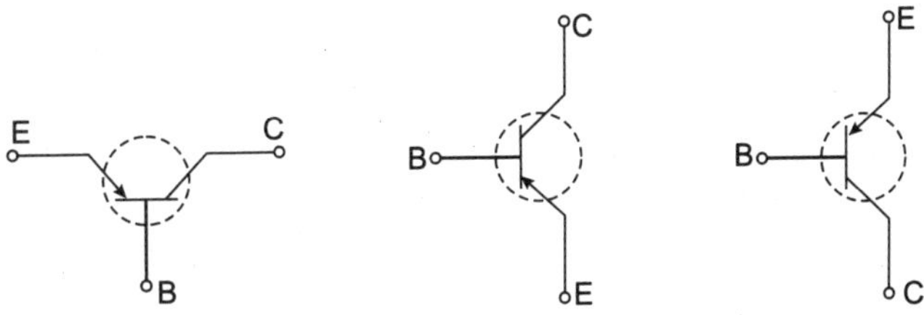

Figure 5.8 Three configurations of transistor (*p-n-p*)

In Figure 5.8, we have considered a *p-n-p* transistor but we can have all the three configurations for a *n-p-n* transistor also. We shall now discuss the characteristics of all the three configurations.

5.4.2 Common Base Characteristics

Consider Figure 5.9 where a *p-n-p* transistor has been shown connected with its base terminal common to both input and output circuit.

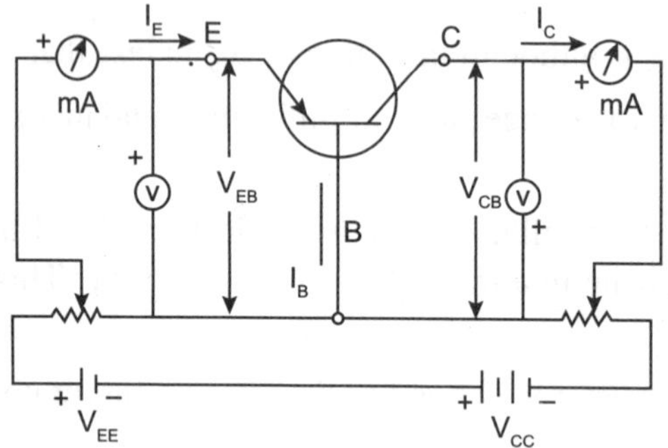

Figure 5.9 Circuit for CB Configuration for *p-n-p* transistor (the same circuit holds good for *n-p-n* transistor with polarity reversed)

Here, emitter and base terminals represent the input side, whereas collector and base terminals represent output side; thus base is common. The EB junction is forward-biased and CB junction is reverse-biased; so polarity of the battery is selected accordingly.

Input characteristics To draw the (V_{EB}, I_E) characteristics, output voltage V_{CB} is maintained constant for a particular curve.

Input voltage V_{EB} is set at several convenient levels and corresponding emitter current I_E is noted at each level. A graph is drawn between V_{EB} and I_E keeping V_{CB} constant for one particular curve (Figure 5.10); for another curve, V_{CB} is set at another value and again a graph is drawn between V_{EB} and I_E.

Since the EB junction is forward-biased, the common base input characteristics are essentially those of a forward-biased *p-n* junction. Figure 5.10 shows that for a given level of input voltage, more input current flows when higher levels of V_{CB} are employed because large V_{CB} voltage (reverse bias) causes the depletion region to enter more into the base region there by its resistance decreases. These characteristics are called emitter characteristics.

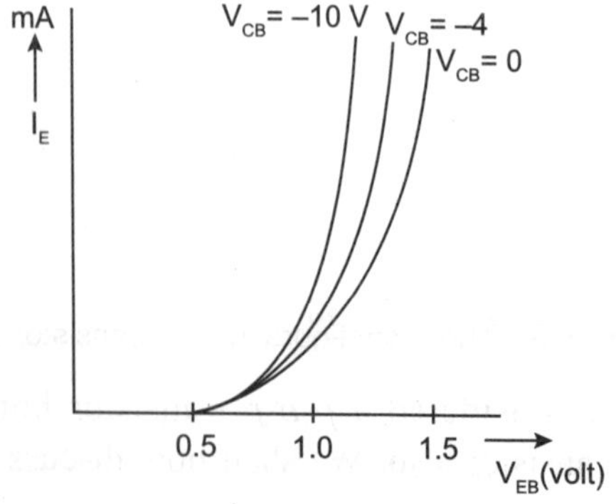

Figure 5.10 Input characteristics of CB configuration

The inverse of slope of any curve at a particular point represents input impedance,

$$r_i = \left[\frac{\Delta V_{EB}}{\Delta I_E}\right]_{V_{CB}\ \text{const}}$$

If we change the point or select a different curve, then the value of r_i changes.

Output characteristics To draw the (V_{CB}, I_C) characteristics, input current I_E is maintained constant for one particular curve.

Output voltage V_{CB} is set at several convenient levels and the corresponding collector current I_C is noted at each level. A graph is drawn between V_{CB} and I_C, keeping I_E constant for one particular curve (Figure 5.11). For another curve, I_E is set at another value and again a curve is drawn between V_{CB} and I_C (Figure 5.11).

The graph shows that for each level of fixed I_C, collector current I_C is almost equal to I_E and it appears constant when the value of V_{CB} is increased. All such curves for different values of I_E are almost parallel. The constant value of I_C for increasing value of V_{CB} is because an increase in V_{CB} widens the CB depletion region and thus shortens the gap between two depletion regions. Similarly, with I_E held constant, the increase in I_C is so small that it is not noticeable and the curve appears nearly horizontal.

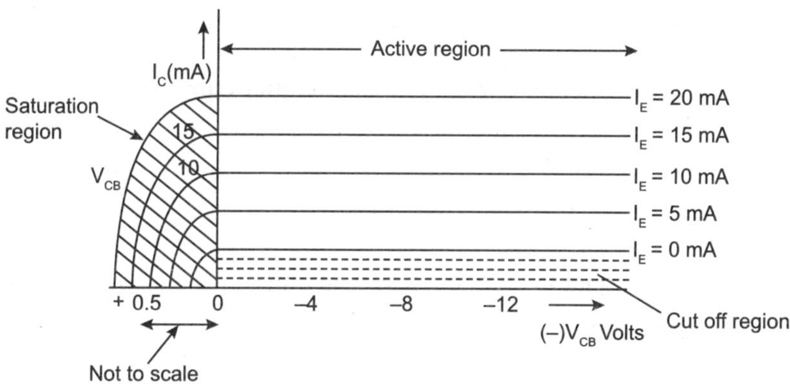

Figure 5.11 (a) Output characteristics

When the value of externally applied bias V_{CB} voltage is zero, there is still a barrier potential existing across CB junction and this potential helps the current flowing. Thus collector current I_C flows as shown in the Figure 5.11(a). The charge carriers which makeup I_C constitute minority carries as they cross CB junction. Thus, reverse-bias V_{CB} and (unbiased) CB barrier potential assist flow of I_C to stop the flow of charge carriers, the CB junction has to be forward-biased. Consequently, I_C is reduced to zero only when V_{CB} is made positive. The region for forward-biased CB is known as **saturation region**, and the region for which CB is reverse-biased is known as **active region**. This is the region in which the transistor is normally operated. The region between the curve for $I_E = 0$ and X-axis is called **cutoff region**.

If reverse-bias voltage (V_{CB}) across the CB junction is increased over the maximum safe limit specified, then current I_C will no longer remain constant and it increases suddenly, and we say that breakdown of the device has occurred. In such situations, the depletion region at CB junction penetrates more in the base region and it makes contact with the depletion region of EB junction. This condition is known as **punch through** and large current flows through the device, possibly destroying the device.

5.5 Dynamic Output Resistance (r_o)

It is defined as the ratio of change in collector base voltage (ΔV_{CB}) to the change in collector current (ΔI_C) for a given value of emitter current (I_E):

$$r_o = \left[\frac{\Delta V_{CB}}{\Delta I_C}\right]_{I_E \text{ const}}$$

Its value is very large of the order of mega ohms since change in collector current ΔI_C is very small. This is obvious from the output characteristics which are nearly horizontal, i.e. parallel to X-axis (Figure 5.11(a)).

5.6 Current Amplification Factor (α_{dc})

It is defined as the ratio of collector current (I_C) and emitter current (I_E) for a given value of V_{CB}.

$$\alpha_{\text{dc}} = \frac{I_C}{I_E}$$

Its value is very nearly equal to unity but less than unity.

Similarly, for dynamic conditions,

$$\alpha_{\text{ac}} = \left.\frac{\Delta I_C}{\Delta I_E}\right|_{V_{CB}=\text{const}}$$

5.6.1 Current Gain Characteristics

These characteristics are also referred to as forward transfer characteristics. Experimentally, they can be drawn by holding V_{CB} constant at a convenient value and recording the I_C level measured for various settings of emitter current I_E. The slope of such characteristics is very nearly equal to unity, i.e. $I_E \approx I_C$ (Figure 5.11(b)).

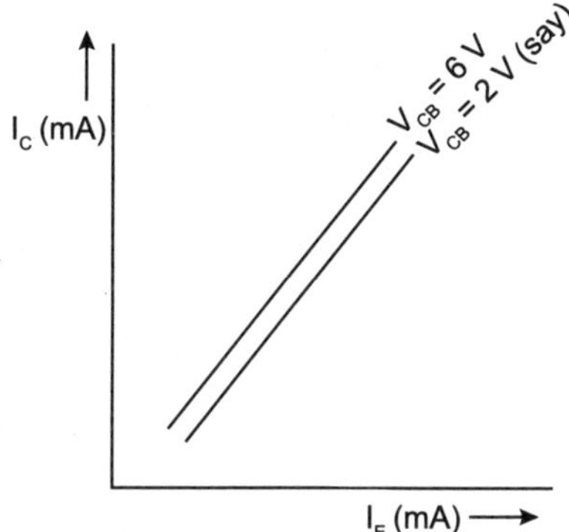

Figure 5.11 (b) Current gain characteristic

5.6.2 Common Emitter Characteristics

The circuit showing the characteristics of a *p-n-p* transistor in CE configuration is shown in Figure 5.12.

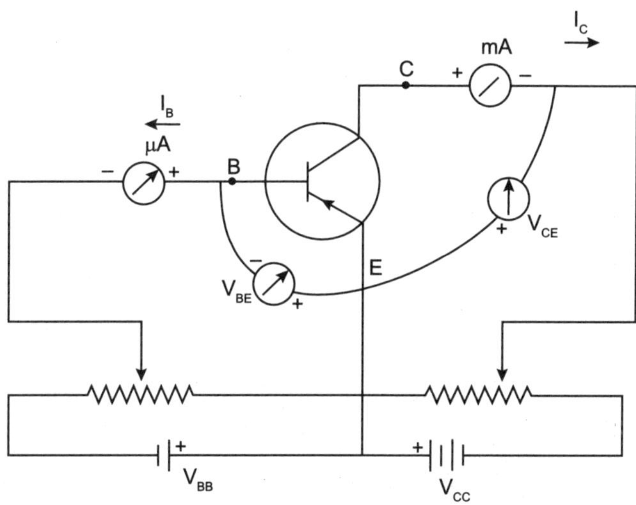

Figure 5.12 CE configuration for *p-n-p* transistor (the same circuit holds for n-p-n a transistor but with polarity changed)

Here the base and emitter terminals represent the input side, whereas collector and emitter represent the output side. Thus, emitter is common to both. The EB junction is forward-biased and CE junction is reverse-biased; so polarity of the battery is selected accordingly.

Input characteristics To draw the (V_{BE}, I_B) characteristics, output voltage V_{CE} is maintained constant for a particular curve.

Input voltage V_{BE} is set at several convenient levels and the corresponding base current I_B is noted at each level and a graph is drawn between the various values of V_{BE} and I_B keeping V_{CB} constant (Figure 5.13). For another curve, V_{CE} is set at another value and again a graph is plotted between V_{BE} and I_B. Thus, a family of curves is drawn.

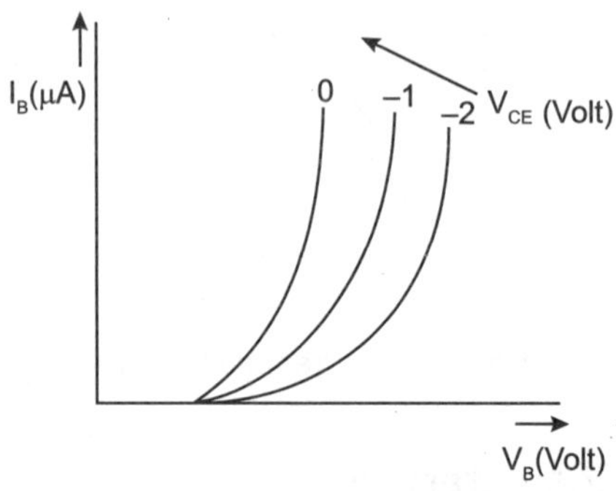

Figure 5.13 Input characteristics for CE configuration

The input characteristics are essentially for a forward-biased *p-n* diode. Figure 5.13 shows that for a given level of input voltage, more input current flows. When a higher level of V_{CE} voltage is employed, they shift to the left.

The inverse of the slope of a curve at a particular point gives input resistance,

$$r_i = \left.\frac{\Delta V_{BE}}{\Delta I_B}\right|_{V_{CE}\ \text{const}}$$

If we change the point or select a different curve, then the value of r_i changes.

Output characteristics To draw (V_{CE}, I_C) characteristics input current I_B is maintained constant for a particular curve.

Output voltage V_{CE} is set at several convenient levels and the corresponding collector current I_C is noted at each level. A graph is drawn between V_{CE} and I_C keeping base current I_B constant for a particular curve.

For another curve, base current I_B is set at another level and again a curve in drawn between V_{CE} and I_C. Thus, a family of curves is drawn. From the curves in Figure 5.14, we find that for a given value of I_B collector current I_C initially increases sharply for a slight increase in voltage V_{CE}, but later on increasing V_{CE} current I_C increases slowly.

On increasing value of I_B, collector current increases for the same value of V_{CE}. But here, the family of curves is not parallel like in the case of common base configuration, and also on decreasing voltage V_{CE}, collector current I_C reduces. Finally, when V_{CE} reduces to zero, current I_C also reduces to zero because V_{CE} is the sum of V_{CB} and V_{BE} and at the knee of the characteristic, collector base junction bias V_{CB} has been reduced to zero. So, a further reduction of V_{CE} causes the collector base junction to be forward-biased. The forward-bias repels the minority charge carriers which constitute I_C, and so I_C reduces to zero, which is unlike the companion base output characteristics.

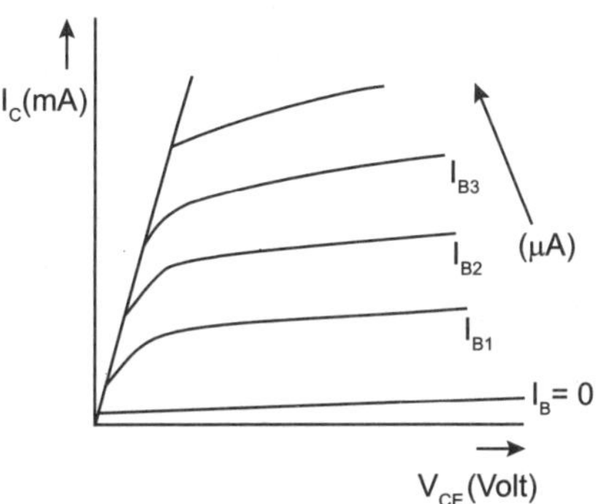

Figure 5.14 Output characteristics for CE configuration

Further, it is to be noted that the reverse bias. V_{CC} causes holes (majority carriers) to be injected from emitter to collector in addition to holes injected due to forward-bias of base emitter junction. Consequently, a fraction of total holes reaching the collector increases, i.e. α_{dc} increases and in turn β_{dc} increases (Eqn. 6). This increase in α_{dc} and β_{dc} accounts for upward (positive) slope of characteristic, which is more pronounced than in case of common base output characteristics. The cutoff region, active region and saturation region are as shown in Figure 5.13.

When reverse-biased voltage V_{CE} crosses the safe level, collector current I_C increases rapidly and it may destroy the device. This is largely due to the **punchthrough effect** as in the case of common base configuration.

5.6.3 Current Gain Characteristics

These characteristics are simply I_C vs. I_B for various values of V_{CE}. Like common base current gain characteristics, they are also obtained experimentally or determined from output characteristics. The curves are just straight lines but they are not parallel to each other as is in the case of CB configuration.

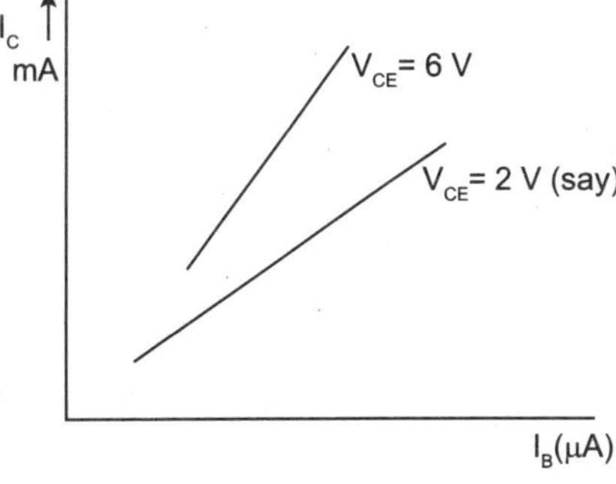

Figure 5.15

5.6.4 Common Collector Characteristics

Figure 5.16 shows the circuit for CC characteristics. This circuit is similar to that of CE (Figure 5.12) except that here the collector is common to both input and output side. So using the circuit (Figure 5.16), we can draw input, output and current gain characteristics.

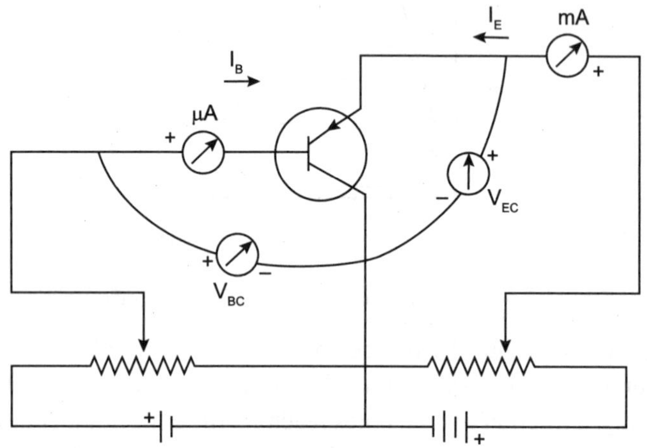

Figure 5.16 Circuit for common collector configuration *p-n-p* transistor

The output characteristics are shown in Figure 5.17.

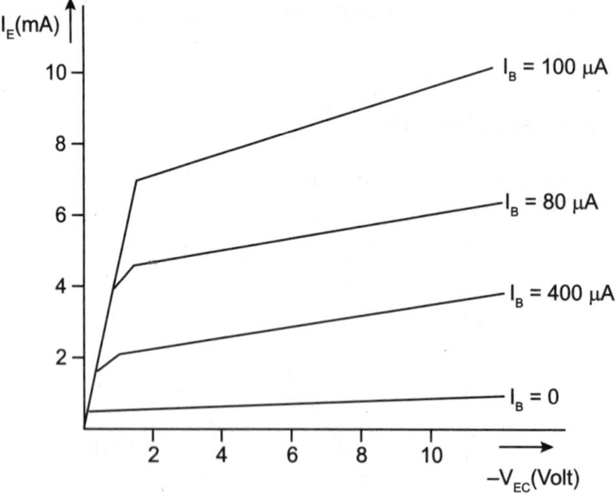

Figure 5.17 CC configuration output characteristics

The common collector characteristics are similar to that in CE configuration. Here emitter current, I_E, is plotted versus V_{EC} for several fixed values of base current, I_B.

The input characteristics are shown in Figure 5.18. They are quite different from either common base or common emitter configuration input characteristics.

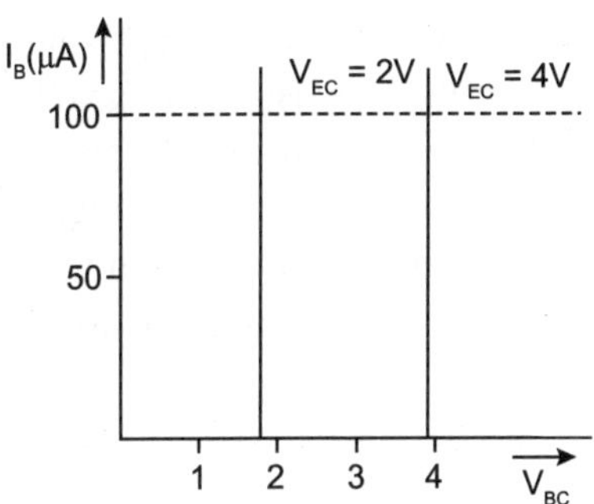

Figure 5.18 Common collector input characteristics

We see that the characteristics are just vertical lines, i.e. base current increasing very suddenly for a given V_{BC}.

The difference is due to the fact that input voltage V_{BC} is largely determined by the level of E_C voltage. This is because when the transistor is biased ON, V_{BC} remains around 0.7 V for silicon (0.3V for germanium) and V_{EC} may be much greater than 0.7 V.

5.6.5 Current Relation

We know that

$$I_E = I_B + I_C \quad \text{(i)}$$

and

$$I_C = \alpha_{dc} I_E + I_{CBO} \quad \text{(ii)}$$

Substituting for I_C in (i) from (ii),

$$I_E = I_B + \alpha_{dc} I_E + I_{CBO}$$

or

$$I_E - \alpha_{dc} I_E = I_B + I_{CBO}$$

$$\therefore I_E = \frac{I_B}{1-\alpha_{dc}} + \frac{I_{CBO}}{1-\alpha_{dc}}$$

But $\dfrac{1}{1-\alpha_{dc}} = \beta_{dc} + 1$

So

$$I_E = (\beta_{dc} + 1) I_B + I_{CBO}(\beta_{dc} + 1)$$

assuming I_{CBO} to be negligible

Hence $I_E = (\beta_{de} + 1)\, I_B$

5.7 CB, CE and CC Configurations of Transistor and their Equivalent Circuits

A transistor is basically a three-terminal device comprising of base, emitter and collector. While studying the characteristics of the transistor, we have kept one terminal common to the input and output side. In this respect we can consider the transistor as four-terminal or two-port device keeping one common to both input and output (Figure 5.19). Consecutively, we have three configurations where one terminal is kept common turn by turn. The configurations are (a) common base, (b) common emitter, and (c) common collector.

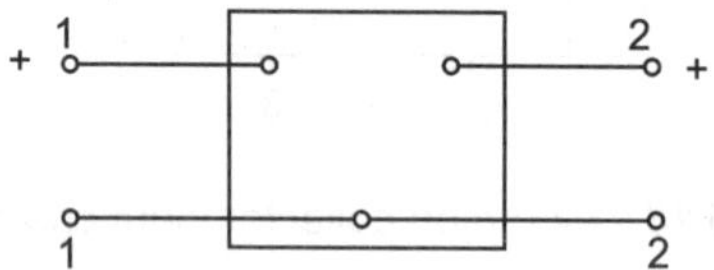

Figure 5.19 Two-port network

5.7.1 CB Configuration

Here, the base terminal is kept common to the input and output side (Figure 5.20).

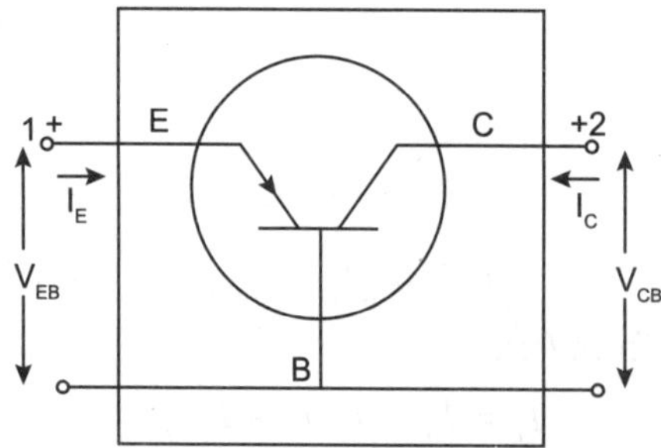

Figure 5.20

In such a two-port network, we assume the upper terminals, both input and output side as positive. So current will enter the black box from both the terminals and there by it is taken as positive. However, current may come out of the terminal in actual situation. In this case, current will be taken as negative.

As shown in Figure 5.20, emitter current I_E enters the box and taken as positive since emitter base junction is forward-biased and emitter being *p*-type is connected to the positive terminal of battery and so current flows into the box. Similarly, collector current I_C will come out of the box since the collector is reverse-biased, i.e. it is connected to the negative terminal of battery and as such I_C is taken negative although by convention upper terminal's output side is shown positive. However, in case of *n-p-n* transistor, the situation is different as shown in Figure 5.21.

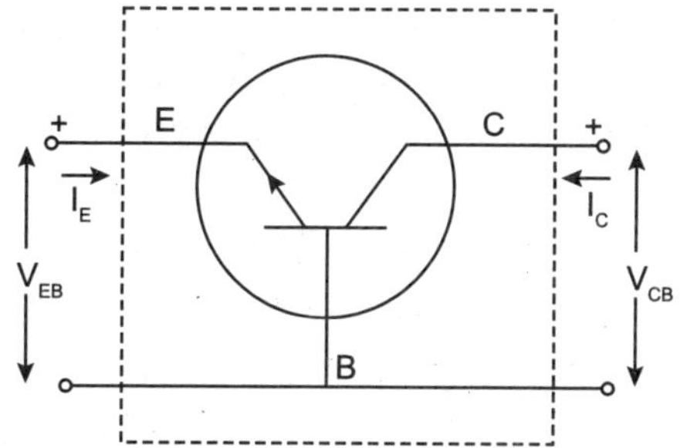

Figure 5.21

Here, I_E is actually negative and I_C is positive. Assuming the above situation, we can have the following two more configurations: (a) common emitter, (b) common collector.

5.7.2 CE Configuration

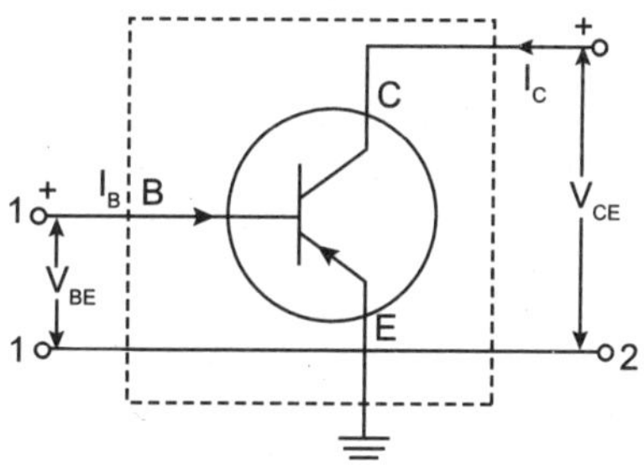

Figure 5.22

Here in case of *p-n-p* transistor, both I_B and I_C are actually negative. But in case of *n-p-n* transistor, both I_B and I_C are positive.

5.7.3 CC Configuration

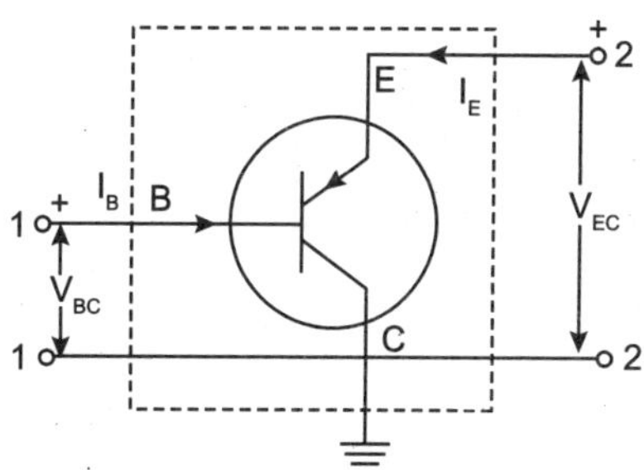

Figure 5.23

Here in case of *p-n-p* transistor, both I_B and I_E are actually positive. But in case of *n-p-n* transistor, both I_B and I_E are negative. A summary of common base, common collector and common emitter configuration circuit characteristics is given in Figure 5.24.

Common Base	
Input impedance	Lowest (approx. 50 Ω)
Output impedance	Highest (approx. 1 MΩ)
Current gain	No (less than 1)
Voltage gain	Yes (500–800)
Power gain	Yes
Input-output phase voltage	In phase
Application	RF amplifier
Configuration diagram The configuration is named after the transistor connection that is common to both input (source) and output (load). Another explanation: A transisor connection that does not provide an input or output to the circuit.	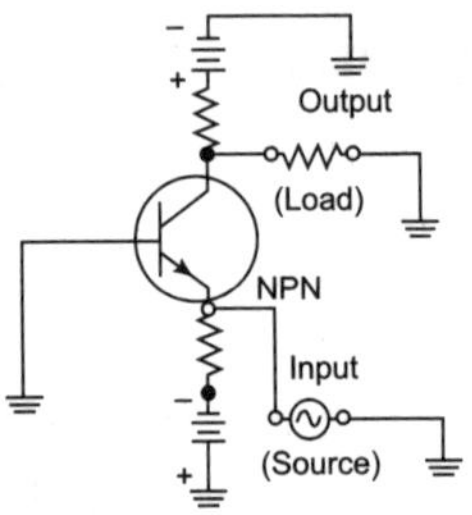 Output (Load) NPN Input (Source)

Common Collector	
Input impedance	Highest (approx. 300 kΩ)
Output impedance	Lowest (approx. 300 kΩ)
Current gain	Yes (30–100)
Voltage gain	No (less than 1)
Power gain	Yes
Input-output phase voltage	In phase
Application	Isolation amplifier emitter follower
Configuration diagram The configuration is named after the transistor connection that is common to both input (source) and output (load). Another explanation: A transisor connection that does not provide an input or output to the circuit.	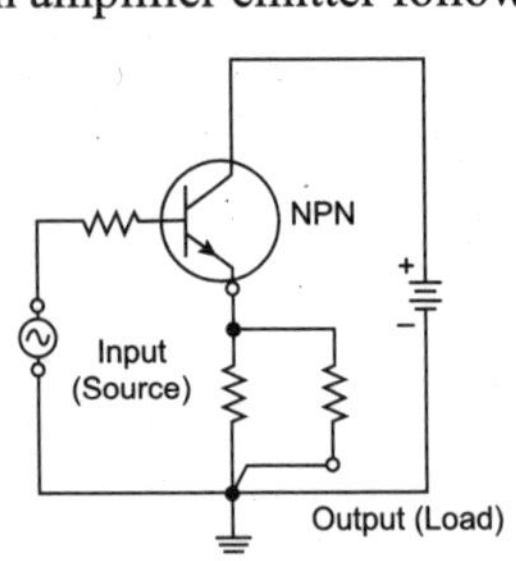 NPN Input (Source) Output (Load)

Common Emitter	
Input impedance	Medium (approx. 1 kΩ)
Output impedance	Medium (approx. 5 kΩ)
Current gain	Yes (30–100)
Voltage gain	Yes (300–600)
Power gain	Yes (Highest)
Input-output phase voltage	180° out of phase
Application	Universal
Configuration diagram The configuration is named after the transistor connection that is common to both input (source) and output (load). Another explanation: A transisor connection that does not provide an input or output to the circuit.	NPN Input (Source) Output (Load)

Figure 5.24 Summary of circuit characteristics

5.8 Small Signal Model of Transistor using *h* Parameters

5.8.1 CB Configuration

As shown in Figure 5.20, input voltage is V_{eb} and input current is i_e so also output voltage is v_{cb} and output current is i_c. Now we can write

$$v_{eb} = h_{ib}i_e + h_{rb}v_{cb}$$
$$i_c = h_{fb}i_e + h_{ob}v_{cb}$$

Using Kirchofff's law, the above equations represent the following AC (small signal) equivalent circuit (Figure 5.25) of a transistor in CB configuration (Figure 5.20).

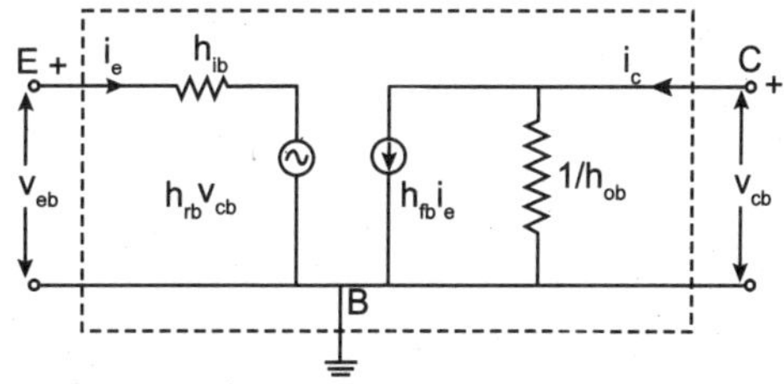

Figure 5.25 AC equivalent circuit of CB transistor using *h* parameters

5.8.2 CE Configuration

As shown in Figure 5.21, input voltage is v_{be} and input current is i_b, so also output voltage is v_{ce} and output current is i_c. Now we can write

$$v_{be} = h_{ie} i_b + h_{re} v_{ce}$$
$$i_c = h_{fe} i_b + h_{oe} v_{ce}$$

Using Kirchhoff's law, the above equations represent the following AC (small signal) equivalent circuit (Figure 5.26) of a transistor in CE configuration.

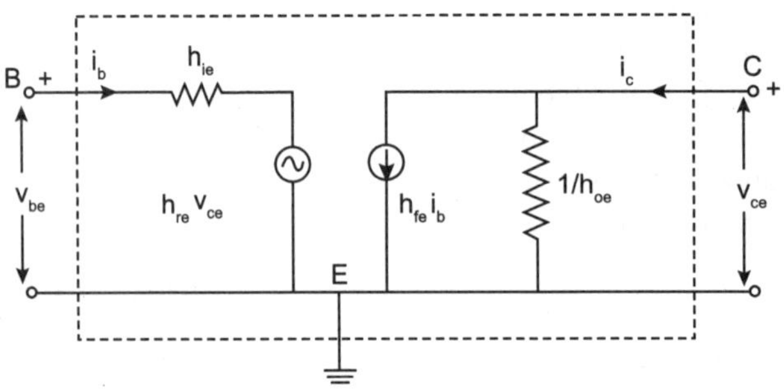

Figure 5.26 AC (small signal) equivalent circuit of a transistor in CE configuration

5.8.3 CC Configuration

As shown is Figure 5.22, input voltage is v_{bc} and input current is i_b, so also output voltage is v_{ec} and output current is i_e. We can write

$$v_{bc} = h_{ic} i_b + h_{rc} v_{ec}$$
$$i_e = h_{fc} i_b + h_{ob} v_{ec}$$

Using Kirchhoff's law, the above equations represent the following AC (small signal) equivalent circuit (Figure 5.27) of a transistor in CC configuration.

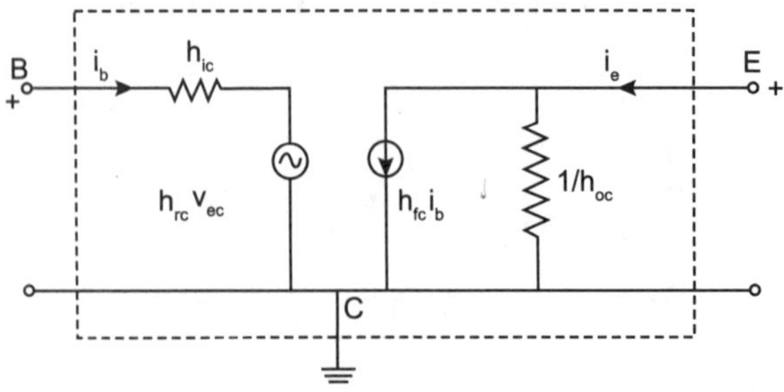

Figure 5.27 AC equivalent circuit of CC transistor using *h* parameters

5.9 Unipolar Junction Transistors

5.9.1 Field Effect Transistor (FET)

While discussing function of BJT we observed that there both type of charge carriers (electron and holes) take part in conduction and it is current controlled device. We have another device known as Field Effect Transistor in which only one type of charge carrier (majority charge carrier either electron or hole) take part in conduction. It is a semiconductor device in which current is controlled by the electric field produced hence the nomenclature Field Effect Transistor. Since current is constituted of one type of charge carriers, there by it is a unipolar junction transistor. It comprises of two major groups.

(a) Junction field effect transistor (JFET)

(b) Insulated gate field effect transistor (IJFET), more commonly known as metal oxide semiconductor (MOS FET) transistor.

A field effect transistor differs from a bipolar transistor in the following important features:

(i) Its operation depends on majority (one type) carriers only. It is therefore a unipolar device.

(ii) It is a voltage operated device like a vacuum tube.

(iii) It exhibits high input resistance.

(iv) It exhibits no offset voltage at zero drain current.

(v) It is simple to fabricate and occupies less space in an integrated circuit.

(vi) It is thermally more stable and less noisy.

(vii) It is a linear device which follows square law.

(viii) Here conduction is through drift of charge carriers which is faster than diffusion so it can be operated at high frequencies of the order of MHz.

Junction field effect transistor Its structure is quite simple. An n-type (or p-type) semiconductor bar is taken and contacts are made at its two ends along its length, which are called source (S) and drain (D), respectively (Figure 5.28). In the lightly doped n-type bar SD, the charge carriers are electrons and they would flow from S to D when the two ends are properly biased by battery V_{DD}. In a p-type for the battery connections are reversed i.e. S is made positive and D is made negative (Figure 5.28(b)).

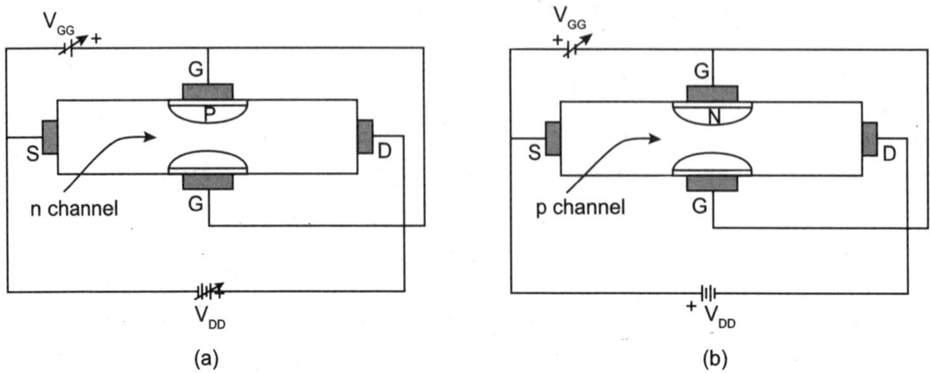

Figure 5.28 FET: (a) *n*-type channel, (b) *p*-type channel

The other two faces of the bar are diffused with a heavily doped *p*-type semiconductor and metallic contacts are made at *G*. These two terminals are internally connected and called gate *G*. The *p*-type gate *G* and *n*-type bar form a *p-n* junction on its two faces. If we take a *p*-type bar, then gate *G* is *n*-type (Figure 5.28(b)); so here also a *p-n* junction is formed on the two faces. In both cases, gate *G* is reverse-biased by battery V_{GG} with respect to source. This reverse-bias controls the flow of drain current I_D (majority carriers) from source to drain.

The symbols for the two types of FET are shown in Figure 5.29.

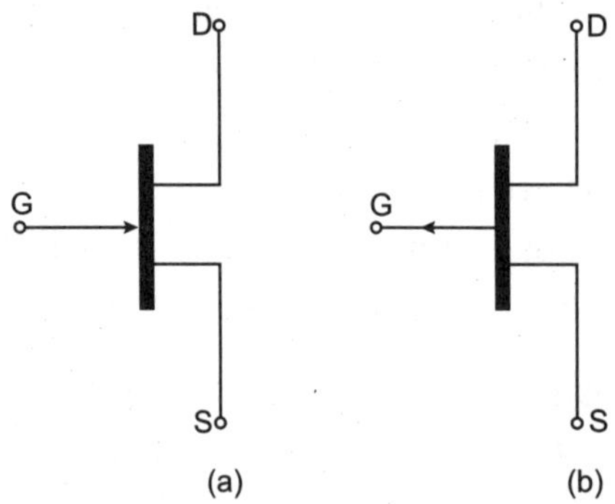

Figure 5.29 FET symbol: (a) n channel FET, (b) p channel FET

5.10 Characteristics of FET

5.10.1 Drain Characteristics

To draw the characteristic curves of FET, the circuit connections are made as shown in Figure 5.30. Here we have considered an *n* channel FET.

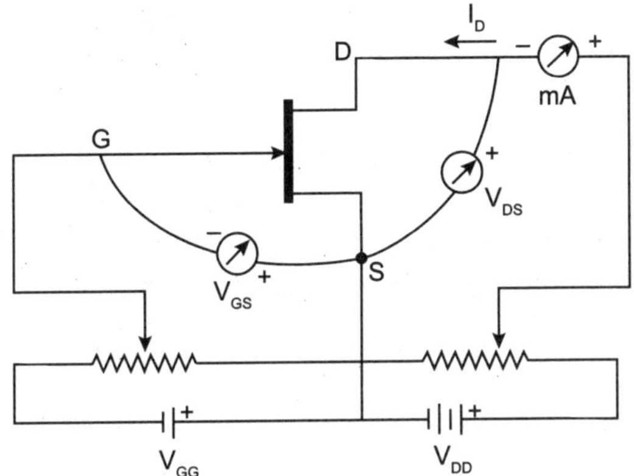

Figure 5.30 Circuit to draw the characteristics of *n*-channel FET

On applying a proper varying voltage V_{DS} across Source (*S*) and Drain (*D*) of N-channel FET, keeping gate voltage $V_{GS} = 0$, the electrons will flow from source to drain through the channel, and the resulting current I_D is read using a milliammeter. As voltage V_{DS} is increased drain current I_D increases linearly with voltage, showing that the channel behaves like a resistor of almost constant value. As voltage V_{DS} is further increased, the increment in current I_D is not linear, and it attains a saturation value after $V_{DS} = V_P$ which is due to the fact that voltage drop across the channel bar becomes large enough to cause a considerable penetration of the depletion region formed at gate (Figure 5.34).

Drain current at this point with V_{GS} at zero is referred to as drain-source saturation current I_{DSS} (Figure 5.31).

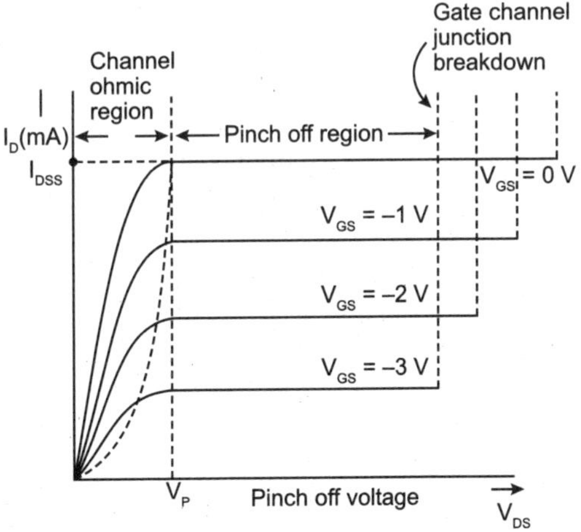

Figure 5.31 Drain characteristic curves of JFET (N-channel)

Drain source voltage V_{DS} at which drain current attains a fairly constant value, i.e. saturated, is called pinch off voltage, $V_{p.}$ However, if external reverse-bias voltage V_{GS} is increased, i.e.

it is made negative, and I_D vs. V_{DS} curve is again drawn, then the shape of the curve remains nearly same (Figure 5.31), but the increase in drain current is much less and saturation value is attained at a much lower value of V_{DS}.

By applying several values of external reverse-bias voltage V_{GS}, a family of drain characteristics (I_D vs. V_{DS}) is obtained as shown in Figure 5.31. In this respect the (V_{DS}, I_D) curve shifts downwards. The broken line connects the points on the different curves at which drain current I_D attains saturation at each level of gate voltage.

5.10.2 Transfer Characteristics

To draw the transfer characteristic of FET, we make use of the circuit shown in Figure 5.30. Drain source voltage V_{DS} is maintained at a constant level and V_{GS} is adjusted in convenient steps. At each step of V_{GS}, the corresponding drain current I_D is noted and a graph is plotted between V_{GS} and I_D as shown in Figure 5.32(a).

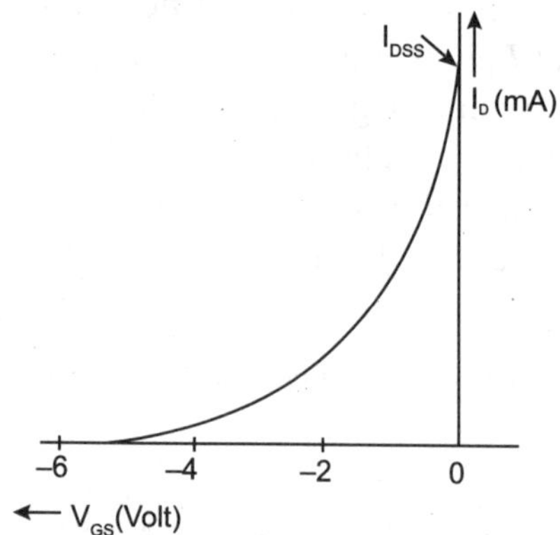

Figure 5.32 (a) Transfer characteristic for an *n*-channel FET

The resulting curve is known as transfer characteristic. It is evident from the curve that, as V_{GS} is made more and more negative (in case of *N* channel), drain curve I_D progressively reduces and attains zero value when V_{GS} is equal to V_P.

The FET transfer characteristic approximately follows the equation

$$I_D = I_{DSS}\left[1-\frac{V_{GS}}{V_p}\right]^2$$

According to the above equation, when $V_{GS} = 0$, drain current $I_D = I_{DSS}$ as shown in Figure 5.31.

In the above discussion, we have considered a *n*-channel FET. However, if we take a *p*-channel FET, then drain source voltage V_{DS} is so applied that source is positive and the drain is negative. Thus, drain current I_D flows (in the conventional direction) from source to drain.

To reverse-bias the gate, we make G positive with respect to p-type channel. Therefore, bias voltage is such that gate is positive and source is negative.

By knowing the characteristics of a FET, we can define the following parameters of the device.

DC resistance R_{DS} of channel For small values of V_{DS} (ohmic region), resistance of the channel is defined as

$$R_{DS} = \frac{V_{DS}}{I_D}$$

Its unit is ohm. This value of R_{DS} is linearly related to the applied V_{GS} and it makes the device useful as a voltage variable resister (VVR).

AC resistance of channel Beyond $V_{DS} = V_p$, the (I_D vs. V_{DS}) curve is nearly horizontal and this region is called pinch off region (Figure 5.31). In this region, resistance of the channel is defined as the ratio of variation in drain source voltage to drain current, keeping gate voltage constant

or

$$r_d = \left.\frac{\Delta V_{DS}}{\Delta I_D}\right|_{V_{GS=\text{const}}}$$

Its unit is ohm. It is to be noted that the slope of drain characteristics is nearly zero, and r_d is not easily determined from the characteristic. This value of r_d is not be confused with drain source ON resistance R_{DS}. While r_d is a dynamic (or AC) quantity, R_{DS} is DC resistance of channel when the depletion region is nearly absent, i.e. when the device is biased on in the channel ohmic region.

Trans conductance (g_m) It is defined as the ratio of variation in drain current to variation in gate-source voltage, keeping drain source voltage.

$$g_m = \left.\frac{\Delta I_D}{\Delta V_{GS}}\right|_{V_{DS}=\text{const}}$$

It can be measured as a slope of transfer characteristics. It defines how drain current varies in relation to gate-source voltage.

It unit is microsiemens (μs) or microampere per volt (μA/V).

Amplification factor (μ) It is defined as the ratio of variation in drain to source voltage to variation in gate to source voltage, keeping drain current constant.

$$\mu = \left.\frac{\Delta V_{DS}}{\Delta V_{GS}}\right|_{I_D=\text{const}}$$

$$= \frac{\Delta V_{DS}}{\Delta I_D} \times \frac{\Delta I_D}{\Delta V_{GS}} = r_d \times g_m$$

Principle of operation of FET

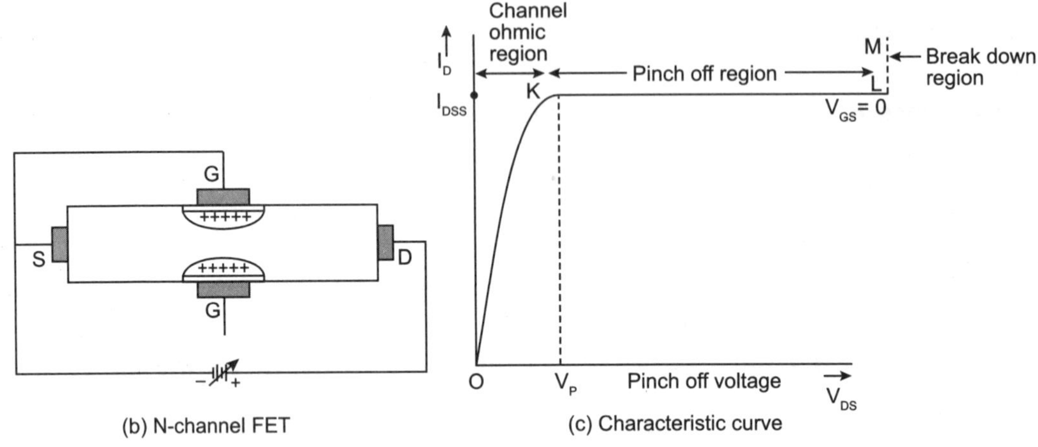

(b) N-channel FET

(c) Characteristic curve

Figure 5.32

5.10.3 Effect of V_{DS} on Characteristic Curve

The drain characteristic (V_{DS}, I_D) at $V_{GS} = 0$ is shown in Figure 5.32(c) when $V_{DS} = 0$, $I_D = 0$ and voltage between gate and all points in the channel is also zero. When V_{DS} is increased by a small amount, a small drain current I_D flows through the channel, causing some voltage drop along the channel. This reverse biases the gate-channel junction by a small amount, causing a little penetration of the depletion region (Figure 5.34) and having negligible effect on channel resistance. With further increase in V_{DS}, drain current increases linearly and so the channel behaves like a resistance of nearly constant value. The channel continues to behave like a constant resistance in *OK* part of curve (Figure 5.32(c)) until voltage drop V_P along the channel becomes large enough to cause considerable penetration of the depletion region at gate-channel junction. A further increase in V_{DS} produces a smaller increase in current I_D; and a saturation level of current I_D is noted; part *KL* of the curve known as pinch off region where any increase in V_{DS} causes only a very minor increase in current.

When drain current saturation level is reached, the shape of depletion region on the two sides of channel is such that they appear to pinch off the channel. For this region, the drain to source voltage at which drain current I_D levels off, i.e. nearly constant, is called **pinch off voltage** V_p as shown in Figure 5.32(c).

The region of characteristic between $V_{DS} = 0$ and $V_{DS} = V_p$ is called **channel ohmic region** because in this region the channel behaves like a resistor.

With continued increase in V_{DS}, a voltage will be reached at which the gate-channel junction breaks down. This is as a result of charge carriers, which make up reverse saturation current at the gate-channel junction, being accelerated to high velocity and producing an avalanche effect. At this point, drain current increases very rapidly and the device may get destroyed. This is called breakdown region. The normal operating region of the device is pinch off region also called active region.

5.10.4 Pinch Off Voltage

At zero bias, there are two sets of depletion region, one on each side of the *n*-type bar near gate terminals which are internally connected. The two sets are alike.

Let W_n be the width of depletion region in the *n* type bar which is lightly doped, and W_p be that in the *p*-type gate which is heavily doped, at a distance *x* from source terminal. If N_A is the concentration of acceptor ions in the *p*-type gate and N_D is that of donor ions in the *n*-type channel, then

$$N_A W_P = N_D W_n$$

and $W_p \ll W_n$ since $N_A \gg N_D$.

The total width of depletion region is $W = W_p + W_n \approx W_n$ since W_p is much less.

Here *W* is the half-width of depletion layer (region) since we have considered only one side of the bar.

Moving perpendicular to the *X* direction, i.e. along *Y* direction, Poisson's equation for the depletion region in *n*-type bar is

$$\frac{d^2V}{dy^2} = (-)\frac{e}{\in} N_D \qquad \text{(i)}$$

where e = electronic charge

$\in$ = permittivity of the bar material

The boundary condition is that at $y = W_n = W$, barrier electric field

$$F = (-)\frac{dV}{dy} = 0$$

Integrating Eqn. (i) and applying boundary condition,

$$\frac{dV}{dy} = (-)\frac{eN_D}{\in}(y - W) = (-)F$$

Again integrating, we have

$$\int_0^V dV = -\frac{eN_D}{\in}\int_0^Y (y - W)dy$$

or

$$V(y) = (-)\frac{eN_D}{2\in}[y^2 - 2Wy] \qquad \text{(ii)}$$

At $y = W$, we have $Y(y) = |V(x)|$

which is the applied voltage across space-charge region at X. We have neglected the built in potential at channel-gate junction.

$V(x)$ is negative because of reverse-bias for an n-channel bar JFET. Let 'a' be the half width of the bar and $b(x)$ halfwidth of the conducting channel (Figure 5.33).

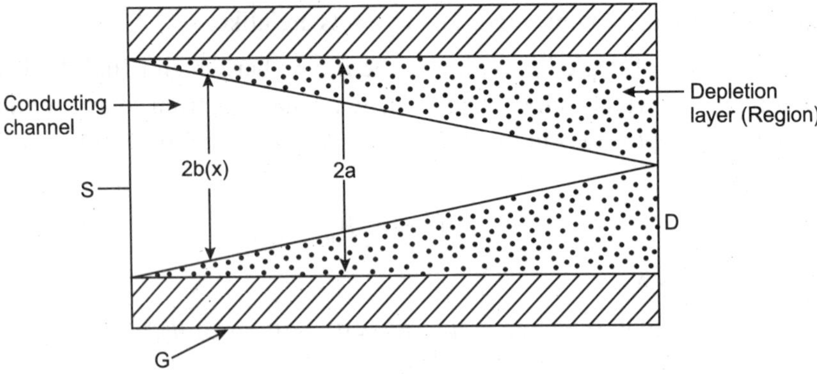

Figure 5.33

The depletion layer halfwidth at x can be determined from Eqn. (ii) putting $y = W(x)$,

$$W_x = a - b(x) = \left[\frac{2\in}{eN_D}|V(x)|\right]^{\frac{1}{2}}$$

Putting $c = \frac{eN_D}{2\in}$, we have

$$|V(x)| = c[a - b(x)]^2 \quad \text{(iii)}$$

The magnitude of voltage across space-charge region for which the channel is just pinched off near drain (D) is called pinch off voltage denoted by

$$V_p = V_{D(sat)} - V_{GS} \quad \text{(iv)}$$

where V_{GS} is either zero or negative.

Eqn. (iii) shows that pinch off takes place from a combination of gate to source voltage and drain to source voltage.

It is to be noted that pinch off is reached at a lower value of $N_{D(sat)}$ when a negative gate bias is applied.

At pinch off, $b(x) = 0$ at drain end (Figure 5.33). Substituting this value in Eqn. (iii)

we have

$$V_p = ca^2 = \frac{eN_D}{2\in}a^2 \quad \text{(v)}$$

Note: For a p-channel JFET, N_D in Eqn. (v) be replaced by N_A which represents concentration of acceptor atoms.

If $b(x)$ decreases linearly with x such that $b(x) = a - kx$ where k is constant, then from Eqn. (iii) and (v) we have

$$V(x) = V_p\left[1 - \frac{b(x)}{a}\right]^2 = V_p \frac{K^2x^2}{a^2} \tag{vi}$$

5.10.5 Effect of V_{GS} on Characteristic Curve

In Figure 5.31, it is noted that drain characteristics are also affected by varying V_{GS} which is reverse-biasing the gate, i.e. it is negative if we have *n*-channel and positive if there is *p*-channel.

With a small reverse-bias voltage, $V_{GS} = (-)1$ V, assuming *n*-channel, the depletion region is quite narrow (Figure 5.34(a)) where we find only immobile positive ions near the channel. Due to the depletion region on both sides of the channel, the channel width is quite large and conduction through the channel is easier on applying V_{DS} and large I_D current is observed.

As reverse-bias is increased to (–)3 and (–)6 V, the depletion region further widens (Figure 5.34(b, c) reaching further into the channel thereby channel width reduces which restricts the passage of electrons into the *n*-type channel (or holes into *p*-type channel), i.e. its resistance increases resulting in lesser drain current (Figure 5.31) for increasing V_{GS}, i.e. more and more negative.

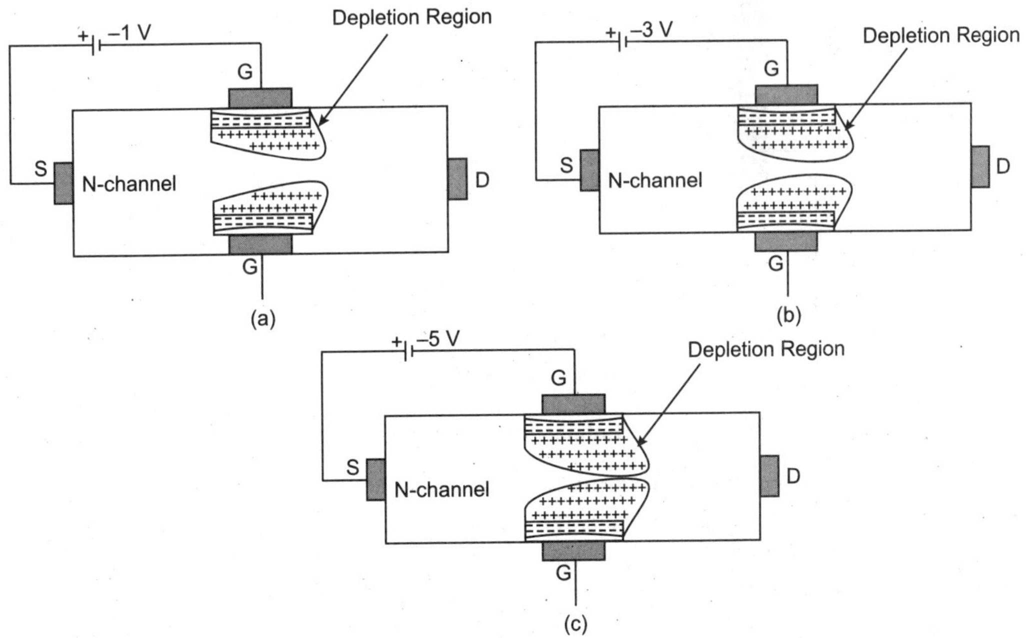

Figure 5.34 Effect of reverse-bias on width of *N* channel (for *p*-channel battery connections are reversed) V_{DS} battery has not been shown.

However, it is to be noted that the depletion region is not symmetrical around the *p-n* junction. It is wider as one moves from source to drain, i.e. the channel width goes on becoming smaller and smaller.

5.11 MOSFET

MOSFET is also called Insulated Gate Field Effect Transistor (IGFET). It can be of *n*-channel or *p*-channel type. Each type is further divided into two classes namely (i) Enhancement MOSFET, (ii) Depletion MOSFET. Its construction and operation differ from usual JFET.

5.11.1 Enhancement MOSFET (N-channel)

Its structure is shown in Figure 5.35.

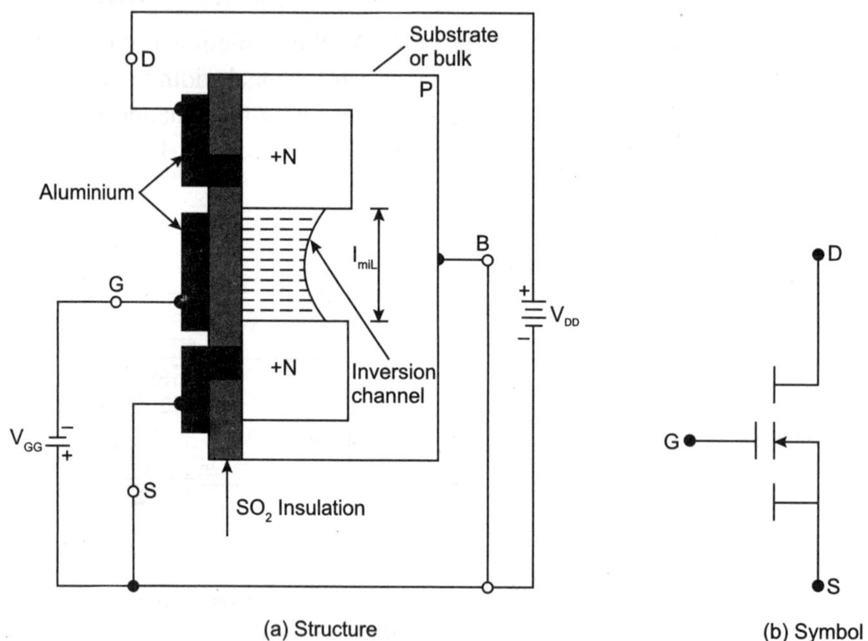

Figure 5.35 Enhancement MOSFET (N-channel)

Here, we have the usual three terminals: drain (*D*), source (*S*) and gate (*G*). The fourth terminal (*B*) connected to the *p*-type substrate is extra (Figure 5.35(a)) but when the substrate is connected internally to the source it is not shown. Figure 5.35(b) shows the symbol for MOSFET. Here a line representing the channel appears as broken into three segments to show that the channel does not exist until a gate voltage is applied, i.e. to show that the device operates in enhancement mode.

5.11.2 Construction and Operation

Its usual construction can be understood in the following steps.

Assume that *p*-type substrate is highly resistive and two blocks of heavily doped *n*-type material are diffused into the substrate and then the surface is coated with a fine layer of silicon dioxide (SiO_2). Gaps are cut through the silicon dioxide layer to make contact with *n*-type blocks. The metal is deposited through the holes to form drain (*D*) and source (*S*) terminals (Figure 5.36(a)) and on the remaining surface area between *D* and *S*, a thin metallic layer is deposited. This metal film will act as gate (*G*). It is to be noted that gate *G* is insulated from the channel by a thin layer of SiO_2 as such there is no *p-n* junction between the gate and channel which is usual in JFET.

Operation A very simplified version of its working is as described below. Assume that drain *D* is positive with respect to source *S* and no potential is applied across gate *G*. The two *n*-blocks and *p*-type substrate from back to back *p-n* junctions connected by a resistor formed due to *p*-substrate (Figure 5.36(b)).

Both junctions cannot be forward-biased at the same time, so only extremely small current (reverse leakage current) flows.

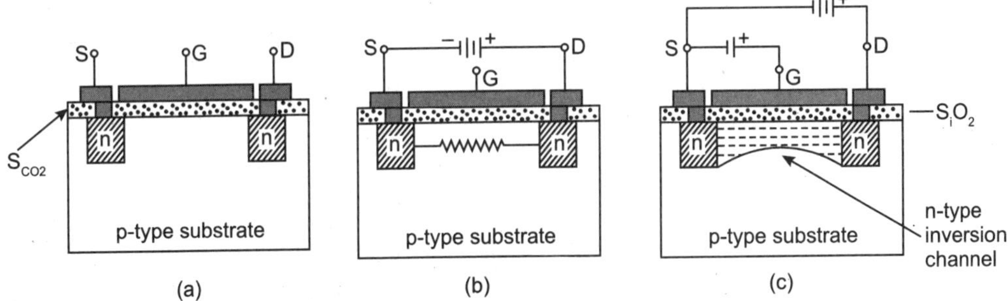

Figure 5.36 (a) n-channel enchancement MOSFET, (b) equivalent circuit when no drain source voltage applied (negative bias), (c) effect of positive gate bias

When gate (*G*) is made positive with respect to source and substrate (Figure 5.35(c)) negative (minority) charge carriers within the substrate are attracted towards the positive gate. These charge carriers (electrons) accumulate close to the upper surface of substrate near the SiO_2 layer below the metallic terminal connected to gate. Accumulation of electrons is subject to a minimum potential on gate called **Threshold Voltage**.

As gate potential V_{GS} is increased, more and more electrons accumulate under the gate (metallic layer) below SiO_2 layer. These electrons do not reach the metallic terminal due to silicon dioxide layer in between substrate and metal, so they just accumulate on upper surface of substrate just below the metallic layer (*G*). These minority charge carriers stretch between *S* and *D* and constitute an *n*-type inversion channel between the source (*S*) and drain (*D*)

(Figure 5.36(c)) so drain current flows from source to drain through the channel. Its magnitude depends upon channel resistance, which in turn depends upon the number of minority charge carriers collected there. Their number is controlled by gate potential, so conductivity (number of charge carriers) of the channel is enhanced by gate potential; so the device is known as enhancement mode MOSFET. The minimum V_{GS} that creates n-type inversion layer is called threshold voltage V_{GS}(Th). Typical drain and transfer characteristics are shown in Figure 5.37.

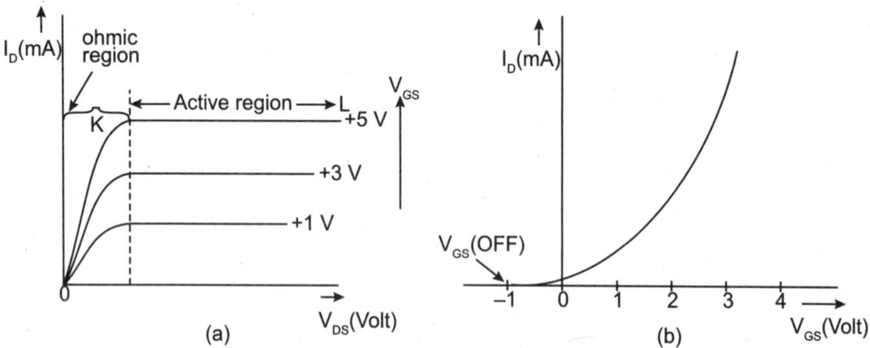

Figure 5.37 N-channel MOSFET (a) Drain characteristics, (b) transfer characteristic

Note that drain current increases with positively increasing gate-source voltage V_{GS}. As the gate of MOSFET is insulated from channel, no leakage current is involved. This gives the device very high input resistance of the order of $10^{10} - 10^{15}$ ohms. Trans conductance value for a MOSFET typically varies from 100 μs to 200 μs or 1 to 2 mA/V.

It is to be noted that the linear region *OK* of drain characteristic (Figure 5.37(a)) is the **ohmic region** and almost horizontal part *KL* is called **active region**. When the device is biased in ohmic region, it behaves like a resistance, and when biased in active region it behaves like current source. Its main use is ohmic region, when it is used as a switching device mainly in computers.

Maximum gate source voltage MOSFET have a thin layer of SiO_2 (Figure 5.35). It is an insulating layer which prevents the flow of gate current from *G* to *S* through the inversion layer. The SiO_2 layer is kept as thin as possible to give the gate voltage maximum control over drain current. This thin insulating layer may be destroyed on applying large gate to source voltage. So one has be cautious while applying V_{GS}. For this season one has to handle the device very carefully. One should never connect or disconnect the device while the power is ON, and before picking up the device, one should ground their body by touching the chassis of the equipment.

5.11.3 Depletion MOSFET

Its structure is shown in Figure 5.38(a). Here too we have the usual three terminals: drain (*D*), source (*S*) and gate (*G*). The fourth terminal *B* connected to substrate is extra. However, when the substrate is connected to a source internally, the extra lead is not shown.

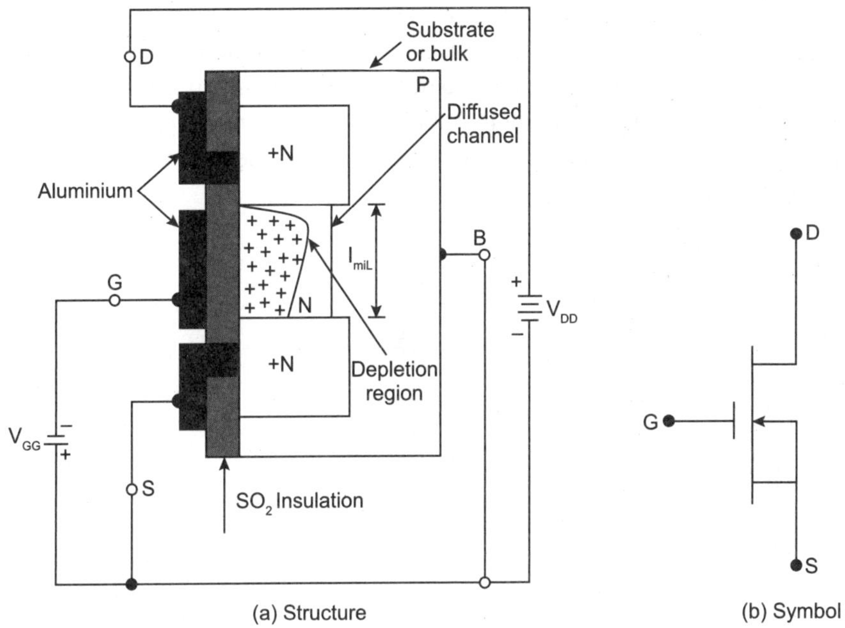

Figure 5.38 N-channel depletion MOSFET

5.11.4 Construction and Operation

Its construction is very similar to that of enhancement MOSFET, except that a highly doped *n*-type channel has been introduced between two heavily doped source (*S*) and drain (*D*) blocks (Figure 5.39). Drain is made positive with respect to source.

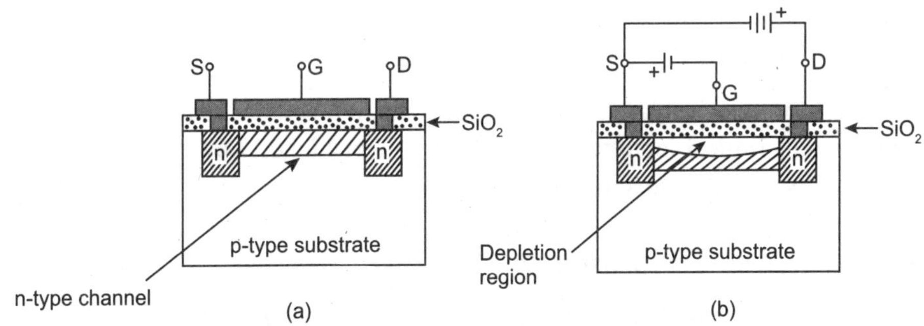

Figure 5.39 Structure of *n* channel MOSFET: (a) n channel depletion MOSFET with no gate bias, (b) depletion mode operation with gate bias

Operation When drain (*D*) is made positive with respect to source (*S*), substantial amount of drain current flows, even with zero gate bias. If the gate is made negative with respect to substrate (Figure 5.39(b)) some of the negative charge carriers are repelled from gate and driven out of

n-type channel. This creates a depletion region in the channel (Figure 5.39(b)) and causes channel resistance to increase, thereby drain current decreases and the action is similar to n-channel JFET. Since the action of negative bias on gate is to deplete the channel of free n-type charge carriers, the device is referred to as depletion-mode MOSFET. The drain characteristics (Figure 5.40(a)) are drawn for various negative gate voltage; the curves obtained are very similar to those of an n-channel JFET (Figure 5.37(a)).

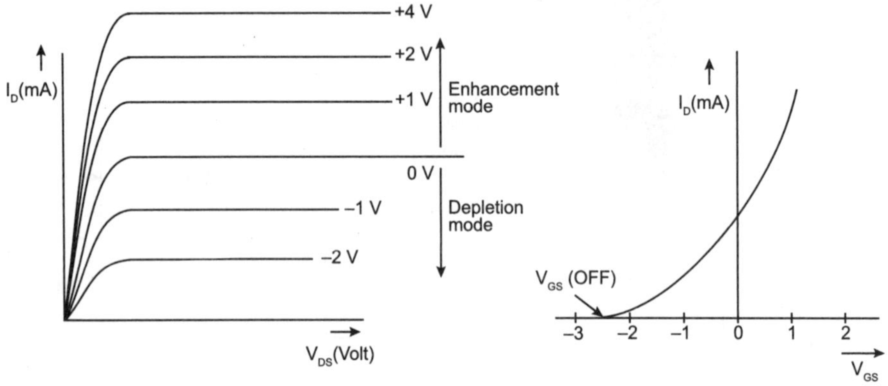

Figure 5.40 Characteristic of depletion MOSFET

However, if gate is made positive with respect to substrate, then additional n-type charge carriers are attracted from p-type substrate into the channel, so channel resistance decreases and I_D flows from source to drain, which increases on increasing gate potential (Figure 5.39(a)). Therefore, depletion-mode MOSFET is capable of operating in enhancement mode also. The resultant characteristics are as shown in Figure 5.40(a). The transfer characteristic is shown in Figure 5.40(b).

5.11.5 Ohmic Region

Although EMOSFET can be biased in the active region, it is seldom done because it is primarily a switching device. Typical input voltage is either low or higher. Low voltage is 0 V and high voltage is V_{GS}(ON), a value specified on data sheets.

5.11.6 Drain Source ON Resistance

When an EMOSFET is biased in the ohmic region, it is equivalent to resistance R_{DS}(ON). There is a Q_{test} point in the ohmic region of $V_{GS} = V_{GS}$ (ON) as shown in Figure 5.41.

$$R_{DS} = \frac{V_{DS}(\text{ON})}{I_D(\text{ON})}$$

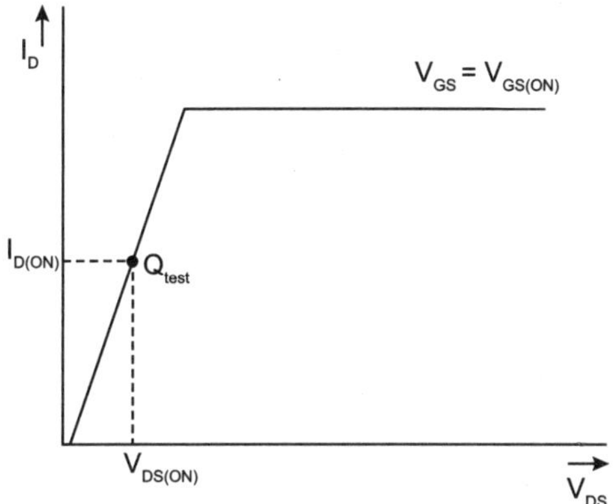

Figure 5.41

5.11.7 Biasing in the Ohmic Region

Consider Figure 5.42(a) where saturation drain current,

$$I_{D(\text{Sat})} = \frac{V_{DD}}{R_D}$$

Drain cutoff voltage is V_{DD}.

Figure 5.42(b) shows the DC load line.

When $V_{GS} = 0$, the Q point is at the lower end of load line.

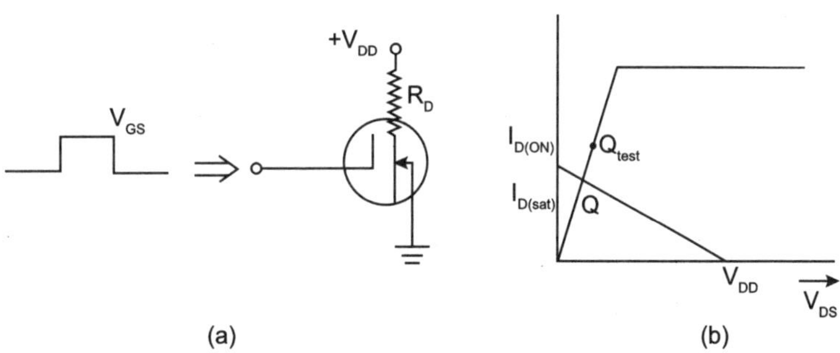

Figure 5.42

When $V_{GS} = V_{GS}$(ON), the Q point is at the upper end of load line. When Q point is below Q_{test} point as shown in Figure 5.41(b), the device is biased in the ohmic region. Stated another way, that EMOSFET is biased in the ohmic region when

$$I_{D(\text{Sat})} < I_{D(\text{ON})} \text{ at } V_{GS} = V_{GS(\text{ON})} \tag{1}$$

So Eqn. (1) defines the state of EMOSFET for it to be working in the ohmic region.

5.11.8 Applications

Digital switching When gate voltage is above threshold voltage, the device switches from cut off to saturation. This OFF-ON action of EMOSFET has made it an important component in any computer, where millions of EMOSFET are used.

5.11.9 Analogue, Digital and Switching Circuits

When we speak of analogue signal, we are talking about signals that change continuously like a sinusoidal wave. The signal need not always be sinusoidal. As long as there are no sudden jumps (changes) between two distinct levels, it is referred as analogue signal.

Digital means discontinuous signal. It means the signal jumps between two distinct levels like a pulse. Such signals are common in a computer. These signals are computer codes that represent numbers, letters or any other symbols.

Switching Switching circuits include digital as a subset. In other words, switching can also refer to circuits that turn on motors, lamps, heaters and other heavy current devices.

Switching is of two types: (i) passive load switching and (ii) active load switching.

Passive load switching Figure 5.43 shows an EMOSFET with passive load R_D. The word passive refers to ordinary resistors like R_D.

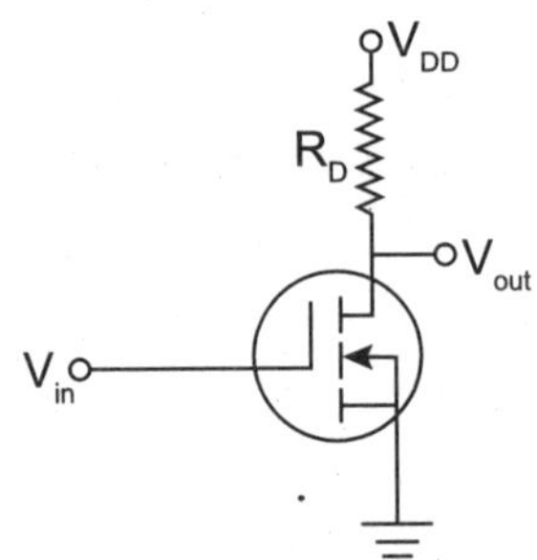

Figure 5.43

In this circuit, V_{in} may be either low or high.

When V_{in} is low, the MOSFET is cut off and V_{out} equals supply voltage V_{DD}, i.e. it is high. When V_{in} is high, the MOSFET saturates and V_{out} drops to a low value.

For the circuit to work properly, the drain saturation current has to be less than $I_{D(\text{ON})}$ when input voltage is equal to or greater than $V_{GS(\text{ON})}$. This is equivalent to saying that resistance $R_{DS(\text{ON})}$ in the ohmic region has to be much smaller than passive drain resistance R_D.

$$R_{DS(\text{ON})} \ll R_D$$

5.12 CMOS (Complimentary MOSFET)

Here the IC designer combines *n*-channel and *p*-channel MOSFETS. It is symbolically represented in Figure 5.44.

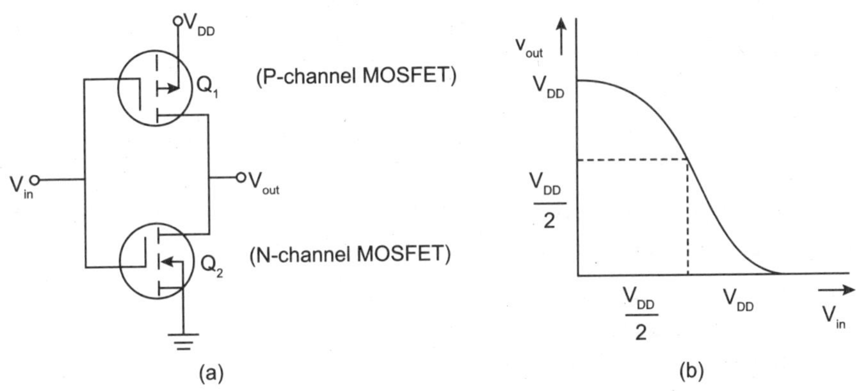

Figure 5.44 (a) CMOS inverter, (b) input-output graph

As shown in the figure, Q_1 is a *p*-channel MOSFET and Q_2 is an *n*-channel MOSFET. These two devices are complimentary, that is they have equal and opposite values of $V_{GS(\text{Th})}$, $V_{GS(\text{ON})}$, $I_{D(\text{ON})}$ and so on. The above circuit is similar to a class B pushpull amplifier. When one MOSFET conducts, the other is in nonconducting state.

5.12.1 Basic Action

When a CMOS circuit like that shown in Figure 5.44 is used in switching applications (like in computers), the input voltage is either high ($+V_{DD}$) or low (0 V).

When the input voltage is high, Q_1 is OFF and Q_2 is ON. In this situation, the shorted Q_2 pulls output voltage down to ground (i.e. it attains a low value). On the other hand, when input voltage is low, Q_1 is ON and Q_2 is OFF. In this situation, the shorted Q_1 pulls output voltage up to $+ V_{DD}$.

Since the output voltage is inverted in both situations, the circuit is called a CMOS inverter.

Summary

1. A bipolar junction transistor consists of a very thin region called base, sandwiched between two regions called emitter and collector. Their base material is silicon or germanium.
2. Transistors are either *p-n-p* or *n-p-n* type. The first letter designates the emitter material, the middle letter the base and last letter, the collector material.
3. The base region is most lightly doped while the emitter and collector region is heavily doped.

4. Junction transistors have two junctions, emitter-base (EB) junction and collector-base (CB) junction.
5. For the purpose of biasing a transistor for normal operation, the E-B junction is forward-biased while CB junction is reverse-biased.
6. The emitter is treated as source of majority charge carriers, i.e. it injects majority charge carriers most of which are received by the collector. The base controls the flow of charge carriers and the three currents are related as $I_E = I_B + I_C$.
7. For *n*-type emitter the majority charge carriers are electrons, while that for *p*-type emitter the majority carriers are holes.
8. There are minority carriers also. Small minority carrier current is in collector base region with the emitter open; the current is known as I_{CBO}.
9. Transistors may be connected in three configurations, i.e. common emitter (CE), common base (CB) and common collector (CC). Here, the term common indicates that this terminal is common to both input side and output side.
10. The term "transistor" comes from "transfer + resistor", i.e. as the signal is applied on the input side, it is transferred from low resistance to high resistance.
11. Since a transistor has input side and output side it has two types of characteristics in each configuration, namely input characteristics and output characteristics.
12. The slope of the characteristics is an indication of resistance offered by the device in that configuration.
13. The slope of output characteristics in CB configuration is practically very small; it indicates that output resistance is very high and the slope in case of CE configuration is finite.
14. In all the three configurations, the transistor has one terminal common and the other two refer to input and output side. Thus, it has four terminals (one common) or it is represented like a two-port network.
15. The equivalent circuit using two-port networks in all the configurations of a transistor can be represented by *h* parameter.
16. The static (*DC*) forward current transfer ratio of a transistor in CE configuration is designated by $\beta = \frac{I_C}{I_B}$ and represents current gain of the transistor. It is always more than 1; in terms of *h* parameters, it is represented by h_{FE}.
17. In case of CB configuration, the ratio of collector current to emitter current is designated as $\alpha = \frac{I_C}{I_E}$; its value is always less than 1; then $\beta = \frac{\alpha}{1-\alpha}$.
18. Unipolar transistors are called field effect transistors (FET) because there is only one type of charge carriers in the channel for their operation. These charge carriers are majority charge carriers as per the nature of channel (*n*-type or *p*-type).

19. There are two types of FETs. They are JFET and metal oxide FET (MOSFET).
20. A uniploar transistor or FFT has three terminals: source (*S*), gate (*F*) and drain (*D*) which emitter (*E*), base (*B*) and collector (*C*) in case of BJT. But here we have only one junction between gate (*G*) and channel.
21. FET is a voltage controlled device while BJT is a current controlled device.
22. In JFET, the gate is negatively biased with respect to channel (source) which may be *n*-type or *p*-type.
23. The family of drain characteristics of a JFET reveals that:
    (i) I_{DSS} is the maximum drain current the transistor can carry in normal operation.
    (ii) As the reverse-bias of gate (V_{GS}) increases, drain current (I_D) decreases, until at $V_{GS(\text{off})}$ it is cut off.
24. The input impedance of JFET is very high.
25. The two types of MOSFET are (i) depletion and (ii) enhancement.
26. The gate of MOSFET is insulated from substrate by a thin layer of SiO_2. The device is therefore called insulated gate field effect transistor (IGFET).
27. Enhancement type MOSFET has no physical channel between drain and source similar as in JFET. Here the channel, consisting of charge carriers, is electrically induced in the substrate by biasing the gate. It is also called Inversion Layer. Its gate is reverse (negatively) biased.
28. Depletion type MOSFET (also called depletion-enhancement type) contains a channel. Its gate can be positively biased or negatively biased or zero biased.
29. When the gate of *N*-channel depletion MOSFET is biased positively, it operates in enhancement mode, and when the gate is biased negatively it operates in depletion mode.
30. MOSFETs are very fragile, since its SiO_2 layer being very thin gets punctured. To protect the layer from transient voltage (static charge), a metallic ring is put around all the terminals when not in use.

Review Questions

1. Draw a block diagram of (i) *p-n-p* and (ii) *n-p-n* transistor. Show the depletion region and barrier potential in them under zero bias. Explain briefly their formation.
2. How does the depletion region width and barrier potential change when a *p-n-p* transistor is properly biased?
3. What are the different current components in the three segments of a *p-n-p* transistor when it is properly biased? Depict the various current components in the block diagram of the transistor.

4. Sketch and explain the shape of input and output characteristics of a transistor in CB configuration.
5. Sketch and explain the shape of input and output characteristics of a *n-p-n* transistor in CE configuration. How these characteristics can be drawn experimentally? Draw necessary circuit.
6. Define α and β for a transistor and derive the necessary relation between them.
7. Explain how current amplification is obtained in a transistor?
8. Explain how a transistor may be used for voltage amplification?
9. Sketch the '*h*' parameter equivalent circuit for a transistor in CE configuration. Identify each component of the circuit and discuss its origin.
10. Define '*h*' parameters of a transistor. Sketch the circuit of a transistor in CB configuration using '*h*' parameters. Identify each component in the circuit.
11. Define '*h*' parameters in case of CE configuration. Sketch a circuit that could be used for determining their values. Write proper units for them.
12. What do the terms unipolar and bipolar transistor refer to? Explain the working of unipolar transistor. Draw its drain characteristics.
13. Draw drain characteristics of a *p*-type JFET. Explain the shape of the characteristics. How do they depend upon V_{DS} and V_{GS}? Explain.
14. Define the following terms is case of JFET: (i) pinch off voltage (ii) channel ohmic region (iii) drain resistance (iv) transconductance.
15. Draw drain characteristics of a *n*-type JFET. Depict (i) pinch off region (ii) I_{DSS} (iii) channel ohmic region (iv) pinch off voltage on them.
16. Explain the phenomenon of narrowing the channel in a JFET with (i) V_{DS} (ii) V_{GS}. How the pinching off current in channel occurs?
17. Explain the working of MOSFET in (i) enhancement mode (ii) depletion mode. Draw necessary block diagram showing the structure of MOSFET.
18. Why the input impedance of a JFET is very high. Mention its order. Where we require such high impedance?
19. Give the merits and demerits of JFET over a BJT.
20. How does the construction of MOSFET differ from that of JFET? Sketch the two devices. In what respect a MOSFET is better. What precaution is taken while using it?
21. Sketch a typical family of drain characteristics for a *p*-channel JFET, for various values of V_{GS}. Explain the effect of variation of V_{GS}. Also sketch the transfer characteristics of the device. Define transconductance and mention its unit.
22. Draw the circuit of JFET to show its drain characteristics. Define (i) drain resistance (ii) transconductance. Mention their unit.

23. Draw a complete equivalent circuit for a JFET. Explain the origin of each component. How this circuit gets simplified for low frequency operation?
24. Write an equation for calculation of drain current (I_D) for various levels of V_{GS}. How the equation matches the characteristic.
25. Explain the operation of a *n*-channel depletion mode MOSFET. Sketch the drain characteristics for the device and explain their shape.

Short Answer Questions

1. What is β_{dc} in case of BJT. Is it a constant quantity?

   Ans: β_{dc} of BJT represents current gain. It is the ratio of collector current to base current. It is not a constant quantity. It depends upon three factors: (i) transistor (ii) collector current drawn (ii) temperature of the device.

2. How many regions are there in the output characteristics of a transistor? Name them.

   Ans: There are four regions in the output characteristics of a transistor. They are (i) saturation region (ii) active region (iii) cutoff region (iv) breakdown region.

3. Why the base region in a BJT is narrow and lightly doped?

   Ans: The base region in a BJT is the narrowest or smallest region so that the maximum percentage of majority carriers injected from emitter reach the collector region. Light doping the base region helps in lesser recombination of majority carriers entering from emitter when they reach there.

4. Why EB junction is forward-biased and CB junction is reverse-biased in a BJT?

   Ans: For a BJT to operate as an amplifier, the operating point must remain in active region. It is possible only when EB junction is forward-biased and CB junction is reverse-biased.

5. In which region the value of current gain β_{dc} of a transistor is minimum.

   Ans: Current gain β_{dc} of a transistor has minimum value when the transistor operates in saturation region.

6. Why the collector current in CB configuration attains nearly a constant value?

   Ans: The collector current is CB configuration attains nearly a constant value, i.e. curve is flat, because on increasing reverse-bias V_{CB}, the depletion region in base-collector region widens more in the base region, thereby chances of recombination of majority carriers from emitter are minimized and the entire charge carriers reach the collector region, i.e. collector current remains constant.

7. Can a JFET behave like a resistor?

   Ans: Yes, when it operates below pinch off voltage in the ohmic region.

8. Can a JFET behave like current source?

   Ans: Yes, when it operates beyond pinch off voltage. Drain current is nearly constant with increasing drain to source voltage (V_{DS}).

9. Why drain current in a JFET attains nearly a constant value?

   Ans: Drain current in a JFET attains nearly a constant value because on increasing V_{DS} the depletion region formed at gate on the two sides of FET widens and the two nearly touch each other, thereby increase in drain current is stopped or it is pinched off.

10. How an EMOSFET operates? Mention the condition.

    Ans: An EMOSFET operates only when an inversion layer of charge carriers is formed between source and drain. It is formed when gate voltage is equal to the threshold voltage of the device.

11. Why a thin metal wire ring is put around the leads of a MOSFET during shipment?

    Ans: A thin metal wire ring is put around the leads to save the thin SiO_2 layer from breaking up due to electrostatic charge.

12. What does the flat part in the CB output characteristic represent?

    Ans: The flat part in (V_{CB}, I_E) characteristic in CB configuration represents that output resistance r_o of the device in this configuration is very high. Moreover, the device in this region behaves like current source.

13. What does the term "transistor" suggest?

    Ans: It suggests that there is "transfer of resistance", i.e., when a signal is applied across the device, it undergoes from a region of low resistance to a region of high resistance.

14. What does the term "field effect transistor" refer to?

    Ans: It suggests that drain current in a FET is affected by the field created by ions present in the depletion region formed at the gate.

15. Why is the collector base junction in a CE configuration thought of as leaky diode?

    Ans: In CE configuration, the collector-base junction is forward-biased for minority carriers but the emitter is forward-biased by the reverse-bias of CB of the junction. So majority carriers from the emitter cross the base-collector junction and in this respect it behaves like a leaky diode.

16. In a BJT, which current is maximum and which current is minimum?

    Ans: In a BJT under normal operation, emitter current is maximum and base current is minimum.

17. Can a transistor act as voltage amplifier in CB configuration? Can it also work as current amplifier in this configuration.

    Ans: Yes, a transistor can function as voltage amplifier in CB configuration. But it cannot function as current amplifier.

18. In CB configuration, emitter at zero bias. Can we have collector current? Explain.

    Ans: We have collector current in CB configuration even if I_E is zero due to emitter being at zero bias. The collector current is made up of minority carriers across CB junction since it is reverse-biased. Its value is equal to reverse saturation current for the diode.

19. What does h_{FE} represent using a hybrid parameter for a bipolar transistor?

    Ans: It represents current gain in CE configuration. It is defined as the ratio of collector current to base current. For AC circuit, it represents the ratio of change in collector current to change in base current.

20. What does the term I_{CBO} represent in a transistor?

    Ans: It represents reverse-bias saturation current through collector-base junction when emitter-base junction is open.

21. What is being represented by the value of β in case of transistor?

    Ans: The value of β in case of transistor represents its ability to handle change in collector current for a given change in base current.

Problems

1. In a transistor calculate I_C and I_E if $\alpha_{dc} = 0.77$ and $I_{CBO} = 10$ μA, I_B is measured as 50 μA.

   [**Ans:** 1.95 mA, 2.02 mA]

2. For a certain transistor, $I_C = 5.255$ mA, $I_B = 100$ μA and $I_{CBO} = 5$ μA. Calculate α_{dc}, β_{dc} and I_E. Calculate the new level of I_B required to make $I_C = 15$ mA.

   [**Ans:** (a) 0.98 (b) 49 (c) 5.4 mA]

3. For a given transistor, collector current is 51 mA when base current is 0.4 mA. If $\beta_{dc} = 12$, calculate cutoff current I_{CBO}. [**Ans:** 1 mA]

4. The cutoff current I_{CBO} for a transistor is 14 μA at room temperature and its $\beta_{dc} = 50$. Calculate its collector current when $I_B = 0.2$ mA. [**Ans:** 10.71 mA]

5. In a transistor current gain is 125 and base current is 30 μA. Calculate its collector current. If we assume $I_{CBO} = 10$ μA. How does collector current change?

   [**Ans:** 3.75 mA, 1.25 mA]

6. A JEFT has $V_p = 6$ V and $I_{DSS} = 100$ mA. Calculate its channel ohmic resistance and gate-source cutoff voltage. [**Ans:** (i) 60 Ω (ii) (–) 6v]

7. Calculate transconductance of a JFET if its $V_p = 8$ V; assume $I_{DSS} = 1.6$ mA and $V_{GS} = (-)$ 1 V.

   $$\left[\text{Hint: } g_m = \frac{2I_{DSS}}{V_p}\left(1 - \frac{V_{GS}}{V_P}\right)\right]$$

   What will be its value if V_{GS} changes to (–) 2 V.

   [**Ans:** 3500 μs]

Objective Questions

1. In a bipolar transistor and a unipolar transistor, the number of junctions are
   (a) 3, 3 (b) 3, 2 (c) 2, 2 (d) 2, 1 **[Ans: (d)]**
2. In a bipolar transistor, the number of doped regions are
   (a) 1 (b) 2 (c) 3 (d) 4 **[Ans: (c)]**
3. In a *n-p-n* transistor, the majority carriers are
   (a) electrons (b) holes
   (c) both holes and electrons (d) neither holes nor electrons **[Ans: (a)]**
4. Current gain in a BJT in CE configuration is defined as
   (a) I_C/I_E (b) I_C/I_B (c) I_E/I_B (d) I_E/I_C **[Ans: (b)]**
5. Base current in a BJT is 50 μA. If current gain is 125, collector current is closest in value to
   (a) 6.25 mA (b) 175 μA (c) 0.04 μA (d) 0.04 mA **[Ans: (a)]**
6. The current of gain in a junction transistor in CE configuration is 200 and collector current is 100 mA. Now, base current is
   (a) 0.5 mA (b) 2 mA (c) 0.2 mA (d) 5 mA **[Ans: (a)]**
7. The relation between α and β of a transistor is
   (a) $\alpha = \beta$ (b) $\alpha = \frac{\beta}{\beta - 1}$ (c) $\beta = \frac{\alpha}{1+\alpha}$ (d) $\beta = \frac{\alpha}{1-\alpha}$ **[Ans: (d)]**
8. In a BJT, collector current (I_C), emitter current I_E and base current (I_B) are related as
   (a) $I_C = I_E + I_B$ (b) $I_E = I_C + I_B$ (c) $I_B = I_E + I_C$ (d) $I_B = I_C - I_B$ **[Ans: (b)]**
9. In a *n-p-n* transistor, the conventional flow of current is
   (a) 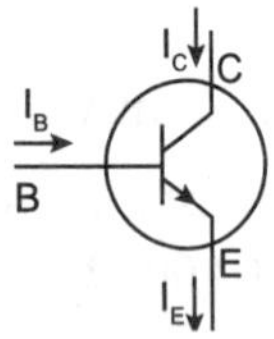 (b) 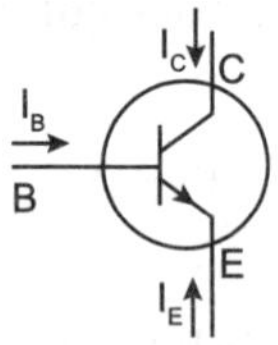
   (c) 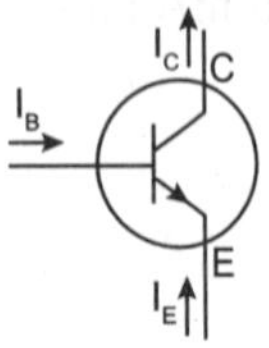 (d) 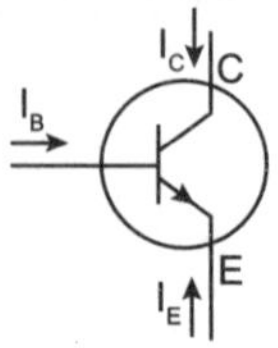 **[Ans: (a)]**
10. For the normal operation of a bipolar junction transistor
    (a) EB junction is forward-biased and CB junction is reverse-biased
    (b) EB junction is reverse-biased and CB junction is forward-biased
    (c) Both EB and CB junctions are forward-biased
    (d) Both EB and CB junctions are reverse biased **[Ans: (a)]**

11. Base emitter voltage of an ideal silicon transistor is
    (a) 0 V (b) 0.7 V (c) 0.3 V (d) 1.0 V **[Ans:** (b)]
12. In a *p-n-p* or *n-p-n* transistor, the percentage of doping is
    (a) all the three regions are equally doped
    (b) emitter is most heavily doped but base region and collector region are lightly doped
    (c) emitter is lightly doped but base and collector regions are heavily doped
    (d) emitter region > collector region > base region **[Ans:** (d)]
13. For high voltage gain, BJT is normally used in
    (a) CE configuration (b) CB configuration
    (c) CC configuration (d) all the three configurations **[Ans:** (b)]
14. Power dissipated by a transistor is approximately equal to collector current times:
    (a) base-emitter voltage (b) collector-emitter voltage
    (c) base-collector voltage (d) collector-supply voltage **[Ans:** (b)]
15. For a transistor to operate as amplifier, it is to be used in
    (a) active region (b) cut-off region
    (c) saturation region (d) break-down region **[Ans:** (a)]
16. Charge carriers in a JFET are
    (a) electrons only
    (b) holes only
    (c) both electrons and holes together
    (d) both electrons and holes but one at a time **[Ans:** (d)]
17. Compared to a bipolar transistor, a JFET has much higher
    (a) voltage gain (b) current gain
    (c) input impedance (d) output impedance **[Ans:** (c)]
18. The source to drain resistance (R_{DS}) in a JFET is given by
    (a) $R_{DS} = \frac{V_p}{I_D}$ (b) $R_{DS} = \frac{V_{DS}}{I_D}$
    (c) $R_{DS} = \frac{V_p}{I_{DSS}}$ (d) $R_{DS} = \frac{V_{DS}}{I_{DSS}}$ **[Ans:** (d)]
19. Trans conductance of a JFET is measured in
    (a) ohms (b) siemens
    (c) volt (d) ampere **[Ans:** (b)]
20. When a JFET is cutoff, the depletion layers on two sides are
    (a) far apart (b) close together
    (c) conducting (d) break down **[Ans:** (b)]

Mark the Following Statements True/False.

1. The value of 'α' for a transistor is always less than unity. [**Ans:** True]
2. The value of β_{de} for a transistor is less than its α_{dc} value. [**Ans:** False]
3. JFET can be used as a variable resistor below pinch off voltage. [**Ans:** True]
4. A depletion region is formed in a MOSFET. [**Ans:** True]
5. MOSFET terminals are shorted by a metallic ring while not in use to save its SiO_2 layer from being punctured. [**Ans:** True]
6. JFET is a unipolar device but MOSFET is a bipolar device. [**Ans:** False]
7. In CB configuration, the collector-base junction is forward-biased in the saturation region. [**Ans:** False]
8. Collector current in CE configuration is given by $I_C = \beta I_B + (\beta + 1)I_{CO}$ [**Ans:** True]
9. There is no cutoff region in the output characteristics of a transistor in CB configuration. [**Ans:** False]
10. The slope of output characteristics in the active region of a transistor in CE configuration is higher than those in CB configuration. [**Ans:** True]

Fill in the Blanks.

1. The emitter region of a transistor is ____________ doped while the base region is ____________ doped. [**Ans:** heavily, lightly]
2. The base region is ____________ in comparison with emitter region. [**Ans:** thinner/narrower]
3. In CB configuration, resistance of a transistor is ____________ while output resistance is ____________. [**Ans:** low, high]
4. Current gain in CE configuration is always ____________ than 1 while it is ____________ in ____________ configuration. [**Ans:** more, less, common base]
5. Current gain in CE configuration is the ratio of ____________ and ____________. [**Ans:** collector current, base current]
6. Modern transistors are usually made up of ____________. [**Ans:** silicon]
7. In a *p-n-p* transistor, the emitter is kept ____________ relative to base while the collector is kept ____________ relative to base. [**Ans:** positive, negative]
8. The arrow head on the emitter in the symbol of transistor shows ____________ of conventional current which is ____________ to electron flow in the circuit. [**Ans:** direction, opposite]

Chapter

6 Thermal Stability

6.1 Introduction

A semiconductor device like diode or transistor can be seriously affected by temperature. In a way, they are temperature-sensitive devices. In Chapter 3, we have discussed that the characteristics of both forward-bias and reverse-bias shift on rise in temperature. So also the characteristics of a transistor are affected and thus for smooth operation of transistors, we must ensure they are thermally stable.

6.2 Factors Contributing to Thermal Stability

Two of the most important temperature sensitive quantities are base emitter voltage (V_{BE}) and collector base reverse saturation current (I_{CBO}).

To make any use of a transistor, it must be DC biased (forward-biased as well as reverse-biased). The purpose of DC biasing is to obtain certain collector current at certain collector voltage. These values of current and voltage define its operating point.

6.2.1 Operating Point

To understand the concept of operating point, we consider Figure 6.1. It represents a transistor in CE configuration with proper DC bias and load R_L.

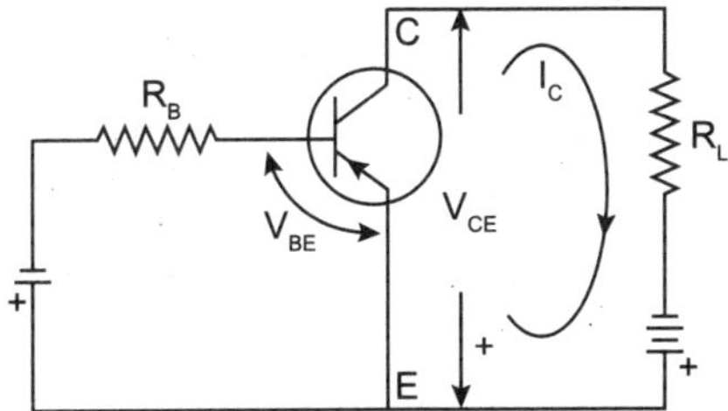

Figure 6.1 Fixed bias CE configuration

With biasing as shown, let collector current I_C be flowing through the output circuit.

Applying KVL we have

$$V_{CC} = V_{CE} + I_C R_L$$

or
$$I_C = (-)\frac{1}{R_L}V_{CE} + \frac{V_{CC}}{R_L} \tag{1}$$

Equation (1) represents the equation of straight line called **load line**. It can be represented on the output (V_{CE}, I_C) characteristic (Figure 6.2) of the transistor.

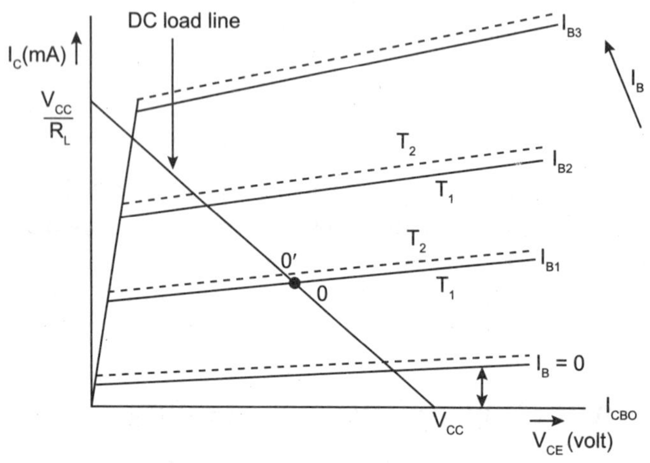

Figure 6.2 Load line CE configuration

With biasing applied to the transistor, suppose base current I_{B1} is flowing; then the intersection of load line with (V_{CE}, I_C) curve for base current I_{B1} represents the operating point *O*.

Collector current I_C flowing through a transistor in CE configuration is given by Eqn. (2).

$$I_C = \beta I_B + (\beta + 1)I_{CBO} \tag{2}$$

The above equation represents that collector current I_C depends upon three variables: (i) β (or h_{FE}), (ii) I_B, (iii) I_{CBO} which in turn depends upon temperature.

As collector current flows through the device, its temperature rises due to thermal generation of charge carriers. This is turn rises I_{CBO} and consequently collector current further rises. This process is regenerative, and so at a certain point, the device may be damaged, which we call thermal runaway.

It has been observed that I_{CBO} approximately doubles for every 10°C rise in temperature.

From Figure 6.2, it is clear that as collector current changes due to temperature for the same base current the entire family of characteristic shifts due to this shift; of characteristic, operating point *O* shifts to *O*′ for the same load. If the shift is large, then the operating point may reach the saturation region, resulting in the clipping and distortion of output signal even if the input signal remains the same.

We conclude that an acceptable operating point at one fix temperature may become poor at increased temperature. To stop or minimize this tendency of shifting of operating point, we make some circuit modification (Figure 6.1) to decrease the base current so that the operating point O' will shift back near to the original operating point O.

A decrease in I_B shall decrease the first quantity βI_B in Eqn. (2) which will compensate the increase in second quality, i.e $(\beta + 1)\ I_{CBO}$. As a result, collector current shall remain nearly constant even on increase in temperature.

Changes in V_{BE} may also produce significant changes in I_C, and thereby the DC operating point shifts. However, because of the possibility of thermal runaway, the changes in I_{CBO} are by far most important. Thermal stability of a circuit is assessed by deriving a stability factor S.

6.2.2 Stability Factor (S)

It is defined as the ratio of rate of change of collector current to the rate of change of I_{CBO} keeping β and I_B (or V_{BE}) constant. Accordingly,

$$S = \left.\frac{\Delta I_C}{\Delta I_{CBO}}\right|_{\beta, I_B \text{ constant}} \tag{3}$$

The process of making the operating point independent of temperature changes or variation in transistor parameters is known as stabilization. Once stabilization is achieved, the operating point remains practically fixed.

Differentiating Eqn. (2) w.r.t. I_{CBO}, assuming β and V_{BE} (I_B) as constant, we have

$$S = \frac{\Delta I_C}{\Delta I_{CBO}} = \beta + 1 \tag{4}$$

For a typical transistor, β has a value 100; so the value of $S = 101$ which means that collector current changes by a factor 101 times as much as any change in I_{CBO} is concerned.

The fact that S is very large in fixed bias circuit (Figure 6.1) indicates that such circuits are not suitable because they give poor thermal stability. In CB configuration with fixed bias, we have

$$I_C = \alpha I_E + I_{CEO}$$

Now, stability factor $S = \dfrac{\Delta I_C}{\Delta I_{CEO}} = 1$ (keeping α and I_E constant)

It represents that the value of S is 1, which is ideal. Hence in common base configuration, we do not require any modification in the circuit to keep it thermally stable.

6.3 Biasing Circuits for Thermal Stabilization (CE Configuration)

The most common methods for obtaining transistor biasing using single source supply (V_{CC}) are

(i) Collector to base bias

(ii) Emitter bias

(iii) Voltage divider with emitter bias

In all the above methods, the following requirements should be considered:

(i) It should ensure proper zero signal collector current.

(ii) It should ensure that V_{CE} does not fall below 0.5 V for germanium transistor and 1.0 V for silicon transistor.

(iii) It should ensure stabilization of operating point.

6.3.1 Collector-to-Base Bias

Here, one end of resistor R_B is connected to collector (*C*) and the other to base (*B*) as shown in Figure 6.3. Here we use only one supply V_{CC}.

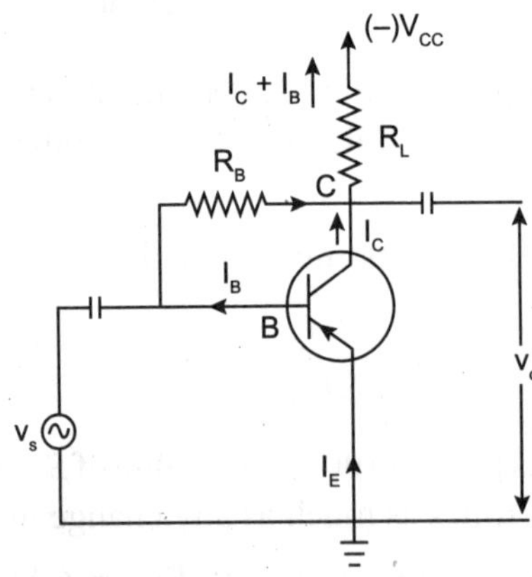

Figure 6.3 Collected-to-base bias

The base-emitter junction is forward-biased by V_{CB} which also ensures zero signal base current I_B. If collector current I_C tends to increase due to increase in temperature or change of β (change of transistor), potential drop across load R_L will increase and it results in lowering the collector to emitter voltage $[V_{CE}]_Q$. This is turn reduces base current I_B, and the operating point O' shifts back to O (Figure 6.2). However, it is to be kept in mind that decrease in base current I_B does not compensate exactly the increase in I_C; so O' does not shift exactly to O.

Circuit analysis The required value of resistor R_B needed for zero signal collector current I_C can be determined by applying KVL to the input side of circuit (Figure 6.3).

$$V_{CC} = (I_C + I_B)R_L + I_B R_B + V_{BE}$$

$$I_B = \frac{V_{CC} - V_{BE} - I_C R_L}{R_L + R_B}$$

Also $$I_C = \beta I_B + (\beta + 1) I_{CBO} \tag{5}$$

Substituting for I_B from Eqn. (5),

$$I_C \left[1 + \frac{\beta R_L}{R_B + R_L} \right] = \frac{\beta (V_{CC} - V_{BE})}{R_B + R_L} + (\beta + 1) I_{CBO}$$

Differentiating the above equation with respect to I_{CBO},

$$\text{Stability factor, } S = \frac{\partial I_C}{\partial I_{CBO}} = \frac{\beta + 1}{1 + \dfrac{\beta R_L}{R_B + R_L}} \tag{6}$$

Equation (6) suggests that the value of stability factor S has reduced much from $(\beta + 1)$ given by Eqn. (4).

However, this circuit has two drawbacks.

(i) The circuit does not provide good stabilization. It is because the operating point does not shift to its original position.

(ii) The circuit provides a path for feedback to AC signal and thereby it reduces the voltage gain of the amplifier. This drawback can be partially overcome if R_B is split into two equal parts and a capacitor is connected between the midpoint of R_B and ground. The value of C is so selected that its reactance is smaller than $R_B/2$.

The other two stability factors S' and S'' are given by

$$S' = \frac{\partial I_C}{\partial V_{BE}} = \frac{\beta}{R_B + R_L(\beta + 1)} \tag{7}$$

$$S'' = \frac{\partial I_C}{\partial \beta} = \frac{V_{CC} - V_{BE} - I_C R_L + (R_B + R_L) I_{CO}}{R_B + R_L(\beta + 1)}$$

$$= \frac{V_{CC} - I_C R_L + (R_B + R_L) I_{CO}}{R_B + R_L(\beta + 1)} \quad \text{since } V_{BE} \ll V_{CC} \tag{8}$$

6.3.2 Emitters Bias

The collector to base bias circuit does not give good stabilization or has low stability factor in cases where the DC load R_L is quite low (of the order of 5–10 Ω). In such cases, if it can be so arranged that, when collector current I_C increases, the forward bias across base-emitter junction is made to decrease simultaneously, resulting in a decrease of base current I_B and subsequently, the operating point shifts back to the original position along load line. The above mechanism is achieved by placing a resistor R_E in series with emitter (Figure 6.4). When emitter current flows through it, voltage across it provides a negative feedback to the input which reduces the signal strength and base current reduces. It is called current feedback since voltage feedback is proportional to emitter current.

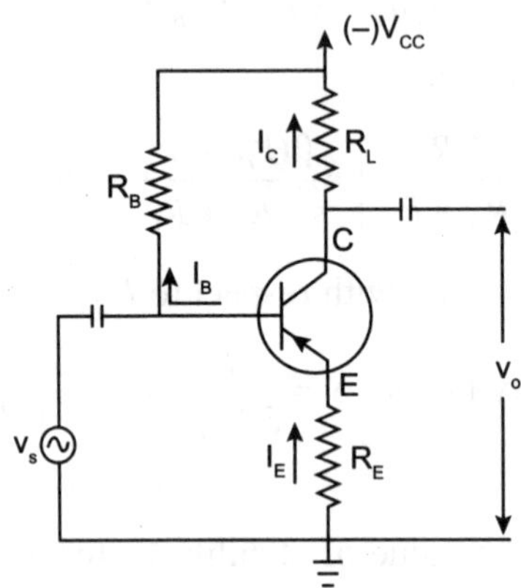

Figure 6.4 Emitter bias circuit

Circuit analysis Applying KVL to the input circuit, we have

$$V_{CC} = I_B R_B + V_{BE} + I_E R_E \tag{9}$$

$$I_B = \frac{V_{CC} - V_{BE} - I_E R_E}{R_B} \tag{9A}$$

$$= \frac{V_{CC} - I_E R_E}{R_B}, \text{ neglecting } V_{BE} \text{ since } V_{BE} \ll V_{CC}$$

$$= \frac{V_{CC} - I_C R_E}{R_B} \text{ since } I_E \approx I_C \tag{10}$$

From the above relation, it is clear that any increase in I_C due to rise in temperature will decrease the R.H.S. Since I_C increases, the value of I_B will also decrease which in turn offsets the rise in I_C.

Also
$$I_E = I_C + I_B$$

Substituting for I_E in Eqn. (9A), we have

$$V_{CC} = I_B R_B + V_{BE} + (I_C + I_B) R_E$$

or
$$I_C R_E = V_{CC} - V_{BE} - I_B (R_B + R_E) \tag{11}$$

From Eqn. (2), $I_B = \dfrac{I_C - (\beta + 1) I_{CO}}{\beta}$.

Substituting for I_B in Eqn. (11), we have

$$I_C\left[R_E+\frac{R_E+R_B}{\beta}\right]=V_{CC}+\frac{\beta+1}{\beta}I_{CO}\ (R_E+R_B)$$

Differentiating with respect to I_{CO}, we have

$$S=\frac{\partial I_C}{\partial I_{CO}}=\frac{\beta+1}{1+\dfrac{\beta R_E}{R_E+R_B}} \tag{12}$$

In Eqn. (12), if $\dfrac{R_B}{R_E}\ll 1$

then
$$S=\frac{\beta+1}{\beta+1}=1 \tag{13}$$

Thus, the stability factor approaches unity which is ideal. Closer the value of S to unity, lesser the variation of operating point with temperature.

However, if $\dfrac{R_B}{R_E}\gg 1$

then the value of S approaches $\beta + 1$. (14)

This value is very high and so a compromise is made between the values of R_B and R_E.

Other stability factors are given as

$$S'=\frac{\partial I_C}{\partial\beta}=\frac{\Delta I_C}{\Delta\beta} \tag{15}$$

which is defined as the rate of change of I_C with β, keeping I_{CO} and V_{BE} constant.

$$S''=\frac{\partial I_C}{\partial V_{BE}}=\frac{\Delta I_C}{\Delta V_{BE}} \tag{16}$$

which is defined as the rate of change of I_C with V_{BE}, keeping I_{CO} and β constant.

6.3.3 Voltage Divider Bias with Emitter Bias

The variation of operating point due to change in β of transistor cannot be offset by the above circuits. The circuit shown in Figure 6.5 is most commonly used to compensate variation of operating point both due to change in β and change in temperature.

Here, an additional resistance R_2 is connected between the base and ground. The term voltage divider comes as a voltage divider is formed by resistors R_1 and R_2.

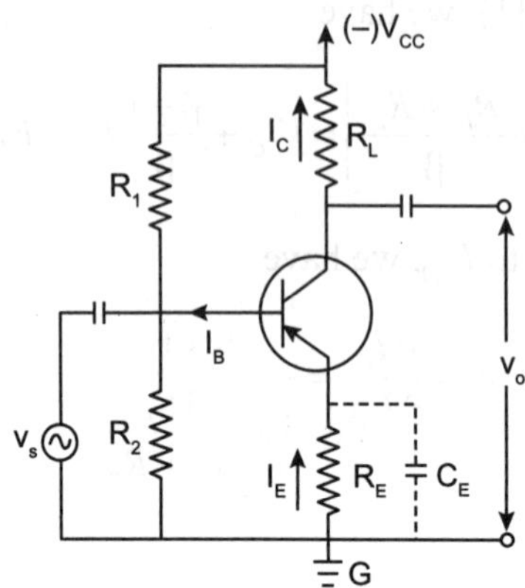

Figure 6.5 Voltage divider bias with emitter bias

This voltage dividing biasing circuit is redrawn in Figure 6.6. Here, Figures 6.6 (a, b and c) are all equivalent.

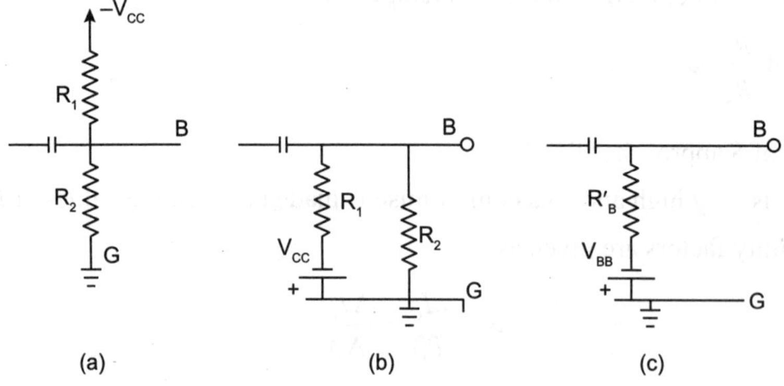

Figure 6.6

Figure 6.6(c) is Thevenin's equivalent of Figure 6.6(a) looking back from B and G.

Applying Thevenin's theorem at terminals B and G we have (Figure 6.6(b))

$$V_{BB} = \frac{R_2}{R_1 + R_2} V_{CC}$$

and $R'_B = \dfrac{R_1 R_2}{R_1 + R_2}$ as shown in Figure 6.6 (c).

Applying KVL to the input side,

$$V_{BB} = I_B R'_B + V_{BE} + I_E R_E + I_E R_E \tag{17}$$

Now $I_C = \beta I_B + (\beta + 1) I_{CO}$

and $I_C = I_E + I_B$,

We can prove that stability factor,

$$S = \frac{\partial I_C}{\partial I_{CO}} = \frac{\beta + 1}{1 + \dfrac{\beta R_E}{R_E + R'_B}} \tag{18}$$

From Eqn. (18) it is clear that R'_B represents the parallel combination of R_1 an R_2; so its value is less than R_B (Figure 6.4). Thus, stability factor S (Eqn. 18) is further reduced in comparison with Eqn. (12). Stabilization due to the addition of R_E in series with emitter has been achieved due to negative feedback. When the circuit is used as an amplifier circuit, AC signal will also pass through R_E. So, to avoid degenerative AC feedback through R_E, a large electrolytic capacitor C_E is joined across R_E whose reactance at lowest frequency is less than resistance R_E; so AC is by passed through C_E.

Effects of changes in V_{BE} In the circuit (Figure 6.5), voltage across R_E is $V_E = V_B - V_{BE}$ where V_B is voltage across R_2

and
$$I_C \approx I_E = \frac{V_E}{R_E}$$

Substituting for V_E,

$$I_C = \frac{V_B - V_{BE}}{R_E} \tag{19}$$

Eqn. (19) shows that any change in V_{BE} affects the value of I_C. The simplest way to counteract the effect of changes in V_{BE} is to make V_B quite large, i.e. V_E should be quite large.

However, for circuits that must retain externally stable bias conditions, it is sometimes necessary to use a compensating technique, where we connect a diode in series base and R_2. In such a circuit, base voltage is

$$V_B = V_{R2} + V_D$$

where V_D is voltage drop across the diode

Substituting this value in Eqn. (19), we get

$$I_C = \frac{V_{R2} + V_D - V_{BE}}{R_E} \tag{20}$$

So when V_{BE} changes, by say ΔV_{BE}, due to change in temperature (its temperature coefficient for silicon is $\dfrac{\Delta V_{BE}}{\Delta I} = (-)1.8$ mV/C°), V_D changes by ΔV_D since $\Delta V_{BE} \approx \Delta V_D$; these changes tend to cancel each other out and finally collector $I_C = \dfrac{V_{R2}}{R_E}$ remains very nearly constant.

6.4 Biasing Transistor Switching Circuits

So far, we have discussed biasing of transistor when used as an amplifier, i.e. Q point should not shift to either saturation region or cutoff region (Figure 6.8).

When a transistor is used as switch, it is either biased OFF (cutoff region) or biased ON to its maximum collector current level.

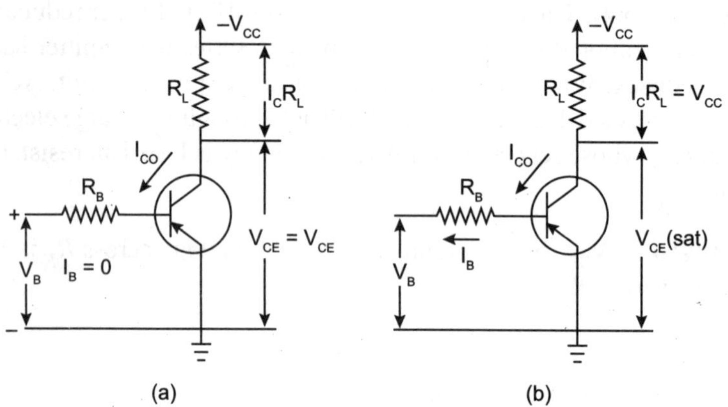

Figure 6.7 Direct coupled transistor switching

The circuit in Figure 6.7 represents the two conditions. In Figure 6.7(a), the base input voltage polarity is such that the transistor is biased off. In this situation, no base current or collector current flows and only current flowing is collector to base leakage current I_{CO}.

From the circuit, we have

$$V_{CE} = V_{CC} - I_C R_L \tag{21}$$

(i) At cutoff,

$$V_{CE} = V_{CC} - I_{CO} R_L = V_{CC} \tag{22}$$

since I_{CO} is very small.

(ii) In Figure 6.7(b), V_B biases the transistor in ON condition to the maximum possible collector current, which is limited by supply voltage V_{CC}. Thus,

$$I_C R_L = V_{CC}$$

$$V_{CE} = V_{CC} - I_C R_L$$

$$= 0 \tag{23}$$

since $I_C R_L = V_{CC}$.

The output characteristic of transistor in CE configuration and load line is shown in Figure 6.8.

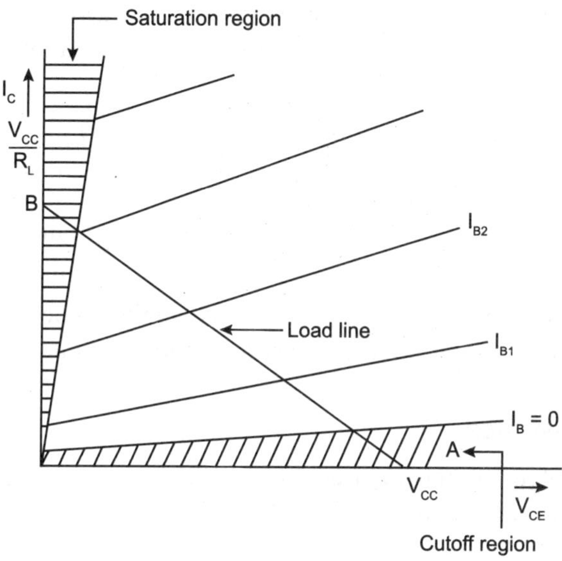

Figure 6.8

6.5 Summary of Stabilization Circuit

The various S factors have been derived in all the circuits where we have a bias of $I_B = 0.3$ mA.

One of the circuits has been more flexible in obtaining the bias point, where the bias point was obtained by adding a resistor from base to ground. In another circuit, an emitter resistor could be added (using collector to base biasing) to increase the stability further.

In general, reducing the value of S will improve thermal stability due to all the effects.

The total change in collector current as a function of all three stability factors is given by

$$\Delta I_C = S\Delta I_{CO} + S'\Delta V_{BE} + S''\Delta\beta \tag{24}$$

Due to smaller dependence on I_{CO} for a silicon transistor, manufacturers specify the average range operation for germanium transistor as –65°C to +75°C and for a silicon transistor as –65°C to +175°C.

6.6 Transistor Dissipation

The maximum power that a transistor is capable of dissipating may vary from a few milliwatts to several hundred watts, depending upon its construction and heat sinks. The power handling capability for a given transistor depends upon; (i) the maximum temperature at which the junction of the device is permitted to operate; (ii) the actual ambient temperature expected in the operation.

However, if ambient temperature exceeds 25°C in a particular environment, then a power derating curve is used to determine the maximum dissipation capability at elevated temperature.

For a given load line, point A represents $I_C = O$, $V_{CE} = V_{CC}$; and point B represents $I_C = \frac{V_{CC}}{R_L}, V_{CE} = 0$. When $I_B = O$, collector current $I_C = I_{CO}$ and the transistor is said to be in cut off region. When $I_B = I_{max}$ (say), then $V_{CE} = V_{CE}$ (sat) and $I_C = \frac{V_{CC}}{R_L}$; the transistor is said to be in saturation region. The region between the two said regions is called active region in which the transistor is biased for amplification. V_{CE}(sat) is the minimum possible collector to emitter voltage for the device, and is referred to as transistor saturation voltage.

6.6.1 Thermal Resistance

Consider a transistor (Figure 6.9) being used in a circuit where temperature of surrounding air is T_A °C and temperature of collector-base junction of transistor is T_J°C. Due to heating within transistor, $T_J > T_A$.

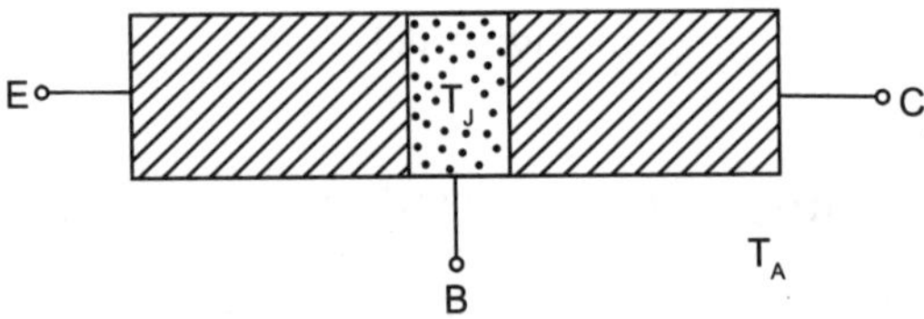

Figure 6.9 Schematic diagram of transistor

Now, it is obvious that greater the power dissipated, P_d, in the transistor, the greater will be the temperature difference between junction temperature (T_J) and ambient temperature (T_A).

So, we can say that temperature rise at collector base junction above ambient temperature is proportional to power dissipated.

$$T_J - T_A \; \alpha \; P_d \tag{25}$$

or

$$T_J - T_A = \theta P_d \tag{26}$$

where θ is a constant of proportionality called thermal resistance.

Its unit of measurement can be found by rearranging equation (26).

So

$$\theta = \frac{T_j - T_A}{P_d} \Rightarrow \frac{°\text{C}}{\text{Watt}}$$

Hence θ is measured in °C per watt. Is typical value may be as low as 0.2°C/watt for small signal, low power transistors that have no cooling provision.

Eqn. (26) may be rearranged as

$$T_J = T_A + \theta P_d \tag{27}$$

Eqn. (27) states that collector base junction temperature T_J will be higher than ambient temperature T_A by an amount equal to the product of θ (thermal resistance) and P_d (power dissipated in the transistor).

Example 1 A transistor having thermal resistance 10°C/watt is operated in an ambient temperature of 25°C. What will be its collector-base junction temperature if it dissipates 2 watts.

Solution

$$\begin{aligned} T_J &= T_A + \theta P_d \\ &= 25 + 10 \times 2 \\ &= 45^\circ\text{C} \end{aligned}$$

θ = 10°C/watt means for every watt consumed its temperature will rise by 10°C.

While using a transistor, we must keep in mind that it must not reach a condition called thermal runaway. The heat released at collector base junction must not exceed the rate at which heat can be dissipated under steady state. For this,

$$\frac{\partial P_d}{\partial T_J} < \frac{1}{\theta} \tag{28}$$

This is the relation for thermal stability.

6.6.2 Junction to Case Thermal Resistance

It has been assumed that θ represents total thermal resistance of the transistor. It should be represented by $\theta_{J\text{–}A}$, i.e. resistance from junction to ambient air. Often this value may not be given in the manual, especially for power transistors. For these transistors, we have thermal resistance from junction to case, i.e. $\theta_{J\text{–}C}$. In this case,

$$T_J = T_C + \theta_{J-C} P_d \tag{29}$$

and for ambient air,

$$T_J = T_A + \theta_{J-A} P_d \tag{30}$$

Thus, if thermal resistance from junction to air is considered, then it consists of two parts; (i) thermal resistance from junction to case, (ii) thermal resistance from case to air.

$$\theta_{J-A} = \theta_{J-C} + \theta_{C-A} \tag{31}$$

It can be expressed that heat dissipated in the junction must make its way to the sorrounding air through two series paths: (i) from junction to case, (ii) case in air.

6.7 Heat-Sink Theory

How we may increase the amount of power, P_d, that a transistor can safely handle? For Eqn. (30),

$$P_d = \frac{T_J - T_A}{\theta_{J-A}}$$

Making use of Eqn. (31), we can write

$$P_d = \frac{T_J - T_A}{\theta_{J-C} + \theta_{C-A}} \tag{32}$$

Here θ_{J-C} is determined by the manufacturer and θ_{C-A} is determined by the surface area of case (cover) or flange and its contact with the surrounding air; thus, it is a variable quantity since effective surface area of the transistor could be increased and resistance to heat flow or θ_{C-A} would be decreased. This can be achieved using a heat sink.

A heat-sink is a relatively large, finned heat conducting device placed in close contact with the transistor case or flange as shown in Figure 6.10.

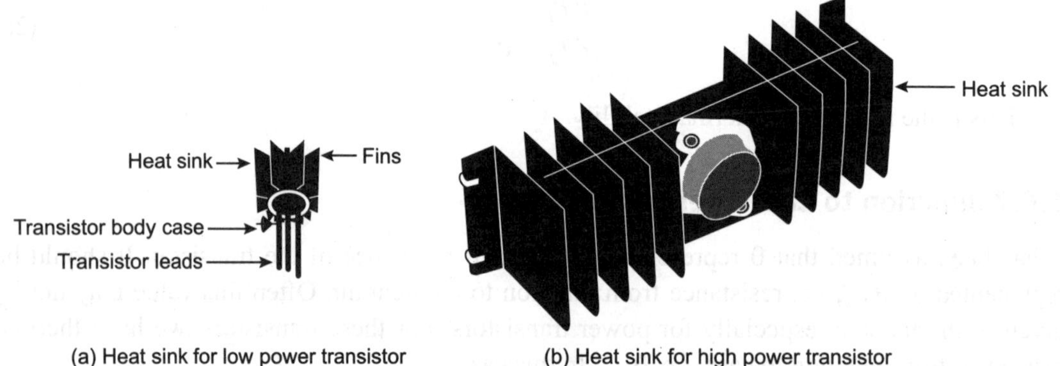

Figure 6.10 Heat sink used to dissipate heat generated in a transistor

When we use a heat-sink, heat dissipated in the junction will make its way to the surrounding air through three series paths: (i) junction to case, (ii) case to sink, (iii) sink to air (Figure 6.11(a)).

We have

$$\theta_{J-A} = \theta_{J-C} + \theta_{C-S} + \theta_{S-A}$$

In this situation, we have

$$P_d = \frac{T_J - T_A}{\theta_{J-C} + \theta_{C-S} + \theta_{S-A}} \tag{33}$$

Eqn. (22) can be represented by thermal resistance equivalent circuit (Figure 6.11(b)).

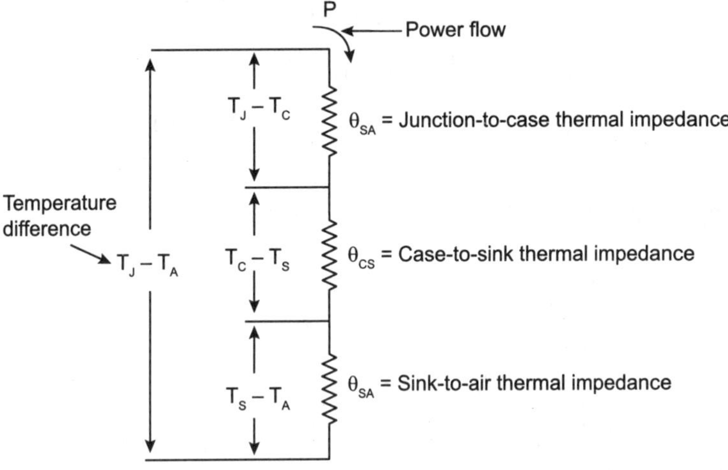

Figure 6.11 Paths of heat dissipation

Example 2 Determine the dissipation capacity of a transistor at an ambient (case) temperature of 50°C, if its maximum dissipation is 165 mW at an ambient temperature of 25°C. Given maximum permissible operating temperature is 71°C.

Solution

Draw a vertical line from temperature axis at 50°C to cut the 71°C line. Join the point of intersection *P* through a horizontal line at *Y*-axis. The point where it intersects *Y*-axis gives the value of permissible dissipation equal to 45% of maximum rating.

The dissipation capability at 50°C ambient is $\left(\frac{45}{100}\times 165\right) = 74$ mW.

Its value ranges from $0.2\frac{°\text{C}}{\text{Watt}}$ to $1000\frac{°\text{C}}{\text{Watt}}$ for a transistor that has an efficient heat sink.

From $T_J - T_A = \theta P_d$ we have

$$T_J = T_A + \theta P_d \tag{34}$$

The above equation reflects that collector-junction temperature T_J of the transistor will be higher than ambient temperature T_A by an amount equal to the product of θ and P_d (power dissipated in the transistor).

Example 3 For a fixed current bias circuit, $R_B = 200$ kΩ, $R_L = 2$ kΩ and $V_{CC} = 10$ V. Assume $V_{BE} = 0.7$. Find the operating point when β = 50. Also determine the change of operating point when β = 60.

Solution

$$I_B = \frac{V_{CC} - V_{BE}}{R_B} = \frac{10 - .7}{200 \times 10^3} = 46.5 \ \mu\text{A}$$

$$I_C = \beta I_B = 50 \times 46.5 \ \mu\text{A} = 2.33 \ \text{mA}$$

$$V_{CE} = V_{CC} - I_C R_L = 10 - 2.33 \times 2000 \times 10^{-3}$$
$$= 5.35 \ \text{V}$$

The operating point coordinates are [5.35 V, 2.33 mA].

If β = 60,

$$I_C = 60 \times 46.5 \ \mu\text{A} = 2.79 \ \text{mA}$$

$$V_{CE} = V_{CC} - I_C R_L = 10 - 2.79 \times 2 \times 1000 \times 10^{-3}$$
$$= 4.43 \ \text{V}$$

The new operating point coordinates are [4.42 V, 2.79 mA].

Example 4 In a collector to base bias circuit, $R_L = 3.3$ kΩ, $R_B = 330$ kΩ and $V_{CC} = 15$ V. Assume β – 60 and $V_{BE} = 07$ V. Calculate V_{CE}.

Solution

$$V_{CC} = I_B R_B + V_{BE}$$

$$I_B = \frac{V_{CC} - V_{BE}}{R_B} = \frac{15 - .7}{330 \times 1000} = \frac{14.3}{33} \times 10^{-4} = 43 \ \mu\text{A}$$

$$I_B = \beta I_B$$
$$= 60 \times 43 \times 10^{-6}$$
$$= 2.58 \ \text{mA}$$

$$V_{CC} = [I_C + I_B]R_L + V_{CE}$$

or

$$V_{CE} = V_{CC} - [I_C - I_B]R_L$$
$$= 15 - [2.58 + 0.43] \times 10^{-3} \times 3.3 \times 10^3$$
$$= 15 - [3.01] \times 3.3$$
$$= 15 - 9.9$$
$$= 5.1 \ \text{V}$$

6.8 Derating Curve

A typical power temperature derating curve chart is shown in Figure 6.9. It consists of a graph of percent permissible dissipation against transistor case temperature.

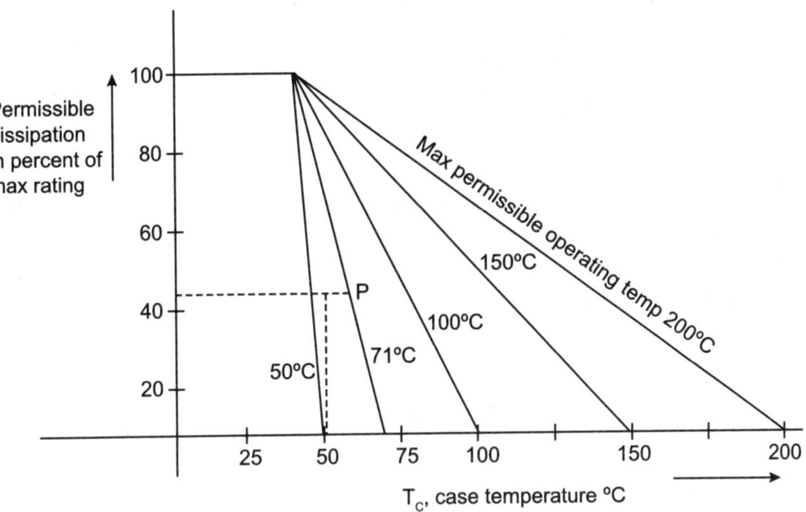

Figure 6.12 Typical power-temperature derating curve. The use a of derating curve can be understood with Example 2.

Summary

1. The variation of DC collector current with temperature in CE configuration of amplifier depends upon β, I_B and I_{CO}.
2. A shift in operating point may cause clipping and distortion in the output signal or it may cause thermal runaway.
3. Stability factor S is defined as the rate of change of collector current I_C with respect to leakage current I_{CO} keeping β and V_{BE} constant. This change in I_C depends upon biasing of the circuit.
4. Stability factor S for CE amplifier with fixed bias is (β + 1), which is quite high.
5. A common base circuit has stability factor $S = 1$, which is most stable.
6. Stability methods involve the reduction of forward-bias with temperature increase to reduce I_B and hence I_C by some means of negative feedback in various bias circuits.
7. For a collector to base bias, a resistor is connected from collector base and

$$S = \frac{\beta + 1}{1 + \beta R_L / R_L + R_B}$$

It may not provide much overall increase in stability in silicon transistor circuit.

8. Emitter bias consists of the addition of a resistor R_E in series with emitter lead bypassed by a large capacitor CE; in this case

$$S = \frac{\beta + 1}{1 + \dfrac{\beta R_E}{R_E + R_B}}$$

9. Maximum reduction of S with emitter bias is accomplished by the use of minimum value of $\frac{R_B}{R_E}$.

10. Voltage divider bias with emitter bias will permit the selection of any operating point with any desired value of S.

    In this case, $S = \dfrac{\beta + 1}{1 + \dfrac{\beta R_E}{R_E + R_B'}}$ where $R_B' = \dfrac{R_1 R_2}{R_1 + R_2}$.

11. Additional stability factors are S' and S'' and defined as

    $S' = \dfrac{\Delta I_C}{\Delta V_{BE}}$ and $S'' = \dfrac{\Delta I_C}{\Delta \beta}$.

12. Total change in collector current as a function of all three stability factors is given by $\Delta I_C = S \Delta I_{CO} + S' \Delta V_{BE} + S'' \Delta\beta$.

13. Every transistor has the maximum power dissipation capability. This capability at elevated ambient temperature is reduced by an amount depending upon case temperature and maximum permissible junction operating temperature. This is usually determined by power-temperature derating curve chart.

14. Thermal resistance of a transistor is defined by the ratio of power dissipation to difference of junction temperature and ambient temperature. Its unit is °C per watt.

15. A set of derating curve represents variation of power dissipation with case temperature.

Review Questions

1. Why a transistor used as amplifier needs biasing?
2. What type of biasing is preferred in a transistor amplifier? Explain its action.
3. Define the term stability factor S. Why would it seem more reasonable to call it "instability factor"? Which type of biasing in transistor amplifier has the highest stability factor?
4. What do you mean by "stability factor"? What is its ideal value? In which configuration of transistor amplifier can we achieve this value? Derive a relation for stability factor in CE configuration using voltage divider biasing and emitter biasing.

5. Define "operating point" as used in an amplifier? Why do we select operating point in the active region of output characteristics of CE amplifier? How this point changes on rise of temperature of the device? How the shifting of operating point is compensated.
6. How does collector to base bias circuit stabilize the operating point of an amplifier? Explain what are its demerits? How they are removed?
7. How will you bias a transistor when used as switch?
8. In what respect a transistor is a temperature-sensitive device? How its input and output characteristics in CE configuration change? Explain their variation.
9. How do you define the various types of stability factors? How they affect the overall change in collector current? Which one is most important and why? Explain.
10. What do you mean by the term "thermal resistance"? Explain. Mention its unit.
11. How the power dissipation capability of a transistor is regulated by using the derating curve? Explain.
12. What is represented by a derating curve? Explain its use by a suitable example.
13. Explain the purpose of load line for a transistor circuit and discuss the bias point.
14. Explain what is meant by AC degeneration in a transistor circuit. Show how AC degeneration may be eliminated in (i) collector to base bias circuit (ii) emitter current bias circuit.
15. Show how an emitter current bias circuit may be compensated for changes in V_{BE} due to change in temperature. Derive an equation for collector current.
16. Define the stability factor for a transistor. Compare the three basic bias circuits with regard to stability factor.
17. Sketch an emitter current bias circuit using (a) an *n-p-n* transistor, (b) *p-n-p* transistor. In each case, show the polarity of V_{CC}, V_{BE} and V_{CE}. Also show the direction of I_C and I_B.
18. Compare the three basic bias circuits with regard to stability of collector current I_C versus changes in transistor current gain β.
19. For an emitter current bias circuit, derive an exact equation for collector current I_C in terms of supply voltage, circuit resistors and transistor current gain factor β.

Short Answer Questions

1. Which of the three configurations of a transistor has ideal stability factor? What is its value?

   Ans: A transistor amplifier in common base configuration has the ideal stability factor. Its value is unity.

2. Which bias is most preferred in CE amplifier? Explain.

   Ans: The most preferred bias in CE configuration is emitter bias since in this case stability factor can be reduced to a suitable value by using current feedback through R_E.

3. What factors other than I_{CBO} contribute to shift in operating point?

   Ans: There are two factors which affect the shift of operating point of an amplifier. The first factor is change in β value due to change of transistor it self ; the second factor is change in V_{BE} due to change in temperature.

4. What is the unit of thermal resistance?

   Ans: The unit of thermal resistance is °C/watt.

5. What is represented by a derating curve?

   Ans: A derating curve represents the permissible dissipation percent of maximum rating with increase in case temperature.

6. What do you mean by the term thermal runaway?

   Ans: Any increase in I_{CBO} of a transistor increases collector current I_C which in turn increases the *C–B* junction temperature resulting into more minority carriers, i.e. $I_{CBO.}$ This increase in I_{CBO} further increases I_C. This effect is cumulative and is known as thermal runaway. It may lead to destruction of the transistor.

7. What is the value of stability factor in case of CB amplifier? Is the amplifier most stable?

   Ans: The value of stability factors in case of CB amplifier is 1 (unity); yes the amplifier is most stable since the value of *S* is ideal.

8. What is the value of stability factor in case of fixed bias circuit employed in a CE amplifier? Is it satisfactory.

   Ans: The value of stability factor *S* in case CE amplifier having fixed bias circuit is β + 1. It is the least satisfactory since it is the maximum value, which is not desired.

9. How we can achieve maximum reduction in the value of stability factor *S* in case of CE amplifier having R_E in emitter lead and R_B for base biasing.

   Ans: Maximum reduction in *S* is achieved by keeping the minimum value of R_B/R_E that is practical.

10. Which stability factor is most prominent in case of CE amplifier?

    Ans: Stability factor $S = \dfrac{\Delta I_C}{\Delta I_{CBO}}$ is most prominent in case of CE amplifier since reduction in its value would also cover $S' = \dfrac{\Delta I_C}{\Delta V_{BE}}$ and $S'' = \dfrac{\Delta I_C}{\Delta h_{FE}}$.

11. Is collector to base bias method the most suitable for achieving suitable value of *S* in case of CE amplifier.

    Ans: No, since by employing collector base bias in CE amplifier the value of *S* remains higher and does not attain ideal value.

12. What do you mean by thermal runaway of transistor?

    Ans: With increase in temperature of transistor, I_{CBO} increases and it results into increase in I_C which is (β + 1) times that of I_{CBO}. Increase in I_C further increases transistor collector junction temperature and consequently I_{CBO} increases. Thus, the action is cumulative and amounts to thermal runaway in which situation the transistor burns out or causes distortion in output.

13. How we can increase the amount of power dissipation capability of a transistor?

    Ans: The power dissipation capability of a transistor is increased by the use of heat sink.

14. What is a heat sink?

    Ans: A heat sink is a metallic heat conducting device placed in close contact with a transistor to increase the power dissipation capability of the transistor.

Objective Questions

1. The ideal value of stability factor S is

   (a) 100 (b) 10

   (c) 1 (d) 0 **[Ans:** (c)**]**

2. The ideal value of stability factor S can be achieved in

   (a) common base amplifier (b) common emitter amplifier

   (c) common collector amplifier (d) in all the three configurations **[Ans:** (a)**]**

3. Under temperature variation, the most stable circuit is

   (a) common base amplifier (b) CE amplifier with collector base bias

   (c) CE amplifier with emitter bias (d) CE amplifier with fixed bias **[Ans:** (a)**]**

4. Stability factor S is defined as

   (a) $S = \dfrac{\Delta I_C}{\Delta I_{CBO}}$ (b) $S = \dfrac{\Delta I_C}{\Delta I_B}$

   (c) $S = \dfrac{\Delta I_C}{\Delta h_{FE}}$ (d) $S = \dfrac{\Delta I_C}{\Delta V_{BE}}$ **[Ans:** (a)**]**

5. Variation in $h_{FE}(\beta)$ of a transistor affects

   (a) S (b) S'

   (c) S'' (d) all the three **[Ans:** (a)**]**

6. The unit of thermal resistance is

   (a) Ω (b) °C/watt

   (c) watt/°C (d) Ω/°C **[Ans:** (b)**]**

7. Thermal resistance is given by the relation

   (a) $\dfrac{T_J - T_A}{P_d}$ (b) $\dfrac{T_J - T_C}{P_d}$

   (c) $\dfrac{T_C - T_A}{P_d}$ (d) $\dfrac{T_C - T_A}{P_d}$ **[Ans:** (a)**]**

Mark the Following Statement True/False.

1. The ideal value of stability factor S is 1. **[Ans:** True]
2. The collector to base bias circuit gives ideal stabilization. **[Ans:** False]
3. The unit of thermal resistance is °C/watt. **[Ans:** True]
4. The larger the temperature difference between transistor base-collector junction and ambient temperature, the higher the power dissipation capability of the transistor. **[Ans:** True]
5. The collector to base bias in a CE amplifier makes the circuit completely insensitive to variation in h_{FE} of transistor. **[Ans:** False]
6. The value of stability factor 'S' in case of fixed bias circuit in a amplifier is $[h_{FE} + 1]$ or $(\beta + 1)$. **[Ans:** True]
7. The thermal resistance of heat sink is constant. **[Ans:** True]
8. Heat sinks are used to increase the power dissipation capability of a transistor. **[Ans:** True]

Fill in the Blanks.

1. The unit of thermal resistance is ______________. **[Ans:** $\frac{°C}{watt}$]
2. Thermal resistance is the ratio of ______________ and ______________.
   [Ans: $\frac{T_J - T_A}{P_d}$; temperature difference between junction temperature-ambient temperature to power dissipation.]
3. Stability factor S is given by the ratio of change in ______________ to change in ______________.
   [Ans: collector current (I_C), collector to base leakage current (I_{CBO})]
4. Stability factor S'' is given by the ratio of change in ______________ to change in ______________. **[Ans:** collector current (I_O), DC current gain (β)]
5. Ideal value ______________ for S with emitter bias is attained by keeping the value of $\frac{R_B}{R_E}$ less than ______________.
   [Ans: unity, unity]
6. The maximum power which a transistor can handle is determined by ______________ and its ______________. **[Ans:** temperature difference $[T_J - T_A]$, thermal resistance]

Chapter 7

Amplifiers (Small Signal Amplifiers)

7.1 Introduction

Electronic circuits which perform the function of amplifying the voltage/current or power level of a given electrical signal are called amplifiers. In real sense, power gain is involved for the circuit so that it may be used as an amplifying circuit. In most of the circuits vacuum Tube (Triode) was used in early days, with the invention of transistor, the triode has been replaced.

When the output signal is more than input signal then ratio of the output signal (voltage/current) to the input signal is defined as voltage gain or current gain and denoted as

(i) Voltage gain $A_v = \frac{V_o}{V_i}$

(ii) Current gain $A_i = \frac{I_o}{I_i}$

This gain either in voltage or current signal is achieved by DC power (Battery) employed in the circuit. From the signal strength point amplifiers are classified as (i) small signal amplifier (ii) large signal amplifier. In this chapter we shall confine our study to small signal amplifier.

7.1.1 Small Signal Amplifier

Small signal means that input signal (voltage/current) is so small that the device (transistor) conducts for the entire cycle (assuming sinusoidal signal) of the signal i.e. (0 – 360°) and amplified output signal (wave) is also sinusoidal without any distortion.

Such a amplifier using fixed bias in CE configuration is shown in (Figure 7.1(a)).

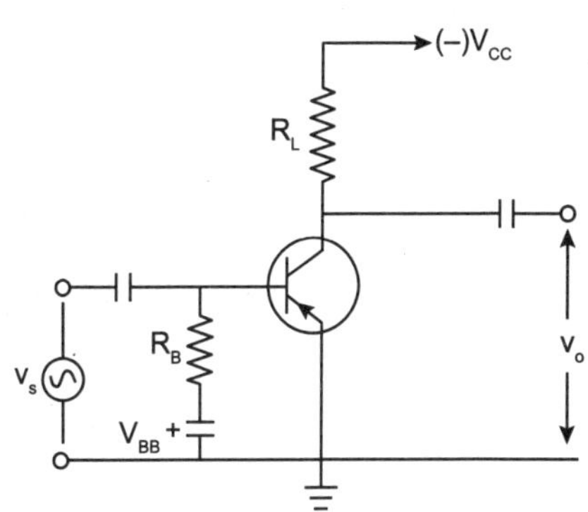

Figure 7.1 (a) Transistor amplifier (general case) using fixed bias (shown as CE configuration)

7.2 Transistor Amplifier Performance

Assuming input signal v_s to be small and having internal resistance R_S, the AC equivalent circuit (Figure 7.1(a)) of the amplifier having load R_L using '*h*' parameters is shown in Figure 7.1(b). The frequency of the signal is such that the reactance of reactive elements present can be neglected.

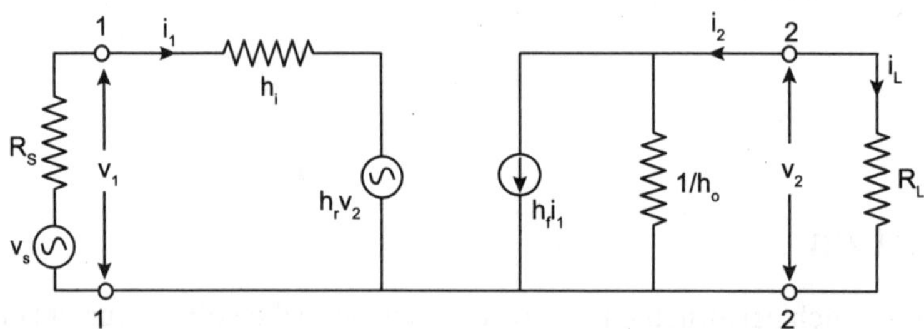

Figure 7.1 (b) AC equivalent circuit of transistor amplifier (general case)

The above circuit represents a transistor in any configuration (CE, CB or CC). A particular configuration can be represented by adding the proper suffix to the various parameters (i.e. h_{ie}, h_{ib} or h_{ic} for CE, CB and CC configuration, respectively).

Using Kirchhoff's law to the circuit, we have

$$v_1 = h_i i_1 + h_r v_2 \tag{1}$$

$$i_2 = h_f i_1 + h_o v_2 \tag{2}$$

and

$$v_2 = (-)\, i_2 R_L \tag{3}$$

Here the negative sign signifies the direction of current i_2 is opposite to the usual convention.

7.2.1 Current Gain (A_i)

It is defined as the ratio of output (load) current to the input current.

$$A_i = \frac{i_L}{i_1} = \frac{(-)i_2}{i_1} \quad \text{where } i_L = (-)i_2$$

Substituting for v_2 in Eqn. (2) from (3),

$$i_2 = h_f\, i_1 + h_o[-i_2\, R_L]$$

or

$$i_2[1 + h_o\, R_L] = h_f\, i_1$$

$$\frac{i_2}{i_1} = \frac{h_f}{1 + h_o\, R_L}$$

$$\therefore \text{ Current gain } A_i = \frac{i_L}{i_1} = (-)\frac{i_2}{i_1} = (-)\frac{h_f}{1+h_o R_L} \tag{4}$$

Normally the value of h_o is very small that the quantity $h_o R_L$ is neglected and the maximum value of current gain

$$[A_i]_{max} = (-)h_f \tag{5}$$

Eqn. (5) represents that the maximum current gain of a transistor amplifier approaches the value h_f which is dependent on the particular transistor used.

Input resistance (R_i) It is defined as the ratio of input voltage to input current flowing through the device as seen at terminal (1).

$$R_i = \frac{v_1}{i_1}$$

Substituting for v_2 from Eqn. (3), we have

$$v_1 = h_i i_1 + h_r(-i_2 R_L)$$

Dividing both sides by i_1, we have

$$\frac{v_1}{i_1} = h_i - \frac{i_2}{i_1} h_r R_L$$

By definition of current gain, $A_i = (-)\frac{i_2}{i_1}$

So we have $$R_i = \frac{v_1}{i_1} = h_i + A_i \, h_r \, R_L \tag{6}$$

Substituting for A_i from Eqn. (4), we have

$$R_i = h_i + \left(-\frac{h_f}{1+h_o R_L}\right) h_r \, R_L$$

$$= \frac{h_i[1+h_o R_L] - h_f \, h_r R_L}{1+h_o R_L}$$

$$= \frac{h_i + [h_i \, h_o - h_r \, h_f] R_L}{1+h_o \, R_L}$$

$$= \frac{h_i + \Delta h \, R_L}{1+h_o \, R_L} \approx h_i \text{ where } \Delta h = (h_i \, h_o - h_r \, h_f) \tag{7}$$

Since Δh and h_o are very small, Eqn. (7) suggests that the input resistance of an amplifier is nearly equal to h_i of the transistor used in a particular configuration.

7.2.2 Voltage Gain (A_v)

It is defined as the ratio of output voltage to input voltage,

$$A_v = \frac{v_2}{v_1}$$

We know that $v_2 = (-)i_2 R_L$ and $i_2 = (-)A_i\, i_1$

Substituting for i_2, we have

$$v_2 = A_i\, i_1\, R_L$$

Dividing both sides by v_1, we have

$$\frac{v_2}{v_1} = A_i\, R_L \frac{i_1}{v_1}$$

$\therefore$ Voltage gain $A_v = \dfrac{v_2}{v_1} = A_i\, R_L / R_i$ since $R_i = \dfrac{v_1}{i_1}$

Making use of Eqn. (4) and (7), we have

$$\text{Voltage gain, } A_v = \frac{(-)\left[\dfrac{h_f}{1+h_o R_L}\right] R_L}{\dfrac{h_i + (h_i h_o - h_r h_f)R_L}{1+h_o R_L}}$$

$$= (-)\frac{h_f R_L}{h_i + \Delta h R_L} \quad (8)$$

If $h_i \gg \Delta h R_L$, then

$$A_v = (-)\frac{h_f}{h_i} R_L \quad (8b)$$

Output resistance It is defined as the ratio of output voltage to output current at terminals (2, 2).

$$R_o = \frac{v_2}{i_2}$$

(Remember, this value is not equal to R_L as given by Eqn. (3). Here R_S is taken as zero and R_L as $\propto$)

Output conductance is defined as $G_o = \dfrac{i_2}{v_2} = \dfrac{1}{R_o}$

From Eqn. (2),

$$G_o = h_f \frac{i_1}{v_2} + h_o \quad (9)$$

Assuming $v_s = 0$ and applying KVL to the input side, we have

$$R_S i_1 + h_i i_1 + h_r v_2 = 0$$

or

$$\frac{i_1}{v_2} = (-)\frac{h_r}{R_s + h_i}$$

Substituting this value in Eqn. (9), we have

$$G_o = h_o - \frac{h_r h_f}{R_s + h_i} \quad (10)$$

So the value of R_o can be found by making use of Eqn. (10).

7.3 Transistor Amplifier Circuits in Different Configurations

A transistor can be used as an amplifier in the following three configurations:

(i) Common emitter

(ii) Common base

(iii) Common collector

We have discussed general case of transistor amplifier using hybrid parameters. Now we shall discuss a simplified version in all the three configurations.

7.3.1 CE Transistor Amplifier

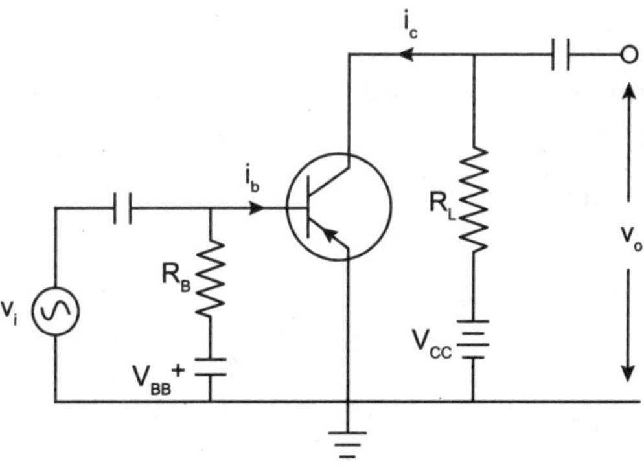

Figure 7.2 (a) CE transistor amplifier

A basic circuit of the amplifier is shown in Figure 7.2 using two battery connections. Its AC equivalent circuit using '*h*' parameters, in midrange frequencies of the input signal, has been shown in Figure 7.2(b).

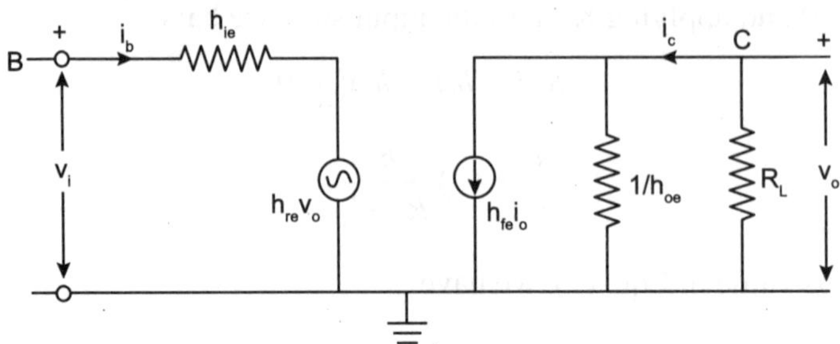

Figure 7.2 (b) AC equivalent circuit of a CE amplifier

When drawing the above equivalent circuit, we have assumed that the input signal frequency range is such that the reactance of various capacitors present is neglected.

Circuit analysis By definition, current gain $A_i = \frac{i_c}{i_b}$.

From the circuit (Figure 7.2(b)),

$$i_c = h_{fe}i_b + h_{oe}v_o$$
$$= h_{fe}i_b - h_{oe}i_cR_L$$

because $v_o = (-)i_cR_L$

or
$$i_c[1 + h_{oe}R_L] = h_{fe}\, i_b$$

$\therefore$ Current gain,
$$A_{ie} = \frac{i_c}{i_b} = \frac{h_{fe}}{1 + h_{oe}R_L} \tag{11}$$

$$= h_{fe} \tag{12}$$

assuming $1 \gg h_{oe}\, R_L$ being small.

From Eqn. (12), we note that the maximum value of current gain is h_{fe} which depends upon the transistor used in a circuit.

By definition, voltage gain $A_v = \frac{v_o}{v_i}$.

From the circuit (Figure 7.2(b)),

$$v_i = i_bh_{ie} + h_{re}v_o$$
$$= i_bh_{ie} - h_{ve}i_cR_L$$

$$\therefore \text{ Voltage gain, } A_{ve} = \frac{(-)i_cR_L}{i_bh_{ie} - h_{re}i_cR_L} = \frac{(-)\left(\frac{i_c}{i_b}\right)R_L}{h_{ie} - \left(\frac{i_c}{i_b}\right)h_{re}R_L}$$

Now $$A_{ve} = \frac{(-)A_{ie}R_L}{h_{ie} - A_{ie}h_{re}R_L} \text{ since } \frac{i_c}{i_b} = A_{ie}$$

Substituting for A_{ie} from Eqn. (54), we have

$$A_{ve} = \frac{(-)\left[\dfrac{h_{fe}}{1+h_{oe}R_L}\right]R_L}{h_{ie} - \left[\dfrac{h_{fe}}{1+h_{oe}R_L}\right]h_{re}R_L}$$

$$= \frac{(-)h_{fe}R_L}{h_{ie} + h_{ie}h_{oe}R_L - h_{re}h_{fe}R_L}$$

$$= \frac{(-)h_{fe}R_L}{h_{ie} + [h_{ie}h_{oe} - h_{re}h_{fe}]R_L}$$

$$= \frac{(-)h_{fe}R_L}{h_{ie} + \Delta h R_L} \tag{13}$$

$$= (-)\frac{h_{fe}}{h_{ie}}R_L \tag{13b}$$

where $\Delta h = \begin{vmatrix} h_{ie} & h_{re} \\ h_{fe} & h_{oe} \end{vmatrix}$ is neglected, it being small.

Eqn. (13) shows that voltage gain depends upon load R_L connected and h_{fe} of the transistor used. Further, the negative sign represents the phase difference of π between output voltage and input voltage.

Input resistance (R_i) By definition,

$$R_i = \frac{v_i}{i_b} = \frac{i_b h_{ie} + h_{re}v_o}{i_b}$$

$$= \frac{i_b h_{ie} - h_{re}i_c R_L}{i_b}$$

$$= h_{ie} - A_{ie}h_{re}R_L$$

$$= h_{ie} - \frac{h_{fe}h_{re}R_L}{1+h_{oe}R_L}$$

$$= h_{ie} + h_{ie}h_{oe}R_L - h_{fe}h_{re}R_L$$

$$= h_{ie} + [h_{ie}h_{oe} - h_{fe}h_{re}]R_L$$

$$= h_{ie} + \Delta h R_L \tag{14}$$

$$\therefore R_{ie} = h_{ie} \tag{15}$$

neglecting $\Delta h R_L$, it being small.

Hence the input resistance attains a value 'h_{ie}' which depends on the transistor used.

Output resistance (R_o) To determine the value of R_o, we assume input signal v_i to be zero and imagine that a voltage source v_o is replaced by L_o and R_L.

By definition,
$$R_{oe} = \frac{v_o}{i_c}$$

but
$$i_c = h_{fe} i_b + h_{oe} v_o$$

and
$$i_b h_{ie} + h_{ve} v_o = o$$

or
$$i_b = (-)\frac{h_{re}}{h_{ic}} v_o$$

Substituting for i_b, we have

$$i_c = (-)h_{fe}\frac{h_{re}}{h_{ie}} v_o + h_{oe} v_o$$

$$= v_o \left[\frac{h_{oe} h_{ie} - h_{fe} h_{re}}{h_{ie}}\right]$$

$$= \frac{v_o \Delta h}{h_{ie}}$$

$$\therefore \text{Output resistance, } R_{oe} = \frac{v_o}{i_c} = \frac{h_{ie}}{\Delta h} \tag{16}$$

Note: In the above, the suffix '*e*' for various parameters shows that the transistor is in common emitter configuration.

Simplified version In the AC equivalent circuit (Figure 7.2(b)) the following approximations can be made. The value of $h_{re} v_o$ can be neglected in comparison with v_i because it is very small. The value of h_{oe} for a transistor is of the order of 10^{-5} mhos, and if R_L is of the order of kΩ then $1/h_{oe}$ can be taken as an open circuit. Therefore, the circuit takes the form as shown in Figure 7.3.

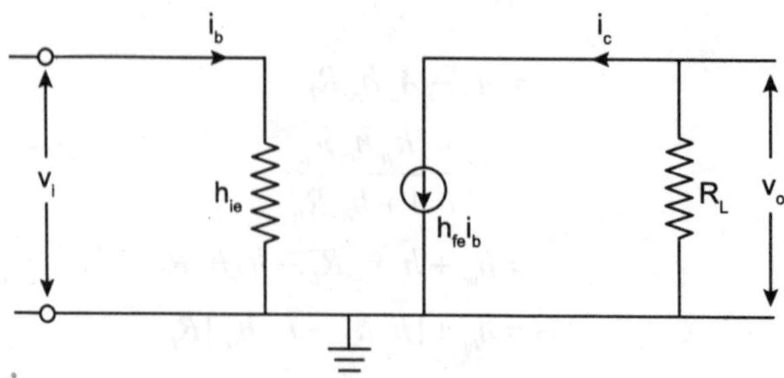

Figure 7.3 Simplified AC equivalent circuit of a CE amplifier

$$i_c = h_{fe} i_b$$

$\therefore$ Current gain $$A_{ie} = \frac{i_c}{i_b} = h_{fe} \tag{17}$$

which is same as Eqn. (55).

Voltage gain, $A_{ve} = \dfrac{v_o}{v_i}$

$$= \frac{(-)i_e R_L}{i_b h_{ie}}$$

$$= (-)\, A_{ie} \frac{R_L}{h_{ie}}$$

Substituting for A_{ie},

$$A_{ve} = (-)\frac{h_{fe} R_L}{h_{ie}} \tag{18}$$

$$= (-) g_m R_L \quad \text{since} \quad \frac{h_{fe}}{h_{ie}} = g_m$$

which is same as Eqn. (13b).

Input resistance

$$R_i = \frac{v_i}{i_b} = h_{ie} \tag{19}$$

which is same as Eqn. (20).

Output resistance

$$R_o = \frac{v_o}{i_c} = \frac{1}{h_{oe}} \Rightarrow \infty \tag{20}$$

Because v_i is assumed to be zero, so $i_b = 0$.

Power gain, $$A_p = [\text{Voltage gain}]\ [\text{Current gain}]$$

$$= |A_v||A_i|$$

$$= \frac{h_{fe} R_L}{h_{ie}} \times h_{fe} = \frac{h_{fc}^2 R_L}{h_{ie}} \tag{21}$$

Trans conductance model By definition, $h_{fe} = \dfrac{i_o}{i_b}$ and $h_{ie} = \dfrac{v_{be}}{i_b}$

So,
$$\frac{h_{fe}}{h_{ie}} = \frac{i_o / l_b}{v_{be} / i_b} = \frac{i_c}{v_{be}}$$

The dimension of R.H.S. is like that of conductance. So, this quantity (L.H.S.) is called transconductance and represented by g_m.

$$g_m = \frac{h_{fe}}{h_{ie}}$$

Eqn. (21) can be expressed as

$$A_p = g_m h_{fe} R_L \tag{22}$$

7.3.2 Common Base Amplifier

The basic circuit of an amplifier is shown in Figure 7.4(a). Its AC equivalent circuit is shown in Figure 7.4(b) using '*h*' parameters and assuming a small signal in the frequency range where reactance of the capacitors is negligible.

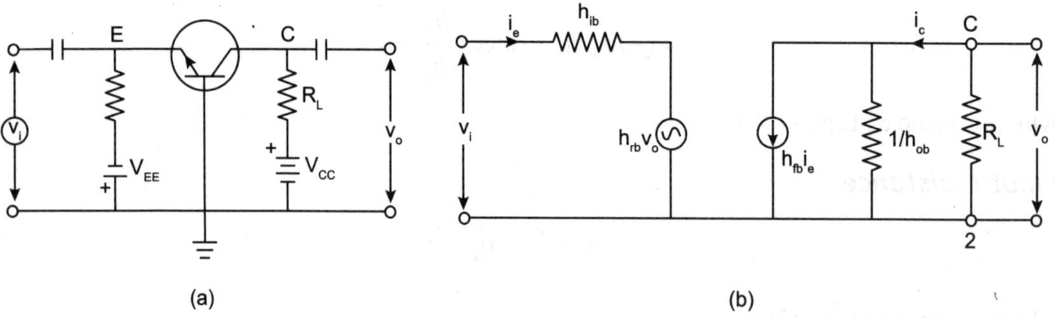

Figure 7.4 (a) CB transistor amplifier circuit, (b) AC equivalent of CB amplifier in mid frequency range

By definition, current gain $A_{ib} = \dfrac{i_c}{i_e}$

$$i_c = h_{fb} i_e + h_{ob} v_o$$
$$= h_{fb} i_e - i_c h_{ob} R_L \quad \text{since } v_o = (-) i_c R_L$$

or
$$i_c = \frac{h_{fb} i_e}{1 + h_{ob} R_L}$$

$$\therefore A_{ib} = \frac{i_c}{i_e} = \frac{h_{fb}}{1 + h_{ob} R_L} \tag{23}$$

Voltage gain (A_{vb}) By definition,

$$A_{vb} = \frac{v_c}{v_i}$$

$$= \frac{(-)i_c R_L}{h_{ib} i_e + h_{rb} v_o}$$

$$= \frac{(-)i_c R_L}{h_{ib}\, i_e - i_c h_{rb} R_L}$$

$$= \frac{-\frac{i_c}{i_e} R_L}{h_{ib} - \frac{i_c}{i_e} h_{rb} R_L}$$

$$= \frac{(-)\, A_{ib} R_L}{h_{ib} - A_{ib} h_{rb} R_L}$$

Substituting for A_{ib},

$$A_{vb} = \frac{(-)\left[\frac{h_{fb}}{1 + h_{ob} R_L}\right] R_L}{h_{ib} - \left[\frac{h_{fb}}{1 + h_{ob} R_L}\right] h_{rb} R_L}$$

$$= \frac{(-)\, h_{fb} R_L}{h_{ib} + [h_{ib} h_{ob} - h_{fb} h_{rb}] R_L}$$

$$= \frac{(-) h_{fb} R}{(h_{ib} + \Delta h R_L)} \tag{24}$$

Input resistance (R_i) By definition,

$$R_{ib} = \frac{v_i}{i_e}$$

$$= \frac{h_{ib} i_e + h_{rb} v_o}{i_e}$$

$$= h_{ib} - \frac{i_c}{i_e} h_{rb} R_L$$

$$= h_{ib} - A_{ib} h_{rb} R_L$$

$$= h_{ib} - \left[\frac{h_{fb}}{1 + h_{ob}R_L}\right] h_{rb} R_L$$

$$= h_{ib} - \frac{h_{fb}h_{rb}}{1 + h_{ob}R_L} R_L$$

$$= h_{ib} + [h_{ib} - h_{ob} - h_{fb}h_{rb}]R_L$$

$$= h_{ib} + \Delta h R_L \tag{25}$$

$$\therefore R_{ib} = h_{ib} \tag{25a}$$

ΔhR_L is neglected, it being small.

Output impedance (R_o) By definition, $R_{ob} = \frac{v_o}{i_c}$ assuming $v_i = 0$ and replacing load R_L by a voltage source v_o,

$$i_c = h_{ib}i_e + h_{ob}v_o$$

and

$$h_{ib}i_e + h_{rb}v_o = 0$$

or

$$i_e = (-)\frac{h_{rb}}{h_{ib}} v_o$$

Substituting for i_c we have

$$i_c = (-)\frac{h_{ib}h_{rb}}{h_{ib}} v_o + h_{ob}v_o$$

$$= \frac{[h_{ob}h_{ib} - h_{ib}h_{rb}]v_o}{h_{ib}}$$

$$\therefore i_c = v_o \frac{\Delta h}{h_{ib}}$$

Substituting the value of i_c, we have

$$R_{ob} = \frac{v_o}{v_o \frac{\Delta h}{h_{ib}}}$$

$$= \frac{h_{ib}}{\Delta h} \tag{26}$$

The AC equivalent circuit (Figure 7.4(b)) can be now simplified (Figure 7.4(c)) to look like a CE amplifier.

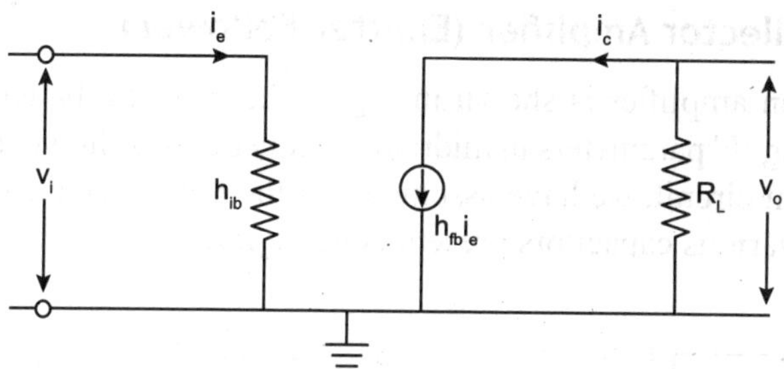

Figure 7.4 (c) AC equivalent of CB amplifier (simplified version)

Current gain,
$$A_{ib} = \frac{i_c}{i_e} = \frac{h_{fb}\, i_e}{i_e} = h_{fb} \tag{27}$$

Voltage gain, $A_{vb} = \dfrac{v_o}{v_i}$

Here $v_o = (-)i_c\, R_L$ and $v_i = i_e\, h_{ib}$

Substituting for v_o and v_i, we have

$$A_{vb} = \frac{(-)i_c R_L}{i_e\, h_{ib}}$$

$$= (-)A_{ib} \cdot \frac{R_L}{h_{ib}}$$

Substituting for A_{ib}.

$$A_{vb} = (-)\frac{h_{fb}}{h_{ib}} R_L \tag{28}$$

Input resistance (R_i) By definition,

$$R_{ib} = \frac{v_i}{i_e} = h_{ib} \tag{29}$$

Output resistance (R_o) By definition,

$$R_{ob} = \frac{v_o}{i_c} = \frac{1}{h_{ob}} \Rightarrow \infty \tag{30}$$

Power gain,
$$A_p = |A_i||A_v| = \left[\frac{h_{fb}^2 R_L}{h_{ib}}\right] \tag{31}$$

(**Note:** In the above, suffix '*b*' for the various parameters shows that the transistor is in common base configuration.)

7.3.3 Common Collector Amplifier (Emitter Follower)

The basic circuit of an amplifier is shown in Figure 7.5(a) using battery connections. It AC equivalent circuit using '*h*' parameters in midrange frequencies is shown in Figure 7.5(b). When drawing the equivalent circuit, we have assumed that the input signal frequency range is such that the reactance of various capacitors present is negligible.

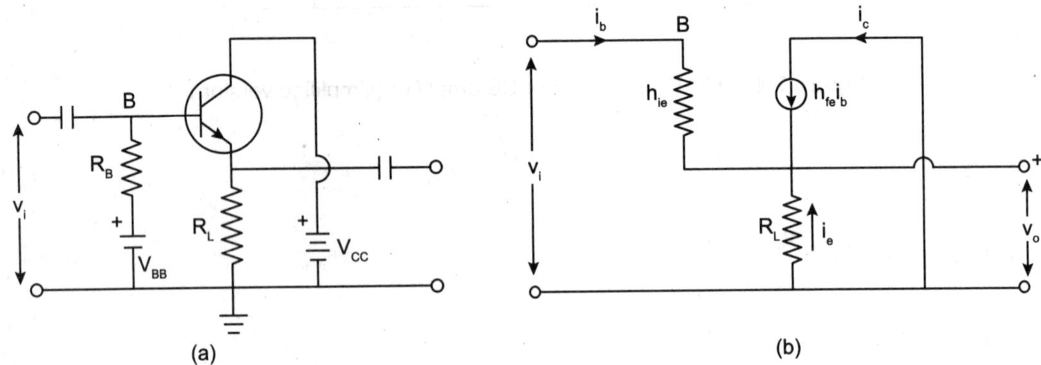

Figure 7.5 (a) Common collector amplifier circuit, (b) AC equivalent circuit of *CC* amplifier using '*h*' parameters in midfrequency range

Current gain (A_i) By definition, current gain

$$A_{ic} = \frac{i_e}{i_b}$$

$$= (-)\frac{i_b + i_c}{i_b}$$

$$= -\left[1 + \frac{i_c}{i_b}\right] = (-)\left[1 + \frac{h_{fc}i_b}{i_b}\right]$$

$$= (-)(1 + h_{fe}) \quad (32)$$

$$= (-)h_{fc} \quad (33)$$

since $h_{fe} >> 1$.

Voltage gain (A_v) By definition, voltage gain $(A_{vc}) = \frac{v_c}{v_i}$

From the circuit, (Figure 7.5(b))

$$v_o = (-)i_e R_L = (i_b + i_c)R_L$$

$$= (+)i_b[1 + h_{fe}]R_L$$

(It is to be noted that the direction of i_e is opposite to that of i_b and i_c, and so the negative sign is deleted.)

$$\text{also} \qquad v_i = i_b\, h_{ie} + (i_b + i_c)R_L$$
$$= i_b\, h_{ie} + i_b(1 + h_{fe})R_L$$
$$= i_b[h_{ie} + (i + h_{fe})R_L]$$

$$\text{Voltage gain, } A_{vc} = \frac{v_o}{v_i}$$
$$= \frac{(+)i_b[1 + h_{fe}]R_L}{i_b[h_{ie} + (1 + h_{fe})R_L]}$$
$$= \frac{1}{1 + \dfrac{h_{ie}}{(1 + h_{fe})R_L}} \tag{34}$$
$$= 1 \tag{35}$$
$$\text{since } 1 \gg \frac{h_{ie}}{(1 + h_{fe})R_L}$$

Eqn. (35) represents that the maximum voltage gain in case of *CC* amplifier is unity.

Eqn. (34) can also be written as

$$A_{vc} = \frac{g_m R_L}{1 + g_m R_L} \tag{36}$$

Assuming

(i) $\dfrac{R_L}{h_{ie}}$ to be small and

(ii) $\dfrac{h_{fe}}{h_{ie}} = g_m$

Eqn. (34) shows that there is no phase change in the output and input voltage.

Input resistance (R_i) By definition,

$$R_{ic} = \frac{v_i}{i_b} = h_{ie} + (1 + h_{fe})R_L$$
$$= h_{ie} + h_{fe}R_L \quad \text{neglecting 1}$$
$$= h_{ie}(1 + g_m R_L) \tag{37}$$

Eqn. (37) represents that the value of input resistance is very high.

Output resistance (R_o) Assuming input signal to be zero, the circuit takes the form as shown in Figure 7.6.

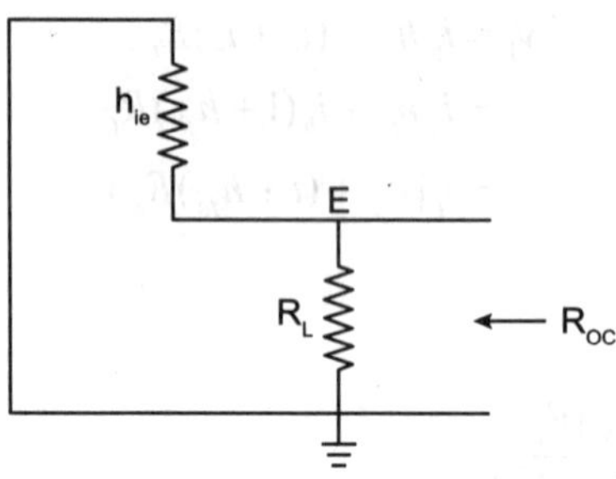

Figure 7.6

Here h_{ie} and R_L form a parallel combination so that

$$R_{oc} = \frac{h_{ie} R_L}{h_{ie} + R_L} \tag{38}$$

From Eqn. (38), it is clear that the value of output resistance is very small. From the above discussion, it is clear that the common collector amplifier (or emitter follower) has very high input impedance (resistance) and low output impedance (resistance).

(**Note:** In the above, the suffix '*c*' for the various parameters shows that the transistor is in common collector configuration.)

7.4 T Model (Eber-Moll Model) or π Model

7.4.1 π Model

When an AC input signal drives a transistor amplifier, an AC base-emitter voltage v_{be} is across the emitter diode as shown in Figure 7.7.

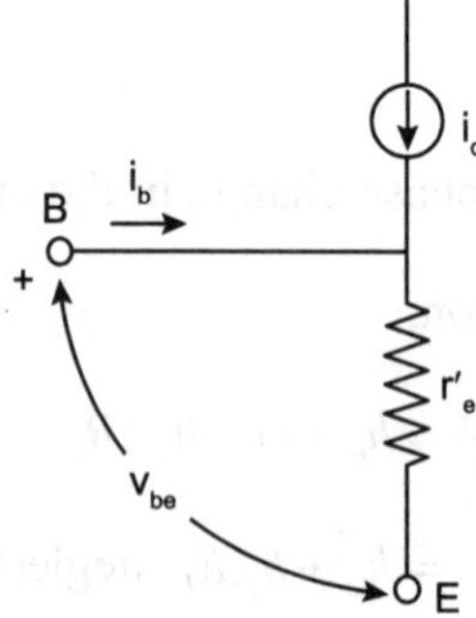

Figure 7.7

This produces an AC base current i_b. The AC voltage has to supply the AC base current, so that the transistor amplifier will work. Stated another way, the AC voltage source is loaded by input impedance r_e' of the base.

7.4.2 AC Resistance of Emitter Diode

Normally, the base emitter junction of a transistor is forward-biased and so current I_E always flows through the junction due to applied voltage V_{BE}. The graph between (V_{BE}, I_E) is as shown in Figure 7.8.

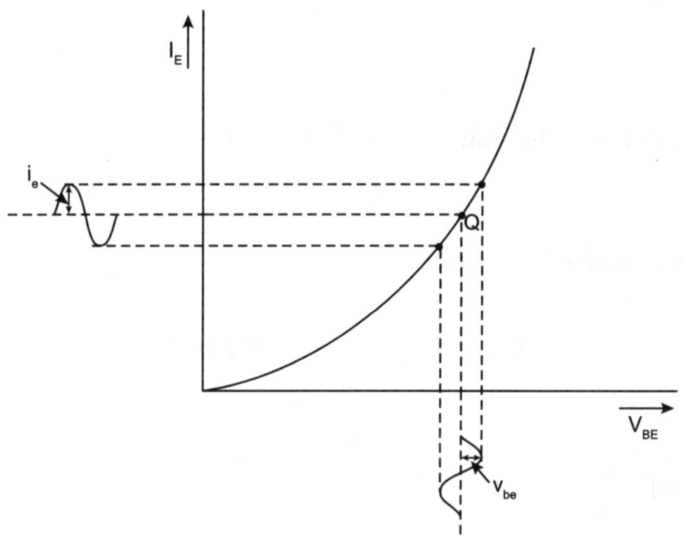

Figure 7.8

When an AC signal v_{be} is applied across the emitter-base diode, it produces an AC emitter current i_e. The magnitude of i_e depends on the location of Q point.

The total emitter current flowing through the junction is

$$I_E = I_{EQ} + i_e$$

and the total base emitter voltage is

$$V_{BE} = V_{BEQ} + v_{be}$$

Thus, a sinusoidal variation in V_{BE} produces a sinusoidal variation in current, provided the signal is small. Because of the curvature in the graph, a fixed v_{be} is produced as more emitter current i_e as the Q point is biased higher up on the curve (where we have a large slope). Stated another way, the AC resistance of the emitter-diode decreases when DC emitter current increases.

The AC resistance of the emitter diode is defined as

$$r_e' = \frac{v_{be}}{i_e}$$

Here r_e' indicates that resistance is inside the transistor. We can calculate the value of r_e' of a transistor by using the following formula:

$$r'_e = \frac{25 \text{ mV}}{I_E} \tag{39}$$

From Figure 7.7, the AC voltage source sees an input impedance Z_{in}(base). At low frequencies, this impedance is purely resistive and defined as

$$Z_{\text{in(base)}} = \frac{v_{be}}{i_b}$$

Applying Ohm's law to the emitter diode (Figure 7.7),

$$v_{be} = i_e r'_e$$

Substituting the value of v_{be}, we have

$$Z_{\text{in (base)}} = \frac{i_e r'_e}{i_b} = \left[\frac{i_c}{i_b}\right] r'_e \text{ since } i_e \approx i_c$$

$$= \beta r'_e \tag{40}$$

since by definition $\beta = \frac{i_c}{i_b}$.

The above equation tells us that an input impedance of base is equal to AC current gain (β) times the AC resistance (r'_e) of the emitter diode.

Taking into consideration the AC emitter diode, the resistance of a transistor amplifier can be shown as in Figure 7.9.

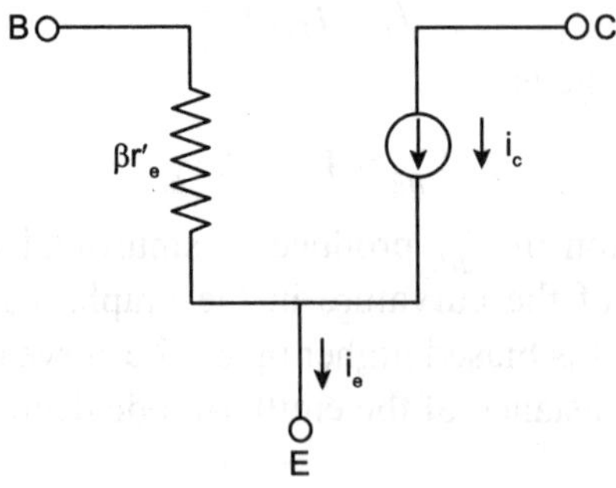

Figure 7.9

The above model suggests that an input impedance of $\beta r'_e$ will load the AC source (signal) to drive the base.

Thus, the AC equivalent circuit of an RC amplifier (Figure 7.3) can be shown as in Figure 7.10.

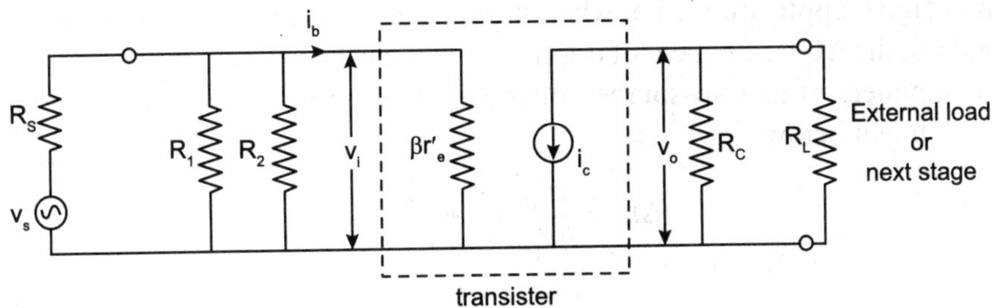

Figure 7.10 AC equivalent circuit of a CE amplifier (using π model)

Circuit analysis

$$v_{in} = i_b \beta r'_e$$

In the output circuit,

$$v_o = (-)i_c [R_C \,||^{\text{rl}}\, R_L]$$

here sign is due to sign convention.

Substituting for $i_c = \beta i_b$,

$$v_o = (-)\beta i_b [R_C \,||^{\text{rl}}\, R_L]$$

$$\therefore \text{ Voltage gain, } A_v = \frac{v_o}{v_{in}} = \frac{(-)\beta i_b R'_L}{i_b \beta r'_e}$$

$$= (-)\frac{R'_L}{r'_e} \qquad (41)$$

The negative sign represents a phase change of 180° between input and output voltage.

Note: In h parameter, h_{ie} represents the equivalent input impedance, and as per the definition

$$r'_{ie} = \frac{h_{ie}}{h_{fe}} = \frac{h_{ie}}{\beta}$$

$$\therefore h_{ie} = \beta r'_e$$

7.5 FET Amplifier

We have discussed the use of BJT as an amplifier. We can also use JFET as an amplifier. When we discussed the characteristics of FET (Chapter 5), it was observed that drain current (I_d) depends upon gate-to-source voltage (V_{GS}) and drain-to-source voltage (V_{DS}). We can write

$$I_d = f(V_{GS}, V_{DS})$$

For small signal applications, i.e. when these quantities vary by a small amount about the quiescent values, the AC component of drain current i_d can be expressed as a linear combination of the AC component of gate-to-source voltage v_{gs} and the AC component of drain-to-source voltage v_{ds} as shown below:

$$\Delta I_d = \frac{\partial I_d}{\partial V_{GS}} v_{gs} + \frac{\partial I_d}{\partial V_{DS}} v_{ds}$$

$$\text{or} \quad i_d = g_m v_{gs} + \frac{1}{r_d} v_{ds} \tag{42}$$

where $g_m = \dfrac{\partial I_d}{\partial V_{GS}}$ at V_{ds}, constant is called transconductance of FET, and $r_d = \dfrac{\partial V_{DS}}{\partial I_D}$ at v_{GS}, constant is called output resistance of FET. Applying Kirchoff's current law, Eqn. (42) is represented by the circuit shown in Figure 7.11.

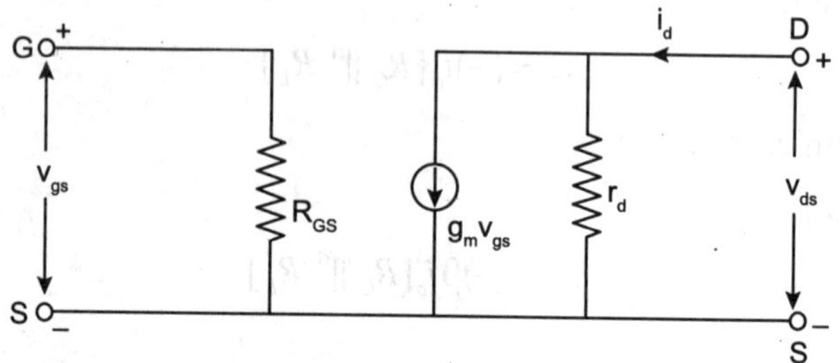

Figure 7.11 Current source equivalent circuit of FET (low frequency)

Here v_{gs} is the AC (signal) voltage applied across gate and source S. The input signal will see the reverse gate source resistance R_{GS}.

Eqn. (42) can also be written as

$$v_{ds} = i_d v_d - \mu v_{gs} \tag{43}$$

where $\mu = g_m v_d$

By applying Kirchoff's voltage law, Eqn. (43) can be represented by the following circuit (Figure 7.12). The circuit consists of a voltage source μv_{gs} called a small signal voltage source.

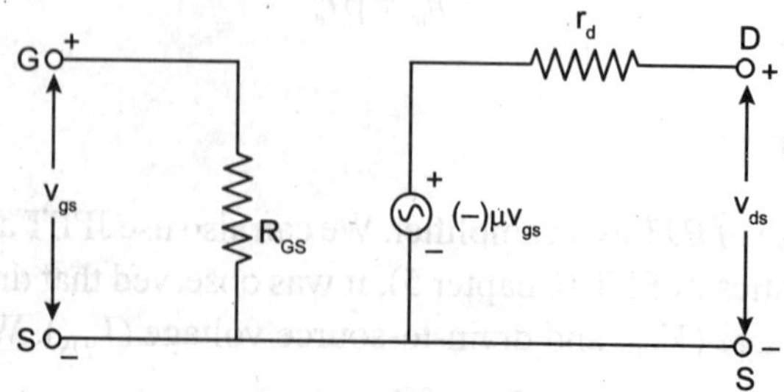

Figure 7.12 Voltage source equivalent circuit of FET (low frequency)

Comparing the circuit (Figure 7.12) with that of a hybrid AC equivalent circuit of CE amplifier (Figure 7.2), we see that current source $h_{fe}\, i_b$ depends upon input current i_b, while current source $g_m v_{gs}$ depends upon input voltage v_{gs}. In the FET model, no feedback takes place from the output to the input as in the case of BJT where h_{ve} provides such a feedback. In addition, FET has a very high input resistance, whereas for a CE amplifier, it is quite small and is of the order of 1 kΩ. Thus, FET gives a better lowfrequency amplifier than a conventional transistor amplifier.

Just as in a transistor, here too we have two useful configurations: (i) common source and (ii) common drain, which are similar to the CC and CE configurations in transistors. We shall discuss the two configurations of a FET amplifier.

7.5.1 Common Source FET Amplifier

A common source FET amplifier circuit with biasing batteries is shown in Figure (7.45). The AC signal v_{gs} to be amplified is in series with gate-battery V_{GG}.

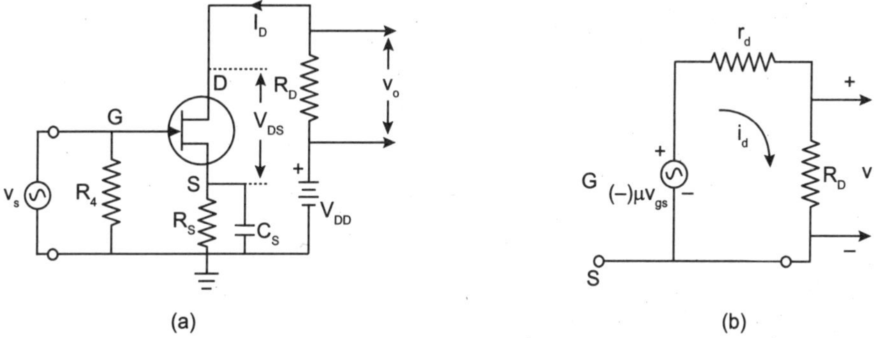

Figure 7.13 (a) Circuit common source in channel FET amplifier, (b) AC equivalent circuit of common source FET amplifier

As shown in Figure 7.13(a), load $R_L = R_D$ is connected between battery V_{DD} and drain *D*. The negative terminal of the battery (supply) is connected to the source since we have used a *N* channel FET (if *P* channel is used then battery V_{DD} connections are reversed and the source is connected to the positive terminal of battery). Gate *G* is negatively biased because of the voltage drop across R_S due to drain current I_D flowing through it. Input signal v_s (assumed to be small) is connected across R_G as shown. R_G is connected for DC return path.

7.5.2 Action of FET as Amplifier

When a small input voltage v_{gs} is applied and if the signal goes positive, then gate-source bias decreases (less negative) and thereby drain current I_D increases. The increase in drain current causes voltage drop across load R_L to increase. This causes an voltage drop between drain and source. Thus, drain *D* becomes less positive w.r.t. the source *S*. A small change in gate input voltage causes a large change in voltage across load, i.e. we can say there is voltage amplification.

Circuit analysis Consider the AC equivalent circuit (Figure 7.45(b)).

$$(-)\mu v_{gs} = i_d [R_D + r_d] \tag{44}$$

Output voltage (i.e. change in voltage across drain D and source S) is

$$v_o = v_{ds} = i_d R_D \tag{45}$$

Substituting for i_d from Eqn. (44), we have

$$v_o = (-)\frac{\mu v_{gs}}{R_D + r_d}(R_D)$$

$$\text{Voltage gain, } A_v = \frac{v_o}{v_{gs}} = (-)\frac{\mu R_D}{R_D + r_d} = (-)\frac{\mu R_L}{R_L + r_d} \tag{46}$$

where $R_D = R_L$ called load.

The negative sign represents a phase change of π between output and input voltage.

If $r_d \gg R_L$,

$$A_v = (-)\frac{\mu}{r_d}[R_L]$$

$$\therefore A_v = (-)g_m R_L \tag{47}$$

since by definition $\mu = g_m r_d$.

Eqn. (46) can also be written as

$$A_v = (-)\frac{\mu}{1 + r_d/R_L} \tag{48}$$

Eqn. (48) represents that voltage gain (in case of a common source FET amplifier) depends upon

(i) Amplification factor (μ) of FET and

(ii) Load connected

If R_L goes to infinity in Eqn. (48), the maximum value of voltage amplification is

$$|A_v|_{\max} = \mu \tag{49}$$

i.e. it attains the constant value μ.

The variation of gain with load R_L is shown in Figure 7.14.

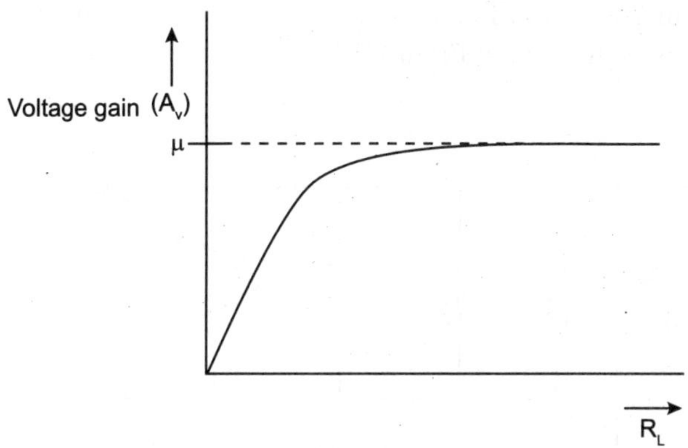

Figure 7.14 Variation of voltage gain with load

The (V_{DS}, I_D) characteristic of N-channel FET is shown in Figure 7.15.

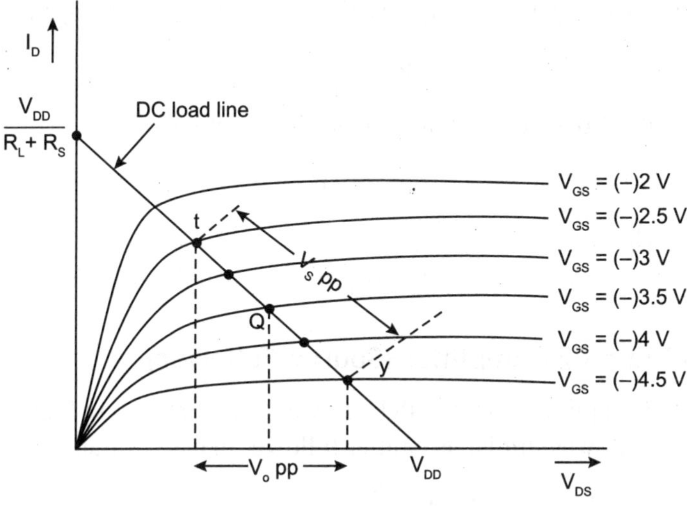

Figure 7.15 [V_{DS}, I_D] characteristic curve with DC load line

From Figure 7.13(a), applying K_{VL} to the output side,

$$V_{DD} = I_D R_D + V_{DS}$$

or

$$I_D = (-)\frac{1}{R_D}V_{DS} + \frac{V_{DD}}{R_D} \tag{50}$$

Eqn. (50) represents the equation of a straight line $y = mx + c$ drawn on V_{DS} vs. I_D axis having a slope $(-)\frac{1}{R_D}$. The equation represents the equation of DC load line (Figure 7.15) just like that of a CE transistor amplifier.

Just like the CE transistor amplifier's using '*h*' parameter circuit (Figure 7.2(b)), the FET amplifier can also be represented as in Figure 7.16.

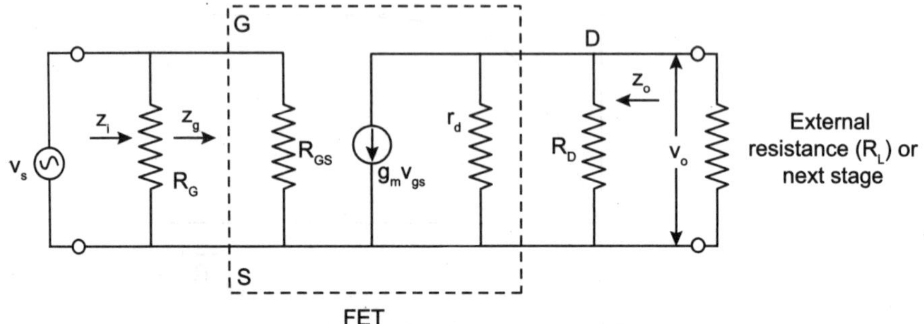

Figure 7.16 AC equivalent circuit (complete) of a common source FET amplifier

Input impedance Looking into the gate *G* and source *S* terminal (Figure 7.16), the resistance seen is gate source leakage resistance, R_{GS}. Hence the device's input impedance is

$$Z_g = R_{GS} \tag{51}$$

Output impedance Consider the drain and source terminals.

$$Z_o = r_d \parallel R_D$$

Generally $r_d >> R_D$.

$$\text{So, } Z_o = R_D \tag{52}$$

7.5.3 Common Drain FET Amplifier (Source Follower)

A common drain FET amplifier circuit with battery biasing is shown in Figure 7.17. The circuit is also called source follower similar as emitter follower in case of transistor amplifier.

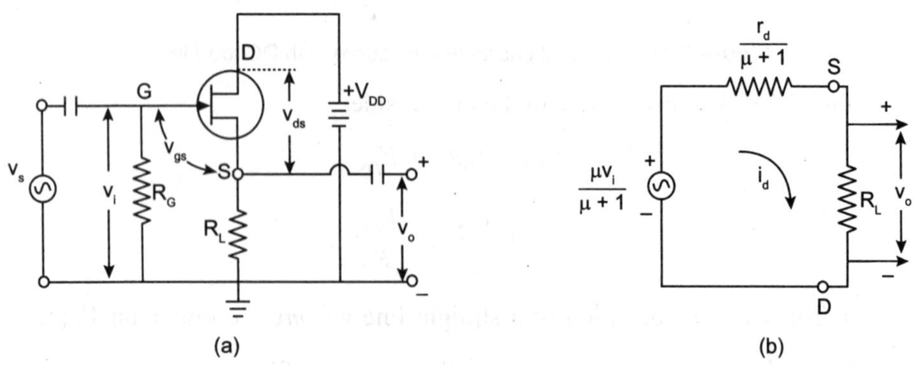

Figure 7.17 Common drain N-channel amplifier circuit: (a) circuit, (b) voltage source equivalent circuit of common drain FET amplifier

7.5.4 Action of FET as Amplifier

The complete AC equivalent circuit of an amplifier is shown in Figure 7.18.

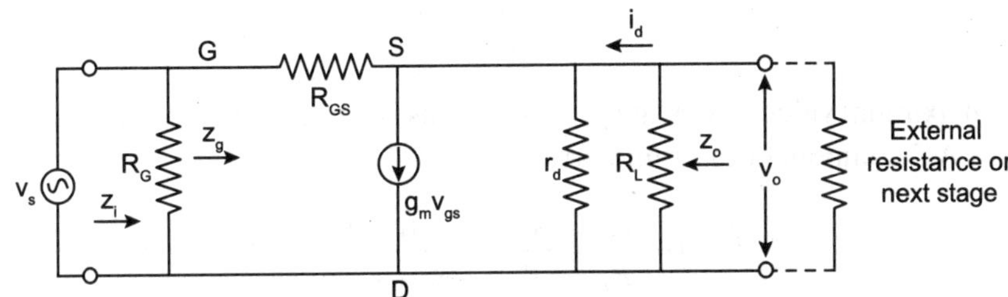

Figure 7.18 AC equivalent circuit (complete) of common drain FET amplifier

Circuit analysis Consider AC conditions (neglecting V_{DD} DC supply). The small signal condition gives us

$$i_d = g_m v_{gs} + \frac{v_{ds}}{v_d} \tag{53}$$

For the AC equivalent component in output circuit, we have $v_{ds} + i_d R_L = 0$

The AC component of gate source voltage is

$$v_{gs} = v_i - i_d R_L = v_i + v_{ds}$$

since change in voltage across R_L is opposite to that in v_{ds}.

Substituting for i_d from Eqn. (i),

$$v_{ds} + \left[g_m v_{gs} + \frac{v_{ds}}{r_d} \right] R_L = 0$$

$$v_{ds} + \left[g_m (v_i + v_{ds}) + \frac{v_{ds}}{r_d} \right] R_L = 0$$

$$\text{or } v_{ds} \left[1 + g_m R_L + \frac{R_L}{r_d} \right] + g_m R_L v_i = 0$$

also $v_o = (-)v_{ds}$

$$\text{Voltage gain } A_v = \frac{v_o}{v_i} = (-)\frac{v_{ds}}{v_i} = \frac{g_m R_L r_d}{r_d + g_m r_d R_L + R_L}$$

$$\text{or } A_v = \frac{\dfrac{\mu R_L}{\mu + 1}}{\dfrac{R_L + r_d}{\mu + 1}}$$

since $\mu = r_d g_m$.

Thus, in case of common drain FET amplifier, there is no phase change between output and input voltage. When $R_L \gg \frac{r_d}{\mu+1}$ (which is normally observed),

$$A_v = \frac{\mu}{\mu+1} \approx 1$$

Thus, the maximum value of voltage gain is unity as is the case in common collector amplifier (Eqn. 35). The AC component of drain current is

$$i_d = (-)\frac{v_{ds}}{R_L} = \frac{A_v v_i}{R_L} = \frac{\mu v_i / \mu+1}{R_L + r_d / \mu+1}$$

$$\text{or } \frac{\mu}{\mu+1} v_i = i_d \frac{r_d}{\mu+1} + i_d R_L \tag{54}$$

The small signal voltage source circuit of common drain amplifier given by Eqn. (54) is shown in Figure 7.17(b).

Input impedance Looking into the gate and drain terminal (Figure 7.18), the common drain equation for Z_g can be derived and given by

$$Z_g = R_{GS}[1 + g_m R_L]$$

which gives an extremely large value for Z_g.

Thus, input impedance

$$R_i = R_Q \,||^{rl}\, Zg$$
$$\approx R_G$$

Output impedance To determine its value, the input signal is assumed to be zero and current through load is I_d. The output voltage is derived between R_{GS} and R_G.

So,
$$v_{gs} = v_o \times \frac{R_{GS}}{R_{GS} + R_G} = v_o \text{ since } R_{GS} \gg R_G$$

$$I_d = g_m v_{gs}$$

$$Z_s = \frac{v_o}{Z_d} = \frac{v_{gs}}{g_m v_{gs}} = \frac{1}{g_m}$$

Actually, r_d is in parallel with $(1/g_m)$ but $r_d \gg (1/g_m)$; so it is neglected.

Circuit output impedance,

$$Z_o = R_L \,||^{rl}\, (1/g_m)$$

Example 1 Calculate the lowest frequency for which the coupling in figure is good?

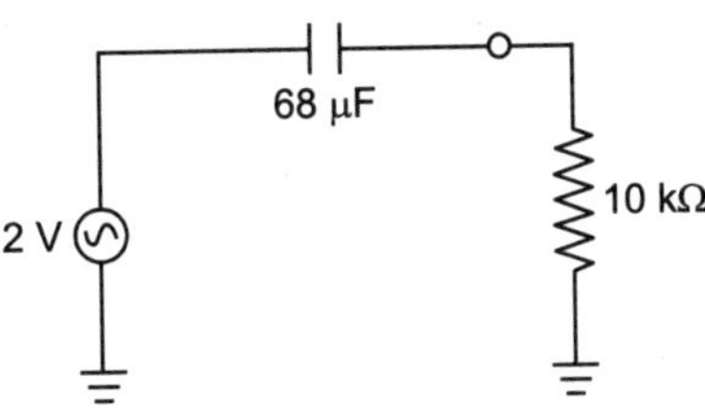

Figure 7.19

Solution For good coupling,

$$X_C \leq 0.1\text{ R}$$

$$\frac{1}{2\pi fc} = 0.1 \times 10 \times 1000$$

$$\therefore \text{Lowest frequency}, f = \frac{1}{2\pi \times 68 \times 10^{-6} \times 0.1 \times 10 \times 1000}$$

$$= 2.34$$

$$= 3\text{ Hz}$$

Example 2 Calculate the lowest frequency for which we have good by passing in the figure.

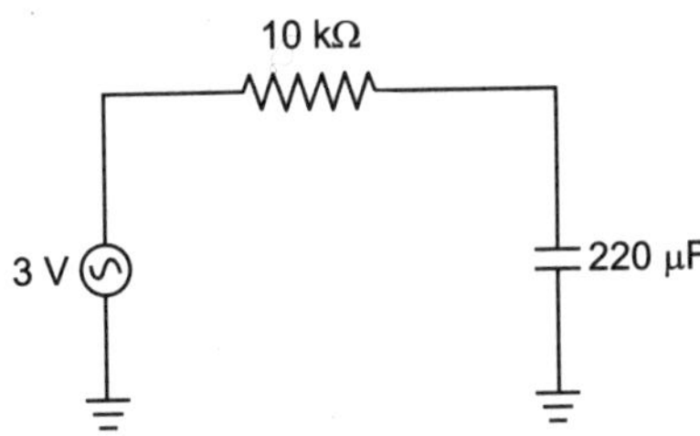

Figure 7.20

Solution For good bypassing, $X_C \leq 0.1$ R.

$$\frac{1}{2\pi fc} = 0.1 \times 1000$$

$$\therefore \text{Lowest frequency}, f = \frac{1}{2\pi \times 220 \times 10^{-6} \times 0.1 \times 1000}$$

$$= 0.72$$

$$= 1.0\text{ Hz}$$

Example 3 Calculate (i) maximum allowable emitter current so that operation of the amplifier is considered a small signal amplifier, (ii) AC emitter diode resistance, (iii) input impedance of the base if $\beta = 250$.

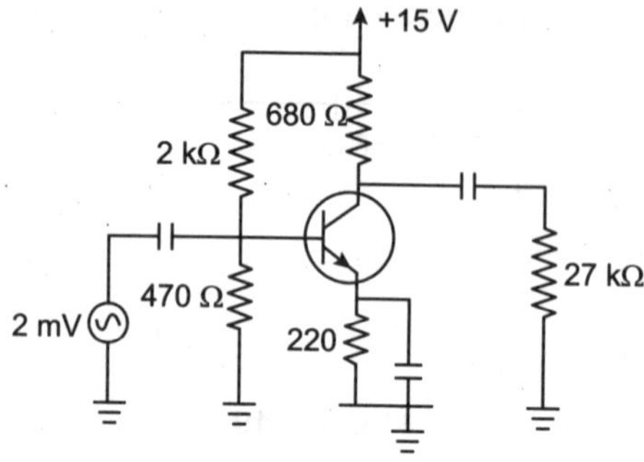

Figure 7.21

Solution

(i) Base voltage $V_B = \dfrac{V_{CC}}{R_1 + R_2} \times R_2$

$$= \frac{15}{2000 + 470} \times 470$$

$$= \frac{15}{2470} \times 470 = 2.85 \text{ V}$$

Voltage at emitter end, $V_E = V_B - 0.7 = 2.85 - 0.7 = 2.15$ V.

Emitter current, $I_E = \dfrac{V_E}{R_E} = \dfrac{2.15}{220} = .97 \times 10^{-2}$ A

For small signal operation, $i_e \le 0.1\, I_E = 0.1 \times 9.7 \times 10^{-3}$

$$= 0.97 \text{ mA}$$

(ii) AC emitter diode resistance $= \dfrac{25 \text{ mV}}{I_E} = \dfrac{25}{.97} = 26\ \Omega$

(iii) $Z'_{vm} = \beta r'_e = 250 \times 26 = 6500\ \Omega$

Example 4 In the circuit, calculate the output voltage. Assume $\beta = 100$.

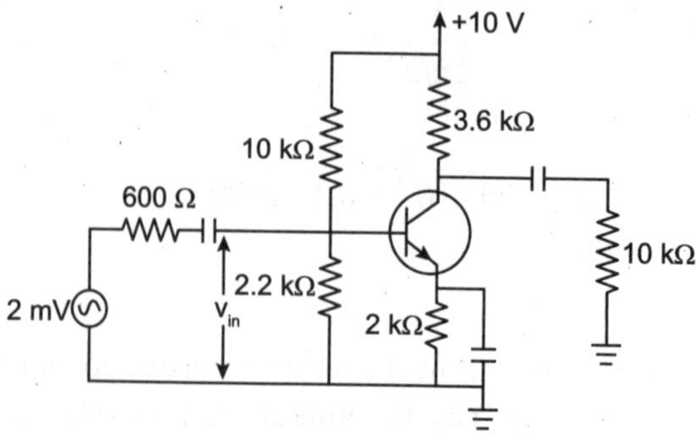

Figure 7.22

Solution For DC calculation, neglecting capacitance

$$\text{Base voltage, } V_B = \frac{2.2\text{ k}\Omega}{(10+2.2)\text{ k}\Omega} \times 10$$
$$= 1.8\text{ V}$$

Emitter voltage = $V_B - 0.7 = 1.8 - 0.7 = 1.1$ V

$$\text{Emitter current, } I_E = \frac{1.1}{2000} = 0.55\text{ mA}$$

$$\text{Emitter diode resistance, } r_e = \frac{25\text{ mV}}{I_E} = \frac{25}{0.55} = 45\ \Omega$$

$$A = \frac{r_c}{r_e} = \frac{(3.6 \,||\, 10)\text{ k}\Omega}{45} = \frac{2.65 \times 1000}{45} = 53$$

$$V_{in} = \frac{Z_{in}\text{ stage}}{R_S + Z_{in}\text{ stage}} \times v_s$$

$$Z_{in}\text{ stage} = 10 \,||\, 2.2 \,||\, \beta r_e'$$
$$= 10 \,||\, 2.2 \,||\, 100 \times 53$$
$$= [10 \,||\, 2.2 \,||\, 5.3]$$
$$= 1.4\text{ k}\Omega$$

$$\therefore\ V_{in} = \frac{1.4 \times 1000}{600 + 1.4 \times 1000} \times 1\text{ mV} = 0.7\text{ mV}$$

$$\text{Gain} = \frac{v_{out}}{v_{in}}$$

$$\therefore v_{out} = \text{gain} \times v_{in}$$
$$= 53 \times 0.7$$
$$= 37.1\text{ mV}$$

Example 5 A RC-coupled CE amplifier has collector load of 10 kΩ. The input signal applied is 0.02 V. The internal resistance of voltage signal applied is 500 Ω. Calculate the voltage gain if h_{ie} = 1 kΩ and h_{fe} = 50 for the transistor used. Also calculate output voltage.

Solution

$$\text{Voltage gain, } |A_v| = \frac{h_{fe} R_L}{R_S + h_{ie}}$$
$$= \frac{50 \times 10 \times 1000}{500 + 1000} = 333.3$$

$$\text{Output voltage} = A_v \times v_s$$
$$= 333.3 \times 0.002$$
$$= 6.66 \text{ V}$$

Example 6 An audio amplifier has an input resistance of 10 kΩ. An input signal of 0.5 V gives an output of 5 V across a load of 10 Ω. Calculate the power gain of the amplifier. What will be the power gain if input resistance is equal to output resistance?

Solution

$$\text{Voltage gain} = \frac{\text{Output voltage}}{\text{Inut voltage}} = \frac{5}{0.5} = 10$$

$$\text{Input power, } P_i = \frac{V_i^2}{R_i}$$

$$\text{Output voltage, } P_o = \frac{V_o^2}{R_o}$$

$$\text{Power gain (in decibel)} = 10\log\frac{P_o}{P_i} = 10\log\frac{V_o^2/R_o}{V_i^2/R_i} = 10\log\left[\frac{V_o}{V_i}\right]^2 + 10\log\frac{R_i}{R_o}$$

$$= 20\log 10 + 10\log\frac{10\times 10^3}{10} = 20 + 30$$

$$= 50 \text{ dB}$$

When $R_i = R_o$, power gain = 20 dB

Example 7 A RC-coupled CE amplifier has a load of 2 KΩ. It is coupled to the next stage of input resistance 500 Ω through a capacitor of 2 μF. Calculate the lower cut-off frequency of the amplifier.

Solution

$$\text{Lower cut-off frequency, } f_1 = \frac{1}{2\pi C_C[R_L + R_i]}$$

$$= \frac{1}{2\times 3.14\times 2\times 10^{-6}[2500]}$$

$$= 33 \text{ Hz}$$

Example 8 A RC-coupled amplifier has (–)3 dB frequencies at 33 Hz and 33 KHz, respectively. Calculate the value of coupling capacitor and shunting capacitor if collector load is 20 KΩ and input resistance of next stage 5 KΩ.

Solution

$$\text{Lower cut-off frequency}, f_1 = \frac{1}{2\pi C_C (R_L + R_i)}$$

$$\therefore \text{ Coupling capacitor, } C_C = \frac{1}{2\pi f_1 (R_L + R_i)} = \frac{1}{2\pi \times 33(25000)}$$

$$= \frac{1}{200 \times 25000} = \frac{1}{5} \times 10^{-6}$$

$$= 0.2 \ \mu\text{F}$$

$$\text{Upper cut-off frequency}, f_2 = \frac{1}{2\pi C_S \,\text{Req}} = \frac{1}{2\pi C_S \left[\dfrac{R_L R_i}{R_L + R_i}\right]}$$

$$\therefore \text{ Shunting capacitance}, f_1 = \frac{1}{2\pi C_C (R_L + R_i)}$$

$$= \frac{1}{2\pi \times 33000 \times \dfrac{20000 \times 5000}{25000}}$$

$$= 1250 \text{ pF}$$

Example 9 A FET amplifier used in common source configuration has $R_D = 3$ KΩ and external resistance $R_L = 5$ kΩ coupled with 0.2 μF capacitor. The input signal is 100 mV. Calculate output voltage, voltage gain assuming $g_m = 4500$ μMhos.

Solution In the mid-frequency, range reactance of the capacitor is neglected. Thus, effective value of drain resistance,

$$R'_L = \frac{R_D R_L}{R_O + R_L} = \frac{3 \times 5}{3 + 5} \text{ k}\Omega = 1.88 \text{ k}\Omega$$

$$\text{Voltage amplification}, A_v = (-) g_m R'_L$$

$$= (-)4500 \times 10^{-6} \times 1.88 \times 10^3$$

$$= (-)8.46$$

$$= 9 \angle 180^\circ$$

$$\text{Output voltage}, v_o = |A_v| v_i$$

$$= 9 \times 100 \times 10^{-3}$$

$$= 0.9 \text{ V}$$

7.5.5 Small Signal High Frequency Amplifier

We understand that the common base amplifier circuit has very low input impedance as well as output impedance similar to that of a CE amplifier circuit. The voltage gain of CB amplifier is also of the same range as that of CE amplifier, with the important exception that there is no phase inversion between input and output signal. Because of the absence of phase inversion, there is no Miller effect (amplification of input capacitance) in case of a CB circuit. Consequently, a CB circuit can be used to operate at much higher frequencies than by a CE circuit. A practical CB circuit is illustrated in Figure 7.23.

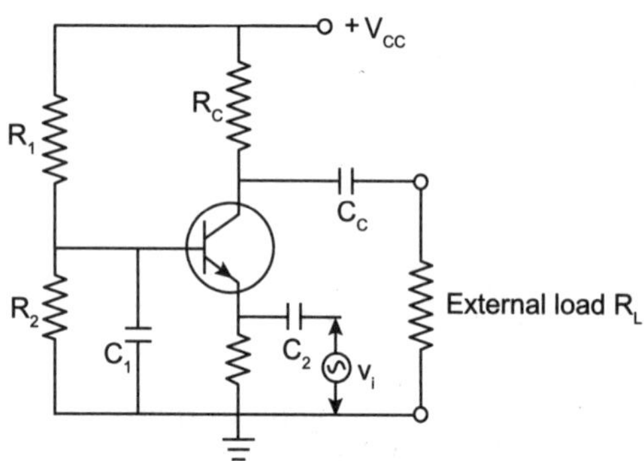

Figure 7.23 CB amplifier circuit

When the AC equivalent circuit of the above is drawn, impedance in series with capacitor C_2 is that seen when looking into the emitter of the transistor, i.e.

$$h_{ib} = \frac{h_{ie}}{1+h_{fe}}$$

This low input impedance makes C_2 the largest capacitor in the circuit. Consecutively, C_2 is used to set the lower 3 dB frequency f_1, and is calculated similar as CC in the case of CE amplifier.

Impedance in series with capacitor C_1 is the impedance h_{ie} seen when looking into the base of the transistor. For the circuit set at lower (3 dB) frequency, X_{C1} should be very much smaller than h_{ie}. This is similar to the calculation of C_1 in a CE circuit:

$$X_{C1} = \frac{h_{ie}}{10}$$

7.5.6 Cascade Amplifier

The cascade amplifier (Figure 7.24) is one approach to solve low input impedance problems of a CB amplifier circuit. Transistor Q_1 and its associated components operate as CE input stage, while the circuit of Q_2 functions as a CB amplifier output stage. This arrangement gives the cascade the high input impedance of a CE amplifier as well as the moderate voltage gain and high frequency performance of a CB amplifier.

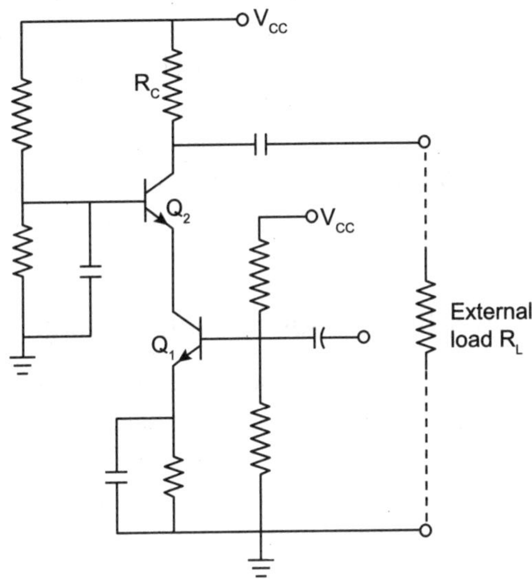

Figure 7.24 Cascade amplifier

7.5.7 RC-Coupled CE Amplifier

It is one of the most common amplifiers (Figure 7.25) used. Here, the best method for biasing is the voltage divider method with emitter bias, with which the stability factor could be kept quite low.

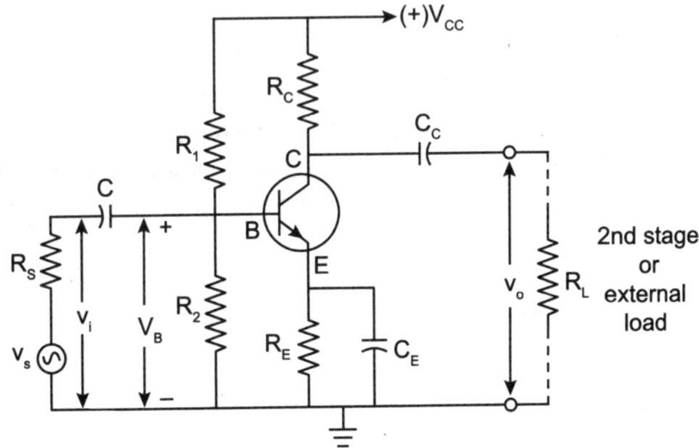

Figure 7.25 Single-stage CE amplifier

The signal source, v_s, with internal resistance R_s is connected to transistor base via coupling capacitor C.

Here the capacitor C_C couples the amplifier to the next stage, if any. Otherwise, we get only AC voltage as output voltage, v_o. Capacitor C_E bypasses AC, which otherwise would have been the feedback to input as regards the AC signal is concerned. It is important to note the polarity of capacitors. The curve inside the symbol of capacitor represents the negative terminal; a straight line indicates the positive terminal. Each capacitor must be connected with its negative terminal at a DC voltage level lower than (or equal) the DC level of the positive terminal. For example, C_1 in Figure 7.3 has its negative terminal at DC ground level, because it is grounded through signal source resistance, R_S. The positive terminal of C is at potential $+V_B$ w.r.t. the ground. If incorrectly connected, the polarized capacitor can result in changes in transistor bias conditions and poor AC performance.

7.5.8 AC Analysis of CE Amplifier (Mid-frequency Range)

The first step in AC analysis is to draw AC equivalent circuit. This is done by replacing the capacitors with short circuits in the frequency range where the reactance of capacitor is not effective. Also, since supply voltage V_{CC} remains constant, it must be treated as another short circuit. This means that the $+V_{CC}$ terminal (in case of NPN transistor) is at the same AC potential as ground. Using the above, AC consideration in the mid-frequency range (Figure 7.25) can be redrawn as shown in Figure 7.26.

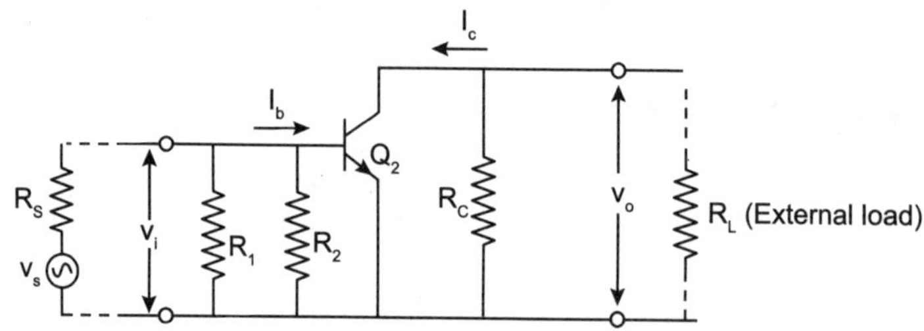

Figure 7.26 AC equivalent of CE amplifier (mid-frequency range)

In the AC equivalent circuit (Figure 7.4), R_1 is shown connected between transistor base and ground. R_1 and R_2 are thus in parallel. Similarly, R_C appears across transistor collector terminal and ground. When an external load R_L is present, R_C and R_L are in parallel for AC consideration. The circuit of Figure 7.26 can be redrawn (Figure 7.27) using 'h' parameters.

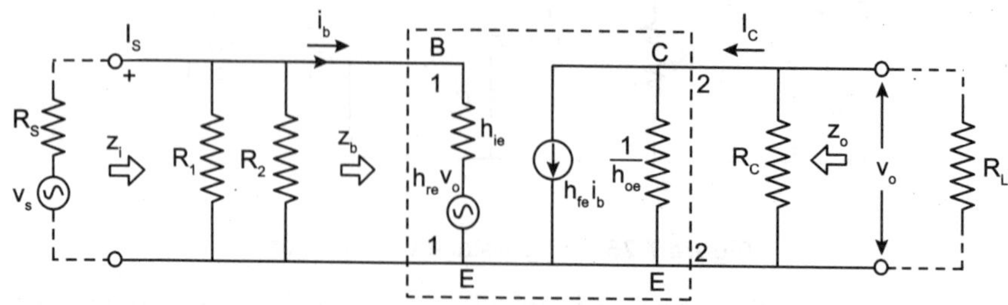

Figure 7.27 CE amplifier equivalent circuit using 'h' parameters (mid-frequency range)

The circuit of Figure 7.5 can be further simplified using the fact that $\frac{1}{h_{oe}}$ is very large (open end) and h_{re} is very small; so $h_{re}v_o$ can be neglected in Figure 7.27.

Such a circuit is drawn is Figure 7.28.

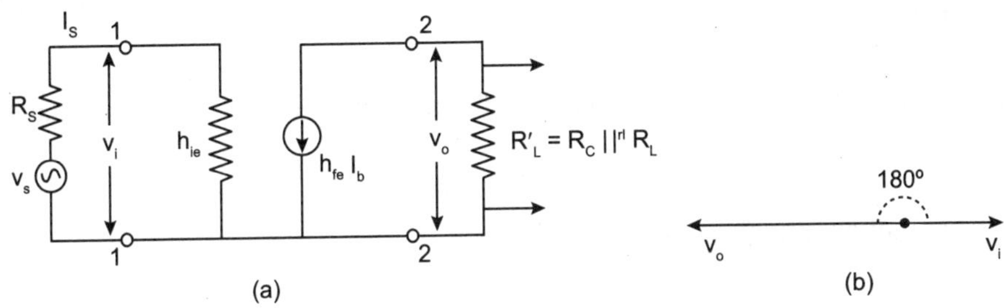

Figure 7.28 (a) Simplified CE amplifier equivalent circuit using '*h*' parameter (mid-frequency range), (b) Phasor diagram

7.6 Bias Consideration

For consideration of bias conditions, the capacitors must be treated as open-circuit element. If the source signal was to be directly connected to transistor base, then source resistance R_S would be directly parallel with resistor R_2. This would reduce bias voltage at transistor base and, consequently, alter the collector current. Similarly, if external load R_L was directly connected to the transistor collector, the DC levels of V_C and V_{CE} would be affected. Therefore, coupling capacitors C and C_C function as open circuit to the DC and short circuit for AC. Similarly, bypass capacitor C_E effectively ties the emitter of transistor to the ground level for AC consideration. Thus, the entire input voltage v_i is developed across the transistor base-emitter function.

7.7 Transistor Action

By applying an AC signal, voltage V_{BE} alternately increases and decreases as per the signal strength. This change in V_{BE} causes the I_B level to change. The change in I_B produces a variation in collector current I_C, which alters voltage drop across collector resistor R_C. This variation in voltage across resistor R_C constitutes the AC output voltage v_o.

The phase relationship between circuit input voltage v_i and output voltage v_o can be determined by considering the effect of a positive going input signal. When v_i increases in positive direction, it increases base-emitter voltage V_{BE} (Figure 7.1). An increase in V_{BE} raises the level of collector current I_C, thereby increasing the voltage drop across resistor R_C.

Hence voltage at collector

$$V_C = V_{CC} - I_C R_C \tag{55}$$

The increase in I_C increases the second term, resulting in an overall decrease in collector voltage V_c, or the output voltage goes negative. Thus, as V_i goes in the positive direction,

V_o goes in the negative direction. Similarly, when v_i goes negative, the resultant decrease in V_{BE} reduces the level of I_C, in turn reducing voltage drop across resistor R_C, i.e. 2nd term in Eqn. (55) reduces, which produces a positive going output voltage. Thus, v_i going negative causes V_o to go positive.

Hence we conclude that output voltage v_o is 180° but in phase with input voltage v_i [Figure 7.28(b)].

Circuit analysis Applying KVL to output and input circuit (Figure 7.27), we have

$$v_o = (-)h_{fe} i_b R'_L \tag{56}$$

where R'_L is parallel combination of R_C and R_L

Also

$$i_b = \frac{v_s}{R_s + h_{ie}} \tag{57}$$

Substituting for i_b in Eqn. (56), we have

$$v_o = (-)h_{fe}\left[\frac{v_s}{R_s + h_{ie}}\right]R'_L$$

Hence voltage gain

$$A_{vm} = \frac{v_o}{v_s} = (-)\frac{h_{fe} R'_L}{R_S + h_{ie}} \tag{58}$$

If $R_S \ll h_{ie}$,

$$A_{vm} = (-)\frac{h_{fe}}{h_{ie}} R'_L \tag{59}$$

Here the negative sign represents phase change of 180° between input and output voltage (Figure 7.28(b)), so also the gain is independent of frequency.

Eqn. (59) can also be written as

$$A_{vm} = \frac{h_{fe}}{R_S + h_{ie}} R'_L \angle 180° \tag{60}$$

If there is no external load or no second stage, then $R'_L = R_C$

and

$$A_{vm} = (-)\frac{h_{fe}}{h_{ie}} R_L' = \frac{h_{fe}}{h_{ie}} R_C \angle 180° \tag{61}$$

Equation (61) is similar to Eqn. (8b), which was derived for a general configuration.

The phasor diagram between output voltage (v_o) and input voltage (v_s) is shown in Figure 7.28(b).

In eqn. (58), we consider the magnitude part; then

$$|A_{vm}| = \frac{h_{fe}R_L'}{R_S + h_{ie}}$$

We know that $R_L' = \dfrac{R_C R_L}{R_C + R_L}$, where R_L is external load.

If we do not have an external load but a second stage of the amplifier is connected then R_L represents the input resistance of 2nd stage.

$R_L = h_{ie}$ so $R_L' = \dfrac{R_C h_{ie}}{R_C + h_{ie}}$ (since R_C and h_{ie} appear in parallel, Figure 7.27)

and
$$|A_{vm}| = \frac{\dfrac{h_{fe}R_C h_{ie}}{R_C + h_{ie}}}{R_S + h_{ie}} = \frac{h_{fe}R_C h_{ie}}{(R_C + h_{ie})h_{ie}} \quad \text{when } R_S << h_{ie}$$

$$|A_{vm}| = \frac{h_{fe}}{1 + \dfrac{h_{ie}}{R_C}} \tag{61a}$$

When R_C is infinity,

$$|A_{vm}|_{\max} = h_{fc} \tag{61b}$$

Eqn. (61b) suggests that the maximum value of voltage gain in mid-frequency range (for high collector resistor) is constant when there is a next stage, and is equal to h_{fe} of the transistor used and independent of frequency.

7.7.1 Low Frequency Range

In this range, the reactance of coupling capacitor C_C as well as emitter bypass capacitor C_E is effective. This range is divided in two parts: (i) where only C_E is effective, (ii) where only C_C is effective.

(a) Effect of C_E While considering the effect of C_E it is assumed that C_C is sufficiently large enough so that its reactance does not have any effect on the frequency response of the amplifier. The AC equivalent diagram of the amplifier using '*h*' parameters is shown in Figure 7.29.

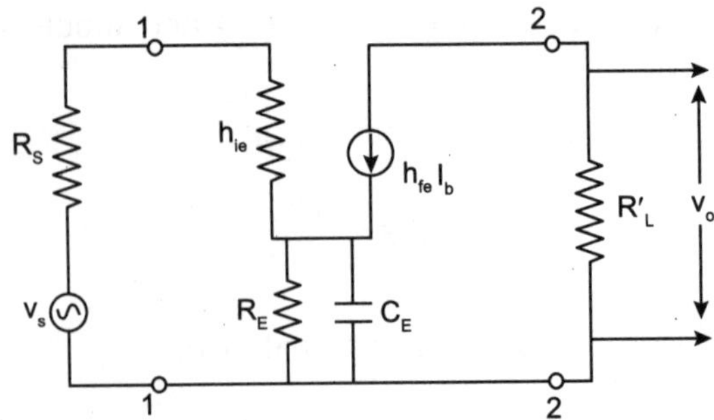

Figure 7.29 Equivalent circuit of CE amplifier in low frequency range (effect of C_E)

Applying KVL to the output and input circuit, we have

$$v_o = (-)h_{fe} i_b R'_L \tag{62}$$

and

$$i_b = \frac{v_s}{R_S + R_i} \tag{63}$$

where R_i represents input resistance between the terminals (1, 1)

$$R_i = h_{ie} + (1 + h_{fe}) Z_E \tag{64}$$

where

$$Z_E = \frac{R_E}{1 + j\omega C_E R_E}$$

is a parallel combination of R_E and C_E.

Substituting the value of R_i in Eqn. (63),

$$i_b = \frac{v_s}{R_S + h_{ie} + (1 + h_{fe})\left[\dfrac{R_E}{1 + j\omega C_E R_E}\right]}$$

Substituting the value of i_b in Eqn. (62),

$$v_o = (-)h_{fe} R'_L \left[\frac{v_s}{R_S + h_{ie} + \dfrac{(1 + h_{fe})R_E}{1 + j\omega C_E R_E}}\right]$$

Hence voltage gain,

$$A_{vL} = \frac{v_o}{v_s} = (-)h_{fe} R'_L \left[\frac{1}{R_S + h_{ie} + \dfrac{(1 + h_{fe})R_E}{1 + j\omega C_E R_E}}\right] \tag{65}$$

(b) Effect of coupling capacitor (C_C) Here the frequency range is so assumed that the reactance of coupling capacitor is effective while those of other capacitors is neglected. The equivalent circuit is as shown in Figure 7.29.

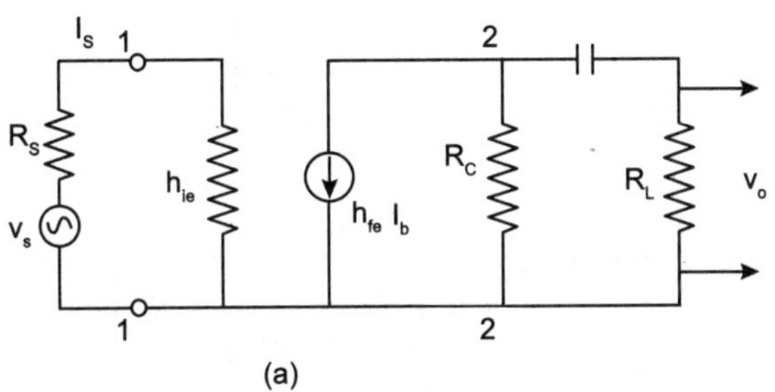

(a)

Figure 7.30 (a) Equivalent circuit of CE amplifier in low frequency range (effect of C_C)

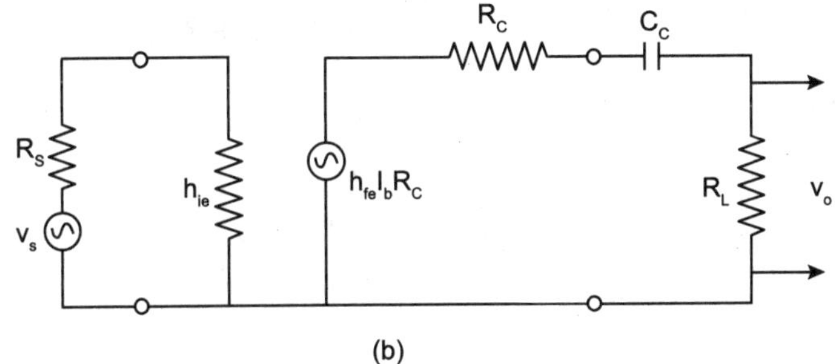

(b)

Figure 7.30 (b) Thevenin equivalent circuit of Figure 7.30 (a)

The Thevenin equivalent circuit (Figure 7.30) is drawn by applying Thevenin's theorem across points (2, 2) of Figure 7.29.

Current flowing through the output circuit across external load R_L is

$$I = (-)\frac{h_{fe}\, i_b\, R_C}{R_C + \dfrac{1}{j\omega C_C} + R_C}$$

Output voltage v_o across external load R_L is

$$\begin{aligned} v_o &= IR_L \\ &= (-)\left[\frac{h_{fe} i_b R_C}{R_C + 1/j\omega C_C + R_L}\right] R_L \end{aligned} \qquad (66)$$

Substituting for i_b from input circuit,

$$\left[i_b = \frac{v_s}{R_S + h_{ic}}\right]$$

Output voltage, $v_o = (-)\dfrac{h_{fe}R_C R_L}{R_C + \dfrac{1}{j\omega C_C} + R_L}\left[\dfrac{v_s}{R_S + h_{ie}}\right]$

$$\therefore \text{Voltage gain, } A_{vL} = \frac{v_o}{v_s} = (-)\frac{h_{fe}R_C R_L}{(R_S + h_{ie})\left[R_C + \dfrac{1}{j\omega C_C} + R_L\right]}$$

$$= (-)\frac{h_{fe}\left[\dfrac{R_C R_L}{R_C + R_L}\right]}{R_S + h_{ie}}\left\{\frac{1}{1 + \dfrac{1}{j\omega C_C(R_C + R_L)}}\right\}$$

Using Eqn. (60), we have

$$A_{vL} = A_{vm}\left[\frac{1}{1 + \dfrac{1}{j\omega C_C(R_C + R_L)}}\right]$$

$$= A_{vm}\left[\frac{1}{1 - j\dfrac{1}{\omega C_C(R_C + R_L)}}\right]$$

$$= A_{vm}\left[\frac{1}{1 - j\dfrac{\omega_1}{\omega}}\right] \qquad (67)$$

where $\omega_1 = \dfrac{1}{C_C(R_C + R_L)}$

or

$$A_{vL} = A_{vm}\left[\frac{1}{\sqrt{1 + \left(\dfrac{\omega_1}{\omega}\right)^2}}\right]\angle\theta_1 \qquad (68)$$

where

$$\tan\theta_1 = \frac{\omega_1}{\omega}$$

and $(180 + \theta_1)$ represents the phase change between output and input voltage (Figure 7.34).

Equation (68) suggests that voltage gain in low frequency range increases on increasing the frequency (Figure 7.32). When applied signal frequency $\omega = \omega_1$, the gain is 0.707 times the midfrequency gain.

$$\therefore [A_{vL}] = \frac{A_{vm}}{\sqrt{2}} = 0.707[A_{vm}] \tag{68a}$$

This frequency $\omega = \omega_1$ is called lower half power frequency since power gain is half of midfrequency gain. Now, the lower half power frequency

$$f_1 = \frac{1}{2\pi C_C[R_C + R_L]} \tag{69}$$

depends upon the value of C_C, i.e. any change in C_C inversely affects the lower half power frequency as shown in Figure 7.33.

7.7.2 High Frequency Range

In the mid frequency range discussed above, the reactance of coupling capacitor C_c is not effective, but the reactance of shunt capacitance C_s does have an effect so it is to be considered. Shunt capacitance constitutes the transistor collector base capacitance C_{bc} in parallel with wiring capacitance C_w so the combination is shown as shunt capacitance C_S as shown in Figure 7.31, which represents the AC equivalent circuit of the amplifier.

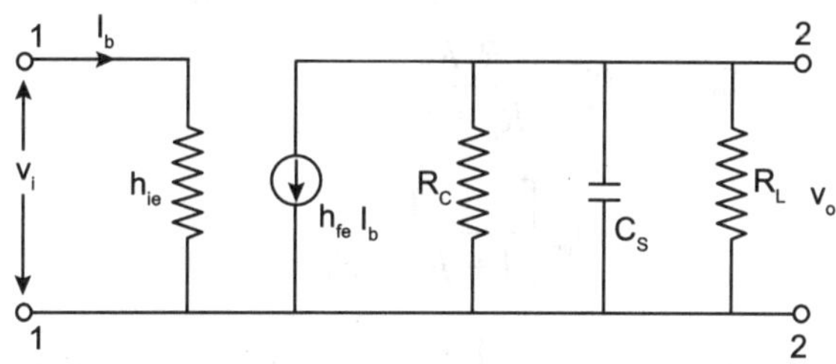

Figure 7.31 Equivalent circuit of CE amplifier in high frequency range

In the output circuit, the equivalent load Z_L is a parallel combination of R_C, C_S and R_L (external load or next stage if any). It is given by.

$$\frac{1}{Z_L} = \frac{1}{R_C} + \frac{1}{R_L} + j\omega C_S$$

$$\text{or} \quad Z_L = \frac{R_C R_L}{R_C + R_L + j\omega C_S R_C R_L}$$

Output voltage $v_o = (-)h_{fe} i_b Z_L$

Substituting for i_b, from input circuit $i_b = \dfrac{v_s}{R_S + h_{ie}}$

$$v_o = -h_{fe}\frac{v_s Z_L}{R_s + h_{ie}}$$

Hence voltage gain, $A_{vh} = \frac{v_o}{v_s} = (-)\, h_{fe} \frac{Z_L}{R_S + h_{ie}}$ (70)

Substituting for Z_L, we have

$$A_{vh} = (-)\frac{h_{fe}}{R_S + h_{ie}}\left[\frac{R_C R_L}{R_C + R_L + j\omega C_S R_C R_L}\right]$$

However, if $R_S \ll h_{ie}$ which is generally the case, then

$$A_{vh} = (-)\frac{h_{fe}\left(\frac{R_C R_L}{R_L + R_C}\right)}{h_{ie}\left[1 + j\omega C_S \frac{R_C R_L}{R_L + R_C}\right]}$$

$$\text{or} \quad A_{vh} = (-)\frac{h_{fe} R_L'}{h_{ie}}\left[\frac{1}{1 + j\omega C_S R_L'}\right]$$

using equation (60) where $R_L' = \frac{R_L R_C}{R_L + R_C}$

$$A_{vh} = A_{vm}\left[\frac{1}{1 + j\omega C_S R_L'}\right]$$

where $\omega_2 = \frac{1}{C_S R_L'}$ is called the upper half power frequency.

$$= A_{vm}\left[\frac{1}{1 + j\frac{\omega}{\omega_2}}\right]$$

In magnitude

$$|A_{vh}| = \frac{|A_{vm}|}{\sqrt{1 + \left(\frac{\omega}{\omega_2}\right)^2}} \angle -\theta_2 \qquad (71)$$

where $\tan\theta_2 = \frac{\omega}{\omega_2}$ and $[180 - \theta_2]$ represents phase change between output and input voltage (Figure 7.34)

If $\omega = \omega_2$, then

$$\therefore\ |A_{vh}| = \frac{|A_{vm}|}{\sqrt{2}} = 0.707\, A_{vm} \qquad (71(a))$$

At upper half power frequency gain is 0.707 times the gain in mid-frequency range.

This frequency ω_2 is called upper half power frequency since gain at this frequency is $\frac{1}{2}$ of mid-frequency gain.

Half power frequencies f_1 and f_2 are also called (–)3 dB frequencies.

The upper half power frequency ω_2 depends upon the value of C_S i.e. any change in C_S inversely affects the upper half frequency as shown in Figure 7.32.

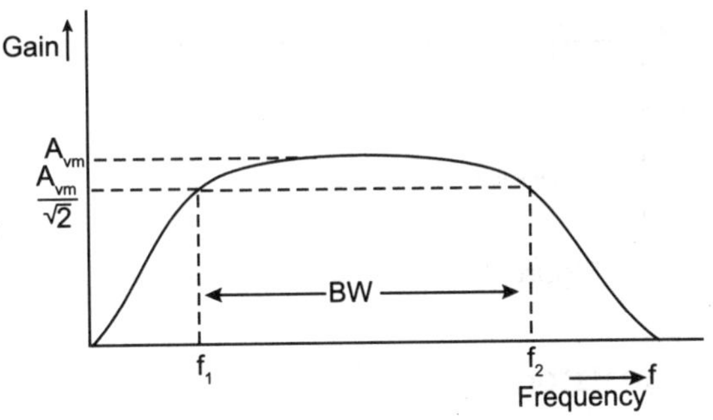

Figure 7.32 Frequency response curve

The frequency curve depicts a graph of gain vs frequency. In Figure 7.32, f_1 represents lower half frequency and f_2 represents upper half power frequency.

The difference of the two half power frequencies is called **Bandwidth** (BW) of the amplifier.

$$\text{BW} = f_2 - f_1 \tag{72}$$

The Bandwidth of the amplifier changes by changing either the value of C_C or the value of C_S. As is clear from Figure 7.33, the value of lower half power frequency f_1 decreases on increasing the value of C_C and the value of upper half power frequency f_2 increases on decreasing the value of C_S.

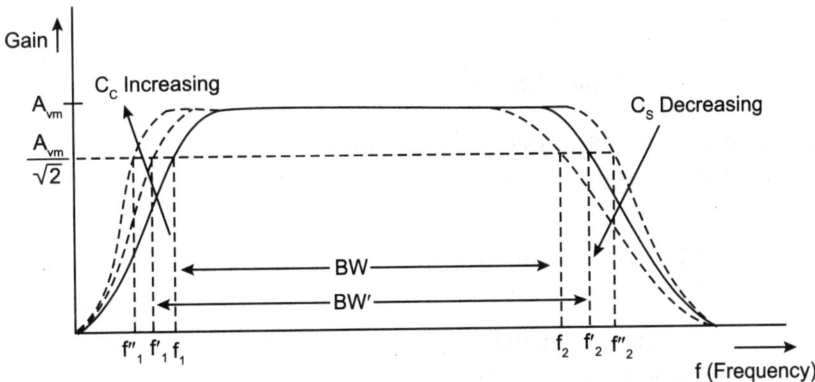

Figure 7.33 Variation of f_1 and f_2 with C_C and C_S

Figure 7.34 represents the phase change between output voltage and input voltage. It is clear from the figure that phase difference increases in the lower frequency range and decreases in the high frequency range with respect to the midfrequency where phase change is 180°. Further, it is to be noted that total phase difference at f_1 is 225° and at f_2 is 135°.

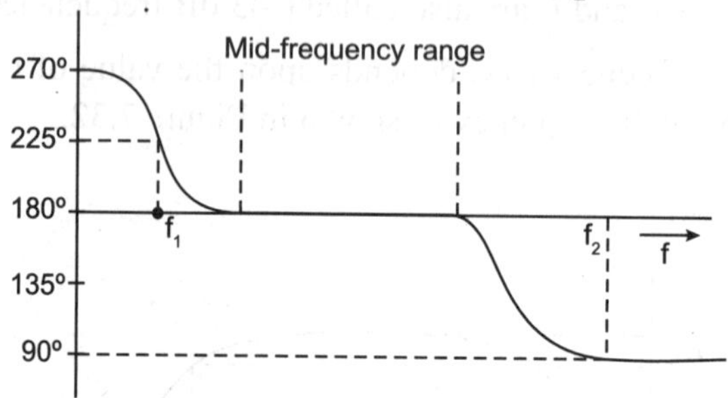

Figure 7.34 Phase vs. frequency

7.7.3 Limitations of Transistor Amplifier in High Frequency Range

When discussing transistor voltage amplifier in the high frequency range, we had shown shunting capacitor C_S (Figure 7.31) in parallel with the load. This shunting capacitance was shown in lieu of collector capacitor and extra wire capacitance. However, we must consider that fact the hybrid π model of transistor takes care of different internal components of the transistor. An hybrid π model of transistor is shown in Figure 7.35.

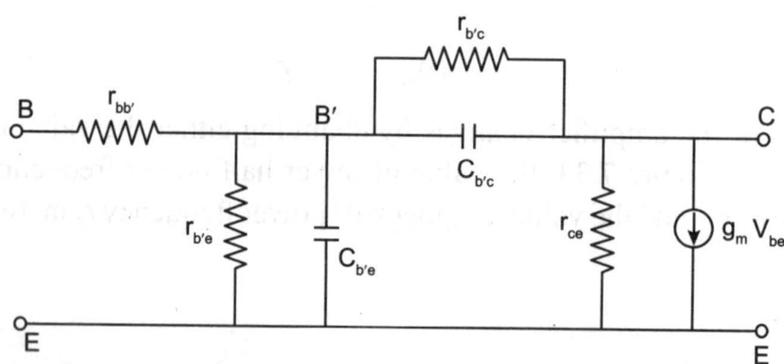

Figure 7.35 Hybrid π model of transistor

The internal node B' is not physically accessible. Ohmic base spreading resistance r'_{bb} is represented as lumped resistance between B and B'. The following components are represented in Figure 7.35.

$C_{b'e}$ – Diffusion capacitance between base and emitter when it is forward-biased.

$C_{bc'}$ – Diffusion capacitance between collector and base when collector-base junction is reverse-biased.

$r_{b'c}$ and $r_{b'e}$ – These two resistors are shown in the equivalent circuit of amplifier.

$r_{b'c}$ – Resistor between base and collector, but it is normally taken as open circuit in the equivalent circuit diagram since its value is very high, of the order of 4 MΩ.

r_{ce} – Output resistance of the transistor; whenever there is load R_C along with collector, then its effect is negligible.

$g_m v_{be}$ – For small changes in voltage $v_{b'e}$ across the emitter junction, the excess minority carrier concentration injected into the base is proportional to $v_{b'e}$, and it results in a small signal collector current generator.

Assuming load R_C is short circuit and approximations made, the hybrid equivalent circuit (Figure 7.35) reduces to Figure 7.36.

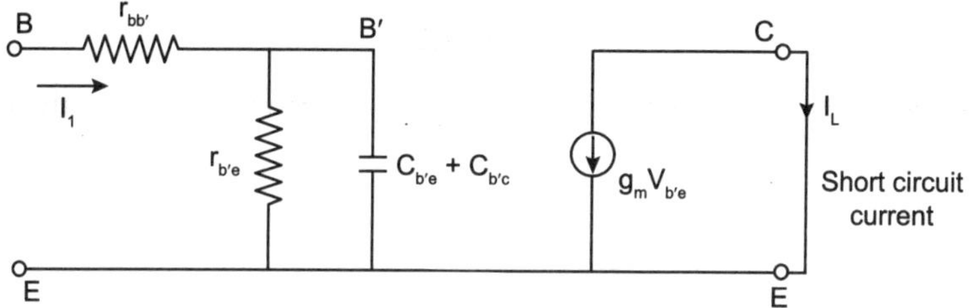

Figure 7.36 Simplified hybrid π model assuming short circuit between *C* and *E*

Current gain in a short-circuited circuit,

$$[A_i]_{\text{short circuit}} = \frac{I_L}{I_1} = (-)\frac{h_{fe}}{1 + j\left(\frac{f}{f_\beta}\right)} \tag{73}$$

$$\text{where } f_\beta = \frac{1}{2\pi r_{r'e}(C_{b'e} + C_{b'c})} \tag{74}$$

If we again connect the collector load R_C, the circuit takes the form as shown in Figure 7.37.

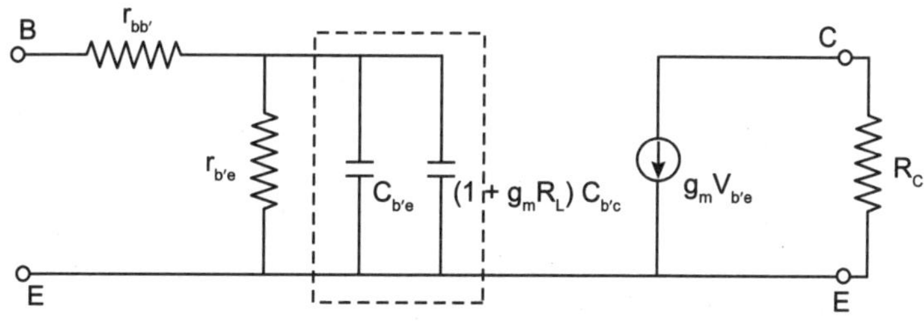

Figure 7.37

In Eqn. (74) frequency f_β represents the high frequency handling capacity of the transistor, up to which the midfrequency gain does not fall much.

Similarly, there is another characteristic frequency f_t of the transistor where the gain of the short circuited transistor gain attains unity value, as shown in Figure 7.38.

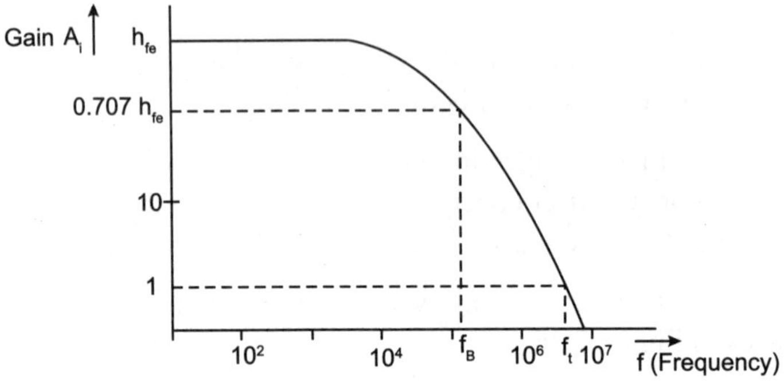

Figure 7.38

Both the frequencies have a specific meaning for a transistor and are always shown in the data sheet.

Normally $f_t \gg f_\beta$

So
$$1 = \frac{h_{fe}}{\left(\frac{f_t}{f_\beta}\right)}$$

or
$$\frac{f_t}{f_\beta} = h_{fe}$$

$$\therefore\ f_t = h_{fe}\,(f_\beta) \tag{75}$$

Eqn. (75) holds good for CE configuration. However, for CB configuration

$$f_t = h_{fb}[f_\alpha]$$

Substituting for f_β from Eqn. (74) in eqn. (75), we have

$$f_t = h_{fe}\left[\frac{1}{2\pi r_{b'e}[C_{b'e} + C_{B'C}]}\right]$$

$$= \left(\frac{h_{fe}}{r_{b'e}}\right)\left[\frac{1}{2\pi(C_{b'e} + C_{b'c})}\right]$$

or
$$f_t = \frac{g_m}{2\pi(C_{b'e} + C_{b'c})} \tag{76}$$

where
$$g_m = \frac{h_{fe}}{r_{b'e}}$$

From Eqn. (7.73), the magnitude of current gain of short circuited amplifier is

$$|A_i| = \frac{h_{fe}}{\sqrt{1+\left(\frac{f}{f_\beta}\right)^2}} \tag{77}$$

As per the definition, when applied frequency attains the value f_t, current gain reduces to unity, and thus Eqn. (77). can be written as

$$|A_i| = 1 = \frac{h_{fe}}{\sqrt{1+\left(\frac{f_t}{f_\beta}\right)^2}} \tag{78}$$

7.8 Curves of Response vs. Frequency

Figure 7.32 shows the gain versus frequency curve, known as frequency response curve. This curve has been drawn approximately as per the effect of coupling capacitor C_C and shunting capacitor C_S.

It is to be noted that amplitude and phase angle in the midrange (Eqn. 61) are independent of frequency, and midrange gain is useful as a normalizing standard in the comparative study of low and high frequency gains as in eqn. (68) and (71). When these equations are written in decibel units, they exhibit useful properties. For instance consider eqn. (68).

$$\frac{A_{vL}}{A_{vm}} = \frac{1}{\sqrt{1+\left(\frac{\omega_1}{\omega}\right)^2}}$$

$$\text{dB}\left|\frac{A_{vL}}{A_{vm}}\right| = 20\log\left[\frac{1}{1+\left(\frac{\omega_1}{\omega}\right)^2}\right]^{1/2}$$

or

$$\text{dB}\left|\frac{A_{vL}}{A_{vm}}\right| = (-)10\log\left[1+\left(\frac{\omega_1}{\omega}\right)^2\right]$$

$$= (-)10\log\left(\frac{\omega_1}{\omega}\right)^2 = (-)20\log\frac{\omega_1}{\omega}$$

or

$$\text{dB}\left|\frac{A_{vL}}{A_{vm}}\right| = (-)20\log\omega_1 + 20\log\omega$$

or

$$\text{dB}\left|\frac{A_{vL}}{A_{vm}}\right| = 20\log\omega - 20\log\omega_1 \tag{79}$$

Equation (79) represents the equation of a straight line on dB vs log frequency graph ($y = mx + c$) having a positive slope (Figure 7.39). The slope $m = 20$ dB/decade.

Similarly, considering Eqn. (71),

$$\text{dB} = \left|\frac{A_{vh}}{A_{vm}}\right| = 20\log\left[\frac{1}{1+(\omega/\omega_2)^2}\right]^{1/2}$$

$$= (-)10\log\left[1+\left(\frac{\omega}{\omega_2}\right)^2\right]$$

or

$$\text{dB}\left|\frac{A_{vh}}{A_{vm}}\right| = (-)20\log\left(\frac{\omega}{\omega^2}\right)$$

or

$$\text{dB}\left|\frac{A_{vh}}{A_{vm}}\right| = (-)20\log\omega + 20\log\omega_2 \tag{80}$$

Equation (80) represents the equation of straight line $y = mx + c$ on dB vs log frequency graph passing through the corner point at 0 dB (Figure 7.39) and having a negative slope (–)20 dB/decade, i.e. it falls with a slope of (–)20 dB per frequency decade, which is equivalent to (–)6 dB per frequency octave. Figure 7.39 shows the asymptote representing the actual gain curve for $\frac{\omega}{\omega_2} \approx 3$.

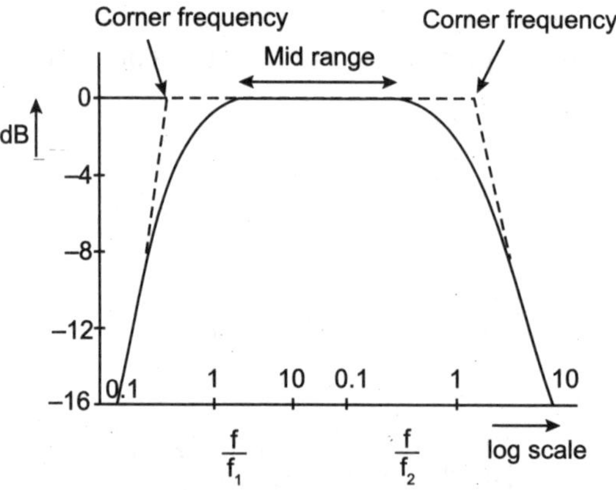

Figure 7.39 Generalized gain curve

Gain-frequency can be easily plotted by drawing an asymptote from the corner point with appropriate slope, locating the (–)3 dB point.

7.8.1 RC Amplifier (Current Gain)

So far we have discussed voltage gain due to an RC amplifier whenever a voltage signal is applied across its input. However, we may have a current source as an input and determine the current gain due to RC amplifier.

Mid-frequency range In case of a current source, its internal resistance R_S appears in parallel (Figure 7.40).

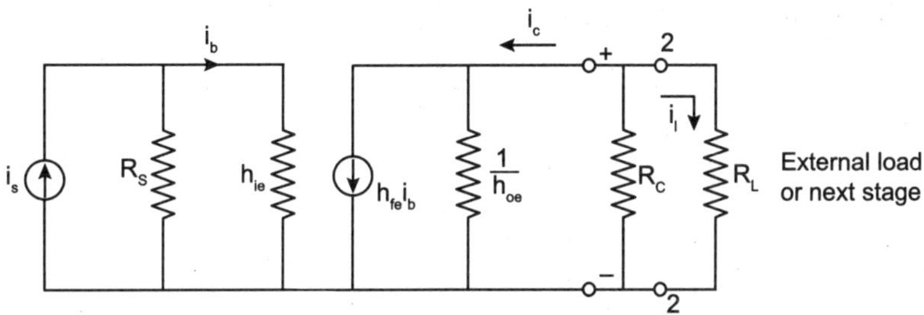

Figure 7.40 AC equivalent circuit of RC amplifier (mid-frequency range)

Circuit analysis For an ideal current source, internal resistance R_S is taken as infinity and in its output $\dfrac{1}{h_{oe}}$ as open.

By definition, current gain

$$A_i = \frac{\text{Output (load) current}}{\text{Input current}} = \frac{i_L}{i_S} = (-)\frac{i_c}{i_b} = (-)\frac{h_{fe} i_b}{i_b}$$

$$= (-)h_{fe} \tag{81}$$

The maximum value of current gain can be h_{fe} (i.e., which depends upon the transistor used).

However, if we consider R_S and $1/h_{oe}$ also, effective load is R_L' and input current is i_s.

In this case, current gain $A_i' = (-)\dfrac{i_c}{i_s}$.

Actually, input current is i_b,

$$i_b = \frac{i_s R_S}{R_S + h_{ie}}$$

$$A_i' = (-)\frac{h_{fe}}{1 + h_{oe} R_L'} \cdot \frac{R_S}{R_S + h_{ie}} \tag{82}$$

Figure 7.41 Thevenin's equivalent of output circuit of CE amplifier

Figure 7.40 has been redrawn in Figure 7.41 by applying Thevenin's theorem across terminal (2, 2). Here R_{Lo} represents equivalent parallel combination of $1/h_{oe}$ and R_C and acts as internal resistance of voltage source $\frac{h_{fe}\, i_b}{h_{oe} + 1/R_C}$.

$$\text{Current gain, } A_{im} = \frac{i_o}{i_b} = \frac{(-)\left[\dfrac{h_{fe} i_b}{h_{oe} + \dfrac{1}{R_C}}\right]}{i_b\,(R_{LO} + R_L)}$$

$$\text{or } A_{im} = (-)\frac{\dfrac{h_{fe}}{h_{oe} + \dfrac{1}{R_C}}}{(R_{LO} + R_L)}$$

Now

$$A_{im} = (-)\frac{h_{fe}\, R_C}{R_{LO} + R_L} = (-)\frac{h_{fe}\, R_C}{R_L} \tag{83}$$

since $R_L \gg R_{LO}$

If we have the next stage, then $R_L = R_i$ (input resistance of next stage)

So $R_L = R'_L$

$$\text{Hence current gain, } A_{im} = (-)\frac{h_{fe} R_C}{R_i} = (-)\frac{h_{fe} R_C}{h_{ie}} \tag{84}$$

$$= (-) g_m\, R_C \tag{85}$$

where $g_m = \frac{h_{fe}}{h_{ie}}$

If $R_C = h_{ie}$, current gain,

$$A_{im} = (-)h_{fe} \quad \text{which is maximum value} \tag{85a}$$

Here –ve sign denotes the direction of load current is opposite to i_c.

Low frequency range (Effect of C_C) The AC equivalent circuit of amplifier is shown in Figure 7.42.

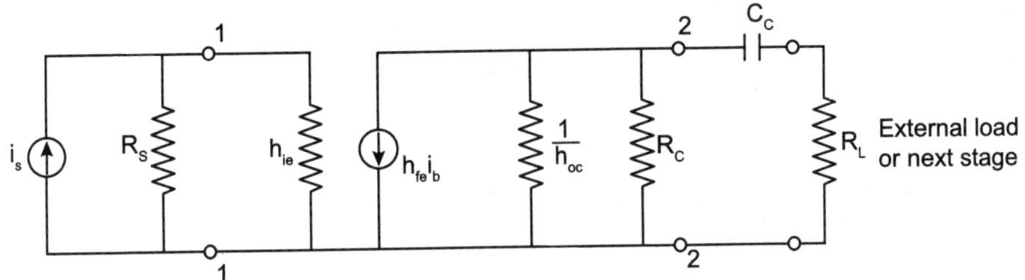

Figure 7.42 AC equivalent circuit of RC amplifier (low frequency range)

Applying Thevenin's theorem across the terminal (2, 2), we can redraw Figure 7.42 as shown in Figure 7.43. Here R_{Lo} is internal resistance of voltage source.

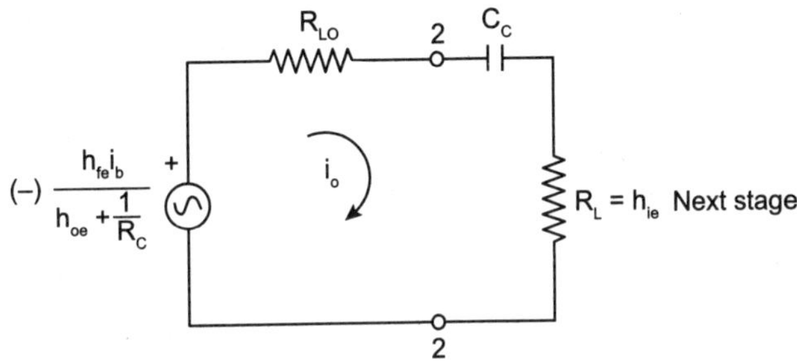

Figure 7.43 Equivalent output circuit

$$\text{Output current, } i_o = (-)\frac{h_{fe}\dfrac{i_b}{(h_{oe}+1/R_C)}}{R_{LO}+\dfrac{1}{j\omega C_C}+h_{ie}}$$

$$\text{Current gain, } A_{iL} = \frac{\text{Output current}}{\text{Input current}} = \frac{i_o}{i_b}$$

$$= (-)\frac{\dfrac{h_{fe}}{(h_{oe}+1/R_C)}}{R_{LO}+h_{ie}+\dfrac{1}{j\omega C_C}}$$

Now,

$$A_{iL} = (-)\frac{h_{fe}R_C}{(R_C+h_{ie})\left[1+\dfrac{1}{j\omega C_C(R_C+h_{ie})}\right]}$$

Making use of Eqn. (83),

$$A_{iL} = A_{im} \frac{1}{\left[1 - j\frac{\omega_1}{\omega}\right]} \tag{86}$$

where $\omega_1 = \dfrac{1}{C_C(R_C + h_{ie})}$ is called lower half power frequency.

Taking magnitude, we have

$$A_{iL} = A_{im} \frac{1}{\sqrt{1 + \left(\frac{\omega_1}{\omega}\right)^2}} \tag{87}$$

If $\omega = \omega_1$,

$$A_{iL} = \frac{A_{im}}{\sqrt{2}}$$

which is similar to that in a voltage source amplifier.

High frequency range The AC equivalent circuit is shown in Figure 7.44.

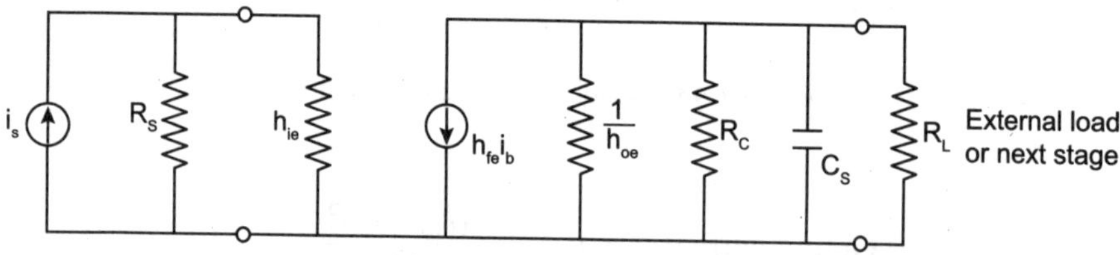

Figure 7.44 AC equivalent circuit of RC amplifier (high frequency range)

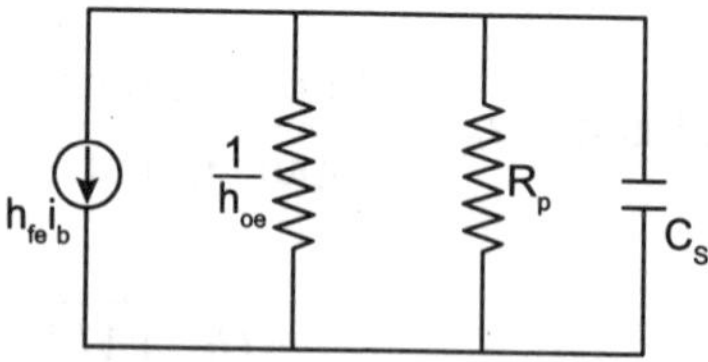

Figure 7.45 AC equivalent circuit of output

Here R_p is a parallel combination of R_C and R_L, and we can neglect $1/h_{oe}$ taken as open.

Current in $R_p = h_{fe}\, i_b \left[\dfrac{1/j\omega C_S}{R_p + 1/j\omega C_S}\right] = i$

Current in external load ($R_L = h_{ie}$),

$$i_o = i\frac{R_C}{R_C + h_{ie}}$$

Substituting for *i*, we have

$$i_o = \frac{h_{fe}\, i_b}{1 + j\omega R_p C_S}\left[\frac{R_C}{R_C + h_{ic}}\right]$$

Taking upper terminal as positive, the direction of current is negative and so we can write

$$i_o = (-)\frac{h_{fe} i_b}{(1 + j\omega R_p C_S)}\left[\frac{R_C}{R_C + h_{ie}}\right]$$

$$\text{Current gain, } A_{ih} = \frac{i_o}{i_b} = (-)\left[\frac{h_{fe} R_C}{R_C + h_{ie}}\right]\left[\frac{1}{1 + j\omega R_p C_S}\right]$$

Making use of Eqn. (83),

$$A_{ih} = A_{im}\left[\frac{1}{1 + j\dfrac{\omega}{\omega_2}}\right] \tag{88}$$

where $\omega_2 = \dfrac{1}{R_p C_S}$ is called upper half power frequency.

Taking magnitude, we have

$$A_{ih} = A_{im}\frac{1}{\sqrt{1 + \left(\dfrac{\omega}{\omega_2}\right)^2}} \tag{89}$$

If $\omega = \omega_2$, $A_{ih} = \dfrac{A_{im}}{\sqrt{2}}$ which is just like voltage source.

In case of current gain, gain frequency response curve is similar to voltage gain frequency response curve (Figure 7.32).

7.9 Design of CE Amplifier

We have discussed bias circuit design from stability point of view. To design such a circuit, it is necessary to work from a specification which might state the following: (i) supply voltage, (ii) desired voltage gain, (iii) frequency response along with (–)3 dB frequencies, (iv) signal source internal resistance and (v) load.

7.9.1 Selection of I_c, R_c and R_E

When designing for a particular voltage gain, it requires the use of an AC negative feedback to stabilize the gain. The circuit may not have any provision for negative feedback, thus it is designed to achieve the largest (maximum) possible gain given by

$$A_{vm} = \frac{(-)h_{fe}[R_C \parallel^{rl} R_L]}{h_{ie}}$$

From the above equation, it is clear that gain is directly proportional to $R_C \parallel^{rl} R_L$. It normally requires the selection of the largest possible value of collector resistance R_C. However, an extremely large value of R_C may make collector current I_C too small for satisfactory transistor operation. A good minimum value of I_C is 1 mA.

The transistor collector current might be selected to have the largest value for h_{fe}, again to achieve the maximum value of A_v. However, a high level of I_c results in a small value of R_C [for a given value of voltage drop (V_{RC}) across it]. So a high value of I_c may actually produce a lower voltage gain although h_{fe} may be large. To make V_{RC} as large as possible, V_{CE} and V_E should be kept to a minimum, so that transistor operators proceed in the active region.

Another consideration for selecting R_C is that there is nothing to be gained by making R_C larger than external load R_L. In fact, $R_C \ll R_L$ so that R_L does not have much effect on voltage gain. For good bias stability, emitter resistor voltage drop V_E should be much larger than V_{BE}.

This is because $V_E = V_B - V_{BE}$ and when $V_E \gg V_{BE}$, any variation in V_{BE} has very slight effect on V_E. Consequently, I_E and I_c remain fairly stable at $I_C \approx I_E = \frac{V_E}{R_E}$. In most cases, a minimum of $V_E \gg 5$ V gives good stability. But, where supply voltage is near 10 V, the value of V_E may reduce to 3 V.

Once V_E, V_{CE} and I_c are decided upon, V_{RC} can be determined

$$V_{RC} = V_{CC} - V_{CE} - V_E$$

Then, R_C and R_E are calculated.

$$R_C = \frac{V_{RC}}{I_C} \text{ and } R_E = \frac{V_E}{I_C}$$

7.9.2 Bias Resistors

To select these resistors, potential divider current I_2 as $I_C/10$ gives good stability and high input resistance. But, when input resistance is not the consideration, then $I_2 \approx I_C$.

$$R_1 = \frac{V_B}{I_2} \text{ and } R_2 = \frac{V_{CC} - V_B}{I_2 + I_B}$$

7.9.3 Bypass Capacitor (C_E)

Bypass and coupling capacitors should be chosen to have the smallest possible capacitance both from economy and space point of view, since large electrolytic capacitors occupy large space.

Each capacitor has its highest reactance at lowest frequency, and so capacitor values are calculated at lowest operating frequency which the circuit is to amplify. This frequency is the lower cut-off frequency 'f_1'. We should have

$$\frac{1}{\omega_1 C_E} \ll R_E$$

Knowing $\omega_1 = 2\pi f_1$, we can calculate C_E.

7.9.4 Coupling Capacitor (C_C)

Coupling capacitors should have very little effect on amplifier frequency response. It is known that X_{CC} and Z_i (or R_L) constitute a potential divider. If X_{CC} is too large, v_i will be smaller than v_s and will increase with increasing applied (input) frequency. Similarly, X_{CC} and R'_L potentially divide the AC collector voltage. To minimize the effect of X_{CC}, the reactance of each coupling capacitor at lowest operating frequency for the circuit is selected to be approximately equal to 1/10 of impedance in series with it.

$$X_{CC1} = \frac{Z_i}{10}$$

and

$$X_{CC} = \frac{R'_L}{10}$$

7.9.5 Shunting Capacitor

Many a times, an amplifier is required to have a particular upper cut-off frequency (f_2). Normally a transistor must operate satisfactorily to a frequency much higher than f_2, i.e. transistor (selected) cut-off frequency must exceed the circuit upper cut-off frequency (f_2). It can be proved that at f_2 reactance, $X_{CS} = R'_L = R_C \,||^{rl}\, R_L$.

7.10 Multistage Amplifier (Cascading Amplifier)

So far we have studied about single-stage amplifiers only. But, many a times we do not get the desired level of output voltage specially when the input signal is weak (small). In such a situation, we use more than one stage so that final output voltage is of the desired level.

A capacitor-coupled, two-stage amplifier circuit is shown in Figure 7.46.

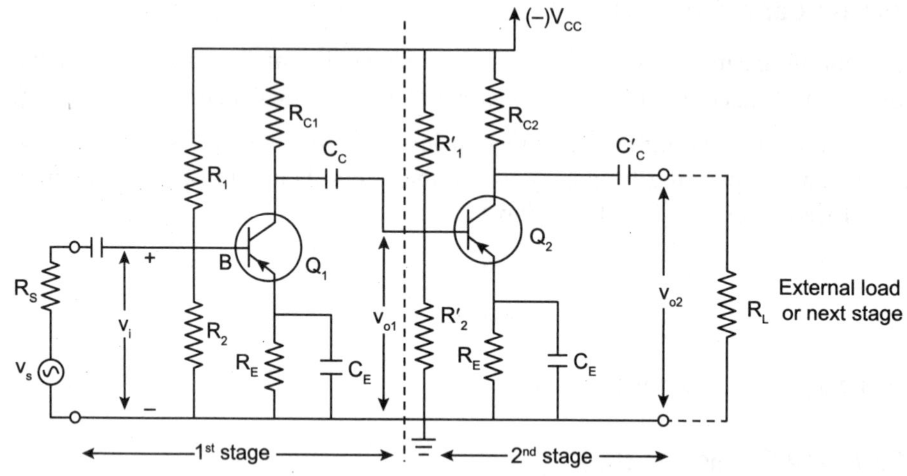

Figure 7.46 Two-stage, RC-coupled CE amplifier circuit

As shown in the circuit diagram, each stage is made identical, i.e. the component $R_1 = R_1', R_2 = R_2'\ R_{C1} = R_{C2}$ but $C_C \neq C_C'$. It is to be noted that C_C is calculated in terms of input resistance of stage and C_C' is determined as per the conditions given.

$$X_C = \frac{1}{\omega C_C'} \ll R_L \text{ (external load)}$$

One way in which the design of a two-stage circuit differs from that of single stage is in the calculation of C_E.

Consider the equation

$$X_{CE} = \frac{h_{ie}}{1 + h_{fe}}$$

This equation was derived to make the voltage gain of stage 1 circuit 3 dB down from midfrequency gain at f_1 (lower cut-off frequency). If the above equation is used to calculate C_E for both stages, it is found that both the gains of stage 1st and 2nd are down by 3 dB at f_1. This means that the overall gain of the entire circuit is down by a total of 6 dB at f_1. For a 3 dB total reduction in overall voltage gain at f_1, the bypass capacitors must be calculated to give a 1.5 dB reduction in gain at each stage. A 1.5 dB reduction in stage gain at f_1 is achieved by making the reactance of C_E equal to 0.65 times the impedance in series with the capacitor.

$$X_{CE} = 0.65 \frac{h_{ie}}{1 + h_{fe}}$$

7.10.1 AC Analysis

The '*h*' parameter equivalent circuit of a two-stage amplifier is shown in Figure 7.47.

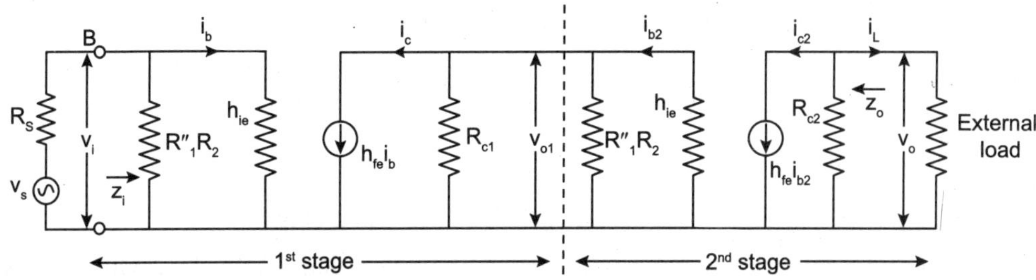

Figure 7.47 AC equivalent of a two-stage (RC) amplifier

The above AC equivalent circuit has been produced by using '*h*' parameters and assuming that capacitors are short circuited.

Voltage polarities show a 180° phase shift between v_i and v_{o1} and a further 180° phase shift from v_{o1} to v_o.

Input impedance as before is

$$Z_i = R_1 \,||\, R_2 \,||\, h_{ie}$$

Output impedance is

$$Z_o = R_{C2}$$

Voltage gain equation for the 2nd stage is the same as for the 1st stage. For capacitor-coupled impedance at the collector terminal of the 1st stage and input impedance for the 2nd stage, the corresponding voltage gain equations are

$$A_{v1} = \frac{(-)h_{fe}(R_{C1} \,||^{l}\, Z_{i2})}{h_{ie}} \tag{90}$$

$$A_{v2} = \frac{(-)h_{fe}(R_{C2} \,||^{l}\, R_L)}{h_{ie}} \tag{91}$$

The over all voltage gain is

$$A_v = A_{v1} \times A_{v2}$$

We have discussed a two-stage amplifier, but consider we have a number of such stages coupled with the preceding stage (Figure 7.48).

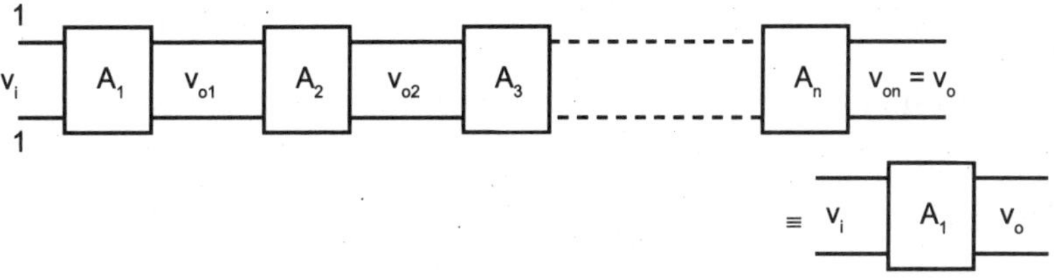

Figure 7.48 Block diagram of multistage amplifier

From Figure 7.48, where each block represents a stage, it is clear that

$$v_{o_1} = A_1 v_i \text{ and } v_{o2} = A_2 v_{o1},\ v_{o3} = A_3 v_{o2}$$

For the last stage, $v_{on} = A_n[v_{o(n-1)}]$.

Here $A_1, A_2, A_3, \ldots A_n$ represent voltage gain of a particular stage.

If we can represent an equivalent overall stage having voltage gain A_{OV}, which gives an output v_o [$= v_{on}$] for the input signal v_i, then

$$v_o = A_{ov}[v_i]$$

$$\therefore A_{ov} = \frac{v_o}{v_i} = \frac{v_{on}}{v_i} = \frac{v_{o1}}{v_i} \times \frac{v_{o2}}{v_{o1}} \times \frac{v_{o3}}{v_{o2}} \times \ldots \frac{v_{on}}{v_{o(n-1)}}$$

In the R.H.S. of the above equation, each fraction represents the respective voltage gain of the particular shape.

$$\therefore A_{ov} = A_1 \times A_2 \times A_3 \times \ldots A_n \tag{92}$$

Thus, the above equation represents that over all voltage gain of cascading stages, and is equal to the product of voltage gain of each stage.

If voltage gain of each stage is equal, then

$$A_1 = A_2 = A_3 = \ldots A_n = A$$

So Eqn. (92) can be written as

$$A_{ov} = [A]^n \tag{93}$$

If the above equation is expressed in dB, then it is to be kept in mind that Eqn. (37 or 38) holds good for voltage gain only and not for current gain since the output of 1st stage is not the input current for 2nd stage. Before reaching the 2nd stage, output current is distributed, i.e. it value diminishes.

$$\begin{aligned} \text{dB} &= 20\log_e[A_{ov}] \\ &= 20\log_e[A_1 \times A_2 \times A_3 \times \ldots A_n] \\ &= 20\log_e A_1 + 20\log_e A_2 + 20\log A_3 + \ldots 20\log_e A_n \end{aligned} \tag{94}$$

The frequency response curve of n coupled stage amplifier is shown in Figure 7.49.

From Figure 7.49, it is clear that the overall gain has increased from A to A^n for n stages. But the (–)3 dB frequencies (cut-off frequencies) have also changed. The over all bandwidth has reduced to $\text{BW}' = [f_{2n} - f_{1n}]$ from $\text{BW} = (f_2 - f_1)$.

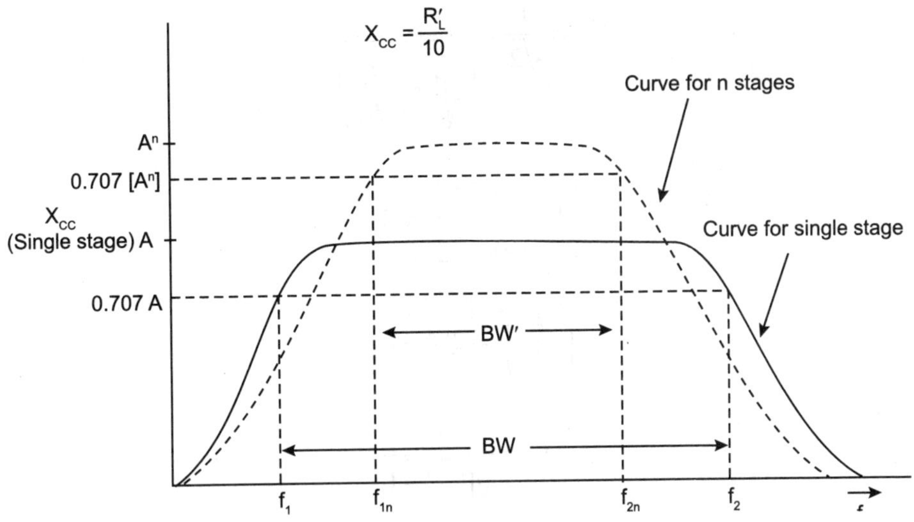

Figure 7.49 Frequency response for *n* coupled amplifier stages

Hence, (*n* stages) BW′ < BW (single stage).

Thus, in cascading the amplifier, gain increases but bandwidth decreases. The resulting upper cut-off frequency (f_{2n}) and lower cut-off frequency (f_{1n}) can be derived as follows:

For a single-stage amplifier, gain in low frequency range is

$$A_{vL} = \frac{A_{vm}}{\sqrt{1+\left(\frac{f_1}{f}\right)^2}}$$

or

$$\frac{A_{vL}}{A_{vm}} = \frac{1}{\sqrt{1+\left(\frac{f_1}{f}\right)^2}}$$

We know that for *n* identical stages,

$$[A_{ov}] = [A]^n$$

so

$$\frac{A_{VLo}}{A_{vmo}} = \left[\frac{A_{VL}}{A_{vm}}\right]^n = \left[\frac{1}{\sqrt{1+\left(\frac{f_1}{f}\right)^2}}\right]^n$$

If the lower cut-off frequency for the overall stage is f_{1n}, then at this frequency

$$\frac{A_{VLo}}{A_{vmo}} = \frac{1}{\sqrt{2}} = \left[\frac{1}{\sqrt{1+\left(\frac{f_1}{f_{1n}}\right)^2}}\right]^n$$

or

$$\left[1+\left(\frac{f_1}{f_{1n}}\right)^2\right]^n = 2$$

or

$$\left(\frac{f_1}{f_{1n}}\right)^2 = 2^{1/n} - 1$$

$$\therefore f_{1n} = \frac{f_1}{\sqrt{[2^{1/n} - 1}} \qquad (95)$$

From Eqn. (96), it is clear that $f_{1n} > f_1$ as was represented in Figure 7.28.

If we assume a two-stage amplifier, then $n = 2$.

So,

$$[f_1]_2 = \frac{f_1}{\sqrt{1.44 - 1}} = \frac{f_1}{\sqrt{0.44}} = (1.6) f_1 \qquad (96)$$

Similarly, in the high frequency range,

$$\frac{A_{vho}}{A_{vmo}} = \left[\frac{A_{vh}}{A_{vm}}\right]^n = \left[\frac{1}{\sqrt{1+\left(\frac{f}{f_2}\right)^2}}\right]^n$$

If the upper (half power) cut-off frequency for the overall stage is f_{2n}, then at this frequency

$$\frac{A_{vho}}{A_{vmo}} = \left[\frac{A_{vh}}{A_{vm}}\right]^n = \left[\frac{1}{\sqrt{1+\left(\frac{f}{f_2}\right)^2}}\right]^n$$

If the upper (half power) cut-off frequency for the overall stage is f_{2n}, then at this frequency

$$\frac{A_{vho}}{A_{vmo}} = \frac{1}{\sqrt{2}} = \left[\frac{1}{\sqrt{1+\left(\frac{f_{2n}}{f_2}\right)^2}}\right]^n$$

or
$$\left[1+\left(\frac{f_{2n}}{f_2}\right)^2\right]^n = 2$$

or
$$\left[\frac{f_{2n}}{f_2}\right]^2 = \frac{1}{2^n} - 1$$

$$\therefore\ f_{2n} = f_2\left[\sqrt{2^{1/n}-1}\right] \tag{97}$$

From Eqn. (97), it is clear that $f_{2n} < f_2$ as shown in Figure 7.28.

For a two-stage amplifier, $n = 2$.

$$\therefore\ [f_2]_2 = [0.643]f_2 \tag{98}$$

It can be proved that the gain-bandwidth product of an amplifier is constant since if one increases the other decreases.

For a single-stage amplifier,

Midfrequency gain, $A_{vm} = g_m R'_L$

Bandwidth, $f_2 - f_1 = f_2 = \dfrac{1}{2\pi C_S R'_L}$ since $f_2 \gg f_L$.

Substituting for A_{vm} and f_2, we have

$$\text{Gain bandwidth} = [A][\text{BW}] = [A_{vm}][f_2]$$
$$= [g_m R'_L]\left[\frac{1}{2\pi C_S R'_L}\right]$$

or $$[A]\ [\text{BW}] = \frac{g_m}{2\pi C_S} = \text{const} \tag{99}$$

The R.H.S. of the equation is fixed for a specific amplifier. Thus, we can say gain-bandwidth product of an amplifier is constant.

7.11 Transformer Coupled Amplifier

A transformer may be used in an amplifier to couple: (i) input signal source to the first stage, (ii) the output of one stage to the input of following stage, (iii) the output of last stage to the load.

The interstage transformer isolates the DC voltages and currents similar as capacitor coupling. Further, it can be also step-up or step-down AC voltage/current. The DC resistance of transformer windings is very small, and so there is no effect on transistor bias conditions. Furthermore, collector supply voltage V_{CC} is higher than quiescent collector voltage since the DC resistance of primary is quite small. Further, by using transformer coupling, we may adjust its windings ratio so that output impedance of the first stage matches with the input impedance of the next stage, resulting into maximum power transfer from one stage to the next stage. The only disadvantage in using a transformer is its cost and space requirement. Another disadvantage is that stray magnetic fields can interlink the different transformers used in a cascaded amplifier and it may lead to complications.

The circuit diagram of a transformer-coupled CE amplifier is shown in Figure 7.50.

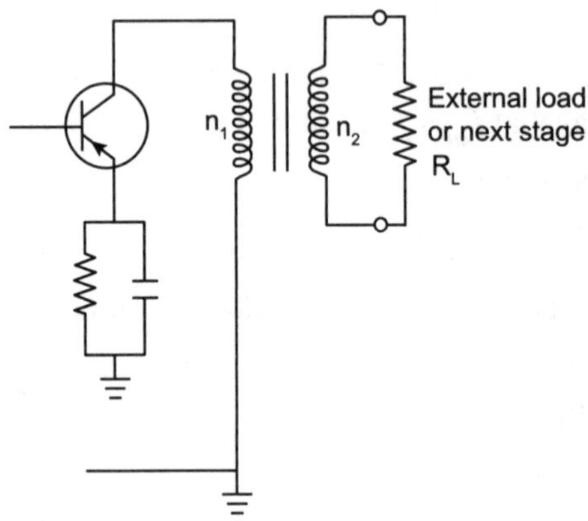

Figure 7.50

The turn ratio 'a' $\left(=\dfrac{N_1}{N_2}\right)$ of transformer windings is so selected that maximum power is transferred to the load connected. If actual load connected is R_L, then load as seen towards the primary of transformer, i.e. reflected resistance, is $R_L' = a^2 R_L$ and in this respect we can select 'a'. It is to be remembered that a is not selected for voltage rise, instead it is selected for matching of impedance.

Further, it is to be noted that the primary coil is directly connected to power supply V_{CC}. The resistance of the primary coil is negligible and we may assume effective DC load to be zero. In this situation, the slope of DC load line is given by $\tan\theta \Rightarrow \infty$, i.e. load line is the vertical line shown in Figure 7.51.

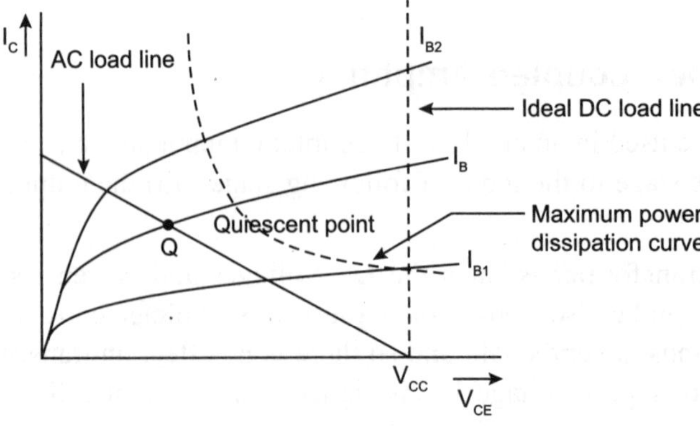

Figure 7.51

The AC load line has a slope $\tan\theta = (-)\dfrac{1}{R'_L}$.

Before analyzing the circuit, it is necessary to understand the AC equivalent circuit (Figure 7.52) of iron core transformer used in the circuit.

The AC equivalent circuit of iron core transformer is shown in Figure 7.52.

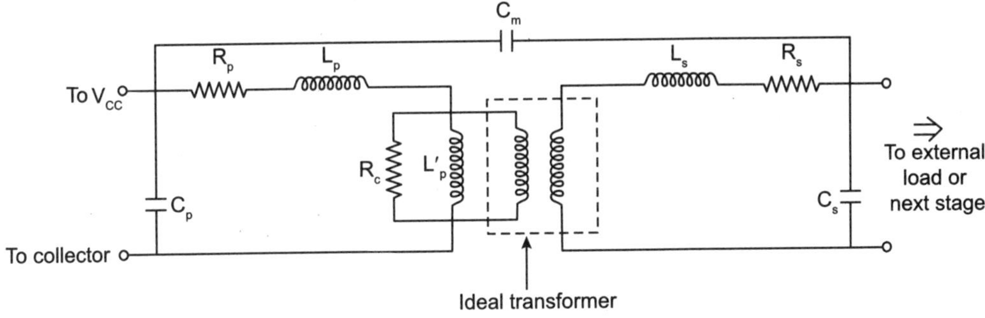

Figure 7.52 AC equivalent circuit of iron core transformer

Here,

(i) C_P and C_S represent the distributed capacitance of primary and secondary coils, respectively.

(ii) L_P and L_S represent the leakage inductance of primary and secondary coils, respectively.

(iii) R_P and R_S represent series resistance of primary and secondary coils, respectively.

(iv) R_C represents the losses of core.

Analyzing such a complicated circuit is very difficult. But, we can simplify it as frequency range of the applied signal. Just like RC amplifier, here too we have three ranges of frequency: (i) low (ii) mid and (iii) high.

Mid frequency range In this range, both capacitive and inductive reactance are negligible and the AC equivalent circuit of the amplifier using '*h*' parameters is shown in Figure 7.53.

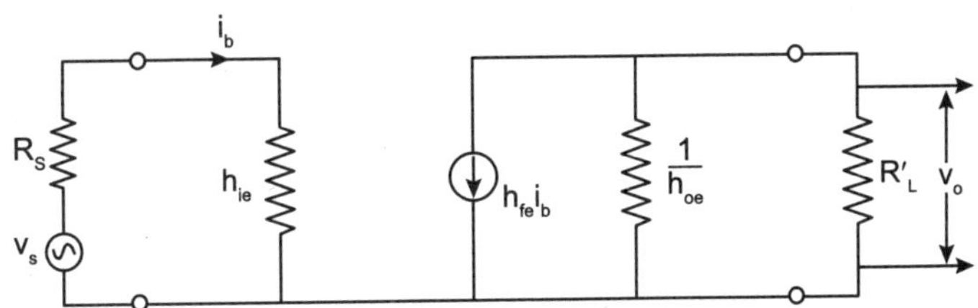

Figure 7.53 AC equivalent circuit of transformer-coupled CE amplifier (mid-frequency range)

From output circuit,

Output voltage, $v_o = (-)h_{fe}i_o(R)$ (100)

where R represents a parallel combination of $1/h_{oe}$ and R'_L.

From input circuit,

$$i_b = \frac{v_s}{R_s + h_{ie}}$$

$$= \frac{v_s}{h_{ie}}$$

assuming R_S to be very small.

Substituting the value of i_b in Eqn. (100),

$$v_o = (-)h_{fc}\left[\frac{v_s}{h_{ie}}\right]R$$

Now, voltage gain,

$$A_{vm} = \frac{v_o}{v_s} = (-)\frac{h_{fe}}{h_{ie}}R$$

The negative sign represents that output voltage is out of phase with input voltage.

$$\therefore |A_{vm}| = \left[\frac{h_{fe}}{h_{ie}}\right]R \quad (101)$$

Equation (101) suggests that voltage gain in midfrequency range for a given load is constant just like an RC amplifier.

Low frequency range In this range, the reactance of various capacitors C_p, C_s and C_m is not effective. Similarly, the values of resistors R_p, R_s and R_c have been taken as negligible since they are very small.

Only leakage inductance whose reactance is ωL_P is effective. It is taken in parallel with reflected load R'_L (Figure 7.54)

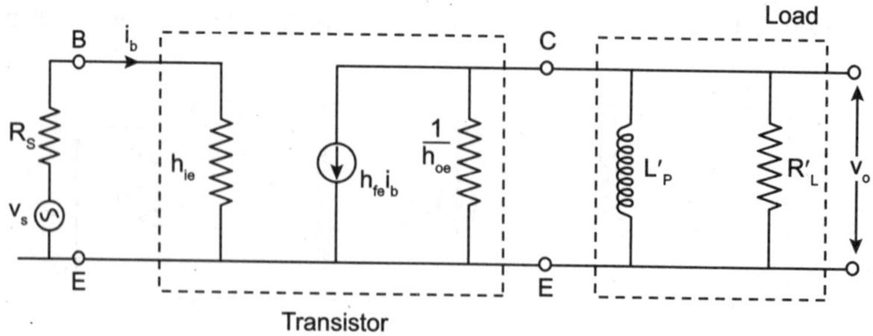

Figure 7.54 AC equivalent circuit of transformer-coupled CE amplifier (low frequency range)

From Figure 7.54, effective load Z_L is a parallel combination of $\frac{1}{h_{oe}}$, L'_P and R'_L.

Thus, $\dfrac{1}{Z_L} = h_{oe} + \dfrac{1}{j\omega L_P} + \dfrac{1}{R'_L}$. Normally $1/h_{oe}$ can be taken as open circuit and in that case Z_L is a parallel combination of L_P and R'_L only.

From output circuit,

Output voltage, $v_o = (-)(h_{fe}\, i_b)|Z_L|$ (102)

From input circuit,

$$i_b = \frac{v_s}{R_s + h_{ie}} = \frac{v_s}{h_{ie}} \text{ assuming } R_S \ll h_{ie}$$

Substituting for i_b in Eqn. (102),

$$v_o = (-)h_{fe}\left[\frac{v_s}{h_{ie}}\right]|Z_L|$$

Hence voltage gain, $A_{vL} = \dfrac{v_o}{v_s} = (-)\dfrac{h_{fe}}{h_{ie}}|Z_L|$

Voltage gain (in magnitude), $|A_{vL}| = \dfrac{h_{fe}}{h_{ie}}|Z_L|$ (103)

Making use of Eqn. (101),

$$A_{vL} = A_{vm}\frac{|Z_L|}{R} \quad (103(a))$$

Eqn. (103) suggests that gain in the low frequency range increases as frequency increases and attains a maximum value A_{vm}.

The external load in the amplifier is not R_L but another stage of an amplifier. In this case, the value of R_L is equal to input resistance of amplifier stage, which is equal to h_{ie2}. The value of Z_L is given by

$$\frac{1}{Z_L} = h_{oe} + \frac{1}{j\omega L_P} + \frac{1}{\left(\dfrac{N_1}{N_2}\right)^2 h_{ei2}}$$

This value of Z_L should be substituted in Eqn. (103).

High frequency range At high frequencies, the shunting action of inductor L_P can be ignored, but series leakage inductance and distributed capacitance of primary (C_P) and secondary (C_S) windings play an important role, thereby the equivalent circuit of amplifier takes the form as shown in Figure 7.55.

As shown in Figure 7.55(a), leakage inductance of the primary, L_S, is the leakage inductance of secondary, and C_P and C_S are lumped capacitances representing the primary and secondary distributed capacitances, respectively.

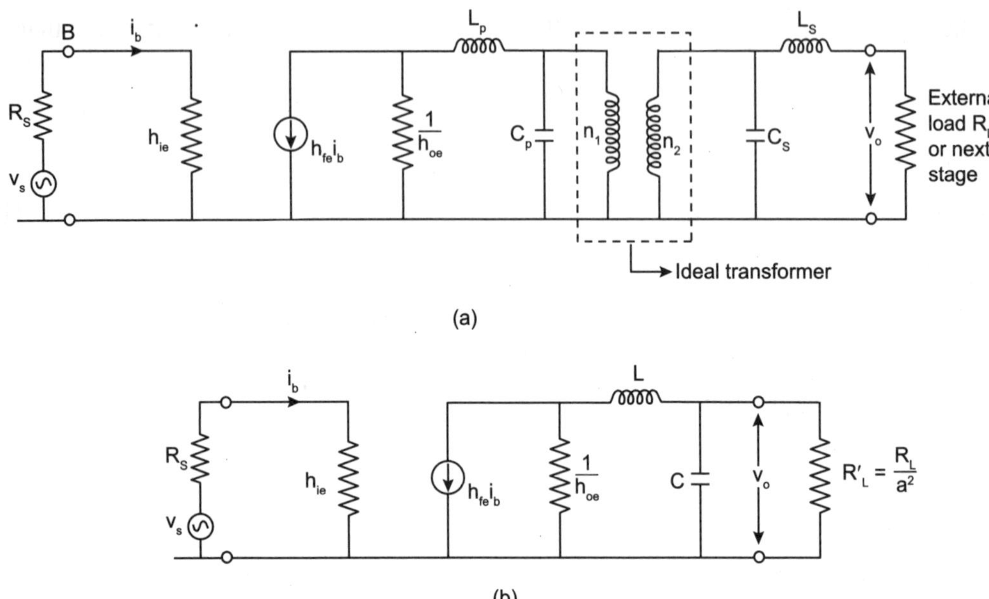

Figure 7.55 AC equivalent circuit of transformer-coupled CE amplifier (high frequency range)

In the simplified equivalent circuit (Figure 7.55(a)), the resistance, inductance and capacitance on the secondary side of transformer (Figure 7.55(b)) are reflected into the primary side where their respective values are

$$L = L_{1p} + a^2 L_s \text{ and } C = C_p + \frac{1}{a^2} C_S$$

Due to LC combination in Figure 7.55(b), there is series resonance at a high frequency when $\omega_0 L = \frac{1}{\omega_0 C}$ where ω_o is resonant frequency.

At a frequency below about 1/10 of the resonant frequency ω_o, the effect of their reactance on voltage gain can be ignored. This frequency then roughly demarcates the mid-frequency range and high frequency range. Thus, the mid-frequency range is given by

$$\frac{5}{h_{oe} L_p} < \omega < \frac{1}{10\sqrt{LC}}$$

In an approximate treatment, resistance R'_L is parallel with C and can be ignored at high frequencies. Transforming the current source $h_{fe}\, i_b$ (Figure 7.55(a)) in shunt with $1/h_{oe}$ into a equivalent voltage source, we obtain the approximate equivalent circuit (Figure 7.56) from Figure 7.55(b).

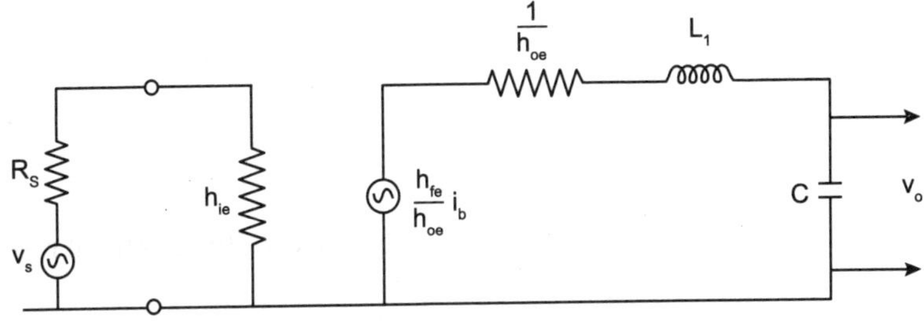

Figure 7.56 Output voltage across capacitor C

$$v_o = (-)\frac{\left[h_{fe}\dfrac{i_b}{h_{oe}}\right]\dfrac{1}{j\omega C}}{\dfrac{1}{h_{oe}} + j\left(\omega L - \dfrac{1}{\omega C}\right)} \tag{104}$$

Ignoring R_s (being small) in the input side, input current $i_b = \dfrac{v_s}{h_{ie}}$.

Substituting the value of i_b in Eqn. (104), we have $v_o = (-)\dfrac{\dfrac{h_{fe}v_s}{h_{oe}h_{ie}} \times \dfrac{1}{j\omega C}}{\dfrac{1}{h_{oe}} + j\left(\omega L - \dfrac{1}{\omega C}\right)}$

Hence voltage gain,

$$A_{vh} = \frac{v_o}{v_s} = (-)\frac{\dfrac{h_{fe}}{h_{oe}h_{ie}} \times \dfrac{1}{j\omega C}}{\dfrac{1}{h_{oe}} + j\left(\omega L - \dfrac{1}{\omega C}\right)} \tag{105}$$

Since the reflected value of external load R_L' has been ignored in the above circuit, the mid-frequency gain would be twice the value given in Eqn. (101).

Thus, $A_{vh} = (-)\, a^2 h_{fe} = (-)\dfrac{h_{fe}}{h_{oe}\, h_{ie}}$ under matched condition; so

$$A_{vh} = \frac{A_{vm}\dfrac{1}{j\omega C}}{\dfrac{1}{h_{oe}} + j\left(\omega L - \dfrac{1}{\omega C}\right)}$$

or

$$\frac{A_{vh}}{A_{vm}} = \frac{\dfrac{1}{j\omega C}}{\dfrac{1}{h_{oc}} + j\left(\omega L - \dfrac{1}{\omega C}\right)} \tag{106}$$

The series resonant frequency $\omega_o = \dfrac{1}{\sqrt{LC}}$ and effective Q of the circuit at resonant frequency is given by $Q = \omega_o L h_{oe}$. Equation (106) can also be written as

$$\frac{A_{vh}}{A_{vm}} = \frac{1}{\left[1 - \dfrac{\omega^2}{\omega_o^2}\right] + j\dfrac{\omega}{\omega_o Q}} = \frac{1}{\sqrt{\left[\left(1 - \dfrac{\omega^2}{\omega_o}\right)^2 + \dfrac{\omega^2}{\omega_o^2 Q^2}\right]}} \angle\theta$$

where

$$\tan\theta = \frac{\dfrac{1}{Q}}{\dfrac{\omega_o}{\omega} - \dfrac{\omega}{\omega_o}}$$

At $\omega = \omega_o$,

$$|A_{vh}| = Q[A_{vm}] \tag{107}$$

Because of the small value of h_{oe}, the overall Q of the circuit is rather low; its order is 0.5 to 2. For a high value of Q of the order of 1 and above, the voltage gain curve (frequency response curve) (Figure 7.57) shows a peak at frequency slightly below ω_o. As Q increases, the peak becomes more and more sharp and it shifts towards ω_o.

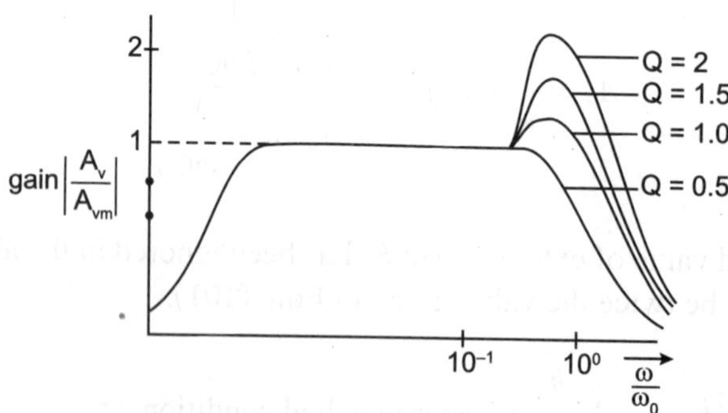

Figure 7.57 Frequency response of transformer-coupled CE amplifier

The drop of voltage gain at high frequencies ($\omega > \omega_0$) is essentially due to the shunting effect of capacitance C, which show that the shunting effect of C is stronger at higher frequencies due to the decreasing value of reactance of C. To avoid an appreciable peak in voltage gain at high frequencies near ω_0, the Q of the circuit is not allowed to be larger than unity.

7.12 Direct Coupling between Stages

7.12.1 Direct Coupled Circuit

The use of direct coupling (Figure 7.58) between stages is one way of eliminating certain components for economical reasons. In Figure 7.58, the base of transistor Q_2 is directly coupled to the collector of Q_1.

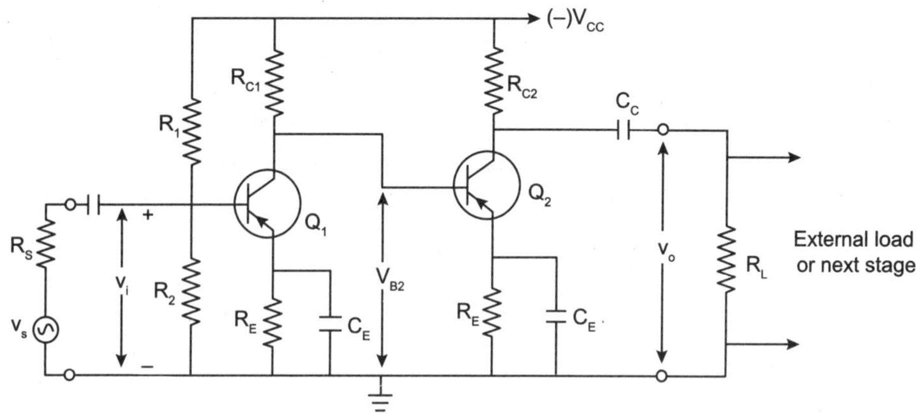

Figure 7.58 Direct coupled amplifier

Comparing Figure 7.58 with Figure 7.25, we find that biasing resistor R_1 and R_2 for 2nd stage and coupling C_C has been eliminated.

7.13 Use of Complimentary Transistors

A direct-coupled two-stage circuit (Figure 7.59) is similar to that seen in (Figure 7.46) except that if Q_1 is *p-n-p* type then the transistor Q_2 is *n-p-n* device. Q_1 and Q_2 are usually so selected that their characteristics and parameters are similar. Such transistors are known as complimentary transistors.

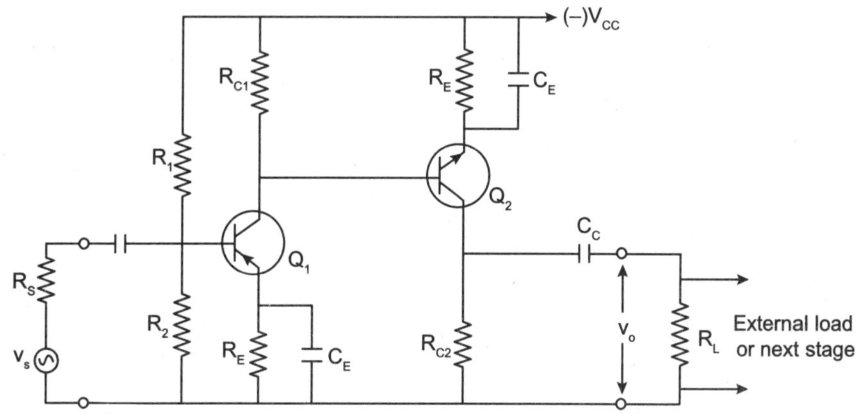

Figure 7.59

Summary

1. Small signal transistor voltage amplifiers are used in three configurations: (a) common emitter, (b) common base, (c) common collector. Voltage gain is minimum in case of common collector amplifier and its maximum value is unity.
2. The most common CE amplifier is RC-coupled amplifier since both voltage gain and current gain have finite value.
3. In low frequency range, both C_E and coupling capacitor C_C are effective; and in high frequency range, interelectrode capacitors of the transistor are effective. In mid-frequency range, neither series nor shunt capacitors are effective since their reactances can be neglected.
4. The operating point of an amplifier is selected in the middle of load line.
5. In mid-frequency range, voltage gain is expressed by

$$A_{vm} = \frac{(-)h_{fe}R'_L}{R_s + h_{ie}}$$

   Here the negative sign indicates that there is a phase change of π between input and output signals. The maximum value of gain is h_{fe}.
6. In low frequency range, voltage gain is expressed by

$$A_{vL} = A_{vm}\left[\frac{1}{1+\dfrac{1}{j\omega C_C(R_D + R_i)}}\right] = \frac{A_{vm}}{1 - j\dfrac{\omega_1}{\omega}} = \frac{A_{vm}}{\sqrt{1+\left(\dfrac{\omega_1}{\omega}\right)^2}}$$

   where ω_1 is called lower half power frequency since power gain here reduces to half of the mid-frequency range.
7. In high frequency range, voltage gain is expressed by

$$A_{vh} = A_{vm}\left[\frac{1}{1 + j\omega C_S \mathrm{R}_{eq}}\right] = \frac{A_{vm}}{1 + j\dfrac{\omega}{\omega_2}} = \frac{A_{vm}}{\sqrt{1+\left(\dfrac{\omega}{\omega_2}\right)^2}}$$

   where ω_2 is called upper half power frequency since here also power gain reduces to half of the mid-frequency range.
8. The bandwidth of an amplifier is defined as $\omega_2 - \omega_1$ or $f_2 - f_1$. The two frequencies are known as (–)3 dB frequencies.
9. The gain bandwidth product of an amplifier is always constant.
10. To increase voltage gain which has a maximum fixed value 'h_{fc}' for a single stage, we cascade two or three amplifiers as required. The overall voltage gain of a cascaded or multistage amplifier having *n* stages is given by

$$A_{ov} = A_1 \times A_2 \times A_3 \ldots \times A_n$$

If voltage gain of each stage is equal to 'A', then overall gain for the equivalent amplifier is $A_{ov} = [A]^n$

In this case,

Lower half power frequency, $[f_1]_{ov} = \dfrac{f_1}{\sqrt{2^{\frac{1}{n}} - 1}}$ and

Upper half power frequency, $[f_2]_{ov} = f_2\left[f_2\sqrt{2^{1/2} - 1}\right]$

Thus, the bandwidth of equivalent amplifier reduces and gain increases, which is in consonance with the fact that gain bandwidth of an amplifier is constant.

11. For radio frequencies we prefer transformer coupling instead of RC coupling. In transformer coupling DC load is zero but AC load is the reflected value towards primary of the actual load. R_L given by $R'_L = \dfrac{R_L}{a^2}$ where $a = \dfrac{n_1}{n_2}$.
12. In addition to the transistor, a FET is also used as voltage amplifier. A FET can also be used in three configurations; (i) common source, (ii) common gate, (iii) common drain.
13. The most common configuration is common source configuration when we get appropriate gain whose value R'_L.
14. FET amplifier is preferred where we require high input impedance.
15. The frequency response curve of FET amplifier is similar to that of transistor amplifier. The maximum voltage gain is observed in mid-frequency range and it falls both in the low frequency and high frequency range, i.e., it falls at the rate of 20 dB or 6 dB per diode.

    The maximum value of voltage gain in case of common drain configuration is 1, similar as in a common collector transistor amplifier.

Review Questions

1. Discuss the procedures for determining I_C, R_C, R_E in a single-stage CE amplifier. Also discuss the selection of bias resistor R_1 and R_2.
2. Derive the equation for calculating the value of emitter bypass capacitor and coupling capacitor for a RC common emitter transistor amplifier.
3. Sketch the circuit of a two-stage capacitor-coupled CE amplifier. Sketch its 'h' parameter equivalent circuit and write equations for input impedance, output impedance and voltage gain of each stage.
4. Prove that the cut-off frequencies of multi-stage amplifiers are different from those of individual stages. What is the relation between f_{1n} and f_1?
5. Prove that gain bandwidth product of an amplifier is constant. How does the bandwidth change for a stage amplifier?

6. Prove the following relations:

$$f_{1n} = \frac{f_1}{\left[\sqrt{2^{1/n}-1}\right]}$$

$$f_{2n} = f_2\left[\sqrt{2^{1/n}-1}\right]$$

where f_{1n} and f_{2n} represent the cut-off frequencies of n stage amplifier and f_1, f_2 represent the cut-off frequencies of single stage amplifier.

7. How do you define the mid-range band of a CE amplifier? Derive a relation for voltage gain in the mid-range frequencies of CE amplifier.

8. What do you mean by "operating point" of an amplifier? How is it selected? Show the same for a CE amplifier using appropriate diagram.

9. What do you mean by (i) DC load line, (ii) AC load line of an amplifier. Represent the same in case of RC-coupled amplifier.

10. Define the following terms as applied to an amplifier:

    (i) Input resistance (ii) Output resistance

    (iii) Voltage gain (iv) Current gain

11. Draw the circuit diagram of a common collector amplifier showing DC bias. Why it is called emitter follower? Derive a relation for (i) voltage gain (ii) input impedance.

12. Draw the circuit diagram of an emitter follower showing DC bias. Derive the relation for (i) voltage gain (ii) input impedance. What is the maximum value of voltage gain? Where such an amplifier is used?

13. Why does the voltage gain of a CE transistor amplifier fall in the low as well as high frequency range? What is the rate of fall of gain?

14. Prove that voltage gain of an amplifier in high frequency range or low frequency range falls at 20 dB/decade.

15. What are the various methods of cascading amplifiers? When do we use them? Give reasons for your choice of the method used.

16. Define amplification factor, transconductance and drain resistance in case of field effect transistor. Draw mid-frequency AC equivalent circuit for a given FET.

17. Draw the complete circuit of a common drain FET amplifier. Derive the relation for (i) voltage gain, (ii) input impedance, (iii) output impedance.

18. Sketch a practical common source FET (n channel) amplifier using fixed bias voltage. Explain its action as amplifier and derive the relation for voltage gain.

Problems

1. An amplifier has a mid-band voltage gain of 1000. The cut-off frequencies are f_1 =100 Hz and f_2 = 100 kHz. Calculate voltage gain at input frequency f = 20 Hz and f = 300 kHz.

   [**Ans:** 196, 316]

2. A single-stage CE amplifier with emitter current bias is to be designed using 2N 3903 transistor. The power supply is 15 V and external load is 70 kΩ. Assume h_{fe} = 100, f_1 = 60 Hz, f_2 = 60 kHz, h_{ie} = 1 kΩ.

   [**Ans:** C_C = 0.39 μF, C_S = 43 μF, C_E = 56 μF, R_C = 7 kΩ]

3. Design a single-stage CE amplifier having emitter current bias. The circuit is to use 2N 3904 transistor, supply voltage V_{CC} = 24 V, external load R_L = 120 kΩ and have cut-off frequencies f_1 = 100 Hz and f_2 = 50 kHz (Assume h_{ie} = 1 kΩ, h_{fe} = 100).
   [**Ans:** R_C = 12 kΩ, R_E = 3.75 kΩ, R_L = 43 kΩ, R_1 = 125 kΩ, C_E = 159 μF, C_C = 0.13 μF, C_S = 292 pF]

4. Calculate suitable resistor values for a stage, direct-coupled CE amplifier. The circuit is to use 2N 3903 transistors, the supply voltage V_{CC} = 14 V and load resistor = 50 kΩ. (Assume h_{fe} = 50, h_{ie} = 0.5 kΩ, f_1 = 75 Hz, V_{E1} = 5 V, V_{CE1} = 3 V)
   [**Ans:** R_{C2} = 4 kΩ, R_{E2} = 7.7 kΩ, R_{C1} = 6 kΩ, R_{E1} = 4.7 kΩ, R_2 = 53.3 kΩ, R_1 = 68 kΩ, C_{E1} = C_{E2} = 330 μF, C_C = 0.53 μF].

5. In a transformer-coupled CE amplifier, load connected to the secondary is 100 Ω. Calculate its mid-frequency voltage gain. Assume $N_p : N_S$ = 10 : 1, h_{fe} = 100, h_{ie} = 10 kΩ.

   [**Ans:** 100]

6. Calculate the lower half power frequency of a CE amplifier having R_C = 6.8 kΩ, C_C = 1 μF and external load 1 kΩ. [**Ans:** 20.4 Hz, 32.64 Hz]

7. In a cascaded amplifier, voltage gain of 1st stage, 2nd stage and 3rd stage are 100, 200 and 400, respectively. Calculate the overall voltage gain. Also find its value in dB.

   [**Ans:** 8×10^6, 138 dB]

8. The voltage gain of an amplifier is 65 dB. The input voltage is 0.5 mV. Calculate the output voltage. [**Ans:** 0.89 V]

9. The voltage gain of a two-stage amplifier is 100 dB. The voltage gain of 2nd stage is 100. Calculate the voltage gain of 1st stage. Express the same in dB also. [**Ans:** 1000, 60 dB]

10. In a CE amplifier mid-frequency voltage gain is 200. It falls to 141.4 at 50 Hz and also at 10 kHz. Calculate the (–)3 dB frequencies of the amplifier. What is its bandwidth?

    [**Ans:** 50 Hz, 10 kHz, 9950 Hz]

11. An audio amplifier has an input impedance of 160 kΩ. The external load connected to the amplifier is 8 Ω. On applying an input signal of 0.5 V, we get an output of 5 V across the load. Calculate power gain in dB. [**Ans:** 63 dB]

12. In an RC-coupled amplifier, load R_C is 1.5 kΩ and the value of emitter lead resistor $R_E = 350\ \Omega$ which is bypassed by a capacitor of 50 μF. If the capacitor goes open, what is the value of DC load with amplifier. **[Ans:** 1850 Ω]
13. In a CE amplifier, load R_C is 10 kΩ and input impedance of next stage is 10 kΩ. Calculate the output voltage if input signal of 1 mV is applied. Assume $h_{ie} = 2.5$ kΩ, $h_{fe} = 100$. **[Ans:** 200 mV]

Short Answer Questions

1. What is the role of (i) coupling capacitor (ii) shunt (bypass) capacitor in an amplifier?

   Ans: When a capacitor is so connected in a circuit that it couples or transmits AC signal to a resistor, it is called a coupling capacitor. They are important in an amplifier because they transmit AC signal without disturbing the Q point of the amplifier. They are so selected at lowest frequency $X_c < 0.1R$.

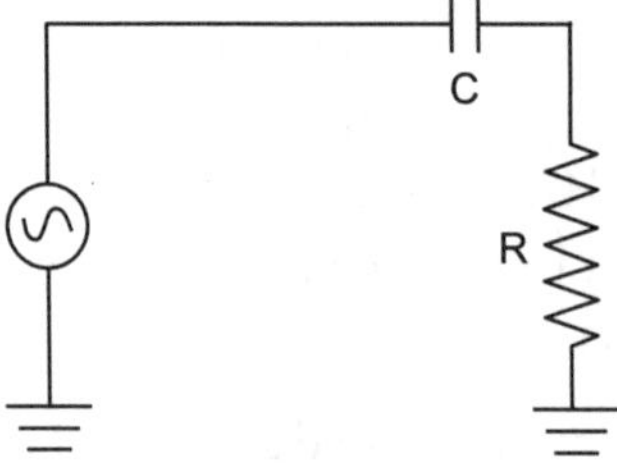

When a capacitor is so connected in a circuit that all the AC source voltage appears across the resistor, that is, point E is effectively shorted to ground. In this respect, the capacitor is called a bypass capacitor because it bypasses or shorts point E to ground without disturbing the Q point.

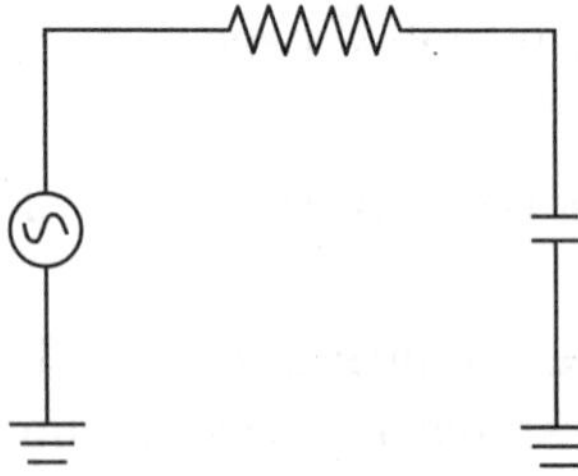

2. What do you mean by small signal operation in an amplifier?

   Ans: When an AC voltage sinusoidal source is coupled into the base of a transistor, the variation in base voltage produces a sinusoidal variation in base current resulting into sinusoidal variation in collector current and collector voltage of the amplifier. Operation with such a voltage signal is called small signal operation and there is no distortion in the output.

3. What do you mean by AC resistance of emitter diode?

   Ans: It is defined as the ratio of change in base to emitter voltage $\Delta V_{BE} = (v_{be})$ to the corresponding change in emitter current $\Delta I_E = (i_e)$, i.e.,

$$r_e' = \frac{v_{be}}{i_e} = \frac{\text{AC base emitter voltage}}{\text{AC emitter current}}$$

4. Why the Q point is selected in the middle of AC load line of an amplifier?

   Ans: The Q point is selected in the middle of AC load line because on applying AC signal base current varies and so the position of Q point varies. If it goes much down, it may be in the cut-off region; or if it goes much up, then its may be in the saturation region, which is not desirable for smooth operation of the amplifier.

5. Which capacitance affects the gain in high frequency ranger?

   Ans: In CE stage, gain in high frequency range is affected by (i) stray wiring capacitance and (ii) collector to base capacitance.

6. How does gain in a CE amplifier roll off in high frequency range in a bode plot?

   Ans: The gain in a CE amplifier breaks at upper cut-off frequency and then rolls off at a rate of 20 dB/decade, which is equivalent to 6 dB/octave.

7. What do you mean by cut-off frequencies in case of CE amplifier?

   Ans: The frequencies at which the gain is 0.707 times the maximum gain are called cut-off frequencies. There are two cut-off frequencies: (i) lower cut-off frequency (ii) upper cut-off frequency.

8. How is the mid-band of frequencies defined in an amplifier?

   Ans: Mid-band of an amplifier is defined as the band of frequencies between $10\,f_1$, and $0.1\,f_2$ where f_1 is lower cut-off frequency and f_2 is upper cut-off frequency.

9. Why is a DC amplifier called so?

   Ans: Many times, we have direct coupling between two amplifier stages. This allows the circuit to amplify all the way down to zero hertz. (lowest frequency).

10. In which systems do we require the matching of source resistance (R_S) and input resistance of amplifier as well as matching of output resistance of amplifier and external load. Why we require such matching.

    Ans: Impedance matching is generally required in microwave, television and telephone systems. It is required so that maximum power is transferred.

11. What is ideal bode plot in case of an amplifier?

    Ans: Ideal bode plots are approximations that allow us to draw the frequency response of an amplifier quickly and easily. In such a plot there is a break at cut-off frequency. There are two cut-off frequencies, namely (i) lower cut-off frequency, (ii) upper cut-off frequency. The gain is a roll off at 20 dB/decade before the lower cut-off frequency (f_1) and after the upper cut-off frequency (f_2) as shown.

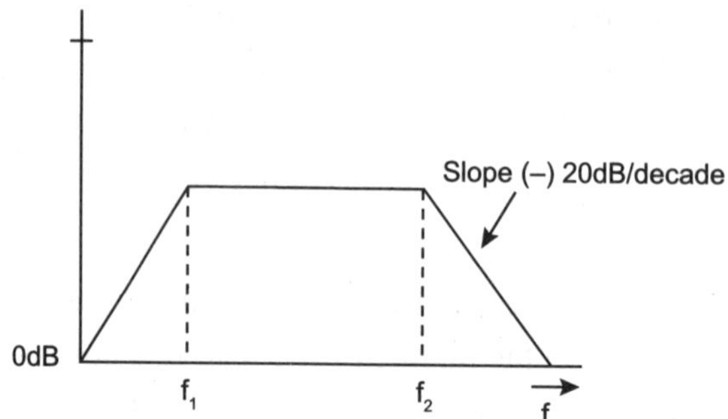

12. Why do we get a peak in the frequency response curve of a transformer-coupled amplifier? In which range the peak is located.

    Ans: In case of a transformer-coupled amplifier, we get a peak in the high frequency range in the frequency response curve because of series resonance due to inductance and capacitance of the transformer being effective.

13. Why the DC load line in case of an ideal transformer-coupled amplifier is a vertical line?

    Ans: This load line is vertical because the slope of the load line is given by $\tan\theta = (-)\frac{1}{R_L}$ and here the resistance of primary of transformer connected to the power supply is negligible, i.e. $R_L = 0$. So $\tan\theta = \alpha$ and hence $\theta = 90°$ or it is a vertical line drawn on (V_{CE}, I_C) characteristic.

14. Why the maximum value of voltage gain in case of common collector amplifier is unity?

    Ans: Voltage gain in case of a CC amplifier is given by $A = \frac{(1+h_{fe})R_E}{h_{ie} + (1+h_{fe}R_E)}$. Neglecting h_{ie} being small, gain attains a value of unity which is the maximum value.

15. Why a transformer is not an amplifier?

    Ans: A transformer increases the voltage a level of signal but does not increase the power level. Obviously, a transformer is not an amplifier.

Objective Questions

1. Frequency response of an amplifier is a graph between voltage gain and
   (a) load (b) frequency
   (c) input voltage (d) output voltage **[Ans: (b)]**
2. The stray wiring capacitance in an amplifier has an effect on
   (a) lower cutoff frequency (b) mid-band frequencies
   (c) upper cutoff frequency (d) output resistance **[Ans: (c)]**

3. Two stages of an amplifier have decibel voltage gain of 20 and 40 dB each. The overall ordinary voltage of an amplifier is
   (a) 1 (b) 10
   (c) 100 (d) 1000 **[Ans:** (d)]
4. Two stages of an amplifier have gain of 100 and 200 each. The overall decibel gain of the amplifier is
   (a) 46 dB (b) 86 dB
   (c) 66 dB (d) 106 dB **[Ans:** (b)]
5. The voltage gain of an amplifier decreases 20 dB per decade above 2 kHz. If the mid-band voltage gain is 86 dB, the ordinary voltage gain at 20 MHz is
   (a) 20 (b) 200
   (c) 2000 (d) 20000 **[Ans:** (d)]
6. The AC load line in an amplifier is
   (a) horizontal (b) vertical
   (c) steeper than DC load line (d) less steeper than DC load line **[Ans:** (c)]
7. The power gain of an amplifier is
   (a) equal to its voltage gain (b) smaller than voltage gain
   (c) square of the voltage gain (d) square root of the voltage gain **[Ans:** (c)]
8. The frequency response curve of DC amplifier is same as
   (a) RC amplifier (b) TC amplifier
   (c) OP amplifier (d) tuned amplifier **[Ans:** (c)]
9. The ordinary power gain of an amplifier is expressed as
   (a) log G (b) 10 log G
   (c) 20 log G (d) 100 log G **[Ans:** (b)]
10. In a multistage amplifier, the AC load of the first stage includes
    (a) load resistance connected in the collector branch
    (b) input impedance of first stage
    (c) emitter resistance of first stage
    (d) input resistance of next stage **[Ans:** (d)]
11. In a CE amplifier if collector resistance gets open, the AC output voltage will
    (a) remain unchanged (b) decrease
    (c) increase (d) go to zero **[Ans:** (d)]
12. In a CE amplifier if the coupling capacitor gets open, the AC output voltage will
    (a) remain unchanged (b) decrease
    (c) increase (d) go to zero **[Ans:** (d)]

13. In a CE amplifier the external load gets open, the AC output voltage will
    (a) Remain unchanged (b) Decreases
    (c) Increase (d) Go to zero [**Ans:** (c)]
14. In a CC amplifier, the output voltage is
    (a) amplified and is out of phase w.r.t. input signal
    (b) amplified and is in phase with input signal
    (c) not amplified and is out of phase w.r.t. input signal
    (d) not amplified and is in phase with input signal [**Ans:** (d)]

Mark the Following Statement True/False.

1. The operating point must be in the middle of load line. [**Ans:** True]
2. The Q point of an emitter-biased circuit is immune to changes in current gain. [**Ans:** True]
3. The capacitor which produces an AC ground is called coupling capacitor. [**Ans:** False]
4. The overall current gain in a multistage amplifier having n identical stages is n times the current gain of single stage. [**Ans:** False]
5. If a capacitor connected in the emitter lead of a CE amplifier gets open, the voltage gain remains unchanged. [**Ans:** False]
6. The output voltage of a two-stage CE amplifier is in phase with the input signal. [**Ans:** True]
7. JFET is referred to as square law device. [**Ans:** True]
8. The output voltage in common source FET amplifier is in phase with the input signal. [**Ans:** False]
9. The DC load line in a transformer-coupled amplifier is steeper than AC load line. [**Ans:** True]
10. In a transformer-coupled amplifier, a resistive load behaves like a resonant load in the high frequency range. [**Ans:** True]

Fill in the Blanks.

1. In a p-channel FET amplifier, the source is connected to the ____________ terminal of the battery and gate is kept ____________ with respect to source. [**Ans:** positive, negative]
2. The slope of DC load line in a transformer-coupled amplifier is ____________. [**Ans:** infinity]
3. The maximum value of voltage gain in a CE transistor amplifier is ____________, but in case of CC amplifier it is ____________. [**Ans:** h_{fe}, unity]

4. The input resistance of FET common source amplifier is ____________ than that of CE transistor amplifier. [**Ans:** greater]
5. The gain of a multistage amplifier is ____________ and its bandwidth is ____________ than that of single-stage amplifier. [**Ans:** more, less]
6. For low frequency amplification, ____________ amplifier is generally used.
   [**Ans:** direct coupled]
7. In RC transistor amplifier the coupling capacitor affects the ____________ frequency whereas the shunting capacitor affects the ____________ frequency.
   [**Ans:** lower half power, upper half power]
8. Gain-bandwidth product of an amplifier is ____________. [**Ans:** constant]
9. FET is ____________ operated and transistor is ____________ operated.
   [**Ans:** voltage, current]

Chapter

8 Feedback Amplifiers

8.1 Concept of Feedback

The transfer of energy from the output of an electrical network to its input is called feedback. When such a transfer of energy in the form of voltage or current is made in an amplifier, the working of the amplifier is affected in many respects. Such an amplifier is called feedback amplifier. There are of two types: (i) positive feedback and (ii) negative feedback.

8.1.1 Positive Feedback

When the feedback signal from the output and input (initial) are in same phase in an amplifier, the net input signal to the amplifier increases. Such a feedback is called positive or regenerative feedback. The gain of such an amplifier increases and this type of feedback is generally employed in oscillators.

8.1.2 Negative Feedback

When the feedback signal from the output and input (initial) in an amplifier are in opposite phase, the net input signal to the amplifier decreases and the gain of such an amplifier decreases. This type of feedback is called negative feedback. Although the gain of such amplifiers decreases, there are many advantages such as (i) distortion and noise in output decreases, (ii) improvement in stability and in frequency response (iii) increment in bandwidth and (iv) change in input and output impendence of the amplifier. Such amplifiers are used in many devices.

Negative feedback can be defined as feeding a portion of an amplifier output signal back to the input. The signal is fed back in such a way that it is 180° out of phase with the input signal. This feedback reduces the gain of the amplifier, but it also reduces the amount of distortion in the output signal.

8.2 General Theory of Feedback

The block diagram of normal amplifier and feedback amplifier is shown in Figure 8.1.

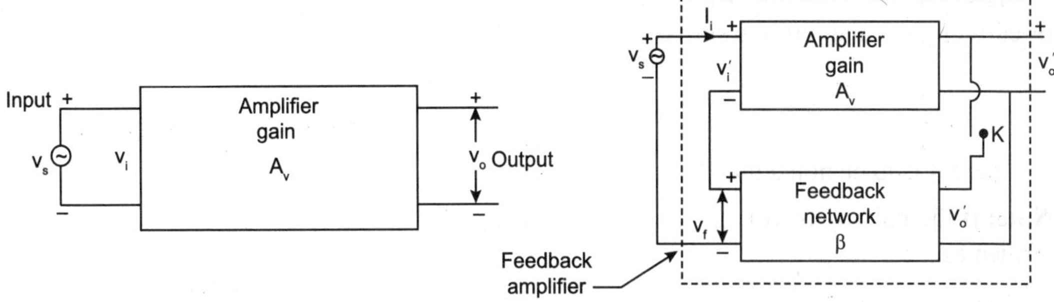

Figure 8.1 (a) Normal amplifier, (b) feedback amplifier

Internal gain of normal amplifier (Figure 8.1a) is defined as,

$$A_v = \frac{\text{Output voltage } (v_o)}{\text{Input voltage } (v_s)} = \frac{v_o}{v_s} = \frac{v_o}{v_i} \tag{1}$$

When key *k* (Figure 8.1(b)) is pressed, output voltage is applied as input across the feedback network. A fraction (v_f) of the output voltage is fed back to the input side of the amplifier where it is fed in series with input signal (v_s) as shown in Figure 8.1(b).

Let the feedback factor of feedback network be β. It is defined as

$$\beta = \frac{\text{Voltage feedback}}{\text{Output voltage}} = \frac{v_f}{v_o'} \tag{2}$$

In this situation, effective input voltage.

$$v_i' = v_s - v_f \tag{3}$$

assuming the polarity of voltages as shown in Figure 8.1(b). [However, if the polarity of feedback voltage is opposite, then $v_i' = v_s + v_f$.]

Substituting for v_f in equation (3) from Eqn. (2), we have

$$v_i' = v_s - \beta v_o'$$

This input voltage is amplified by the amplifier, and so

$$v_o' = A_v\, v_o' = A_v[v_s - \beta v_o']$$

or $$v_o' + \beta A_v\, v_o' = A_v v_s$$

or $$v_o'[1 + \beta A_v] = A_v v_s$$

or $$\frac{v_o'}{v_s} = \frac{A_v}{1 + \beta A_v}$$

Hence voltage gain after feedback, by definition, is

$$A_{vf} = \frac{v_o'}{v_s} = \frac{A_v}{1 + \beta A_v} \tag{4}$$

Here we have used suffix v to A to show voltage gain. But, in general, we may have voltage or current gain, and thus the suffix v is dropped. We can write Eqn. (4) as a general case

$$A_f = \frac{A}{1+\beta A} \quad (5)$$

Here it is to be noted that β and A are complex.

Note: If the polarity of voltage feedback, v_f, is reversed in Figure 8.1(b), then Eqn. (5) can be written as

$$A_f = \frac{A}{1-\beta A}$$

However, in general, we have $A_f = \frac{A}{1-\beta A}$ where both β and A are complex quantities and expressed as vector.

8.2.1 Positive Feedback

(i) $|1+\beta A|$ less than unity, i.e. $|(1+\beta A)| < 1$: This is possible only when β is negative, then the value of A_f increases. Hence

$$A_f > A$$

(The polarity of v_f connection is opposite to that shown Figure 8.1(b).)

In this case, feedback is called positive feedback and the amplifier tends to be unstable.

8.2.2 Negative Feedback

(i) $|1+\beta A|$ greater than unity, i.e. $|(1+\beta A)| > 1$: This is possible only when β is positive. In this situation, the value of A_f decreases. Hence

$$A_f < A$$

In this case, feedback is called negative feedback.

(ii) $|1+\beta A| = 0$:

Then the value of A_f in Eqn. (5) goes to infinity.

$$A_f \Rightarrow \infty \quad (6)$$

Equation (6) represents that in this situation we are getting output without input signal, i.e. applied input signal, v_s, is absent or zero.

This is the condition for a feedback oscillator and the condition is known as Barkhausen criterion for oscillation.

(iii) $\beta A \gg 1$: Then from Eqn. (5),

$$A_f = \left|\frac{A}{\beta A}\right| = \left|\frac{1}{\beta}\right| \tag{7}$$

Equation (7) represents that gain after feedback does not depend upon any parameter of amplifier but depends only upon β, i.e., the feedback network comprising passive parameters. For satisfactory operation of the feedback amplifier, the open loop gain '*A*' must always be greater than the closed loop gain. In general, for A_f to be stable to within 1% when *A* changes by 50%, *A* should be equal or greater than 100 A_f.

In equation (5), βA is called return ratio or open loop gain. In most amplifier circuits, both β and *A* are complex and they dependent on frequency.

In equation (5), if $\beta = 0$. Then $A_f = A$ in this situation. *A* is called internal gain of amplifier or gain without feedback.

Consider $\beta \neq 0$. Then feedback loop is present and so A_f is called closed loop gain and $|1+\beta A|$ in case of negative feedback or $|1-\beta A|$ in case of positive feedback is generally called return difference.

(iv) The amount of feedback employed in an amplifier can be expressed in. It is defined as the ratio of output power with feedback to output power without feedback.

We can write the amount of feedback as

$$N\,\text{dB} = 10\log\frac{P_o'}{P_o}$$

$$= 10\log\frac{(V_o')^2/R}{(V_o)^2/R}$$

$$= 20\log\frac{V_o}{V_o'}$$

where P_o' – output power with feedback

P_o – output power without feedback

Dividing both numerator and denominator by V_s, we have by definition

$$N\,\text{dB} = 20\log\frac{V_o'/V_s}{V_o/V_s} = 20\log\frac{A_f}{A} \tag{8}$$

From Eqn. (5), we have $\dfrac{A_f}{A} = \dfrac{1}{1+\beta A}$

where βA is negative for negative feedback

$$N\,\text{dB} = 20\log\left[\frac{1}{1+\beta A}\right] \tag{8a}$$

$$\therefore N\,\text{dB} = (-)20\log[1+\beta A]$$

$$= (-)20\log(\text{return difference})$$

Thus, for negative feedback, N will be a negative number.

Eqn. (5) representing voltage gain with feedback depends upon the following assumptions:

(i) The input signal is transmitted through amplifier circuit only and not through feedback network.

(ii) The feedback signal transmitted from output to input is through feedback network only and not through amplifier.

(iii) The feedback factor β does not depend upon load and source (signal) resistance.

It has been mentioned that the effect of negative feedback in an amplifier is to:

(i) improve stabilization

(ii) minimize distortion and noise

(iii) improve frequency response and increase in bandwidth

(iv) change (increase/decrease) input impedance and output impedance

8.2.3 Stabilization

The gain of an amplifier depends on temperature variation in power supply, voltage, changes in the parameters (constants) of transistor or change of transistor itself. These effects are studied by stabilizer factor. It is defined as ratio of change in gain (either without feedback or with feedback) to the original gain. So we have

(i) Stabilization without feedback, $S = \dfrac{dA}{A}$

(ii) Stabilization with feedback, $S_f = \dfrac{dA_f}{A_f}$

The gain of an amplifier with negative feedback is given by Eqn. (5)

$$A_f = \frac{A}{1+\beta A}$$

Differentiating the above equation, we have

$$\frac{dA_f}{dA} = \frac{1+\beta A-\beta A}{(1+\beta A)^2} = \frac{1}{(1+\beta A)^2}$$

Using Eqn. (5), we have

$$\frac{dA_f}{dA} = \frac{1}{(1+\beta A)}\left[\frac{A_f}{A}\right]$$

$$\text{or} \quad \frac{dA_f}{A_f} = \frac{1}{(1+\beta A)} \frac{dA}{A}$$

which by defination leads to

$$S_f = \frac{S}{1+\beta A} \tag{9}$$

Equation (9) suggests that $s_f < s$.

Gain sensitivity may be defined as

$$\frac{S_f}{S} = \frac{dA_f / A_f}{dA / A} = \frac{1}{1+\beta A} \tag{9a}$$

For stabilization, it is necessary that the value of stability factor S_f be minimum. This can be achieved by increasing the value of $[1 + \beta A]$.

There are two possibilities:

(i) If feedback be negative, then only the value of return difference $|1+\beta A|$ will be greater than unity.

(ii) If the value of gain A without feedback be quite high. So the value of return ratio $\beta A \gg 1$.

In this case, we get

$$A_f = \frac{1}{\beta}$$

Thus, closed loop gain A_f depends only on the feedback factor and does not depend on the internal gain of amplifier A.

With negative feedback, the gain value reduces but there is improvement in stabilization or variation as shown in Figure 8.2.

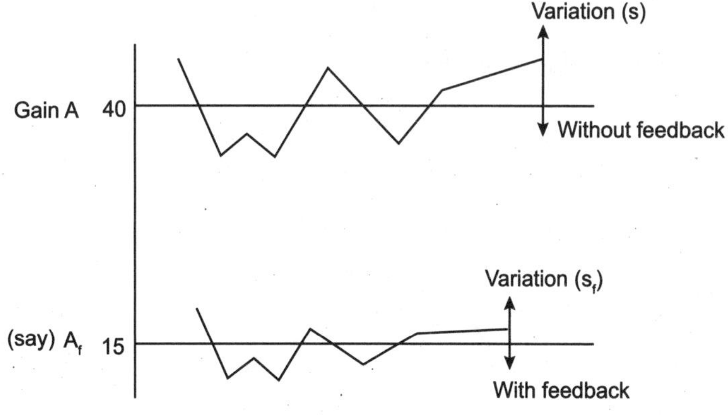

Figure 8.2 Variation of gain (a) with (b) without feedback

8.2.4 Reduction in Non-Linear Distortion and Noise

When a large signal is applied across the input of an amplifier, the output signal is distorted due to the curvature present in the transfer characteristic of the amplifier. The distorted output signal comprises the fundamental frequency of the input signal as well as harmonics. In addition to this type of distortion, output signal is also distorted due to thermal noise signal produced by the transistor, and so on.

Distortion due to nonlinearity in the characteristics and noise can be expressed as voltage source v_D, which is in series of the output of amplifier. In an amplifier, if output voltage is v_o and input voltage is v_i, then

$$v_o = Av_i + v_D$$

The above expression is represented in Figure 8.3(a).

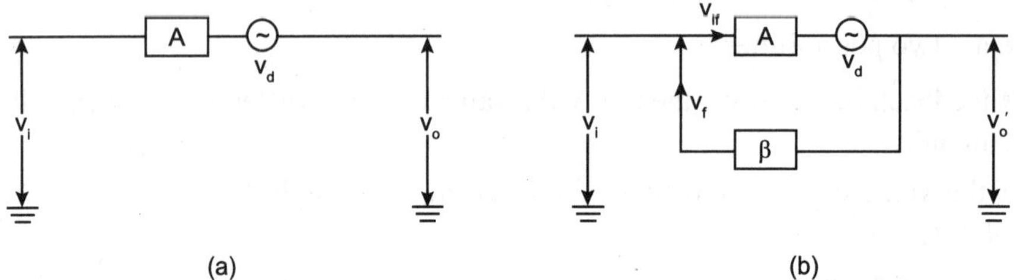

Figure 8.3

Output voltage without distortion is

$$v_{os} = Av_i$$

If we adjust the input voltage with feedback such that the output voltage remains unchanged, i.e. v_{os}, then in this situation input voltage is different from v_i. Suppose this input voltage is v_{if}, then as per Figure 8.3(b), we have

$$v'_o = Av_{if} + v_D$$

Considering negative feedback, we have

$$v'_o = A(v_i - v_f) + v_d$$

or $$v'_o = A(v_i - \beta v'_o) + v_D$$

or $$v'_o + \beta A v'_o = Av_i + v_D$$

or $$v'_o[1 + \beta A] = Av_i + v_D$$

$$\therefore\ v'_o = \frac{A}{1+\beta A} v'_i + \frac{v_D}{1+\beta A} \qquad (10)$$

Now we know that

$$v_{os} = Av_i \text{ (without feedback)}$$

and $$v'_o = \frac{A}{1+\beta A} v'_i$$

hence if $$v'_i = (1+\beta A) v_i \tag{11}$$

then $$v_{os} = v'_{os} \tag{12}$$

Thus if, with feedback, we have input voltage v'_o equal to $(1+\beta A)v_i$, then output voltage remains unchanged but distortion voltage v_d changes to

$$v'_d = \frac{v_d}{1+\beta A} \tag{13}$$

Clearly, the new value $v'_d < v_d$ due to negative feedback, i.e. distortion and noise, reduces by a factor $\frac{1}{1+\beta A}$, but output voltage remains unchanged.

This expression applies for small distortion only. Figure 8.4 clearly shows the reduction in distortion by employing negative feedback.

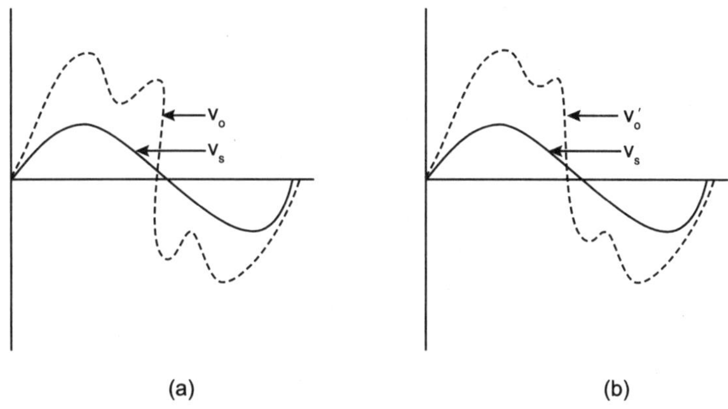

Figure 8.4 (a) Without feedback, (b) with feedback

Reduction in noise Suppose we have noise expressed by voltage V_n in an ordinary amplifier of gain A, then the output is

$$v_{no} = A v_n \tag{14}$$

If we employ negative feedback and get output noise as v'_{no}, then a part, $\beta v'_{no}$, is fed back and net input is

$$v'_n = v_n - \beta v'_{no} \tag{15}$$

Output in this situation is

$$v'_{no} = A v'_n = A[v_n - \beta v'_{no}]$$

or
$$Av_n = v'_{no} + \beta A v'_{no} \tag{16}$$

L.H.S. represents output v_{no} without feedback.

So
$$v_{no} = v'_{no}[1 + \beta A]$$

$$\therefore v'_{no} = \frac{v_{no}}{1 + \beta A} \tag{17}$$

If we represent v'_{no} as N_f and V_{no} as N, then

$$N_f = \frac{N}{1 + \beta A} \tag{18}$$

Expression (18) represents that $N_f < N$, i.e. noise in output after negative feedback is less than noise in output without feedback in case of negative feedback.

Improvement in frequency response-increase in bandwidth When discussing the theory of RC-coupled amplifiers, we studied the frequency response of amplifier under three heads:

(i) Mid-frequency range (A_m)
(ii) Low frequency range (A_l)
(iii) High frequency range (A_h)

and derived the respective expressions for voltage gain:

$$A_l = \frac{A_m}{1 + j\dfrac{f_l}{f}} \tag{19}$$

$$A_h = \frac{A_m}{1 + j\dfrac{f}{f_h}} \tag{20}$$

Here f_e and f_h represent the lower and upper half power respectively, where gain reduces by a factor $\frac{1}{\sqrt{2}}$. The above expressions represent gain without feedback. When we employ negative feedback, these expressions take the form of (5a) i.e.

$$A_{lf} = \frac{A_l}{1 + \beta A_l} \tag{21}$$

$$A_{hf} = \frac{A_h}{1 + \beta A_h} \tag{22}$$

Substituting the value of A_l from Eqn. (19) in Eqn. (21), we have

$$A_{lf} = \frac{\dfrac{A_m}{\left(1 - j\dfrac{f_l}{f}\right)}}{1 + \beta\dfrac{A_m}{\left(1 - j\dfrac{f_l}{f}\right)}}$$

$$= \frac{A_m}{1 + \beta A_m - j\dfrac{f_l}{f}} = \frac{\dfrac{A_m}{1 + \beta A_m}}{1 - j\dfrac{f_1}{f(1 + \beta A_m)}}$$

Using Eqn. (5a), the above expression can be written as

$$A_{lf} = \frac{A_{mf}}{1 - j\dfrac{f_{lt}}{f}} \tag{23}$$

where
$$A_{mf} = \frac{A_m}{1 + \beta A_m}$$

and
$$f_{lf} = \frac{f_l}{1 + \beta A_m}$$

In case of negative feedback, $|1 + \beta A_m|$ is greater than 1.

$$\therefore\ f_{lf} < f_l \tag{24}$$

So, lower half power frequency, f_{lf}, after feedback is less than lower half power frequency fl without feedback as shown in Figure 8.5.

Similarly, for high frequency range, we have

$$A_{hf} = \frac{A_h}{1 + \beta A_h} = \frac{\dfrac{A_m}{\left(1 + j\dfrac{f}{f_h}\right)}}{1 + \beta\dfrac{A_m}{\left(1 + j\dfrac{f}{f_h}\right)}}$$

Simplifying, we have

$$A_{hf} = \frac{A_{mf}}{\left(1 + j\dfrac{f}{f_{hf}}\right)}$$

Here $$f_{hf} > f_h \tag{25}$$

where $f_{hf} = f_h\,(1 + \beta A)$ represents that upper half power frequency with feedback is greater than upper half power frequency without feedback as shown in Figure 8.5.

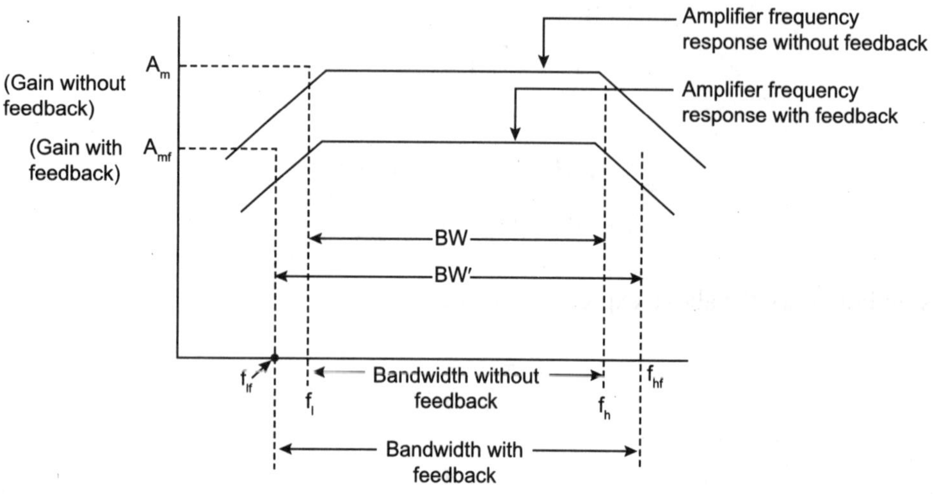

Figure 8.5 Amplifier frequency response with and without feedback

In Figure 8.5, BW′ represents bandwidth $(f_{hf} - f_{lf})$ with feedback, and BW represents bandwidth $(f_h - f_l)$ without feedback. From Figure 8.5, we note that

$$\text{BW}' > \text{BW} \tag{26}$$

Hence we say bandwidth increases with negative feedback but the respective gain decreases.

(i) Change in input impedance The input impedance of an amplifier (without feedback) is defined as impedance z_i seen when "looking into" the base of the transistor employed.

From Figure 8.1(b), input impedance

$$R_i = \frac{v_i}{I_i}$$

or $$I_i = \frac{v_i}{R_i} \tag{27}$$

After employing negative feedback, input impedance of feedback amplifier is

$$R_i' = \frac{v_s}{I_i} \tag{28}$$

Substituting for I_i from Eqn. (27),

$$R_i' = \frac{v_s}{v_i} R_i$$

Also from Figure 8.1(b), we have

$$v_i = v_s - v_f = v_s - \beta v_o' = v_s - \beta A v_i$$

or
$$v_i[1+\beta A] = v_s \tag{29}$$

Substituting for v_s in Eqn. (28),

$$R_i' = \frac{v_i}{I_i}(1+\beta A)$$

Using Eqn. (27), we have

$$R_i' = R_i(1+\beta A) \tag{30}$$

Thus, in this situation called voltage series negative feedback, input impedance of the amplifier increases by a factor $(1 + \beta A)$.

$$\therefore R_i' > R_i$$

(We can have another voltage shunt feedback situation where $R_i' < R_i$.)

(ii) Change in output impedance The output impedance of an amplifier (without feedback) is defined as impedance seen between the output terminals looking into the amplifier assuming input signal as zero.

For an amplifier with feedback shown in Figure 8.6(b), feedback voltage v_f acts as input voltage $(-)\beta V_o'$. The negative sign shows that it is negative feedback.

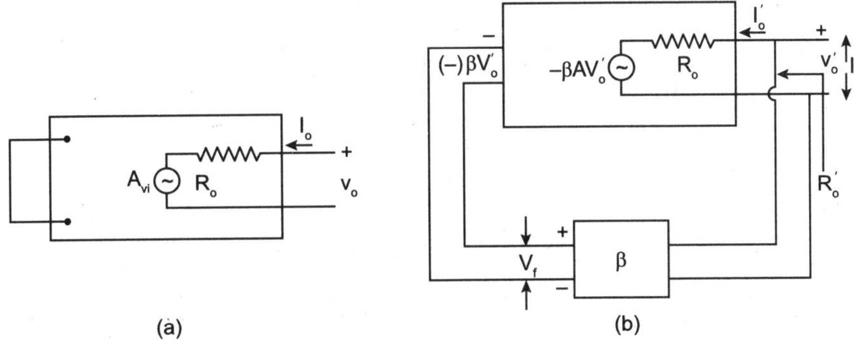

Figure 8.6

Here it has been assumed that the feedback network does not act as extra load in the output, i.e. it draws negligible current.

Applying KVL to the output terminals, we have

$$V_o' = I_o' R_o - \beta A V_o'$$

or
$$V_o'[1+\beta A] = I_o' R_o$$

or
$$\frac{V_o'}{I_o'} = \frac{R_o}{1+\beta A} \tag{31}$$

L.H.S. in Eqn. (31) represents output impedance of the amplifier with feedback. So,

$$R'_o = \frac{R_o}{1+\beta A} \tag{32}$$

From Eqn. (32), it is clear that with voltage series negative feedback, output impedance of the amplifier reduces by a factor $\left[\frac{1}{1+\beta A}\right]$.

Hence $R'_o < R_o$

However, we can have another situation (current series negative feedback) where $R'_o > R_o$.

Example 1 The voltage gain of an amplifier is 120. A signal of 50 mV (rms) can be used as input voltage across it. We employ negative feedback circuit having feedback factor 0.1. Calculate (i) input signal, (ii) output voltage, (iii) voltage gain, (iv) feedback in dB after feedback is employed.

Solution

$$\text{Output voltage without feedback} = A\, v_i$$
$$= 120 \times 50 \times 10^{-3}$$
$$= 6 \text{ V}$$

$$\text{Input signal, } v_s = v_i - v_f$$
$$= v_i - \beta v_o$$
$$= 50 \times 10^{-3} + 0.1 \times 6$$
$$= 0.65 \text{ V}$$

(positive sign because of negative feedback)

$$\text{Gain after feedback} = \frac{A}{1+\beta A}$$
$$= \frac{120}{1+0.1\times 120} = \frac{120}{13}$$
$$= 9.2$$

$$\text{Feedback (dB)} = (-)20\log[1-\beta A]$$
$$= (-)\log[1+\beta A] \text{ (since } \beta \text{ is negative)}$$
$$= (-)20\log\,[1+12]$$
$$= (-)20\log 13$$
$$= (-)20\times 1.1139$$
$$= (-)22.278$$

Example 2 The bandwidth of an amplifier is 6 MHz. The feedback factor is (–)4. Calculate the bandwidth when negative feedback is employed.

Solution

$$\text{Bandwidth after feedback BW}' = \text{BW}(1-\beta A)$$
$$= 6(1+4)\,\text{MHz (since } \beta \text{ is negative)}$$
$$= 30\,\text{MHz}$$

Example 3 For an amplifier, open loop gain is 60000 and closed loop gain is 10000. Calculate negative feedback factor and also find βA.

Solution

$$\text{Gain after negative feedback } A_f = \frac{A}{1+\beta A}$$
$$10000 = \frac{60000}{1+\beta(1000)}$$

or

$$10000\,\beta + 1 = 6$$
$$\therefore \text{ Feedback factor } \beta = \frac{5}{10000}$$
$$= 0.0005$$
$$\therefore \ \beta A = 0.00055 \times 60000$$
$$= \frac{5}{10000} \times 60000$$
$$= 30$$

Example 4 An amplifier has a voltage gain of 200 and produces 8% harmonic distortion. How much feedback is employed so that harmonic distortion reduces to 0.5%. What will be the voltage gain after feedback?

Solution

$$\text{Distortion after negative feedback, } D_f = \frac{D}{1+\beta A}$$
$$D_f = \frac{0.5}{100} = \frac{8/100}{1+\beta A}$$

or
$$1 + \beta A = 16$$

or
$$\beta A = 15$$

$$\therefore \text{ Feedback factor, } \beta = \frac{15}{200} = 0.075$$

$$\text{Gain after negative feedback, } A_f = \frac{A}{1+\beta A} = \frac{200}{1+15} = 12.5$$

Example 5 The voltage gain of an amplifier is 100. To reduce the distortion produced in the amplifier, negative feedback of 10% is introduced. Calculate voltage gain after feedback and decrement in distortion.

Solution

$$\text{Gain after negative feedback, } A_f = \frac{A}{1+\beta A}$$

$$\text{Distortion after feedback, } D_f = \frac{D}{1+\beta A}$$

$$\text{Decrement in distortion, } D - D_f = D - \frac{D}{1+\beta A} = \frac{D\beta A}{1+\beta A}$$

$$A_f = \frac{100}{1+\frac{10}{100}\times 100} = 9.1$$

$$\beta A = 10$$

$$\text{Percentage change in distortion} \left[\frac{D-D_f}{D}\right] = \frac{\beta A}{1+\beta A}\times 100 = \frac{10}{11}\times 100 = 91\%$$

∴ Reduction is 9%.

Example 6 An amplifier has a voltage gain of 50. 10% negative feedback is employed to reduce distortion.

(a) What will be the voltage gain with feedback?

(b) If 5% negative feedback is employed, will the new voltage gain be double of the previous case? How much it differs from the expected value of gain?

(c) How much feedback should be introduced so that voltage is double of case (a).

Solution

(a) Voltage gain after negative feedback,

$$A_f = \frac{A}{1+\beta A}$$

$$= \frac{50}{1+\frac{10}{100}\times 50} = 8.33$$

(b)

$$A_f = \frac{A}{1+\beta A}$$

$$= \frac{50}{1+\frac{5}{100}\times 50} = \frac{50}{1+\frac{5}{2}}$$

$$= \frac{100}{7} = 14.3$$

So the new gain is not double the previous case.

The difference between expected value and actual value of gain obtained is

$$= 2\times 8.33 - 14.3$$

$$= 2.36$$

(c) To have double the gain of case (a), i.e. 16.66, let the feedback introduced be β (assuming negative feedback)

$$16.66 = \frac{100}{6} = \frac{50}{1+\beta(50)}$$

or

$$\beta\left(\frac{5000}{6}\right) = 50 - \frac{100}{6}$$

$$\therefore \beta = \frac{2}{50} = 0.04$$

8.3 Feedback Technique

In a feedback amplifier we have the following networks in addition to the amplifier circuit:

(i) Sampling network (S) which selects the voltage or current to be fed back

(ii) Feedback network (β) which feeds a fraction of the output voltage or current to the input

(iii) Mixing network (Σ) which mixes the feedback voltage or current with input signal.

The above arrangement has been shown in Figure 8.7.

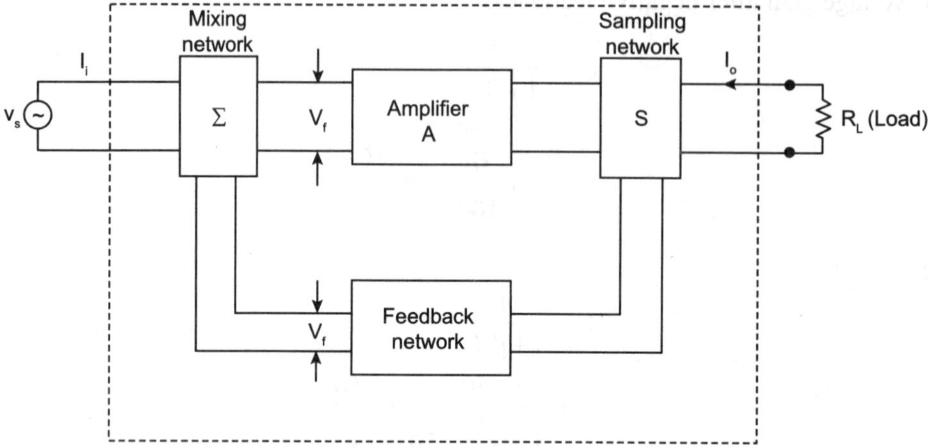

Figure 8.7 Feedback amplifier

In all feedback amplifiers the parameter to be fed is of two types: (i) voltage (ii) current. So they are classified into (a) voltage feedback amplifiers and (b) current feedback amplifier.

8.3.1 Voltage Feedback

In this arrangement, the output voltage is applied across the feedback network which transfers a fraction of the voltage or current to be fed to the input of the amplifier as shown in Figure 8.8.

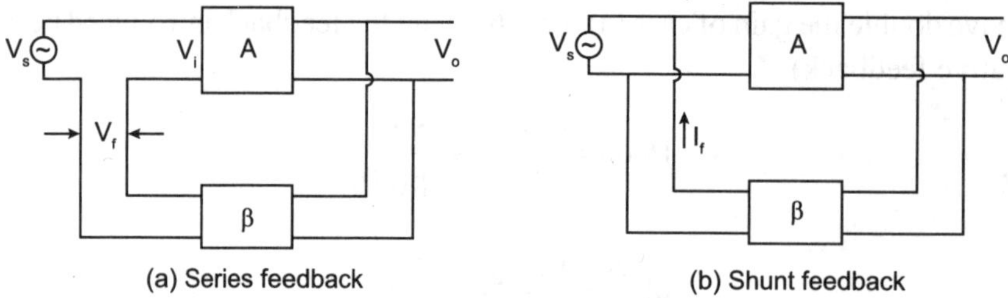

Figure 8.8 Voltage feedback

The connection for the feedback signal to the input can be either in series (Figure 8.8(a)) or in parallel (shunt) (Figure 8.8(b)). The two circuits are known as voltage series feedback circuit and voltage shunt feedback circuit, respectively.

Here voltage feedback is βV_o and current feedback is $I_f = \beta' V_o$.

8.3.2 Current Feedback

In this arrangement, the feedback network is connected in series with the output of amplifier as shown in Figure 8.9. Here the feedback network transfers a fraction of output current to the input of amplifier.

The connection for feedback signal to the input can be either in series (Figure 8.9(a)) or in parallel (or shunt) (Figure 8.9(b)). The two circuits are known as (i) current series feedback and (ii) current shunt feedback, respectively.

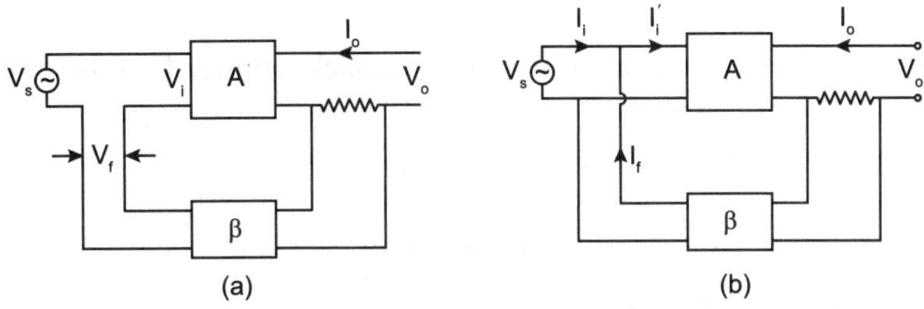

Figure 8.9 (a) Current series feedback, (b) current shunt feedback

(a) Voltage series feedback (Emitter follower) The circuit of emitter follower of common collector amplifier is shown in Figure 8.10. This circuit is an example of voltage series feedback (Figure 8.8(a)) and can be redrawn as shown in Figure 8.11(a) to match Figure 8.8(a).

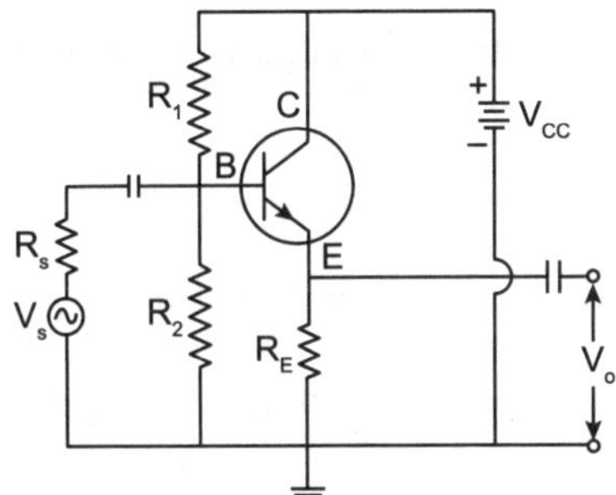

Figure 8.10 Emitter follower (common collector amplifier)

Its AC equivalent circuit is shown in Figure 8.11(a)

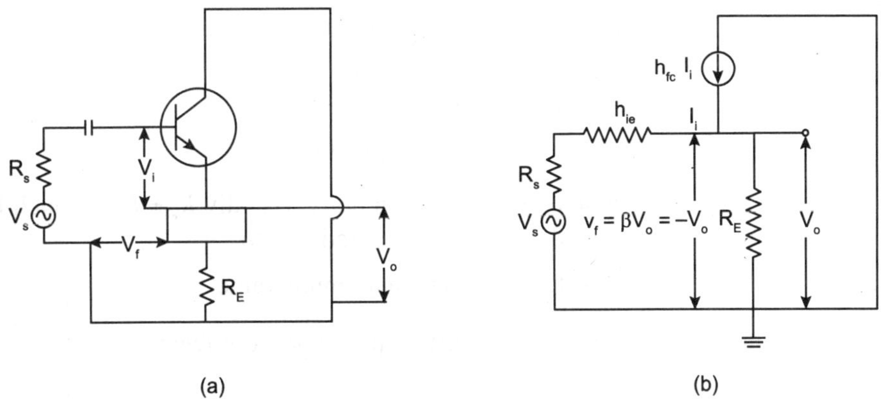

Figure 8.11 (a) Emitter follower circuit without DC bias, (b) AC equivalent circuit using '*h*' parameters for emitter follower

In drawing the circuit using 'h' parameters in Figure 8.11(b), we have neglected h_{re} and $\frac{1}{h_{oe}}$. The circuit (Figure 8.11) resembles voltage series feedback circuit and its theory can be discussed as under.

(i) Voltage gain To find voltage gain without feedback, emitter E of the circuit is assumed to be grounded. Thus,

Input voltage, $$V_s = (R_s + h_{ie})I_i \tag{33}$$

Output voltage, $$V_o = (1 + h_{fe})I_i R_E \tag{34}$$

Substituting the value of I_i from (33) in (34),

$$V_o = (1 + h_{fe})\left(\frac{V_s}{R_s + h_{ie}}\right)R_E$$

By definition, voltage gain, $$A = \frac{V_o}{V_s} = \frac{(1 + h_{fe})R_E}{(R_S + h_{ie})} \tag{35}$$

But in Figure 8.11(b) the entire output voltage v_o is being fed back. Thus,

Feedback voltage, $v_f = \beta V_o = (-)V_o$

Hence $\beta = (-)1$

To calculate voltage gain after feedback, we use Eqn. (5)

$$A_f = \frac{A}{1 + \beta A}; \text{ here } \beta \text{ is negative}$$

Substituting for A and β, we have

$$A_f = \frac{(1 + h_{fe})R_E / R_{S+hie}}{1 + \dfrac{(1 + h_{fe})R_E}{R_S + h_{ie}}}$$

$$\therefore\ A_f = \frac{(1 + h_{fe})R_E}{R_S + h_{ie} + (1 + h_{fe})R_E} \tag{36}$$

Eqn. (36) reflects that the value of A_f is less than unity. Usually $R_s + h_{ie}$ is much less than $(1 + h_{fe})R_E$, and so it can be neglected. Now Eqn. (36) reduces to

$$A_f = 1, \text{ which is its maximum value}$$

(ii) Input impedance Input impedance of the amplifier without feedback is

$$R_i = \frac{V_s}{I_i} = R_S + h_{ie}$$

Output voltage, $V_o = (1 + h_{fe}) I_i \, R_E$

and gain $A = \dfrac{(1 + h_{ie}) R_E}{R_S + + h_{ie}}$

Feedback factor β = (–)1

We known that input impedance after negative feedback is given by $R_i' = (1 + \beta A) R_i$.

Substituting for β, A and R_i, we have

$$R_i' = (R_S + h_{ie}) \left[1 + \frac{(1 + h_{fe}) R_E}{R_S + h_{ie}} \right]$$

$$= R_S + h_{ie} + (1 + h_{fe}) R_E \quad (37)$$

Eqn. (37) reflects that input impedance with negative feedback increases.

(iii) Output impedance It is defined as impedance seen between the output terminals looking towards the amplifier and assuming input signal to be zero.

Voltage gain of the amplifier without feedback is given by Eqn. (35)

$$A = \frac{h_{fe} R_E}{(R_S + h_{ie})}$$

and feedback factor β = (–) 1.

Hence output impedance with negative feedback is given by Eqn. (32):

$$R_o' = \frac{R_o}{1 + \beta A}$$

$$\text{or } R_o' = \underset{R_E \to \alpha}{Lt} \frac{R_E}{1 + \dfrac{h_{fe} R_E}{R_S + h_{ie}}}$$

[Here R_E acts as load; to calculate output impedance, it is taken as ∞, or the output terminals are assumed to be open.]

or

$$R_o' = \underset{R_E \to \alpha}{\alpha_L} \frac{R_S + h_{ie}}{\dfrac{R_S + h_{ie}}{R_E} + h_{fe}}$$

This first term in the denominator goes to zero as per our assumption, and so

$$R_o' = \frac{R_s + h_{ie}}{h_{fe}} \quad (38)$$

Eqn. (38) reflects that output impedance after feedback reduces a lot since h_{fe} of the transistor is larger than unity.

(b) Voltage shunt feedback (Fixed bias amplifier) The circuit of fixed bias amplifier is shown in Figure 8.12.

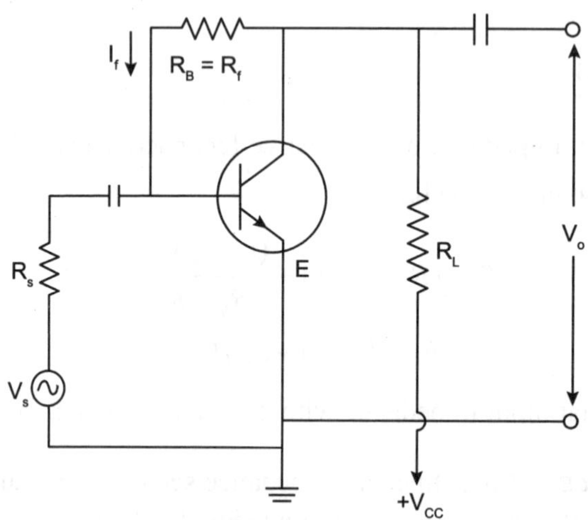

Figure 8.12

The above circuit represents voltage shunt feedback circuit (Figure 8.8(b)) and can be redrawn without supply as in Figure 8.13(a).

The AC equivalent circuit using '*h*' parameters is shown in Figure 8.13(b). The circuit theory can be discussed as under.

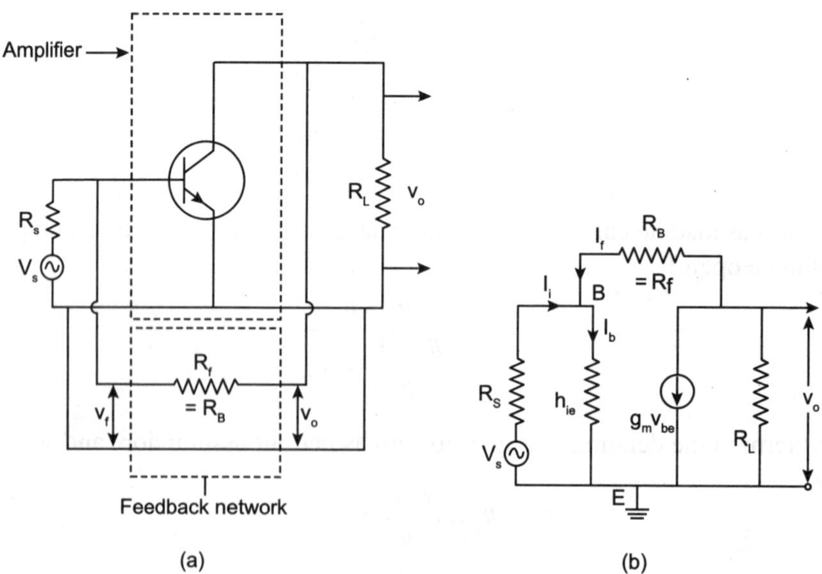

Figure 8.13 (a), (b) AC equivalent circuit for fixed bias amplifier using 'h' parameters

(i) Voltage gain From the circuit in Figure 8.13(b), at junction B we have

$$I_i + I_f = I_B$$

If internal voltage of the amplifier is quite high, then base current I_B can be assumed to be zero and so

$$I_i = (-)I_f \tag{39}$$

Also from the circuit,

$$I_i = \frac{V_s - V_{be}}{R_s} = \frac{V_s}{R_s} \tag{i}$$

and

$$I_f = \frac{V_o - V_{be}}{R_f} = \frac{V_o}{R_f} \tag{ii}$$

since V_{BE} is quite low.

From Eqn. (ii) it is clear that $I_f \alpha V_o$ (output voltage) and this current flows through R_f (feedback network) and that is why it is called voltage shunt feedback.

Also $V_{Be} = I_i h_{ie}$.

Assuming I_f to be quite small, we have from Figure 8.13(b)

$$V_{be} = I_i h_{ie}$$

$$\text{or } V_{be} = \frac{h_{ie} V_s}{R_s + h_{ie}}$$

Also output voltage $V_o = (-)[g_m v_{be}]R_L$.

Substituting for V_{be}, we have

$$V_o = (-)g_m \frac{h_{ie} V_s}{R_S + h_{ie}} R_L$$

By definition, $h_{fe} = g_m h_{ie}$; so

$$V_o = (-)\frac{h_{fe} V_s}{R_S + h_{ie}} R_L$$

Hence voltage gain without feedback is

$$A = \frac{V_o}{V_s} = (-)\frac{h_{fe} R_L}{R_S + h_{ie}} \tag{40}$$

Feedback factor,

$$\beta = \frac{\text{Voltage feedback}}{\text{Output voltage}} = (-)\frac{R_S I_i}{R_f I_f}$$

And from Eqn. (39) (assuming negative feedback), $I_i = (-)I_f$

so $$\beta = \frac{R_s}{R_f} \tag{41}$$

Hence loop gain, $$\beta A = (-)\frac{h_{fe}R_L\,R_s}{(R_s + h_{ie})R_f} \tag{42}$$

We know that with negative feedback voltage gain,

$$A_f = \frac{A}{1+\beta A}$$

Substituting from Eqn. (40) and (41) for A and β, we have

$$A_f = (-)\frac{h_{fe}\dfrac{R_L}{R_s + h_{ie}}}{1 + \dfrac{h_{fe}R_L\,R_s}{(R_S + h_{ie})R_f}}$$

If $h_{fe}\dfrac{R_L\,R_S}{(R_S + h_{ie})R_f} \gg 1$, then

$$A_f = \frac{(-)h_{fe}\dfrac{R_L}{R_S + h_{ie}}}{h_{fe}\dfrac{R_L\,R_S}{(R_S + h_{ic})R_f}} = (-)\frac{R_f}{R_S} \tag{43}$$

(ii) Input impedance Input impedance after negative feedback is given by

$$R_i' = \frac{R_i}{1+\beta A}$$

For an amplifier without feedback, $R_f = \infty$

So input impedance, $R_i = R_S + h_{ie}$

Substituting of R_i, β and A, we have

$$R_i' = \frac{R_S + h_{ie}}{1 + h_{fe}\dfrac{R_L R_S}{(R_S + h_{ie})R_f}} \tag{44}$$

Eqn. (44) reflects that input impedance after feedback reduces.

(iii) Output impedance We know that for a CE amplifier with load R_L, output impedance is given by $R_o = \dfrac{R_L}{1 + h_{oe}R_L}$.

Normally $h_{oe} R_L \ll 1$; so

$$R_o = R_L$$

As per Eqn. output impedance after feedback (Eqn. 32), we have

$$R'_o = \frac{R_o}{1+\beta A}$$

Substituting for R_o, β and A, we have

$$R'_o = \underset{R_L \to \infty}{Lt} \frac{R_L}{1+\dfrac{h_{fe} R_L R_S}{(R_S + h_{ie})R_f}}$$

$$= \frac{(R_S + h_{ie})R_f}{h_{fe} R_S}$$

However, if $R_S \gg h_{ie}$ which usually is the case,

$$R'_o = \frac{R_f}{h_{fe}} \tag{45}$$

Thus, output impedance after feedback reduces a lot since h_{fe} is quite large.

(c) Current series feedback (Emitter bias amplifier) The circuit of emitter bias amplifier is shown in Figure 8.14.

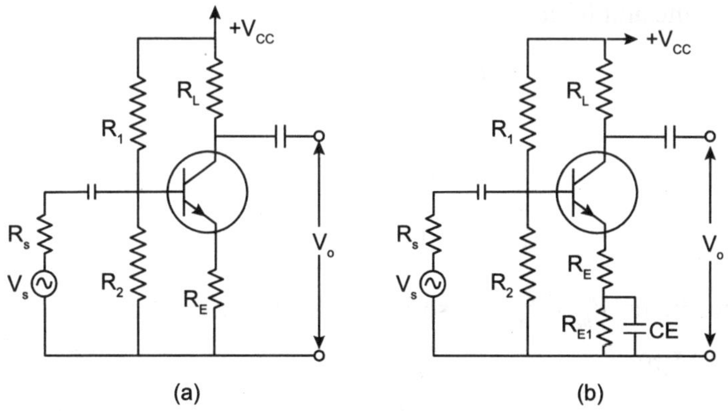

Figure 8.14 Emitter bias amplifier

In Figure 8.14(b), resistance R_{E1} is by passed by capacitor C_E and so it is not effective.

The above circuit represents current series feedback (Figure 8.9a) as can be seen from Figure 8.15(a) where it has been redrawn.

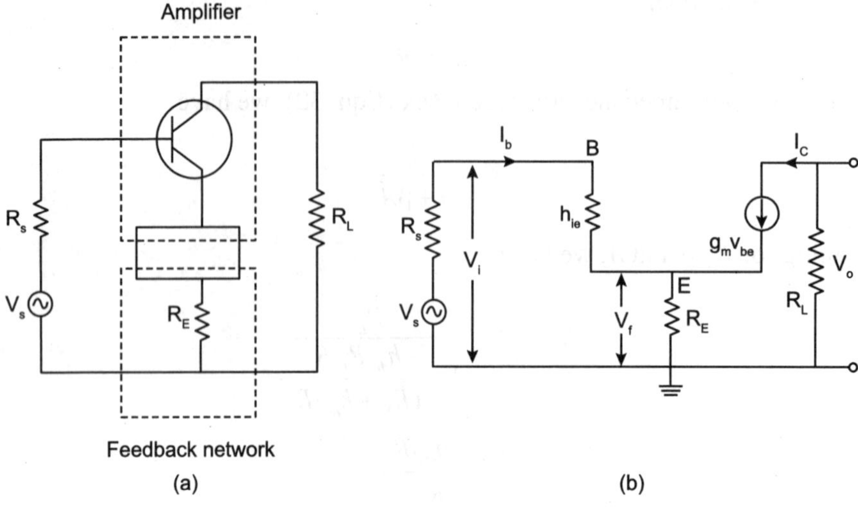

Figure 8.15 ac equivalent circuit of emitter bias amplifier using 'h' parameters.

From Figure 8.15(b),

$$V_f = (I_b + I_c)R_E$$

or

$$V_f = I_C R_E \left(1 + \frac{I_b}{I_c}\right)$$

Current gain of the amplifier,

$$A_i = \frac{I_c}{I_b} = h_{fe}, \text{ by definition}$$

$$\therefore v_f = \left\{\left[\frac{1 + h_{fe}}{h_{fe}}\right] R_E\right\} I_C$$

or

Also from the circuit, $V_{be} = V_i - V_f$; so V_f represents series feedback.

From Figure 8.15(b),

$$v_o = (-)g_m V_{be} R_L$$

Substituting for V_{be}, we have

$$V_{be} = \frac{h_{ie}}{h_{ie} + R_E} V_i$$

Voltage gain, $A = \frac{V_o}{V_i} = (-)\frac{g_m R_L}{1+\frac{R_E}{h_{ie}}}$ (46)

The negative sign represents phase difference of π between v_o and v_i

$$\text{Feedback factor, } \beta = \frac{V_f}{V_o} = \frac{(I_b + I_c)R_E}{g_m V_{bc} R_L}$$

$$= \frac{I_b(1+h_{fe})R_E}{\frac{h_{fe}}{h_{ie}}(h_{ie}I_b)R_L} = \left[\frac{1+h_{fe}}{h_{fe}}\right]\frac{R_E}{R_L}$$

$$= \frac{R_E}{R_L} \quad (47)$$

since $h_{fe} \gg 1$

Voltage gain after feedback (negative feedback) is

$$A_f = \frac{A}{1+\beta A}$$

Substituting for A, β, we have

$$A_f = \frac{(-)g_m \frac{R_L}{1+\frac{R_E}{h_{ie}}}}{1+\left(\frac{g_m R_L}{1+\frac{R_E}{h_{ie}}}\right)\frac{R_E}{R_L}} = (-)\frac{g_m R_L}{1+\frac{R_E}{h_{ie}}+g_m R_E} \quad (48)$$

Normally, $1+g_m R_E \gg \frac{R_E}{h_{ie}}$

So $$A_f = (-)\frac{R_L}{R_E} \quad (49)$$

Input impedance Input impedance without feedback is

$$R_i = R_E + h_{ie} = h_{ie}\left(1+\frac{R_E}{h_{ie}}\right)$$

Input impedance with feedback is

$$R_i' = R_i(1+\beta A)$$

Substituting for R_i, β and A, we have

$$R_i' = h_{ie}\left(1+\frac{R_E}{h_{ie}}\right)\left[1+\frac{\dfrac{g_m R_E}{1+\dfrac{R_E}{h_{ie}}}}{}\right]$$

$$= h_{ie}\left[1+\frac{R_E}{h_{ie}}+g_m R_E\right]$$

$$= h_{ie} + R_E[1+h_{fe}] \text{ since } h_{fe} = g_m h_{ie} \tag{50}$$

$$= h_{fe} R_E$$

Since h_{ie} is very small, R_E is also small.

Hence input impedance after feedback increases.

Output impedance Output impedance of the amplifier without feedback, i.e. there is no R_E in the circuit, is given by

$$R_o = \left[\frac{R_L}{1+h_{oe}R_L}\right]$$

Input impedance with feedback is

$$R_o' = R_o(1+\beta A)$$

Substituting for R_o, β and A, we have

$$R_o' = \underset{R_L\to\infty}{Lt}\left[\frac{R_L}{1+h_{oe}R_L}\left\{1+\frac{R_E}{R_L}\left[\frac{g_m R_L}{1+\dfrac{R_E}{h_{ie}}}\right]\right\}\right]$$

$$= \left[1+\frac{g_m R_E}{1+\dfrac{R_E}{h_{ie}}}\right] \tag{51}$$

Hence output impedance after feedback increases.

The results for various cases of negative feedback are summarized in Table 8.1.

Table 8.1

Characteristics of amplifier	Voltage feedback		Current feedback	
	Series	Shunt	Series	Shunt
Gain	Decreases	Decreases	Decreases	Decreases
Gain stability	Increases	Increases	Increases	Increases
Harmonic distortion	Less	Less	Less	Less
Noise	Less	Less	Less	Less
Bandwidth	Increases	Increases	Increases	Increases
Input impedance	Increases	Decreases	Increases	Decreases
Output impedance	Decreases	Decreases	Increases	Increases

We have discussed the theory of negative feedback amplifiers using transistor, which we have represented as two-port networks (four terminals, Figure 8.1). But the same theory applies for operational amplifiers, which are normally represented as three-terminal networks where we have two terminals for input and one for output. Such an Op-Amp having feedback network is shown in Figure 8.16.

8.4 OP Amplifiers Having Feedback

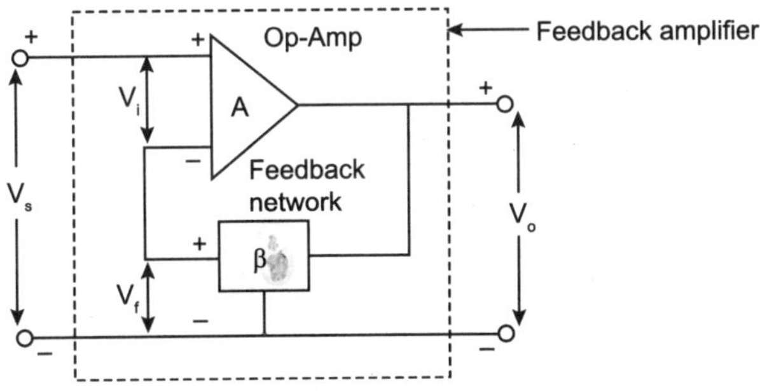

Figure 8.16 Feedback amplifier using Op-Amp

From the polarity of V_f shown in Figure 8.16, it represents negative feedback amplifier and

$$V_i = V_s - V_f$$

Eqn. (5) for gain after negative feedback, $A_f = \frac{A}{1+\beta A}$, applies here too; and similarly other relations (30) and (32) apply for input and output impedance, respectively.

With the above set up, an expression for input impedance and output impedance can be easily derived.

Input impedance

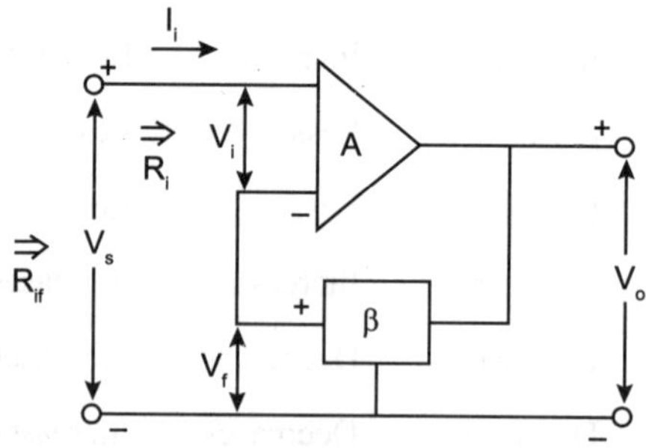

Figure 8.17 Series voltage negative feedback

Input impedance (resistance) without feedback is

$$R_i = \frac{v_i}{I_i} \quad \text{(i)}$$

$$\text{or } I_i = \frac{v_i}{R_i}$$

Input impedance with feedback is

$$R_{if} = \frac{v_s}{I_i}$$

Substituting for I_i, we have

$$R_{if} = \frac{v_s}{v_i} R_i \quad \text{(ii)}$$

Also from figure,

$$v_i = v_s - v_f = v_s - \beta v_o = v_s - \beta A v_i$$

or

$$v_i(1+\beta A) = v_s$$

or

$$\frac{v_i}{v_s} = \frac{1}{1+\beta A} \quad \text{(iii)}$$

Substituting in (ii), we have

$$R_{if} = R_i(1+\beta A) \quad \text{(iv)}$$

The above relation is same as that given by Eqn. (30).

Output impedance

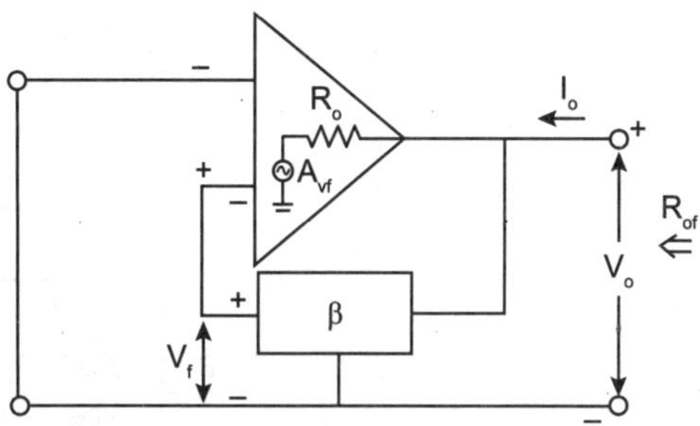

Figure 8.18

Here the input terminals are short-circuited so that feedback voltage v_f is now input voltage to the amplifier. Let output resistance of the amplifier be R_o; when output voltage is changed by v_o, feedback voltage appears as amplifier input, producing a voltage (–) Av_f in series with R_o. Assuming instantaneous positive polarity for v_o as shown, instantaneous current I_o would occur in the direction shown.

From the circuit, applying KVL at output,

$$v_o = I_o R_o - Av_f$$

or

$$I_o R_o = v_o + Av_f$$

$$= v_o + A\beta v_o$$

or

$$R_o = \frac{v_o}{I_o}(1+\beta A) \qquad \text{(i)}$$

Also

$$R_{of} = \frac{v_o}{I_o}$$

Substituting for $\frac{V_o}{I_o}$ from (i), we have

$$R_{of} = \frac{R_o}{1+\beta A} \qquad \text{(ii)}$$

This equation resembles Eqn. (32).

8.5 Darlington Connection

In many applications, we require very high input impedance for instruments like multimeter and voltmeter. This is desirable in many cases especially when a high gain (≈ 50,000) Op-Amp is used, for very high input impedance for its operation in analog applications.

A common collector or emitter follower has 100% negative feedback, thereby it has largest input impedance and lowest output impedance of all three amplifiers.

A darlington connection can be considered as two cascaded emitter followers, with the first having infinite emitter resistance as shown in Figure 8.19.

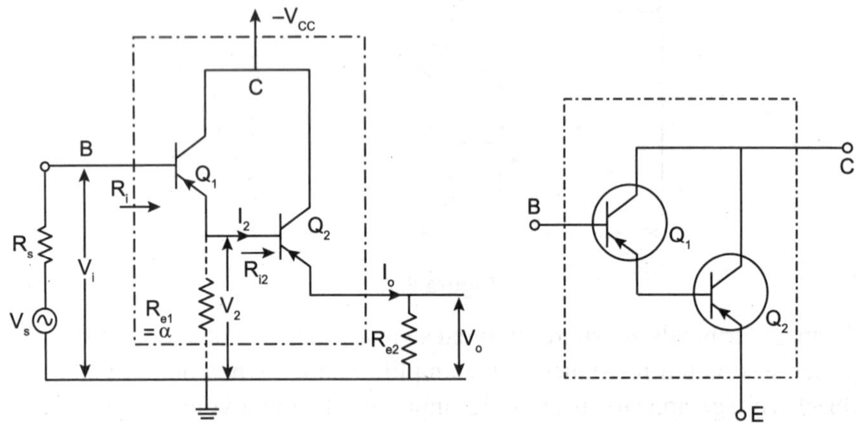

Figure 8.19 High input impedance amplifier utilizing darlington connection (biasing resistors have not been shown)

Assuming the operating point for transistor Q_1 to be correct by employing biasing resistors (not shown), we have two transistors Q_1 and Q_2 cascaded as shown.

Since there is no connection to external circuitry from emitter of Q_1 or base of Q_2, these two transistors can be fabricated as a single device and denoted as darlington pair with three terminals B, C and E as shown such that the pair has voltage capability of $V_{CE} = 60$ V and h_{fe} between 7000 and 70,000 at collector current 100 mA. Its power capability is 500 mW.

Resistance R_{e2} connected to the emitter of transistor Q_2 acts as load for it.

Current gain for this conventional emitter follower is

$$A_{i2} = \frac{I_o}{I_2} = \frac{1+h_{fe}}{1+h_{oe}R_{e2}} = 1 + h_{fe} \tag{52}$$

since $h_{oe}R_{e2} \ll 1$

Input impedance,

$$R_{iz} = h_{ie} + (1+h_{fe})R_{e2}$$
$$= (1+h_{fe})R_{e2} \tag{53}$$

assuming h_{ie} to be small enough.

But R_{i2} is the effective load resistance for first transistor Q_1 so current gain at this stage is

$$A_{i1} = \frac{I_2}{I_1} = \frac{1+h_{fe}}{1+h_{oe}R_{i2}} \tag{54}$$

Substituting for R_{i2} from Eqn. (53), we have

$$A_{i1} = \frac{1+h_{fe}}{1+h_{oe}(1+h_{fe})R_{e2}} = \frac{1+h_{fe}}{1+h_{oe}h_{fe}R_{e2}} \tag{55}$$

assuming $h_{oe}\,R_{e2}$ to be small.

The over all current gain is

$$A_i = \frac{I_o}{I_1} = \frac{I_o}{I_2} \times \frac{I_2}{I_1}$$
$$= A_{i2} \times A_{i1}$$

Substituting for A_{i2} and A_{i1}, we have

$$A_i = (1+h_{fe})\left[\frac{1+h_{fe}}{1+h_{oe}h_{fe}R_{e2}}\right]$$

$$A_i = \frac{(1+h_{fe})^2}{1+h_{oe}h_{fe}R_{e2}} \tag{56}$$

Input impedance of the device is

$$R_i = \frac{A_i R_{e2}}{A_v} = A_i R_{e2} \text{ since } A_v = 1$$

Substituting for A_i, we have

$$R_i = \frac{(1+h_{fe})^2 R_{e2}}{1+h_{oe}h_{fe} R_{e2}} \tag{57}$$

The overall voltage gain for a darlington circuit is less than unity, since it consists of two emitter followers in cascade as shown in Figure 8.19, each having A_v less than unity.

$$A_v = \frac{V_o}{V_i} \approx 1 - \frac{h_{ie}}{R_{i2}} \tag{58}$$

Output impedance is

$$R_o = \frac{R_S + h_{ie}}{(1+h_{fe})^2} + \frac{h_{ie}}{1+h_{fe}} \tag{59}$$

Eqn. (59) represents that output impedance is very low for darlington connections.

8.5.1 Stability

We know that if $|1+A\beta|<1$, the feedback is termed as positive or regenerative. Under this condition the resultant transfer gain A_f is greater than internal gain A, which is the nominal gain without feedback.

Thus, regeneration is an effective means of increasing the amplification of an amplifier, as was first suggested by Armstrong. However, because of reduced stability of an amplifier with positive feedback, this method is seldom used. To illustrate instability in an amplifier with positive feedback, we consider the following situation:

Let there be no external signal v_s, but because of transient disturbance, a signal x_o appears at output. A fraction of it (–) βx_o is fed back to input, and it will appear in output as an amplified signal $-A\beta x_o$. If this term just equals x_o,

then $$(-)A\beta x_o = x_o$$

or $$(-)A\beta = 1$$

Therefore, the amplifier will oscillate since $A_f \Rightarrow \infty$ in this situation.

Thus, if an attempt was made to obtain a large gain by making $|\beta A|$ almost equal to 1, there is a possibility of the amplifier to break out and result into spontaneous oscillations.

In the second situation, the amplifier is designed to have negative feedback in a particular frequency range, but breaks out into oscillations at some other (low or high) frequency. Hence in designing a feedback amplifier, it must be ascertained that the circuit is stable at all frequencies and not merely over the frequency range of interest. A system can be stable if transient disturbance results in a response that dies out.

Therefore, in order for a system to be stable, the poles of the transfer function must all lie in the left half of the complex frequency plane. If the system without feedback is stable, the poles of A do lie in the left half plane. It follows from Eqn. (5), therefore, that stability condition requires that zeros of $[1 + \beta A]$ all lie in the left of the complex plane.

8.6 The Nyquist Criterion

Nyquist obtained an alternative but equivalent condition for stability, which may be expressed in terms of study state or frequency response characteristic. It is given here without proof.

Since the product βA is a complex number, it may be represented as a point in complex plane, the real component being plotted along X-axis and the j component (imaginary component) along Y-axis. Furthermore, βA is a function of frequency. Consequently, points in the complex plane are obtained for the values of βA corresponding to all values of f from (–)∞ to +∞. The locus of all such points forms a closed curve.

The criterion of Nyquist is that the amplifier is unstable if this curve enclose the point (–)1 $+ j_o$, and the amplifier is stable if the curve did not enclose this point.

The criterion for positive or negative feedback may also be represented in the complex plane. We see that when $|1+\beta A| = 1$, a critical condition exists. It may be written as $(1-a)^2 + b^2 = 1$.

This equation represents a circle of unit radius, with its centre at $-1 + jo$ (Figure 8.20). If for any frequency, vector βA extends out side the circle, the feedback is negative; then $|1+\beta A| > 1$

and amplifier is stable. If, however, βA lies within the circle (Figure 8.20(a)) then $|1+\beta A|<1$ and feedback is positive. In the latter case, the system is unstable and will not oscillate unless Nyquist criterion is satisfied.

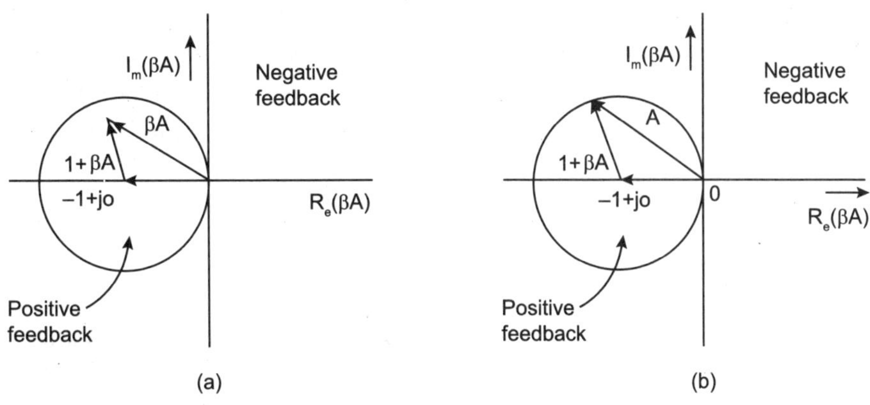

Figure 8.20 (a) Vector βA ends in circle, (b) Vector βA touches circle

Summary

1. If an amplifier having voltage gain A has some of the output voltage fed back to the input, then voltage gain with feedback is given by $A_f = \dfrac{A}{1-\beta A}$ where $\beta(=v_f/v_o)$ is called feedback factor, which is negative for negative feedback and positive for positive feedback.
2. Although negative feedback reduces the gain, the result is stabilization. Since $A_f = \dfrac{1}{\beta}$ when βA is very large than unity, the gain simply depends upon feedback factor and not on any parameter of the amplifier. The product βA is called open loop gain and the term $[1-\beta A]$ is called return difference.
3. Feedback in dB (decibel) is defined as $\text{dB} = 20\log\dfrac{A_f}{A} = (-)20\log(1-\beta A)$. The advantages of negative feedback on input and output impedance of the amplifier depends upon the type of negative feedback employed: (i) voltage series feedback, (ii) voltage shunt feedback, (iii) current series feedback, (iv) current shunt feedback.
4. A unbypassed resistor in the emitter lead of CE amplifier is an example of current series feedback that increases both input and output resistance of amplifier, and feedback factor

$$-\beta = \frac{R_E}{R_L}$$

5. A common collector or emitter follower is an example of voltage series feedback that increases input resistance but decreases output resistance.
6. A voltage shunt feedback, such as a resistor connected from collector to base in CE amplifier, reduces both input and output impedance of the amplifier.
7. A darlington connection involves tying together the collectors of two transistors and emitter of first to the base of second to provide a three-terminal device with very high current gain.
8. Input resistance and current gain of darlington amplifier is very high, its output resistance is very low, and its voltage gain is approximately one.

Review Questions

1. Explain the working of a feedback amplifier and prove that voltage gain after feedback is $A_f = \dfrac{A}{1-\beta A}$.
2. How do you differentiate between positive feedback and negative feedback in an amplifier. Illustrate your answer with a suitable diagram.
3. What is effective negative feedback on nonlinear distortion in an amplifier? Device appropriate relation.
4. What are the advantages of employing negative feedback in an amplifier? How it affects noise in an amplifier?
5. Discuss the various types of feedback employed in an amplifier? How input impedance and output impedance of the amplifier is affected by employing those techniques?
6. What do you mean by feedback? Define (i) open loop gain (ii) closed loop gain (iii) return difference. How do we define feedback in dB.
7. Derive a relation for voltage amplification after feedback in an amplifier. How does the negative feedback affect the satiability of the amplifier? Explain.
8. Draw the circuit of an emitter follower. Using the principle of feedback, derive a relation for its voltage gain, input impedance and output impedance.
9. Prove that input impedance increases and output impedance decreases on employing negative voltage series feedback in an amplifier.
10. Draw a circuit of an amplifier employing current series feedback. How its input impedance and voltage gain is affected? Derive a necessary relation.
11. What type of feedback is employed in an emitter follower? Explain and sketch the circuit showing feedback network and amplifier circuit? Derive a necessary relation for its voltage gain.

12. Describe the working of an emitter bias amplifier. Draw the necessary circuit and analyze it mathematically. What type of feedback is employed there?
13. Draw the circuit of a fixed bias amplifier. What type of feedback is employed there? Give necessary theory and derive relation for (i) voltage amplification (ii) input and output impedance.
14. Draw the circuit of an amplifier employing voltage shunt feedback. Give its necessary theory and derive relation for (i) voltage gain (ii) input impedance.
15. How is the (i) upper half power frequency and (ii) lower half power frequency of an amplifier affected by employing negative feedback in an amplifier. Show the effect by drawing frequency response of the amplifier; (a) with feedback (b) without feedback. What is the new bandwidth of the amplifier?
16. (a) What does a Darlington connection consist of essentially? (b) What are the primary features of a Darlington amplifier?
17. (a) How is current series feedback introduced in a CE amplifier? (b) What is an approximate expression for feedback factor? (c) Derive expressions for (i) voltage gain (ii) input resistance (iii) output resistance. How these values differ from that of a CE amplifier?
18. Mention the various types of negative feedback that can be utilized in an amplifier. (a) Which type always increases input impedance? (b) Which type increases input resistance but decreases output resistance.
19. (a) Why is voltage gain of an amplifier with negative feedback smaller than that with no feedback? Explain. (b) Does certain percentage of feedback produce a fixed percentage of reduction in gain? If yes, why?

Problems

1. The internal voltage gain of an amplifier can vary ± 10%. Calculate its variation when 20 dB feedback is employed in it. [**Ans:** ±1%]
2. An amplifier having gain of 200 produces a distortion of 8%. Calculate the feedback factor of the feedback network so that distortion reduces to 0.5%. Also calculate its gain after feedback. [**Ans:** (a) 0.075, (b) 12.5]
3. The voltage gain of an amplifier is 100 and its bandwidth is 2 MHz. Negative feedback by feedback network having feed ratio 1/20 is employed in it. Calculate its voltage gain and bandwidth. [**Ans:** (a) 16.67, (b) 12 MHz]
4. The voltage gain of an amplifier is 200 and its bandwidth is 2 MHz. Voltage gain reduces to 100 by employing a negative feedback. Calculate its bandwidth and feedback ratio. [**Ans:** (a) 4 MHz, (b) 0.005]

5. An amplifier with negative feedback has $\beta = \dfrac{1}{105}$ and produces 2 V output when a signal of 20 mV is employed as input. Distortion content in the output is 1%. If negative feedback circuit is disconnected, find the new level of signal to give same output. Also calculate the new distortion percentage.

   [Ans: (a) 0.95 mV, (b) 21%]

6. An amplifier has voltage gain of 100. A technician decides that 10% negative feedback be employed to reduce distortion. (a) What is the voltage gain after feedback? (b) Finding the voltage gain to be excessively low, he decides to introduce negative feedback of 5%? Will the new gain be double now?

   [Ans: 9.1, No, 16.6]

7. An amplifier has bandwidth of 200 KHz and voltage gain of 100. If a 5% negative feedback is introduced, what will be the (a) new gain (b) new bandwidth. (c) What should be the amount if feedback of the bandwidth is to be restricted to 1 MHz.

   [Ans: (a) 16.7, (b) 1.2 MHz, (c) (–) 4%]

8. An amplifier with 2.2 kΩ input impedance and 40 kΩ output impedance has a voltage gain of 80. The amplifier is modified to provide 10% negative voltage series feedback. Calculate: (a) voltage gain with feedback (b) input impedance (c) output impedance. (d) If we introduce negative feedback as current series feedback, which of the above quantities will change.

   [Ans: (a) 8.89, (b) 19.8 kΩ (c) 4.4 kΩ]

9. An amplifier has (–)40 dB of feedback introduced. (a) What is the ratio of voltage gain before feedback to gain with feedback? (b) Determine the value of closed loop gain. (c) If $\beta = (-)5\%$, what is the gain before and after feedback?

   [Ans: (a) 100, (b) (–)99, (c) 1980, 19.8]

10. An amplifier has a voltage gain of 40 and has (–)14 dB of feedback introduced. Calculate (i) voltage gain with feedback (ii) closed loop gain (iii) percentages of feedback.

    [Ans: (i) 8, (ii) 4 (iii) 1/10]

11. A transistor with $h_{ie} = 1.5$ kΩ and $h_{fe} = 75$ is used in a emitter follower circuit having $R_E = 860$ Ω and biasing resistors $R_1 \| R_L = 20$ kΩ. Calculate: (a) A_i (b) R_i (c) R_i' (d) A_v (e) R_o (f), R_o' assuming input voltage signal source resistance $R_s = 1$ kΩ.

    [Ans: (a) 76, (b) 66.9 kΩ, (c) 15.4 kΩ, (d) 0.98, (e) 32.9 Ω, (f) 31.7 Ω]

12. An amplifier employs collector to base biasing to provide stabilization of operating point and negative feedback results. Assume $R_f = R_B = 80$ kΩ and load $R_C = R_L = 1$ kΩ, $h_{ie} = 2$ kΩ and $h_{fe} = 75$. Assume input signal source resistance $R_s = 1$ kΩ. Calculate (a) R_o' (b) A_v (c) R_{of} (d) R_{of}'.

    [Ans: (a) 1.02 kΩ, (b) (–)37, (c) 3200 Ω, (d) 760 Ω]

13. An amplifier utilizes current series feedback. The transistor used has $h_{fe} = 100$, $h_{ie} = 2\ \text{k}\Omega$ $R_L = R_C = 470\ \Omega$ and R_E consists of two 100 Ω resistors in series, so that both or one may be bypassed by a capacitor C_E which has negligible reactance. Assuming biasing resistors $R_2 = 5.6\ \text{k}\Omega$ and $R_1 = 15\ \text{k}\Omega$, Calculate (a) voltage gain (b) R_i when complete R_E is bypassed. (c) If one R_E is only bypassed by C_E, determine feedback factor β (d) feedback in dB.

[**Ans:** (a) 23.5, (b) 2 kΩ, (c) $\beta = 0.213$, (d) 15.6 dB]

Short Answer Questions

1. What is the effect of negative feedback on the bandwidth of an amplifier?

   Ans: The bandwidth of an amplifier increases on employing negative feedback.

2. Does the gain bandwidth product remain constant in a feedback amplifier?

   Ans: Yes.

3. What is the affect of negative feedback on nonlinear distortion in an amplifier?

   Ans: Non linear distortion existing in the output of an amplifier reduces on employing a negative feedback.

4. Illustrate an example of negative feedback where input impedance increases and output impedance decreases? And where both increase?

   Ans: When we employ voltage series feedback, input impedance increases and output impedance decreases. On employing current series feedback, both input and output impedance increase.

5. What type of feedback is employed in emitter follower? What is the maximum value of voltage amplification in it?

   Ans: If emitter follower voltage series feedback is employed, the maximum value of voltage amplification is unity.

6. What is the value of feedback in dB?

   Ans: Feedback in dB is defined as

$$\text{NdB} = 20\log\frac{A}{A_f} = 20\log(1-\beta A)$$

7. Can the gain after feedback be more than the without feedback? If yes under what condition it is possible?

   Ans: Yes, it is possible by employing positive feedback in an amplifier.

8. How is the (i) input impedance (ii) output impedance of an amplifier affected by employing current series negative feedback?

   Ans: Both input impedance and output impedance of an amplifier increase on employing current series negative feedback.

9. What is the relation between $\frac{dA_f}{A_f}$ and $\frac{dA}{A}$ in case of negative feedback amplifier?

Ans: The two are related as

$$\frac{dA_f}{A_f} = \left[\frac{1}{1+\beta A}\right]\frac{dA}{A}$$

10. What is the nature of A and β in a feedback amplifier?

Ans: Both A and β are complex in a feedback amplifier. They depend upon the frequency employed.

11. What does βA represent in a feedback amplifier?

Ans: βA represents return ratio in a feedback amplifier.

12. What is closed loop gain and open loop gain in a feedback amplifier?

Ans: Gain after feedback A_f is defined as closed loop gain and gain without feedback A is defined as open loop gain.

Objective Questions

1. The bandwidth and gain of an amplifier using negative feedback
   (a) decreases, increases respectively
   (b) decreases, decreases respectively
   (c) increase, decreases respectively
   (d) increases, increases respectively **[Ans: (c)]**

2. Emitter follower is an amplifier employing
   (a) voltage series feedback
   (b) current series feedback
   (c) voltage shunt feedback
   (d) current shunt feedback **[Ans: (a)]**

3. The gain of an amplifier is 600; it reduces to 100 on employing negative feedback. The feedback factor is
   (a) (+) 0.008 (b) (–) 0.008 (c) (+) 0.011 (d) (–) 0.011

   [Ans: (b)]

4. The gain of an amplifier is 500; it reduces to 100 on employing negative feedback. The value of feedback employed in dB is
   (a) (–) 20 $\log_{10}5$ (b) (–) 20 $\log_e 5$ (c) (–) 20 $\log_{10}4$ (d) (–) 20 $\log_e 4$

   [Ans: (a)]

5. If return ratio in a feedback amplifier is quite large, then voltage gain of feedback amplifier is

   (a) $\dfrac{A}{1+\beta A}$ (b) $\dfrac{A}{1-\beta A}$ (c) $\dfrac{1}{\beta}$ (d) Zero

   [Ans: (c)]

6. The gain of an amplifier is 100. If return ratio in the amplifier employing negative feedback is also 100, then feed factor is

   (a) 1 (b) $\dfrac{1}{10}$ (c) $\dfrac{1}{100}$ (d) $\dfrac{1}{10000}$

   [Ans: (a)]

7. Voltage gain and stability of an amplifier is expressed as A and S, respectively. On employing negative feedback, their respective values are A_f and S_f. Their relation is expressed as

   (a) $A_f < A, S_f < S$ (b) $A_f < A, S_f > S$ (c) $A_f > A, S_f < S$ (d) $A_f > A, S_f > S$

   [Ans: (b)]

8. On employing negative feedback in an amplifier, the value of return difference is

   (a) 1 (b) more than 1 (c) less than 1 (d) zero

   [Ans: (b)]

9. On employing negative feedback in an amplifier, the lower and upper power frequencies are respectively given by

   (a) $f_1' = \dfrac{f_1}{1+\beta A}, f_2 = \dfrac{f_2}{1+\beta A}$ (b) $f_1' = f_1(1+\beta A),\ f_2' = \dfrac{f_2}{1+\beta A}$

   (c) $f_1' = \dfrac{f_1}{1+\beta A}, f_2' = \dfrac{f_2}{1+\beta A}$ (d) $f_1' = f_1(1+\beta A),\ f_2' = f_2(1+\beta A)$ **[Ans:** (c)]

10. In an emitter follower, the maximum value of voltage gain is

    (a) 1 (b) less than 1

    (c) more than 1 (d) any value **[Ans:** (a)]

Mark the Following Statements True/False.

1. The maximum value of voltage gain of emitter follower is 1. **[Ans:** True]
2. Voltage gain in a negative feedback amplifier employing large return ratio depends only one feedback network. **[Ans:** True]
3. Input impedance increases in an amplifier both in employing current series negative feedback and current shunt feedback. **[Ans:** False]
4. On employing negative feedback in an amplifier, both its gain and bandwidth decrease.

   [Ans: False]

5. On employing negative feedback in an amplifier, its lower half frequency decreases but upper half power frequency increases. [**Ans:** True]
6. The value of feedback ratio β in a feedback amplifier is generally less than unity. [**Ans:** True]
7. To increases bandwidth of an amplifier, we employ positive feedback in it. [**Ans:** False]
8. Both stability and harmonic distortion in an amplifier decrease in a negative feedback amplifier. [**Ans:** False]
9. On employing negative feedback in an amplifier, its harmonic distortion decreases but noise remains constant. [**Ans:** False]
10. Input impedance and output impedance of emitter follower are very low. [**Ans:** False]
11. Emitter follower has highest input impedance and lowest output impedance among various negative feedback amplifiers. [**Ans:** True]

Fill in the Blanks.

1. The maximum value of voltage gain of emitter follower is ____________. [**Ans:** unity]
2. On employing negative feedback in an amplifier, its bandwidth ____________ but gain ____________. [**Ans:** increases, decreases]
3. On employing negative feedback in an amplifier, its stability ____________ but harmonic distortion ____________. [**Ans:** increases, reduces]
4. Input impedance of emitter follower is ____________ and its output impedance is ____________. [**Ans:** high, low]
5. The gain of an amplifier employing feedback is independent of amplifier ____________ when ____________ is ____________. [**Ans:** internal gain, return ratio, large]
6. In an emitter follower, feedback voltage is ____________ to output voltage. [**Ans:** equal]
7. In an amplifier employing negative feedback, the lower half power frequency ____________ but the upper half power frequency ____________. [**Ans:** decreases, increases]
8. Input impedance of a feedback amplifier increases in ____________ feedback but decreases in ____________. [**Ans:** voltage series, voltage shunt]
9. Output impedance of a feedback amplifier ____________ both in ____________ feedback as well as ____________. [**Ans:** decreases, voltage series, voltage shunt]

Chapter 9

Negative Feedback Amplifier (Using Op-Amp)

9.1 Introduction

Harold Black invented the negative feedback amplifier which can stabilize voltage gain, increase input impedance and decrease output impedance. With the advent of Op-Amps, more types of negative feedback became available.

The input to a negative feedback amplifier can be either voltage or current signal. Also, the output of the amplifier can be either voltage or current signal. In this way, we can have four types of negative feedback as follows:

(i) Voltage-controlled voltage source (VCVS)
(ii) Current-controlled voltage source (ICVS)
(iii) Voltage-controlled current source (VCIS)
(iv) Current-controlled current source (ICIS)

9.2 Voltage-Controlled Voltage Source

Here, the input is voltage signal and output is also voltage signal. It is an ideal voltage amplifier because it has stabilized voltage gain, infinite input impedance and zero output impedance. For practical purposes, input impedance is high and output impedance is low (Figure 9.1).

9.3 Current-Controlled Voltage Source

Here, the input current controls output voltage. It is sometimes known as transresistance amplifier. The word resistance is used because the ratio V_{out}/I_{in} has the unit ohms. The prefix "trans" refers to the ratio of an output quantity to input quantity. Here input impedance is low and output impedance is also low (Figure 9.1(b)).

9.4 Voltage-Controlled Current Source

Here, input voltage controls output current. It is sometimes called transconductance amplifier. The term conductance is used because the ratio I_{out}/V_{in} has the unit of Mhos. Here, input impedance is high and output impedance is also high (Figure 9.1(c)).

9.5 Current-Controlled Current Source

Here, input current is amplified to get a large output current. It is an ideal current amplifier because of stabilized current gain, zero input impedance and infinite output impedance. For practical purposes, input impedance is low and output impedance is high (Figure 9.1(d)).

9.6 Converters

Referring to VCVS and ICIS circuits as amplifiers, makes sense because the first is a voltage amplifier and the second is current amplifier. The combination of the term amplifier with transresistance or conductance is not appropriate because both input and output quantities are different. So many prefer to use the term "converter" with it.

In this respect, VCIS is called voltage-to-current converter and ICVS is called current-to-voltage converter.

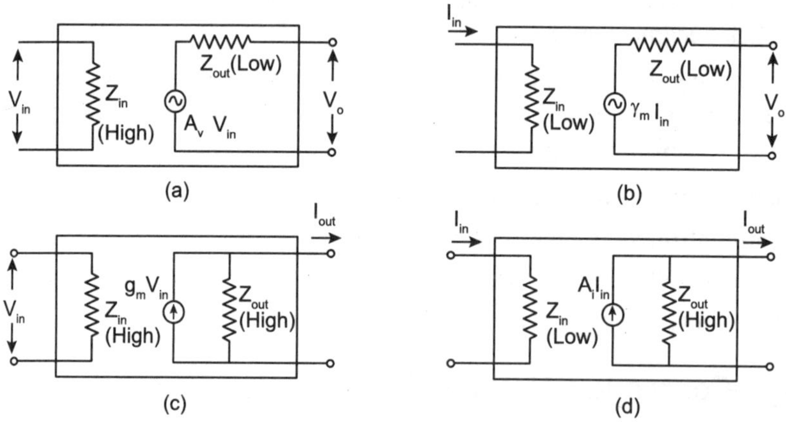

Figure 9.1 (a) VCVS, (b) ICVS, (c) VCIS, (d) ICIS

As mentioned earlier, there are four types of negative feedback: (i) VCVS, (ii) ICVS, (iii) VCIS, (iv) ICIS. Of these, VCVS and VCIS are controlled by an input voltage signal and ICVS and ICIS are controlled by an input current signal. The output side of VCVS and ICVS acts like voltage source and the output side of VCIS and ICIS acts like current source.

9.6.1 VCVS Amplifier

For this amplifier, loop gain is the voltage gain of forward and feedback paths. In any practical design, loop gain should be very high. As a result, the closed loop voltage gain is ultrastable because it no longer depends upon the characteristics of the amplifier, but it depends only upon the characteristics of external resistors which are very precise.

In VCVS, negative feedback has a curative effect on the flow of an amplifier because it stabilizes voltage gain, increases input impedance, decreases output impedance and decreases harmonic distortion.

9.6.2 ICVS Amplifier

This is a transresistance amplifier, also known as current-to-voltage converter. Because of virtual ground, it ideally has zero input impedance. The input current produces a precise value of output voltage.

9.6.3 VCIS Amplifier

This is a transconductance amplifier, also known as voltage-to-current converter. It ideally has infinite input impedance. The input voltage produces a precise value of output current. The output impedance ideally approaches infinity.

9.6.4 ICIS Amplifier

Because of heavy negative feedback, the ICIS amplifier approaches a perfect current amplifier, having zero input impedance and infinite output impedance.

Negative feedback increases the bandwidth of an amplifier because roll off in open loop voltage gain means less voltage is fed back, which produces more input voltage as compensation. Because of this, the closed loop cut-off frequency is higher than the open loop cut-off frequency.

9.7 Voltage-Controlled Voltage Sources

9.7.1 Voltage Gain

The circuit of a VCVS is shown in Figure 9.2. As seen, the circuit represents a non-inverting amplifier.

Such an Op-Amp has open loop voltage gain of the order of 100,000. The feedback voltage, V_f, which is fed to the inverting input. The feedback fraction for the circuit is given by

$$\beta = \frac{v_f}{V_o} = \frac{R_1}{R_1 + R_2}$$

Feedback fraction is also called feedback attenuation factor, as it indicates how much out put voltage is attenuated before feedback.

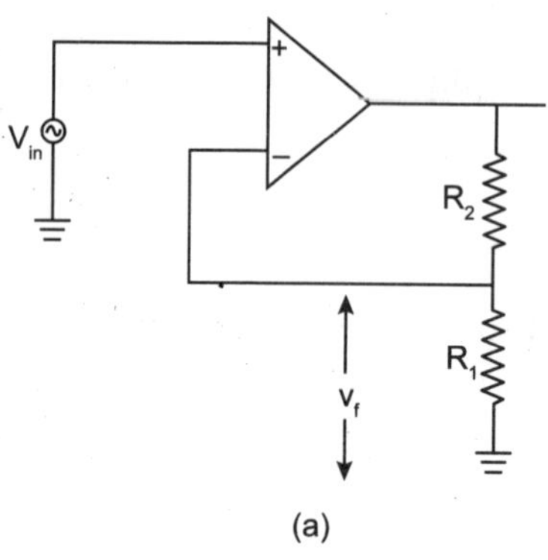

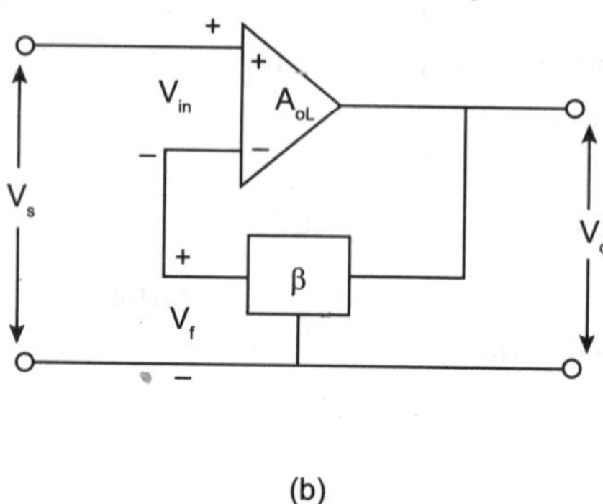

Figure 9.2 Voltage-controlled voltage source (non-inverting amplifier)

Looking at Figure 9.2(b), the closed loop voltage gain can be derived as

$$v_o = A_{\text{OL}}\ v_{\text{in}} \tag{1}$$

where A_{OL} is open loop voltage gain and $v_{\text{in}} = V_S - V_f$ with polarity shown

or
$$V_{\text{in}} = V_S - \beta V_o \tag{2}$$

where β represents a fraction with which output V_o has been attenuated.

Substituting for v_{in} from (2) in (1),

$$V_o = A_{\text{OL}}(V_S - \beta V_o)$$

or
$$V_o[1 + \beta A_{\text{OL}}] = A_{\text{OL}}\ V_S$$

Hence, closed loop voltage gain (also called feed back voltage gain) is

$$A_{\text{CL}} = \frac{v_o}{v_s} = \frac{A_{\text{OL}}}{1 + \beta A_{\text{OL}}} \tag{3}$$

Eqn. (3) shows that $A_{\text{CL}} < A_{\text{OL}}$.

Here, the term $\beta[A_{\text{OL}}]$ in Eqn. (3) is called loop gain because it is voltage gain of the forward and feedback paths.

If $\beta\, A_{\text{OL}}$ is quite high (which is usual in circuit design), then Eqn. (3) can be written as

$$A_{\text{CL}} = \frac{1}{\beta} = \frac{R_1 + R_2}{R_1} = 1 + \frac{R_2}{R_1} \tag{4}$$

This equation is familiar for non-inverting amplifier and it suggests that the gain does not depend upon the Op-Amp, rather it depends only on external parameters. However, the exact closed loop voltage gain will be slightly less than the ideal closed loop voltage gain given by Eqn. (4). Percent error between the two is given by

$$\%\,\text{Error} = \frac{100\%}{1 + \beta A_{\text{OL}}}$$

Negative feed-back has a *curative effect* on the flow or short-comings of an amplifier in the same way as how open loop gain may show wide variations from one Op-Amp to the other of the same make. In such a case, negative feedback stabilizes the voltage gain, i.e. it almost eliminates internal Op-Amp variations and makes closed loop voltage gain dependent primarily on external resistances as given by Eqn. (4). These resistances can be precision resistances with very low temperature coefficient, which results into ultrastable closed loop voltage gain.

Input impedance Negative feedback in a VCVS amplifier increases input impedance.

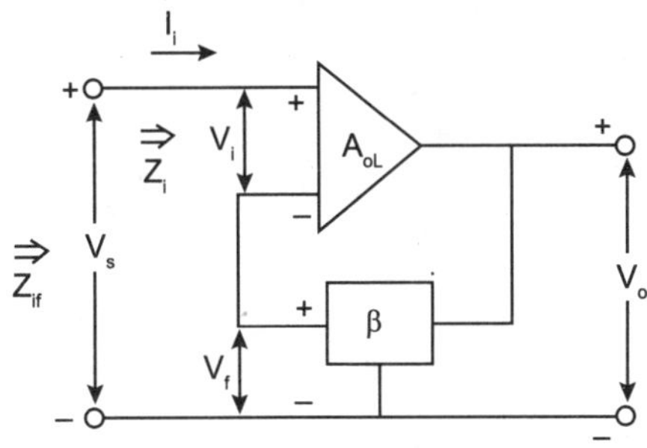

Figure 9.3

The input impedance without feedback is usually impedance Z_i seen in an Op-Amp. From Figure 9.3, we have

$$Z_i = \frac{V_i}{I_i}$$

or

$$I_i = \frac{V_i}{Z_i} \tag{5}$$

With negative feedback present, input impedance is

$$Z_{if} = \frac{V_s}{I_i}$$

Substituting for I_i,

$$Z_{if} = \frac{V_s Z_i}{V_i} \tag{6}$$

From the circuit (Figure 9.3), we have

$$V_i = V_s - V_f$$
$$= V_s - \beta v_o$$

where β is the feedback factor

or
$$V_i = V_s - A_{OL}\,\beta V_i$$

or
$$V_i[1 + \beta A_{OL}] = V_S$$

Substituting the value of V_s in Eqn. (6),

$$Z_{if} = \frac{V_i[1+\beta A_{OL}]}{V_i} Z_i$$

$$= (1+\beta A_{OL}) Z_i \qquad (7)$$

Eqn. (7) shows that after negative feedback, input impedance increases.

Output impedance Negative feedback in a VCVS amplifier decreases output impedance.

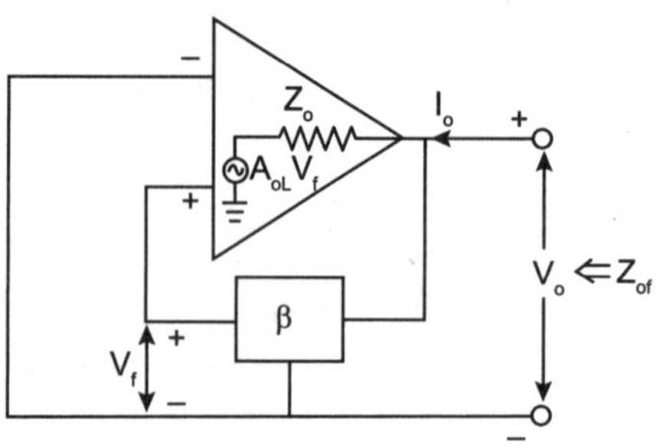

Figure 9.4

To determine output impedance, here the input terminals are shorted (Figure 9.4) so that the input voltage is equal to v_f, the feedback voltage. This input voltage appears as output voltage, $A_{OL}\, v_f$, in the amplifier having an output impedance of Z_o.

From the circuit (Figure 9.4),

$$v_o = i_o Z_o - A_{OL}\, v_f$$

$$\text{or} \quad i_o Z_o = v_o + A_{OL}\, v_f$$

$$= v_o + A_{OL}\, \beta v_o$$

$$= v_o[1+\beta A_{OL}]$$

Output impedance after feedback is given by

$$Z_{of} = \frac{v_o}{i_o} = \frac{Z_O}{1+\beta A_{OL}} \qquad (8)$$

Eqn. (8) shows that output impedance of a VCVS amplifier decreases by a factor $[1 + \beta A_{OL}]$. This is the same factor by which voltage gain (Eqn. 3) reduces and input impedance increases.

9.7.2 Non-Linear Distortion

In the later stages of an amplifier, non-linear distortion occurs with large input signal because the input/output response of the amplifying device becomes non-linear. For instance, the nonlinear graph of base emitter diode distorts large signals by elongating the positive half cycle and compressing the negative half cycle of input signal (Figure 9.5).

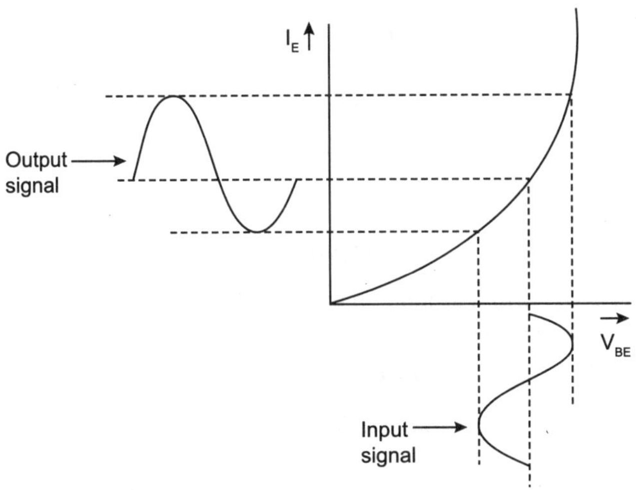

Figure 9.5 Nonlinear distortion

Nonlinear distortion shows that the output is not a pure sine wave as in the input signal. It implies that, in the output we have harmonics of the input signal. Because of this, nonlinear distortion is called harmonic distortion. By employing negative feedback, harmonic distortion can be reduced.

Such a distortion can be measured using a distortion analyzer. We get percent Total Harmonic Distortion (THD), which is defined as

$$\text{THD} = \frac{\text{Total harmonic voltage}}{\text{Fundamental voltage}} \times 100\%$$

It may be proved that

$$\text{THD}_{\text{CL}} = \frac{\text{THD}_{\text{OL}}}{1+\beta A_{\text{OL}}} \tag{9}$$

where THD_{OL} is open loop harmonic distortion

THD_{CL} is closed loop harmonic distortion.

9.8 Current-Controlled Voltage Source Amplifier

It is also known as current-to-voltage converter because it has zero input impedance and zero output impedance under ideal conditions.

Figure 9.9 shows a transresistance amplifier. It has current source as input, and in output we get voltage.

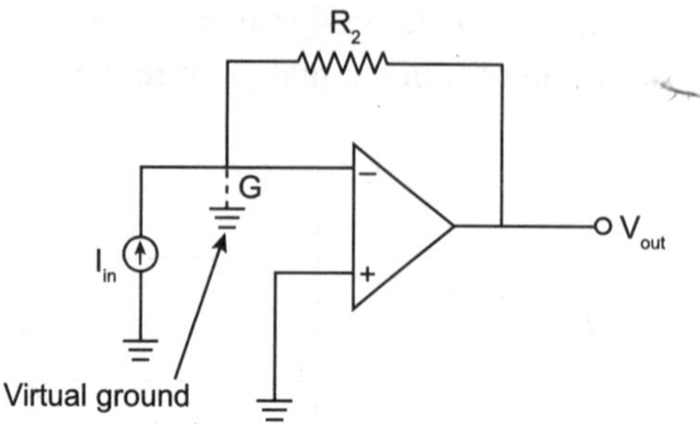

Figure 9.9

Junction G behaves like virtual ground on the input terminal. Thus, the entire current reaching G flows through the feedback resistor R_2. Hence

$$V_{\text{out}} = I_{\text{in}} R_2 \text{ (in magnitude)} \tag{10}$$

The exact equation for output voltage is given by

$$V_{\text{out}} = I_{\text{in}} R_2 \left[\frac{A_{\text{OL}}}{1+\beta A_{\text{OL}}} \right] = I_{\text{in}} R_2 \frac{A_{\text{OL}}}{1+A_{\text{OL}}} \tag{11}$$

since $\beta = 1$.

Normally A_{OL} of an Op-Amp is very high. So Eqn. (11) reduces to

$$V_{\text{out}} = I_{\text{in}} R_2$$

which is same as Eqn. 10.

Eqn. (10) justifies the nomenclature current-to-voltage converter since by selecting different values of feedback resistor R_2, we can get different values for voltage:

(i) If $R_2 = 15\ \Omega$ for an input of 1 mA, we get

$$V_{\text{out}} = 1 \text{ V}$$

(ii) If $R_2 = 10\ \Omega$, then for the same input, i.e. 1 mA,

$$V_{\text{out}} = 10 \text{ V}$$

9.9 Input and Output Impedance

The exact equations for closed loop input and output impedance can be derived as

$$Z_{\text{inCL}} = Z_{if} = \frac{R_2}{1+A_{\text{OL}}} \tag{12}$$

and

$$Z_{\text{outCL}} = Z_{of} = \frac{R_{\text{out}}}{1+A_{\text{OL}}} \tag{13}$$

Thus, in this case, both input and output impedance decrease.

9.10 Inverting Amplifier

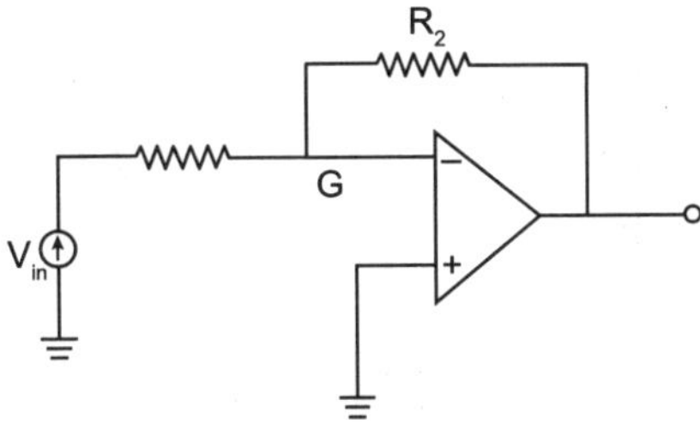

Figure 9.10

Suppose we have a voltage source in place of current source (Figure 9.10) and connect it to the inverting terminal of Op-Amp through resistance R_1 (Figure 9.10). This circuit is known as inverting Op-Amp or transresistance amplifier.

This type of amplifier (Figure 9.10) uses ICVS negative feedback. Since junction G behaves like a virtual ground,

$$I_{in} = \frac{V_{in}}{R_1}$$

and the entire current flows through feedback resistor R_2.

Thus, output voltage $$v_o = (-)\, I_{in} R_2 \qquad (14)$$

Substituting for I_{in}, we have

$$v_o = (-)\frac{V_{in}}{R_1} R_2 \qquad (15)$$

and closed loop voltage gain is $$A_{CL} = \frac{v_o}{V_{in}} = (-)\frac{R_2}{R_1} \qquad (16)$$

In magnitude, its value is

$$A_{CL} = \left|\frac{R_2}{R_1}\right| \qquad (17)$$

Also, the value of transresistance is given by

$$\frac{v_o}{I_{in}} = R_2 = r_m$$

We call it transresistance because it is the ratio of output voltage to input current.

9.11 Voltage-Controlled Current Source Amplifier (Voltage-Current Converter)

With VCIS, the input voltage controls output current. Because of heavy negative feed back in such an amplifier, the input voltage is converted to a precise value of output current.

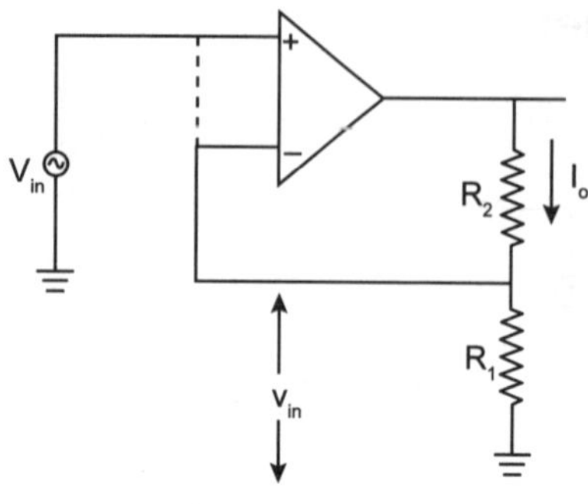

Figure 9.11

The circuit of Figure 9.11 represents such an amplifier. Because of virtual short between the two input terminals of Op-Amp, voltage across R_1 is V_{in}.

So, current through R_1 is $\frac{V_{in}}{R_1}$.

This current also flows through R_2, which is in series with R_1 and there is no other path for it. No current the enters negative input terminal of Op-Amp. Thus,

$$I_{out} = \frac{V_{in}}{R_1} = g_m V_{in} \tag{18}$$

Here
$$g_m = \frac{1}{R_1}$$

Eqn. (18) represents that output current is proportional to input voltage and the above circuit is also called voltage-to-current converter. We can select different values of R_1 to get different conversion factors (trans conductances). For instance,

(i) if $R_1 = 1$ kΩ, an input voltage of 1 V produces an output current of 1 mA.

(ii) if $R_1 = 100$ kΩ, an input voltage of 1 V produces an output current of 10 mA.

Input impedance The input side of the above circuit (Figure 9.11) is same as that of VCVS amplifier circuit (Figure 9.2) and so the closed loop input impedance of the amplifier is same as given by Eqn. (7).

$$Z_{in(CL)} = Z_{if} = (1 + \beta A_{OL}) Z_{in} \tag{19}$$

Thus, input impedance increases.

Output impedance The stabilized output current sees a closed loop output impedance given by

$$Z_{out(CL)} = (1 + A_{OL}) R_1 \tag{20}$$

Hence, output impedance increases. Thus, in this case, input and output impedance increases.

9.12 Current-Controlled Current Source Amplifier

An ICIS amplifier amplifies input current just as how VCVS amplifies input voltage. It is a perfect current amplifier because of heavy negative feedback. It has very low input impedance and very high output impedance.

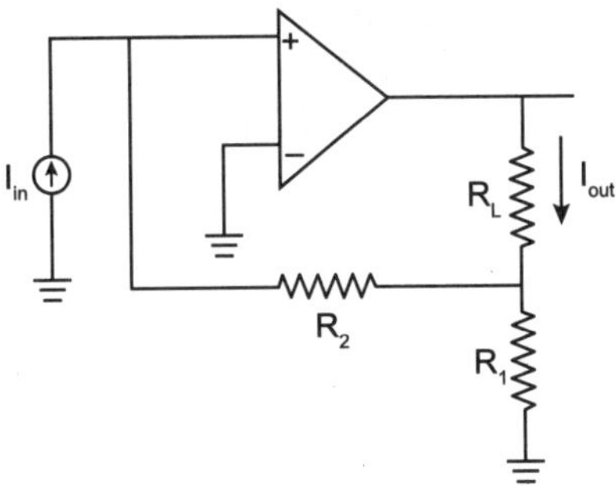

Figure 9.12 ICIS amplifier

Figure 9.12 shows the circuit of an ICIS amplifier. It represents a current inverting amplifier just like the VCVS amplifier; here the closed loop current gain is given by

$$[Ai]_{\text{CL}} = \frac{A_{\text{OL}}(R_1 + R_2)}{R_L + A_{\text{OL}}\, R_1}$$

The second term in the denominator is usually very large; so we can neglect R_L and

$$[Ai]_{CL} = \frac{A_{\text{OL}}(R_1 + R_2)}{A_{\text{OL}}\, R_1} = \frac{R_1 + R_2}{R_1}$$

$$= 1 + \frac{R_2}{R_1} \tag{21}$$

Input impedance The closed loop input impedance is given by

$$Z_{if} = Z_{\text{in(CL)}} = \frac{R_2}{1 + \beta A_{\text{OL}}} \tag{22}$$

where $\beta = \frac{R_1}{R_1 + R_2}$ is called feedback factor.

Output impedance The stabilized output current sees closed loop output impedance given by

$$Z_{of} = Z_{\text{out(CL)}} = (1 + A_{\text{OL}})\, R_1 \tag{23}$$

Thus, we find that a large A_{OL} gives small input impedance (Eqn. (2)) and large output impedance.

Because of the above characteristic, the ICIS circuit is considered a perfect current amplifier.

9.13 Bandwidth

The open loop bandwidth of an Op-Amp is very low of the order of 10 Hz because of internal compensating capacitor. When negative feedback is employed, the overall bandwidth increases because when input frequency increases and is greater than f_{2OL}, the open loop gain A_{OL} decreases by 20 dB/decade. When V_{out} tries to decrease, less opposing voltage is fed back to the input, resulting into an increase in input voltage and it compensates for the decrease in A_{OL}. Because of this closed loop, gain A_{CL} breaks at higher frequency than f_{2OL}. Thus, greater the negative feedback, higher the closed loop cut-off frequency f_{2CL}. Stated another way, the smaller the A_{CL}, the greater the f_{2CL}.

$$f_{2(CL)} = \frac{f_{unity}}{A_{CL} + 1} \tag{24}$$

$$= \frac{f_{unity}}{A_{CL}} \quad \text{since } A_{CL} \gg 1$$

or

$$[f_{2(CL)} A_{CL}] = f_{unity}$$

or

$$A_{CL}\ f_{2(CL)} = \text{constant} \tag{25}$$

since f_{unity} is fixed for a device.

Thus, we say that gain bandwidth for the product of VCVS is constant.

For VCVS, the following equations also hold good.

$$f_{2(CL)} = [1 + \beta A_{OL}]\ f_{2(OL)}$$

or

$$f_{2(CL)} = \frac{A_{OL}}{A_{CL}} [f_{2OL}] \tag{26}$$

since $A_{CL} = \dfrac{A_{OL}}{1 + \beta A_{OL}}$

Because gain-bandwidth product is constant for a given Op-Amp, a designer has to trade off gain for bandwidth. The lesser the gain used, the more the bandwidth that results. Conversely, if the designer needs more gain, they should settle for lesser bandwidth.

The only way to improve is to use an Op-Amp device having a high value of gain-bandwidth product equal to f_{unity}, which is quite high.

9.13.1 Bandwidth and Slew Rate Distortion

Although negative feedback reduces distortion in the later stages of the amplifier, it has absolutely no effect on slew rate distortion. Therefore, after you calculate closed loop bandwidth, you can calculate power bandwidth. For undistorted output over the entire closed loop bandwidth, the closed loop cut-off frequency must be less than the power bandwidth.

We know that $f_{2(CL)} < f_{unity}$

$< f_{max}$

This means that the peak value of output should be less than

$$V_{p(max)} = \frac{S_R}{2\pi f_{2(CL)}}$$

Example 1 The A_{OL} and $f_{2(OL)}$ of a VCVS amplifier are 250,000 and 160 Hz, respectively. Calculate the closed loop bandwidth for $A_{CL} = 50$.

Solution

$$f_{2(CL)} = \frac{A_{OL}}{A_{CL}} f_{2(OL)}$$

$$= \frac{250000}{50} \times 160 = 800 \text{ kHz}$$

Example 2 An ICVS amplifier has $A_{OL} = 50000$ and $f_{2OL} = 14$ Hz. Calculate the closed loop bandwidth.

Solution

$$f_{2(CL)} = (1 + A_{OL}) f_{2(OL)}$$

$$= (1 + 50000)14$$

$$= 700 \text{ kHz}$$

Example 3 The open loop cut-off frequency of an ICIS amplifier is 20 Hz. The value of $(1 + A_{OL}\beta)$ is 2500. Calculate the closed loop bandwidth.

Solution

$$f_{2(CL)} = (1 + \beta A_{OL}) f_{2OL}$$

$$= 2500 \times 20$$

$$= 50 \text{ kHz}$$

Example 4 A VCVS amplifier has $f_{unity} = 1$ MHz, and $S_R = 0.5$ V/μs. Its closed loop gain is 10. Calculate the closed loop bandwidth and largest undistorted peak output voltage.

Solution

(i) $f_{2CL} = \dfrac{f_{unity}}{A_{CL}} = \dfrac{1 \text{ MHz}}{10} = 100 \text{ KHz}$

(ii) $V_{P(max)} = \dfrac{S_R}{2\pi f_{2(CL)}} = \dfrac{0.5 \text{ V}/\mu s}{2 \times 3.14 \times (100 \text{ KHz})} = 0.795 \text{ V}$

Example 5 Calculate feedback fraction, ideal closed loop gain and exact voltage gain in VCVS amplifier circuit (assume $A_{OL} = 88$ dB) shown.

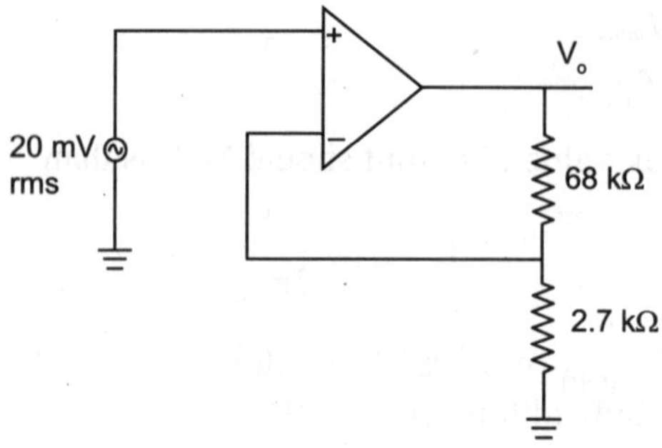

Solution

Feedback fraction,

$$\beta = \frac{R_1}{R_1 + R_2}$$

$$= \frac{2.7 \times 10^3}{2.7 \times 10^3 + 68 \times 10^3} = 0.038$$

Ideal closed loop gain,

$$A_{CL}\,(\text{Ideal}) = \frac{1}{\beta} = \frac{1}{0.038}$$

$$= 26.32$$

The exact closed loop voltage gain or feedback voltage gain is

$$A_{CL} = \frac{A_{OL}}{1 + \beta A_{OL}}$$

Decibel A_{OL} = 88 dB = 20 $\log_{10}x$ ($\log_{10}x = 4.4$)

Example 6 The given VCVS amplifier circuit has R_{in} = 3 MΩ and R_{CM} = 500 MΩ. What is the closed loop input impedance? Assume A_{OL} = 200,000

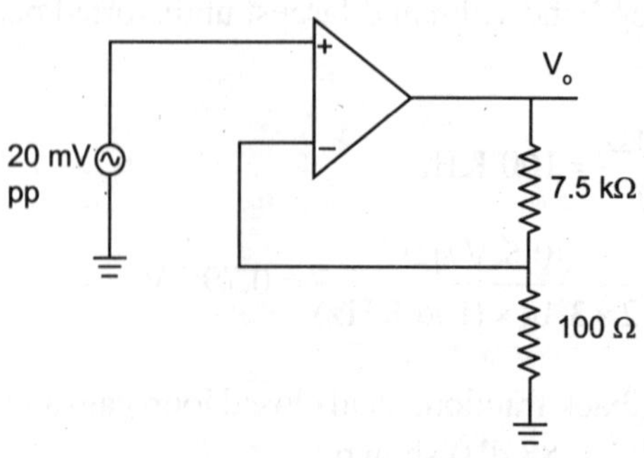

Solution

$$\text{Feedback fraction} = \beta = \frac{R_1}{R_1 + R_2} = \frac{100}{100 + 7500} = \frac{1}{76}$$

Closed loop input impedance,

$$Z_{\text{in CL}} = (1 + \beta A_{\text{OL}}) R_{\text{in}} \parallel R_{\text{out}}$$

$$= \left(1 + \frac{1}{76} \times 20000\right)\left(\frac{3 \times 500}{3 + 500}\right) \times 10^6$$

$$= 470 \text{ M}\Omega$$

Example 7 Calculate load current and power dissipated in the load 1 Ω connected in an ICIS amplifier circuit.

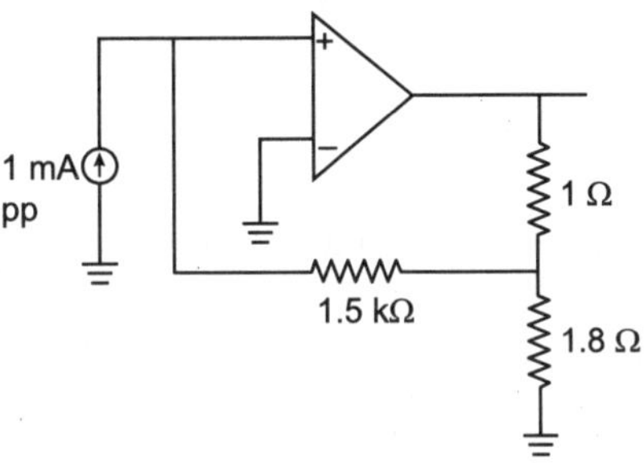

Solution

$$\text{Current gain, } A_i = 1 + \frac{R_2}{R_1}$$

$$= 1 + \frac{1.5 \text{ k}\Omega}{1.8} = 834$$

$$\text{Output current or load current, } I_L = A_i I_{\text{in}}$$

$$= 834 \times 1 \text{ mA}$$

$$= 834 \text{ mA pp}$$

Example 8 Calculate output current and load power in a VCIS amplifier circuit. What will be load power if load resistance is changed to 2 Ω?

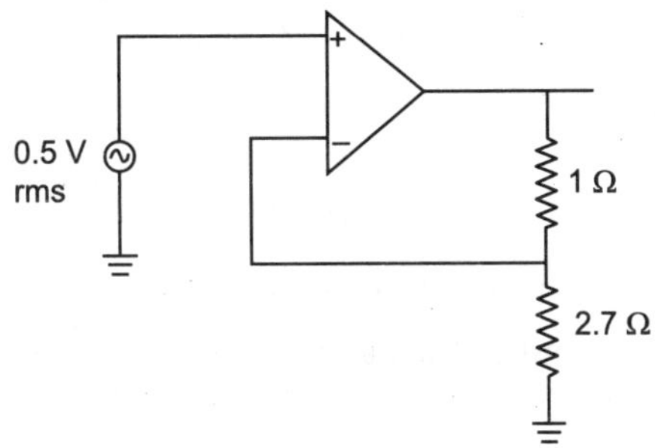

Solution

(i) Output current $= \frac{V_{\text{in}}}{R_1}$

$= \frac{0.5}{2.7} = 185 \text{ mA}$

(ii) Load power $= (I_{\text{rms}})^2 R_2$

$= (185 \times 10^{-3})^2 1$

$= 34.2 \text{ mW}$

(iii) If load is changed to 2 Ω, load current remains the same.

$$\text{Load power} = (185 \times 10^{-3})^2 \times 2$$
$$= 68.4 \text{ mW}$$

Summary

1. With the introduction of Op-Amp, we have four types of negative feedback amplifiers, namely
   (a) Voltage-controlled voltage source (VCVS)
   (b) Current-controlled voltage source (ICVS)
   (c) Voltage-controlled current source (VCIS)
   (d) Current-controlled current source (ICIS)

2. (a) In VCVS, input is a voltage signal and output is also a voltage signal. It is a perfect voltage amplifier.
   (i) Voltage gain of such an amplifier is

$$A_{\text{CL}} = \frac{A_{\text{OL}}}{1 + \beta A_{\text{OL}}}$$

   where A_{OL} is open loop gain
   β is feedback fraction

   (ii) Input impedance is

$$Z_{if} = (1 + \beta A_{\text{OL}})Z_i \text{ or } Z_{if} > Z_i$$

   (iii) Output impedance

$$Z_{of} = Z_o / 1 + \beta A_{\text{OL}} \text{ or } Z_{of} < Z_o$$

   (b) ICVS – It is also known as current-to-voltage converter.
   Ideally it has zero input impedance and zero output impedance.

$$Z_{if} = \frac{R_2}{1 + A_{\text{OL}}}$$

$$Z_{of} = \frac{R_{\text{out}}}{1 + A_{\text{OL}}}$$

$$A_{\text{CL}} = (-)R_2 / R_1$$

(c) VCIS – It is also known as voltage-to-current converter. Here, input voltage controls output current.

Input impedance

$$Z_{if} = (1 + \beta A_{\text{OL}})Z_i \text{ or } Z_{if} > Z_i$$

Output impedance

$$Z_{of} = (1 + A_{\text{OL}})R_1$$

(d) ICIS – It amplifiers the input current just as how VCVS amplifies the input voltage. It is a perfect current amplifier.

$$[Ai]_{\text{CL}} = 1 + \frac{R_2}{R_1}$$

Input impedance

$$[Z_{if}] = \frac{R_2}{1 + \beta A_{\text{OL}}}$$

Z_{if} decreases with large A_{OL}.

Output impedance

$$[Z_{of}] = [1 + A_{\text{OL}}]R_1$$

Z_{of} increases with large A_{OL}.

3. The gain bandwidth product of VCVS is constant and given by $A_{\text{CL}}[f_2]_{\text{CL}} = f_{\text{unity}}$.

Review Questions

1. How many types of negative feedback amplifiers we have? Mention the characteristics of each.
2. Prove that the closed loop voltage gain of voltage-controlled voltage source amplifier reduces and its input resistance increases.
3. Justify the nomenclature "current-to-voltage converter" for current-controlled voltage source amplifier. Derive the necessary relation.
4. Justify the nomenclature "voltage-to-current converter" for voltage-controlled current source amplifier. Derive the necessary relation.

5. Can we call current-controlled current source amplifier as current amplifier? Justify the statement. How its input impedance and output impedance change after feedback?

6. How does the bandwidth of an amplifier employing negative feedback change? What can be its maximum value?

Short Answer Questions

1. How many types of negative feedback amplifiers using Op-Amp we have? Give their nomenclature.

   Ans: We have four types of negative feedback amplifiers. They are:

   (i) voltage-controlled voltage source (ii) current-controlled current source

   (iii) voltage-controlled current source (iv) current-controlled voltage source

2. Can we call voltage-controlled voltage source amplifier an ideal voltage amplifier? Justify your answer.

   Ans: Yes. Since it has stabilized voltage gain, infinite input impedance and zero output impedance.

3. Mention the characteristics of current-controlled current source amplifier.

   Ans: It is an ideal current amplifier since it has

   (i) stabilized current gain

   (ii) zero input impedance

   (iii) infinite output impedance

4. Can voltage-controlled current source amplifier be called voltage-to-current converter? Justify your answer.

   Ans: Yes, since it is a transconductance amplifier.

5. Can we call current controlled voltage source amplifier as current-to-voltage converter? Justify your answer.

   Ans: Yes, since it is a transresistance amplifier.

6. What is the effect of negative feedback on amplifier characteristics?

   Ans:

	Types of feedback			
Characteristic	**Current series**	**Voltage series**	**Voltage shunt**	**Current shunt**
Voltage gain	Decrease	Decrease	Decrease	Decrease
Bandwidth	Increase	Increase	Increase	Increase
Harmonic distortion	Decrease	Decrease	Decrease	Decrease
Noise	Decrease	Decrease	Decrease	Decrease
Input resistance	Increase	Increase	Increase	Increase
Output resistance	Increase	Decrease	Decrease	Increase

Problems

1. A VCVS amplifier has $R_i = 2$ MΩ and $R_{CM} = 200$ MΩ. Calculate its closed loop input impedance. Assume $A_{OL} = 100{,}000$. [**Ans:** 192 MΩ]

2. A VCVS has $R_1 = 100\ \Omega$ and $R_2 = 3.9$ kΩ. Calculate the closed loop total harmonic distortion if open loop total harmonic distortion is 7.5%. Assume $A_{OL} = 100{,}000$. [**Ans:** 0.003%]

3. Calculate output voltage in the given circuit. Also find closed loop (i) input impedance (ii) output impedance.

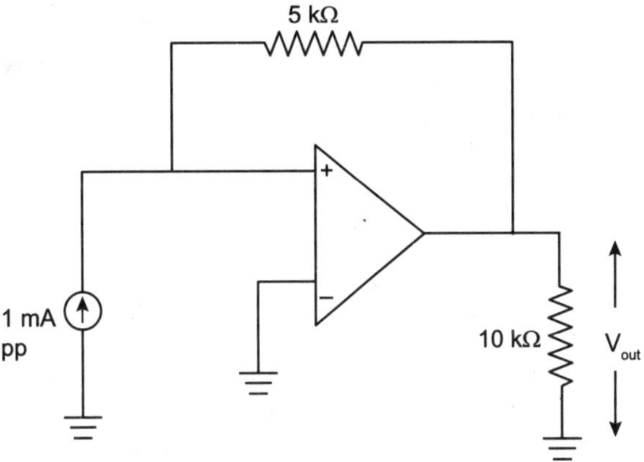

Assume $A_{OL} = 100{,}000$

[**Ans:** 5 Vpp, 0.05 Ω, 0.00075 Ω]

4. In a voltage-to-current converter circuit, load is 2 Ω. Calculate load current and load power. What will be the load current if load is changed to 4 Ω.

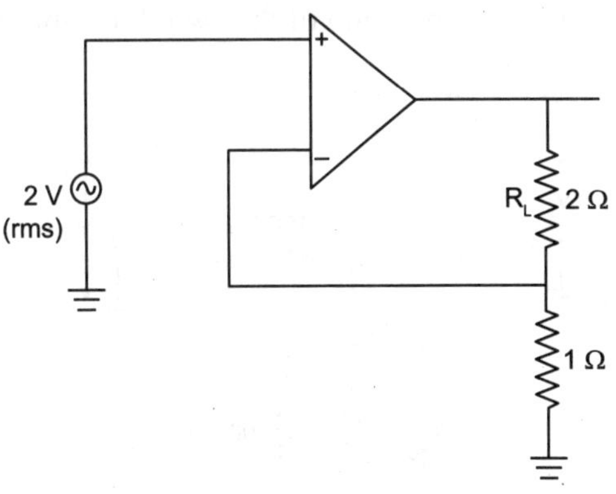

[**Ans:** 2 A(rms), 8 W, 2 A(rms)]

5. In an ICIS amplifier circuit, load is 1 Ω. Calculate the load current of the circuit and find power dissipated in the load.

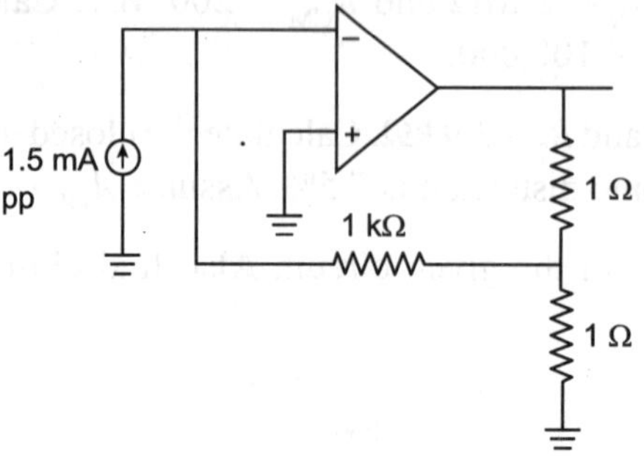

[**Ans:** 1.5 A(rms), 2.25 W]

6. Calculate output voltage of the circuit.

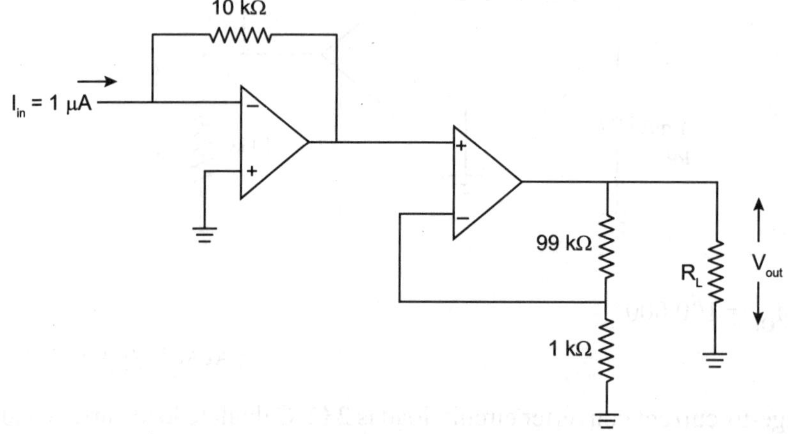

[**Ans:** 1 V]

7. Calculate output voltage for each position of the switch if input voltage is 1 mV.

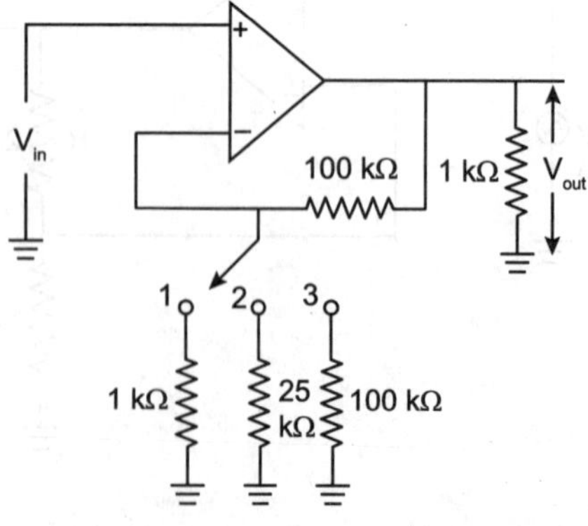

[**Ans:** (1) 101 mV, (2) 5 mW, (3) 2 mV]

Objective Questions

1. The closed loop gain βA_{OL} in a negative feedback amplifier is __________
   (a) between 0 and 1 (b) equal to 1
   (c) much greater than 1 (d) much less than 1 and negative **[Ans:** (d)]
2. The feedback ratio β in a negative feedback amplifier is __________.
   (a) always less than 1 (b) equal to 1
   (c) always greater than 1 (d) much less than 1 and negative **[Ans:** (a)]
3. A VCVS amplifier approximates ideal __________
   (a) voltage amplifier (b) current-to-voltage converter
   (c) voltage-to-current converter (d) current amplifier **[Ans:** (a)]
4. An ICVS amplifier circuit approximates an ideal __________
   (a) voltage amplifier (b) current-amplifier
   (c) current-to-voltage converter (d) voltage-to-current converter **[Ans:** (c)]
5. The transresistance of an amplifier is the ratio of __________
   (a) output current to input voltage (b) input voltage to output current
   (c) output voltage to input current (d) input current to output voltage **[Ans:** (c)]
6. The closed loop input impedance in an ideal VCVS amplifier is __________
   (a) usually larger than open loop input impedance
   (b) usually equal to open loop input impedance
   (c) ideally zero
   (d) ideally infinity **[Ans:** (d)]

Fill in the Blanks.

1. Loop gain is the __________ gain of forward and feedback paths in a VCVS amplifier. **[Ans:** voltage]
2. ICVS is a __________ amplifier. **[Ans:** transresistance]
3. Because of heavy negative feedback, the ICIS amplifier approaches a perfect __________ amplifier. **[Ans:** current]
4. Voltage-controlled voltage source Op-Amp has an open loop voltage gain of the order of __________. **[Ans:** 100,000]

Chapter

10 Operational Amplifiers

10.1 Introduction

An operational amplifier is a very high-gain amplifier with two high impedance input terminals and one low impedance output. It is a direct-coupled differential amplifier whose gain is controlled by external negative feedback circuitry. Using an operational amplifier, various operations like addition, subtraction, differentiation, etc., can be performed. Figure 10.1 shows a block diagram of an Op-Amp.

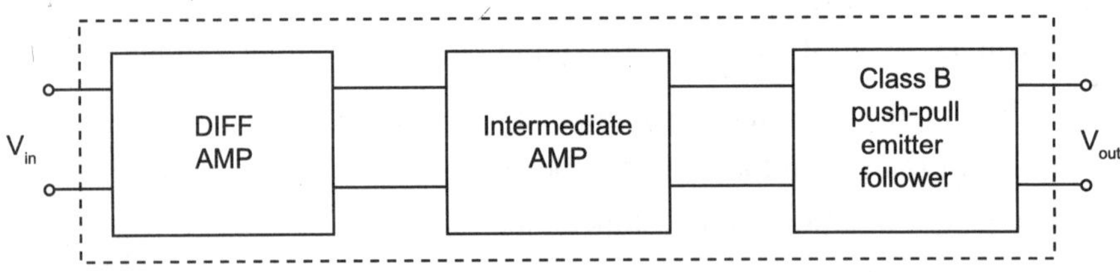

Figure 10.1 Op-Amp comprising sub-system

At the input stage is a differential amplifier followed by an intermediate amplifier having more than one stages of gain and a class B push-pull emitter follower which has low output impedance. Because a differential amplifier is the first stage, it determines the input characteristics of Op-Amp. In most amplifiers, the output is single ended as shown in Figure 10.29(a) with positive and negative supplies; the single ended output is designed to have a quiescent value of zero. This way, zero input voltage ideally results in zero output voltage.

Not all Op-Amps are designed like that shown in Figure 10.1. For instance, some do not use a class B push-pull output, and others may have double-ended output. The internal design of a monolith Op-Amp is very complicated, using dozens of transistors as current mirrors, active load and other components that are not possible in discrete designs.

Figure 10.2(a) shows the schematic symbol of Op-Amp. It has two inputs, a non-inverting and an inverting input, with a single-ended output. Figure 10.2(b) represents the equivalent circuit of Op-Amp. Figure 10.2, (ii) shows a dual-in-line package and numbering diagram.

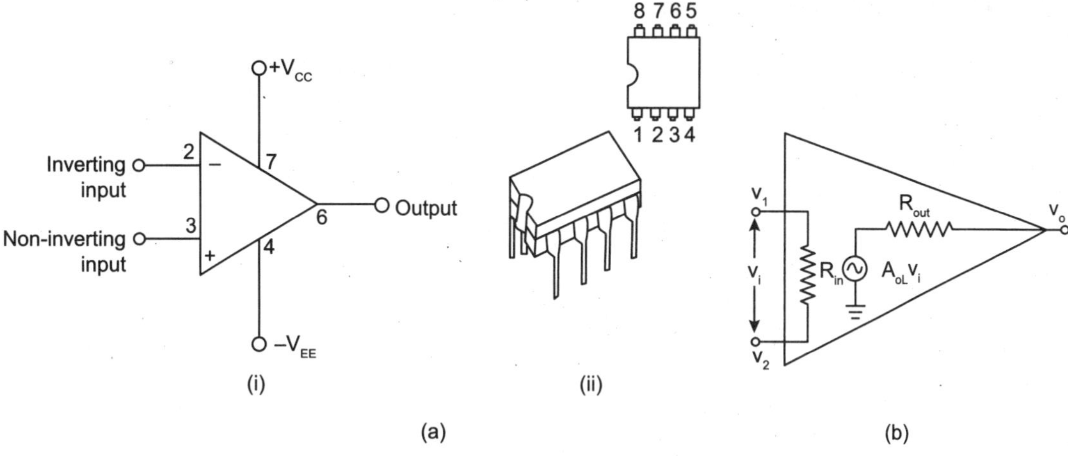

Figure 10.2 (a) Operational amplifier circuit symbol, dual-in-line package and numbering system, (b) equivalent circuit of Op-Amp

An ideal Op-Amp has the following characteristics:

(i) Infinite voltage gain

(ii) Infinite bandwidth

(iii) Infinite input impedance

(iv) Zero output impedance

(v) Perfect balance in the circuit so that output is zero for equal inputs.

(vi) Its characteristics do not change with change in temperature or DC voltage.

Such an ideal Op-Amp represents a perfect voltage amplifier and is often referred to as voltage-controlled voltage source (VCVS).

Some of the important characteristics of Op-Amp are listed in Table 10.1.

Table 10.1 Op-Amp characteristics

Parameter	**Symbol**	**Ideal value**	**Practical value (for 741 C)**
Open loop voltage gain	A_{OL}	∞	100,000
Unity gain frequency	f_{unity}	∞	1 MHz
Input resistance	R_{in}	∞	2 MΩ
Output resistance	R_{out}	Zero	75 Ω
Common mode rejection ratio	CMRR	∞	90 dB

When no feedback path (or loop) is used, voltage gain is maximum and called open loop voltage gain (A_{OL}). Normally its value is very high of the order of 100,000.

10.2 Differential Amplifier

A differential amplifier is used in the input stage of an Op-Amp. Figure 10.3 shows a differential amplifier. It has two stages in parallel with a common emitter resistor. Although it has two input voltages (v_1 and v_2) and two collector voltages (v_{c1} and v_{c2}), the overall circuit is considered to be a single stage. Because no coupling or bypass capacitors are present, there is no lower cutoff (half power) frequency.

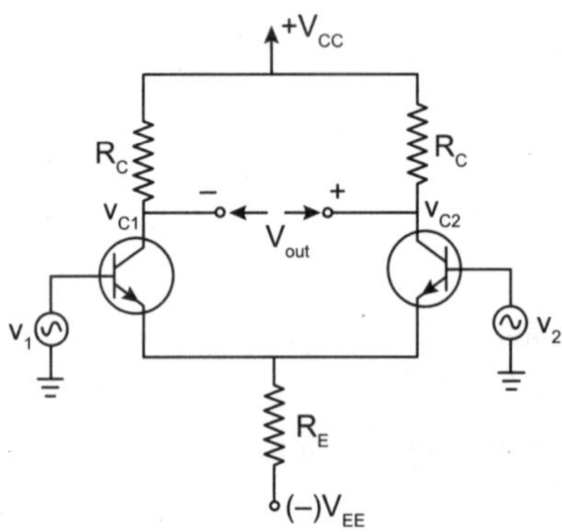

Figure 10.3 Differential amplifier (differential input and differential output for the polarity shown in output $v_1 > v_2$)

Ideally the circuit has two identical transistors and equal collector resistance (R_C) in each branch. The two input voltages are v_1 and v_2; when the two are equal and in perfect symmetry, $v_{out} = (v_{c2} - v_{c1}) = 0$.

When v_1 is greater than v_2, the output voltage has polarity as shown in Figure 10.2 (i.e. $v_{c2} > v_{c1}$). However, when v_2 is greater than v_1, the output has reverse polarity (i.e. $v_{c1} > v_{c2}$).

In the above configuration, input v_1 is called non-inverting input because v_{out} is in phase with v_1. On the other hand, v_2 is called inverting input because v_{out} is 180° out of phase with v_2.

When both non-inverting (v_1) and inverting (v_2) inputs are present, the net input is called differential input, because output voltage equals the voltage gain times the difference of two input voltages. So in this case, output voltage

$$v_{out} = A\,(v_1 - v_2) \tag{1}$$

Note: By convention, v_1 is called non-inverting and v_2 is called inverting input. We can have the following different configurations for differential amplifiers.

(a) Differential input and single-ended output as shown in Figure 10.4(a).

(b) Non-inverting input and differential output as shown in Figure 10.4(b).

(c) Inverting input and differential output as shown in Figure 10.4(c).

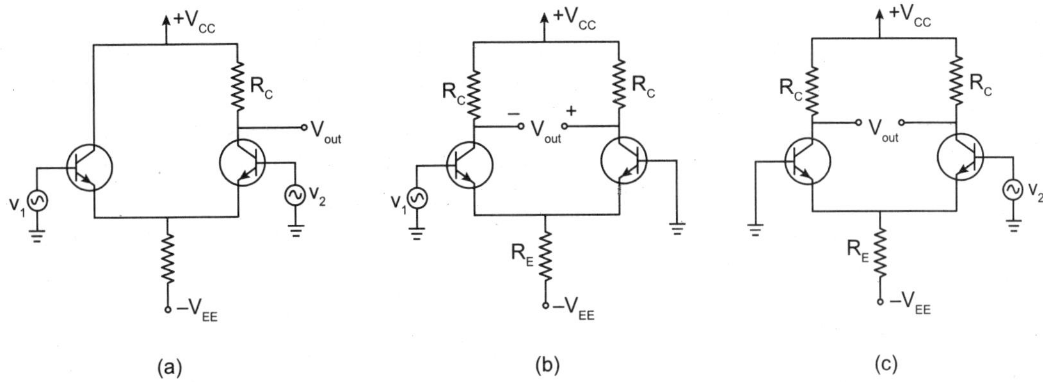

Figure 10.4 Differential amplifier: (a) differential input and single-ended output, (b) non-inverting input and differential output, (c) inverting input and differential output

For circuit shown in Figure 10.3(a), we have from Eqn. (1)

$$v_{out} = A\ (v_1 - v_2) \text{ assuming } v_1 > v_2$$

For the circuit shown in Figure 10.3(b), we have

$$v_{out} = A\ v_1 \tag{2}$$

For the circuit shown in Figure 10.3(c), we have

$$v_{out} = (-)\ A\ v_2 \tag{3}$$

10.2.1 AC Analysis of Differential Amplifier

We shall start with the simplest configuration, having non-inverting input and single-ended output (Figure 10.4(a)). With a large R_E, tail current is almost constant when a small AC signal is present. Because of this, the two halves of a differential amplifier respond in a complimentary manner to the non-inverting input. In other words, an increase in emitter current of transistor Q_1 produces a decrease in emitter current of transistor Q_2 and vice versa.

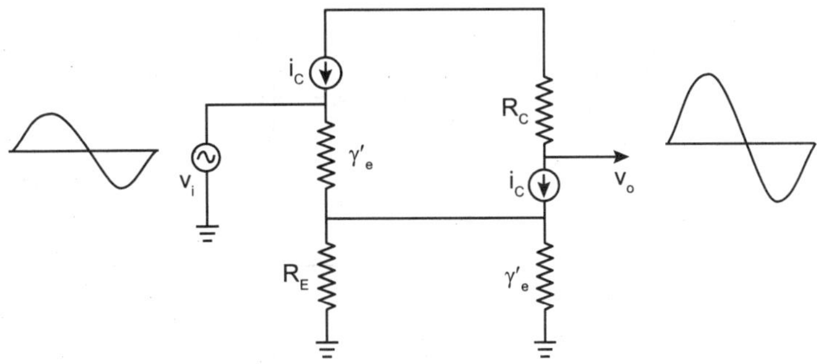

Figure 10.5 (a) AC equivalent circuit of (non-inverting input, single-ended output) differential amplifier.

Transistor Q_1 acts like an emitter follower that provides an AC voltage across emitter resistance R_E. This AC voltage is half of the applied input voltage as can be proved by analyzing the AC equivalent circuit of the amplifier (Figure 10.5(a)).

On the positive half cycle of input voltage, Q_1 emitter current increases, Q_2 emitter current decreases and Q_2 collector voltage increases. Similarly, on the negative half, Q_2 emitter current increases, and collector voltage decreases. This is why the amplified output sine wave is in phase with input voltage as shown in figure, and hence these terminals are called non-inverting terminals.

Output gain As shown in Figure 10.5, each transistor has emitter diode resistance r_e'. Also, biasing resistor R_E is in parallel with r_e' of the right transistor. Normally the value of R_E is very large in comparison with r_e', and so we can neglect R_E in our analysis and consider only r_e'. From Figure 10.5, it is to be noted that input v_i is across the first r_e' in series with second r_e' (neglecting R_E, being very large); since the two r_e' resistors are equal, voltage across each r_e' is half the input voltage.

AC output voltage $v_{\text{out}} = i_c R_C$

Input voltage $v_{\text{in}} = i_e(r_e' + r_e')$

Hence voltage gain (for single-ended output of amplifier) is

$$A = \frac{v_o}{v_i} = \frac{i_e R_C}{i_e(2r_e')}$$

or

$$A = \frac{R_C}{2r_e'} \tag{4}$$

since

$$i_c \approx i_e$$

10.2.2 Non-Inverting Input and Differential Output

The AC equivalent circuit of Figure 10.4(b) is shown in Figure 10.5(b).

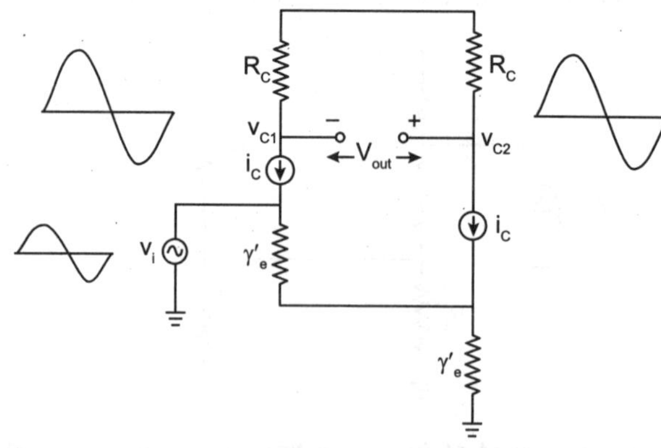

Figure 10.5 (b) AC equivalent circuit of (non-inverting input and differential output) amplifier

$$v_{out} = [v_{c_2} - v_{c_1}] = i_c\ R_c - (-i_c R_c) = 2i_c R_C$$

The negative sign in second term is because v_{c_1} is negative, i.e. 180° out of phase with v_{c_2}.

Now input voltage $v_i = 2i_e\ r_e'$

Hence voltage gain, $$A = \frac{v_o}{v_i} = \frac{2i_c R_C}{2i_e r_e'} = \frac{R_C}{r_e'} \quad (5)$$

Common mode gain

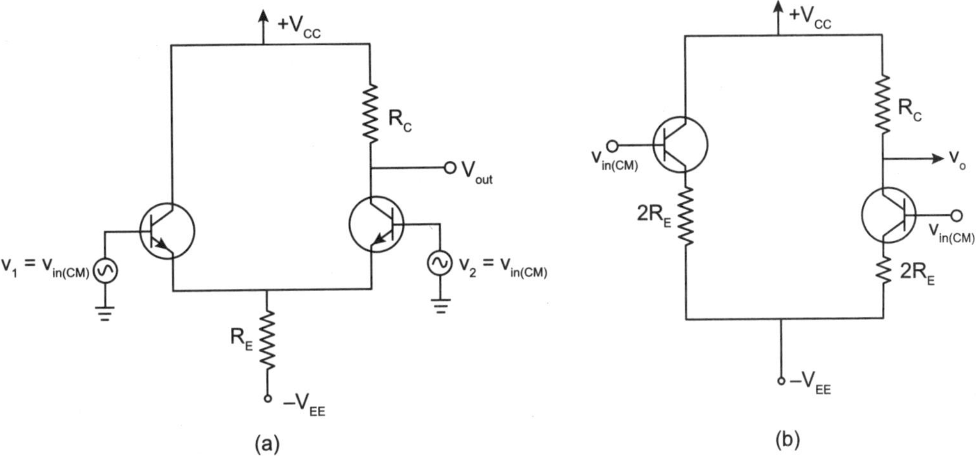

Figure 10.6 Common mode differential input, single-ended output amplifier

Figure 10.6 shows a differential amplifier having same input $v_{i(CM)}$ to both input terminals, i.e. there is differential input with a single-ended output.

The input voltage being equal is called common mode signal. If differential amplifier is perfectly symmetrical, then the AC analysis can be simplified using superposition theorem. For non-inverting input, $V_{out} = A[v_1]$, and for inverting input, $V_{out} = A[v_1 - v_2] = 0$. Since $v_1 = v_2 = v_{CM}$, we do not have any output and such amplifiers are seldom used except to eliminate most static and other undesirable signals. However, if the amplifier is non-symmetrical, then there is some AC output.

DC equivalent of Figure 10.6(a) is shown in Figure 10.6(b). Since the two input voltages are equal, there will be no current through a wire joined between the two emitter leads.

With common mode signal, the right side of the circuit is equivalent to having a swamped C_E amplifier. Normally resistor R_E is quite large than r_e'. The two inputs are not only equal but they are in phase too. Thus, the two transistors are working in parallel and total emitter current changes when flowing through R_E.

$$\Delta I_E = \frac{v_{in}}{R_E}$$

If current change through each emitter of transistor is ΔI_{E1}, then

$$\Delta I_E = 2\Delta I_{E1}$$

Let the corresponding change in collector current be ΔI_C;

then $$\Delta I_C = \Delta I_{E1}$$

Now, change in collector voltage is

$$\begin{aligned}\Delta V_c &= (+)\Delta I_C R_C \\ &= (+)\Delta I_{E1} R_C \\ &= (+)\left[\frac{\Delta I_E}{2}\right] R_C \\ &= (+)\frac{v_{\text{in}}}{2R_E}\cdot R_C\end{aligned}$$

Hence common mode gain, $$A_{\text{CM}} = \frac{\Delta v_c}{\Delta v_i} = (+)\frac{R_C}{2R_E} \tag{6}$$

With typical values of R_C and R_E, the value of A_{CM} is usually less than unity and ideally its value is zero since $v_1 = v_2$.

10.2.3 Common Mode Rejection Ratio (CMRR)

The common mode rejection ratio is defined as voltage gain '*A*' either of single input or differential input divided by common mode voltage gain, A_{CM}.

$$\text{CMRR} = \frac{A_d}{A_{\text{CM}}}$$

Data sheets usually express CMRR in decibels. In decibels, it is expressed as

$$\text{CMRR (dB)} = 20\log_{10}\text{CMRR} = 20\log_{10}\left|\frac{A}{A_{\text{CM}}}\right|$$

In a way, it determines the ability to reject common mode signal and amplify wanted signal. The value of CMRR is generally very high, e.g. if $A = 200$ and $A_{\text{CM}} = 0.5$,

$$\text{CMRR} = \frac{200}{0.5} = 400$$

$$\text{CMRR(dB)} = 20\log_{10} 400 = 20\times 2.6 = 52\text{ dB}$$

Example 1 Calculate the value of common mode voltage gain when R_C = 1 MΩ and R_E = 2 MΩ. Also find output voltage if input voltage is 1 mV.

Solution

Common mode voltage gain, $A_{CM} = \frac{R_C}{R_E} = \frac{1\times10^6}{2\times10^6} = 0.5$

Output voltage, $v_{out} = A_{CM}\, v_{in} = 0.5\times1$ mV

$= 0.5$ mV

Thus, a differential amplifier in common mode attenuates the input signal rather than amplifying it.

Example 2 In a differential amplifier, $A = 150$ and $A_{CM} = 0.5$ and $v_{in} = 1$ mV. If base leads are picking up a common mode signal of 1 mV, calculate output voltage.

Solution

Amplified output voltage,

$$V_{out(1)} = A\, v_{in}$$
$$= 150\times1 \text{ mV}$$
$$= 150 \text{ mV}$$

Attenuated output voltage (common mode),

$$V_{out(2)} = A_{CM}\, v_{in}$$
$$= 0.5\times1 \text{ mV}$$
$$= 0.5 \text{ mV}$$

Input characteristics of Op-Amp There are three characteristics to an Op-Amp. They are:

(i) Input bias current

(ii) Input OFF set current

(iii) Input OFF set voltage

Input bias current In an integrated Op-Amp, the value of β_{dc} for each transistor is slightly different, which means base current for the two transistors is also slightly different.

So, I_{in}(bias) is the average of two base currents.

$$I_{in(bias)} = \frac{I_{B1} + I_{B2}}{2} \quad \text{(i)}$$

Input OFF set current It is defined as the difference of DC base currents.

$$I_{in(OFF)} = I_{B1} - I_{B2} \text{ (2) assuming } I_{B1} > I_{B2} \quad \text{(ii)}$$

From Eqn. (i) and (ii) we have,

$$I_{B1} = I_{\text{in(bias)}} + \frac{I_{\text{in(OFF)}}}{2}$$

$$I_{B2} = I_{\text{in(bias)}} - \frac{I_{\text{in(OFF)}}}{2}$$

10.2.4 Equivalent Circuit of an Ideal Op-Amp

The characteristics of an ideal Op-Amp are as under:

(i) **Its voltage gain is infinite:** For a constant finite output voltage, input voltage between inverting and non-inverting terminal is zero.

(ii) **Its input impedance is infinite:** Current entering through either inverting or non-inverting terminal is zero, i.e. $I_1 = I_2 = 0$.

(iii) **Its output impedance is zero:** Current through a load connected to the output terminal can be of any value, but its output voltage remains constant for any value of current drawn.

Keeping in view the above characteristics, its equivalent circuit is shown in Figure 10.7.

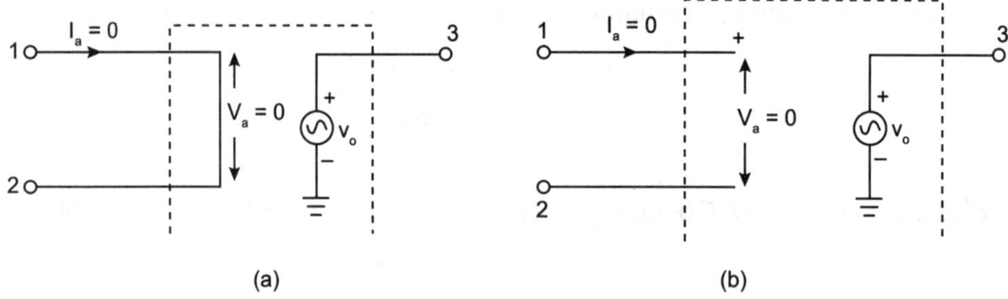

Figure 10.7 Equivalent circuit of an ideal Op-Amp

10.2.5 Operations of an Op-Amp

Inverting mode amplifier It is the most basic Op-Amp circuit. It uses negative feedback to stabilize the overall voltage gain. The reason we need to stabilize the overall voltage gain is that A_{OL} (open loop gain) is too high and unstable for use without some form of negative feedback. For example, 741C has a minimum A_{OL} of 20,000 to a maximum value of 200,000. Thus, an unpredictable voltage gain of this magnitude and variation is useless.

In this mode, a feedback resistance R_f is connected between output terminal (3) and inverting terminal (2) as shown in Figure 10.8(a).

Input signal v_i is connected between terminal (2) and ground. This voltage drives the inverting input through resistor R_1. This will result in an inverting input voltage v_2. The input voltage is amplified by open loop gain to produce an inverted output voltage (Figure 10.8(a)). The output voltage is fed back to input through a feedback resistor R_f. This will result in a negative feedback because the output is 180°C out of phase with input. In other words, any change in input v_2 produced by input voltage v_i is opposed by output voltage v_o.

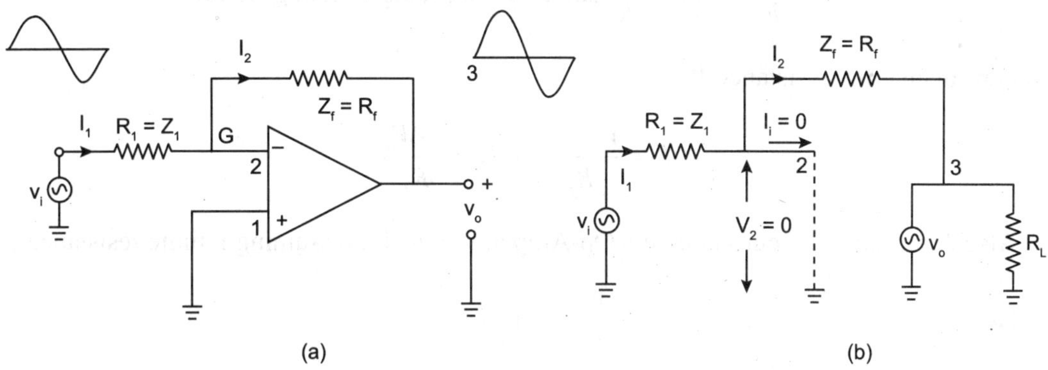

Figure 10.8 Inverting mode Op-Amp

Virtual ground When we connect a point in a circuit to ground through a piece of conducting wire, the voltage of the point connected becomes zero, since the conducting wire provides a path for current to flow to ground. The point in circuit connected this way is called mechanical ground to both voltage and current.

However, a virtual ground is different. The concept of virtual ground is based on an ideal Op-Amp whose characteristics are as under:

(i) It has infinite voltage gain (A_{OL}).

(ii) It has infinite input resistance (R_{in}).

Because of these characteristics, we can deduce the following ideal properties for inverting Op-Amp (Figure 10.8).

Since R_{in} is infinite, current entering terminal 2 is zero.

Since A_{OL} is infinite, voltage v_2 is zero because by definition $A_{OL} = \frac{v_{out}}{v_{in}} = \frac{v_{out}}{v_2}$, and for an ideal Op-Amp, A_{OL} tends to be infinity only when v_2 is zero.

When V_2 is zero (Figure 10.8), current entering the feedback resistor R_f must be equal to input current through R_1 ($I_1 = I_2$). Furthermore, since v_2 is zero, virtual ground G as shown in Figure 10.8 means the inverting input acts like a ground for voltage but open for current.

Thus, a virtual ground is half of a mechanical ground because it is short for voltage but open for current as shown by dashed line in Figure 10.8.

With a virtual ground, the analysis of an inverting Op-Amp and related circuits becomes incredibly easy. Although it is an ideal approximation, it gives accurate answers when used with heavy feedback.

Voltage gain Current entering resistance R_1 is

$$I_1 = \frac{V_i - V_2}{R_1} \text{ or } I_1 = \frac{V_i}{R_1} \text{ since } V_2 = 0 \text{ (being virtual ground)}$$

and current entering resistance R_f is

$$I_2 = \frac{V_2 - V_o}{R_f} \text{ or } I_2 = \frac{(-)V_o}{R_f}$$

Also $I_1 = I_2$ since no current enters Op-Amp at terminal 2 (assuming infinite resistance).

Hence $$\frac{V_i}{R_i} = (-)\frac{V_o}{R_f}$$

Therefore, voltage gain $A_{CL} = \frac{V_o}{V_i} = (-)\frac{R_f}{R_1} = (-)\frac{R_2}{R}$ where $R_2 = R_f$ (7)

Hence A_{CL} is called closed loop gain because it is voltage gain when there is a feedback path between output and input. Further, it is to be noted that closed loop gain A_{CL} is always less than open loop gain A_{OL}.

Input impedance From Figure 10.8, it is to be noted that the right end G of resistances $R_1(Z_1)$ is virtual ground to closed loop input impedance.

$$Z_{in(CL)} = Z_1 + \frac{Z_f}{A}$$

or $$Z_{in(CL)} = Z_1 = R_1 \quad (8)$$

Since A is large, the second term is neglected.

Output impedance It can be proved that output impedance with feedback decreases than without feedback $Z_{o(CL)} = Z_o$.

Applications of inverting Op-Amp

(a) Inverter

If $R_f = R_1$ in Figure 10.8, then from Eqn. (7), closed loop voltage gain

$$A_{CL} = \frac{v_o}{v_i} = (-)\frac{R_f}{R_1} = (-)1$$

Hence output voltage, $$v_o = (-)v_i \quad (9)$$

It represents that output voltage is numerically equal to input voltage but out of phase with it. Thus, in the above situation, it does not act like an amplifier but acts like an invertor.

(b) Scale changer

If $R_f = kR_1$ where k is a positive number, then from Eqn. (7), closed loop voltage gain

$$A_{CL} = \frac{v_o}{v_i} = (-)\frac{k R_1}{R_1} = (-)k$$

Hence output voltage, $$v_o = (-)k v_i \tag{10}$$

Eqn. (10) represents that output voltage is a multiple of input voltage with phase change of π. So Op-Amp in this situation is known as scale changer.

(c) Phase shifter

When R_1 and R_f are not pure resistors, then

$$Z_1 = |Z_1| \angle\theta_1 \text{ and } Z_f = |Z_f| \angle\theta_2 \text{ and if } |Z_1| = |Z_f|$$

Now from Eqn. (7), voltage gain, $$A_{CL} = \frac{v_o}{v_i} = (-)\frac{|Z_1|\angle\theta_2}{|Z_1|\angle\theta_1}$$

$$= (-)1\angle\theta_2 - \theta_1 \tag{11}$$

Eqn. (11) shows that out put voltage is equal to input voltage but not in phase with input voltage. Its phase can be changed from 0 to 180°. In this respect, Op-Amp acts like a phase shifter.

10.3 Non-Inverting Op-Amp

Non-inverting amplifier is another basic version of Op-Amp. It uses negative feedback to the stabilize the overall voltage gain. It also increases input impedance and decreases output impedance. Its circuit is shown in Figure 10.9(a). Figure 10.9(b) shows another version of Figure 10.9(a).

The input voltage v_i drives the non-inverting terminal (1). Here output v_o is in phase with input voltage v_i. A part of output voltage v_o is fed back (Figure 10.9(b)) to input through the voltage divider formed by the combination of R_f and R_1. So voltage across R_1 is fed back and it is fed to the inverting terminal.

This feedback voltage v_f is almost equal to input voltage v_i but it opposes the input voltage. So there is negative feedback which stabilizes the overall voltage gain. If open loop gain A_{OL} increases for any reason, output voltage will increase and so also feedback voltage increases. This opposing feedback voltage reduces the net input voltage. Therefore, even though A_{OL}

increases, the net input voltage decreases and the final output voltage increases by much less than it would without feedback.

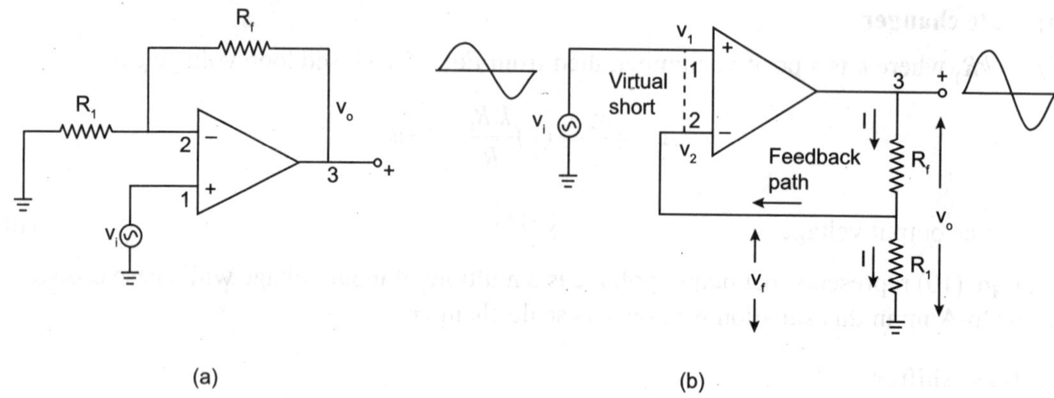

Figure 10.9 Non-inverting Op-Amp

10.3.1 Virtual Short

We know that a wire connected between two points in a circuit is a mechanical short and it is short for both voltage and current.

However, a virtual short is different. It uses two properties of Op-Amp:

(i) Input impedance of an ideal Op-Amp is infinite, and so both input current entering terminal 1 and 2 respectively is zero.

(ii) For an ideal Op-Amp, open loop gain A_{OL} is infinite, i.e. $v_{in} = v_1 - v_2$ is zero.

Figure 10.9(b) shows a virtual short between the two input terminals of Op-Amp. It is short for voltage but not for current, i.e. no current flows through it unlike in a mechanical short.

Because of virtual short, inverting input voltage (v_2) follows non-inverting input voltage (v_1). If non-inverting input voltage increases or decreases, then inverting input voltage also increases or decreases immediately. This action of following the leader is called bootstrapping action.

10.3.2 Voltage Gain

We know that voltage $v_2 = v_1$ because of virtual short between the two input terminals, and so $v_1 = v_2$ and $v_2 = v_f$ (Figure 10.9(b)).

Hence input voltage $v_i = v_f$.

If I is current flowing through the combination of R_f and R_1 (Figure 10.9 (b)), then $v_f = IR_1$ and v_f is also equal to v_i; so $v_i = IR_1$.

Output voltage, $v_o = I_1(R_1 + R_2)$

$$\text{Hence voltage gain, } A_{\text{CL}} = \frac{v_o}{v_i} = \frac{I(R_1 + R_f)}{IR_1}$$

$$= \left(1 + \frac{R_f}{R_1}\right) = \left[1 + \frac{R_2}{R_1}\right] \text{ where } R_2 = R_f \tag{12}$$

Eqn. (12) shows that voltage gain of non-inverting Op-Amp is always more than unity and there is no phase change between output and input voltage.

Input impedance Circuit in Figure 10.9(b) can also be shown as in Figure 10.10 where we have a separate feedback network which provides feedback voltage $v_f = \beta v_o$ to input.

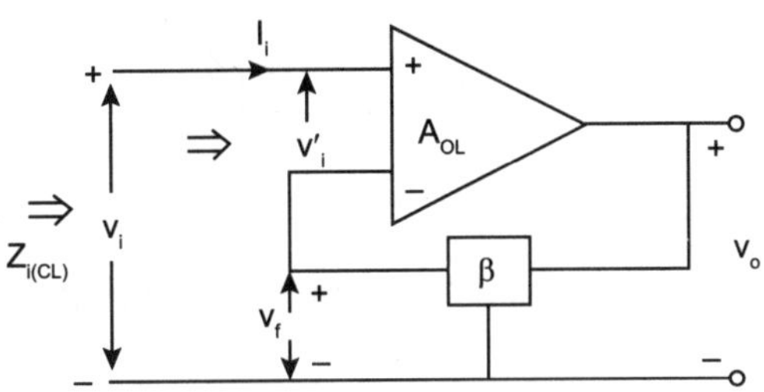

Figure 10.10

From Figure 10.10, input impedance of amplifier without feedback is

$$Z_i = \frac{v_i'}{I_i}$$

or

$$I_i = \frac{v_i'}{Z_i}$$

Input impedance of amplifier with feedback is

$$Z_{i(\text{CL})} = \frac{v_i}{I_i}$$

Substituting for I_i, we have

$$Z_{i(\text{CL})} = \frac{v_i}{\left[\dfrac{v_i'}{z_i}\right]} = z_i\left[\frac{v_i}{v_i'}\right] \tag{1}$$

Also from the circuit in Figure 10.10, we have

$$v_i' = v_i - v_f \tag{1a}$$

where v_f is feedback voltage

and $v_f = \beta v_o = \beta[A_{\text{OL}}\ v_i']$ where β is a fraction of output voltage feedback.

Substituting for v_f in (i),

$$v_i' = v_i - \beta A_{\text{OL}}\, v_i'$$

or
$$v_i'[1 + \beta A_{\text{OL}}] = v_i$$

or
$$v_i' = \frac{v_i}{1 + \beta A_{\text{OL}}} \tag{2}$$

Substituting for v_i' in Eqn. (1),

$$Z_{i(\text{CL})} = \left[\frac{v_i}{1 + \beta A_{\text{OL}}}\right] Z_i$$

$$= Z_i[1 + \beta A_{\text{OL}}] \tag{13}$$

Eqn. (13) represents that input impedance of Non-inverting Op-Amp increases by a factor of $[1 + \beta A_{\text{OL}}]$.

Output impedance It can be proved that output impedance with feedback decreases larger than output impedance without feedback.

$$Z_{o(\text{CL})} = \frac{Z_o}{1 + \dfrac{AZ_1}{Z_1 + Z_f}} \tag{14}$$

10.4 Applications

10.4.1 Unity Gain Buffer (Voltage Follower)

The use of Op-Amp as unity gain buffer is universal; its circuit is shown in Figure 10.11.

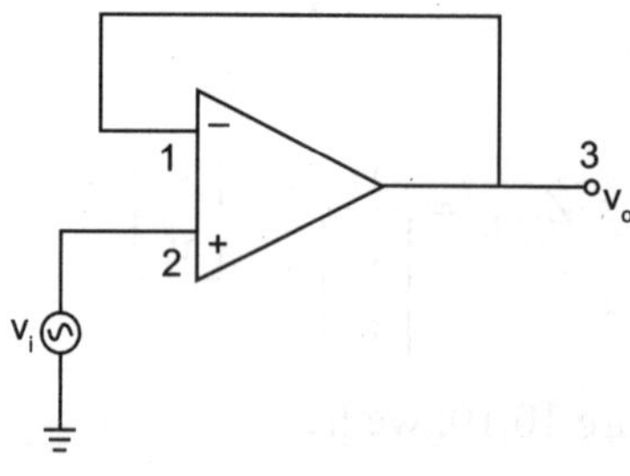

Figure 10.11 Unity gain buffer

Turning to Figure 10.9(a) for non-inverting Op-Amp, the above circuit matches with it provided $R_1 = \alpha$ and $R_f = 0$. Here we have maximum feedback and the entire output is fed back to inverting input terminal. Substituting these values we find the gain of Op-Amp.

Making use of Eqn. (12),

$$A = 1 + \frac{R_f}{R_1}$$

Substituting for $R_f = 0$, we have

$$A = 1 \tag{15}$$

$\therefore v_o = v_i$

So the above circuit is also known as voltage follower.

Matching device In the above situation, gain reduces to unity. Such a circuit is mostly used as buffer or impedance matching device between a source of high impedance and a load of low impedance since the input impedance of Op-Amp is very high and its output impedance is low. In addition, it is also used as an isolating device.

10.4.2 Adder (Summing Amplifier)

Whenever we need to combine two or more analog signals into a single output, its circuit is designed as shown in Figure 10.12(a).

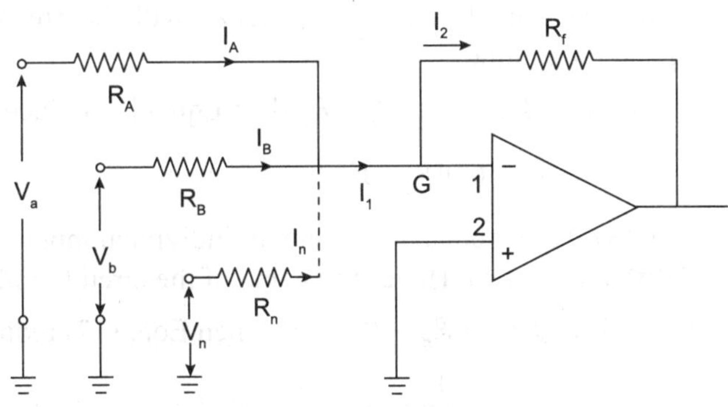

Figure 10.12 (a) Adder (summing amplifier)

As we know, terminal 1 (point *G*) behaves as virtual ground. Applying Kirchhoff's current law to the circuit, we have

$$I_A + I_B + ...I_n = I_1 = I_2 \tag{16}$$

since no current enters terminal 1.

From the circuit, we note that

$$I_A = \frac{V_A}{R_A}, I_B = \frac{V_B}{R_B}, \ldots I_n = \frac{V_n}{R_n}$$

(since terminal G is virtual ground)

where V_A, $V_B \ldots V_n$ are analogue signals to be combined.

Also $$I_2 = (-)\frac{V_o}{R_f}$$

Substituting the values of I_A, $I_B \ldots I_n$ and I_2 in Eqn. (16), we have

$$\frac{V_A}{R_A} + \frac{V_B}{R_B} + \ldots + \frac{V_n}{R_n} = (-)\frac{V_o}{R_f}$$

or $$V_o = (-)R_f\left[\frac{V_A}{R_A} + \frac{V_B}{R_B} + \ldots + \frac{V_n}{R_n}\right]$$

$$\therefore\ V_o = (-)\left[\frac{R_f}{R_A}V_A + \frac{R_f}{R_B}V_B + \ldots + \frac{R_f}{R_N}V_n\right] \tag{17a}$$

Making use of Eqn. (7), we have

$$V_o = [A_{\text{CL}_1}V_A + A_{\text{CL}_2}V_B + \ldots + A_{\text{CL}_3}V_c + \ldots A_{\text{CL}_\text{N}}V_n] \tag{17b}$$

Eqn. (17) represents that the sum of input voltage signals with their respective weight factor is equal to output voltage in magnitude.

If in the circuit we have $R_A = R_B = \ldots = R_n = R_f$, then Eqn. (17) reduces to

$$V_o = (-)[V_A + V_B + \ldots + V_n] \tag{18}$$

In this situation, output voltage is equal to the sum of individual input signals with negative sign (i.e. it has a phase difference of π). Hence the name of the circuit 'adder' goes with it.

If we have n input signals and $R_A = R_B \equiv R_n = nR_f$, then Eqn. (17) reduces to

$$V_o = (-)\frac{1}{n}[V_A + V_B + \ldots + V_n] \tag{19}$$

Eqn. (19) shows that output voltage is the average of individual input voltages. Thus, such a circuit can also be used as an averager.

10.4.3 Summing Amplifier Using Both Inputs

Some times we may have a circuit as shown in Figure 10.12(b), where both inputs of Op-Amp are being used.

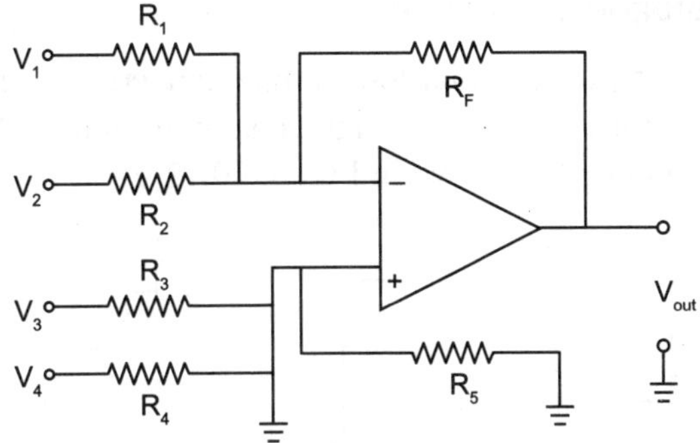

Figure 10.12(b)

The inverting amplifier has two input channels, and the non-inverting side also has two input channels. The total gain of the amplifier is the superposition of individual channel gains.

In the inverting mode, the gain of each channel (Eqn. 7) is the ratio of feedback resistor R_F to input channel resistance, either R_L or R_2.

$$A_1 = (-)\frac{R_F}{R_1}$$

$$A_2 = (-)\frac{R_F}{R_2}$$

In the non-inverting mode, the gain of each channel is $\left[1+\frac{R_F}{R_1 \| R_2}\right]$ reduced by the voltage divider factor, either $\frac{R_4 \| R_5}{R_3 + R_4 \| R_5}$ or $\frac{R_3 \| R_5}{R_4 + R_3 \| R_5}$

Thus,

$$A_3 = \left[1+\frac{R_F}{R_1 \| R_2}\right]\left[\frac{R_4 \| R_5}{R_3 + R_4 \| R_5}\right] \tag{20}$$

and

$$A_4 = \left[1+\frac{R_F}{R_1 \| R_2}\right]\left[\frac{R_3 \| R_5}{R_3 + R_4 \| R_5}\right] \tag{21}$$

After knowing the gain of each channel, we can calculate total output voltage by superposition theorem.

$$V_{\text{out}} = A_1 v_1 + A_2 v_2 + A_3 v_3 + A_4 v_4 \tag{22}$$

10.4.4 Digital-to-Analogue Converter

In digital electronics, we mostly use a digital-to-analogue converter which is a weighted summing circuit that produces an output equal to the weighted sum of inputs. The weight is same as the gain of the channel. Such a circuit is shown in Figure 10.12(c).

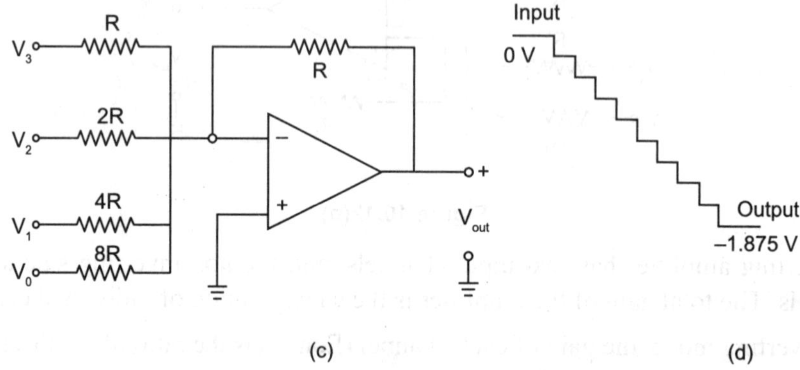

Figure 10.12 (c) Weighted summing circuit

Making use of Eqn. (7), the respective channel gain is

$$A_3 = (-)\frac{R}{R} = (-)1$$

$$A_2 = (-)\frac{R}{2R} = (-)0.5$$

$$A_1 = (-)\frac{R}{4R} = (-)0.25$$

$$A_0 = (-)\frac{R}{8R} = (-)0.125$$

Using the superposition theorem, output voltage is

$$\begin{aligned} V_{\text{out}} &= (-)[A_3 v_3 + A_2 v_2 + A_1 v_1 + A_0 v_0] \\ &= (-)[1 + 0.5 + 0.25 + 0.125]V_i \\ &= (-)1.875\ V_i \end{aligned}$$

where all inputs are equal to V_i.

10.5 Subtractor (Difference Amplifier)

If we apply two separate voltages as input voltage at the two terminals, their output voltage is proportional to the difference of two applied voltages. In this respect, the circuit which is a combination of inverting and non-inverting amplifier acts like a subtractor.

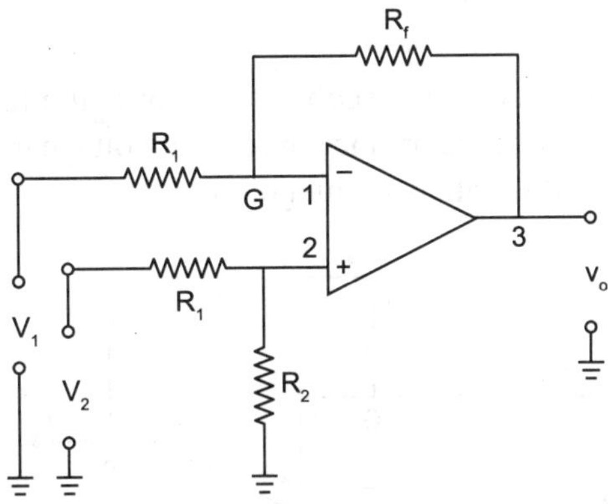

Figure 10.13 Subtractor (difference amplifier) circuit

Suppose voltages V_1 and V_2 are impressed at the positive and negative terminals 1 and 2, respectively, of an Op-Amp through equal resistances R_1. The feedback resistor R_f is normally larger than R_1.

Voltage gain in inverting mode (Eqn. 7) is

$$A_1 = (-)\frac{R_f}{R_1} \tag{i}$$

Voltage gain in non-inverting mode (Eqn. 12) is

$$A_2 = 1 + \frac{R_f}{R_1} = \frac{R_f}{R_1} \text{ if } \frac{R_f}{R_1} \gg 1 \tag{ii}$$

If output voltage due to V_1 at inverting terminal is V_{01} (assuming $V_2 = 0$),

$$V_{01} = (-)V_1 \frac{R_f}{R}$$

Similarly, output voltage due to V_2 at non-inverting terminal is

$$V_{o2} = \frac{R_f}{R_1} V_2 \text{ (assuming } v_1 = 0\text{) (making use of Eqn. ii)}$$

By superposition theorem, total output voltage is

$$V_o = V_{01} + V_{02} = (-)\frac{R_f}{R_1} V_1 + \frac{R_f}{R_1} V_2 \tag{23}$$

$$\therefore\ V_0 = \frac{R_f}{R_1}(V_2 - V_1) \tag{24}$$

Eqn. (24) shows that output voltage is proportional to the difference of two input voltages.

10.6 Integrator

If we connect a capacitor C in place of the feedback resistor R_f in Figure 10.8, then the circuit in Figure 10.14(a) behaves like an integrator since the output voltage is proportional to the integral of input voltages. Such a circuit is called an integrator.

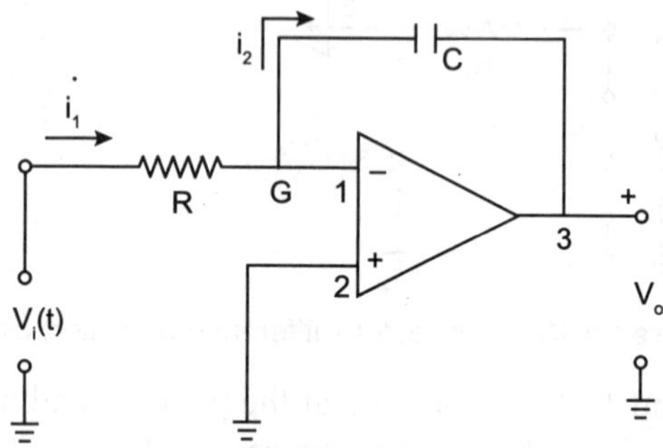

Figure 10.14(a) Integrator

Junction G is behaving like a virtual ground; so current flowing through R_1 is

$$i_1 = \frac{V_i(t)}{R}; \text{ here } V_i(t) \text{ represents voltage at instant } t$$

The same current flows through capacitor C since no current flows in terminal 1 virtual ground. So

$$i_2 = i_1$$

Now

$$i_2 = \frac{dq}{dt} = \frac{d}{dt}[-CV_o(t)]$$

where $V_o(t)$ is output voltage at instant t

or

$$i_2 = -C\frac{d[V_o(t)]}{dt}$$

Equating this value to $i_1 = \frac{V_i(t)}{R}$, we have

$$-C\frac{d}{dt}(V_{o(t)}) = \frac{V_i(t)}{R}$$

Integrating, we have

$$V_{o(t)} = (-)\frac{1}{RC}\int V_{i(t)}dt \tag{25}$$

Eqn. (25) represents that output voltage is proportional to the integral of input voltages.

Thus, if we have a square wave signal as input, then output wave is triangular as shown in Figure 10.14(b).

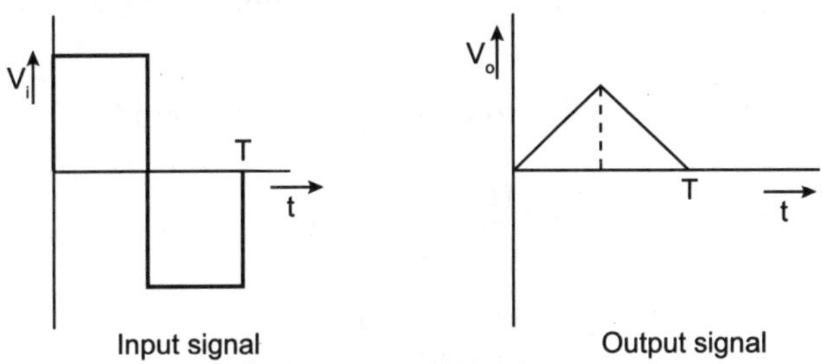

Figure 10.14(b)

10.7 Differentiator

Here the positive terminals of capacitor (*C*) and resistor (*R*) are interchanged in Figure 10.14(a) and the circuit takes the form as shown in Figure 10.15.

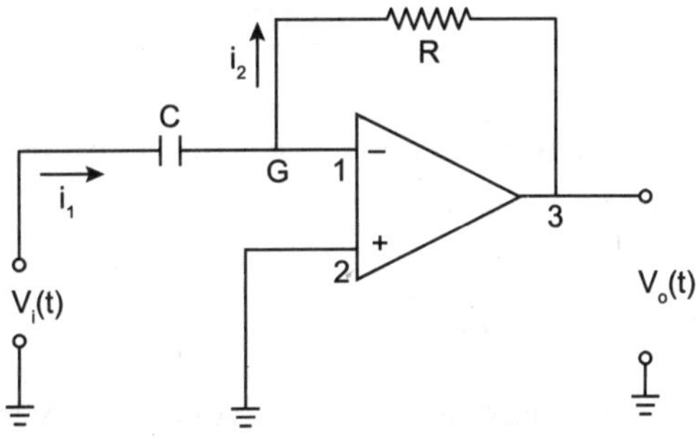

Figure 10.15 Differentiator

Since junction *G* behaves like a virtual ground, the entire input voltage $v_i(t)$ appears across capacitor *C*. If *q* is charge stored by the capacitor, then

$$v_i(t) = \frac{q(t)}{C}$$

where $q(t)$ is instantaneous value of charge on *C*.

Current i_1 through capacitor is given by

$$i_1(t) = \frac{d[q(t)]}{dt}$$

Substituting for $q(t)$, we have

$$i_1(t) = \frac{d}{dt}[C v_i(t)]$$

Current flowing through resistor R is

$$i_2(t) = (-)\frac{v_o(t)}{R} \text{ where } v_o(t) \text{ is output voltage}$$

So $$i_2(t) = i_1(t) \text{ due to } G \text{ being virtual ground}$$

or $$(-)\frac{v_o(t)}{R} = \frac{d}{dt}[C v_i(t)]$$

or $$v_o(t) = (-)RC\frac{d}{dt}[v_1(t)] \tag{26}$$

Eqn. (26) represents that output voltage is proportional to the rate of change (or differential) of input voltage.

If we assume that $RC = 1$ or $\begin{bmatrix} R = 1\ \text{M}\Omega \\ C = 1\ \mu\text{F} \end{bmatrix}$, then Eqn. (26) takes the form

$$v_o(t) = (-)\frac{d[v_i(t)]}{dt} \tag{27}$$

In this situation, the shape of output voltage wave is just a differential of input voltage wave and so the above circuit is called a differentiator.

10.8 Current-to-Voltage Converter

Circuit as shown in Figure 10.16 is used as current-to-voltage converter.

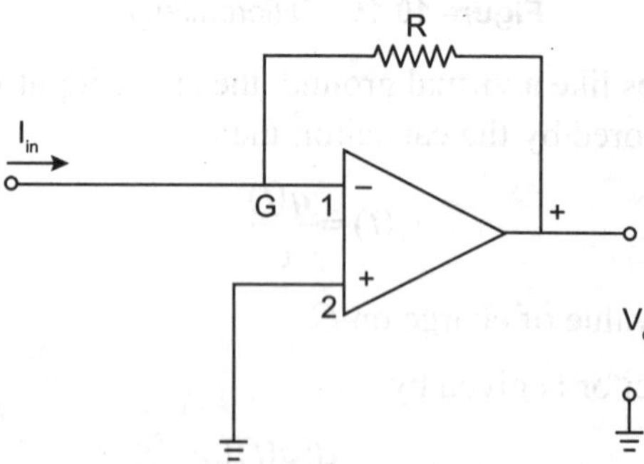

Figure 10.16 Current-to-voltage converter

Junction G is a virtual ground and no current enters terminal 1 since the input impedance of an ideal Op-Amp is assumed to be infinity. So, active input current I_{in} enters the feedback resistor R.

Voltage appears as output voltage and so

$$v_o = (-)I_{\text{in}}(R) \tag{28}$$

Assuming R to be constant, Eqn. (28) represents that output voltage is proportional to input current. We can select different values of R to get different conversion factors, e.g. if $R = 1$ kΩ, then an input of 1 mA gives an output of 1 V. However, if $R = 10$ kΩ, the same input gives the output of 10 V. Such a circuit is commonly used in electronic ammeters. The main characteristics of these ammeters is that their input resistance is very low. When such ammeters are connected in series with a circuit, then current entering the circuit is not much affected by such measuring instruments since its resistance is very low.

10.9 Voltage-to-Current Converter

The circuit shown in Figure 10.17 is used as voltage-to-current converter.

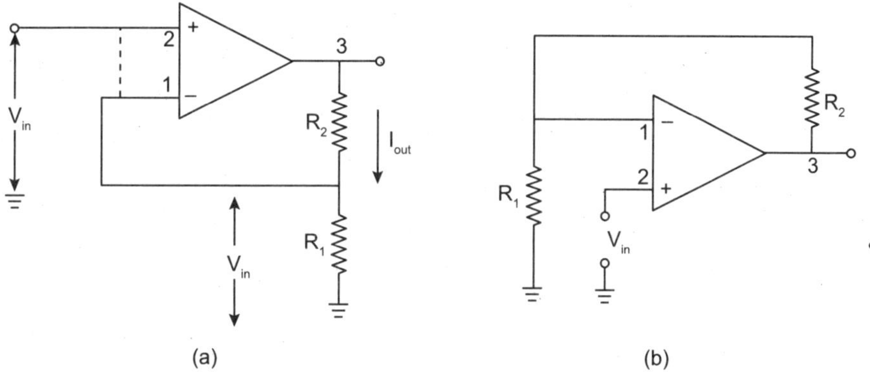

Figure 10.17 Voltage-to-current converter; both circuits are equivalent

Here an input voltage controls an output current. Because of heavy negative feedback in this type of amplifier, the input voltage is converted to a precise value of output current.

From the circuit, it is to be noted that because of virtual short between the two input terminals, v_{in} appears across R_1.

No current from the junction of R_2 and R_1 can flow to the negative input terminal.

So, $$i_{\text{out}} = \frac{v_{\text{in}}}{R_1}$$

The above equation can sometimes be written as

$$i_{\text{out}} = g_m v_{\text{in}} \text{ where } g_m = \frac{1}{R_1}$$

We can select different values of R_1 to get different conversion factors, e.g. if $R_1 = 1\ \text{k}\Omega$, an input voltage of 1 V produces an output current of 1 mA. However, if $R_1 = 1\ \text{k}\Omega$, the same input voltage produces an output current of 10 mA.

The above circuit is a perfect voltage-to-current converter because it has very high input and output impedance.

10.10 Single Supply Operation of Op-Amp

Using dual DC supplies is a typical way to power Op-Amps. Often this is undesirable in some applications. Both inverting and non-inverting amplifiers are discussed using a single supply.

10.10.1 Inverting Amplifier

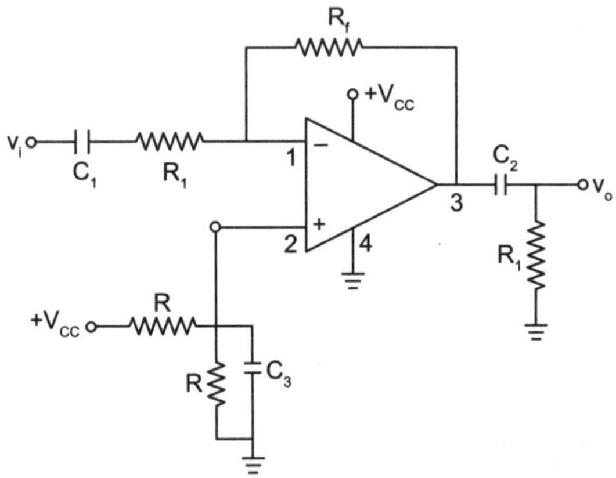

Figure 10.18 Inverting amplifier circuit using a single supply

Figure 10.18 shows a single supply inverting amplifier that can be used with AC singles. The usual V_{EE} supply (pin 4) has been grounded, and the voltage divider applies $1/2V_{CC}$ to the non-inverting input terminal. Because both input terminals are virtually shorted, the inverting terminal 1 has quiescent voltage of approximately $1/2V_{CC}$. Input off sets are minimized because of unity voltage gain.

In the AC equivalent circuit, all capacitors are shorted and the circuit is an inverting amplifier. Thus, its voltage gain is

$$A = (-)\frac{R_f}{R_1}$$

A bypass capacitor C_3 has been shown with non-inverting input; this reduces the ripple of power supply and noise also.

10.10.2 Non-Inverting Amplifier

Figure 10.19 shows a single supply non-inverting amplifier. The usual V_{EE} supply pin (4) has been grounded. The voltage divider applies 1/2 V_{CC} to the non-inverting input terminal. Because of negative feedback through R_2, the inverting input is also boot strapped to the same value 1/2 V_{CC}.

In the DC equivalent, all capacitors are open and the circuit has a voltage gain of unity, which minimizes the output offset voltage. The DC output voltage of Op-Amp is (+)1/2V_{CC} but is blocked from load R_L by capacitor C_2.

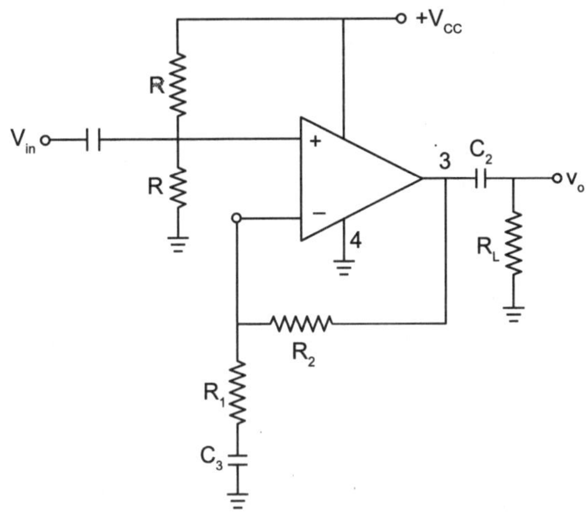

Figure 10.19 Non-inverting amplifier circuit using a single supply

In the AC equivalent, all capacitors are shorted and the circuit behaves like a non-inverting amplifier, and now the gain is

$$A = 1 + \frac{R_2}{R_1}$$

10.11 The 741 Op-Amp

Among the existing Op-Amps, the most popular one is 741 Op-Amp, whose 741C is the least expensive and most widely used. Here C represents the commercial grade.

Figure 10.20 shows a simplified circuit of 741C Op-Amp.

A detailed study of the circuit is not possible at this stage. However, we shall give a brief discussion of its main stages as shown in Figure 10.17.

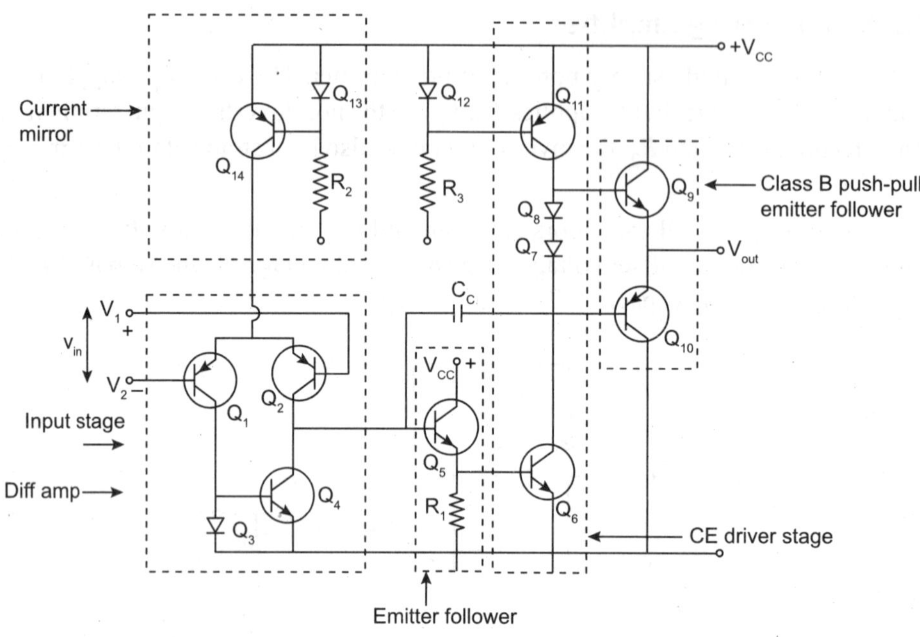

Figure 10.20 Circuit diagram of 741C Op-Amp

10.11.1 Input Differential Amplifier

The input stage is a differential amplifier comprising of transistors for Q_1 and Q_2. Transistor Q_{14} represents a current source that replaces the tail resistor of amplifier. R_2, Q_{13} and Q_{14} represent a current mirror that produces the tail current for Q_1 and Q_2. Instead of using an ordinary collector resistor for the differential amplifier here we make use of an active load resistor.

The active load Q_4 acts like a current source with extremely high impedance. Because of this, voltage gain of differential amplifier is much higher than with a passive load resistor.

The amplifier signal from differential amplifier drives the base of transistor Q_5 which is used as an emitter follower. This stage steps up the impedance level to avoid loading down the differential amplifier. The out signal of emitter follower goes to the base of Q_6. Diodes Q_7 and Q_8 are part of the biasing for final stage.

Q_{11} is an active load resistor for Q_6. Therefore, Q_6 and Q_{11} act like a CE driver stage with a very high voltage gain.

Final stage The amplified signal out of the CE driver stage (Q_6) goes to the final stage, which is a class B push-pull emitter follower (Q_9 and Q_{10}). Because of split supply (equal positive V_{CC} and negative V_{EE}), the quiescent output is ideally 0 V when input voltage is zero. Any deviation from 0 V is called output error voltage.

When V_1 is greater than V_2, the input voltage v_{in} produces a positive output voltage V_{out}, but when V_1 is less than V_2, the input voltage v_{in} produces a negative output voltage.

Ideally the (+) output can attain $+V_{CC}$ and (–) output can attain $(-)V_{EE}$ before clipping occurs. The output swing is within 1–2 V of each supply voltage because of voltage drop within the circuit; capacitor C_C is a compensating capacitor. Because of Miller effect, this small capacitor behaves like $C_{in} = (A + 1)\ C_C$. The entire circuit is prepared on a single chip of IC.

10.11.2 Frequency Response of Op-Amp

Figure 10.21 shows an ideal bode-plot of open loop voltage gain versus frequency.

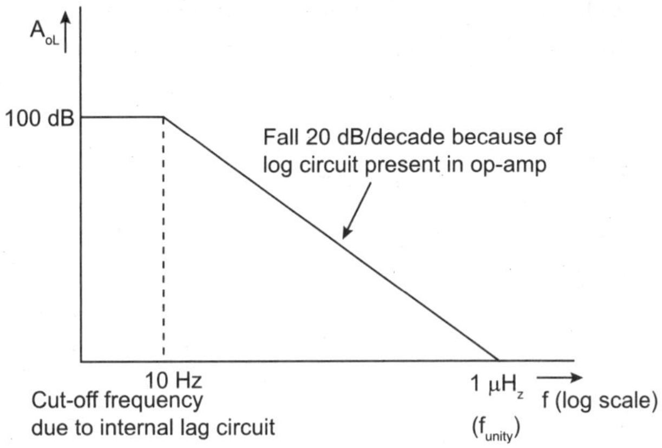

Figure 10.21 Frequency response curve (Ideal bode plot)

A typical value of voltage gain of the Op-Amp is 100000. Here this value is expressed in decibel; it is 100 dB as shown in Figure 10.21. Since the open loop cut-off frequency is 10 Hz, the above gain breaks at this value of frequency and then rolls off at 20 dB/decade until its value goes to 0 dB at 1 MHz, which is known as f_{unity}. (Here we have shown 1st order response only.)

10.12 Inverter/Non-Inverter Circuits

We have so far discussed inverting and non-inverting amplifiers separately. Now we shall discuss circuits in which the input signal drives both the input terminals of Op-Amp simultaneously. When an input signal drives both the input terminals, we achieve amplification for both at the same time, and the output is a superposition of both amplified signals since total voltage gain is

$$A_{over\ all} = A_{inv} + A_{non\text{-}inv}$$

10.12.1 Switchable Inverter/Non-Inverter

Figure 10.22 shows a circuit that can function either as a inverter or non-inverter.

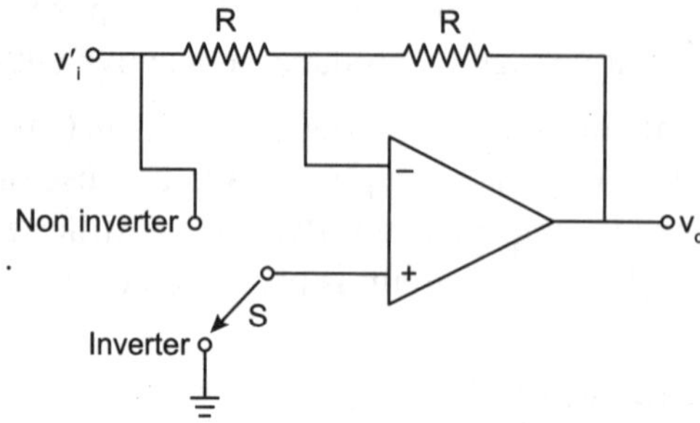

Figure 10.22 Reversible voltage gain circuit or switchable inverter/non-inverter circuit

(i) With switch *S* is in lower position, the non-inverting terminal is grounded and the circuit behaves like an inverting amplifier.

So voltage gain, $A_v = (-)\dfrac{R}{R} = -1$

$= |1|$ with a phase change of π

(ii) With switch *S* is in the upper position, the signal drives both inverting and non-inverting terminals simultaneously. So

$$A_{\text{in}} = (-)\frac{R}{R} = (-)1 \text{ and } A_{\text{non}} = 1 + \frac{R}{R} = 2$$

Total voltage gain is

$$A_{\text{over}} = A_{\text{in}} + A_{\text{non}} = -1 + 2 = (+1)$$

Hence there is no phase change.

Thus, the above circuit can have either (–)1 or (+)1 depending upon the position of the switch.

10.12.2 Inverter with Adjustable Gain

Figure 10.23 shows a circuit where we have variable gain $\left[0 \text{ to } \dfrac{R_f}{R_1}\right]$.

(i) When variable resistor R_f is zero, the non-inverting input is grounded and the circuit behaves like an inverting amplifier. So voltage gain,

$$A_{\text{in}} = (-)\frac{R_f}{R_1}$$

(ii) When variable resistor R_f is increased to full, the same input voltage v_i drives both the input terminals, i.e. we have a common mode signal and because of the common mode rejection, net output voltage is zero (gain is zero).

The above circuit behaves like an inverter with adjustable gain $\left[0 \text{ to } \dfrac{R_f}{R_1}\right]$.

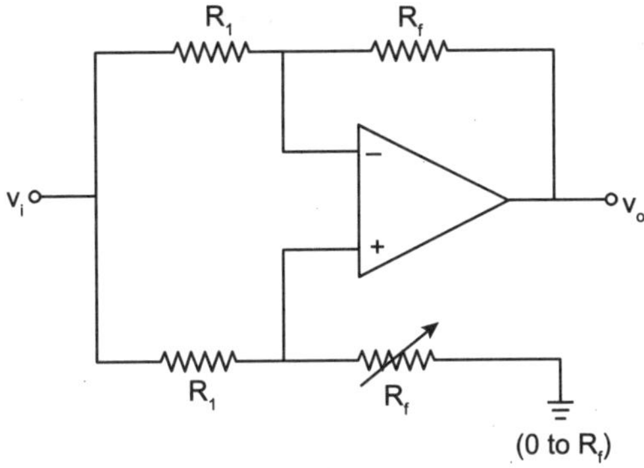

Figure 10.23 Inverter with adjustable gain circuit

10.12.3 Sign Changer

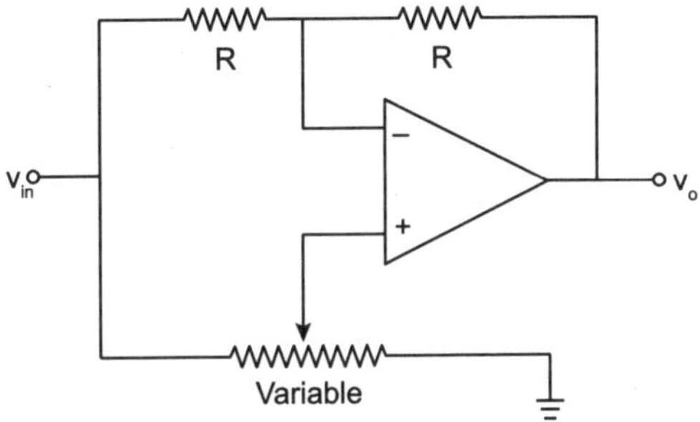

Figure 10.24 Sign changer circuit

The circuit shown in Figure 10.24 is called a sign changer. Here the gain varies from –1 to +1 (like that of switchable inverter/non-inverter circuit; Figure (10.22)). The circuit is very similar to that shown in Figure 10.23.

(i) When a wiper of variable resistor is all the way to the right end, the non-inverting input terminal is grounded. Hence the circuit behaves like an inverting amplifier and gain

$$A_{\text{in}} = (-)\frac{R}{R} = (-)1$$

(ii) When a wiper of variable resistor is all the way moved to the left end, the input signal drives the inverting input as well as non-inverting input. In this case, the total gain is obtained by applying super-position.

$$A_{ov} = A_{\text{in}} + A_{\text{non}} = (-)1 + 2 = +1$$

Thus, when a wiper is moved from right end to left end, gain continuously varies from (–)1 to (+)1.

(iii) When a wiper is just at the center of variable resistor, the common mode signal drives the Op-Amp and output is ideally zero. Hence the gain is zero.

10.12.4 Phase Shifter (Variable Phase Changer)

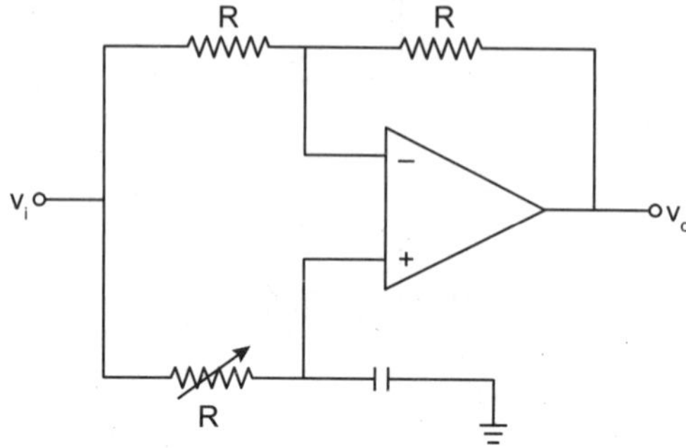

Figure 10.25 Phase shifter circuit

The circuit (Figure 10.25) shows a phase shifter circuit where the phase output voltage can vary continuously with respect to input voltage. In case of a suitable inverter/non-inverter circuit, the phase of output voltage was zero or 180° fixed.

Here, the non-inverting channel has an RC lag circuit, and the inverting channel has equal resistors; so voltage gain in the inverting mode is always unity with a phase change of π.

But voltage gain in the non-inverting mode depends upon the cut-off frequency of RC circuit.

(i) When frequency of applied signal v_i is much less than cut off frequency of RC circuit ($f \ll f_c$), the capacitor behaves like an open circuit, and the gain of non-inverting channel is

$$A_{\text{non}} = 1 + \frac{R}{R} = 2$$

and $A_{\text{in}} = (-)1$

Hence the overall gain is

$$A_{ov} = A_{\text{Non}} + A_{\text{in}} = 2 - 1 = 1$$

It shows that output signal is equal to input and it is in phase with input signal.

(ii) When the frequency of applied signal v_i is much more than the cut-off frequency of RC circuit ($f \gg f_c$), the capacitor appears shorted and the gain of non-inverting channel is zero because of common mode signal

$$A_{\text{non}} = 0$$

The over-all gain is

$$\begin{aligned} A_{ov} &= A_{\text{Non}} + A_{\text{in}} \\ &= A_{\text{in}} \\ &= (-)1 \end{aligned}$$

Thus, the output signal is 180° out of phase with input signal.

In between the two extreme cases, there is a phase change of A depending upon frequency. Thus, a phase shifter produces an output voltage with same magnitude as input signal but with a phase angle which can be varied continuously between 0° and 180°.

10.13 Operational Amplifier Circuit Stability

10.13.1 Inverting Mode

Consider the inverting amplifier circuit and wave forms shown in Figure 10.26(a).

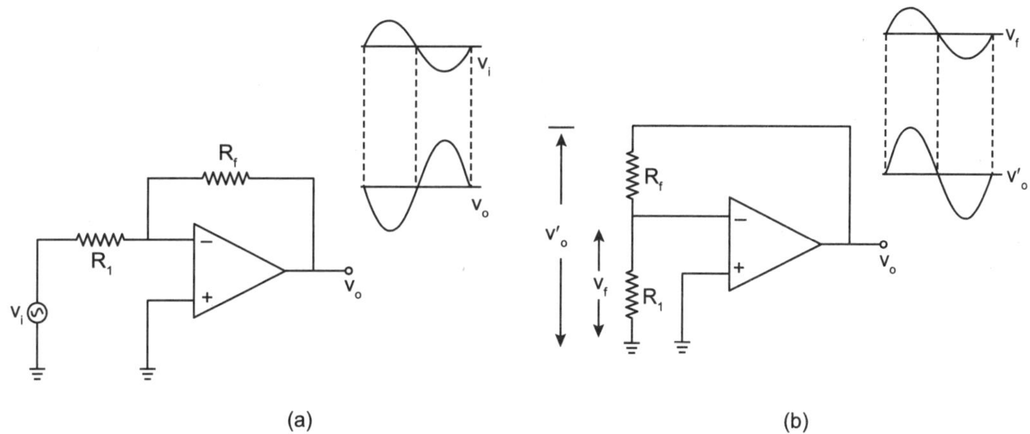

Figure 10.26 (a) Inverting amplifier (b) Inverting amplifier oscillating

The signal v_i is amplified by a factor $\frac{R_f}{R_1}$ with its phase shifted through 180° (Figure 10.26(a)). The entire circuit is redrawn in Figure 10.26(b) to show that output voltage v_o' is potentially divided by the feedback network to produce a feedback voltage v_f (Figure 10.26(b)). Whether or not an input signal is present, any AC voltage in the Op-Amp circuit at the inverting terminal, is amplified by a factor of open loop gain A_{OL} to produce an output voltage v_o'.

Since an AC voltage v_f occurs at the Op-Amp inverting terminal, the amplified v_f is $v_o' = (A_{\text{OL}}\ v_f)$, and the potentially divided $v_f = \beta v_o'$ (where $\beta = \frac{R_1}{R_1 + R_f}$). If $\beta v_o'$ is exactly equal to and in phase with the original v_f, the circuit is supplying its own input and a state of continuous oscillations occurs.

10.13.2 Non-Inverting Mode

Consider the non-inverting amplifier circuit and its wave forms as shown in Figure 10.27(a).

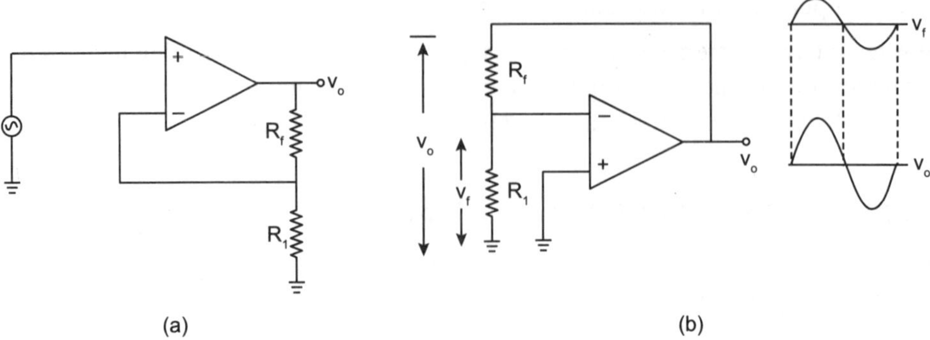

Figure 10.27 (a) Non-inverting amplifier, (b) non-inverting amplifier oscillating

The signal v_i is amplified by a factor $\left[1+\frac{R_f}{R_1}\right]$ without a phase change (Figure 10.27(a)). The entire circuit is redrawn in Figure 10.27(b) to show that output voltage v_o' is potentially divided by the feed back network to produce a feedback voltage v_f which is applied to the amplifier inverting terminal. Whether or not an input signal is present an AC voltage in the Op-Amp at inverting terminal is amplified by a factor of open loop gain A_{OL} to produce an output voltage v_o'.

Since an AC voltage v_f occurs at the Op-Amp inverting terminal, the amplified v_f is $v_o' = A_{\text{OL}}\ v_f$ and potentially divided $v_f = \beta v_o'$ where $\beta = \frac{R_1}{R_1 + R_f}$. If $\beta v_o'$ is exactly equal and in phase with the original v_f, the circuit is supplying its own input and the state of continuous oscillations occurs.

Hence we conclude that because of the feedback network, virtually every Op-Amp circuit is a potential oscillator. For a circuit to oscillate, two conditions normally have to be fulfilled: (i) loop gain should be equal or greater than unity; (ii) loop phase shift should be 0° or 360°. Here loop gain is gain around the loop from inverting input terminal to amplifier output and back to input via feedback network.

It is to be noted that loop gain from input to output is A_{OL}, and for feedback network, gain from output to input is actually an alternation, i.e. feedback factor β. Hence

$$\text{Loop gain} = (\text{Amplifier gain}) \times (\text{Feed-back factor})$$
$$= A\beta$$

10.14 Frequency Compensation

Measures taken to prevent oscillations or to stabilize an Op-Amp are referred to as frequency compensation. Among these measures, the prominent ones are: (i) phase lag and (ii) phase lead compensation. In each case, a resistance capacitance network is connected into the circuit so that it is a part of the loop. Details of the circuit are beyond the scope of this chapter.

Example 3 In an Op-Amp operating circuit in inverting mode, $R_1 = 10\ \text{k}\Omega$, $R_f = 50\ \text{k}\Omega$, and its internal gain varies from 5000 to 10,000. Calculate its output voltage in both extreme shapes of gain when input voltage is 10 mV. If its $R_0 = 5000\ \Omega$, calculate its R'_o.

Solution

In the inverting mode,

$$A' = \frac{V_o}{V_i} = (-)\frac{R_f}{R_1}\left[1 + \frac{1}{A}\left(1 + \frac{R_f}{R}\right)\right]^{-1}$$

(a) When gain is 5000,

$$V_o = 10 \times 10^{-3}\left[\frac{\frac{50}{10}}{1 + \frac{1}{5000}\left(1 + \frac{50}{10}\right)}\right]$$

$$= 10 \times 10^{-3}\left[\frac{5}{1 + \frac{6}{5000}}\right]$$

$$= 10 \times 10^{-3} \times \frac{5 \times 10000}{50006}$$

$$= 49.90\ \text{mV}$$

(b) When gain is 10000,

$$V_o = 10 \times 10^{-3}\left[\frac{\frac{50}{10}}{1 + \frac{1}{10000}\left(1 + \frac{50}{10}\right)}\right]$$

$$= 10 \times 10^{-3} \times \frac{5 \times 10000}{10006}$$

$$= 49.95\ \text{mV}$$

(c) $R'_o = \dfrac{R_o}{1 + A\dfrac{R_1}{R_f}} = 5000$

When $A = 5000$,

$$R'_o = \frac{5000}{1 + 5000 \times \frac{10}{50}} = 4.994\ \Omega$$

When $A = 10000$,

$$R'_o = \frac{5000}{1 + 10000 \times \frac{10}{50}} = 2.49\ \Omega$$

Example 4 In an Op-Amp used in inverting mode, $R_f = 75$ kΩ and $R_1 = 1.5$ kΩ. Calculate closed loop voltage gain and bandwidth. What is its output voltage at 1 kHz if input signal is 10 mV? Assuming $f_{\text{unity}} = 1$ MHz pp.

Solution

Closed loop voltage gain,

$$A_{\text{CL}} = (-)\frac{R_f}{R_1} = (-)\frac{75000}{1500} = (-)50$$

$$= 50 \text{ in magnitude}$$

Closed loop bandwidth,

$$f_{2\text{CL}} = \frac{f_{\text{unity}}}{|A_{\text{CL}}|} = \frac{1 \times 10^6}{50} = 20 \text{ kHz}$$

$$V_{\text{out}} = A(V_i) = 50 \times 10 \text{ mV pp} = 500 \text{ mV pp}$$

Example 5 Calculate minimum and maximum closed loop voltage gain.

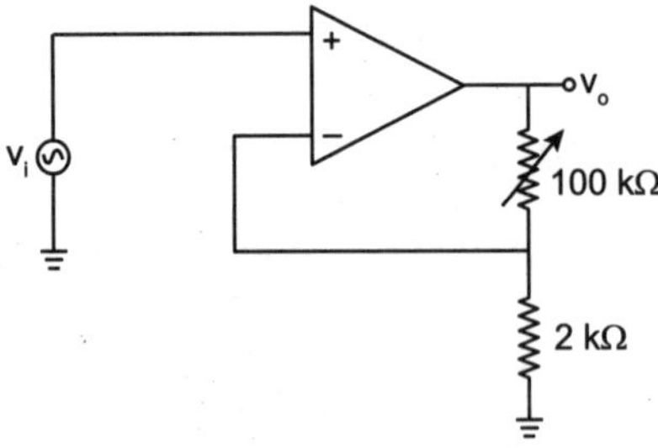

Figure 10.28

Solution

(i) When resistor 100 kΩ is in zero position,

$$\text{Gain} = 1 + \frac{R_f}{R_1} = 1 + \frac{0}{2} = 1$$

(ii) When resistor 100 kΩ is in maximum position,

$$\text{Gain} = 1 + \frac{100}{2} = 51$$

Example 6 In an inverting Op-Amp, R_1 = 50 kΩ and R_f = 300 kΩ. The internal voltage gain is 10,000. Calculate the output voltage when input signal is 1 V. If the same input voltage is applied is non-inverting mode, then what is the output voltage.

Solution

In the inverting mode, voltage gain is

$$A' = \frac{V_o}{V_i} = (-)\frac{R_f}{R_1}\left[1 + \frac{1}{A}\left(1 + \frac{R_f}{R_1}\right)\right]^{-1}$$

Here R_f = 300 kΩ, R_1 = 50 kΩ, A = 10000, V_i = 1 V

$$V_o = (-)1\left\{\frac{\frac{300}{50}}{1 + \frac{1}{1000}\left(1 + \frac{300}{50}\right)}\right\}$$

$$= (-)\frac{6}{1 + \frac{7}{10000}} = (-)\frac{60000}{10007}$$

$$= (-)5.996 \text{ V}$$

Example 7 An Op-Amp has slew rate 15 V/μs. What is its power bandwidth for a peak output voltage of 10 V?

Solution

$$\text{Power bandwidth, } f_{\text{max}} = \frac{S_R}{2\pi V_p} = \frac{15 \text{ V}/\mu\text{S}}{2 \times 3.14 \times (10 \text{ V})} = 239 \text{ KHz}$$

Example 8 The voltage gain of an Op-Amp used in unity gain buffer circuit is 5000, its input resistance is 10,000 Ω and output resistance is 100 Ω. Calculate the input and output resistance of buffer circuit.

Solution

Input resistance of buffer circuit,

$$R_{\text{in}} = \frac{AR_i}{1 + \frac{R_f}{R_i}}; \text{ here } R_f = 0$$

$$\therefore R_{\text{in}} = AR_i = 5000 \times 10000 = 5 \times 10^7\ \Omega$$

Output resistance of buffer circuit,

$$R_{\text{out}} = \frac{R_o}{A}\left[1 + \frac{R_f}{R_i}\right]$$

$$= \frac{R_o}{A} \text{ since } R_f = 0$$

$$= \frac{100}{1000} = 0.02\ \Omega$$

Summary

An Op-Amp amplifier is a high gain, direct coupled differential amplifier whose gain is controlled by an external negative feedback circuit. It is generally denoted by two input terminals, called inverting and non-inverting, and one output terminal. An ideal Op-Amp has infinite voltage gain, infinite input impedance, infinite bandwidth and zero output impedance.

It functions as (i) an inverting amplifier where output is 180° out of phase with input or (ii) a non-inverting amplifier where output is in phase with input. The voltage gain in inverting mode is $(-)\frac{R_f}{R_1}$, which explains that gain is controlled by external parameters only. The voltage gain in non-inverting mode is $1 + \frac{R_f}{R_1}$. Here gain is always more than unity.

It has many applications including as (i) scale changer, (ii) summing amplifier, (iii) differentiator, (iv) integrator, etc.

CMRR of an Op-Amp is defined as the ratio of differential voltage gain to common mode voltage gain. Its value is quite high, and slew rate is defined as the maximum rate at which the output voltage can change. Its typical value is 0.5 V/μs.

Power band-width is the highest undistorted frequency that can be amplified without slew rate distortion.

The internal gain of an amplifier is called open loop gain, while that of the overall circuit employing negative feedback is called closed loop gain.

Negative feedback increases input impedance and decreases output impedance of the Op-Amp. It also increases the upper cutoff frequency up to 1 MHz, but the gain bandwidth product remains constant.

Review Questions

1. Sketch the circuit of a basic differential amplifier. Briefly explain its working.
2. For a differential amplifier, show that gain

$$A = \frac{R_C}{2r_e'}$$

   Also derive an equation for output voltage at each collector when signal v_s is applied to one input terminal.
3. Sketch the circuit of a differential amplifier with an emitter follower output. Identify inverting and non-inverting input terminals.
4. What is an operational amplifier? What are the characteristics of an ideal Op-Amp? Give typical values for them in a particular case.
5. Sketch the circuit of an Op-Amp used as an inverting amplifier. Derive the relation for (i) gain, (ii) input impedance, (iii) output impedance.
6. Sketch the circuit of an Op-Amp used as a non-inverting amplifier. Derive the relation for (i) gain, (ii) input impedance, (iii) output impedance.
7. Sketch the circuit of an Op-Amp summing amplifier. Derive a relation for output voltage in terms of inputs.
8. Explain how an Op-Amp in inverting mode is used as (i) inverter, (ii) scale changer, (iii) phase shifter.
9. How an Op-Amp can be used as a subtractor? Explain and derive necessary relations.
10. How an Op-Amp can be used as (i) Integrator (ii) Differentiator? Explain and derive necessary relations. Give appropriate circuit.
11. Can we use an Op-Amp as (i) current-to-voltage converter, (ii) voltage-to-current converter? If yes, necessary circuit and explain its working in brief.
12. What is an Op-Amp? What substages does it consist of? Draw a block diagram to show the various stages.
13. Draw the circuit diagram of 741C Op-Amp. Identify the various stages it consists of.
14. Draw any differential amplifier that has voltage gain $\frac{R_c}{r_e'}$. Now draw any other differential amplifier that has voltage gain $\frac{R_c}{2r_e'}$.
15. Show how feedback in an inverting amplifier can produce instability? What measures are taken to stabilize the Op-Amp?
16. Show how feedback in a non-inverting amplifier circuit can produce instability? What measures are taken to stabilize the Op-Amp?
17. Prove that gain of an Op-Amp (in common mode) with a common signal is zero.

Short Answer Questions

1. Define the term CMRR. What it shows?

   Ans: CMRR (common mode rejection ratio) is defined as the ratio of voltage gain (for single input or double input) to common mode voltage gain:

$$\text{CMRR} = \frac{A_d}{A_{\text{CM}}}$$

   A high CMRR means that the differential amplifier is amplifying the wanted signal and rejecting the common mode signal.

2. What is virtual ground in an Op-Amp?

   Ans: It is like a half mechanical ground because it is short for voltage but open for current.

3. What is virtual short in an Op-Amp?

   Ans: It is like a half mechanical short because it is short for voltage but open for current. Because of this, the inverting input voltage follows the non-inverting input voltage.

4. What is bootstrapping action in an Op-Amp?

   Ans: In bootstrapping, one voltage follows the other voltage. In an Op-Amp, when inverting input voltage follows non-inverting input voltage, the action is called bootstrapping.

5. What are the characteristics of an ideal Op-Amp?

   Ans: An ideal Op-Amp has (i) infinite voltage gain, (ii) infinite input impedance, (iii) infinite bandwidth and (iv) zero output impedance.

6. Do we have two bandwidths in case of an Op-Amp?

   Ans: Yes, we have two bandwidth in an Op-Amp. They are (i) Small signal bandwidth determined by first order response, (ii) large signal or power bandwidth determined by slew rate.

7. What do you mean by the term "slew rate" in an Op-Amp?

   Ans: If we apply a voltage step across a capacitor, then it gets charged exponentially rather than a sudden change in voltage. The initial slope of the exponential rise of wave form is called slew rate.

8. Define the term "input offset current" in case of differential amplifier.

   Ans: It is defined as the difference of DC base currents,

$$I_{\text{in(OFF)}} = I_{B_1} - I_{B_2}$$

   It indicates how closely the two transistors match. If the two transistors are identical, its value is zero.

9. Define the term "input offset voltage" in case of differential amplifier.

   Ans: It is defined as the input voltage that would produce the same output error voltage in a perfect differential amplifier. It is given by

$$V_{\text{in(OFF)}} = \frac{V_{\text{eror}}}{A}$$

10. What is power-bandwidth of an Op-Amp?

    Ans: It is equal to the highest frequency that can be amplified by an Op-Amp without slew rate distortion.

Problems

1. The slew rate of LM675 is 8 V/μs. What is its power bandwidth for a peak output voltage of 5 V. [**Ans:** 256 KHz]
2. Calculate the closed loop voltage gain of an Op-Amp having $R_f = 3$ kΩ, $R_1 = 150$ Ω and input signal is 25 mV operating in inverting mode. Also calculate output voltage at 1 KHz assuming $f_{\text{unity}} = 20$ MHz. Find its bandwidth. [**Ans:** 20,500 mVpp, 1 MHz]
3. Calculate the closed loop voltage gain of an Op-Amp operating in non-inverting mode having $R_f = 3.9$ kΩ, $R_1 = 100$ Ω and input signal is 50 mV pp. Find its bandwidth and output voltage at 1 KHz assuming $f_{\text{unity}} = 1$ MHz. [**Ans:** 40, 25 KHz, 2 V]
4. In an Op-Amp having three channels, the feedback resistor is 100 KΩ. The respective channels having resistors 20 kΩ, 10 kΩ and 50 kΩ are connected to 100 mV, 200 mV and 300 mV signal. Calculate the output voltage. [**Ans:** 3.1 V]
5. An AC voltage source of 10 mV with an internal resistance of 100 kΩ drives a voltage follower. The load connected is 1 kΩ. Calculate output voltage and bandwidth if 741C Op-Amp is used. [**Hint:** $f_{2\text{CL}} = 1$ MHz] (10 mV, 1 MHz)
6. What is the closed loop voltage gain for each position of switch in the circuit given:

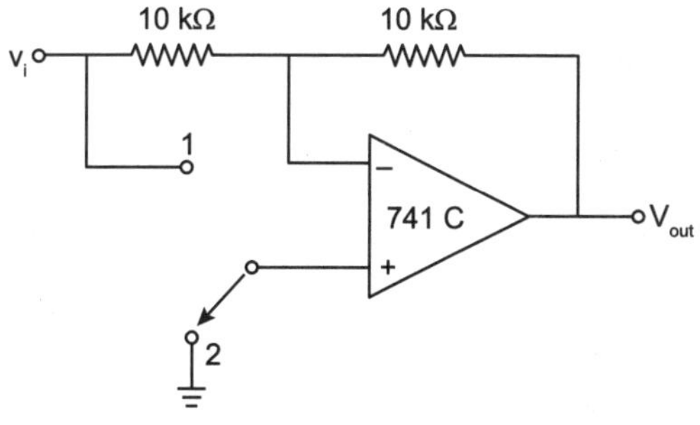

Figure 10.29

(i) When switch connected to ground $A_{CL} = 1$

(ii) When switch connected to terminal $A_{CL} = 2$

7. What is the AC output voltage of the circuit?

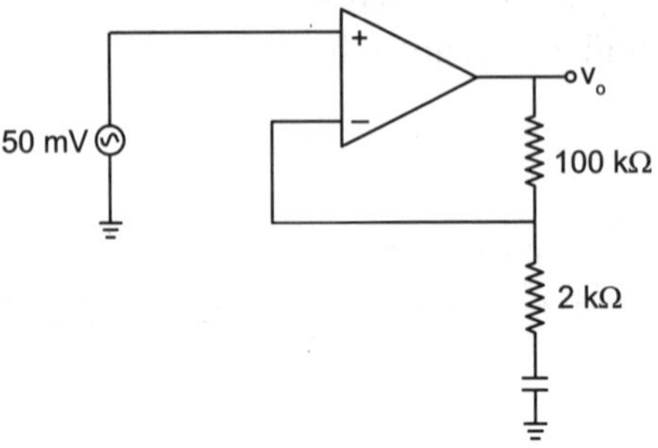

Figure 10.30

[**Ans:** 2.55 V]

8. In the circuit, the technician leaves the ground off to a 6 kΩ resistor. What is the closed loop gain in each position of switch?

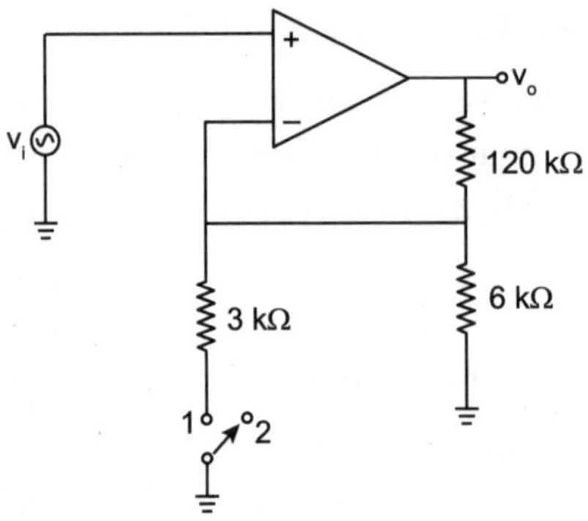

Figure 10.31

(i) When switch is in contact with terminal 1

$A_{CL} = 41$

(ii) When switch is in contact with terminal 2

$A_{CL} = 1$

9. In a differentiator circuit using Op-Amp having R = 100 kΩ and C = 1 μF, a sinusoidal signal of peak value 5 mV at 1000 Hz is impressed upon. What is the output voltage.

[**Ans:** $-\pi \cos(2000\ \pi t)$ V]

10. In an integrated circuit using Op-Amp having $R = 10^6\ \Omega$ and C = 0.5 μF, a [0.4 t] V signal is impressed. Find the output voltage. [**Ans:** $-\ 0.4\ t^2$ V]

11. The voltage gain of an Op-Amp used in a unity gain buffer is 10000. Op-Amp input resistance is 50000 Ω and output resistance is 200 Ω. Calculate the input and output resistance of unity gain buffer circuit. [**Ans:** $5 \times 10^8\ \Omega$, 0.02 Ω]

12. Calculate the closed loop gain and bandwidth of the circuit given. Calculate output voltage at 1 KHz. Assume $f_{unity} = 20$ MHz.

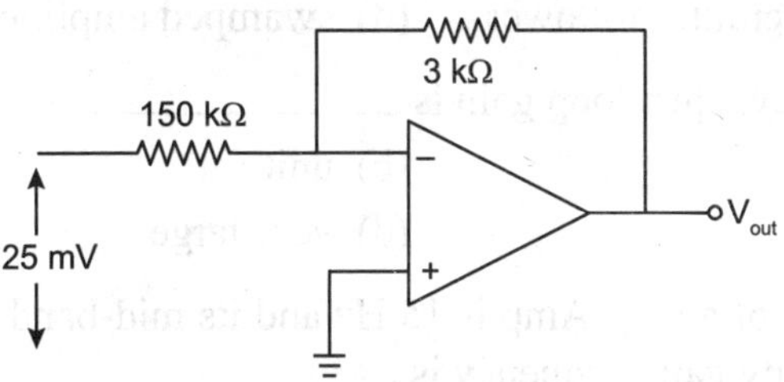

Figure 10.32

[**Hint:** $f_{unity} = A_{CL} f_{2CL}$] [20, 1 MHz, 0.2 V]

13. Calculate the closed loop voltage gain and bandwidth of the circuit given. Find output voltage at 1 KHz. What will be the output voltage if V_{in} is reduced to zero?
Assume $f_{unity} = 20$ MHz.

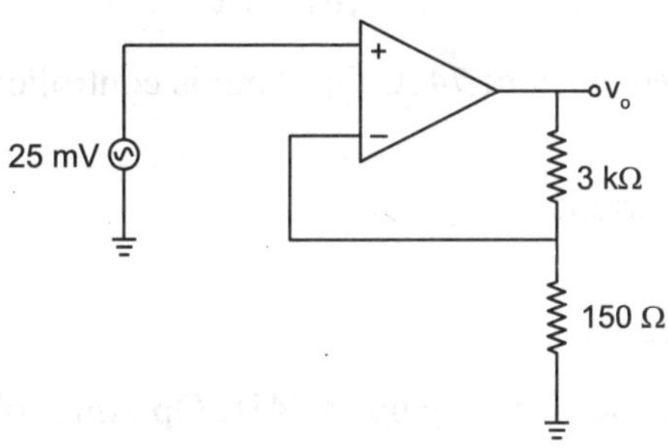

Figure 10.33

[**Ans:** 21, 9.5×10^5 Hz, 525 mV]

Objective Questions

1. The voltage gain of an Op-Amp with a single-ended output is R_C divided by
   (a) r'_e (b) $r'_{e/2}$
   (c) $2r'_e$ (d) $2\,R_E$ [**Ans:** (c)]
2. The common mode rejection ratio of an Op-Amp is ______________.
   (a) $\dfrac{A_{CM}}{A}$ (b) $\dfrac{A}{A_{CM}}$
   (c) $A - A_{CM}$ (d) $A_{CM} - A$ [**Ans:** (b)]
3. The common mode voltage gain of a differential amplifier is R_C divided by
   (a) r'_e (b) $r'_{e/2}$
   (c) $2r'_e$ (d) $2\,R_E$ [**Ans:** (d)]

4. The input stage of an Op-Amp usually is ____________.
   (a) differential amplifier (b) class B push-pull amplifier
   (c) class B push-pull emitter follower (d) swamped amplifier **[Ans:** (a)]
5. At unity gain frequency, open loop gain is ____________.
   (a) zero (b) unity
   (c) infinity (d) very large **[Ans:** (b)]
6. The cut-off frequency of an Op-Amp is 15 Hz and its mid-band open loop voltage gain is 1000,000, then the unity gain frequency is
   (a) 25 Hz (b) 1.5 MHz
   (c) 1 MHz (d) 15 MHz **[Ans:** (d)]
7. The two supply voltages in an Op-Amp are ± 15 V. The MPP value of the Op-Amp is close to
   (a) 0 (b) (–) 15 V
   (c) + 15 V (d) 30 V **[Ans:** (d)]
8. The open loop cut-off frequency of 741C Op-Amp is controlled by
   (a) a coupling capacitor
   (b) output short circuit current
   (c) power bandwidth
   (d) a compensating capacitor **[Ans:** (a)]
9. Above cut-off frequency, the voltage gain of 741C Op-Amp rolls down
   (a) 10 dB per decade (b) 20 dB per octave
   (c) 10 dB per octave (d) 20 dB per decade **[Ans:** (d)]
10. Op-Amp when used as voltage follower has ____________.
    (a) closed loop voltage gain of unity
    (b) open loop voltage gain of unity
    (c) closed loop bandwidth infinity
    (d) open loop bandwidth infinity **[Ans:** (a)]
11. An ideal Op-Amp draws ____________.
    (a) zero current (b) sum of two input currents
    (c) difference of two input currents (d) average of two input currents **[Ans:** (a)]
12. For an Op-Amp used as scale changer ____________.
    (a) $R_f = R_1$ (b) $R_f = k\,R_1$
    (c) $R_f = R_1 = 0$ (d) $R_f = 0, R_1 = \alpha$ **[Ans:** (b)]

Mark the Following Statements True/False.

1. The voltage gain of a single-ended Op-Amp is greater than that of Op-Amp having differential output. **[Ans:** True]
2. The cut-off frequency of an Op-Amp is generally of the order of hundreds. **[Ans:** False]
3. When the initial slope of a sine wave is greater than the slew rate, the function of Op-Amp is at its best. **[Ans:** False]
4. Power bandwidth of an Op-Amp increases if our application can accept a peak output voltage of 1 V. **[Ans:** True]
5. The unity gain frequency equals the product of closed loop voltage gain and closed loop cut-off frequency. **[Ans:** True]
6. The open loop cut-off frequency of an Op-Amp cannot be controlled by the compensating capacitor present. **[Ans:** True]
7. A summing amplifier can have no more than two input signals. **[Ans:** False]
8. When slew rate distortion of a sine wave occurs, the output appears as a triangular wave. **[Ans:** True]
9. Gain-bandwidth product of an Op-Amp does not remain constant. **[Ans:** False]
10. Virtual ground exists in an Op-Amp. **[Ans:** True]

Fill in the Blanks.

1. A differential amplifier has ____________ transistors and ____________ inputs. **[Ans:** two, two]
2. The differential output voltage of an Op-Amp is ____________ than the single-ended output voltage for differential input in both cases. **[Ans:** less]
3. CMRR of an Op-Amp is ____________ of differential voltage gain and ____________. **[Ans:** ratio, common mode gain]
4. Input offset current is ____________ of ____________ DC base currents of Op-Amp. **[Ans:** difference, two]
5. Common mode voltage gain of an ____________ is ____________ than its usual voltage gain. **[Ans:** Op-Amp, less]
6. Unity gain frequency of an Op-Amp is referred as ____________. **[Ans:** gain bandwidth product]
7. The output signal of Op-Amp in inverting mode has a phase difference of ____________ with input signal. **[Ans:** 180°]
8. The output impedance of an Op-Amp is ____________ and its input impedance is ____________. **[Ans:** low, high]

Chapter

11 Power Amplifiers

11.1 Introduction

We have studied small signal amplifiers (Chapter 7) where the input signal was considered to be small, i.e. only the linear portion of the characteristic curve of the device is involved so that the output signal is quite similar in shape to the input signal and there is no distortion. In such cases, the primary aim is to amplify the input voltage or current level and we can use simple mathematics since the device behaves like a linear device.

However, when the input signal is large, a large portion of the characteristic is covered, which includes the non-linear part also, and so the shape of the output signal is distorted and mathematical solving is quite complex. In such cases, we discuss the behaviour of the device by making use of graphical methods.

In case of a large input signal, it may be possible that the device may not conduct through the entire cycle of input signal, and accordingly amplifiers are classified as per the angle or time period of the signal for which the device conducts. It is to be noted that power amplifiers are not like simple voltage/current amplifiers; it actually draws DC power from the battery/power supply connected in the circuit and converts it into useful AC power.

A comparison of voltage and power amplifiers is shown in Table 11.1.

Table 11.1 Comparison of power and voltage amplifiers

S. No.	Parameters	Power amplifier	Voltage amplifier
1.	Coupling	Invariably transformer coupling	R–C coupling
2.	Power output	Very high	Low
3.	Collector current	Very high ($\approx$ 100 mA)	Low (10 mA)
4.	Output impedance	Low (100 to 200 Ω)	High ($\simeq$ 10 kΩ)
5.	Input voltage	High (2 to 5 V)	Very low (2 to 5 mV)
6.	Current amplification factor (β)	Low (25 to 50)	High ($\approx$ 100)
7.	Collector load (R_C)	Low (0 to 5 Ω)	Quite high (5 to 10 kΩ)

11.2 Classification of Power Amplifiers

With respect to the period or angle of conduction of the device, amplifiers have been classified mainly into three categories:

1. Class A angle of conduction 2π
2. Class B angle of conduction π
3. Class C angle of conduction less than π

One more class exists which is known as class AB.

11.2.1 Class A Operation

Here, the operating point is kept in the middle of the dynamic transfer characteristic (Figure 11.1(a)) so that the device conducts for the entire period of input signal, or output current flows for 0–360°. In this condition, voltage amplification is prominent and distortion in the output signal is zero or minimum since the input signal is small. Its conversion efficiency is minimum in comparison with other classes, which is nearly 25% and the maximum value is 50%.

11.2.2 Class B Operation

Here, the operating point is not kept in the middle of the dynamic transfer characteristic but it is so DC biased that the operating point Q is kept at cutoff point 0, i.e. both Q and 0 coincide (Figure 11.1(b)) so that the device conducts for the half period of current input signal, or output current flows for 0 to π and 2π to 3π, etc., i.e. flows for positive half cycle only and no current flows during negative half cycle (π to 2π). In this situation, distortion is much more than class A operation. Its conversion efficiency is quite high and its value is nearly 78%.

11.2.3 Class C Operation

Here, the DC biasing of the amplifier is such that the operating point Q is kept much beyond cutoff point 0 [Figure 11.1(c)] so that the device conducts for less than the half period or 180° of the input signal, or output current flows for a very short period lasting from θ_1 to θ_2 [Figure 11.1(c)] in positive half cycle. In this situation, distortion is much more than class A or class B operation. Its conversion efficiency is maximum among the three classes and ranges from 80–100%.

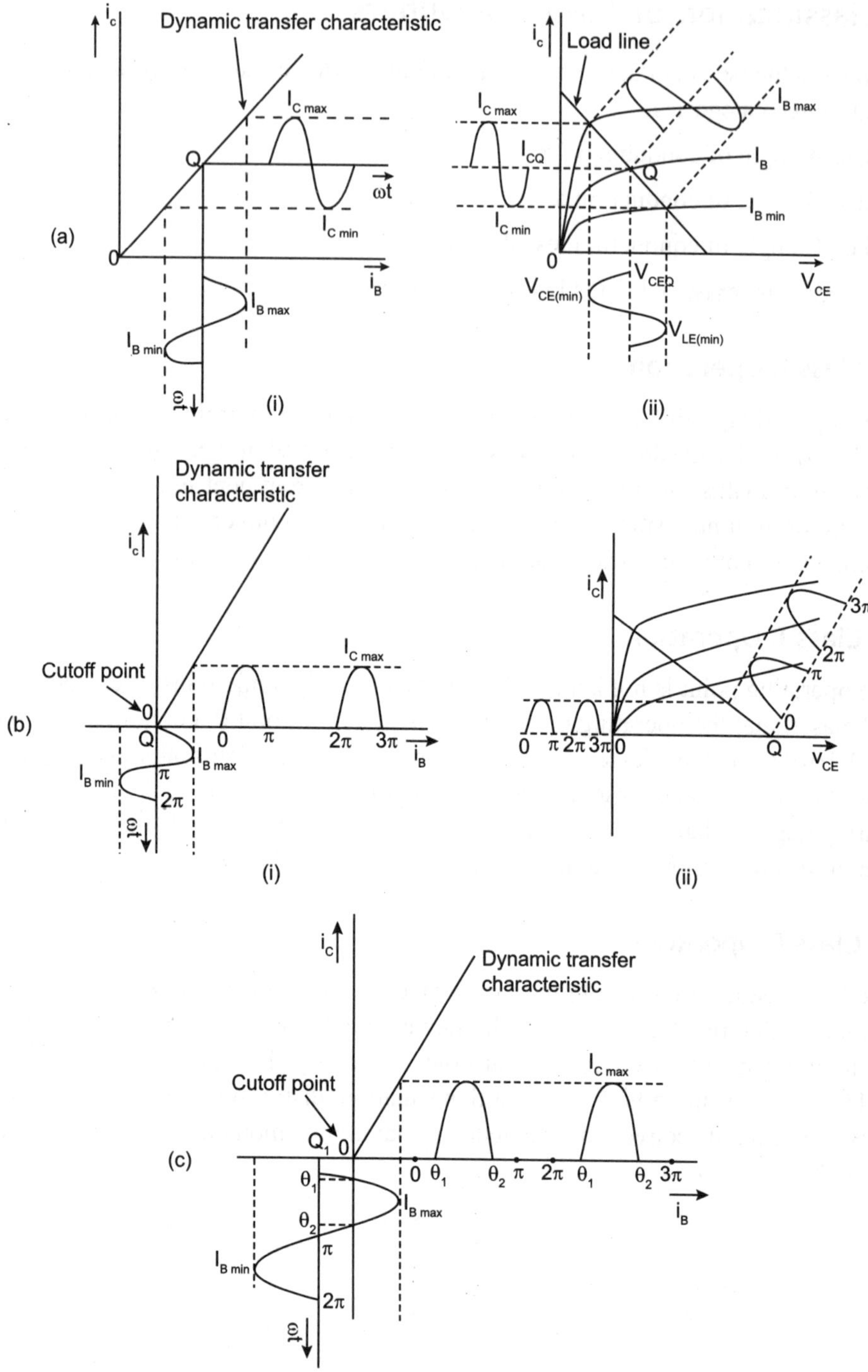

Dynamic transfer characteristic
i_c
$I_{C\,max}$
Q
ωt
$I_{C\,min}$
(a)
0
i_B
$I_{B\,max}$
$I_{B\,min}$
ωt
(i)
Load line
$I_{B\,max}$
I_{CQ}
I_B
$I_{B\,min}$
V_{CEQ}
$V_{CE(min)}$
V_{CE}
$V_{LE(min)}$
(ii)
Dynamic transfer characteristic
Cutoff point
(b)
$I_{C\,max}$
$0\ \pi\ 2\pi\ 3\pi$
$I_{B\,max}$
$I_{B\,min}$
(c)
Q_1
θ_1
θ_2

Figure 11.1

11.3 Conversion Efficiency of Class A Amplifier

We shall calculate the conversion efficiency of different types of amplifiers operating under class A.

The circuit of direct-coupled CE amplifier is shown in Figure 11.2(a). The operating point Q on load line (Figure 11.2(b)) has been chosen in the middle so that the operation is under class A, i.e. output current flows for the entire period of input signal (0–360°). Let the DC voltage and current flowing be V_{CEQ} and I_{CQ}, respectively, before applying the signal.

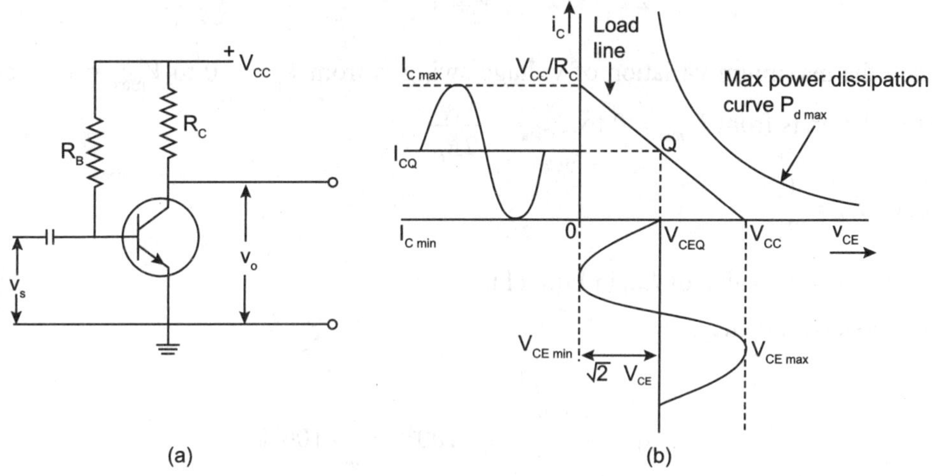

Figure 11.2 (a) Direct coupled amplifier circuit, (b) variation of voltage and current

As the input signal V_s is applied to the base, current varies (not shown in diagram) which causes collector current I_{CQ} to vary thereby voltage across the load varies. The maximum variation of voltage V_{CEQ} can be $V_{CE\min} = 0$ and $V_{CE\max} = V_{CC}$. So, the peak value of voltage variation is $\sqrt{2}\,V_{CE}$.

The maximum AC power obtained across the load is

$$p_{\text{ac(max)}} = P_o = I_{c(rms)}\,V_{C(rms)}$$

$$= \frac{I_C(p-p)}{2\sqrt{2}} \times \frac{V_C(p-p)}{2\sqrt{2}}$$

From the graph (Figure 11.1(b)), $I_{C(p-p)} = V_{CC}/R_C$ and $V_{Cp-p} = V_{CC}$. Substituting, we have

$$P_{\text{ac(max)}} = P_o = \frac{\left(\dfrac{V_{CC}}{R_C}\right)V_{CC}}{8} = \frac{V_{CC}^2}{8R_C}$$

DC power drawn from the power supply (battery) is

$$P_{i(\text{dc})} = V_{CC}\ I_{CQ}$$

By definition,

$$\text{Conversion efficiency, } \eta = \frac{\text{Maximum AC power from load}}{\text{DC power given to device}}$$

$$= \frac{V_{CC}^2/8R_C}{V_{CC}\ I_{CQ}} \qquad (1)$$

Since the maximum variation of voltage swing is from $V_{\min} = 0$ to $V_{\max} = V_{CC}$, collector current variation is from $I_{C\min} = 0$ to $I_{C\max} = \dfrac{V_{CC}}{2R_C}$

Hence $I_{CQ} = \dfrac{V_{CC}}{2R_C}$

Substituting the value of I_{CQ} in Eqn. (1),

Conversion efficiency,

$$\eta = \frac{\dfrac{V_{CC}^2}{8R_C}}{V_{CC}\left[\dfrac{V_{CC}}{2R_C}\right]} \times 100\% = \frac{1}{4} \times 100\%$$

$$= 25\% \qquad (2)$$

From the graph, it is to be noted that the total power received from power supply (battery) is

$$P_{in} = P_{\text{dc}} = V_{CC} \times I_{CQ} = 2V_{CEQ} \cdot I_{CQ}$$

This power is distributed in two parts:

(i) DC power to load $[P_L]_{\text{dc}} = I_{CQ}^2 R_C$

This entire power is dissipated in the form of heat.

(ii) DC power to device (transistor) $[P_{Tr}]_{\text{dc}}$

So $[P_{Tr}]_{\text{dc}} = P_{in} - P_L$

This power is divided into two parts:

(a) One part appears as AC power in load.

$$[P_o]_{\text{ac}} = [I_{rms}]^2 R_L$$

(b) Another part is dissipated in the form of heat in transistor. This is expressed as $[P_C]_{dc}$. As per the energy conservation law, we have

$$[P_{in}]_{dc} = [P_L]_{dc} + [P_O]_{ac} + [P_C]_{dc} \tag{3}$$

11.4 Transformer-Coupled Amplifier

Normally in a voltage amplifier, the load is directly connected to the battery (power supply). So, it draws current directly from the battery and power is dissipated in it in the form of heat. Moreover, the load connected may not match with the device; so the power transferred may not be maximum.

For the above reasons, we generally use a transformer as a coupling device in the circuit (Figure 11.3(a)); so the DC power lost is minimized and we can also match the load for maximum power transfer by adjusting the turn ratio in the transformer.

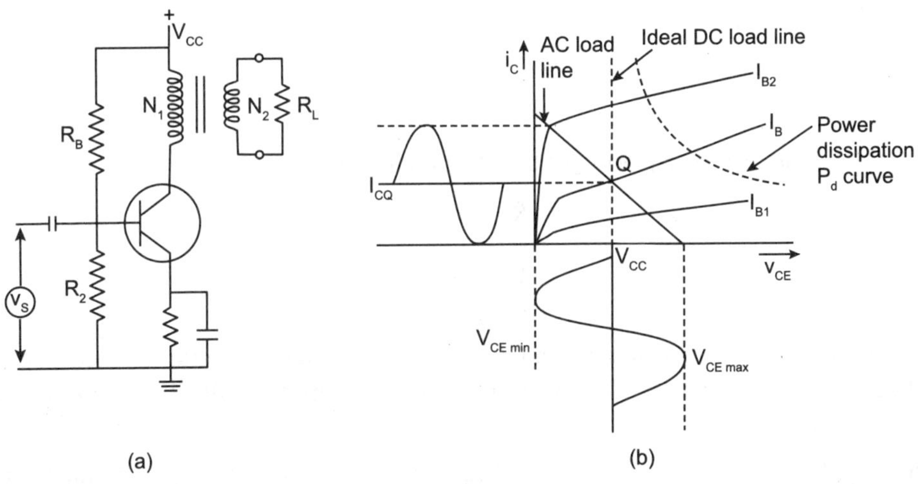

Figure 11.3 (a) Transformer-coupled amplifier, (b) variation of voltage and current in TC amplifier

For an ideal transformer, the DC load connected is zero. Otherwise, it is low resistance of the primary coil which is connected to the battery (Figure 11.3(a)). Here, we assume the operating point Q is in the middle of AC load line (Figure 11.3(b)) so that the operation of amplifier is under class A. It is also the circuit in which maximum possible variation of V_{CE} is achieved, hence the variation on two sides is equal (Figure 11.3(b)).

In an ideal transformer, N_1 is the number of turn in primary coil and N_2 the number of turns in secondary coil. Let V_1 and I_1 be the primary AC voltage and current, respectively, while V_2 and I_2 be the induced secondary voltage and current respectively.

Now, actual load $R_L = \dfrac{V_2}{I_2}$

Let the AC load that could be measured across the primary terminals be designated as R'_L; then $R'_L = \frac{V_1}{I_1}$ (also called reflected load).

We know that for a transformer

$$\frac{V_1}{V_2} = \frac{N_1}{N_2}$$

and

$$\frac{I_1}{I_2} = \frac{N_2}{N_1}$$

so

$$V_1 = \frac{N_1}{N_2} V_2$$

$$I_1 = \frac{N_2}{N_1} I_2$$

Substituting these values, we have

$$R'_L = \frac{(N_1/N_2)V_2}{(N_2/N_1)I_2} = \left(\frac{N_1}{N_2}\right)^2 \frac{V_2}{I_2}$$

Therefore, load as seen by primary or reflected load is

$$R'_L = a^2\ R_L \tag{4}$$

where

$$a = \frac{N_1}{N_2} \text{(turn ratio)}$$

From Eqn. (4), it is clear that we can select a transformer turn ratio 'a' such that Eqn. (4) is satisfied and there is transfer of maximum power as desired, resulting into minimum power loss and increased efficiency.

The (V_{CE}, I_C) characteristic is shown in Figure 11.3(b). Assuming an ideal transformer, (DC load is zero) load line is a vertical line drawn at V_{CC} and passing through the operating point Q, AC load line is drawn through the operating point Q such that its slope with V_{CE} is given by $\tan\theta = (-)\frac{1}{R'_L}$. On applying signal V_S across the base, the base current varies. Now let the collector current flowing through primary be I_C and the rms value of voltage across primary be V_C AC power.

$$[P_o]_{ac} = I_{Crms}\ V_{Crms}$$

$$= \frac{I_{Cpp}}{2\sqrt{2}}\ \frac{V_{Cpp}}{2\sqrt{2}} \text{ (from Figure 3(b))}$$

where I_{Cpp} and V_{Cpp} represent the peak to peak value of current and voltage. From Figure 11.3(b), we have

$$P_o = \frac{\left[\dfrac{2V_{CC}}{R'_L}\right][2V_{CC}]}{8} = \frac{V_{CC}^2}{2R'_L} \tag{5}$$

$$[P_i]_{dc} = V_{CC}\, I_{CQ} = V_{CC} \cdot \frac{V_{CC}}{R'_L} = \frac{V_{CC}^2}{R'_L} \tag{6}$$

By definition, conversion efficiency $\eta = \dfrac{\text{Maximum useful AC power from load}}{\text{DC power given to device}}$

$$\therefore\ \eta = \frac{V_{CC}^2 / 2R'_L}{V_{CC}^2 / R'_L} = \frac{1}{2} \times 100\%$$

$$= 50\% \tag{7}$$

So the maximum value of conversion efficiency in case of transformer-coupled amplifier under class A operation is 50%.

11.4.1 Characteristic of Class A Amplifier

The device conducts (output current flows) throughout the entire cycle of input signal. The input signal is confined across the linear region of dynamic curve, and so the shape of output signal is similar to that of input signal.

The operating point is selected in the middle of load line so that there is no clipping of peak of either output voltage or current wave whenever there is a variation of base current on applying the input signal. The AC power output is less in this case; consequently efficiency is less. Further, it is to be noted that in the absence of input signal, the entire DC power supplied is lost in the form of heat.

11.5 Transformer-Coupled Amplifier Design

The design of a transformer-coupled amplifier commences with a specification of load R_L and power to be delivered to it. A signal voltage amplitude may be specified, as well as the lower and upper cut-off frequencies. If a supply voltage is given, the design must determine the specification for the transformer.

Power delivered to the transformer primary, P'_o, is determined using output power and transformer efficiency.

$$P'_o = \frac{P_o}{\eta}$$

Also, power delivered to primary can be expressed as

$$P'_o = \frac{(V_{rms})^2}{R'_L}$$

where V_{rms} is primary voltage

R'_L is reflected load towards primary

Using peak voltages, the above relation can be expressed as

$$P'_o = \frac{(V_P/\sqrt{2})^2}{R'_L} = \frac{V_P^2}{2R'_L}$$

$$\therefore R'_L = \frac{V_P^2}{2P'_o}$$

Also, the reflected load towards primary in a transformer is $R'_L = \left(\frac{N_1}{N_2}\right)^2 R_L$ for an usual transformer.

Impedance seen when looking into the whole winding of a transformer with centre tapping is

$$R''_L = \left(\frac{2N_1}{N_2}\right)^2 R_L = 4\left(\frac{N_1}{N_2}\right)^2 R_L = 4R'_L$$

Referring to Figure 11.10, we note that when Q_1 is OFF and Q_2 is ON, a voltage with a peak value of approximately $+V_{CC}$ is induced in half of the transformer primary winding connected to Q_1. Induced voltage occurs because the other half of primary has $+V_{CC}$ applied to it via Q_2 and because both primary half windings are on same core. This induced voltage is superimposed upon the supply so that voltage that appears at the collector of Q_1 is now

$$V_{CE(\max)} = V_{CC} + V_{CC}$$
$$= 2\,V_{CC}$$

Example 1 In RC-coupled amplifier, the various components used are as under: – $V_{CC} = 15$ V, biasing resistors $R_1 = 2$ kΩ and $R_2 = 470$ Ω. The load connected in the collector $R_C = 680$ Ω and external coupled load $R_L = 2.7$ kΩ. The resistor connected in emitter lead, $R_E = 220$ Ω. Calculate (i) $I_{\text{dc(sat)}}$, (ii) DC and AC load resistance, (iii) I_{CQ}, (iv) V_{CEQ}, (v) efficiency of amplifier.

Solution

(i) $$I_{\text{dc(sat)}} = \frac{V_{CC}}{R_C + R_E}$$

$$= \frac{15}{680+220} = \frac{15}{900} = 16.7 \text{ mA}$$

(ii) DC load = 680 Ω

(iii) AC load $R'_L = \frac{R_C R_L}{R_C + R_L} = \frac{680 \times 2700}{680+2700} = \frac{1836000}{3380} = 543\ \Omega$

(iv) $I_{CQ} = \dfrac{V_B - 0.7}{R_E}$

$$V_B = \frac{R_2}{R_1 + R_2} V_{CC} = \frac{470}{2470} \times 15 = 2.89$$

$$\therefore I_{CQ} = \frac{2.89 - 0.7}{220} = \frac{2.19}{220} = 9.9 \text{ mA}$$

(v) $V_{CEQ} = V_C - V_{R_E}$

$$V_C = V_{CC} - I_C \times R_C = 15 - (9.9 \times 10^{-3} \times 680) = 8.2\,\text{V}$$

$$\therefore V_{CEQ} = 8.2 - 9.9 \times 10^{-3} \times 220 = 6 \text{ V}$$

$$\eta = \frac{P_{\text{ac}}}{P_{\text{dc}}} = \frac{\dfrac{V^2}{8R_L}}{V_{CC} I_{\text{dc}}} = \frac{2V_{CE}^2}{8R_L} \frac{12 \times \dfrac{12}{8} \times 2700}{15 \times 9.9 \times 10^{-3}} \approx 4\%$$

11.6 Conversion Efficiency of Class B Amplifier

Here, the device conducts only during the positive half cycle of input signal, i.e. output current flows for 0–180°. For this, operating point is set at the cut off point of dynamic transfer characteristic (Figure 11.1(b)) or at the end of AC load line. Since the device is biased at the cutoff point, no current flows through it in the absence of a signal, thereby no DC power is dissipated in the device till a signal is applied. But, on applying the signal, the device conducts and draws current during positive half of signal, which increases with increase in signal amplitude, so average current flows through the device. Thus, during this interval, some power is dissipated into the device, but in case of class A operation, power is dissipated through the device even if a signal is absent. The average value of current under class B operation is less than that of class A operation, and so power dissipated is also much less.

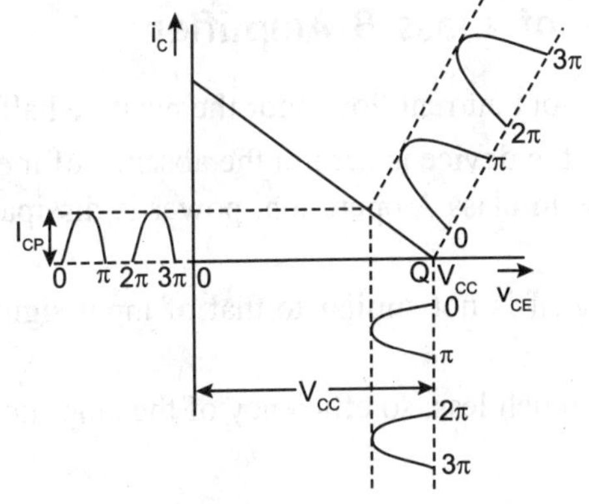

Figure 11.4

From Figure 11.4, it is clear that during the positive half cycle of input signal (base current variation), collector current flows. As collector current flows, potential drop across the collector and emitter V_{CE} drops. The maximum swing of V_{CE} can be equal to V_{CC}; let the peak value of collector current be I_{CP}.

$$\text{Output power } P_{o(\text{ac})} = V_{RMS}\, I_{RMS}$$

$$= \left[\frac{V_{CC}}{\sqrt{2}}\right]\left[\frac{I_{CP}}{\sqrt{2}}\right]$$

$$= \frac{1}{2} V_{CC}\, I_{CP} \tag{8}$$

$$\text{Input power, } P_{i(\text{dc})} = V_{CC}\, I_{\text{dc}} \tag{9}$$

By Fourier analysis, $I_{\text{dc}} = \dfrac{2I_{CP}}{\pi}$ since we get only half sinusoidal pulse in output.

$$\text{So, } P_{i(\text{dc})} = V_{CC}\left[\frac{2I_{CP}}{\pi}\right]$$

$$\text{Conversion efficiency, } \eta = \frac{P_{o(\text{ac})}}{P_{i(\text{dc})}} \times 100\%$$

$$= \frac{\frac{1}{2} V_{CC} I_{CP}}{\frac{2V_{CC} I_C}{\pi}} \times 100\%$$

$$= \frac{\pi}{4} \times 100\%$$

$$= 78.5 \tag{10}$$

11.7 Characteristics of Class B Amplifier

1. The device conducts (output current flows) for the positive half cycle of input signal.
2. DC power dissipated in the device is zero in the absence of input signal. It increases with signal strength, whereas in class A operation, power is dissipated even in the absence of signal.
3. The shape of output signal is not similar to that of input signal, but it is half sinusoidal pulse.
4. DC power dissipated is much less; so efficiency of the amplifier is much higher than class A operation.

11.8 Conversion Efficiency of Class C Operation

Here, the operating point *Q* is set much beyond the cut off point of dynamic transfer characteristic (Figure 11.5), and so the device conducts, i.e. output current flows only for a small interval of positive half cycle; now i.e. conduction angle $\theta = [\theta_2 - \theta_1]$ which is much less than 180°. Consequently, output current is in the form of pulses and not half-sinusoidal or sinusoidal as in case of class B or class A operation.

No current flows through the device in the absence of input signal. Even in the presence of a signal, current exists for a very small interval of positive half cycle.

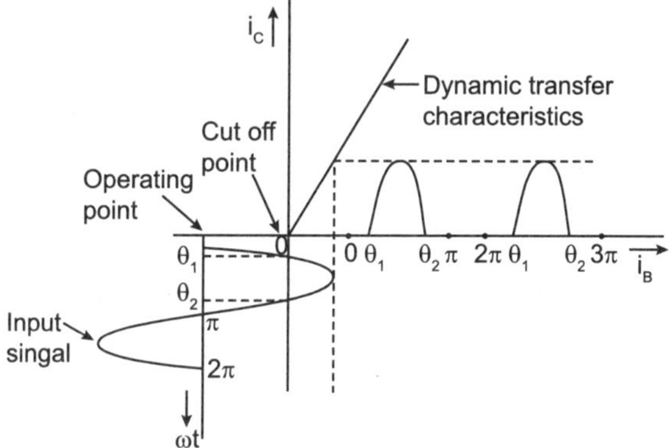

Figure 11.5

So, current pulse exists for $\theta = [\theta_2 - \theta_1] < 180°$.

Mathematical derivation of the expression for efficiency is not possible, but by graphical analysis, we can prove that efficiency reaches 100%.

11.9 Distortion due to Non-Linear Characteristics

When discussing the working of different amplifiers, we have assumed the dynamic transfer characteristic to be linear, so distortion in the output signal under class A operation is negligible. But actually, the transfer characteristic is not linear and it is curved; so distortion appears in the output signal. To calculate distortion, we make use of graphical analysis.

Consider the characteristic curve (Figure 11.6) relating output and input quantities, which can be used to predict the output wave form resulting from arbitrary driving functions.

The input signal which is sinusoidal is applied at the operating point *Q*. The output signal obtained is not sinusoidal. Here, the positive half cycle is magnified but the negative half cycle is rather reduced due to curvature present in the characteristic. Distortion in the output signal represents that there are harmonics in addition to fundamental frequency. To calculate the amount of distortion, we make use of Tylor series around the operating point of dynamic curve. The relation between current *i* and applied voltage *e* is given by

$$i = a_o + a_1(e - e_Q) + a_2(e - e_Q)^2 + a_3(e - e_Q)^3 + \cdots \quad (11)$$

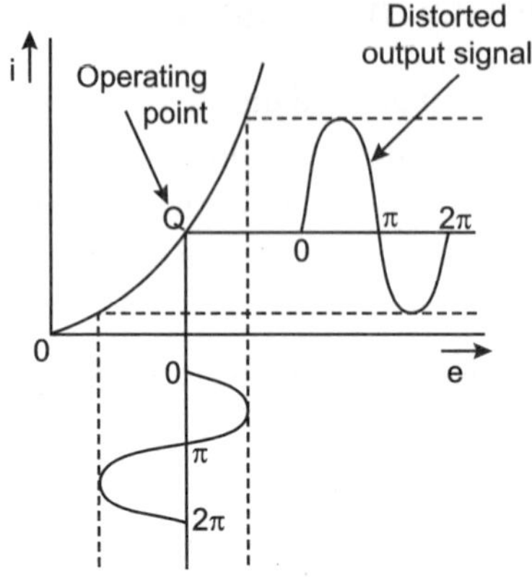

Figure 11.6

Let the input signal be

$$e - e_Q = E_m \cos \omega t$$

Substituting this value in Eqn. (11), we have for transistor output current

$$i_c = a_0 + a_1 E_m \cos \omega t + a_2 E_m^2 \cos^2 \omega t + a_3 E_m^3 \cos^3 \omega t + \cdots + a_4 \cos^4 \omega t + \cdots \quad (12)$$

Eqn. (12) can be expressed as

$$i_c = a_o + A_0 + A_1 \cos \omega t + A_2 \cos 2\omega t + A_3 \cos 3\omega t + A_4 \cos 4\omega t + \cdots \quad (13)$$

where a_0, A_0, A_1 and A_2 are constants to be determined. If input signal is absent, then current flowing through the device (transistor) is

$$i = I_C$$

So the value of $a_o = I_c$

The first two terms in Eqn. (13) represent average current (DC current) with signal present.

So, $a_0 + A_o = I_C + A_o = I_{Cs}$

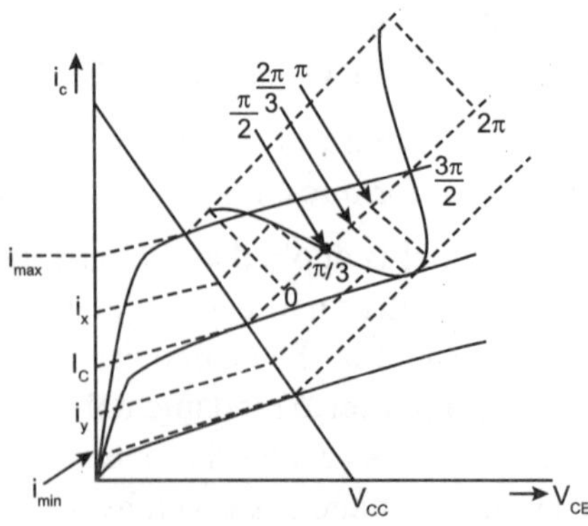

Figure 11.7

The amplitude of distortion components can be determined by the evaluation of A coefficients.

We can determine the values of A_0, A_1, A_2, A_3, A_4 by considering five points on load line. These five points are selected at the following angles (ωt):

(i) $\omega t = 0$, current is maximum at this instant

(ii) $\omega t = \frac{\pi}{3}$, current is i_x

(iii) $\omega t = \frac{\pi}{2}$, current is I_C since input signal is zero

(iv) $\omega t = \frac{2\pi}{3}$, current is i_y

(v) $\omega t = \pi$, current is minimum at this instant.

Substituting these values one by one in Eqn. (13), we have

$$\text{at } \omega t = 0,\ i_{\max} = I_C + A_0 + A_1 + A_2 + A_3 + A_4 \tag{a}$$

$$\text{at } \omega t = 60^\circ,\ i_x = I_C + A_0 + \frac{A_1}{2} - \frac{A_2}{2} - A_3 - \frac{A_4}{2} \tag{b}$$

$$\text{at } \omega t = 90^\circ,\ i_y = I_C + A_0 - A_2 + A_4 \tag{c}$$

$$\text{at } \omega t = 120^\circ,\ i_y = I_C + A_0 - \frac{A_1}{2} - \frac{A_2}{2} + A_3 - \frac{A_4}{2} \tag{d}$$

$$\text{at } \omega t = 180^\circ,\ i_{\min} = I_C + A_0 - A_1 + A_2 + A_3 - A_4 \tag{e}$$

Using the above equations, the values of various coefficients can be determined and we have

$$A_0 = \frac{1}{6}(i_{\max} + i_{\min}) + \frac{1}{3}(i_x + i_y) - I_C \tag{a}$$

$$A_1 = \frac{1}{3}(i_{\max} - i_{\min}) + \frac{1}{3}(i_x - i_y) \tag{b}$$

$$A_2 = \frac{1}{4}(i_{\max} + i_{\min}) - \frac{1}{2}I_C \tag{c}$$

$$A_3 = \frac{1}{6}(i_{\max} - i_{\min}) - \frac{1}{3}(i_x - i_y) \tag{d}$$

$$A_4 = \frac{1}{12}(i_{\max} + i_{\min}) - \frac{1}{3}(i_x + i_y) + \frac{1}{2}I_C \tag{e}$$

It is possible to define percentage distortion due to a particular harmonic as the ratio of harmonic amplitude to that of fundamental times 100 percent.

Distortion at a particular harmonic is usually expressed as a percentage. They are defined as

$$2^{nd} \text{ harmonic distortion, } D_2 = \frac{A_2}{A_1} \times 100\%$$

$$3^{rd} \text{ harmonic distortion, } D_3 = \frac{A_3}{A_1} \times 100\%$$

$$4^{th} \text{ harmonic distortion, } D_4 = \frac{A_4}{A_1} \times 100\% \tag{14}$$

Total harmonic distortion, D, is expressed as

$$D = \sqrt{D_2^2 + D_3^2 + D_4^2}$$

$$= \frac{\sqrt{A_2^2 + A_3^2 + A_4^2}}{A_1} \times 100\% \tag{15}$$

11.9.1 Brief Analysis

If we select three points instead of five, then our discussion to evaluate distortion can be simplified. Accordingly Eqn., (13) can be written as

$$i_C = I_C + C_0 + C_1 \cos \omega t + C_2 \cos 2\omega t \tag{16}$$

Now, the value of constants (three amplitudes) respectively is

$$C_0 = \frac{1}{4}(i_{max} + i_{max}) - \frac{1}{2} I_C$$

$$C_1 = \frac{1}{2}(i_{max} - i_{min})$$

$$C_2 = \frac{1}{4}(i_{max} + i_{min}) - \frac{1}{2} I_C \tag{17}$$

Knowing these amplitude constants, we can determine 2nd harmonic, and by this we delivered at fundamental frequency,

$$P_\omega = \frac{I_\omega^2 R_C}{2} = \frac{A_1^2 R_C}{2}$$

Power delivered at 2nd harmonic,

$$P_{2\omega} = \frac{I_{2\omega}^2 R_C}{2} = \frac{A_2^2 R_C}{2}$$

Total power delivered to load,

$$P_0 = [I_\omega^2 + I_{2\omega}^2 + I_{3\omega}^2]\frac{R_C}{2}$$

$$= \left[1 + \left[\frac{I_{2\omega}}{I_\omega}\right]^2 + \left[\frac{I_{3\omega}}{I_\omega}\right]^2\right] I_\omega^2 \frac{R_C}{2}$$

$$= [1 + D_2^2 + D_3^2 + \cdots]P_\omega$$

$$= [1 + D^2]P_\omega \tag{18}$$

11.10 Push-Pull Amplifier

It is clear from the previous discussion that distortion is always present in the output signal due to curvature present in the characteristic curve.

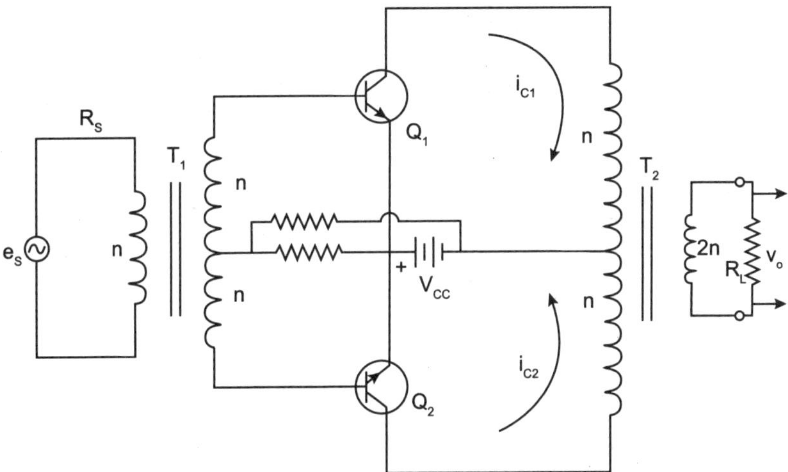

Figure 11.8

To eliminate such distortion (Figure 11.6), we make use two identical devices (transistors) connected as shown in Figure 11.8. Such an arrangement is called push pull. With such an arrangement, power delivered will also be enhanced.

11.10.1 Class A Push-Pull Amplifier

The circuit of this amplifier is shown in Figure 11.8. The two transistors Q_1 and Q_2 are identical. Two transformers T_1 and T_2 are employed in which secondary of T_1 and primary of T_2 is centre tapped.

The number of turns in the primary of T_2 is $2n$ and the number of turns in its secondary is n. The input signal at the bases of transistors Q_1 and Q_2 is e_{S1} and e_{S2}, respectively, and the strength of each signal is equal to input signal e_S connected in the primary of transformer T_1.

Battery current I_{C1} and I_{C2} flowing in the upper half and lower half of the primary of transformer T_2 is also equal, but their direction is opposite to each other; so the effect of magnetic field produced is nullified and there will be no saturation in the iron core. Here, we do not connect any resistor R_E in the emitter lead of transistors as is usual in other class A (CE) amplifiers. The characteristic curves of the transistors are similar since the two transistors are identical.

The magnitude of input signals e_{S1} and e_{s2} to the two transistors is equal, but they have phase difference π, i.e. one is opposite to the other.

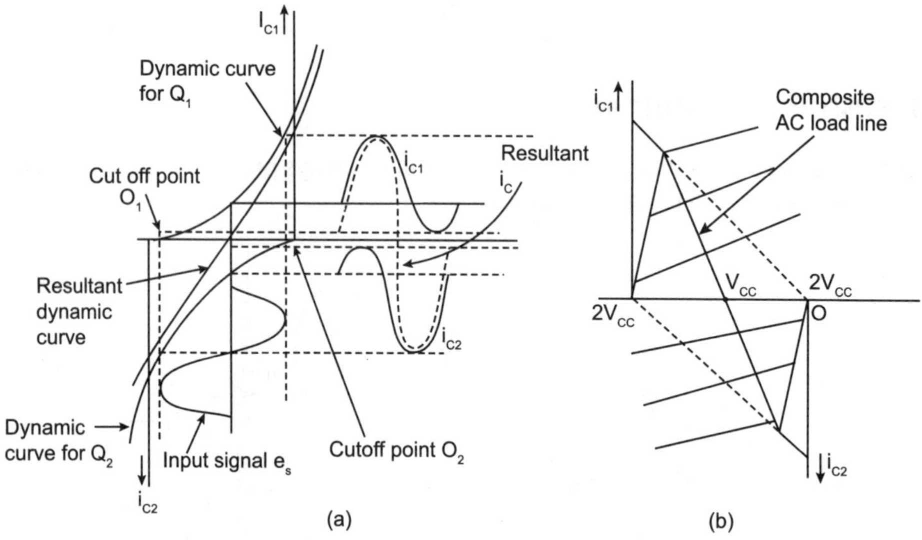

Figure 11.9

$$e_{S1} = E\sin\omega t \text{ and } e_{S2} = E\sin[\omega t + \pi]$$

The relation between output current and input signal is given by Eqn. (12).

For transistor Q_1,

$$i_{C1} = I_{O1} + a_1 E\sin\omega t + a_2 E^2 \sin^2 \omega t + a_3 E^3 \sin^3 \omega t \tag{19}$$

For transistor Q_2,

$$i_{C2} = I_{O2} + a_1 E\sin(\omega t + \pi) + a_2 E^2 \sin^2(\omega t + \pi) + a_3 E^3 \sin^3(\omega t + \pi) \tag{20}$$

Equations (19) and (20) can be written as

$$i_{C1} = I_{O1} + B_0 + B_1 \sin\omega t - B_2 \cos 2\omega t + B_3 \sin 3(\omega t) - B_4 \cos 4\omega t \tag{21}$$

$$i_{C2} = I_{O2} + B_0 + B_1 \sin(\omega t + \pi) - B_2 \cos 2(\omega t + \pi) + B_3 \sin 3(\omega t + \pi) - B_4 \cos 4(\omega t + \pi) \tag{22}$$

But $\sin(\omega t + \pi) = (-)\sin\omega t$ and $\cos 2(\omega t + \pi) = \cos 2\omega t$

So Eqn. (22) can be written as

$$i_{C2} = I_{O2} + B_0 - B_1 \sin\omega t - B_2 \cos\omega t - B_3 \sin 3\omega t - B_4 \sin 4\omega t \tag{23}$$

From the centre point C of primary of transformer T_2, the two collector currents i_{c1} and i_{c2} flow in opposite directions; and with increase in input signal e_s, if total current I_{c1} increases then I_{C2} decreases; it is for this reason the two characteristic curves have been shown as one facing up and the other upside down (Figure 11.9(a)).

These varying currents induce secondary voltage e_o, which is proportional to their difference since they flow in opposite directions in the primary.

$$e_o \alpha [i_{c1} - i_{c2}] \tag{24}$$

Similarly, when input signal e_s decreases, then collector current I_{C1} decreases but I_{C2} increases, and the induced emf in secondary is proportional to their difference. In this way, the action of two transistors is push-pull and output voltage value remains as in the previous case.

From Eqn. (22) and (23),

$$i_{c1} - i_{c2} = 2B_1 \sin \omega t + 2B_3 \sin 3\ \omega t + 2B_5 \sin 5\ \omega t \tag{25}$$

We can say output voltage,

$$\begin{aligned} e_o &= K[i_{c1} - i_{c2}], \text{ where } K \text{ is constant} \\ &= 2K[B_1 \sin \omega t + B_3 \sin 3\ \omega t + B_5 \sin 5\ \omega t] \end{aligned} \tag{26}$$

From Eqn. (26), it is clear that 2nd harmonic is not present, and as such distortion is much less since only higher harmonic 3ω, 5ω are present and they do not affect much the output (wave) signal shape because their amplitude is much less than fundamental. The resultant current shape is shown in Figure 11.9(a), which is very nearly a sinusoidal wave as per the input signal. Since the resultant dynamic characteristic is nearly a linear or straight line the input signal is so applied that the operation is limited to class A, i.e. swing of input signal is that current flows for the entire period of input cycle. When Q_1 conducts, then Q_2 is OFF, and when Q_1 is OFF, then Q_2 is ON.

To determine the composite load line, the (V_{CE}, I_C) characteristic of the two transistors (which are identical) are drawn so that one is up and another is down and so adjusted that the load line for one transistor is drawn between $2V_{CC}$ and I_{cm} point, and V_{CC} for both is overlapping. The composite load line passes through V_{CC} and meets the maximum I_{cm} points and both sides. As shown in Figure 11.9(b), the slope of load line is such that $\tan\theta = \frac{V_{CC}}{I_{cm}}$ and we can determine effective load R'_L. This value is arrived at by adjusting the turn ratio of transformer T_2, which satisfies the equation $R'_L = a^2 R_L$ where R_L is actual load connected across the secondary of transformer (Figure 11.8).

11.10.2 Class B Push-Pull Amplifier

The low efficiency of class A amplifiers is largely due to transistor bias conditions where it draws current from the battery even in the absence of input signal. In class B amplifiers, the transistor is biased to cut-off, so that the transistor does not conduct in the absence of input signal and so there is no power dissipation in the absence of input signal. This gives class B amplifiers much greater efficiency than class A amplifiers.

A basic class B circuit is shown in Figure 11.10. Transformer T_2 couples load resistor R_L to the collector circuits of two transistors, Q_1 and Q_2. Note that the primary of the transformer has a centre tap to which DC supply V_{CC} is connected. Transistors Q_1 and Q_2 have grounded emitters and both are biased off via resistors R_1 and R_2, which ground the bases of two transistors.

Input signals applied to the bases are equal but are in anti-phase. Thus, when e_{s1} is going positive, e_{s2} is negative so that transistor Q_2 is biased further off when Q_1 is biased ON. As collector current in Q_1 increases from zero, it produces a half sine wave of voltage across the upper half of primary winding of transformer T_2. When the positive half cycle of input signal to Q_1 begins to go negative, the signal at Q_2 base is commencing to go to positive. Thus, Q_1 becomes biased OFF again and Q_2 is biased ON and a half cycle of voltage signal is generated across the lower half primary of transformer T_2.

The effect of two half cycles in separate halves of transformer primary is to produce negative flux in the iron core, which first flows in one direction and then in the opposite direction. The flux links with transformer secondary and generates a complete cycle of sine wave across load (R_L) connected to the secondary.

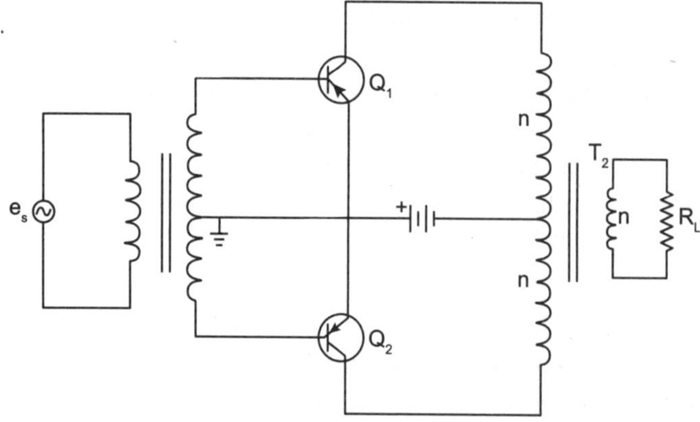

Figure 11.10 Class B push-pull amplifier

The circuit of class B push-pull amplifier is shown in Figure 11.10. The circuit is very similar to that of class A push-pull amplifier except that biasing is such that it operates under class B where the transistor is biased at cut-off of dynamic characteristic and it does not conduct in the absence of signal. The DC bias applied across the base of each transistor is zero.

Working and circuit analysis In class B, the dynamic transfer characteristic of two transistors are taken as one up side down and their cut-off point concides as shown in Figure 11.11(a).

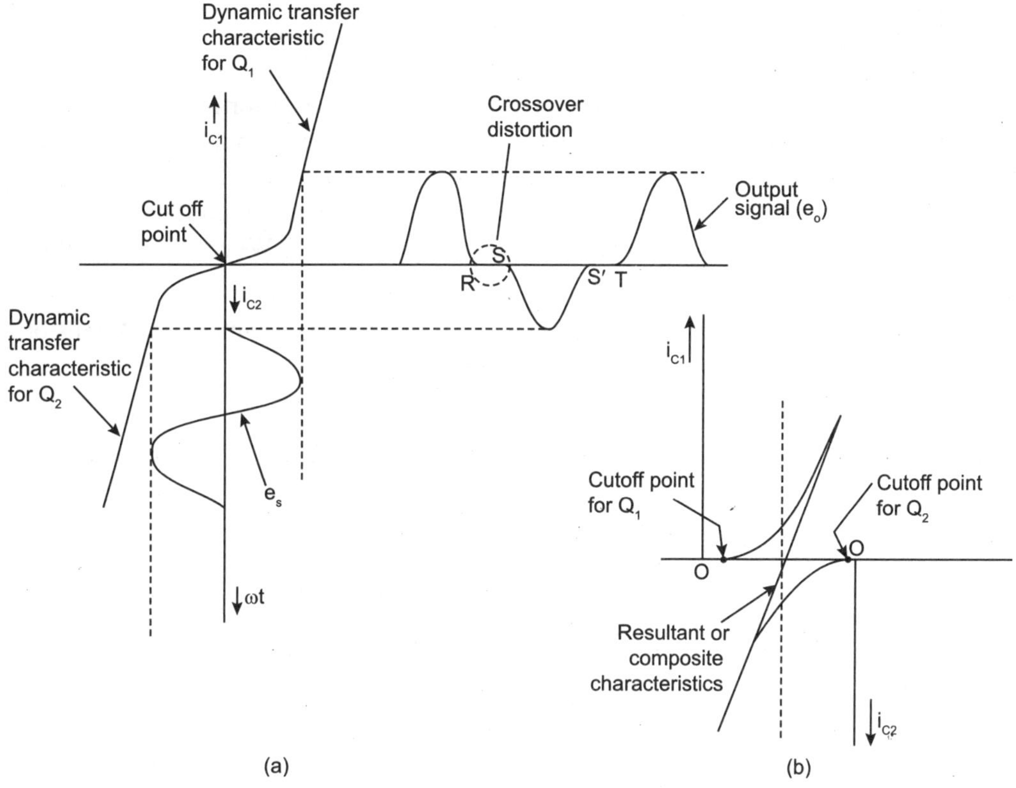

Figure 11.11

The transfer characteristic of transistor Q_2 has been shown as inverted. From the characteristic, we see that output current i_c flows for the entire cycle of input signal e_s as is clear from the fact that transistor Q_1 conducts during the positive half cycle of input signal, while transistor Q_2 conducts during the negative half cycle of input cycle.

From Figure 11.11(a), as the signal attains zero value, current does not attain zero value instantaneously, but it attains zero value slowly as shown by portion *RS* and *S′T* of the current cycle. In this respect, for low voltage input signal we may not even get (collector) output current because of the curvature of transfer characteristic. But, in case of high voltage input signal (large input signal), there is output current. Output current under such condition is called stepped output current. This type of distortion is called cross over distortion.

This distortion can be eliminated by keeping a provision for forward-bias in the circuit. This means locating point *Q* slightly above cut-off (Figure 11.11(b)) and in this situation output current will have the same shape as input signal, i.e. sinusoidal. For this, the dynamic transfer characteristic of Q_2 is shifted towards right; so the cut-off points that were coinciding earlier will be separated and we can think of a composite (or resultant) characteristic that is a straight line (Figure 11.11(b)) thereby the output current will remain throughout the entire input cycle. This type of operation is no more a class B operation, but it is called class AB operation.

11.11 Class AB

We know that the wave form delivered to output transformer T_2 primary and the resultant output are not perfectly sinusoidal in a class B circuit due to cross-over distortion (Figure 11.11(a)). It exists due to the fact that transistors do not begin to turn ON first on the beginning of input sinusoidal voltage. The transistor turns ON when the input base-emitter voltage reaches about 0.5 V for silicon device and 0.15 V for germanium device. To eliminate this effect, the transistors may be partially biased ON instead of being biased completely OFF. With this modification, the circuit becomes a class AB amplifier.

Figure 11.11(c) shows a class AB transformer-coupled output stage with a class A transformer-coupled driver stage.

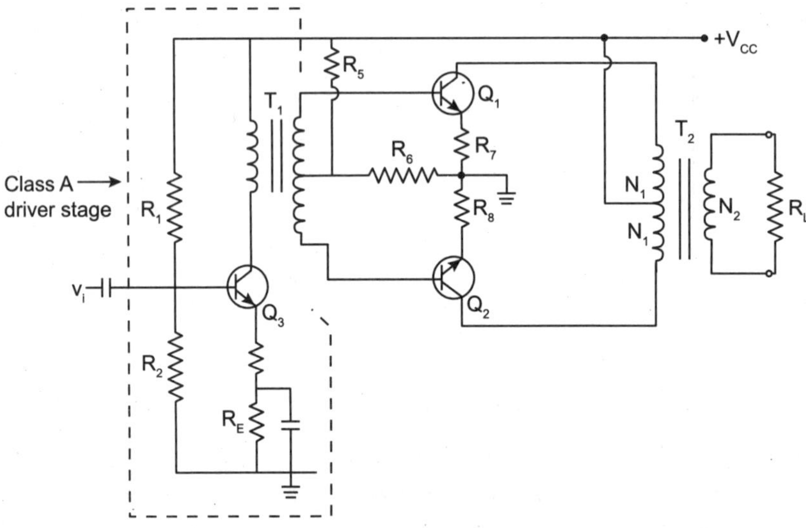

Figure 11.11 (c) Class AB output stage

As seen in the figure, output transformer T_2 has a centretapped primary winding, each half of which forms a load for one of the transistors, Q_1 and Q_2. Resistors R_5 and R_6 bias Q_1 and Q_2 partially ON. Resistors R_7 and R_8 limit emitter (and collector) currents to the desired bias level. Transformer T_1 and transistor Q_3 form the driving stage. The secondary of transformer T_1 is centre tapped to provide necessary anti-phase signals to Q_1 and Q_2. As for transistors Q_1 and Q_2, one transistor is ON and the other is OFF. The output functions exactly as for a class B circuit, except that each transistor commences to conduct just before the signal to its base becomes positive. This eliminates delay in transistor turn ON, which creates crossover distortion in a class B amplifier.

The class A circuit in figure is referred to as a driver stage, simply because it provides input signals to drive the class AB output stage.

11.12 Efficiency of Class B Push-Pull Amplifier

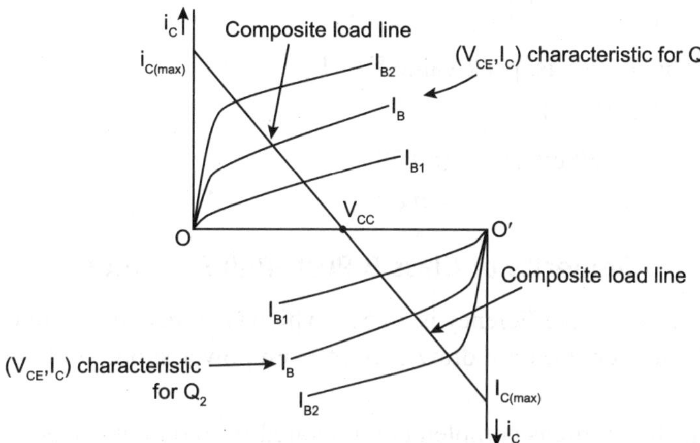

Figure 11.12 (V_{CE}, I_C) characteristic, and composite load line

The (V_{CE}, I_C) characteristic for transistor Q_1 has been drawn as usual, but the characteristic for transistor Q_2 has been drawn upside down. The AC load lines for Q_1 and Q_2 together form a straight line since both meet at V_{CC}. For the composite load line of the amplifier, slope is $\frac{I_{C\max}}{V_{CC}}$.

Average value of collector current = I_C

Peak value of collector current = $I_P = I_{CM}$

Average value of output voltage = V_C

Peak value of voltage = $V_P = V_{CM}$

AC power transferred to load $= P_{oac} = V_C I_C = \frac{V_{cm}}{\sqrt{2}} \times \frac{I_{cm}}{\sqrt{2}} = \frac{V_p I_p}{2}$

$$\text{Power dissipated in the two transistors} = 2[P_{Tr\,dc}] = P_{in\,dc} - P_{o\,ac}$$
$$= V_{dc} I_{dc} - P_{o\,ac}$$
$$= V_{CC}\frac{2I_p}{\pi} - \frac{V_p I_p}{2}$$
$$= 2I_p\left[\frac{V_{CC}}{\pi} - \frac{V_p}{4}\right]$$

$$\text{Conversion efficiency, } \eta = \frac{p_{oac}}{p_{in\,dc}} = \frac{\frac{V_{cm} I_{cm}}{2}}{V_{CC}\frac{2I_{CM}}{\pi}}$$

$$= \frac{\pi}{4}\left[\frac{V_{cm}}{V_{cc}}\right] = 0.785\,\frac{V_{cm}}{V_{cc}} \tag{27}$$

When output voltage swings, peak value $V_p = V_{cc}$

In this situation, $V_p = V_{cm} = V_{cc}$

Maximum conversion efficiency, $\eta = 0.785$

$= 78.5\%$

11.12.1 Merits and Demerits of Class B Push-Pull Amplifier

1. The maximum conversion efficiency is 78.5%, which is more than that of class A operation. It is because the device does not draw current from power supply in the absence of input signal.
2. Here, the second harmonic is completely eliminated from output current.
3. Since the second harmonic is absent, output power per transistor is much more for any distortion.
4. There is no DC component in the output current.
5. There is no chance of saturation of iron core in the transformer since the DC current flowing in upper half primary is opposite to that flowing in lower half primary of output transformer.
6. The main disadvantage of the amplifier is the use of bulky transformers. However, newer designs have eliminated the need for transformers in most applications.

11.13 Transformer-Less Push-Pull Amplifier

It is known that transformer T_1 produces two equal but out-of-phase input signals for transistors Q_1 and Q_2 respectively. In this respect, the transformer acts like a phase splitter.

In the circuit (Figure 11.13), we do not use any transformer but it functions on the principle of push-pull, thereby it has all the characteristics of a push-pull amplifier.

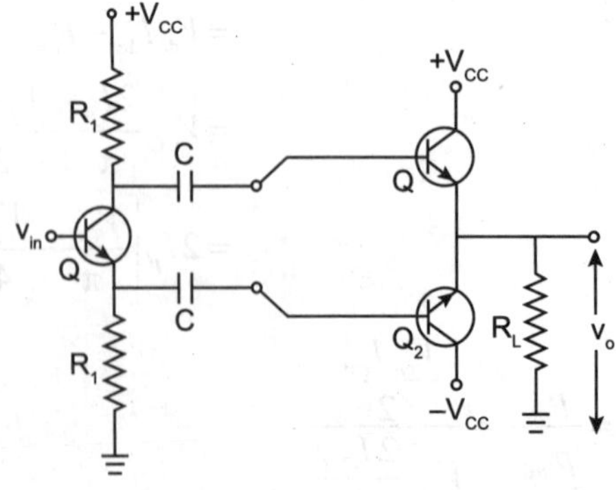

Figure 11.13 Transformer-less push-pull amplifier

The first part of the circuit acts like a phase splitter, which was the function of transformer T_1 in push-pull amplifier. Here, two equal resistors, R_1, are connected each along with collector and emitter, respectively, of transistor Q. The output signal at emitter is in phase with input signal, but the output signal at collector has a phase difference of 180° with input signal v_{in}. Thus, we get two equal and opposite because signals from transistor Q. Here too, we do not use any output transformer T_2 (Figure 11.10), we do not require any matching for the given load for maximum power transfer, since we have specific power transistors, Q_1 and Q_2, with output impedance that matches with load R_L up to 8 Ω.

The use of positive and negative power supply V_{cc} along with transistors Q_1 and Q_2 is that in output we do not have any DC component because load R_L is directly connected in common to the emitter.

11.13.1 Complimentary Symmetry

In complimentary symmetry, we do not require any transformer but still have the push-pull actions. As seen in Figure 11.14, we use two similar transistors, Q_1 and Q_2, of which one is *p-n-p* type and another is *n-p-n* type.

A common input v_s is applied to the two bases of transistors Q_1 and Q_2, respectively.

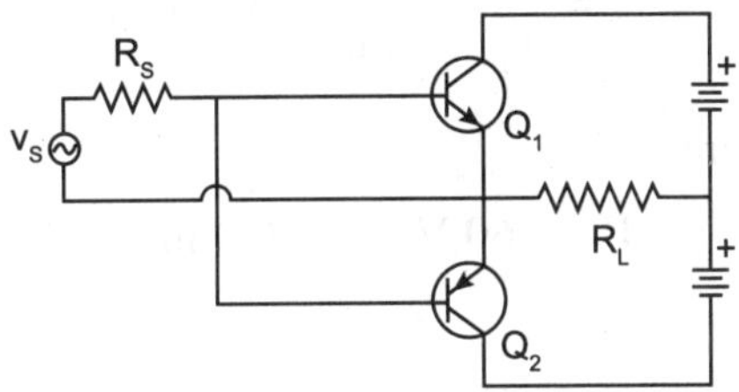

Figure 11.14

But, the action of two transistors is opposite to each other, i.e. if base current increases in one, it decreases in the other transistor because in the positive half cycle of input signal v_s, transistor Q_1 is forward-biased but transistor Q_2 is reverse-biased, thereby collector current increases in one and decreases in the other transistor. The entire action is reversed in the negative half cycle of input signal. Thus, in one complete cycle, the action is a push-pull one. These two transistors together act like a phase splitter. To avoid transitional distortion, one transistor is slightly forward-biased as in the case of class AB operation.

Example 2 A class B transformer-coupled amplifier is to supply 4 W power to a 16 Ω load connected across the secondary. The available power supply is $V_{CC} = 30$ V. Specify the output of transformer and transistor. Assume transformer efficiency of 80%.

Solution

Effective power to be transferred is

$$P_o' = \frac{P_o}{\eta} = \frac{4}{0.8} = 5\,\text{W}$$

Impedance seen when "looking into" the whole winding of centretapped transformer,

$$V_p = V_{CC} = 30 \text{ V}$$

Effective load, $R'_L = \dfrac{V_p^2}{2P'_o} = \dfrac{(30)^2}{2\times 5} = 90\ \Omega$

$$R''_L = 4R'_L$$

$$= 4\times 90 = 360\ \Omega$$

Transformer specifications are

$$P_o = 4\text{ W}, R_L = 16\ \Omega, R''_L = 360\ \Omega$$

Maximum transistor voltage,

$$V_{CE(\max)} = 2V_{CC} = 2\times 30 = 60 \text{ V}$$

Maximum transistor current,

$$I_{C(\max)} = I_p$$

$$I_p = \frac{2P'_o}{V_p} = \frac{2\times 5}{30} = 330 \text{ mA}$$

Transistor specifications are

$$V_{CE} = 60 \text{ V}, I_C = 330 \text{ mA}$$

11.14 Darlington Pair

It consists of two transistors connected in such a way that its overall current gain equals the product of individual current gains.

$$\text{i.e. } \beta_{\text{overall}} = \beta_1\beta_2$$

Such a connection is shown in Figure 11.15(a).

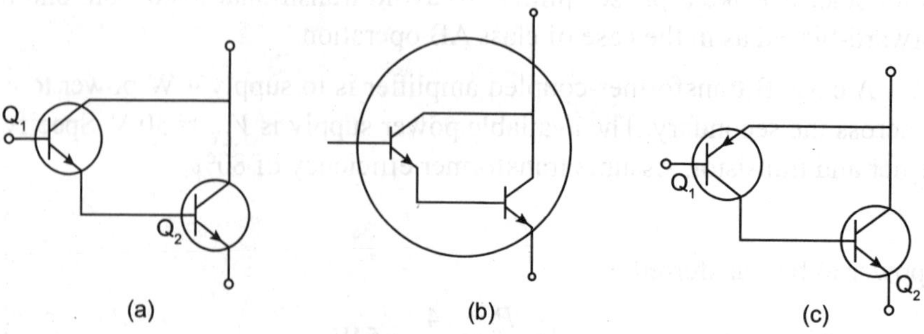

Figure 11.15 Darlington connection

The two transistors are put into a single case as shown in Figure 11.15(b). The device is known as Darlington transistor and its works as a single transistor with very high current gain. For instance, 2N6725 is a Darlington transistor having a current gain of 25000 at 200 mA.

11.14.1 Complimentary Darlington

Figure 11.15(c) shows another Darlington connection called complimentary Darlington. Here, the collector current of one (Q_1) acts like the base current of another (Q_2). It acts like a single *p-n-p* transistor of current gain

$$\beta_{\text{overall}} = \beta_1\beta_2$$

Thus, a Darlington connection has very high current gain. Since its current gain is very high, it has very high input impedance and can produce very large output currents.

Such a device is used in voltage regulators and in class B push-pull power amplifiers.

11.15 Class B Push-Pull Emitter Follower

We have discussed emitter follower (common collector amplifier) in detail in a previous chapter. The circuit for class B emitter follower is shown in Figure 11.16. Here, we use two complimentary transistors.

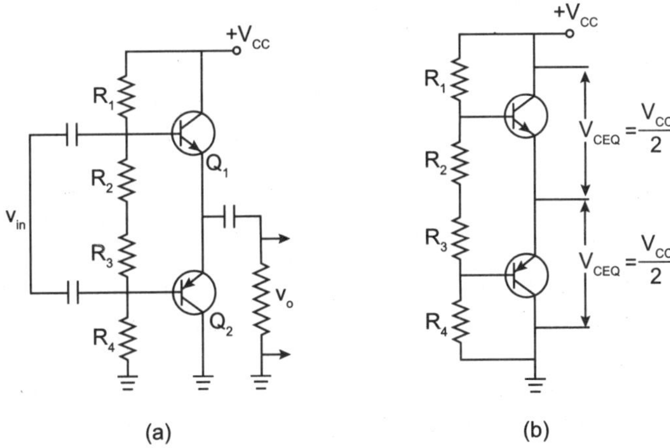

Figure 11.16 (a) Class B push-pull emitter follower circuit, (b) DC equivalent circuit

Transistor (Q_1) is *n-p-n* and other (Q_2) is *p-n-p*, which are so connected that we have an *n-p-n* emitter follower and a *p-n-p* emitter follower connected in push-pull arrangement.

11.15.1 Circuit Analysis

The DC equivalent circuit of the amplifier is shown in Figure 11.16(b). We select a biasing resistor (R_1, R_2, R_3 and R_4) such that the Q point is at cutoff. This will bias the emitter diode of each transistor between 0.6 and 0.7 volt so that it is on the verge of conduction and ideally

$$I_{CQ} = 0$$

Because the biasing resistors are equal, each emitter diode is biased with the same bias voltage, and as a result half the supply voltage V_{CC} is dropped across each transistor's collector-emitter voltage, i.e.

$$V_{CEQ} = \frac{V_{CC}}{2}$$

DC load line Since there is no DC resistance either in the collector or emitter circuit and DC load line is vertical at cutoff point, DC saturation current is infinite (Figure 11.17).

AC load line

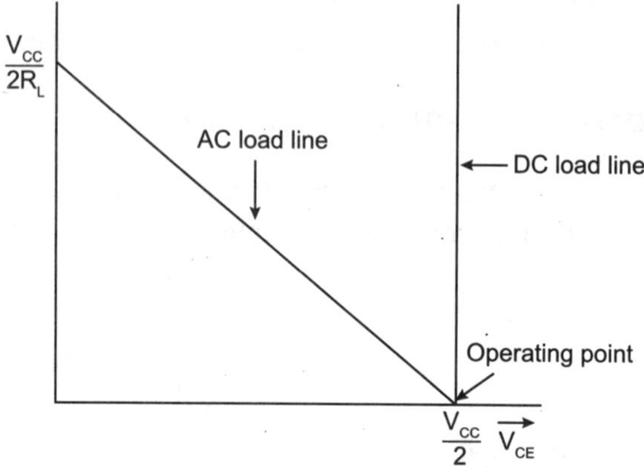

Figure 11.17 DC and AC load line

Figure 11.17 shows the DC and AC load line. When either transistor is conducting, its operating point moves up and down along the AC load line. The voltage swing of the conducting transistor can go all the way from cutoff to saturation. On the alternate half cycle, the other transistor does the same thing; so the maximum peak-to-peak output is

$$M_{PP} = V_{CC}$$

AC analysis As mentioned earlier, the maximum voltage gain of emitter follower is

$$A = 1$$

and input impedance of base is

$$Z_{i(\text{base})} = \beta R_L$$

On the positive half cycle, upper transistor Q_1 conducts and lower transistor Q_2 is cutoff. The upper transistor acts like an ordinary emitter follower and so output voltage is nearly equal to input voltage since $A = 1$. On the negative half cycle, upper transistor Q_1 is cut off and lower transistor Q_2 conducts and acts as an ordinary emitter follower. Thus, again output voltage is nearly equal to input voltage. During either half cycle of input voltage, the source sees high input base impedance.

Thus, a class B push-pull amplifier using complimentary transistors is a large signal amplifier having unity voltage gain and high input impedance.

Here too, we find crossover distortion, since the incoming signal has to rise about 0.7 V to overcome the barrier potential of emitter diodes in either half of the input signal. Now conduction angle is less than 180°. Output voltage across the load appears as shown in Figure 11.18.

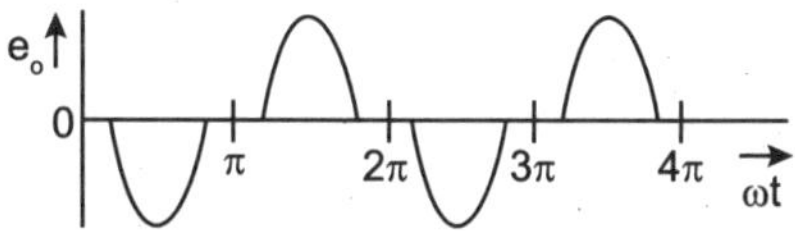

Figure 11.18 Output voltage wave shape

From the wave shape, it is clear that output voltage is distorted, since clipping occurs between the time one transistor cutsoff and other one is ON. We call it crossover distortion. To eliminate it, we need to apply a slight forward-bias to each emitter diode, so that operating point Q is slightly raised. By raising the operating point, conduction angle is now slightly more than 180°. In a strict sense, we call it class AB operation where conduction is greater than 180° but less than 360°.

11.15.2 Zener Follower

Emitter follower is largely used in voltage regulator circuits, but when it is used in conjuction with a zener dioxide, we not only have regulated output voltage but also have larger output current. Such a circuit is called zener follower.

The circuit for zener follower is shown in Figure 11.19.

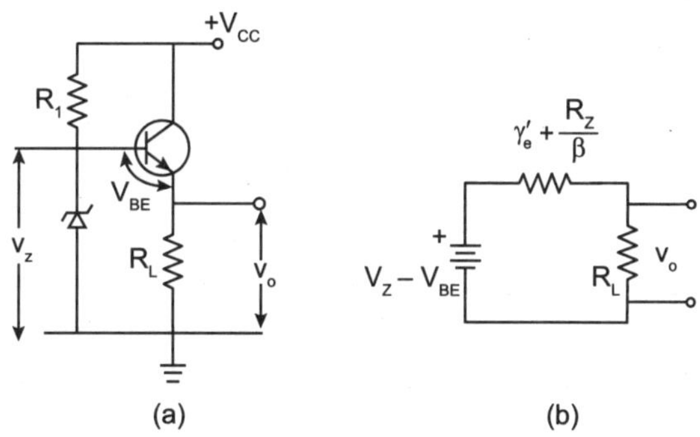

Figure 11.19 (a) Zener follower circuit, (b) equivalent circuit

Working Zener voltage V_Z is the input to the base of emitter follower. The DC output voltage is

$$V_{\text{out}} = V_Z - V_{BE}$$

This output voltage is fixed since R.H.S. is constant. If supply voltage changes, zener voltage still remains approximately constant, so does the output voltage. Thus, the entire circuit acts like a voltage regulator.

It has two advantages over an ordinary voltage regulator circuit:

(i) Firstly, the zener diode has to produce a load current of value

$$I_B = \frac{I_{\text{out}}}{\beta}$$

Since base current I_B is usually much smaller than output current, we can use a smaller zener diode whose current carrying capacity is much small. When we are to supply several amperes to load resistor, an ordinary zener regulator requires a zener diode capable of handing such large current. On the other hand, zener needs to handle only tens of milliampere (requirement of base current).

(ii) Its another advantage is that zener follower has low output impedance. Here, its value is

$$Z_{\text{out}} = re' + \frac{R_z}{\beta}$$

whereas for an ordinary voltage regulator,

$$Z_{\text{out}} = R_z \text{ (zener impedance)}$$

Further, we note that Z_{out} is usually much less than load R_L; so an emitter follower can hold the DC output voltage almost constant because the source is much stiff.

Thus, a zener follower provides voltage regulation of zener diode and increased current heating capability of emitter follower.

11.16 Tuned Amplifier (Class C Amplifier)

When we use a push-pull arrangement in class B function, almost all class B amplifiers are push-pull amplifiers. With class C, we need to use a resonant circuit as load. This is why almost all class C amplifiers are tuned amplifiers.

DC biasing in class C operation is such that no collector current flows when signal is absent; collector current flows in the form of pulses for less than a half cycle of input signal when it is applied. A parallel resonant circuit connected to the collector arm filters the pulses of collector current and produces a pure since wave in the output voltage; so in class C operation we have a tuned circuit as load. The main application for class C amplifiers is with RF amplifiers, whose efficiency goes upto 100%.

11.16.1 RF Amplifiers

These amplifiers are also known as band-pass amplifiers, since they amplify a select band of frequencies in the range 10 MHz to 30 at a centre frequency (f_c) of 100 MHz.

Centre frequency is generally the resonant frequency of tuned (resonant) circuit as shown in Figure 11.20.

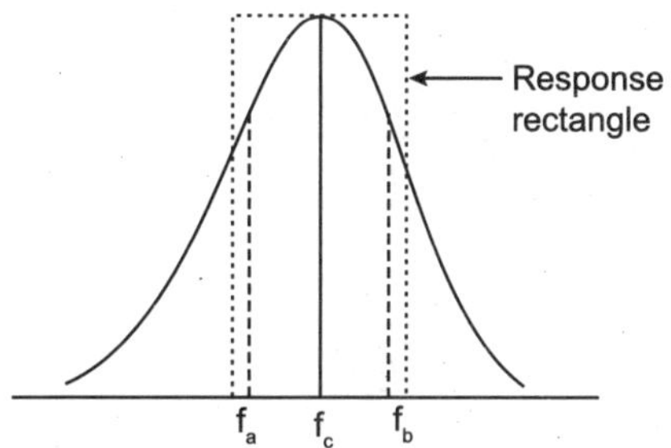

Figure 11.20 Response of a narrow-band RF amplifier

As shown in the response curve (Figure 11.20), f_c is centre frequency (say 100 MHz) and bandwidth is $(f_b - f_a) \approx 2\,\Delta f = 10$ to 30 MHz. In a similar fashion, an IF (intermediate frequency) amplifier has a bandwidth of few KHz at a centre frequency to few 100 KHz for voice modulated signals; such amplifiers in television receivers have a bandwidth of 6 MHz at a centre frequency of 40 MHz. These systems in which pass band ranges from a few percent to 30% of centre frequency are called broad band amplifiers. It is in the scope of this chapter to discuss narrow band amplifiers (tuned amplifiers) only. Another basic requirement of an RF amplifier is usually "tunability", or that it should be possible to shift or tune the response rectangle (Figure 11.20) over a considerable range of frequencies, which is usually the case in radio receivers or television receivers. The tunability requirement is most easily met using a parallel resonant circuit with a variable capacitor to vary the resonant frequency. Thus, these amplifiers have a parallel resonant circuit as their load.

We shall now discuss and analyze a resonant circuit.

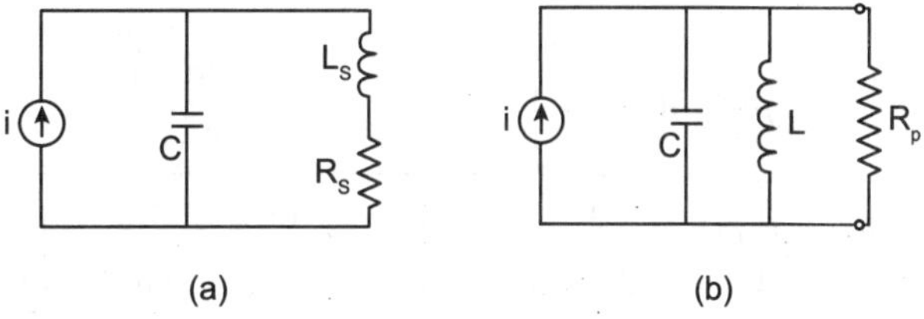

Figure 11.21 (a) Series resonant form, (b) parallel resonant form

A parallel resonant circuit can be shown in two forms (Figure 11.21):

(i) Series resonant form

(ii) Parallel resonant form

Q of a resonant circuit is defined as

$$Q = \frac{2\pi \times \text{Maximum energy stored per cycle}}{\text{Energy dissipated per cycle}}$$

It indicates relative efficiency of the circuit.

If I_m is peak current in the inductor connected in the circuit (series resonant form), then by definition

$$Q = \frac{2\pi L_s I_m^2}{\frac{I_m^2 R_s}{f_o}}$$

$$= \frac{\omega_o L_s}{R_s} \tag{28}$$

where (i) f_o is resonant frequency (or center frequency)

$$\omega_o = 2\pi f_o$$

(ii) R_S is series resistance of inductor (L_s)

Then, anti resonant frequency or resonant frequency of a parallel circuit is given by

$$f_t = \frac{1}{2\pi}\sqrt{\frac{1}{L_S C} - \frac{R_S^2}{L_S^2}} = \frac{1}{2\pi}\sqrt{\frac{1}{L_S C}\left[\frac{Q_2}{1+Q^2}\right]} \tag{29}$$

The value of Q at radio frequencies is usually more than 10 and may reach 100 or more. So, for high Q values, undamped resonant frequency is

$$f_o = \frac{\omega_o}{2\pi} = \frac{1}{2\pi}\sqrt{\frac{1}{L_S C}} \tag{30}$$

Impedance of the circuit (Figure 11.20(a)) is

$$Z = \frac{(R_S + j\omega L_S)\left[\frac{1}{j\omega C}\right]}{R_S + j\left[\omega L_S - \frac{1}{\omega C}\right]} = \frac{\frac{L_S}{C}\left[1 - \left(\frac{jR_S}{\omega L_S}\right)\right]}{R_S\left[1 + j\left(\frac{\omega L_S}{R_S} - \frac{1}{\omega C R_S}\right)\right]} \tag{31}$$

The reactive term in the denominator [Eqn. (31)] involves the difference of two terms which are nearly equal at resonant frequency.

For greater accuracy in computation, the expression for Z can be written by defining

$$\delta = \frac{\omega - \omega_o}{\omega_o} \tag{32}$$

Then

$$\frac{\omega L_S}{R_S} = (1+\delta)Q_o \text{ and } Q_o = \frac{1}{R_S}\sqrt{\frac{L_S}{C}} \tag{33}$$

So Eqn. (33) can be written as

$$Z = \frac{\left[\frac{L_S}{R_S C}\right]\left\{1 - \frac{j}{[Q_o(1+\delta)]}\right\}}{1 + jQ_o\left[1 + \delta - \frac{1}{(1+\delta)}\right]}$$

When the value of Q is high, i.e. $Q_o^2 \gg 1$, impedance

$$Z = \frac{R_S Q_o^2}{1 + j\left[Q_o \frac{\delta(2+\delta)}{1+\delta}\right]} \tag{34}$$

At resonance, $\delta = 0$ since $\omega = \omega_o$.

Impedance of the circuit (Eqn. 34) can be expressed as

$$Z_o = R_S Q_o^2 = \frac{L_S}{C R_S} \tag{35}$$

From Eqn. (35), we find that impedance is a function of $\frac{L_S}{C}$

Circuit impedance (Z) may be normalized against impedance at resonance (Z_o); then Eqn. (34) and (35) can be combined as

$$\frac{Z}{Z_o} = \frac{Z}{R_s Q_o^2} = \frac{1}{1 + j\left[Q_o \delta\left(\frac{2+\delta}{1+\delta}\right)\right]} \tag{36}$$

The above equation has been represented by the curve shown in Figure 11.22.

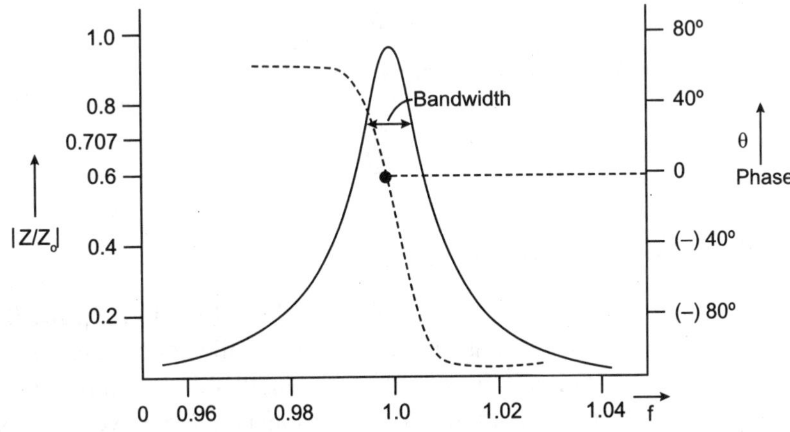

Figure 11.22 Impedance of a parallel resonant circuit (Q = 100 assumed) and phase variation with frequency

An important factor in parallel resonant circuit is variation of phase angle θ (Figure 11.22), which changes from inductive below resonant frequency f_o to capacitive above resonant frequency f_o.

So far, we have discussed the series resonant form of parallel resonant circuit. However, we may find useful the parallel resonant form often (Figure 11.21(b)), where R_p may be readily combined with other amplifier circuit resistance. Here, L and R_p branches are equivalent to L_S and R_S series of Figure 11.21(a), and its admittance is

$$Y = \frac{1}{R_s + j\omega L_s} = \frac{1}{R_s}\left(\frac{1}{1+\dfrac{\omega^2 L_s^2}{R_s^2}}\right) - \frac{j}{\omega L_s}\left(1+\frac{1}{\dfrac{R_s^2}{\omega^2 L_s^2}}\right) \tag{37}$$

For high Q, i.e. $Q \gg 10$, parallel resistance and inductance in Figure 11.21(b) are

$$R_p = \frac{\omega_o^2 L_s^2}{R_s} = Q_o \omega L_s \text{ and } L \approx L_s \tag{38}$$

Here, Q_o is quality factor at resonance of the coil only.

By the definition of Q and with E_m as the peak value of voltage across the circuit, we can write

$$Q_o = \frac{2\pi C E_m^2}{\dfrac{E_m^2}{R_p f_o}} = \omega_o C R_p \tag{39}$$

It represents Q of the circuit, with R_p equivalent to R_s. In effect, this Q value is equal to

$$Q = \frac{\omega_o L_s}{R_s}$$

Bandwidth of the response curve is given by

$$\text{BW} = \frac{f_o}{Q} \tag{40}$$

11.17 Single Tuned Amplifier

A single tuned CE amplifier circuit is shown in Figure 11.23. Resistors R_1, R_2, R_E and supply V_{CC} establish the operating point, and biasing is such that no collector current flows when signal is absent. The parallel resonant circuit tuned to the frequency of desired input signal acts as the AC load of the amplifier having maximum value at resonance. Here, we have used coupling capacitor C_C to couple it with the next stage, which will prevent the succeeding stage from interacting with this stage. Due to the presence of C_C, it is termed as single tuned amplifier. At frequencies to be amplified, the reactances of C_L and C_E are negligible. Capacitor C_1 couples the v_s source AC to the base of the transistor, but it is a short circuit for AC signal. Since the bandwidth of the amplifier is assumed to be narrow, the variation of h_{fe} with frequency can be ignored and its value is assumed constant at frequency f_o (resonant frequency of tank circuit) for practical consideration.

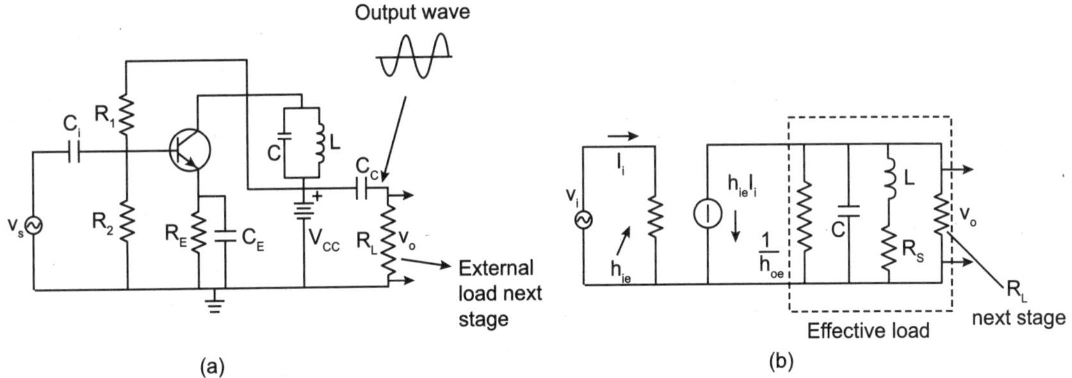

Figure 11.23

The AC equivalent circuit is shown in Figure 11.23(b).

As explained earlier, tank circuit (tuned circuit) impedance as given by Eqn. (34) is

$$Z = \frac{Q_o^2 R_s}{1 + j\left[Q_o \delta \left(\frac{2+\delta}{1+\delta}\right)\right]}$$

It has a maximum value at $\delta = 0$ (or when $\omega = \omega_o$)

$$Z_o = Q_o^2 R_S = R_o \text{ (say)} \tag{40a}$$

and $\dfrac{Z}{Z_o} = \dfrac{1}{1 + jQ_o\delta\left(\dfrac{\delta+2}{\delta+1}\right)}$ (called normalized impedance)

Putting $\dfrac{1}{h_{oe}} = r_o$, output voltage (Figure 11.23(b)).

$$v_o = (-)h_{fe} I_i Z_i$$

The negative sign is due to the direction of current. Here, Z_i is the equivalent impedance of parallel combination r_o, Z and R_L.

$$\frac{1}{Z_i} = \frac{1}{r_o} + \frac{1 + jQ_o\delta\left(\dfrac{\delta+2}{\delta+1}\right)}{R_o} + \frac{1}{R_L} \tag{41}$$

Assuming $\frac{1}{R_p} = \frac{1}{r_o} + \frac{1}{R_o} + \frac{1}{R_L}$ and substituting for Z_i, we have output voltage

$$v_o = \frac{(-)h_{fe} I_i}{\frac{1}{R_p} + \frac{jQ_o\delta\left(\frac{\delta+2}{\delta+1}\right)}{R_o}} \tag{42}$$

Effective Q of the circuit is defined as

$$Q_e = \frac{Q_o R_p}{R_o}$$

We also have $R_p < R_o$; so $Q_e < Q_o$.

Hence output voltage from Eqn. (42) is

$$v_o = \frac{(-)h_{fe} I_i R_p}{1 + jQ_e\delta\left[\frac{\delta+2}{\delta+1}\right]} \tag{43}$$

Now, input voltage $v_i = h_{ie} I_i$. Substituting it, we have

$$\text{Voltage gain, } A_v = \frac{v_o}{v_i} = \frac{(-)\left(\frac{h_{fe}}{h_{ie}}\right)R_p}{1 + jQ_e\delta\left(\frac{\delta+2}{\delta+1}\right)}$$

$$= \frac{(-)g_m R_p}{1 + jQ_e\delta\left(\frac{\delta+2}{\delta+1}\right)} \tag{44}$$

at $\delta = 0$ or at resonant frequency $\omega = \omega_0$.

Maximum voltage gain,

$$[A_v]_{\max} = (-)g_m R_p \tag{45}$$

Making use of Eqns. (44) and (45), we have

Now, normalized gain

$$\frac{A_v}{(A_v)_{\max}} = \frac{1}{1 + jQ_e\delta\left(\frac{\delta+2}{\delta+1}\right)} \tag{46}$$

Eqn. (46) has the same form as Eqn. (36), except that Q_e replaces Q_o. The normalized amplitude response is shown in Figure 11.24.

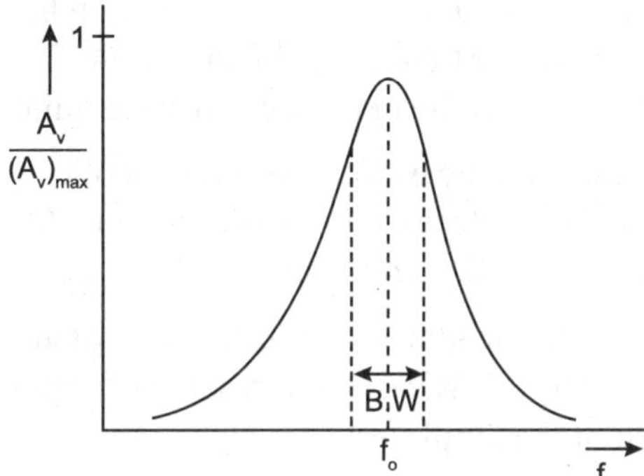

Figure 11.24 Normalized amplitude response of single tuned amplifier

Its shape is very similar to that for $\frac{Z}{Z_o}$ (Figure 11.22).

Bandwidth of the amplifier is given by

$$\text{B.W} = \frac{\omega_o}{Q_e}$$

and gain-bandwidth product is

$$[Av]_{\max}[\text{B.W}] = g_m R_p \frac{\omega_o}{Q_e} \tag{47}$$

$$= \frac{g_m}{C} \tag{48}$$

(which is constant)

where $\frac{R_p \omega_o}{Q_e} = \frac{R_p \omega_o}{(Q_o R_p / R_o)} = \frac{\omega_o R_o}{Q_o} = \omega_o Q_o R$

$$= \frac{\omega_o R}{\omega_o C R} = \frac{1}{C}$$

We have shown that there is voltage amplification. Collector current flows in the form of pulses (Figure 11.1(c)) since the amplifier operates under class C, and the maximum amplitude of current pulse is $\frac{V_c}{R_p}$ where R_p is equivalent resistance of combination of r_o, R_o and R_L.

The Output current flowing in the form of pulses is non-sinusoidal and rich in harmonics (multiple of input frequency). The resonant tank circuit has maximum impedance (Eqn. (40a)) only at resonant frequency (f_o) which is equal to input signal frequency. This produces a maximum output voltage, hence voltage gain is also maximum (Figure 11.24). On the other hand, the impedance of tank circuit is much less at other frequencies or higher harmonics. This is why voltage across the tank almost looks like pure sine wave (Figure 11.23(a)) since all harmonics are filtered by the tank circuit.

It is to be noted that the tank circuit under class C operation has a high Q ($Q > 10$). It means bandwidth is less than 10% of resonant frequency (Figure 11.24). For this reason, these amplifiers are called narrow band amplifiers and the resonance curve is quite sharp.

When the tank circuit is resonant, the AC load as seen by collector current source is maximum and purely resistive. Therefore DC collector current is minimum. One way to tune the tank circuit is to look for decrease (minimum) DC current supplied by V_{CC}.

Example 3 In a tuned amplifier, the tank circuit comprises of an inductor of inductance 2 μH and capacitor of capacitance 220 pF. What is the resonant frequency? If Q of the inductor is 125, calculate the bandwidth of the amplifier.

Solution

$$\text{Resonant frequency, } f_o = \frac{1}{2\pi\sqrt{LC}}$$

$$= \frac{1}{2\pi\sqrt{2\times10^{-6}\times220\times10^{-12}}}$$

$$= 7.9 \text{ MHz}$$

$$\text{Bandwidth, BW} = \frac{f_o}{Q} = \frac{7.9\times10^6}{125} \simeq 63 \text{ KHz}$$

Example 4 In a RC-coupled amplifier, the value of components used is $-R_C = 3.6$ kΩ, $R_E = 680$ Ω, $R_i = 10$ kΩ. Calculate transistor power dissipation. What will it be if R_C is replaced by a tank circuit?

Solution

(a) Bias voltage, $V_B = \dfrac{R_2}{R_1 + R_2} V_{CC} = \dfrac{2.2}{12.2} \times 10 = 1.8\text{V}$

Emitter current, I_E, flowing through $R_E = \dfrac{V_B - 0.7}{R_E} = \dfrac{1.8 - 0.7}{680} = 1.62$ mA

$$\text{DC collector voltage, } V_C = V_{CC} - I_C R_C = V_{CC} - I_E R_C$$

$$= 10 - (1.62\times10^{-3}\times3.6\times10^3)$$

$$= 4.17 \text{ V}$$

$$\text{DC collector to emitter voltage, } V_{CEQ} = V_C - I_E R_E$$

$$= 4.17 - 1.1$$

$$= 3.07 \text{ V}$$

Hence transistor power dissipation, $P_D = V_{CEQ} \times I_{CQ} = 3.07 \times 1.62 \times 10^{-3} = 4.97$ mW.

(b) If collector resistance, R_C, is replaced by tank circuit, there is no DC voltage drop across it.

So

DC collector voltage = $V_{CC} = 10$

DC collector to emitter voltage, $V_{CEQ} = V_{CC} - I_E R_E$

$$= 10 - 1.1$$

$$= 8.9 \text{ V}$$

Thus, transistor power dissipation $= V_{CEQ} \times I_{CQ}$

$$= 8.9 \times 1.62 \times 10^{-3}$$

$$= 14.4 \text{ mW}$$

Example 5 In a tuned class 'C' amplifier, input signal voltage is 5 V rms. What is its peak input voltage? Calculate base-to-ground voltage.

Solution

The peak value (maximum amplitude) of input signal

$$= 5\sqrt{2} \text{ volt}$$

$$= 7.06 \text{ volt}$$

Now, peak-to-peak value of input voltage

$$= 2 \times 7.06 = 14.12 \text{ V}$$

Here DC biasing (class C operation) is such that no collector current flows in the absence of signal and the base in negatively clipped.

So, base-to-ground voltage = (–) [7.06 – 0.7] = (–) 6.36 V

Here, 0.7 V is the base to emitter drop.

11.18 Double Tuned Amplifier

The circuit of a double tuned amplifier is shown in Figure 11.25.

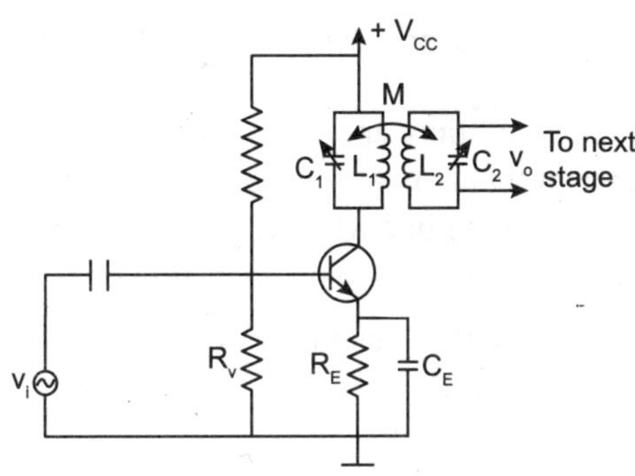

Figure 11.25 Double tuned amplifier circuit

As shown in Figure 11.25, there are two parallel resonant circuits comprising $L_1 - C_1$ and $L_2 - C_2$ tuned to the same frequency and they have the same Q value. The first (primary) tuned circuit $L_1 - C_1$ is connected to the collector of transistor and it is magnetically coupled with the secondary tuned circuit $L_2 - C_2$; M represents the coefficient of mutual inductance between the two coupled circuits. Its value depends upon the coefficient of coupling, K, between the two resonant circuits and given by

$$M = k\sqrt{L_1, L_2}$$

Here, the two turned circuits L_1 and L_2 represent the inductance of two coils but they have their own series resistance R_1 and R_2 as usual (not shown).

A detailed analysis of the circuit is beyond the scope of this chapter since it is arrived using the theory of coupled circuits. However, if Q_e is the effective quality factor of each tuned circuit, then the coefficient of critical coupling $K_C = \frac{1}{Q_e}$. The gain of the amplifier at resonance is also a function of K and is maximum at $K = K_c$.

A plot of normalized gain with frequency f for different values of coefficient coupling K is shown in Figure 11.26.

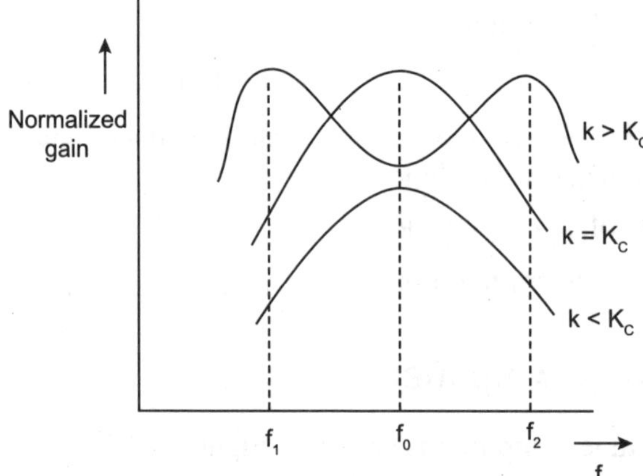

Figure 11.26 Frequency response of double tuned amplifier

When the coils are so placed that they are undercoupled ($K \leq K_C$), the response resembles the usual resonance curve, except that at near resonance $f = f_o$, the top of the curve is somewhat flat. The flatness of the curve is maximum when $K = K_C$; when the coils are overcoupled i.e. $K > K_C$, the response curve has two peaks one on each side of $f = f_o$.

Since such an amplifier has a flatter response near resonance and the response curve is steeper on two sides, such amplifiers are used as IF (intermediate frequency) amplifiers in super heterodyne receivers.

Summary

1. A power amplifier does not amplify power, but rather it draws DC power from the supply and converts it into useful AC signal power at the output.
2. Conversion efficiency of the amplifier is defined as

$$\eta = \frac{\text{AC output power}}{\text{DC input power}}$$

   Its maximum value:
   (i) in class A direct-coupled amplifier is 25%
   (ii) in class A transformer-coupled amplifier is 50%
   (iii) in class B amplifier (push-pull) is 78.5%
   (iv) in class C amplifier is 100%
3. Amplifiers are broadly classified as:
   (i) **Class A** where conduction of the device, i.e. output current, flows for 360° (entire period of input signal)
   (ii) **Class B** where conduction of the device (i.e. output current) flows for 180° (during positive half input signal)
   (iii) **Class C** where conduction of the device (i.e. output current flows) for $\theta < 180°$.
4. In most power amplifiers, input signal is large, i.e. covers a large part of dynamic transfer characteristics and thereby curved part is also covered. Due to the curvature of the characteristic, output signal is not the same shape as input signal and we say the output is distorted.
5. Total harmonic distortion is defined as

$$D = \sqrt{D_2^2 + D_3^2 + D_4^2 + \cdots}$$

   where D_2, D_3, D_4, etc. are called 2nd harmonic, 3rd harmonic, 4th harmonic percentage distortion.
6. For a single ended transformer-coupled amplifier, effective load or load seen by the primary is $R'_L = a^2 R_L$ where a represents turn ratio (n_1/n_2) and R_L is actual load connected across the secondary and power delivered to load is $\frac{1}{2} I_C^2 R'_L$.
7. In a push-pull class B amplifier, we employ two similar transistors, each conducting for 180°, i.e. one in positive half and other during negative half input signal, and the operation is such that even harmonics are cancelled without a change in fundamental thereby distortion is much reduced.

   However, under class A operation, we get two equal and opposite input signals for both transistors of which each conducts for 360°.
8. In an usual push pull arrangement, we employ two transformers, one at input side and the other at output side, thereby the whole arrangement is quite bulky and costly. These days

we have transformer-less push-pull arrangements, where in one arrangement, it is a phase splitting device and in another arrangement we employ complimentary symmetry.

9. Tuned voltage amplifiers consist of a tank circuit or tuned circuit comprising a parallel resonate circuit as load. The value of load Z varies with frequency and is maximum at resonance $Z_o = \frac{L}{RC}$; the resonant frequency is given by $f_o = \frac{1}{2\pi}\sqrt{\frac{1}{LC} - \left(\frac{R}{L}\right)^2}$ which reduces to $f_o = \frac{1}{2\pi}\sqrt{\frac{1}{LC}}$ for high values of quality factor Q, whose value is given by $Q = \frac{\omega L}{R}$. The tuned circuit comprises the load of the amplifier. We may have both primary and secondary tuned circuits called double tuned amplifiers.

10. The maximum voltage gain is obtained at resonant frequency. The response of gain with frequency in such amplifiers is different from usual voltage amplifier (R–C amplifier) and it is more or less like a resonance curve of tuned circuit, so such amplifiers are also called tuned amplifiers or narrow band amplifiers, since they amplify those input frequencies which are near resonant frequency.

Review Questions

1. Classify power amplifiers. Give their main features. Which type of amplifier has maximum efficiency?
2. What is a power amplifier? How is it different from a voltage amplifier? Explain.
3. What do you mean by small signal and large signal amplifiers? How do they differ from each other? Give their main characteristics.
4. Derive an expression for conversion efficiency of direct coupled class A amplifier. Draw necessary characteristics of CE amplifier and show the variation of voltage and current when an input AC signal is applied.
5. Sketch the circuit of a transformer-coupled class A transistor amplifier. Briefly explain the circuit operation.
6. Derive the equation for the AC load as seen by the primary of a transformer-coupled class A amplifier. Write the equations for DC and AC load resistances. Show these load lines on appropriate characteristics of the amplifier.
7. Sketch the (V_{CE}, I_C) characteristics and AC load line for a class A amplifier. Draw the wave forms to show transistor output voltage variation with change in base current.
8. For a class A transformer-coupled amplifier, write the equations for DC input power, AC power delivered to the load and circuit efficiency. Show that the maximum theoretical efficiency of such amplifier is 50%.

9. Sketch the circuit of a push-pull class A amplifier. Briefly explain the operation of the circuit. Show composite load line on appropriate (V_{CE}, I_C) characteristics.
10. Sketch approximate composite characteristics for a class B push-pull amplifier. Show the AC load line also. Do we find crossover distortion. If yes, how can it be eliminated?
11. What do you mean by the term "distortion" in an amplifier. How it is present? Derive a relation for total harmonic distortion assuming five-point analysis.
12. Define percent harmonic distortion in an amplifier. Derive the relation for total harmonic distortion. What measures you would employ to minimize it? Draw the necessary circuit diagram.
13. Draw the circuit of a class B push-pull amplifier. How much load is seen by the primary of the output transformer? Derive the necessary equation. Derive the necessary expression for its efficiency. What is its maximum value?
14. How distortions in the output wave of a class A push-pull amplifier are minimized? Explain by drawing necessary graphical representation. Briefly describe the action of amplifier circuit.
15. Can we have a transformer-less push-pull arrangement in an amplifier? If yes, draw necessary circuit for such an amplifier. Briefly describe its action.
16. What is complimentary symmetry? How the same is used in a push-pull amplifier. Explain briefly its working.
17. Draw the circuit diagram of a class B push-pull amplifier. Briefly describe its working. Mention the merits and demerits of such a circuit.
18. What is a Darlington pair? Where such a device is used?
19. Briefly describe the working of class B emitter follower. Draw the necessary circuit diagram.
20. What is a zener follower? Where such a circuit is used? Mention its merits.
21. Describe the working of a single tuned amplifier. Derive the necessary relation for voltage amplification at resonance. Prove that gain-bandwidth of the amplifier is constant.
22. A tuned circuit employed in a transistor amplifier acts as a parallel resonant circuit. Explain. What is the maximum value impedance? Show the variation of impedance with frequency.
23. Prove that a single tuned amplifier is a narrow band amplifier? Draw the frequency response curve for such an amplifier.
24. Describe briefly the working of a double tuned amplifier. How does the frequency response curve of the amplifier vary with the coupling of two coils?

Short Answer Questions

1. What is the value of load in a power amplifier if load R_L is connected across the secondary of transformer having turn ratio unity?

   Ans: The effective value of load reflected towards primary is $a^2R_L = R_L$ since $a = 1$.

2. Why do we get harmonics in a power amplifier?

   Ans: In a power amplifier, harmonics are produced in the output because of large input signal which covers the curved part of dynamic transfer characteristic.

3. Under which class of operation an amplifier has maximum efficiency? What is its value?

   Ans: The maximum value of efficiency is obtained in class C operation. Its maximum value is 100%.

4. In which class of operation conversion efficiency is the lowest? Give reason.

   Ans: Class A operation has the minimum value of efficiency. The value may approach 50%. The reason for this being that current is drawn from the power supply throughout and power is dissipated in the device.

5. Where the DC input power is dissipated in a power amplifier?

   Ans: The DC input power is dissipated in three parts:

   (i) load

   (ii) device

   (iii) output power, which is AC

6. What are complimentary transistors?

   Ans: Complimentary transistors are similar to each other but one is *n-p-n* type and other is *p-n-p* type.

7. What is crossover distortion in a push-pull amplifier?

   Ans: For large input signals, the output current does not start and end instantly but rather it starts and ends slowly due to the curvature present in the characteristic. Thus, in one complete cycle, there is interval during which practically we do not have any output current. This flat region of output characteristic is known as cross over distortion.

8. What is composite load line in a push-pull amplifier?

   Ans: It is the resultant load line obtained from individual load lines for the two transistors employed. On (V_{CE}, I_C) characteristic, it passes through $(V_{CC}, 0)$ and has a slope I_{cm}/V_{CC}.

9. How do you define total harmonic distortion in a power amplifier?

   Ans: Total harmonic distortion in a power amplifier is defined as $D = \sqrt{D_2^2 + D_3^2 + D_4^2 + \cdots}$ where D_2, D_3 and D_4 are 1st, 2nd, 3rd harmonic distortions.

10. How do we get a sine wave in the output of a tuned amplifier operating under class C operation?

    Ans: Under class C operation, the output consists of pulses. These pulses consists of harmonics of input signal frequency. The tuned circuit of the amplifier at resonance acts as filter for harmonics and only the fundamental is passed. Thus, the output is a sine wave.

Problems

1. The AC load line of a class B push-pull amplifier has a cut off voltage of 12 V. What is its peak to peak voltage? [**Ans:** 24 V]
2. In a class C power amplifier, output power is 11 mW and input power is 50 μw. What is the power gain? [**Ans:** 220]
3. An amplifier has peak to peak output voltage of 15 V across a load of resistance 1 KΩ. What is the power gain if input power is 400 μW. [**Ans:** 70]
4. A loud speaker of resistance 16 Ω is connected in a power amplifier having effective output impedance of 10 kΩ. Find the turn ratio of a transformer so that maximum power is transferred to the load? [**Ans:** 25 : 1]
5. The input signal is 5 V rms in a class 'C' amplifier. What is its peak to peak input voltage? [**Ans:** 14.14 V]
6. An inductor of 1 μH and a capacitor of capacitance 220 pF form the tuned circuit of a class C amplifier. Calculate the resonant frequency. How the same will change if inductor value is doubled. If the Q of the inductor is 125, calculate the bandwidth of the amplifier. [**Ans:** (i) 11 MHz and (ii) 7.8 MHz (iii) 8.8 MHz]
7. In a amplifier having tank circuit connected to a power supply of 30 V, an external load of 10 kΩ is connected to its collector through a capacitor. Calculate the maximum output power. [**Ans:** 45 mw]

Objective Questions

1. A transformer in the output of a power amplifier is used for
   (a) amplifying voltage
   (b) getting high voltage
   (c) matching the load connected in secondary
   (d) safety of transistor [**Ans:** (c)]
2. The maximum value of conversion efficiency in a class B amplifier is
   (a) 100% (b) 78.5%
   (c) 50% (d) 25% [**Ans:** (b)]

3. The operating point in class B amplifier is
   (a) coincident with cutoff point of its dynamic transfer characteristic
   (b) beyond cutoff point of its dynamic transfer characteristic
   (c) within origin and cutoff point of its dynamic transfer characteristic
   (d) kept anywhere on its dynamic transfer characteristic [**Ans:** (a)]

4. An amplifier is working under class 'C' operation; the output current flows for
   (a) full cycle of input signal
   (b) positive half cycle of input signal
   (c) less than positive half cycle of input signal
   (d) negative half cycle of input signal [**Ans:** (c)]

5. The conduction angle 'θ' of device under class A and C operation respectively is
   (a) $\theta = 2\pi, \theta = \pi$ (b) $\theta = 2\pi, \theta > \pi$
   (c) $\theta = 2\pi, \theta < \pi$ (d) $\theta = 2\pi,\ \theta < \pi$ [**Ans:** (d)]

6. In the output of a push-pull class A amplifier, we get
   (a) all harmonics (b) only odd harmonics
   (c) only even harmonics (d) only 2nd harmonic is absent. [**Ans:** (b)]

7. The push-pull action of an amplifier is maintained by
   (a) input transformer (b) output transformer
   (c) both inputs and output transformer (d) transformer has no role [**Ans:** (b)]

8. In class A push-pull amplifier, load R_L is connected to the secondary of output transformer. Its effective value as seen by device is
   (a) R_L (b) $R_L/2$
   (c) $R_L/4$ (d) $2R_L$ [**Ans:** (c)]

9. Conversion efficiency of a direct coupled and a transformer coupled amplifier working under class A, respectively, is
   (a) 25%, 25% (b) 50%, 50% (c) 50%, 78.5% (d) 25%, 50% [**Ans:** (a)]

10. Total harmonic distortion in a power amplifier is defined as
   (a) $D = \sqrt{D_2 + D_3 + D_4 + \cdots}$ (b) $D = \sqrt{D_2^2 + D_3^2 + D_4^2 + \cdots}$
   (c) $D = D_2 + D_3 + D_4 + \cdots$ (d) $D = \sqrt{\dfrac{D_2^2 + D_3^2 + D_4^2}{n}}$ [**Ans:** (b)]

11. In a class B push-pull amplifier, the load is connected in secondary of output transformer R_L. The load as seen by each transistor is
   (a) R_L (b) $\dfrac{R_L}{2}$
   (c) $\dfrac{R_L}{4}$ (d) $n^2 R_L$ [**Ans:** (c)]

12. The angle for which output current flows through each transistor in a class B amplifier is
    (a) $\theta = \pi$ (b) $\theta = 2\pi$
    (c) $\theta < \pi$ (d) $\theta > \pi$ [**Ans:** (a)]
13. The gain-bandwidth product is constant only in
    (a) RC amplifier (b) transformer-coupled amplifier
    (c) power amplifier (d) all of the above [**Ans:** (d)]
14. In a tuned amplifier, the tank circuit is tuned so that its resonant frequency is
    (a) equal to input signal frequency
    (b) less than input signal frequency
    (c) greater than input signal frequency
    (d) has no relation with input signal frequency [**Ans:** (a)]
15. In a single tuned amplifier, its frequency response curve is
    (a) similar to RC amplifier
    (b) similar to resonance curve of parallel circuit
    (c) similar to transformer coupled amplifier
    (d) similar to resonance curve of series circuit [**Ans:** (d)]

Mark the Following Statements True/False.

1. A high Q circuit has smaller bandwidth and high selectivity. [**Ans:** True]
2. An amplifier under class C operation has minimum efficiency. [**Ans:** False]
3. The efficiency of a class B push-pull amplifier is less than that of a class A push-pull amplifier. [**Ans:** False]
4. In class A push-pull amplifier, both the transistors conduct simultaneously for 360°. [**Ans:** False]
5. A tuned circuit in an amplifier acts as a parallel resonant circuit. [**Ans:** True]
6. The impedance of a tuned circuit connected in an amplifier is maximum at resonance. [**Ans:** True]
7. A double tuned amplifier has lower selectivity than a single tuned amplifier. [**Ans:** True]
8. We can replace both the transformers in a push-pull amplifier and still have push-pull action. [**Ans:** True]
9. The composite load line in a push-pull amplifier is the same as load line for a simple transformer-coupled amplifier. [**Ans:** False]
10. Tuned amplifiers are basically power amplifiers. [**Ans:** True]
11. The output signal is in the form of pulses in a class C tuned amplifier. [**Ans:** False]

Fill in the Blanks.

1. Output transformer in a power amplifier is ____________ transformer. [**Ans:** iron core]
2. The maximum efficiency of a class B amplifier is ____________. [**Ans:** 78.5]
3. Maximum power dissipation occurs in ____________ in class A amplifier during ____________ signal condition. [**Ans:** device/transistor, no]
4. In a single tuned amplifier, load at resonant frequency is ____________ and voltage gain is ____________. [**Ans:** max, max]
5. ____________ tuned amplifier is used as I.F amplifier in ____________.

   [**Ans:** double, radio receiver]
6. The composite load line in a push-pull class B amplifier passes through ____________ point on (V_{CE}, I_C) characteristic. [**Ans:** V_{CC}, O]
7. Tuned amplifiers are generally used in ____________ range. [**Ans:** radio frequency]
8. Tuned amplifiers are basically ____________ amplifiers. [**Ans:** narrow band]
9. In class B push-pull amplifier, each transistor is ____________ biased to avoid ____________. [**Ans:** slightly forward, cross over distortion]
10. In a push-pull class A amplifier, we use ____________ signal which is ____________ distributed and fed to the ____________ transistors. [**Ans:** large, equally, two]

Appendix: MOSFET Power Amplifier

Power MOSFET have several advantages over bipolar transistors as regards their use as power amplifiers is concerned. One of their most important advantages is that the transfer characteristic of the device is more linear than that of bipolar transistor. This property enables MOSFET power amplifiers to have much less distortion in the output than those of bipolar transistor power amplifiers. In addition, the problem of thermals run away does not occur with power MOSFETs. Thus, the emitter resistors used with a bipolar output stage are not required, thereby eliminating DC power dissipation in emitter resistors.

Power MOSFETs can readily be operated in parallel to reduce total channel resistance and increase the output current level. Unlike bipolar transistors operated in parallel, there is no need for use of resistors to equalize current distribution between parallely connected MOSFETs. For class AB operation, the MOSFET gate-source should be biased to threshold voltage V_{th} for the device to ensure that it is conducting at a low level.

A simple power amplifier circuit using MOSFETs is shown in Figure A.1. Here, output transistors Q_3 and Q_4 operate as common source amplifiers. This permits the peak output voltage to approach the supply level. The differential amplifier input stage allows the use of DC negative feedback to stabilize the DC output voltage level, and AC negative feed-back to set off the AC voltage gain.

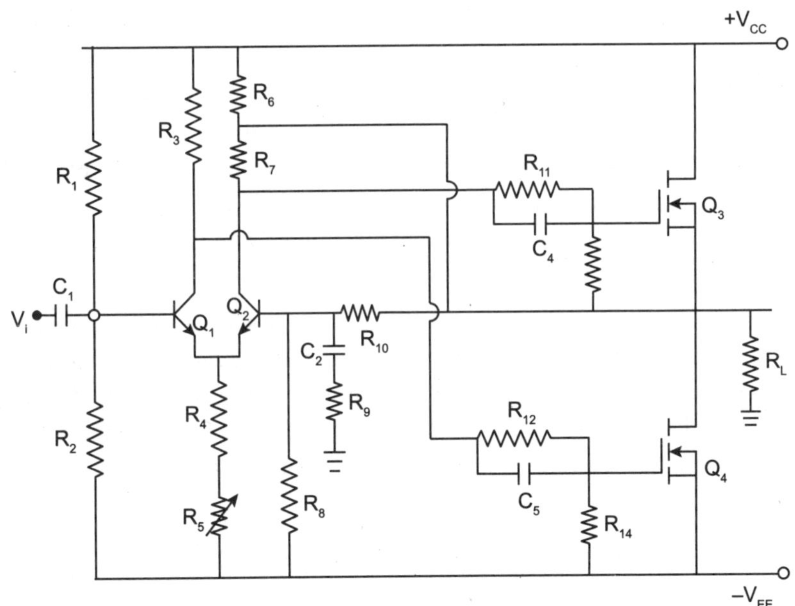

Figure A.1 Power amplifier using MOSFET output stage

Output transistors Q_3 and Q_4 are a similar type of MOSFETs and both operated as common source amplifiers, rather than as source followers. Resistors R_{12} and R_{14} potentially divide the collector voltage of Q_1 to bias Q_4 gate to the threshold level for MOSFET. Similarly, the bias voltage for Q_3 is derived from the collector of Q_2 via R_{11} and R_{12}. Capacitors C_4 and C_5 short circuit R_{11} and R_{12} respectively at signal frequencies, so that all of the AC voltage from the first stage is applied to the FET gate terminals. When a positive going voltage from Q_1 collector is applied to the gate of Q_4, it is clear that the voltage is developed across gate-source terminals. Consequently, Q_4 turns on harder and causes drain current I_{D4} to flow through R_L in a negative direction. Q_4 is thus operating as common source amplifier with a voltage gain of $g_m R_L$. During the time that Q_4 is conducting, Q_3 has a negative going voltage as applied to its gate to turn OFF.

If capacitor C_3 were not present, then transistor Q_3 would simply operate as source follower, passing on the AC voltage from Q_2 collector to the lead, and its voltage gain will be unity.

Consequently, its performance would not match with that of Q_4. Capacitor C_3 is termed as bootstrapping capacitor, and by behaving as a short circuit at signal frequencies, it effectively ties the function of R_6 and R_7 to the source terminal of Q_3. This means that all of the voltage developed across R_7 is applied to the gate and source terminals of Q_3 and so it has the same amplification as common source circuit. When the signal from Q_2 collector is positive going, Q_3 conducts, providing drain current I_{D3} in a positive direction through R_L. At this time, Q_4 has a negative going voltage from Q_1 collector, so Q_4 is OFF.

Note that in the circuit, the base of Q_1 is not biased to ground level, but instead has a base voltage between zero and (–) V_{EE}. Since the DC output voltage must be biased to ground, resistors R_8 and R_{10} are required to set V_{B2} equal to V_{B1}. The reason for this is that voltage V_{CE2} must be large in order to handle change in V_{R7} and peak output voltage swing transferred via C_3.

Adjustable resistor R_5 provides control over total emitter current for Q_1 and Q_2. If R_5 is reduced in order to increase I_{E1} and I_{E2}, then I_{C1} and I_{C2} are also increased. Thus, voltage drops across R_3, R_6 and R_7 are increased and collector voltage of Q_1 and Q_2 is decreased. The decrease in V_{C1}, and V_{C2} causes a reduction in the gate source bias voltage of output transistors and consequently reduces the quiescent drain current through Q_3 and Q_4. It is seen that R_5 facilitates the adjustment of bias condition of Q_3 and Q_4.

In designing a B class MOSFET power amplifier, the gate service voltage of output transistors should be biased to the specified threshhold voltage for the devices. Peak output voltages and currents are determined in the usual way and the supply voltage is calculated as $V_p + I_p R_{DS(ON)}$ where $R_{DS(ON)}$ is the channel resistance of MOSFET when it is fully turned ON. The selected MOSFETs must handle the maximum drain current, approximately equal to 1.1 I_p, and must survive the supply voltage of 2 V_{CC}.

Chapter

12 Oscillators

12.1 Introduction

An oscillator is an amplifier circuit that uses a positive feedback to provide output waveforms without an input (external) signal. Since an oscillator has no input, it can be considered a signal generator. In other words, an oscillator converts the DC used to bias the circuit to an AC output signal.

Oscillators provide different waveforms for a variety of equipment. With the use of solid-state components and circuitry, oscillators are used to create stable waveforms for use in control circuitry. Oscillators are also used in the control circuitry of variable-speed motor controls and other output devices.

12.2 LC Oscillator (Sinusoidal Oscillator)

The simplest circuit to generate electrical oscillations is an LC circuit, popularly known as tank circuit.

Suppose we have an LC combination as shown in Figure 12.1. Let capacitor C be charged by some astray DC source (Figure 12.1(a)). On closing the key K, the capacitor will get discharged through inductor L, i.e. electrons from, negative plate begin to flow through the inductor (Figure 12.1(b)) towards the positive plate, and we say the capacitor begins to discharge. This flow of electrons amounts to a flow of current in the inductor where it builds up a magnetic field around inductor L shown by the magnetic lines of force. This field goes on increasing and attains a maximum value (Figure 12.1(c)) till the capacitor is fully discharged. In a way, electrostatic energy stored in the capacitor gets converted into magnetic energy in the inductor.

Now, the magnetic field around the inductor gets decayed (since there is no flow of electrons from the capacitor) and magnetic flux changes (Figure 12.1(d)), which induces an e.m.f. across the capacitor, i.e. it gets charged but now in the opposite direction, and the upper plate of capacitor is negatively charged. The process goes on untill the entire magnetic flux reduces to zero and the capacitor is fully charged (Figure 12.1(e)) in opposite sense. Now, we can say that magnetic energy has converted into electrostatic energy. The capacitor again begins to discharge (Figure 12.1(f)) and the cycle is repeated in other direction. In this way, charge and discharge of the capacitor (or creation and decay of magnetic field) goes on. This interchange of electrostatic energy to magnetic and vice versa between L and C is repeated again and again, resulting into electromagnetic oscillations (Figure 12.2(a)). In actual practice, the amplitude of oscillations (EM wave) goes on decreasing which we call damped oscillations (Figure 12.2(b)) due to the small resistance of inductor, i.e. energy is partly lost into heat energy developed in the resistance of coil.

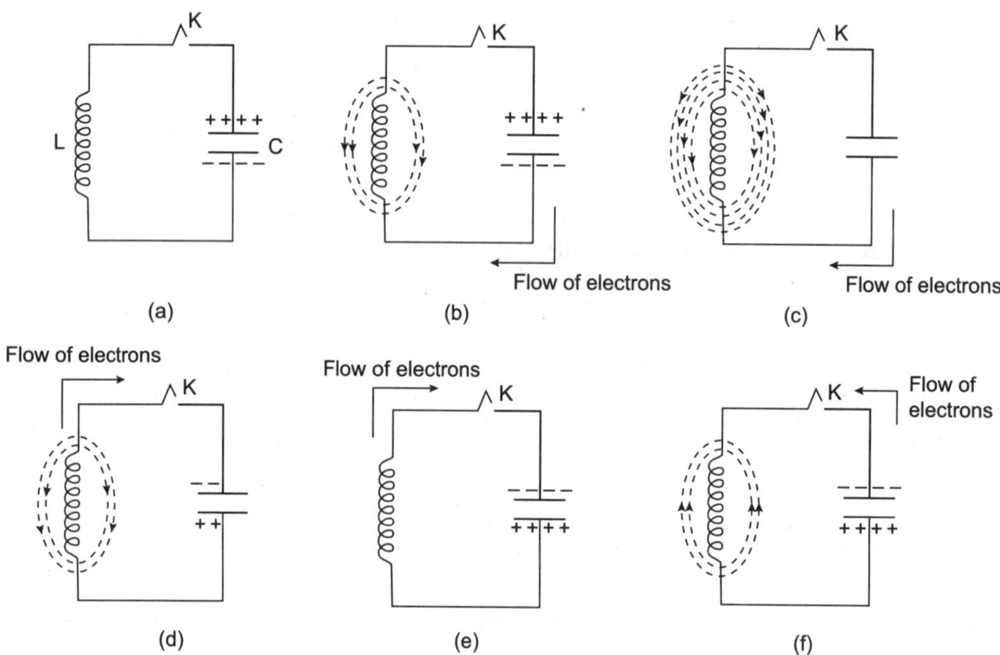

Figure 12.1 Charge and discharge of capacitor through an inductor

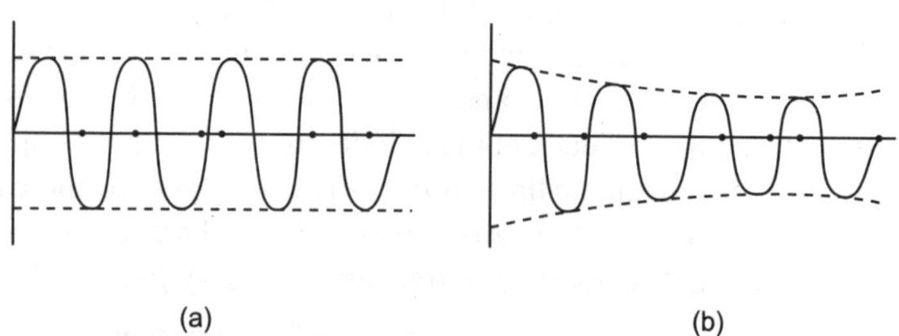

Figure 12.2 (a) Ideal oscillations, (b) practical oscillations

The frequency of oscillations (ideal) is given by $f = \dfrac{1}{2\pi\sqrt{LC}}$.

For practical applications, we require undamped oscillations (Figure 12.2(a)) (i.e. sustained oscillations). For this, we require a device to feed proper amount of energy at proper time. To achieve sustained oscillations in an LC combination, we make use of the principle of positive feedback.

The oscillator can have different frequency ranges by varying the values of RC or LC components. Oscillators that have this capability are called variable frequency oscillators (VFOs). One method of frequency control is the lead-lag network. VFOs are sometimes built using circuits, with continuously variable inductors or capacitors to adjust the frequency as required. Varactors are particularly suited for use in LC tuned circuits.

12.3 Principle of Feedback Oscillator

In a previous chapter, we discussed the effect of feeding back some of the output voltage to the input in such a way that feedback voltage reduces the input. The result was a reduction is overall voltage gain, although numerous beneficial effects were cited. It was termed as negative feedback. We now examine the case when feedback is positive such that the magnitude of return difference $|1-\beta A|$ is zero or $|\beta A|=1$. In such a situation, gain A_f after feedback increases to ∞ (Eqn. 2), i.e. we have output even without an external input signal and the result is that oscillation occurs in the output. Positive feedback must be in phase with the original (sine) wave so that it adds the needed energy to overcome and maintain signal loss. Another name for this type of feedback is *regenerative feedback*.

The required positive feedback effect can be accomplished only if the feedback is greater than the signal loss and if the feedback is in phase. Such an arrangement or device is called 'oscillator' with feedback.

12.4 Barkhausen Criterion

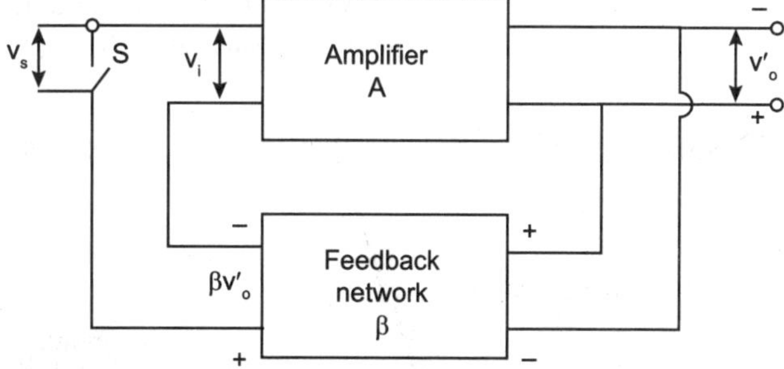

Figure 12.3 Block diagram of an oscillator in which both *A* and β are both negative

For a feedback amplifier, we have overall gain,

$$A_f = \frac{A}{1-\beta A} \tag{1}$$

If in Eqn. (1), $$|\beta A| = 1 \tag{2}$$

then $$|A_f| \Rightarrow \alpha \tag{2a}$$

It means we have output even when there is no external input. The above condition (Eqn. (2)) is called Barkhausen criterion for oscillation. Clearly, this criterion expresses the relationship between circuit feedback factor and voltage gain of the circuit for proper oscillator operation. This means that the product of feedback and gain must be exactly equal to 1 in order to maintain oscillation at a constant level. If $\beta A < 1$, oscillations will fade within cycles. If $\beta A > 1$, oscillations increase each cycle and will eventually drive the oscillator into clipping and saturation. Thus, for oscillations to be sustained, we must have (i) loop gain of the circuit be equal to (or greater than) 1 and (ii) total phase shift around the circuit be zero or 360°. For this condition to be met, both β and A must be either positive or negative. When this is satisfied, switch S be closed and external signal V_s may be removed, and still there will be output, i.e. oscillations are sustained since

$$|v_i| = |\beta v_o'|, \text{ in general}$$

$$v_i = v_f = (-)KV_o' \tag{3}$$

If the requirement for oscillation $\beta A \geq 1$ is met at only one frequency (loop gain phase shift is zero or integral multiple of 360°), then sinusoidal oscillations will result, and we have a set up which we call feedback sinusoidal oscillator; further amplitude of oscillations is limited by the ON set of nonlinearity. If the above requirement is met over a band of frequencies, then the output may be nonsinusoidal such as a square wave.

LC oscillators based on the principle of feedback have three features in common:

(i) There should be an active device that works as amplifier and introduces a phase shift of 180° between input and output signal.

(ii) There must be a feedback network that provides positive feedback, i.e. it introduces an additional phase shift of 180°, so that the total phase shift is 360° or integral multiple of 2π.

(iii) The amount of feedback must be sufficient to overcome circuit losses so that oscillations (e.m.f. wave) have equal amplitude.

12.5 Oscillator Characteristics

Oscillators can be sensitive to stray or line capacitance. This is capacitance that comes from the skin effect of wires and interelectrode capacitance of transistor junctions. These excess capacitance sources can cause false readings or increase lag time at very high frequencies. Unknown feedback paths can be developed in complicated circuits because of the unwanted capacitance and inductance.

Another impact on oscillator stability is power supply voltage. An increase or decrease in power supply voltage can cause the biasing of circuit to become unbalanced. For example, cut off points could be reached earlier or saturation levels exceeded. Other effects are lowering or increasing the need for feedback from the output. In a crystal oscillator, the increased voltage will vary the amplitude of frequency.

12.5.1 General Form of Feedback Oscillator Circuit

Most oscillators based on the principle of feedback may be represented by a block diagram (Figure 12.4). It uses an amplifier circuit with Z_1, Z_2 and Z_3 impedances in the output.

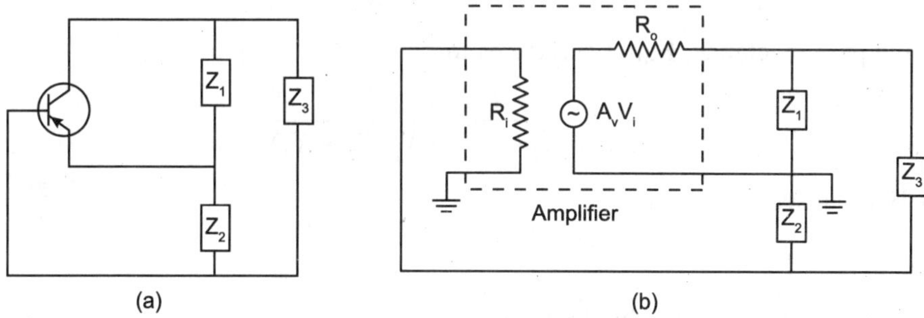

Figure 12.4 (a) Generalized block diagram for resonant circuit oscillators, (b) equivalent circuit

As shown in Figure 12.4, load Z_L of the amplifier circuit consists of impedance Z_1 in parallel with series combination of Z_2 and Z_3 impedance arranged so that a part of output voltage across Z_2 is fed back, i.e. it acts as a feedback network. As seen from Figure 12.4(a), feedback factor

$$\beta = \frac{Z_2}{Z_2 + Z_3} \tag{4}$$

Gain of the amplifier without feedback [Figure 12.4(b)] is

$$A = \frac{A_v Z_L}{Z_L + R_o} \tag{5}$$

where R_o represents output resistance of amplifying device, Z_L is effective load and A_v is internal gain of amplifier.

$$\text{Effective load, } Z_L = \frac{Z_1(Z_2 + Z_3)}{Z_1 + Z_2 + Z_3}$$

Figure 12.4(b) shows the AC equivalent circuit of resonant circuit oscillator using feedback. Here A_v represents internal gain of amplifier.

Substituting the values of β and A, we have

$$\beta A = \left[\frac{Z_2}{Z_2 + Z_3}\right]\left\{\frac{A_v \dfrac{Z_1(Z_2 + Z_3)}{Z_1 + Z_2 + Z_3}}{\dfrac{Z_1(Z_2 + Z_3)}{Z_1 + Z_2 + Z_3} + R_o}\right\} \tag{6}$$

or
$$\beta A = \frac{A_v Z_1 Z_2}{R_o(Z_1+Z_2+Z_3)+Z_1(Z_2+Z_3)} \tag{7}$$

If Z_1, Z_2 and Z_3 comprise a pure reactance, then

$$Z_1 = jX_1,\ Z_2 = jX_2 \text{ and } Z_3 = jX_3$$

$$\left(\text{For an inductor } X = \omega L; \text{and for a capacitor } X = \frac{1}{\omega C}\right)$$

Assuming pure reactance, Eqn. (7) reduces to

$$\beta A = \frac{A_v X_1 X_2}{j(X_1+X_2+X_3)-X_1(X_2+X_3)} \tag{8}$$

As per Barkhausen criterion,

Loop gain, $\beta A = 1 + jo$ (8a)

i.e., here the imaginary term should be zero.

Applying this condition in Eqn. (8), we have

$$X_1 + X_2 + X_3 = 0 \tag{9}$$

and Eqn. (8) can now be written as

$$\beta A = (-)\frac{A_v X_2}{X_2 + X_3}$$

Using Eqn. (9), we have also

$$\beta A = (+)\frac{A_v X_2}{X_1} \tag{10}$$

From Eqn. (10), we find that the nature of X_2 and X_1 should be same, i.e. either both inductive or both capacitive taking A_v should be positive since βA must be positive and equal to unity for maintaining Barkhausen criterion.

From Eqn. (9),

$$X_3 = (-)(X_1 + X_2) \tag{11}$$

The negative sign represents that if

(i) both X_1 and X_2 are inductive, then X_3 should be capacitive

(ii) both X_1 and X_2 are capacitive, then X_3 should be inductive

A circuit making use of the first criterion is called Hartley oscillator (Figure 12.5) and a circuit making use of the second criterion is called Colpitts oscillator (Figure 12.6). The Hartley and Colpitts oscillators are very similar in design. The major difference is how the feedback circuit is produced.

12.5.2 Hartley Oscillator

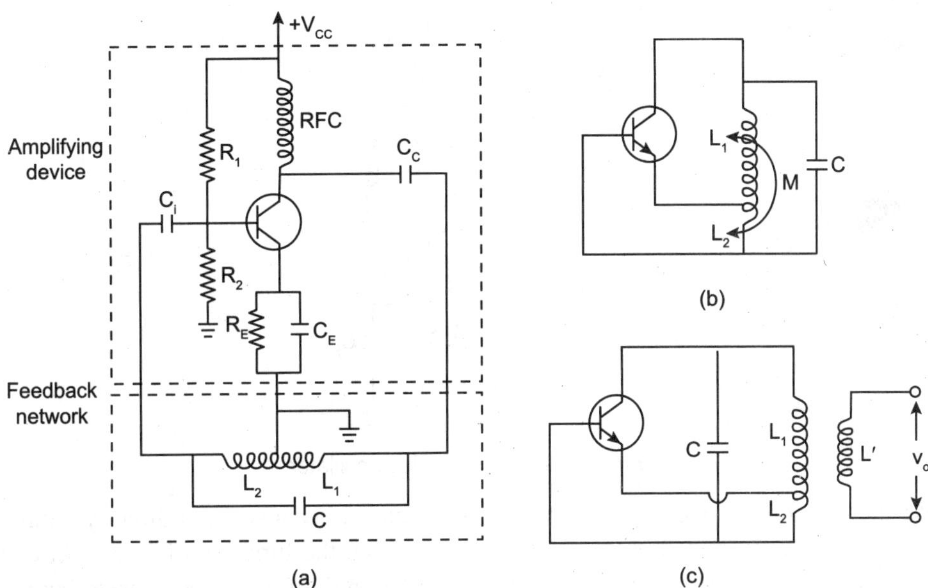

Figure 12.5 (a, b, c) Hartley oscillator

As shown in Figure 12.5(a), we use a tapped inductor which constitutes the two inductors L_1 and L_2. The entire inductor is connected across capacitor C. So it forms an LC circuit. A part of it acts as feedback network as shown. Coil L_1 is inductively coupled to coil L_2 and the combination functions as an autotransformer. The presence of coupling capacitor C_C does not allow DC to flow through the tank circuit. A radio frequency coil is connected in series with supply V_{CC} to the collector. It permits the flow of DC to the collector but does not allow any AC. Capacitor C_i allows the feedback voltage to input.

In this case, total Inductance $L = L_1 + L_2 + 2M$

where M is mutual inductance between L_1 and L_2.

It is an example of shunt feed oscillator, so called because the RF path is in parallel with DC path. This type of circuit is frequently used as a local oscillator in super heterodyne receivers. Capacitor C_C and inductor L_1 form a path for RF current in collector to ground circuit. RF current through L_1 induces a voltage in L_2 of proper phase and amplitude to sustain oscillations. The position of emitter lead on inductor L determines how much signal is fed back to base circuit. Capacitor C and autotransformers L_1 and L_2 make up a resonant circuit which determines the resonant frequency. Capacitor C can be made variable for tuning the oscillator to various frequencies. The RF choke in the collector circuit keeps off the RF signal to flow through supply V_{CC}. Another version of the circuit is shown in Figure 12.5(d).

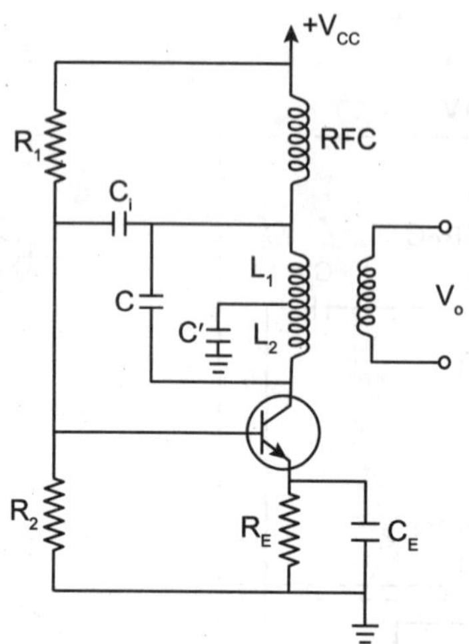

Figure 12.5 (d) Hartley oscillator

Here the coupling capacitor C_C is no longer required because L_2 is directly connected to the amplifier. However, because of the direct connection, the function of L_1 and L_2 cannot be directly grounded, so another capacitor C' is connected and grounded. The use of capacitor C' does not allow any DC to flow to ground.

Circuit analysis Looking at Figure 12.5(b), we identify

$$Z_1 = jX_1 = j\omega(L_1 + M)$$
$$Z_2 = jX_2 = j\omega(L_2 + M)$$
$$Z_3 = jX_3 = (-)\frac{j}{\omega c}$$

Substituting these in Eqn. (8) and using Eqn. (8a) to equate real and imaginary quantities separately, we have the condition for sustained oscillation as

$$h_{fe} \geq \frac{L_l + M}{L_1 + M} \tag{12}$$

And frequency of oscillation,

$$\omega_o = \frac{1}{\sqrt{C(L_1 + L_2 + 2M)}}$$

$$= \frac{1}{\sqrt{CL_T}}$$

where L_T is inductance of the entire coil and is equal to $L_1 + L_2 + 2M$

The output is observed across a secondary coil L' inductively coupled with coil L as shown in Figure 12.5(c).

12.5.3 Colpitt Oscillator

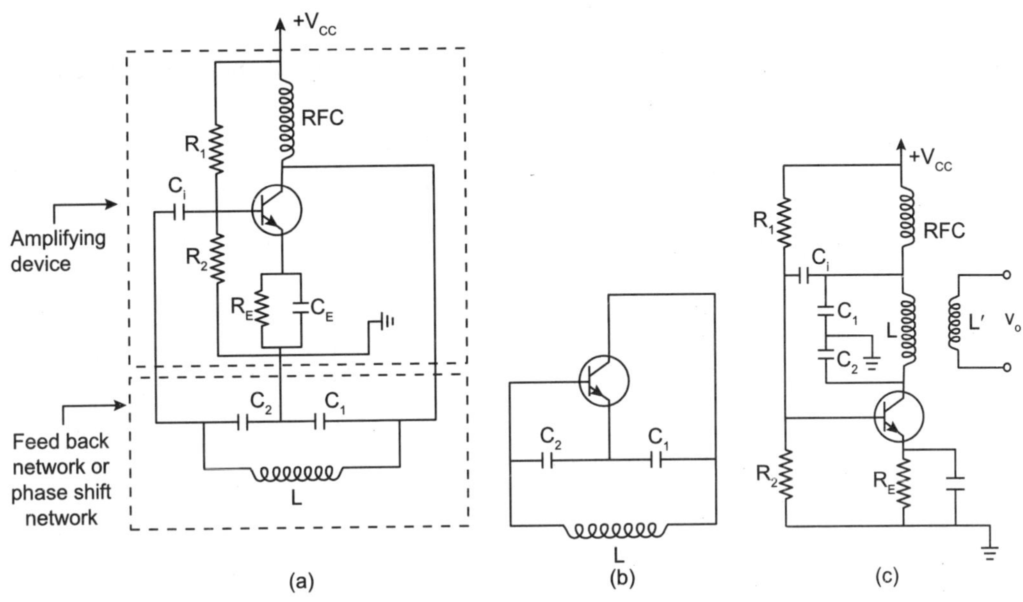

Figure 12.6 (a, b, c) Colpitt oscillator

As shown in Figure 12.6(a), a radio frequency coil is connected in series with supply V_{CC} to collector. It allows DC to flow through the collector but offers very high impedance to AC at the oscillating frequency of LC tank circuit (C_1, C_2 and L), which provides necessary phase shift between the output and feedback voltage. The right end of L is AC isolated due to RFC, from V_{CC} and ground. The output of feedback voltage across C_2 is fed to base at input through capacitor C_i. Output voltage v_o (Figure 12.6(c)) is derived from a secondary winding L' coupled to inductor L.

The amount of feedback depends upon the relative capacitance of C_2 and C_1. Capacitors C_1 and C_2 act as simple AC voltage divider so that the two extreme points of capacitor combination are 180° out of phase. This is in addition to the phase change of 180° introduced by the amplifier; so there is a total phase change of 360° as required for positive feedback. It is widely used to generate frequency greater than 1 MHz.

Another version of the oscillator is shown in Figure 12.6(c). The output of phase shift network is coupled via C_i from the junction of L and C_1 to the base of transistor. As in the case of phase shift amplifier, there too transistor current gain in h_{FE} is important.

Circuit analysis Looking at Figure 12.5(b), we identify

$$Z_1 = jX_1 = (-)\frac{j}{\omega C_1}$$
$$Z_2 = jX_2 = (-)j/\omega c_2$$
$$Z_3 = jX_3 = j\omega L$$

Substituting these in Eqn. (8) and using Eqn. (8a) to equate real and imaginary quantities separately, we have the condition for oscillation as

$$h_{fe} \geq \frac{C_1}{C_2} + h_{ie}\,h_{oe}\frac{C_2}{C_1} + \frac{h_{oe}}{h_{ie}}\frac{L}{C_2} \quad (12)$$

and frequency of oscillation,

$$\omega_o = \frac{1}{\sqrt{L\left(\frac{1}{C_1} + \frac{1}{C_2}\right) + \frac{h_{oe}}{h_{ie}} \times \frac{1}{C_1 C_2}}} = \frac{1}{\sqrt{L\left(\frac{C_1 C_2}{C_1 + C_2}\right)}} \quad (13)$$

Since $\frac{h_{oe}}{h_{ie}}$ being very small, it can be neglected.

12.5.4 Clapp Oscillator

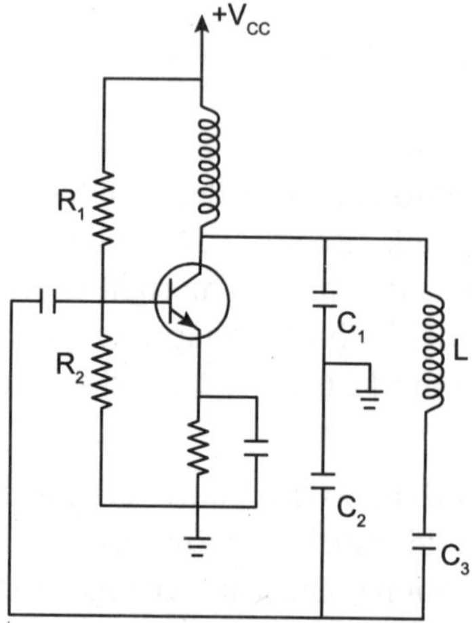

Figure 12.6 (d) Clapp oscillator

Its circuit is shown in Figure 12.6(d). It is a refinement of Colpitts oscillator. The capacitive voltage divider produces feedback signal as explained in the case of Colpitts oscillator. An additional capacitor C_3 is put in series with inductor L. Since the circulating current flows through C_1, C_2 and C_3 in series, its equivalent capacitance is

$$C = \frac{1}{\frac{1}{C_1} + \frac{1}{C_2} + \frac{1}{C_3}} \approx C_3$$

since normally C_3 is much less than both C_1 and C_2.

Frequency of oscillation is

$$f_r = \frac{1}{2\pi\sqrt{\text{LC}}} = \frac{1}{2\pi\sqrt{\text{LC}_3}}$$

Feedback factor, $\beta = \frac{C_1}{C_2}$.

Minimum voltage gain to start oscillations $A_{\min} = \frac{C_2}{C_1}$

so that $\beta A = 1$ is satisfied. Here, stray capacitors formed due to transistor and wiring has no effect, and frequency is more stable.

Example 1 Capacitors C_1 and C_2 each of value 0.001 μF and 0.01 μF are used in a Colpitts oscillator along with inductor L of 5 μH. What should be the minimum value of h_{fe} of transistor used in the circuit so that the oscillator vibrates. Calculate the frequency of oscillations produced.

Solution

To maintain vibrations in a colpitts oscillator,

$$h_{fe} \geq \frac{C_2}{C_1}$$

$$= \frac{0.01 \times 10^{-6}}{0.001 \times 10^{-6}} = 10$$

So the value of h_{fe} of transistor used must be greater than 10.

Frequency of oscillations produced,

$$f = \sqrt{\frac{1}{L}\left(\frac{1}{C_1} + \frac{1}{C_2}\right)}$$

$$= \sqrt{\frac{1}{5}\left(\frac{10^6}{0.001} + \frac{10^6}{0.01}\right)}$$

$$= 1.5 \times 10^4 \text{ Hz}$$

Example 2 In a phase shift oscillator, $R_L = 3.3\ \text{k}\Omega$, $R = 5.6\ \Omega$ and $C = 0.001\ \mu\text{F}$. What should be the value of h_{fe} of transistor used so that the oscillator vibrates? Calculate the frequency of oscillations produced.

Solution

For oscillations to be maintained in a RC oscillator,

$$h_{fe} = \left[23 + 29\frac{R}{R_L} + 4\frac{R_L}{R}\right]$$

$$\therefore h_{fe} = \left[23 + \frac{29 \times 5.6 \times 10^3}{3.3 \times 10^3} + \frac{4 \times 3.3 \times 10^3}{5.6 \times 10^3}\right]$$

$$= 74.6$$

$$\approx 75$$

Frequency of oscillations,

$$f = \frac{1}{2\pi C\sqrt{4RR_L + 6R_2}}$$

$$= \frac{1}{2 \times 3.14 \times 10^{-8}\sqrt{[4 \times 5.6 \times 3.3 + 6 \times 5.6 \times 5.6) \times 10^6}}$$

$$= 983.6\,\text{Hz}$$

Example 3 The electrical parameters of a crystal used in an oscillator are $L = 0.33$ H, $C = 0.065$ pF and $C_m = 1.0$ pF, $R = 0.55\ \text{k}\Omega$. Calculate the series resonant frequency of oscillations produced and Q of the crystal.

Solution

$$\text{Series resonant frequency, } f_s = \frac{1}{2\pi\sqrt{\text{LC}}}$$

$$= \frac{1}{2 \times 3.14\sqrt{0.33 \times 0.065 \times 10^{-12}}}$$

$$= 1.09\ \text{MHz}$$

$$Q \text{ of the crystal} = \frac{2\pi f_s L}{R}$$

$$= \frac{2\pi \times 1.09 \times 10^6 \times 0.33}{0.55 \times 10^3}$$

$$= 3800$$

12.6 RC Coupled Oscillators

Oscillator circuits described so far have two drawbacks:

(i) They suffer from frequency instability and poor waveforms.

(ii) They generate very high frequency and are too much bulky and expensive too.

Excellent frequency stability and waveform with desired low frequency range can be obtained from oscillators based on positive feedback by employing resistive and capacitive elements called RC coupled oscillators. RC oscillators are generally used only below a frequency of 1 MHz. RC circuits are used to produce the required phase shift for an oscillator to function. An RC oscillator is made of three basic sections: the amplifier, the phase-shift network, and the feedback loop.

Among RC coupled amplifiers, two are very popular:

(i) Phase shift oscillator

(ii) Wien bridge oscillator

12.6.1 Phase Shift Oscillator

Principle We know that a single stage CE transistor amplifier amplifies the input signal at the same time there is a phase difference of 180° between the output and input signal. To provide the rest of 180° phase shift, we use a RC network and then a part of output signal may be fed to the input.

Figure 12.7(a) represents a RC network across AC voltage V_i. Voltage V_o across R is in phase with current through it but it leads input voltage by an angle ϕ as shown in Figure 12.7(b). Thus, we find a phase shift of ϕ between the input and output voltage given by

$$\tan\phi\frac{V_C}{V_O}=\frac{IX_C}{IR}=\frac{X_C}{R} \tag{14}$$

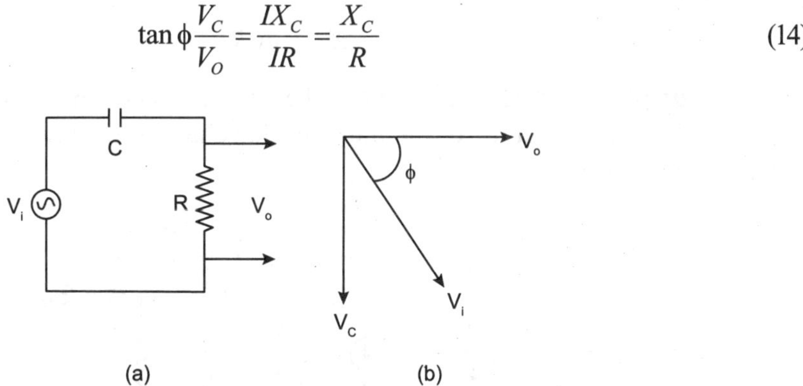

Figure 12.7 RC network

where I is current flowing in the circuit and X_C is reactance of capacitor at a particular frequency. If we manage X_C and R such that $\phi = 60°$ for one RC section, then three such sections in sequence will produce a phase shift of $3 \times 60 = 180°$.

Therefore, the voltage after three RC sections, which is fed back to the input of amplifier, would be in phase with the input signal since the voltage fed back has a total phase shift of 360° (180° due to transistor amplifier and another 180° due to three RC sections).

A phase oscillator circuit employing the above principle is shown in Figure 12.8.

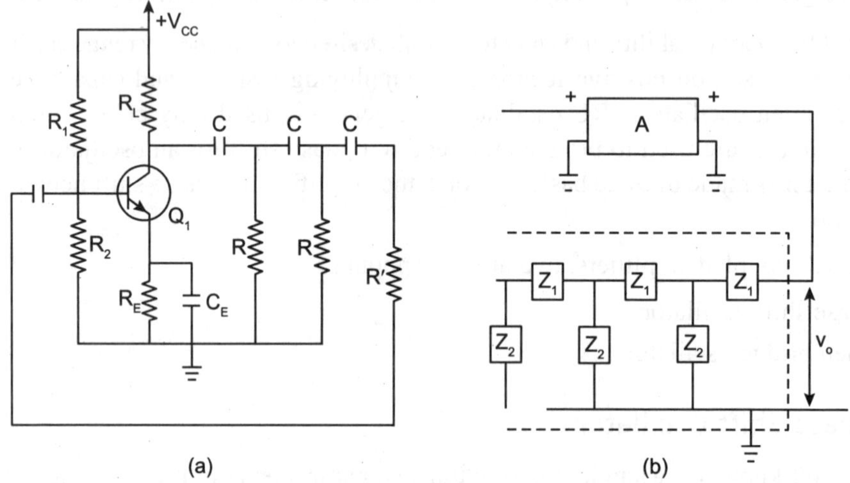

Figure 12.8 Phase shift oscillator

Q_1 is the transistor used in CE amplifier configuration having proper bias by employing biasing resistors R_1 and R_2. The three RC sections are connected to the amplifier as shown.

The frequency of oscillator output depends upon capacitor and resistance values employed in the phase shift network. Using basic CR circuit analysis techniques, it can be shown that network phase shift is 180° when

$$X_C = \sqrt{6}R \tag{15}$$

Analysis of phase shift oscillator The AC equivalent circuit of oscillator is shown in Figure 12.9. Here h_{re} and h_{oe} of the transistor have not been shown since their values being small have been neglected.

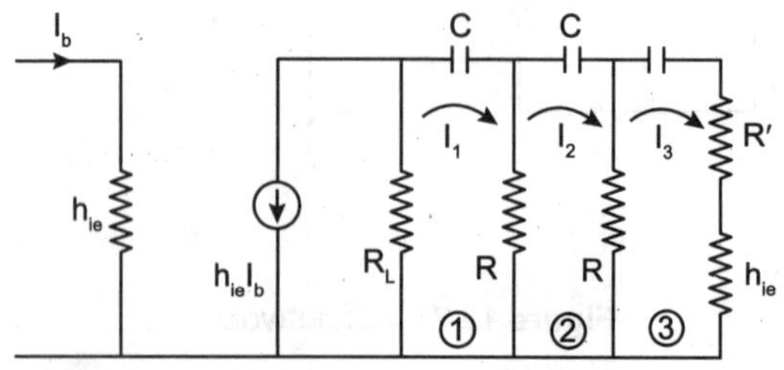

Figure 12.9 AC equivalent circuit of phase shift oscillator

Assuming the three loops carry different currents I_1, I_2 and I_3, respectively, we can write the following equations (applying KVL) for each loop.

$$\left(R_L + R + \frac{1}{j\omega c}\right)I_1 - RI_2 + h_{fe}R_L I_3 = 0$$

$$(-)RI_1 + \left[2R + \frac{1}{j\omega c}\right]I_2 - RI_3 = 0$$

$$(-)RI_2 + \left[2R + \frac{1}{j\omega c}\right]I_3 = 0$$

The above equations can be written in matrix form as

$$\begin{vmatrix} R_L + R - \dfrac{j}{\omega c} & -R & h_{fe}R_L \\ -R & \left(2R - \dfrac{j}{\omega c}\right) & -R \\ 0 & -R & \left(2R - \dfrac{j}{\omega c}\right) \end{vmatrix} \begin{vmatrix} I_1 \\ I_2 \\ I_3 \end{vmatrix} = 0$$

The above relation gives oscillating frequency,

$$f = \frac{1}{2\pi \mathrm{RC}\sqrt{6}} \tag{16}$$

as well as phase shift in the RC network attenuates the amplifier output.

Network analysis shows that, when 180° phase shift is achieved, the network attenuates the output voltage by a factor of 29. It means that the amplifier must have a voltage gain of 29 for the circuit to oscillate.

The amplifier gain of 29 and network attenuation $\beta = \frac{1}{29}$ gives a loop gain of as required by Barkhausen criterion. Also, the amplifier phase shift of 180° combined with the network phase shift of 180° gives a loop shift of 360° or 0°.

If amplifier gain is much greater than 29, the oscillator output waveform is likely to be distorted. The amplitude of output wave depends upon power supply V_{CC} and amplifier bias conditions.

From the above discussion, it would appear that a CE transistor amplifier with a voltage gain of 29 or more simply has to be used in the circuit (Figure 12.8(a)). An attempt to design such a circuit reveals that the amplifier output is overloaded by phase shift network or else the phase shift network is overloaded by amplifier input impedance. The problem can be solved by introducing an emitter follower in the circuit. However, the circuit can be made to function if the transistor in Figure 12.8(a) is treated as current amplifier rather than as voltage amplifier. In this case, the analysis of the circuit gives

$$f = \frac{1}{2\pi RC\sqrt{6 + 4R_L/R}} \tag{17}$$

and

$$h_{fe(\min)} = 23 + \frac{29R}{R_L} + \frac{4R_L}{R} \tag{18}$$

If transistor current gain is less than the minimum value determined above, the circuit will not oscillate. If $R_L = R_C$, then h_{fe} (minimum) must be greater than 56 to sustain oscillations.

Although a phase shift oscillator is the easiest to understand, it is rarely used because it is not stable and is difficult to tune.

12.6.2 Wien-Bridge Oscillator

The Wien-Bridge is the standard common RC oscillator circuit for low to moderate frequencies in the range of 5 Hz to 100 KHz. It is mostly used in commercial audiogenerators and also preferred for other low frequency applications. This oscillator provides no phase shift at resonant frequency. The result of this "non-phase shifting" characteristic is that neither the amplifier nor the feedback network produces a phase shift. This oscillator uses two feedback circuits, one positive and one negative. The positive feedback circuit is used to control the operating frequency of the oscillator. The negative feedback circuit is used to control the gain of the circuit.

Before we discuss its principle of working, we should be familiar with lag circuit and lead circuit which play an important role in its working.

Lag circuit

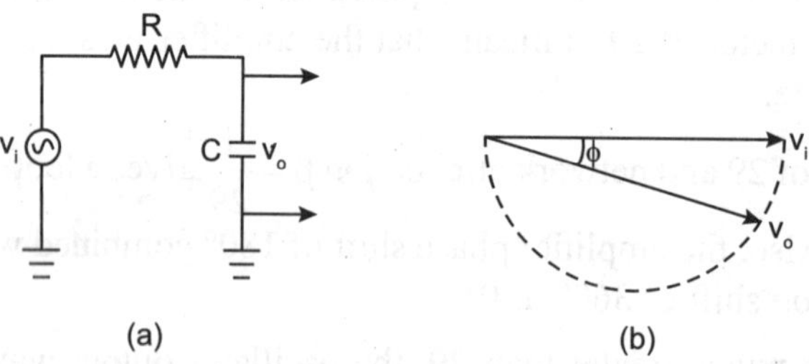

Figure 12.10 (a) By pass capacitor (lag circuit), (b) phase diagram of lag circuit

A lag circuit is shown in Figure 12.10(a). As seen in figure, v_i is applied across RC combination and output v_o is observed across the capacitor. It lends to

$$\frac{v_o}{v_i} = \frac{X_C}{\sqrt{R^2 + X_c^2}}$$

and phase angle between the output and input is given by $\tan\phi = (-)\frac{R}{X_C}$. Notice the negative sign; it shows that the output lags behind input voltage and because of this the bypass circuit is also known as lag circuit. The phase angle lies between 0° and (–) 90°.

Lead circuit

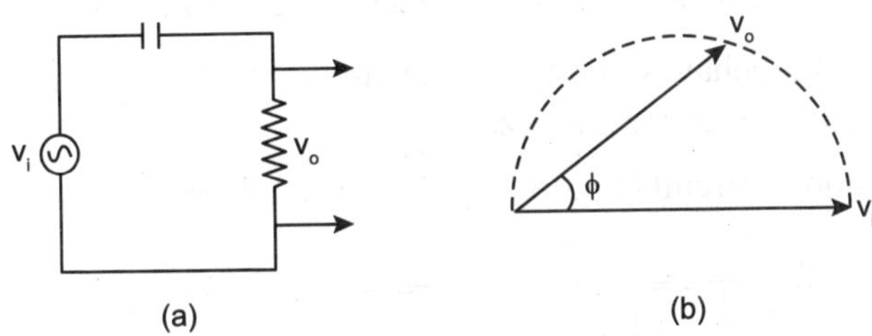

Figure 12.11 (a) Lead circuit (coupling circuit), (b) phasor diagram of lead circuit

Figure 12.11 shows a coupling circuit. Here, input voltage v_i is applied across RC combination and output v_o is observed across the resistor. It leads to

$$\frac{v_o}{v_i} = \frac{R}{\sqrt{R^2 + X_C^2}}$$

and phase angle ϕ between the output and input is given by

$$\tan\phi = \frac{X_C}{R}$$

Notice here that the phase angle is positive, i.e. the output is leading the input. Because of this, a coupling circuit is also called a lead circuit. The output phasor can lead through angle 0° to 90°.

Coupling and bypass circuits are examples of phase shifting circuits. A sinusoidal oscillator always uses some kind of phase shifting circuit to produce oscillations at some frequency.

12.6.3 Principle of Working of Wien-Bridge Oscillator

A Wien-Bridge oscillator uses a resonant feedback circuit called lead-lag circuit (Figure 12.12).

At very low frequencies, the series capacitor appears open to input signal v_i and there by we do not observe any output. At very high frequencies, the shunt capacitor looks shorted and again there is no output. In between these extreme frequencies, the output voltage exists and it attains a maximum value (Figure 12.13). The frequency at which maxima occurs is called resonant frequency f_r. At this frequency, feedback factor β attains a value 1/3.

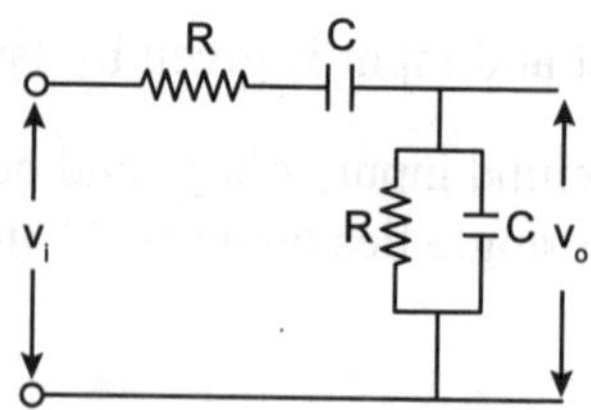

Figure 12.12 Lead-lag circuit (phase shift circuit)

Thus, a lead-lag circuit behaves like a resonant circuit in which the feedback fraction attains a maximum value 1/3, having phase angle zero.

By analyzing the above circuit (Figure 12.12), we can derive that

$$\frac{v_o}{v_i} = \beta = \frac{1}{\sqrt{9+\left(\frac{1/\omega c}{R} - \frac{R}{1/\omega c}\right)^2}} = \frac{1}{\sqrt{9+\left(\frac{1}{\omega cR} - \omega cR\right)^2}}$$

and
$$\tan\phi = \frac{\frac{1/\omega c}{R} - \frac{R}{1/\omega c}}{3} = \frac{\left(\frac{1}{\omega cR} - \omega cR\right)}{3}$$

These two equations produce Figure 12.13(a) and (b).

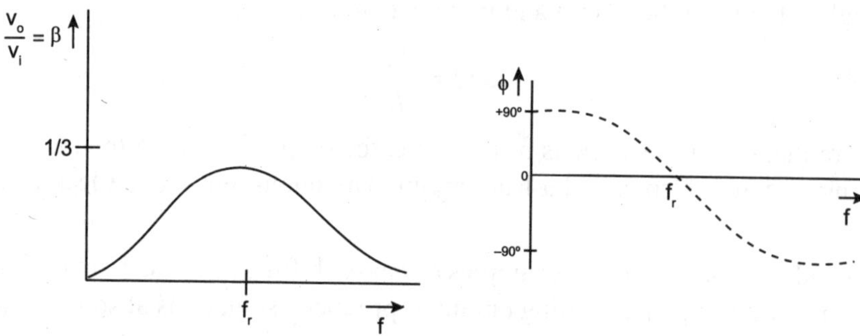

Figure 12.13 (a) Graph for voltage feedback factor vs. frequency
(b) Phase response of lead-lag circuit

The feedback fraction has a maximum value at resonance where $\frac{X_C}{R} = \frac{R}{X_C}$.

At resonant frequency, (1) $\beta = \frac{1}{3}$, (2) $X_C = R$

or
$$\frac{1}{2\pi f_r C} = R$$

or
$$f_r = \frac{1}{2\pi RC}$$

Wien bridge oscillator circuit It primarily consists of a lead-lag network with a two-stage amplifying circuit. In the lead-lag network, the phase shift is positive at low frequencies and negative at high frequencies. So, the phase shift may attain a value zero at a particular frequency f_o called resonant frequency.

Voltage gain of the lead-lag network attains a maximum value of 1/3 at resonant frequency $f_o = \frac{1}{2\pi RC}$. A tungsten lamp is used in a voltage divider circuit to stabilize the output voltage. Its output levels off when lamp resistance, which is variable, attains a value $\frac{R_3}{2}$. At this point, the voltage divider has a gain of 1/3.

The circuit of Wien-Bridge oscillator is shown in Figure 12.14(a). It represents the bridge version of:

(a) a Wien-Bridge oscillator using a two-stage RC amplifier
(b) a Wien-Bridge circuit using Op-Amp

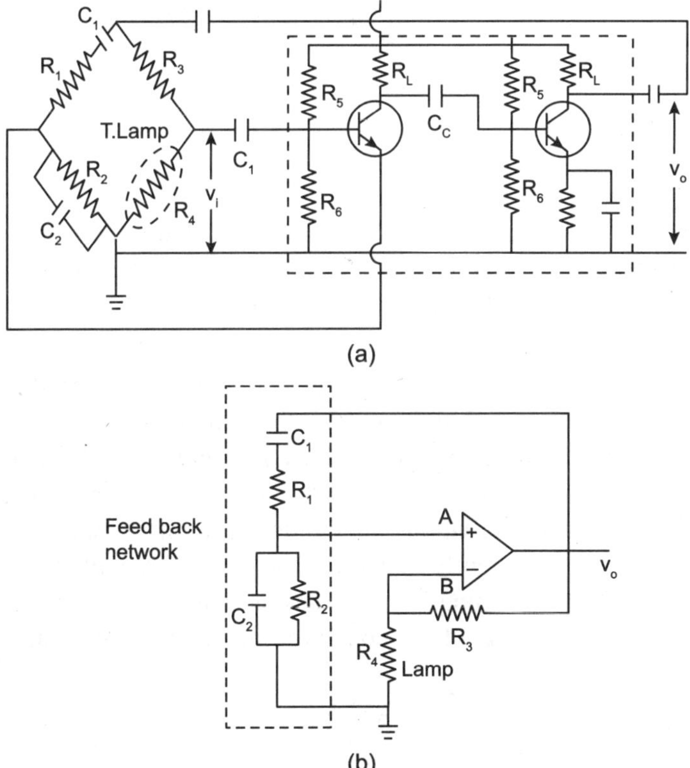

Figure 12.14 (a) Wien-Bridge oscillator using a two-stage RC amplifier, (b) Wien-Bridge oscillator using Op-Amp

Resistor R_4 in the circuit (Figure 12.14(a)) is a tungsten lamp whose resistance varies; as the lamp gets heated up, its resistance increases. Initially the tungsten lamp has a low resistance and there is no output; so negative feedback is less. For this reason, loop gain at resonant frequency βA is greater than 1. As oscillations grow up, the tungsten lamp gets heated up and resistance increases. At some high temperature (not enough to glow lamp), lamp resistance is equal to half of R_3 as can be seen by the following equation when the bridge is balanced.

$$\frac{R_3}{R_4} = \frac{R_1}{R_2} + \frac{C_1}{C_2}$$

Resonant frequency, $$f_r = \frac{1}{2\pi\sqrt{R_1C_1R_2C_2}}$$

When $$R_1 = R_2 \text{ and } C_1 = C_2$$

so $$f_r = \frac{1}{2\pi RC}$$

Under the above condition,

$$\frac{R_3}{R_4} = 2$$

Resistance of lamp, $$R_4 = \frac{R_3}{2}$$

Alternative

Its another version using Op-Amp is shown in Figure 12.14(b). It uses both positive and negative feedback because there are two paths for feedback. The path for positive feedback is from the output through lead-lag circuit to the non-inverting input terminal A. The path for negative feedback is from the voltage divider to the inverting input terminal B.

Working When the circuit is initially turned ON, there is more positive feedback than negative feedback. This allows oscillations to build up. After the output signal reaches a desired level, the negative feedback becomes large enough to reduce loop gain $\beta A = 1$, because at power up, tungsten lamp (R_4) has a low resistance and negative feedback is low. For this reason, loop gain βA is greater than unity and oscillations can buildup at resonant frequency. As oscillations grow, the tungsten lamp heats up slightly (but not to the extent that it glows) and its resistance increases. At some high input level, tungsten lamp has resistance R' which is just half of R_3. At this point, the closed loop voltage gain from non-inverting input to output decreases to

$$A_{CL} = \frac{R_3}{R'} + 1$$

$$= \frac{R_3}{R_3/2} + 1 = 3$$

The lead-lag circuit has $\beta = 1/3$.

$\beta A = \frac{1}{3} \times 3 = 1$ and Barkhausen criterion is satisfied.

When power is first turned ON, the resistance of tungsten lamp is less than R'. As a result, the closed loop voltage gain from non-inverting to output is greater than 3 and so

$$\beta A > 1$$

As oscillations build up, the peak-to-peak output becomes large enough to increase the resistance of tungsten lamp. When its resistance is R', loop gain βA is exactly equal to unity. At this point, oscillations become stable and output voltage has a constant peak to peak value.

12.6.4 Merits and Demerits of RC Oscillators

Merits

1. They do not require any transformer or inductor, thereby the entire size of device is much reduced and is less expensive.
2. They are quite useful in the low frequency range where tank circuit (LC) can be used.
3. They provide constant output and good frequency stability.

Demerits

1. It is difficult to start oscillations.
2. The circuit requires a large number of components.
3. They cannot generate high frequency oscillations and are unstable as variable frequency generators.
4. The values of components used vary with temperature, so the frequency is not constant as temperature varies.

12.7 Crystal Oscillators

It is known that circuits do not provide frequency stability. To provide frequency stability, i.e. maintain frequency within a few hertz per mega hertz, crystal oscillators making use of the piezoelectric effect are generally used in place RC or LC oscillators.

Crystal oscillators use a quartz crystal to control operational frequency. Due to the piezoelectric effect, the crystal under compression produces an electric charge on its surface.

The reverse is also true. When an electric charge is put across the crystal, it expands and contracts. The frequency of this expansion and contraction depends on the physical dimensions of the crystal.

There are three types of crystals used in oscillators: quartz, tourmaline and rochelle salt. Rochelle salt is the best from performance standpoint, but it breaks easily. Tourmaline is the most rugged, but it is the least stable. Quartz is the best crystal for oscillator work. Quartz crystals are made of silicon dioxide. This is the same material used in the insulating layer of MOSFET gate.

The resonant frequency of a crystal is determined by its thickness. The thinner the crystal, the higher the frequency. Crystals also have a very high Q (quality) because of their natural frequency of vibrations. This high Q is important because the higher the Q, the higher the frequency stability. That is, with a high Q, the oscillator will operate at resonance within a very narrow frequency band.

In crystal oscillators, the frequency changes by less than 0.1% due to changes in temperature. In most applications, quartz crystal is used.

12.7.1 Principle

In such crystals, the alternating potential applied across one pair of faces of the crystal causes it to vibrate. For each crystal, there is a natural frequency of resonance at which maximum continuous oscillations can be made to occur. The frequency of oscillations depends upon crystal dimensions and is quite stable. Quartz crystal is the most preferred among other crystals showing piezoelectric effect.

12.7.2 Preparation of the Crystal

The preparation of electronic crystals begins by cutting a natural or cultured quartz crystal into sections. In its natural form, the crystal is approximately in the form of hexagonal prism Figure 12.15(a).

The Z-axis, passing through the ends of the prism, is known as optical axis. No piezoelectric effect is produced by electrical or mechanical stresses along the Z-axis. The Z-axis, which passes through the corners (Figure 12.15(b)) of the hexagonal cross section and are perpendicular to the Z-axis, are termed as electrical axis.

The mechanical axis, i.e. Y axis passing through the faces (Figure 12.15(b)) of the hexagonal prism, perpendicular to both Z-axis and the edge. Mechanical stress along Y-axis produces a voltage along the perpendicular of X-axis.

A variety of crystal cuts are possible and each has its own characteristics. Figure 12.15(c) shows a x-cut and a y-cut separately. Mechanical stress applied to the edges of an x-cut crystal generates an electrical potential across the flat sides. Similarly, mechanical stress on the flat sides of y-cut crystal generates an electrical potential across its edges. The change of resonant frequency with temperature is termed as crystal temperature coefficient. Sections cut out from a crystal are called *blanks*. Each blank must be carefully grinded to accurate dimensions to achieve the desired resonance frequency. After grinding, silver or gold electrodes are plated on to the opposite faces of the blank to form electrical connections. The crystal is then mounted inside a vacuum sealed glass envelope or in a hermetically sealed metal can. Fundamental frequency of the crystal is given by

$$f = \frac{k}{t}$$

t

Crystal blank

where t is thickness of crystal chip (blank) and K is a constant.

Since fundamental frequency is inversely proportional to thickness, there is a limit to the highest fundamental frequency since the thinner the crystal, the more fragile the crystal chip and likely to break up when in vibration, vibrating. It works well up to 10 MHz on fundamental frequency. One can reach higher frequencies up to100 MHz. Occasionally, the more expensive but stronger and rigid chip of tourmaline crystal is used at such higher frequencies.

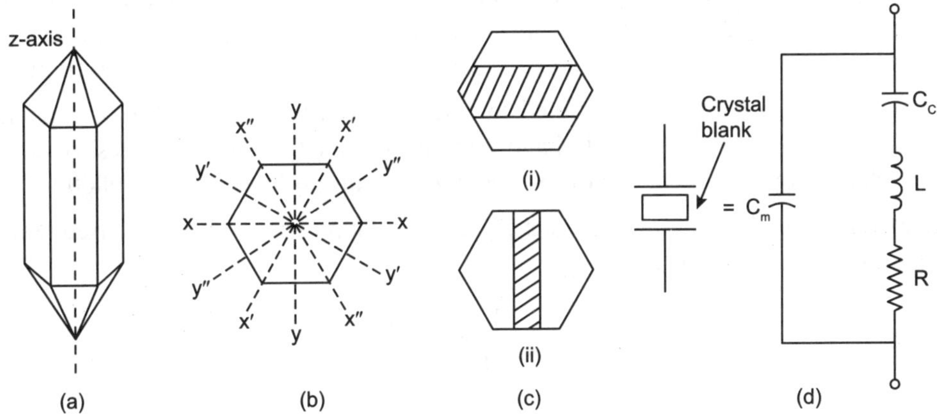

Figure 12.15 (a) *Z*-axis, (b) *X*-axis are electrical, *Y*-axis are mechanical, (c) (i) *Y*-cut, (ii) *X*-cut, (d) Crystal equivalent electrical circuit

12.7.3 Electrical Version of a Crystal

The electrical equivalent circuit for a crystal is shown in Figure 12.16.

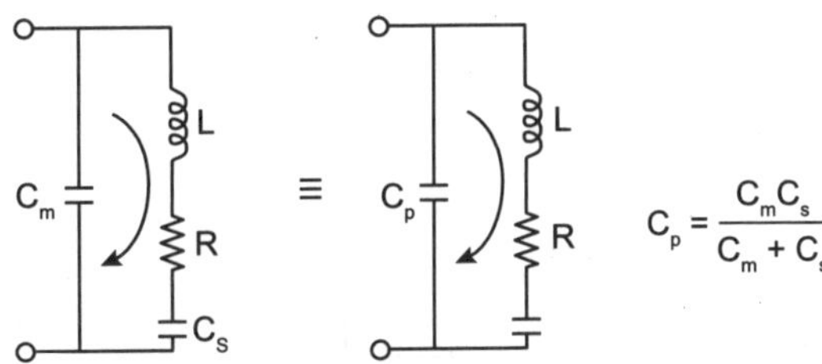

Figure 12.16 Electrical equivalent circuit of a crystal

Physical properties of a crystal and their equivalent electrical parameters are given in Table 12.1.

Table 12.1

Physical property	Electrical parameters
Mass	Inductance
Friction	Resistance
Elastance	L/C

When the crystal is not vibrating, it is equivalent to a capacitor of capacitance C_m because there are two metal plates between which the blank (crystal) is mounted, which acts as a dielectric.

When the crystal is vibrating, it acts like a tuned circuit (tank circuit). Figure 12.16 represents its AC equivalent circuit. Typical values of L are in Henry, C_2 in fraction of picofarads, R in hundreds of ohm, and C_m in picofarad. For instance, $L = 3$ H, $C_S = 0.05$ pF, $R = 2$ kΩ, $C_m = 10$ pF.

The two capacitors can be combined to form an equivalent capacitor of capacitor C_p (Figure 12.16(b)) so that the crystals have incredibly high Q. For the values given above, $Q = 4000$ which is quite high. It can even have higher values of the order of 10000.

The extremely high value of Q means crystal oscillators have very stable frequency. The exact equation of frequency is given by

$$f_r = \frac{1}{2\pi\sqrt{LC}}\sqrt{\frac{Q_2}{Q^2+1}}$$

When Q approaches infinity, frequency f_r approaches an ideal value $\dfrac{1}{2\pi\sqrt{LC}}$.

Because of the presence of C_m, the crystal has low resonant frequencies (Figure 12.17). One of these is series resonant frequency (f_s) where $2\pi f L = \dfrac{1}{2\pi f C}$. In this situation, crystal impedance is very low. The other resonant frequency is f_p, it is slightly higher than f_s and is due to parallel resonance of capacitance C_m and reactance of the series circuit. At f_p, impedance of the crystal is very high.

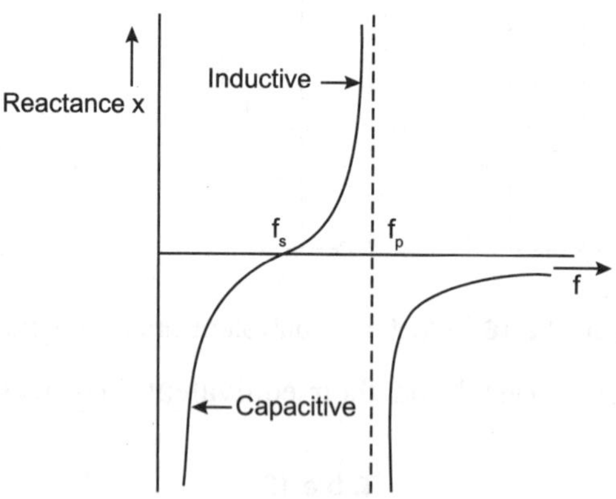

Figure 12.17

12.7.4 Crystal Stability

The frequency of an oscillator tends to change slightly with time. This drift is produced due to temperature, aging and many other causes. In a crystal oscillator, drift is very small typically less than 1 part in 10^6 per day. Stability of this kind is important in electronic watches.

12.7.5 Crystal Oscillator Circuits

Figure 12.18 shows a Colpitts crystal oscillator.

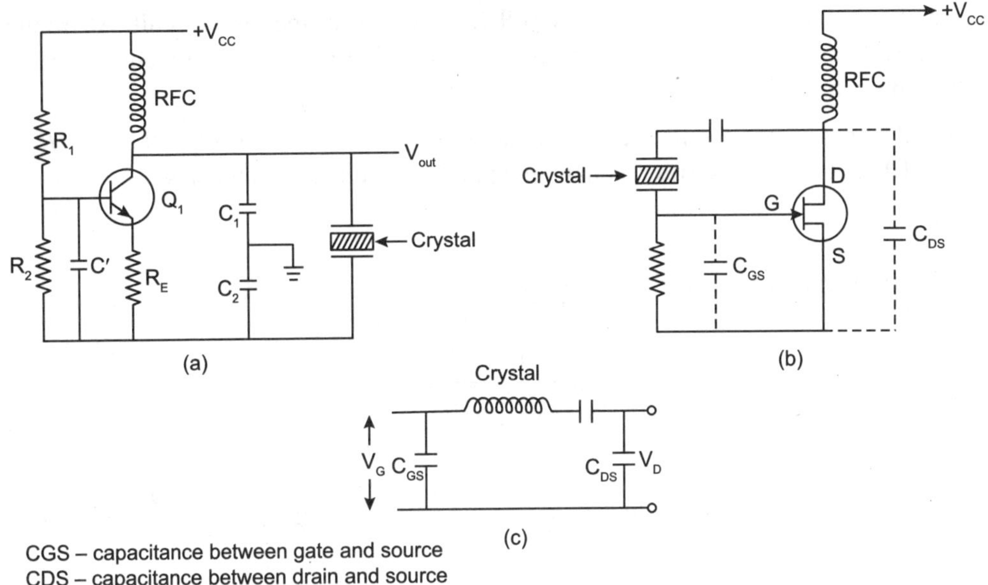

Figure 12.18 (a) Colpitts crystal oscillator with crystal mounting in parallel resonance, (b) pierce crystal FET oscillator with crystal operating in series resonance, (c) phase shift network of pierce oscillator

As explained earlier, transistor Q_1 combined with R_1, R_2, RFC and R_E (Figure 12.18(a)) constitute a common base circuit. Capacitors C_1 and C_2 form the voltage divider which returns a portion of output voltage to the base of Q_1.

As output voltage increases positively, emitter voltage also increases, and since base voltage is fixed, base-emitter voltage is reduced. The reduction in V_{BE} causes I_C to decrease, and this in turn causes collector voltage V_C to increase positively. Thus, the circuit is supplying its own input and there exists a state of oscillations. The crystal in parallel with C_1 and C_2 permits maximum voltage feedback from collector to emitter when its impedance is very high, i.e. at parallel resonant frequency. At other frequencies, crystal impedance is low and low impedance causes the feedback voltage to be two small to sustain oscillations. Thus, oscillation frequency is stabilized at parallel resonant frequency.

12.8 Pierce Oscillator

A pierce oscillator (Figure 12.18(b)) operates at series resonant frequency of the crystal. The FET drain-source capacitor C_{DS} and gate-source capacitor C_{GS} are an important part of phase shift network as can be seen from Figure 12.18(c). It is very similar to a colpitts oscillator and is very simple to construct. In a crystal oscillator, frequency drift with aging is very small, typically less than 1 part in 10^6 per day.

Crystal oscillators can operate in parallel or in series. At resonance, the series (crystal) circuit will have minimum resistance and allow full feedback from the tank circuit. At frequencies other than resonance, the crystal will act as band-stop filter circuit and prevent feedback from returning to the transistor's base.

Example 3 In a pierce oscillator, the crystal constants are $L = 3$ H, $C_S = 0.05$ pF, $R = 2$ kΩ and $C_m = 10$ pF. Calculate series resonant and parallel resonant frequencies.

Solution

$$\text{Series resonant frequency, } f_s = \frac{1}{2\pi\sqrt{LC_S}}$$

$$= \frac{1}{2\pi\sqrt{3\times 0.05\times 10^{-12}}}$$

$$= 411 \text{ KHz}$$

$$\text{The equivalent parallel capacitance, } C_p = \frac{C_m \times C_s}{C_m + C_s} = \frac{10\times 10^{-12}\times 0.05\times 10^{-12}}{10.05\times 10^{-12}} = 0.0498\,\text{pF}$$

$$\text{Parallel resonant frequency, } f_p = \frac{1}{2\pi\sqrt{L\times C_p}} = \frac{1}{2\pi\sqrt{3\times 0.0498\times 10^{-12}}}$$

$$= 412 \text{ KHz}$$

Example 4 A crystal has thickness 't'. If its thickness is reduced by 1%, what happens to the fundamental frequency of oscillation of the crystal.

Solution

Fundamental frequency of oscillation of crystal,

$$f_r = \frac{K}{t}$$

Let new thickness of the crystal be t'.

$$\frac{f_r'}{f_r} = \frac{t'}{t} = \frac{99}{100}$$

So, new frequency $f_r' = \dfrac{k}{t'}$

or
$$f_r' = \frac{99}{100}\times f_r$$

or reduction in frequency,

$$f_r - f_r' = f_r - \frac{99}{100} f_r$$

$$= f_r \left[\frac{1}{100} \right]$$

or

$$\frac{f_r - f_r'}{f_r} = \frac{1}{100}$$

$\therefore f_r'$ reduces by 1%.

12.9 Multivibrators (Non-Sinusoidal Oscillators)

There are three groups of multivibrators:

1. Astable (free running)
2. Monostable
3. Bistable (flip flop)

12.9.1 Astable or Free Running Multivibrator

An astable multivibrator is basically a RC coupled oscillator. It has two quasi stable states. It switches back and forth from one state to the other, remaining in each for a time dependent upon the discharging of capacitive circuit. It is then an oscillating circuit since it requires no external triggering. (It does of course need a source of power). Its circuit is shown in Figure 12.19.

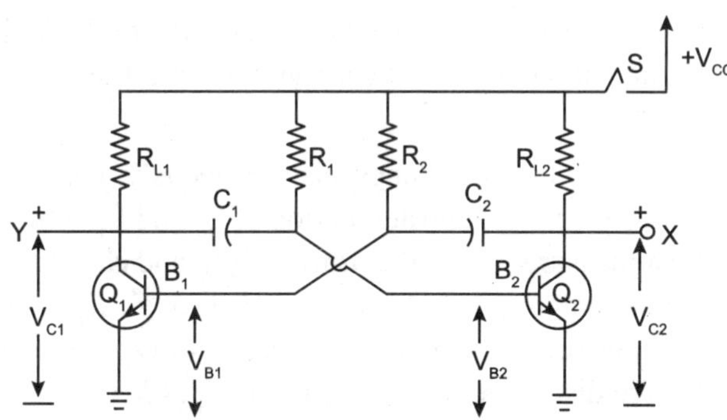

Figure 12.19 Astable multivibrator circuit

As shown in Figure 12.19, it consists of two-terminal RC coupled transistors Q_1 and Q_2 connected so that the output of Q_1 is fed to Q_2 and the output of Q_2 is fed to Q_1. Components R_1, R_{L1} and C_1 in one half of the circuit are considered to be equal to R_2, R_{L2} and C_2 connected in the other half of the circuit. It is clear that the output of one transistor is fed to the other as input through the capacitor.

To analyze the circuit operation, let us assume that the two halves of the circuit are symmetrical. When switch S is closed, power is applied to the two transistors and current flows through the collectors of both transistors.

(a) Since it is not conceivable that both transistors although of the same type draw exactly equal amount of current. Due to certain circuit imbalance, assume that the collector of Q_1 draws more current than that of Q_2. So the voltage of collector of Q_1 goes more negative (less positive) than the collector voltage of Q_2.

(b) This negative voltage change at collector of Q_1 will be coupled by capacitor C_1 to the base of Q_2, reducing the forwardbias on Q_2. This will result in a decrease in collector current of Q_2; so collector voltage of Q_2 will increase or become more positive.

(c) The positive voltage change at collector of Q_2 is coupled by capacitor C_2 to the base of Q_1, increasing the forwardbias on Q_1. This will result in an increase in collector current of Q_1. So the voltage at collector will decrease or become more negative, driving the base of Q_2 more negative, thereby reducing collector current Q_2 further; thus driving collector voltage of Q_2 becomes even more positive. In a very short time, this regenerative feedback from the collector of one transistor to the base of next will cause Q_1 collector current to reach saturation while Q_2 collector current is cut off. The entire action is practically instantaneous.

How long does Q_1 remain ON and Q_2 OFF? The time constant C_1R_1 determines this, as explained below.

(d) Assume (i) $V_{CC} = 12$ V, and (ii) that after power is applied and Q_2 saturates, collector voltage V_{C1} of Q_1 drops practically to zero, i.e., collector to emitter resistance of Q_1 approaches zero. This 12-volt negative change at collector looks like a negative battery of (–) 12 V to C_1, and at the instant of change, the circuit to C_1 looks like that in Figure 12.19(a).

Since voltage across C_1 cannot change instantaneously, the entire 24 V must appear across R_1, and voltage w.r.t. ground at junction of C_1 and R_1, which is point B_2 or base of transistor Q_2, must measure (–) 12 V cutting off Q_2, and capacitor C_1 starts discharging through emitter-collector resistance of Q_1 and through R_L toward V_{CC} as shown in Figure 12.20(a). Since emitter-collector resistance of Q_1 (which undergoes saturation) is very low, effective discharge time constant of C_1 is R_1C_1.

(e) Voltage V_{B2} at base of Q_2 will change in the manner shown in Figure 12.21 as capacitor C_1 discharges.

(f) When voltage at base of Q_2 has increased from 12 V to 0 V to a low positive voltage, Q_2 is brought out of cutoff to conduction. Q_2 base current and collector current starts to flow.

(g) Voltage V_{C2} at collector of Q_2 drops and this negative voltage change is coupled by C_2 to base of Q_1, reducing base current and collector current in Q_1.

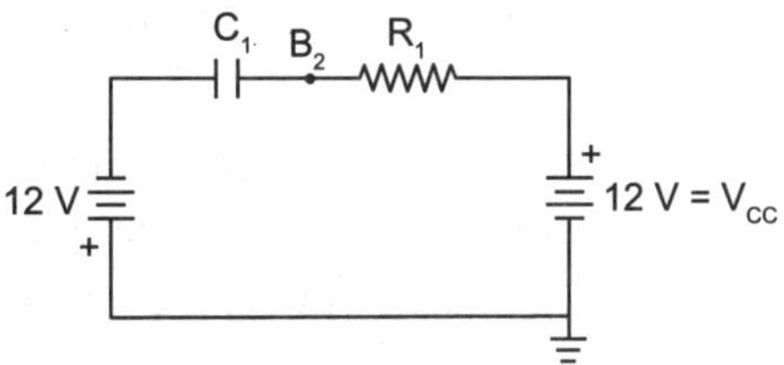

Figure 12.19(a)

(h) As collector current in Q_1 is reduced, collector voltage V_{C1} of Q_1 rises. The positive change at collector of Q_1 is coupled by C_1 to base of Q_2, increasing base current and collector current in Q_2.

(i) Now voltage V_{C1} at collector of Q_2 reduces or goes more negative driving the base of Q_1 more negative. Almost instantaneously as Q_2 starts conducting, the regenerative feedback action turns Q_2 ON and turns Q_1 OFF. Q_2 remains ON as long as Q_1 is OFF. The length of time Q_1 remains OFF can be determined by time constant R_1C_2, for voltage V_{B1} at base of Q_1 is determined by discharge of C_2 through emitter-collector resistance of Q_2 and through R_2 as shown in Figure 12.20(c).

(j) When Q_1 starts to conduct, it switches Q_2 OFF. When Q_2 starts to conduct, it switches Q_1 OFF. The length of time for which Q_1 conducts is determined by the length of time Q_2 is OFF as shown by the waveform of V_{B2} (Figure 12.21). Similarly, the length of time for which Q_2 conducts is determined by the length of time Q_1 is OFF as shown by the waveform of V_{B1} (Figure 12.21).

(k) The idealized waveforms (Figure 12.21) merit some discussion. We see that as base voltage V_{B2} is driven 12 V negative during interval t_1, collector voltage V_{C2} of Q_2 goes 12 V positive i.e. up to V_{CC}. We might expect that voltage at collector or Q_2 would rise to V_{CC} the moment Q_2 is OFF. However, this is not so, since capacitor C_2 cannot charge up instantaneously to V_{CC}. Therefore, V_{C2} rises to V_{CC} at the rate determined by the charging of C_2 through R_{L2} in series with emitter base resistance of Q_1 (Figure 12.20(d)). The exponential rise of voltage V_{C2} (Figure 12.21) represents the charge of C_2. At the end of interval t_1, when V_{B1} has risen from cut off condition to conducting state, the voltage at collector of Q_2 drops from (+) V_{CC} to a very low value (close to zero). At that instant, Q_1 is switched OFF and voltage V_{C1} at collector of Q_1 starts rising from 0 to (+) V_{CC} exponentially. The exponential rise in V_{C1} (Figure 12.21) is the result of C_1 charging towards (+)V_{CC} through R_{L1} [Figure 12.20(b)]. At the end of period t_2, when V_{B1} has risen from (–) 12 V to slightly positive voltage, Q_1 is switched ON and Q_2 OFF and the cycle is repeated.

It is to be noted that ON-OFF time ($t_1 = t_2$) of each transistor is equal, since the circuit is symmetrical. The waveforms at collectors of Q_1 and Q_2 are 180° out of phase. Similarly, the waveforms at bases of Q_1 and Q_2 are also 180° out of phase.

To analyze the circuit when switch S is closed, it is assumed that it is oscillating and justify that it will continue to do so.

(a) When the supply is ON, one transistor will conduct more than the other due to certain circuit imbalance. Assume that Q_1 is conducting and Q_2 is cutoff.

Then the output of Q_1 is approximately 0 V, and $V_{C2} = V_{CC}$. (assume V_{CC} = 12 V). We can say that voltage at collector of Q_1 is more negative than collector voltage of Q_2.

(b) This will permit capacitor C_2 to charge up through R_{L2} and base-emitter junction resistance of Q_1. The voltage will be approximately equal to V_{CC} (= 12 V).

(c) Now consider capacitor C_1 which had previously charged up to V_{CC} (when Q_2 was conducting) through R_{L1}. When Q_1 conducts, a path is provided for C_1 to discharge through Q_1 and R_1 in the manner shown in Figure 12.20(a).

(d) The initial pulse of discharge current from C_1 through R_1 makes the base of Q_2 suddenly very negative (–) V_{CC} as shown by the wave form of V_{B2} in Figure 12.21. This causes Q_2 to be cutoff as assumed.

(e) Q_1 is kept in conducting state by base current provided from V_{CC} through R_2; so V_{B2} is slightly positive. The charging current of C_2 through R_{L2} has now ceased.

(f) The length of time for which Q_2 is held off is determined by the time constant of discharging of C_1 through Q_1 and R_1. After approximately five time constants, Q_2 is no longer cutoff but starts to conduct.

(g) This allows C_1 to be recharged very quickly through R_{L1} and Q_2 (Figure 12.20(b)), causing a momentary increase in positive voltage at the base of Q_2 and aiding the turning ON of Q_2. The result is fast-falling edge in the waveform of V_{C2} as shown in Figure 12.21(c).

(h) At the same time, C_2 is allowed to discharge through Q_2 and R_2 coupling a large negative pulse to the base of Q_1 (due to voltage developed across R_2 upon the initial discharge of C_2) (Figure 12.20(c)). This turns Q_1 off quickly and collector voltage V_{C1} rises towards $+V_{CC}$. However, voltage at the collector of Q_1 is somewhat less than V_{CC} at this instant due to a voltage drop across R_{L1}, occurring while C_1 is recharging. This is relatively a shorter time, but it causes a rounding of the rising edge in V_{C1} (Figure 12.21(a)).

(i) Q_1 is held in a nonconducting state for an interval dependent upon R_2C_2 time constant, after which Q_1 begins to conduct. C_2 discharges through Q_1 and C_2 recharges through Q_1, C_1 discharges through Q_1 and C_2 recharges through Q_1, etc.

(j) It will be noted that ON-OFF time for each transistor is equal since the circuit is symmetrical, i.e. time duration of positive and negative alterations are equal.

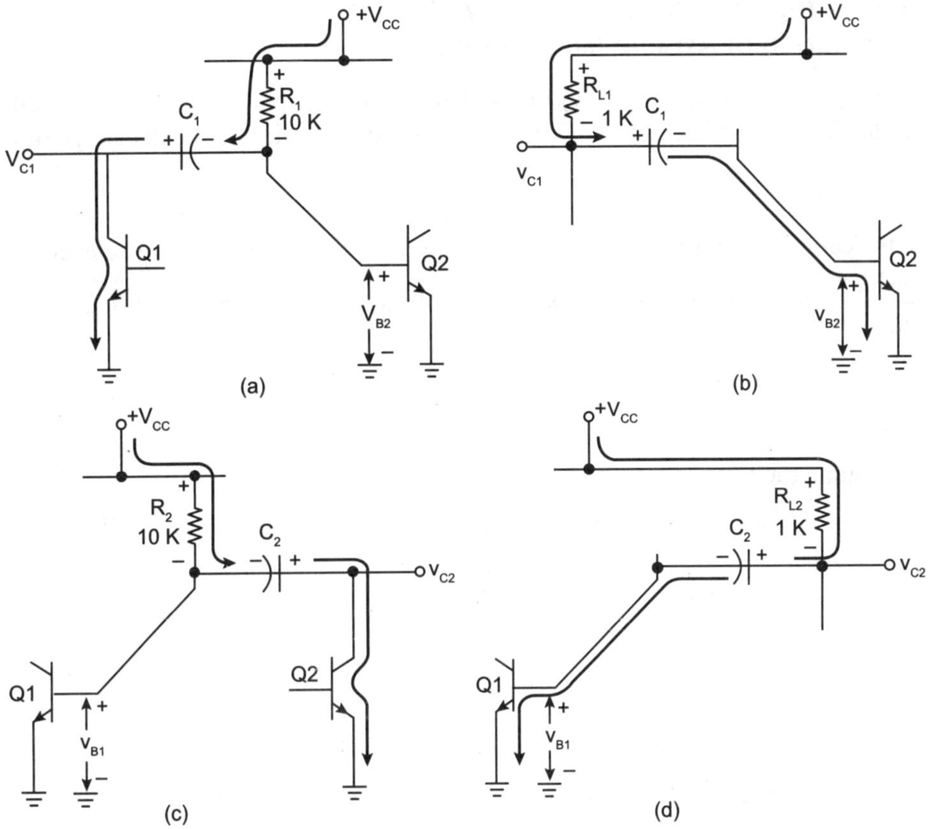

Figure 12.20 (a) Discharging path for C_1 through Q_1 and R_1, (b) recharging path for C_1 through Q_2 and R_{L1}, (c) discharging path for C_2 through Q_2 and R_2, (d) recharging path for C_2 through Q_1 and R_{L2}

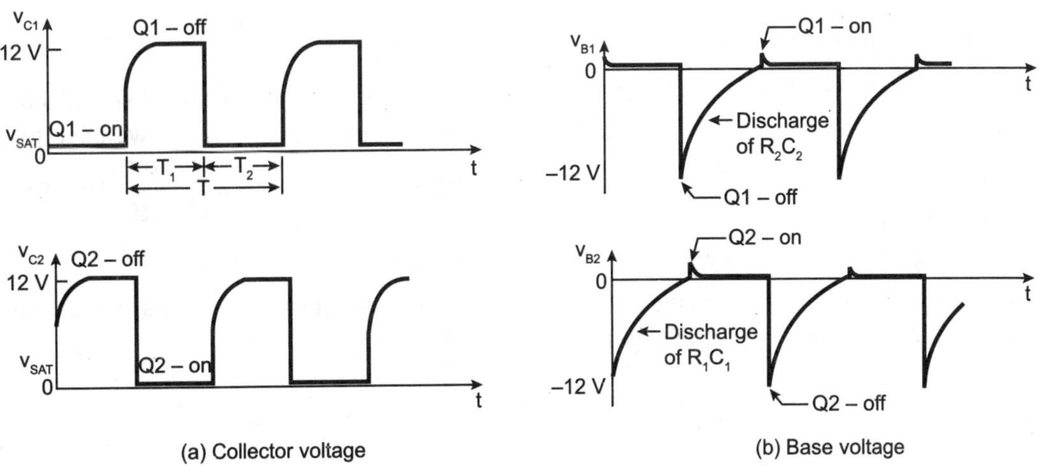

Figure 12.21 Wave forms at the base and collector of Q_1 and Q_2

12.9.2 Switching Times and Frequency of Oscillation of Astable Multivibrator

To obtain an expression for the frequency of oscillations, consider that the period of wave depends only upon the discharge of capacitors C_1 and C_2. Consider V_{B2} during the discharge of C_2.

$$V_{B_2} = V_{CC} - i_{c1}R_1$$

Since the capacitor charges up to approximately V_{CC}, the initial discharge current value of $i_{C1} = \frac{V_{CC} + V_{CC}}{R_1}$, and this current decays exponentially with a time constant of R_1C_1.

$$\therefore \; V_{B_2} = V_{CC} - 2V_{CC}\exp\left[\frac{-t}{R_1C_1}\right]$$

Let us assume the transistor will switch when $V_{B2} = 0\ V$.

(Actually, +0.5 V for silicon transistor and 0.2 V for germanium, is quite small in comparison with $V_{CC} = 12$ V.)

$$\therefore \; 0 = V_{CC}\left[1 - 2\exp\left(\frac{-t}{R_1C_1}\right)\right]$$

or

$$\exp\left[\frac{t}{R_1C_1}\right] = 2$$

from which the OF time for Q_2 (and the ON time for Q_1) is T_2.

$$t = T_2 = R_1C_1\log_e^2 = 0.694\ R_1C_1 \approx 0.7\ R_1C_1$$

Similarly, the OFF time for Q_1 is T_1.

$$T_1 = 0.694\ R_2C_2 \approx 0.7\ R_2C_2$$

Hence the total period of waveform is $T = T_1 + T_2 = 0.7\ [R_1\ C_1 + R_2\ C_2]$.

If $R_1 = R_2$ and $C_1 = C_2$, we have a symmetrical mutivibrator, with outputs at the two collectors having same waveforms but out of phase with each other. Since $R_1C_1 = R_2C_2$, we have time period of wave $T = T_1 + T_2 = 1.4$ RC sec, and frequency of oscillation $f = \frac{1}{T} = \frac{0.7}{\text{RC}}$ Hz if R is in ohms and C in farads.

Assuming RC = 200 μs, the value of frequency of oscillations is 3.5 kHz.

The frequency of oscillations may be varied by using a separate variable voltage source for R_1 and R_2. If this voltage is V_{BB}, then period of oscillations is

$$T = 2\text{RC}\log_e\left(1 + \frac{V_{CC}}{V_{BB}}\right)$$

where $R = R_1 = R_2$.

This furnishes an effective way of controlling the frequency or time period of wave. Unsymmetrical waves may be obtained making different time constants, i.e. $R_1C_1 \neq R_2C_2$. The output may be observed at X and Y.

Minimum value of h_{FE} for oscillations To ensure oscillation, the transistors must saturate so that R_1 and R_2 must be chosen with the h_{FE}, the DC current gain of transistor.

Assume Q_2 is saturated, so that its collector to emitter voltage is approximately equal to base to emitter voltage. Then, voltage drop across R_1 must equal to the drop across R_{L2}.

$$I_{B2}\ R_1 = I_{C2}\ R_{L2}$$

$$\text{or } R_1 = \frac{I_{C2}}{I_{B2}} R_{L2} \text{ but } \frac{I_{C2}}{I_{B2}} = h_{FE}$$

$$= h_{FE}\, R_{L2}$$

So, if $R_1 \leq h_{FE}\, R_{L2}$, then transistor will be saturated

Hence
$$h_{Fe} \geq \frac{R}{R_L}$$

where $R = R_1 = R_2$ and $R_L = R_{L1} = R_{L2}$

The above condition must be satisfied for oscillations to be maintained.

12.10 UJT Relaxation Oscillator

It is a nonsinusoidal oscillator where a unijunction transistor (UJT) is used along with a RC circuit. A relaxation oscillator is a circuit that uses the charge/discharge cycle of a capacitor or inductor to produce a pulse output. The output pulse is normally used to fire or trigger an SCR or a triac. The output pulse waveform is either sawtooth or rectangular. Both the waveforms can be used as output, depending on where the output is taken. Here, we make use of the (V_{EB}, I_E) characteristic of UJT as shown in Figure 12.22.

As shown in figure, it has a negative resistance region between I_P and I_V where I_p is called peak current. Upto this point the device is operating in cutoff region. When current increases beyond I_p, the device enters the negative resistance region in which resistance r_{B1} falls very rapidly and applied potential V_{EB} falls to valley voltage V_v where current is I_V. A further increase in I_E causes the device to enter into the saturation region. So, it is to be noted that the device operates in the negative resistance region.

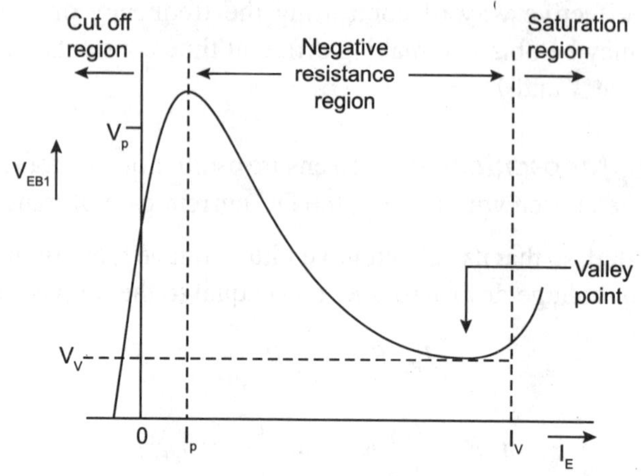

Figure 12.22

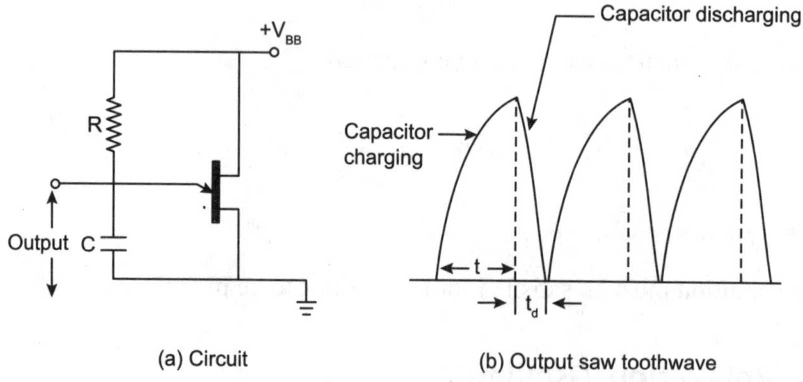

(a) Circuit

(b) Output saw toothwave

Figure 12.23 UJT relaxation oscillator

The circuit of the oscillator is shown in Figure 12.23(a). It comprises of a UJT and capacitor C which is charged through resistor R_E.

12.10.1 Function

Capacitor C gets charged by supply V_{BB} through R when the capacitor voltage reaches V_p (peak voltage of UJT, Figure 12.22). The UJT fires and the capacitor gets discharged rapidly to V_{EB1}, (sat). The device then cutsoff and the capacitor commences charging again as shown by the output wave (Figure 12.23(b)). As soon as voltages reache V_p, the UJT again fires and the capacitor discharges. This cycle of charging and discharging is repeated continuously and we have a sawtooth wave in output across capacitor C. The time for the capacitor to charge from V_{EB} (sat) to V_p can be calculated, and the frequency of oscillations is $\frac{1}{t}$ (Figure 12.23(b)). Discharge time t_d is very small and can be neglected.

The general equation for charging time of capacitor is given by

$$t = CR\frac{E - E_o}{E - e_c}$$

where C – capacity in farads

R – ohms

E – supply voltage (V_{BB})

e_c – capacitor voltage at time t

E_o – initial voltage of capacitor

Summary

1. An LC circuit known as tank circuit may be excited in oscillations. These oscillations are damped due to energy loss in resistance present in the inductor. If energy loss is replaced from time to time, then oscillations that are sinusoidal will continue.
2. Oscillators generate AC output without external AC input using noise AC signal generated in switching. DC power from V_{CC} or V_{DD} is converted to AC power. The range of frequency signals generated by the circuit can be high and output power is small.
3. Special arrangements of feedback circuits using a transistor in conjunction with LC circuit, may supply the energy required to over come resistance losses in tank circuits and oscillations can be sustained.
4. To sustain oscillations after feedback, Barkhausen criterion

$$\beta A = 1 + jo$$

should be satisfied.

5. To satisfy Barkhausen criterion, we employ three reactive elements X_1, X_2 and X_3 in the amplifying circuit. These reactive elements must satisfy

$$X_1 + X_2 + X_3 = 0 \text{ for oscillation.}$$

Under the above condition, we have

$$\beta A = \frac{AX_2}{X_1}$$

So the nature of X_2 and X_1 should be same either inductive or capacitive. But X_3 must be opposite in nature so that $X_3 = (-)\,(X_1 + X_2)$.

(i) If X_1 and X_2 are inductive, then X_3 is capacitive.

(ii) If X_1 and X_2 are capacitive, then X_3 is inductive. An Hartley oscillator satisfies the first condition and a Colpitts oscillator satisfies the second condition.

6. (a) In Hartely oscillator using transistors, oscillations are sustained only when

$$h_{fe} \geq \frac{L_2 + M}{L_1 + M}$$

and frequency of oscillations is

$$\omega_o = \frac{1}{\sqrt{C(L_1 + L_2 + 2M)}}$$

(b) In Colpitts oscillator using transistors, oscillations are sustained when

$$h_{fe} \geq \frac{C_1}{C_2} + h_{ie}\, h_{oe} \frac{C_2}{C_1} + \frac{h_{oe}}{h_{ie}} \frac{L}{C_2}$$

and frequency of oscillations is

$$\omega_o = \frac{1}{\sqrt{L\left(\frac{C_1 C_2}{C_1 + C_2}\right)}}$$

7. To produce electrical oscillations, we also use RC oscillators. Among them, the popular ones are
   (i) Phase shift oscillator
   (ii) Wien-bridge oscillator

   A phase shift oscillator consists of an amplifying device with a feedback circuit comprising of RC sections such that feedback from output to input is in phase to sustain oscillations.
8. Frequency stability of an usual oscillator is poor. To have frequency stability, we use a crystal oscillator. A crystal oscillator uses a crystal which show one piezoelectric effect. Among such crystals showing piezoelectric effect, the most common one is quartz. A crystal behaves like a LC circuit, but it has two resonant frequencies, one series resonant and another parallel resonant. The two are very close to each other. The most common oscillators are (i) Colpitts crystal oscillator and (ii) Pierce oscillator.
9. Multivibrators are RC coupled oscillators which generate a nonsinusoidal wave form.

   The most common multivibrator is free running type or astable type, which is self excited. Here in the output we get nearly a square wave when the multivibrator is symmetrical. For such multivibrators, the frequency of oscillation is $f = \frac{0.7}{\text{RC}}$, and the minimum value of h_{fe} of the transistor used must satisfy

$$h_{FE} \geq \frac{R}{R_L}$$

10. An UJT oscillator is used for having a sawtooth wave.

Review Questions

1. State and mention the Barkhausen criterion for sustained oscillations in a feedback oscillator. Why it must be fulfilled to maintain oscillations? Explain.
2. Draw the basic block diagram for a feedback oscillator. Explain its working. What is the condition for sustained oscillations?
3. Sketch the circuit of a phase shift oscillator. Explain, how it operates to maintain oscillation? Derive the necessary relation for frequency and feedback factor.
4. Sketch the circuit of Hartley oscillator. Explain its working and derive the necessary relation for frequency generated. Mention the condition for maintaining oscillations.
5. Sketch the circuit of Colpitts oscillator. Explain its working and derive the necessary relation for frequency generated. Mention the condition for maintaining oscillations.
6. How the Barkhausen criterion is fulfilled in Colpitts oscillator? Draw the circuit of the oscillator. What is the order of frequency generated?
7. Prove that in a Hartley oscillator the value of h_{fe} of the transistor used must satisfy

$$h_{fe} \geq \frac{L_1 + M}{L_2 + M}$$

to sustain oscillations produced, where symbols have usual meaning. What is the order of frequency generated?

8. Prove that the frequency generated in a phase shift oscillator is

$$f = \frac{1}{2\pi C\sqrt{4RR_L + 6R^2}}$$

where symbols have usual meaning.

9. Prove that in a RC oscillator

$$h_{fe} \geq 23 + \frac{29}{K} + 4K \text{ where } K = \frac{R_L}{R}$$

must be achieved to produce sustained oscillations.

Also show that the minimum value of

$$K = 2.875 \text{ and } h_{fe} = 45$$

10. Draw the circuit of Wien-Bridge oscillator. Explain its working. Prove that to maintain oscillations the gain of the amplifier employed be more than 3.
11. What is the nature and role of feedback voltage in an oscillator? Illustrate your answer by taking a specific example. Derive the necessary relation to maintain oscillations in a specific oscillator.
12. For the Colpitts oscillator, write relations for oscillating frequency, phase shift network attenuation and amplifier gain. Discuss the design approach for the circuit.

13. Compare the phase shift networks of Colpitts and Hartley oscillators.
14. What do you mean by lead and lag network? Explain their use in Wien-Bridge oscillator and discuss the working of the oscillator. What is the range of frequencies generated by the oscillator.
15. How many types of multivibrator we have? Draw the circuit of a free running multivibrator and explain its working. Derive the necessary relation for its frequency. Draw the wave shape produced.
16. Explain the working of a astable multivibrator. Draw its circuit diagram and show the wave shape obtained. Derive the relation of the frequency of oscillations produced.
17. How can we have a square wave generated in an oscillator? Draw the circuit diagram of such an oscillator. Describe its working and show the shape of output waveform.

Short Answer Questions

1. What causes the output of an oscillator to be a sine wave at a particular frequency?

   Ans: In a tank circuit, the charge and discharge of a capacitor through inductor causes the wave shape to be a sine wave where the damping of oscillations is powered at fixed intervals by the amplifying device having a DC supply.

2. How does the phase shift oscillator gets its name?

   Ans: The phase shift oscillator gets its name due to the fact that, in a RC network there is a phase difference between output across R and input across RC combination. The phase shift between the two depends on their respective values. So, if we have a phase shift of 60° in an RC network, such networks will produce a total phase shift of 180° necessary for oscillations.

3. What is the Barkhausen criterion for oscillation?

   Ans: In a feedback amplifier, the gain after feedback is given by $A_f = \dfrac{A}{1-\beta A}$. If $|1-\beta A|$ is zero, then A_f goes to infinity and we say there is output without external inputs or the amplifier oscillates. This condition $\beta A = 1 + jo$ is called Barkhausen criterion of oscillation.

4. What should be the minimum gain of the amplifier in a phase shift oscillator?

   Ans: The minimum gain of the amplifier in a phase shift oscillator should be 29.

5. How does a lead-lag circuit in Wien-Bridge oscillator act as a resonant circuit?

   Ans: At low frequencies in the series RC circuit, C behaves like an open circuit, so there is no output, and at high frequencies a parallel RC circuit behaves like a short circuit and there is no output. So in between these two extreme situations, there is maximum output which is same as resonant circuit.

6. What does the half period of oscillation of a symmetrical free running multivibrator depend upon?

   Ans: The half period of a symmetrical free running multivibrator depends upon the value of R and C.

7. What causes the rounding of leading edge of V_C in the output wave form of a multivibrator?

   Ans: When a transistor turns OFF its collector voltage V_C rises towards supply voltage V_{CC} but it is slightly less than V_{CC} due to voltage drop across load R_L while the capacitor is recharging. This is relatively a shorter time interval, but it causes a rounding of the leading edge in V_C.

8. How feedback is employed in a multivibrator?

   Ans: When one transistor say, Q_2, is turned ON, the voltage of its collector falls, thus falling voltage permits the discharge of capacitor which drives transistor Q_1 into OFF.

9. Is the feedback in multivibrator positive or negative?

   Ans: The feedback is positive since the process is regenerative.

10. What is the order of frequency of oscillations in a multivibrator?

    Ans: The frequency of oscillations in a multivibrator is the order of hundreds.

Problems

1. In a colpitts oscillator, $C_1 = 0.001\ \mu F$, $C_2 = 0.01\ \mu F$ and $L = 15\ \mu H$. Calculate the frequency of oscillation. What is the feedback fraction? How much voltage gain is required to start the oscillations? [**Ans:** 1.36 MHz, 0.1, 10]

2. In a crystal oscillator, the crystal has the following values; L = 3 H, C_S = 0.05 pF, C_m = 10 pF and R = 2 kΩ. Calculate the series and parallel resonant frequencies of the crystal.

   [**Ans:** 411 KHz, 412 KHz]

3. In a colpitts oscillator, $C_1 = 0.001\ \mu F$, $C_2 = 0.01\ \mu F$ and $L = 5\ \mu H$. Calculate (i) approximate frequency of oscillator, (ii) feedback factor, (iii) maximum value of voltage gain to start oscillator, (iv) if the value of L is doubled, what is the new frequency of oscillator.

   [**Ans:** (iv) 1.67 MHz]

Objective Questions

1. Barkhausen criterion for a feedback oscillator is

   (a) $\beta A = 1$ (b) $\beta A = -1$

   (c) $\beta A = 1 + jo$ (d) $\beta A = 1 + jo$ [**Ans:** (c)]

2. In a feedback oscillator, we employ

   (a) an external input signal (b) positive feedback signal

   (c) negative feedback signal (d) zero signal [**Ans:** (b)]

3. The best device to get a signal of 1 KHz is
   (a) hartley oscillator
   (b) colpitts oscillator
   (c) crystal oscillator
   (d) wein-bridge oscillator **[Ans:** (d)]
4. In a crystal oscillator,
   (a) there is no external tank circuit
   (b) there is no feedback net work
   (c) output can be obtained at two frequencies
   (d) Q value is very high **[Ans:** (d)]
5. Stability of frequency is fulfilled in
   (a) crystal oscillator
   (b) hartley oscillator
   (c) phase shift oscillator
   (d) weib-bridge oscillator **[Ans:** (a)]
6. The total phase difference in a closed loop in a feedback oscillator is
   (a) 90°
   (b) 180°
   (c) 360°
   (d) $180 < \phi > 360°$ **[Ans:** (c)]
7. A split type capacitor in the tank circuit is used in
   (a) hartley oscillator
   (b) colpitts oscillator
   (c) weib-bridge oscillator
   (d) phase shift oscillator **[Ans:** (b)]
8. The relation between reactive elements X_1, $X_2$and X_3 connected in a feedback network of an oscillator is
   (a) (i) $X_1 + X_2 + X_3 = 0$ (ii) $\frac{X_1}{X_2} = (+)^{ive}$
   (b) (i) $X_1 = X_2 + X_3$ (ii) $\frac{X_1}{X_2} = (+)^{ive}$
   (c) (i) $X_1 + X_2 + X_3 = 0$ (ii) $\frac{X_1}{X_2} = (-)^{ive}$
   (d) (i) $X_1 = X_2 + X_3$ (ii) $\frac{X_1}{X_2} = (-)^{ive}$ **[Ans:** (a)]
9. In crystal oscillator, the order Q and frequency generated is
   (a) 10,000, 10 MHz
   (b) 1000, 10 KHz
   (c) 100, 10 MHz
   (d) 10000, 10 KHz **[Ans:** (a)]
10. In a phase shift oscillator,
    (a) $h_{fe} > 56$, $\beta = 1$
    (b) $h_{fe} > 56, \beta = \frac{1}{29}$
    (c) $h_{fe} \approx 10, \beta = \frac{1}{29}$
    (d) $h_{fe} \approx 10\ \beta = 1$ **[Ans:** (b)]

11. In crystal oscillator, the crystal maintains
    (a) vibration only at its series resonant frequency
    (b) vibration only at its parallel resonant frequency
    (c) vibration between its series and parallel resonant frequency
    (d) vibration either at series or parallel resonant frequency [**Ans:** (c)]
12. A wien bridge oscillator uses
    (a) positive feedback (b) negative feedback
    (c) both types of feedback (d) no feedback [**Ans:** (c)]
13. A phase shift oscillator usually has
    (a) two lead-lag circuits (b) three lead-lag circuits
    (c) a lead-lag circuit (d) a twin T filler [**Ans:** (b)]
14. For oscillation to start in a feedback oscillator, the loop gain must be greater than unity when phase shift around the loop is
    (a) 90° (b) 180° (c) 210° (d) 360° [**Ans:** (d)]
15. The range of frequency of oscillations developed in a colpitts oscillator is
    (a) hundred hertz (b) kilo hertz
    (c) mega hertz (d) billion hertz [**Ans:** (c)]
16. The material that shows piezoelectric effect is
    (a) quartz (b) tourmaline
    (c) rochell salt (d) all of the above [**Ans:** (d)]
17. When Q decreases in a colpitts oscillator the frequency of oscillation.
    (a) decreases (b) increases
    (c) remains constant (d) becomes erratic [**Ans:** (c)]

Mark the Following Statements True/False.

1. In a phase shift oscillator, we can have a variable frequency output. [**Ans:** False]
2. A *p-n-p* or *n-p-n* transistor can be used in a phase shift oscillator. [**Ans:** True]
3. A sine wave can be generated from a free running multivibrator. [**Ans:** False]
4. If $R_1 \neq R_2$ then the time period for which two transistors in free running multivibrator remain off is also different. [**Ans:** True]
5. If $R_1 C_1 \neq R_2 C_2$ in a free running multivibrator, then shape of output wave produced is not same as when $R_1 C_1 = R_2 C_2$. [**Ans:** True]
6. In a feedback oscillator, the total phase shift through the loop is 0°. [**Ans:** True]
7. A Wien-Bridge oscillator has resonant circuit. [**Ans:** False]
8. The starting condition for a feedback oscillator is $BA = 1$. [**Ans:** False]

9. A crystal oscillator has only one frequency of resonance. [**Ans:** False]

10. An astable multivibrator is also called free running multivibrator. [**Ans:** True]

Fill in the Blanks.

1. The purpose of RF choke in a Hartley oscillator is to keep the ____________ out of the ____________. [**Ans:** alternating current, power supply]

2. The voltage measured from base to emitter (in Hartley oscillator) when oscillator is ON will be ____________. [**Ans:** negative]

3. In the first section of RC phase shift network in a phase shift oscillator circuit, signal v_1 across R is ____________ the applied signal. [**Ans:** leading]

4. In a phase shift oscillator, the feedback network introduces an additional ____________ phase shift, so that signal feedback from collector to base is ____________ with the signal on base. [**Ans:** 180°, in phase]

5. In a symmetrical astable multivibrator, $C_1 = 0.02$ μF and $R_2 = 100$ kΩ. The time period of one complete cycle is ____________ sec and frequency of oscillation is ____________ Hz. [**Ans:** 2.8×10^{-3}, 357]

6. In a symmetrical astable multivibrator resistor, R_2 is changed to 1.5 times its present value. The time for which transistor Q_1 is ____________ is ____________ from that of transistor Q_2. [**Ans:** off, different]

7. At resonant frequency, the phase shift of lead-lag network in Wien-Bridge oscillator is ____________. [**Ans:** zero]

8. When power is ON in a Wien-Bridge oscillator, the resistance of tungsten lamp is ____________ and gain of the voltage divider is less than ____________.

   [**Ans:** low, 1/3]

Chapter 13

Ionosphere and Communication

13.1 Introduction

The ionized region of the upper atmosphere (ionosphere) extends from about 40 km to a height of about 400 km above earth's surface (Figure 13.1(a)). It is made up of electrons, positive and negative ions in the background of neutral particles, comprising of molecules of oxygen and nitrogen gas. The concentration of neutral particles and gas molecules is much greater than that of charged particles. In this respect, we can say that the ionosphere is a weakly ionized medium. Here, cosmic rays and ultraviolet rays, α, β and γ radiations, coming from the sun and interstellar space interact directly with gas molecules and produce electrons and positive ions, simultaneously. A few of these electrons recombine with positive ions and generate neutral particles. In this way, there is continuous generation and annihilation of electrons and ions in the ionosphere, and it remains stratified rather than regularly distributed. However, it is to be noted that the concentration of electrons and negative ions taken together equals that of positive ions; so the entire region is electrically neutral.

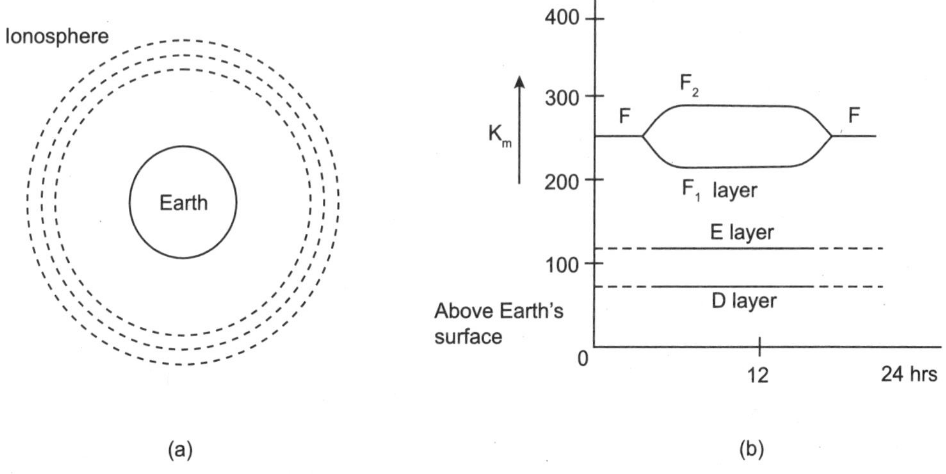

Figure 13.1 Ionospheric layers

The over all result is a range of four main layers, *D*, *E*, F_1 and F_2, as shown in Figure 13.1(b). The last two combine at night to form a single layer.

In view of their low mass and high mobility, electrons in the ionosphere greatly influence the propagation of radio waves through it. The variation of electron concentration with height in the ionosphere must be known. The maximum electron concentration in a layer is called peak electron concentration of the layer (Figure 13.2). The height at which peak concentration occurs is called peak height of the layer, as shown in Figure 13.2(a) and Figure 13.2(b). Figure 13.2(a) refers to day time and Figure 13.2(b) refers to night time.

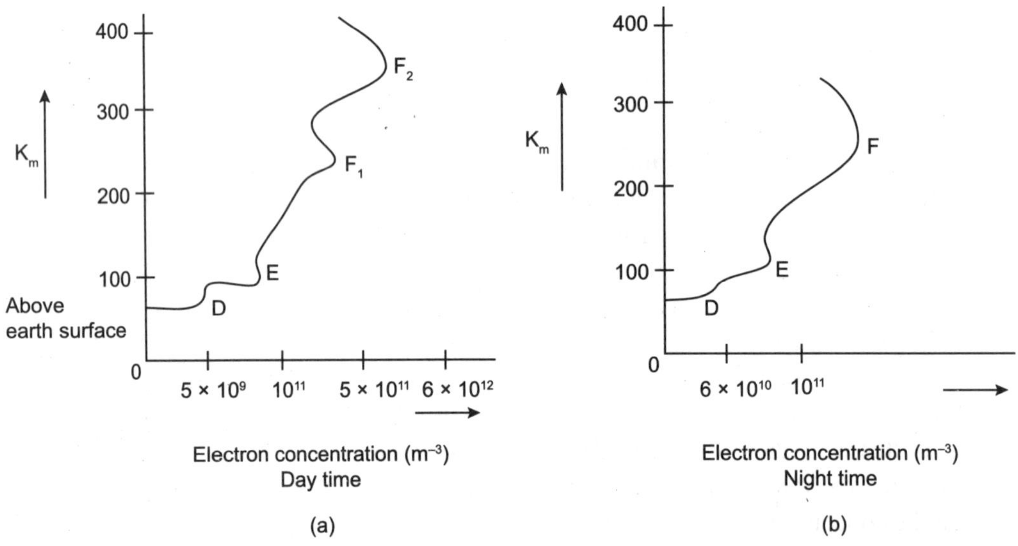

Figure 13.2 Height distribution of electron concentration in different layers

As shown in Figure 13.2(a), four layers exist in the day time. Electron concentration in '*F*' layer diminishes rapidly with height. At night, '*D*' layer practically disappears completely, and F_1 and F_2 layers coalesce to produce a single layer known as *F* layer (Figure 13.2 (b)). These variations are regular and predictable. The variation of F_2 layer is quite fast and greater than *E* and F_1 layers. The *E*, F_1 and F_2 layers are exploited to reflect radio waves back to the earth in sky wave propagation. *D* layer attenuates the radio wave.

13.2 Propagation of Radio Wave through Ionosphere

Whenever a radio wave propagates through the ionosphere, its propagation is unlike through free space since it is affected by electrons and ions present there. The effect of electrons on propagation is much greater than that by ions since the mass of electrons is much less than that of ions. As such, for simplicity, we may neglect the effect of ions and consider that of electrons only in the following analysis.

Let N be the concentration of electrons per m^3 at heigh h in the ionosphere; electric field associated with the radio wave of angular frequency ω be E cos ωt. Force experienced by an electron due to the electric field of radio wave is

$$F = (-)\, eE \cos \omega t$$

Here, the negative sign is for the charge of electron e.

Let v be instantaneous velocity of electrons due to the force experienced. Ignoring collision among electrons and ions, the equation of motion of electrons is

$$m\frac{dv}{dt} = (-)eE \cos\omega t, \text{where } m - \text{mass of electron}$$

Integrating the above equation,

$$v = (-)\frac{eE}{m\omega}\sin\omega t + K \tag{1}$$

Here: (i) (–) sign shows that the movement of electrons is opposite to that of field direction; (ii) K is a constant of integration.

Now, $v = 0$ when $E = 0$.

So $K = 0$

or
$$v = (-)\frac{eE}{m\omega}\sin\omega h \tag{2}$$

The movement of electrons produces current. So, current density produced by N electrons is

$$J_e = (-)Nev$$

Substituting for v from Eqn. (1), we have

$$J_e = \frac{Ne^2}{m\omega}E\sin\omega t \tag{3}$$

Eqn. (3) shows that current density lags behind electric field by 90°. Since the ionosphere is a weakly ionized medium, its permittivity is nearly equal to that of free space. So displacement current density is approximately given by

$$J_d = \frac{d}{dt}(\varepsilon_0 E\cos\omega t) = (-)\varepsilon_0 E\sin\omega t \tag{4}$$

where ε_0 is permittivity of free space.

Eqn. (4) shows that displacement current leads electric field by 90°. Total current density of ionosphere is

$$J = J_e + J_D = (-)\omega\left[\varepsilon_0 - \frac{Ne^2}{m\omega^2}\right]E\sin\omega t \tag{5}$$

Assuming total current density as displacement current, Eqn. (5) predicts that effective permittivity of the ionized region of ionosphere is

$$\varepsilon = \epsilon_0 \left[1 - \frac{Ne^2}{m \epsilon_0 \omega^2}\right] \tag{6}$$

In SI units, we have $\varepsilon_0 = 8.854 \times 10^{-12}$ F/m, $e = 1.6 \times 10^{-19}$ C and $m = 9.1 \times 10^{-31}$ kg. Substituting these values, Eqn. (6) reduces to

$$\epsilon = \epsilon_0 \left[1 - \frac{81\text{ N}}{f^2}\right] \tag{7}$$

where frequency of radio wave, $f = \dfrac{\omega}{2\pi}$.

Phase velocity of a plane wave propagating in a medium of permittivity ϵ and permeability μ is

$$v_p = \frac{1}{\sqrt{\epsilon \mu}}$$

However, in free space, $\epsilon = \epsilon_0$ and $\mu = \mu_0 = 4\pi \times 10^{-7}$ H/m. Hence, phase velocity of radio wave in free space is

$$c = \frac{1}{\sqrt{\epsilon_0 \mu_0}} = 3 \times 10^8 \text{ m/s}$$

which is also the velocity of light. However, in an ionized medium, $\mu = \mu_0$ but ϵ is less than ϵ_0 as predicted by Eqn. (7); so phase velocity of a radio wave in an ionized medium (ionosphere) exceeds that of light in free space, but group velocity of a radio wave cannot exceed c.

The refractive index of ionized media relative to free space is

$$n = \frac{c}{v_p} = \sqrt{\frac{\epsilon}{\epsilon_0}} \tag{8}$$

13.3 Refraction of Radio Wave in Ionosphere

Let a radio wave propagate through the ionosphere from the underlying unionized region. As explained earlier, phase velocity is greater in ionosphere, thereby the corresponding refractive index of ionosphere is less than a unionized medium, and so it behaves like a rarer medium. Whenever a ray enters a rarer medium from a denser medium, it bends away from normal at the interface as shown in Figure 13.3.

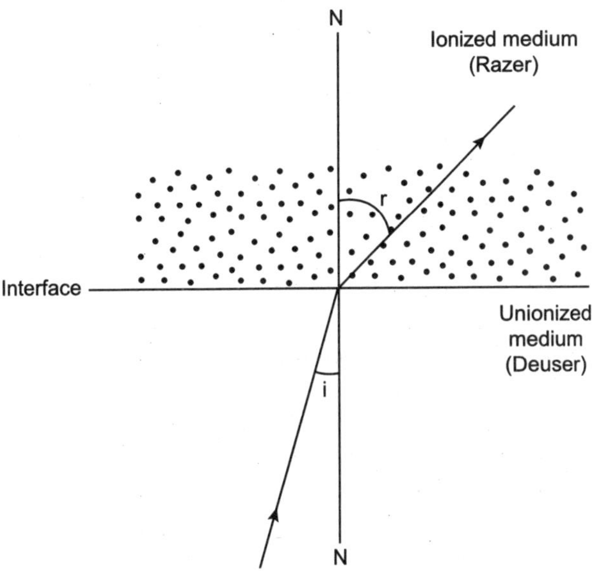

Figure 13.3 Refraction of a radio wave

Snell's law governs the refraction of wave according to its refractive index

$$n = \frac{\sin i}{\sin r}$$

$$= \sqrt{\frac{\epsilon}{\epsilon_0}}$$

$$= \sqrt{1 - \frac{81\,\mathrm{N}}{f^2}} \qquad (9)$$

As a wave travels through the ionosphere which extends to a certain height, its electron density gradually increases with height, thereby the refractive index of various layers goes on decreasing and the refracted ray bends more and more away from normal (Figure 13.4(a)), i.e. angle of refraction goes on increasing. At a certain height, i.e. at top point M (Figure 13.4(a)), refraction angle may touch 90°. In this situation, refractive index

$$n = \frac{\sin' i'}{\sin r} = \frac{\sin i}{\sin 90^\circ} = \sin i$$

If N_r is electron density at the point, then

$$n = \sin i = \left[1 - \frac{81 N_r}{f^2}\right]^{1/2} \qquad (10)$$

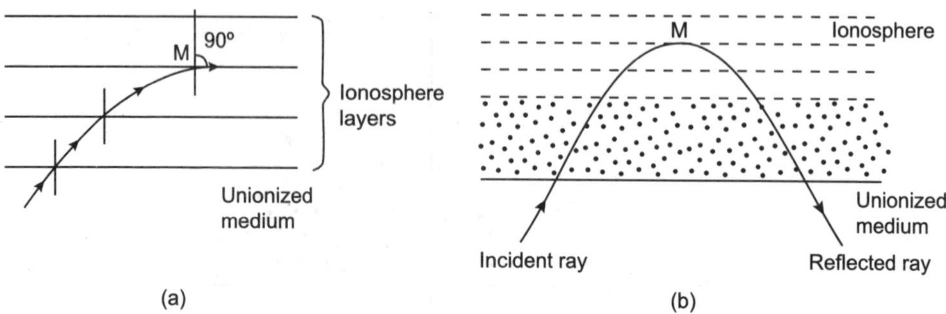

Figure 13.4 Ray bending in ionosphere and reflection in ionosphere

At point *m*, the refracted ray is just parallel to a particular layer (parallel to earth's surface) and thereafter it undergoes total reflection (Figure 13.4(b)) and returns to earth. In this respect, point *M* is known as the point of reflection.

Eqn. (10) shows that as the angle of incidence increases, refractive index also increases and lower is the value of N_r ray nos. 3, 4, 5, 6, the point of reflection is at lower height and electron concentration decreases with decreasing height. This situation is illustrated in Figure 13.5. The situation is depicted by ray nos. 3 through 5.

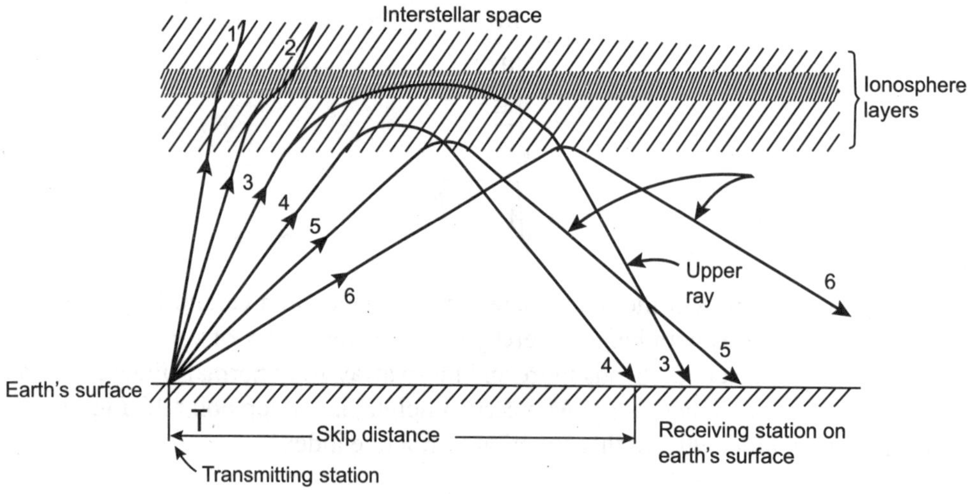

Figure 13.5 Trajectories of radio waves incident at different angles through the ionosphere

For a sufficiently small angle of incidence *i*, Eqn. (10) is not satisfied even with the largest value of electron concentration in the layer. The ray then penetrates the ionosphere layers and goes into the interstellar space (ray nos. 1, 2).

When the radio wave is incident vertically, $i = 0$ and so $n = 0$ (Eqn. 10). So for vertical incidence, the maximum frequency of wave reflected towards earth from a ionospheric layer is referred as critical frequency (f_c) of that layer. Eqn. (10) shows that critical frequency (f_c) related to peak electron concentration N_p of a particular layer is given by

$$f_c = 9\sqrt{N_p} \tag{11}$$

So, in terms of f_c, Eqn. (10) can be written as

$$n = \sin i = \left[1 - \left(\frac{f_c}{f}\right)^2\right]^{1/2} \tag{12}$$

Eqn. (12) shows that, a frequency higher than f_c can be reflected back by the layer for oblique incidence. Eqn. (12) gives

$$\left(\frac{f_c}{f}\right)^2 = 1 - \sin^2 i = \cos^2 i$$

or

$$f = f_c \sec i \tag{13}$$

Eqn. (13) is known as Secant law.

It relates frequency f of the reflected wave for angle of incidence i to critical frequency f_c. When a ray (wave) is launched tangentially (Figure 13.6) to earth's surface, the angle of incidence i attains the maximum value i_m. If R is the radius of earth, H is the height of ionized layer (i.e. point M is the top point of trajectory), then from Figure 13.6, we can write

$$\sin i_m = \frac{R}{R+H} \tag{14}$$

So maximum frequency f_m corresponding to this angle i_m can be obtained from Eqn. (13).

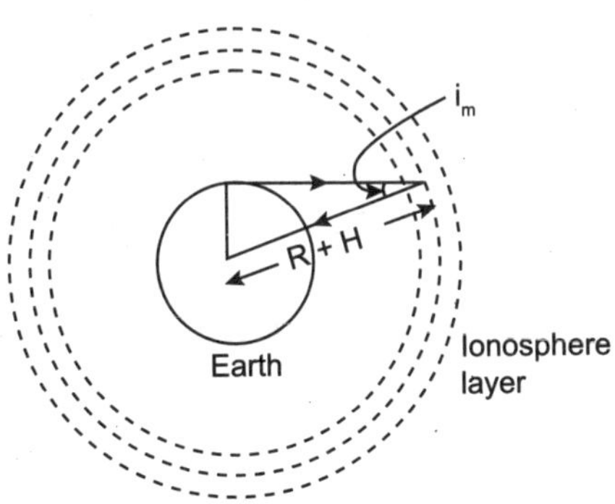

Figure 13.6

Critical frequencies for E and F layers of ionosphere are nearly 2.5 MHz and 9 MHz, respectively. The maximum frequency f_m of radio wave that can be reflected from these layers is about 12 MHz and 33 MHz, respectively. These values vary since their height and peak electron concentration of layers are subject to diurnal, seasonal and the sun spot cycle variations.

The peak electron concentration of a layer decreases at night. So from Eqn. (11), critical frequency of the layer at night time is less than that at day time. So, lower carrier frequencies are employed for sky wave transmission at night than those during day time. Reflections from E layer cover a distance of about 200 km over the earth's surface. F layer reflections are used for transmission around the globe (Figure 13.7). For carrier frequencies above 50 MHz, communication via sky wave is not possible because radio waves with these frequencies are not reflected back to the earth by the ionosphere.

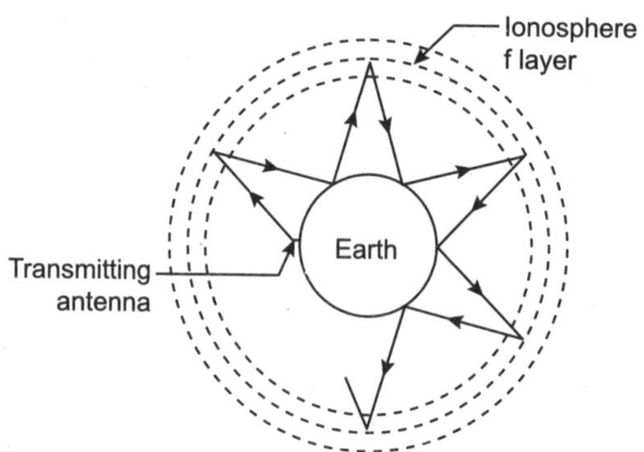

Figure 13.7 Round-the-globe communication through sky wave

While discussing sky wave propagation through the ionosphere, it was mentioned that the reflection of the wave due to ionosphere layer depends upon the angle of incidence (Figure 13.5). It is found that as the angle of incidence decreases, the distance between transmitting and receiving stations on the surface of earth first diminishes (rays 5, 4 and 3) and then increases (ray 2); and for an angle of incidence less than the critical value, the ray passes through the layer (ray 1). Compared to other rays, ray 2 moves through a longer horizontal path in the ionospheric layer. The reason for this is unnknown. For the ray to be received at earth's surface, there is a minimum distance between transmitting and receiving stations for a given frequency. This minimum distance between transmitting and first receiving stations is called skip distance.

Skip distance increases with the frequency of incident wave. For a given frequency, skip distance depends upon the height of reflecting layer. It increases with increasing height of ionosphere layer. As the height of the layer is least at noon and it increases in the pre-noon and afternoon hours, skip distance is minimum at noon.

The frequency of radio wave, for which the distance on earth's surface equals skip distance is referred as Maximum Usable Frequency (MUF). If the frequency of transmitted radio wave is higher than MUF, the sky wave will not reach the receiving station. The MUF too has diurnal variation' it is maximum at noon and falls on either side of noon hour. For sky wave communication between two points, optimum frequency of the wave is taken about 85% of MUF for two given points.

13.4 Propagation of Radio Waves

Radio waves are electromagnetic (EM) waves whose frequency ranges from 10 KHz to 300 GHz. These waves are further subdivided into groups as per their frequency and use. These waves are radiated (transmitted) from a transmitting station and are not markedly affected by the atmosphere or surrounding. They readily pass through nonmetallic obstacles but markedly attenuate while crossing a metallic surface. They travel at the speed of light ($C = 3 \times 10^8$ m/s) through the atmosphere or free space. Their classification is shown in Table 13.1.

Table 13.1 Radio waves spectrum

Band	Frequency range	Wavelength range	Principal applications
Very low frequency (VLF)	10 to 30 KHz	30 to 10 km	Direct long range communication
Low frequency (LF)	30 to 300 KHz	10 to 1 km	Marine navigational aids
Medium frequency	300 KHz to 3 MHz	1 km to 100 m	Broadcasting (radio)
High frequency	3 to 30 MHz	10 to 1 m	Broadcasting and all types of communication
Very high frequency (VHF)	300 MHz to 3 GHz	1 m to 10 cm	Microwave communication, radar
Super high frequency	3 to 30 GHz	10 cm to 1 cm	Radio, relay, navigation, radar, satellite communication
Extremely high frequency	30 to 300 GHz	1 cm to 1 mm	Experimental

On the basis of mode of propagation, radio waves have been broadly classified as

(i) Ground or surface wave
(ii) Space or troposphere wave
(iii) Sky wave

1. Ground wave In ground wave propagation, radio waves are guided by the earth's surface and as such they propagate along earth's curved surface from the transmitter to the receiver. As waves propagate along the ground, they are strongly influenced by the electric and magnetic properties of the surface. As high frequency EM waves are strongly absorbed by the ground, ground wave propagation is useful for low frequency radio waves only.

Ground waves below 500 KHz can be used for communication within a distance of 1500 km from the transmitting station. AM radio broadcast in the medium frequency range covers local areas, and so it is mainly through ground wave. It is quite reliable irrespective of weather conditions around the transmitter.

2. Space wave In space wave propagation, radio waves propagate through the earth's troposphere within a height of about 15 km from earth's surface. It has two components: (a) direct or line-of-sight wave from the transmitting station to the receiving station (b) ground-reflected wave traversing from the transmitting station antenna to the ground (earth's surface) and reflected towards the receiving station antenna. Figure 13.8 illustrates the two components.

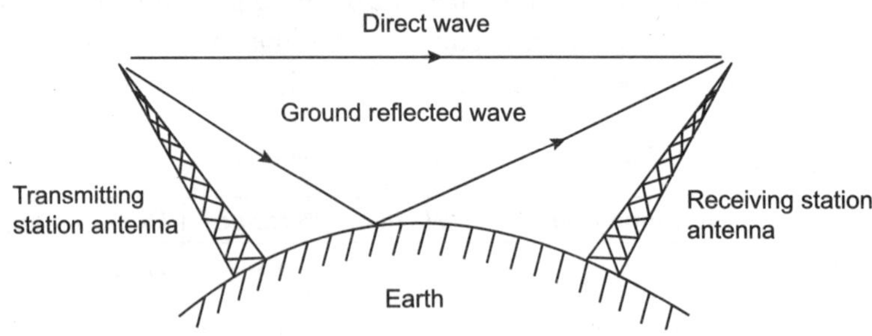

Figure 13.8 Space wave communication

As shown in Figure 13.8, a ground-reflected wave suffers a phase change of 180° after being reflected from ground points. Therefore, when path lengths of direct wave and ground-reflected wave from the transmitting station to the receiving station antenna are such that they reach in opposite phase, they cancel each other. But when the two path lengths are such that they differ by half wavelength, they are in phase so they reinforce each other. By changing the height of antenna, their path lengths can be changed. In this way, alternate enhancing or reducing of signal strength can be achieved. This phenomenon is known as selective fading.

The space wave is not continuously absorbed by the earth's surface, thereby it can cover a greater range than a ground wave. FM and TV programmes in VHF band are propagated by direct waves with contribution from ground-reflected waves. In UHF band, transmission occurs entirely by direct wave.

3. Sky wave In sky wave propagation, radio waves propagated from the transmitting antenna reach the receiving antenna after being reflected from the ionosphere (Figure 13.7). Short wave transmission around the globe is possible through sky wave via successive reflections at the ionosphere (Figure 13.7).

13.5 Satellite Communication

Satellites offers a reliable communication link over long distances.

We know that for sky wave propagation, usually the frequency band of 3 to 30 MHz is employed. Radio links for longer distances over the earth's surface can be established by multi-loop transmission. However, reliable communication by sky wave is hampered due to many problems like fading, ionospheric disturbances, etc. These can be overcome with the help of artificial satellites. An artificial satellite is launched into space in a circular orbit above earth in the equatorial plane at a height of 36000 Km. The period of revolution of these satellites around the earth is 24 hours. To an observer on earth's surface, the satellite at this height appears to be stationary. Such satellites are referred to as geostationary satellites and the orbit is called synchronous orbit. The transmitter established at a particular radiostation radiates information in the form of EM wave to the satellite, where it is processed, amplified and retransmitted towards a receiving station on the surface of the earth. The satellite being geostationary, the transmitting and receiving antennas on earth's surface can be oriented in specific directions. The angle between the direction of propagation of the EM wave radiated from the transmitting station antenna and horizontal is termed the angle of elevation. The horizontal pointing angle of the antenna as measured from true North, is called Azimuth.

To orient an earth station antenna towards a particular satellite, elevation angle and azimuth must be known. These angles are referred to as look angles. For smaller angles of elevation, the radiated EM wave travels a longer distance through earth's atmosphere, and thus suffers more absorption and greater noise contamination. To avoid excessive deterioration of the wave, the angle of elevation is kept larger than 5°. The radiation pattern of a satellite antenna can cover nearly one-third of the earth's surface. Thus, three equally spaced geostationary satellites cover the entire earth's surface except for the unpopulated areas of North and South Poles.

Satellites operating at or near the same frequency are separated by 3° to 6° to prevent mutual interference. Usually, carrier frequency for the wave transmitted from earth station to satellite, called uplink frequency, is 6 GHz, and carrier frequency of the wave from satellite to earth's receiving station, called downlink frequency, is 4 GHz. An uplink frequency of 14 GHz and a down link frequency of 12 GHz are also employed in satellite communication. The available bandwidth is several MHz. Different uplink and down link antenna frequencies are used to avoid unwanted feedback. Frequency band separation also permits the same antenna to be employed for both transmission and reception, simplifying satellite hardware. Downlink frequency is lower than uplink frequency. As absorption increases with increasing frequency, the signal received by earth station suffers less attenuation.

For the carrier frequency employed in satellite communication, the antenna size is not inconveniently large. FM or special modulation techniques for digital signals are employed in satellite communication. The use of 6/4 GHz band offers the following advantages – (i) relatively in expensive microwave equipment can be employed; (ii) attenuation due to rainfall and fog is less than 14/12 GHz band; (iii) sky background noise due to galactica, solar, and terrestrial noise sources is in significant.

The 14/12 GHz band used as 6/4 GHz band coincides with that used for terrestrial microwave systems. In the 14/12 GHz band, smaller and hence less expensive antenna can be employed.

A satellite system (Figure 13.9(a)) essentially consists of three primary sections: (i) an uplink, (ii) a satellite transponder and (iii) a downlink. The uplink section consists of a transmitter situated at transmitting station, which radiates EM wave of particular frequency to the satellite. The satellite transponder consists of a frequency translator which converts uplink frequency to downlink frequency by feeding the incoming wave (uplink frequency) and the signal from a

2 GHz shift oscillator in a mixer. The signal is amplified and transmitted to earth station receivers. The various circuits on board of the satellite are powered by solar cells mounted on the panels attached to satellite. The downlink station consists of earth receiver which recovers information from modulated signal.

Communication between satellites is possible via crosslinks or inter-satellite links (ISLS) shown in Figure 13.9(b). Conventional speech telephone signals, data (in analogue or digital form) and television picture (video) and sound (Audio) are conveyed through satellite systems.

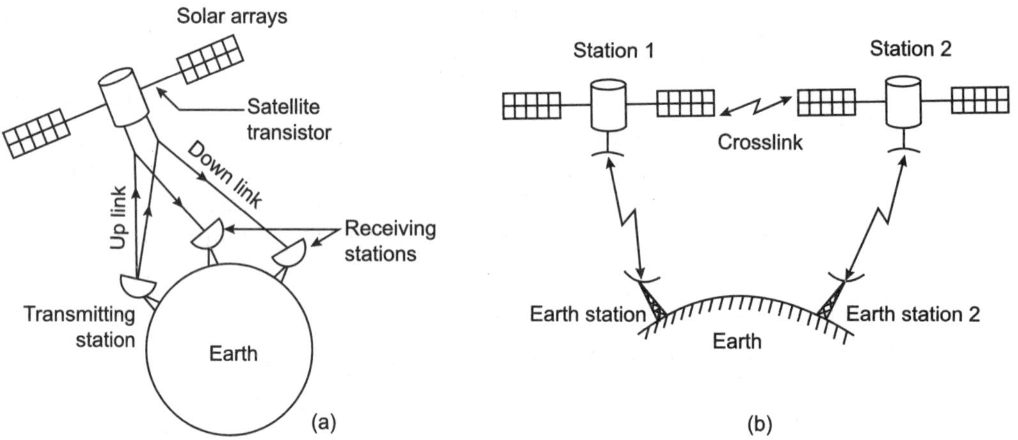

Figure 13.9 (a) A satellite system, (b) inter-satellite link

13.5.1 Merits of Satellite Communication

(i) The quality of reception is quite good and the system is most reliable.

(ii) The bandwidth is large.

(iii) A geostationary satellite can cover approximately onethird of earth's surface.

(iv) A geostationary satellite appears stationary to an observer on earth, so expensive tracking equipment is not required.

(v) The wide band go-any-where service has opened up possibilities for satellite systems to handle a large number of signals and data.

13.5.2 Demerits of Satellite System

(i) The installation cost is quite high. Geostationary satellites require sophisticated and heavy propulsion devices on board to keep them in fixed orbit.

(ii) The high altitude of geostationary satellite introduces long propagation times. The round-the-trip propagation delay between two earth stations via satellite is 500 to 600 mS.

(iii) The system requires high output power (several kilowatts) of earth station transmitters and sensitive receivers.

13.6 Digital Communication

In an analogue communication system the information-bearing signal varies continuously in amplitude and time, and it is employed directly to modify some features (such as amplitude, frequency or phase) of the sinusoidal carrier wave. On the other hand, a digital communication system processes the information-bearing signal in a way to represent it by a sequence of discrete messages. The impact of computers as a data source and as a tool for communication, the availability of wide band channels provided by optical fibres, and the growth of solid state devices and ICs to accommodate system complexities in an economic way led to the development of digital communication.

The merits of the system are:

(i) Digital communication is more immune to channel noise.

(ii) Accumulation of noise along the path can be prevented with repeaters.

(iii) The system permits the use of computers, LSI circuits and can accommodate various sources of information in flexible and compatible manner.

(iv) The system is highly reliable and gives high fidelity, security and low error rate.

(v) A number of digital signals can be easily and efficiently multiplexed into the system.

A detailed discussion of the system is beyond the scope of this chapter. However, basic ideas of the system are as under:

(i) Information such as human voice, music or TV picture, etc. can be transformed into an electrical wave with the help of a transducer. The electrical form is referred to as message signal or base band signal. Both amplitude and time have discrete values in digital signal. Computer data are examples of digital signals.

(ii) The basic operations namely sampling, quantizing and encoding are combined to convert an analogue signal into digital form as shown in Figure 13.10.

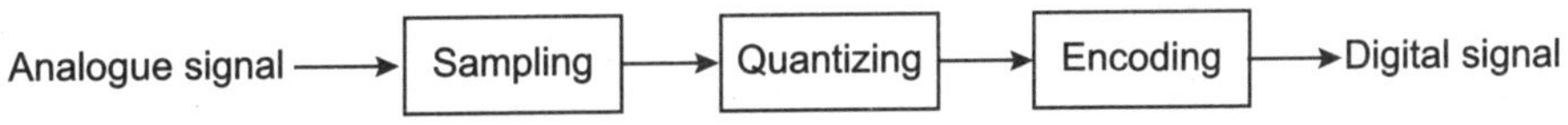

Figure 13.10 A/D conversion in a digital system

(a) The sampling process retains only sample values of analogue signals at uniformly spaced discrete instants of time.

(b) The quantizing system rounds off each sample value to the nearest level which is a member of finite set of discrete levels.

(c) The encoding process represents the selected level by a code word formed with a prescribed number of code elements.

Summary

1. There is a layer of ionized particles and gas molecules surrounding the earth's atmosphere at a height of nearly 100 to 400 km, which is called the ionosphere. It comprises of mainly four layers, D, E, F_1 and F_2, in the ascending order. The last two layers combine at night to form a single layer.
2. When radio waves transmitted from a radio station reach the ionosphere, some of them depending on the angle of incidence get refracted through the ionosphere and traverse to interstellar space but other waves undergo the phenomenon of total reflection and return to earth at far distant point from the transmitting station.
3. The ionosphere behaves like a dielectric media for radio waves; its permittivity is given by

$$\in = \in_0 \left[1 - \frac{81\,\mathrm{N}}{f_2}\right]$$

4. There is always a minimum distance from the transmitting station, called skip-distance, up to which no radio wave reaches after reflection from the ionosphere. It increases with frequency of incident wave.
5. There is also a frequency of radio wave called critical frequency f_c above which the frequency of wave getting reflected from the ionosphere is related to the angle of incidence 'i'. It follows the secant law

$$f = f_c \sec i$$

6. Radio waves have been classified as (i) ground wave, (ii) space wave and (iii) sky wave.
   (i) Communication along the surface of earth is possible through ground wave up to a distance of 1500 km from the transmitting station.
   (ii) Communication through space wave is possible along the line of sight from the transmitting station, which is usually large than the skip distance.
   (iii) Communication through sky wave is possible after reflection from the ionosphere. Communication round the globe takes places only via this mode of propagation. The frequency band of this wave is 3 to 30 MHz.
7. The most common mode of communication at present is through satellite which is geostationary. It is very reliable. Here, the signal called uplink first traverses to the satellite and then the satellite converts the uplink frequency; the converted frequency wave called downlink reaches the receiving station.
8. Here, the frequency band of the radio wave is quite large and its bandwidth is in GHz range.

Review Questions

1. What do you mean by the following terms with reference to ionosphere: (a) virtual height, (b) critical frequency, (c) maximum usable frequency, (d) skip distance?
2. What are the common modes of radio wave propagation? Explain briefly the mechanism of propagation. Mention their limitations.
3. What is "ionosphere"? Describe its constituents and various layers. How do these layers change their position during day and night? Show by appropriate figure.
4. How does the presence of ionosphere affect the propagation of radio waves through it? Show the effect by appropriate diagram and explain the paths of radio wave.
5. Mention the nomenclature of various layers in ionosphere. Show their relative height from the surface of earth using an appropriate diagram. How do they change during day and night?
6. Describe the mechanism with which radio waves transmitted from a radio station are totally reflected by ionosphere to the earth.
7. Prove that the ionosphere acts as a media of dielectric constant $\in = \in_0 \left[1 - \frac{81\,N}{f^2}\right]$ where N represents electron number density and f represents frequency of the wave reaching ionosphere.
8. Discuss the mode of propagation of radio waves via ionosphere.
9. Derive the relation for Secant law as related to the propagation of radio waves.
10. Classify frequency band of radio waves. Mention the uses of various groups of radio waves.
11. How do you distinguish between space wave and sky waves?
12. Describe satellite communication in brief. Give its merits.
13. What do you mean by the term "digital communication"? Explain.

Short Answer Questions

1. What are the constituents of ionosphere. Is it electrically neutral?

   Ans: The constituents of the ionosphere are electrons, ions and molecules of gases like nitrogen, oxygen, etc. They are distributed throughout in the form of layers and are of varying densities. The entire region is electrically neutral.
2. Give the nomenclature of layers of ionosphere and their approximate height from the surface of the earth.

   Ans: The nomenclature of layers of ionosphere is D, E, F_1 and F_2. D layer is the inner most layer at a height of 50–70 km E layer is at 100 km. F_1 and F_2 layers are at height of 200 to 300 km and are seperated from each other.

3. How do the layers of ionosphere change their position during day and night?

   Ans: In the day, we have four layers in ionosphere, namely D, E, F_1 and F_2. F_1 and F_2 merge and form a single F layer. So effectively during night, we observe E and F layers.

4. How is the dielectric constant of ionosphere related with the frequency of incoming radio wave?

   Ans: Dielectric constant of ionosphere is given by

$$\in = \in_0 \left[1 - \frac{81\,\text{N}}{f^2}\right]$$

   $\in_0$ – dielectric constant of free space = 8.85×10^{-12} F/m

   f – frequency of incoming radio wave

   $N \rightarrow$ electron density.

5. Do radio waves radiated from a transmitter reach the earth via ionosphere?

   Ans: All waves radiated from the antenna of a transmitting station do not reach the earth via the ionosphere. (1) Radio waves of very high frequency of the order of GHz get refracted through the ionosphere and travel through free space above it. (2) Radio waves having frequency of the order of MHz and are radiated at certain angle do reach the earth's surface after getting totally reflected from the ionosphere.

6. What is Secant law?

   Ans: Secant law relates the frequency (f) of radio wave with the angle of incidence (i) and critical frequency (f_c). It is expressed as

$$f = f_c \sec i$$

7. Which layer of ionosphere helps radio wave transmission throughout the globe?

   Ans: Reflections of radio wave from E layer of ionosphere make it reach throughout the globe.

8. Classify radio waves as per their mode of transmission.

   Ans: Radio waves can be classified as

   (a) **Ground wave** – Radio waves travel that along the surface of earth.

   (b) **Space wave** – Radio waves that travel through earth's troposphere within a height of 15 km from earth surface and reach the receiving station directly from transmitter station.

   (c) **Sky wave** – Radio waves transmitted from a transmitter that reach the receiving station after reflection from the ionosphere.

Objective Questions

1. Radio waves having frequency in the UHF range propagate as ____________.
   (a) ground waves (b) sky waves
   (c) surface wave (d) space wave **[Ans:** (b)]
2. As radio waves travel in free space, they undergo ____________.
   (a) absorption (b) attenuation
   (c) refraction (d) reflection **[Ans:** (b)]
3. The absorption of radio waves through atmosphere depends upon ____________.
   (a) their frequency (b) their distance from transmitter
   (c) their degree of polarization (d) their intensity **[Ans:** (a)]
4. During night, the layer of ionosphere which disappears is ____________.
   (a) D layer (b) E layer
   (c) F layer (d) surface wave **[Ans:** (a)]
5. Communication of radio waves around the globe is through ____________.
   (a) ground wave (b) space wave
   (c) sky wave (d) surface wave **[Ans:** (c)]
6. The mechanism through which sky wave reaches earth via ionosphere is ____________.
   (a) reflection (b) refraction
   (c) interface (d) difference **[Ans:** (b)]

Mark the Following Statements True/False.

1. During night, both F_1 and F_2 layers of ionosphere merge. **[Ans:** True]
2. Communication of radio waves around the globe is through space waves. **[Ans:** False]
3. The most prominent layers of ionosphere during night are E and F layers. **[Ans:** True]
4. Skip distance is the minimum distance from the radio station where radio wave reaches through ground wave. **[Ans:** False]
5. Maximum usable frequency of radio wave is always more than critical frequency. **[Ans:** True]
6. Ionosphere behaves like a dielectric media for radio waves. **[Ans:** True]
7. Ionosphere is basically a neutral media. **[Ans:** True]

Fill in the Blanks.

1. The height of ionosphere from the surface of earth is between ____________ and ____________ km. **[Ans:** 50, 500]

2. Ionosphere consists of ____________ layers of ____________ of ____________ density. **[Ans:** conducting, electrons, varying]

3. The prominent layers of ionosphere during ____________ are ____________ and ____________. **[Ans:** night, *E*, *F*]

4. The minimum distance called ____________ is the distance from transmitter to a far distance place up to which ____________ wave does not reach after reflection from ____________.

   [Ans: skip distance, radio, ionosphere]

5. Dielectric constant of ionosphere depends upon the ____________ of radio waves reaching there. **[Ans:** frequency]

6. The innermost layer of ionosphere is called ____________ and outermost layer is called ____________ layer during day time. **[Ans:** D, F_2]

7. The radio wave going to a satellite is called ____________ and radio wave relayed by the satellite is called ____________. **[Ans:** uplink, downlink]

Chapter 14 Optical Fibres and Communication

14.1 Introduction

Fibre optics is the technology used in telecommunications as well as in cable TV. This technology is related to transportation of optical energy, i.e. light energy in guiding media specifically glass fibres. Optical fibres are mostly very thin and usually made from glass or plastic are far better than the normal telephone lines. Optical fibres are usually about 100 micrometers in diameter, used to carry signals in the form of pulses of light over distances upto 50 km without any need for repeaters. These signals may be coded voice communications or computer data. The more general term for an optical fibre is *light wave guide* and for fibre optics communication as *light wave communication*. However, the name *photonics* for fibre optics communications is becoming popular day by day. *Photonics* is very much helping the technology for fibre optic communication to enhance its performance and become more cost effective. Today, optical fibres are finding use in large number of areas, e.g. sensors, display systems, copying machines, defence services due to high privacy, etc. In telecommunications, modulating a beam of light (LED or LASER) as a source of light and using optical fibre waveguides have the inherent advantage of very large bandwidth, small dimensions, low losses bandwidth and insensitivity to electrical interference.

Optical fibres are dielectric waveguides, which are fabricated from glass or plastic. A fibre is as thin as human hair designed to guide light wave along their length. They are operated on optical frequencies. Usually an optical fibre is a cylindrical waveguide system through which the optical wave can propagate. The principle by which light wave travels through the fibre is 'total internal reflection' without loss of incident intensity.

An important structure used in optical systems is layered structure or waveguide structure. These structures are used to confine optical waves in a well defined region and guide their propagation. One can make layered structures from non-crystalline or crystalline materials. For example, glass is used to produce optical fibres for use in optical communication, whereas semiconductor waveguides are used in semiconductor lasers.

An optical fibre has three principal sections: (i) the core (ii) the cladding and (iii) the jacket (Figure 14.1). The innermost section of the fibre is referred to as cladding. The optical properties of cladding are different from those of the core. The outermost section of the fibre is known as jacket. Jacket is made of plastic or polymer and protects the fibre structure from moisture, abrasion, mechanical shocks and other environmental hazards.

The core forms the actual working structure of optical fibre. An incident (wave) ray of light entering the core suffers a number of total internal reflections at the core-cladding interface. Obviously the interface between core and cladding acts as a mirror at which total internal reflection of the transmitted light takes place. The structure of a single optical fibre is shown in Figure 14.1.

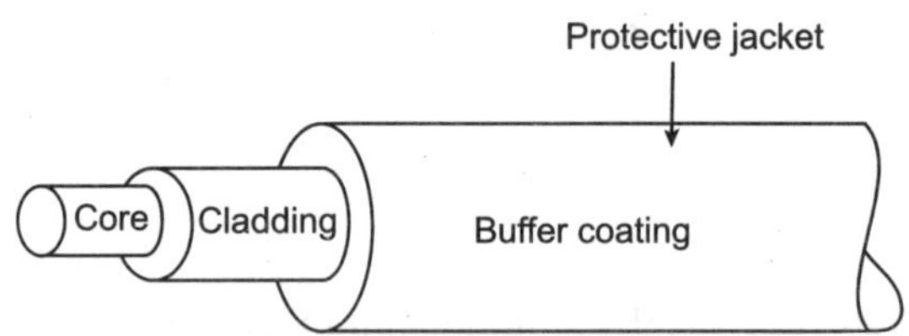

Figure 14.1 A single fibre structure

(i) Core Core is the innermost region of the fibre. Core has specific property of conducting an optical beam. Core is usually made of glass ($n_1 = 1.5$) or plastic. Glass is used as core material for low and medium loss fibres, whereas plastic core is used for relatively higher loss fibres. Quartz ($n_1 = 1.46$) is also used for core material. The diameter of core varies from 8 to 200 μm. We may note that the core is the actual working structure of the optical fibre, which is covered with another layer of glass or plastic having slightly different chemical composition known as cladding.

(ii) Cladding The core is surrounded by a solid dielectric cladding having refractive index n_2 which is less than n_1 of core ($n_2 < n_1$). The cladding material is either glass or plastic. Although cladding is not mandatory for light propagation along the core, however, it has the following advantages:

(a) It reduces scattering loss arising due to dielectric discontinuities at the boundary of core surface and air ($n = 1$).

(b) It adds mechanical strength to the fibre.

(c) It protects the core from absorbing surface contaminants.

We can have three types of fibres for core and cladding.

(i) Plastic core and cladding

(ii) Glass (silica) core and cladding

(iii) Glass core with plastic cladding

Although glass has low loss, plastic provides more flexibility and ruggedness to fibre, is less expensive, is 60% less heavy than glass and is stress tolerant. Hence plastic fibre is preferred for small distance communication. The optical fibre may have an abrupt boundary between core or cladding, i.e. they may exhibit a gradual change in material between them.

(iii) Jacket or Sheat The outermost section of optical fibre is made up of plastic or special kind of polymer and other material usually opaque in nature. This is known as jacket or sheat. A typical single fibre cable is shown in Figure 14.2.

Jacket protects the core from abrasion, interaction with environment, crushing, absorption, moisture and all other adversities of the terrestrial atmosphere. Obviously, jacket enhances tensile strength of fibre.

Apart from its use as communication channel, optical fibres find wide uses in many other areas. Sensors for detecting electrical, mechanical and thermal energies are made using optical fibres. Copier machines, simple display systems, medical diagnostics, fibroscopes, etc. also utilize fibre optics.

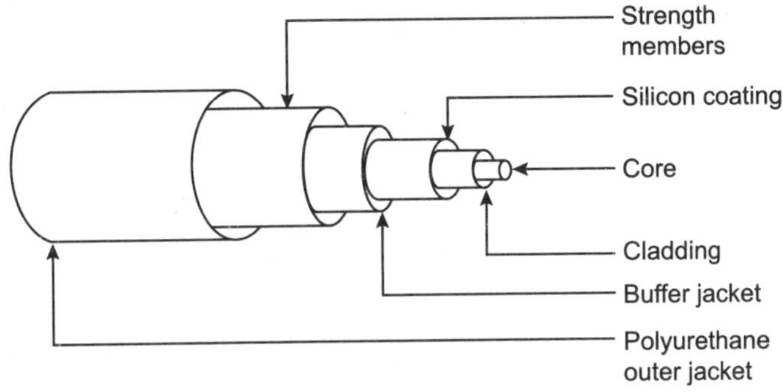

Figure 14.2 A schematic representation of typical fibre cable

14.2 Principle of Optical Guides

Light is guided along the fibre due to its "total internal reflection" property when it travels from one dielectric material to the other. The fibre material is basically a dielectric having dielectric constant ε. The refractive index n of a dielectric material depends on its dielectric constant $\varepsilon (n = \sqrt{\varepsilon})$. The refractive index of some common materials is given in Table 14.1. When a light ray encounters a boundary separating two different dielectric media, a part of the ray is reflected back and the remainder is refracted as it enters the second medium. This is shown in Figure 14.3 in which a light ray is entering from core material (n_1) to cladding (n_2). The bending is caused because of change in the speed of light ($v = c/n$) in the medium. The ray is bending away from normal as shown in Figure 14.3 if $n_1 > n_2$, i.e. ray travels from denser to rarer material. The ray will bend towards normal in case $n_1 < n_2$, i.e. if ray travels from rarer to denser materials.

Table 14.1 Typical indexes of refraction (at 589 nm)

Medium	Index of refraction (*n*)
Vacuum	1.0
Air	1.0003 (= 1.0)
Water	1.33
Ethyl alcohol	1.36
Fused quarts	1.46
Glass fibre	1.5–1.9
Diamond	2.0–2.42
Silicon	3.4
Gallium arsenide	3.6

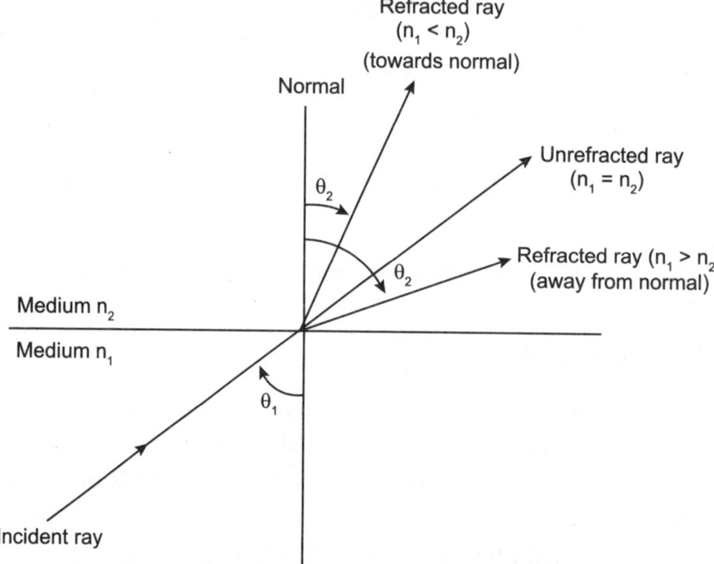

Figure 14.3 Refractive model for Snell's law

14.2.1 Snell's Law

The relationship between incident ray and refracted ray is governed by Snell's law:

$$n_1 \sin\theta_1 = n_2 \sin\theta_2 \quad 1(a)$$

or

$$n_1 \cos\phi_1 = n_2 \cos\phi_2 \quad 1(b)$$

where θ_1 and θ_2 are angle of incidence and angle of refraction, respectively, with reference to normal, whereas ϕ_1 and ϕ_2 are angled with reference to axis of fibre, as shown in Figure 14.4(a). As the angle of incidence θ_1 increases, refracted angle θ_2 also increases and ray is further bent away from normal and moves towards boundary. As the angle of incidence θ_1 further increases, the refracted ray approaches parallel to boundary, i.e. refracted angle θ_2 tends towards 90°. The angle of incidence for which the refracted ray becomes parallel to boundary (θ_2 becomes 90°) is known as critical angle of incidence, denoted by θ_c. This is shown in Figure 14.4(b). From Eqn. 1(a), we have a relation for critical angle as

$$n_1 \sin\theta_c = n_2, (\sin\theta_2 = \sin 90° = 1)$$

or

$$\theta_c = \sin^{-1}\left(\frac{n_2}{n_1}\right) \quad (2(a))$$

Similarly, if the angle of incidence is measured with respect to the axis of fibre, then Eqn. 1(b) yields

$$\theta_c = \cos^{-1}\left(\frac{n_2}{n_1}\right) \quad (2(b))$$

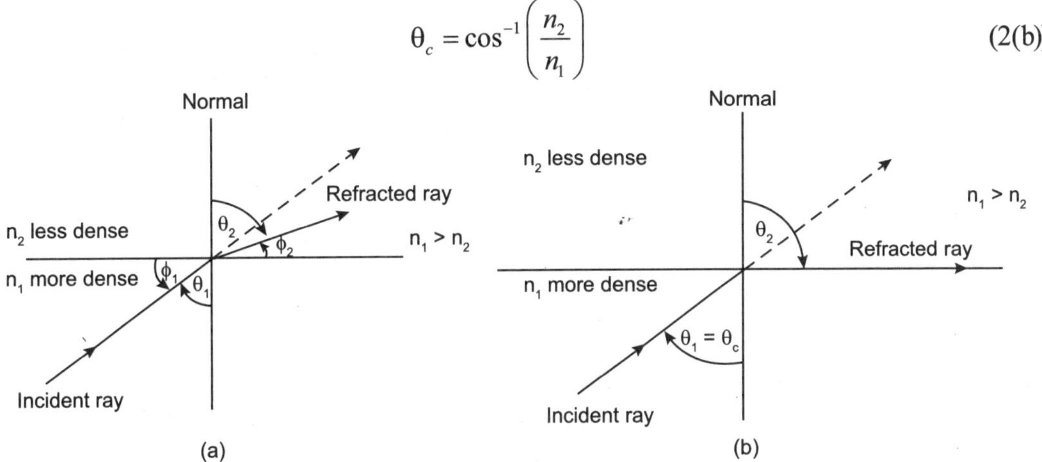

Figure 14.4 (a) Light ray refracted away, (b) critical angle from normal

When the angle of incidence θ_1 is further increased and becomes greater than critical angle θ_c, the condition of total reflection occurs and the entire light is reflected back to core medium (n_1) and no light is refracted in cladding (n_2) in ideal condition. This total reflection is shown in Figure 14.5(a). In this way, whenever a light ray strikes the core-cladding boundary at critical angle, it is reflected back towards the core and so light travels along the guide and remains confined inside the core, as shown in Figure 14.5(b).

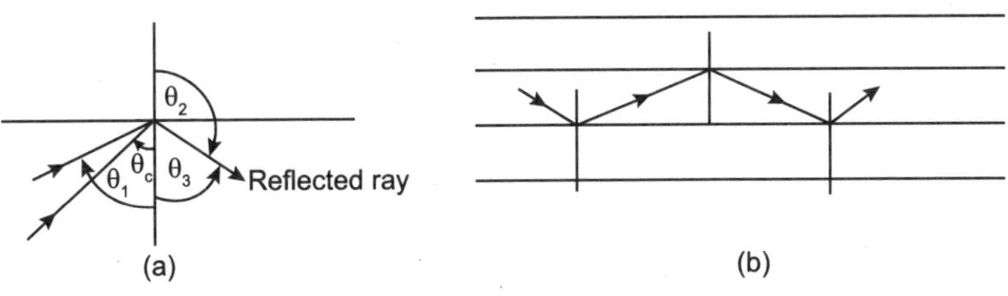

Figure 14.5 (a) Total reflection θ_1 = angle of incidence, θ_2 = angle of refraction, $\theta_3 = \theta_1$ = angle of reflection, (b) light confinement in the core

14.2.2 Acceptance Angle and Acceptance Cone

Acceptance angle represents the light gathering capability of a fibre. This defines the ability to couple light from source into the fibre cable. Consider a fibre section shown in Figure 14.6.

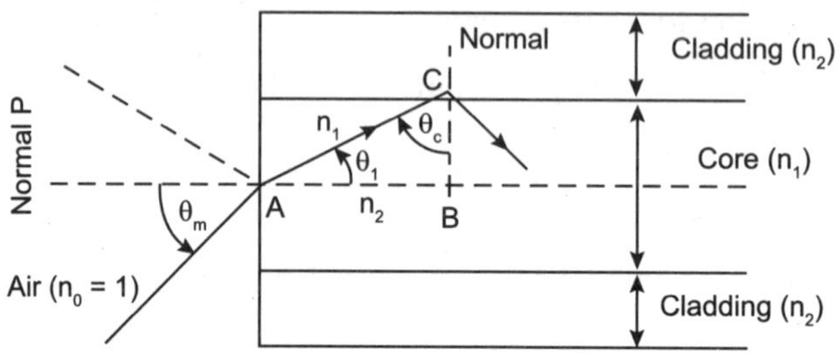

Figure 14.6 Ray propagation into an optical fibre cable

At the interface of core and cladding, consider point C (Figure 14.6) for the critical angle θ_c

$$n_1 \sin\theta_c = n_2 \sin 90° = n_2$$

or

$$\sin\theta_c = \left(\frac{n_2}{n_1}\right)$$

Therefore, in ΔABC, $\sin\theta_c = \dfrac{\text{AB}}{\text{AC}} = \dfrac{n_2}{n_1}$

Assume $AB = n_2$ (in magnitude), then AC $= n_1$ (magnitude). Using Pythagoras theorem in ΔABC, we get

$$BC = \sqrt{AC^2 - AB^2} = \sqrt{n_1^2 - n_2^2}$$

Now, consider point A at the air-core interface on axis of the cable, (with reference to normal P)

$$n_o \sin\theta_{\text{in}} = n_1 \sin\theta_1$$

since $n_0 = 1$ (for air)

or,

$$\sin\theta_{in} = n_1 \sin\theta_1 = n_1 \frac{BC}{AC} \text{ (from } \Delta ABC)$$

Putting values of BC and AC, we get

$$\sin\theta_{\text{in}} = n_1 \frac{\sqrt{n_1^2 - n_2^2}}{n_1}$$

$$\sin\theta_{\text{in}} = \sqrt{n_1^2 - n_2^2} \qquad (3)$$

or

$$\theta_{\text{in}} = \sin^{-1}\sqrt{n_1^2 - n_2^2}$$

θ_{in} is known as **acceptance angle** or **acceptance cone half angle**, which defines the maximum angle in which external light rays couple at air-fibre interface and travel down the fibre with a response maximum 10 dB down the peak value. Rotating the acceptance angle around the fibre axis, we get the acceptance cone of the fibre as shown in Figure 14.7.

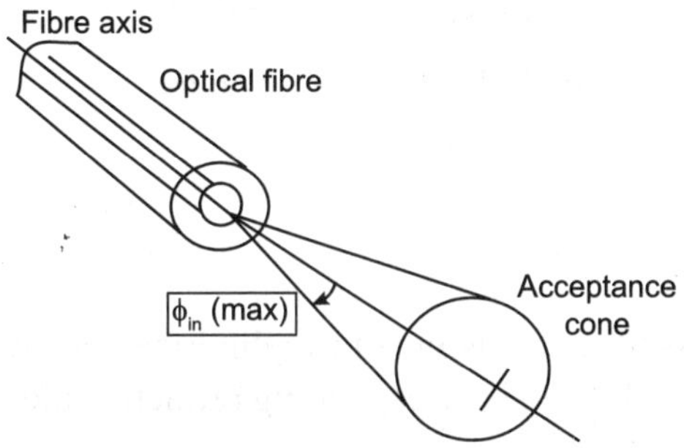

Figure 14.7 Acceptance cone of a fibre cable

14.2.3 Numerical Aperture (NA)

This is the parameter which determines the quantity of light coupled from the source to the fibre. NA is related with θ_{in} (acceptance angle or acceptance cone half angle) as follows:

From Eqn. (3),

$$\sin\theta_{in} = \sqrt{n_1^2 - n_2^2} = \text{NA}$$

or
$$\text{NA} = \sqrt{n_1^2 - n_2^2} = n_1\sqrt{2\Delta} \tag{4}$$

where the parameter D is called index difference

and
$$\theta_{in} = \sin^{-1}(\text{NA}) \tag{5}$$

Thus, NA represents a figure of merit used to find the light gathering capability of a fibre. Higher the NA, higher is θ_{in} and higher the amount of light coupled to the fibre. Obviously, NA is a measure of the ability of the fibre to accept light for transmission. NA is effectively depending only on the refractive indices of the core and cladding of materials and is not a function of fibre dimension.

NA is a dimension-less quantity lying typically between 0.14 and 0.50. The maximum value of NA for step index fibre occurs for air cladding, i.e. without any solid cladding, which can be determined from Figure 14.8

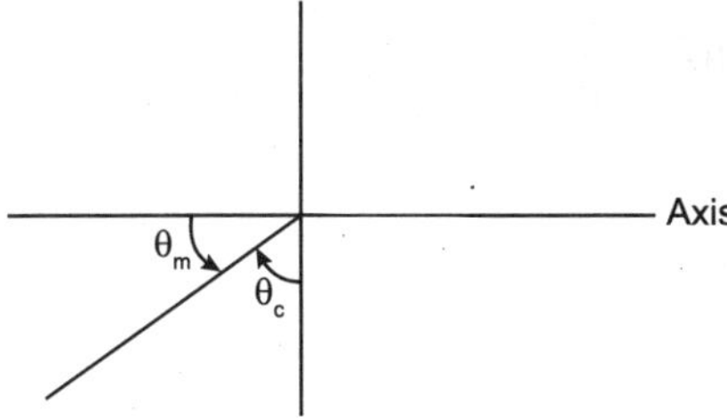

Figure 14.8 Acceptance angle for air cladding

For air cladding:

Acceptance angle, $\theta_m = 90 - \theta_c$

$$\text{NA} = \sin\theta_{in} = \sin(90 - \theta_c) \tag{6}$$

Acceptance angle θ_{in} for grade index fibre is given as

$$\text{NA} = \sin\theta_c \tag{7}$$

14.3 Optical Fibre

Figure 14.9(a) shows a glass fibre consisting of a (cylindrical) central core cladded by a material of slightly lower refractive index. The corresponding refractive index distribution (in transverse direction) is given by

$$\begin{aligned} n(r) &= n_1 \quad 0 < r < a \quad \text{core} \\ &= n_2 \quad r < a \quad \text{cladding} \end{aligned} \tag{8}$$

and is as shown in Figure 14.9(b). Obviously, n_1 is the refractive index of core of a fibre, and n_2 is that of cladding ($n_1 > n_2$). A ray of light *OA* incident on one end-face of the core is refracted into the core along *AB*. If n_0 is the refractive index of air, we have from Snell's law

$$\frac{\sin i}{\sin\theta} = \frac{n_1}{n_0} \tag{9}$$

The ray *AB* is incident on the core-cladding interface at angle $\phi = (90 - \theta)$. We can see that as *i* decreases, θ also decreases so that ϕ increases. When this ray has to suffer total internal reflection at core-cladding interface,

$$\sin\phi\ (= \cos\theta) > \frac{n_2}{n_1} \tag{10}$$

Thus,

$$\sin\theta < \left[1 - \left(\frac{n_2}{n_1}\right)^2\right]^{\frac{1}{2}} \tag{11}$$

and one obtains

$$\sin i_m < \frac{n_1}{n_0}\left[1 - \left(\frac{n_2}{n_1}\right)^2\right]^{\frac{1}{2}} = \left[\frac{(n_1^2 - n_2^2)}{n_0^2}\right]^{1/2}$$

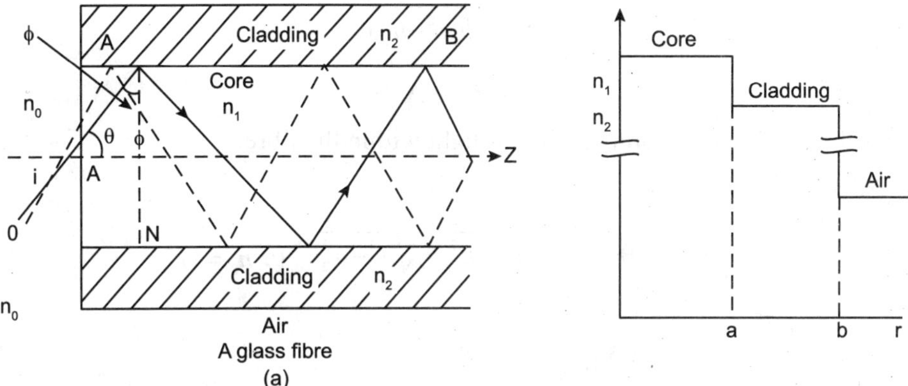

Figure 14.9 (a) Refraction of a ray of light into core, (b) distribution of refractive index of a cladded optical fibre which consists of a cylindrical glass structure surrounded by a material of slightly lower refractive index

If $(n_1^2 - n_2^2) \geq n_0^2$, then for all values of i, total internal reflection will occur. Assuming $n_0 = 1$, one obtains the maximum value of sin i for a ray to be guided as

$$\begin{aligned} \sin i_m &= (n_1^2 - n_2^2)^{1/2} \quad \text{when} \quad n_1^2 < n_2^2 + 1 \\ &= 1 \quad \text{when} \quad n_1^2 > n_2^2 + 1 \end{aligned} \tag{12}$$

Obviously, if a core of light is incident on one end of the fibre, it will be guided through it provided the semi-angle of the core is less than i_m. This angle is a measure of light gathering power of the fibre. The core of the semi vertical angle θ_m is termed as **acceptance core** and angle θ_m.

$$\theta_m = \sin^{-1}\left(n_2^1 - n_2^2\right)^{1/2}$$

is called the **acceptance angle** or the **acceptance cone half angle**.

Acceptance angle is defined as the maximum angle that a ray can have relative to the axis of the fibre so that it may propagate down the fibre. It may also be defined as the maximum angle from the fibre axis at which light may enter the fibre so that it will propagate in the core through total internal reflection. We may note that:

(i) The larger is the diameter of the core, the larger is the acceptance angle.

(ii) The large the difference in the refractive indices of the core and cladding, the larger will be the acceptance angle.

14.3.1 Condition for Propagation

If i is the angle of incidence of an incident ray, then the ray will be able to propagate, if

$$i < i_m$$

or

$$\sin i < \sin i_m$$

or
$$\sin i < \sqrt{(n_1^2 - n_2^2)}$$

or
$$\sin i < \text{NA} \tag{13}$$

This is the condition for propagation of light within the fibre.

The NA is defined as

$$\text{NA} = \sin\theta_m = \frac{1}{n_0}\sqrt{n_1^2 - n_2^2} = \sqrt{n_1^2 - n_2^2} \quad (\because n_0 = 1) \tag{14}$$

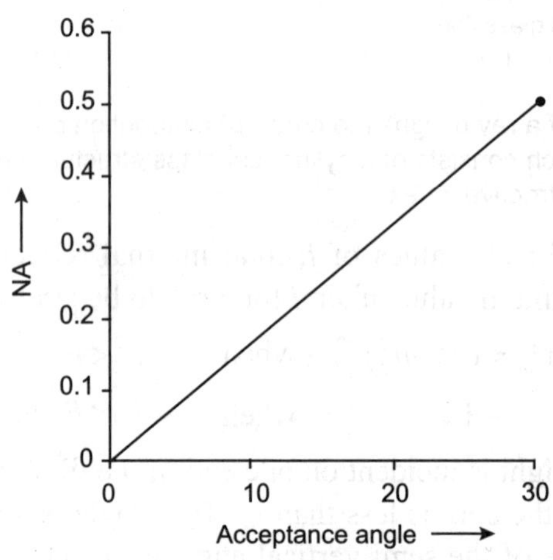

Figure 14.10 Variation of NA with acceptance angle

Numerical aperture (NA) determines the light gathering ability of the fibre. So, it is a measure of the amount of light that can be accepted by the fibre. Figure 14.10 shows the variation of NA with acceptance angle.

It is obvious from the figure that for fibres used in short distance communication. NA ranges from 0.4 to 0.5, whereas for long distance communications NA ranges from 0.1 to 0.3.

Usually $(n_1 - n_2)$ is not large and, therefore, one can have

$$\left(\text{writing } n_1 - n_2 = n_1\left(1 - \frac{n_2}{n_1}\right) = n_1\sqrt{2\Delta}\right)$$

$$\text{NA} = n_1\sqrt{2\Delta} \tag{15}$$

where

$$\Delta = \frac{\delta_n}{n_1} = \frac{n_1^2 - n_2^2}{2n_1^2} = \frac{(n_1 - n_2)(n_1 + n_2)}{2n_1^2} = \frac{n_1 - n_2}{n_1} = \frac{(\text{NA})^2}{2n_1^2} \tag{16}$$

The quantity Δ is usually referred to as the fractional difference of refractive indices of the core and cladding. Δ is also called *relative refractive index difference* when $n_1 \neq n_2$. Typically, Δ is of the order of 0.01. For $n_1 = 1.48$ and $n_2 = 1.46$, one obtains NA ≈ 0.24. This gives $i_m \approx 14°$. Obviously, acceptance angle i_m is less than 30°, so that NA < 0.5. We can easily see that if NA is too small, it is difficult to launch power into the fibre. The range of NA is 0.13 – 0.50. Larger the value of NA means more is the amount of light accepted by the fibre from the source.

We may note that NA is effectively dependent only on refractive indices of core and cladding material. Moreover, NA is not a function of fibre dimension.

To prevent energy losses via absorption and scattering, the cladding should be at least a few wavelengths thick. One can see that a typical value for n_2/n_1 is 0.99. For this value, one finds the critical angle as 0.142 radius or 8.11°. Thus, rays travelling at an angle less than 8.11° relative to the reference axis will be totally internally reflected and guided by the fibre. Higher angle rays will enter the cladding and be lost due to high levels of scattering and absorption.

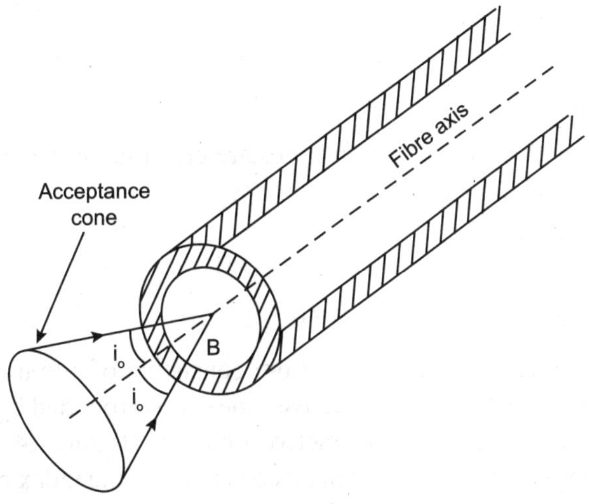

Figure 14.11

We may note that dielectric cladding on glass core reduces scattering loss, protects core from absorbing external optical disturbances and provides mechanical strength to main core glass fibre. Sometimes, there is buffer coating over cladding which adds further strength to main fibre and protects it from mechanical vibrations and impact.

If ray *AB* is rotated around the fibre axis keeping i_0 same, it describes a conical surface as shown in Figure 14.11. Now only those rays that are funneled into the fibre within this cone having a full angle $2i_0$ will only be total internally reflected and thus confined within the fibre for propagation, i.e., only rays within the cone are accepted. Therefore, the cone is called acceptance cone. Light incident at an angle beyond i_m will be refracted though the cladding and the corresponding optical energy is lost.

14.3.2 Modes in Optical Fibre Waveguides

Like metallic waveguides, propagating light along the fibre can be described in terms of a set of guided electromagnetic waves called **modes**. Each guided mode is a pattern of electric and magnetic field lines that repeat along the fibre at intervals equal to wavelength. Only certain discrete number of modes can propagate along the guide. In fibre optics terminology, mode simply means path through which light waves can propagate down the fibre.

14.4 Classification of Fibres

Optical fibres are classified into two major categories based on

(i) number of modes and
(ii) refractive index

On the basis of number of modes of propagation, optical fibres are classified into two types:

(i) Single mode fibre (SMF)
(ii) Multimode fibre (MMF)

On the basis of refractive index, optical fibres are classified into two types:

(i) Stepped index optical fibre
(ii) Graded index optical fibre

14.4.1 Stepped Index Multimode Fibre

Stepped index is the simplest type of optical fibre consisting of a thin cylindrical structure of transparent glassy material of uniform refractive index n_1 surrounded by a cladding of another material of uniform but slightly lower refractive index n_2 (Figure 14.9(a)). These fibres are referred to as stepped index fibres due to step discontinuity of the index profile at core-cladding interface. Light rays incident on the end-face of the core at angles of incidence within the acceptance angle are guided along the core in different paths. It is observed that light rays of different colours of frequencies are also separated as they travel inside the core. This indicates that multimode propagation is possible in the fibre as shown in Figure 14.12

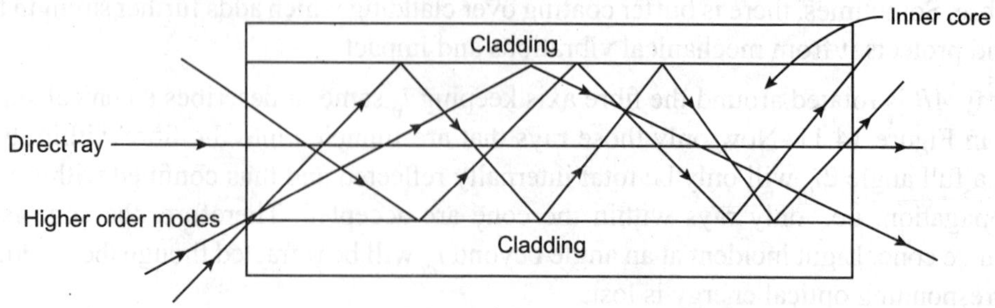

Figure 14.12 Multimode propagation in a fibre

From Figure 14.12, we see that time taken by different light rays propagating along the paths of different lengths upto the destination is different. This reveals that variations of successive pulses of light may overlap, thus distorting the information being carried. Such distortions are usually termed as **transit-time dispersions**.

One can show that time taken by a ray is a function of angle θ made by the ray with z-axis which leads to pulse dispersion and given by (Figure 14.9)

$$\tau = \frac{n_1 L}{c\cos\theta} \tag{17}$$

where $L = AB$. If we assume that all rays lying between 0 and θ_c (critical angle) are present, then time taken by the rays corresponding to $\theta = \theta_c \approx \cos^{-1}\left(\frac{n_2}{n_1}\right)$ would be given by

$$\tau_{\min} = \frac{n_1 L}{c} \text{ and } \tau_{\max} = \frac{n_1^2 L}{n_2 c} \tag{18}$$

Obviously, if all the input rays were excited simultaneously, the rays would occupy a time interval at the output end of duration.

$$\Delta\tau = \tau_{\max} - \tau_{\min} = \frac{n_1 L}{c}\left(\frac{n_1}{n_2} - 1\right) \tag{19}$$

For a typical fibre, one can assume $n_2 = 1.46$ (pure silica).

$$\Delta = \frac{n_1 - n_2}{n_1} \Rightarrow n_1 \approx 1.4746$$

One obtains

$$\Delta\tau \cong 50 \times 10^{-9}\text{ s/km}$$

$$\cong 50\text{ ns/km}$$

This reveals that an impulse after traversing through the fibre of length 1 km broadens to a pulse dispersion. For this purpose, two alternative solutions exist, one involving the use of graded index fibres and the other involving single mode fibres.

The advantages of step-index multimode fibre are:

(i) They are relatively easy to manufacture.

(ii) They are cheaper than other types of optical fibres.

(iii) They have large core of diameter (50–100 mμ and higher).

(iv) They have longer life than laser diodes.

The disadvantages are:

(i) They have lower bandwidth.

(ii) They have higher dispersion.

(iii) They have smearing of single pulse.

14.4.2 Stepped Index Monomode Fibre

One can make core very thin, i.e. core diameter is almost of the same order as wavelength for the light to be propagated. This helps in minimizing transit time dispersion. Figure 14.13 shows propagation in a stepped index monomode optical fibre. Step index monomode fibre has low attenuation, small numerical aperture (NA) and wide bandwidth.

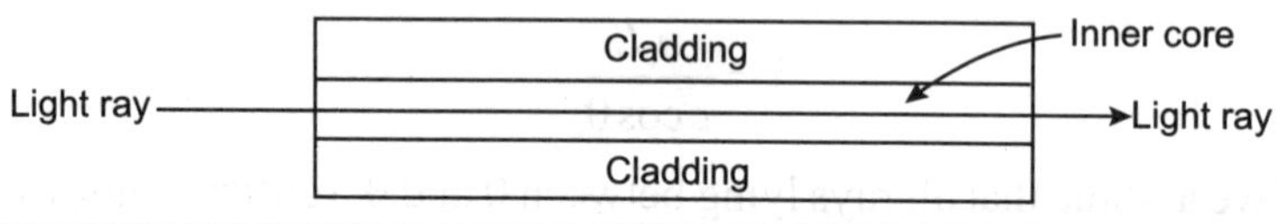

Figure 14.13 Propagation in stepped index monomode optical fibre

One can show that when normalization frequency, $V\left(=\dfrac{2\pi}{\lambda}a(\text{NA}), \text{ where } a \text{ is core radius}\right)$ <2.405, only the LP_{01} mode will propagate, i.e. the fibre becomes single mode fibre. Theoretically, LP_{01} mode will propagate no matter how small the V value is, i.e. how small the core radius a.

$$V=\frac{2\pi}{\lambda}a(\text{NA})$$

$$a=\frac{V\lambda}{2\pi(\text{NA})}$$

When $V<2.405$ then

$$a<\frac{2.405\lambda}{2\pi(\text{NA})}$$

or

$$a<\frac{2.405\lambda}{2\pi\sqrt{n_1^2-n_2^2}}$$

for single mode fibre.

In general, in single mode fibre, the cores tend to be only a few microns in diameter. However, it is advantageous if core can be made as large as possible. One can easily see that this may be achieved by reducing the NA value, i.e. by making core and cladding very close to each other. One of the major advantages of using single mode fibre is that intermodel dispersion no longer exists, and thus they are capable of carrying very large bandwidth signals over long distances.

14.4.3 Graded Index Multimode Optical Fibre

This fibre is quite cost effective and offers a less expensive method of overcoming the problem of transit time dispersion. It is observed that in such fibres, the refractive index of core falls of gradually from the axis towards the core-cladding interface. This means that the individual light rays are gradually refracted in the core itself. Figure 14.14 shows wave (ray) propagation at different angles of incidence, travelling different distances from the central axis prior to suffering reflections to recross the central axis.

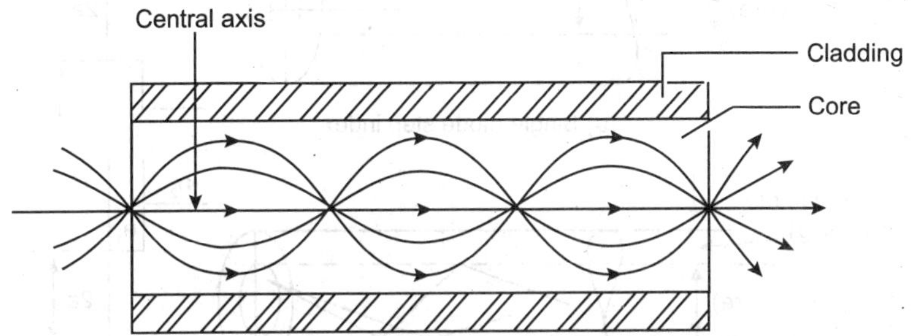

Figure 14.14 Propagation in a graded index multimode optical fibre

We may note that the number of guided modes is dependent upon physical parameters, i.e. refractive index difference between core and cladding given by NA $= \sqrt{n_1^2 - n_2^2}$, core radius of the fibre a and wavelength λ of transmitted light, included in the normalized frequently V for the fibre, where $V = (2\pi/\lambda)a$ (NA). One finds that the total number of guided modes for step index fibre is related to V by the following approximate expression:

$$M_s = \frac{V^2}{2}$$

From the figure, it is obvious that light rays with large angles of incidence transverse greater paths than those with smaller angles of incidence. However, the velocity of propagation increases with a drop of refractive index and thus all the rays recross the central axis at nearly the same instant of time. This helps in reducing transit time dispersion considerably.

Fibre optic systems use optical fibres to transmit information in the form of coded pulses or fragmented images. Over moderate distances they are used in telecommunications for which purpose they are becoming competitive with electric cables. In fibre communication, two bands are extensively used: 800 nm to 900 nm and 1200 nm to 1400 nm. The losses in fibres for the said two bands are extremely low. Lasers and LED's serve as light sources for fibres. Photodetectors are employed for processing the signal at the receiver end of optical fibre link. Photodetectors convert light signals into electrical waveforms. Optical fibres are also used in medical instruments (fibrescopes) to examine internal body cavities, such as stomach and bladder.

Obviously, graded index multimode fibre provides larger bandwidth as compared to multimode step index fibre but is relatively more complex to fabricate. The question of intermodal dispersion does not arise in single mode fibres (as it has only one mode) and hence they can provide higher bandwidth. Although, theoretically we can have four combinations, practically there are only three main classes of optical fibres as shown in Figure 14.15 (a, b, c).

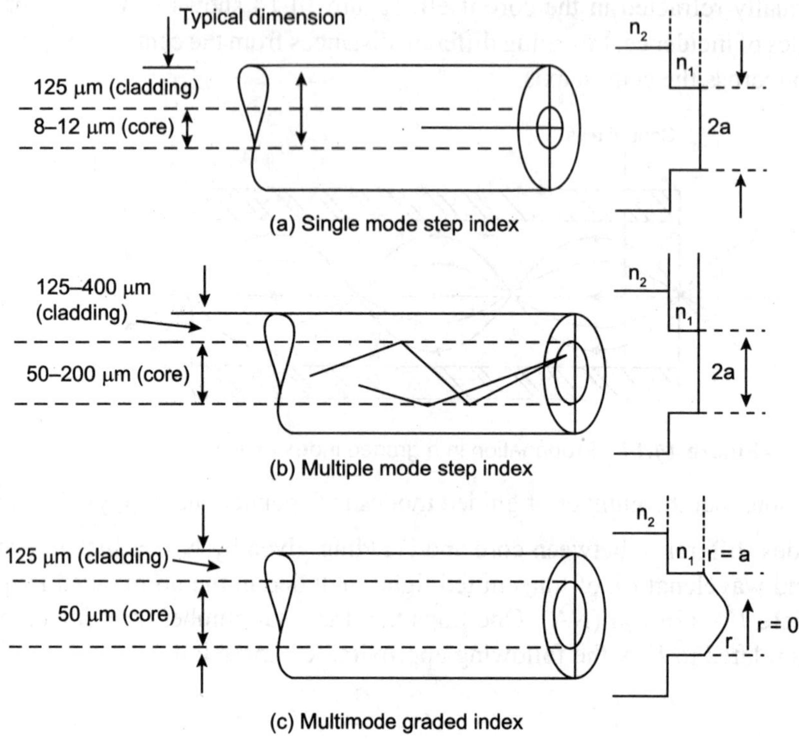

Figure 14.15 (a, b, c) Classification of optical fibres

The acceptance angle and NA in the graded index fibre decrease with radial distance from the axis. NA of the graded index multimode fibre is obtained as

$$\text{NA} = n_1 \sqrt{2\Delta\left[1 - \left(\frac{r}{a}\right)^2\right]}$$

where r is the varying radius and a is the total radius of the core.

Advantages of graded index multimode fibre are

(i) Dispersion is low

(ii) Bandwidth is greater

(iii) Easy to couple with optical sources

Disadvantages of graded index multimode fibre are

(i) Expensive

(ii) Very difficult to manufacture

A comparison of step index and graded index fibres are given in Table 14.2

Table 14.2

Parameter	Step index fibre	Graded index fibre
Data rate	Slow	Higher
Efficiency of coupling with fibre	Higher	Lower
Path of the ray	Light ray follow total internal reflection	Light ray travels in oscillatory fashion
Index variation	$\Delta = 1 - \frac{n_2}{n_1} = \frac{n_1 - n_2}{n_1}$	$\Delta = \frac{1}{2} - \frac{n_2^2}{2n_1^2} = \frac{n_1^2 - n_2^2}{2n_1^2}$
Material used	Usually glass or plastic is preferred	Only glass is preferred
Numerical aperature (NA)	Remain same	NA changes continuously with distance from fibre axis.
Pulse spreading	more (by fibre length)	less
Typical light source	LED	LED, Lasers
Attenuation of light	Less (typically 0.34 dB/km at 1.3 μm)	More (0.6 to 1 dB/km at 1.3 μm)
Applications	Local network communication (subscriber)	Local as well as wide area networks

V-number or V-parameter An optical fibre is characterized by yet another important parameter called V-number. This is more generally called *normalized frequency* or *cut off perameter* or V-parameter of the optical fibre and given by the relation

$$V = \frac{2\pi a}{\lambda}\sqrt{n_1^2 - n_2^2} \tag{20}$$

where a is the radius of core and λ is free space wavelength. Using

$$\text{NA} = \sqrt{n_1^2 - n_2^2} = n_1\sqrt{2\Delta}$$

one obtains

$$V = \frac{2\pi a}{\lambda}(\text{NA}) \tag{21}$$

$$V = \frac{2\pi a}{\lambda} n_1 \sqrt{2\Delta} \tag{22}$$

Equation (22) can also be expressed in terms of fractional refractive index change (Δ) as

$$V = n_1 \sqrt{2\Delta}\ ka \tag{22a}$$

where $k = 2\pi/\lambda$ is the radial wave vector.

The maximum bar of modes N_m supported by a stepped index (SI) fibre is determined by

$$N_m \cong \frac{1}{2} V^2 \tag{23}$$

Thus for $V = 10$, $N_m = 50$.

For $V < 2.405$, the optical fibre can support only one mode and is classified as SMF. MMFs can have values $V > 2.405$ and can support many modes simultaneously.

The wavelength corresponding to the value of $V = 2.405$ is known as *cutoff wavelength* (λ_c) of the optical fibre. We have

$$\lambda_c = \frac{\lambda V}{2.405} \tag{24}$$

From Eqn. (22), one can deduce that single mode properties can be realized by decreasing core diameter and or decreasing Δ such that $V > 2.405$. We may note that either a larger core and/or a larger Δ will result in multimode properties.

In case of GRIN fibres, for large values of V, we have

$$N_m \cong \frac{V^2}{4} \tag{25}$$

V-parameter depends only on (i) The wavelength of light propagating through fibre, and (ii) Characteristics of fibre.

The total number of modes can be obtained as

$$N = \int V\, dV = \frac{V^2}{2}$$

$$= \frac{1}{2}\left(\text{NA} \times \frac{2\pi a}{\lambda} \right)^2$$

We may note that mode refers to the number of paths for the light rays in the fibre cable. In single mode, light follows single path through core, whereas in multimode, the light follows several paths through the core.

14.4.4 Mode Calculation in Step Index Fibres

The total number of modes M entering the fibre depends on wavelength (λ), radius of fibre (r) and refractive indices (n_1, n_2) and is given by the relation.

$$M = \frac{2\pi^2 r^2}{\lambda^2}(n_1^2 - n_2^2) = \frac{1}{2}\left[\frac{\pi d}{\lambda}\text{NA}\right]^2 \tag{26}$$

Where d is the core diameter ($d = 2r$ or $2a$)

An important parameter connected with cut off conditions for fibre modes is normalized frequency V (or V-number or V-parameter) given by the relation

$$V^2 = 2M \tag{27}$$

A model is referred to as being cut off when it is no longer bound to the core of the fibre. V-number is a dimensionless number that determines how many modes a fibre can support. The percentage of power flow in cladding depends on M and hence V as given by the relation

$$\frac{P_{\text{clad}}}{P} = \frac{4}{3}M^{\frac{1}{2}} \tag{28}$$

14.4.5 Signal Degradation

When a light signal modulated by analogue or digital information propagates through a fibre, it is degraded at other end in the sense that power loss and shape distortions takes place in the signal. Loss or signal attenuation weakens the strength of light signal after travelling a certain distance. A repeater is required at a distance where the signal is not severely attenuated. Similarly, due to distortion, optical signal pulses (in digital communication) are broadened as they travel along the fibre. The excessive broadening may cause overlapping with neighbouring pulses, thereby creating error ('1' may be replaced by '0' or vice versa) in the receiver output. Obviously, signal degradation is mainly due to two reasons: (i) attenuation (loss) and (ii) distortion (broadening of pulses).

Fibre loss and distortion depends on the material used and structural defects during fibre manufacturing. Materials used in making the fibre are glasses and plastics. The most common optically transparent glasses from which fibre is made is silica which has refractive index $n_1 = 1.458$ at 850 nm. The cladding material (having slightly lower index n_2) is made by adding 'dopants' such as B_2O_3, GeO_2 or P_2O_5 in silica.

14.4.6 Fibre Losses (Attenuation)

The basic mechanisms responsible for fibre losses are:

1. **Absorption** Absorption depends on fibre material and wavelength. Glass introduces less absorption loss as compared to plastic. Absorption of energy may be caused by atomic defects, impurity atom or basic constituent atoms of fibre material. This loss includes ultraviolet absorption, infrared absorption and ion-resonance absorption. Absorption loss due to impurities is the major source of loss in optical fibre. There are two important types

of impurities: (i) the transition metal ions and hydroxyl (OH) ions. The transition metal impurities, i.e. Cu, Fe, Co, V, Cr, Ni and Mn absorb strongly in the region of interest. This means the fibre be free from these impurities in the electronic absorption. The trapped OH^- ions in the fibre material absorb at 0.95 μm, 1.23 μm and 1.39 μm. These regions corresponds to region of interest and hence the presence of OH^- ion should be minimized and as far as possible should be kept below 0.01 part per million (ppm). There are certain wavelength ranges at which attenuation is minimum. Such a range of wavelength is called as **optical window** or **transmission window**. These windows are quite suitable for transmission of information.

2. **Scattering losses** Light scattering (Rayleigh scattering) is caused by structural imperfection in the guided mode besides the fibre material. The order of losses introduced per kilometer due to scattering is 2.5 dB at 820 nm, 0.24 dB at 1300 nm, and 0.012 dB at 1550 nm. This reveals that loss reduces with increasing wavelength. This loss arises due to microscopic irregularities in material density, structural inhomogeneities or defects during manufacture process. Light is refracted by imperfections and some of it is spread out to escape through the cladding. For Rayleigh scattering $\propto \lambda^{-4}$, this sets a lower limit on wavelength that can be transmitted through a glass fibre to about 0.8 μm. Below this wavelength, scattering loss is quite high.
3. **Radiative losses** These originate from perturbations in fibre geometry, e.g. bends (microscopic or macroscopic) of finite radius. Bends can arise when cable turns a corner or when fibre is incorporated in cable.
4. **Coupling losses (splicing losses)** These losses occurs at junctions connecting source to fibre, fibre to fibre and fibre to photodetector. Losses at the junction of two fibre sections are caused mainly due to misalignment as shown in Figure 14.16. The figure shows four types of misalignment, namely lateral misalignment, gap displacement, angular misalignment and imperfact finish of surface.
5. **Loss due to Geometrical Effects** During manufacture and/or installation, bending (microscopic or macroscopic) causes loss of power. Due to small cracks in glass, microbending is caused. When the cracks extend to a large distance along the length of the optical fibre, microbending occurs. However, coupling can occur due to microbending but microbending obstructs more propagation.

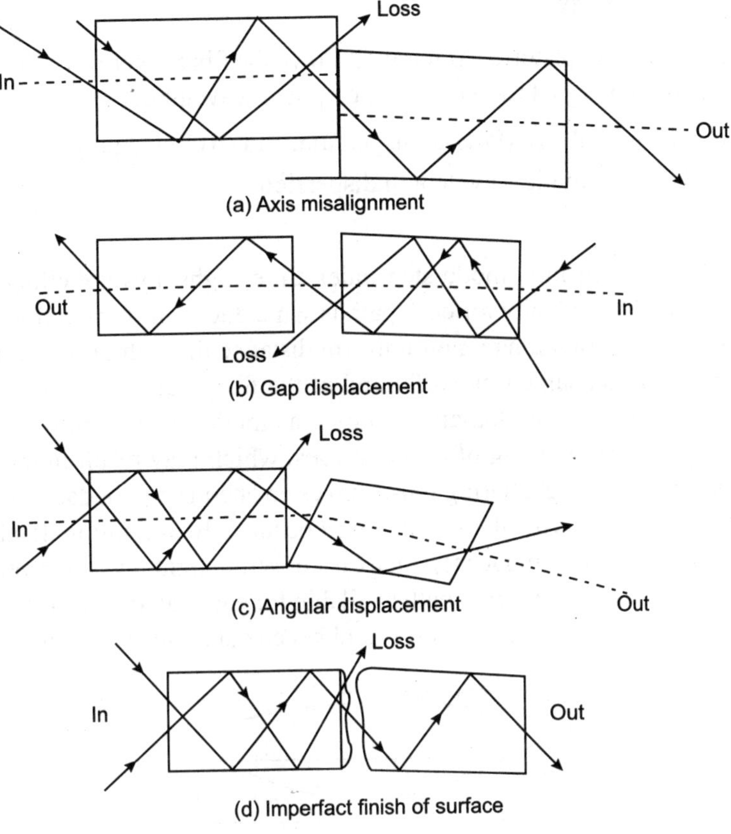

Figure 14.16 (a, b, c, d) Fibre coupling losses

14.4.7 Units for Measuring Fibre Loss

Fibre loss is defined as the ratio of output power P_{out} from a fibre of length L (in km) to optical input power P_{in}. The attenuation factor α is commonly expressed in decibels per kilometer, and denotes power loss per km of fibre or fibre attenuation, given as

$$\alpha = \frac{10\log_{10}\left(\frac{P_{out}}{P_{in}}\right)}{L} \tag{29}$$

In ideal case, $P_{in} = P_{out}$ and $\alpha = 0 \frac{\text{dB}}{\text{km}}$ or o Np/km. However, in practice, an attenuation upto 3 dB/km is considered within permissible limit. We may note that attenuation depends on the wavelength of the signal.

14.4.8 Signal Distortions

An optical pulse can also get distorted as it travels down the fibre. There are mainly three sources of distortion as the digital signal travels over the optical waveguide:

1. Modal or intermodal: (i) Material dispersion, and (ii) Waveguide dispersion
2. Intramodal or Chromatic wavelength dispersion
3. Intermodal delay

1. Modal dispersion In multimode step index fibre, light rays travelling down the fibre follow different paths. The velocity for each path is same due to uniform refractive index (n) of core material. Hence rays travelling a non-uniform distance through different paths with same velocity do not reach at the same time at far end of the fibre, causing a spread in the pulse of light energy. This spread in time domain is known as modal dispersion or modal spreading. Modal spreading causes broadening of digital pulses, which may result in the overlapping of pulses. The overlapping of neighbouring pulses gives erroneous output data due to intersymbol interference. Obviously, this type of spreading will occur only in multimode fibres and can be entirely eliminated using single mode step index fibre. While light rays are propagating down a single mode fibre, there is only a single path available for each ray due to small diameter of the fibre. There is no pulse stretching in this case, and hence data rate is tremendously high.

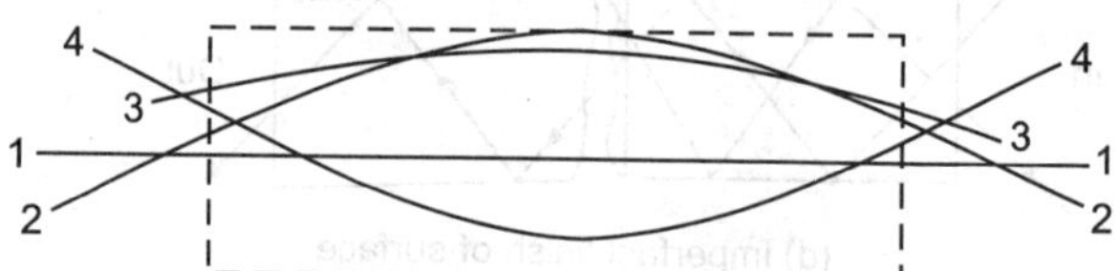

Figure 14.17 Propagation in graded index fibre

Figure 14.17 shows the propagation of light along multimode graded index fibre. In this case, different rays follow different paths, but with different velocity. As the ray moves towards cladding, its velocity increases (as n reduces). Therefore, rays travelling through longer paths move with greater velocity than rays travelling shorter paths near the axis of fibres. Hence all the rays reach at the far-end almost simultaneously. The velocity varies because the refractive index of fibre decreases with radial distance from the centre. Therefore, this fibre has much less pulse spreading as compared to multimode step index fibre, but of course, higher than single mode fibre where dispersion is almost nil.

2. Chromatic dispersion or wavelength dispersion This type of dispersion is pulse spreading that occurs in a fibre as a result of the fact that group velocity is a function of wavelength. This is because refractive index n of fibre material is wavelength dependent. Dependence on λ is often referred as **chromatic dispersion.** Its effect increases with the spectral width of the source. Spectral width means the band of wavelength over which the source emits light. For LEDs, the RMS spectral width is about 5% of central wavelength. Therefore, an LED source emitting peak value of light at central wavelength of 850 nm will have a typical value for spectral width of 40 nm, i.e. source emits most of its optical power in the 830 to 870 nm wavelength band as shown in Figure 14.18(a). Light rays in this band are emitted simultaneously from an LED, but their velocity is different. Hence, while propagating along the fibre they do not arrive at the other end at the same time causing pulse broadening or spreading as shown in Figure 14.18(b). This causes erroneous data and is known as chromatic distortion.

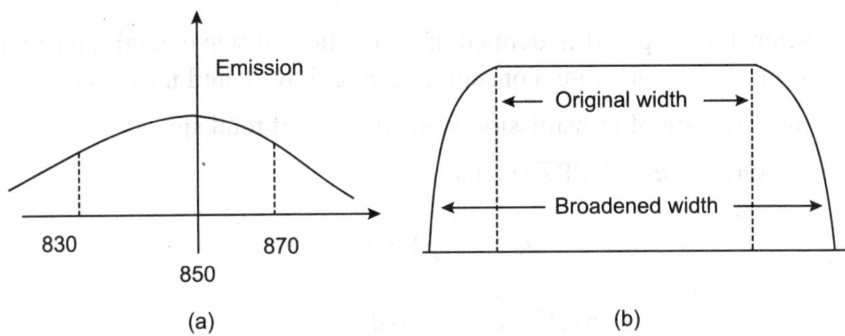

Figure 14.18 (a) Spectral width, (b) pulse spreading

The laser diode is a monochromatic source and has much narrower spectral width, typically 1 to 2 nm. Hence, chromatic dispersion is much less in case where a laser diode is used.

Causes of chromatic dispersion

Material dispersion This type of dispersion arises when the refractive index of core material is wavelength dependent. Due to this, group velocity of a given mode varies with λ causing dispersion of the pulse at the other end of the guide.

Waveguide dispersion This type of dispersion occurs because modal propagation constant β is a function of (a/λ), i.e. core radius a relative to wavelength λ.

3. Intermodal delay This delay occurs when each mode has a different group velocity at a given frequency. Group velocity is the velocity at which energy propagates in the guide. Intermodal delay also causes pulse spreading.

14.4.9 Information Capacity

Pulse spreading due to dispersion may cause overlapping between neighbouring pulses causing error at receiver end and this limits the information capacity of the fibre.

A measure of information of an optical waveguide is usually referred by bandwidth-distance product in MHz-km. For step index fibre, a typical value is 20 MHz-km, whereas in graded index profile this value is typically 2.5 GHz-km.

14.4.10 Pulse Spreading due to Dispersion

Dispersion D is defined as

$$D = \frac{1}{L}\frac{dt_g}{d\lambda} \tag{30}$$

where L is distance travelled by optical pulse and $\frac{1}{L}\frac{dt_g}{d\lambda}$ is variation of group delay in spectral width per kilometer. Pulse spread is defined as a function of wavelength and is measured in n sec per km per nm. Pulse spreading constant is denoted by Δt and measured in n s/km.

The maximum data rate of transmission is the inverse of total spread.

For unipolar return to zero (UPRZ) signal:

$$f_b = \frac{1}{\Delta t.L}\text{ Mb/s} \tag{31a}$$

For unipolar non-return to zero (UPNRZ) signal:

$$f_b = \frac{1}{2\Delta.L}\text{ Mb/s} \tag{31b}$$

where Δt is spreading constant in n s/km and L is the length of fibre in km over which light travels.

14.5 Collection of Light from Diffuse Source by Fibre

Let a small diffuse source be placed on fibre axis near one face as shown in Figure 14.19(a). Now, consider light radiated from diffuse source in the form of a cone with solid angle $d\omega$ in a direction θ w.r.t. fibre axis. Power radiated into small solid angle $d\omega$ in direction θ to normal of end face surface (fibre axis) is

$$dP_r = I(\theta)\, d\omega = I_0 \cos\theta\, d\omega \tag{32}$$

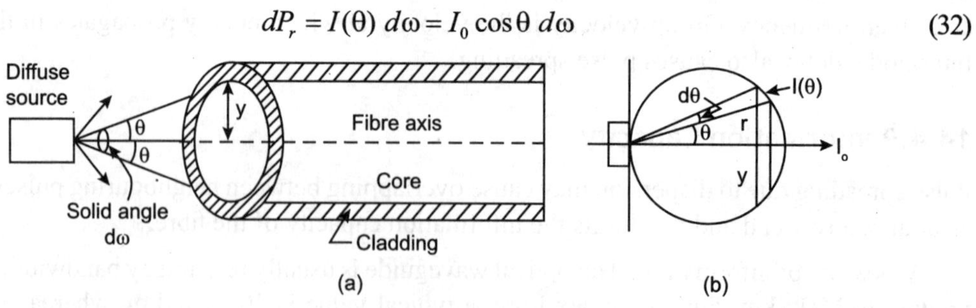

Figure 14.19

where I_0 is intensity of light along fibre axis.

Let us calculate $d\omega$. (see Figure 14.19 (b))

$$d\omega = \frac{\text{Surface area of angular ring}}{(\text{Radius}^2)} = \frac{(2\pi y)(rd\theta)}{r^2}$$

$$= \frac{(2\pi r \sin\theta)(rd\theta)}{r^2} = 2\pi \sin\theta \; d\theta \quad (33)$$

($\because y = r\sin\theta$ from figure)

Substituting the value of $d\omega$ from Eqn. (33) in Eqn. (32), one obtains

$$dP_r = I_0 \cos\theta \times 2\pi \sin\theta d\theta = 2\pi I_0 \sin\theta \cos\theta \; d\theta$$

$$= \pi I_0 \sin 2\theta \; d\theta \quad (\because \sin 2\theta = 2\sin\theta\cos\theta) \quad (34)$$

Total power radiated by diffuse source can be obtained by integrating Eqn. (34) within the limit of $\theta = 0$ to $\theta = \pi/2$. Hence

$$\int dP_r = \int_0^{\pi/2} \pi I_0 \sin 2\theta \; d\theta = \pi \; I_0 \int_0^{\pi/2} \sin 2\theta \; d\theta$$

or

$$P_r = \pi I_0 \left[\frac{-\cos 2\theta}{2}\right]_0^{\pi/2} = \pi I_0 \left[\frac{\cos 0 - \cos \pi}{2}\right]$$

$$= \pi I_0 \left[\frac{1-(-1)}{2}\right] = \pi I_0 \quad (35)$$

Suppose the diameter of core is greater than the diameter of diffuse source. Now power collected by the fibre through an angle $\theta = i_0$ (maximum value of angle of acceptance) is given by

$$P_c = \int_0^{i_0} \pi I_0 \sin 2\theta \; d\theta = \pi I_0 \left[\frac{-\cos 2\theta}{2}\right]_0^{i_0}$$

$$= \frac{\pi I_0}{2}[\cos 0 - \cos 2i_0] = \frac{\pi I_0}{2}[1 - \{1 - 2\sin^2 i_0\}]$$

$$\therefore \; P_c = \pi I_0 \sin^2 i_0 \quad (36)$$

Dividing Eqn. (36) by Eqn. (35), we have

$$\frac{P_C}{p_r} = \frac{\pi I_0 \sin^2 i_0}{\pi I_0} = \sin^2 i_0$$

$$= (\text{NA})^2 \; (\therefore \; \sin i_o = \text{NA})$$

$$\therefore \; P_C = P_r \, (\text{NA})^2 \quad (37)$$

So, to collect more rays, NA should be large.

14.5.1 Cut off Wave Length

We may note that single mode operation only occurs above a theoretical cutoff wavelength λ_c given by

$$\lambda_c = \frac{2\pi a n_1}{V_c}\sqrt{2\Delta} \tag{i}$$

Where V_c is cutoff normalized frequency. Obviously, λ_c is the wavelength above which a particular fibre becomes single moded. We have

$$\lambda = \frac{2\pi}{V} a n_1 \sqrt{2\Delta} \tag{ii}$$

Dividing (i) by (ii), we obtain the inverse relationship for the same fibre

$$\frac{\lambda_c}{\lambda} = \frac{V}{V_c} \tag{iii}$$

But for step index fibre where $V_c = 2.405$, one obtains the cutoff wavelength as

$$\lambda_c = \frac{V\lambda}{2.405} \tag{iv}$$

An effective cutoff wavelength has been defined which is obtained from a 2 m length of optical fibre containing a single 14 cm radius loop. Practical transmission systems are generally operated close to the effective cutoff wavelength in order to enhance the fundamental mode confinement, but sufficiently distant from cutoff so that no power is transmitted in the second-order LP_{11} mode.

14.5.2 Advantages of Optical Fibres

Fibre optic systems offer several advantages. Few of them are as follows:

(i) Optical fibres are light in weight and occupy less space than coaxial (bundles of twisted pair) cables.

(ii) Optical fibres are composed of dielectric materials and therefore these are totally immune to extraneous interfering electromagnetic signals.

(iii) There is virtually no signal leakage from optical fibres and hence cross-talks is not possible.

(iv) Attenuation of signals in optical fibres is much smaller than in coaxial cables or twisted pair.

(v) Optical fibres are immune to electromagnetic interference and do not pick up line currents. Obviously, these can be safely used in high voltage environments, e.g. power station, and can be laid alongside metallic power cables.

(vi) The basic raw material used in the fabrication of low loss fibres is silica, which is abundantly available in nature (although in impure form), whereas copper constitutes the basic raw material in coaxial cables.

(vii) Optical fibres can withstand environmental hazards in a better manner and also have longer life compared to copper cables.

(viii) Optical fibres are nonconductive and noninductive. This reveals that there is no interference with other communication systems.

(ix) The speed of transmission through optical fibres is very high because the signals are carried by light.

(x) The channel capacity of optical fibres is very large as compared to the audioband (5 to 10 Hz); the frequency of carrier light wave is very high ($\sim 10^{14}$ Hz). Obviously, one can accommodate a large number of speech signal channels in fibres whose bandwidth is larger than that of a wire transmission line.

14.5.3 Disadvantages of Optical Fibre Systems

(i) Optical fibre systems are virtually of less use unless they are connected to standard electronic facilities which often require expensive interfaces.

(ii) Optical fibres have significantly lower tensile strength than coaxial cables. This can be improved by coating the fibre with standard Kevlar and a protective jacket of PVC.

(iii) Occasionally it is necessary to provide electrical power to remote interface or regenerating equipment. Thus, additional metallic cables have to be included in the cable assembly.

(iv) Optical fibres require special tools to splice and repair cables and special test equipment to make routine measurements. Repairing fibre cables is also difficult and expensive and working on optical fibre cables also requires special skills and training.

14.6 Types of Rays

The plane containing the central axis of the fibre is called meridian plane and meridional rays remain in the meridian plane which can be divided into two categories (Figure 14.20(a,b)): (a) bound rays that propagate along axis and unbound rays that are not bound along axis and are refracted out of core and absorbed in cladding or buffer.

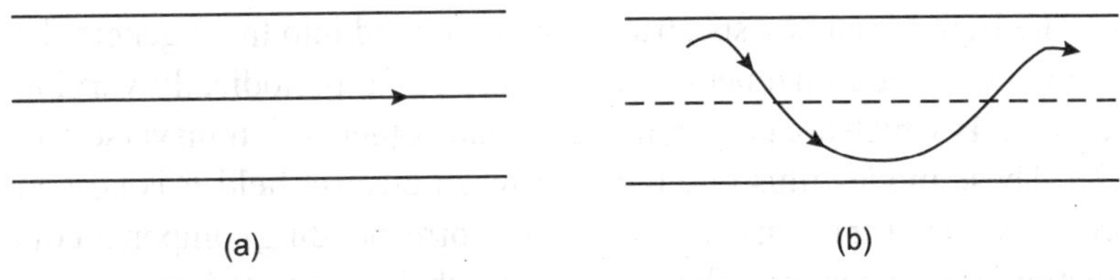

Figure 14.20 Meridional rays propagating through optical fibre cable

14.6.1 Skew Rays

These are rays that enter the optical fibre obliquely and not confined to a plane and follow a helical path around the axis (Figure 14.21). These rays are soon scattered out of core at bends and irregularities and thus do not contribute significantly in optical communication.

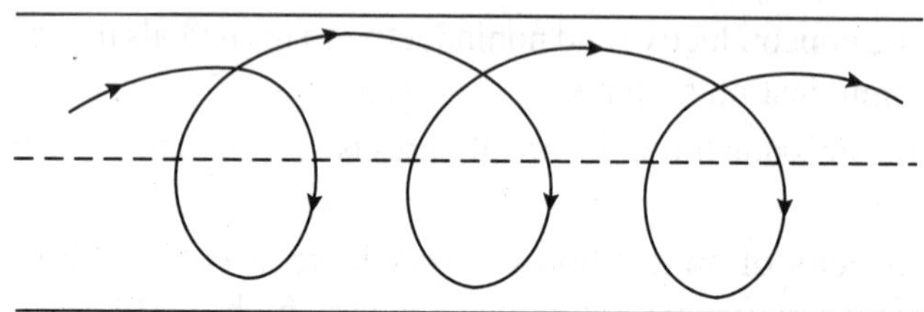

Figure 14.21 Skewed ray propagation through optical fibre cable

14.7 Modes of Propagation in Optical Fibre

Light propagation through straight and curve step index optical fibre with cladding is shown in Figure 14.22.

Here we have shown only one ray and, therefore, this is called ray diagram. However, in reality, an infinite number of rays (Figure 14.23 (a, b)) all slightly displaced from each other can be propagated at a time. At the time of propagation, they will follow different paths in the optical fibre (Figure 14.23).

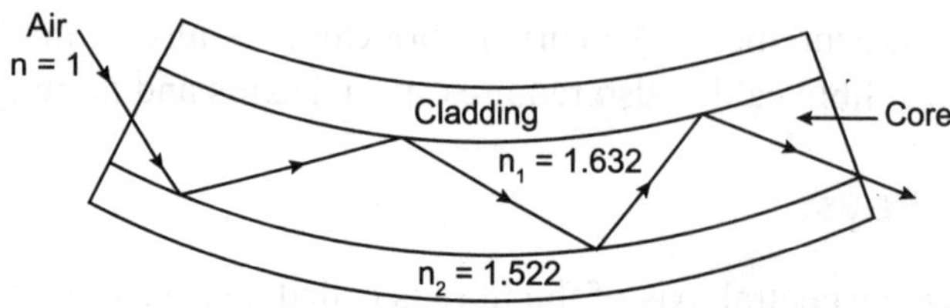

Figure 14.22 Propagation of light through straight and curve step index optical fibre with cladding

We know that light can be described as an electromagnetic wave. It can be seen that the frequency spectrum of electromagnetic waves extends from subsonic (a few *Hz*) to cosmic rays (10^{22} *Hz*). The light frequency spectrum can be divided into three general bands: infrared, visible and ultraviolet. Electromagnetic waves consist of a periodically varying electric field **E** and magnetic field **H** which are perpendicular to each other. The transverse modes are shown in Figure 14.23. These modes illustrate the case when electric field E is perpendicular to the direction of propagation (Z) and hence $E_z = 0$, but the corresponding component of magnetic field H is in the direction of propagation. These modes are therefore called transverse electric (TE_m) modes. Similarly, when a component of electric field E is in the direction of propagation, but

$H_z = 0$, the modes so formed are called transverse magnetic (TM_m) modes. We have incorporated mode number m into this nomenclature by referring to TE_m or TM_m modes. We may note that each value of m is associated with a distinct wave pattern or mode as shown in Figure 14.24; within the waveguide they follow different paths. We also find that for each value of m there is a value of $\gamma(\gamma_m)$, γ_m can take values in the range γ_c to $\pi/2$, which implies that the smaller the value of γ_m, i.e. γ_m closer to γ_c, higher the mode number and vice versa. We can see from Figure 14.24 that higher order modes carry along with them a higher portion of mode energy within cladding. This part of the wave that is carried in cladding is generally referred to as "evanescent wave". For free space communication, the total electromagnetic wave field lies in the transverse plane, where $E_z = 0$ and $H_z = 0$. This is called *transverse electromagnetic wave* (TEM). However, TEM wave occurs in metallic conductors like coaxial cable but not in optical waveguides. For planar waveguide, generally $TE(E_z = 0)$ and $TM(H_z = 0)$ modes are obtained. In optical fibres the cylindrical waveguide is bounded in two dimensions, rather than one as in plane waveguide. Obviously two integers l and m are necessary in order to specify the modes, in contrast to a single integer (m) required for planar guide. Thus, for cylindrical waveguide we refer TE_{lm} and TM_{lm} modes. TE_{lm} and TM_{lm} modes correspond to meridional rays (see Figure 14.20 a, b) which are travelling within the optical fibre. However, we may note that hybrid modes where E_z and H_z are non-zero also occur within the cylindrical waveguide.

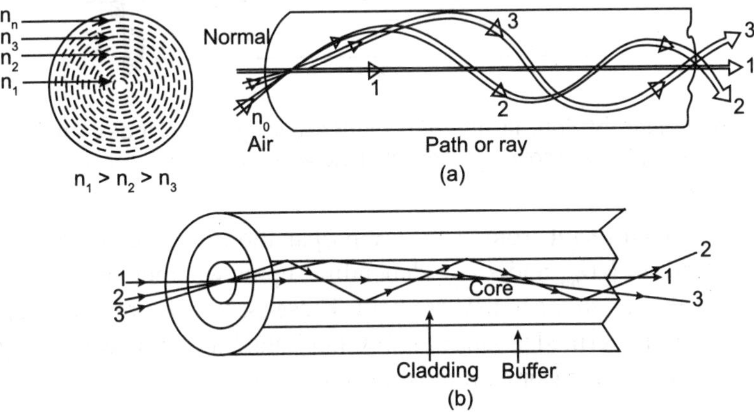

Figure 14.23 Propagation of light in multimode graded index and step index fibre

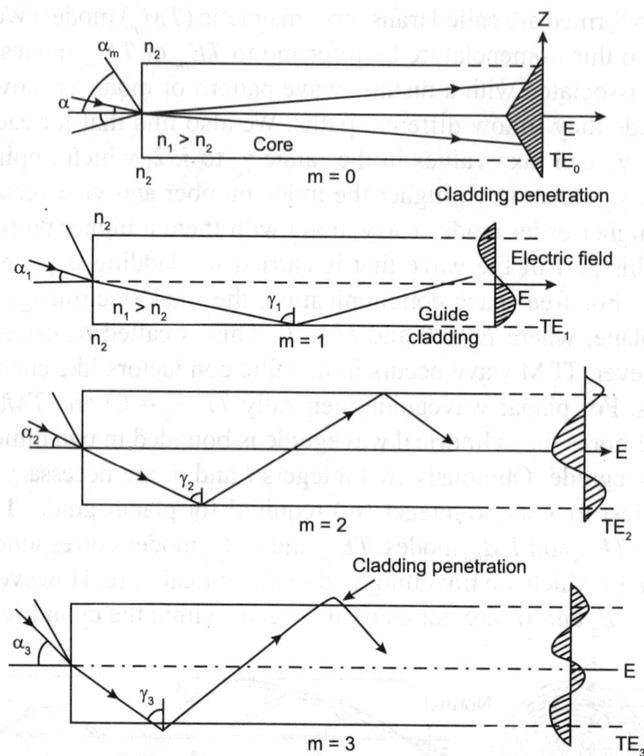

Figure 14.24 Propagation of light in different modes through step index optical fibre, where α_m = acceptance angle, r_c = critical angle and $\pi/2 > r_1 > r_2 > r_3 > \ldots r_c$, whereas $\alpha_0 < \alpha_1 < \alpha_2 < \alpha_3 \ldots < \alpha_m$.

These modes, which result from skew ray propagation (Figure 14.25) within the optical fibre, are designated as HE_{lm} and EH_{lm} depending upon whether the components of H or E make larger contribution to the transverse field. These skew rays are known as *leaky rays* and corresponding modes are termed as *leaky modes,* not bound modes because they can propagate only by appreciable distance prior being lost from the optical fibre.

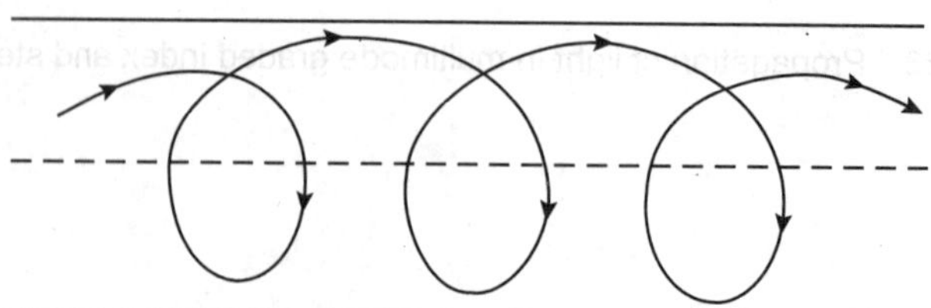

Figure 14.25 Propagation of skew rays through optical fibre cable

In most practical waveguides, it is found that the refractive indices of core and cladding differ from each other by only a few per cent and in that case the full set of modes (i.e. *EH*, *HE* and *TE*) can be easily approximated by a single set called linearly polarized (LP_{lm}) **modes**. An LP_{lm} mode in general has m field maxima along a radius vector and $2l$ field maxima **a round** a circumference (Figure 14.26).

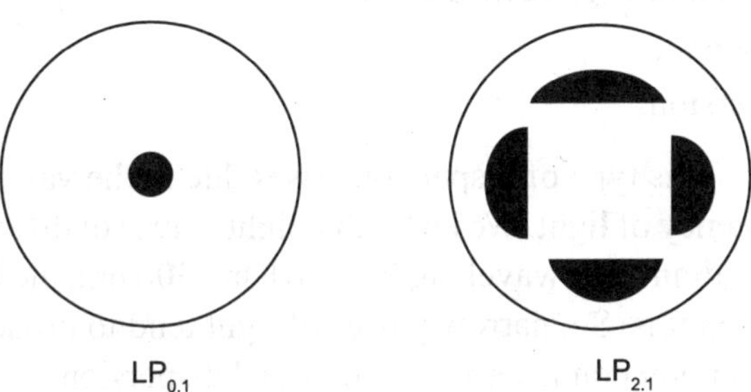

Figure 14.26 Linearly polarised (LP_{tm}) modes

14.8 Dispersion and Dispersion Losses

14.8.1 Dispersion

When data is being sent along the communication fibre, information is generally contained in the form of pulses of intensity of light. Figure 14.27 shows the input and output pulses across a fibre. It is observed that the output pulse is wider than the input pulse, and moreover the amplitude of output pulse also decreases. This shows that the pulse becomes distorted as it propagates through the fibre. The reason of this distortion is dispersion. The term dispersion describes the pulse broadening effect in optical fibres. Let us consider an input pulse of width W_1 and output pulse of width W_2, where $W_2 > W_1$. Then fibre dispersion D is defined as

$$D = \sqrt{(W_2^2 - W_1^2)}$$

Figure 14.27

Dispersion is measured in the unit of time either in nanoseconds (10^{-9} s) or picoseconds (10^{-12} s). Total dispersion produced by the fibre depends directly on its length. So, dispersion is typically measured in nanoseconds per kilometre (ns/km).

14.8.2 Dispersion Losses

Following are the three mechanisms which cause distortion to the light pulse in optical fibres:

1. Material dispersion or chromatic dispersion,
2. Waveguide dispersion
3. Intermodal dispersion

Material dispersion This type of dispersion arises due to the variation of refractive index with wavelength or frequency of light. We know that light waves of different wavelengths travel at different speeds in a medium. For wavelengths less than 1300 nm, the long wavelengths travel faster than shorter wavelengths. So, narrow pulses of light tend to broaden as they travel down the optical fibre. This phenomenon is termed as material dispersion.

However, above 1300 nm, the short wavelengths travel faster than longer wavelengths. At 1300 nm, the difference in speed is minimised and hence this wavelength is often called **zero dispersion wavelength**. Material dispersion is very low in single mode fibre due to axial light propagation path. This can be reduced using a more monochromatic source in true sense.

Waveguide dispersion This type of dispersion arises from the guiding properties of optical fibre. The effective refractive index for any mode varies with wavelength. This causes pulse spreading similar to variation in refractive index. This is known as waveguide dispersion. Waveguide dispersion is present in both single and multimode optical fibres. It should be remembered that higher the frequency bandwidth of transmitted pulse, the higher the waveguide dispersion.

Intermodal dispersion We have seen that light wave propagates through a fibre in different modes. Each mode in a multimode fibre travels at a slightly different velocity. So, there is difference in propagation time for different modes. This causes the pulse to spread in time as it travels along the fibre. This is known as intermodal dispersion. This kind of dispersion is completely absent in single mode optical fibres. It also does not depend on spectral width of the source. The pulse broadening is entirely due to intermodal dispersion. However, single mode fibres exhibit the least pulse broadening.

(a) For multimode stepped index fibre When the ray of light of different modes follows different paths for example, *AA′* propagates along the optical fibre axis and some (say CC′ or BB′) propagate along the trajectories of oblique rays (Figure 14.28) there will be difference in time taken for different modes to travel a distance *l*. We know that refractive index (RI) of a medium is simply a measure of velocity of propagation v of light in the medium. Thus, we can write

$$\text{RI}(n) - \frac{v_a}{v_m} \quad \text{or} \quad v_m = \frac{v_a}{n} = \frac{c}{n}$$

where $c(= v_a)$ is the velocity in air and v_m is the velocity of light in medium.

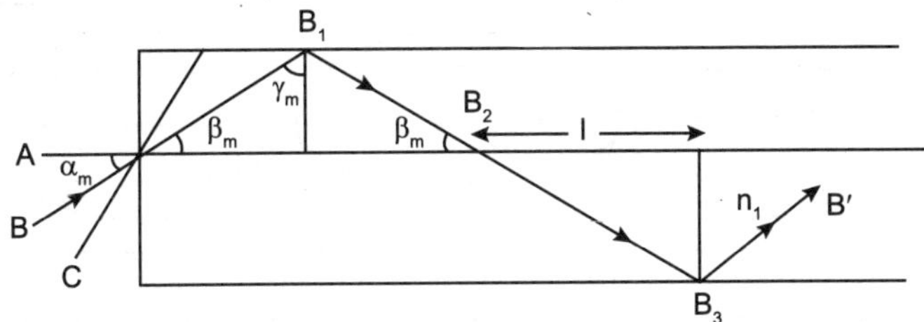

Figure 14.28 Multipath time dispersion for multimode step index optical fibre

The axial ray AA' travels an axial distance l along the optical fibre in time t_1, whereas the oblique ray BB' ravels the same axial distance l along the path B_2B_3 in the time

$$t_2 = \frac{n_1}{c}\frac{1}{\cos\beta_m} = \frac{n_1}{c}\frac{1}{\cos(90-\gamma_c)} \qquad \left(\because \sin\gamma_c = \frac{n_2}{n_1}\right)$$

$$= \frac{n_1}{c}\frac{1}{\sin\gamma_c} = \frac{n_1^2 l}{n_2 c}$$

The above results reveal that if two rays (AA' and BB') are launched together they will be separated on arrival at the other end by ΔT, where

$$\Delta T = t_2 - t_1 = \frac{n_1 l}{c}\left(\frac{n_1}{n_2} - 1\right)$$

$$= \frac{n_1 l}{n_2}\cdot\frac{l}{c}\Delta n \quad (\text{where } \Delta n = n_1 - n_2)$$

Obviously, a pulse containing rays at all possible angles spreads out during propagation by an amount given by

$$\frac{\Delta T}{l} = \frac{n_1}{n_2}\cdot\frac{\Delta n}{c}$$

This is called **multipath time dispersion** of step index fibre and this causes pulse broadening.

(b) For graded index multimode fibre Different modes follow different paths sinusoidally. We may note that rays making larger angles (Figure 14.29) traverse longer path but they do so in lower refractive index area and hence they travel with higher speed of propagation. However, refractive index changes gradually in graded index optical fibre, so there is a self-focusing mechanism that leads to a smaller value of pulse dispersion than that of step index optical fibre. This shows that for long distance communication, graded index optical fibre will be more useful than step index optical fibre which causes less time dispersion.

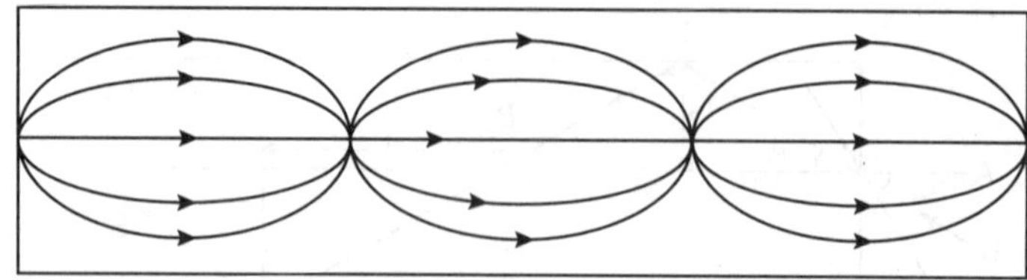

Figure 14.29 Less multiple time dispersion for graded index optical fibre

14.9 Losses in Optical Fibre Cable

Losses in optical fibre, i.e. attenuation, are mainly caused due absorption and scattering of rays. We may note that

(i) Absorption of rays is basically a material property and it occurs at all wavelengths when electronic transitions take place within the material and are followed by non-radiative relaxation processes. This causes an increase of thermal energy in the material.

(ii) Scattering is also partly a material property, but in optical fibres it is caused by imperfections in their structures. Scattering occurs when the mode of propagation of light is changed, such that some of the optical energy leaves the optical fibre (Figure 14.30).

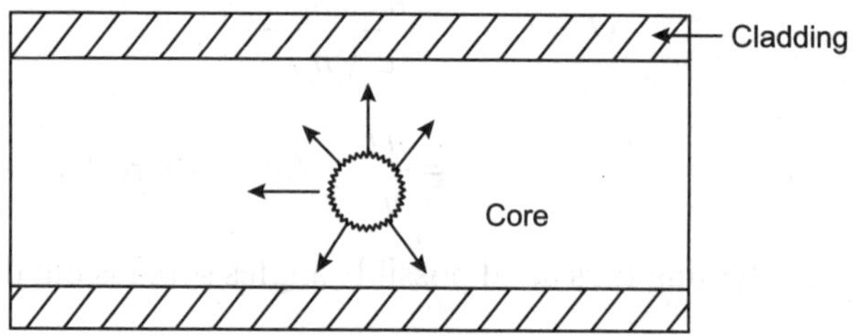

Figure 14.30

(iii) **Losses due to tight bending of fibres** Tight bends of the optical fibre cause some of the light not to be internally reflected but to propagate into the cladding and be lost (Figure 14.18).

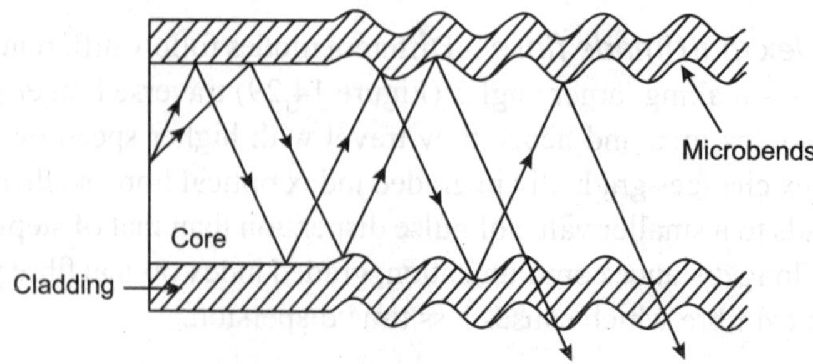

Figure 14.31

(iv) **Loss due to imperfections in fibre** The presence of imperfections in optical fibre material causes power losses from the optical fibre cable.

(v) **Losses due to connectors and splices** Power losses in long distance communication cables are caused due to connectors and splices which are required to connect two optical fibres temporarily.

Table 14.3 summarizes some standard optical fibres.

Table 14.3 Standard optical fibres

Optical fibre (type)	Cladding diameter (μm)	Core diameter (μm)	Δ	Applications
Single mode (8/125)	125	8	0.1% to 0.2%	(i) Long distance (ii) High data rate
Multimode (50/125)	125	50	1% to 2%	(i) Short distance (ii) Low data rate
Multimode (62.5/125)	125	62.5	1% to 2%	LAN
Multimode (100/140)	140	100	1% to 2%	LAN

14.10 Optical Fibre Communication System

Figure 14.32 shows a simplified block diagram of an optical fibre communication link. The three primary building blocks of the link are the transmitter, the receiver, and the fibre guide.

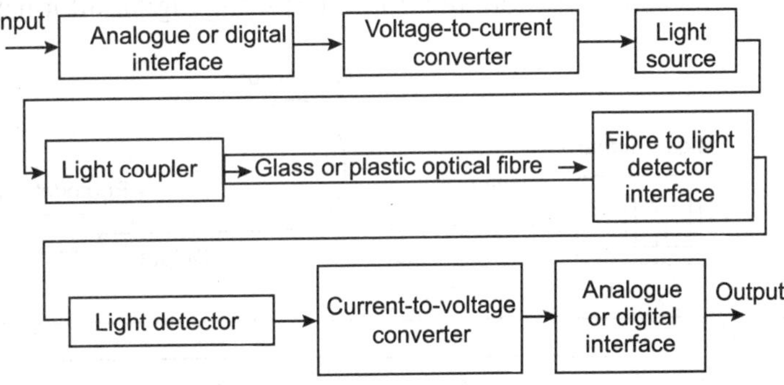

Figure 14.32 Simplified schematic diagram of fibre optic communication link

The transmitter consists of an analogue or digital interface, a voltage-to-current converter, a light source, and a source-to-fibre light coupler. The fibre guide is either an ultrapure glass or plastic cable. The receiver includes a fibre-to-light detector coupling device, a photodetector, a current-to-voltage converter, an amplifier and an analogue or digital interface. In an optical fibre transmitter, the light source can be modulated by a digital or analogue signal. For analogue modulation, the input interface matches impedances and limits input signal amplitude. For digital modulation, the original source may already be in digital form or, if in analogue form, it must be converted to a digital pulse stream. For the latter case, an analogue-to-digital converter must be included in the interface.

The voltage-to-current converter serves as an electrical interface between the input circuitry and light source. The light source is either a light emitting diode (LED) or an injection laser diode (ILD). The amount of light emitted by either an LED or ILD is proportional to the amount of drive current. Obviously, the voltage-to-current converter converts an input signal voltage-to-current that is used to drive the light source.

The source-to-optical fibre coupler (such as a lens) is a mechanical interface. Its function is to couple light emitted by the source into the optical fibre cable. The fibre-to-light detector coupling device is also a mechanical coupler. Its function is to couple as much light as possible from the fibre cable into the light detector.

The light detector is very often either a PIN (P-type-intrinsic-N-type) diode or an APD (avalanche photodiode). Both APD and PIN diodes convert light energy into current. Consequently, a current-to-voltage converter is required. The current-to-voltage converter transforms detector current to output (signal) voltage.

The analogue-to-digital interface at the receiver output is also an electrical interface. If analogue modulation is used, the interface matches impedances and signal levels to the output circuitry. If digital modulation is used, the interface must include a digital-to-analogue converter.

An optical communication system is quite similar in concept to an electronic (microwave) communication system (10^9 Hz), except that the frequency range is different (10^{15} Hz) and optical fibre is used instead of metallic fibre.

Figure 14.33 shows an optical fibre link for transmitting digital information.

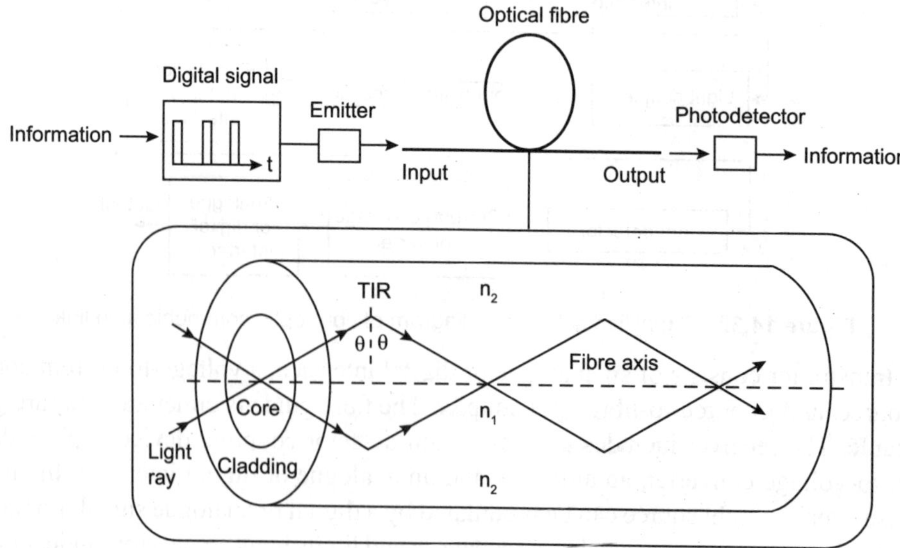

Figure 14.33 Optical fibre link for transmitting digital information

Optical fibre communication has the following advantages:

(i) This system has far more information carrying capacity.

(ii) Smaller size and weight of the system makes optical fibre quite suitable in space and aeronautical applications.

(iii) The raw material of optical fibre, i.e. glass (silica), is available abundantly on earth.

(iv) The per unit length cable cost of optical fibre is lower in comparison to its metallic counterpart.

(v) Optical fibres (plastic or glass) are insulators. No electric current flows through them, due to either transmitted signal or internal radiation striking the optical fibre. Moreover, light wave within the fibre is trapped, so nothing leaks out during transmission or interfere with signals in other optical fibres.

(vi) Optical fibres will not pick up or propagate electromagnetic pulses caused by nuclear explosion.

(vii) Light cannot couple into the optical fibre from its sides. Obviously, an optical fibre is well protected from interference and coupling.

(viii) A wire communication link could short circuit with high voltage lines falling across them, causing considerable damage. Optical fibres are free from such problems.

(ix) Optical fibres do not radiate energy within them and hence it is quite difficult for an intruder to detect the signal being transmitted. This is why optical fibres offer high degree of security and privacy.

(x) Optical fibers can withstand extreme temperatures prior to deteriorating. Glass fibres remain unaffected even at 800°C.

(xi) In comparison to copper, corrosion due to water or chemicals is less severe in optical fibres.

(xii) No equipment is needed for grounding optical fibres.

14.11 Photodetector

The receiving end of an optical communication system requires an optical detector. The function of the optical detector is to receive optical signal and convert it into electrical signal. This is achieved with the help of a photodetector which works on the principle of photoelectric effect. Some important photo detectors used in optical communication systems are photodiodes, PIN photodiode and avalanche photodiode.

The most common semiconductor photodetector is the photodiode. A photodetector detects the optical power incident upon it and converts the variation of this optical power into a correspondingly varying electric current. The most commonly used system in optical fibre communication is 1M/SS.

14.11.1 Photodiode Mechanism

The basic structure of an extrinsic P-N junction silicon photodiode is shown in Figure 14.34(a). In normal operation, a sufficiently large reverse-bias voltage is applied across the diode, so that depletion region (D.R.) as well as barrier potential will be large. Now, light incident on the reverse-biased P-N junction diode causes a photon ($h\upsilon$) to be absorbed. If the energy of photon is more than band gap energy (E_g), i.e. $h\upsilon > E_g$, then the absorbed photon excites an electron from valence band (VB) to conduction band (CB) and leaving a hole in VB. The electron-hole pair created due to incident light that is absorbed in the depletion region of photodiode is called photocarrier.

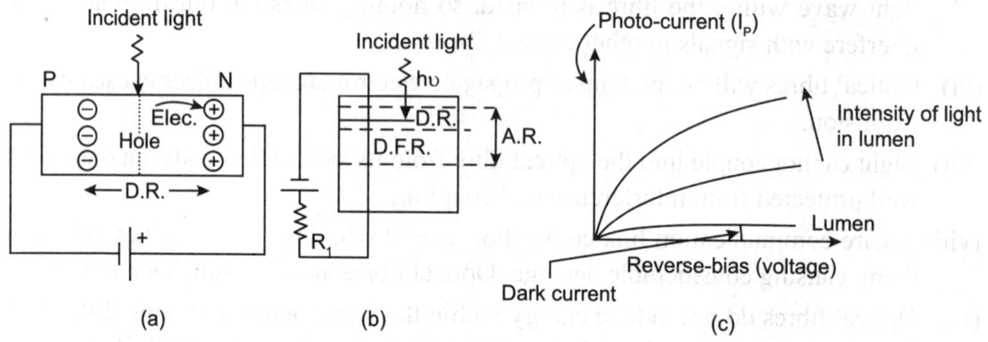

Figure 14.34 Mechanism of P-N junction silicon photodiode

Under reverse-bias voltage, electron and hole flow in the depletion region in opposite directions, some of the electron-hole pairs will recombine and hence lost. The distance they will traverse prior to their recombination is called **diffusion length** (L_n and L_p for electrons and holes, respectively), and the time it takes for an electron and hole to recombine is known as carrier life time (T_n and T_p). The relation between carrier life time and diffusion length can be expressed as

$$L = (D_n T_n)^{1/2}, \quad L = (D_p T_p)^{1/2}$$

where D_n and D_p represent electron and hole diffusion coefficients, respectively, which are in cm^2/sec.

We may note that photons can be absorbed in depletion region as well as diffusion region too, which is usually indicated as absorption region (AR) (Figure 14.34(b)). Obviously, this region becomes a void of free photocarriers. Now, within the depletion region only those pairs of electron-hole remain, which cannot recombine because of the existence of main electrostatic field due to depletion region voltage. Any photon-induced electron-hole pair there in will be separated and drawn in opposite directions by the combined effect of equivalent depletion region voltage as well as reverse bias voltage. We may note that polarities of these two voltages are in the same direction. This effect leads to a flow of photocurrent in the external circuit (Figure 14.34(b)), as photo carriers drift away across the depletion region. Figure 14.34(c) shows the typical output characteristics for reverse bias P-N photodiode. Different operating conditions are marked from no light to high level light. When there is no incident light, then also a small reverse current will be there. This is called *dark current*.

If electron-hole pairs are produced outside the absorption region, then there will be higher probability available and in that case no photocurrent will be available. Obviously, this means we have to make the top layer as thin as possible and effective depletion region as wide as possible to maximize the efficiency of photodiode as photodetector. This is termed as *quantum efficiency*. One can increase the quantum efficiency of photodiode if one goes from ordinary photodiode to PIN photodiode or avalanche photodiode (APD) or *reach through avalanche photodiode (RAPD)*.

Usually, two different structures for photodiodes are possible (Figure 14.35). Figure 14.35(a) shows a photodiode with front illuminated operation at 0.8 to 0.9 μm range. Figure 14.35(b) shows the side illuminated photodiode structure operating at higher wavelength (1.09 μm) range.

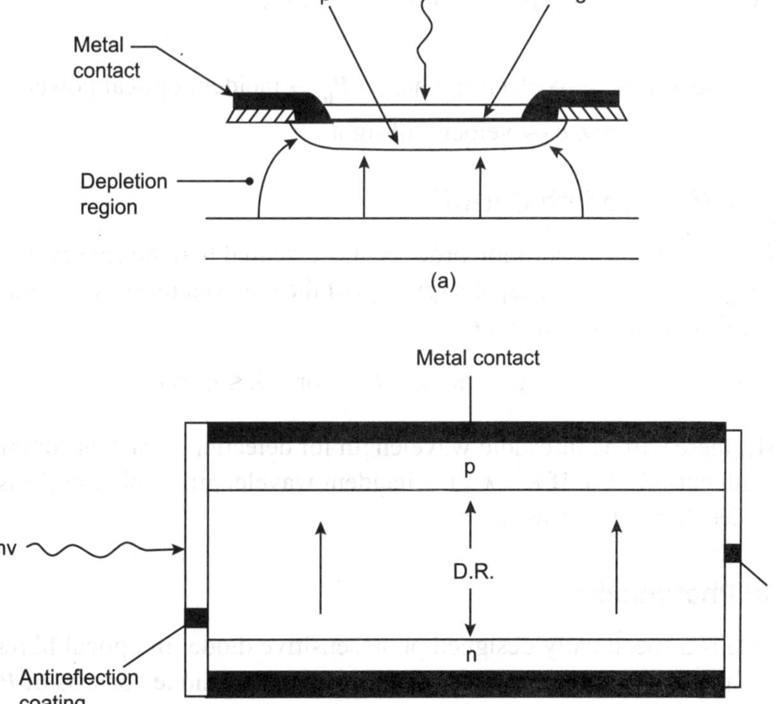

Figure 14.35 Two different possible structures of photodiode

14.11.2 Quantum Efficiency (η) and Responsivity (*R*)

For a photodetector η and *R* are two important characteristics. These two parameters depend on material band gap, operating wavelength, percentage of doping and thickness of P and N regions. In order to achieve high quantum efficiency, the depletion region must be thick enough to permit a large fraction of incident light to be absorbed. This is why PIN or APD photodiode is better than ordinary photodiode. One can define quantum efficiency and responsivity as follows:

$$\text{Quantum efficiency, } \eta = \frac{\text{Rate of generation of electron-hole pair (electron/sec.)}}{\text{Rate of photons incident (photons/sec.)}}$$

or

$$\eta = \frac{I_p / e}{P_0 / h\upsilon}$$

$$\text{Responsivity,} \qquad R = I_p / P_0 = \eta e / h\upsilon = \frac{\eta e\lambda}{hc} \text{ A/watt}$$

where $I_p \rightarrow$ photocurrent, $e \rightarrow$ electron charge, $P_0 \rightarrow$ incident optical power
$h\upsilon \rightarrow$ photon energy, $c \rightarrow$ velocity of light.

14.11.3 Long Wave Length Cut-off

When we consider intrinsic absorption process, it is essential that the energy of incident photon ($h\upsilon$) should be greater than band gap energy (E_g) of the material used as photodiode; otherwise electron-hole pair will not generate, i.e.

$$h\upsilon \geq E_g \quad \text{or} \quad hc/\lambda \geq E_g \quad \text{or} \quad \lambda \leq \lambda_c \leq hc/E_g$$

Obviously, there will be threshold wavelength for detection, which is commonly known as long wavelength cut-off (λ_c). If $\lambda > \lambda_c$, i.e. incident wavelength of photon (λ) is more than λ_c, then the photodetector will not work.

14.11.4 PIN Photodiode

PIN photodiode is a specifically designed photosensitive diode. In optical fibres systems, PIN diode is one of the most commonly used photodetector. PIN diode consists of *P* and *N* regions separated by a neutral layer of semiconductor known as intrinsic layer (*I*) (Figure 14.36). We may note that PIN diode is operated in reverse-bias mode with depletion region exposed to incident light.

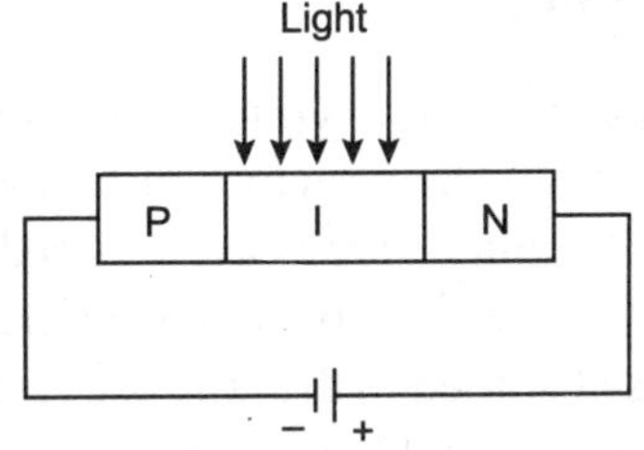

Figure 14.36

The intrinsic layer increases the size of depletion region. This leads to an increased light-sensitive area. The increased area allows the detector to pick up a wide range of incoming light. As a result, the amount of current increases with the amount of light absorbed. The P region of photodiode is made very thin so that it is transparent to the incoming light. On the other hand, the depletion region is made relatively thick so that it may absorb all the incident light. Photons incident on the depletion region produce electron-hole pairs (mobile charges). They are separated by high electric field present in the depletion region. Positive charges move towards P region while negative charges move towards N region. This gives rise to photocurrent to flow in the external circuit.

The performance of a PIN photodiode can be judged by two parameters: (i) quantum efficiency (η) and (ii) responsivity (R).

(i) Quantum efficiency (η) is defined as the number of electron-hole pairs generated by the number of incident photons, i.e.

$$\eta = \frac{\text{Number of electron-hole pair generated}}{\text{Number of incident photons}}$$

$$= \frac{i/e}{P/h\upsilon}$$

where, i = photocurrent developed, e = charge on electron

P = incident optical power, $h\upsilon$ = energy of incident photon

(ii) Responsivity (R) of a photodiode is defined as photocurrent generated per unit optical power, i.e.,

$$R = \frac{i}{P} = \frac{\eta e}{h\upsilon}$$

So, a PIN photodiode acts as a linear device, i.e.

$$i = RP$$

Obviously, photocurrent is directly proportional to optical power incident on the photodiode.

14.11.5 Avalanche Photodiode (APD)

This diode is another major type of optical communication detector having sophisticated structure to create more photocurrent. The structure of an avalanche photodiode is shown in Figure 14.37. It has four regions. N^+ and P^+ are heavily doped semiconductors and hence have very low resistances. The I region is very lightly doped and hence nearly intrinsic. Light enters the diode through P^+ region and is absorbed in the intrinsic region. Now, electron-hole pairs are generated which are separated by electric field in intrinsic region I. The photogenerated electrons drift to PN^+ junction through intrinsic region. Here high electric field exists. This high electric field region is also known as avalanche region or multiplication region. Here photo generated electrons are accelerated and collide with the bound electrons of valence band. They release more number of electrons as free or conduction electrons. This effect is known as **avalanche effect**. A single photogenerated electron can liberate 1100 free electrons in APD. This makes APD a highly sensitive detector.

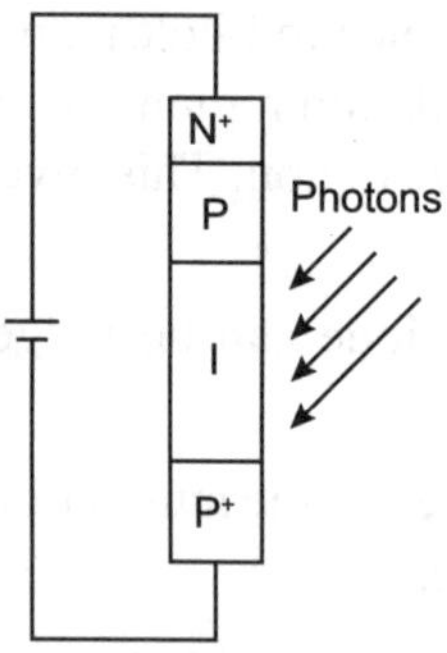

Figure 14.37 Avalanche photodiode

Avalanche multiplication M is given by

$$M = \frac{\text{Total (multiplied) output current diode } (I_m)}{\text{Primary photo current } (I_p)}$$

The value of M can be greater than 50.

Responsivity R is given by

$$R = \frac{\eta e}{h\upsilon} \times M$$

Due to avalanche multiplication, avalanche photodiodes have high sensitivity as well as responsivity.

The advantages of APD are as follows:

(i) APD have good sensitivity.

(ii) Current gain is very high (30 – 100).

(iii) Response time is very fast (2.5×10^{-10} s).

(iv) The output from APD is peaked at one of the better optical fibre wavelengths. This makes APD an ideal detector.

14.12 Connectors

This is a device by means of which an optical cable is usually connected to another optical cable or to peripheral equipment. Optical connectors may also be defined as components that allow direct transition of optical power from one core to another. Optical fibre connectors perform the same job as that of electric plugs and jacks. A good connector is one that connects fibre ends as close as possible. If fibre ends are not connected properly, light will escape from the junction to result in power loss.

A variety of optical fibre connectors have been developed and these are mainly consist of:

(i) a central alignment section, and

(ii) two identical insertion sections attached to two connecting fibre ends.

Optical fibre connectors are mainly of two types:

(a) butt joined connectors, and

(b) expanded beam connectors

Butt joined connectors may also be classified into three categories:

(a) basic ferrule connectors

(b) jewelled ferrule connectors, and

(c) ceramic capillary ferrule connectors

14.12.1 Basic Ferrule Connector

This is the simplest optical fibre connector. This is also known as concentric sleeve connector. Figure 14.38 shows the butt joined alignment type connector. It can be used in both multimode and single mode fibre. First of all, the coating of two fibre ends is removed for a desired length. Each fibre end is molded with a metal, ceramic or plastic called ferrule. The ferrule and fibre surfaces are grounded and polished. Then the two ferrules are brought into contact inside a precisely guiding cylindrical sleeve. The cylindrical sleeve allows fibre ends to be butt joined.

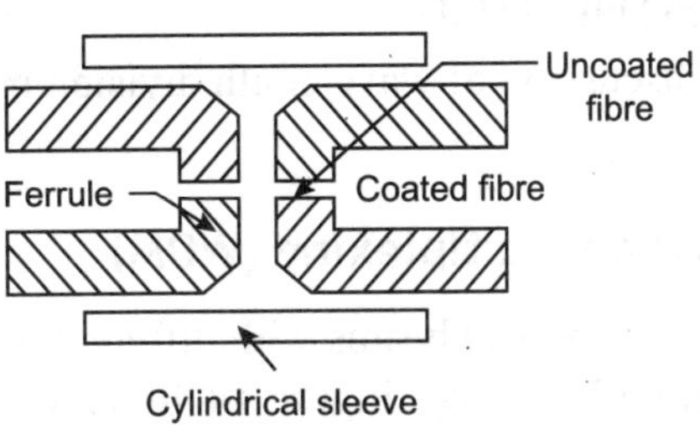

Figure 14.38 Ferrule connector

14.12.2 Expanded Beam Connector

An expanded beam connector (Figure 14.39) utilizes collimating lens at the end of transmitting fibre and focussing lens at the entrance end of receiving fibre in order to expand the beam between fibre ends. The use of lenses makes the achievement of lateral alignment much less critical than a butt joined connector. The collimating lens converts light from the fibre into a parallel beam of light (because fibre to lens distance is equal to the focal length of the lens). Further, the focussing lens converts the parallel beam of light into a focussed beam of light.

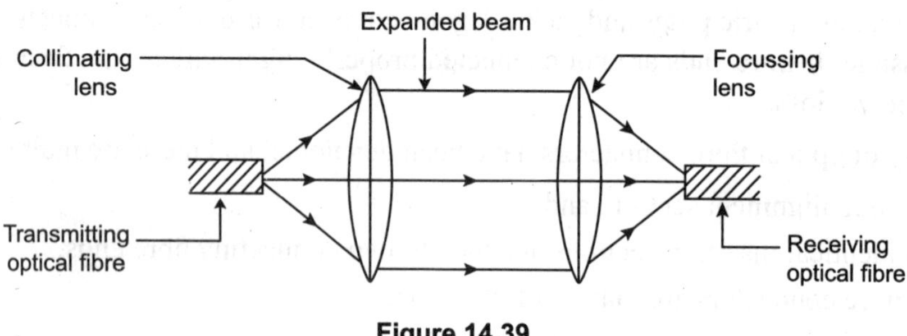

Figure 14.39

This beam of light falls on the core of receiving fibre. The lenses are antireflection-coated spherical microlenses. These connectors are useful for multifibre connection and edge connection for printed circuit boards where longitudinal and lateral alignment are difficult to achieve.

14.13 Multiplexers

These are devices to combine optical signals. Multiplexing is a process of simultaneous transmission of several messages over a single channel in a way that they do not interfere with each other. At the receiving end, individual signals can be extracted from the multiplexed signal by a process called demultiplexing. Telephone calls are multiplexed so that one fibre or wire can carry multiple conversations. Similarly, many video channels are multiplexed on a cable television network. There are three types of multiplexing:

1. Frequency division multiplexing (FDM)
2. Wavelength division multiplexing (WDM)
3. Time division multiplexing (TDM)

Here we shall restrict to discussion of wavelength division multiplexing and time division multiplexing.

14.13.1 Wavelength Division Multiplexing (WDM)

The simultaneous transmission of optical beams with different wavelengths over a single fibre is called wavelength division multiplexing. Figure 14.40 shows the basic principle of WDM technique. First of all, different wavelengths carrying separate signals are multiplexed with the help of a multiplexer. After this they are transmitted through a single fibre. Finally, the output of optical fibre is demultiplexed with the help of a demultiplexer to obtain different wavelengths. These different wavelengths are passed to separate receivers. From the receiver side also, signals can be transmitted in the same way. It is important to mention that a demultiplexer can also be used as a multiplexer.

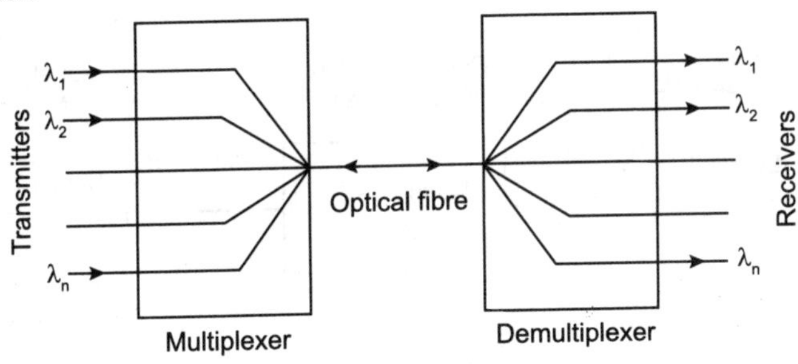

Figure 14.40

There are two types of wavelength division multiplexers:

(i) Angularly dispersive devices such as prisms or gratings.

(ii) Interference filter based devices such as multilayer thin film interference filters.

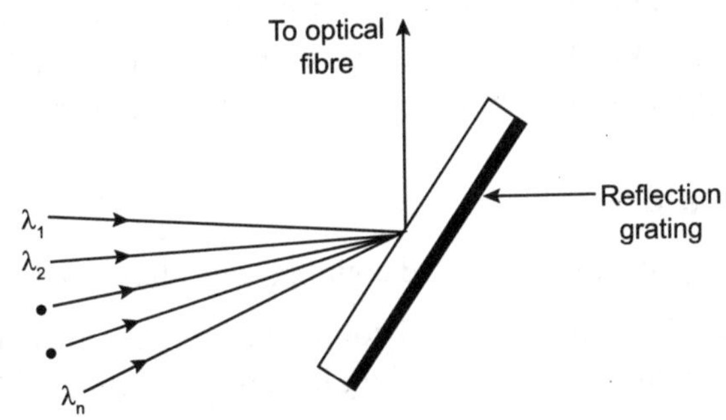

Figure 14.41

Let us consider the use of plane diffraction grating in wavelength division multiplexer (Figure 14.41). Different signals having different wavelengths $\lambda_1, \lambda_2, \ldots\lambda_n$ are collimated by a convex lens. Then they are allowed to incident upon a plane reflection grating. The reflected light is a composite light or multiplexed light. If the direction of light is changed, then the grating multiplexer can also be used as demultiplexer.

14.13.2 Time Division Multiplexing (TDM)

TDM technique is based on the principle of dividing time frame into slots with one slot for each message signal (Figure 14.42).

Three input signals are shown in the Figure (14.42). These signals are sequentially sampled at the transmitter with the help of a rotating switch known as commutator. The rotating switch extracts one sample from each input during each revolution. The output of the switch containing samples of input signals are periodically interlaced in time. At the receiver, individual signals are separated and distributed by another rotating switch. Commutators are electronic devices that are carefully synchronised. Finally, samples from each channel are filtered and then they produce original message signal.

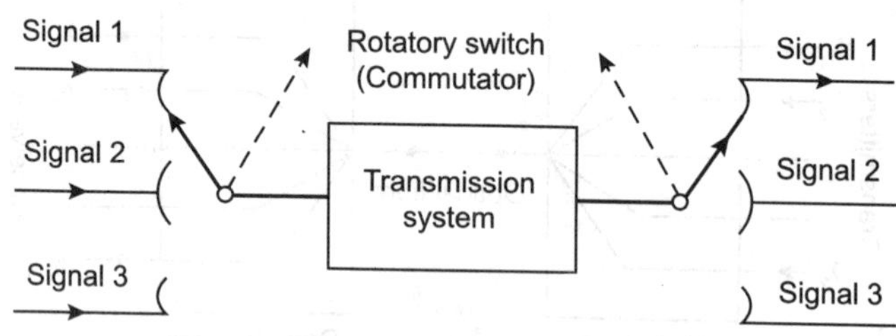

Figure 14.42

14.14 Couplers

In order to distribute optical signals from one to many or many to one, some special devices called as couplers are required. A coupler is a device that distributes light from a main fibre into one or more branch fibres. Couplers connect two or more fibres together without terminating the fibres. According to the physical mechanism involved, couplers can be classified into three main types: (i) diffusion type, (ii) area splitting type and (iii) beam splitting type.

14.14.1 Diffusion Couplers

Diffusion couplers use either evanescent wave coupling or radiative coupling.

Evanescent wave coupling In this type of coupling, the fibres are placed sufficiently close to each other for a finite length known as coupling length C_L (Figure 14.43).

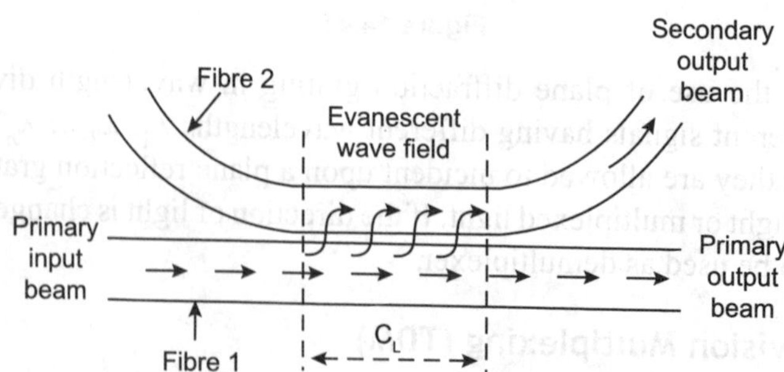

Figure 14.43 Coupler with evanescent wave field.

Here, evanescent field from the primary fibre builds up a propagation field in the secondary fibre to provide two outputs. This cannot readily be used in multimode fibres.

Radiative coupling This is suitable for multimode fibres. In this coupling, the power of input fibre is made to radiate towards output fibre. Thus, the bent fibres are coupled together through radiative coupling as shown in Figure 14.44.

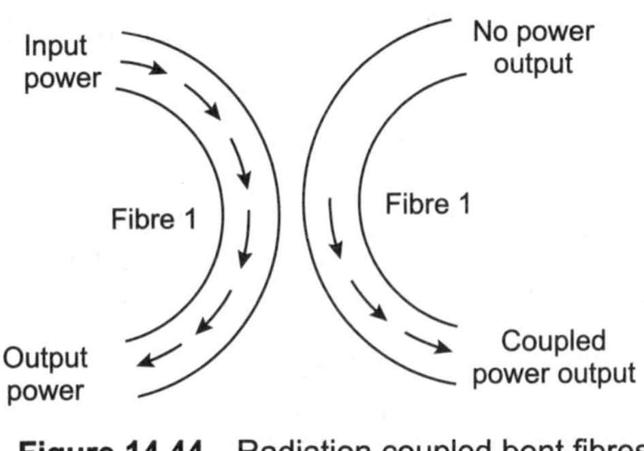

Figure 14.44 Radiation coupled bent fibres

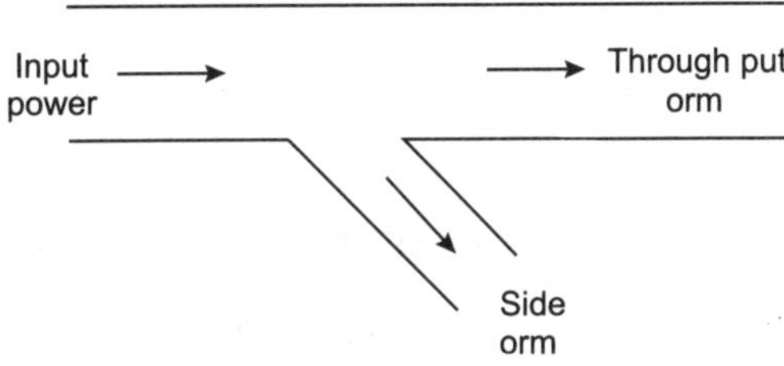

Figure 14.45 Twisted pair coupler

Another important example of radiative coupling is twisted pair coupler or fused biconical taper coupler (Figure 14.45). This coupler is made from a pair of fibres that are twisted. They are fused and elongated into biconical tapers. Light entering the coupler along one fibre core is radiated out of the core due to down taper. This power is trapped in the cladding. Some of the trapped radiation reenters the core of each output fibre in the up taper region.

14.14.2 Area Splitting Coupler

This coupler is shown in Figure 14.46. Single input fibre core is situated between the cores of two output fibres. Obviously, light from input fibre is coupled to output fibres. Here, power splitting is not equal but depends upon the surface area of two fibres.

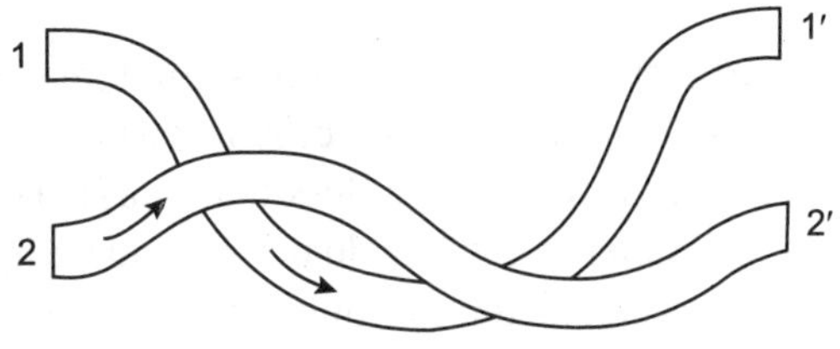

Figure 14.46 Area splitting coupler

14.14.3 Beam Splitting Coupler

This coupler is based on the principle of partial reflection of the main beam by means of a partially reflecting surface as shown in Figure 14.47. The partially reflecting surface can directly face incident beam and the face of a fibre polished at an angle or can be in the form of a separate partially reflecting mirror.

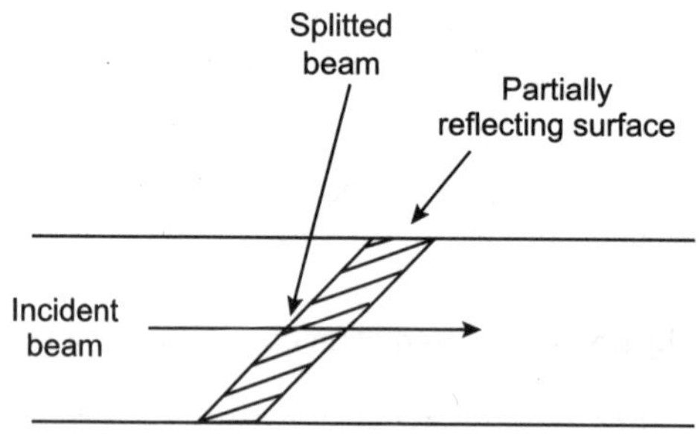

Figure 14.47 Principle of beam splitter coupler

14.15 Optical Fibre Cable Construction

A single optical fibre can transmit light energy very effectively. However, in order to transmit an image of an object, one requires an assembly or bundles of fine glass fibres or plastic fibres because each fibre transmits rays from a very small region of the object (Figure 14.48). Quality of the image depends upon the diameter of optical fibre. It can be as small as 10^{-6} m. A bundle of optical fibre may consist of thousands of individual optical fibres.

Essentially, three varieties of optical fibres are available constructed of either glass, plastic or a combination of glass and plastic. These are:

(i) Plastic core and cladding

(ii) Glass core with plastic cladding (often called plastic clad silica, PCS)

(iii) Glass core and glass cladding (often called silica clad silica, SCS)

There is also a fourth variety that uses a nonsilicate substance (zinc chloride). Investigations have shown that fibres made of this substance will be as much as 1000 times as effective as glass.

Plastic fibres have several advantages over glass fibres. First, plastic fibres are more flexible and consequently more rugged than glass. They are easy to install, can better withstand stress, are less expensive, and weigh approximately 60% less than glass. The disadvantage of plastic fibres is their high attenuation characteristic. They do not propagate light as efficiently as glass. Consequently plastic fibres are restricted for use only within a single building or a building complex.

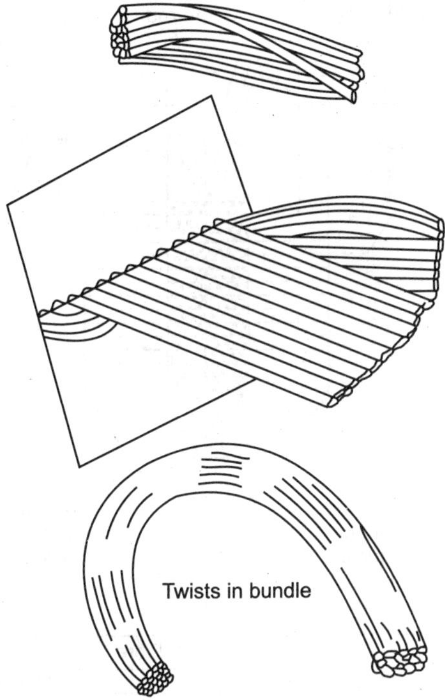

Figure 14.48 Bundle of optical fibre

Fibres with glass core exhibit low attenuation characteristics; however, PCS fibres are slightly better than SCS fibres. PCS fibres are also less affected by radiation and, therefore, are more attractive to military applications. SCS fibres have the best propagation characteristics and they are easier to terminate than PCS fibres. Unfortunately, SCS cables are the least rugged, and they are more susceptible to increase in attenuation when exposed to radiation.

The selection of fibre for a given application depends on specific system requirements.

14.15.1 Cable Construction

Figure 14.48 shows examples of different fibre optic cable configurations available. Depending upon the configuration, the cable may include a core, a cladding, a protective tube, buffer strength members and one or more protective jackets.

There are a number of techniques available to manufacture glass fibers. Double crucible method is a well known technique used for manufacture of glass fibres (Figure 14.49). It consists of two concentric platinum crucibles having very thin orifices at the bottom. The raw material for core (glass) is placed in the inner crucible while the raw material for cladding is fed into the outer crucible. The arrangement is vertically placed between electric furnace. The temperature of furnace is so maintained to transform the raw materials into molten state.

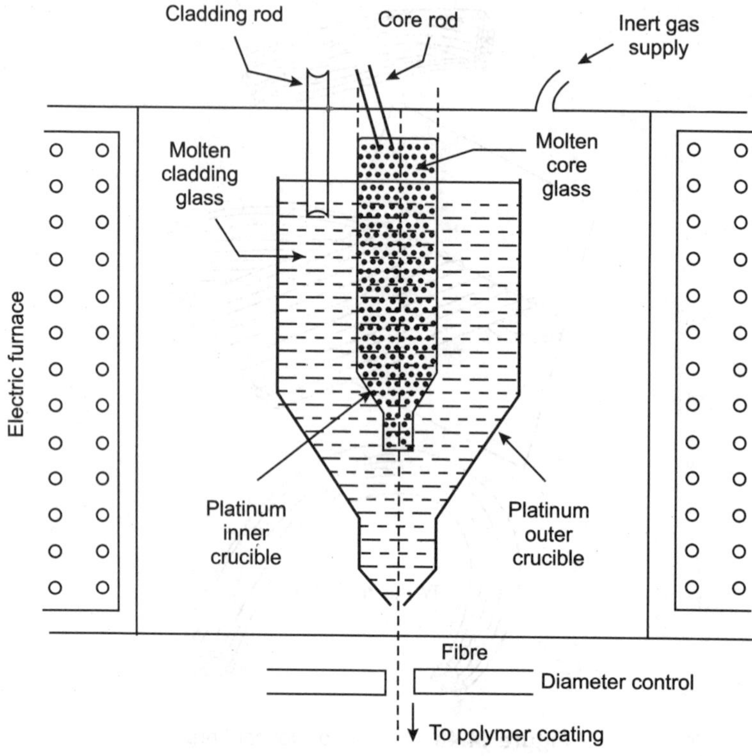

Figure 14.49 Double crucible technique

When the two molten glasses come together at the base of outer container, they form a glass-cladded core. Thickness of the fibre is also monitored. Finally, the fibre is coated with polymer. It appears that this technique can produce only step index fibres. To produce graded fibres the core and cladding glasses are allowed to interdiffuse after they come together. Diffusion causes a gradual change of refractive index between core glass and cladding glass.

14.16 Fibre Optic Sensor

It is a transducer that can convert various input variables (physical quantity) into electrical signals in a measurable form. In comparison to traditional sensors, optical fibre sensors are less costly and find wider applications. Optical fibre sensors are used in measuring and sensing various parameters, e.g., pressure, temperature, magnetic field, electric field, humidity, acoustic vibrations, current, etc.

Optical fibres have the following major advantages in sensing applications:

(i) Optical fibre sensors are light in weight and small.

(ii) Optical fibre sensors have good geometrical flexibility.

(iii) Optical fibre sensors are made of silica and therefore are free from the risk of fire and sparks.

(iv) Optical fibre sensors are immune to electromagnetic interference, i.e. they are electrically passive and they do not distort the surrounding electrical and magnetic fields.

(v) Optical fibre sensors have larger bandwidth and are highly sensitive.

(vi) Chemical and environmental ruggedness is high for optical fibre sensors.

Optical sensors are non contact and generally high-accuracy devices and systems. In optical fibre sensors, the optical wave acts as information carrier.

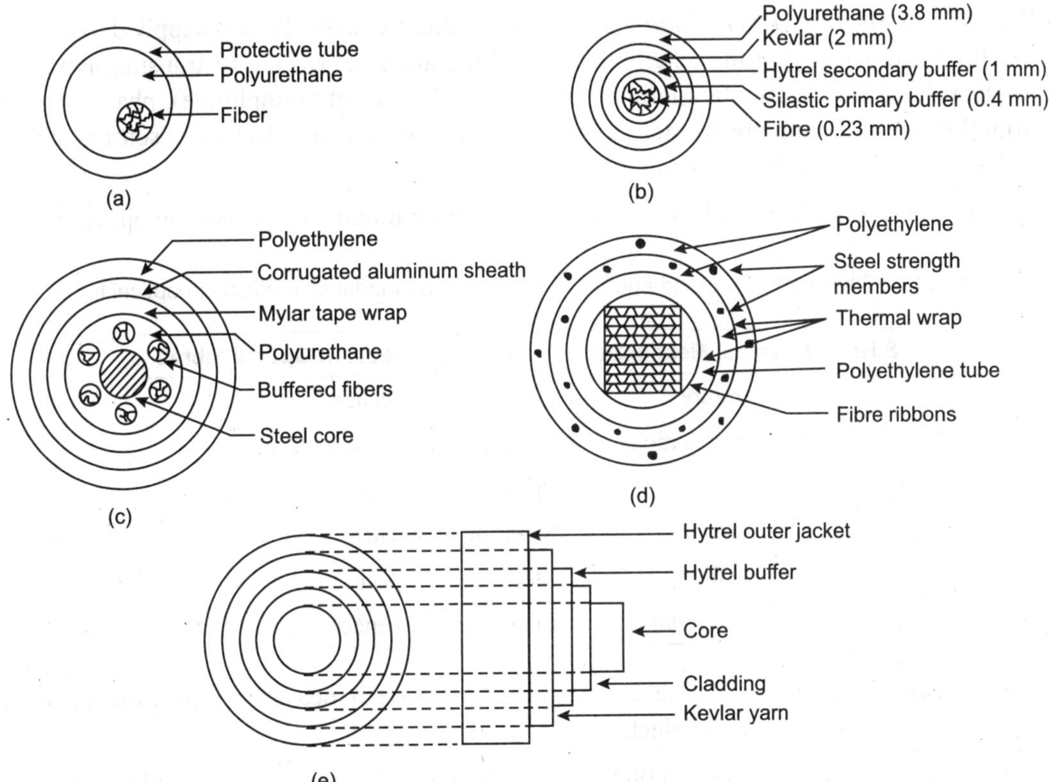

Figure 14.50 Fibre optic cable configuration: (a) loose tube construction, (b) constrained fibre, (c) multiple strands, (d) telephone cable, (e) plastic-clad silica cable

When optical sensors are used for measuring a physical parameter, any one of the characteristics like phase, amplitude, intensity, frequency, polarisation and direction of propagation of the wave gets modulated by the measured quantity. However, the detected quantity has intensity and thus on demodulation it results in change of intensity relating to the measured quantity.

14.16.1 Types of Optical Fibre Sensors

Optical fibre sensors are of two types:

(a) Extrinsic or hybrid sensors (passive sensors) In these sensors, the interaction between light and quantity under measurement takes place outside the optical fibre.

In these sensors, the fibre acts merely as a wave guide. This type of sensors has a sensor head, and the sensed optical signal is transferred to the measurement point with low attenuation and increased mechanical stability for signal processing. Extrinsic sensors can be used in the measurement of voltages, current, pressure, temperature, force, displacement, etc.

(i) Intrinsic or all-fibre sensors (active sensors) In these sensors, light is supplied to one end of the fibre and detected at the other end of a continuous fibre. During its transmission along the optical fibre the radiation is modified in terms of its intensity (amplitude), phase, spectral distribution or polarisation by the measuring parameter. Examples of this type of sensors include liquid level sensor, Faraday sensor and gyroscope.

Table 14.4 gives a few of physical measures and their modulation effects on optical fibres.

Table 14.4 Physical measurand and their corresponding modulation effects in optical fibres

S.No.	Physical Measurand	Modulation Effects in Optical Fibres
1.	Temperature	Thermoluminescence
2.	Nuclear radiations	Radiation induced luminescence
3.	Pressure	Piezoelectric effect
4.	Mechanical force	Stress birefringence
5.	Electric field	Electro optic effect
6.	Electric current	Electroluminescence

(ii) Interferometric sensors These sensors are widely used to measure temperature, pressure, displacement, magnetic field and electric field.

These sensors are mainly based on the principle of interference patterns produced due to a phase shift on superimposing a reference beam and a sensing beam.

(b) Displacement sensor

(i) Based on intensity measurement To measure displacement, the intensity modulation of transmitted light is used.

The arrangement of a displacement sensor is shown in Figure 14.51. One end of the transmitting fibre is coupled to a laser source. The other free end is made to face a target located at a distance *d* metre. Similarly, one end of the receiving fibre is connected to a detector and its free end is made to face the same target.

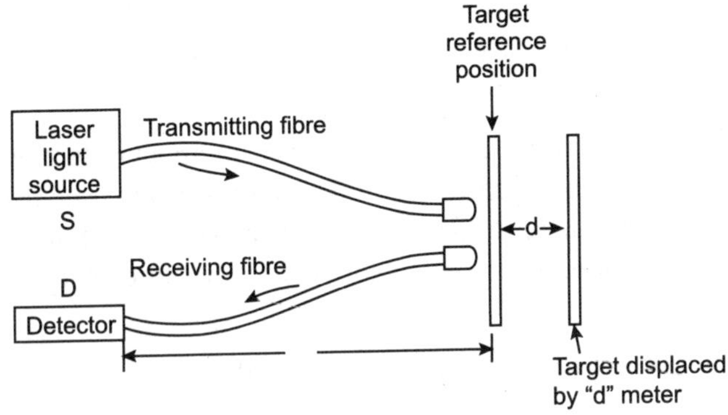

Figure 14.51 Displacement sensor

A light beam is allowed to incident over the target at a distance *d* meter from the transmitting end of the fibre. After being reflected from the target, light is collected by the detecting fibre and detected by detector *D*. Obviously, displacement of the object is related to the intensity on light reaching at the detector. Clearly, there will be either an increase or decrease in intensity with respect to the decrease or increase in distance of the target. Obviously, we can measure the position of the target using the optical fibre detector.

In order to enhance the density of the sensor, the axes of the transmitting fibre and receiving fibre should be placed at a specific angle with respect to each other and also to the target.

(ii) Displacement measurement with the help of diffraction pattern The arrangement of an optical fibre displacement sensor based on phase modulation is shown in Figure 14.52. A collimated laser light beam is allowed to fall over butted grating arrangement (Figure 14.52). Light beam after getting transmitted through the gratings is received by a detector. Clearly, when any one of the gratings is transversely displaced, it causes the diffraction angle of light beam to vary.

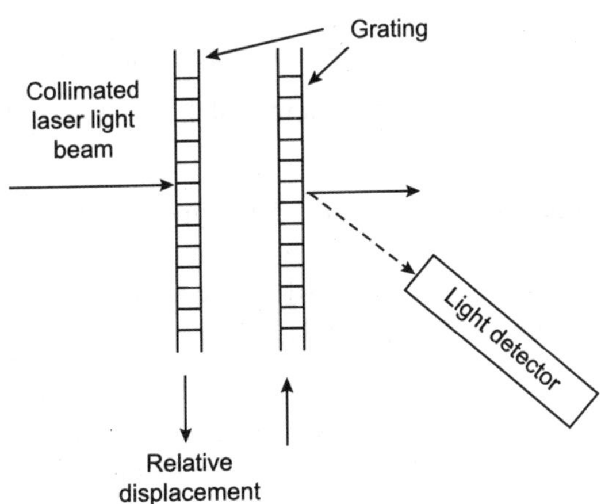

Figure 14.52 Displacement measurement based on phase modulation

Obviously, the detector senses this as an intensity variation. Clearly, the displacement of gratings results in a high or low intensity beam on the digital detector, thereby providing a direct measurement of displacement.

(c) Pressure sensor Optical fibre sensors for the measurement of temperature and pressure have been developed for down hole measurement. These sensors are used in different fields like process control, oceanography, meteorology, hydrology, aerospace, laboratory instrumentation and energy pressure exploration. These are also used to measure blood gas parameters and dosage levels.

There are different types of following pressure sensors:

(i) Microbending sensor This type of sensor is based on the measurement of light losses caused by controlled microbending.

The basic construction of a microbending sensor, which can be used for measuring pressure, temperature and displacement is shown in Figure 14.53.

A fibre is mounted inbetween a pair of plates containing parallel grooves. When we apply pressure over the plates, it causes microbends in the fibre and therefore results in the leakage of light thereby reducing the power of transmitted light. Thus, the measurement of output power provides the pressure that acted over the fibre.

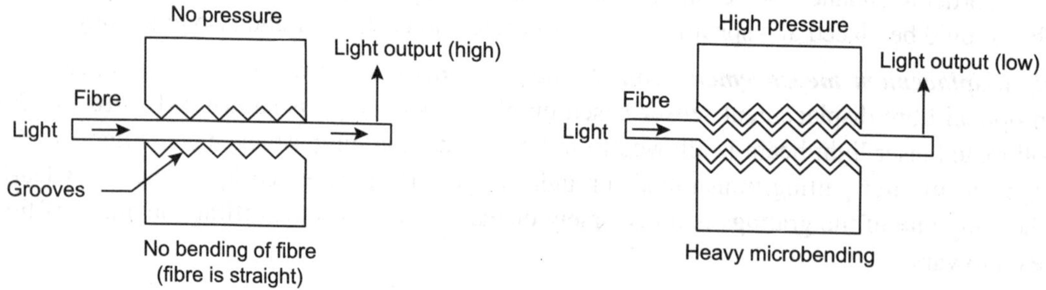

Figure 14.53 Schematic of microbending sensor

Microbending sensors require an environment which causes the fibre to bend under certain conditions.

(ii) Phase modulation based pressure sensor This type of sensor based on the principle of interference patterns produced due to phase shift between signals from the reference fibre and the test fibre in the measuring environment. The basic construction of interferometric arrangement to measure pressure is illustrated in Figure 14.54.

Light from a source of laser beam (S) is allowed to pass through a beam splitter. This beam splitter is a glass plate G, which is inclined at an angle of 45° to the direction of laser beam. Obviously, beam splitter G divides a single beam incident on it into two parts travelling at right angles to each other, marked by 1, 1 and 2, 2.

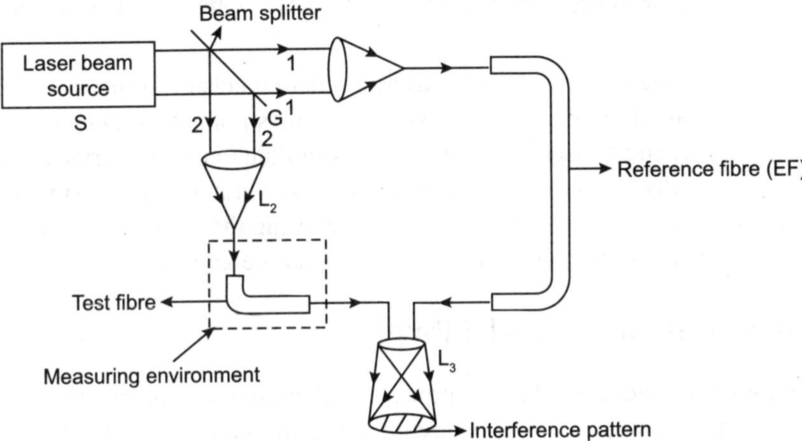

Figure 14.54 Schematic of pressure sensor based on phase modulation

The main or direct beam 1,1 passes through lens L_1 and then allowed through a reference fibre which is isolated from any influence of external environment. Finally, the beam from the other end of this reference fibre is allowed to pass through lens L_3.

The splitted beam 2,2 at right angle to the main beam is allowed to pass through lens L_2 and then through a test fibre which is kept in a varying environment to be measured. Clearly, when pressure or temperature acts over this test fibre, a path difference is produced due to the change in fibre length, core diameter, and refractive index with respect to the beam from reference fibre. Finally, this beam after getting transmitted through the test fibre is allowed to pass through lens L_3.

Due to phase difference between the two beams from reference fibre, interference pattern is produced. With the help of this sensor, very accurate measurement of pressure or temperature may be obtained from this interference pattern.

(iii) Photoelastic pressure sensor based on polarisation In this type of pressure sensor, the phenomenon of photoelastic effect is used to measure pressure.

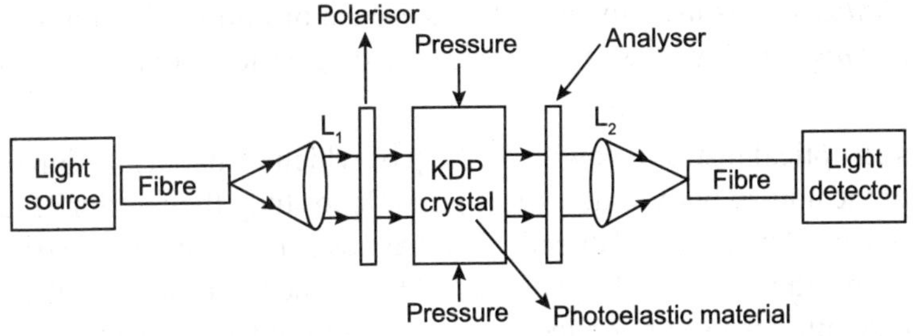

Figure 14.55 Photoelastic pressure sensor based on polarisation

Figure 14.55 shows the schematic of a multimode fibre optic sensor used for measuring pressure.

In this sensor, the photoelastic effect induced as a result of mechanical stress on a photoelastic material like KDP crystal which is placed between two cross polarisers. Due to applied stress over the crystal upon passing a laser beam through it simultaneously, the crystal material gets transformed into an optically anisotropic material. This causes variation in intensity of transmitted light through the optical fibre sensor. Obviously, by measuring the intensity of transmitted light beam, pressure that acted over the photoelastic material can be known.

14.17 Applications of Optical Fibres

Fibre optic technology has developed into a powerful alternative for information transport and sensing. Fibre communication has large bandwidth, and the system is capable of handling a large number of channels. Few important applications of optical fibres are:

(i) **Communication** Optical fibres can carry signals in optical communications due to their unique features, such as large channel capacity, wide bandwidth, good electrical isolation, no crosstalks, etc. At present, telephone, television, satellite and computer links of advanced communication systems rely heavily on this technology.

(ii) **Scanning** Optical fibres can be used to measure the distribution of light intensity over an illuminated area. For this purpose, the input end of the optical fibre scans the area, and light output from optical fibres is fed into a photodetector. The variation of photodetector output reveals light intensity distribution.

(iii) **Medical Instruments** Optical fibres are used in medical instruments e.g. fibroscopes to examine internal body cavities such as stomach and bladder. Optical fibres are also useful in the study of tissues and blood vessels. For this purpose, a bundle of fibres is enclosed in a hypodermic needle. These flexible fibre bundles are used as endoscope. Endoscopes are used to make observations along a curved path. Such types of fibre bundles are said to be passive, i.e. they merely convey light from one end to the other end.

Bronchoscope is used to check the presence of foreign bodies, infections in trachea and larger air ways. *Cystoscope* is used to check the presence of tumors, inflammation and stones in urinary bladder. *Gastroscope* is used to check tumors, gastric ulcer and gastritis in stomach and so on.

Endoscope is an optical instrument with the help of which visual inspection or photography of internal parts of the human body is done. It is possible to study the interior of lungs and other parts of the body that cannot be viewed directly. For diagnostic purposes, endoscopy is extensively used. In an endoscope, there are two fibre bundles. One is used to illuminate the inner structure of the object and the other is used to collect the reflected light from desired area.

An endoscope (Figure 14.56) mainly consists of the following four parts:

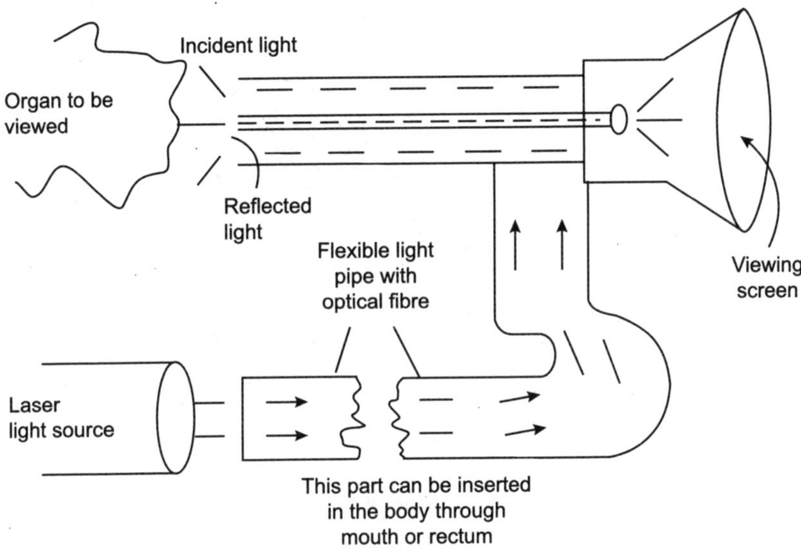

Figure 14.56 Endoscope

(i) A light guide (LG)
(ii) A coherent bundle (CB)
(iii) An instrumental channel (IC_1)
(iv) An irrigation channel (IC_2)

The light guide (LG) consists of a large number of optical fibres of large numerical aperture. They are bundled together and fused at the ends. They are free at all other parts. The bundle as a whole is enclosed in a thin plastic sleeve for protection. Now this unit acts as light guide. The input end of light guide is illuminated by high intensity light source. Light is transmitted through the light guide and emerges from the other end. This light illuminates the part or organ of the body.

The coherent bundle (CB) consists of a bundle of closely packed optical fibres in a perfect order. These fibres are packed in a light sleeve. A tiny lens is fixed to one end of the bundle. The reflected light from the body part is brought to focus by the lens at the end of the plane of the bundle. Now each part picks up a part of the picture. Thus, the array of fibres in the bundle collects the picture bit by bit and transmits them in order. As a result, the picture as a whole is reproduced at the other end of the bundle.

Instrument channel (IC_1) provides a path for insertion of surgical instruments to carry out an operation. The irrigation channel (IC_2) is provided to blow air or send some transparent liquids to clear the blood at operating region for better view. The whole process is shown in Figure 14.57.

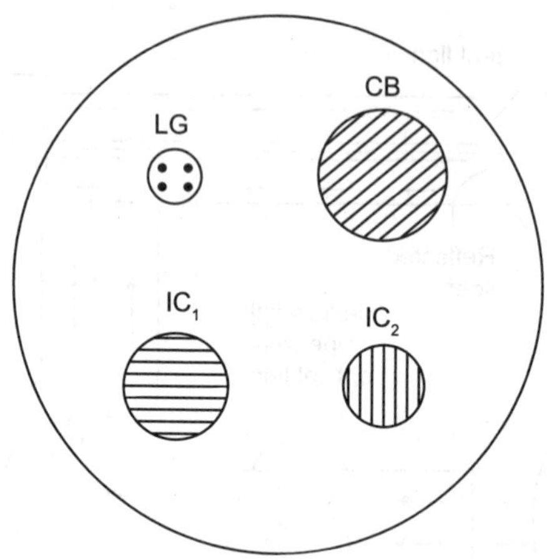

Figure 14.57

Optical fibres are also used in bloodless surgery, laser ophthalmoscope for retinal welding, etc.

(iv) **Coupler** Using the optical fibre, one may couple two circuits without introducing a direct electrical link. An LED forms a load in the primary circuit, so that a light output is obtained. Light guide collects the light and transmits to a photodetector in the secondary circuit. The secondary circuit gives an electrical signal in this circuit corresponding to that in the first.

(v) **Display and illumination** Optical fibres are employed to carry light to the display units. They are also used to illuminate dials when measuring instruments are used in dim light.

(vi) **Transfer of energy** Optical fibres are employed in the ultraviolet region for spectroscopic work and also used to transfer infrared energy from the source to the point of application of heat.

(vii) **Defence** The optical fibre system is widely used in defence services because these helps in maintaining high privacy. The military applications include communications, command and control links on ships and aircrafts, and data links for satellite earth stations.

(viii) Optical fibres are particularly suited for the transmission of digital data such as that generated by computers.

(ix) **Optical fibre sensors** These are used to measure temperature, pressure, rotary and linear position and liquid levels. Gyroscopes measure rotational motion. Hydrophones are used to measure acoustic disturbance in water.

(x) Optical fibres are also used in cable telephone, space vehicles, ships, and submarine cable.

(xi) Optical fibres have wide applications in security and alarm systems, electric instrumentation systems, industrial automation and process control.

Illustrative Examples

Example 1 For a 3 μm diameter optical fibre with core and cladding indexes of refraction of 1.545 and 1.510, respectively, determine the cut off wavelength.

Solution

$$\lambda_c = \frac{2\pi a n_1 \sqrt{2\Delta}}{2.405}$$

where

λ_c - cut off wavelength (μm)

n_1 - core index of refraction (unit-less)

n_2 - cladding index of refraction (unit-less)

a - core radius (μm)

$$\Delta = \frac{n_1 - n_2}{n_1}$$

$$\Delta = \frac{1.545 - 1.510}{1.545} = 0.023$$

$$a = \frac{3}{2} = 1.5\ \mu\text{m}$$

$$\lambda_c = \frac{(2\pi)(1.5)(1.545)\sqrt{2(0.023)}}{2.405}$$

$$= 1.29\ \mu\text{m}$$

Example 2 For a multimode index fibre $n_1 = 1.53$, $n_2 = 1.50$ and $\lambda = 1$ μm. If core radius is 50 μm, calculate the normalized frequency of the fibre (v) and the number of guided mode.

Solution

$$\text{Normalized frequency, } v = \frac{2\pi a(\text{NA})}{\lambda}$$

$$= \frac{2 \times 3.14 \times 50 \times 10^{-6} \times \sqrt{(1.53)^2 - (1.50)^2}}{1 \times 60^{-6}}$$

$$= 94.72$$

$$\text{Total number of guided mode, } m_s = \frac{\upsilon^2}{2} = \frac{(94.72)^2}{2} = 4486$$

Example 3 For a multimode stepped index fibre with a glass core ($n_1 = 1.5$) and a fused quartz cladding ($n_2 = 1.46$), determine θ_c, θ_m and NA. The source to fibre medium is air.

Solution

$$\text{Critical angle}, \theta_c = \sin^{-1}\frac{n_2}{n_1} = \sin^{-1}\frac{1.46}{1.5} = 76.7°$$

$$\theta_m = \sin^{-1}\frac{\sqrt{n_1^2 - n_2^2}}{n_0}$$

and

$$\theta_{m(\max)} = \sin^{-1}\sqrt{n_1^2 - n_2^2}$$

$$= \sin^{-1}\sqrt{(1.5)^2 - (1.46^2)}$$

$$= 20.2°$$

$$\text{NA} = \sin\theta_m = \sin 20.2 = 0.344$$

The maximum diameter a single-mode optical fibre can have is proportional to the wavelength of light ray entering the cable and numerical aperture of the fibre. Mathematically, the maximum radius of core of a single-mode fibre is

$$r_{\max} = \frac{0.383\text{Å}}{\text{NA}}$$

where $r_{\max}$ is maximum core radius

NA → Numerical Aperture (unit-less)

λ → light ray wavelength

Example 4 The numerical aperture of an optical fibre is 0.5 and core refractive index 1.54. Determine (i) Refractive Index (RI) of cladding, (ii) change in core cladding refractive index per unit RI of the core.

Solution

(i) $\text{NA} = \sin\theta_m = \sqrt{n_1^2 - n_2^2}$

$\therefore\ (0.5)^2 = (1.54)^2 - n_2^2$ or $n_2 = 1.456$

(ii) RI of the core $= \dfrac{n_1 - n_2}{n_1} = \dfrac{1.54 - 1.456}{1.54} = 0.0542$

Example 5 Determine the numerical aperture of a step index fibre when core refractive index $n_1 = 1.5$ and cladding refractive index $n_2 = 1.48$. Find the maximum angle for entrance of light if the fibre is placed in air. [B. Tech.]

Solution

(i) $\text{NA} = \sqrt{(n_1^2 - n_2^2)} = \sqrt{(1.5)^2 - (1.48)^2}$

$= \sqrt{(2.98 \times 0.02)} = 0.24413$

(ii) The maximum entrance angle i_0 can be found from

$$i_0 = \sin^{-1}\left(\frac{\text{NA}}{n}\right) = \sin^{-1}\left(\frac{0.24413}{1}\right) = 14.13°$$

Example 6 An optical fibre has NA of 0.20 and cladding refractive index of 1.59. Determine acceptance angle for the fibre in water with refractive index 1.33. [B. Tech.]

Solution

$$\text{NA} = \frac{\sqrt{(n_1^2 - n_2^2)}}{n_0}$$

When the fibre is in air, $n_0 = 1$, then

$$\text{NA} = \sqrt{(n_1^2 - n_2^2)} = 0.20$$

$$\therefore\ n_0 = \sqrt{(\text{NA})^2 + n_2^2} = \sqrt{(0.20)^2 + (1.59)^2} = 1.6025$$

When the fibre is in water, $n_0 = 1.33$.

Now
$$\text{NA} = \frac{\sqrt{(n_1^2 - n_2^2)}}{n_0}$$

$$= \frac{\sqrt{(1.6025)^2 + (1.59)^2}}{1.33} = 0.15$$

$$\therefore\ i_0 = \sin^{-1}(\text{NA}) = \sin^{-1}(0.15) = 8.6°$$

Example 7 Calculate the refractive indices of core and cladding material of a fibre from the following data: NA = 0.22 and $\Delta = 0.012$. [B. Tech.]

Solution

$$\text{NA} = n_1\sqrt{2\Delta}$$

$$\therefore\ n_1 = \frac{NA}{\sqrt{2\Delta}} = \frac{0.22}{\sqrt{2 \times 0.012}} = 1.42$$

$$\because\ \Delta = \frac{n_1 - n_2}{n_1}$$

$$\therefore\ 0.012 = \frac{1.42 - n_2}{1.42} \text{ or } n_2 = 1.40$$

Example 8 Calculate numerical aperture, acceptance angle and critical angle of an optical fibre, having refractive index of core 1.52 and refractive index of cladding 1.46.

[B. Tech.]

Solution

We have, $$\Delta = \frac{n_1 - n_2}{n_1} = \frac{1.52 - 1.46}{1.52} = 0.0395$$

Now $$\text{NA} = n_1\sqrt{(2\Delta)} = 1.52 \times \sqrt{(2 \times 0.0395)}$$

$$= 1.52 \times 0.281 = 0.427$$

Acceptance angle, $$i_0 = \sin^{-1}(\text{NA}) = \sin^{-1}(0.427)$$

Critical angle, $$\theta_C = \sin^{-1}(n_2 / n_1)$$

$$= \sin^{-1}(1.46/1.52) = \sin^{-1}(0.96)$$

Example 9 A step index fibre in air has NA of 0.16, core refractive index of 1.45 and core diameter of 60 cm. Determine the normalized frequency for the fibre when light at a wavelength of 0.9 μm is transmitted.

[B. Tech.]

Solution

$$V = \frac{\pi d}{\lambda}\sqrt{(n_1^2 - n_2^2)} = \frac{\pi d}{\lambda}(\text{NA})$$

Substituting the given values, one obtains

$$V = \frac{3.14 \times 0.60 \text{ m}}{9 \times 10^{-7} \text{ m}} \times 0.16$$

$$= 335103 = 3.35 \times 10^5$$

Example 10 Find the core radius necessary for single mode operation at 850 μm in SI fibre with $n_1 = 1.480$ and $n_2 = 1.47$. Find numerical aperture and maximum acceptance angle of this fibre.

[B. Tech.]

Solution

For single mode propagation, V - number is 2.405.

Now, $$V = \frac{\pi d}{\lambda}\sqrt{(n_1^2 - n_2^2)}$$

$$\therefore\ 2.405 = \frac{\pi d}{850 \times 10^{-9}} \times \sqrt{[(1.480)^2 - (1.47)^2]}$$

or
$$2.405 = \frac{3.15 \times d}{850 \times 10^{-9}} \times 0.1717$$

$$\therefore\ d = \frac{2.405 \times (850 \times 10^{-9})}{3.143 \times 0.1717} \times m = 3.79\ \mu\text{m}$$

So $r = 1.89\ \mu$m

Further,
$$\text{NA} = \sqrt{(n_1^2 - n_2^2)} = \sqrt{[(1.480)^2 - (1.47)^2]}$$
$$= 0.1717$$

$$\sin i_0 = \text{NA} = 0.1717$$

$$i_0 = \sin^{-1}(0.1717) = 9°53'2''$$

Example 11 An optical signal has lost 85% of its power after traversing 500 m of fibre. Find the loss in dB/km of this fibre. [B. Tech.]

Solution

We know that,
$$\alpha = \frac{10}{L}\log\left(\frac{P_{in}}{P_{out}}\right)$$

Here, $P_{in} = 1$ mW, $P_{out} = 85\%$, $P_{in} = 0.85$ mW and $L = 500$ m $= 0.5$ km

$$\alpha = \frac{10}{0.5}\log\left(\frac{1}{0.85}\right) = 1.41\ \text{dB/km}$$

Example 12 A communication system uses a 10 km fibre having a loss of 2.5 db/km. Calculate output power if input power is 500 μW. [B. Tech.]

Solution

Loss in the fibre $= \alpha L = -2.5 \times 10 = -25$ dB

$$\text{Loss, } \alpha = 10\log\left(\frac{P_{out}}{P_{in}}\right)$$

or
$$-25 = 10\log\left(\frac{P_{out}}{P_{in}}\right) \text{ or } \log\left(\frac{P_{out}}{P_{in}}\right) = -2.5$$

$$\therefore\ \frac{P_{out}}{P_{in}} = \text{Antilog } (-2.5) = 0.00316$$

or
$$P_{out} = P_{in} \times 0.00316$$

$$\therefore\ P_{out} = 500 \times 0.00316 = 1.58\ \mu\text{W}$$

Example 13 A GRIN optical fibre with parabolic refractive index profile core has a refractive index at the core axis of 1.5 and relative index difference at 1%. Show that the maximum core diameter that allows signal mode operation at $\lambda = 1.3\ \mu m$ is 6.6 μm. [B. Tech.]

Solution

Given, $\Delta = 1\% = 0.01$, $\lambda = 1.3\ \mu m$
$= 1.3 \times 10^{-6}$ m and $n = 1.5$

We have for a GRIN, maximum value of normalized frequency for single mode operation is given by

$$V = 2.4\left(1+\frac{2}{\alpha}\right)^{1/2}$$

$$= 2.4\left(1+\frac{2}{2}\right)^{1/2} = 2.4\sqrt{2}$$

For maximum core radiation, we have

$$r = \frac{V\lambda}{2\pi n_1\sqrt{2\Delta}}$$

$$= \frac{24\sqrt{2}\times 1.3\times 10^{-6}}{2\pi\times 1.5\times\sqrt{0.02}} \doteq 3.3\ \mu m$$

∴ Maximum core diameter which permit single mode operation $= 2r = 6.6\ \mu m$.

Example 14 A multimode step index optical fibre having relative refractive index difference 1.5% and core refractive index 1.48 is used for single mode operation. The operating wavelength is 0.85. Calculate the maximum core diameter. [B. Tech.]

Solution

Given: $\Delta = 1.5\% = 0.015$, $n_1 = 1.48$ and $\lambda = 0.85\ \mu m = 0.85 \times 10^{-6}$ m
The maximum V value for a fibre which gives single mode operation is 2.4.

We have
$$V = \frac{2\pi}{\lambda} r(\text{NA})$$

$$= \frac{2\pi}{\lambda} r n_1\sqrt{2\Delta}$$

∴
$$r = \frac{V\lambda}{2\pi n_1\sqrt{2\Delta}}$$

$$r = \frac{2.4 \times (0.85 \times 10^{-6})}{2\pi \times (1.48) \times \sqrt{0.03}} = 1.3\ \mu\text{m}$$

Thus, the maximum core diameter for single mode operation is 2.6 μm.

Example 15 An optical fibre has an attenuation 3.5 dB/km. If 0.5 mW of optical power is initially launched into the optical fibre, show that its power level after 4 km is 19.9 mW.

[B. Tech.]

Solution

Given: $\alpha = 3.5$ dB/km,
$P_i = 0.5$ mW
$L = 4$ km

The attenuation of an optical fibre is given by

$$\alpha = \frac{10}{L}\log\frac{P_i}{P_o}$$

or
$$3.5 = \frac{10}{4}\log\frac{0.5}{P_o}$$

or
$$1.4 = \log\frac{0.5}{P_o}$$

$$\therefore \quad P_o = \log\left(\frac{0.5}{10^{1.4}}\right) = 19.9\ \text{mW}$$

Example 16 The core refractive index and core radius for step index fibre in single mode operation are 1.46 and 4.5 μm respectively and the relative index difference is 0.25%. Estimate cut-off wavelength. [B. Tech.]

Solution

Given: $n_1 = 1.46$, $r = 4.5$ μm and $\Delta = 0.25\% = 0.0025$
For cut-off wavelength, $V_C = 2.405$
We have, cut-off wavelenth expression

$$\lambda_c = \frac{2\pi r n_1\sqrt{2\Delta}}{V_c}$$

$$= \frac{2\pi \times 4.5 \times 1.46\sqrt{0.005}}{2.405} = 1.214\ \mu\text{m}$$

Summary

1. Fibre optics is a technology related to the transportation of optical energy (light energy) in guiding media, specifically glass fibres.
2. In case of optical fibres, there must be very little absorption of light as it travels through a long distances inside the fibre. One can achieve this by purification and special preparation of the materials used.
3. Fibres that are used in optical communication are wave guides made of transparent dielectrics. Its function is to guide visible and infrared light over long distances.
4. Optical fibres are classified into two categories based on: (i) number of modes and, (ii) refractive index.
5. On the basis of number of modes of propagation, optical fibres are classified into two types: (i) single mode fibre (SMF) and (ii) multimode fibre (MMF).
6. On the basis of refractive index, fibres can be classified into two types: (i) step-index optical fibre and (ii) granded-index optical fibre.
7. The condition for propagation of light within the optical fibre is

$$\sin i < \text{NA}$$

8. V-number is an important parameter for optical fibre and is generally called normalized frequency of the fibre.

$$V = \frac{2\pi a}{\lambda}(\text{NA}) = \frac{2\pi a}{\lambda} n_1 \sqrt{2\Delta}$$

where $\Delta = \frac{n_1 - n_2}{n_1}$ is fractional refractive index change.

The wavelength corresponding to the value of $V = 2.405$ is known as cutoff wavelength.

9. Losses occurring in optical fibres may be mainly due to the following mechanisms: (i) absorption, (ii) scattering and (iii) fibre bends.
10. Skew rays does not pass through optical fibre axis. However, skew rays travel in zig-zag fashion.
11. Axis rays always travel through the axis of optical fibre.
12. Meridian rays cross the axis of optical fibre.
13. The index of refraction in graded index fibres varies gradually from the axis of fibre.
14. Mode of fibre refers to the number of paths the light ray can travels down the optical fibre.
15. The single mode propagation of LP_{01} mode in step index optical fibre is possible over the region $0 < V < 2.405$.

16. In a single mode operation for a graded index optical fibre, the cut-off value of normalized frequency (V_c) is as follows:

$$V_c = 2.405\left(1+\frac{2}{\alpha}\right)^{1/2}$$

and cut off wavelength

$$\therefore \qquad \lambda_c = \frac{2\pi a\, n_1}{V_c}\sqrt{(2\Delta)}$$

Review Questions

1. Define an optical fibre system. What is numerical aperture of an optical fibre? What is its significance?
2. Obtain the expression for NA is terms of RI of the core and cladding of an optical fibre.
3. Outline the primary building blocks of a fibre optic system.
4. Draw a block diagram for optical communication through fibre optic cables. Mention the advantages of optical communication.
5. Explain the working of monomode and multimode optical fibres. For long distance communication which optical fibre is preferred and why?
6. Contrast glass and plastic fibre cables.
7. Explain the principle and working of fibre optic sensors.
8. Explain the working of displacement sensor based on: (i) intensity measurement and (ii) diffraction pattern.
9. Explain the working of a pressure sensor based on: (i) phase modulation and (ii) photoelastic effect.
10. Briefly explain the principle and working of different pressure sensors.
11. Explain the principle and working of microbending sensor.
12. Explain the principle and working of photoelastic pressure sensor.
13. Write short notes on
    (i) Fibre optic cable
    (ii) Step index fibre
    (iii) Graded index fibre
    (iv) Fibre optic communication system
    (v) Fibre optic sensors

Problems

1. A multimode step index fibre has $n_1 = 1.53$, $n_2 = 1.50$ and $\lambda = 1$ μm. In order to make this fibre single mode, what will be core radius?

$$\left[\text{Hint: } a \le \frac{2.405\lambda}{2\pi(\text{NA})},\ \text{NA} = \sqrt{(1.53)^2 - (1.50)^2}\right]$$

$$\alpha \le \frac{2.405 \times 1 \times 10^{-6}}{2 \times 3.14 \times \sqrt{(1.53)^2 - (1.50)^2}},\ \therefore\ a < 1.27\ \mu\text{m}$$

2. Numerical aperture of a optical fibre is 0.5 and core refractive index is 1.48. Show that cladding refractive index is 1.393 and acceptance angle is 30°.

[Hint: $\text{NA} = \sqrt{n_1^2 - n_2^2} = 0.5$, RI of core = n_1 = 1.48; that is n_2 = 1.393; RI of cladding = n_2 = 1.393; Acceptance angle $\theta = \sin^{-1}$ (NA) – 30°]

3. A step index fibre in air has NA of 0.16, core refractive index of 1.45 and core diameter of 60 cm. Show that the normalized frequency for the fibre when light at a wavelength of 0.9 μm transmitted is 3.35 × 10^5. [B. Tech.]

$$\left[\text{Hint: } v = \frac{\pi d}{\lambda_0}\sqrt{n_1^2 - n_2^2} = \frac{3.143 \times 0.6\ \text{m}}{9 \times 10^{-7}\ \text{m}} \times 0.16 = 3.35 \times 10^5\right]$$

4. An optical fibre has NA of 0.20 and cladding refractive index of 1.59. Show that the acceptance angle for the fibre in water which has a refractive index of 1.33 is 8.6°.

$$\left[\text{Hint: NA} = \frac{\sqrt{(n_1^2 - n_2^2)}}{n_0}\right]$$

When the fibre is in air, $n_o = 1$ and $\text{NA} = \sqrt{n_1^2 - n_2^2} = 0.20$

$$\therefore\ n_1 = \sqrt{(\text{NA})^2 + n_2^2} = \sqrt{(0.20)^2 + (1.59)^2} = 1.6025$$

When the fibre is in water, $n_o = 1.33$

$$\text{NA} = \frac{\sqrt{n_1^2 - n_2^2}}{n_0} = \frac{\sqrt{(1.6025)^2 - (1.59)^2}}{1.33} = 0.15$$

$$\theta_0(\max) = \sin^{-1}(\text{NA}) = \sin^{-1}(0.15) = 8.6°$$

5. Show that the core radius necessary for single mode operation at 850 nm Sl fibre with $n_1 = 1.480$ and $n_2 = 1.47$ is 1.89 μm. Also show that NA and maximum acceptance angle of this fibre are 0.1717and 9° 53′ 12″ respectively.

[Hint:

(i) $\nu = \frac{\pi d}{\lambda_0}\sqrt{n_1^2 - n_2^2}$

$$2.405 = \frac{\pi d}{450 \times 10^{-9}\ \text{m}} \times 0.1717$$

or $d = 3.79$ mm

$\therefore r = 1.89$ mm

(ii) $\text{NA} = \sqrt{n_1^2 - n_2^2} = \sqrt{(1.48)^2 - (1.47)^2} = 0.1717$

(iii) $\sin\theta_0(\max) = \text{NA}$

$$\therefore \theta_0(\max) = \sin^{-1}(\text{NA}) = (n_1^2 - n_2^2)^{1/2}$$

$$= \sin^{-1}(0.1717) = 9^\circ\ 53'12'']$$

6. A fibre cable has index of refraction core 1.52 and cladding 1.31. Show that numerical aperture is 0.77.

7. A 5 km long optical fibre has pulse spreading constant of 10 ns/km. Determine maximum digital transmission rate if the transmission of data is (i) UPRZ, (ii) UPNRZ

   (i) $f_b = \frac{1}{\Delta t L} = \frac{1}{10\ \text{ns/km} \times 5\ \text{km}} = 20\ \text{Mbps}$

   (ii) $f_b = \frac{1}{2\Delta L} = \frac{1}{2 \times 10\ \text{ns/km} \times 5\ \text{km}} = 20\ \text{Mbps}$

8. For a multimode step index fibre with glass core ($n_1 = 1.5$), acceptance angle is required to be 20.2°. Find numerical aperture and refractive index of cladding (n_2).

   [Hint: NA = sin (θ_{in})

   Here $\theta_{in} = 20.2°$

   Hence NA = sin 20.2° = 0.344

   $$\text{NA} = \sqrt{n_1^2 - n_2^2}$$

   Hence, $0.344 = \sqrt{1.5^2 - n_2^2} = \sqrt{2.25 - n^2}$

   By solving, we get $N_2 = 1.46$]

9. A light ray travels from a medium of refractive index n_1=1.5 to another with refractive index n_2 = 1.46. Calculate the critical angle of incidence measured from normal at the boundary. What will be this angle if the second medium is air?

[Hint:
For $n_1 = 1.5$ and $n_2 = 1.46$,

Critical angle, $\theta_c = \text{arc sin}\left(\frac{n_2}{n_1}\right) = \text{arc sin}\left(\frac{1.46}{1.5}\right) = \text{arc sin}(0.9733) = 76.7°$

When second medium is air, $n_2 = 1.0003$

$$\theta_c = \text{arc sin}\left(\frac{1.0003}{1.5}\right) = 41.8]$$

Short Answer Questions

1. To prevent energy losses via absorption and scattering, what should be the order of cladding in an optical fibre?

   Ans: At least a few wavelengths thick.

2. How numerical aperture (NA) of an optical fibre is defined?

   Ans: $NA = \sqrt{n_1^2 - n_2^2}$ where n_1 is refractive index of the core of a fibre and n_2 is that of cladding ($n_1 > n_2$): One can define the numerical aperture of a fibre as the physical area that light must enter in order to propagate through the fibre. The numerical apertures for light sources can be calculated in terms of angle involved with the output.

3. How will you define acceptance angle (θ_A) for an optical fibre?

   Ans: $\theta_A = \sin^{-1}\sqrt{n_1^2 - n_2^2}$.

4. What is major economic benefit offered by fibre optics in information technology?

   Ans: Very high information transmission rate at low cost per circuit-km.

5. What type of reflection is expected from an optical fibre?

   Ans: Total internal reflection

6. How is an optical fibre fabricated?

   Ans: In the fabrication of optical fibres one can use either different glasses with different refractive indices for core and cladding or use fused silica (quartz, glass) whose refractive index is modified by doping. The type of glass used will also vary with application, but purity (lack of foreign materials as well as imperfections such as cracks or bubbles) must also be kept very high.

7. What is kevlar?

   Ans: It is a yarn type material having high tensile strength. It is used in fibre construction to provide additional strength to the cable.

8. Explain the relationship between information capacity and bandwidth.

   Ans: The information carrying capacity of a communication system is directly proportional to its bandwidth, i.e. higher the bandwidth, the greater its information carrying capacity.

9. Explain the purpose of cladding in optical fibre.

   Ans: Due to its lower refractive index than core, cladding helps in retaining light with the core. Moreover, it provides mechanical strength, reduces scattering loss and also protects the core from absorbing surface contaminants with which it could come in contact.

10. Mention the advantages of optical fibre communication over metallic cable communication.

    Ans: (i) Extremely wide bandwidth
    (ii) Light in weight and small in size
    (iii) Immune to electrostatic interference
    (iv) More secured, greater safety and no crosstalk

11. Can light propagate through an optical fibre cable with the angle of incidence at entrance end greater than acceptance angle?

    Ans: No. Only when light enters at an angle less than the acceptance angle, light can propagate through the cable by undergoing total internal reflection.

12. What is an optical fibre sensor?

    Ans: It is a transducer which can convert various input variables (physical quantity) into an electrical signal in measurable form.

13. What are the advantages of optical sensors in sensing applications?

    Ans: (i) They are electrically passive.
    (ii) They have good geometrical flexibility.
    (iii) They are light in weight and small in size.
    (iv) Chemical and environmental ruggedness is high.

14. What are the characteristics of wave that may get modulated in fibre optic sensors?

    Ans: Amplitude or intensity, phase, polarisation, frequency and direction of propagation may get modulated.

15. Which type of optical fibre (single mode or multimode) tends to have lower loss and produces less singal distortion?

    Ans: Single mode fibre.

16. Which properties of optical fibre reduce connection loss in short distance systems?

    Ans: (i) Higher fibre numerical aperature, and (ii) Larger fibre core.

17. Which basic optical material property is considered relevant for optical fibre light communication?

    Ans: Index of refraction.

18. Out of core or cladding, which optical fibre has a higher index of refraction?

    Ans: Core.

19. Multimode step index fibres have a core and cladding of constant refractive index n_1 and n_2 respectively. Out of these, which refractive index core or cladding is lower?

    Ans: Cladding.

20. What is the benefit of having large core diameters and large NA for multimode step index fibres?

    Ans: This makes it easier to couple light from an LED into the optical fibre.

21. Meridional rays are classified as either bound or unbound rays. While propagating through the fibre, the bound rays follow what property of light?

    Ans: Total internal reflection.

22. What are the advantages of optical fibre over copper wire conductors?

    Ans: Optical fibre transmission has following advantages over convention copper wire system:

    (i) More data can be sent over the long distances through optical fibre due to its lower transmission losses.

    (ii) The light weight and the small thin hair-sized dimension of the optical fibres are quite advantageous, e.g. in aircraft.

    (iii) Dielectric nature of optical fibre provides optical immunity to the electromagnetic interferences.

    (iv) The principal material for the optical fibre, i.e. silica, whose source is sand, is abundantly available.

Objective Questions

1. In fibre-optic communication, the propagation of light through the fibre is based on the principle of

   (a) diffraction of light (b) interference of light
   (c) total internal reflection (d) polarization of light **[Ans: (c)]**

2. The refractive index in a graded index fibre

   (a) increases throughout the core
   (b) remains constant throughout the core
   (c) decreases parabolically throughout the core
   (d) None of the above **[Ans: (c)]**

3. The acceptance angle or acceptance cone half angle for the optical fibre is

   (a) $\theta_m = \sin^{-1}(n_1 - n_2)$ (b) $\theta_m = \sin^{-1}\sqrt{n_1^2 - n_2^2}$
   (c) $\theta_m = \sqrt{n_1^2 - n_2^2}$ (d) $\theta_m = n_1^2/n_2^2$ **[Ans: (b)]**

4. Numerical aperature of step index optical fibre is

   (a) $\text{NA} = n_1\sqrt{2\Delta}$ (b) $\text{NA} = n_1^2\sqrt{2\Delta}$
   (c) $\text{NA} = n_1^2 - n_2^2$ (d) $\text{NA} = n_1\, 2\Delta$ **[Ans: (b)]**

5. The jacket of an optical fibre enables
   (a) to prevent from moisture trapping
   (b) to prevent from mechanical abrasions
   (c) to prevent from interaction with internal atmosphere
   (d) all of the above **[Ans:** (d)]
6. When V parameter is less than 2.405, then optical fibre will support
   (a) three modes (b) two modes
   (c) one mode (d) none of the above **[Ans:** (c)]
7. The optical fibres are made of
   (a) dielectric material
   (b) metallic material
   (c) magnetic material
   (d) plastic doped with metallic impurities **[Ans:** (a)]
8. Pulse broadening can be minimised by making use of
   (a) multimode graded index fibre (b) single mode index fibre
   (c) multimode step index fibre (d) none of the above **[Ans:** (a)]
9. The total number of modes in a multimode fibre is given by

   (a) $M = \frac{1}{2}\left[\frac{\pi d}{\lambda} NA\right]^2$ (b) $M = \left[\frac{2\pi d}{\lambda} NA\right]$

   (c) $M = \frac{1}{2}\left[\frac{\pi d}{\lambda} NA\right]^{1/2}$ (d) $M = \frac{1}{2}\left[\frac{\pi d}{\lambda} NA\right]^3$ **[Ans:** (a)]

10. Attenuation is an optical fibre is measured by

    (a) $\text{loss} = 10 \log_{10}(P_i/P_o^2)$ (b) $\text{loss} = -10 \log_{10}\left(\frac{P_o}{P_i}\right)$

    (c) $\text{loss} = 10 \log(P_i^2/P_o)$ (d) $\text{loss} = -10 \log\left(\sqrt{P_i/P_o}\right)$ **[Ans:** (b)]

11. Information signals are transmitted from optical fibre in the form of
    (a) analogue signals (b) digital signals
    (c) binary and hexadecimal bits (d) none of the above **[Ans:** (b)]
12. Dispersion in optical fibres is due to
    (a) chromatic dispersion (b) material dispersion
    (c) intermodal dispersion (d) all of the above **[Ans:** (d)]

13. The propagation of light along the waveguide is decided by the
    (a) interference (b) dispersion
    (c) modes of the waveguides (d) none of the above **[Ans:** (c)]
14. In multimode propagation the light propagates the fibre in
    (a) straight line (b) zig zag fashion
    (c) circular path (d) eliptical path **[Ans:** (b)]
15. In order that mode remain guided, the propagation factor β must satisfy
    (a) $n_2 < \beta < n_1$ (b) $n_2 \frac{2\pi}{\lambda} < \beta < n_1 \frac{2\pi}{\lambda}$
    (c) $n_2 < \beta = n_1$ (d) $n_2 = \beta < n_1$ **[Ans:** (b)]

Fill in the Blanks.

1. Optical fibre is ____________ carrier and immune to ____________ interference. **[Ans:** non-electrical, electromagnetic]
2. Optical fibre does not emit any ____________ that might interfere with signals. **[Ans:** noise]
3. Light has much ____________ frequency than radio waves or modulated signals in copper wire. **[Ans:** higher]
4. Optical fibre does not rust or corrode and is much ____________ and ____________ than copper cable. **[Ans:** smaller, lighter]
5. Scattering mechanism causes the transfer of ____________ from one propagating mode into a different mode. **[Ans:** optical power]
6. Major implementation of optical fibre transmission involves some form of digital ____________. **[Ans:** modulation]
7. Optical fibres are ____________ pipes that guide light through them on the principle of total ____________. **[Ans:** light, internal reflection]
8. The refractive index of ____________ region is always less than that of the core region. **[Ans:** cladding]
9. Optical fibres are divided into different groups based on their ____________ profile. **[Ans:** modulation]

Chapter

15 Communication Systems

15.1 Introduction

Communication is the transfer of information (intelligence) from one point (source) to another (destination). In electrical sense, communication refers to (i) sending, (ii) receiving and (iii) processing of information by electrical means. Information to be transmitted can be in the form of audio or video signals, or it could even be digital data. Figure 15.1(a) illustrates the basic transfer of information from source to destination. Signals can be transmitted from the source to destination by two means:

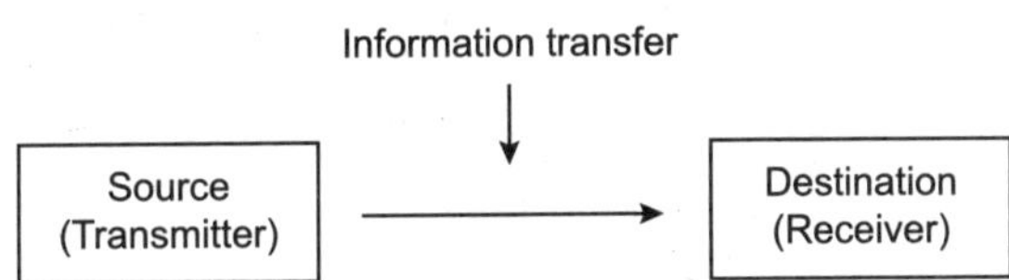

Figure 15.1 (a) Transfer of information

(i) *Wire communication*, where an antenna is used as a transmitting device

(ii) *Line communication*, where waveguides, transmission lines or optical fibers are used as transmitting devices.

An oscillator can be used to produce a high frequency signal. When the output of this oscillator is fed to a transmitting antenna, it converts the high frequency alternating current to a radio wave. A radio wave travels at approximately the speed of light ($= 3 \times 10^8$ m/s).

When this radio wave is incident on another antenna, a high frequency current is induced in it. This induced current is of smaller strength but identical in shape to current in the transmitting antenna. Thus, transfer of electrical energy from one point to the other without the use of wires (wireless communication) is accomplished.

Wireless or radio communication, made possible by the invention of triode tube, was greatly stimulated by the work done during 1940–45. It subsequently became even more widely used and refined through the invention of transistors, integrated circuits and other semiconductor devices. More recently, the use of satellites and fibre optics has made communication even more faster, with increasing emphasis on computers and other data communication equipment.

There are two types of radio communication: (i) radio telegraphy and (ii) radio telephony.

In radio telegraphy, a message is transmitted by means of telegraphic code that periodically interrupts continuous radio waves. In radio telephony, speech is broadcasted with the help of radio waves modulated by the speech. It was also known as wireless since wires are not used for the transmission of information.

There are three basic types of communication systems based on the characteristics of the carrier that is alternated.

(i) Systems that use amplitude-modulated (AM) carriers
(ii) Systems that use angle-modulated modulation, which in turn comprise two variations:
 (a) Frequency modulated (FM) carriers
 (b) Phase modulated (PM) carriers
(iii) Systems that use digital techniques, normally referred to as pulse modulation.

Figure 15.1(b) shows a block diagram of basic communication system.

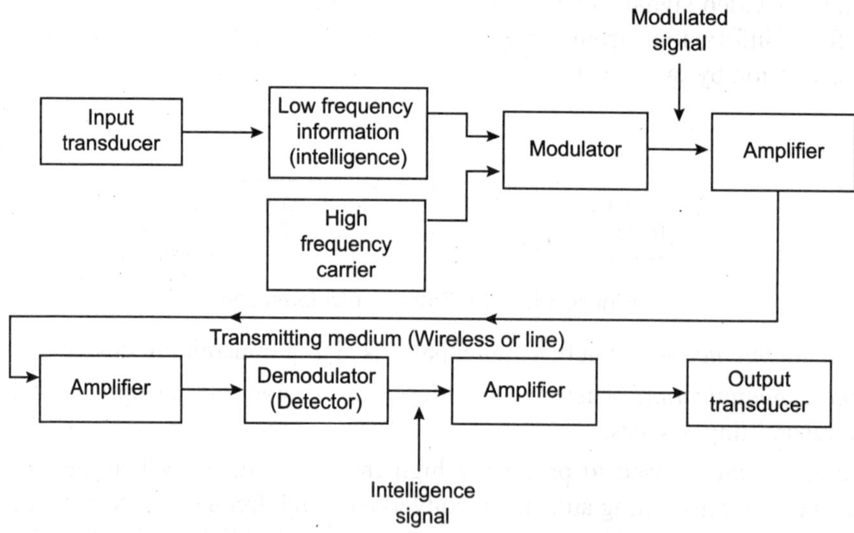

Figure 15.1 (b) Basic communication system

Transmitting end An input transducer (such as a microphone or a camera) converts the information into an electrical form. This electrical information is the modulating signal. The modulating signal and a high-frequency signal (called the carrier) are injected into the modulator, which produces a modulated signal. The modulated signal is then amplified.

Receiving end The receiver picks up the signal and amplifies it to compensate for attenuation that occurred during the transmission. The output of the amplifier is fed to a demodulator or detector where information is extracted from the carrier. The demodulated signal is then fed into another amplifier, which suitably amplifies it so that it can be fed into an output transducer such as a speaker (audio signals) or monitor (video signals) or computer (digital signals). The output transducer converts the information in its electrical form to a physical form such as sound or picture.

15.2 Radio Communication

The basic elements of radio communication (Figure 15.1(c)) are:

(a) *Microphone* that amplifies human voice/speech and converts it into electrical signals which modulate the radio waves generated in the transmitter in accordance with the information or intelligence to be transmitted.

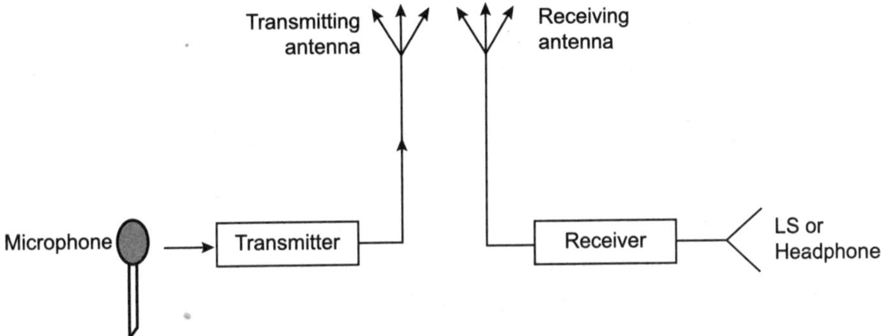

Figure 15.1 (c) Black diagram of radio communication system

(b) *Transmitter* generates radio waves.

(c) *Transmitting antenna* radiates the radio waves in all directions and in a specified direction if required.

(d) *Receiving antenna* receives a part of the incoming radiated radio wave.

(e) *Receiver* that selects, amplifies and detects the desired information or intelligence.

(f) *Loudspeaker or head phone* converts the detected electrical signal/information into sound.

15.3 Radio Transmitter

In a radio transmitter, a fixed frequency radio wave referred to as carrier wave is generated. Information in the electrical form known as modulating wave to be broadcasted is superimposed on the carrier wave. It modulates one parameter, e.g. amplitude or frequency or phase, of the carrier wave. The resulting wave known as modulated wave (amplitude modulated or frequency modulated or phase modulated) is transmitted by the transmitting antenna into space in the form of modulated electromagnetic wave. We shall confine our study to amplitude modulation and frequency modulated only.

15.3.1 Amplitude-Modulated Transmitter

Figure 15.2 depicts the basic subsystem of an amplitude modulated radio telephone transmitter.

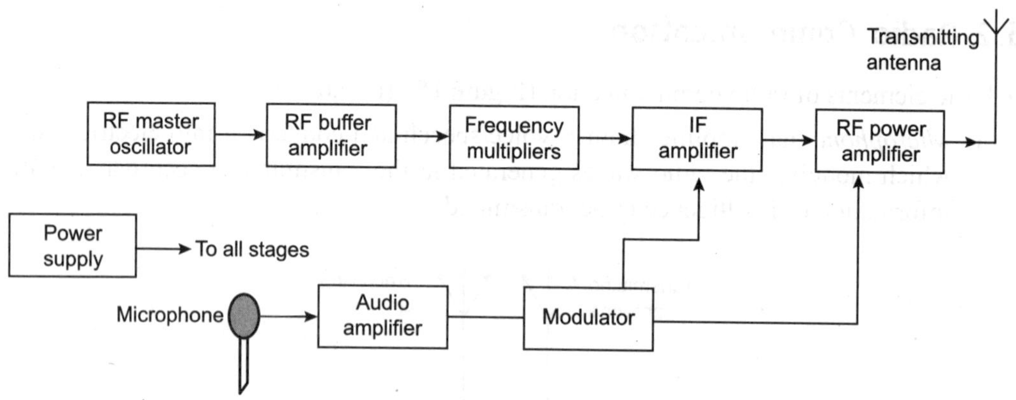

Figure 15.2 Block diagram of amplitude-modulated radio-telephone transmitter

(i) **RF master oscillator** A crystal-controlled master oscillator is used to operate the transmitter at desirable fixed radio frequency. The power output of master oscillator being not sufficiently large is amplified in stages to the desired level.

(ii) **RF buffer amplifier** It isolates the master oscillator from succeeding stages so that variations of coupling and antenna loading do not affect oscillator frequency generated by it.

(iii) **Frequency multipliers** Crystals have limitation in the frequency generated and as such we cannot get a carrier wave of very high frequencies. To obtain high-frequency carrier waves, frequency multipliers are used to multiply the frequency of carrier wave generated by the oscillator.

(iv) **Audio amplifier** It is used to amplify the audio signal output of microphone which converts sound wave into electrical (vibrations) wave.

(v) **Modulator** Its function is to modulate the amplitude of RF carrier wave in accordance with the amplified audio signal.

(vi) **Intermediate RF amplifier** The modulated carrier is fed into this stage for final amplification before being carried by the antenna for high-level modulation.

(vii) **Power supply** It is common to all stages and supplies the required power.

15.4 Amplitude Modulation (AM)

In an AM system, the amplitude of the carrier is altered by the information.

Here, the amplitude (E_C) of carrier wave is varied by the modulating wave having amplitude E_m and whose frequency (f_m) is invariably much lower than carrier wave frequency (f_c). The amplitude of carrier wave is made proportional to instantaneous amplitude (E_m) of modulating wave as shown in Figure 15.3.

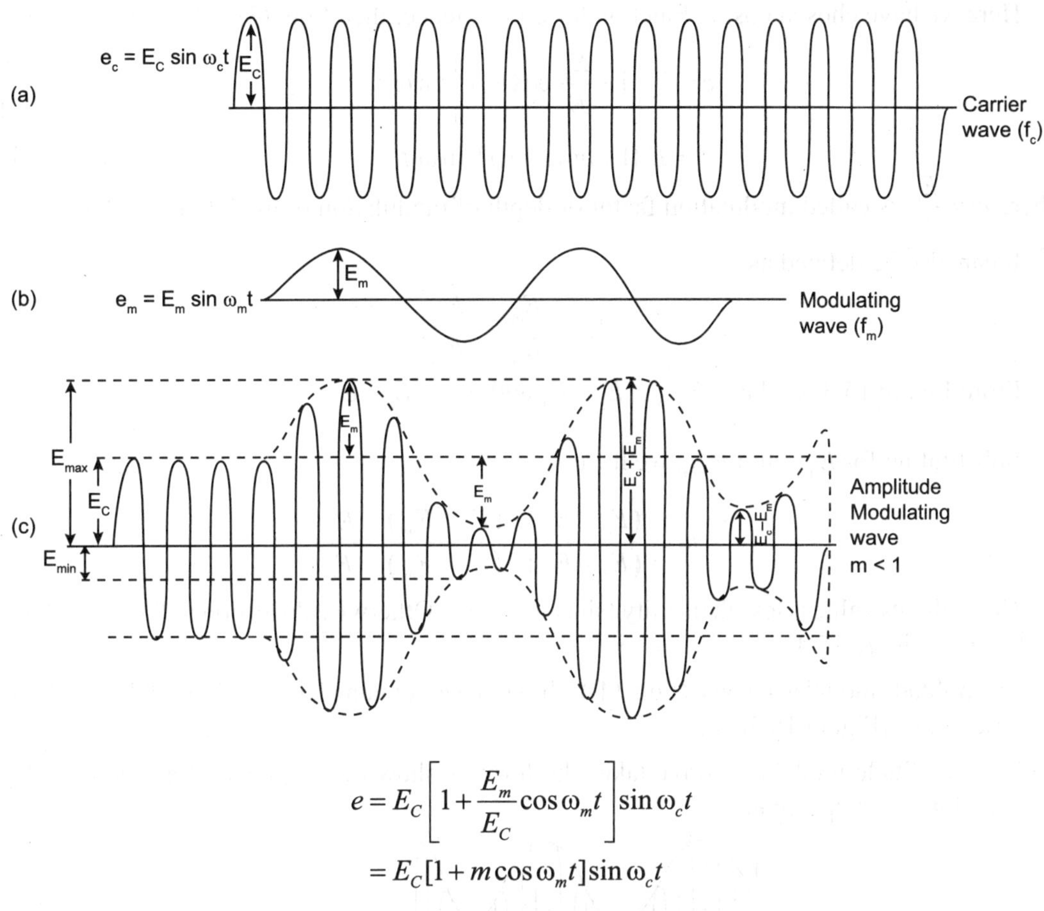

Figure 15.3 Amplitude modulation

Let the carrier wave be represented by

$$e_c = E_c \sin \omega_c t \tag{1}$$

and modulating wave be represented by

$$e_m = E_m f(t) \tag{2}$$

where e_m is the instantaneous value of voltage which is a certain function of time. In amplitude modulation, amplitude E_C of the carrier wave is varied in accordance with the amplitude of modulating wave. By superposition of the two, the modulated wave is given by

$$e = E_c \left[1 + \frac{E_m}{E_c} f(t) \right] \sin \omega_c t \tag{3}$$

The modulating signal is also represented by a sinusoidal wave and given by

$$e_m = E_m \cos \omega_m t$$

Here we have chosen cosine function for convenience; then Eqn. (3) reduces to

$$e = E_c \left[1 + \frac{E_m}{E_c} \cos \omega_m t \right] \sin \omega_c t \tag{4}$$

$$= E_c [1 + m \cos \omega_m t] \sin \omega_c t \tag{4a}$$

where $m = \frac{E_m}{E_c}$ is called modulation factor or depth of modulation or modulation index.

It can also be defined as

$$m = \frac{E_{max} - E_{min}}{E_{max} + E_{min}}$$

From Figure 15.3, we have $E_{max} = E_c + E_m$ and $E_{min} = E_c - E_m$

Substituting for E_{max} and E_{min}, we have

$$m = \frac{(E_c + E_m) - (E_c - E_m)}{(E_c + E_m) + (E_c - E_m)} = \frac{E_m}{E_c}$$

Normally its value is less than unity (since $E_m < E_c$). Otherwise distortion will be introduced as shown in Figure 15.4.

In amplitude modulated wave (Eqn. (4)), the signal is contained in variations of the amplitude of carrier wave (Figure 15.3(c)).

The amplitude modulated wave takes the form as shown in Figure 15.4(a) and (b) when $m = 1$ and $m > 1$, respectively.

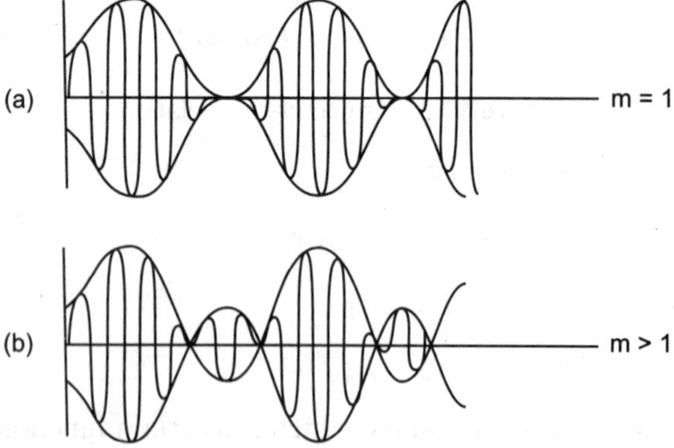

Figure 15.4

Eqn. (4a) can be expressed as

$$e = E_c \sin \omega_c t + \frac{mE_c}{2} \sin[\omega_c + \omega_m]t + \frac{mE_c}{2} \sin(\omega_c - \omega_m)t \tag{5}$$

From Eqn. (5), we conclude that the amplitude modulated wave is equivalent to the sum of three sinusoidal waves:

(i) first having amplitude E_C and frequency f_c

(ii) second having amplitude $\frac{mE_c}{2}$ and frequency $(f_c + f_m)$

(iii) third having amplitude $\frac{mE_c}{2}$ and frequency $(f_c - f_m)$

Thus, in amplitude modulated wave, two more frequencies have crept in, namely $(f_c + f_m)$ and $(f_c - f_m)$, which are called side band frequencies. The first one is called upper side band (USB) and the second one is called lower side band (LSB) as shown in Figure 15.5.

The bandwidth is $[(f_c + f_m) - (f_c - f_m)] = 2f_m$, i.e. it is twice that of modulating signal frequency.

It is of much importance because of the fact that the tuned amplifier used in the receiver must have the required bandwidth to include these side bands.

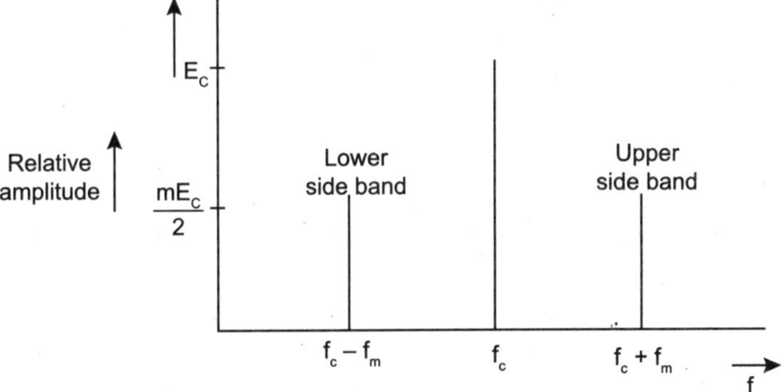

Figure 15.5

15.4.1 Power in AM Wave

Power carried by a voltage signal is proportional to the square of its amplitude. The r.m.s. value of amplitude of the three components of AM wave in Eqn. (5) is $\frac{E_c}{\sqrt{2}}, \frac{mE_c}{2\sqrt{2}}, \frac{mE_c}{2\sqrt{2}}$, respectively

Total power p_t carried by the AM wave is the sum of the power of carrier component, power of upper side band and power of lower side band.

It is proportional to the square of amplitude.

So,

$$p_t \ \alpha \left(\frac{E_C}{\sqrt{2}}\right)^2 + \left(\frac{mE_C}{2\sqrt{2}}\right)^2 + \left(\frac{mE_C}{2\sqrt{2}}\right)^2$$

$$\alpha \ \frac{E_C^2}{2} + \frac{m^2 E_C^2}{8} + \frac{m^2 E_C^2}{8}$$

or $$p_t \;\alpha\; \frac{E_C^2}{2}\left[1+\frac{m^2}{2}\right] \tag{6}$$

Power of carrier component,

$$p_c \;\alpha\; \frac{E_C^2}{2} \tag{7}$$

Eqn. (6) can be written as

$$p_t = p_c\left[1+\frac{m^2}{2}\right] \tag{8}$$

Power in side bands,

$$p_s \;\alpha\; \frac{m^2E_C^2}{8}+\frac{m^2E_C^2}{8}\left(=\frac{E_C^2}{2}\left[\frac{m^2}{2}\right]\right) \tag{9}$$

Making use of Eqn. (7), we have

$$p_s = \frac{m^2}{2}P_C \tag{10}$$

So power in side bands of modulated wave is some fraction of power in carrier wave.

If $m = 1$, then Eqn. (10) gives

$$p_s = \frac{P_C}{2} \tag{11}$$

Thus, in this special case, power in side bands is half that of power in carrier wave.

15.5 Methods for Amplitude Modulation

Normally vacuum tubes are used for amplitude modulation due to high power involved in the process. Presently, high-power transistors are also available which can bear high power (current). However, vacuum tubes are still in use in radio stations since they are capable of bearing high power.

15.5.1 Principle

The process of modulation can be understood mathematically as follows:

Every device like triode, diode or transistor is a not a linear device as revealed by their (e, I) characteristic. In general, the relation between i and e is expressed as

$$i = a_0 + a_1 e_i + a_2 e_i^2 + a_3 e_i^3 + \ldots \tag{12}$$

where, a_0, a_1, a_2 are constants.

Let input voltage e_i comprise

(i) carrier wave $e_c = E_C \sin \omega_c t$

(ii) modulating wave $e_m = E_m \sin \omega_m t$

Substituting for e_i which comprises two frequencies in Eqn. (2), we have

$$i = a_0 + a_1[E_C \sin \omega_c t + E_m \sin \omega_m t]$$
$$+a_2[(E_C \sin \omega_c t + E_m \sin \omega_m t)^2]$$
$$+a_3[(E_C \sin \omega_c t + E_m \sin \omega_m t)^3] + \ldots$$

On simplification, we have (neglecting higher order terms)

$$i = a_0 + a_1[E_c \sin \omega_C t + E_m \sin \omega_m t] + a_2 E_C^2 \sin^2 \omega_c t + a_2 E_m^2 \sin^2 \omega_m t$$
$$+2a_2[(E_C \sin \omega_c t)(E_m \sin \omega_m t)] + \ldots$$

or

$$i = a_0 + a_1 E_C \left[1 + 2\frac{a_2}{a_1} \sin \omega_m t\right] \sin \omega_c t + \frac{1}{2} a_2 E_C^2 [1 - \cos 2\omega_c t]$$

$$+\frac{1}{2} a_2 E_m^2 (1 - \cos 2\omega_m t) + a_1 E_m \sin \omega_m t + \ldots \qquad (13)$$

On comparing the 2nd term $a_1 E_C \left[1 + 2\frac{a_2}{a_1} \sin \omega_c t\right] \sin \omega_c t$ with Eqn. (4a), we observe that it represents an amplitude modulated wave, having modulation factor or modulation index,

$$m = 2\frac{a_2}{a_1} \qquad (14)$$

The frequency of other signals represented by Eqn. (13) is much different from carrier frequency so they can be easily filtered out.

A transistor can be used in two ways for modulation:

(i) collector modulated transistor amplifier

(ii) emitter modulated transistor amplifier

We also have plate modulated triode amplifier.

15.5.2 Collector Modulated Transistor Amplifier

The circuit of such a amplifier is shown in Figure 15.6.

Such circuits are used where low power transmitters are required. Here, the carrier wave is fed to two transistors Q_2 and Q_3 through transistor Q_1. Transistors Q_2 and Q_3 are connected in parallel and they operate under class C conditions. The radio frequency chokes LC_1 and LC_2 stop the carrier wave from mixing directly with the modulating wave and it gets mixed only in the collector circuit. Capacitor c_b does not allow carrier wave a direct path to earth.

The carrier wave signal gets amplified by the push pull action of Q_2 and Q_3, and after amplification it gets mixed by modulating signal (e_m) within the collector circuit, thereby in the output we get an amplitude modulated wave as expressed by Eqn. (14).

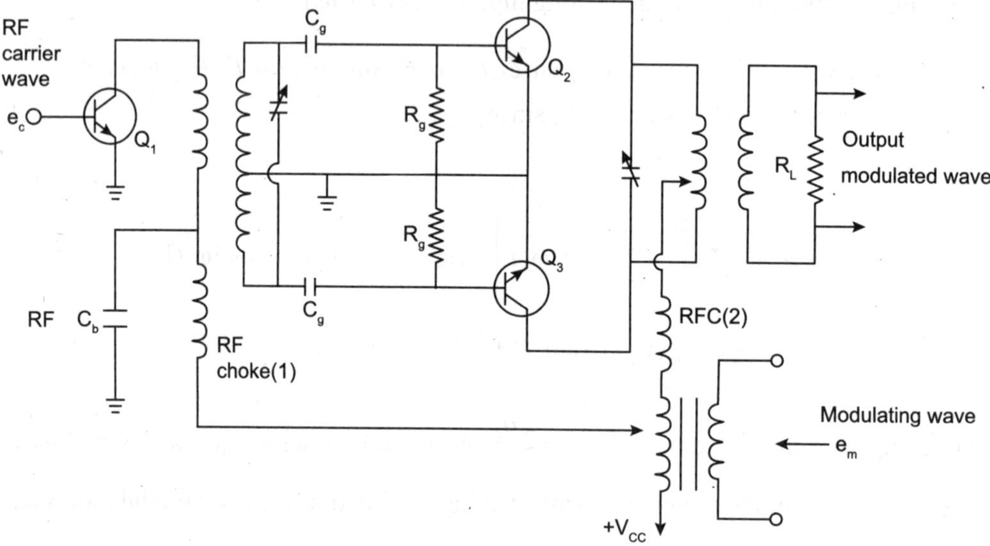

Figure 15.6 Collector modulated transistor amplifier

15.5.3 Plate Modulated Class C Triode Amplifier

It is the most widely used amplified for obtaining amplitude modulated wave for broadcasting and other high-power applications.

The RF carrier signal is amplified by triode Q_1 functioning as class C amplifier. The AF modulating signal (e_m) is amplified by the pushpull amplifier comprising of triodes Q_2 and Q_3. After being amplified, it gets mixed with amplified carrier signal in the plate circuit of triode Q_1 and we get an amplitude modulated wave in the output across load connected in the tank circuit. Here, the radio frequency choke (RFC) is connected in series with modulating transformer so that the amplified carrier signal does not reach directly to transformer T_3. However, if some signal reaches then it is grounded through capacitor C_b. But capacitor C_b does not allow the modulating signal to reach the ground since it offers very high reactance to the signal.

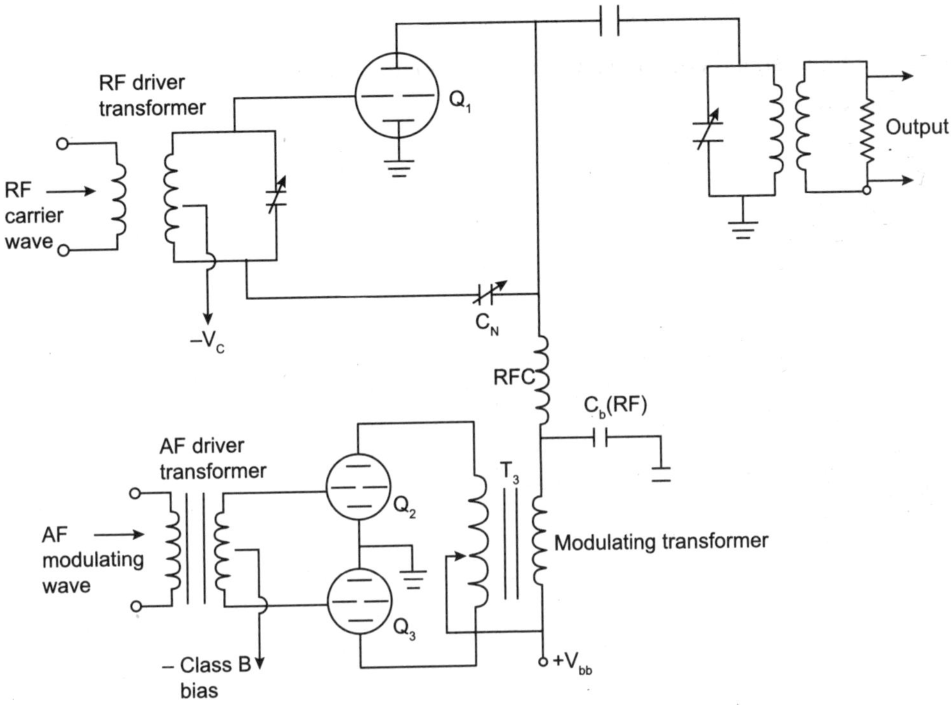

Figure 15.7 Plate modulated class C triode amplifier

15.5.4 Single Side Band (SSB) Transmission

In the wave transmitted from the transmitter, the input power is distributed among carrier wave and two side bands (Figure 15.5) as per Eqn. (6). Most part of the power goes waste since no information/intelligence is carried by the carrier wave component. Also, power carried by one of the side bands goes waste. So, if only one side band which carries useful information is transmitted, the bandwidth of the spectrum (Figure 15.5) has to be halved and also the carrier wave component to be eliminated, it would result into saving of power and at the same time useful information will be transmitted to the destination.

Principle Two separate side bands are generally employed following two methods: (i) phase method, (ii) filter method. In both the methods, a balanced modulator is used so that the carrier wave component is completely eliminated.

15.5.5 Balanced Modulator

A simple circuit of balanced modulator is shown in Figure 15.8.

The function of the above circuit depends upon the (e, i) characteristic of the device (transistor or tube) used. The relation between e and i of the device is generally expressed as:

$$i = a_o + a_1 e_i + a_2 e_i^2 + a_3 e_i^3 + \cdots \quad (15)$$

Here, the input signal comprises carrier and modulating signals.

For transistor Q_1,

$$e_i = E_m \sin \omega_m t + E_c \sin \omega_c t \quad (16)$$

and for transistor Q_2,

$$e_i = (-)E_m \sin \omega_m t + E_c \sin \omega_c t \quad (17)$$

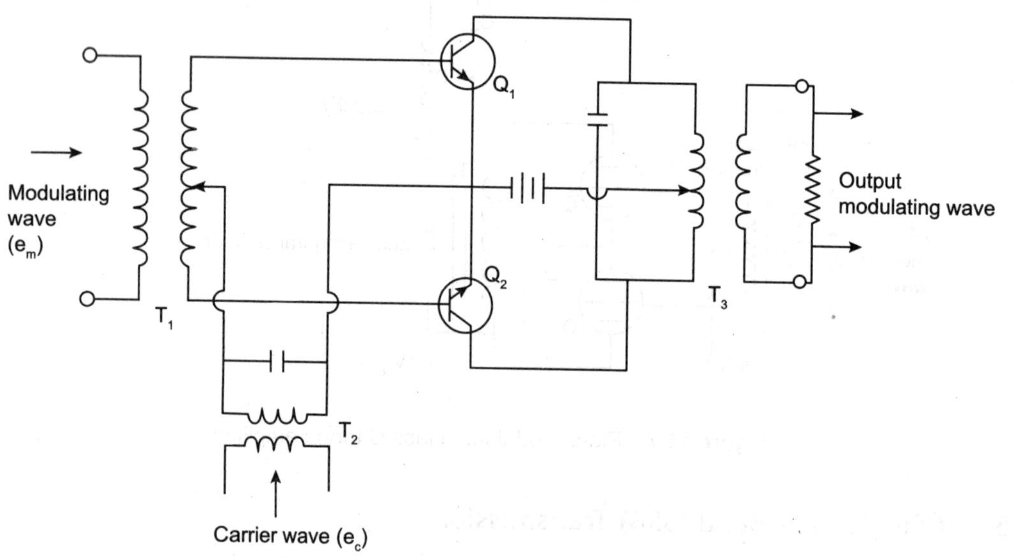

Figure 15.8 Balanced modulator

Using Eqn. (15), output current in transistor Q_1 is given by

$$i_1 = a_o + a_1 (E_m \sin \omega_m t + E_c \sin \omega_c t) + a_2 [E_m \sin \omega_m t + E_c \sin \omega_c t]^2 + \ldots$$

Neglecting higher order terms,

$$i_1 = a_0 + a_1 E_m \sin \omega_m t + a_1 E_c \sin \omega_c t + a_2 E_m^2 \sin^2 \omega_m t + 2 a_2 E_m E_c \sin \omega_m t \sin \omega_c t + a_2 E_c^2 \sin^2 \omega_c t + \cdots$$

Similarly, substituting Eqn. (17) in (15), we have output current in transistor Q_2 as

$$i_2 = a_0 - a_1 E_m \sin \omega_m t + a_1 E_c \sin \omega_c t + a_2 E_m^2 \sin^2 \omega_m t - 2 a_2 E_m E_c \sin \omega_m t \sin \omega_c t + a_2 E_c^2 \sin^2 \omega_c t + \cdots$$

These two currents flow in opposite directions of the primary of transformer T_3 as such the induced e.m.f. in its secondary is proportional to the difference of two currents.

So $$e_o = K(i_1 - i_2)$$

Substituting for i_1 and i_r, we have

$$e_o = 2k[a_1 E_m \sin\omega_m t + 2a_2 E_m E_c \sin\omega_m t \sin\omega_c t + \cdots] \tag{18}$$

where k is constant is timed for carrier frequency (ω_c) and as such modulating frequency (ω_c) is automatically eliminated since $\omega_c >> \omega_c$. A tuned (L–C) circuit resonant to the carrier frequency ω_c so it will reject the modulation frequency ω_m provided $\omega_c >> \omega_m$. It will also reduce the effect of any components at multiples of ω_c due to higher even order terms is connected in the collector circuit. So the output is

$$e_o = 2E_m E_c K\, a_2[\cos(\omega_c + \omega_m)t - \cos(\omega_c - \omega_m)t] \tag{19}$$

From Eqn. (19), we find that the output consists of two side bands only. To get one side band, we can use either of the two methods: (i) SSB by phasing method (ii) SSB by filter method. In the above circuit, transistors have been used; however we can use diode or FET, in which case also the square law in Eqn. (15) applies.

SSB by phasing method:

To get a single side band in phasing method, we use two balanced modulator circuits and two phase shift circuits. The block diagram (Figure 15.3) represents the schematic of entire circuit employed.

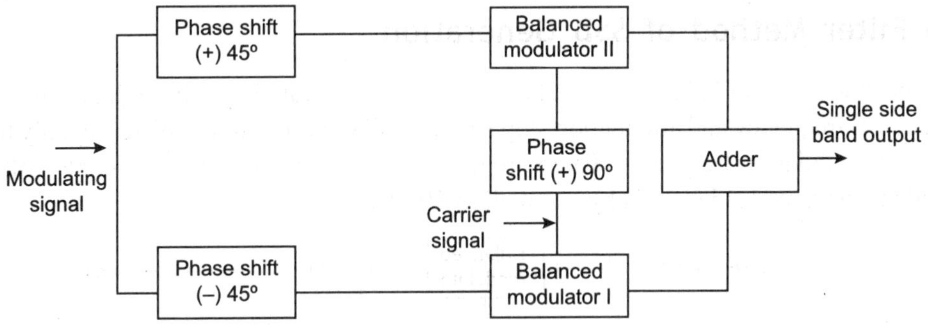

Figure 15.9 Phase method for SSB

One circuit produces a phase shift of 90° in modulating wave and the other also produces a phase shift of 90° with respect to the carrier wave. For producing a phase shift of 90°, each branch of circuit introduces a phase shift of (+) 45° and (–) 45°, thereby resulting into a phase shift of 90°.

In the modulating circuit, square law applies; so for the 1st modulating circuit, the input signal is

$$e_{s1} = E_m \sin\omega_m t + E_c \sin\omega_c t$$

and

$$e_{s2} = (-)E_m \sin\omega_m t + E_c \sin\omega_c t$$

thereby output signal is given by Eqn. (19):

$$e_{o1} = 2k\, a_2 E_m E_c[\cos(\omega_c + \omega_m)t - \cos(\omega_c - \omega_m)t] \tag{20}$$

For the 2nd modulating circuit, input signal is

$$e'_{s1} = E_m \cos \omega_m t + E_c \cos \omega_c t$$

$$e'_{s1} = E_m \cos \omega_m t + E_c \cos \omega_c t$$

and output signal is

$$e'_{s2} = (-)E_m \sin \omega_m t + E_c \cos \omega_c t$$

Using Eqn. (19), we have output

$$e_{o2} = 2k\, a_2 E_m E_c[\cos(\omega_c + \omega_m)t + \cos(\omega_c - \omega_m)t] \tag{21}$$

Hence the output at adder circuit [making use of Eqn. (20) and (21)] is

$$e_o = e_{o1} + e_{o2}$$

$$= k'a_2\ E_m E_c\ \cos(\omega_c + \omega_m)t \tag{22}$$

From Eqn. (22) we find that the output consists of upper side band (comprising $(\omega_c + \omega_m)$ frequency) only. However, if we subtract the two output signals given by Eqn. (20) and (21), then in the output, the lower side band (comprising of $(\omega_c - \omega_m)$ frequency) is obtained and as expressed by $e_o = e_{o1} - e_{o2} = k'a_2 E_m E_c \cos(\omega_c - \omega_m)t$. Thus by this method, we can have either upper band or lower band.

15.6 Filter Method of SSB Generation

When the side bands of balanced modulator output are separated by a filter, this network must have a cutoff frequency and attenuation slope which will separate signals differing only by twice the lowest modulation frequency. There are quartz crystal or mechanical resonator filters that will reduce unwanted side bands by as much as 50 dB.

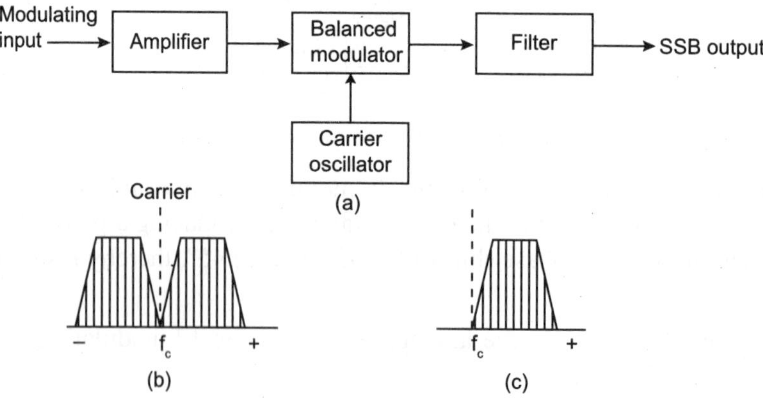

Figure 15.10 (a) Filter method of SSB generation, (b) balanced modulated output, (c) output after filtering single side band

The block diagram of filter method is shown in Figure 15.10. Without infinitely steep cutoff slope, it is customary to place filter cut off so that the carrier frequency would be about 20 dB below the wanted side band level.

Key circuits in the transmitter are balanced modulator and side band suppression filter.

Basically, such a filter must have a flat pass band and extremely high attenuation outside the pass band. There is no limit on this. Higher attenuation is better. In radio communication system, the frequency range used for voice is 300 to about 2800 Hz in most cases. If it is necessary to suppress the lower band and if the transmitting frequency is f, then the lowest frequency the filter must pass without attenuation is $[f + 300]$ Hz, whereas the highest frequency that must be fully attenuated is $[f - 300]$ Hz. In other words, the filter's response must change from zero attention to full attenuation over a range of 600 Hz; if the transmitting frequency is much above 10 MHz, this is virtually impossible.

Looking at the situation from other end, we find that there must be an upper frequency limit for any type of filter used. It has been found, for instance, that multistage LC filters cannot be used for RF values much greater than 100 KHz. Above this frequency, attenuation outside the pass band is insufficient. However, mechanical filters have been used at frequencies upto 500 KHz and crystal ceramic filters upto about 20 MHz.

Of the three major types of SS filters in use, the mechanical filter seems to be the one with best properties, including small size, good pass band, good attenuation characteristics and adequate upper frequency limit.

Crystal or ceramic filters may be cheaper, but are technically preferable only at frequencies in excess of 1 MHz.

15.7 Systems Comparison

Single side band modulation is a form of modulation in which the carrier has been cancelled out with a balanced modulator and one of the side bands has been filtered/removed. SSB is made possible by the fact that in ordinary AM the carrier contains no information, while the two side bands are mirror images of each other. From information-carrying point of view, one of the side bands is superfluous. The main advantages of SSB over DSB are that it requires much less power to transmit and occupies only half the bandwidth. Its main disadvantage is that it requires higher system frequency stability for satisfactory operation and this feature is particularly demanding on receivers.

15.8 Frequency Modulation

Here, the frequency of carrier wave (Figure 15.11(b)) is changed in accordance with the amplitude of modulating wave (Figure 15.11(a)) and the amplitude of modulated wave remains constant as that of carrier wave.

Until recently, amplitude modulation has been employed in radio transmission. At present, frequency modulation is employed for audio signal and amplitude modulation for video signal.

Here, angular frequency ω_c of the carrier wave is made to vary in accordance with the amplitude of modulating wave as

$$\omega(t) = \omega_c + k_f E_m \cos\omega_m t$$

where

$$\omega(t) = \omega_c + k_f E_m \cos\omega_m t$$

where

(i) $f_c = \dfrac{\omega_c}{2\pi}$ is called centre frequency

(ii) ω_m is called modulating frequency and it is assumed that $\omega_m < \omega_c$

(iii) $k_f E_m$ is the degree of frequency variation proportional to modulation amplitude.

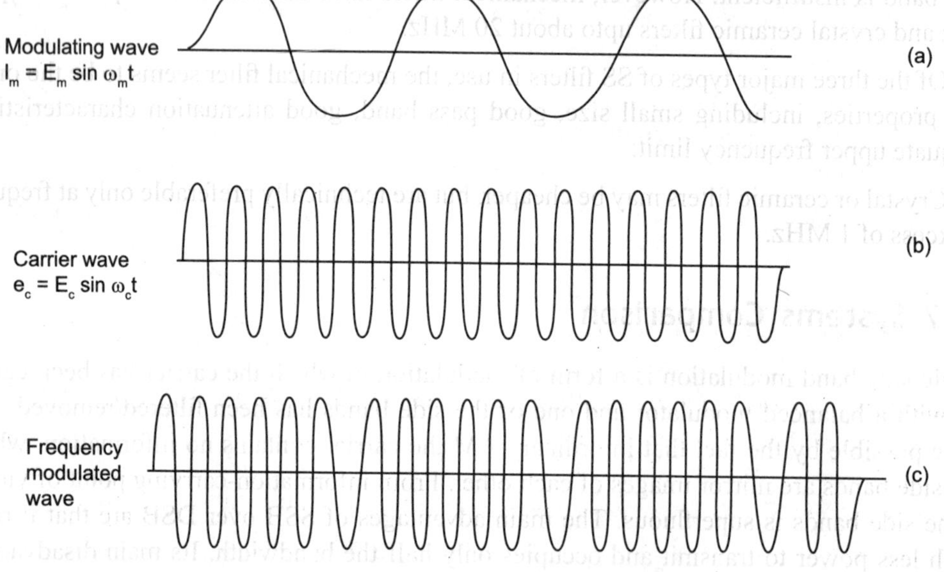

Figure 15.11 Frequency modulation

The equation of modulated wave (Figure 15.7(c)) can be expressed as $e = E_c \sin \int \omega dt + \theta_o$, where θ_o is another constant

Substituting for ω, we have

$$e = E_c \sin \int (\omega_c + k_f E_m \cos\omega_m t)dt + \theta_o$$

Integrating, we have

$$e = E_c \sin\left(\omega_c t + \frac{k_f E_m}{\omega_m}\sin\omega_m t\right); \text{ here } \theta_o \text{ is dropped} \quad (23)$$

The maximum value of $\frac{k_f E_m}{\omega_m}$ is called frequency deviation and it is generally denoted by $(\Delta f)_{max}$. The ratio of Δf (frequency deviation) to maximum modulating frequency $(f_m)_{max}$ is called deviation ratio of the system or frequency modulation index and denoted by m_f or δ.

$$\delta = m_f = \frac{(\Delta f)_{max}}{(f_m)_{max}} = \frac{K_f E_m}{(\omega_m)_{max}}$$

Hence the equation of frequency modulated wave (Eqn. (23)) can be written as

$$e = E_c \sin[\omega_c t + \delta \sin \omega_m t] \tag{24}$$

Such a wave is shown in Figure 15.11(c).

It is to be noted that for simplicity we have assumed modulating signal to be sinusoidal. This signal has two parameters: (i) amplitude and (ii) frequency; these parameters must be represented by modulation. It is assumed that phase relation of a complex modulation signal will be preserved.

By definition of frequency modulation, the amount by which carrier frequency is varied from its unmodulated value, called deviation, is made proportional to the instantaneous value (amplitude) of modulating wave (Figure 15.11(c)). The rate at which this frequency variation takes place is naturally equal to modulating frequency. In this situation, all modulating signals having the same amplitude will deviate carrier frequency to the same amount no matter what their frequency is.

Similarly, all modulating signals of same frequency will deviate carrier frequency at the same rate, no matter what their individual amplitude is.

The above expression [Eqn. (24)] is of the form sin $(a + b)$ and it can be expanded as

$$\sin[\omega_c t + \delta \sin \omega_m t] = \sin \omega_c t \cos(\delta \sin \omega_m t) + \cos \omega_c t \sin(\delta \sin \omega_m t) \tag{25}$$

These functions can be evaluated by Bessel coefficients where

$$\cos(\delta \sin \omega_m t) = J_o(\delta) + 2J_2(\delta)\cos 2\omega_m t + 2J_4(\delta)\cos 4\omega_m t + \cdots$$

$$\sin(\delta \sin \omega_m t) = 2J_1(\delta)\sin \omega_m t + 2J_3(\delta)\sin 3\omega_m t + 2J_5(\delta)\sin 5\omega_m t + \cdots$$

where J_n' S are Bessel functions of the first kind and order n and are defined by

$$J_n(s) = \sum_{\delta=0}^{\alpha} \frac{(-1)^K}{\lfloor K \lfloor K+n} \left(\frac{\delta}{2}\right)^{n+2k}$$

The values of $J_n(\delta)$ are tabulated from different reference tables available elsewhere.

$$\sin A \cos B = \frac{1}{2}[\sin(A+B) + \sin(A-B)]$$

$$\cos A \sin B = \frac{1}{2}[\sin(A+B) - \sin(A-B)]$$

The frequency of modulated wave in Eqn. (24) becomes

$$\begin{aligned} e = E_o[&J_0(\delta)\sin\omega_o t + J_1(\delta)[\sin(\omega_o - 2\omega_m)t - \sin(\omega_o - \omega_m)t] \\ &+J_2(\delta)[\sin(\omega_o + 2\omega_m)t + \sin(\omega_o - 2\omega_m)t] \\ &+J_3(\delta)[\sin(\omega_o + 3\omega_m)t - \sin(\omega_o - 3\omega_m)t] \\ &+J_4(\delta)[\sin(\omega_o + 4\omega_n)t + \sin(\omega_o - 4\omega_m)t] \end{aligned} \tag{26}$$

Eqn. (26) is a reasonable approximation at $\delta = 20$ and above. For small deviation ratio, $\delta < 0.6$. The spectrum reduces to

$$e = E_i J_o(\delta)\cos\omega_c t + J_1(\delta)[\sin(\omega_o + \omega_m)t - \sin(\omega_o - \omega_m)t]$$

which is similar to the double side band AM wave given by Eqn. (5) but with a phase shifted carrier. The bandwidth is $\text{BW} = 2f_{m(\max)}$ and the system is termed narrow band FM. For $\delta > 0.6$ the system is considered wide band.

For FM broadcasting, Δf range is usually 75 kHz and receivers having bandwidths are designed for ±100 kHz. The signal band would then be near 200 kHz and would fit within the receiver design band. The function becomes zero or has roots at $\delta = 2.40, 5.52, 8.65, 11.79, 14.93\ldots$ as shown in Figure 15.12. At those values, centre frequency does not exist.

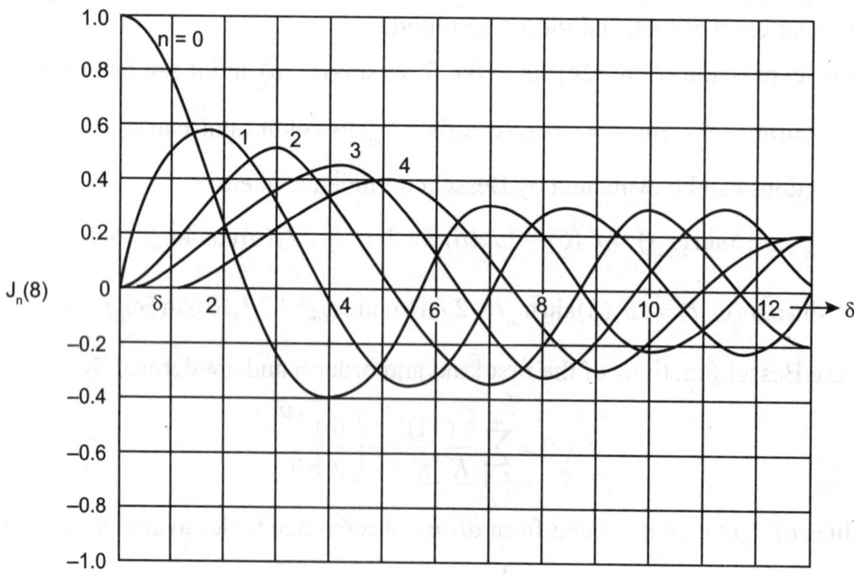

Figure 15.12 Bessel functions of the first kind

15.9 Methods for Frequency Modulation

Variation of reactance in a tuned circuit of oscillator at modulation frequency will modulate oscillator frequency. It is possible to employ a tube or transistor as variable reactance, as that shown in Figure 15.13.

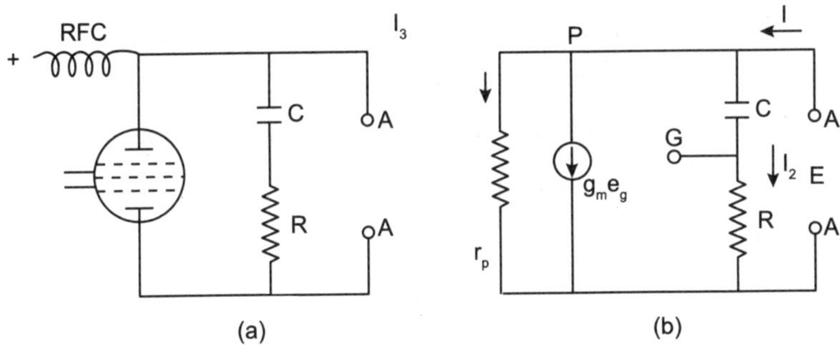

Figure 15.13 (a) Reactance tube circuit, (b) equivalent circuit of reactance tube

The variable reactance acts across terminals *A*, when these terminals are shorted across the oscillator tank circuit as shown in Figure 15.14.

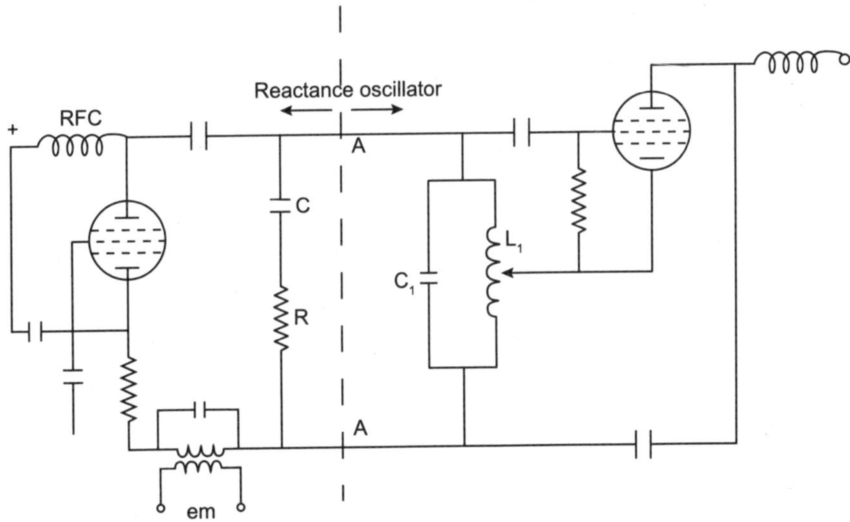

Figure 15.14 Use of a reactance tube

The reactance of capacitor *C* (Figure 15.13(b)) must be large as compared to resistor *R* so that current I_2 will be in quadrature to *E*. Neglecting r_p of the pentode it being parallel and usually quite high, current *I* is given by

$$I = I_2 + g_m E_g$$

$$= j\omega CE + g_m E_g \tag{27}$$

Voltage at the grid, $E_g = \approx j\omega$ CER so that

$$I \cong j\omega c(1+g_m R)E \tag{28}$$

Hence the circuit represents an effective capacitance

$$c_{eff} = c(1+g_m R) \approx g_m RC \tag{29}$$

The capacitance is controlled by transconductance g_m of the tube used, which as a function of the grid, bias, and g_m can be varied by modulating the signal as in Figure 15.14.

Linearity of frequency of bias change is small. The value of g_m under modulation is

$$g_m = g_e(1+k_f E_m \cos\omega_m t] \tag{30}$$

And oscillator frequency

$$\omega = \sqrt{\frac{1}{L_1(c_1 + RCg_e(1+k_f E_m \cos\omega_m t)}} \tag{31}$$

The centre frequency with zero modulation will be

$$\omega_o = \sqrt{\frac{1}{L_1(C_1 + RCg_e)}} \tag{32}$$

The shift due to modulation is

$$\frac{\omega}{\omega_o} = \sqrt{\frac{1}{1+\dfrac{K_f E_m \cos\omega_m t}{1+\dfrac{C_1}{R_C g_e}}}} \tag{33}$$

The frequency shift is small and the above expression may be expanded by binomial theorem. The first two terms give

$$f \approx f_o\left[1 - \frac{k_f}{2(1+C_1/RCg_e)} E_m \cos\omega_m t\right] \tag{34}$$

which is a frequency modulated wave with

$$\delta = \frac{K_f E_m}{2(1+C_1/RCg_e)} \tag{35}$$

15.10 Diode FM Generation

Semiconductor diode (varactor diode) capacitance is a function of reverse voltage, so the diode forms a suitable means of frequency modulation of an oscillator wave as shown in Figure 15.15.

A polarizing voltage V_R maintains the reverse-bias in series with modulating voltage e_m. The choke RFC keeps polarizing circuits from affecting higher frequency oscillator currents. The capacitance of p-n junction is proportional to $\frac{1}{\sqrt{V_B}}$ and oscillator frequency is given by

$$\omega = \frac{1}{L(C + K\frac{1}{\sqrt{V_B}})} \tag{36}$$

Figure 15.15 Diode FM generator

In order to achieve linearity of frequency deviation with VB, it is necessary to keep the effect of $\left[K\frac{1}{\sqrt{VB}}\right]$ small.

15.10.1 FM Generation

A simpler way to generate FM signals is by using a voltage-controlled oscillator. Figure 15.15(a) shows a simplified version of voltage-controlled oscillator.

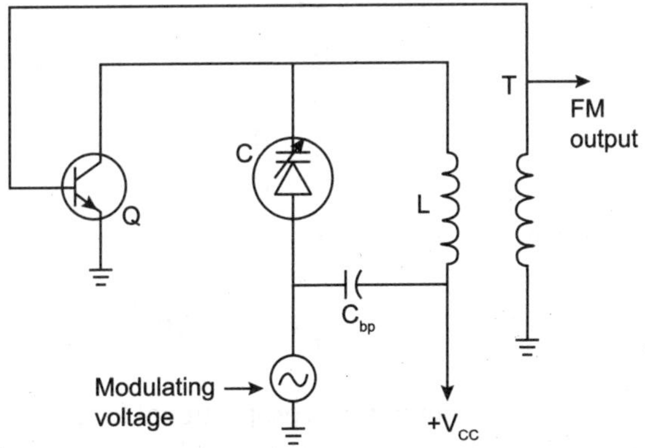

Figure 15.15 (a) FM generator

Here we employ a varactor diode *D* to convert the instantaneous value of modulating signal to change in frequency. A parallel resonant circuit consisting of varactor diode (capacitor) *C* and C_{bp} is used in the collector circuit of transistor *Q*, thereby forming a tuned collector oscillator.

In the absence of a modulating voltage, the centre frequency of oscillator is given by $f_c = \frac{1}{2\pi\sqrt{LC}}$ where L is the inductance of primary winding of transformer T_1 and C is the capacitance of varactor diode capacitor. C_{bp} is quite large and hence its reactance is negligible.

On applying modulating voltage, it changes the reverse-bias of varactor diode and its capacitance changes. Capacitance C *is* used for the value of a given reverse-bias V_{CC} and given by

$$C = \frac{K}{V_{CC}^n}$$

where K is a constant, n is 1/2 for an abrupt junction varactor diode.

If ΔC is change in capacitance due to modulating voltage ΔV, the new frequency of oscillation is given by

$$f = \frac{1}{2\pi\sqrt{L(C - \Delta C)}} = f_c\left[1 - \frac{\Delta C}{2C}\right] \text{ since } \Delta C \ll C$$

So we can use bionomial theorem

and
$$f_c = \frac{1}{2\pi\sqrt{Le}}$$

Hence

$$(C - \Delta C) = \frac{K}{(V_{CC} + \Delta V)^n} = \frac{K}{V_{cc}^n}\left[1 + \frac{\Delta V}{V_{CC}}\right]^{-n}$$

$$= C\left[1 - n\frac{\Delta V}{V_{CC}}\right]$$

$$\therefore \quad \frac{\Delta C}{C} = n\frac{\Delta V}{V_{CC}}$$

Therefore, frequency change

$$\Delta f = f - f_c\frac{\Delta C}{2C} = \frac{n f_c}{2V_{CC}}\Delta V \quad \text{(i)}$$

Eqn. (i) represents that frequency deviation is proportional to ΔV (modulating signal voltage). In this respect the output is a frequency-modulated signal.

15.10.2 FM Systems

The block diagram of a complete FM transmitter is shown in Figure 15.16.

The stability of center frequency is dependent on the stability of oscillator and reactance tube, and does not usually meet channel tolerances which usually allow ±2 kHz at 100 MHz. It is customary to employ a frequency stabilizing system, in which a portion of the output is sampled out and applied to the circuit called frequency discriminator, which gives a DC output voltage proportional in magnitude and polarity to the instantaneous shift of modulated wave from center frequency. This DC voltage is averaged and applied as additional bias to the reactance tube so as to shift average frequency in a direction to correct the drift from true center.

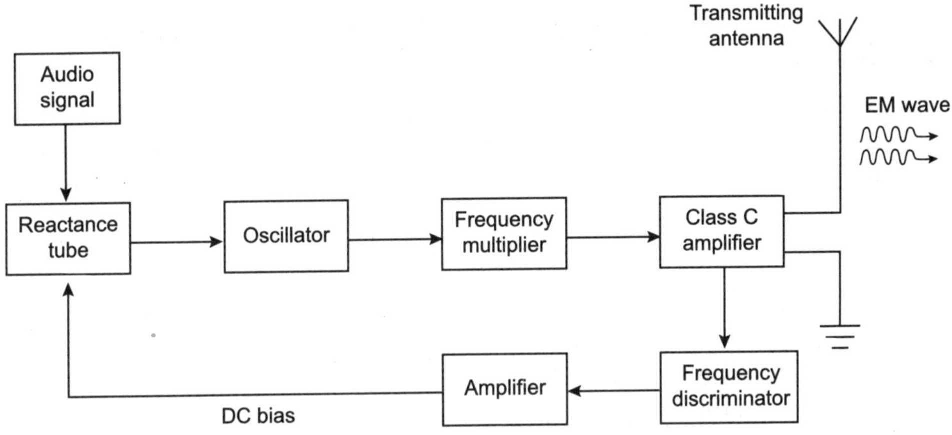

Figure 15.16 FM transmitter

The above method employs a tuned circuit in the discriminator as a standard against which frequency is compared. In other equipment, the output frequency is compared against a crystal oscillator, with the difference signal being rectified and averaged for use as correction bias for reactance tube modulator.

The Armstrong system (Figure 15.17) employs a crystal oscillator with stable center frequency. The system operates on the audio signal with a weighing network so that the modulating signal amplitude becomes inversely proportional to frequency. By phase modulation with this signal, it is possible to generate a synthetic FM wave.

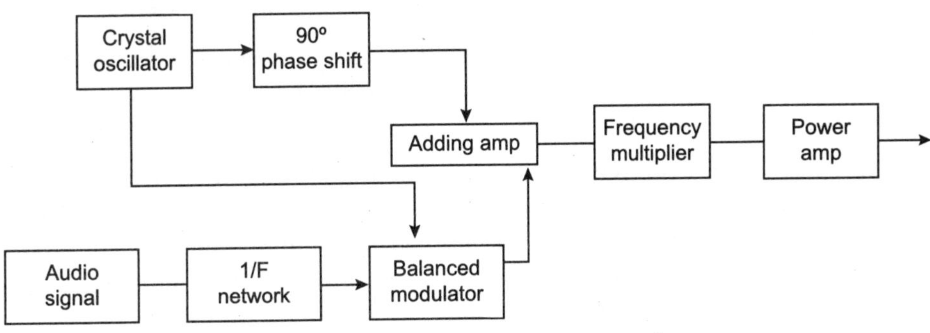

Figure 15.17 Armstrong system

15.11 Comparison of Amplitude Modulation and Frequency Modulation

S.No.	Amplitude modulation	Frequency modulation
1.	The amplitude of modulated wave changes as per modulation wave but its frequency remains constant.	The frequency of modulated wave changes as per amplitude of modulating wave but its amplitude remains constant.
2.	The power of modulated wave depends on modulation index.	The power of modulated wave depends on modulation index.
3.	Such transmitters have low efficiency since a large amount of power is dissipated in carrier wave.	Such transmitters have high efficiency since low energy level modulation is possible and thereafter their power could be increased by using class C amplifiers which have high efficiency.
4.	The total input power is utilized since a large part of it is carried by carrier wave and it goes as waste.	The entire input power is utilized and there is not waste of power.
5.	Unwanted audio signal is always present due to variation in amplitude.	These receivers can be connected to amplitude limiters, so unwanted audio signal remains minimum.
6.	No standard frequencies are used and thereby no guard band of frequencies is kept between two neighboring stations. Moreover, reception is poor due to interface between frequencies generated.	Only standard frequencies are used which have been in use internationally. A guard ring can be put between two neighboring stations and so there is no interference between frequencies generated.
7.	These do not work at high and medium frequencies, so noise is always present.	These work at very high and ultra high frequencies, so there is no noise signal.
8.	Bandwidth of frequencies transmitted is very small, so it is very advantageous.	Bandwidth of frequencies transmitted is quite large, nearly ten times that of AM wave. It is one of the biggest disadvantages.
9.	The entire arrangement of transmitter comprising various subsystems is quite simple, so its maintenance is cheaper.	The entire arrangement of transmitter comprising various subsystems is very complex and elaborate, thereby its maintenance is costly.
10.	The area covered for receiving stations can be quite large since the signal from transmitting station can reach the receiving station both directly along the line of sight and after reflection from ionosphere. This is a big advantage.	The area covered for receiving stations is quite limited since the signal from transmitter can reach there directly along the line of sight only.

15.12 AM vs. FM with Respect to Interference

It is known that noise signals caused by static or manmade electrical discharge create reception problems for radio signals. We shall now discuss interference due to noise signal in amplitude modulation and frequency modulation.

(i) If a receiver is designed for the reception of FM signals with wide frequency bandwidth and is made insensitive to amplitude variation of the signal, such random interference can be largely eliminated. Thus, FM has a major advantage over AM for noisy channel use particularly in the mobile service.

(ii) Interference between two simultaneously received signals on the same or adjacent channels is a major difficulty in radio reception. In this respect, FM also has an edge over AM. In Figure 15.14(a), one signal is represented by phasor $\vec{A}$ rotating about point O. A second signal of amplitude $\vec{B}$ is received so that the sum voltage of signal applied to the receiver is $\vec{R}$. As phasor $\vec{B}$ rotates through its cycle, the amplitude of $\vec{R}$ will change greatly, thus input to the receiver is varied in amplitude by signal $\vec{B}$, and in an AM receiver signal $\vec{B}$ may cause appreciable amplitude interference with $\vec{A}$ even though $\vec{B}$ is quite small.

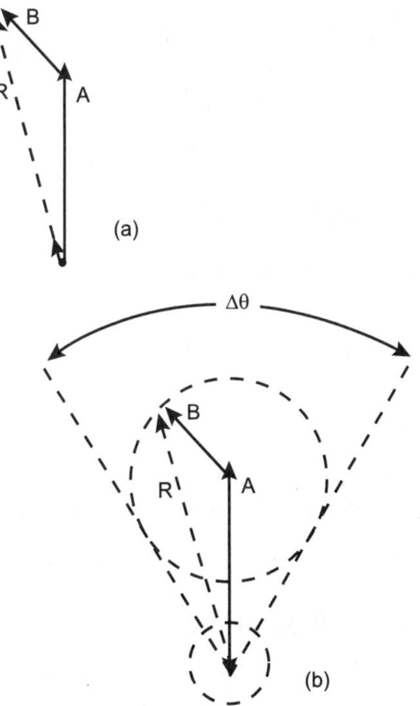

Figure 15.18 Effect of interference between two FM signals

(iii) If signal $\vec{A}$ is frequency modulated as in Figure 15.18(b) and an interfering FM signal $\vec{B}$ is added, the resultant input to receiver will be $\vec{R}$. Phasor $\vec{B}$ will rotate about the end of

$\vec{A}$ and resultant $\vec{R}$ will vary in amplitude and also in relative phase angle. For amplitudes shown, the variation $\Delta\theta$ (Figure 15.18(b)), approximates 1 radian as shown by the circle locus. Even if the two signals are equal in amplitude, phase angle variation cannot be much greater, although amplitude variation would then be so severe that AM reception would be impossible.

(iv) An FM receiver can be made insensitive to amplitude variations by clipping all signals to a common low level. Amplitude variations, due either to amplitude modulation on the signal or amplitude variation of $\vec{R}$ caused by the interfering FM signal, can be removed. Voltage effective on the receiver would then be fixed by a dashed circle, but varying about $\pm\frac{1}{2}$ radian in phase with respect to phase $\vec{A}$. If the original frequency modulation causes its phase angle to vary through $\frac{2\pi\Delta f}{f_m}$ radian per hertz or 10π radians for $\Delta f = 75000$ hertz and $f_m = 15000$ hertz, it can be seen that phase variation of $\pm\frac{1}{2}$ radian due to interference is quite small. If $f_m = 150$ hertz for the same value of Δf, then the desired signal phase variation would by $500\ \pi$ radians, still further reducing the effect of interference.

(v) The effect of interfering signal can be reduced by widening the frequency band of the signal or increasing Δf, sufficiently strong signals being assumed. E.H. Armstrong is credited with demonstrating that increased bandwidth of the signal does reduce interference when the signal is defiantly above the noise level.

15.13 Demodulation (Detection)

It is now possible to discuss the process of demodulation or detection, whereby a (AM or FM) modulated wave is translated to recover the original information. The diode is usually employed for the purpose. We shall first discuss AM detection.

15.13.1 AM Demodulator

Demodulators for AM are large, signal or linear envelope detectors or of small-signal square law form. The latler has a level of distortion which makes its use undesirable, so the envelope detector is preferred.

15.13.2 Linear Envelope Demodulation

A diode in series with large load resistor (Figure 15.19(a)) will have linear (e, i) characteristics as in Figure 15.19(b).

An AM wave (voltage signal) is applied across the diode circuit (Figure 15.19(a)). It is a half wave rectifier and rectification will occur; so the original modulation can be recovered as a variation of the average value of output. When averaged over a time long enough with respect to carrier cycle time period and short with respect to modulation cycle time period, average current over a cycle is

$$I_{av} = \frac{1}{2\pi} \int_{-\pi/2}^{\pi/2} \frac{E_C}{r_d + R}(1 + m_a \cos\omega_m t)\cos\omega_c t \, d(\omega_c t) \tag{37}$$

where r_d is diode internal resistance and limits are $-\pi/2$ to $+\pi/2$ since we get current only for positive input cycle in case of a diode. Output voltage across R

$$e_d = I_{av}(R)$$

substituting for I_{av},

$$e_d = \frac{E_C R}{\pi(r_d + R)} + \frac{K_a E_m R \cos\omega_m t}{\pi(r_d + R)} \tag{38}$$

In Eqn. (38) R.H.S, the first term represents a DC component proportional to E_C and the second term represents modulation components.

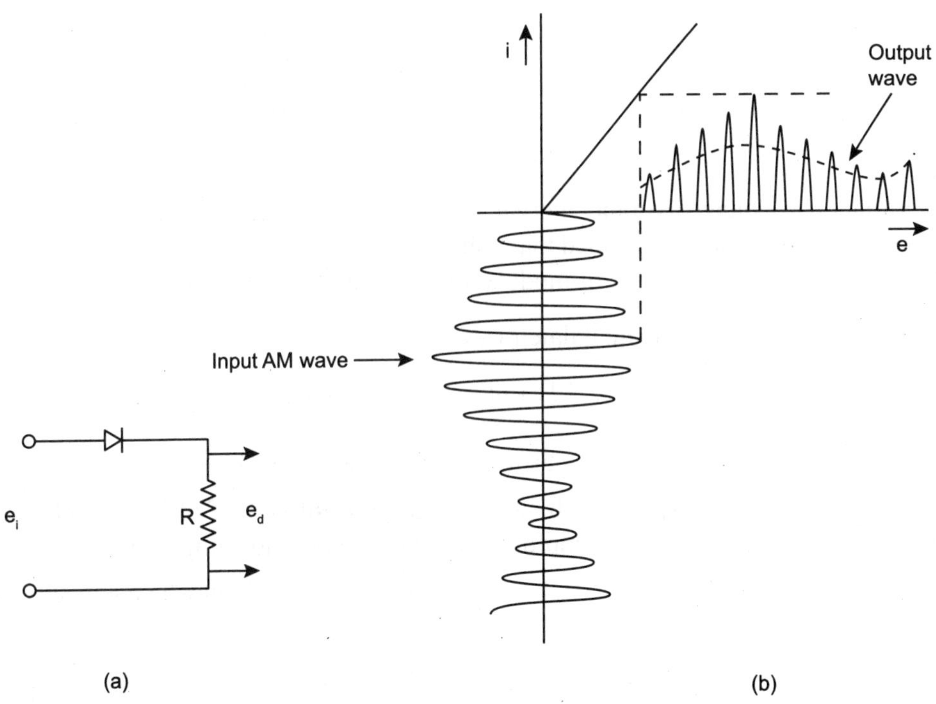

Figure 15.19 (a) Linear diode detectors circuit, (b) (*e*, *i*) characteristic

This simple circuit (Figure 15.19(a)) is insufficient because the $1/\pi$ coefficient in Eqn. (38) limits the output to about 1/3 of the input. To improve efficiency, a filter capacitor can be added (Figure 15.20(a)).

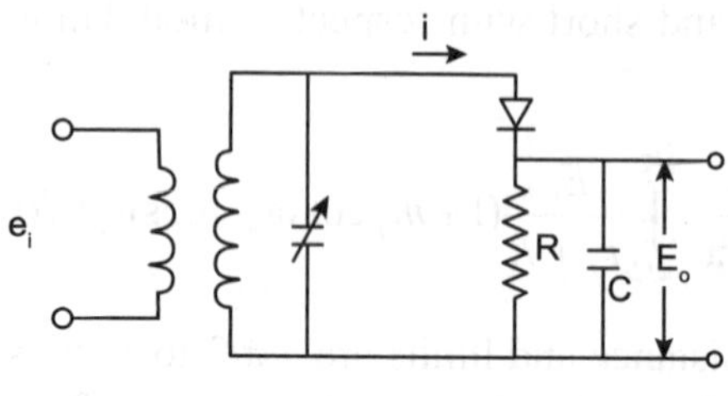

Figure 15.20(a)

Diode detector circuit with RC filter During the positive half cycle the capacitor charges along path (1, 2) up to the peak value of carrier cycle and discharges through resistor R along path (2, 3) during negative half cycle (Figure 15.20(b)). Voltage across C and load tends to follow the envelope of modulated wave.

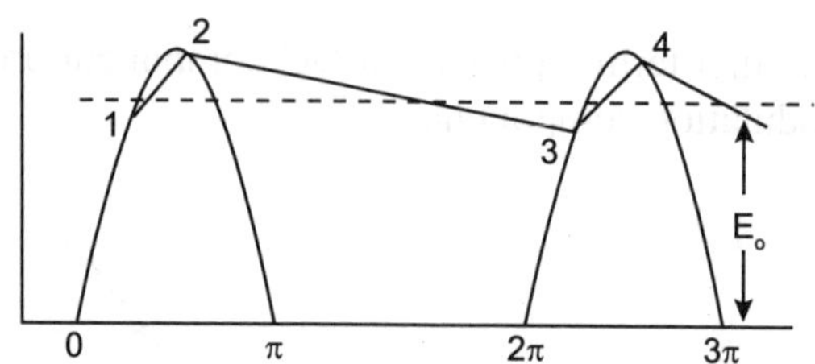

Figure 15.20(b)

The variation of voltage over the interval [1 – 2] and [3 – 4] .... is at carrier frequency. The variation is small and it can be assumed that voltage across C is constant (E_O) over a few cycles of carrier, but variable over a cycle of modulation frequency, since $f_m \ll f_c$.

On employing an amplitude modulated wave,

$$e_r = E_C[1 + m_a \cos\omega_m t]\cos\omega_c t$$

Diode voltage will be biased by E_O as indicated in Figure 15.21.

Current is in the form of pulses ($-\theta_1$ to $+\theta_1$) during positive half of input cycle. But, the method of analysis is similar to that employed for class B operation amplifier.

Voltage across diode $e_d = E_C(1 + m_a \cos\omega_m t)\cos\omega_C t - E_o$ which may be compactly written for one cycle as

$$e_d = E'\cos\theta - E_o \tag{39}$$

Since the term E' is practically constant over a few cycles of f_c (carrier wave), current may be written as

$$i_d = \frac{e_d}{r_d}$$

and averaged as

$$I_o = \frac{1}{\pi r_d}[E' \sin\theta_1 - E_o\theta_1] \tag{40}$$

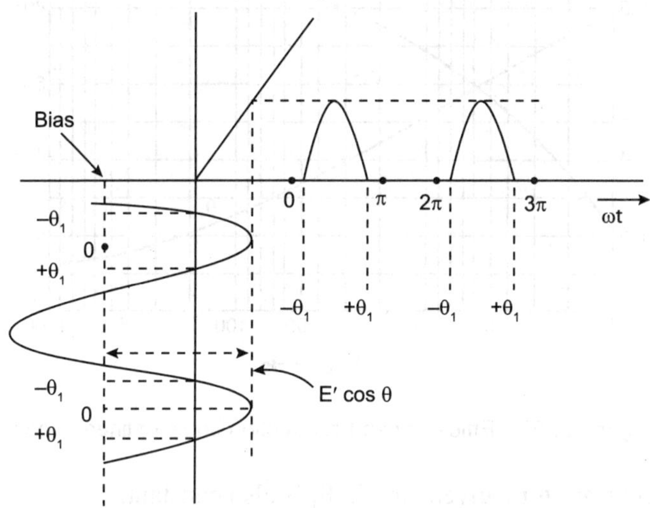

Figure 15.21 Operation of biased linear diode

Noting that conduction limits are $-\theta_1$ to $+\theta_1$ from the figure, it is apparent that

$$\cos\theta_1 = \frac{E_o}{E'}$$

and so

$$E_o = R I_o = \frac{E'R}{\pi r_d}(\sin\theta_1 - \theta_1 \cos\theta_1) \tag{41}$$

So $$E_o = \frac{E_C R}{\pi r_d}[(\sin\theta_1 - \theta_1\cos\theta_1) + m_a(\sin\theta_1 + \theta_1\cos\theta_1)\cos\omega_m t] \tag{42}$$

in view of definition of E'.

The first term of R.H.S. of Eqn. (42) is DC voltage with magnitude dependent on E_C. The second term has amplitude proportional to $m_a E_c = E_m$ and varies at modulation frequency ω_m. This represents recovery of original modulation, without distortion.

Eqn. (41) may be transformed to

$$\frac{R}{r_d} = \frac{\pi}{\tan\theta_1 - \theta_1} \tag{43}$$

This ratio is a function of design choice, and θ_1 may be determined from Figure 15.22.

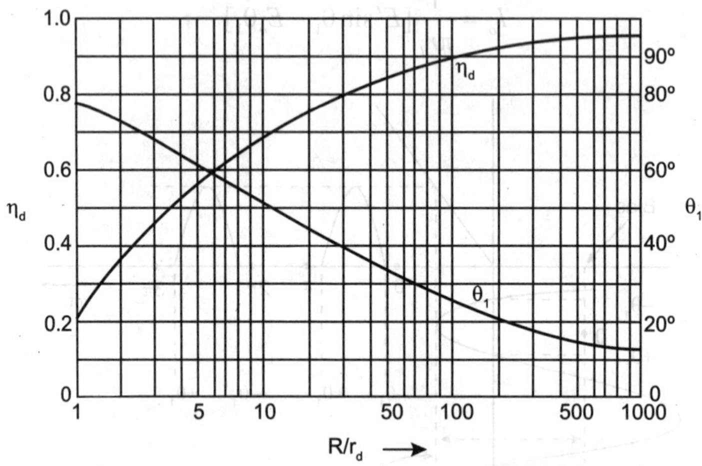

Figure 15.22 Efficiency and conduction angle of linear diode

Since R/r_d is a constant for a given circuit, θ_1 is also constant.

If load R were infinite, then $E_o = E'$ and this output could not be exceeded. The ratio of E_o actually obtained to the value E' which could be obtained in theory is defined as detection efficiency η_d.

So, $$\eta_d = \frac{E_o}{E'} = \frac{R}{\pi r_d}[\sin\theta_1 - \theta_1 \cos\theta_1] \tag{44}$$

The value of efficiency η_d as a function of R/rd is plotted in Figure 15.22. Its value also varies as the value of ωRC changes. The plot shown in Figure 15.23 is for $\omega RC = 50$; it approaches the maximum value of unity for $\omega RC = \infty$ at $\frac{R}{r_d} = 1000$.

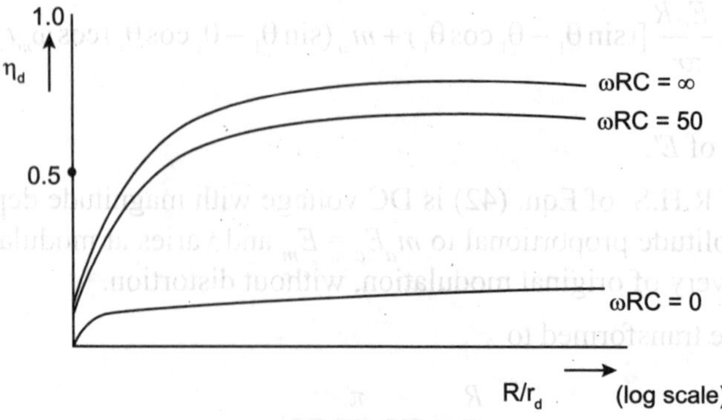

Figure 15.23

15.14 Input Impendence of Envelope Detector

The usual detector input is from a tuned circuit and the shunting effect of diode and load reduce Q of the tuned circuit. It is thus desirable to determine the magnitude of loading effect. The average input power to diode and load is

$$P_{in} = \frac{1}{2\pi}\int_{-\pi}^{\pi} ei\,dt = \frac{(E')^2}{2\pi r_d}(\theta_1 - \sin\theta_1 \cos\theta_1) \tag{45}$$

The input resistance of diode and load is equal to the resistance which would dissipate. P_{in} with the same voltage applied. Since E' is a peak value. So

$$R_i = \frac{(E')^2}{2P_{in}} = \frac{\pi r_d}{\theta_1 - \sin\theta_1 \cos\theta_1} \tag{46}$$

Eqn. (46) gives input impedance of the detector.

Making use of Eqn. (43) we have

$$\frac{R_i}{R} = \frac{\tan\theta - \theta_1}{\theta_1 - \sin\theta_1 \cos\theta_1} \tag{47}$$

This function is plotted in Figure (15.24) with usual circuits employing R/r_d values 50 or more. The equivalent load resistance of diode tends to 0.5 R but in linear diode it tends to 2 R.

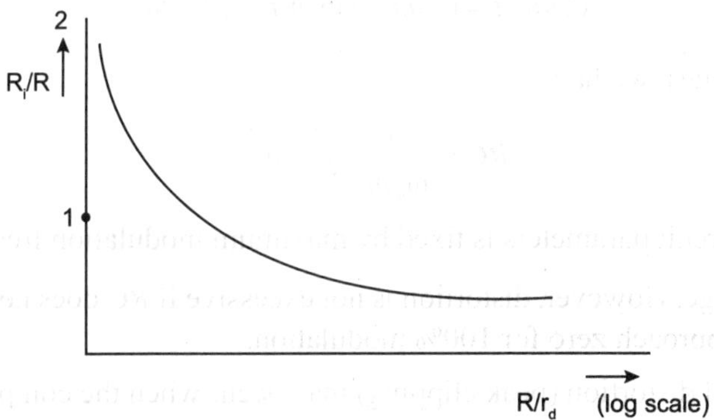

Figure 15.24

15.14.1 Distortion in Envelope Detectors

The R–C time constant of diode load determines the rapidity with which load voltage can change. If voltage cannot change fast enough to follow the modulation envelope, then amplitude distortion will occur. This occurs only as modulation envelope is decreasing, since C must discharge through R and time constant is large. On the increasing portion of the envelope, the capacitor is being charged through r_d, which is much smaller than R, and time constant is limiting.

The rate of change of modulation envelope is

$$\frac{de}{dt} = \frac{d}{dt}[E_C(1 + m_d \cos\omega_m t)] = (-)E_C\omega_m m_\alpha \sin\omega_m t \tag{48}$$

The capacitor discharges according to

$$e_c = E_C \exp(-t/RC) \text{ so that}$$

$$\frac{de_c}{dt} = (-)\frac{e_c}{RC} = (-)\frac{E_C(1 + m_a \cos\omega_m t)}{RC} \tag{49}$$

In order that voltage across C may follow variations in the modulation envelope, the rate of fall of voltage across C must be potentially greater than the rate of fall of signal envelope. The critical condition that will just permit changes in envelope voltage can be found by setting Eqn. (48) equal to Eqn. (49).

This leads to

$$RC \le \frac{1}{\omega_m}\left[\frac{1 + m_a \cos\omega_m t}{m_a \sin\omega_m t}\right] \tag{50}$$

Maximizing the R.H.S. of Eqn. (50) w.r.t. time '*t*', we have

$$\cos\omega_m t = (-)m_a,\ \sin m_a t = \sqrt{1 - m_a^2}$$

Putting these values, we have

$$RC \le \frac{1}{\omega_m m_a}\sqrt{1 - m_a^2} \tag{51}$$

The choice of circuit parameters is fixed by maximum modulation frequency and expected modulation percentage. However, distortion is not excessive if RC does not exceed $\frac{1}{\omega_m m_a}$ even though RC should approach zero for 100% modulation.

A second form of distortion (peak clipping) may occur when the complete circuit of Figure 15.25(a) is used.

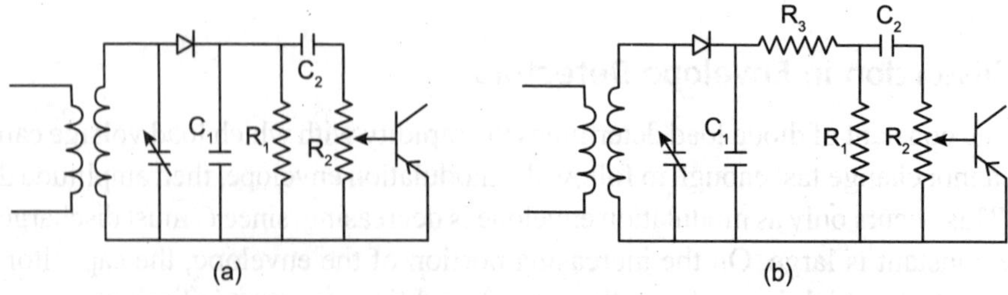

Figure 15.25 Distortion reduction in envelope detector

Here C_2 blocks DC component and R_2 serves as gain control. The DC and AC load impedances are different, the AC resistance being due to parallel combination of R_1 and R_2, the reactance of C being small. The AC resistance is smaller than R_{dc} and AC current might exceed the DC component. However, negative currents are impossible in a diode, and so negative peak would be clipped at zero, and it represents distortion.

Eqn. (42) can also be written as

$$\frac{\text{peak } E_{ac}}{E_{dc}} = \frac{(\text{peak } I_{ac})R_{ac}}{R_{dc} I_{dc}} = m_a$$

If clipping is to be avoided, the maximum value of peak $\frac{I_{ac}}{I_{dc}}$ is unity and so the maximum value of m_a follows as $[m_a]_{max} = \frac{R_{ac}}{R_{dc}}$

In the circuit (Figure 15.25(a)), this becomes $\left[\frac{R_2}{R_1 + R_2}\right]$ which makes it desirable to lower R_1 to permit $m_a = 1$, but in doing so detection efficiency decreases. Employing the circuit of Figure 15.25(b), we can try to equalize DC and AC load by introducing resistor R_3 in series with AC load. This would sacrifice some output voltage.

With a large input signal and large R/r_d to linearize the diode curve, it is possible to hold distortions to one or two percent in envelope detectors.

15.15 Detection of FM Wave

Here we detect the modulated signal on to the carrier wave via frequency variations. It changes the frequency derivations of modulated wave into amplitude variation corresponding to the modulating signal. The transformation of frequency variation into amplitude variation should be linear and efficient.

15.15.1 Foster-Seeley Discriminator

It is one of the most common means of detection of FM waves using minimum circuitry. It is also known as phase discriminator. Its circuit is depicted in Figure (15.26). It consists of two tank circuits L_1C_1 and $(L_1 + L_2)C_2$ tuned to carrier frequency f_c. Capacitors C_C, C_3 and C_4 act as shortcircuit at this frequency. The following features apply:

1. Carrier wave voltage V_1 applies across L_4 since reactance of C_C and C_4 is negligible at frequency f_c.
2. Voltage induced in the secondary $(L_2 + L_3)$ of transformer is 180° out of phase with V_1. Due to the resistance of secondary coil, voltage v_s across capacitor C_2 will differ in phase from the induced secondary voltage depending on the frequency. At frequency equal to f_c, voltage leads V_1 by 90°.

3. The total voltage V_4 applied across diode D_1, C_3 and R_1 network will be the phasor sum of V_1 (across L_4) and V_2 as shown in Figure 15.28.

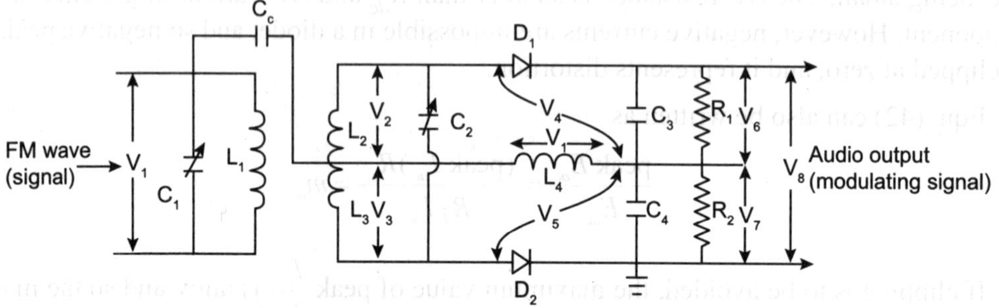

Figure 15.26 Foster-Seeley discriminator

Similarly, voltage V_5 is the phasor sum of V_1 (across L_4) and V_3 as shown in Figure 15.28.

The frequency response curve of discriminator is shown in Figure 15.27. Its shape is approximately *S* type. The output varies linearly over a limited range of frequency called useful range. Beyond this frequency range, the output is not linear.

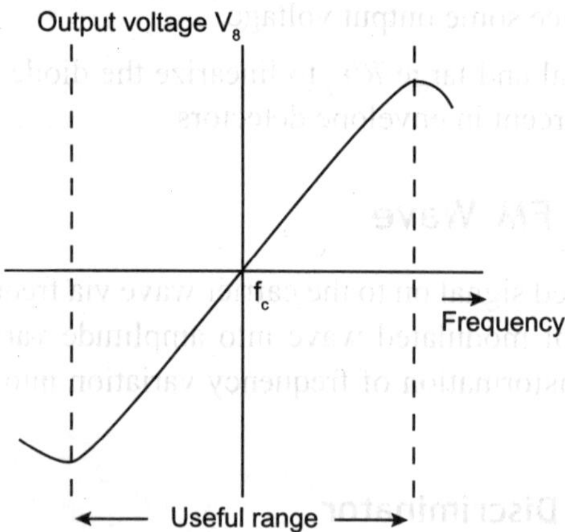

Figure 15.27 Frequency responses of discriminator

The linearity of discriminator is an advantage but its disadvantage is that it does not provide any amplitude limiting action. To achieve it, the circuit must be preceded by a limiter which limits the variation of incoming signal so that reception becomes less distorted and less noisy.

Since V_2 and V_3 are taken from the center tap, they are 180° out of phase with each other and their magnitude is $\left[\frac{V_S}{2}\right]$ where V_s is induced voltage across the secondary. Also, the magnitude of voltages across the two diodes is equal.

4. The magnitude of V_6 is proportional to V_4 and magnitude of V_7 is proportional to V_5. The output voltage V_8 is equal to the difference of V_6 and V_7 and it would be zero at resonance frequency f_c as shown in Figure 15.28(a). Voltage V_6 proportional to V_4 is now greater than V_7, i.e. V_5, and so output V_8 will be positive.

When frequency of FM wave is less than f_c, the capacitive reactance of tank circuit predominates and V_S leads by an angle more than 90° [Figure 15.28(c)].

The voltage proportional to V_5 is now greater than V_6, i.e. V_4, and so output V_8 will be negative.

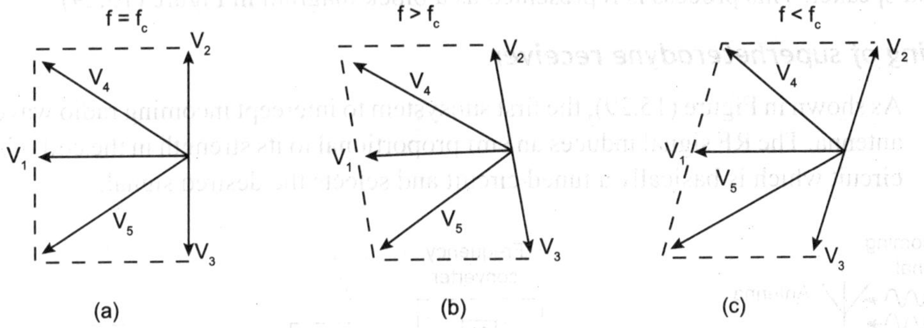

Figure 15.28 Phasor diagram of discriminator

15.16 Radio Receivers

It is a device to detect information contained in the modulated wave reaching from the transmitter. We shall first confine our study to receivers used in the detection of AM waves. These are called AM receivers.

15.16.1 Principle

For satisfactory radio reception, the signal applied to the linear detector must be of the order of a few millivolts. Thus, considerable amplification of the incoming radio signal is required before detection. The receiver must have the following characteristics:

1. It must be capable of receiving the large number of incoming radio signals; for this, an antenna is required.
2. The desired audio signal must be detected; for this we have a detector circuit.
3. The audio signal must be of a particular strength so that it is audible; for this we have an AF power amplifier and loud speaker.

Radio receivers are classified as: (i) straight radio receivers and (ii) super hetro dyne receivers. The former suffers from a large number of demerits and so they have been abandoned.

15.17 AM Receiver (Super Heterodyne Receiver)

15.17.1 Principle

Here, the desired incoming signal after proper amplification is superimposed on a locally produced RF signal which is always kept higher than the incoming signal. The resulting signal called IF signal is again amplified and fed to a detector which separaters audio signal from IF signal which is a modulated signal. The audio signal is fed to an AF amplifier and then fed to the loud speaker. This process is represented as a block diagram in Figure (15.29).

Working of superheterodyne receiver

1. As shown in Figure (15.29), the first subsystem to intercept incoming radio waves is the antenna. The RF signal induces an emf proportional to its strength in the coil of selector circuit which is basically a tuned circuit and selects the desired signal.

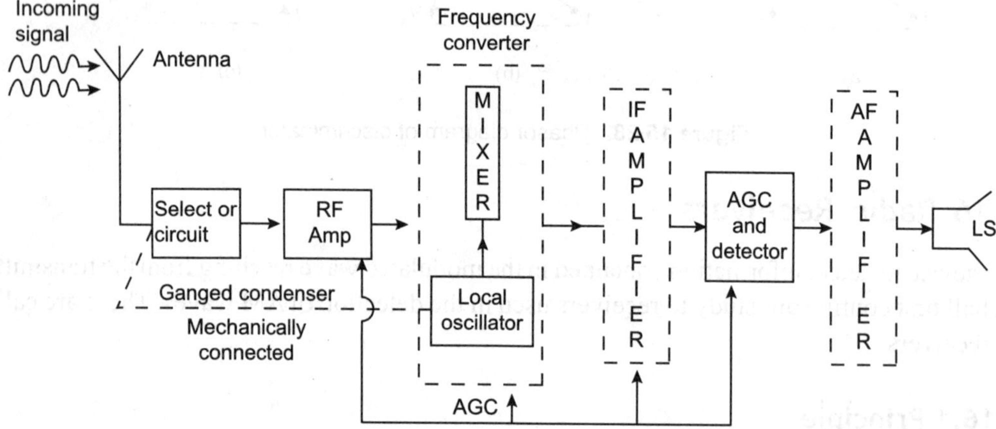

Figure 15.29 Simplified block diagram of super heterodyne receiver

2. The selected signal is fed to RF amplifier. The amplifier amplifies the received RF signal without introducing any extra signal.
3. The output of RF amplifier is connected to the frequency converter circuit. It comprises a mixer circuit and a local oscillator whose frequency can be changed as per incoming signal at selector circuit since both are mechanically connected through a ganged condenser; so if one changes as per signal, the other (local oscillator) automatically changes to produce the necessary signal (f_o). This signal frequency (f_o) is always higher than incoming signal frequency (f_s) and the two when superimposed in a mixer circuit produce a new signal called intermediate frequency signal (IF signal) which is also known as beat frequency signal having fixed value 455 kHz ($f_o - f_s = 455$ kHz $= f_i$).
4. *IF amplifier* The beat frequency signal (f_i) is fed to intermediate frequency amplifier for proper amplification.

5. *Detector* The output of IF amplifier is fed to detector stage which primarily consists of diode circuit (Figure 15.20). It filters the desired incoming (AF) signal from the IF signal and transmits the AF signal to AF amplifier.
6. *AF amplifier* The detected signal is amplified to a proper level so that it has proper power to run the loud speaker and it produces the sound (information) concerned with the incoming RF signal intercepted by the antenna.

Because of unintentional positive feed-back, there will be limit to the amount of stable gain achievable on any one frequency. By shifting the frequency band of the signal, it is possible to approach this limiting stable gain in each of several frequency bands in cascade.

Thus high gain and controllable selectivity through fixed frequency transformer design are advantages of super heterodyne system.

The various frequency bands have been standardised internationally. Primarily, we have three bands in a receiver:

(i) Medium Band (M) having range 500 kHz to 1600 kHz
(ii) Short Band I (S_1) having range 2.5 MHz to 7.5 MHz
(iii) Short Band II (S_2) having range 7.5 MHz to 25 MHz

The above two short bands have been further divided into four, five or seven bands in highly selective radio receivers.

15.18 Image Frequency

In a radio receiver the incoming (desired) signal is selected by the selector circuit. This signal after amplification is superimposed over another signal produced by the local oscillator, resulting into an **Intermediate Frequency** (IF) **signal** better known as beat frequency signal, whose frequency is always kept constant, whatever be the frequency of incoming signal it is achieved correspondingly by changing the frequency of oscillator.

Let the frequency of desired incoming signal be 'f_s' and the frequency of local oscillator be f_o ($f_o > f_s$).

Resultant Beat frequency or Intermediate frequency

$$f_i = f_o - f_s \qquad \text{(i)}$$

However, if another signal f_{si} reaches the oscillator and the resultant beat frequency or Intermediate frequency is again f_i

$$f_i = f_{si} - f_o \qquad \text{(here } f_{si} > f_o\text{)} \qquad \text{(ii)}$$

This IF produced by undesired signal f_{si} would also be amplified and f_{si} will be detected, there by it would interfere with desired signal. f_s which is not desired. Such undesired signal f_{si} satisfying equation (ii) is called Image Frequency.

Substituting for f_o from relation (i) in (ii)

$$f_i = f_{si} - (f_i + f_s)$$

or
$$f_{si} = f_s + 2f_i \qquad \text{(iii)}$$

Thus the undesired signal known as image frequency must satisfy relation (iii), i.e. its value is $2f_i$ more than desired signal as illustrated by the following example.

Let the desired signal frequency $f_s = 545$ kHz

Beat frequency (IF) $f_i = 455$ kHz (fixed)

The oscillator frequency $f_o = f_s + f_i$

$= 545 + 455$

$= 1000$ kHz

with the given oscillator frequency $f_o = 1000$ kHz, another undesired signal

$f_{si} = 1455$ kHz would also produce beat frequency

$f_i = f_{si} - f_o$

$= 1455 - 1000$

$= 455$ kHz

This undesired signal of frequency 1455 kHz is known as image frequency since it produces the same IF frequency 455 kHz with given oscillator.

Difference of undesired and desired signal frequency

$f_{si} - f_s = 1455 - 545$

$= 910$

$= 2(455)$ kHz

Which is double of IF.

To avoid the undesired signal (image frequency) we introduce additional selectivity at selector circuit, thereby the strength of undesired signal is lowered and by use of higher beat frequency (IF) the undesired signal is separated from the desired signal. Thus it is possible to suppress the undesired signal (Image Frequency).

15.19 Communication Receiver

The purpose of a communication receiver is the reception of signals used for communication (for scientific purpose) rather than entertainment. It is a radio receiver designed to perform the task of low and high frequency reception. Its block diagram is shown in Figure 15.30. The discussion of this receiver is beyond the scope of this chapter.

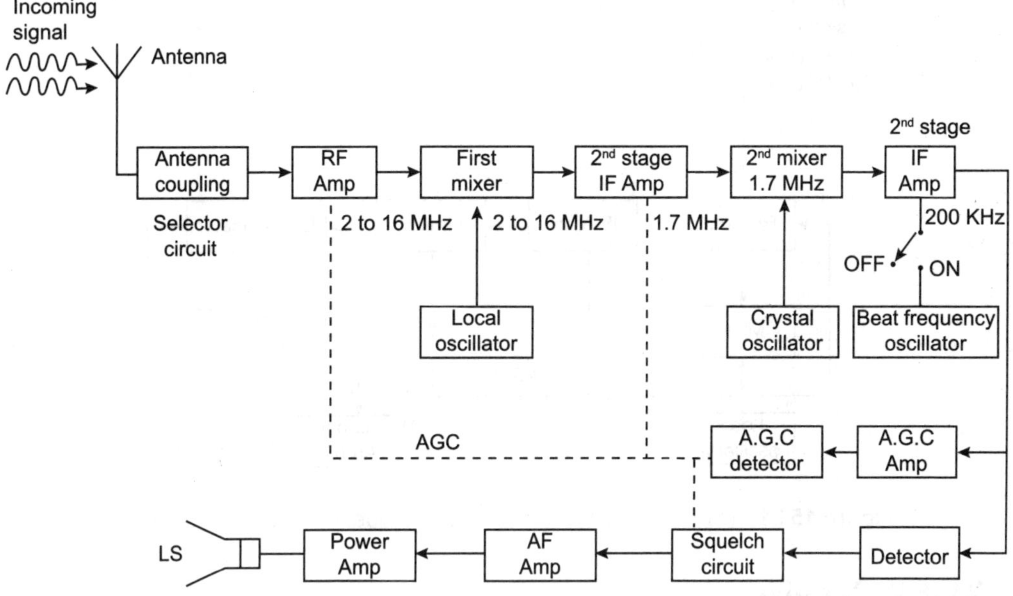

Figure 15.30 Communication receiver

15.20 FM Receivers

It is also a super heterodyne receiver. Its block diagram is shown in Figure 15.31 which is very similar to an AM receiver (Figure 15.29).

The basic differences are:

1. Much higher frequencies than AM
2. Need for limiter and de-emphasis circuits
3. Totally different method of detection
4. Different method for AGC

As shown in the block diagram (Figure 15.31(a)), the FM receiver has limiter circuits, discriminator circuits and de-emphasis network which mainly differs from the AM receiver. Moreover, its frequency band is also much higher than that of AM receiver. It also employs an AGC circuit just like in the AM receiver. The last stage is power amplifier whose output is fed to a loud speaker.

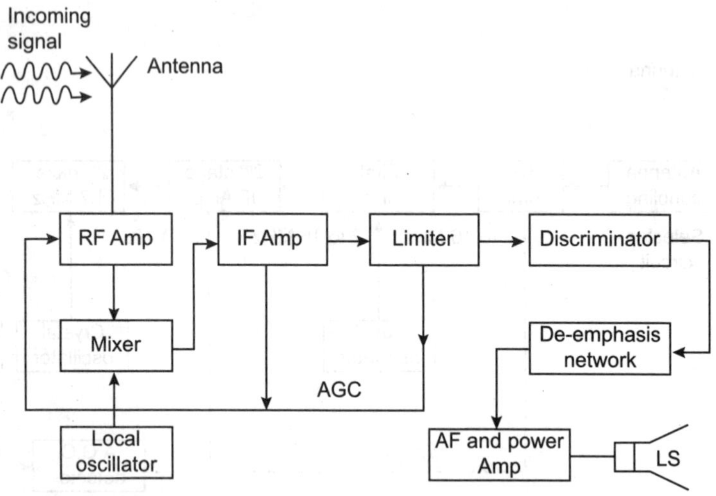

Figure 15.31 (a) Simplified block diagram of frequency modulated receiver

15.21 FM Receiver

FM receivers employ the superheterodyne circuit just like AM receivers, since it provides a large gain to enhance the amplitude of weak signals. The gain is sufficient so that the signal after amplification can be held constant by the limiter employed in the receiver circuit. The constant amplitude of the signal ensures low noise feature of frequency modulation.

Figure 15.31(b) shows the block diagram of an FM superheterodyne receiver. It resembles Figure (15.29) of AM receiver except that the conventional detector of the AM receiver is replaced by limiter and discriminator circuits.

If a ratio detector is used in place of the discriminator, the limiter stage can be avoided. The limiter discriminator or ratio-detector maintains the amplitude of carrier wave constant and also transforms the wave into equivalent amplitude (audio) variations.

Another important difference is that the FM receiver requires a wide band IF amplifier. It is because the bandwidth of broadcast FM signal is 150 kHz or more which is 10 times that of AM signal. The IF amplifier in FM receiver must have a wide band frequency response to pass the 150 kHz band. The IF frequency for FM is usually 10.7 MHz since the operating FM frequency is about 100 MHz.

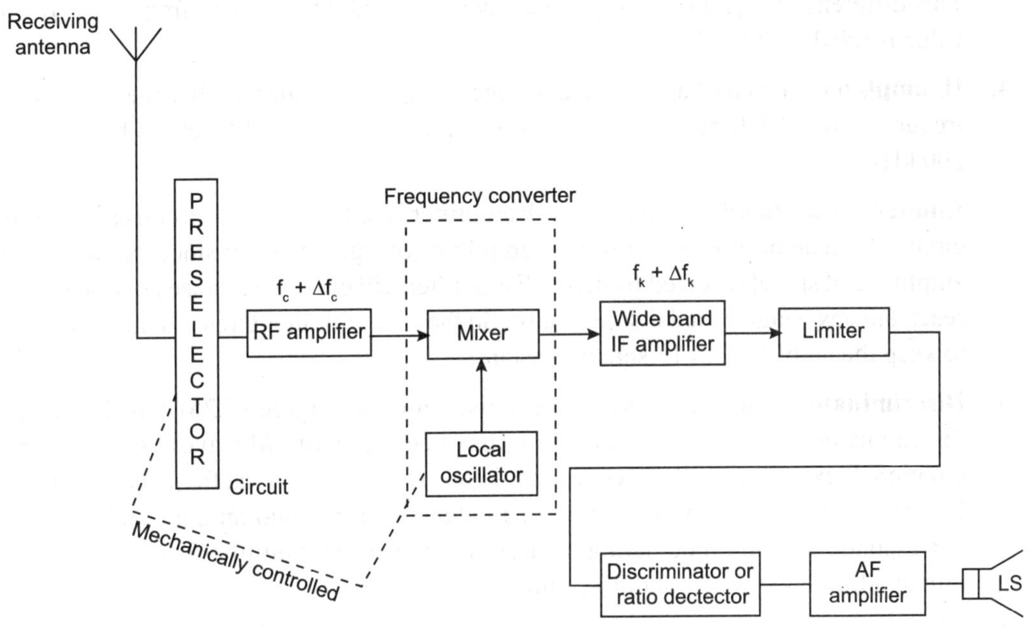

Figure 15.31 (b) FM superheterodyne receiver

The functions of various subsystems employed in the FM superheterodyne receiver are as follows:

1. **Antenna** Its main function is to catch/intercept the desired RF signal transmitted by the transmitter and existing in the space. This RF signal induces an emf proportional to the signal in the selector coil shown as preselector tuned circuit, which selects the desired signal and rejects all other signals present in space. Here, the incoming signals for FM receivers are kept between 88 and 108 MHz.

2. **RF amplifier** The selected signal is fed to the RF amplifier as its input signal. The amplifier amplifies the input signal (desired frequency) without introducting any unwanted signal (frequency). This amplified signal is fed to the next stage called mixer.

3. **Frequency converter** It actually consists of a mixer circuit and a local oscillator which generates a signal (frequency) which is always more than the incoming signal frequency, and changes according to the received signal so that the difference between two frequencies is always kept constant. For this, the local oscillator circuit is coupled mechanically with the selector coil attached to the antenna thereby as frequency (f_s) of the incoming signal changes, then frequency f_o of the signal generated by oscillator automatically changes to the extent that difference between the two remains constant.

This difference frequency or beat frequency is called intermediate frequency (IF); its value is fixed at 10.7 MHz.

4. **IF amplifier** It may have more than one stage of amplifier so that the IF signal of frequency 10.7 MHz received is amplified upto a certain fixed level. Its bandwidth is 200 kHz.

5. **Limiter** The amplified signal from IF amplifier is fed to a special circuit called limiter circuit. Its main purpose is to limit the amplitude of signal fed to it since variation in the amplitude of signal received from the IF amplifier will other wise cause noise when they reach the next stage called discriminator. So the role of limiter circuit is most important to keep the amplitude of IF signal constant.

6. **Discriminator** Here the Foster-Seeley discriminator (Figure 15.26), which is a phase discriminator, is used. It is a popular means of detection of FM waves using minimum circuitry. It extracts the intelligence modulated on to the carrier via frequency variations. It changes the frequency derivation of modulated signal into an amplitude variation corresponding to the modulating signal. This transformation of frequency variations into amplitude variations should be linear and efficient.

   We can also use a ratio detector in place of Foster-Seeley discriminator which also normaly responds to undesired variations in amplitude and so there was need of limiter circuit with it. But in the case of a ratio detector, the limiter circuit can be dispensed with.

7. **AF amplifier** After the desired signal is detected by the discriminator or ratio detector, it is fed to the AF amplifier which is a wide band amplifier. The AF amplifier amplifies the signal to a level which is audible when fed to a loud speaker.

15.21.1 Pre-Emphasis and De-Emphasis

All electronic circuits produce noise (unwanted signal). It is low in the low AF range and high in high frequency range. For information signals with uniform signal level, a non-uniform signal-to-noise ratio is obtained. Higher modulating frequencies have lower signal-to-noise ratio than lower modulating frequencies. To overcome this partially, the high-frequency modulating signals are emphasized or boosted in amplitude in the FM transmitter before modulation, i.e. before noise signal gets into. This process is known as *pre-emphasis*.

To compensate for pre-emphasis, higher frequency signals are attenuated or de-emphasised in the FM receiver after demodulation. This process is known as *de-emphasis*.

In this way, de-emphasis is the reciprocal of pre-emphasis and the de-emphasis circuit restores the original amplitude-frequency characteristic of the information signal.

Thus, by employing the two processes, we get more or less a uniform signal-to-noise ratio over the entire audio frequency range.

15.22 Automatic Gain Control (AGC)

15.22.1 Simple AGC

In radio reception, the strength (amplitude) of incoming signal varies from time to time or station to station, thereby the output of radio will vary accordingly and at times it may not be suitable for audio reception, i.e. either it is too loud or too weak to hear; so it is desirable to have the gain of receiver nominally constant (suitable for hearing). For this to be accomplished, we have to automatically vary the gain of receiver in inverse proportion of incoming signal strength. This is achieved using automatic gain control (AGC) or better called automatic volume control (AVC) circuits.

A measure of the strength of received carrier is available in the DC output of the detector diode (Eqn. 23),

$$e_{dc} = \frac{E_C R}{\pi r_d}[\sin\theta_1 - \theta_1 \cos\theta_1]$$

A low pass RC filter, with time constant long enough w.r.t. the lowest modulation cycle, as R_3C_3 in Figure 15.32, will separate this DC voltage from modulated wave. This DC voltage may then be applied as a control bias to vary the gain of amplifier ahead of detector.

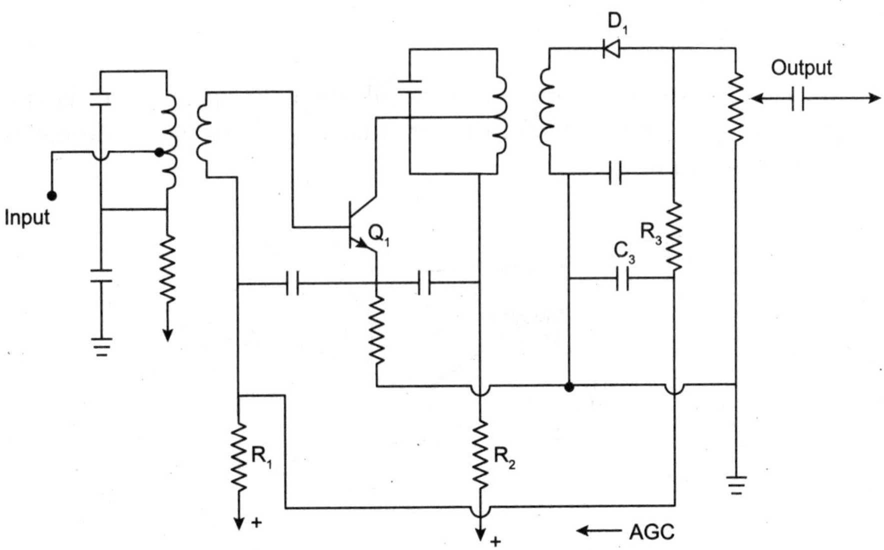

Figure 15.32 AGC system

In transistor amplifiers, several currents might be varied to change its gain, but the most suitable and desirable is I_E. The variation of gain with I_E is shown in Figure 15.33 for a typical transistor.

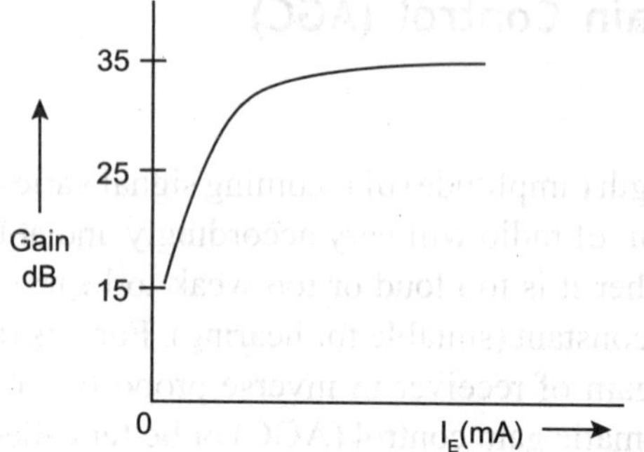

Figure 15.33 AGC system

In Figure 15.32, negative voltage from diode D_1 changes the base bias of transistor Q_1 and reduces the emitter current with increasing signal strength.

As emitter current is reduced, h_{fe} of the transistor falls and so gain of amplifier is reduced; h_{ie} and h_{oe} also change and impedance mismatches, which results in further reducing the gain of the amplifier.

Thus, in a simple AGC system, the over all gain of radio receiver is varied automatically with changing strength of incoming signal and we may have sufficient audible sound.

15.22.2 Delayed AGC

A simple AGC is an improvement on No AGC at all where the gain of receiver is reduced to audible strength for very strong signals. But unfortunately, weak signals do not escape this reduction and so they are not audible.

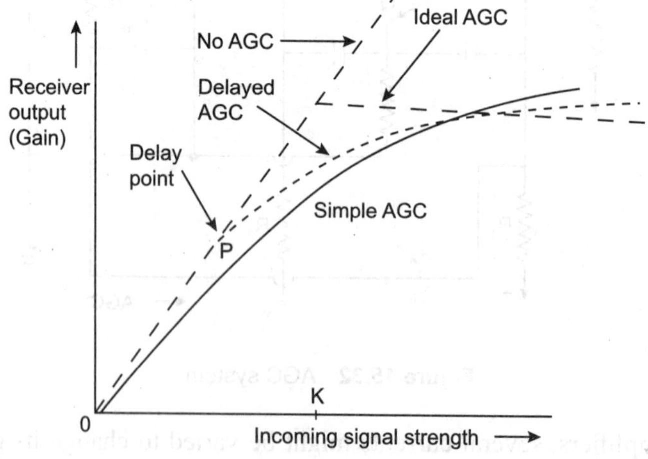

Figure 15.34

To overcome the above shortcomings, we have two options:

(i) **Ideal AGC** Here no AGC would be applicable till the incoming signal strength is adequate shown as OK (Figure 15.34) and after that AGC functions resulting into constant average output, no matter how much more signal strength increases.

(ii) **Delayed AGC** Here AGC bias is not applied until the signal strength has reached a predetermined level OP (Figure 15.34) after which bias is applied as with normal AGC, but more strongly. As incoming signal strength rises, receiver output also rises but relatively slightly. Thus, the problem of reduced gain for weak signals has been avoided as in the case of an ideal AGC.

A simple circuit employing delayed AGC is shown in Figure 15.35.

As shown, here we use two diodes D_1 and D_2, the first one is the usual detector diode, the second diode D_2 is the AGC detector. A positive bias is applied to the cathode of D_2 to prevent its conduction till a predetermined signal level reaches (shown by P in Figure 15.34). A manual control is often provided through a variable resistor. If weak stations are most likely to be received, the delay control setting may be quite high (i.e. no AGC until the signal level is fairly, high). Nevertheless, it should be made as low as possible to prevent overloading of the last IF amplifier by unexpected strong signals.

As mentioned earlier, we need the voltage gain to change automatically when input signal strength changes. Specifically we need voltage gain to decrease when input voltage increases. In a way we have audio AGC as described below.

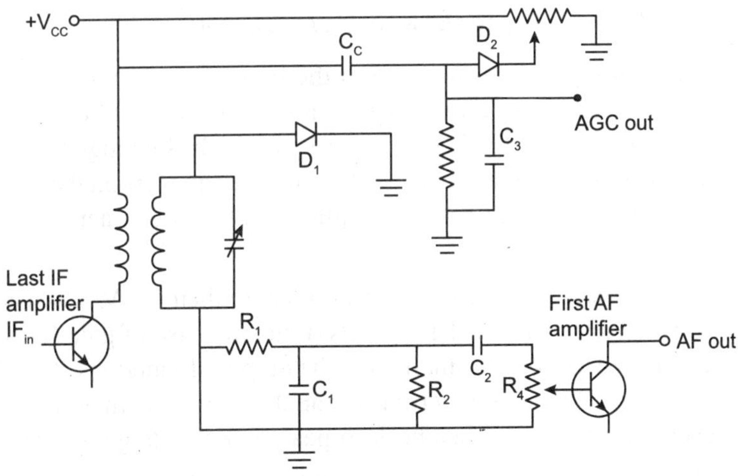

Figure 15.35 Delayed AGC circuit

15.22.3 Audio AGC

Figure 15.36 showns an audio AGC circuit. As shown, Q_1 is JFET used as voltage controlled resistor. For small signal operations with drain (D) voltage to be nearly zero, the JFET operates in the ohmic region and offers resistance r_{ds} to AC signals. This resistances can be controlled by gate (G) voltage. The more the gate voltage, the larger the r_{ds}. With a JFET like 2N4861, r_{ds} can vary from 100 Ω, to 10 Mr Ω so resistor R_3 and Q_1 act as voltage divider. This output terminal is connected to the non-inverting terminal of Op-Amp. Thus, the input to non-inverting terminal changes as output of junction R_3 and D changes. The input voltage of inverting terminal is $\left[\frac{R_2}{R_1}+1\right]$ times the input voltage V_i.

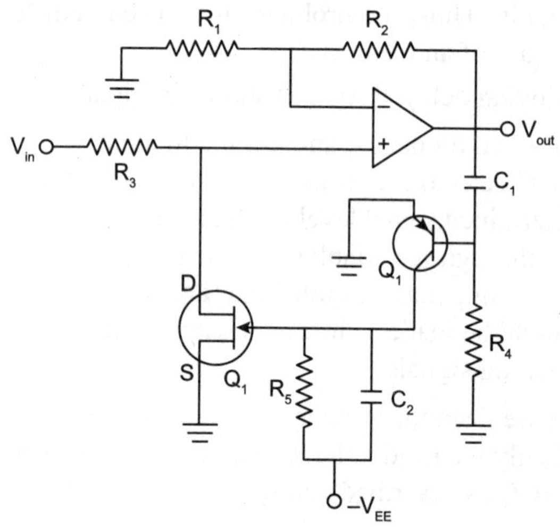

Figure 15.36 Audio AGC circuit

The output voltage of Op-Amp is coupled to the base of transistor Q_2. For peak-to-peak output less than 1.4 V, transistor Q_2 is cutoff because there is no bias on it. With Q_2 OFF, capacitor C_2 remains uncharged and gate of FET Q_1 is at (–) V_{EE}, which is enough to make FET cutoff. This means that maximum input voltage reaches the non-inverting terminal of Op-Amp. In other words, the output voltage is less than 1.4 V. It implies that the circuit acts like a non-inverting voltage amplifier with maximum input voltage.

When output peak-to-peak voltage of Op-Amp is more than 1.4 V, transistor Q_2 conducts and thereby capacitor C_2 gets charged. This results in the increase of gate (G) voltage thereby resistance r_{ds} of FET decreases. With a smaller r_{ds}, the output of junctions R_3 and Q_1 decreases, so there is less input voltage to the non-inverting terminal of Op-Amp. In other words, the overall voltage gain of the circuit decreases when peak-to-peak output voltage is greater than 1.4 V.

We conclude that output voltage increases much less for a large increase in input signals.

15.23 Advanced Communication Systems

In the past few decades, several new technologies have come up and they are extensively used for communication purposes. As a result, *satellite communication*, *optical communication* and *mobile communication* have become very popular. A brief account of these systems is as follows:

15.23.1 Satellite Communication

This is a mode of communication of signal between a transmitter and receivers through a satellite. This is similar to extension of line-of-sight *microwave communication*. In the microwave frequency range (> 1 GHz), communication is neither possible through ionosphere (since the signal penetrates the ionosphere) nor through earth's surface (very high attenuation at such high frequencies). Obviously, the only possible communication mode is space wave propagation which is line-of-sight communication. Due to the curvature of earth, the maximum permissible distance between the transmitter and receiver is around 50 km, depending on various factors such as antenna height, transmitter output power, terrain, atmospheric condition, receiver noise threshold, etc. If communication is to be established between two distant points, then a suitable number of repeater stations has to be installed. Figure 15.37 shows a typical microwave communication system. The microwave transmitter transmits the signal which is received at the first microwave repeater. It consists of a receiver, amplifier-reshaper and transmitter. The signal passes through all the repeaters and then reaches the receiver. The microwave systems can be either analogue (such as FM/FDM) or digital (such as PCM-PSK/TDM).

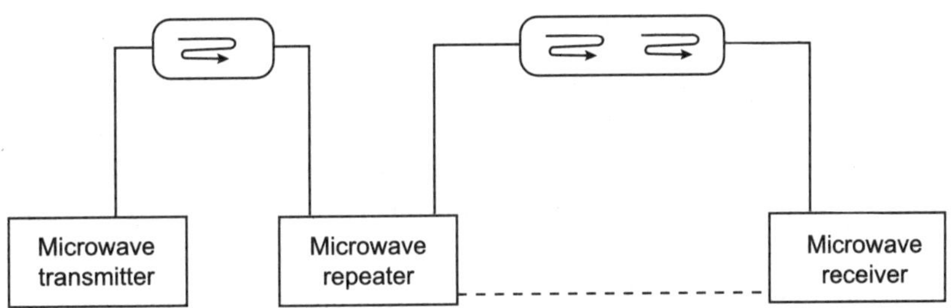

Figure 15.37 Microwave communication system

When there is necessity to cover very long distance, the arrangement shown in Figure 15.37 becomes inconvenient due to the use of a large number of repeaters. Further, when transoceanic communication or worldwide communication is to be established in microwave frequency range, practically it becomes impossible to achieve it. This problem may be solved if a repeater is set up in the sky. With the successful launching of satellite in 1957, the idea of having a repeater in the sky became feasible and communication via satellite became a reality. This gave rise to a new field of communication called **satellite communication**. Since satellite communication is achieved through space, it is also known as space communication. Satellite communication is achieved through two types of satellites: (i) **Passive satellites** and (ii) **Active satellites**. The field of satellite communication become very popular due to its immense potential and the ever lasting communication needs of the world.

15.23.2 Satellite Communication System

Basically a communication satellite acts as a big microwave repeater in the sky. The satellite communication system can take on any form as shown in Figure 15.38.

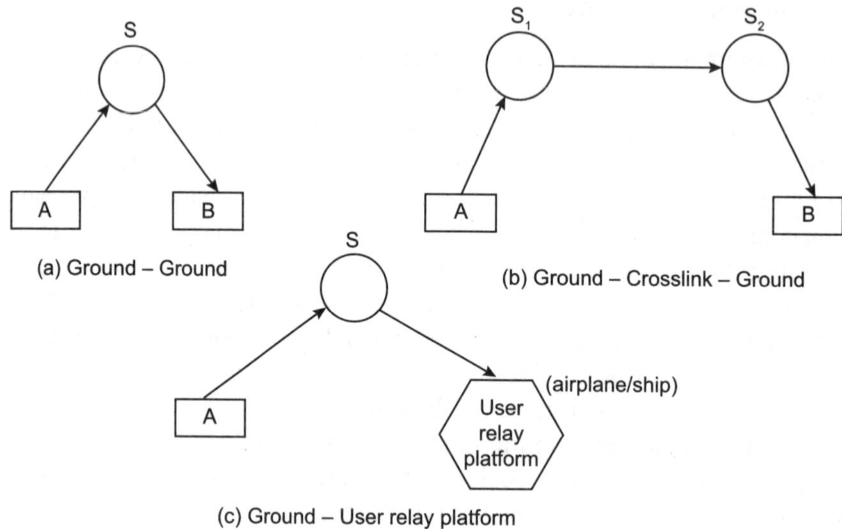

Figure 15.38 Different forms of satellite communication systems

Figure 15.38 shows three schemes of satellite communication systems. In all these schemes there can be more than one transmitting and receiving earth stations and user ready platforms. The ground-ground arrangement shown in Figure 15.38(a) will take the form shown in Figure 15.39. In this figure, A, A', A'', … are transmitting earth and B, B', B'',… are receiving earth stations. A single satellite can be used by different transmitting and receiving earth stations with the help of multiple access technique.

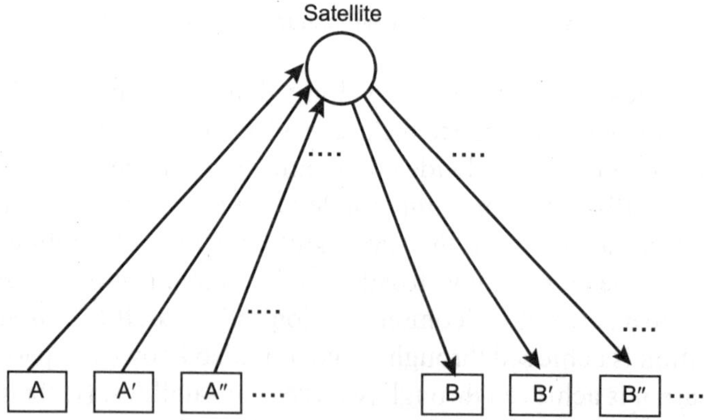

Figure 15.39 More than one transmitting and receiving earth stations

Figure 15.40 shows a satellite communication system in which a satellite links the earth station transmitter and earth station receiver. Normally, one satellite is connected to the earth station. At earth stations, a user interface between the user and terrestrial system, and a terrestrial interface between the terrestrial system should be used.

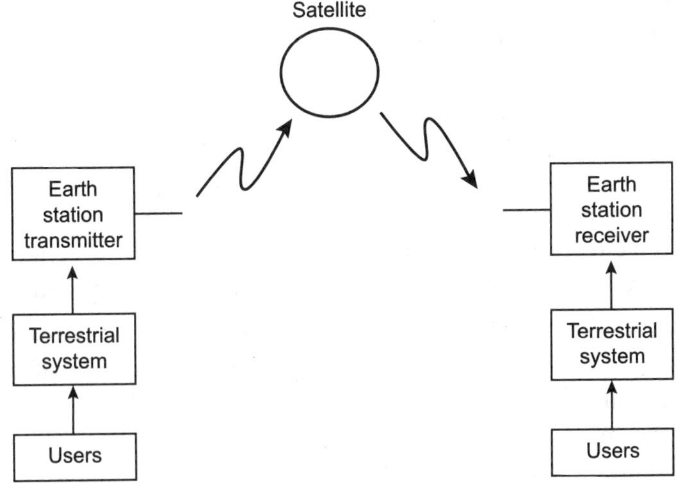

Figure 15.40 Schematic diagram of a satellite communication system

A satellite used for communication have transponder power supply system, arrangements to control its position, fuel, etc. Transponder is used for carrying the electronic processing. The earth station transmitter and the earth station receiver are referred as **uplink section** and **downlink section** respectively.

Several transmitting and receiving earth stations connected via a single communication satellite are shown in Figure 15.39. System may work properly, one has to ensure: (i) there should be no mixing of signals from different earth stations, and (ii) at earth station receiver signals can be separated. One can achieve this by making use of the concept of multiplexing resulting in various multiple access technique.

The transponder capacity of a satellite must be quite high to handle the traffic as several earth stations are communicating via a single satellite. In reality, the capacity of a transponder is quite higher in comparison to the traffic provided by a single earth station. This is why, a single earth station will under use transponder's capacity whereas multiple access seems to make best use of full capacity of transponder.

Four domains, e.g. frequency, time, code and space are used for achieving multiple access. The respective multiple access techniques are: Frequency Division Access or Frequency Domain Multiple Access (FDMA), Time Division Multiple Access or Time Domain Multiple Access (TDMA), Code Division Multiple Access or Code Domain Multiple Access (CDMA), and (Space Division Multiple Access or Space Domain Multiple Access (SDMA).

An earth station is given access to a satellite transponder through one of the following ways:

(i) Fixed Assigned Multiple Access or Pre-Assigned Multiple Access (FAMA or PAMA).

(ii) Demand Assigned Multiple Access (DAMA)

(iii) Random Multiple Access (RMA)

Launching of a satellite is a very complex process. A geostationary satellite is launched with the help of **Geostationary Satellite Launch Vehicle**, whereas a polar or inclined satellite is launched by **Polar Satellite Launch Vehicle** (PSLV). Satellite launch vehicle can be expendable type or reusable type. The expendable type launch vehicle carries one or more satellite within it. The vehicle takes the satellite upto a certain altitude in space and then leaves satellite to spin and proceed. When all the satellites placed within it are released in space, the satellite launch vehicle gets destroyed in space. In case of reusable type satellite launch vehicle, the satellite is placed over the vehicle. Vehicle returns to earth after leaving the satellite a certain altitude in space. Obviously, it can be reused for subsequent launches as and when required, e.g. space shuttle.

An major aspect of any satellite is its **altitude stabilization**. It is always desired that the satellite antenna should point towards uplink and downlink antennas of earth stations. This is **satellite stabilization** (orientation). One can achieve it by the technique of spinning. In spin stabilization, the entire satellite is kept to physically spin at 30 to 100 rpm about the axis perpendicular to the orbital plane in order to maintain a fixed altitude axis. Thus, all points in space are in a fixed direction relative to that axis. To achieve this, it is necessary to "despin" the antenna and solar panels so that they should continuously point towards the desired direction. One can achieve despinning in two ways: (a) by placing the antennas and solar panels on platforms that are spun in positive direction to that of the satellite spin, and (b) by using multiple elements that are phased in such a way that only the element in proper direction is activated at any time. However, there are some problems with spin stabilization, i.e. spinning of satellite causes motion of liquid fuel, which is about one-third weight of satellite, and disturbs the stability of inertia axis; the quality of liquid fuel degrades with age of satellite and hence the disturbance varies accordingly; and the despin operation becomes more difficult with increase in antenna and solar panel weight. Due to these problems, the spin stabilization method is usually avoided except for some satellites where this method is still used with some modifications.

15.23.3 Optical Communication

Optical communication systems employ light waves to transmit information from one place to another. This is also analogues to any electrical communication system except that in this system optical wave (10^{14} to 10^{18} Hz) serve as carrier.

Similar to radio frequency spectrum, two transmission media can be used for optical communication, e.g. (i) an atmosphere channel and (ii) guided wave channel. The first mode is used in applications where the use of optical waveguides is not possible such as laser radar and earth to space communication. The tremendous system and installation cost, and the limitations imposed on the atmospheric channel by rain, fog, snow and dust make this system unattractive. On the other hand, guided wave channels provide much more reliable and versatile optical channel than the open atmospheric channel. The media used for guided channels in the optical range are non-metallic guides popularly known as **optical fibre**.

The optical link used for communication engineering applications has definite inherent advantages over conventional copper cable and radio links.

(i) **Tremendous bandwidth** The optical system operating in the frequency range of the order of 10^{14} Hz provides a very large bandwidth as compared to copper wire and radio channels. A single optical fibre can handle as many voice channels as can be handled by 1500 copper cables. The bandwidth of copper cable is limited by inherent capacitance and inductance causing low pass filtering effect.

(ii) **Light weight and small diameter** Light weight and hair-thin dimension of the optical fibre makes it quite suitable for aircraft, automotive and military applications. This helps in physical portability to the optical fibre.

(iii) **Low loss** Losses offered by the optical fibre cable are very small as compared to that of copper cable and radio links. Typical comparative loss values are: copper = 50 dB/km, satellite = 20 dB/km, optical = 0.1 dB/km.

(iv) **Less number of repeaters** Due to low loss, the spacing of repeaters ranges from 35 to 80 km for optical fibres links as compared to 1–1.5 km for wire communication. This reduces quite considerably the number of repeaters needed. The spacing is further increased from 100 to 1000 km with modern technologies using fibres.

(v) **Immune to electromagnetic interference (EMI)** A significant feature of optical fibre relates to their dielectric nature. This makes them immune to interference from lighting, cross-talk, and electromagnetic radiations generated from electrical appliances. Optical fibres neither produce EMI nor pick it up.

(vi) One can obtain a high degree of data security from fibres due to the fact that optical signal is well confined within the fibre waveguide. Thus, tapping is almost impossible.

(vii) Silica is the principal material from which optical fibre is made. The main source of silica is sand which is abundantly available and is quite inexpensive.

(viii) Noise is negligibley small in optical communication systems.

(ix) Controlled dispersion yield low error rate of the order of 10^{-9} to 10^{-12}, i.e. 1 bit error in 10^9 to 10^{12} bits.

(x) Transmission cost per bit is quite low.

Optical communication has the following disadvantages:

(i) Initial installation cost is high and therefore not desirable for low density traffic.

(ii) Launching of light, coupling and splicing of cable is quite difficult and costly. Obviously, maintenance and repair is costly and cumbersome.

(iii) Laying cable in inhospitable terrains is also cumbersome.

However, the above disadvantages of optical fibre communication are becoming insignificant due to advances in technology.

Figure 15.41 presents an schematic illustration of the elements of an optical fibre link.

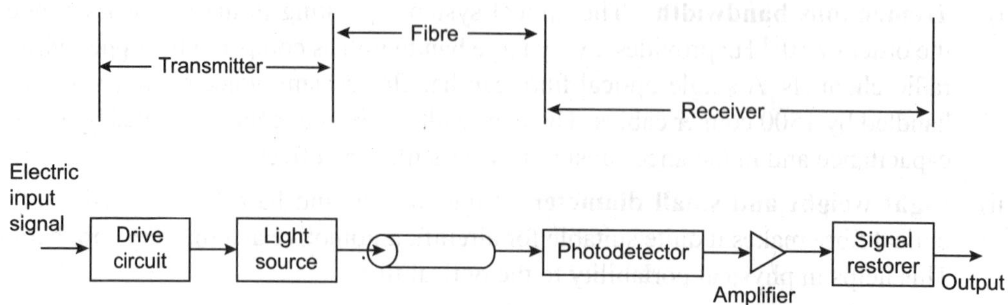

Figure 15.41 Elements of an optical fibre link

Wavelength-Division Multiplexing Technology (WDM technology) enables us to utilize a significant portion of the available fibre bandwidth by allowing several independent signals to be transmitted simultaneously in one optical fibre. In addition, high bandwidth routing can also be facilitated by means of a multi-user network. One can route and detect independently with the wavelength determining the communication path by acting as the origin, destination or routing.

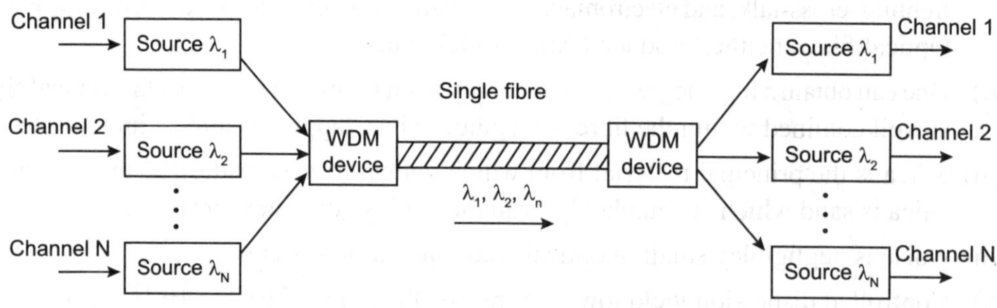

Figure 15.42 A schematic illustration of unidirectional WDM scheme

Optical fibre communication is being widely used in telephone communication, data communication, multimedia transmission, computer networks, etc. Now, the technology has been developed for wideband high-speed transmission of information. The spectral bandwidth of optical fibre is being improved in dispersion shifted and dispersion compensated optical fibres. Optical amplifiers have been developed to replace electrical repeaters. The information capacity of optical link is enhanced by optical multiplexing such as Wavelength Division Multiplexing (WDM) and Angular Division Multiplexing (ADM), in which multiple light beams are transmitted over a single fibre. Each optical beam carries millions of telephone channels by electrical multiplexing. WDM or DWDM (Dense Wavelength Division Multiplexing) and ADM techniques enhance the channel capacity of a optical fibre manifold. This may reduce the number of optical fibre cables to be laid between long distances or overseas. Figures 15.42 and 15.43 illustrate the principle of WDM and ADM over a fibre, respectively.

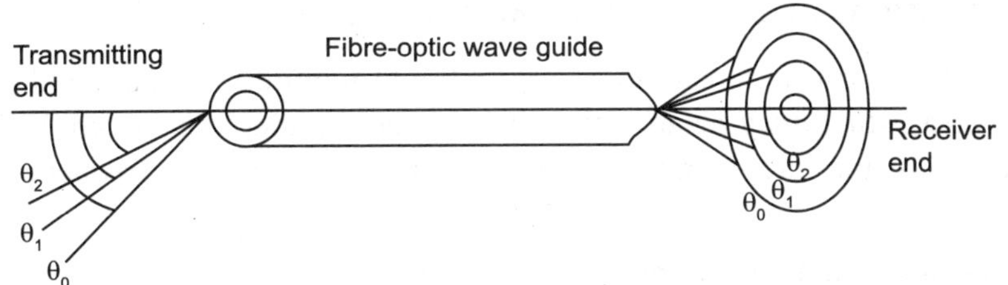

Figure 15.43 An angular division multiplexing (ADM) scheme

In a WDM channel, millions of telephone channels modulate a light source having wavelength λ_1. Similarly, channels 2, 3,... n modulate light sources having wavelength λ_2, λ_3,... λ_n, respectively. This WDM signal is transmitted in a single optical fibre. At the receiver, optical beam corresponding to each wavelength is separated by optical filters and demodulated so that it may give the respective electrical baseband signal.

In angular division multiplexing (ADM), acceptance angle of the optical fibre is divided into smaller angular segments as shown in Figure 15.43. Each angular segment in used to transmit one channel. At the receiver side, we find a series of annual cones corresponding to each channel.

15.24 Mobile Communication

Communication has developed so rapidly that today one can hear the buzz of a bee across a pane of glass in USA to somewhere remote in any part to the world. Celluar mobile communication system began just a few years back has advanced to such a stage that today it is used for transmission of all sorts of information, e.g. data, voice, video, internet and several value added services.

The concept of cellular technology addresses the problem of limited spectrum. This is allocated by the government for radio/mobile communications. However, there is a need to have more channels in a limited area rather than having a large coverage area. To achieve this, the cellular technology divide the geographical region into small regions of uniform shape and area, referred to as **cells**. The cells are formed to achieve an objective of **frequency reuse**. Each cell is provided with a base station of its own so that it may take care of the coverage in that particular cell. Obviously, the region to be covered is small and therefore, the antennas used for the transmission are usually low powered. Some portion of the spectrum is allocated to each cell to meet the needs of the cellular subscribers in that cell. Moreover, the base stations in the neighbouring cells are allocated a different group of channels that fall in another portion. This prevents interference between the mobiles in adjacent cells.

The cellular system is facilitated with *hand-off capability* so that call is not dropped when mobile unit moves from one cell to other cells. Each cell has a base station consisting of low-power transmitters so that its radiation is restricted to the respective cell boundary. All the base stations are connected to a Master Telephone Switching Office (MTSO) which monitors and controls the entire call processing.

15.24.1 Limitations of Conventional Mobile System

Due to use of high power transmitters, conventional mobile systems could cover a large geographical area, which may be an entire city or zone. Further, this system had no hand-off facility. Due to possible interference by large power transmitters, the frequency-reuse seems not feasible.

The limitations of conventional mobile systems are as follows:

(a) Inefficient spectrum utilization One can define the spectrum utilization as

$$N = \frac{\text{Number of subscribers}}{\text{Number of channels}}$$

i.e. the maximum number of subscribers that could be served per channel in the busy hour. In conventional mobile systems, the typical value of N is limited to 37 to 53. We may note that in the conventional mobile system, each channel can serve only one customer at a time in the entire area. A major problem faced by wireless communication industry is the availability of limited RF spectrum. To utilize RF spectrum efficiently, the major approaches are:

- Single Side Band (SSB), Quadrature Amplitude Modulation (QAM) and similar other modulation techniques which require less bandwidth per user.
- Code Division Multiple Access (CDMA) system in which many users can use the same spectrum at the same time and the user is distinguished by a distinct code allotted to them.
- Cellular system which reuses the allocated spectrum in different geographical locations located beyond radiation coverage of each other. This cellular concept solved a major problem faced by mobile telephone system worldwide, i.e. spectrum scarcity.

(b) Poor service Due to nonavailability of frequency reuse technique, the number of customers allotted per channel in conventional mobile system was quite large (~37 or 53). This created a high blocking probability (50% or more) during busy hour. During busy hour, large number of calls did not mature, causing deterioration in the service performance. With the introduction of cellular concept problem is resolved.

(c) Limited service capability There was no provision for hand-off in conventional system. Due to this the call was dropped when a mobile user moved from one geographical zone to other and the user had to initiate the call again. Use of hand-off mechanism in cellular mobile system helped to overcome this difficulty.

15.24.2 Cellular Mobile System

In cellular mobile communication, the user carries the mobile instrument with them and obviously, it is location and time independent. With the concept of cellular phone and satellite phone, mobile communication is possible for any distance worldwide, i.e. it is distance independent. One can interface the digital mobile communication system with any device, e.g. fax machine, telephone equipment, laptop, etc. through which information can be exchanged in any form obviously, this is device independent.

Requirements for mobile communication

1. **Terminal mobility** A portable subscriber set alongwith a battery, so that users can carry the set anywhere they move.
2. **Wireless connectivity** The user's set and mobile telephone base station must have wireless connectivity for free mobility.
3. **Personal mobility** The users should be able to use the same telephone number every where, whether in coverage area or outside coverage area with roaming facility.
4. **Value addition** Modern mobile communication system is not restricted to voice only i.e., it should be supported with all types of facilities to exchange information including voice, data, image, video and multimedia.
5. **Fast and efficient call processing**.

15.24.3 Concept of a Cell

The cellular technology divides the geographical region into small regions of uniform shape and area. These are referred to as **cells**. The diameter of a cell ranges from 2 to 50 km. Each cell is provided with a base station of its own that care of coverage in that particular cell. The antennas used for transmission are generally low powered as the region to be covered is small. The transmitters of adjacent cells operate on different sets of frequency. Cells located at far distances can reuse the same set of carrier frequencies without causing interferences from any co-channel. This is termed as the cellular concept of frequency reuse. With the concept of cells and reuse of frequency, the channels that are allocated in a certain cell are reused in a cell that does not fall in the neighbourhood of the cell. The **spectrum scarcity** is the main bolleneck for limited capacity. This can be solved by spectrum conservation, i.e. to make frequence reuse maximum possible.

Frequency reuse concept Following are the few techniques that can be used for frequency or spectrum reuse:

(i) TDMA

(ii) CDMA

(iii) Sectoral concept (frequency reuse direction)

(iv) Cellular concept (frequency reuse distance)

TDMA technique is used for spectrum reuse in time-domain utilizing the same spectrum in different time slots, whereas in CDMA, same frequency spectrum is used by several users all a time without interference. Here, each user is assigned a distinct code to identify at the receiver.

The sectoral concept and cellular concept provide frequency reuse in space-domain, i.e. one can reuse the same frequency in radiation-limited regions. In sectoral concept, different antenna lobes radiating in different geographical sectors (angular coverage) will not create interference even if they reuse same spectrum as shown in Figure 15.44. Normally, neighbouring sectors avoid use of the same spectrum to so that co-channel interference may not happen. Distant sectors can reuse the same spectrum.

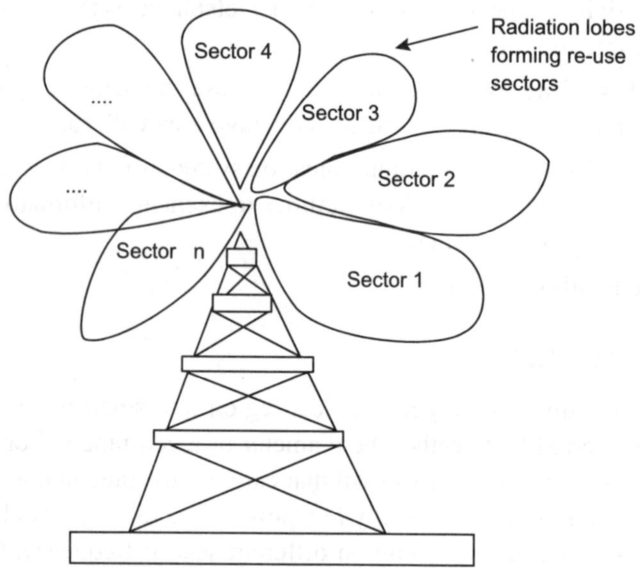

Figure 15.44 Illustration of sectoral concept of frequency reuse

Cellular concept of frequency reuse This concept helps frequency reuse in space-domain. Frequency domain can be divided into following two categories:

(i) Same frequency spectrum is assigned in two different geographical areas separated by large distance, such as high-power AM or FM radio stations using same frequencies in different cities. This can be considered large size cells.

(ii) Same frequency is repeatedly used in same geographical area, divided to form several small sized (few km) co-channel cells. This system utilizes low-power transmitters, for making the spectrum reused after a gap of few cells.

Frequency reuse distance The fundamental principle of frequency reuse in cellular systems is that the radiation field of transmitting antenna diminishes with distance, and after a distance D (termed as frequency reuse distance) the field becomes so weak that it does not interfere with any radiation from other antenna radiating at same frequency (Figure 15.45).

Each cell has coverage radius equal to R, and uses the same frequency spectrum f_1, and hence termed *co-channel cell*. D is the distance between co-channel cells where frequency can be reused.

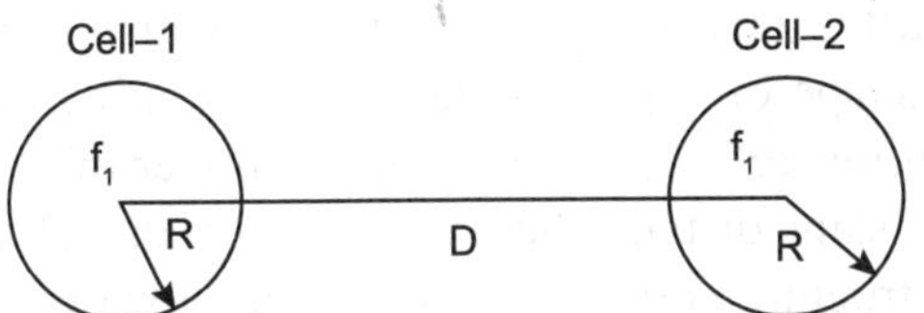

Figure 15.45 Frequency reuse distance

The minimum frequency reuse distance D is given by

$$D = \sqrt{3K} \cdot R \quad \text{or} \quad \frac{D}{R} = \sqrt{3K}$$

where K is a constant depending on *frequency reuse* pattern, and denotes the number of cells in a *cluster* of cells. Here, cluster denotes a group of neighbouring cells in which each cell uses a different frequency spectrum.

The space between co-channel cells is filled by creating similar other cells and making use of different frequency spectrum. As the frequency of these neighbouring cells is different, theoretically, we can say that they have no mutual interference. However, in practice, there may be some inter-channel interference. *Inter-channel interference* is interference between the neighbouring cells of a cluster using different frequency spectrum. Figure 15.46 represents a cluster of four cells, i.e. $K = 4$ for this pattern.

The cells shown are circular, and resembles the most efficient radiation pattern of an isotropic antenna. However, this cluster of circular cells leaves some overlapping area (hole), which is not possible to cover by any cell. One can be eliminate the hole by choosing a hexagonal shape for cells. Any other regular shape can be chosen (e.g. square). However, the hexagonal shape avoids hole and provides most effective transmission approximately closer to a circular pattern. The hexagonal shaped cells fitted together look like a honeycomb pattern (Figure 15.47). The size of a cell and number of cells per system is left to the service provider.

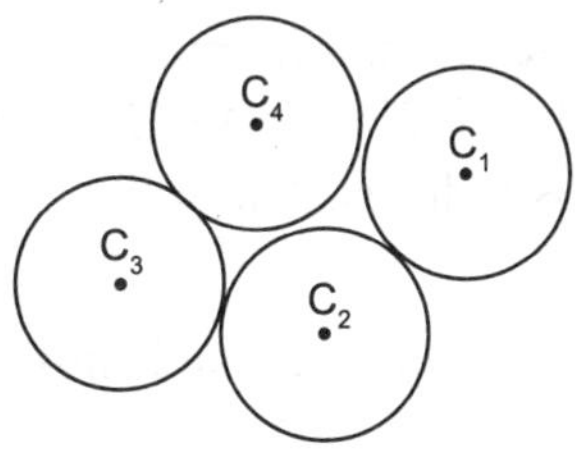

Figure 15.46 A cluster of four cells ($K = 4$)

Figure 15.47 shows three clusters, each cluster having a group of four cells. Each cell is allotted a spectrum depending on the number of channels and total available bandwidth. The four cells of a cluster make use of different sets of frequencies for their base stations. However, cells marked '1' in all the three clusters reuse the same frequency for their base stations. Similarly, cells marked '2' in each cluster reuse the same frequency which is different from cells marked '1' or '3'. The hexagonal shape of cells provide 6 neighbouring equidistant cells. The angle between adjacent neighbouring cells is 60°. A limited number of cluster size and cell layout is possible. Obviously, for the same distance between the base station and the farthest point in the foot print, the hexagonal structure covers a greater area. However, the actual cell size might differ in practice. In order to cover the entire system coverage area and to minimize any gaps of overlaps between adjacent clusters, K can only have values which satisfy the equation.

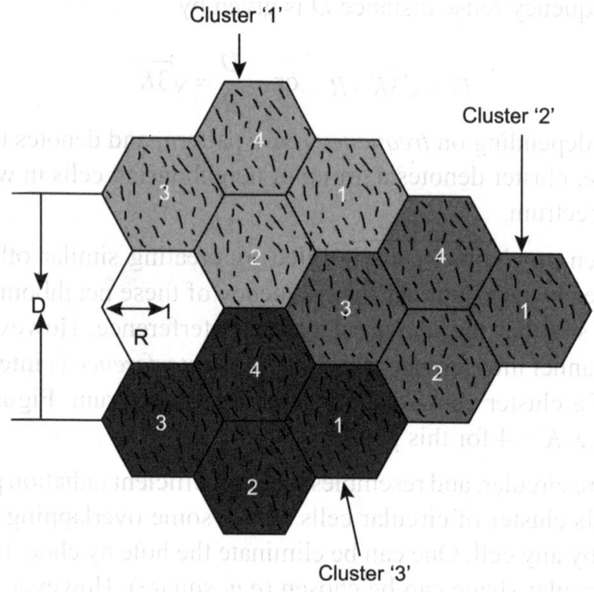

Figure 15.47 Cell pattern for $K = 4$

$$K = i^2 + j^2 + ij$$

where i and j are non-negative integers. The minimum value of K comes out to be 3, for $i = 1$ and $j = 1$. Theoretically, a large value of K is desired, but if it is too large then the number of channels assigned to each of K cells reduces. The following method is useful for locating the nearest co-channel cell.

(i) Count i cells leaving the reference cell and move through the centre of successive cells.

(ii) Turn 60° counter-clockwise from the centre of ith cell.

(iii) Move j cells to reach the nearest co-channel cell.

The co-channel interference can be determined by a factor called frequency reuse ratio given as

$$Q = D/R$$

Smaller the Q, larger will be the channel capacity of cell pattern. This increases the co-channel interference. The above equation yield a compromised value for D/R. Accordingly, one finds, minimum frequency reuse distance D for different K-pattern as follows:

For

$$K = 4; \quad D = 3.46R$$

$$K = 7; \quad D = 4.6R$$

$$K = 12; \quad D = 6R$$

$$K = 19; \quad D = 7.55R$$

Each mobile service area is divided into clusters. Each cluster has n disjoint cells, and each channel has G number of full duplex channels. However, if a cluster is duplicated m times in a mobile service area, the total number of channels (N) in the service area can be obtained from

$$N = mGn$$

Physical size of cell This varies depending on user density and calling pattern. Broadly speaking, one can classify the physical size in three categories:

(i) *Macrocell* It has a radius of 2–24 km and base station transmitter power lies between 1 and 6 W.

(ii) *Microcell* These cells are used in high density areas like large cities. The radius of such cells is 500 m or less, and base station power is 0.1–1 W.

(iii) *Picocell* Such cells are smaller than micro cells and used in well-shielded area with high interference where normal signal is very weak, e.g. underground malls.

We may note that smaller cells require more hand-offs and hence more processing.

Cell splitting The macro-cells of cellular system can be split into micro and picocells. The main purpose of cell-splitting is to enhance the channel capacity of cellular system and improve its availability and reliability by reducing blocking for congested cells. However, cell-splitting has certain demerits. It requires more hand-off per cell. Reduction of cell radius by a factor of 4 gives 10-fold increase in hand-off rate per subscriber. Obviously, this increases the processing load per subscriber and moreover the cell-splitting increases the cost of the system.

Hand-off This refers to switching the channel that the mobile is currently using, when the mobile moves from one cell to another cell. As the mobile unit moves away, the base station of cell–1 goes on decreasing and when it drops to below a pre-decided threshold level, the MTSO (Mobile Telephone Switching Office) locates the base station of cell–2 receiving stronger signals from mobile unit and transfers the mobile unit under the control of base station of cell–2. The call does not drop and remains continuous. The hand-off process can be accomplished in four stages.

(i) Initiation

(ii) Resource reservation

(iii) Execution

(iv) Completion (resources are relinquished).

Smaller cells require frequent hand-off. The analogue cellular systems use hardware circuits for accomplishing hand-offs. The digital cellular systems can use soft hand-off techniques which are quite flexible and adaptive. When the mobile unit moves from one service area to another service area (may be from other company with prior agreement), the process is called **roaming**.

A basic cellular network Figure 15.48 shows a basic cellular network. It has the following components:

(i) Mobile unit

(ii) Cell-site (base station)

(iii) MTSO

(iv) System interconnects

(v) Communication protocol

Mobile unit This is a portable, low-weight handset carried by the user which has wireless connectivity with the nearest base station. A simple mobile unit contains a control unit, a radio transreceiver and an antenna system. The modern mobile unit has camera and also other value addition. The transreceiver consists of a transmitter and receiver for two-way telephonic conversation. It consists of a frequency synthesizer for timing the designated channels. The control unit houses all the user interfaces.

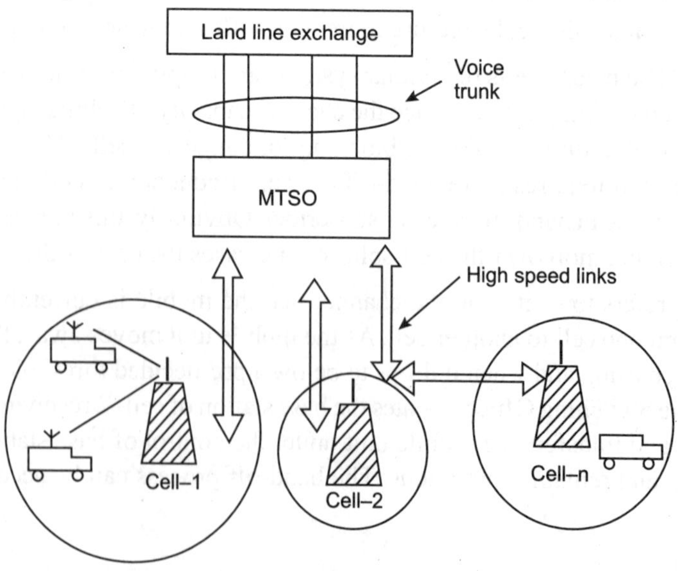

Figure 15.48 A basic cellular system

Cell site This provides an interface between the mobile unit and MTSO. It is also called as base station and consists of control unit (base station controller), radio cabinets, antennas, power plant and data terminals. It is connected with MTSO on one side using wideband media, and on the other side it has wireless connectivity with the mobile unit.

MTSO (Mobile Telephone Switching Office) MTSO is considered the heart of the mobile system. Its processor provides cellular administration and central coordination. It can also be connected with landline telephone network. It contains the cellular processor and cellular switch.

The cellular switch is an analogue or digital telephone exchange which controls switching between landline subscriber unit to base station for landline-to-mobile connectivity and vice versa. It controls switching between a base station and another base station for mobile-to-mobile conversation.

System interconnect Radio connectivity, voice-grade four-wire line connectivity, optical fibre connectivity, microwave link, data link, etc. may be used for connecting mobile unit, base stations, MTSO and public switched telephone network (PSTN). Each mobile unit may use only one channel at a time for communication link with a base station. But this channel is not fixed. It can be assigned to any channel from the entire band allocated for the service area. For second conversation, some other channel may also be assigned. Microwave link or T-carriers (wire-line) are used to carry both data and voice between MTSO and base station. Voice-trunks are used to connect MTSO with PSTN. Optical fibres can be used for wideband data and information.

Communication protocols The protocols govern the process of call connection and disconnection at the end of conversation. Several layers of protocols may used and each company can have different protocols for its cellular network, e.g. IS-54 and IS-95, GSM, etc.

Operation of a cellular system Besides a hand-off procedure, the operation of a cellular system can be divided into following parts:

1. ***Mobile unit initialization*** Out of the total radio channels (say 416) available for a cellular coverage area, few channels (say 21) are designated for setting up connections, and are termed *set-up-channels*. When a subscriber activates his mobile unit by switching on the power, its receiver scans the set-up channels, selects the strongest one (nearest cell-site) and locks on it for a certain time. This is termed as self-location scheme. This limitation can be eliminated by *registration* process.
2. ***Mobile unit originated call*** The mobile user enters the called number and presses the send button. The request goes to the cell-site through a set-up channel. The cell-site then sends the request via a high-speed link to MTSO for allocating a *voice-channel*. The MTSO allots a suitable free voice channel and cell-site links the two subscribers for conversation.
3. ***Land-line originated call*** When a land-line subscriber dials a mobile unit number, the respective telephone exchange transfers it to MTSO via voice-grade trunk lines. The MTSO then sends this information on relevant cell-sites along with a search algorithm. Each

cell-site uses its set-up channel to transmit information to its mobile units. The concerned mobile unit recognizes its identification and locks into it.

4. ***Call-termination*** When the mobile user switches off its transmitter, a signaling tone is sent to the respective cell-site and both sides make the voice channel free.

15.25 Personal Communication System (PCS)

This is a combination of cellular telephone network and intelligent network (IN). IN is an entity that distinguishes the physical components of a network from the services provided by the network. The main aim of IN is to decouple the functions required for call and connection control from the function required for service control. Obviously, it provides two sets of functions placed on different platforms. The key attributes of IN-based networks are extensive use of information processing techniques, efficient use of network resource, service independent interfaces, modular service creation and flexible allocation of network functions. PCS is a North American implementation of European GSM standard. Some companies define PCS to compare with standard telephone systems but offer enhanced services and better quality, may be by using new standards or simply extension of existing standards. The extension of standard includes: (i) extend the frequency range (e.g. from 800 MHz band to 1900 MHz), (ii) smaller cell-size, (iii) all digital systems, (iv) extra features like short messaging service (SMS), e-mail and limited web browsing, (v) coverage extension for roaming, (vi) lower tariff. The future communication network aims at becoming all-wireless and mobile communication supporting PCS with a single unique identification number through a personal communication network for *global roaming*.

15.26 Standards of Mobile and Personal Communication Systems

Developments in mobile communication systems were based on improvements in technologies (analogue-to-digital), access techniques (FDMA, TDMA, CDMA, SDMA), operating frequencies, bandwidth availability, roaming distance, traffic density, type of service handled (voice, data, fax, video, multimedia, internet, etc.), information security, and many other parameters. Mobile and public communication systems (PCS) standardized in every generation of development by various agencies. The first (analogue systems) and second generation (digital systems) have already been implemented. Further improvement in second generations (2G) has been operative as 2.5G or 2.75G. The third generation (3G) has also been implemented in several countries. The draft of fourth generation (4G) is in the final stage.

First generation (G) The first analogue cellular system introduced in Chicago in 1983. This was based on the **trunking principle**. The analogue mobile systems fall in the category of **first-generation** mobile systems. The first generation cellular system had capacity bottleneck and incomplete standardization due to which wider use of cellular systems was impossible. These led the development of second generation cellular systems with higher capacity and improved compatibility.

Second generation (2G) The second generation (2G) of cellular mobile systems is digital in nature and uses TDMA/CDMA for multiple access. Advanced digital modulation techniques like QPSK and GMSK are used for bandwidth conservation.

Third generation (3G) The third generation mobile and PCS are aiming for improvement over previous 1G and 2G generations in the following respects:

(i) Improved data communication for wireless connectivity with high speed applications such as multimode or video telephony
(ii) Greater capacity for voice and data communication (huge transmission bandwidth).
(iii) Greater standardization
(iv) Availability to mobile and fixed operations
(v) Interoperatibility with 2G systems like IS-95 and W-CDMA
(vi) Extended roaming facility with low handover time
(vii) High security
(viii) Complemented with portable internet technologies like Wi-Max.

Fourth generation (4G) The 4G network promises speed up to 100 Mbps in cell phone, and up to 1 Gbps in local Wi-Fi network. It will be compatible with Bluetooth, WCDMA, etc. It can support interactive multimedia, internet and other wideband services. Some of the advantages of 4G deployment are:

(i) Very high data rate (say 50 times more than that of 3G)
(ii) Enhanced GPS (Global Positioning System)
(iii) Evaluation to provide subscriber usage and analysis, visitors usage and analysis, network usage.
(iv) GPRS roaming through GRX (GPRS roaming exchange).

Example 1 A transmitter supplies 10 kW power to an aerial when unmodulated. Show that the power radiated will be 10.45 kW when modulated to 30%.

Solution

Given $p_c = 10$; $m = 0.30\% = 0.3$

$$\text{Total power} = \text{Carrier power}\left(1+\frac{m^2}{2}\right)$$

or

$$P_T = P_C\left(1+\frac{m^2}{2}\right)$$

$$= 10\left(1+\frac{(0.3)^2}{2}\right)$$

$$= 10.45 \text{ kW}$$

Example 2 The unmodulated carrier current to the aerial of a transmitter is 100 A. What will be the increase in current which will result from the application of 80% modulation.

Solution

Given $I_C = 100$; $m = 0.8$

We know $I_T = I_C\sqrt{1+\dfrac{m^2}{2}}$

$$= 100\sqrt{1+\frac{(0.8)^2}{2}}$$

$$= 114.9\,A$$

Therefore, increase in current due to modulation = 114.9 – 100 = 14.9 A

Example 3 A 100 V, 10 kHz carrier is modulated with the help of 5 V, 50 Hz signal. Determine (a) modulation factor, (b) amplitude of each side band, (c) frequency of each side band and (d) bandwidth of modulated wave.

Solution

(a) Modulation factor,

$$m = \frac{E_m}{E_c} = \frac{5}{100} = 5\%$$

(b) Amplitude of each side band $= \dfrac{mE_c}{2}$

$$= \frac{0.05 \times 100}{2} = 2.5$$

(c) Frequency of sidebands:

$$\text{USB} = f_c + f_m = 50 + 10000 = 10050\,\text{Hz}$$
$$\text{LSB} = f_c + f_m = 10000 - 50 = 9950\ \text{Hz}$$

(d) Bandwidth of modulated wave

$$\text{BW} = 2f_m = 2 \times 50 = 100\ \text{Hz}$$

Example 4 An AM wave has peak-to-peak voltage of 600 V and valley-to-valley voltage of 100 V. Show that the percentage depth of modulation is 70%. [Diploma]

Solution

From the figure, we have

$$m = \frac{V_{\max} - V_{\min}}{V_{\max} + V_{\min}}$$

$$= \frac{600-100}{600+100} = \frac{500}{700} = 71.4\%$$

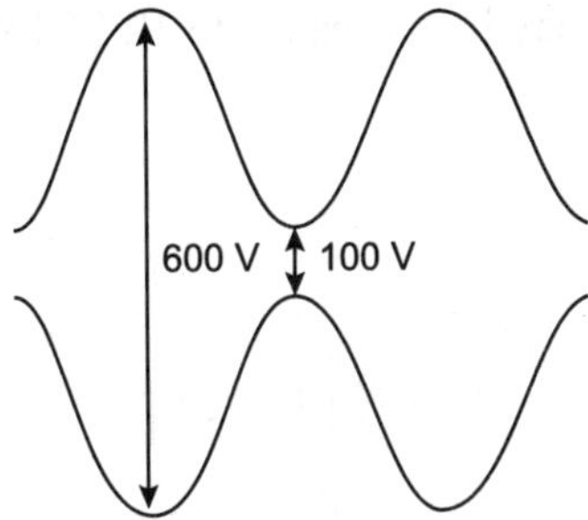

Figure 15.49

Example 5 Equation $e = 20(1+0.7\sin 6280t)\sin 628000t$ represents an AM wave. Find (i) modulation factor (m), (ii) carrier amplitude (E_C), (iii) signal frequency (f_m), (iv) carrier frequency (ω_c), (v) maximum amplitude of AM wave (E_{max}) and (vi) minimum amplitude of AM wave and (vii) bandwidth.

Solution

The standard equation of AM wave is

$$e = E_C(1+m\sin\omega_m t)\sin\omega_c t \qquad \text{(i)}$$

Given equation is

$$e = 20(1+0.7\sin 6280t)\sin 628000t \qquad \text{(ii)}$$

Comparing (i) and (ii), one obtains

(i) Modulation factor, $m = 0.7$

(ii) Carrier amplitude, $E_C = 20$ V

(iii) $\omega_m = 6280$

$\therefore$ Signal frequency, $f_m = \frac{\omega_m}{2\pi} = \frac{6280}{2\pi} = 1$ kHz

(iv) $f_c = \frac{\omega_c}{2\pi} = \frac{628{,}000}{2\pi} = 100$ kHz

(v) $E_{max} = E_c + mE_C$
$= 20 + (0.7\times 20) = 34$ V

(vi) $E_{min} = E_C - mE_C = 20 - mE_C$
$= 20 - (0.7\times 20) = 6$ V

(vii) Bandwidth $= 2f_m = 2\times 1 = 2$ kHz

Example 6 A radio transmitter using amplitude modulation has unmodulated power output of 10 kW and can be modulated to a maximum depth of 90% by sinusoidal voltage. Show that the power of modulated wave is 14 kW. If maximum permitted modulation is 40%, then show that unmodulated carrier power may be increased to 12.96 kW. [B. Tech]

Solution

Given $P_C = 10$; $m = 0.9$

We have

$$P_T = P_C\left(1+\frac{m^2}{2}\right)$$

$$= 10\left(1+\frac{(0.9)^2}{2}\right)$$

$$= 14 \text{ kW}$$

This will be the maximum power handled by the transmitter. Now, increased unmodulated carrier power can be obtained by

$$14 = P_C\left[1+\frac{(0.4)^2}{2}\right] \Bigg\| m = 40\% = 0.4$$

or

$$P_C = 12.96 \text{ kW}$$

We may note that original power of the carrier was 10 kW.

Example 7 A 100 V, 100 kHz carrier is modulated with the help of a 10 V, 1 kHz signal to the extent of 50%. Show that the equation for the AM wave is

$$E = 100 \sin 628000\, t + 25 \cos 621720\, t - 25 \cos 634280\, t$$

[Diploma]

Solution

Given

$$m = 50\% = 0.5$$

$$E_C = 100 \text{ V};\ E_m = 10 \text{ V}$$

$$f_c = 100 \text{ kHz}$$

$$f_m = 1 \text{ kHz}$$

$$\omega_c = 2\pi f_c = 2\times 3.14\times 100\times 10^7$$

$$= 628000$$

and

$$\omega_m = 2\pi f_m = 2\times 3.14\times 1\times 10^3$$

$$= 6280$$

Now, putting these values in the standard equation for modulated voltage wave,

$$e = E_C \sin \omega_c t + \frac{m E_C}{2} \cos(\omega_c - \omega_m)t - \frac{m E_C}{2} \cos(\omega_c + \omega_m)t$$

one obtains

$$e = 100 \sin 628000t + \frac{0.5 \times 100}{2} \cos(628000 - 6280)t$$

$$- \frac{0.5 \times 100}{2} \cos(628000 + 6280)t$$

$$= 100 \sin 628000t + 25 \cos 621720t - 25 \cos + 634280t$$

This is the desired equation of AM.

Summary

1. Communication is the transfer of meaningful information from one point (source) to another (destination). The information to be transmitted can be in the form of audio or video signals, or it could even be digital data.
2. Signals can be transmitted from source to destination by two means: (i) wireless communication, and (ii) line communication.
3. Radio communication comprises: (i) radio telegraphy and (ii) radio telephony.

   Radio telephony or wireless communication has two major subsystems: (a) transmitting system or transmitter. (b) receiving system or receiver.

   For transmission, we make use of variation of some characteristics, say amplitude or frequency or phase of radiowave, and mix it with a locally produced wave known as carrier wave. This process is known as modulation.

   (i) Modulation is defined as the process by which some characteristics of a signal called carrier is varied in accordance with the instantaneous value of another signal called modulating signal.

   (ii) Signals containing information to be transmitted are called modulating signal. This information-bearing signal is also called *baseband signal.*

   (iii) The carrier frequency is much greater than modulating frequencies. The signal resulting from the process of modulation is called modulated signal.

   (iv) When the carrier wave is continuous in nature, the modulation process is known as continuous wave (CW) or analogue modulation. The carrier wave is usually a sinusoidal signal. Amplitude modulation and angle modulation are examples of this type of modulation.

   (v) When the carrier wave is a pulse type waveform, the modulation process is known as pulse modulation. PAM, PWM, PCM, DM, etc. are examples of this type of modulation.

(vi) Modulation is needed in communication systems to achieve the basic needs: (i) multiplexing, (ii) practicability of antennas.

(vii) Simultaneous transmission of multiple messages over a channel is known as multiplexing.

(viii) When free space is used as communication media, messages are transmitted and received with the help of antenna. The message signal is radiated by an antenna at the transmitter. The antenna radiates effectively when its size is of the order of wavelength of the signal being transmitted. For example, if an audio frequency is translated to a radio frequency carrier of frequency 10 MHz, the antenna height required will be

$$\frac{\lambda}{2} = \frac{3\times10^8}{10^7} = 15\,\text{m}.$$

4. This antenna height can be practically achieved. We have three types of modulation: (i) amplitude modulation, (ii) frequency modulation, (iii) phase modulation.

   (i) In amplitude modulation, the amplitude of carrier wave changes as per the amplitude of desired signal to be transmitted. However, the variation in amplitude of modulated wave is at the same rate as that of carrier wave.

   (ii) In frequency modulation, the frequency of carrier wave changes as per the amplitude of desired signal (modulated wave) to be transmitted.

   The equation of amplitude modulated wave is

$$e = E_C[1 + m\cos\omega_m t]\sin\omega_c t$$

   Here E_C – amplitude of carrier wave; of E_m – amplitude of modulating wave; $E_c m = \frac{E_m}{E_c}$ is called modulation factor;

   ω_m - Angular frequency of modulating wave;

   ω_c - Angular frequency of carrier wave ($\omega_c \gg \omega_m$).

   Here the modulated wave consists of two side bands $(f_c + f_m)$ and $(f_c - f_m)$ known as upper and lower side bands, respectively.

   The equation of frequency modulated wave is given by

$$e = E_C\sin(\omega_c t + \delta\sin\omega_m t)$$

   where $\delta = m_f = \frac{(\Delta f)_{\max}}{f_{\max}}$ is frequency modulation index.

5. Demodulation is the process in which the detector extracts information or audio signal from the modulated wave at the receiving station.

6. For the detection of audio signal from AM wave, we generally employ a diode, thereby the carrier wave is filtered out by employing suitable filters. This is linear detection.

$$\text{Detection efficiency, } \eta = \frac{E_m}{mE_C}$$

7. A radio receiver is a device which reproduces the audio signal from modulated radio waves interfaced by its antenna.
8. The basis of the cellular technology is frequency reuse. The same frequency channels can be reused at a distance far enough to keep the interference within tolerable limits.
9. With the concept of cells and frequency reuse, the channels that are allocated in a certain cell are reused in a cell that does not fall in the neighbourhood of the cell.
10. The actual physical area covered by a cell is termed a footpoint. N cells among which the total allocated bandwidth is shared are together called a **cluster**. $1/N$ is termed as frequency reuse ratio, where N can only take values that satisfy the equation $N = i^2 + j^2 + ij$ for integer values of i and j.

Review Questions

1. What do you mean by the term modulation? How many types of modulation you are familiar with? Describe each in brief and represent the respective modulated wave.
2. Which of the modulation processes for radio wave is commonly used? Mention its merits and demerits vis-a-vis other processes.
3. Derive an expression of amplitude modulated wave. Depict the respective expression for AM modulation for (i) $m < 1$, (ii) $m = 1$, (iii) $m > 1$. Explain the various parameters used.
4. An amplitude modulated wave is expressed as

$$e = E_c(1 + m\cos\omega_m t)\sin\omega_c t$$

   Derive an expression for power carried by the wave. Prove that power carried by side bands is half the total power for $m = 1$.
5. Prove that in a balanced modulator, the output consists of side band frequencies only. Give its merits over phasing method.
6. What is a balanced modulation process? Draw appropriate circuit diagram explaining the process. Why we prefer single side band method for modulation? Explain.
7. Describe the filter method for generation of single side band. Give its merits over phasing method.
8. What is frequency modulation? Draw a frequency modulated wave given its modulating wave and carrier wave. Write the equation of FM wave.
9. Draw a block diagram of FM transmitter. Explain briefly the function of each block.

10. What is the principle of linear envelope detector? Draw a circuit diagram for such a detector. Describe its working and draw appropriate related waves.
11. Define "detection efficiency". Derive the nccessary relation for efficiency?
12. What is AGC? Where and how it is employed? Discuss its function. How the same is improved upon by employing delayed AGC? Draw the necessary curves.
13. Draw the circuit diagram of a delayed AGC employing FET. Describe its working.
14. What do you mean by the term "superheterodyne". Explain the working of a superheterodyne receiver by drawing the necessary block diagram.
15. Why we prefer SSB transmission? Draw the necessary block diagram and explain its function.
16. How an amplitude modulated wave is demodulated? Describe the working of the method employed. Give necessary circuit diagram.
17. Describe phasing method for generating single side band. Draw necessary block diagram. Derive the necessary relation.
18. Draw the block diagram of
    (i) AM transmitter
    (ii) FM transmitter
    Explain briefly the function of each block.
19. What is image frequency in superheterodyne receiver? How the same is eliminated?
20. Compare amplitude modulation and frequency modulation.
21. Draw a block diagram showing the various subsystems used in frequency modulated receiver. How it differs from that of AM receiver?

Short Answer Questions

1. What is amplitude modulation?

   Ans: In amplitude modulation, the amplitude of carrier wave is varied according to the variation of amplitude of modulating wave.
2. How power is distributed in an amplitude modulated wave?

   Ans: The power in AM wave is distributed among three components, namely (i) carrier wave (ii) upper and lower side bands. The distribution of power among the components depends upon the modulation factor.
3. Why SSB is preferred in amplitude modulation?

   Ans: In amplitude modulation, SSB is preferred because the information component is carried only by side bands. So one side band is suppressed and carrier wave is completely eliminated since it does not carry any useful information.

4. How carrier component is eliminated in AM wave?

   Ans: The carrier component in AM wave is eliminated by using a balanced modulater.

5. Which type of modulation completely uses the input power?

   Ans: Frequency modulation uses the input power completely and in amplitude modulation a large part of input power goes waste in carrier component since it does not contain information.

6. What is detection efficiency? What is its maximum value?

   Ans: Detection efficiency is defined as the ratio of amplitude of modulating wave (E_m) to m times amplitude of carrier wave.

$$\eta = \frac{E_m}{m E_C}$$

   Its maximum value is equal to $\frac{1}{\pi}$.

7. Why frequency modulation is preferred over amplitude modulation?

   Ans: Frequency modulation is preferred over amplitude modulation because here the entire input power is utilized, whereas in amplitude modulation a large part of input power goes waste in the carrier component.

   Moreover, in frequency modulation, noise signal is minimum since it operates at high frequencies, thereby $\left[\frac{S}{N}\right]$ is large.

8. What is the principle of working of a superheterodyne receiver?

   Ans: In a superheterodyne receiver, the incoming radio wave signal (f_s) is mixed with radio frequency waves of frequency f_o produced by a local oscillator (keeping $f_o > f_s$) so we get a beat frequency or IF signal. This signal is amplified and the audio signal is detected from it by a detector which is further fed into the loud speaker after proper amplification.

9. What is the value of beat frequency signal in radio?

   Ans: The beat frequency is 455 kHz.

10. What is image frequency?

    Ans: It is unwanted incoming signal frequency which reaches the mixer and produces the same beat frequency (≈ 4555 kHz) when mixed with the signal of local oscillator. Its value is equal to the sum of desired signal frequency and twice the beat frequency.

11. What is automatic gain control? Where it is used?

    Ans: It is a system used in radio receivers, by means of which the overall gain of the receiver is varied automatically with the strength of incoming signal to keep the output at audible level. Generally, we used delayed AGC so that the incoming weak signal is not affected and we can also bear it.

12. What are passive satellites?

    Ans: A passive satellite is a metallic sphere or a metal coated plastic balloon, that works as a passive reflector. Microwave signals are reflected from it from one region to another region.

13. What are active satellites?

    Ans: An active satellite works as an active microwave repeater or transponder in the sky and carries an antenna system, a transmitter, a receiver and a power supply. The microwave signals emitted by the transmitter on the earth's surface directly towards the satellite are received by the transponder on the active satellite. After amplification, signals are sent to the receiving station on the earth either instantaneously or after storing in the memory until the satellite faces receiving station on earth's surface.

14. Can the moon, an natural satellite of earth act as a passive satellite?

    Ans: yes.

Problems

1 A 400 W carrier wave is modulated to a depth of 75%. Calculate the total power of modulated wave. **[Ans:** 512.5 W]

2. A radio transmitter radiates wave of strength 100 kW when modulation index is 60%. How much of this is carrier power? **[Ans:** 8.47 kW]

3. The antenna current of an amplitude modulated transmitter is 8 ampere when only the carrier is transmitted, but it increases to 8.93 ampere when the carrier is modulated by a single modulating wave. Find the percentage modulation. Determine antenna current when $m = 0.8$. **[Ans:** 70.1%; 9.19 A]

4. A certain transmitter radiates 9 kW with unmodulated carrier wave and 10.125 kW when the carrier is modulated. Calculate modulation index. If another wave corresponding to 40% modulation is also simultaneously transmitted, determine the total radiated power.

   [Hint: final $m_f = \sqrt{m_2^1 + m_2^2}$] **[Ans:** 0.5, 10.84 kW]

5. A 1000 kHz carrier is simultaneously modulated with 300 Hz, 800 Hz and 2 kHz sine waves. What frequencies will be present in the output?

   [Ans: 9980.0, 999.2, 999.7, 1000.3, 1000.8, 1000.2 kHz]

6. A transistor class C amplifier has maximum permissible dissipation of 20 watt and collector efficiency of 75%. Calculate (i) maximum unmodulated carrier power and (ii) side band power. **[Ans:** 42.7 W, 17.3 W]

7. Calculate modulation index in case of AM wave: when (i) modulating wave amplitude is zero and amplitude of carrier wave is E_C, (ii) modulating wave amplitude is equal to carrier wave amplitude.

   $$\left[\text{Hint: } m = \frac{E_{\max} - E_{\min}}{E_{\max} + E_{\min}}\right] (m = 0, m = 100\%)$$

8. When a broadcast AM transmitter is 50% modulated, its antenna current is 12 ampere. What will be its current when modulation depth is increased to 0.9.

   [**Ans:** 13.4 ampere]

9. In a radio receiver the beat frequency is 455 kHz and the desired signal frequency is 1000 kHz. Calculate the image frequency. [**Ans:** 545 kHz]

10. Prove that image frequency is greater than desired frequency signal by twice the beat frequency.

Objective Questions

1. The carrier of a 100% modulated AM wave is suppressed; the percentage power saving is:

   (a) 50 (b) 66.66 (c) 100 (d) NIL

   [**Ans:** (b)]

2. Modulation index of an AM wave is changed from 0 to 1. The transmitted power

   (a) is doubled (b) is halved
   (c) is tripled (d) increases by 50% [**Ans:** (d)]

3. A carrier wave is simultaneously modulated by two sine waves with modulation index 0.3 and 0.4, respectively. The total modulated index is

   (a) 1 (b) 0.7 (c) 0.5 (d) zero

   [**Ans:** (c)]

4. The plate supply voltage for a plate modulated class C amplifier is 10 V. The maximum plate cathode voltage could almost be as high as

   (a) 40 V (b) 30 V (c) 20 V (d) 10 V

   [**Ans:** (a)]

5. An AM receiver uses a diode detector for demodulation. This enables it to satisfactorily receive

   (a) single side band, suppressed carrier
   (b) single side band, reduced carrier
   (c) single side band, full carrier
   (d) double side band, full carrier [**Ans:** (c)]

6. In a super heterodyne receiver, the local oscillator generates a signal of frequency
   (a) 455 kHz
   (b) 455 kHz above incoming signal
   (c) 455 kHz below incoming signal
   (d) double of the incoming signal **[Ans:** (b)]
7. In a super heterodyne receiver, beat frequency is
   (a) 455 kHz
   (b) 455 kHz above incoming signal
   (c) 455 kHz below incoming signal
   (d) depends upon incoming signal **[Ans:** (a)]
8. In an AM wave, useful power is carried by:
   (a) side bands (b) carrier wave
   (c) IF wave (d) None of the above **[Ans:** (a)]
9. In a super heterodyne receiver, the deserved incoming signal is 3.5 MHz; the image frequency is ____________.
   (a) 4.41 MHz (b) 4.5 MHz (c) 3.5 MHz (d) 4.4 MHz **[Ans:** (a)]
10. The propagation time from one earth station to antoher via a geostationary satellite is approximately
    (a) 125 ms (b) 125 μs (c) 300 ms (d) 25 μs **[Ans:** (c)]
11. The approximate height of geostationary orbits from earth's surface is
    (a) 3500 km (b) 6400 km (c) 35,800 km (d) 45,600 km **[Ans:** (c)]
12. The order of optical frequencies is
    (a) THz (b) GHz (c) MHz (d) kHz **[Ans:** (a)]
13. The second-generation mobile communication system is
    (a) GSM (b) IMJ-2000 (c) AMPS (d) none **[Ans:** (a)]

Mark the Following Statements True/False.

1. An amplitude modulated wave consists of carrier signal and one side band only. **[Ans:** False]
2. For zero modulation index, the total power carried by AM wave is equal to power carried by carrier wave. **[Ans:** True]
3. A balanced modulator is used for suppression of carrier AM wave. **[Ans:** True]
4. In AM wave, the value of modulation index is always kept greater than 1. **[Ans:** False]
5. For SSB in phase shift method, no balanced modulator is required. **[Ans:** False]

6. In a superheterodyne receiver, beat frequency is fixed and equal to 455 kHz. [**Ans:** True]
7. All super heterodyne receivers use delayed AGC. [**Ans:** True]
8. A varactor diode is used to producer a frequency modulated wave. [**Ans:** True]

Fill in the Blanks.

1. An FM transmitter consists of __________ which gets DC bias from __________. [**Ans:** reactance tube, frequency discriminator]
2. The two side bands in an A.M wave are called __________ and __________. [**Ans:** lower side band, upper side band]
3. The frequency of modulating wave is always __________ than the frequency of carrier wave. [**Ans:** less]
4. A GC in __________ is used to keep the sound __________. [**Ans:** superhetrodyne receiver, audible]
5. In a frequency modulated wave, amplitude remains __________ but frequency __________. [**Ans:** constant, changes]
6. Noise in FM receiver is __________ because its frequency range is __________. [**Ans:** minimum, ultra high]
7. Uplink frequency is __________ downlink frequency in satellite communication. [**Ans:** greater than]
8. In CDMA, __________ can be used all the times. [**Ans:** entire bandwidth]
9. The figure of merit for a satellite transmitter is __________. [**Ans:** G/T ratio]
10. The main objective in a cellular mobile system is __________. [**Ans:** frequency reverse]

Appendix: Pulse Modulation

A signal may be represented by a train of pulses, and the pulses can be transmitted with

(a) varying amplitude known as pulse amplitude modulation (PAM)
(b) varying widths or time duration known as pulse duration modulation (PDM)
(c) varying positions known as pulse position modulation (PPM)
(d) coded group of pulses known as pulse code modulation (PCM)

Here we shall briefly describe the principle of each technique.

Pulse Amplitude Modulation (PAM)

Here the amplitude of each pulse is made proportional to the instantaneous amplitude of modulating signal (Figure 15.30a). The largest pulse *PQ* (Figure A.1(b)) represents the greatest positive signal amplitude *OA* sampled while the smallest pulse *RS* represents the greatest negative signal *BC*. The time duration of each pulse is quite small but equal to Δt for every pulse large or small, and the time interval between neighbouring pulses is quite long.

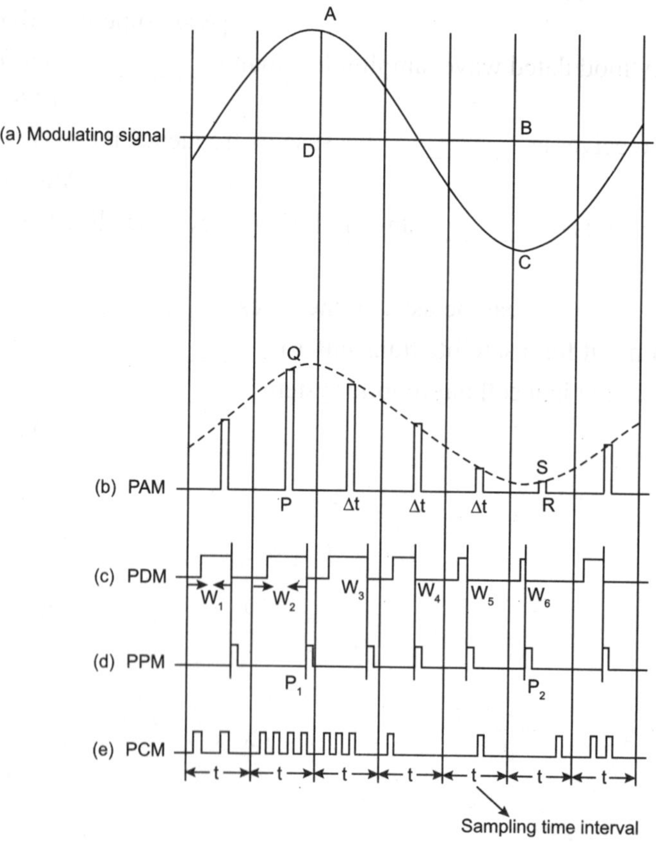

Figure A.1 (a) Modulating signal, (b) PAM, (c) PDM, (d) PPM, and (e) PCM

If a radio frequency wave is pulse amplitude modulated instead of simply being amplitude modulated, much less power is required for transmission of information because the transmitter is actually switched off between two neighbouring pulses. Less power requirement is one of the advantages of pulse modulation.

Pulse Duration Modulation (PDM)

It is also called pulse width modulation. Here the pulses have equal amplitude but variable width W_1, W_2, W_3, W_4, etc., i.e. time duration. The time duration (or width) of each pulse is proportional to the instantaneous amplitude of modulating signal (Figure A.1(c)). In this case, the narrowest pulse W_6 represents the most negative sample BC of the signal and the widest pulse W_2 represents the most positive sample OA of the signal. When PDM is applied to radio transmission, the carrier frequency has a constant amplitude and the transmitter ON time is carefully controlled (limited to pulse duration or width) as in case of magnetic tape recording where pulse widths can be recorded and reproduced with less error than pulse amplitudes.

Pulse Position Modulation (PPM)

In this type of modulation, all pulses have the same amplitude and width which is quite narrow. Here the position in time of pulses is made to vary in proportion to the amplitude of the modulating signal (Figure A.1(d)). It is to be noted in the figure that each PPM pulse occurs just at the end of PDM pulse (Figure A.1(c)) so pulse P_1 near the right side of sampling time interval 't' represents the largest positive signal OA of sample and those, i.e. P_2, towards the left side correspond to the most negative sample BC of original signal.

PPM uses less power than PDM and essentially has all the advantages of PDM. Thus, it is more efficient than both PAM and PDM for radio transmission.

One disadvantage of PPM is that demodulation process to recover the original signal is more difficult than PDM.

Pulse Code Modulation (PCM)

It is the most complicated type of modulation. However, it is the most accurate and efficient of all other types employed under certain conditions.

Here, each amplitude sample of the signal is converted from analogue to digital. The sample amplitude is then represented by a group of pulses.

Pulse Decoding and Demodulation

The various techniques employed are beyond the scope of book. However, for PCM signals they are first processed through a TDM decoding system to separate bits into different channels; then DAC circuits are employed to convert digital signals to analogue signals.

Chapter

16 Television and Radar

16.1 Introduction

The word "Television" means to see a scene/object which is a distant place. For this purpose telescope is generally used but it too has limitations. Moreover we have a still picture. So to view a live picture (moving object) along with sound associated with the scene/object television or commonly known as TV set is used. Here the picture (video signal) is generated by a TV camera and sound (audio signal) is generated by a microphone and both signals are simultaneously received by TV set installed at home. Thus, the TV set is essentially a receiving set. The mechanism to transmit still pictures was devised by Blackwell in 1847, but the credit for inventing the television goes to American scientist J.L. Baird.

16.2 Television System

Television system mainly comprises of two processes: (i) **Transmission** (ii) **Reception**. The first one is known as **Transmitter** and the second one is called a **Receiver**. We are familiar with Black and White TV and Colour TV which is very popular these days. Both types of TV employ the two process.

16.2.1 Black and White TV Transmitter

Block diagram of Black and White TV transmitter is shown in Figure (16.1).

Different subsystems employed in the transmitter are:

(i) **TV camera** It develops the image (picture) of the object in the form of electrical signals by focusing light incident from the object. Electrical signal called video is developed by a process known as scanning and synchronizing.

(ii) **Scanning and synchronizing system** In this system, we have horizontal and vertical sweep oscillators for scanning. It also develops the synchronizing signal separately to be fed to the TV camera.

(iii) **Video amplifier** It amplifies the video signal, which is the output of TV camera. It is basically a wide band amplifier which can amplify RF signals having bandwidth 7 MHz.

(iv) **AM video transmitter** Here, the video signal is converted into amplitude modulated signal by mixing it with a carrier signal of appropriate frequency. This AM signal comprising the video signal is fed to the antenna through a combinational circuit where it gets mixed with the audio signal.

(v) **Microphone** It converts sound into electric signal which is generally called AF (Audio Frequency) signal.

(vi) **AF amplifier** It amplifies the AF signal received from the microphone.

(vii) **FM audio transmitter** Here, the amplified audio signal is converted into frequency modulated signal by mixing it with a RF carrier wave. The FM modulated signal is fed to the antenna through combinational network.

(viii) **Antenna** It transmits (radiates) both AM signal (video signal) and FM signal (audio signal) into the space or to a satellite in the form of Electromagnetic Waves.

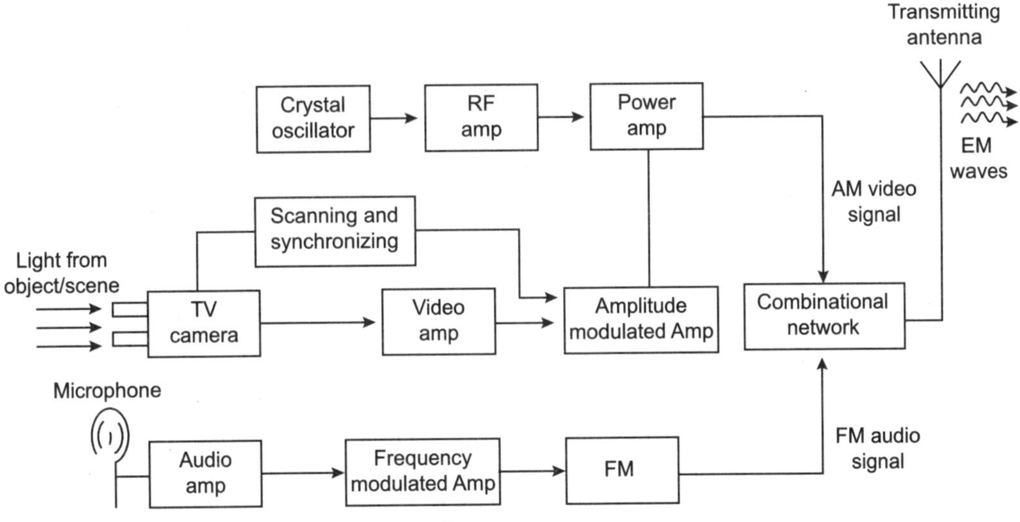

Figure 16.1 Block diagram of black and white TV transmitter

16.2.2 TV Transmission Frequency Range

We know that radio waves used in radio transmission have a frequency range distributed in three bands:

(i) Medium band (500 KHz to 1600 KHz)

(ii) Short band I (2.5 MHz to 7.5 MHz)

(iii) Short band II (7.5 MHz to 25 MHz)

For TV transmission, the bandwidth of signal must be 6 to 7 MHz (See Figure 16.8). Thus, none of the bands used for radio wave transmission are suitable here. In the short band range, we can have at the most three transmitting stations or three channels since the entire band has a bandwidth of 22.5 MHz. The limit of three stations is too small for worldwide transmission. Hence for effective working of TV transmitting stations, VHF (very high frequency) and UHF (ultra high frequency) bands (Table 16.1) are used. These bands have been standardized by CSIR.

Table 16.1

Frequency band	Channel no.	Channel (frequency) range
(a) VHF (30 to 300 MHz)	1	40 to 47 MHz
Here we have two ranges (i) 40 to 82 MHz	2	47 to 54 MHz
(ii) 174 to 216 MHz	3	54 to 61 MHz
	4 DD I	61 to 68 MHz
	5	68 to 75 MHz
	6	75 to 82 MHz
	7 DD II	174 to 1814 MHz
	8	181 to 188 MHz
	9	188 to 195 MHz
	10	195 to 202 MHz
	11	202 to 209 MHz
	12	209 to 216 MHz
(b) UHF (470 to 890 MHz)	Here we can have 60 channels	

Looking at the table for frequency range of TV transmission, the range is quite higher than that for radio transmission. In radio transmission, signals from the transmitting station can reach the destination via (i) ground wave or (ii) by reflection from the ionosphere. However, both these modes are not suitable for TV transmission since high frequency signals do not get reflected by the ionosphere; rather they are refracted through it and lost into space as shown in Figure 16.2. Thus, TV transmission is possible only through direct wave propagation. The transmission range for direct wave from the transmitting station having antenna of normal height is limited to a distance *TR* of 80 km from it. This limitation is on account of curvature of earth's surface (Figure 16.2).

It is clear from Figure 16.2 that places R_1 and R_2, etc. on earth, which are at a distance more than 80 km from the transmitting station, do not receive any direct wave from the transmitter. As such for places far away from the transmitting station, we can employ the following methods so that they receive signal from the transmitting station:

(i) **High TV antenna** Here we employ High TV Antenna. But the physical height of an antenna cannot be increased beyond a certain limit because of practical engineering limitations. However, its effective height can be increased by installing it on a nearby hill. So, the direct wave transmitted from such an antenna can reach a station that is located at a distance much more than 80 km. But again, there is limitation and also we cannot increase the effective height of antenna due to the limited height of hill and its distance from the transmitting station. We cannot go far away from the station also.

(ii) **Use of communication satellites** With successful installation of geostationary satellites, can utilize its place (≈3500 km above) in space for transmitting TV signals to all places situated in an area covered by it. In this respect, three satellites installed at equidistant all round the earth shall cover the entire surface of earth.

For a country whose boundary is well defined, we can have its own geostationary satellite. So signals from its TV transmitting station are first received by the satellite and thereafter they are relayed back to reach every place within the country. INSAT I-B is such a satellite for India.

(iii) **Microwave link** In this method, aerials are installed at every 25–30 km from the transmitting station, which relay signal to the desired place.

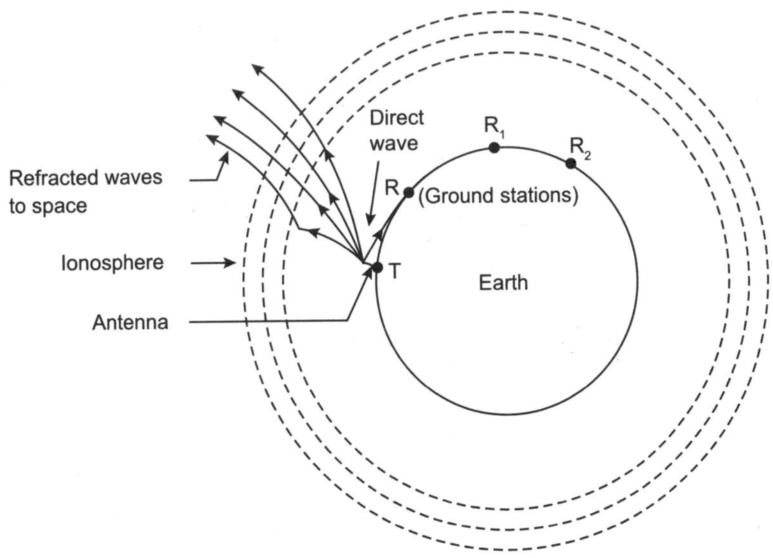

Figure 16.2 TV transmission range

16.3 TV Camera

The TV camera (Figure 16.1) is the first subsystem to collect light from the object or scene and convert it into video signal. We shall now discuss the construction and working of TV camera used in black and white TV (better known as monochrome TV).

Principle When we observe an object or a scene through a magnifying glass, we find that it is made up of infinite number of minute point centers. The intensity of light scattered from each centre is varying. These centers having different light intensity are called picture elements. The device which converts picture (light) scene of an object into an electrical signal as per the intensity of various picture center is called TV camera. It is the heart of TV transmission. The TV camera is of three types:

(i) Iconoscope

(ii) Image orthicon tube

(iii) Vidicon tube

(i) Iconoscope

Construction It is one of the earliest type of camera tubes seldom in use at present but it is very useful in understanding the working of other types of camera tubes. Its construction is shown in Figure 16.3.

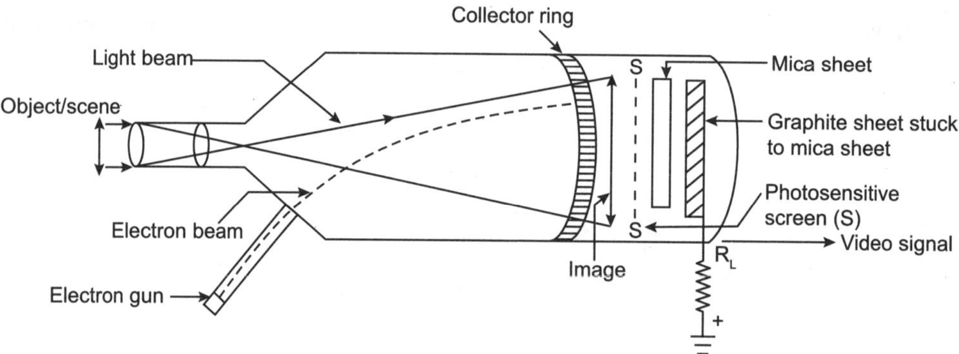

Figure 16.3 Iconoscope

It has three main parts:

(i) **Optical system** Here, two converging lenses are fitted into a tube. The focal length of the combination of lenses is adjustable as per the distance of object/scene; using a rack and pinion arrangement to change the separation between two lenses, thereby the image of the object/scene is formed on the photosensitive screen (S).

(ii) **Photosensitive screen** It works on the principle of photosensitivity. This part is also called mosaic. It is made up of fine particles of silver-cesium oxide stuck in an array of nearly 625 lines on a mica sheet, each line comprising thousands of particles of silver-cesium oxide. A graphite sheet is fixed behind the mica sheet, but in the diagram it is shown separately for clarity.

(iii) **Electron gun** It is a device called cathode ray tube that produces highly accelerated beam of electrons. It comprises a cathode indirectly heated by a filament, which emits electrons. These electrons are accelerated through a series of anodes maintained at very high positive potential. In addition to acceleratory anodes, there is a focusing anode in between them. Each anode has a central hole through it so that the electron beam can pass through it. All these are coaxially aligned so that the beam of electron is well focussed. This beam scans the screen both in vertical and horizontal directions. For scanning there are a set of deflection coils suitably placed around the screen. The whole arrangement is enclosed in a highly evacuated glass tube.

Working Light (photons) beam coming from the object or scene gets focussed at the photosensitive screen and it results into an emission of electrons as per the intensity of light beam. When electrons are emitted from the silver-cesium oxide surface of screen, the surface is positively ionized. This positive surface and graphite sheet behind it form a charged mini capacitor. These charged mini capacitors represent the electrical image of the object/scene. Charged mini capacitors get discharged when electron beam from the electron gun strikes them. The discharging current flows through resistor R_L connected to the screen but kept outside the tube. Voltage developed across it depends upon the discharging current which flows as mini capacitors get discharged. Thus, in output we get a series of varying voltage pulses across the load. The voltage pulse is called video signal. These video signals are suitably amplified by the video amplifier as shown in Figure 16.1 and then properly modulated so that we have an amplitude modulated signal. This modulated signal is fed to the transmitting antenna. The graphite sheet and collector ring, act as anode and thereby the circuit is complete.

(ii) Image orthicon tube This tube works on the principle of photosensitivity but it is more sensitive than the iconoscope since charging in iconoscope depends upon the intensity of incident light from the object to be transmitted. If the intensity of incident light from the object is weak/poor, then ionization of silver-cesium oxide layer will be negligible and thereby leakage current will also be very small. In such cases, a more sensitive arrangement is required, which is available in orthicon tube where we have an electron multiplier tube to increase current.

Construction

(i) **Optical system** Light from the object is focussed at the semitransparent photo-sensitive cathode plate on which a layer of cesium-potassium oxide or sodium oxide (Figure 16.4) is pasted. The paste makes it photosensitive. When light strike the plate, electrons are emitted from it. These electrons are collected by the target made up of glass which further emits secondary electrons and the target acquires positive polarity. The acquisition of positive polarity and its pattern represents the intensity of light received by the cathode from the object; the secondary electrons from the target are attracted by the screen mesh. Ultimately, the number of secondary electrons is proportional to the intensity of incident light.

(ii) **Scanning stage** The scanning of electronic image formed by the camera is done by electrons ejected from the electron gun placed on the other side of the target, thereby the target which was positively charged gets neutralized. The electron beam after neutralizing the target, returns to dianode of the electron multiplier tube and gets intensified by the emission of secondary electrons. This intensified beam of electrons possesses the complete information of the object. In this respect, the orthicon tube is more sensitive than the iconoscope.

(iii) **Multiplier stage** Here, the electron beam returning from the target gets amplified. It is on account of the amplification of electron beam that the tube is more efficient and sensitive. These electrons flow through the load which is earthed, there by a voltage pulse proportional to intensity of light, is produced across the load. These voltage pulses are further amplified by the video amplifier and properly modulated so that we have an amplitude modulated wave, which is fed to the antenna as shown (Figure 16.1).

In addition to the above stage, the device consists of (i) a focussing coil, (ii) an alignment coil and (iii) a horizontal angle vertical deflection coil, which are suitably placed as shown in Figure 16.4.

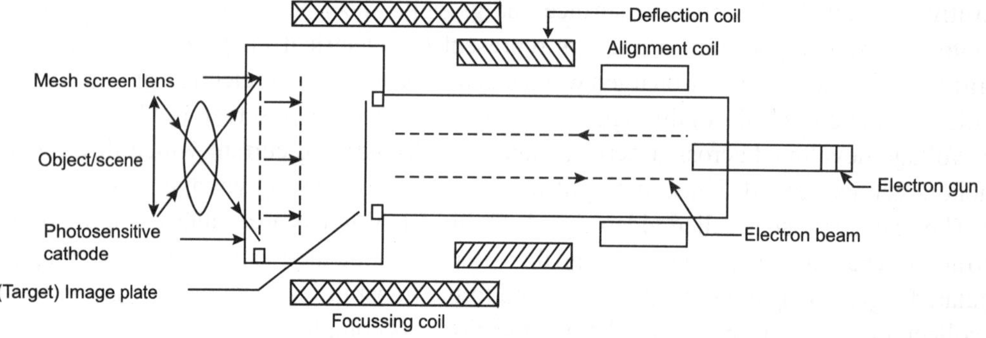

Figure 16.4 Image orthicon tube

(iii) Vidicon tube Its construction is very similar to that of an orthicon tube, but it works on the principle of photoconductivity instead of photosensitivity. Here the target has two surfaces. One surface has a transparent conducting film and the other has a photoconducting film of selenium. The thickness of photo-conducting film is very small. The rest of the arrangement similar to that of the orthicon tube i.e. we have focussing, deflection coil, alignment coil and electron gun.

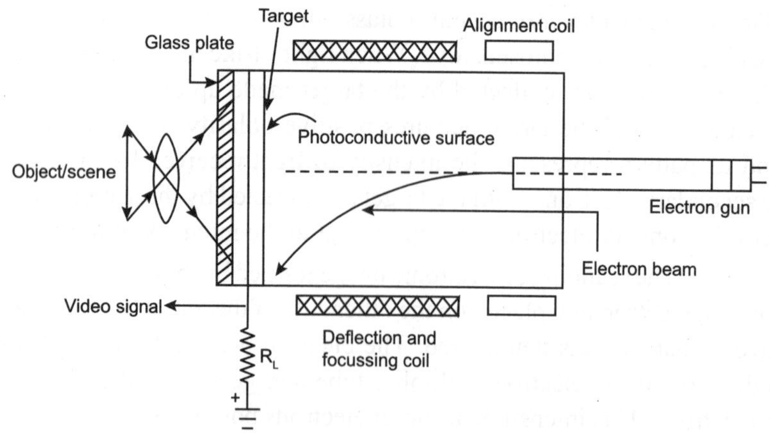

Figure 16.5 Vidicon tube

Working When an image of the scene is focussed at the target, an electronic image is formed on the photoconducting surface. It is scanned on the other side by electron beam coming from the electron gun. Here, the speed of electron beam is kept quite low. For developing the video signal, load R_L is connected to the transparent film of the target. The conductivity of selenium surface, related to the black part (less bright) of the object, is minimum and maximum for bright part of the object. In this way, the corresponding current flowing through, load is different for different brightness of the object. Current flowing through the load develops the video signal of varying voltage, which is fed to the video amplifier (Figure 16.1). The output of video amplifier is amplitude modulated and fed to the antenna after routing it through combinational network. The size of the tube is relatively small and as such suitable for movie camera.

16.4 Scanning

When we read a book, the pupil of our eye moves both horizontally and vertically and we get the complete impression of matter printed on the page. Similarly, in a TV camera tube, electron beam emitted from the electron gun reads the picture elements both horizontally and vertically till the entire picture is covered and the corresponding video signal is generated. The process of reading a picture/frame by electron beam is called scanning. Two methods are employed for scanning: (i) progressive scanning and (ii) interlaced scanning.

Progressive scanning Each picture is divided into 625 equally spaced horizontal lines. Each line consists of thousands of picture elements. For scanning, the electron beam moves from left to right across each line and then to scan the next line it moves downwards. In this way, an entire picture is scanned by the horizontal and vertical motion of electron beam falling over it (Figure 16.6).

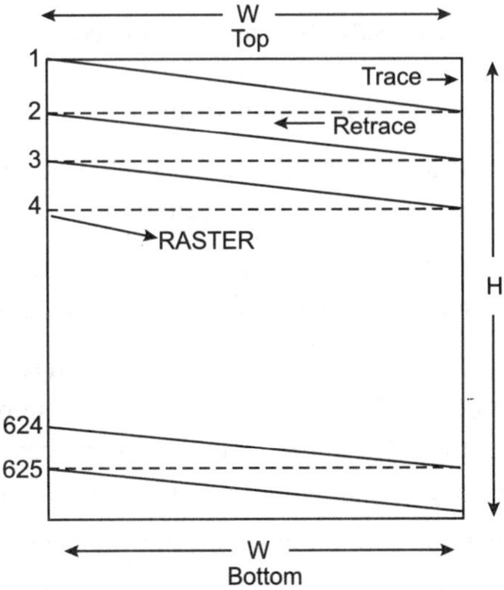

Figure 16.6 (a) Progressive scanning (lines shown not to scale)

Sixteen pictures per sec must pass through the human eye for it to be seen as a moving object with constant velocity. However, for good scanning, we keep 25 photographs per second.

Number of horizontal lines = 625

Number of frames scanned per second = 25

Hence, the frequency of horizontal scanning = 625 × 25

or $f = 15625$ per second

Time period for scanning, $T = \frac{1}{f} = \frac{1}{15625} = 65\ \mu s.$

During this time, the electron beam moves left to right and right to left; then it moves along next line. Both these paths although equal are covered in different time. The total time (t) is divided into tracing time (t_1) and retracting time (t_2) as shown in [Figure 16.6(b)].

The scanning time is divided in two timings as

(i) Tracing time (t_1) is kept as 54 μs

(ii) Retracing time (t_2) is kept as 11 μs

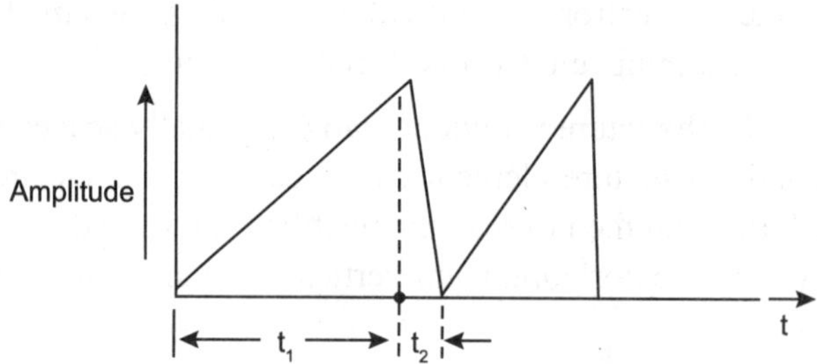

Figure 16.6 (b) Trace and retrace time

The time period for vertical scanning is kept 19 μs.

For horizontal sweeping, the sawtooth wave is generated by a sweep oscillator and its time period is fixed as 65 μs which is equal to scaning time period.

Similarly, for vertical sweeping, the sawtooth wave is generated by a vertical sweep oscillator having its time period as 20 msec. Both these signals are fed to deflection coil of the camera.

Interlaced scanning In this method, every picture is scanned twice since the horizontal lines are divided in two groups (i) odd and (ii) even; so in the first step, a odd group (1, 3, 5, etc.) consisting of 312.5 lines is scanned and in the second step an even group (2, 4, 6, etc.) of 312.5 lines is scanned; every set of lines is scanned alternately (Figure 16.7).

To keep scanning frequency at 15625 per second in this stage, field frequency changes to 50 per second instead of 25 per second as in the case of progressive scanning since in one step only 312.5 lines are scanned.

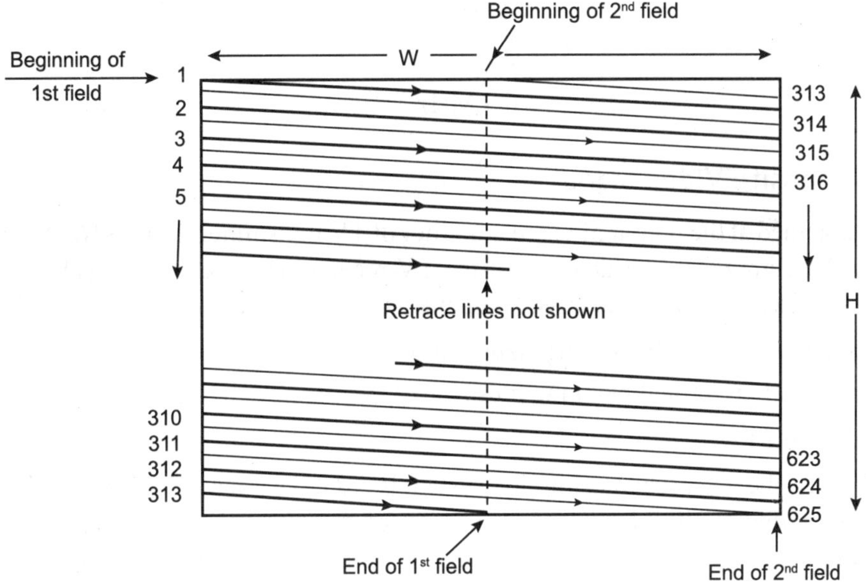

Figure 16.7

So ultimately, we have $312.5 \times 50 = 15625$ per second as in the case of progressive scanning. Here, vertical tracing and retracing is twice that achieved in progressive scanning.

Vertical trace time = 19.75 μs

Vertical retrace time = 0.25 μs

Total time period for sawtooth wave = 20 μs

The main advantage is that flicker effect is eliminated without increasing the scanning rate so there is no need of increasing the bandwidth of channel, i.e. it is still 7 MHz as in case of B/W television.

16.5 Video Signal

The signal we get from the camera after scanning is called video signal. Its frequency is kept 20 MHz to 65 MHz.

$$\text{Frequency of video signal, } f_v = \left\{\frac{4}{3}\left[\frac{(\text{No. of lines})^2 \times (\text{No. of frames per second})}{2}\right]\right\}$$

Here the ratio $\frac{4}{3}$ is called aspect ratio, i.e. width to height ratio of TV screen.

In a frame, there are 625 lines.

Number of frame per second = 25

So $$f_v = \frac{4}{3}\left\{\frac{(625)^2 \times 25}{2}\right\}$$

$$= 6.5 \text{ MHz}$$

16.5.1 Composite Video Signal

The signal obtained at the output of video amplifier of TV transmitter (Figure 16.1) or the signal obtained at the output stage of detector of the TV receiver (Figure 16.9) is called composite signal. It comprises the following three components:

(i) Horizontal or vertical synchronizing pulse

(ii) Horizontal or vertical blanking pulse

(iii) Equalizing pulse

Note: Generally, the negative transmission system is employed in a TV. In this system, voltage pulse corresponding to the black or dark portion of the picture is more than voltage pulse corresponding to the bright portion of the picture. This system has an advantage that external noise signal of a running car or motor or starting ignition signal having frequency in the RF range and where amplitude is constantly varying randomly will not produce any lines on the screen of TV.

16.5.2 Standard Channel Bandwidth

The sound carrier is always positioned at the extremity of fully radiated side band and hence 5.5 MHz away (Figure 16.8) from the picture carrier. This is its logical place since it makes for minimum interference between the two (audio and video) signals. The FM sound signal occupies a spectrum of about ±75 KHz around the sound carrier. However, a guard band of 0.25 MHz (Figure 16.8) is allowed on the sound carrier side of television channel to allow for adequate inter channel separation. Thus, the total channel bandwidth is 7 MHz as shown in Figure (16.8).

16.6 TV Signal Transmission

For TV signal transmission, single side band (SSB) is employed because of the fact that here a bandwidth of 6 MHz is to be kept had DSB been employed, then we would have required a band gap of 12 MHz. As a result, frequency difference between the two neighbouring telecasting stations would be nearly 14 MHz and this would lead to lesser number of telecasting stations on earth. So SSB is preferred in VHF and UHF range. The bandwidth of 7 MHz is kept because audio signal is also to be transmitted along with video signal which itself requires a bandwidth of 6 MHz. While deciding the bandwidth, it has to be kept in mind that attenuation of a signal is not sharp at a particular frequency rather it is gradual (decaying); so there is a slope of 0.5 MHz at both ends of the frequency response curve (Figure 16.8). It is also important to be kept in mind that the eliminated band has the lowest frequency range, but it too cannot be eliminated completely, because in the lowest frequency range there is important information of video signal (visible scene) which would have been lost resulting in picture distortion.

On account of the above reason, only a part of lower frequency band is eliminated. In this way, the transmitted bandwidth comprises carrier frequency, upper side band and the remaining part of lower side band (Figure 16.8).

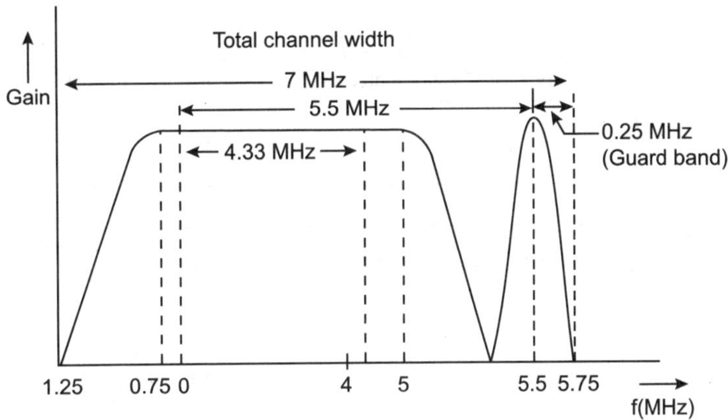

Figure 16.8 Indian TV channel side band

If there is no slope in the frequency response curve in the upper frequency range, then horizontal details are also lost because the modulated video signal consists of horizontal details.

The system in which both video and audio signals are transmitted is called intercarrier system. In this system, the frequency of video carrier wave is 1.25 MHz more than the lower most frequency of the channel concerned, and similarly the frequency of audio carrier wave is 0.25 MHz lesser than the uppermost frequency of the channel concerned.

Accordingly the frequency of audio carrier wave is

$$[7 - (1.25 + 0.25)] = 5.5 \text{ MHz}$$

The audio signal is frequency modulated before transmission; the maximum width of side band is ± 100 KHz since we employ SSB system, so the width of upper side band is 5.25 MHz and that of lower side band is 1.25 MHz.

16.7 Black and White TV Receiver

Its simple block diagram is shown in Figure (16.9).

It consists of the following main stages:

(i) Common video and audio stage

(ii) Audio stage

(iii) Video stage
(iv) Synchronizing and extra HT stage
(v) Rectifier stage

A detailed study of the above stages is beyond the scope of the book. However, we shall consider each stage in brief.

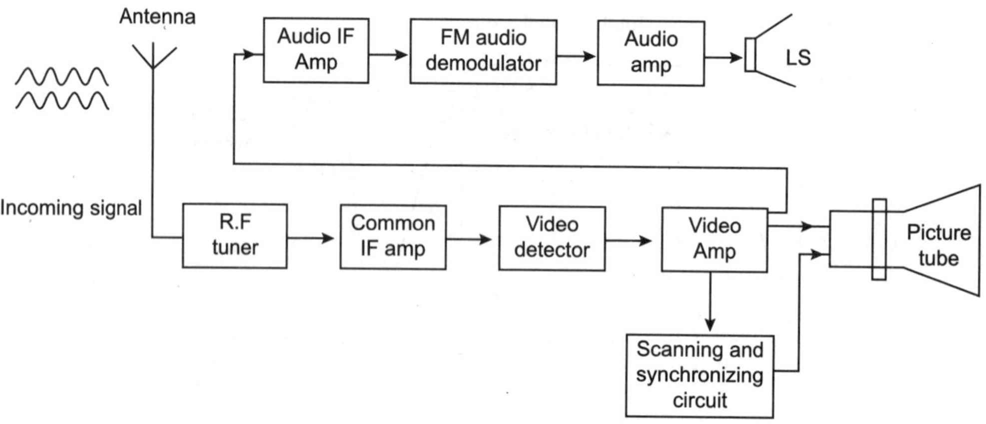

Figure 16.9 Simplified block diagram of B/W TV receiver

16.7.1 Common Video and Audio Stage

This stage consists of dipole antenna, RF amplifier, mixer and local oscillator; combined together it is known as RF tuner stage. A brief discussion is as under.

(a) Dipole antenna It has four main parts as shown in Figure 16.10: (i) dipole, (ii) directors, (iii) reflector and (iv) booster along with feeder line (cable).

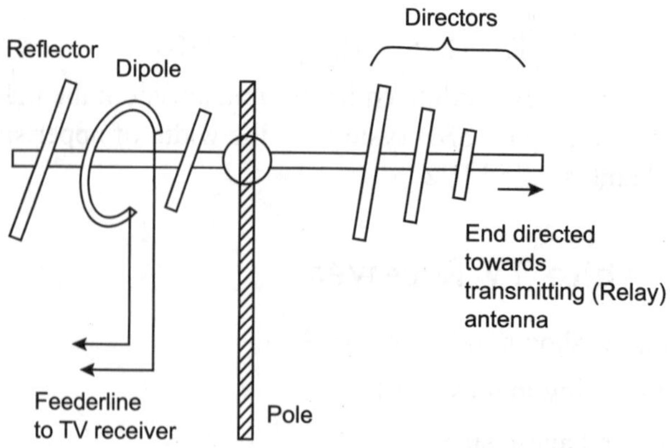

Figure 16.10 Dipole antenna

The feeder line connects the dipole to the receiver. Power loss through the feeder line must be minimum; otherwise the received signal will get further weakened.

(b) RF amplifier It is a radio frequency amplifier having a bandwidth of 9 MHz. It isolates the antenna from mixer stage and amplifies the incoming signal from antenna.

(c) Local oscillator Its function is to heterodyne the incoming RF signal and produce IF signal. The frequency of the oscillator changes according to the incoming signal frequency so that the resulting IF signal frequency remains constant.

(d) Mixer Here the IF signal is produced by mixing RF signal with the signal produced by the local oscillator. Here we have two resultant signals, video and audio, having a frequency difference of 5.5 MHz as explained below.

Let the frequency of incoming signal be within the range of 61 to 65 MHz; so from Figure (16.8),

Video carrier = 61 + 1.25 = 62.25 MHz

Audio signal = 68 – 0.25 = 67.75 MHz

Let the frequency of local oscillator be 101.15 MHz

So,

(i) Resultant video signal frequency = 101.15 – 62.25 = 38.9 MHz

(ii) Resultant audio signal frequency = 101.15 – 67.75 = 33.4 MHz

Thus, the two signals differ by frequency (38.9 – 33.4) = 5.5 MHz.

Note: The frequency of local oscillator is changed as per the incoming signal so that the resultant (beat) frequency of two signals is always 38.9 and 33.4 MHz, respectively.

(e) IF amplifier It has three stages and amplifies the incoming IF signal (video as well as audio) from the mixer stage.

(f) Video detector Here the output of IF amplifier is the input. This stage eliminates the one side band of composite video signal, because it is essential for amplitude modulated video signal. The RF carrier wave is filtered, and by heterodyning a signal frequency of 5.5 MHz is produced. The amplitude of this signal is nearly 3 V.

16.7.2 Audio Stage

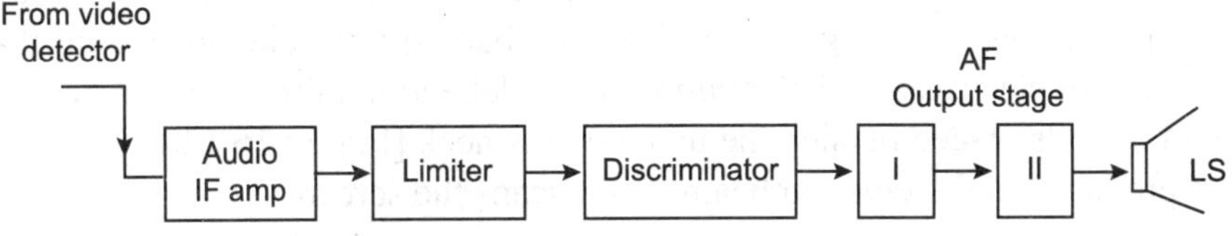

Figure 16.11 Simplified audio stage block diagram

It consists of the following subsystems:

(i) IF amplifier
(ii) Limiter
(iii) Discriminator
(iv) AF driver amplifier (two stage)
(v) Loud speaker

(i) IF amplifier It amplifies the IF signal received from the video detector. Its frequency is equal to the difference of video signal (38.9 MHz) and audio signal (33.4 MHz), i.e. 5.5 MHz.

(ii) Limiter It further amplifies the IF signal but at the same time limits variations in the amplitude of signal so that it is suitable for amplitude detection.

(iii) Discriminator It is also known as FM detector, better known as ratio detector. Here the RF content of modulated wave is separated for FM detector.

(iv) AF output amplifier It is a power amplifying stage consisting of stages for the AF signal.

16.7.3 Video Stage

It consists of the video amplifier and the picture tube. The video amplifier is a wide band amplifier ranging from 20 Hz to 6.5 MHz. Another stage known as synchronizing separator is associated with it, supplying the necessary signal to AGC. Output signal from the video amplifier is nearly (–75 V) and fed to the picture tube.

Picture tube It is similar to a cathode ray tube where the video signal is converted into visible (optical) scene/picture. Here the tube is not as long as in a CRO, also the screen is quite large [Figure 16.12(a)].

The screen of the tube is nearly rectangular whose width to height ratio is kept 4 : 3, i.e. if one side (width) is 40 cm long then the other side (height) is 30 cm. The length of the diagonal of the screen describes the size of the TV receiver. The ratio of width to height is also known as aspect ratio.

The picture tube is a special form of cathode ray tube in which the screen forms the front or face plate [Figure 16.12(a)] of TV. A black and white TV or monochrome picture tube has a single electron gun [Figure 16.12(b)]. The inside of the screen is coated with luminescent material which produces white light when it is bombarded with electrons from the electron gun. Here, instead of electrostatic deflection we have deflection of the electron beam due to the magnetic field of coils placed outside the tube near its neck [Figure 16.12(b)]. It is on account of deflection produced in the electron beam that it scans the screen.

The received video signal from video amplifier (Figure 16.1) drives the electron gun and thus controls the intensity of electron beam at any instant, resulting in variation in brightness of the raster. When the electron beam scans the screen, it results into the black and white picture of the object/scene being televised.

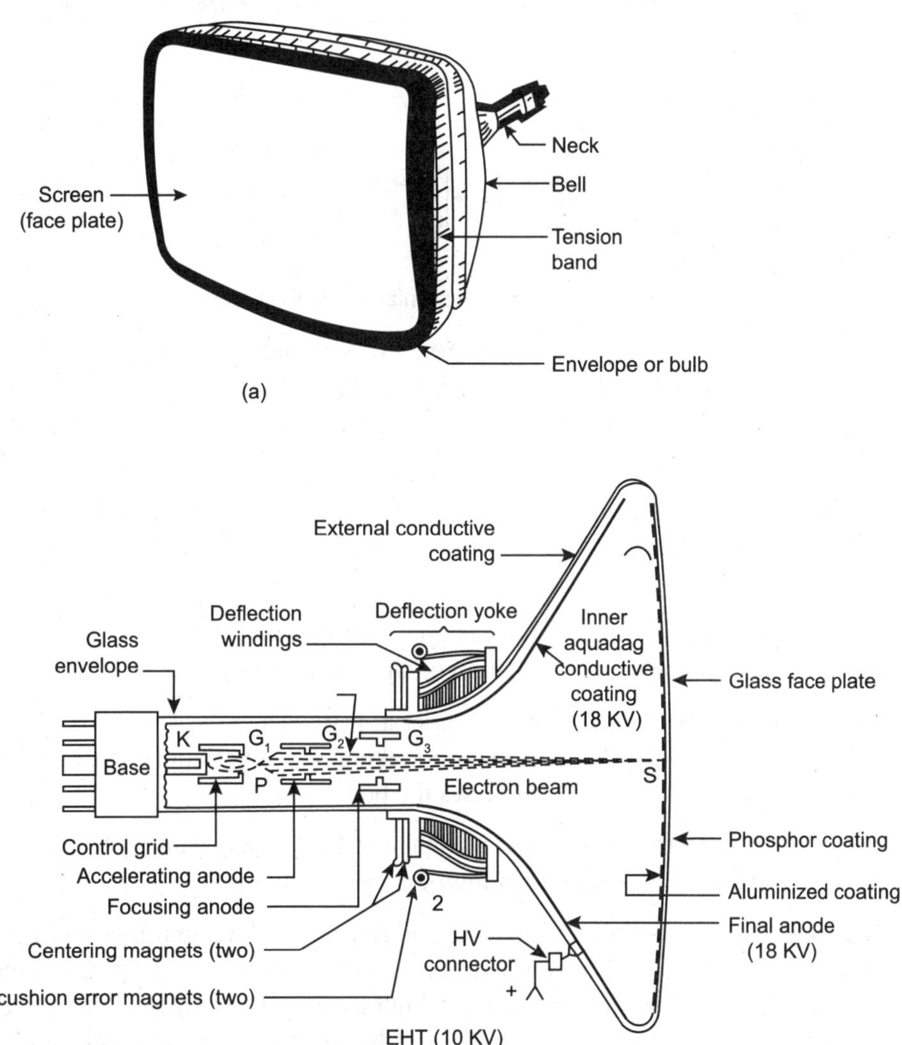

Figure 16.12 Black and white TV picture tube

16.7.4 Synchronizing and EHT Stage

Its simplified block diagram is shown in Figure 16.13. It consists of a synchronizing separator, horizontal oscillator, horizontal or line amplifier, booster, extra high tension (voltage) stage (EHT stage), etc.

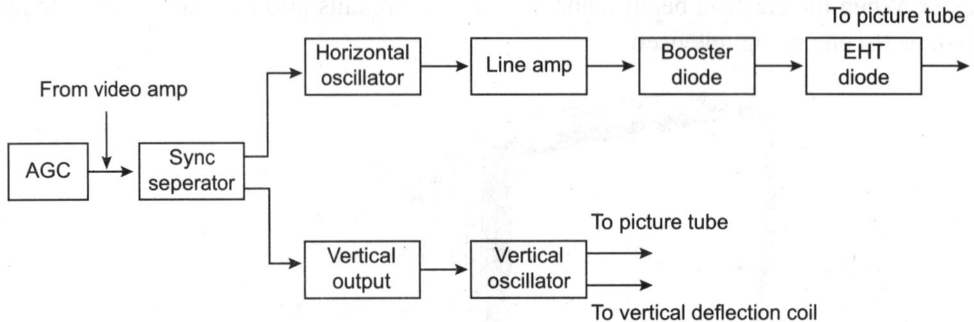

Figure 16.13 Synchronizing and EHT stage

(a) Synchronizing separator or clipper It separates the synchronized signal from composite video signal and further separates horizontal and vertical signal from synchronized stage by making use of differentiating and integrating circuits. These signals are separately fed to horizontal and vertical oscillators.

(b) AGC (Automatic gain control) Signal from the video amplifier is fed to this stage and accordingly bias is developed for RF and IF stage of TV receiver so that the incoming signal is audible.

(c) Horizontal oscillator It generators a signal of frequency 15625 Hz. To keep the frequency constant, it is fed a synchronizing pulse from composite video signal.

(d) Horizontal or line amplifier It works at a frequency of 15625 Hz. The amplified horizontal signal imparts speed to the electron making use of horizontal deflection coil of the picture tube. Simultaneously, it also processes the booster stage.

(e) Booster diode stage It develops a voltage of 1 KV (DC). This high voltage is applied to the anode of picture tube and anode of high voltage amplifier.

(f) Extra high tension diode This stage is run by the line amplifier and it produces a high voltage of the order of 12 to 24 KV depending upon the size of the picture tube. Since it produces a very high voltage, it is called extra high tension. The output of line stage amplifier is a saw tooth wave having frequency 15625 Hz. When the output in each cycle reduces to zero, then this action is known as fly back and it produces a high voltage across the winding of the transformer. The alternating voltage is rectified by a high tension diode, thereby we get a high DC voltage.

(g) Vertical oscillator It is a multivibrator having a frequency of 50 Hz. To keep frequency constant, vertical pulse separated from composite video signal by the synchronizing separator is given to it.

(h) Vertical output amplifier It works on 50 Hz. Its output is fed to the defection coil in picture tube so that the electron beam gets deflected in the vertical direction.

16.7.5 Rectifier Stage

Here, we have a rectifying unit and an assembly of filters so that we get DC output. It provides 230 V DC to the remaining parts of picture tube and other stages of the receiver where LT (low tension) is required.

16.8 Colour Television

The working of colour television is quite similar to that of black and white TV. The only difference is that here we also get information about the colour of the object. Before we go to study the working of a colour we shall study the basics of different colours.

16.8.1 Colour Basics

All colours are constituted of three primary colours: red, green and blue (Figure 16.14). In a way a specific colour of the body as perceived by the eye is a mixture of three basic colours in proper proportion as shown.

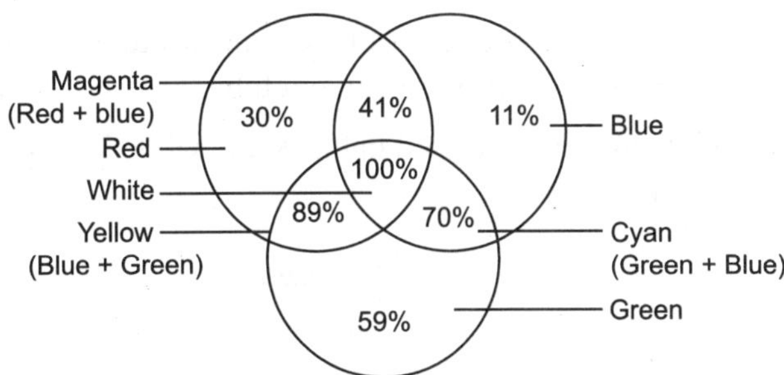

Figure 16.14

Whenever an object is scanned by a colour camera, three video signals corresponding to three primary colours are mixed and the resulting colour is known as complimentary colour:

Primary colours		**Complimentary colour**
(i) Red + Green	⇒	Yellow
(ii) Red + Blue	⇒	Magenta or purplish shade
(iii) Green + Blue	⇒	Cyan or greenish blue shade

When a complimentary colour is mixed in proper proportion with primary colour which is not present in it, then the resulting colour is white (Figure 16.14).

Whenever we view an object, our eyes cannot detect the primary colours with which its colour is made of. However, for a TV system, it is necessary to know the quantum of each primary colour with which it is composed of. For this, we employ CHROMA section in the TV receiver.

To get different colours by mixing the primary colours, the Grassman's Law is useful, which gives the proportion in which different primary colours are mixed to get a specific colour shade.

Accordingly, we have

30% Red + 59% Green	⇒	100% Yellow
30% Red + 11% Blue	⇒	100% Magenta
59% Green + 11% Blue	⇒	100% Cyan
11% Blue + 30% Red + 59% Green	⇒	100% White
70% Cyan + 30% Red	⇒	100% White
89% Yellow + 11% Blue	⇒	100% White
41% Magenta + 59% Green	⇒	100% White

16.8.2 Colour Transmission

Here we employ the same stages as in the case of black and white TV (Figure 16.1). The only difference is in the working and construction of camera tube.

Camera tube It is the vidicon type tube used in black and white TV transmission. Its simple sketch is shown in Figure 16.15.

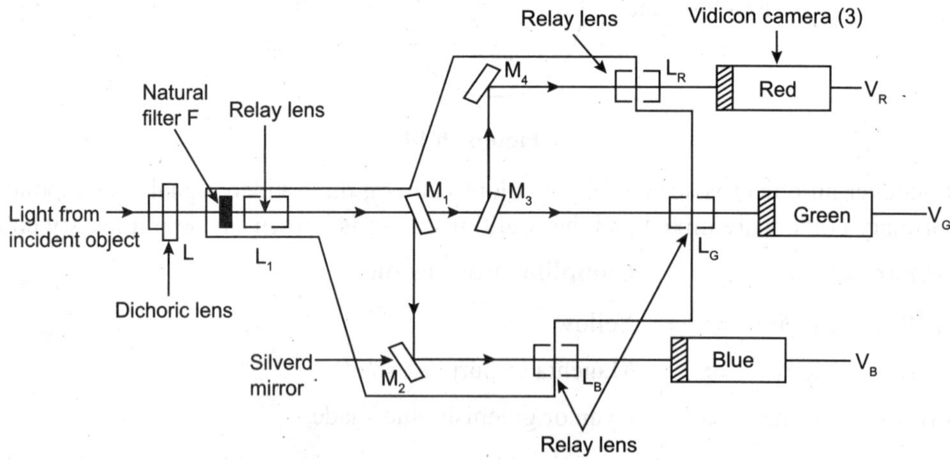

Figure 16.15 Colour camera tube

Here, we use a dichoric lens L to analyze light received from the object into its primary colours. A dichoric lens is used to minimize loss of colours. The incident light falls on a natural filter F and then on dichoric mirror M_1 which reflects blue light and transmits light of rest of the colours. Blue light falls on silvered mirror M_2 which completely reflects it to pass through relay lens L_B, and picked up by vidicon blue camera V_B.

The transmitted beam from mirror M_1 falls on another dichoric mirror M_3 placed in front of it. Mirror M_3 reflects red colour and transmits rest of the colour (green) to pass through relay lens L_G and picked up by vidicon camera V_G. The red colour after being reflected from M_3 falls on mirror M_4 which reflects it to pass through relay lens L_R and picked up by vidicon tube V_R.

In this way, incident light from the object is divided into three primary colours. The output voltage signal of three tubes is proportional to the brightness of three colours received by the respective tubes. These video signals from pickup tubes are fed to such a circuit which mixes them in different proportions and produces Y, Q and I signals respectively. The Y signal represents variation in brightness of the object/scene whereas Q and I signals, represents another class of colour called chromaticity (C). After this, Y and C signals are fed to the adder circuit where they are mixed and one more signal is added to it. The output of adder circuit represents the complete video signal.

The video signal is fed to the video amplifier and subsequent stages (Figure 16.1) of the transmitter of B/W television. In Figure 16.15, we have shown three vidicon tubes for the three primary colours. Presently, we can have one tube to perform the same function. Here, we have stripped the target and the output of such camera tube produces two types of signals: (i) luminance and (ii) chrominance.

16.8.3 Colour TV Receiver

Construction It comprises a number of subsystems as shown in Figure 16.16.

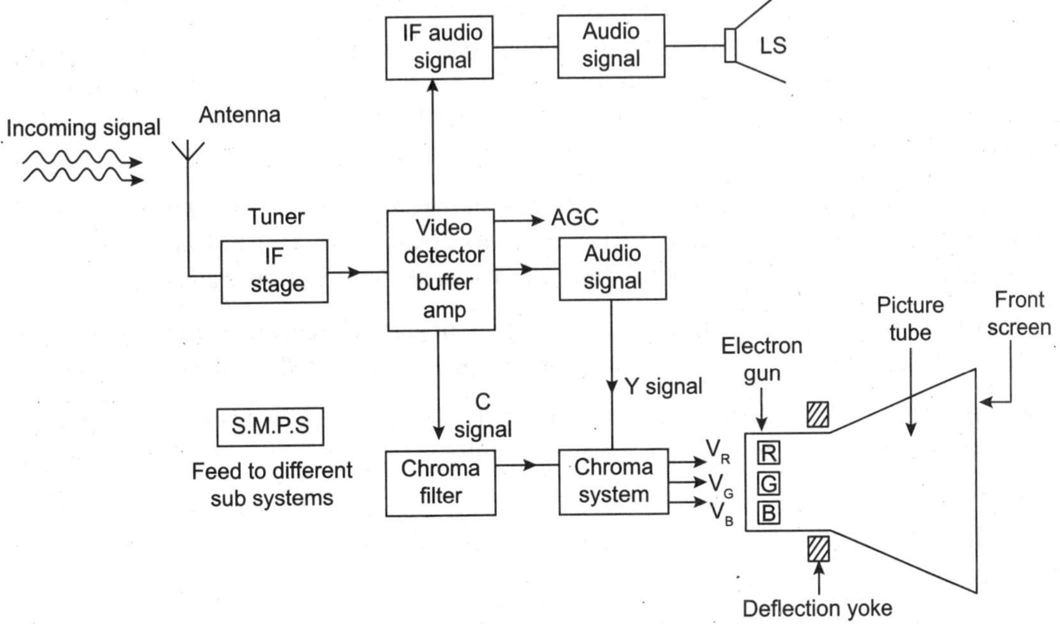

Figure 16.16 Simple sketch of colour TV receiver

It is to be noted that the construction of the picture tube here is quite different since it consists of three (R, G, B) electron guns and so its working is based on a different principle. We shall first discuss the working principle of the receiver.

Working principle Here we have a separate section called "Chroma section", which processes colour information contained in the video signal. The video detector supplies the composite colour signal to both *Y* channel and colour decoder through a buffer amplifier. The luminance signal at the output of *Y* channel and *U* and *V* signals at the output of decoder combine in the matrixing circuit to produce R, G and B colour device signals. The use of delay line in the *Y* channel enables exact time correspondence of colour and brightness information of the picture elements. The automatic frequency tuning circuit is to ensure exact IF output from the tuner. This is necessary to keep the encoded chroma signal, which is located at the upper end of channel within pass band of IF amplifier.

The desired signal from the antenna is fed into the superheterodyne receiver so that video and audio signals are separated. The frequency modulated signals are detected and converted into audio signal which is fed to the LS after amplification.

The video signal is fed to the kinescope, which is the cathode ray tube of the receiver. It is quite similar to a camera tube. The only difference here is that the electron beam is focussed at the fluorescent screen instead of mosaic as in vidicon tube (Figure 16.5). To reconstruct the image of the object/scene, the beam should scan the screen at the same rate as was done in the transmission camera.

This action is furnished by two pairs of coils, one of them scans at a very high frequency and the other scans the same at very low frequency. To produce the original picture, the scanning electron beam of the receiver is synchronized with the transmitter electron beam. The fluorescent effect produced on the screen is related to the video signal and in this way it is indirectly related to the object/scene.

The block diagram of colour TV receiver is shown in Figure 16.16, and comprises the following substages: (i) antenna, (ii) RF tuner, (iii) preamplifier, (iv) FM audio subsystem, (v) chroma subsystem, (vi) R, G, B amplifier, (vii) vertical drive amplifier and (viii) SMPS.

RF tuner It selects the desired signal (channel) out of signals received by the antenna. Here, the selected signal is also processed to develop IF signal which is fed to the preamplifier.

Pre amplifier In a way it is a power amplifier. The IF signal received is a very low power signal, so it is amplified to a proper level (strength). Its output stage is the input to video IF stage.

Video IF subsystem Here the combined video signal is separated into audio (sound) and chromasignal.

FM audio subsystem Here we get frequency modulated audio signal to be fed to the LS after detection.

Chroma subsystem It performs all those functions so as to get R, G and B signals. After receiving the video signal, it first recovers (B-Y) signal; and (R-Y) signal and then by mixing the video signal with Y signal received through Y channel, we get the original V_R, V_G and V_B signals which were prepared in vidicon tube at the time of transmission. The V_G signal is automatically suitably amplified and fed to the picture tube where we have three respective electron guns, R, G and B, in the same glass envelope of picture tube.

16.8.4 Principle of Function of Picture Tube

Here, three electron beams R, G and B generated by three electron guns (Figure 16.17) cross one another at a given distance on a metal sheet called shadow mask from the fluorescent screen and then traverse towards the screen, where a glass plate on which closely spaced groups of phosphor dot trios (R, G and B) are deposited is used as viewing screen. Each trio comprises three phosphor dots. When the respective electron beams strike these dots, then light beam of the colour is emitted. A particular trio in the screen corresponds to a particular hole in the shadow mask. It is to be noted that electron beams R, G and B can hit only the red emitting, green emitting and blue emitting dots, respectively. By separately controlling the number of electrons in each beam, the proportion of three colours can be controlled. Finally, we see the complete colour of the object when the screen is scanned by the electrons.

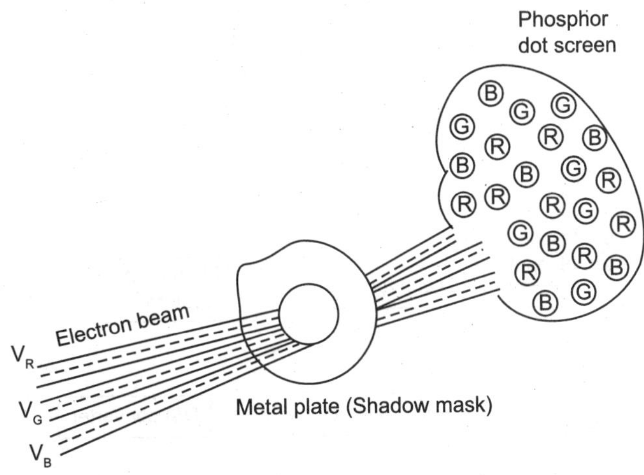

Figure 16.17

16.9 SMPS (Switch Mode Power Supply)

It is a special type of power supply. It gives fixed DC voltages like +5 V or (–)5 V, +12 V or 24 V, as output when input voltage varies from 90 V to 230 V AC. The different power supplies work at 20–30 KHz. Its efficiency is quite high of the order of 70%. (For details see Chapter 4)

16.9.1 Cable TV and Digital TV

(a) Cable TV Presently, the use of cable TV is widespread and increasing especially in urban areas. It has the following subgroups:

(i) Master antenna TV [MATV]

(ii) Closed circuit TV [CCTV]

(iii) Community cable TV [CATV]

The first group is very popular in India and employed in most houses. Here, a very powerful RF signal is fed to the antenna installed at the top of building and this signal is processed by the TV receiver in the home. Other groups are not common in India.

(b) Digital TV Digital Television (DTV) is the transmission of audio and video signals by digital signals incontrast to the analogue signal used by analogue TV. Here the signal is digitized both at transmitting and receiving stage. A simplified version is shown in Figure (16.18). Detailed discussion is beyond the scope at this level.

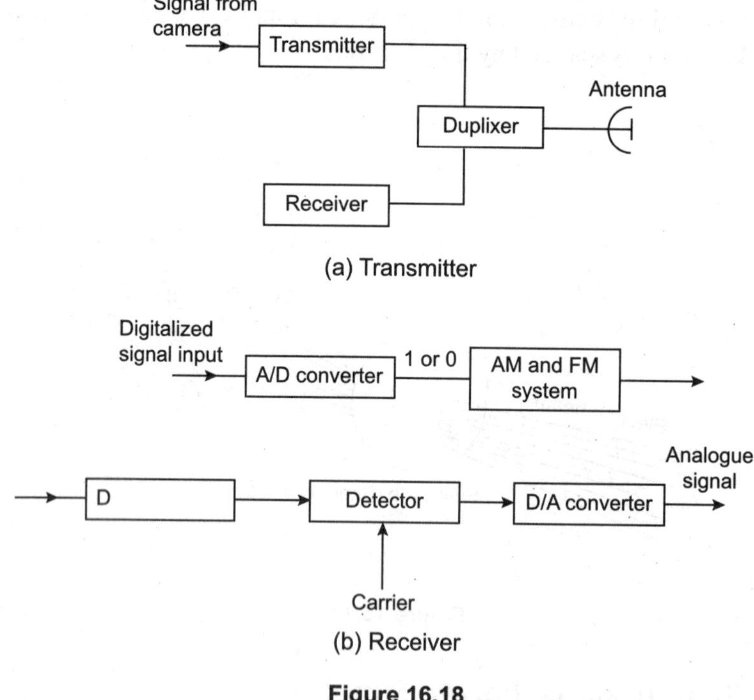

(a) Transmitter

(b) Receiver

Figure 16.18

Digital TV has many merits over analogue TV.

(i) Digital channels take up less band width and the bandwidth needs are continuously variable.

(ii) In a given space one can provide more digital channels.
(iii) The channels provide high-definition television service.
(iv) The channels can also provide non-television services like multimedia or interactivity.

16.10 Radar

It is an electronic device used for the detection and location of objects located far away from the point of observation, better known as ground station. It operates by transmitting a particular type of signal, e.g. a pulse-modulated sine wave and detects the nature of returning signal, better called echo signal.

The word radar is an acronym of "radio detection and ranging". It has been so designed to enable to see through those conditions like fog, snow, smoke, haze, rain, storm, etc. where normal human eye fails. In addition, it has the ability to measure the distance of object location from the point of observation. The only limitation is that it does not give any details of the object, like its colour, structure, size, etc, which is very crucial for an observer.

Basically, a radar system consists of a transmitter and a receiver, each connected to a directional antenna separately through a switching system, better known as duplexer (Figure 16.19).

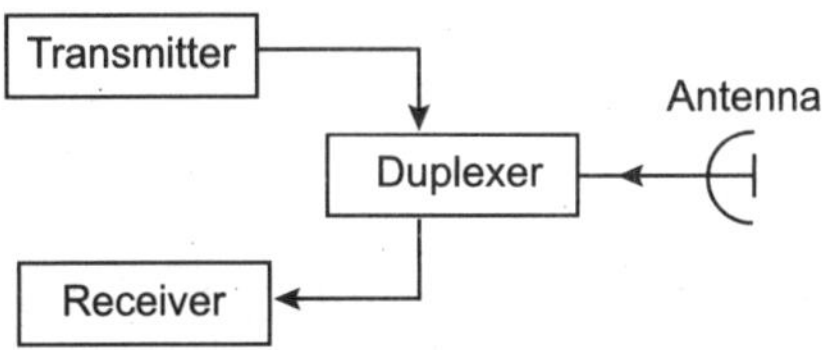

Figure 16.19 Elementary pulsed radar

A small portion of pulse power generated by the transmitter enters the duplexer. The duplexer disconnects the receiver from the antenna and connects the transmitter to the antenna. In most radar systems, the antenna rotates and scans the area to be observed. The scanning speed is quite small compared to the speed of light, so the antenna does not rotate significantly during the time interval taken by the radiated signal to return from normal targets. This is how echoes are received, and the antenna still points in the right direction to collect them.

As soon as the transmitted pulse terminates the duplexer disconnects the transmitter from the antenna and connects the antenna to the receiver. The reflected pulses or echos from the object (target) are received by the antenna and fed to the receiver where they are processed and correctly analysed. The demodulated pulses are then fed to a display device or an indicator for final analysis.

The distance of target is determined by measuring the time taken for the signal to travel to the target from the transmitter and the echo to return from the target to the receiver. The radar provides information on the azimuth (horizontal) and elevation (vertical) direction of the target.

If T (sec) is time taken by the pulse signal to travel to the target and back, then distance travelled by the pulse is CT (meter); here C is the speed of light in vacuum.

Hence the distance of target from the point of observation is

$$d = \frac{CT}{2} \text{ m}$$

Here, the factor 1/2 is multiplied since time taken by the pulse is double the actual time, e.g. if the pulse takes 1/μs to go and return, then the distance of target is

$$d = \frac{3\times10^8 \times1\times10^{-6}}{2} = 150 \text{ m}$$

In a radar system, distance is usually measured in "nautical miles" where 1 nautical mile is 1852 m. If the target is more along the line of sight, then the speed of target is determined by knowing the time interval between two consecutive echoes. In this process, we have limitation in the frequency of pulse generated by the transmitter as explained below.

Once the outgoing pulse is transmitted by the radar, sufficient time must elapse to allow the echo signal to reach and be detected by the receiver there, then only the next pulse may be transmitted, i.e. the width of pulses must be shorter than interpulse time interval. The maximum range for a radar is

$$d_{\max} = \frac{C}{2f_r}$$

where f_r is pulse repetition frequency.

Example If a radar is designed for a maximum range of 30 km, then time between two successive pulses must be

$$t = \frac{2d_{\max}}{C} = \frac{2\times30\times1000}{3\times10^8} = 200 \text{ μs}$$

We must note that pulse repetition frequency must be higher than scanning speed of antenna and, therefore, there is limit on the smallest number of pulses that can be transmitted per second. Usually a radar system transmits a few hundred pulses per second and pulse width is about 1 μs.

The radar antenna must have dimensions of a few wave lengths to produce sharp and narrow beams. It helps us to detect target location precisely. For this, short wave lengths are to be used and so pulse modulated UHF or microwave frequencies are employed in the radar system. The radar power is nearly a few kW.

The frequencies employed by the radar are in upper UHF and microwave ranges. As a result of war time security, many band names came up for various ranges, which are still being used. A few bands are mentioned in Table 16.2.

Table 16.2 Radar bands

Band name	Frequency range (GHz)	Maximum available peak power (mW)
UHF	0.3 – 1.0	5.0
L	1.0 – 1.5	30.0
S	1.5 – 3.9	25.0
C	3.9 – 8.0	15.0
X	8.0 – 12.5	10
Ku	12.5 – 18.0	2.0
K	18.0 – 26.5	0.6
Ka	26.5 – 40.0	0.25
V	40.0 – 80.0	0.12
N	80.0 – 170.0	0.01
A	Above 170	–

16.10.1 Free Space Radar Range

A radar system has a range upto which it can locate objects. We now derive a relation for the maximum range upto which a radar system can detect objects.

Let P_T be peak power value of a transmitter in watts. Assuming an isotropic antenna (which radiates uniformly in all directions), its power density at distance d is

$$\frac{P_T}{4\pi d^2}$$

However, the antenna is directional, so power density at the target to be located at distance d is

$$p = G_T \frac{P_T}{4\pi d^2} \tag{1}$$

where G_T is power gain of the transmitter antenna in the direction of target relative to the isotropic radiator. Let S be effective area of the target; then power received by the target is

$$P_i = pS = \frac{G_T P_T S}{4\pi d^2} \tag{2}$$

This power is incident on the target and gets scattered in all possible directions. A part of this power is reflected back to the radar antenna. So, power density received at antenna

$$P_{rec} = \frac{P_i}{4\pi d^2} = \frac{G_T P_T S}{(4\pi d^2)^2} \tag{3}$$

Power seen by the receiving antenna is proportional to its area. Let the effective area of antenna be A_o; then power received by the receiving antenna is

$$P'_{rec} = A_o P_{rec} = \frac{P_T G_T S\, A_o}{(4\pi d^2)^2} \tag{4}$$

In most radar systems, the transmitting and receiving antenna is one and the same. So we can prove that

$$G_T = \frac{4\pi A_o}{\lambda^2} \tag{5}$$

where λ is wave length of radiations transmitted.

Substituting for G_T,

$$P'_{rec} = \frac{SP_T(4\pi A_o/\lambda^2)A_o}{(4\pi d^2)^2}$$

$$= \frac{SP_T\, A_o^2}{4\pi\lambda^2 d^4} \tag{6}$$

The maximum radar range d_{max} is achieved when received power by the antenna P'_{rec} just equals the minimum detectable signal power (P_{min}) being limited by receiver noise background. From Eqn. (6),

$$d_{max} = \left[\frac{SP_T A_o^2}{4\pi\lambda^2 P_{min}}\right]^{1/4} \tag{7}$$

Using Eqn. (5), we have

$$d_{max} = \frac{SP_T G_T^2 \lambda^2}{(4\pi)^3 P_{min}} \tag{8}$$

Both Eqns. (7) and (8) give a relation for the maximum range of radar system. The above derivation has been made under ideal conditions where the effects of ground or absorption noise and interference have been neglected. However, if we take noise into consideration, we have

$$d_{max} = \left[\frac{P_T A_o^2 S}{4\pi\lambda^2 k T_o \delta_f (F-1)}\right]^{1/4}$$

where

$kT_o\delta_f$ – noise input

k – Boltzmann's constant

T_o – standard ambients temperature = 290 K

δ_f – Bandwidth

16.10.2 Basic Pulsed Radar System

The simple block diagram of a typical high-power pulsed radar system is shown in Figure 16.20.

Transmitter The trigger source, the modulator (pulsers) and the microwave oscillator constitute the radar transmitting system.

Receiver The mixer, the local oscillator, the IF amplifier, the detector and the video amplifier constitute the receiving system. The antenna, the indicator and the duplexer are common to both transmitter and receiver.

Function The trigger source generates pulses of appropriate time and frequency which are fed to the modulator which in turn generate rectangular voltage pulses that are used to supply voltage for the microwave oscillator. These rectangular pulses in fact switch ON and OFF the output tube as per requirement. The output tube may be a magnetron oscillator or an amplifier, e.g. klystron, cross-field amplifier or a travelling wave tube depending upon specific need. The operation of the oscillator by pulses reduces the heating of the tube. The duplexer passes output pulses from the transmitter to the antenna having a parabolic reflector. The duplexer connects the receiver to the antenna when transmission takes place. At the mixer, the received pulse modulated carrier wave is combined with the local oscillator. The main receiver gain is achieved at the IF amplifier stage which has a typical frequency of 30 or 60 MHz. The output of IF amplifier is fed to the detector which recovers the pulse envelope. The video amplifier amplifies the signal received from the detector and then it is fed to the indicator or display unit. In order that the transmitted pulse can be displayed as reference, the trigger source is also connected to the indicator. Angle data from the antenna are also supplied to the indicator. In order to allow passage of narrow pulses, the bandwidths of IF amplifier and video amplifier must be quite large.

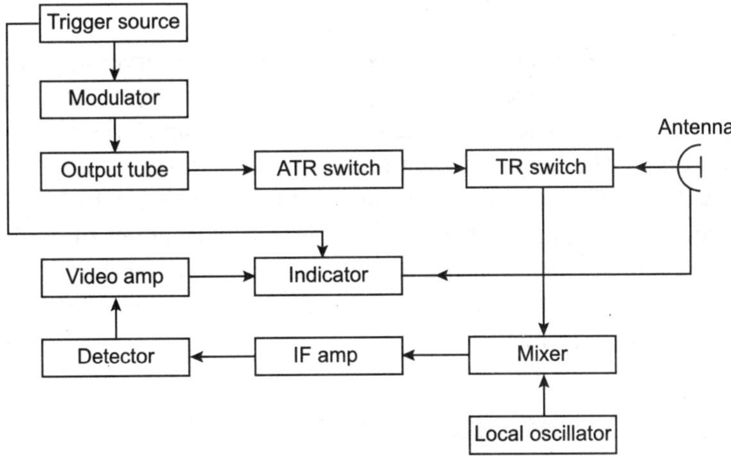

Figure 16.20 Pulse radar

Indicator The output of a radar system may be displayed in a number of ways. The following are the most common forms of display: (i) deflection modulation of a cathode ray tube screen as in the A scope, (ii) intensity modulation of CRT as in the plan position indicator (PPI), (iii) direct feeding to a computer, Additional information such as height, speed or velocity may be shown separately.

Radar displays are mostly two-dimensional, sometimes stereoscopic or three-dimensional, especially with radar systems having 3D information.

A scope In this type of display, the cathode ray tube is deflection modulated. This type of display is similar to an ordinary oscilloscope. The horizontal deflection plates of CRT are fed as sweep wave form, which moves the electron beam slowly from left to right across the screen and then back to initial position. The demodulated receiver output is applied to vertical deflection plates. The linear sawtooth waveform is synchronized with transmitted pulses, so that the width of CRT screen corresponds to the time interval between two pulses. The displacement from left to right on the screen is calibrated so that it gives the range of the target.

A typical a scope display is shown in Figure 16.21.

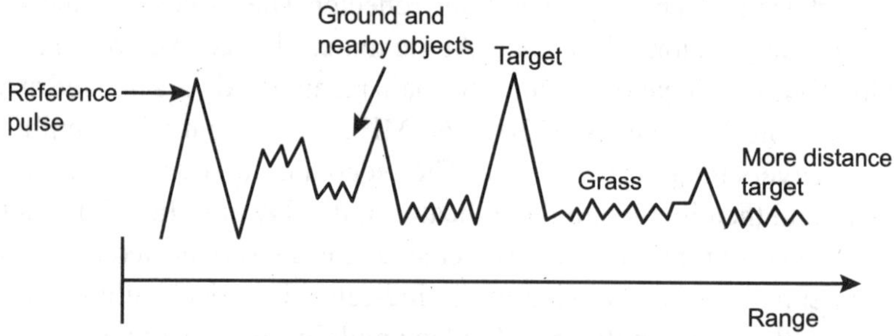

Figure 16.21 A scope display

The first blip on the left is due to the transmitted beam and serves as reference; then come various strong blips due to reflections from ground and near by objects followed by noise which is here called "grass". The various targets then shown up as (ideally) large blips, again interspersed with grass. The height of each blip corresponds to the strength of returned echo, while the distance from reference blip is a measure of range. This is why blips on the right side of screen have been shown smaller than those on the left. For targets of comparable size, one finds that blip for a more distant object/target is smaller. It is interesting to note that it would take a very large object/target at a range of nearly 50 km to produce the same size of echo as normal target nearly 10 km away.

16.11 Plan Position Indicator (PPI)

This type of display is generally employed in search type radars. They are particularly useful when conical scanning is employed. In this type of display, the CRT is intensity modulated so that signal from the receiver after demodulation is applied to the grid of CRT. The CRT is biased slightly beyond cut off, and thus blips corresponding to the targets permit beam current and, therefore, screen brightness. The scanning waveform is now applied to a pair of coils on the opposite side of the neck of the tube, so that magnetic deflection is used and saw tooth current is required. Coils situated in a yoke similar in appearance to that around the neck of a television picture tube are rotated mechanically at the same angular velocity as antenna. Hence the beam is not only deflected radially outward from the centre and neck and then back again rapidly but also rotates continuously around the tube. Brightness at any point on the screen indicates the presence of an object there, with its position corresponding to its physical position and its range being measured radially out from centre. Long persistence phosphors are used to ensure that the face of PPI screen does not flicker.

Summary

1. A television system comprises two processes: (i) transmission and (ii) reception.
2. There are mainly two types of television: (i) black and white and (ii) colour.
3. For black and white television, the transmitter has the following subsystems: (i) TV camera, (ii) scanning and synchronising system, (iii) video amplifier, (iv) AM video transmitter, (v) microphone, (vi) AF amplifier, (vii) FM audio transmitter and (viii) Antenna

   Here, the frequency of radio waves used for transmission is

   (a) Medium band (500 to 1600 KHz)

   (b) Short band I (2.5 to 7.5 MHz)

   (c) Short band II (7.5 to 25 MHz)
4. The most important subsystem of TV is camera. There are three types of camera, namely: (i) iconoscope, (ii) image orthicon tube and (iii) vidicon tube.

   These tubes scan a scene or object to be transmitted. For scanning we employ two methods: (i) progressive scanning and (ii) interlaced scanning.
5. In progressive scanning, we have 25 frames/pictures per second and there are 625 horizontal lines. So the time period for scanning is 65 μ.

   In interlaced scanning, the picture/scene is scanned twice. The horizontal lines are divided in two groups: (i) odd and (ii) even.
6. A TV signal to be transmitted has a total channel width of 7 MHz, which comprises both video and audio signal. The audio wave has a bandwidth of 5.5 MHz.
7. A black and white TV receiver has the following stages: (i) common audio and video stage, (ii) audio stage, (iii) video stage, (iv) synchronizing and extra HT stage, (v) rectifier for all signals to be received that employs Yagi antenna.

8. A colour television works on the principle of primary and complementary colours which are produced by mixing the primary colours on the basis of Grassmann's Law.

   Here the camera tube is completely different from that used in Black and White television. It employs a dichoric lens and relay lenses which, produce the three primary colours separately.

9. A colour TV receiver is completely different from that used in black and white television. Here, the three primary colour beans produced by the electron guns are mixed to produce the actual colour of the object.

Review Questions

1. Draw the block diagram of a monochrome TV transmitter. Explain briefly the function of each block.
2. Draw the block diagram of a monochrome TV receiver explain the function of each block briefly.
3. Describe the construction and working vidicon tube draw a neat labelled diagram. Give its merits over other tubes.
4. What do mean by (i) Progressive scanning (ii) Interlaced scanning? Explain and give their relative merits.
5. What do you understand by interlaced scanning? Explain. Show that it reduces flickering.
6. What do you mean by the term "scanning"? As regards television is concerned.
7. Describe the construction and working of iconoscope employed in black and white TV draw a neat labelled diagram for the same.
8. What do you mean by active and blanking time period in horizontal and vertical scanning in a TV system. Give normal values of active and retrace intervals both for horizontal and vertical scanning in case of BW television.
9. Why we have 625 lines for scanning during transmission in a progressive scanning system? Justify.
10. What do you mean by "Raster" in TV system. Explain.
11. Why TV signals are limited to VHF and UHF regions and not kept in radio transmission region? Explain. Mention the VHF and UHF regions for various channels as decided.
12. Why the bandwidth of a channel is fixed as 7 MHz? Explain. To which side of the frequency response curve we have video signal and audio signal respectively.
13. Name primary colours. What are complementary colours? Give their composition. What is Grass man's Law? Explain how three primary colours are combined to have a particular colour.
14. Give the construction of Yagi antenna. Draw sketch and describe the function of various parts.
15. Give the frequency bands used in TV transmission. Why these bands are preferred. Explain.

16. Draw a labelled diagram of BW television picture tube. Describe the function of various components.
17. Draw a labelled diagram of colour TV transmission camera tube. Describe the function of various components.
18. Explain the terms "Luminescence" and "Chrominance" as referred in a TV system.
19. Explain, what is meant by *Y*, *I* and *Q* signals in a colour TV system. Why are they generated.
20. Discuss the basic principle of a pulsed radar system. Explain its operation. Why are Microwaves used in a radar?
21. Draw the basic block diagram of a pulsed radar system. Draw a relation for the range of a radar. How the maximum range of radar can be increased? Explain.
22. Draw the block diagram of a basic Radar system and explain the essentials of its operation.
23. Derive the basic radar range relation as governed by the minimum receivable echo power p_{min}.
24. What happens in horizontal scanning in TV system? Explain. Giving step by step all events. When beam starts and ends and again repeated.

Short Answer Questions

1. "Radio waves from a transmitting radio station can reach various places on earth but TV signals do not reach". Explain.

   Ans: The frequency range of radio waves is from KHz to a few MHz, so these waves get reflected by the ionosphere and so can reach various places on earth, but for TV signals the frequency range is well above MHz and they do not get reflected by the ionosphere. Rather they are refracted through the ionosphere and get lost, so these signals do not reach various places on earth as well as their direct range from transmitting station is very small, nearly 80 km.

2. What is the range of frequency band used in TV transmission?

   Ans: The frequency band of TV signals is 30 to 300 MHz in VHF range and 470 to 890 MHz in UHF range.

3. How the scanning time period of 65 μs is divided into tracing and retracing time?

   Ans: The scanning time period of 65 μs is divided as under:

   (i) Tracing time = 54 μs

   (ii) Retracing time = 11 μs

4. What is the time period for vertical scanning?

   Ans: The time period for vertical scanning is nearly 20 msec.

5. What does a composite video signal consist of?

   Ans: A composite video signal in TV transmission consists of three types of pulses:

   (i) Horizontal or vertical synchronizing pulse

   (ii) Horizontal or vertical blanking pulse

   (iii) Equalising pulse

6. What is the bandwidth of a particular channel?

   Ans: The bandwidth of a channel is 7 MHz.

7. Why the channel bandwidth in a TV system is kept at 7 MHz?

   Ans: In a TV transmission system, the channel bandwidth has to contain both audio and video signals. So it is confined within 7 MHz.

8. What is the function of synchronizing circuits in a TV system?

   Ans: The function of synchronizing circuits in a TV system is to process the received information in such a way to ensure that vertical and horizontal oscillators in the receiver work at correct frequencies. This task is divided into three specific functions:

   (i) Extraction of sync information from composite waveform

   (ii) Provision of vertical sync pulses

   (iii) Provision of horizontal sync pulses

9. What is the frequency of video signal in a TV system?

   Ans: The frequency of video signal is 6.5 MHz.

10. What is the frequency difference between audio and video signals when each of them is mixed with the signal generated by the local oscillator in TV system?

    Ans: The frequency difference between audio and video signals generated by the mixer in TV system is 5.5 MHz.

11. What is the total number of lines scanned in interlacing?

    Ans: The total number of lines scanned in interlacing is 625, and they are divided into two fields. In the first field 1 to 312.5, and remaining 312.5 to 625 in the second field.

12. Why we require three electron guns in the camera tube of colour TV transmission?

    Ans: Three electron guns are used in the camera tube to separately distinguish the three primary colours present in the object to be telecasted.

13. What is the role of dichoric mirrors in colour TV?

    Ans: Dichoric mirrors are used in colour TV because a simple optical separator causes excessive loss of light intensity, whereas a dichoric mirror reflects completely a particular colour and refracts the remaining for which it is meant.

14. "Direct TV telecast is limited to 80 km on earth's surface." Explain.

    Ans: The limit of 80 km for direct TV telecast is due to the curvature of earth's surface.

15. Which factor limits the range of a radar system?

    Ans: The time interval between two consecutive pulses limits the range of a radar system.

16. What do you mean by the term "grass" in radar system?

    Ans: The term "grass" in radar display is related to noise reflected from nearby objects.

17. What is duplexer in a radar system?

    Ans: It is a device which alternately connects the transmitter and receiver in a radar system so that once a signal is transmitted, its reflection from a far object (echo) is received by the receiver.

Objective Questions

1. In a television, the aspect ratio is ______.
   (a) 2 : 1 (b) 2 : 3 (c) 4 : 3 (d) 4 : 5 **[Ans: (c)]**
2. The number of lines per field in a TV system is ______.
   (a) 312½ (b) 625 (c) 525 (d) 262½ **[Ans: (b)]**
3. The number of frames per second in a TV system is ______.
   (a) 30 (b) 25 (c) 20 (d) 10 **[Ans: (b)]**
4. Line frequency for TV system in India is ______.
   (a) 15625 (b) 15750 (c) 625 (d) 525 **[Ans: (a)]**
5. A signal sent by the TV transmitter to ensure correct scanning in the receiver is called ______.
   (a) sync (b) chroma (c) luminance (d) video **[Ans: (d)]**
6. The three primary colours are ______.
   (a) red, yellow, orange (b) red, blue, yellow
   (c) red, blue, cyan (d) red, blue, green **[Ans: (a)]**
7. After the mixture stage in a transmitter, the beat frequency of video signal and audio signal in MHz respectively is ______.
   (a) 38.9, 33.4 (b) 33.4, 38.5 (c) 38.9 for both (d) 33.4 for both **[Ans: (a)]**
8. Active line period and blanking period during horizontal scanning respectively is ______.
   (a) 50 μs, 10 μs (b) 54 μs, 11 μs (c) 64 μs, 10 μs (d) 64 μs, 20 μs **[Ans: (b)]**
9. The total time period for vertical scanning is ______.
   (a) 64 ms (b) 52 ms (c) 20 ms (d) 18 ms **[Ans: (c)]**
10. The bandwidth of a TV channel in India is ______.
    (a) 7 MHz (b) 6.5 MHz (c) 6 MHz (d) 5.52 MHz **[Ans: (a)]**

11. The location of video signal and audio signal on the frequency response curve is ______________.
    (a) both on the same side (HF range)
    (b) both on the same side (LF range)
    (c) video signal is at lower frequency side and audio signal is at higher frequency side.
    (d) video signal is at higher frequency side and audio signal is at lower frequency side.
    [**Ans:** (c)]
12. The number of electron guns in a colour TV receiver is ______________.
    (a) 1 (b) 2 (c) 3 (d) 4 [**Ans:** (c)]
13. The correct statement is ______________.
    (a) Both orthicon tube and vidicon tube work on the principle of photosensitivity.
    (b) Both orthicon tube and vidicon tube work on the principle of photoconductivity.
    (c) Orthicon tube works on the principle of photosensitivity whereas vidicon tube works on the principle of photoconductivity.
    (d) Orthicon tube works on the principle of photoconductivity whereas vidicon tube works on the principle of photosensitivity. [**Ans:** (c)]

Mark the Following Statements True/False.

1. The aspect ratio of TV screen is height to width ratio.. [**Ans:** True]
2. Yagi antenna is a dipole antenna. [**Ans:** True]
3. There are three electron guns in the picture tube of a colour TV. [**Ans:** True]
4. The frequency for transmission in a TV system ranges from KHz to MHz. [**Ans:** False]
5. Both image orthicon tube and vidicon tube function on the principle of photoconductivity.
   [**Ans:** False]
6. The time period of sawtooth wave for scanning in 65 μs. [**Ans:** True]
7. A composite signal at the output stage of detector consists of three signals. [**Ans:** True]
8. A radar system uses microwave. [**Ans:** True]
9. There is no range for a radar system. [**Ans:** False]
10. Radar system power output is of the order of a few kilowatts. [**Ans:** False]
11. The time period for horizontal scanning is equal to that of vertical scanning. [**Ans:** False]

Fill in the Blanks.

1. A colour TV camera tube uses ____________ mirror and ____________ lens system.
[**Ans:** dichoric, relay]
2. A colour TV picture tube has ____________ electron guns and ____________ screen.
[**Ans:** three, one]
3. ____________ percent Red, ____________ percent Green and ____________ percent Blue make white colour. [**Ans:** 30, 59, 11]
4. The audio signal is ____________ modulated and video signal is ____________ modulated at the transmitting station of colour TV. [**Ans:** frequency, amplitude]
5. In ____________ TV transmission ____________ scanning is preferred to ____________ scanning. [**Ans:** black and white, progressive, interlaced]
6. In scanning, the tracing time is ____________ whereas retrace time is ____________ sec. [**Ans:** 54 μs, 11 μs]
7. The frequency of video signal is ____________ and that of audio signal is ____________.
[**Ans:** 1.25 MHz, 5.5 MHz]
8. The number of frames in a TV system is ____________ per sec and each frame has ____________ lines. [**Ans:** 25, 625]
9. The time interval between two transmitted pulses is ____________ than the time period of echo in radar system. [**Ans:** larger]

Chapter 17

Electronic Instruments

17.1 Cathode Ray Oscilloscope

A cathode ray oscilloscope (CRO) is used for the measurement and analysis of waveforms and other phenomena in electronic circuits. CROs are basically fast XY plotters that display an input signal versus another signal or time.

In addition to voltage signals, it can show a visual representation of various dynamic phenomena by means of transducers that convert current, pressure, temperature, strain, etc. into voltage signal. These observations can be recorded by means of a special camera attached to the device.

17.2 Construction of General CRO

It consists of the following main subsystems as shown in Figure 17.1.

(a) Cathode ray tube

(b) Vertical amplifier

(c) Horizontal amplifier

(d) Delay line

(e) Time-base generator

(f) Trigger circuit

(g) Power supply

  (i) Low voltage supply (LT)

  (ii) High voltage supply (HT)

Schematic block diagram of the above subsystems has been shown in Figure 17.1(b) whereas Figure 17.1(a) represents the front panel of a complete CRO depicting nobs for its various functions.

(a) Cathode ray tube It is the most important sub system of the CRO. It may be called the heart of the device. Its important parts are:

(i) Glass envelope and its base

(ii) Electron gun

(iii) Deflection plates

(iv) Fluorescent screen

Electrodes, deflecting plates, etc. are enclosed in a large, sturdy glass envelope which is conical in shape (Figure 17.2). The electron gun, which produces a fast diverging beam of electrons, consists of a filament *FF* used for heating cathode *K*, which emits electrons. It is surrounded by grid *G* made up of a nickel hollow cylinder with a small centrally located hole coaxial with the tube. Its voltage is kept negative with respect to cathode so it controls the emission of electrons from cathode, i.e. it controls the intensity (nob 3 on front panel, Figure 17.1(a)) of electron beam. Electrons that manage to pass through the small hole of grid together make up the beam current. These electrons are accelerated by the high positive potential applied to the two accelerating anodes A_1 and A_2.

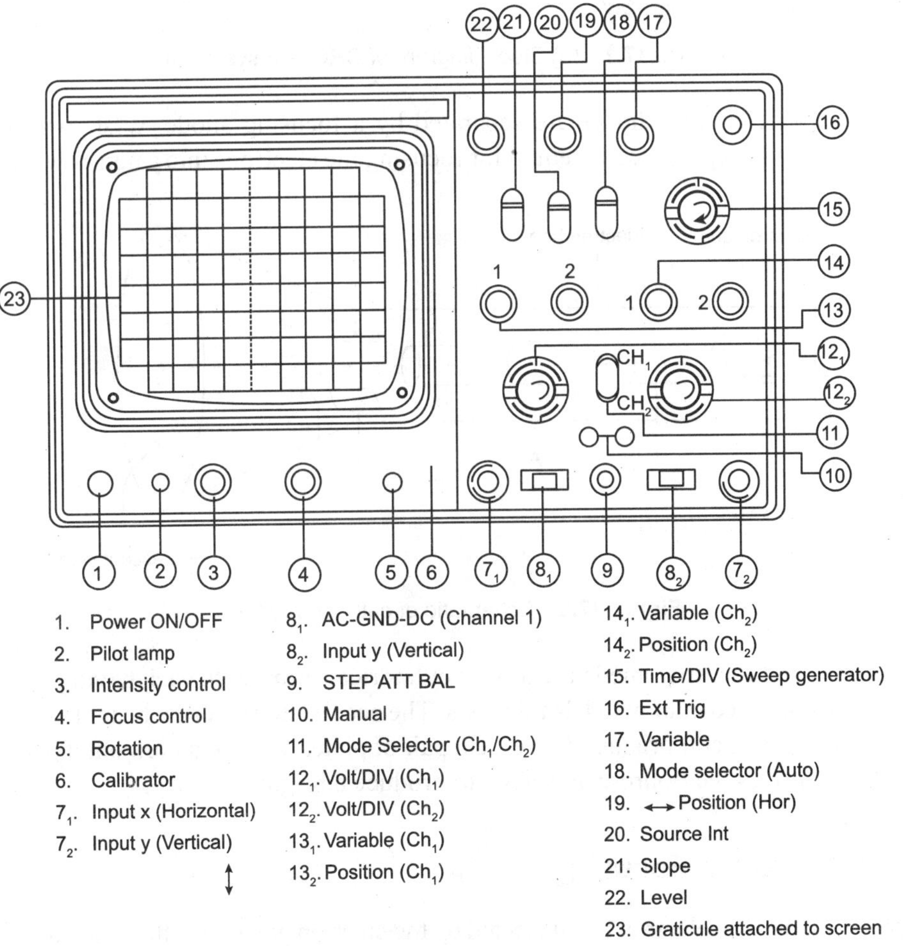

Figure 17.1 (a) Double beam CRO (front panel) typical view

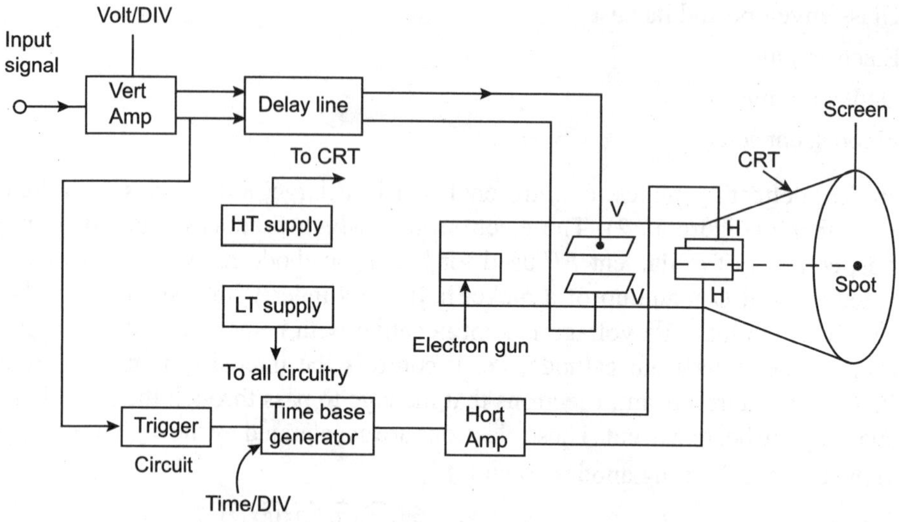

Figure 17.1 (b) Block diagram of CRO (subsystems)

These two accelerating anodes are separated by a focusing anode A_F shown by knob 4 (Figure 17.1(a)), which provides a method for focusing the electrons into a non-row and sharply defined beam.

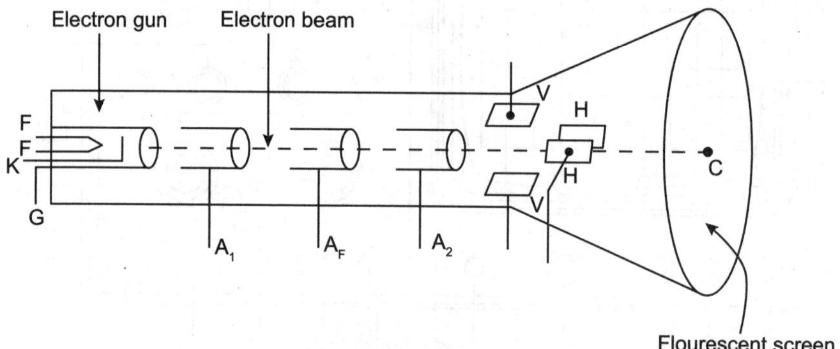

Figure 17.2 Schematic of cathode ray tube

The anodes and focusing anode are also in cylindrical form, with each having a small hole at the centre, which is coaxial with CR tube axis. The accelerated electron beam travels past the deflecting plate system comprising of a vertical pair of plates VV and a horizontal pair of plates HH and finally strikes the flourescent screen to produce a bright spot at *C*.

17.3 Electrostatic Focusing

Principle To focus an electron beam ejected by the electron gun on the fluorescent screen, two techniques are generally employed: (i) electrostatic focusing and (ii) magnetostatic focusing.

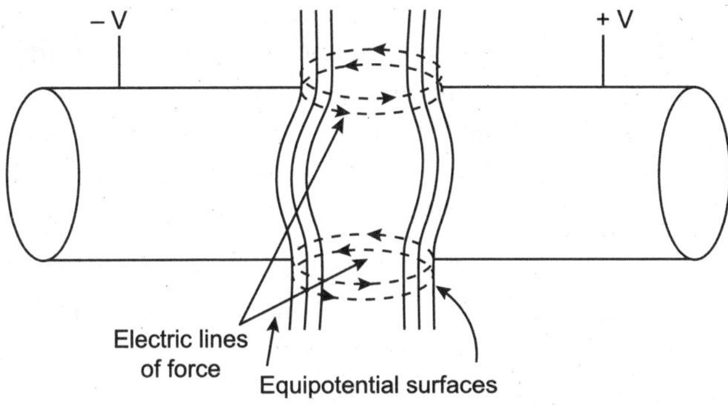

Figure 17.3 Electrostatic focusing

In electrostatic focusing, two similar cylindrical electrodes are placed one before the other (Figure 17.3). They are kept at unequal potentials (–)V and (+)V so that the resultant electric field is not uniform but it will be as shown in Figure (17.3). Here the electric lines of force will repel each other so they will not be parallel to each other, rather they will bulge. Since the density of lines of force is not uniform, the potential surface will also be curved. Consider one such potential surfaces (Figure 17.4).

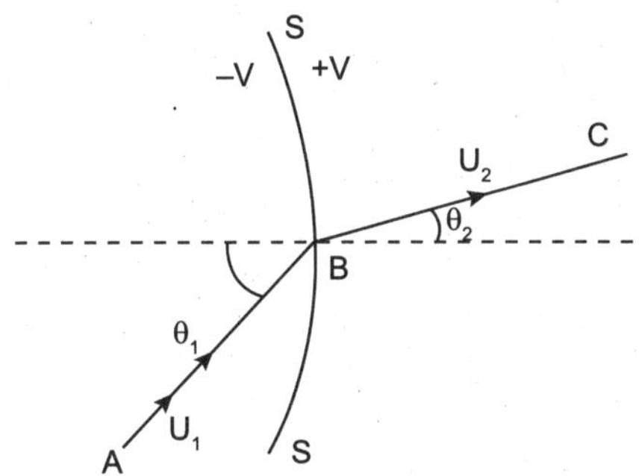

Figure 17.4

Let the potential towards the left side of potential surface *SS* be (–) V and that towards the right side be (+) V. Let an electron moving along path *AB* with velocity U_1 be incident at the surface at point *B*. While moving it is being acted upon by a force. This force is perpendicular to surface *SS*; so as electron crosses the surface, its velocity changes both in direction and magnitude. Let the electron move along *BC* with velocity U_2. It can be proved that transverse velocity of the electron on both side of equipotential surface remains constant. So accordingly,

$$U_1 \sin\theta_1 = U_2 \sin\theta_2 \tag{1}$$

where θ_1 is angle of incidence and θ_2 is angle of refraction

or
$$\frac{\sin\theta_1}{\sin\theta_2} = \frac{U_2}{U_1} \tag{2}$$

Equation (2) is similar to Snell's law for refraction of light ray from one medium to another. So we can assume that, here also the electron ray undergoes a sort of refraction when it crosses a equipotential surface (it moves from one potential field to another potential field). Here the bending of electron ray is used in focusing the beam; this action is very similar to that of lens which focuses the light beam when it refracts through it. Thus, two cylindrical electrodes maintained at different potentials act as electrostatic lens, which focuses the electron beam.

The entire process of focusing is carried out by three electrodes arranged in the cathode ray tube (Figure 17.2). The action of focusing the electron beam is governed by nob no 4 installed on the front of CRO (Figure 17.1(a)). The sharp bright points we see on screen are due to the focusing of electron beam in the cathode ray tube.

17.4 Electrostatic Focusing System

Considered the electrostatic focusing system of CRT. It consists of three elements: (i) pre-accelerating anode, (ii) focusing anode and (iii) accelerating anode (Figure 17.5).

A block diagram of the arrangement (Figure 17.5) explains the entire mechanism of focusing.

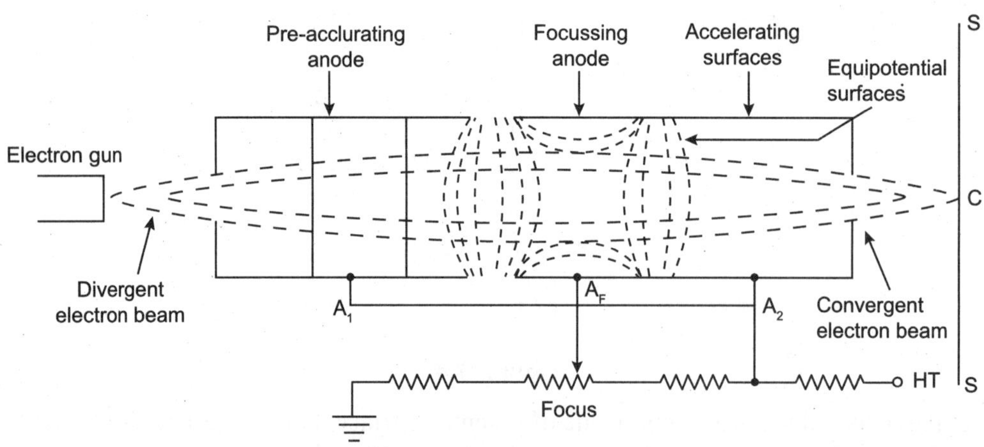

Figure 17.5 Block diagram of electrostatic focusing system of CRT

The three anodes (pre-accelerating (A_1), focusing, accelerating (A_F)) and accelerating anode A_2 form a perfect convergent electrostatic lens, thereby the divergent electron beam emerging from the electron gun becomes a convergent beam while passing through the lens and get focused at C on the fluorescent screen SS. The three electrodes are cylindrical in form and coaxial with tube axis.

The refracting action of lens is explained as thus: those electrons that enter the electric field between pre-accelerating anode and focusing anode at angles other than normal to the equipotential surface are refracted towards the normal. The incident divergent beam of electrons becomes parallel to the CRT axis as shown. The approximately parallel beam of electrons then enter the second lens formed due to focusing anode A_F and accelerating anode A_2 and is refracted again, thereby it becomes convergent and gets focused at *C* on the fluorescent screen *SS*. The focal length of lens systems can be increased or decreased by varying the potential of focusing anode. The potential provides the adjustment of voltage on the focusing anode of CRO (front panel control marked (4) as focus).

17.5 Electrostatic Deflection

While describing the function of CRT, it was said that an electron beam emerging from the electron gun gets deflected by the electric field produced between the deflecting plates. We shall now discuss the mechanism of electrostatic deflection.

An uniform electric field $\vec{E}$ is established between two parallel plates *PP* (Figure 17.6). When an electron travels along the direction of electric field, it experiences a force.

$$F_e = (-)\ eE$$

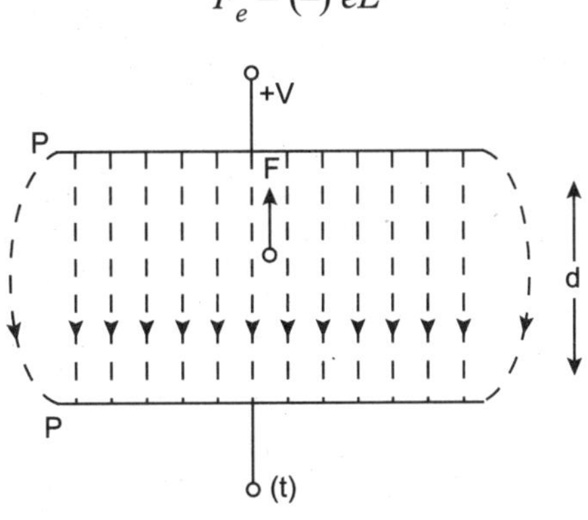

Figure 17.6

This force accelerates the electron, opposite to the direction of electric field since its direction is opposite to the direction of field. Acceleration produced is given by

$$a = \frac{F}{m} = (-)\frac{eE}{m} \qquad (1)$$

However, if the initial motion of electron is not along the direction of electric field, then acceleration produced is also not along the electric field.

Let the electron enter horizontally (along *X* direction) the electric field established along (–) *Y* direction, with velocity V_{OX} (Figure 17.7). This electron experiences a force. Since the field is directed along *Y* axis, there will be no force along either *X* axis or *Z* axis. So acceleration produced in the electron along these directions will also be zero. Zero acceleration means constant velocity. So electron entering along *X* direction as assumed will travel along *X* axis with constant velocity V_{OX}.

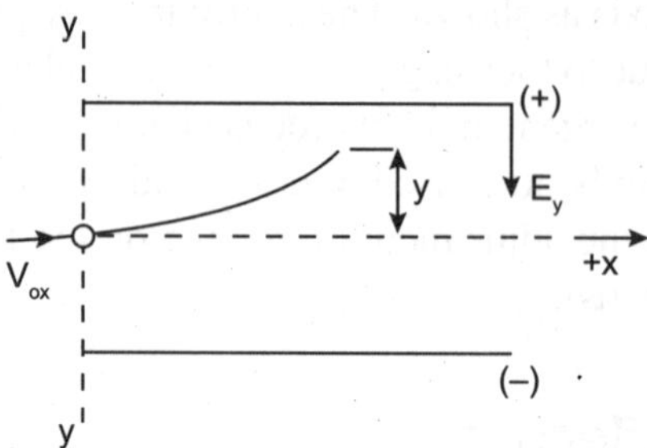

Figure 17.7

Acceleration along Y axis is

$$a_y = \frac{F_y}{m} = (-)\frac{eE_y}{m} \tag{2}$$

Due this acceleration, the electron gets deflected from the original direction of motion. To find displacement due to the accelerating force, we have at instant t, velocity along Y direction $v_y = a_y t$ (since initial velocity along this direction is zero).

Substituting the value of a_y, we have

$$v_y = (-)\frac{eE_y t}{m} \tag{3}$$

During this time, displacement along X direction is $x = v_{ox}t$ (since acceleration along X is zero)

or
$$t = \frac{x}{v_{ox}} \tag{4}$$

During this time, displacement of electron along Y direction is

$$y = \frac{1}{2}a_y t^2$$

(since initial velocity along Y direction is zero)

Substituting for a_y, we have

$$y = (-)\frac{1}{2}\frac{eE_y}{m}t^2$$

Substituting for t, we have

$$y = (-)\left[\frac{eE_y}{2mv_{ox}^2}\right]x^2 \tag{5}$$

Equation (5) shows that the path of an electron entering through an electric field (along *Y* of constant intensity) at right angle (i.e. along *X*) to the direction of field is a parabola in [*X*–*Y*] plane.

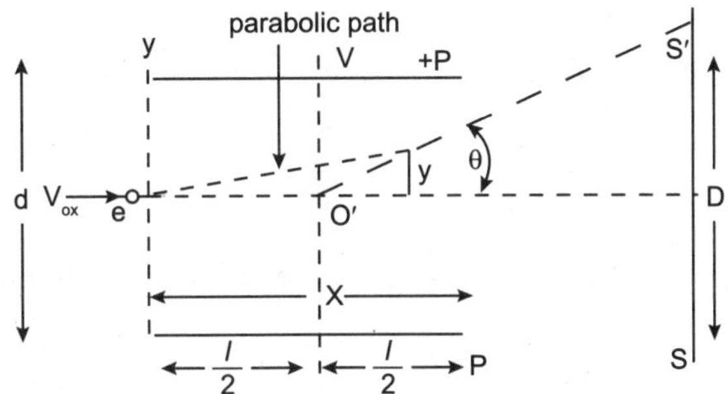

Figure 17.8 Deflection of cathode ray beam

Let the two deflection plates *P*–*P* (Figure 17.8) be at distance *d* apart and are connected to a source so that the potential difference is *V*. Electric field produced along *Y* direction is given by

$$E_y = \frac{V}{d}$$

An electron entering at right angle with initial velocity V_{OX} gets deflected towards the positive plate and follows a parabolic path (Eqn. 5) shown in Figure 17.8. When this electron leaves the electric field region between the plates *PP*, the deflecting force no longer exists and it travels along a straight line striking the screen at *S*′.

The slope of parabola at distance $x = l_d$ is defined as

$$\tan\theta = \frac{dy}{dx}$$

where *y* is given by Eqn. (5).

Differentiating Eqn. (5) w.r.t. *x*, we have

$$\tan\theta = (-)\frac{eE_y l_d}{mv_{ox}^2} \tag{6}$$

The straight line path of the electron after leaving the electric field region is tangent to the parabola at $x = l_d$.

This tangent intersects the *X* axis at point *O*′. The location of this apparent origin *O*′ is given by Eqn. (5) and (6) since

$$x - O' = \frac{Y}{\tan\theta} = \frac{eE_y l_d^2 / 2mv_{ox}^2}{eE_y l_d / mv_{ox}^2} = \frac{l_d}{2} \tag{7}$$

The apparent origin is, therefore, at the center of deflecting plates *PP* and at distance *L* away from screen *S*.

The deflection of electron on the screen is given by

$$D = L \tan \theta$$

Substituting for tan θ from Eqn. (6), we have

$$D = L\frac{eE_y l_d}{mv_{ox}^2} \tag{8}$$

The potential energy of electron entering the electric field with initial velocity v_{ox} is given by

$$\frac{1}{2}mv_{ox}^2 = eV_a$$

where v_a is acclerating voltage of electron

or
$$v_{ox}^2 = \frac{2eV_a}{m} \tag{9}$$

Substituting for E_y and v_{ox} in Eqn. (8), we have

$$D = \frac{LeE_y l_d}{mv_{ox}^2} = \frac{Ll_d V}{2dV_a} \tag{10}$$

where

D – deflection (displacement) on the screen (m)

L – distance from centre of deflection plates to screen (m)

l_d – effective length of plates (m)

d – distance between deflection plates (m)

V – voltage across deflection plates (V)

V_a – accelerating voltage of applied (V)

Equation (10) indicates that for a given accelerating voltage V_a and for the particular dimensions of CRT, the deflection of electron beam on the screen of tube is directly proportional to the potential difference between the deflecting plates.

This direct proportionality of deflection on screen indicates that a CRT can be used as a linear voltage indicating device. However, deflection voltage is usually a varying parameter and the image on screen follows variation of deflection voltage in a linear manner.

17.5.1 Deflection Sensitivity

Deflection sensitivity (*S*) of a CRT is defined as deflection on the screen (in meters) per volt of deflection voltage. Hence we have

$$S = \frac{D}{V} = \frac{Ll_d}{2dV_a} \text{ expressed as } \frac{\text{meter}}{\text{volt}}$$

17.5.2 Deflection Factor

Deflection factor (G) is defined as the reciprocal of deflection sensitivity:

$$G = \frac{1}{S} = \frac{2dV_a}{Ll_d} \tag{11}$$

Equation (10) suggests that sensitivity of CRT is independent of deflection voltage. But it varies inversely with accelerating voltage (V_a).

17.6 Screen of the CRT

When an electron beam strikes the screen of CRT, a spot of light is produced. The material pasted on the inside surface of screen that produces this effect is phosphor. It absorbs the kinetic energy of bombarding electrons and re-emits the same in the form of radiation (light) at lower frequency. The property of some crystalline materials such as phosphor or zinc oxide to emit light when stimulated by radiation is called fluorescence. *Fluorescent* materials have a second characteristic called *phosphorescence*, which refers to the property of the material to continue light emission even after the source of excitation is cut off. The duration of time during which phosphorescence or afterglow exists is called persistence of phosphor. Persistence is usually measured in terms of time required for the CRT image to decay to a certain percentage (usually 10%) of original intensity (brightness).

Table 17.1 Characteristics of some commonly used phosphors

Phosphor type	Fluorescence	Phosphorescence	Relative luminance	Decay time 0.1% (ms)	Use
P_1	Yellow green	Yellow green	50%	95	General purpose
P_2	Blue green	Yellow green	55%	120	Good compromise for high and low speed
P_4	White	White	50%	20	Television display
P_7	Blue	Yellow green	35%	1500	Low speed phenomenon
P_{11}	Purple blue	Purple blue	15%	20	Photographic applications
P_{31}	Yellow green	Yellow green	100%	32	General purpose; brightest available

Note: Luminance is photometric equivalent of brightness.

17.7 Basic Sweep Generator

A CRO usually displays the vertical input signal as a function of time. This requires a horizontal deflection system, which will move or sweep a CRT spot across the screen from left to right with constant velocity and then rapidly (abruptly) returns the spot to its starting point and ready for next sweep. This sweep voltage or time base is produced within the horizontal deflection system by a sweep generator.

The ideal sweep voltage rises at a linear rate from some minimum value to some maximum value and then returns abruptly to the original value (Figure 17.9).

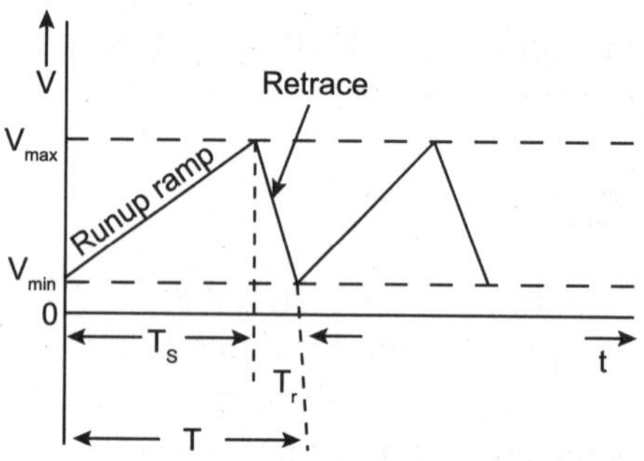

Figure 17.9 Linear sawtooth wave

The linear rising part of sawtooth wave is called runup ramp voltage. During the time interval T_S when it rises from V_{min} to V_{max}, the spot on the screen moves from left to right with constant velocity. In retrace or fly back interval T_r, sweep voltage drops rapidly (ideally abruptly) to its minimum value, so the spot flies back to the starting point on the screen. In all CROs, the electron beam is cut off during retrace interval, so there is no retrace image.

Principle All sweep generators are modifications of a basic *R-C* circuit (Figure 17.10(a)). In this circuit, switch *S* is initially closed and voltage (e_c) across capacitor is zero. When switch *S* is opened, capacitor begins to get charged and its voltage (e_c) rises exponentially (Figure 17.10(b)) to the maximum value *E* (supply voltage). The instantaneous value of voltage is

$$e_c = E\left[1 - \exp\left(-\frac{t}{RC}\right)\right]$$

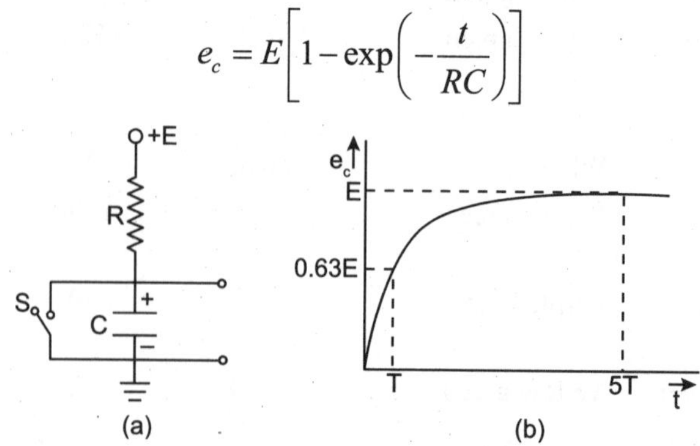

Figure 17.10 (a) *R-C* circuit, (b) charging of capacitor

As shown, the rise of voltage is not linear and it achieves 63% of its final value in one time constant $\left(T = \frac{1}{RC}\right)$. The final value E is achieved in nearly 5 T. So it is clear that this circuit cannot be used for linear sweep voltage. However, linearity can be achieved if rising voltage is terminated at 0.2 T instant.

17.8 Time Base Generator

The signal to be viewed is applied to the vertical amplifier input. The gain of this amplifier is set by a calibrated input attenuator usually marked VOLTS/DIV (nob 12 in front panel). The push-pull output of the amplifier is fed to vertical deflection plates of CRT via a so-called delay line, with sufficient power to drive the CRT spot in vertical direction.

The time base generator or sweep generator, develops a sawtooth wave form that is used as horizontal deflection voltage of CRT. The positive going part of sawtooth wave form is linear, and its rate of rise is set by a front panel nob control (no 15). The sawtooth voltage is fed to the horizontal amplifier. This amplifier includes a phase inverter and produces two waveforms simultaneously: (i) a positive going runup and (ii) a negative going down runup. The positive going sawtooth is applied to the right hand horizontal deflection plate of CRT, and the negative going sawtooth to the left hand horizontal deflection plate. These voltages cause the electron beam to sweep across the CRT screen from left to right and right to left respectively on the screen in time units that are controlled by TIME/DIV control (nob no 15).

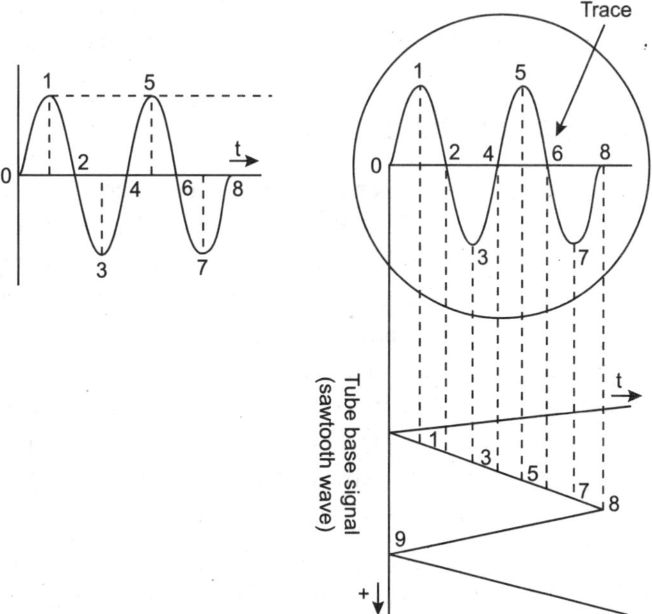

Figure 17.11

The simultaneous application of deflection voltages to both sets of plates causes the spot to trace an image on the screen (Figure 17.11) where a sawtooth voltage is applied to horizontal plates and a sine wave is applied to vertical plates.

Since the horizontal sweep voltage increases linearly, the spot on the screen moves across it from left to right at constant velocity. At the end of sweep when saw tooth voltage suddenly drops from maximum to zero (point 8 to 9), the spot immediately returns to the starting point and remains there till a new sweep starts.

When, simultaneously as with horizontal sweep voltage, an input voltage signal is applied to vertical deflection plates, the electron beam is under the influence of two perpendicular forces (voltages). The spot moves up and down in accordance with magnitude and polarity of the input signal. The resultant motion of electron beam, therefore, produces a CRT display (wave shape) of vertical input as a function of time.

If the vertical input signal is recurrent type, the display on the screen can be made stable, by starting each horizontal sweep at the same point on the signal waveform. To achieve this, a sample of vertical input wave form is fed to the trigger circuit which produces a trigger pulse (Figure 17.11) at some selected time on the input wave form.

The trigger pulse is used to start the time base generator which in turn starts horizontal sweep of CRT spot from left to right.

The purpose of delay line in the circuitry is to retard the arrival of vertical input signal at deflection plates *VV* until the trigger and time base circuits have had a chance to start a sweep of the beam. It introduces a total delay of nearly 0.25 μs in the vertical channel so that the leading edge of wave form can be viewed even though it was used to trigger the sweep.

17.9 Synchronization of the Sweep

The sawtooth generator (Figure 17.9) is free running because there is no external control that starts the generation of each new sweep. The new sweep simply starts as soon as the capacitor has discharged sufficiently to turn the UJT off. It is possible to use a free-running sweep to produce a stable CRT display, provided the frequency of vertical input signal (f_v) is a whole multiple of sweep frequency (f_s), i.e. $f_v = nf_s$. This situation is shown in Figure 17.12, where two cycles of vertical signal sweep frequency (f_s) occur in the same interval as one cycle of sweep voltage. If this exact frequency relationship is not maintained, the CRT display will be unstable, so it drifts across the screen. To produce a stable display, the sweep generator must run synchronously or "in step" with vertical signal, so that both vertical and horizontal signals arrive at the reference point in their cycles at the same time.

In the relaxation oscillator circuit, synchronization of the sweep can be obtained by applying a so-called "sync signal" to the sync input terminal in such a manner that the peak voltage of UJT is lowered. This situation is clarified in Figure 17.13 where a train of negative sync pulses is superimposed on the peak voltage of UJT runup.

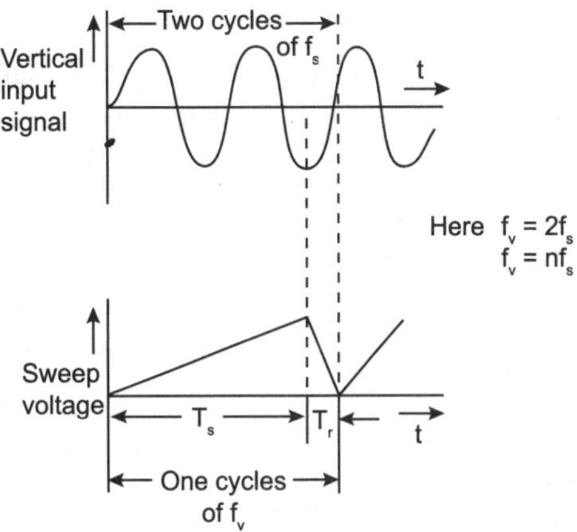

Figure 17.12 Vertcial input signal and a free-running sweep

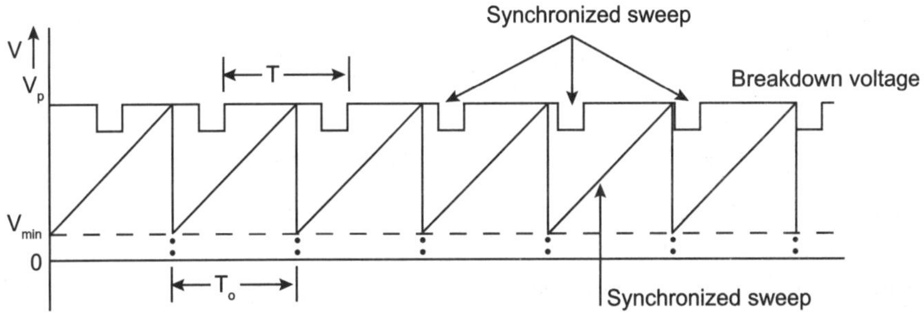

Figure 17.13 Principle of sweep synchronization

As shown in Figure 17.13, first few pulses have no effect on the frequency of sawtooth wave form, and the sweep generator continues to run unsynchronized at its own natural frequency. Eventually, the capacitor charging process is stopped prematurely by a sync pulse that occurs at the exact moment that runup ramp voltage equals the (momentarily) reduced peak voltage of UJT. At this moment the capacitor discharges rapidly through UJT and runup is terminated. When capacitor voltage has dropped to minimum (V_{min}), the voltage required to maintain UJT conducting, the transistor switches off and the capacitor charges again to produce the next ramp.

The synchronizing signal for sweep generator can be obtained from various sources and selected by a CRO front panel control called sync selector. In Figure 17.1, this selector is shown as a three-position switch marked INT-EXT-LINE. On the INT position, a sample of vertical amplifier signal, provided by a voltage divider, is used to develop the input signal under observation. On the EXT position, the sweep generator can be synchronized to an externally applied signal connected to a front panel jack marked EXT. In the LINE position, a sample of power line voltage is applied to the sweep generator, so that the signal under observation is synchronized to the frequency of power line.

17.10 Horizontal Amplifier

In a conventional CRO, the performance requirements (gain/bandwidth) of the horizontal amplifier are of lower order than those of the vertical amplifier. While the vertical amplifier must be able to handle small amplitude, fast rise time signals, the horizontal amplifier has to process a sweep signal with fairly high amplitude and relatively slow rise time. The gain of horizontal amplifier, however, is larger than that of vertical amplifier, because horizontal deflection sensitivity of the CRT is smaller than vertical deflection sensitivity.

Figure 17.14 shows the block diagram of a basic horizontal amplifier that is generally used in conventional CRO.

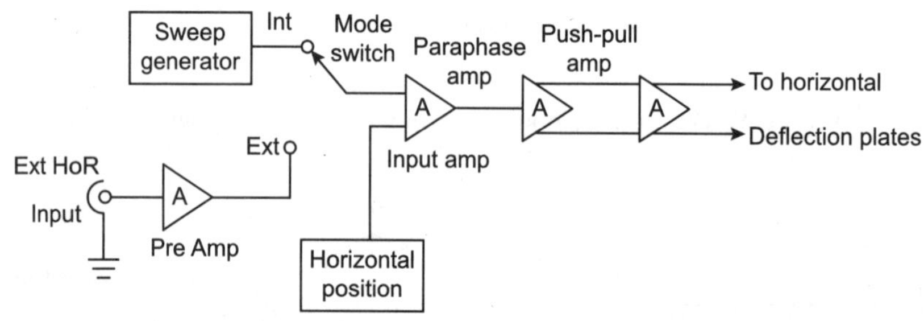

Figure 17.14 Block diagram of horizontal amplifier

As shown, the amplifier consists of three stages: (i) input amp, (ii) paraphrase amplifier and (iii) push-pull output amplifier. In the usual application, the input amplifier receives its signal from the sweep generator, which typically produces a 10-V time base ramp. Together with sweep voltage, it allows horizontal positioning of the spot on the screen. The single ended output is connected to a negative-feedback paraphrase amplifier, which produces two balanced output signals to feed the final stage. The push-pull output stage amplifier provides two ramp voltages, a positive going ramp and a negative going ramp, amplified to the required level, for simultaneous application to horizontal deflection plates.

17.11 Frequency Determination

Among many other uses, frequency determination of a signal is one of the important uses of a CRO. We are familiar with the fact that when two sine waves are superimposed perpendicular to each other, the resultant pattern/curve is known as lissajous figure. The shape of the pattern/curve depends upon the ratio of frequency of two sine waves, phase angle and amplitude. The most common patterns are:

(a) A straight line (Figure (a)) is produced when the phase angle is zero or 180°.

(b) A circle (Figure (b)) is produced when their frequencies are equal and phase angle is 90° or 270° and amplitudes are equal.

(c) An ellipse (Figure (c)) is produced when their amplitudes are unequal and phase angle is 90° or 270°.

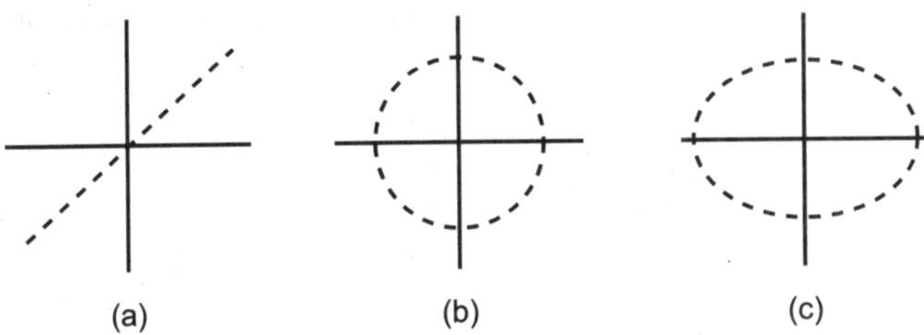

Looking at the figure, the following conclusions can be drawn:

(i) In case of a straight line, the angle formed with horizontal is 45°. When amplitudes are exactly equal, an increase in amplitude of vertical signal shall increase the angle.

(ii) When phase angle is 90° but amplitude of vertical signal is more, an ellipse with vertical major axis is formed and its inclination depends upon the phase angle.

In all the above pattern, the frequency ratio is unity. When frequency ratio is more than unity, other patterns are produced. In these patterns, it is important whether higher frequency or lower frequency signal is applied across the horizontal plates. The most significant consideration, however, is the phase angle of higher frequency signal w.r.t. lower frequency signal. For example, a figure of 8 (eight) resting on its side (side figure) results when both signals start out together. A tangent drawn against the top edge of the pattern would make contact at two points. A tangent drawn against a vertical side would make contact at one point. Evidently the number of horizontal tangencies correspond to the frequency of vertical signal, and number of vertical tangencies correspond to the frequency of horizontal signal. As such the ratio of frequencies in the said pattern is 2 : 1.

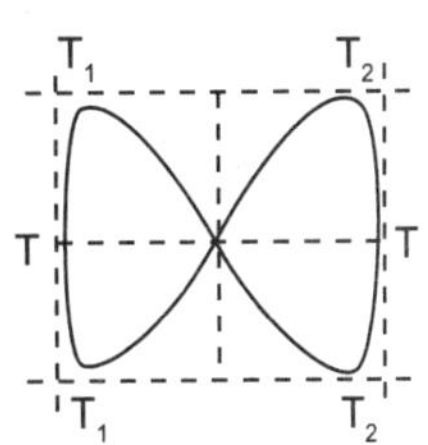

When a double image, such as a parabola (below figure) occurs, a different method of evaluating the frequency ratio is used. In such a case, a tangent drawn against the open end of the pattern is counted as half tangency.

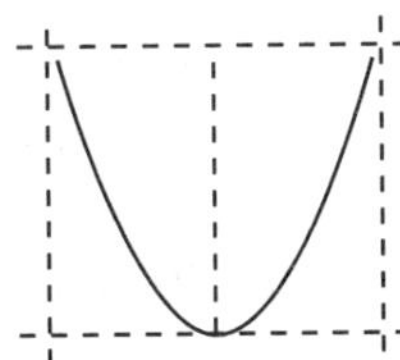

So in this case, a tangent drawn against the top makes two contacts, giving a total of one point. Against the vertical side, there is only one open contact giving a count of one half. So the ratio of vertical frequency to horizontal frequency is 2 : 1.

In addition to the patterns for integral ratios of the two frequencies, there are many other patterns for which the numerator and denominator of the ratio are whole numbers. For example, the pattern (Figure (a)) shows a 3/2 ratio and Figure (b) shows a ratio of 5/3 by applying the method of tangencies.

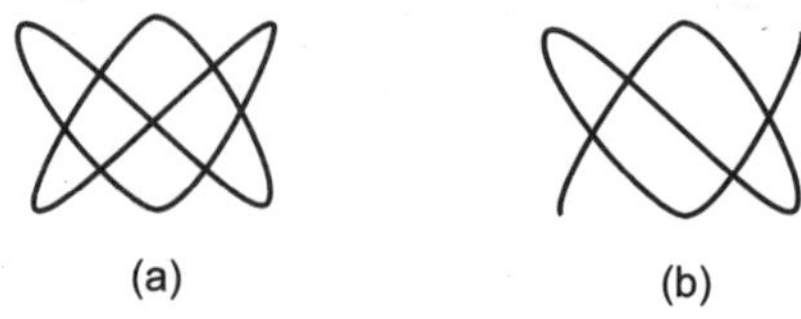

In addition to determining the frequency of an unknown signal, Lissajous patterns are generally used for comparing the frequencies of two signals.

17.11.1 Computation of Phase Angles

Regardless of the relative amplitudes of applied voltage signals, the pattern ellipse provides a simple means of finding the phase difference (angle) between two signals of the same frequency. The method to determine is shown in Figure 17.15. The sine of the angle between two signals of same frequency is equal to the ratio between *Y* axis intercept represented by Y_1 to the maximum vertical deflection represented by Y_2. So

$$\sin\theta = \frac{Y_1}{Y_2}$$

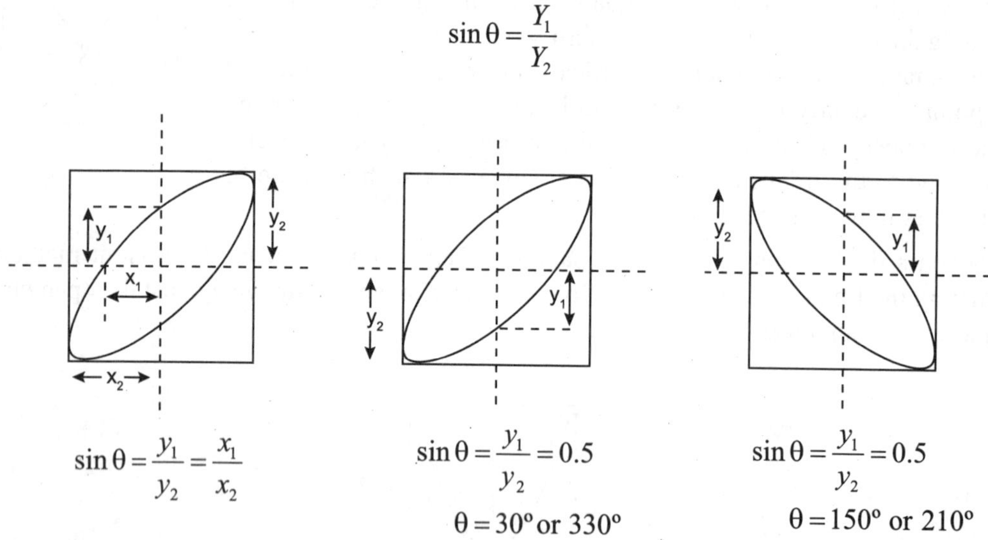

Figure 17.15 Computation of phase angle between two signals of same frequency

17.11.2 Special Purpose Cathode Ray Oscilloscope

(i) Dual trace CRO In conventional CRO, we have a single trace (spot). The single trace capability of CRO can be modified to produce a dual image (two spots) by means of electronic switching of two separate input signals. A simplified block diagram is shown in Figure 17.16.

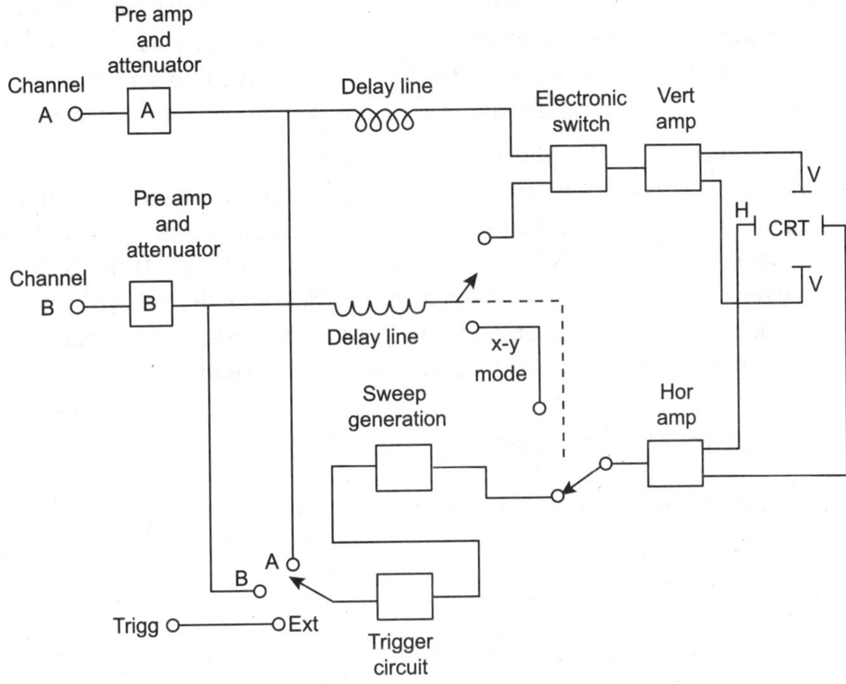

Figure 17.16 Block diagram of dual trace CRO

It has two vertical input circuits, marked channel *A* and channel *B*, with identical pre-amplifiers and delay lines. The *A* and *B* amplifiers' outputs are fed to an electronic switch that alternately connects the input of main vertical amplifier to two input signals. The electronic switch also consists of a circuity to select a variety of display models (not shown). When the display mode selector is in alternate position, the electronic switch alternately connects the main vertical amplifier to channel *A* and to channel *B*. This switching takes place at the start of each new sweep. The switching rate of the switch is synchronized to the sweep rate, so that CRT spot traces channel *A* signal on one sweep and channel *B* signal on the next sweep. Since each vertical amplifier has a calibrated input attenuator (a vertical position control), the amplitudes of input signals can be adjusted individually and two images positioned separately on the screen. This mode of operation is especially useful with relatively fast sweep ratio, when two images appear as one simultaneous stable display.

It is to be noted the sweep trigger signal is available from channel *A* or channel *B*, and that is picked off before electronic switch. This arrangement maintains a correct phase relationship between the two signals.

(ii) Dual-beam CRO A dual-beam CRO accepts two vertical input signals and displays them as two separate images on the screen. Instead of electronically switching the two signals to a single vertical amplifier as in dual trace CRO, the dual beam oscilloscope uses a special CRT that produces two completely separate electron beams, which can be independently deflected in vertical direction. In some dual beam CRTs, the output of a single electron gun is mechanically split into two separate beams (split beam technique), while other CRTs have two separate electron guns, each producing its own beam. The dual beam CRTs have two sets of vertical deflection plates, one set for each channel, and a single set of horizontal deflection plates.

There are a number of useful applications in which a CRO is placed in the so-called X-Y mode of operation. In the X-Y mode, the vertical input signal is applied to the CRO in the usual way, but the horizontal time base is replaced with an external signal that is applied to the horizontal amplifier via pre-amplifier and EXT position on the sweep selector. If the X-Y display must represent a true relationship between horizontal and vertical signals, the two systems must have the same phase delay, deflection factor, and band pass. These requirements place the horizontal amplifier system in the same class as the vertical amplifier system.

17.12 Electronic Multimeter

One of the most versatile general-purpose instruments capable of measuring DC and AC voltage as well as current and resistance is the solid-state electronic multimeter or VOM. It can be classified into (i) analogue type and (ii) digital type.

17.12.1 Analogue Volt-Ohm Meter

Analogue type meters were very popular until a few years ago, but with the advancement of digital electronics, most of them have been replaced by digital multimeters. Although circuit details will vary from one instrument to another, an electronic multimeter generally consists of the following:

(a) Balanced bridge DC amplifier and indicating meter

(b) Input attenuator or range switch to limit the magnitude of input voltage to the desired value

(c) Rectifier section to convert an AC input voltage to a proportional DC value

(d) Internal battery and additional circuitry to provide the capability of resistance measurement

(e) Function switch to select the various measurement functions of the instrument.

In addition, the instrument generally has a built-in power supply for AC line operation and, in most cases, one or more batteries for operation as a portable test instrument.

17.12.2 Voltage Measurement

Figure 17.17 shows the schematic of a balanced bridge DC amplifier using FETs. This circuit also applies to an amplifier with BJTs. The two FETs form the upper arm of the bridge circuit. Source resistors R_1 and R_2 together with zero adjust resistor R_3 from the lower arms of the bridge. The meter (galvanometer) is connected between the source terminals of FETs, representing the two opposite corners of the bridge.

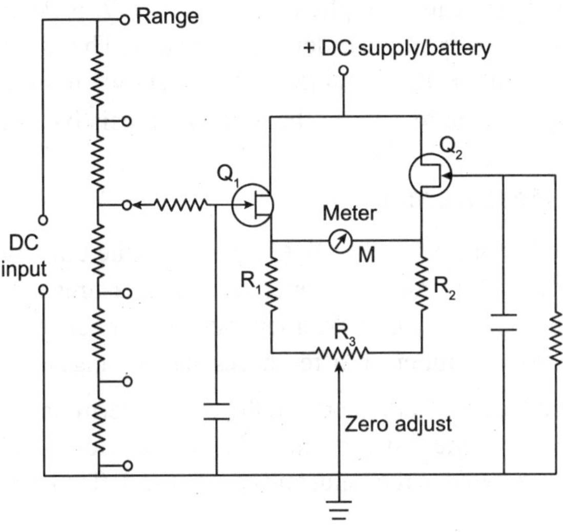

Figure 17.17 Electronic voltmeter

Without an input signal, the gate terminals of FETs are at ground potential and the two FETs operate under identical quiescent conditions. In this situation, the bridge is balanced and the meter shows zero ideally. However, zero can be adjusted by adjusting R_3. When a positive voltage is applied to the gate of input transistor Q_1, its drain current increases which causes the voltage at its source to rise. The resulting inbalance between the voltage of source of Q_1 and Q_2 is indicated by meter M connected between them whose scale has been suitably calibrated to agree with the magnitude of applied voltage.

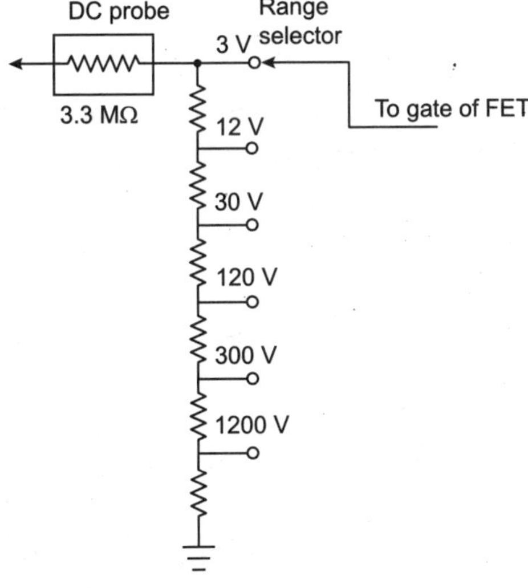

Figure 17.18

The maximum voltage that can be applied to the gate of Q_1 is determined by the operating range of FET used and it is usually of the order of a few volts. The range of input voltage can be easily extended by an input attenuator or range switch as shown in Figure 17.18. The unknown voltage is applied through a large resistor in the probe to a resistive voltage divider.

17.12.3 Resistance Measurement

When the function switch is placed in the OHMS position, the unknown resistor is connected in series with the internal battery and the meter simply measures the voltage drop across it. A typical circuit shown in Figure 17.19, where a separate divider network, used only for resistance measurements, provides for a number of different resistance ranges.

When unknown resistance R_x is connected to the OHMS terminal of multimeter, the battery supplies current through one of the resistors and unknown resistor to ground, so voltage drop V_x across R_x (Figure 17.19) is applied to the input and causes the deflection in the meter proportional to the voltage.

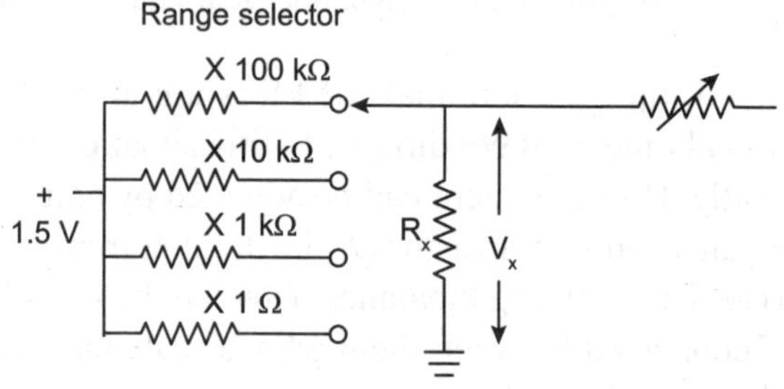

Figure 17.19

17.12.4 Digital Multimeter

A basic digital multimeter is made up of analogue-to-digital converter and circuitry for counting. A block diagram of basic digital multimeter is shown in Figure 17.20(a).

The various subsystems are:

1. Test probes
2. Selector switch
3. Attenuator
4. Compensated attenuator
5. Current-to-voltage converter
6. Constant current source
7. A/D converter
8. Decade counter
9. Digital read out

The test probe is a separate entity; it connects the unknown source or parameter to be measured to the main system. The selector switches S_{1A} and S_{1C} are internally connected mechanically; so if the position of one is changed, then the position of other is automatically changed to the related parameter for which S_{1A} has been set.

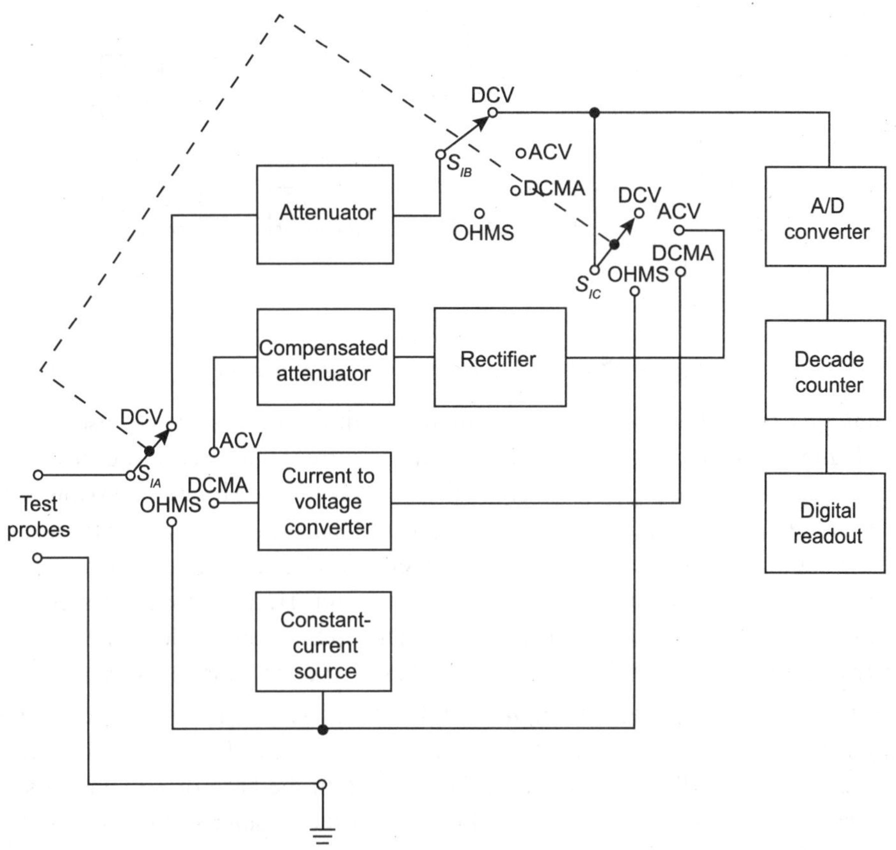

Figure 17.20 (a) Block diagram of digital multimeter

Let us discuss the functions of the various subsystems in brief.

(a) **Attenuator** An attenuator network is always placed between the main circuit and quantity to be measured. Its main function is to attenuate/reduce the size of quantity to be measured. It basically reduces the amplitude of the quantity, say voltage.

(b) **Compensated attenuator** It mainly reduces the amplitude of AC voltage (signals) and these signals are rectified by a rectifying circuit connected to it so that finally we get in output a DC.

(c) **Current-to-voltage converter** Here current to be measured is applied to the summing junction Σ_i at the input of an Op-Amp. Since current at the input of the amplifier is near zero because of high impedance of the amplifier, feedback current is the same as input/incoming current. Voltage drop across the resistor due to the incoming current is the analogue input voltage to the A/D convertor. Its block diagram is shown in Figure 17.20(b).

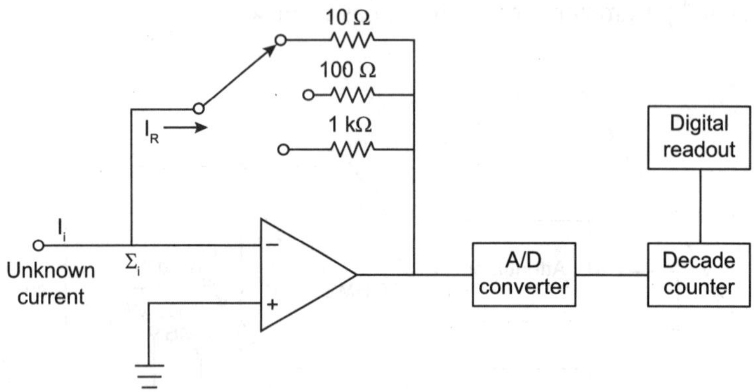

Figure 17.20 (b) Current-to-voltage converter

(d) **Analogue-to-digital converter** Normally digital multimeters are used to measure analogue parameters (quantities) like voltage, current, etc. Therefore, it necessary to convert the analogue signal to an equivalent digital signal. Three conversion techniques are employed. They are (i) single slope, (ii) dual slope and (iii) voltage to frequency. Most of the digital converters use dual slope conversion. The dual slope converter overcomes most of the limitation of the single slope converter. Here a capacitor is charged by a constant current source to provide linear voltage to time conversion. The dual slope not only charges the capacitor but discharges it as well during the measurement cycle.

(e) **Decade counter and digital readout** It behaves like a totalizer by totaling the pulses applied to it during the time interval a gate pulse is present. The decade counter consists of a circuit of flip-flops in cascade which counts in the base of 10 number system. It means there is a sequence of 10 distinct counts the counter progresses through. The counts of the counter are displayed by the digital readout.

The last two subsystems are discussed by taking the example of digital voltmeter.

17.12.5 Digital Voltmeter (DVM)

A digital voltmeter displays the measurement of DC or AC voltage as discrete numerals instead of a pointer deflection on a continuous scale as in an analogue device.

Numerical read out is advantageous in many applications because it:

(i) reduces human reading and interpolation errors,

(ii) eliminates parallax error,

(iii) increases reading speed when a large number of observations are to be recorded.

(iv) often provides outputs in digital form suitable for further processing or recording.

A digital voltmeter is a versatile and accurate instrument used in scientific labs or industry. The size, power consumption and finally cost of the DVM have drastically reduced. Their basic circuitry can be easily supplemented by additional circuitry to measure current, resistance and voltage ratios.

Digital voltmeters are classified into the following broad categories:

(a) Ramp type
(b) Integrating type
(c) Continuous balance type
(d) Successive approximation type

Here, we shall describe the working of ramp-type VOM in brief.

17.12.6 Ramp-Type DVM (Single-Slope Converters)

Principle The operating principle of this type is based on the measurement of time it takes for a linear ramp voltage to rise from 0 V to the level of input voltage or to decrease from the level of input voltage to 0 V. This time interval is measured with an electronic time interval counter and the count is displayed on electronic indicating tubes (LED or LCD).

The conversion from voltage to time interval is illustrated by a wave form diagram (Figure 17.21).

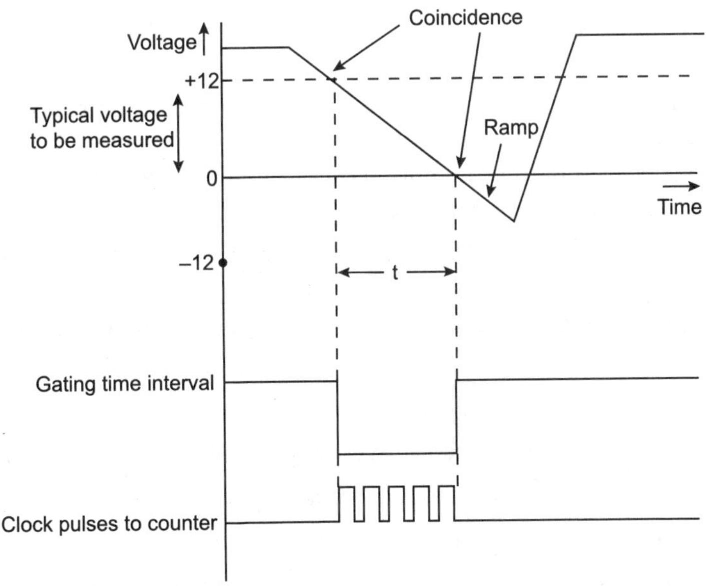

Figure 17.21 Voltage to time conversion using gated clock pulses

At the start of measurement cycle, a ramp voltage is initiated; this voltage can be positive-going or negative-going. The negative-going ramp shown above is continuously compared with an unknown input voltage. At the instant the ramp voltage equals the unknown voltage, a coincidence circuit, or comparator, generates a pulse which opens the gate. This gate is shown in Figure 17.22. The ramp voltage continues to decrease with time until it finally reaches 0 V (or ground potential) and a second comparator generates an output pulse which closes the gate.

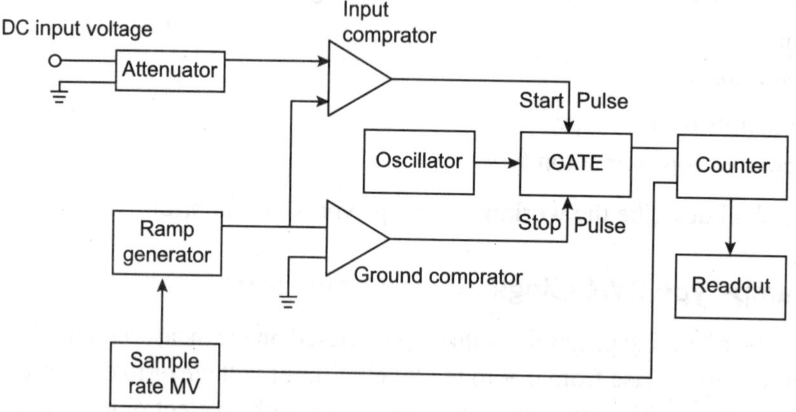

Figure 17.22 Block diagram of ramp-type digital voltmeter

An oscillator generates clock pulses which are allowed to pass through the gate a number of decade counting units (DCUs) which totalize the number of pulses passed through the gate. The decimal number displayed by indicator tubes (LED) associated with DCUs is a measure of the magnitude of unknown input voltage.

The sample rate multivibrator determines the rate at which the measurement cycles are initiated. The oscillations of this multivibrator can usually be adjusted by front panel control from a few cycles per second to as high as 1000 or more per second. The sample rate circuit provides an initiating pulse for the ramp generator to start its next ramp voltage. At the same time, a reset pulse is generated which returns all the DCUs to their zero state, removing the display momentarily from (LED/LCD) display.

Although simplicity is an advantage of the circuit but simplicity is also associated with certain limitations:

1. It can measure voltage of one type of polarity only.
2. Additional circuitry is required for over range condition.
3. The circuit is susceptible to oscillator frequency.
4. The circuit is susceptible to drift in constant current source.
5. Accuracy depends upon the stability of differential voltage that trips the comparator.
6. The converter is very susceptible to noise on analogue voltage.

To overcome most of these limitations, dual slope converters are used. They also use the linear charge concept of single slope converter.

17.13 Dual Slope Converter

Although the dual slope converter uses a capacitor charged by a constant current source to provide voltage to time conversion, the dual slope converter not only charges the capacitor but discharges it as well during measurement cycle. This charge/discharge cycle tends to significantly reduce the long-term drift and stability problems associated with single slope converter. Its schematic block diagram is shown in Figure 17.23.

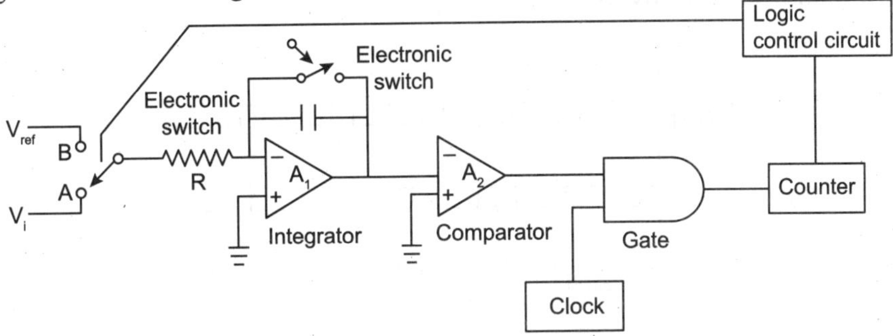

Figure 17.23 Dual slope A/D converter

17.13.1 Decade Counter

A decade counter is made up of flip-flops in cascade which counts in the base 10 number system. This means there is a sequence of 10 distinct counts the counter progresses through. Three flip-flops in cascade progress through eight distinct states (binary number 000 through 111) while four flip-flops in cascade progress through 16 states (binary number 0000 through 1111). Eight distinct states is too few while 16 is too many for a decade counter. This problem can be overcome by using four flip-flops in cascade by resetting the output of each flip-flop, so 0 after the desired 10 counts. Each flip-flop is a binary counter i.e. worth a 1 or 0 according to state high or low. If we start out with our flip-flops all in '0' state and supply a triggering pulse to the first flip-flop, it will change to the state 1. A second pulse will return it to 0 state but in doing so it will trigger second flip-flop to the state 1 and so on with succeeding pulses. Figure 17.24 shows four flip-flops in cascade.

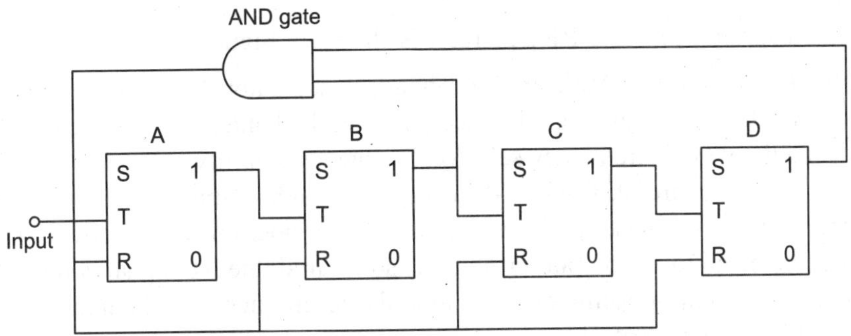

Figure 17.24 Block diagram of a decade counter

After 10 pulses have been applied to the circuit, *B* and *D* output will be high (i.e. binary 1), therefore, output in binary will be 1010. We wish to reset the output 0000 and repeat this counting sequence. This is the purpose of AND gate. When *B* and *D* are equal to 1, there is an output from AND gate that resets all flip-flops to 0.

The wave forms shown in Figure 17.25 show the pulse train applied to the trigger input (*T*) of decade counter and the output of each flip-flop is accordingly obtained as depicted by wave train at *A*, *B*, *C* and *D*, respectively.

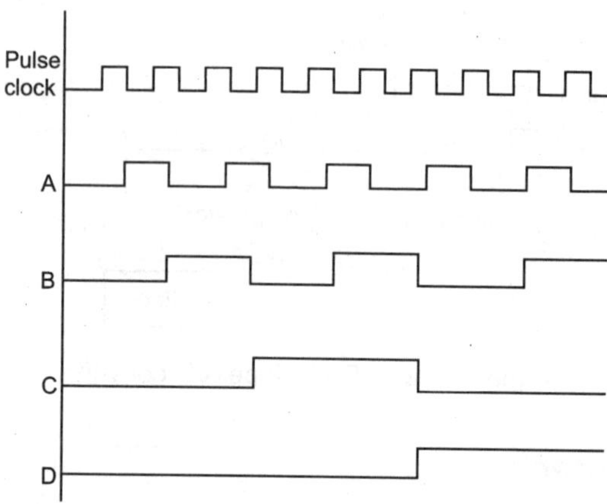

Figure 17.25 Wave form for decade counter

17.13.2 Radio Frequency Generators

The term "generator" is generally used to describe an instrument which is capable of providing a modulated signal, and in this respect it is different from an oscillator which provides only a output signal. RF generators are designed to provide an output signal over a wide range of frequencies from approximately 30 KHz to 3000 MHz.

RF generators contain a precision output attenuator network that permits the selection of output voltages from 1 μV to 3 V nearly in precise steps. The output impedance of RF generator is generally 50 Ω.

The block diagram of a basic RF generator is shown in Figure 17.26.

The circuit is that of a very stable, multiple band RF oscillator. The various blocks shown are self explanatory. The frequency range is selected using the band selector. The exact frequency is selected using the vernier frequency selector. Frequency stability is limited by the frequency stability of LC oscillator circuit. The low-distortion, sinusoidal output of the oscillator is applied to the broadband amplifier that amplifies the signal and provides buffering between the oscillator and any load connected to the output terminal. If so desired, the RF signal is modulated at the amplifier by the modulation oscillator. Both modulation frequency and percent of modulation can be adjusted by vernier control.

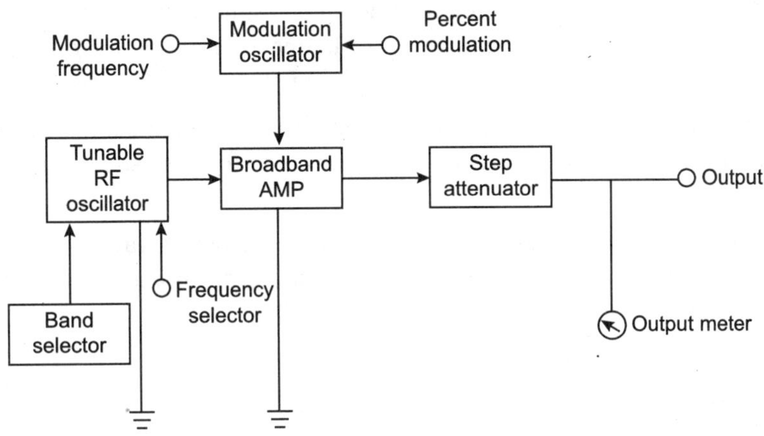

Figure 17.26 Block diagram of RF generator

The amplified output is applied to a step attenuator network. The output of attenuator is monitored by the output-meter which provides the user with an indication of signal level. The entire instrument is contained in a shielded cabinet. Many laboratory-quality RF generators do provide an inbuilt separator shielding for the oscillator circuit.

17.13.3 Function Generators

These instruments provide a variety of output waveforms over a wide frequency range. The most common output wave forms are (i) sine, (ii) square (iii) ramp, (iv) triangular and (v) pulse. The frequency range generally extends from a fraction of hertz to at least several hundred KHz.

Since the function generator provides many types of wave output, any one of these is chosen as primary wave form generated by the instrument. This primary wave form can then be applied to the proper circuitry to generate the remaining wave forms.

Generally, a sine wave is chosen as primary wave form generated by a *R-C* or *L-C* oscillator circuit. Figure 17.27 shows the schematic of a typical function generator.

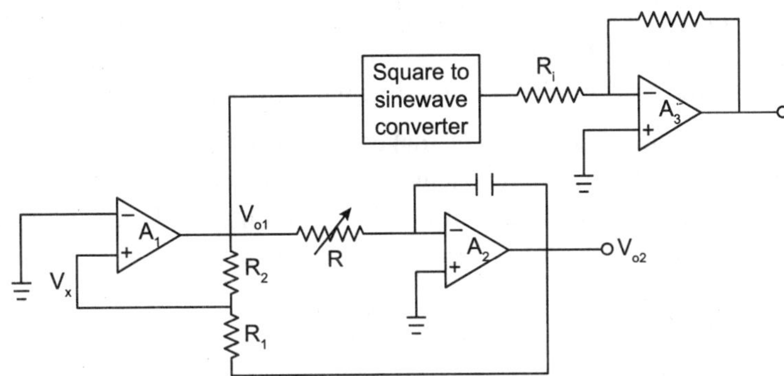

Figure 17.27 Basic circuit of the function generator

Here, the primary wave form in the circuit shown is a square wave. This is because some square wave generator circuits offer significantly better amplitude and frequency stability characteristics with simpler circuits than those required for sine wave forms.

As shown in Figure 17.27, the first stage *A*, which is a voltage comparator, generates a square wave output. Output *A*, is driven to saturation, therefore the square wave is either at $+V_{CC}$ or (–) V_{CC}.

The second stage, A_2, is an integrator that generates a triangular wave. The square wave is applied to a square-to-sine wave converter that filters out odd harmonics making up the square wave while passing on the fundamental sine wave.

17.14 Computer

Computers are available in many forms. There are some computers perform only one task routinely all the time; others perform many tasks either simultaneously or in sequence. For this reason, it is not always easy to give a succinct definition of the computer. A working definition of the computer can be adopted based on its properties:

(i) A computer is an **electronic device.**

(ii) A computer has **internal storage** and can store instructions and data very fast. That is why, the computer can execute programme statements in a sequence at very high speed.

(iii) A computer is capable of using a **stored programme**. A computer internally stores a series of instructions that detail the methods of handling specific operations and the order in which the operations are to be performed. This requires one to 'communicate' with the computer. The languages used in the communication are called **programming languages**.

(iv) A computer allows **programme-execution modification**, i.e., it can make decisions that change the course of a programme's execution, depending on the value of data in storage or on the results of arithmetic or logical operations.

Based on these properties, one can define the computer as: "an electronic device which possesses internal storage capabilities, a stored programme of instructions and also have the capabilities to modify the course of executing the instructions during processing".

Computers are available in many models, makes and sizes, from so-called single task versions (e.g. scanning machine) to the more sophisticated multi-task versions designed to handle jobs of high complexity. A computer system is composed of machines, and programmes for operating the computer. These are called the **hardware** and the **software**.

Hardware It is the physical machinery and a complicated web of electronic connections all interrelated and working in tandem to produce results of computation. In addition, there are also other components that fall under this category, e.g. scanners, printing devices, etc., which are also called **peripheral devices**, or simply peripherals.

Input/Output devices or just I/O devices are the most important parts for a computer system.

Software This is a general name given to the set of instructions that tell the computer system what to do. Sometimes, software is also called **computer programme** e.g. **word processing** and **spreadsheet** programmes. Programming languages enable the computer to communicate with the programmer. Raw data is fed into the computer; the computer then follows software directions for processing the data into useful information.

Computers may be classified according to their size and memory as follows:

(i) **Microcomputer or Personal Computer (PC)**: It is the smallest general-purpose computer system. These computers usually have a 8, 16 or 32-bit microprocessor and can execute programmes to perform a variety of instructions. An 8-bit microprocessor means that it can process 8 bits or 1 byte of data at a single given time.

Microcomputers vary greatly in size, ranging from the **desktop** to the **palm computer**. Desktops are available in two sizes: (i) 'tower' type where the main computer box stands like a tower, (ii) a version in which the display unit sits on top of the main electronic box. Both these versions suffer from the disadvantage that they are cumbersome to carry around. However, flexibility is provided by the '**laptop**' version. A laptop computer is small enough to rest on the user's lap. A specially designed carry bag makes it possible for the user to carry the machine around. Due to its portability, the laptop computer is becoming more popular. Like the desktop, laptop fully satisfies the computer needs of a single user. It is also commonly called as '**notebook**' computer. Still smaller, and of recent introduction into the market, the palmtop is limited in its application. Figure 17.28 shows the various versions of microcomputers.

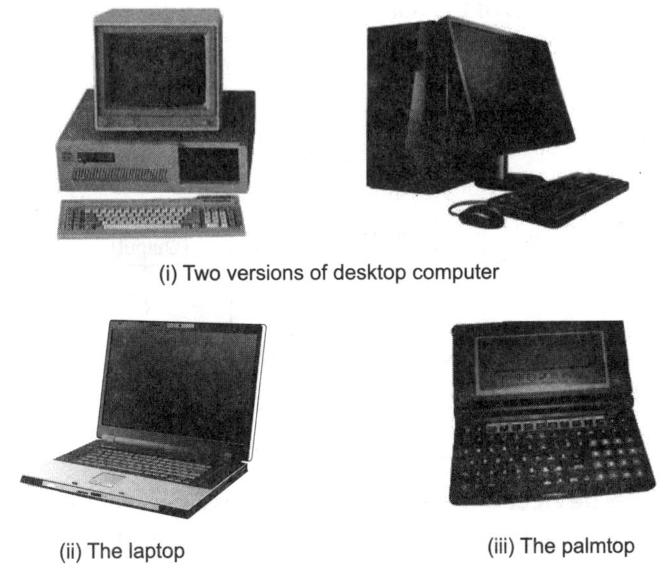

(i) Two versions of desktop computer

(ii) The laptop

(iii) The palmtop

Figure 17.28 Different versions of microcomputer

(ii) **Minicomputer** It is a small general-purpose computer with a 32-bit microprocessor. These computers are very useful in distributed data processing network.

(iii) **Mainframe computers** This computer can process data much faster, as several microprocessors are used in the place of a single one as used in micro and minicomputer systems. In the same unit of time, 2 to 8 bytes can be operated automatically.

(iv) **Super computers** These are the most powerful computers, of importance to research, defence, etc. These are very expensive and time required to execute a single operation may be of the order of nano seconds.

It is also possible to classify computers according to the data they process:

(i) *Analogue computers*: These handle data represented by physical quantities of continuously variable size such as current, voltage, temperature, length etc. It is easy to programme and costs less. It sets up a mathematical analogue of the problem. It has limited memory and is less accurate (1 part in 10^4).

(ii) *Digital computers*: It has very large memory and is highly accurate (1 part in 10^{12} or even higher). It can handle actual numbers expressed in digits and quantities in the problem are represented by discrete numbers. In these computers, pulse circuitry is used. It breaks down the problem into logical arithmetical steps.

17.15 Essential Components of a Computer System

A digital computer is made of the following main units: (a) Input/Output devices, (b) central processing unit (CPU) and (c) memory. The block diagram of a computer system is shown in Figure 17.29. A brief description of the main parts of a computer follows.

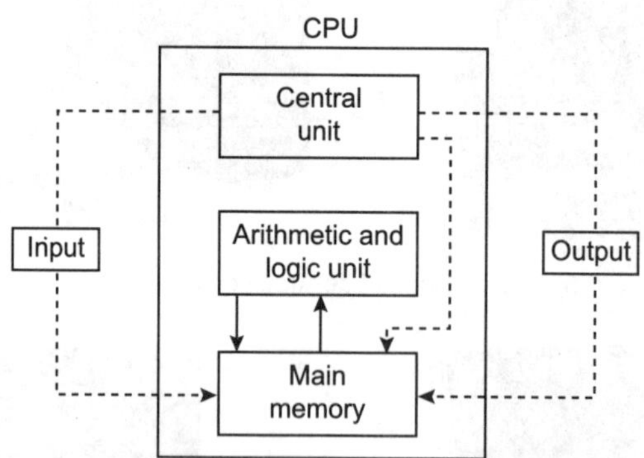

Figure 17.29 Block diagram of a digital computer

(a) Input-Output (I/O) devices These devices are necessary to get decimal and alphanumeric data into and out of the digital computer. The necessary data and instructions are fed to the computer through input devices. Input devices are basically of two types: (a) direct input, and (b) indirect input devices. Examples of input devices are keyboard, data entry terminal, barcode reader, touch screen (visual display unit), light pen, input tablet, mouse, voice drive, magnetic ink character reader and optical character reader. These units transform data and programmes into binary form for processing in the computer.

Output devices take the output generated from the computer in the machine code and either covert it into a form understandable by human beings or store it in some convenient manner. These devices are usually of three types: (i) displayed output, (ii) printed output and (iii) stored output.

(b) CPU It is the nerve centre of computers. Besides data processing, it provides a central control in the operations for computer. It consists of three components: (i) memory or main storage, (ii) arithmetic and logic unit (ALU) and (iii) control unit. All these components of a CPU are electronic circuits. Data processing is conducted by the arithmetic and logic unit, while the control unit directs all operations.

The sequence of operational instructions, known as algorithm, and corresponding data to be processed flow from the memory of CPU. The two units, control and ALU, are usually combined into one single unit fabricated on a semiconductor IC chip, known as *microprocessor*.

(c) File storage It is generally employed in large computer systems where bulk of information is to be stored. It is a permanent storage medium. In early days, magnetic tapes were used as storage medium.

17.16 Computer Programming

A computer does all calculations under the control of a stored programme. The programme consists of a set of instructions which tell the computer how to solve a particular problem step by step including all decisions which are to be made. The different stages in computer programming are as under: (i) analysis of the problem, (ii) algorithm design and (iii) representation of algorithms. For the analysis of a particular problem, one will have to define the problem precisely. One will have to identify the inputs and variables of the problem and desired output. Moreover, all the information required to develop a solution must be known prior to feeding a programme to the computer.

17.17 Computer Languages

Computer languages or programming languages may be placed under four categories:

(i) *Scientific languages*: FORTAN (Formula Translation), BASIC (Beginning All-Purpose Symbolic Instruction Code), ALGOL (Algorithmic Oriented Language), PASCAL and COMAL (Common Algorithmic Language).

(ii) *Commercial language*. COBOL (Common Business Oriented Language) and PL (Programming Language).

(iii) *Special purpose language*: LOGO, PROLOG, COROL and LISP.

(iv) *Multipurpose language*: PL/1, BASIC.

17.18 Microprocessor

It is a ccomputer except for memory and I/O peripherals. Basically, it is a processing unit fabricated on one (or very few) IC chips. Obviously, it contains the arithmetic logic unit, and control logic on a single chip. Presently, not only an entire microprocessor is fabricated on one single chip, but memory and Input/Output (I/O) are also included on the same chip to form an entire microcomputer.

Different types of microprocessor chips are available for a wide range of applications. These are used in simple instruments, calculators, industrial control, data processing acquisition, control and in minicomputers.

The exact structure of a microprocessor depends on the purpose it has to serve. The general structure of a microprocessor is shown in Figure 17.30.

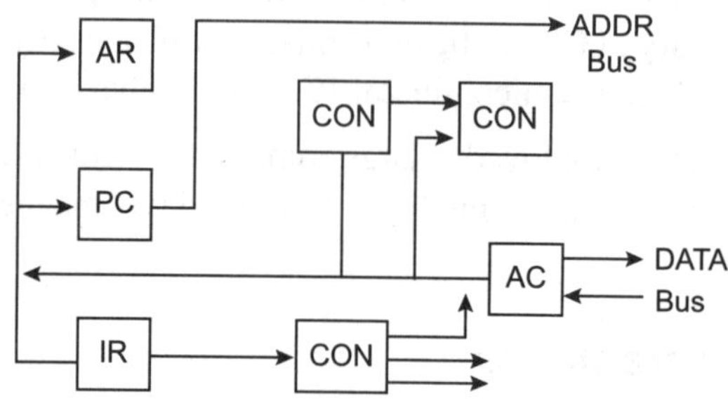

Figure 17.30 General structure of a microprocessor

Accumulator (AR) is a buffer register which can hold intermediate information during computer run. A buffer register can hold one word. Instruction Register (IR) is a part of control unit and its function is to fetch instructions from programme stored in the main memory. Programme Counter's (PC) job is to send to the memory the address of instruction to be fetched, and also keep track of instructions during execution progress.

Microprocessors mainly process three types of information: (i) instruction, (ii) address and (iii) data or operands.

17.18.1 Computer Data

The computer combines a very large number of logic gates, producing 1 when data is present and 0 otherwise. Obviously, computer data comes as a string of zeros and ones. This type of data representation is called **binary system** or **binary code**. Data presented using this code is called bit. Just as representing numbers in binary can lead to very long strings, so also expressing computer data in the corresponding bits is not practical. Computer data can either be in **bits** or **bytes** where 1 byte = 8 bits. One byte of data can be thought of as equivalent to a bank of four switches, each of which can be turned on or off. With huge amounts of data encountered, multiples of byte are commonly used. We have

1 kilobyte (or KB) = 1000 bytes = 10^3 bytes

1 megabyte (or MB) = 1,000,000 bytes = 10^6 bytes

1 gigabyte (or GB) = 1,000,000,000 bytes = 10^9 bytes

Representing the data in bytes offers an easy way of working with the data; it is also less prone to errors.

17.19 Computer Applications

The computer is an indispensable tool for our modern society. It finds wider applications in scientific, industrial and technical activities. The uses of modern computers may be broadly classified into the following three categories: (i) scientific and industrial calculations, (ii) industrial process control and (iii) business data processing.

Industrial process control involves the use of computers to enable automation of industrial operations, e.g. control of electricity generating plants, process involved in petroleum refineries, and automobile assembly plants.

17.20 Microcomputer Hardware

The computer hardware is as a collection of **system unit case**, **input devices**, **output devices**, **data storage devices** and other **peripheral devices**.

The system unit case is the main piece of hardware of the computer system. All devices are connected to the system unit case. Communications between devices go via the system unit case. This is why the unit case is considered the heart of the computer. Figure 17.31 shows the system unit case in the desktop computer.

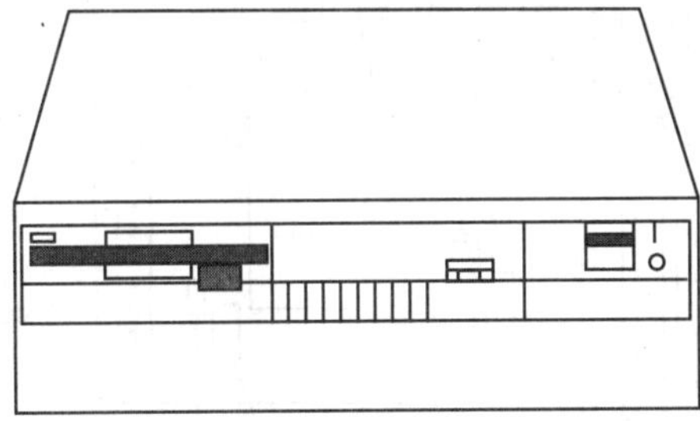

Figure 17.31 System unit case in the desktop computer

At the back of the system unit case are ports for connecting devices to the computer. Ports are of two types: **serial port** and **parallel port**. Normally, the parallel port is used for connecting printers, while the serial port is used for communication devices, e.g. fax and modem. A third type of port called **universal serial bus port** or **USB port** is very popular. Computers equipped with USB ports are generally faster at data communication.

Inside the unit case is power supply unit (PSU), the main circuit board called the **mother board** and internal memory media. The mother board appears to be a very complex electronic component.

It controls all the functions of the computer. All device connections go to the mother board (Figure 17.32). A motherboard has:

(i) **Central processing unit (CPU):** It is the microprocessor.

(ii) **Display chip:** It controls the functions of visual display unit.

(iii) **Memory area** consisting of **RAM (Random Access Memory)** and **ROM (Read Only Memory)**.

(iv) Connections for status indications

(v) Connections for keyboard, mouse, main power, peripheral components and disk drives.

Presently, sound card, modem and network card may also be part of a mother board.

The CPU is a small but very complicated electronic device. From the outside, it looks very much like a square or rectangular box measuring about 7 cm on a side. Its tiny connectors stick out of the sides like leges, allowing the CPU to communicate with other components of the computer. The CPU consists of millions of transistors, organized into groups. Each group carries out a specific class of functions.

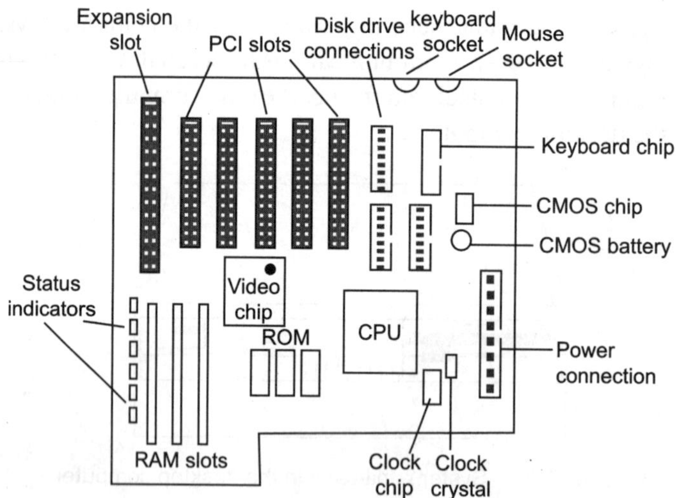

Figure 17.32 Block diagram of mother board in a typical desktop computer (top view)

One can summarize the functions of these groups as follows:

(i) **Programme counter**: This group keeps track of instructions the CPU must carry out.

(ii) Instruction **register** and **decoder**: This class of transistors stores instructions given to the computer and also interprets the instruction.

(iii) **Data register** and **accumulator**, which store information to be processed.

(iv) **Arithmetic** and **logic unit** that performs mathematical and logical operations.

(v) **Address decoder** and **address bus:** The processed information is sent to the bus prior to being sent out as output.

(vi) **Catche**, which holds data and commands that are about to be executed.

The CPU works by comparing pieces of information with each other and making decisions about the information, and storing the results of comparison. The comparisons simply involve determining whether a switch is **on** or turned **off**. This way the microprocessor carries out complex data manipulations and output the results.

One of the major ways of classifying modern computers is with respect to its operating speed. This speed indicates the rate at which the machine processes information. Typical operating speeds for the modern computer are in the range of nanoseconds.

17.21 Input Devices

The computer can process data into information at very high speeds. First, the data must get into the main memory of the computer. Getting data into the computer is called **input**, and the device used to get data into the computer is called **input device**.

An input device enables the user to 'talk' to the computer, and give instructions to it. Some common input devices are **keyboard**, **mouse**, **trackball**, and **joystick** (Figure 17.33).

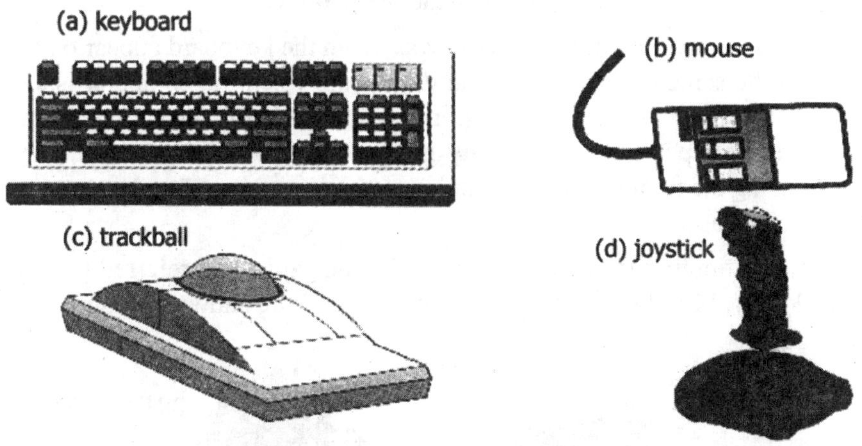

Figure 17.33 Some common input devices

Keyboard

The two major keyboard technologies in use are: (i) **capacitive** technology and (ii) **hard contact** technology. A capacitive keyboard consists of a printed circuit board that completely fills the space under the keys. Above the circuit board and below each key is a capacitor. Current flow in the circuit containing the capacitor is altered each tine a key is pressed and released. This information is sent by the keyboard chip (keyboard controller) to the CPU. The information is then stored in RAM and displayed on the screen. Hard contact keyboards consist of a series of switches that complete the circuit each time a key is pressed. The information is then sent to the computer similar as the capacitive keyboard.

We may note that the keyboard is not always the most efficient way of entering data into a computer. Modem microcomputers are equipped with a '**pointing**' device. This has made computer use very easy. The pointing device places a **cursor** on the computer screen. Taking the cursor to any active area of the screen and clicking issues a command to the computer.

The **mouse** has in it an electronic circuit that communicates with the processor. Through this communication, the user merely has to move the mouse cursor to an item, and a click of the mouse activates the command. Other pointing devices include the **track ball** and the **joystick**, both employ a similar principle of operation.

In addition there are other input devices such as the barcode reader. Using infrared light, this gadget reads the price coding on an item, and feeds the price information directly to the computer. The **electronic pen** is also used for inputting information directly to the computer.

17.22 Output Devices

Computer output can be either temporary or permanent, or can take the form of an action performed by a mechanical device. The most common of output devices are the **monitor**, sometimes called **VDU (visual display unit)** and the **printer**.

VDU gives temporary output. Characters typed from the keyboard appear on the screen of VDU. Output on the screen is temporary and disappears when the computer is switched off. Another temporary output is the **voice response system**. When certain conditions are met, computer systems equipped with this facility give out speech output. Warming systems and elevators announcing the floor level are examples of voice response outputs. Outputs from such devices are described as **soft copy**.

We may note that output from the printer is often on paper, and referred to as **permanent output** or **hard copy**. Monitor and printer (output devices) are shown in Figure 17.34.

Figure 17.34 Output devices

17.23 Data Storage

Computers have the ability to store data. This storage can be programme instructions, raw data before processing, or processed data that would be required later on. Programme statements or data can be stored on a **storage device**. When required for use, the stored programme or data is transferred from the storage unit to the **main memory** of the computer.

During programme execution, the computer stores data in RAM. Data stored in RAM constitutes only a small part of computer memory. RAM consists of **memory modules**, called **single inline memory module (SIMM)** or **double inline memory modules (DIMM)**. These are often colloquially called RAM sticks. Data stored on RAM is only available during programme execution.This data disappears when the computer is turned off. This is why RAM is described as **volatile memory**.

In modern computers, there is an important type of RAM, known as **CMOS** (Complementary Metal Oxide Semiconductor). This semiconductor is powered by a little battery, typically about 3 V. The CMOS chip holds configuration information about the system. For instance, it holds information on the type of monitor, specifications (i.e. cylinders, sectors, tracks, clusters, etc.) of the hard disk drive, and RAM size and allocation. This CMOS chip can be found on the motherboard, usually near its battery. Being battery powered, CMOS does not lose its contents even when the computer is turned off.

At any given point of time, only a small percentage to data need to be available to the computer for use. The rest of the data is stored in external memory devices. This memory is called **secondary memory**. Also called **auxiliary storage**, this memory is cheaper and takes longer to access than RAM. Data stored in secondary devices is permanent and hence called **nonvolatile memory**. It's quicker to access data on secondary storage than to input data all over again.

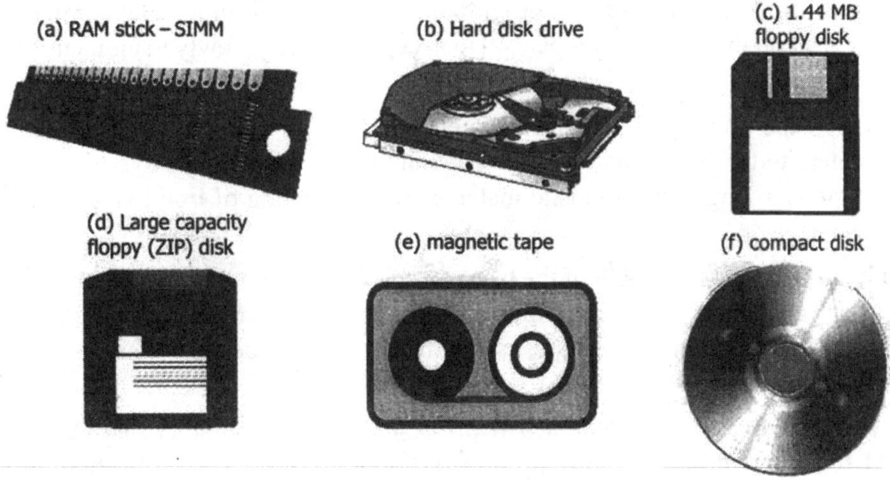

Figure 17.35 Some memory media

Secondary storage media are of many types. Storage media are materials on which data is recorded. Special drives are used to read information from the storage medium into the computer. Some secondary memory media are shown in Figure 17.35.

Several magnetic media are available to store computer data. These can either be in the form of a disk or tape. **Hard disks** are examples of magnetic storage. The computer technology has introduced '**false memory**'. These come as little electronic cards or sticks and are becoming very common at present.

Data on magnetic disks are arranged in **tracks** and **sectors** (Figure 17.36). Sometimes, data in a file can extend from one sector to another to form a cluster.

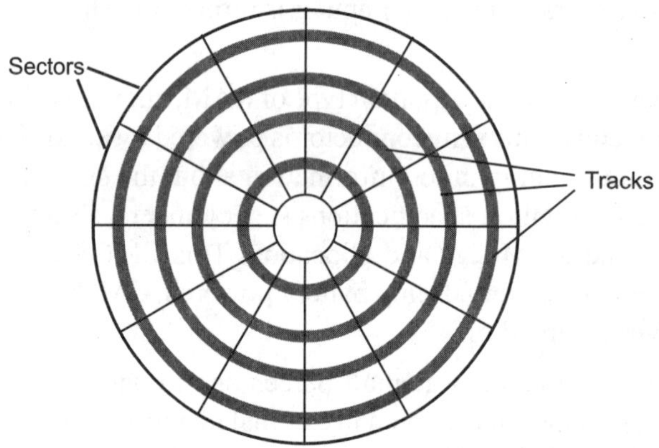

Figure 17.36 Tracks and sectors on magnetic storage disk

Prior to recording, the disk must be electronically prepared to accept data. This process is called **formatting**. A special computer programme carries out this procedure. Formatting erases any existing data on the disk, and prepares it to accept new data. Newly manufactured disks are usually unformatted. However, disks available presently in the market are already formatted.

Magnetic tape storage is the least expensive of high capacity storage media. Most microcomputers today store data on hard disks, but many data backup systems are magnetic tape. The **tape** is a long strip of plastic material, with a coating of iron oxide compound that can be magnetized.

Data is stored along the length of the tape at a specified density, measured in bytes/cm. One byte stores one character. The rate at which data can be retrieved from the tape depends on tape speed. A disadvantage of magnetic tapes is that data access can be very slow, as data is usually accessed sequentially.

The **compact disk** has become a part of the basic microcomputer hardware and called **CD-ROM (compact disk read-only memory)**. Large graphics and application programmes are available for sale on compact disks. One can store upto 700 MB of data on these media. A special drive is used to access data stored on the disk.

DVD (**digital video disk**) is now becoming common. These hold much larger data than the CD-ROM.

17.24 Other Peripheral Devices

Peripheral devices have greatly added to the capabilities of the computer. Figure 17.37 shows some common peripheral devices.

Modem (Modulator-Demodulator) This enables the computer to communicate with other systems over long distances using the telephone line. One can use the modem to send faxes and communicate with other computers over the internet. The modem can be either external or internal. Several computers are equipped with the modem as part of motherboard.

Scanner This is used to transfer documents into the memory of the computer. It works in a manner similar to the photocopier, with one difference that the scanner converts the documents into electronic form and saves it on memory device. This has proved very useful in transferring copies, paintings, maps, legal documents, etc. into the computer.

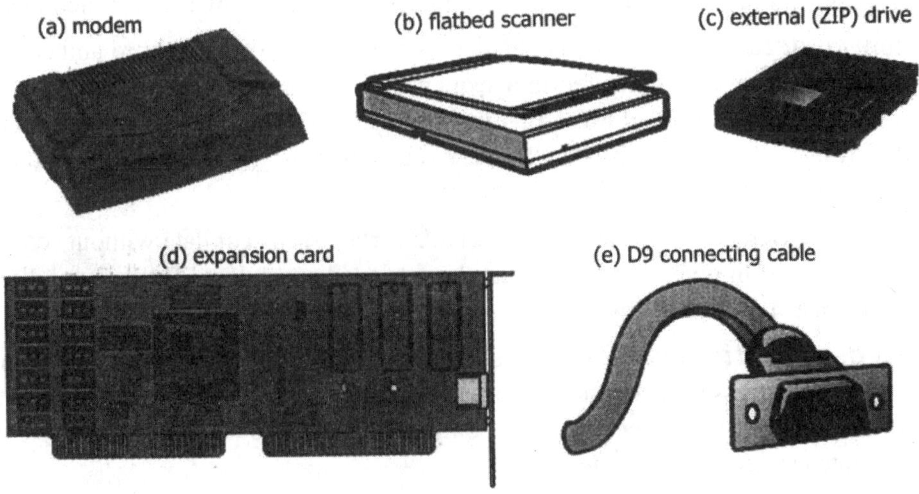

Figure 17.37 Some common peripheral devices and their connectors

There are also a good number of **peripheral devices**, e.g. high capacity hard disk drives, CD writers, digital cameras, etc. Peripheral devices are always available with the following:

(i) A means for connecting them to the host computer. Some devices use **adaptor cards**, while others use a direct cable connection. Figure 17.37(d) shows how an **interface** looks like. This goes into one of the expansion slots in the unit case. There is also direct cable connection, which uses output **ports** at the back of the unit case, e.g. D9 cable having nine connecting pins.

(ii) A device driver for operating the device, usually a programme on a floppy disk or CD-ROM. This device driver alerts the computer of the presence of the peripheral, and allows communication between the device and the computer.

17.25 Organisation of Data

Information in secondary storage is organized into files. The file may be a **word processing file**, a **database file**, or the output of any software programme. Different files have different structures and different characteristics. These differences enable the computer to recognize exactly the file being requested by the user.

The computer can be commanded to perform certain basic operations on data. These operations include **save**, which saves the data onto a secondary memory device; **load**, which loads the data on the memory device for immediate use; **delete**, which erases the data from the file or memory; and **print**, which sends the data in a file to a printer.

The two main ways of accessing data are: (i) **sequential access**, and (ii) **direct access**. **Indexed sequential organisation** represents a hybrid between the two access methods.

(i) Sequential data access In this, the data is retrieved record by record and each record is read. The records are usually stored in a sequence in which they were entered into the file. This access method is also called **First-In-First-Out (FIFO)** method. This type of data access has the disadvantage of being very slow because the system has to scan through all the data in its memory in order to get the one being requested.

(ii) Direct data access This allows access to a certain record or data without sequentially checking every data. In other words, data can be retrieved at random. This is why direct data access is known as **random access**.

Indexed sequential access This is used when a large percentage of the data in the computer is to be processed each time the application is run. Direct or random access is used in situations where only a small percentage of computer data is to be addressed at any one time. The files of interest need to be located quickly.

Indexed sequential accessing is a combination of the two. In this the files are organized such that information in them can be accessed both sequentially and randomly. Files organized for indexed sequential access have a separate index file that lists the key for each record. Some database files run in this way.

17.26 Microcomputer Software

For all its complexity, the hardware by itself is almost useless. The software holds the key to all the marvels of computer performance. Without software, the computer is just an assortment of complicated logic circuits. Software is a programme of instructions written in such a manner that the computer can understand and execute.

A computer can perform many functions depending on the software that drives it. Software programmes are written for specific hardware systems. For instance, the common brands of computers in the market are **Apple** Computer, **IBM** Computer, etc. Some software are written for IBM, while others are written for Macintosh. Fortunately, many software developers do write two versions of the programme for each hardware system.

There are four main categories of computer software: (i) **firmware**, (ii) **operating systems**, (iii) **utility programmes** and (iv) **application** software.

(i) Firmware This is the lowest level of software, considered a hybrid between hardware and software. This is also called **BIOS (Basic Input/Output System)**. This software is coded in the ROM on the motherboard, and is not under user control. BIOS performs non-user functions such as checking the computer's components when the machine is switched on. This programme checks all the components connected on the computer. The testing process is called **POST (Power On Self-Test)**. The programme also tests the integrity of communication ports and all input and output devices connected to the computer.

(ii) Operating systems It is a software programme that controls the computer and allows other programmes to run on the machine. Without the operating system, the computer cannot do anything meaningful. The operating system enables the computer to perform vital operations, e.g. savings files into the memory, reading and writing files, and changing the names of files.

Microsoft designed one of the oldest operating systems, named **MS-DOS (Microsoft Disk Operating System)**. One disadvantage of DOS is that the user needs to learn several commands to use it.

Apple Macintosh introduced the **GUI (Graphic User Interface)** which made computer use very easy. Programmes with GUI display the parts of the software as active cells (called icons) on the computer screen. Using the mouse pointer, the user selects a desired action by moving the mouse pointer to the cell and clicking on the button of the mouse. This clicking action activates the cell, and tells the computer to execute an action such as saving a file, deleting a file, or renaming a file. Microsoft introduced the GUI operating system named **MS Windows**. This has been modified greatly over the years.

(iii) Utility programmes These are programmes that help the user manage computer hardware and software. These programmes are also known as **utilities**. Utility programmes are designed to improve the performance of the computer. Some common utility programmes are hard disk cleanup programmes, file defragmentation programmes, memory management programmes, and file **zipping programmes**.

Disk cleanup programmes erase files from the hard disk that are no longer needed or bits of files left by programmes run previously on the computer.

Defragmentation programmes reorganize the data area such that materials that belong to the same file are kept as close to each other as possible. The computer would run slow if, for instance, one paragraph of a document were in another memory location from the rest of the document.

Memory management programmes set areas of internal memory for specific types of data. A file zipping programme condenses the file-process described in computer language as **zip**. A zipping programme can condense a file to less than $1/10^{th}$ in actual. Complementary to this is the **unzip** programme, which works on a zipped file and opens it to its normal size.

(iv) Application software Most computer users today use **application software**. In this category of software is lumped all programmes that allow the user to do something with the computer.

There is a wide range of application software. The major categories are:

(i) **Word processing** A word processing software allows one to edit and print letters, presentations reports, and manuscripts. The most popular word processing programmes are MS Word and Word Perfect.

(ii) **Electronic spreadsheets** This consists of an arrangement of rows and columns. It is used to do calculations on data entered into its cells. A computer spreadsheet is often known as electronic spreadsheet. Entries into the spreadsheet can be number (e.g. 5.0), text, or formula. A typical spreadsheet showing some columns and rows is shown in Figure 17.38. A formula is entered in cell E3, where cell identified by column E and row 3.

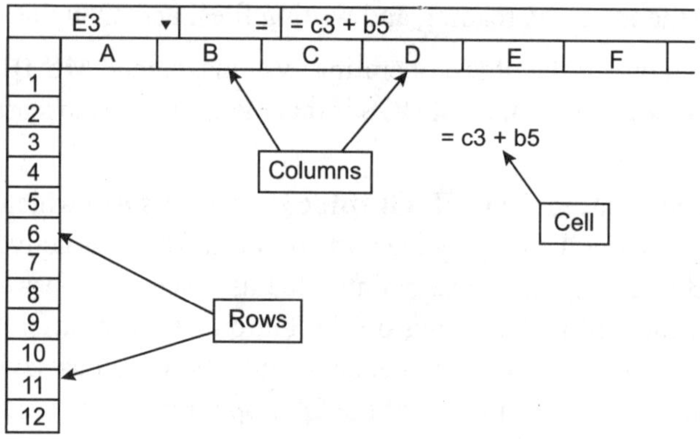

Figure 17.38

(iii) **Data management** Database programmes find a wide range of applications e.g. police department, big business houses keep database of their assets, clientele, etc.

(iv) **Graphic programmes** These allow one to draw, import pictures from peripheral devices such as scanners and digital cameras. For the most part, one uses graphic programmes to produce pictorial representations of work. Modern graphic programmes are often referred to as **CAD (computer aided design)** packages.

(v) **Communication** Programmes that allow one to communicate using the computer are called **communication programmes**. With these programmes one can send and receive faxes, e-mails, etc.

(vi) **Programming languages or development languages** This falls into two categories **low-level** languages, and **high-level** languages.

Also referred to as 'machine code' or assembler language, low-level languages address the basic circuitry of the microprocessor. A microprocessor contains a large collection of transistor switches and logic gates. A low-level language often addresses these switches directly. Microprocessor-based appliances, e.g. washing machines, intruder alarms, etc, are programmed in machine code. Computer users seldom use machine code.

Programming languages for everyday use are high-level languages. These languages are converted into machine code in the process of compilation. **BASIC (Beginners' All-purpose Symbolic Instruction Code**) is the most common of these. Other programming languages include **FORTRAN (Formula Translation)**, PASCAL, etc.

17.27 Software and Hardware Compatibility

Hardware can only accept software of a particular type to work on it. Similarly, software can only work on a computer system for which it was designed. Maching software and hardware to work together is known as **compatibility**.

Each peripheral device comes with its own driver software. This software enables the computer to communicate with the peripheral device.

Compatibility also involves having software to work with other software on the same hardware. Some software tend to conflict with each other. Often it becomes necessary to remove one software off the system in order to use the other.

17.27.1 Coordinating Software and Hardware

Three main factors are to be considered when mixing and maching software and hardware to work on a computer system.

(i) Operating system This is a programme that controls the computer and makes it possible for the user to run other programmes on the computer.

(ii) RAM Every software package needs minimum RAM requirement to run, which is often indicated in the documentation of the software.

(iii) Disk drive Most applications in use today are large programmes, and in operation, they need to make frequent **disk calls**. Disk calls are computer equivalent of 'visits' a programme makes to store and retrieve data to and from the hard disk. Any programme that uses a graphic user interface needs disk space large enough to support innumerable disk calls. In addition to RAM, the user needs to have enough hard disk space for such calls.

17.28 Working of a Computer

Power supply Power supply for the computer is provided either by a standard source, or a battery. The power supply unit converts the AC main voltage to +5 V, –5 V, +12 V and

– 12 V of direct current and distributes it to the motherboard and disk drives through connecting cables. The motherboard then distributes power to the components mounted on it, and to certain peripheral devices as required. Sometimes, power surge can damage the computer.

To avoid damage due to power surge, the use of surge protectors, and voltage regulators is advisable.

Data input Data is input into the computer through an input device such as scanner, keyboard, mouse, modem etc. Each device is assigned a data path and memory location in RAM. An **interrupt** signal is sent to the CPU. This interrupt informs the CPU that data is being communicated from an input device. If the signal makes sense, the computer handles it according to instructions in the BIOS (Basic Input/Output System), operating system or an application programme. Most often, this handling involves storing the input data in a RAM position.

The rate of data transfer from the input device has to match that of retrieval by the CPU. This rate of data transfer is called **band rate**. If a modem is set at 32,000 band and the communication software is set at 32,000 band, problems of reception will arise with the CPU not making sense of data.

Data processing Data and instructions for the CPU are placed in RAM to process. When a programme is to be started, the CPU 'looks' for appropriate RAM locations, reads the instructions it finds there and executes them in the order designated in the programme. To achieve this, the CPU uses the BIOS and disk operating system.

Once the programme is loaded into RAM, the CPU executes its commands. It might, for instance, clear the screen and show the user interface, and wait for input from the keyboard or mouse. Editing a file involves a similar process of reading from the disk, storing the data into RAM, and moving the beginning of the document of the portion of RAM that corresponds to the screen. Text or figure can then appear on the monitor. Each time a key is pressed, the CPU interprets the key code and stores the data in appropriate sections in RAM and on the screen.

Output The CPU sends the data to a memory location dedicated to a specific device. The data is then transferred to the output device through an interface card, serial connector or parallel connector.

On reaching the device, the data is processed again by the circuitry of the output device to give the output in desired form. Indeed, the output system is very complex due to the many activities that need to be coordinated simultaneously. Coordinating the system unit and output devices is critical to successful computing.

17.29 Communication with Computers

One can link any two computers together using telephone lines and the necessary communication equipment and software. Merging communications with computers is the subject known as **telecommunication**. Telecommunication gives people access to a wide range of information and services.

17.29.1 Data Communications

The basic tools needed for data transmission between two locations are: (i) a **sending device**, which can be a modem or a straight wire connection between the computers, (ii) a **communication link**, which may be a telephone line or fibre-optic link, and (iii) a **receiving device** connected to the computer receiving the data. Modem-to-modem communication is normally used for computers in remote locations.

17.29.2 Data Transmission

We know that computers only generate 'digital' data. It would be easy if all components of the communications system work on this principle. However, the telephone cable network on which the system relies was constructed for voice transmission, and hence it is designed for analogue signals. For signals to be sent through analogue lines, the data should be converted into analogue form. An analogue signal is in the form of a transverse wave. Converting this into a digital signal might, for instance, mean designating '1' for maximum displacement and '0' for zero displacement. Converting from digital to analogue signal is referred to as **modulation**, and the reverse process is called **demodulation**. For this the modem is used.

Many computers have modem circuits as part of the motherboard. These computers offer a socket for plugging the telephone line. For two modems in remote computers to communicate with each other, they both must send data at rates not too fast or too slow, so that every bit of data can be received as sent. The word 'handshake' is often used to describe this relationship. The technical term **band rate** is used, which refers to the number of bits of data transmitted per second. Presently modems are available that can transmit at much higher speeds, typically going up to 33,600 bps, 56,000 bps or more.

17.29.3 Communication Links

Some of the hardware necessary to make a communication link are:

(i) Twisted pairs These are inexpensive, and used for data communication over short distances. However, they are susceptible to electrical interference, called noise, which causes distortion in the signal.

(ii) Coaxial cables These are more expensive but are less prone to electrical interferences. A coaxial cable consists of bundles of insulated wires within a shielded enclosure. The radio frequency cable is an example of coaxial cable. Coaxial cables can transmit data much faster than twisted pairs.

(iii) Fibre optic cables Optical fibres are designed to guide light waves along their length. An optical fibre works on the principle of total internal reflection. When light enters from one end of the fibre, it undergoes successive total internal reflections from the side walls and travels down the length of the fibre along a zigzag path. Fibre optic cables have been used extensively in data and telephone communications. In these cables, the electric signal is transformed into light pulses of fixed frequency, and sent down the core of fibre-optic cable. Being non-conducting, these cables are not affected by electrical interferences. The broad bandwidth (i.e. range of frequencies the device can handle) of fibre-optic cables translates into a wide range of multimedia opportunities. These cables are suitable for handling all types of data including voice, pictures, music and video, all at the same time.

(iv) Microwave transmission This uses line-of-sight transmission of data signals through the atmosphere. Microwave transmission uses antennas situated at convenient distances from the microwave relay station. The microwaves received are converted into digital signals and fed to the computer for data transmission.

(v) Satellite transmission Geostationary satellites are used for transmission of data between two earth stations. These satellites are in orbit at a height of about 35,680 km above earth. At this height, the satellite appears stationary from a given point on earth.

Once the communication links are established, there should be a **protocol** to be addressed before two computers can 'talk' to each other. The **ISO (International Standards Organisation)** has defined a set of communication protocols called the **Open Systems Interconnection (OSI) model**. Manufacturers of communication systems adopt this or other standards based on the same principles. That way, a modem can work on a computer system and communicate with other modems even though produced by different manufacturers.

17.29.4 Local Area Network (LAN)

When many computers are to be used in close proximity, it becomes cheaper and more convenient to **network** them. The major advantage of networking is that computers can share **resources**. Networking has made working in groups very easy as each computer can share the peripherals like the printer, scanner, etc. Offices, banks, schools, colleges, etc. already have a LAN (Local Area Network). Few basic requirements of networking computers are (i) hardware and (ii) software.

(i) Hardware On the hardware side, a LAN consists of the following:

(a) **Server** It is a another computer with large hard disk space and a fast CPU.

(b) **Clients** These are computers connected to the server for sharing data and resources. A client can have its own hard disk or be without one. In the latter, the client is referred to as **diskless** station, and would use programmes installed on the server. Clients are also called **work stations**.

(c) **Hub** It contains the essential circuitry enabling the clients to communicate with each other through the server. At the back of the hub are sockets for connecting cables from each computer, and at the front panel are indicators corresponding to each socket being plugged in.

(d) **Connecting cables** In most networks, computers are connected to the hub by eight-core cabling, similar to telephone cables.

(e) **Peripherals** These may include external hard disks, modem, printers, and other devices.

(ii) Software The software control of a network system is more complicated than the hardware. It begins with installing a **network operating system** on the server, e.g. **Windows NT, Netware,** and **Linux**. Then, operating systems compatible with network software are loaded on workstations. Every workstation on the network is identified so that the server 'knows' which client is in which room. All these assignments and many of the details of network operations are in the hands of the network administrator.

17.29.5 Wide Area Network (WAN)

A WAN takes many forms. It covers a large geographical area which spans a country or continent or the whole world. In WAN, communication over long distances is through cabling, microwave, ground radio system or satellite connections. There are two examples of WAN. In first, one uses a single server, sometimes called **host computer**, and many workstations are established over long distances. The computers communicate with each other through the host computer. There is strong similarity between this system and LAN. The other type of WAN arrangement uses many LANs, each with its own server, with all the servers linked to a central server. This is the type of WAN that exists in many educational institutions (Figure 17.39). Often, the main server has a communication device such as modem connected to it.

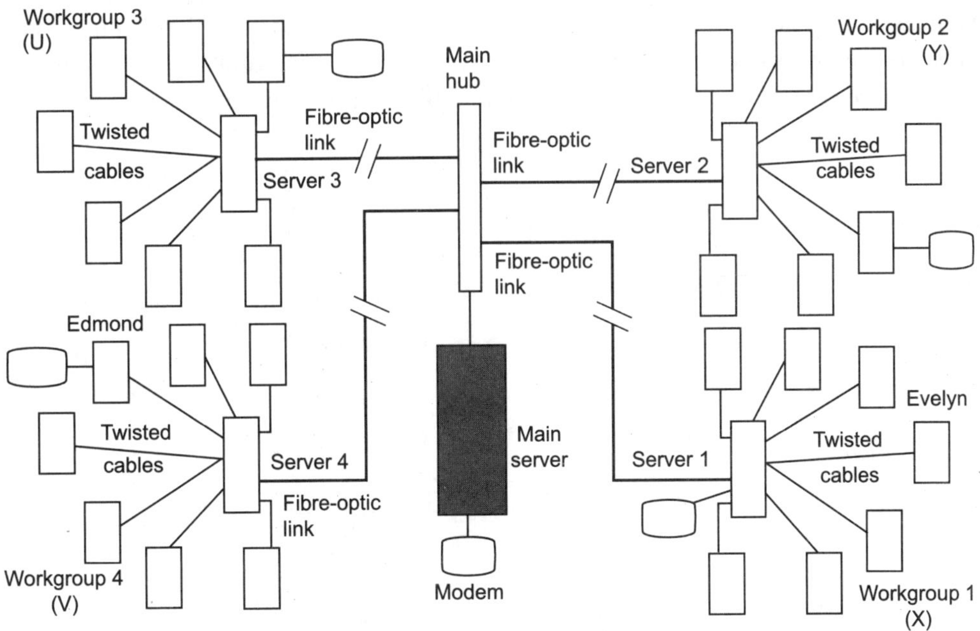

Figure 17.39 A block diagram of wide area network

Managing a WAN is a highly complicated. It requires expertise in computer hardware to put together all the necessary components and get them working properly. Moreover, it also requires skills in software management.

Besides LAN and WAN, there is also Metropolitan Area Network (MAN), which is used to increase connectivity among communities, multiplexes and trunks. MAN interconnects several access subnetworks.

LANs, MANs and WANs are distinguished from each other based on their size, data transmission technology and network topology.

17.29.6 Network Topology

It describes the ways how computers are connected. There are several network topologies but the most important and widely used are **bus, ring, star** and **tree** (Figure 17.40).

In a **bus network**, every node is allowed to receive every transmission on the network. As only one channel is used to serve the complete network, the whole network may go out of operation if this channel fails. In a **ring network**, data flows in a circular fashion in one direction only with one node receiving the data and relaying it to the next node in the ring. In the event of channel failure between two nodes of the ring, the entire network collapses. However, bus and ring network connections are common in LANs. In star network, there is a hub at the centre through which all traffic is routed. In the event of failure of hub computer, the network will collapse. This network is normally used in WANs. Today, the simpler and more common topology is **tree network** or **hierarchical network**. In this configuration, the top-most mode exercises more control on the network and so reliability problems may arise.

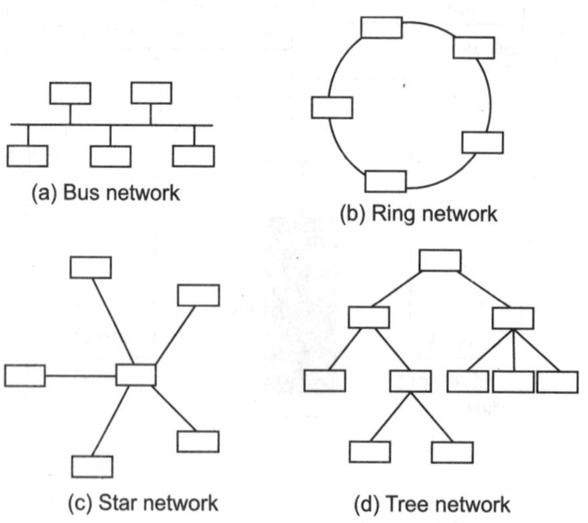

Figure 17.40 Network topologies

17.30 Internet and Web Technology

The internet is considered the ultimate wide area network. With the internet, a computer is linked to any other similar system at any location in the world. Obviously, internet gives its users the power to search information from anywhere in the world. With millions of computers connected together and sharing information, one can get practically any type of information on almost any subject. The interconnection of multiple networks into a single large network is called an internetwork or **internet**. Clearly, internet is a network of networks, formed by using special-purpose computers called **routers** to interconnect the networks.

Figure 17.41 shows the schematic of an Internet. These networks could consists of various types of **LANs** and **ATM** (asynchronous transfer mode) networks of even a single point-to-point link. The internet permits different networks to coexist and interwork effectively. Across an inter-network, the Internet Protocol (IP) has been developed to provide for connectionless transfer of message packet. Packet switches (dedicated special-purpose computers) called **gateways** or **routers** interconnect several computer networks. Routers can interconnect LANs and WANs. The process of selecting a network over which a packet message is sent is termed **routing**. This task is performed by IP routers which transfer message packets across an Internet. After a routing decision is made, the packets are placed in a buffer to wait for transmission over the next network. In effect, packets from different users are statistically multiplexed in these buffers. We may note that the underlying networks are responsible for transferring the packets among routers. The routers obey a routing protocol and choose the most efficient path to reach the destination. A **routing table** is provided to enable the router to choose an optimal path to the destination.

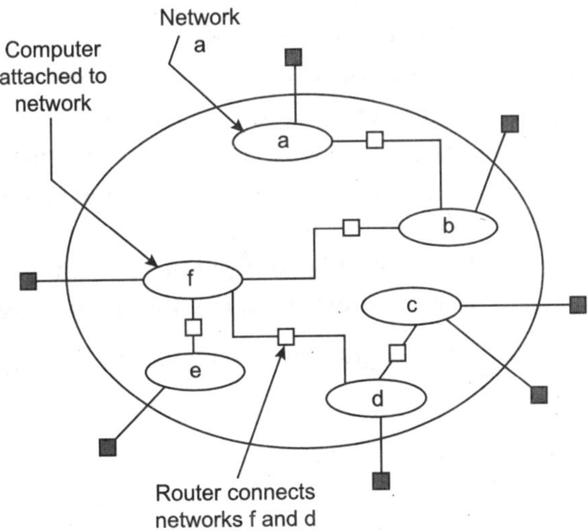

Figure 17.41 Schematic of an internet

Information is transmitted in the form of packets through the internet. Each packet contains the **header** and playload. The header contains the source and destination address, size of data, data type and sequence number, etc., and the advantage of this header information is that wherever the packet resides in the network, the network wherefrom the packet has come and where the packet should go. Obviously, the network can store the packet for quite some time and then forward it to the destination address. This mode of message transmission is termed **store** and **forward** routing. The other important part of the packet is information content in the packet, i.e., **playload**. When a link is disrupted due to any reasons, the incoming packet may be stored and delivered when the link is restored.

17.30.1 Connecting to the Internet

In order to connect a computer to the Internet, one needs the following:

(i) A **modem** to be connected and installed on the computer. This modem connects to the telephone cable.

(ii) **Browser** This performs the job of translating the data from one computer into a form any other computer can access. Examples of browser software are **Netscape Navigator**, and **Microsoft Internet Explorer**.

(iii) **Internet Service Provider (ISP)** A computer user has to apply to an ISP for connecting to the Internet.

17.30.2 Looking up Information

Information in the Internet resides on computers, which in turn are attached to servers. Every server has a **website** address. Users **log on** to this address in order to get the desired information. A website address is also known as **URL (Uniform Resource Locator)**. A typical URL has information on the country in which the server is, the type of server, the name of the server, the Internet service, and the Internet protocol being used. An example of website is http://www.daltron.com.pg. Pg indicates that the server is in Papua (New Guinea). The second segment (com) indicates a commercial service. There are many kinds of Internet services, some are set up by government (.gov), some by organisation (.org), etc. In our example, daltron is the name of a company. The segment **http://www** refers to the internet service; www. stands for the **World Wide Web**, and http stands for **hypertext transfer protocol**. With this protocol, a document is transformed into the language of the Internet (**hypertext markup language** or **html**).

Some of the organizational and geographic domain names are as given below:

au	Australia
jp	Japan
uk	United Kindom
in	India
org	non-profit organisation
edu	educational institutions
com	commercial organisations

gov	government agencies
mil	military organisations

To find information on the internet, one needs to know the URL. Otherwise, the help of **search engines** is taken. Search engines are special websites whose job is to search all the websites for information one needs. Examples of search engines include yahoo!.com, altavista.com, geocities.com, google.com, etc. By 'visiting' any of these search engines, one would be required to type a keyword that identifies the subject one wants to find information on. The search engine then returns all possible results of the search.

Millions of pages of information are stored in a number of servers in the Internet. Internet thus provides user-friendly ways of communicating in the **cyberspace**.

17.30.3 Electronic Mail (e-mail)

E-mail is the most widely used service provided by the Internet. It allows a person (sender) to type a message using one desktop and send it across the Internet to another person (receiver).

An e-mail user must have a **mailbox** to receive e-mail. However, a computer connected to the Internet requires an e-mail programme before a user can send or receive e-mail. This software has some basic functions, e.g. compose a new message, reply to message, notification of arrival of any incoming e-mail message, read the message, forward a message to other users, and maintain an address book.

An e-mail address consists of a string of characters separated into two parts by at symbol (@). The prefix identifies a particular user, and the suffix indicates the domain name of the computer on which the user's mailbox resides, e.g. asian@nda.vsnl.net.in. It means Asian has an e-mail account in the mail server 'nda' of vsnl network belonging to 'in' domain. We may note that Internet e-mail messages can only support ASCII text. However, the **Multipurpose Internet Mail Extensions (MIME)** format was developed to send non-ASCII file as attachment to e-mail messages. We may note that to transfer non-ASCII message, both the sender and receiver require an e-mail software which understands MIME.

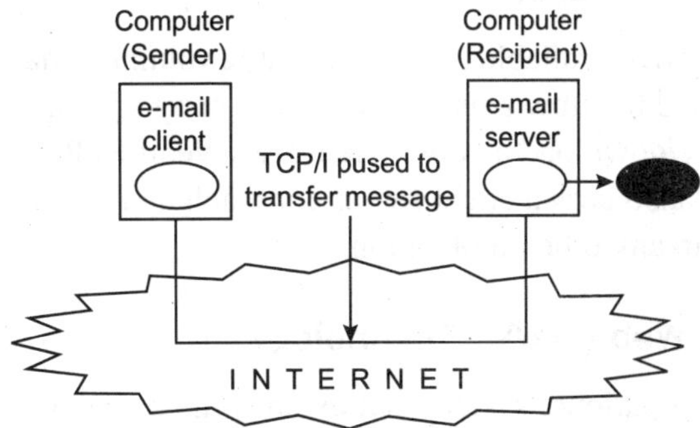

Figure 17.42 Scheme of transfer of e-mail across the internet

When a user sends an e-mail message, software in the sender's computer becomes a **client**. It contacts an e-mail server programme on the computer of the recipient and transfers a copy of the message, which is stored in the mailbox of the recipient. Figure 17.42 illustrates the scheme of working of e-mail.

There are three basic components in an e-mail programme: **User Agent** (UA), **Message Transfer Agent** (MTA) and **Message Store** (MS). UA permits the user to prepare e-mail, MTA sends the message to its destination and MS stores the e-mail until it is received by the recipient. One of the most popular mail systems is simple mail transfer protocol (SMTP), which is supported over Internet's TCP/IP protocol.

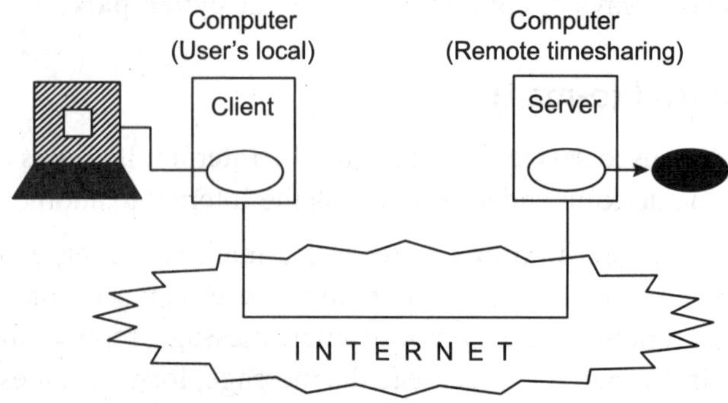

Figure 17.43 Schematic of remote login (TELNET) service across the internet

Another important Internet service is **Remote Login** (telnet). The Internet standard for this service is found in a protocol known as **TELNET**. It specifies how a remote login client and remote login server interact. Using TELNET service, one can remotely login to a machine and work as if they have been directly attached to the machine. Figure 17.43 schematically illustrates this service. This service extends the login concept used by conventional timesharing computer systems to permit access to a remote timesharing system.

17.30.4 File Transfer Protocol (FTP)

FTP is another Internet service available to transfer a large volume of data file from one computer to another. FTP is designed to allow a user to transfer a copy of a file to a remote computer as easily as they can retrieve/download a copy. There are several FTP servers which keep certain files open to public. FTP also uses a client server approach and FTP data transfer causes more traffic on the Internet than any other application.

17.30.5 World Wide Web (WWW) Technology

It is an Internet-based navigational system that extends the concept of hypermedia to include several computers. This is a system for information distribution and management and for linking information stored on several computers by point and click access. The point-and-click access is due to the underlying hypertext or hypermedia approach of web search engine. We may note that searching, retrieving and collection of 'hypertext' documents are the main functions of WWW service. One can access web pages stored in remote computers through the **Hyper Text Transfer Protocol** (HTTP) using hyperlinks and **web browser**. Hypertext refers to computer-based documents in which cross-references are embedded within documents and other entries. Each cross-reference is a pointer to another document. In the document, words, phrases and icons become links that allow one to jump to a new location on the web. If a hypertext contains references to non-textual information, it is referred as **hypermedia system**.

17.30.6 Web Browser

It is a programme that provides access to hypermedia documents on WWW. **Mosaic, Netscape** and **Internet Explorer** are among the prominent web browsers. The first information that one gets from a remote web site is called a **home page**. The home page leads a user to a series of other documents, resources and files that reside on that computer or other web servers anywhere the world.

17.30.7 Hyper Text Markup Language (HTML)

HTML is the language is which web pages are written. We may note that HTML is more of a page description language than a programming language. HTML contains tags which enables the web browser to format text and graphics on the user's screens. HTML uses free-format input because it does not specify exactly how to display the output. A browser has freedom to choose a form that is appropriate for a given computer. HTML is capable to include images such as pictures or graphics as well as text.

17.30.8 Uniform Resource Locator (URL)

An address used to specify a particular page of WWW information is referred to as URL, which represents a link to almost every document, file and resource on the internet. This consists of a short character string that identifies a particular multimedia document provided a valid UTL of that document is known. Web browsers display the URL of a page along with the display of information of that page. Each URL uniquely identifies a page of information by providing the name of a remote computer, a server on that computer, and a specific page of information available from the server.

17.30.9 Electronic Banking

In electronic banking, financial transactions are handled electronically using computer technology. **EFT (Electronic Funds Transfer)** is the most popular electronic banking scheme. The **Automatic Teller Machine** (or ATM) is form of EFT.

17.30.10 Electronic Commerce (e-Commerce)

It is a model of business related to paperless exchange of business information using **Electronic Data Exchange** (EDI), **e-mail, Electronic Bulletin Boards** (EBB), EFT and other technologies based on network. This technology is widely used for on line stock and bond transactions and business-to-business transactions all over the world. This new technology not only automates transactions, but helps organizations to move to a fully electronic environment of operation. This has made possible online shopping with full knowledge of the product. This technology supports the integration of Internet commerce with supply-chain solutions.

17.30.11 Networks and Security

In any remote communication system, it is essential to ascertain that communication is made with the right person at the other end. In computer communication, the use of user names and passwords solves this security problem. In some systems, e.g. banking, the account number and **personal identification number** (or **PIN**) are used to enable to user to use the banking facility effectively.

Summary

1. A cathode ray oscilloscope is an universal instrument in an electronic lab. It comprises many subsystems, mainly a cathode ray tube, power supply (HT and LT), vertical amplifier and horizontal amplifier.
2. There are special types of CRO – (i) dual trace and (ii) dual beam.
3. A CRO is used to measure an unknown signal and its frequency. It is also used to study the wave shape of an unknown signal/pulse.
4. An electronic multimeter is mainly used to measure unknown voltage, current and resistor.
5. In an electronics laboratory, we also have a function generator. It provides a variety of output wave forms over a wide frequency range.
6. A device by which data received in the appropriate form can be manipulated so as to produce a solution to some problem is called the computer. In general, the word 'computer' refers to a *digital computer*, the other basic form, analogue *computer*, being far less versatile.

The digital computer accepts and performs operations on discrete data, i.e. data represented in the form of combinations of characters.

A large general-purpose computer system is called mainframe.

A microcomputer is one in which the central processor is fabricated on a single chip (or a small number of chips) of semiconductor; this processing unit is called a microprocessor.

A modem can convert the digital signals from a computer into an analogueue form for use over an (analogue) telephone system.

Review Questions

1. Draw the block diagram of a general-purpose CRO. Label all the blocks. Briefly describe the function of each block.
2. Describe the working of cathode ray tube. What is the role of grid and focusing anode? Draw a neat diagram of the tube showing the various components and their electrical connection.
3. What is the principle of electrostatic focusing? How it is utilized in the focusing of electron beam in cathode ray tube?
4. How do two electrodes maintained at different potentials act as electrostatic lens? Derive Snell's law of refraction in this case.
5. Derive a relation for electrostatic deflection of electron beam on the screen of cathode ray tube in terms of tube length, effective length of deflecting plates and deflecting voltage, and accelerating voltage. Hence define deflection sensitivity.
6. Give the basic principle of generation of a sawtooth wave. How the same is utilized in time base generator of a CRO and the wave is synchronized?
7. Mention the various applications of a CRO. How will you find the frequency of an unknown signal?
8. What are Lissajous figures? How they are utilized in determining the ratio of frequency of two unknown voltage signals? Can we determine phase difference between two signals? If yes, How?
9. Draw the block diagram of a dual beam CRO. Label each block and describe its function in brief.
10. Draw the circuit diagram of an electronic voltmeter. Describe its function. How it is better than an analogue voltmeter.
11. What is the principle of working of a digital multimeter? Give its main features. Draw its block diagram and describe its working in brief.
12. Draw the block diagram of a digital voltmeter. How it works. Give its limitation.
13. What is a function generator? Give its salient features.
14. Describe the working of dual slope analogue-to-digital converter. Draw its relevant block diagram.
15. What do you mean by hardware and software in a computer? Explain with examples.
16. What is the role of CPU in a computer system?
17. What are the main features of computer communication network?
18. Write the names of three common forms of input devices and three forms of output devices.
19. How many types of computers are there? Describe them briefly.
20. Explain why binary system is used to represent computer data?

21. Distinguish among LAN, MAN and WAN.
22. Define the terms bit, byte, kilobyte, RAM and ROM.
23. What is ISDN? How it works?
24. What do you understand by hard copy and soft copy? Mention the devices that produce each.
25. What is Internet? What are its advantages over other methods of communication?
26. What do you understand by e-mail and e-commerce? Explain clearly.

Short Answer Questions

1. What is the role of delay line in a CRO? Where it is connected?

   Ans: Delay line in a CRO is connected between vertical deflection plates and vertical amplifier. It delays the signal to reach vertical deflection plates since it is essential for the signal to reach both horizontal deflection plates already connected to horizontal amplifier and time base generator and vertical deflection plates.

2. How many power supplies are there in a CRO? What is there approximate range?

   Ans: There are two power supplies in a CRO, known as high tension ($\approx$ 1100 V) and low tension ($\approx$ 300 V) supply.

3. What is time base circuit in a CRO?

   Ans: Time base circuit in a CRO meant to produce a sawtooth wave. Due to generation of run, the spot on the screen moves from left to right and the spot moves from right to left during retrace the path of wave.

4. How sawtooth wave is generated in a CRO?

   Ans: A saw tooth wave is generated by time base circuit which comprises a relaxation oscillator where the capacitor gets charge and discharged. The rising voltage trace across the capacitor is first made linear by chopping exponentially the rising trace by a sync pulse and the discharge of capacitor is through UJT which acts like a closed switch.

5. What is the difference between dual trace CRO and double beam CRO?

   Ans: In dual trace CRO, the single trace capability of CRO is modified into two separate beams by an electronic switch. It has only one vertical amplifier, whereas in dual beam CRO, it accepts two separate vertical input signals which reach to the two vertical amplifiers in the CRO and the output of vertical amplifiers goes to the two sets of vertical deflection plates.

6. How we can know the frequency of an unknown signal using a CRO?

   Ans: The frequency of an unknown signal can be determined by superposing it with a known signal perpendicularly, when Lissajous figures are produced on the screen. The resulting pattern indicates the ratio of the frequencies of two signals. An unknown signal frequency can thus be determined since the frequency of one signal is already known.

7. What comprises the screen in a CRO?

   Ans: The screen in a CRO is a circular flat base of glass envelope of the cathode ray tube. Inside the surface of glass base is painted with phosphor which glows when an electron beam strikes it.

8. What is a CRO probe?

   Ans: It connects the circuit under investigation to the vertical input of CRO. It comprises a probe head, coaxial cable and termination circuitry. The probe head consists of very high resistor (≈ 10 MΩ) shunted by very small capacitor (≈ pF). The termination circuit provides the CRO with source impedance it requires and terminates the cable in its characteristic impedance.

9. How do you define deflection sensitivity of CRT. Is it constant?

   Ans: Deflection capacity (S) of CRT is defined as deflection of spot on the screen in meters per volt. If D (meter) is deflection in screen and E_d (volt) is deflection voltage, then

   $$S = \frac{D}{E_d}\left(\frac{m}{V}\right)$$

   It is not constant and depends upon the geometry of the particular tube and applied accelerating voltage.

10. When we have stable display in a CRO?

    Ans: Stable display of a CRO is obtained when the frequency (f_v) of vertical input signal is a whole multiple of sweep frequency (f_s), i.e. $f_v = nf_s$.

11. Mention the main sections of a computer system.

    Ans: The main sections of a computer system are: (i) input/output devices, (ii) central processing unit (CPU) and (iii) memory.

12. How high-level languages have been classified?

    Ans: High-level languages have been classified as:

    (i) scientific languages (ii) commercial languages
    (iii) special purpose languages (iv) multipurpose languages

13. What do you understand by the statement—Computer data comes as a string of zeros and ones?

    Ans: The computer combines a very large number of logic gates. For data we get a string of zeros and ones.

14. Mention Network Topology?

    Ans: The main network topologies are: (i) bus, (ii) ring, (iii) star and (iv) tree.

15. How a message is transmitted over the Internet?

    Ans: On Internet, the most common communication method is through e-mail.

Objective Questions

1. The sets of vertical deflection and horizontal deflection plates in a laboratory CRO is ______________.
   (a) two, two respectively (b) one, one respectively
   (c) two, one respectively (d) one, two respectively **[Ans: (b)]**
2. The material used to paste the inner surface of the screen in CRO is ______________.
   (a) phodhor (b) fluorescent
   (c) luminiscent (d) none of the above **[Ans: (a)]**
3. The unit of deflection sensitivity in CRO is ______________.
   (a) cm/V (b) cm/mV
   (c) m/V (d) m/mV **[Ans: (c)]**
4. Synchronization of the sweep is possible when the ratio of frequency of vertical input signal and sweep frequency is ______________.
   (a) n (b) $1/n$ (c) n^2 (d) $1/n_2$ **[Ans: (a)]**
5. For synchronization to take place, the time period of synchronizing signal (T) and natural time period (T_o) of sawtooth wave must have the relation.
   (a) $T = T_o$ (b) $T > T_o$
   (c) $T < \frac{1}{T_o}$ (d) $T < T_o$ **[Ans: (d)]**
6. When an inclined straight line is obtained in Lissajous figure, the two signals have:
   (a) equal amplitude, zero phase (b) unequal amplitude, zero phase
   (c) equal amplitude, 90° phase (d) unequal amplitude, 90° phase **[Ans: (a)]**
7. CRO probe consists of ______________.
   (a) high resistor shunted by a very small capacitor
   (b) high resistor shunted by a very high capacitor
   (c) high resistor in series with a very small capacitor
   (d) high resistor in series with a very high capacitor **[Ans: (a)]**

Mark the Following Statements True/False.

1. Deflection sensitivity of a cathode ray tube is constant. **[Ans: False]**
2. The screen of cathode ray tube is pasted with phosphor. **[Ans: True]**
3. Both dual trace CRO and dual beam CRO have two vertical amplifiers. **[Ans: False]**
4. The intensity of electron beam is controlled by voltage on the grid of electron gun. **[Ans: True]**

5. UJT in a relaxation oscillator acts as a switch. [**Ans:** True]
6. The output of relaxation oscillator is a sinusoidal wave. [**Ans:** False]
7. Two accelerating anodes in a CRT acts as electron lens. [**Ans:** True]
8. An electron beam in CRT gets refracted when it crosses through an equipotential surface. [**Ans:** True]
9. High voltage supply is connected to cathode ray tube and low voltage supply is connected to other circuitry in CRO. [**Ans:** True]
10. A microprocessor comprises memory and I/O peripherals. [**Ans:** True]
11. COBOL is a scientific language. [**Ans:** False]
12. RAM is described as volatile memory. [**Ans:** True]
13. A DVD holds much smaller data than a CD-ROM. [**Ans:** False]
14. Modem is a peripheral device. [**Ans:** True]
15. The various sections of a computer know what to do only under the instruction of software. [**Ans:** True]
16. Each type of computer accepts a particular type of software. [**Ans:** False]
17. Microprocessor and microcomputer are one and the same. [**Ans:** True]
18. ALU performs addition, subtraction and digital logic. i.e. AND operation. [**Ans:** True]
19. A ROM is a device where data is permanently stored. [**Ans:** True]

Fill in the Blanks.

1. ______________ is pasted on the ______________ surface of CRT screen. [**Ans:** phosphor, inside]
2. The ______________ of grid in electron gun controls the ______________ of electron beam. [**Ans:** voltage, intensity]
3. In relaxation oscillator, ______________ acts as ______________. [**Ans:** UJT, switch]
4. All electrodes in cathode ray tube are connected to ______________ supply and other circuitry is connected to ______________ supply. [**Ans:** high voltage, low voltage]
5. Dual ______________ CRO has an ______________ switch. [**Ans:** trace, electronic]
6. To determine the______________ of an unknown signal, we make use of ______________ in a CRO. [**Ans:** frequency, Lissajous figure]
7. Dual beam CRO has ______________ sets of ______________ deflection plates. [**Ans:** two, vertical]

8. A relaxation oscillator generates ______________ wave. [**Ans:** sawtooth]
9. An ______________ bit binary number is called a ______________.

   [**Ans:** 8, byte]
10. The computer language is a set of ______________. [**Ans:** binary numbers]
11. Errors in a programme are called ______________. [**Ans:** bugs]

Chapter

18 Optoelectronic Devices

18.1 Introduction

Semiconductor devices can also be designed and fabricated to detect and generate optical signals. These devices are called optoelectronic devices or photonic devices. Optoelectronics is the technology that combines optics and electronics. It deals with the study of all interactions of light with matter for technological applications. It includes the study of lasers, quantum optics, non-linear optics, geometrical optics, waveguide physics and semiconductor physics.

Optoelectronic devices and related emerging technologies have a profound effect on mankind. One can feel the presence of these technologies in areas of computers, home entertainment systems, medical science, telecommunications, etc.

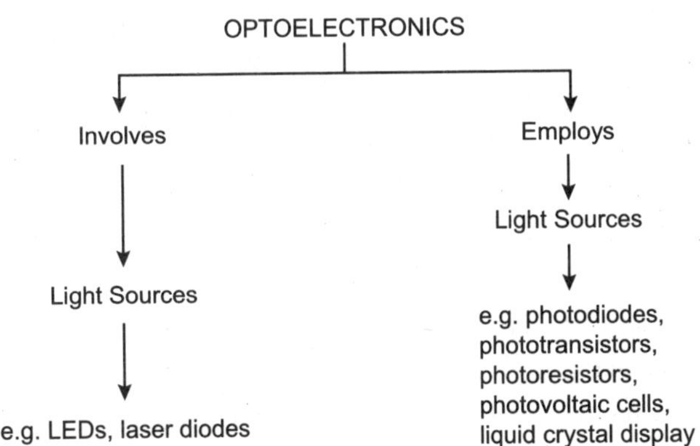

Figure 18.1 Optoelectronic devices

Basically, optoelectronic devices can be placed under two groups (Figure 18.1). Devices that fall under the first type are **light emitting devices**, e.g. light emitting diodes and laser diodes. Devices that fall under the second category work as **light detectors**, e.g. phototransistors, photoresistors, photovoltaic cells and liquid crystal displays (LCDs). There is another device called **optocoupler** or **optoisolator**, which is a combination of two types of devices in a single unit.

In the present chapter, we will mainly concentrate on light emitting diodes, lasers, photodetectors, photodiodes, photoresistors, etc.

There are several semiconductor devices for high frequency applications, which take the advantage of instabilities that occur in the semiconductors. The important type of instability is negative conductance. Devices that work with negative conductances are microwave devices, tunnel diode, IMPATT diode and Gunn diode. These are mainly two-terminal semiconductor devices that can be operated in negative conductance mode to provide amplification or oscillation at microwave frequencies in a proper circuit.

18.2 Optical Spectrum

Light is a form of electromagnetic radiation. If measured in frequency (Hz), the optical light spectrum extends from 30 THz to 3 PHz (Figure 18.2).

The frequencies range from infrared (not visible) to visible light to ultraviolet (not visible).

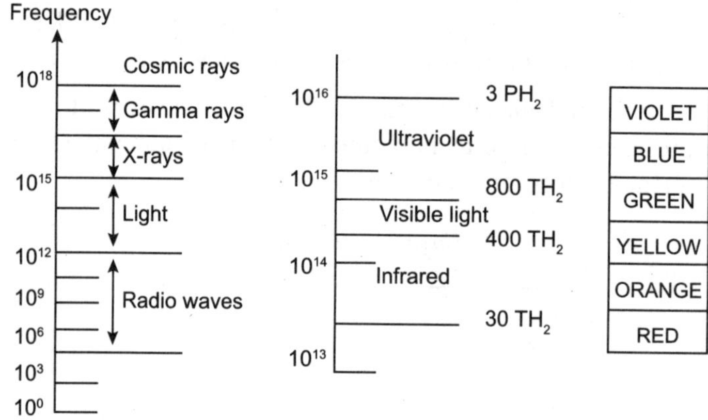

Figure 18.2 The optical spectrum

(i) The infrared band (30 THz to 400 THz) comprises frequencies lower than what the human eye can detect.

(ii) The visible light band (400 THz to 800 THz) consists of frequencies the human eye can detect. This band can be further subdivided into colours of the rainbow, i.e. red, orange, yellow, green, blue and violet.

(iii) The ultraviolet band (800 THz to 3 PHz) is made of frequencies above those the human eye can detect.

We know that an orderly arrangement of radiation according to wavelength or frequency is known as spectrum, and electromagnetic spectrum is such a spectrum (Figure 18.2). The light wave is an electromagnetic wave consisting of periodically varying electric and magnetic fields oscillating at right angles to each other and to the direction of propagation of wave (Figure 18.3).

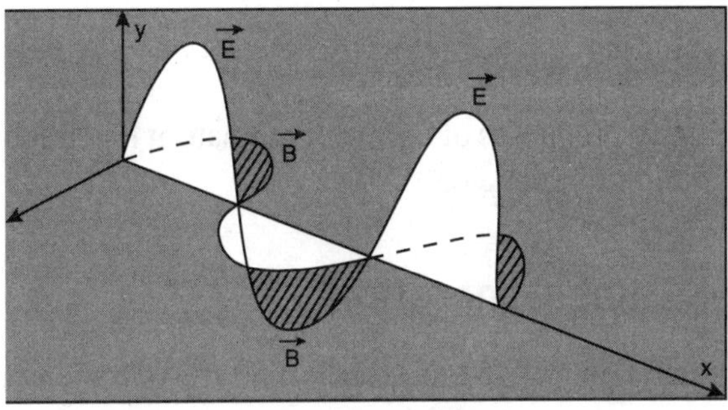

Figure 18.3 Electromagnetic wave

All radiations consisting of the electromagnetic spectrum travel at the same speed of 3×10^8 m/s. This speed is one of the fundamental constants of nature and denoted by *c*. Figure 18.3 shows that one can express frequency as wavelength and vice versa. Wavelength (λ) is the length of one cycle as it travel through space (Figure 18.4). Wavelength is related to frequency by the following relation:

$$\lambda = \frac{c}{\nu} \tag{1}$$

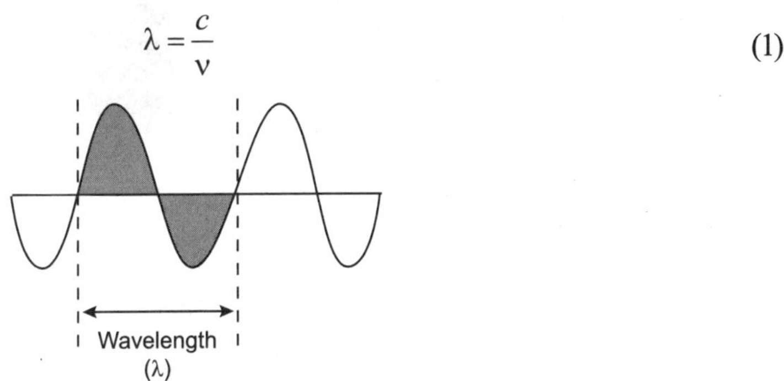

Figure 18.4 One sinusoidal cycle of an electromagnetic wave

where *c* is the speed of light (= 3×10^8 m/s) and υ is frequency in Hertz. For each frequency of the radiation, energy *E* of the wave is given by the relation

$$E = h\nu \tag{2}$$

where *h* is Planck's constant ($h = 6.7 \times 10^{-34}$ Js).

An important characteristic of light is its wave-corpuscle (wave-particle) dualism. On the one hand, light possesses wave properties, manifested in the phenomena of interference, diffraction and polarization. On the other hand, light is a stream of particles, called photons, which have zero rest mass and travel at a velocity equal to that of light in vacuum. Energy (*E*) and momentum (*P*) of a photon for the corresponding electromagnetic wave of frequency υ and wavelength λ in vacuum are given by

$$E = h\upsilon = \frac{hc}{\lambda}$$

and
$$P = \frac{h\upsilon}{c} = \frac{h}{\lambda} \tag{3}$$

We may note that the wave properties of light predominate at low frequencies and corpuscular properties predominate at high frequencies.

18.3 Principles of Light Generation

According to the Bohr model (Figure 18.5), an atom is made up of protons, neutrons and electrons. Protons and neutrons cluster in the center of the atom to form the nucleus of the atom. Electrons orbit the nucleus in somewhat the same way that planets orbit the sun.

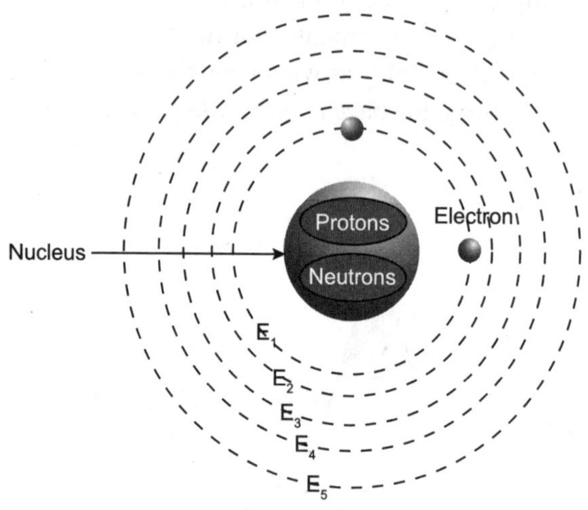

Figure 18.5 Bohr model of atom

Figure 18.5 shows the orbital paths of electrons. Each orbital is associated with a specific energy level. The orbital an electron occupies depends on the amount of energy it possesses. Electrons with lower energy occupy the lower orbits, whereas electrons with higher energy levels occupy the higher orbits. Thus, electrons in orbit E_1 are at lower energy level than electrons in orbit E_2.

The natural state in which an electron exists is said to be its ground state. Energy may be imparted to an atom by heat, electrical force or collision with other particles. Upon absorption of energy, an electron can jump to a higher orbit (Figure 18.6). In this state, the atom is said to be excited. This is an unnatural state, and the electron will remain in this higher orbit for only a very short time before returning to its natural ground state. The electron must shed the excess energy during this transition in the form of electromagnetic radiation. In some materials, this energy is released at frequencies of visible light. The process of energy release is shown in Figure 18.6.

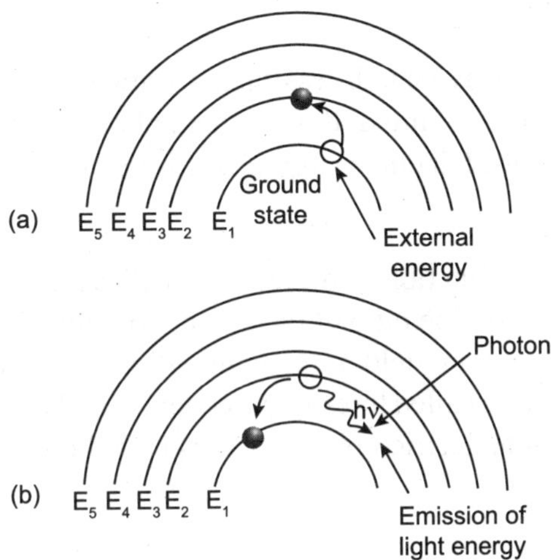

Figure 18.6 Electron absorption and emission of light energy: (a) absorption, (b) emission

The wavelength determines the material to be used in optoelectronic devices. Electric current is generated in a light sensitive material of incident light. Light intensity is measured in terms of 'luminuox flux' falling per unit area of the material. Luminous energy is measured in lumens (lm). The relation between lumen and Watt is

$$1 \text{ lm} = 1.496 \times 10^{-10} \text{ W}$$

Optoelectronic devices are based upon the 'photoelectric effect'. The devices are, therefore, also called 'photoelectric devices'.

18.4 Photoelectric Effect

Photoelectric effect (photoeffect) is the process of emission of electrons from a metal surface when it is illuminated by high-frequency electromagnetic radiation (Figure 18.7). The photoelectric effect is of different forms:

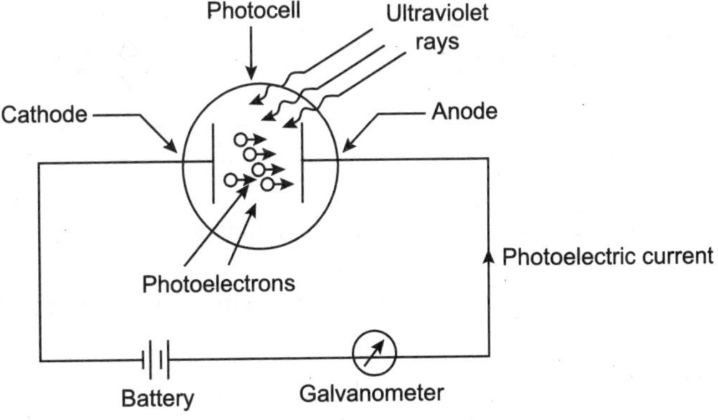

Figure 18.7 Photoelectric effect

18.4.1 Photoemissive Effect

When a radiation (photon) of wavelength less than a critical value is incident upon a metal surface, electrons are found to be emitted; this is called photoemissive or photoelectric effect, e.g. vacuum and gas phototubes. Kinetic energy E of emitted electrons is given by

$$E = h\upsilon - e\Phi \tag{4}$$

where $e\Phi$ is the surface work function. No electrons will be emitted when $h\upsilon < e\Phi$ (or, in terms of wavelength, $\lambda > hc/e\Phi$). The ratio of the number of emitted electrons to the number of absorbed photons is called quantum yield or quantum efficiency. The minimum frequency that can cause a photoelectric effect is called threshold frequency, and the corresponding wavelength of light beyond which the photoelectric effect ceases is called threshold wavelength or cut-off wavelength.

18.4.2 Photoconductive Effect

An electron may be raised from the valence band to the conduction band in a semiconductor by the absorption of a photon of frequency υ provided that

$$h\upsilon \geq E_g \tag{5}$$

where energy gap is Eg, or in terms of wavelength

$$\lambda \leq \frac{hc}{E_g} \tag{6}$$

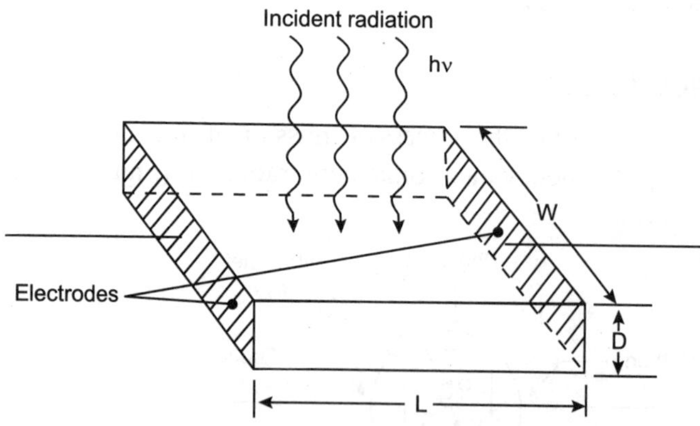

Figure 18.8 Illustration of photoconductive effect

We define bandgap wavelength λ_g as the largest value of wavelength that can cause this transition, i.e.

$$\lambda_g = \frac{hc}{E_g} \tag{7}$$

As long as the electron remains in the conduction band, it will cause an increase in the conductivity of the semiconductor. This is the phenomenon of **photoconductivity**. This is the basic mechanism in the working of photoconductive detectors. Figure 18.8 illustrates the photoconductive effect and a photoconductor bias circuit is shown in Figure 18.8.

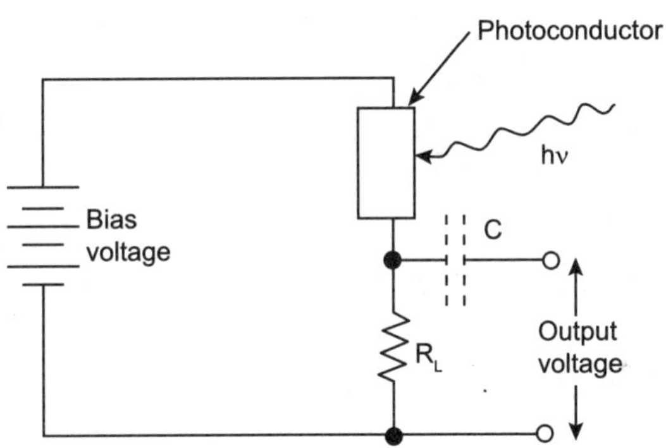

Figure 18.9 Photoconductor bias circuit

Photoconductive mode In this mode, the conductance of the diode changes when light is applied. In this mode, the photodiode is reverse-biased. Figure 18.10 shows a reverse-biased photodiode in the photoconductive mode. The depletion region of the reverse-biased photodiode is very wide and the resistance of the diode is high, and hence there will be only a small reverse current through it. This reverse current that flows through the diode when there is no light being applied is called dark current (I_o).

When light is incident, electron-hole pairs are generated. Electrons are attracted to the positive bias voltage, and holes are attracted to the negative bias voltage. This movement of electrons and holes causes a considerable reverse current to flow through the photodiode. The resistance of the photodiode is extremely low when light is incident. If the intensity of light is increased, resistance decreases and therefore reverse current increases. Current that is passed through the photodiode when light is being applied is called **light current** (I_L).

The conductivity of the photodiode is low when there is no light incident, and conductivity increases as the intensity of light increases. Consequently, the magnitude of dark current is very much smaller than that of light current. Consider the example shown in Figure 18.10. Assume that dark current (I_D) flowing through the diode is 10 nA and light current (I_L) is 100 mA. Output voltage (V_{out}) is dependent on the amount of current flowing in the circuit. With no light present,

$$V_{out} = I_p \times R = 10 \text{ nA} \times 10 \text{ k}\Omega = 100 \text{ }\mu\text{V}$$

With light present,

$$V_{out} = I_L \times R = 100 \text{ }\mu\text{V} \times 10 \text{ k}\Omega = 1 \text{ V}$$

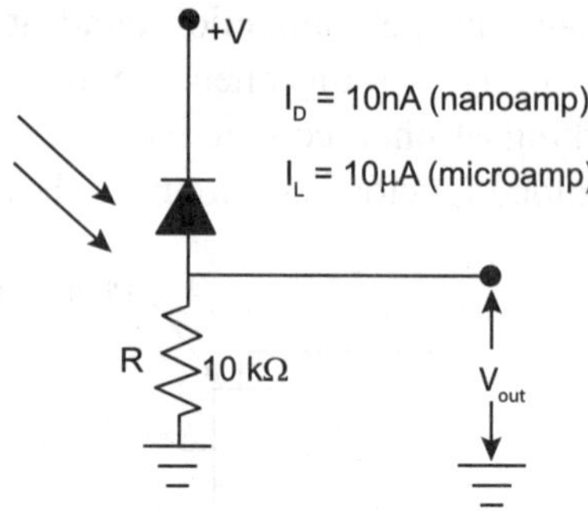

Figure 18.10 Photoconductive mode of operation

18.4.3 Photovoltaic Effect

In photovoltaic effect, an e.m.f. is generated in a PN junction (when reverse-biased) under the influence of incident light.

Photovoltaic mode While in operation in the photovoltaic mode, the photodiode generates a voltage in response to light. The incidence of light on the photodiode creates electron-hole pairs. Electrons generated in the depletion region are attracted to the positively charged ions in the N-type material, and holes are attracted to the negatively charged ions in the P-type material. This creates a separation of charges, and a small voltage drop of about 0.45 V is developed across the diode.

Figure 18.11 shows the photovoltaic mode of operation. In this mode, the photodiode acts as a **solar cell**. If a load resistor is connected across the voltage source, a small current will flow from the cathode to the anode.

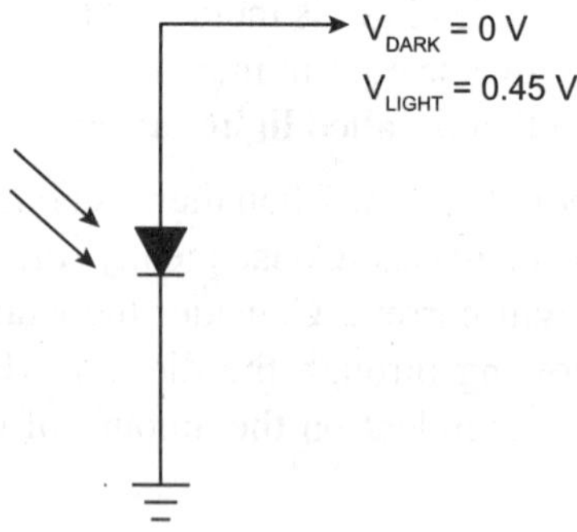

Figure 18.11 Photovoltaic mode of operation

Table 18.1 provides a brief summary of optoelectronic devices and related effect on which they are based.

Table 18.1 Optoelectronic devices

S. No.	Optoelectronic device	Related photoelectric effect
1.	Photoresistor	Photoconductive
2.	Photodiode	Photovoltaic
3.	Phototransistor	Photovoltaic
4.	Photo SCR	Photovoltaic
5.	Solar cell	Photovoltaic
6.	Photomultiplier tube	Photovoltaic
7.	Photo FET	Photoemission

The details of a few materials generally used in optoelectronic devices are given in Table 18.2

Table 18.2 Materials used in optoelectronic devices

S. No.	Materials/compounds	Forbidden energy gap (E_g) (eV)	Emission wavelength (λ) (nm)
1.	Si	1.07	1142
2.	Ge	0.65	1882
3.	GaAs	1.42	910
4.	GaP	2.26	561
5.	SiC	3.2	560
6.	AlSb	1.62	77.7
7.	InSb	0.20	6910

18.5 Optoelectronic Devices

18.5.1 Light Emitting Diodes

An LED is a PN junction diode of special construction. Like other semiconductor diodes, its behaviour is influenced by biasing conditions. Figure 18.12 shows the basic operation of an LED. The PN junction is forward-biased when the cathode (N type region) is negative with respect to the anode. Under forward-bias, electrons from the N type region and holes in the P type material move toward the PN junction.

At the junction, the holes and electrons combine. Electrons are at a higher energy level than are holes. The electrons literally fall into holes as they combine, and in doing so they release the excess amount of energy in the form of light. There is current through the LED in the forward direction, just as in the case of an ordinary diode.

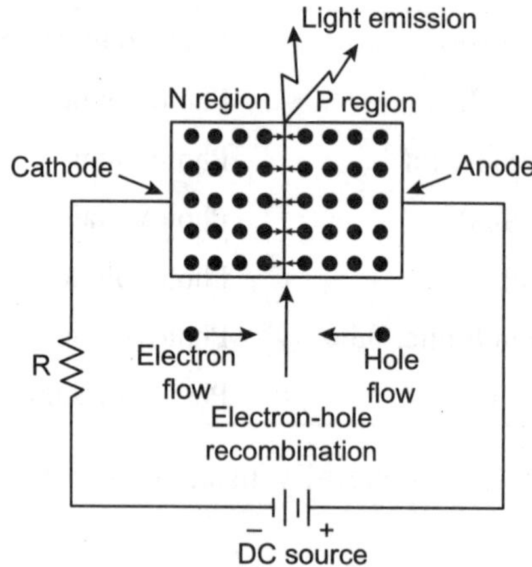

Figure 18.12 Basic operation of a light emitting diode

When the PN junction is reverse-biased, the LED does not conduct and there is no release of light. Hence the LED emits light when forward-biased and does not emit light when reverse-biased. The schematic symbol of LED is shown in Figure 18.13. The LED operates at forward voltages typically ranging from 1.2 V to 4.3 V. The reverse breakdown voltage is in the range of –3 to –10 V.

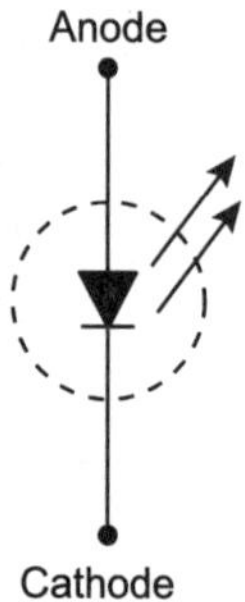

Figure 18.13 Schematic symbol of LED

The main advantages of LEDs are:

(i) LEDs have very long life.

(ii) LEDs are highly durable and reliable.

(iii) The efficiency of LEDs is higher than that of incandescent light bulb and can possibly exceed fluorescence bulbs.

(iv) White-light LED sources can provide significant energy saving in comparison with conventional lighting.

18.5.2 LED Construction and Working

LED emits more light than an ordinary diode because ordinary diodes are made of silicon, which is opaque or impenetrable as far. Energy in ordinary diodes is, therefore, released in the form of heat.

On the other hand, LEDs are made of semiconductor materials that are semitransparent. LEDs are normally made of gallium arsenide phosphide, which emits visible red light. Gallium phosphide produces visible green light. LEDs that produce yellow and blue light are also available.

Figure 18.14 shows an LED chip and a typical LED package. The LED chip is attached to cathode and anode leads through thin wires. The plastic case serves as a lens that conducts light away from the LED and also acts as a magnifier. The entire case may be dyed or tinted with a colour that enhances the on/off contrast of the LED. LED leads are identified by one of the following three ways (Figure 18.15).

1. The leads may have different lengths, and the shorter of the two leads is the cathode Figure 18.15(a).

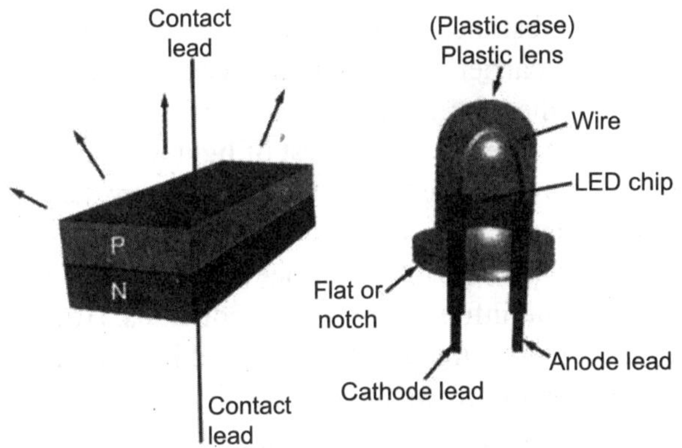

Figure 18.14 The LED: (a) chip, (b) typical package

2. The cathode lead is flattened (Figure 18.15(b)).

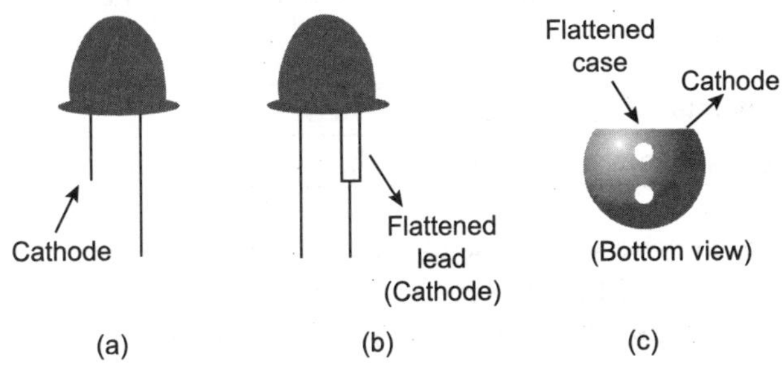

Figure 18.15 Identification of LED leads

3. One side of the case may be flattened. The lead closest to the flattened side is the cathode (Figure 18.15(c)).

A cross-sectional view of a typical diffused LED is shown in Figure 18.16.

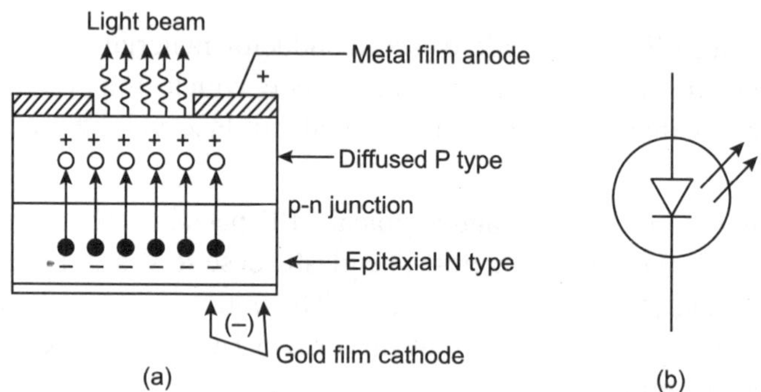

Figure 18.16 (a) Structure of LED, (b) symbol (LED)

The semiconductor material employed may be GaAs, GaAsP or GaP. Here an N type epitaxial layer is grown upon the substrate, and the region is created by diffusion. When the PN junction is forward-biased, charge carrier recombination occurs in the P region, so this region is kept uppermost. The P region, therefore, becomes the surface of the device, and the metal film anode connection must be patterned to allow most of light to be emitted outside it. This is done by making a connection to the outside edges of the P layer or by depositing a comb-shaped pattern in the mid-region of P material. A gold film is applied at the bottom of the substrate to reflect light as much as possible towards the upper surface of the device and to provide a cathode connection. Materials like GaAs emit infrared (invisible) radiations. However, GaAsP provides either red or yellow light, while red or green light is emitted by GaP. Its symbol is shown in Figure 18.16(b) where the direction of arrows indicates emitted light.

Normally a single LED is used to indicate current in the circuit but generally a seven-segment display is used where seven LEDs are combined to show a numerical display (Figure 18.17(a)). Any desired number between 0 and 9 can be displayed. The seven segments are labelled a, b, c, d, e, f, g. We can easily follow that a display for number 2 will be shown by the glow of a, b, g, e, d segments. Similarly, for a display of number 8 all the segments shall glow. In this way we have the following series:

Numerals	0	1	2	3
Display segments	a, b, c, d, e, f	b, c	a, b, g, e, d	a, b, g, c, d

The LED device is usually very small, so to enhance the lighted segments surface, solid plastic light pipes are often used. The LEDs in seven-segment display may be connected in common anode configuration (Figure 18.17(b)) or common cathode configuration (Figure 18.17(c)), so when selecting a display, it is important to choose the configuration.

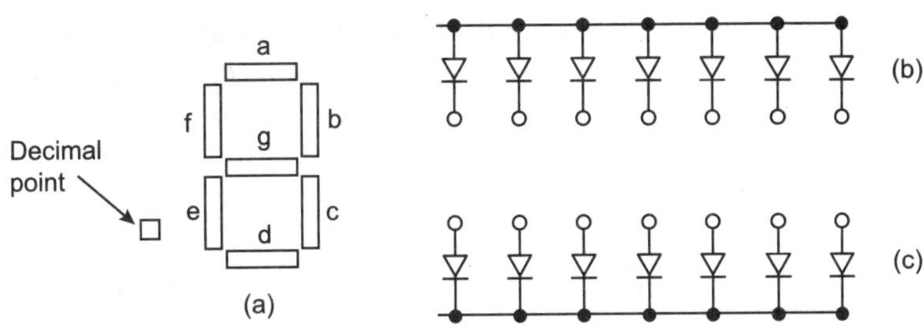

Figure 18.17 (a) Seven-segment display, (b) common anode connection, (c) common cathode connection (arrows have not been shown)

The forward voltage for a LED is typically 1.2 V for forward current 20 mA. The relatively large amounts of current consumed by LEDs are their major disadvantage. Excluding the display, a typical electronic circuitry in LEDs may require as high as 10 mA, necessitating a bulky power supply. However, they have the big advantage of long life and ruggedness. Often LEDs are usually switched ON and OFF by means of a transistor circuit. In such cases, a resistance is connected in series with it.

With the introduction of liquid crystal diodes (LCD), its main disadvantage of requiring high current of the order of 20 mA has been offset. At present, LCDs (Figure 18.24) are very popular and used mostly in measuring meters.

Luminous performance of an LED is related to its optical yield. Optical yield measures the efficiency of LED in the conversion of electric current into output photon. Luminous performance is expressed as the ratio of total output flux in lumens to input power in watts.

The quality of emitted photons reduces by the following three factors:

(i) Absorption within the LED material

(ii) Reflection loss when light passes from semiconductor material to air due to differences in refractive index

(iii) Total internal reflection of light at an angle greater than critical angle (θ_c).

In LEDs, one might apply voltage across the PN junction which results in a diode current, which in turn produces photon and light output. This inverse mechanism is termed as **injection electroluminescence**.

When a voltage is applied across PN junction, electron and holes are injected across the space charge region, where they became excess minority carriers. These excess minority carriers diffuse into regions of neutral semiconductor and recombine with majority carriers. If this recombination is a direct band-to-band process, then photons are emitted. The intensity of output photon is directly proportional to ideal diffusion current.

The forward current voltage characteristics of LED is shown in Figure 18.18, which is same as that of GaAsPN junction. One finds that at low forward voltages, diode current is dominated by non-radiative recombination current mainly due to surface recombination near the perimeter of LED chip. Diode current is dominated by radiative diffusion current at higher voltages. At even higher voltages, the series resistance will limit diode current. The total diode current can be expressed as the sum of saturation current due to diffusion (I_d) and saturation current due to recombination (I_r) as

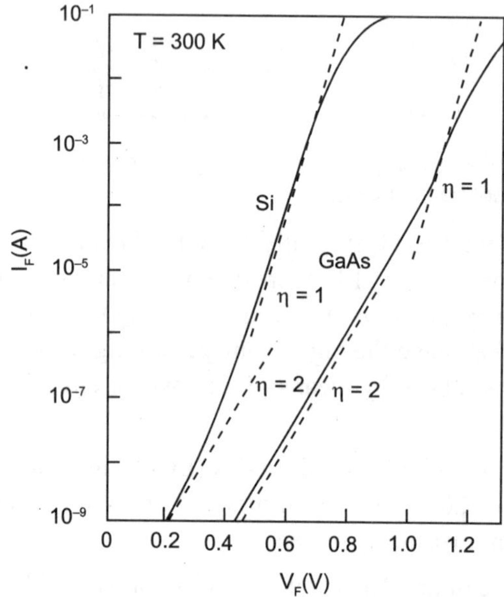

Figure 18.18 Forward-bias current-voltage characteristic of LED

$$I = I_d \exp\left[\frac{q(V - IR_s)}{\text{kT}}\right] + I_r \exp\left[\frac{q(V - IR_s)}{2\text{ kT}}\right] \tag{8}$$

where R_s is device series resistance.

18.5.3 Internal and External Quantum Efficiency of an LED

The internal quantum efficiency of an LED is a function of diode that will produce luminescence and is also a function of injection efficiency as well as of percentage of radiative recombination events compared with the total number of recombination events. The internal quantum efficiency of homo junction LED is influenced by the following factors:

(i) Quality and purity of the material and substrate.

(ii) Existence of a direct energy gap spectrally matched to the desired wavelength

The presence of traps and impurities reduce the internal quantum efficiency of an LED. The internal and external quantum efficiency bears the following relationship.

$$\eta_{\text{ext}} = \eta_{\text{int}}\ C_{\text{ox}} \tag{9}$$

where C_{ox} represents extraction efficiency. Extraction efficiency is defined as the fraction of generated photons that escapes from the device package and emerges from the chip.

Extraction efficiency is determined by taking care of the following issues:

(i) Absorption coefficient of the material at emission wavelength

(ii) Radiation geometry of the LED

(iii) Substrate

We may note that LED is an **incoherent light source** because photons are emitted randomly from the junction in all directions and not in phase with each other. This is why LED is less suitable for the transmission of digital information than lasers.

Recently certain organic semiconductors (e.g. 8-hydroxy-quinolinato aluminium) have been found useful for electroluminescent applications. The organic light emitting diode (OLED) is particularly useful for multicolour, large area, flat-panel display due to its attribute of low power consumption and excellent emissive quality with wide viewing angle.

For designing OLEDs, the criteria are:

(i) Ultrathin layer for low biasing voltage (typically 150 nm)

(ii) Low injection barriers

(iii) Proper band gaps for the required colour

There are two general types of OLED devices: those made with small organic molecules and those made with organic polymers. Low cost of OLEDs is the main advantage over standard LEDs. Since OLEDs are non-crystalline materials, they can be easily deposited. However, there are some problems with OLEDs, e.g. small lifetime of OLEDs; the organic materials used in LEDs are quite sensitive to water and oxygen and therefore they must be deposited on glass substrates and covered with a second sheet of glass. The usage of glass enhances the cost of manufacturing.

Applications LEDs have replaced incandescent light bulbs as display devices. Long life expectancy, ruggedness and low power consumption have contributed to the choice of LEDs over incandescent light bulbs. Different types of liquid crystal displays have now replaced LEDs in some applications because they use much less energy than LEDs.

18.5.4 LED Indicator Circuits

Figure 18.19 shows how an LED can be used as a power level indicator in a computer, CD player, stereo amplifier or other such devices. The LED is connected in parallel to the system's internal power supply along with series resistor R_s. When the power switch is on, the diode is forward-biased and emits a light.

Series resistor R_s limits current to the LED and is often referred to as a "current limiter". Most LEDs have a forward-biased voltage drop between 1.2 V and 2.5 V although high intensity LEDs can have higher forward-biased voltage drops. Note that this is much higher than for a normal silicon diode (0.7 V) or germanium diode (0.3 V). The resistance value and wattage rating of R_s can be calculated using Ohm's law. R_s value is found by subtracting the LED's voltage drop from the source and then dividing by the LED's desired current. To find wattage rating for R_s, simply apply the power formula $P = I^2R$.

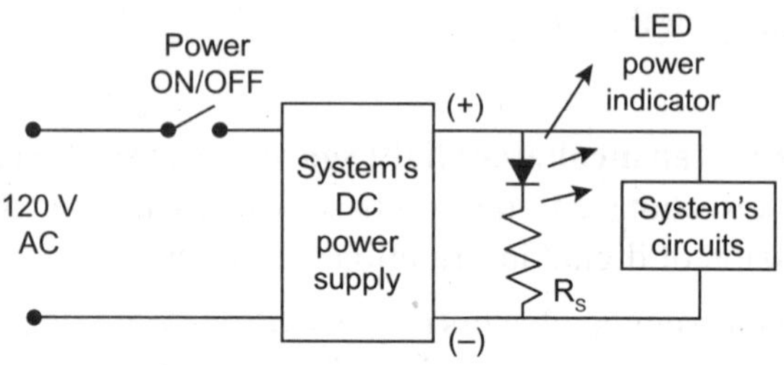

Figure 18.19 LED power indicator

Example 1 What is the resistance and wattage rating of a limiting resistor in a 12 V DC circuit for an LED that requires 2 V forward-bias with 20 mA current rating?

Solution

$$R_s = \frac{V_S - V_D}{I_D} = \frac{12\text{ V} - 2\text{ V}}{20\text{ mA}} = 500\Omega$$

and

$$P = I^2R = (20\text{ mA})^2 \times 500\ \Omega = 200\text{ mW}$$

Therefore, a standard-size 0.25 watt (= 250 mW) resistor is required.

Multicolour LEDs Multicolour LEDs are available that will:

(i) emit one colour when supply voltage is one polarity,

(ii) emit a second colour when polarity is reversed,

(iii) emit a third colour when bias polarity is rapidly switched.

Figure 18.20 shows the schematic of multicolour LEDs. Multicolour LEDs are usually two LEDs connected in antiparallel. That is, the anode of each diode is connected to the cathode of the other. Each LED can emit light only when forward-biased. Thus, when voltage of either polarity is applied, one LED is forward-biased and emits its native colour.

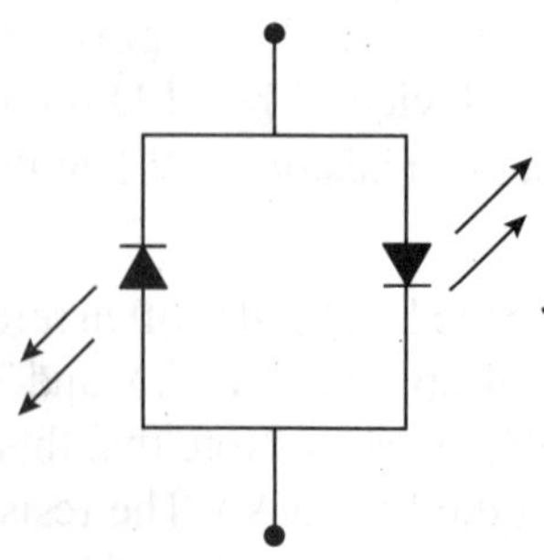

Figure 18.20 Multicolour LEDs

LEDs most commonly used are red and green in colour. The green LED is normally used to indicate whether something is functioning properly, and the red LED is used to indicate that there is a problem. If a multicolour LED is rapidly switched between two polarities, the red/green LED appears to produce a third colour (yellow).

18.5.5 Infrared-Emitting Diode

We know that the infrared band (30 THz to 400 THz) falls below the frequencies the human eye can detect. Diodes made of gallium arsenide release energy by way of heat and infrared light. Such a diode is called infrared-emitting diode (IRED). IREDs are used in remote controls, fiber-optic communication, discriminating organic solvents in the field of medicine and other such applications.

18.5.6 Multisegment LED Displays

LEDs are widely used in multisegment displays. Figure 18.21 shows the most commonly used multisegment display. Its seven segments are labelled a, b, c, d, e, f and g. The LED labelled *dp* is used to display the decimal point. By lighting a combination of different LEDs, any number from 0 to 9 can be displayed.

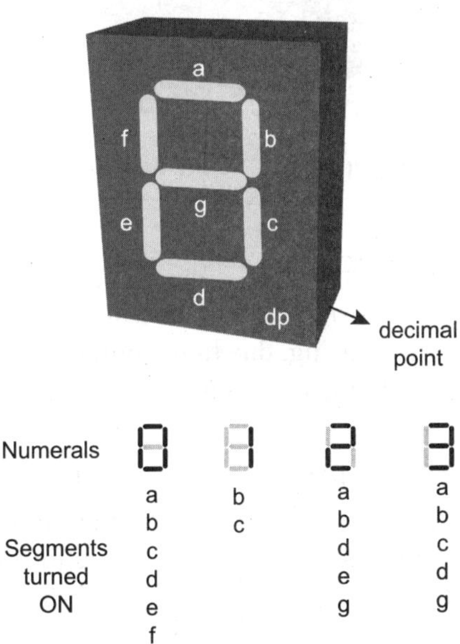

Figure 18.21 Seven-segment display

The seven-segment display cannot be used effectively to display the alphabet, and thus other multisegment displays have been developed. The 16-segment display is illustrated in Figure 18.22, and the 5 × 7 dotmatrix display is shown in Figure 18.23.

Figure 18.22 Sixteen-segment display

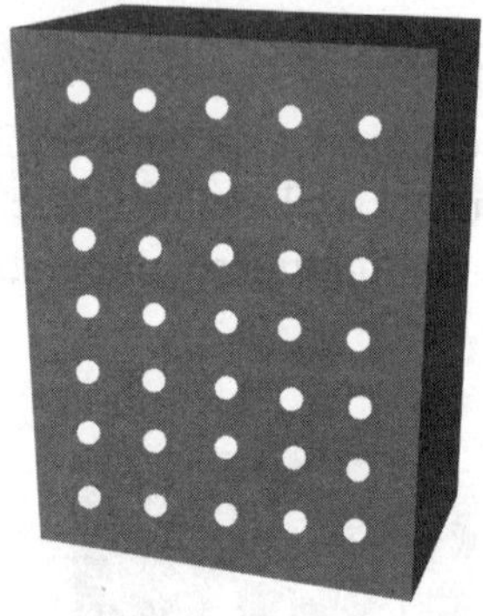

Figure 18.23 5 × 7 dotmatrix

18.5.7 LED Testing

LEDs are usually damaged by excessive current flowing through them. Such LEDs can be identified by discolouration on their casing, due to the burned junction. The LED can be tested with a DMM (digital multimeter). In the diode check position, the DMM supplies about 2.5 V at very low current.

If at this setting the cathode of LED is connected to the negative lead of DMM and the anode of LED is connected to the positive lead of DMM, the supplied 2.5 V should be sufficient to forward-bias the LED and cause it to glow dimly, and the forward breakover voltage should be indicated on the meter display. Reversing the meter leads should, as in conventional dioded, read open (∞).

18.5.8 Liquid Crystal Display (LCD)

Principle Molecules in an ordinary liquid normally have random orientations. On the other hand, molecules in liquid crystals like cholesteryl nonanoate and *p*-azoxyanisole are oriented in a definite crystal pattern as shown in Figure 18.24(a). When an electric field is established in such liquid crystals, the molecules which are approximately cigar shaped tend to orient themselves perpendicularly to the electric field and charge carriers flowing through the liquid disrupt the molecular orientation and cause turbulence in the liquid (Figure 18.24(b)). When not activated, the liquid crystal is transparent. When an electric field is applied, molecular turbulence causes light to be scattered in all directions making the activated area apper bright. This phenomenon is known as **dynamic scattering**.

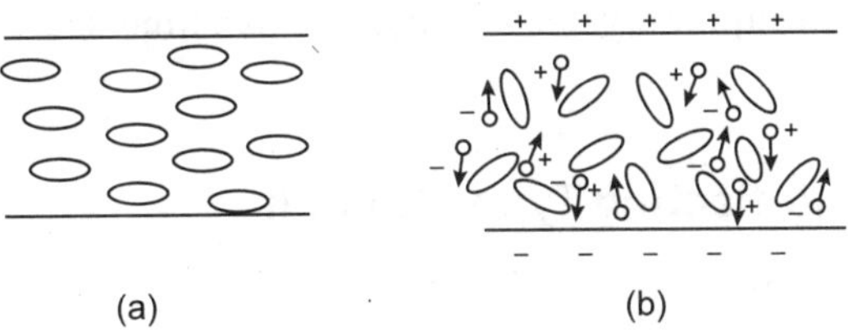

Figure 18.24 (a) Orientation of molecules in liquid crystal, (b) charge carrier flowing through the liquid crystal disturb the orientation of molecules and cause turbulence

18.5.9 Liquid Crystal Cell

A liquid crystal cell consists of a layer of liquid crystal material sandwitched between two glass sheets with transparent metal film electrodes deposited on the inside faces of glass sheets (Figure 18.25). With both glass sheets being transparent, the cell is known as transparent type cell. When only one glass sheet is transparent and other has a reflective coating, the cell is termed reflective type.

When not activated, the transmitive type cell will simply transmit rear or edge light through the cell in straight lines. In this condition, the cell will not appear the bright. When the cell is activated, the incident light is diffusely scattered in forward direction and cell appears quite bright even under high-intensity light conditions.

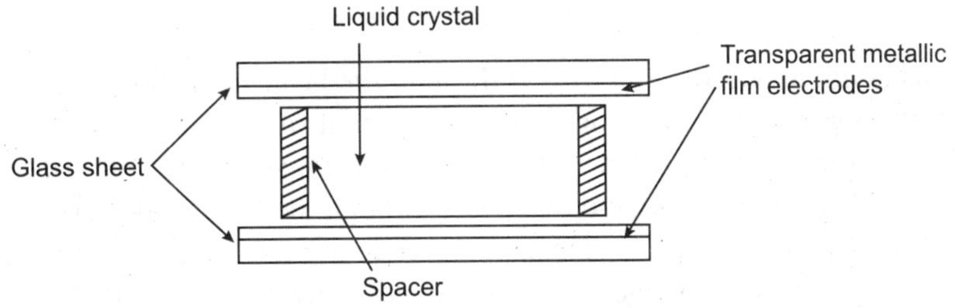

Figure 18.25 Liquid crystal cell

The reflective type cell operates from light incident on its front surface. When it is not activated, light is reflected in the usual way from the mirror surface and the cell does not appear bright. When it is activated, dynamic scattering occurs and the cell appears quite bright.

18.5.10 Electrical Characteristics

Liquid crystals are basically light reflectors or transmitters rather than light generators like LED. Thus, they consume very little power. The only power required is to activate thc liquid crystal material in the cell. The total flow through four small-seven segment displays is typically about 25 μA for field effect cells. However, the LCD requires AC voltage supply either in the form of sine wave or square wave. DC flow would produce plating of cell electrodes, which in turn damages the device.

There are two types of liquid crystal displays: (i) dynamic scattering type, (ii) field effect type.

18.5.11 Field Effect Reflecting Type LCD Cell

The field effect type crystal is shown in Figure 18.26.

Light is incident through a vertical polarizer. Inside the liquid crystal layer, the molecular orientations are graded continuously; the extreme right crystal has vertical orientation and the extreme left crystal has horizontal orientation. The change of orientation from vertical to horizontal occurs progressively as shown by arrows.

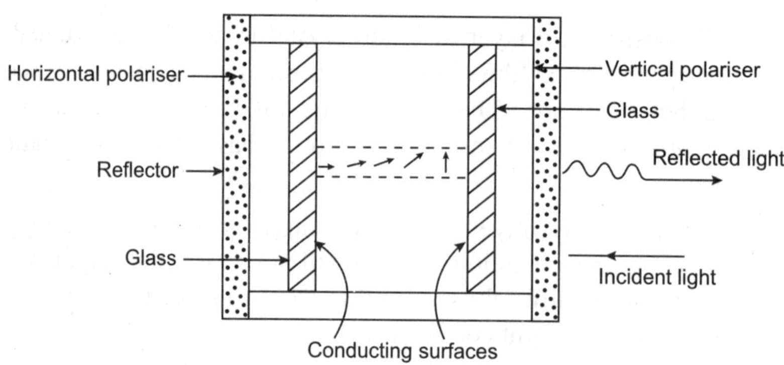

Figure 18.26 Reflecting type LCD

As the vertical component of light travels through the layer of liquid crystal, the direction of polarization alters gradually. The horizontal component of light after passing through the horizontal polarizer gets reflected by the reflector to the right side. This light beam then traces a reverse sequence of transformation and reappears through the vertical polarizer at some point resulting into a bright spot on the LCD. Thus, in presence of light the entire LCD panel appears bright when there is no signal (voltage) across the conducting surfaces of the cell.

Whenever a signal (voltage) is applied across any segment, the resulting electric field disturbs the orientation of molecules of the crystal and light after arriving at the extreme left no longer remains horizontal. It thus fails to pass through the horizontal polarizer and is lost resulting into a dark display of segments. By activating an appropriate combination of segments, a numeric (character) display appears in dark against a bright background.

18.6 Laser

Laser is an acronym for 'light amplification by stimulated emission of radiation'. Laser produces a highly intense, concentrated and parallel beam of monochromatic and coherent light. In the words of C.H. Townes, the inventor of Maser and Laser, "**what the laser does essentially is to give us electronic type control over light. It is a marriage of optics and electronics**".

18.6.1 Principles of Laser Generation

Most light sources are multichromatic, i.e. they comprise many different colours. Each colour of light corresponds to different frequency. Most of these waves are out of phase with one another in an incoherent source of light. Light waves in an incoherent source is shown in Figure 18.27(a). We may note that how these waves (*A*, *B* and *C*) superpose each other and strengthen each other at some points, while opposing and weaking each other at other points. Wave *D* is clearly illustrating this. Light from most ordinary light sources spread or diffuse in various directions. Due to this reason light beam cannot travel very for.

On the other hand laser light sources are **monochromatic**, i.e. in a laser beam all waves are in phase, i.e. laser beam is a coherent beam. Moreover, laser light sources emit light in one direction only. Obviously, laser light is **monochromatic, coherent** and **unidirectional**. Figure 18.27(b) shows waveforms *A*, *B* and *C* to be of one frequency or wavelength and also in phase. We can see that waveform *D* completely agrees with *A*, *B* and *C*. Further, we note that *D* is an constructive superposition of *A*, *B* and *C*.

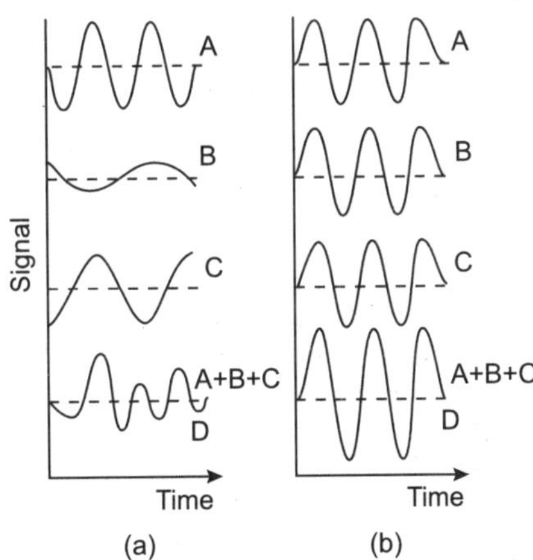

Figure 18.27 Light sources: (a) incoherent, (b) coherent

Let us consider that an electron in an atom undergoes transitions between two energy states or energy levels; it either emits or absorbs a photon, which can be described in terms of a wave of frequency υ, where $\upsilon = \Delta E/h$, ΔE being energy difference between two levels concerned. Let us consider electron transitions which may occur between the two energy levels marked 1 and 2 of an hypothetical atomic system shown in Figure 18.28. If the electron is in the lower level E_1, then in the presence of photons of energy $E_2 - E_1 = h\upsilon_{12}$, it may be excited to the upper level E_2 by **absorbing** a photon (energy $h\upsilon_{12}$). Alternatively, if the electron is in level E_2, it may return to the ground state with the **emission** of a photon. The emission process may take place in two distinct ways: (i) **spontaneous emission** in which the electron falls to the lower level in an entirely random way, and (ii) stimulated emission in which the electron is 'triggered' to undergo transition by the presence of photons of energy $(E_2 - E_1)$. Radiation from **stimulated emission** is monochromatic since each photon has précisely an energy $h\upsilon_{12}$ and is coherent as all emitted photons are in phase. Stimulated emission is a dominating process in laser diodes, whereas absorption is a dominating process in photodetectors and solar cells. Under thermal equilibrium conditions and for $(E_2 - E_1) > 3\,kT$, the population ratio in both excited states may be expressed as:

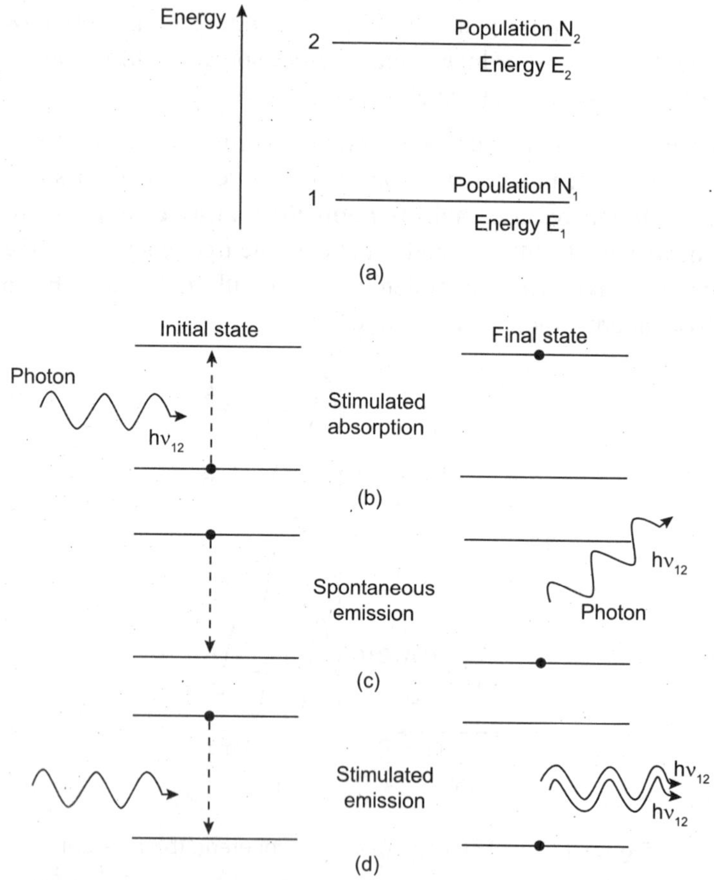

Figure 18.28 Energy level diagram exhibiting (a) two energy level systems, (b) absorption, (c) spontaneous emission and (d) stimulated emission. The black dot in the diagram indicates the state of atom prior and after the transition.

$$\frac{n_1}{n_2} = \exp\left[\frac{-(E_2 - E_1)}{kT}\right]$$
$$= \exp\left[\frac{-h\upsilon_{12}}{kT}\right] \quad (1)$$

where n_1 is the instantaneous population in state E_1 and n_2 is the instantaneous population in state E_2. The negative sign in Eqn. (1) indicates that the excited state E_2 is less populated.

In steady state, to maintain the population in the respective energy states, the stimulated emission rate and spontaneous emission rate must be balanced.

Stimulated emission rate is directly proportional to photon field energy density $\rho(h\upsilon_{12})$ which is defined as the total energy in radiation field per unit volume per unit frequency.

Spontaneous emission rate is proportional to the population of the upper level, i.e.

Stimulated emission rate $= B_{21}n_2\rho(h\upsilon_{12})$

Spontaneous emission rate $= A_{21}n_2$

Here B_{21} and A_{21} are proportionality constants.

Now, rate of absorption is proportional to electron population at the lower state E_1,

i.e. rate of absorption $= B_{12}n_1\rho(h\upsilon_{12})$, where B_{12} is proportionality constant.

Thus, at steady state, we have

Rate of spontaneous emission + Rate of stimulated emission = Rate of absorption

or
$$A_{21}n_2 + B_{21}n_2\rho(h\upsilon_{12}) = B_{12}n_1\rho(h\upsilon_{12}) \quad (2)$$

or
$$\frac{\text{Stimulated emission rate}}{\text{Spontaneous emission rate}} = \frac{B_{21}n_2\rho(h\upsilon_{12})}{A_{21}n_2}$$
$$= \left(\frac{B_{21}}{A_{21}}\right)\rho(h\upsilon_{12}) \quad (3)$$

From Eqn. (3) it is clear that to increase stimulated emission rate over spontaneous emission rate, we have to increase $\rho(h\upsilon_{12})$. One can achieve this by increasing the photon field using an optical resonant cavity. Further,

$$\frac{\text{Stimulated emission rate}}{\text{Absorption rate}} = \left(\frac{B_{21}}{A_{21}}\right)\left(\frac{n_2}{n_1}\right) \quad (4)$$

We also note that stimulated emission rate dominates over absorption rate if the number of atoms is larger in the excited state than that in the ground state. This condition is known as **population inversion**.

A_{21}, B_{21} and B_{12} are called **Einstein coefficients**. One can easily show that

$$g_1 B_{12} = g_2 B_{21} \tag{5}$$

and

$$\frac{A_{21}}{B_{21}} = \frac{8\pi h\upsilon^3}{c^3} \tag{6}$$

where g_1 and g_2 are degeneracies of respective levels. Eqns. (5) and (6) are referred to as Einstein relations. Relation (5) enables us to evaluate the ratio of rate of spontaneous emission to the rate of stimulated emission for a given pair of energy levels. This ratio is given by

$$R = \frac{A_{21}}{\rho(h\upsilon_{12})B_{21}} \tag{7}$$

where

$$R = \exp\,(h\upsilon/\mathrm{kT}) - 1 \tag{8}$$

Obviously, the process of stimulated emission competes with the process of spontaneous emission and absorption. Clearly, if we wish to amplify a beam of light by stimulated emission, then the rate of this process in relation to other two processes must be increased. Eqn. (2) indicates that in order to achieve this for a given pair of energy levels one will have to increase both radiation density and population density N_2 of the upper level in relation to population density N_1 of the lower level. Clearly, to achieve laser action, we must create a condition in which $N_2 > (g_2/g_1)N_1$ even though

$$E_2 > E_1$$

i.e. we must create a so called population inversion. Population inversion can be realized by using a forward-biased, degenerately doped PN junction diode. For a system to lase, the optical gain must exceed all optical losses such that optical field grows with the distance and the field should be resonantly amplified.

18.6.2 Fluorescence and Phosphorescence

A fluorescent material glows under an ultraviolet light source and stops glowing as soon as the source is turned off. A phosphorescent material keeps glowing for a while.

Phosphorescent materials have the ability to store light and release it gradually. One can explain this behaviour by metastable consideration. Whenever molecules of a substance are excited from ground state to metastable state, and if the metastable state can decay slowly back to the ground state via photon emission, phosphorescence occurs.

18.6.3 Classes of Lasers

Since 1960, there has been rapid development in the field of lasers and in the range of materials in which lasing has been shown to occur. The basic requirements which must be satisfied for laser operation are:

(i) There must be an **active medium** which emits radiation in the desired region of electromagnetic spectrum.

(ii) **Population inversion** must be created within the medium. This requires the existence of suitable energy levels associated with lasing transition for pumping.

(iii) For true laser oscillation, there must be **optical feedback** at the ends of the medium to form a **resonant cavity** (the first two conditions can provide light amplification but not highly collimated, monochromatic beam of light). Laser action can be achieved in gases, liquids, solids, flames and ions. The emitted wavelength covers ultraviolet to radio frequency regions. Some lasers emit continuous waves while others in pulses. Characteristics and applications of few types of laser are given in Table 18.3.

Table 18.3 Characteristics and applications of few different types of lasers

Laser	Type	Common wavelengths (μm)	Max. output power (W)	Applications
HeNe	Gas	0.6328, 1.15, 3.39	0.0005–0.05 (CW)	Line-of-sight communication, recording/playback of holograms
CO_2	Gas	9.6, 10.6	500–15,000 (CW)	Heat treating, welding, cutting, scribing, marking
Argon	Gas ion	0.488, 0.5145	0.005–20 (CW)	Surgery, distance measurements, holography
HeCd dye	Metal vapor, liquid	0.441, 0.325 0.38–1.0	0.05–0.1 0.01 (CW) 1×10^6 (P)	Light shows, spectroscopy, pollution detection
Ruby Nd-YAG	Solid state	0.694 1.06	(P) 100(CW) 2 × 108(P)	Pulsed holography, hole piercing, cutting
Nd-glass diode	Solid-state semiconductor	1.06 0.33–40	5×10^{14}(P) 0.6(CW) 100(P)	Pulse welding, hole piercing, barcode reading, CDs and video disks, optical communications.

18.6.4 Semiconductor Lasers

These lasers are not very different in principle from light emitting diodes. A PN junction provides the active medium. Thus, to obtain laser action, we need only to achieve population inversion and optical feedback.

GaAs was the first material to have emitted laser radiation and is related to III-V compound alloys. The other three most important III-V compound alloy systems are $G_{ax}I_{nt\text{-}x}\,A_{SY}\,P_{I\text{-}Y}$, $G_{ax\text{-}ni\text{-}x}\,A_{SY}\,Sb_{I\text{-}Y}$ and $Al_xGa_{i\text{-}x}\,A_{sy}Sb_{i\text{-}y}$ solid solutions. To achieve a heterostructure with negligible face traps, there must be close maching in lattices between two semiconductors.

18.6.5 Population Inversion

To achieve population inversion in a semiconductor laser, it is required that a PN junction or heterojunction form between two degenerated semiconductors. This mean Fermi level on P side should be below the valence band edge, and Fermi level on N side should be above the conduction band edge (Figure 18.29(a)). This will enhance carrier injection across the junction when sufficiently large bias is applied, i.e. large concentration of electrons and holes are injected into the transition region. As a result, the transition region *d* contains a large concentration of electrons in the conduction band and a large concentration of holes in the valence band. This is the required condition to achieve population inversion. The minimum required energy is band gap, E_g. The necessary condition for population inversion is

$$(E_{FC} - E_{FV}) > E_o$$

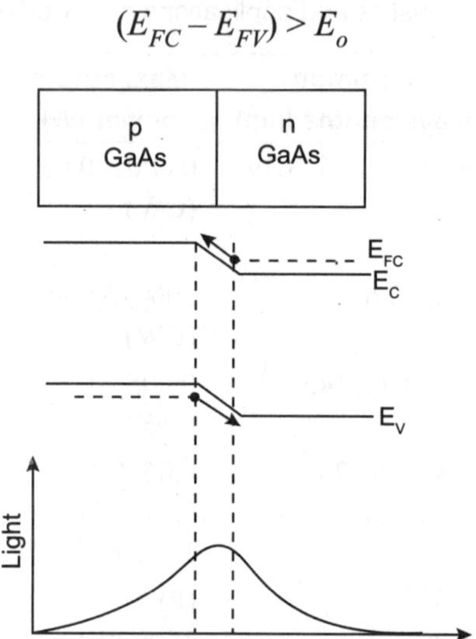

Figure 18.29 Schematic representation of semiconductor laser

We may note that minimum current density and corresponding minimum electron concentration have to be injected into the active region so that one may achieve the threshold and lasing action.

18.6.6 Types of Semiconductor Lasers

Homojunction semiconductor laser, in this type of laser, one can achieve relatively high threshold current densities because both carriers and optical confinement within the active region of device are minimum. However, carriers injected into the active region of device pass through it without any recombination and do not contribute to laser action. It is also seen that optical fields leak into the surrounding inactive layers. Due to these reasons, threshold current density increases and reduces the efficiency of laser action.

The **heterostructure** help to achieve a mean of confining both optical fields and carriers within the active region and reducing threshold current density. For heterostructure, the lattice constants of two materials should match, e.g., GaAs and AlGaAs. The structure of a simple GaAs-AlGa laser is shown in Figure 18.30, which is called a **double heterostructure laser** since it contains two GaAs-AlGaAs interface. Moreover, the energy bandgap of AlGaAs is greater than that of GaAs, and there is energy gap discontinuity between the two materials. GaAs and AlGaAs form a type I heterojunction in which the energy gap discontinuity is equal to the sum of conduction and valence band edge discontinuities.

The presence of conduction and valence band energy discontinuities in the two heterostructures acts to confine the injected carriers within the narrow gap. GaAs layer as shown in Figure 18.30 can result ultimately in a radiative recombination event with subsequent emission of a photon. The heterostructure has lower threshold current density in comparison to the homojunction laser. The double heterostructure also acts to confine optical fields within the GaAs layer due to different refractive indices of semiconductor materials involved.

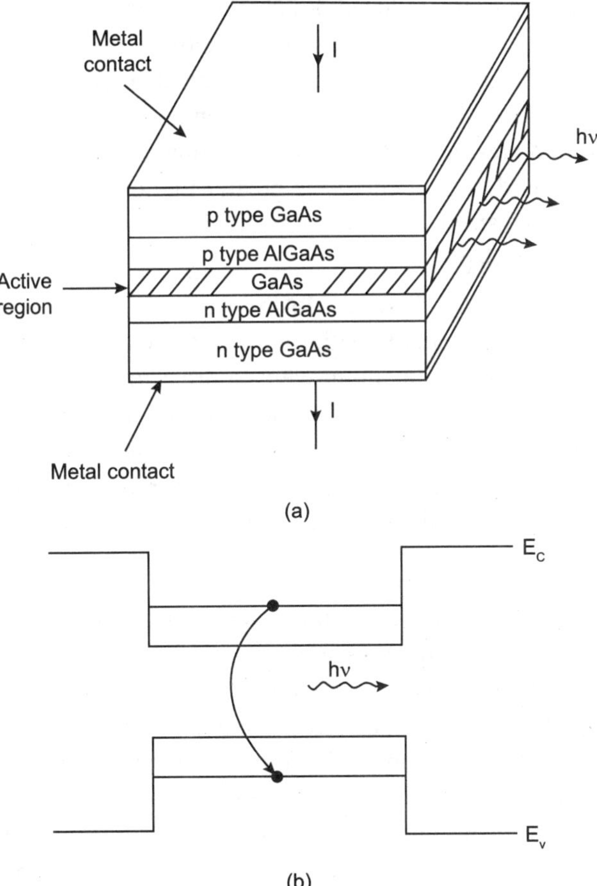

Figure 18.30 Schematic illustration of: (a) Structure of GaAs-AlGa laser, (b) radiative event in GaAs layer

Due to reflection of light from the boundaries of the device, light bounces back and forth. By cleaving the crystal, the reflecting surfaces are produced and the cleaved surface provides for almost complete reflection. Reflected light trigger the stimulated emission within the GaAs layer and coherence is maintained. In case of double heterostructure laser, the emitted light has the energy very close to that of energy gap of narrow gap material. Obviously, the energy of emitted photon depends on the intrinsic properties of active medium.

18.6.7 Quantum Well Lasers

The active layer in the conventional heterostructure lasers is very thin, i.e. between 100 and 200 nm. Obviously, this is thin enough to confine electrons and the optical field. However, the electronic and optical properties remain unchanged.

Quantum well lasers offer an alternative to double heterostructure, i.e. in quantum well lasers the active layer is made thinner than 10 nm. Due to this the electronic and optical properties are altered drastically owing to the reduced dimensionality of the motion of the free electrons, from the three dimensional to two-dimensional nature. Obviously, in quantum well lasers spatial quantization effects occur, which results in the production of energy levels above the conduction band minimum. This means that the minimum electron and hole energies within the well are both above the conduction band and valence band minima, respectively. Now, if electrons and holes recombine, the emitted photon energy is greater that energy gap. It is reported that the energy of quantum states is a strong function of well width. One can tune the energy of emitted photon by adjusting the well depth.

Limitations of quantum well lasers In the well, the cross section for the capture is relatively small. For a radiative recombination event, it is essential for the injected electrons and holes to be captured within the quantum well. In order to confine carriers within the well, they must undergo an inelastic scattering event so that they may lose sufficient energy. In the absence of scattering event within the well width, the injected carriers will traverse the well without being captured, i.e. the injected carriers can pass directly over the well without getting trapped and obviously, do not contribute to the optical output. This means, the threshold current density can be relatively high.

Multiple quantum well lasers One can improve the capture probability of a single quantum well laser by using multiple quantum wells as shown in Figure 18.31. The single quantum well within the active region is replaced by a series of quantum wells. Since more quantum wells are available in the structure, there is a higher probability that the injected carriers will suffer an inelastic scattering event and be captured within a well prior it can completely traverse the device. The injected electrons and holes in multiple quantum well enter the active region from opposite ends and hence electrons and holes are not necessarily captured within the same well which acts to reduce the radiative recombination rate. Due to this drawback, multiple quantum well devices do not have the lowest threshold current density of semiconductor laser.

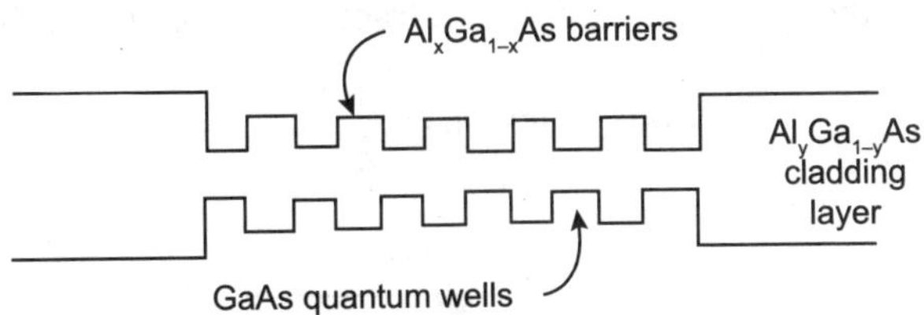

Figure 18.31 Structure of multiple quantum well laser

GRINSCH laser The acronym means graded index separate confinement heterostructures. This has the lowest threshold current density among existing semiconductor lasers. Figure 18.32 shows the structure of Grinsch laser.

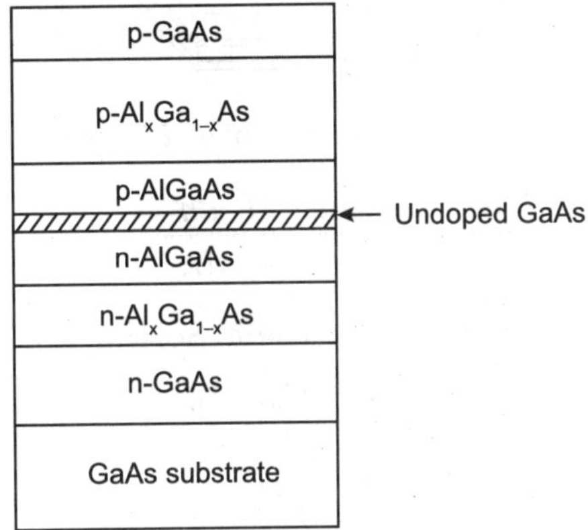

Figure 18.32 Structure of GRINSCH laser

A single quantum well is embedded as shown in figure by graded region. The graded region surrounding the quantum well provides a means by which the injected carriers can be funneled into the well. The bandgap grading induces an electric field that directs the electrons and holes into the same quantum well where they can recombine radiatively. As a result, fewer electrons and holes can traverse the quantum well without being trapped. Thus, the threshold current density of GRINSCH laser is relatively lower in comparison to other lasers.

Vertical cavity surface emitting lasers (VCSEL) In VCSEL laser structure, light is emitted normal to the surface (Figure 18.33). One can achieve a single mode operation due to short cavity length of VCSEL device. This is possible with VCSEL since mode spacing is inversely proportional to a cavity length. VCSEL devices are found better than edge emitting lasers for getting a two-dimensional array. Thus, smaller the cavity length, greater the mode spacing and this leads to single mode operation. Output wavelength is obtained as

$$\lambda_o = \lambda n' \tag{9}$$

where λ is wavelength of incident light and n' is refractive index within the laser.

$$\text{Mode index, } m = \frac{2Ln}{\lambda_o} \tag{10}$$

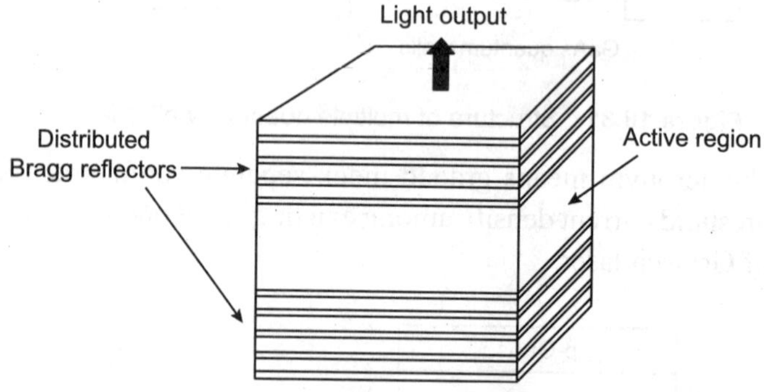

Figure 18.33 Structure of VCSEL laser

Hence
$$\frac{dm}{d\lambda_o} = \frac{-2Ln'}{\lambda_o^2} + \frac{2L}{\lambda_o}\frac{dn'}{d\lambda_o} \tag{11}$$

where $dm/d\lambda_o$ defines mode spacing with wavelength. Solving Eqn. (11), one obtains

$$d\lambda_o = \frac{(\lambda_o / 2Ln')dm}{\left[1 - \frac{\lambda_o}{n'}\frac{dn'}{d\lambda_o}\right]} \tag{12}$$

The value of dm is 1 for adjacent modes.

From Eqn. (12), it is clear that wavelength separation between adjacent modes is inversely proportional to cavity length.

The major disadvantage of VCSEL laser is that cavity length is short as round trip gain of the device is low. The threshold current density of VCSEL laser is significantly higher than that of edge emitting laser.

18.6.8 Laser Diode

In a light emitting diode, the emitted light is incoherent. Obviously, when we require coherent light, LED is not useful. For coherent light, we use a laser beam.

Principle Figure 18.34 illustrates the operating principle of laser light. A PN junction is manufactured with precisely defined length L from GaAs or GaAs combined with other materials. L (junction length) is related to λ (wavelength) to be emitted. The ends of the junction are polished to a mirror surface and possibly may have additional reflective coating so that generated light may reflect back internally into the junction. One end is kept partially reflective so that light can pass through when lasing occurs.

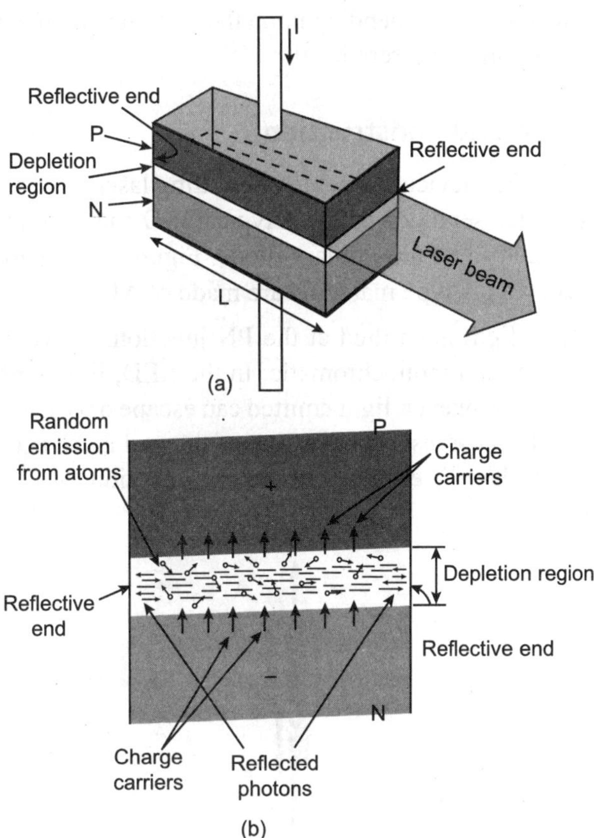

Figure 18.34 (a) Laser diode, (b) Illustration of random emission and laser action within depletion region

Function Let us consider the effect of charge carrier injection into the depletion region when the junction is forward-biased (Figure 18.34(b)). As forward current excites the atoms that strike, first emit photons (radiant energy) of energy randomly, as electrons are raised to a high energy level and then fall back to a lower level. Consequently, several photons strike the reflective ends of the junction perpendicularly so that they are reflected back along their original path. These reflected photons are then again reflected back from the other end of the junction. Obviously, photons have a to-and-fro motion several times and their number goes on increasing as they cause other atoms in their path to emit photons by striking them. This activity of reflection and generation of photons causes amplification of initial reflected photons. A laser beam emerges through the partially reflected surface of the junction. GaAs laser diodes normally require high forward current levels, ranging from 10 mA to tens of amperes. At low current levels, this device emits light similar as in LED. Beyond a threshold current level, light intensity increases sharply and its bandwidth decreases as lasing commences. Due to high energy density, laser beam can be quite harmful to eyes.

Laser diodes operating in a pulsed manner are called **injunction laser diodes**, whereas lasers producing continuous output are called **continuous wave** (CW) laser diodes. Each type of laser emits a particular light wavelength depending upon the material and junction dimensions. Each type of laser has a threshold input current level.

18.6.9 Basic Operation and Construction

Laser diodes are light emitting devices, capable of emitting laser beams. One of the commonly used laser diodes is injection laser diode (ILD). A typical ILD package is shown in Figure 18.35. The laser diode works exactly like an ordinary diode. Figure 18.36 shows the structure of an injection laser. The P type and N type materials are made of AIGaAs.

When forward-biased, light is emitted at the PN junction. However, unlike in the LED, the emitted light is coherent and monochromatic. In the LED, light emitted is scattered in all directions. In the laser diode, however, light emitted can escape only from the end faces because the edges are roughened. In most laser diodes, one of the end faces is coated with a reflective material so that radiation of laser is emitted only in one direction.

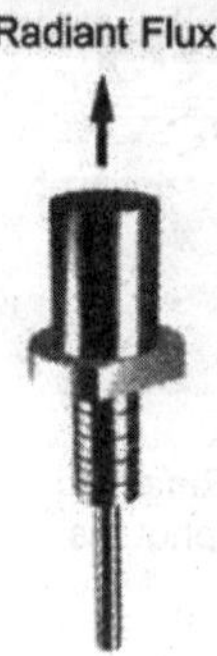

Figure 18.35 A typical injection laser diode (ILD) package

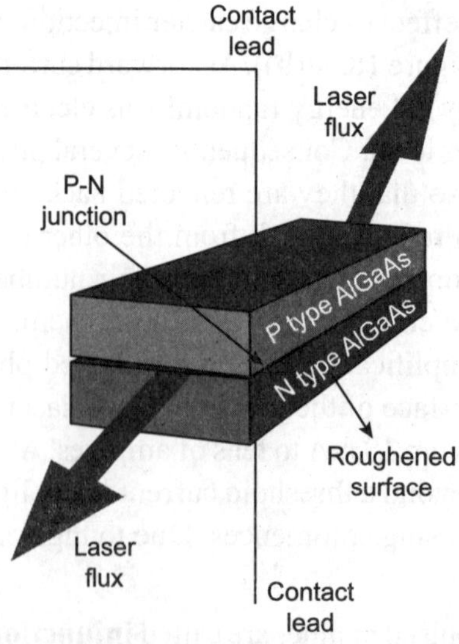

Figure 18.36 Structure of injection laser diode (ILD)

The laser diode emits light when forward-biased and can withstand only relatively small reverse-biased voltages. A high reverse voltage can damage or destroy the laser diode. The symbol for laser diode is shown in Figure 18.37. Note that light emission is shown zigzagged rather than straight, as in the symbol for an ordinary LED.

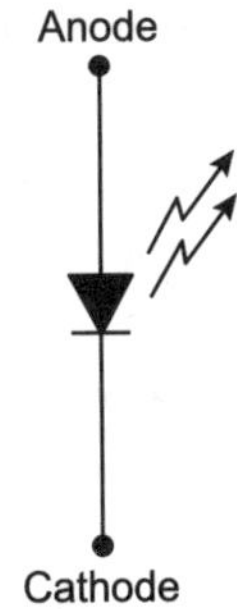

Figure 18.37 Symbol for a laser diode

18.6.10 Applications of Laser Diode

Laser diodes are used primarily in fiber optic communications. It is the only device capable of producing optical energy high enough in concentration to pass through lengthy fiber optic cables. However, it also has some major disadvantages as compared to LEDs.

(i) They cost 10 times more than LEDs.

(ii) Their life expectancy is 10 times shorter than that of LEDs.

(iii) They require elaborate power supplies and consume much more power than LEDs.

Laser diodes do not compete with LEDs on cost-reliability basis. Therefore, LEDs are used as much as possible in fiber optic systems and laser diodes are used only when absolutely necessary.

Some applications of laser diodes are:

(i) Fixed product scanners seen in grocery stores

(ii) Handheld barcode scanners used for inventory control in warehouses

(iii) Handheld laser pointers used in presentations

(iv) Laser projection devices used by surgeons, mechanics, aviators and technicians. Laser scanners project low-power images directly onto the user's retina as an overlay in conjunction with the real-world image.

(v) Land survey range-finding has been revolutionized by the use of laser diodes. Laser transmission from Total Station (reflected from a distant point) is detected, time-delay measured, and converted to an equivalent distance. Figure 18.38 shows a Leica model TC407 Total Station that incorporates laser range-finding.

(vi) Document printers and scanners often incorporate lasers.

(vii) In construction, lasers are used to paint line and angle projections, nearly eliminating the potential for manual measurement errors.

Lasers are categorized into five classes:

(i) **Class I** Laser output between 0.221 and 0.39 milliwatt. Class I lasers are eye safe.

(ii) **Class II** Laser output between 0.39 and 1.0 milliwatt. Class II lasers are eye safe with caution.

(iii) **Class IIIa** Class IIIa applies to visible spectrum laser output between 1.0 and 5.0 milliwatts. These lasers are considered eye safe with danger.

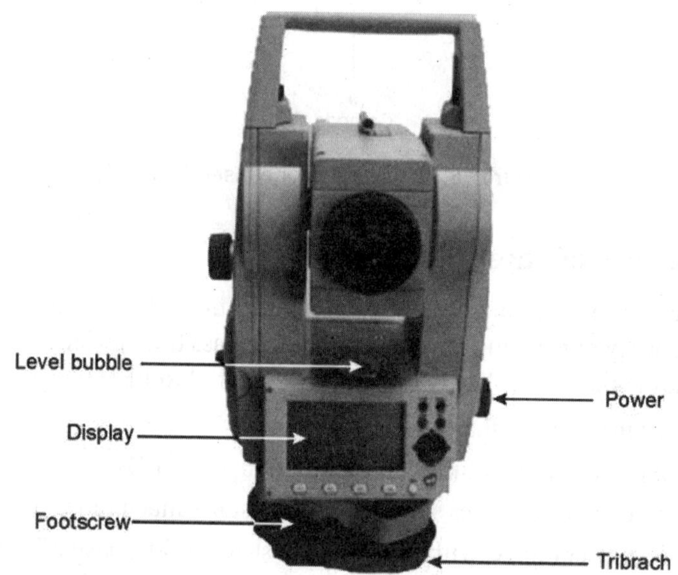

Figure 18.38 Leica model TC407 Total Station

(iv) **Class IIIb** Class IIIb applies to visible spectrum laser output between 5.0 and 500 milli watts, as well as to invisible spectrum laser output of less than 500 milliwatts. These lasers produce sufficient output to cause damage to the retina, and must be labelled "danger".

(v) **Class IV** Class IV applies to visible and invisible lasers of 500 milliwatts and higher. These devices must be labelled "danger".

Laser pointers fall within Class II or Class IIIa. A Class IIIa danger-labelled laser pointer body is shown in Figure 18.39.

Figure 18.39 Class IIIa laser pointer

18.7 Photodiode

18.7.1 Basic Operation and Construction

A photodiode is a light receiving device that contains a semiconductor PN junction. Figure 18.40 shows a typical photodiode package. A glass window or convex lens allows light to enter the case and strike the photodiode mounted within the glass case.

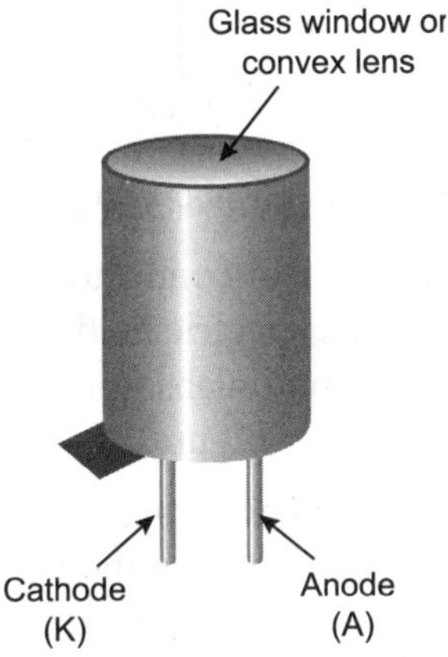

Figure 18.40 Typical package of a photodiode

Figure 18.41 shows the construction of a photodiode. It is constructed basically with a P type region diffused into an N type region. The metal base makes a connection between the cathode terminal and N type region, and the metal ring makes a contact between the anode and P type region. Light enters the photodiode through a hole in the metal ring, which accommodates the glass or convex lens. Figure 18.42 shows the most commonly used symbol of a photodiode. Note that the two arrows point towards the photodiode, indicating that it responds to light. Photodiodes can operate in two modes:

(i) Photovoltaic mode
(ii) Photoconductive mode

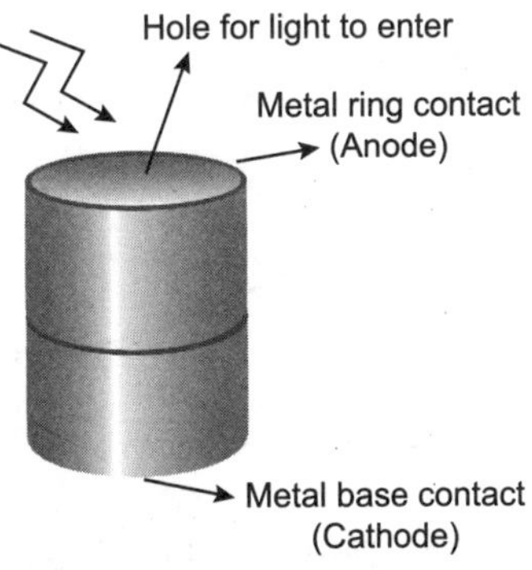

Figure 18.41 Construction of a PN photodiode

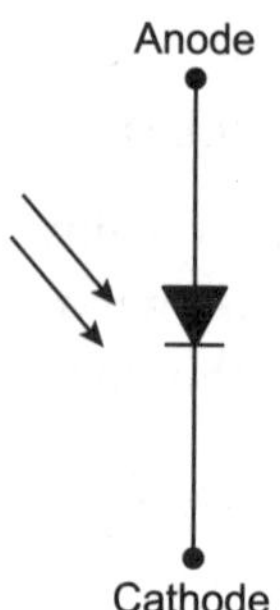

Figure 18.42 Symbol of a photodiode

When a PN junction is formed in the semiconductor material, a region depleted of mobile charge carriers is formed with high internal electric field across it and called as depletion region. If an electron-hole pair is generated by photon absorption within this region, the internal field will cause the electron and hole to separate (Figure 18.43).

One may detect this charge separation in two ways: (i) when the device is left on open circuit, an externally measurable potential appear between the P and N regions, and this is termed as **photovoltaic mode of operation**. (ii) when the device is shortcircuit externally or operating under reverse-bias, then external current flows between P and N regions. This is termed as **photoconductive mode of operations**.

The PN junction will also respond to electron-hole pairs generated away from the depletion region provided they are able to diffuse to the edge of the depletion region prior the recombination takes place. Obviously, only carriers generated within a minority carrier diffusion length or so of the edge of the depletion region are likely to be able to do this; however, this does increase the sensitive volume of the device.

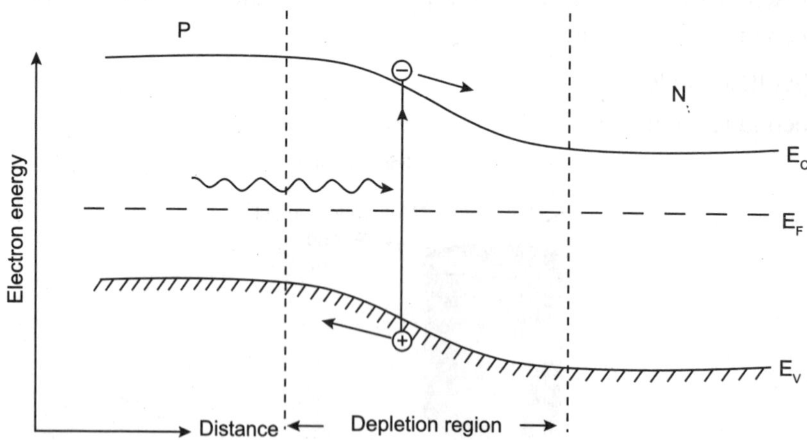

Figure 18.43 Electron energy level diagram illustrating the generation and subsequent separation of an electron-hole pair by photon absorption within the depletion region of a PN junction

In operation, one may represent the photodiode by a constant current generator (current flow i_λ being generated by light absorption) with an ideal diode across it (to stimulate the effect of PN junction) (Figure 18.44). The internal characteristics of the cell may be better modelled by the introduction of a shunt resistor (R_{sh}), a shunt capacitor (C_d) and a series resistor (R_s). If we assume quantum efficiency η for the photon absorption process and also that all the incident radiation is absorbed within the cell, then one may write.

$$i_\lambda = \frac{\eta I_o Ae\lambda}{hc} \tag{1}$$

where I_o is light irradiance falling on a cell of area A.

One can develop expressions for sensitivities of the photodiode when operated in two different modes. For simplicity, we may assume that the operation is at fairly low optical modulation frequencies, so that the effects of shunt capacitance may be neglected (i.e. $i_c = 0$).

Considering current to flow as shown in Figure 18.44, one finds

$$i_\lambda = i_d + i_{sh} + i_{ext} \tag{2}$$

also

$$V_{ext} = V_d - i_{ext} R_s \tag{3}$$

and

$$V_d = i_{sh} R_{sh}$$

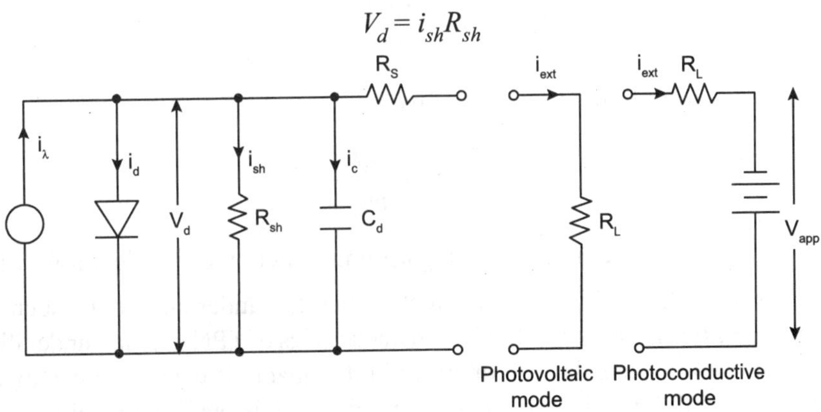

Figure 18.44 Photodiode equivalent circuit

We now take current-voltage behaviour of the diode to be given by

$$i_d = i_o \left[\exp\left(\frac{eV_d}{kT} \right) - 1 \right]$$

where i_o is diode reverse bias leakage current.

In photovoltaic mode, external current flow is very small; hence, taking $i_{ext} = 0$, Eqns. (2) and (3) become $i_\lambda = i_d + i_{sh}$ and $V_{ex} = V_d$, respectively. Substituting for i_d and i_{sh} in the former, one obtains

$$i_d = i_o\left[\exp\left(\frac{eV_d}{kT}\right) - 1\right] + \frac{V_d}{R_{sh}}$$

Rearranging, one obtains

$$\exp\left(\frac{eV_d}{kT}\right) = 1 + \frac{i_\lambda}{i_o} - \frac{V_d}{i_o R_{sh}}$$

Under normal operating conditions, $i_\lambda > i_o$ whereas V_d is usually the same order of magnitude as iR_{sh}. For example, typical values might be $i_o \approx 10^{-8}$ A, $R_{sh} \approx 10^8\ \Omega$ and $V_d \approx 0.6$ V. Thus, to first approximation, one finds

$$\exp\left(\frac{eV_d}{kT}\right) = \frac{i_\lambda}{i_o} = \frac{i_\lambda}{i_o} \tag{4}$$

It then follows that

$$V_d (= V_{ext}) = \frac{kT}{e} \ln\left(\frac{i_\lambda}{i_o}\right)$$

Substituting for i_λ from Eqn. (1), one obtains

$$V_{ext} = \frac{\mathrm{kT}}{e} \ln\left(\frac{\eta I_o e\lambda A}{hci_o}\right) \tag{5}$$

Obviously, external voltage should be a logarithmic function of incident light irradiance.

A solar cell is basically a PN junction detector operated under conditions such that it can deliver power into an external load. The full I-V characteristics of a PN junction under illumination is shown in Figure 18.45. The solar cell operates in the quadrant where *I* is negative and *V* is positive. Most power will be delivered when the product *I-V* is maximum and this determines the **optimum load impedance** across the cell. Presently, most commercially available solar cells are made from silicon which can convert up to 15% of solar radiation incident on them into electrical energy. In bright sunlight, each cell can deliver current densities of the order of 80 AM^{-2} at a voltage of about 0.6.

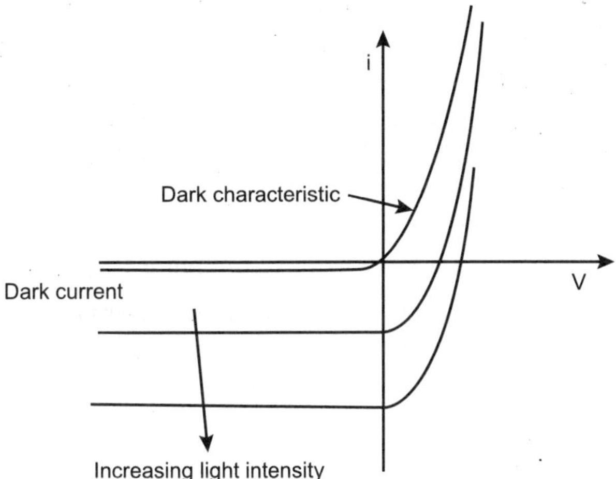

Figure 18.45 Current-voltage characteristic of a PN junction solar cell under various levels of illumination. The dark characteristic is that of an ordinary PN junction diode. Under increasing levels of illumination, the curve is progressively shifted downwards.

V as a means of power generation is not yet found economical except for certain special uses (e.g. low power requirement in remote areas and as power source for satellites). Cheaper processing technology for silicon or development of thin film evapourated cells may well change this situation in the near future.

In photoconductive mode, a relatively small value of reverse bias is applied (i.e. few tenths of a volt). Eqn. (2) can now be written as

$$i_\lambda = i_o + i_{sh} + i_{\text{ext}}$$

Now, $i_o = 10$ nA, while from Eqn (4), $i_{sh} = V_d/R_{sh} \approx 10\ \text{V}/100\ \text{M}\Omega \approx 100$ nA

so that when i_λ is of the order of microamps and above, one may write $i_{\text{ext}} = i_\lambda$.

Substituting from Eqn. (1), one obtains

$$i_{\text{ext}} = \frac{\eta I_o A e \lambda}{hc} \tag{6}$$

Thus, in photoconductive mode, external current flowing is directly proportional to incident light irradiance.

In addition to its inherently linear response, the photoconductive mode usually offers the advantages of **faster response**, **better stability** and **greater dynamic range**. The main drawback of this mode is the presence of dark current ($i_o + i_{sh}$) which as in a photomultiplier gives rise to shot noise and limits the ultimate sensitivity of the device. Both modes of operation are subject to noise generation but recombination noise is absent since charge carriers are separated in the depletion region before they can recombine.

The most common semiconductor material used in photodiodes is silicon. It has an energy gap of 1.14 eV and provides excellent photodiodes with quantum efficiency up to 80% at wavelengths between 0.8–0.9 μm. A typical construction is shown in Figure 18.46. We may note that electrical contact to the semiconductor material is always made via a metal N^+ (or P^+) junction. This is most convenient way of providing an ohmic contact. It is also possible to make detectors that are illuminated from the side (i.e. parallct to the junction). This type of construction results in good sensitivity for wavelengths close to the bandgap limit where optical absorption coefficient is relatively small. Detection efficiency may be in consisting of a $\lambda/2$ coating of SiO_2. Figure 18.47 shows a typical spectral response. Electric field variation within the structure of Figure 18.46 when under reverse biased is illustrated in Figure 18.48.

Let us assume that we have an abrupt junction, having external bias V much larger than internal junction potential and that $N_a \gg N_d$, then depletion layer width can be written as

$$X_N = \left(\frac{2\varepsilon_o \varepsilon_r V}{eN_d} \right)^{\frac{1}{2}} \tag{7}$$

$$X_P = \left(\frac{2\varepsilon_o \varepsilon_r V N_d}{eN^2{}_d} \right)^{\frac{1}{2}}$$

Since we have a P^+ N structure, for which $N_n \gg N_p$. For efficient detection, electron-hole pairs should be generated either inside or within a diffusion length or so of the depletion region. At short wavelengths, where absorption coefficient is relatively high, they will be generated close to the surface. Consequently, to achieve a good short wavelength response, the P^+ region should be made as thin as possible. Conversely, at upper wavelength range of the detector, the absorption coefficient is relatively small and a wide depletion region is necessary for a good response. This shows that large values of reverse-bias voltage are required (Eqn. (8)) which may approach or exceed diode breakdown voltage.

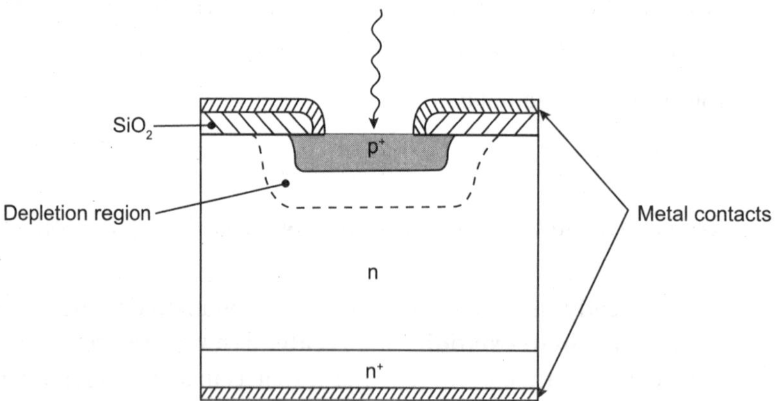

Figure 18.46 Typical silicon photodiode structure for photoconductive operation. A junction is formed between heavily doped P-type material (P^+) and fairly lightly doped N-type material so that the depletion region extends well into the *N* material. The P^+ layer is made fairly thin. Metallic contacts can be made directly to the P^+ material, but to obtain an ohmic contact to the *N* material an intermediate N^+ layer have to be formed.

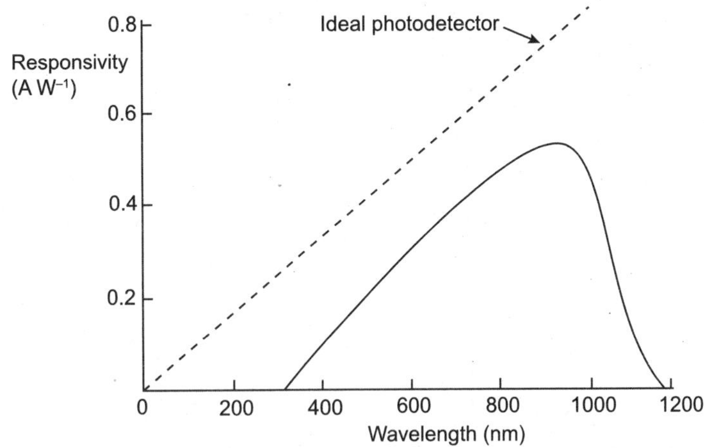

Figure 18.47 Typical current responsivity of a silicon photodiode. Figure also shows the responsivity of an ideal photodiode with unit quantum efficiency

A structure that results in a good long wavelength response with only relatively modest bias levels is the so-called pin (or PIN) structure (Figure 18.49).

The intrinsic (*i*) region has high resistivity (i.e. low values of N_d and N_a) so that only a few volts of reverse-bias are required to cause the depletion region to extend all the way through to the N region and thus provide a large sensitive volume. In practice, the bias is maintained considerably higher than the minimum value and the intrinsic region then remains fully depleted of carriers even at high light levels.

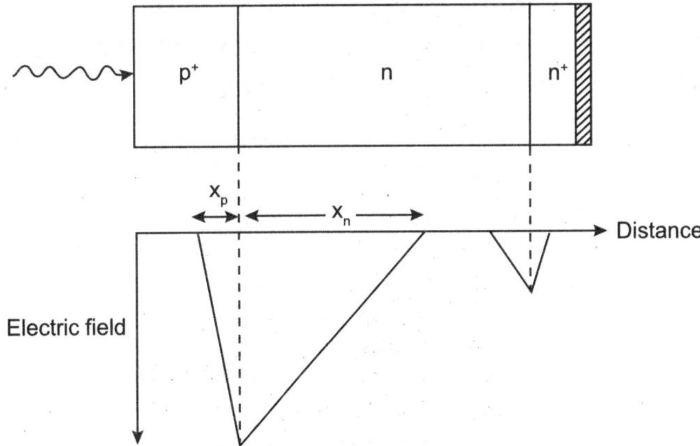

Figure 18.48 Electric field distribution within the P$^+$N junction diode corresponding to Figure 18.46, assuming an abrupt diode structure

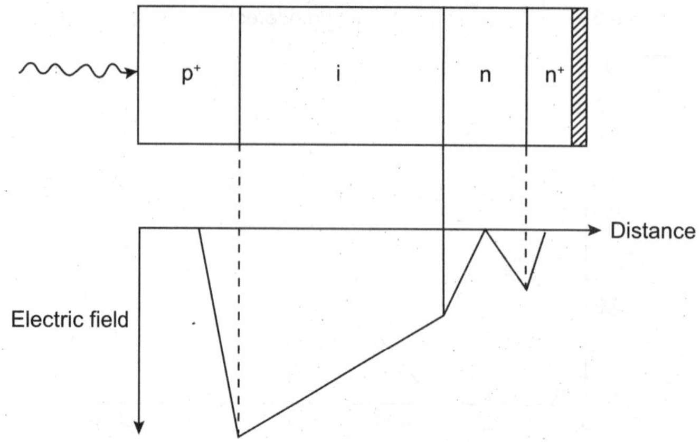

Figure 18.49 Electric field distribution within a PIN structure

18.8 P-I-N Photodiode

P-I-N is one of the most common photodetectors because of its high quantum efficiency and bandwidth. Typically, a P-I-N photodiode is illuminated from P side as shown in Figure 18.50. Since the device is reverse biased, the depletion region contains the intrinsic region, and electric field strength is constant within the intrinsic region. Holes generated near the edge of depletion region on P side usually recombine within the P region or at the contact. In general, electrons can be photodegenerated within the P^+ region and also within the intrinsic region of the diode. Usually, most of the PIN diodes are designed in such a manner that the vast majority of carriers are photogenerated within the intrinsic region. Obviously, intrinsic region acts as a depletion region since N as well as P side is heavily doped. We may note that in a normal photodiode, some of the carriers are generated outside the depletion region.

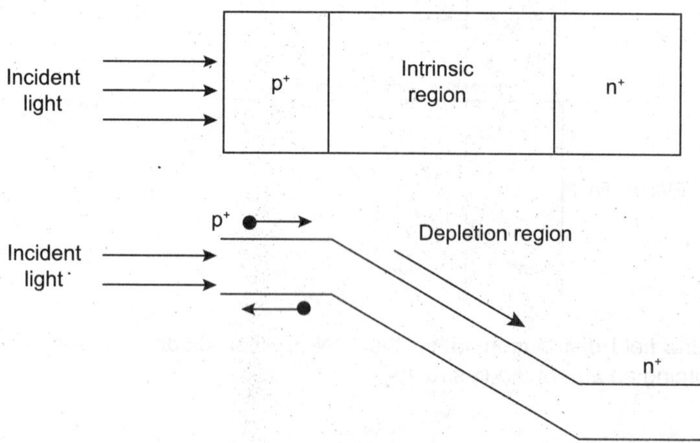

Figure 18.50 Structure and band diagram of P-I-N photodiode

Owing to the relatively high electric field, the current within the intrinsic region flows mainly from drift. Illuminating P side rather than N side, electrons as carries species drift through the depletion region (Figure 18.50). In most PIN photodiodes, the P^+ layer is relatively thin and therefore most of the photogenerated carriers are produced within the intrinsic region. When the applied voltage is such that electric field within the intrinsic region is less than the critical field, then relatively few events takes place in the device.

The **gain** of a photodiode is unity, i.e. there is no carrier concentration or gain in the structure.

When the magnitude of reverse bias is increased to a point at which within the intrinsic region becomes equal or exceeds the critical field, then carrier multiplication happens due to **impact ionization**. Such photodiodes are referred as **avalanche photodiodes** and they have a gain.

Limitation of P-I-N photodiodes The frequency response of a P-I-N photodiode is limited transit time effects or circuit parameters. The most important parameter which due to affects carrier transit time is the width of the intrinsic region. As the width of intrinsic region increases, the total distance travelled by photogenerated carriers also increases and due to this the device bandwidth reduces. However, if intrinsic width is made too small, quantum efficiency is reduced. Obviously, one should select optimal intrinsic width.

Typically, a P-I-N photodiode is designed in such a way that transit time through the intrinsic region is equal to one-half of modulation period of the optical signal,

$$f = \frac{I}{2T_t} \tag{8}$$

where T_t is transit time within the intrinsic region.

18.8.1 Response Time of Photodiodes

Following three main factors limit the speed of response of photodiodes:

Diffusion time of carriers to the depletion region Carrier diffusion is inherently a relatively slow process. Time taken for carriers to diffuse a distance d is found to be

$$T_{\text{diff}} = \frac{d^2}{2D_e} \tag{9}$$

where D_e is minority carrier diffusion coefficient.

To ensure that few carriers are generated outside the depletion region, we require that, at wavelengths being used,

$$W = X_n + X_p \geq \frac{1}{\alpha(\lambda)}$$

where $\alpha(\lambda)$ is **absorption coefficient** at wavelength λ. At wavelengths near the bandgap limit and at fairly low reverse-bias levels, the above inequality may not hold and the speed of detection of some optically generated carriers will then be limited by diffusion. In these circumstances, the response to a narrow optical pulse will be as illustrated in Figure 18.51. The carriers generated within the depletion region respond rapidly, whereas those generated outside give rise to a slow tail. The relative importance of the tail may be decreased by increasing the reverse bias or by using the PIN structure where the depletion region is relatively wide.

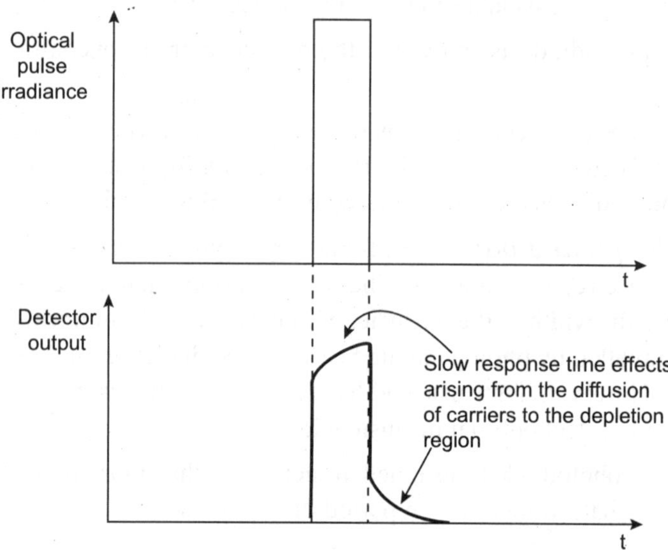

Figure 18.51 Response of a PN junction detector to a narrow optical pulse. The effects arising from the relatively long time taken by carriers generated away from the depletion region to diffuse to the junction is shown.

Drift time of carriers through the depletion region In very high electric fields, drift velocities of carriers in semiconductors tend to saturate. Since the field within the depletion region exceeds the saturation value for most of its length (which is usually the case in practice), then one may assume that carriers move with a constant velocity υ_{sat}. The longest transit time will result when carriers are generated near to one edge of the depletion region. In this case, one of the carriers will have to traverse the full depletion layer width W. The transit time is as follows:

$$T_{drift} = \frac{W}{\upsilon_{sat}} \tag{10}$$

Junction capacitance effects A diode under reverse-bias exhibits a voltage-dependent capacitance caused by variation in stored charge at the junction. For example, an abrupt junction diode has capacitance given by

$$C_1 = \frac{A}{2}\left[\frac{2e\varepsilon_r\varepsilon_o}{(V_o - V)}\left(\frac{N_d N_a}{N_d + N_a}\right)^{1/2}\right]$$

If we assume external bias potential V is large compared to zero bias junction potential V_o and that we have a P^+N junction, then the above expression reduces to

$$C_1 = \frac{A}{2}(2e\varepsilon_r\varepsilon_o N_d)V^{-\frac{1}{2}} \tag{11}$$

In practice, junctions are rarely abrupt. However, it still remains true that capacitance decreases with increasing reverse-bias. For example, in a linearly graded junction, one finds

$$C_1 \ \alpha \ V^{-1/3}$$

One can see from Figure 18.44 that at high frequencies diode capacitance acts as a shunt across the output resistance network and reduces the output.

Usually $R_{sh} > R_L$ and $R_s \ll R_L$, hence as far as diode current i_λ is concerned, diode capacitance C_1 is in parallel with load resistor R_L. One can easily show that output voltage of the device as a function of modulation frequency f is obtained as

$$V_o = \frac{i_\lambda R_1}{(1+4\pi^2 f^2 C_1^2 R_L^2)^{1/2}}$$

Electrical bandwidth Δf_{at} is defined as the frequency range over which the output is above $\sqrt{\frac{1}{2}}$ of its maximum value. Thus, in the present case, we have $4\pi^2 f_{cl}^2 C^2 R_L^2 = 1$, i.e.

$$\Delta f_{et} = \frac{1}{2\pi R_L C_1} \tag{12}$$

The bandwidth may obviously be improved by reducing C_1. Eqn. (11) reveals that this may be achieved by reducing the diode area A, reducing the doping level N_d or increasing the reverse-bias voltage V. To achieve the largest bandwidths, response time associated with diode capacitance is made equal to that associated with the drift of carriers across the depletion region. Silicon photodiodes with bandwidths greater than 1 GHz (i.e. response times less than 1 ns) are available.

The silicon photodiode is found to be one of the most popular all-purpose radiation detectors in the wavelength range 0.4 to 1 μm. It has the virtues of high quantum efficiency, small size, good linearity of response, large bandwidth, simple biasing requirements and relatively low cost.

Photodiodes made from materials other than silicon are also available, but they have not found suitable. Germanium photodiodes can be used up to a wavelength of 1.8 μm but they have rather low quantum efficiencies and somewhat high leakage currents (and hence high noise currents), although cooling to 77 K greatly improves their performance. This is unfortunate since the wavelength region from 1 to 1.6 μm is of great interest from the point of view of optical communications. Recently, diodes made from both ternary (e.g. GaInAs and HgCdTe) and quaternary (e.g. GaInAsP) compounds are gaining importance.

18.8.2 Avalanche Photodiode (AVD)

Most fast photodiodes have been designed for use with 50 Ω load impedance, and voltage output (= $i_{ext}R_L$) often requires considerable amplifications. Useful internal amplification of the photocurrent is achieved in the avalanche photodiode. In this device, a basic PN structure is operated under very high reverse-bias. Carriers traversing the depletion region therefore gain sufficient energy to enable further carriers to be excited across the energy gap by impact excitation. Figure 18.52 shows the process.

An electron having reached point A in the Figure 18.52 has enough energy above the conduction band bottom such that it can collide with an electron from the valence band and raise it to the conduction band (C → D). This generates a new electron-hole pair. In doing so, the electron will of course lose an equivalent amount of energy and move from A to B. These newly generated carriers may both subsequently generate further electron-hole pairs by the same process, although the cross-section for hole impact excitation is much smaller than that for electrons. Excitation cross-sections vary rapidly with electric field E according to a relation of the form exp $(-A/E)$, where A is a constant.

Current gains in excess of 100 are readily obtainable, although, current gain is very sensitive to the value of bias voltage, (Figure 18.53). If bias voltage is made too large, then a self-sustaining avalanche current flows in the absence of any photoexcitation and this sets an upper limit to the voltage that may be used. To ensure a uniformly high avalanche breakdown voltage and uniform gain throughout the diode, one will have to provide a highly uniform field profile across the device. A number of designs have been proposed to achieve this. One design based on guard ring structure is shown in Figure 18.50. The guard ring restricts the avalanche region to the central illuminated part of the cell.

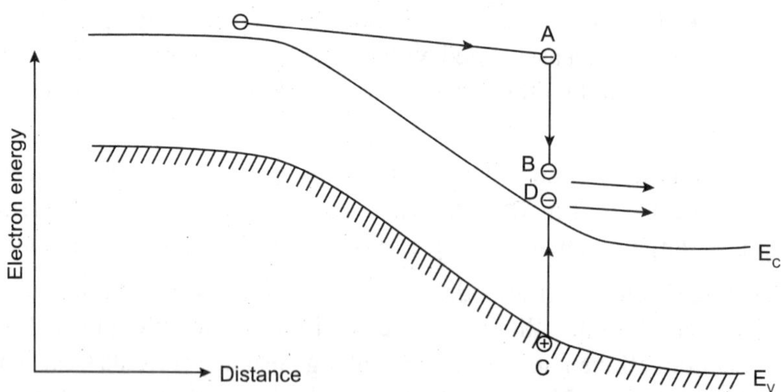

Figure 18.52 Principle of an avalanche photodiode. An electron having reached point A has sufficient energy above the conduction band bottom to enable it to excite an electron from the valence band into the conduction band (C → D). In doing so, it falls from A to B.

While using avalanche photodiode, one will have to take number of precautionary measures: (i) A very stable power supply for maintaining constant gain is required. A typical variation of current gain with reverse bias is shown in Figure 18.53. (ii) Gain is highly dependent on temperature. One will have to overcome this by having two identical diodes. One diode is masked from the incident radiation and operated near breakdown at constant current. A change in temperature causes altering bias voltage at the control diode and one may use this change to regulate the operating bias applied to the signal diode.

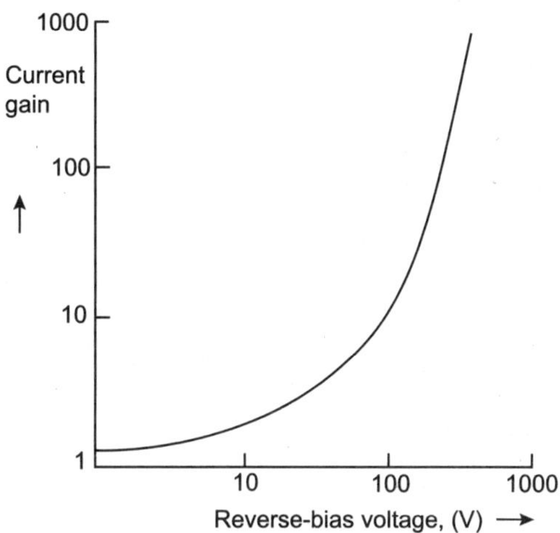

Figure 18.53 Typical variation of current gain with reverse bias voltage for an avalanche photodiode

If each photogenerated carrier always gave rise to the same number of secondary carriers (say M), then, one finds the r.m.s. noise current due to shot noise as

$$\Delta i_f = M(2ie\Delta f)$$

where i is the total current flowing. The multiplication process is a statistical one, however, and any primary carrier may generate more or less than the average value. When this factor is taken into account, then the true r.m.s. noise current is obtained as:

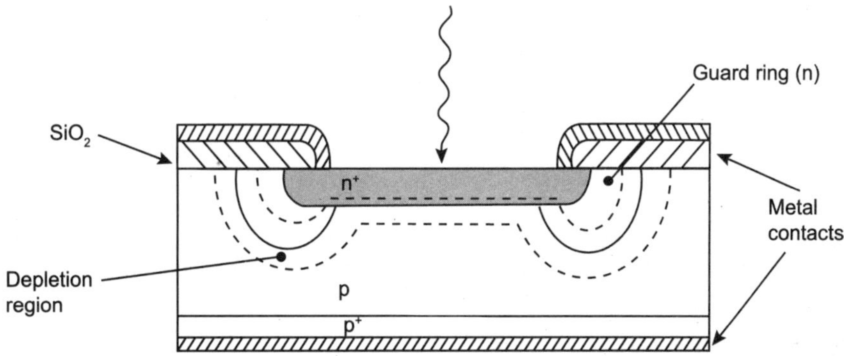

Figure 18.54 Silicon avalanche photo detector with guard ring structure. The guard ring is a region of comparatively two doping, and hence the depletion region extends an appreciable distance into it. Obviously, in the vicinity of the guard ring, the total depletion layer thickness is greater and thus the maximum electric field strengths are less than in the central region.

$$\Delta i_f(\text{avalanche}) = M[2ie\Delta fF(M)]^{1/2}$$

F(*M*), excess noise factor, is obtained as

$$F(M) = M\left[1-\left(1-\frac{1}{r}\right)\left(\frac{M-1}{M}\right)^2\right] \tag{13}$$

where r is the ratio of electron to hole ionization probabilities. In silicon, r tends to have a relatively large value (i.e. about 50), whereas for germanium r is about unity. This reveals that avalanche photodiodes made from germanium are inherently more noisy than those made from silicon.

Avalanche photodiodes are found quite suitable candidates for use in fiber optical communication systems. However, suitable materials giving low noise performance in the region beyond 1.1 μm (limit of silicon detectors) will have be developed.

18.8.3 Schottky Photodiodes

In this photodiode a thin metal coating (usually gold) is applied to an N type silicon substrate (Figure 18.55(a)). The energy band structure in the region of the junction is as shown in Figure 18.55(b). One can see that when an electron-hole pair is generated within the depletion region, the electron and hole will be separated by the action of the internal field as in a PN junction photodiode. The main advantage of this photodiode is that the surface metal layer can be made sufficiently thin to transmit blue and near ultraviolet radiation, thus giving enhanced sensitivity in this region.

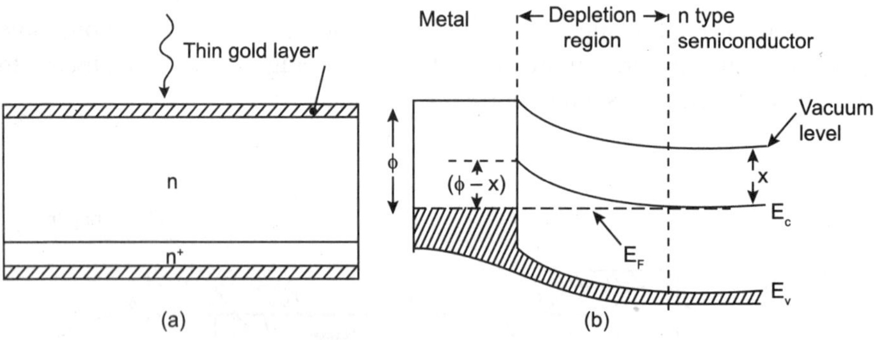

Figure 18.55 Schottky photodiode (a) energy level behaviour in the region of the junction (b); a potential barrier of height ϕ-X is formed between the metal and semiconductor.

18.8.4 Phototransistors

This is another device like the avalanche photodiode, where current flow from a PN junction detector is internally amplified. The construction of phototransistor is basically that of a junction transistor with the base region exposed to incident radiation. Normally, no external connection is made to the base (Figure 18.56(a)). To understand the operation of the device, one will have to consider the external currents as shown in Figure 18.56(b). Base current i_b will be supplied by the photogenerated current. One must have.

$$i_c = i_e - i_b$$

where i_c and i_e are collector and emitter current, respectively. Collector current has two components: (i) the normal diode reverse saturation current i_{co} and (ii) that part of emitter current that manages to cross to the collector. (Current is carried by minority carrier diffusion across the base, and not all minority carriers leaving the emitter will reach the collector.) One can write the latter current as αi_e, where α is slightly less than unity (is known as common base current gain). Thus,

$$i_{co} + \alpha i_e = i_e - i_b \tag{14}$$

or

$$i_e = \frac{i_b + i_{co}}{(1-\alpha)}$$

$$= (i_b + i_{co})\left(\frac{a}{1-\alpha} + 1\right)$$

$$= (i_b + i_{co})(h_{fe} + 1)$$

where $(h_{fe}) = \alpha/(1-\alpha)$ is known as **common emitter current gain of the transistor**. Typical dark values for h_{fe} in phototransistors are about 100. With no incident radiation $i_b = 0$ and current flowing $i_{co}(h_{fe} + 1)$ is dark current of the device. This is obviously larger than for comparable PN junction devices when dark current is just i_{co}.

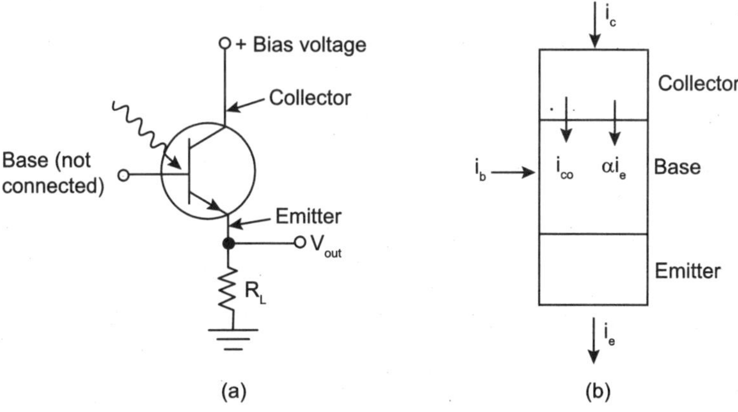

Figure 18.56 NPN phototransistor: (a) Light absorbed within the base region causes emitter current to flow through load resistor R_L and thus a signal voltage will appear across it; (b) currents (assumed to be) flowing in the phototransistor.

When illuminated, there will be base current of magnitude i_λ where, from Eqn. (1) $i_\lambda = \eta(I_o Ae\lambda / hc)$. External current flowing is now $(i_\lambda + i_{co}) \times (1+ h_{fe})$ which, if $i_\lambda > i_{co}$ is equal to $i_\lambda(1+h_{fe})$. Thus, the device gives internal gain, and has responsivity lying between that of a PIN photodiode and an avalanche photodiode. There are two main disadvantages: (i) at low light levels, i.e. low base currents, h_{fc} can drop to quite low values, (ii) frequency response tends to be relatively poor (< 200 kHz), mainly due to time constant associated with the capacitance and resistance of base function.

We can now summarize:

(i) Phototransistor A phototransistor is a light-detecting device that is also called photosensor. It is a transistor whose base current is supplied by carriers generated due to incident or striking light. Collector current of the transistor is thus controlled by the intensity of light. The symbol of a phototransistor is shown in Figure 18.57(a).

Although the phototransistor is a two-terminal device (like the photodiode), it produces a higher output current than the photodiode. The photodiode, however, has a faster response time. Thus, the phototransistor is preferred in high-current applications and the photodiode is used in high speed operations.

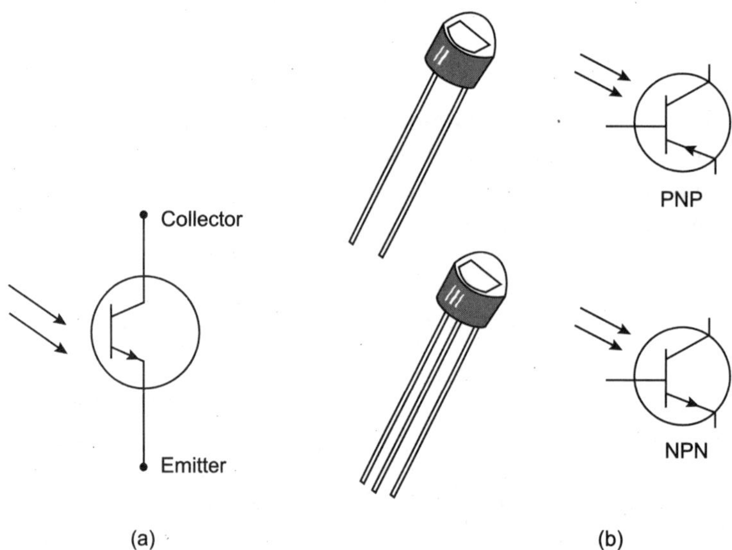

Figure 18.57 (a) Symbol of a phototransistor, (b) various types of phototransistors

(ii) Photodarlington This is a phototransistor packaged with another transistor connected in Darlington configuration. The symbol of photodarlington is shown in Figure 18.58. Due of its large current gain, the photodarlington produces greater output current than either the photodiode or phototransistor. However, the response of the photodarlington is slower than both photodiode and the phototransistor.

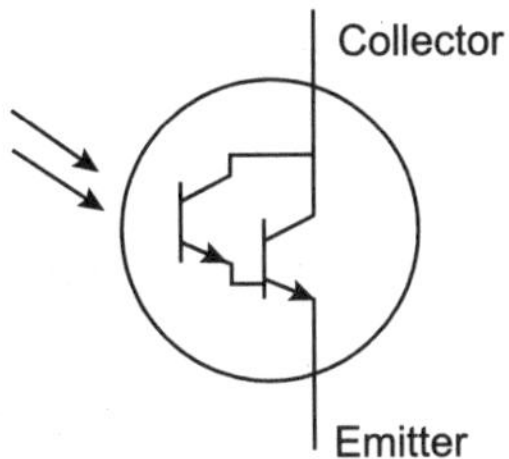

Figure 18.58 Symbol of a photodarlington

(iii) LASCR LASCR is an acronym for light-activated silicon-controlled rectifier. The incoming light strikes the photosensitive surface of the device, which serves as gate signal that triggers LASCR. The LASCR can be used in an optically coupled phase control circuit. Figure 18.59 shows the symbol of LASCR.

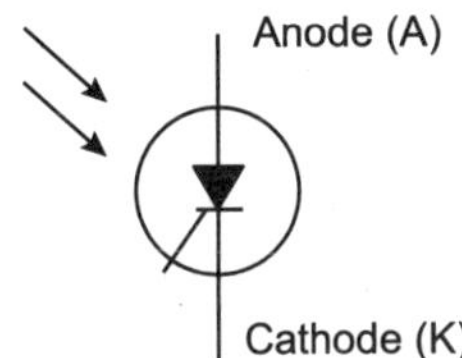

Figure 18.59 Symbol of LASCR

18.8.5 Vidicons and Plumbicons

Vidicon is based on the phenomenon of photoconductivity phenomenon. It converts an optical image into electrical image. A typical structure of vidicon tube is shown in Figure 18.60. An optical image is formed on a tiny target of semiconducting material (usually antimony trisulfide) having a transparent conducting layer (generally SnO_2) on the side on which radiations fall. This conducting layer is connected to a potential of above +50 V above ground via a bias resistor. Similar to CRT, the other side of the semiconducting target is scanned with an electron beam. The target acts rather like a leaky capacitor during operation. Its resistance will be high and charge will accumulate on opposite faces, when not illuminated. The side of the electron beam will charge up to around cathode potential (OV), and the other face around +50 V. Under illumination, the resistivity of the target material will be reduced to much and the capacitor will discharge itself whenever the scanning beam is not incident on the area in question. However, the beam side will get recharge when the beam return to the discharge area and a corresponding amount of opposite charge have to be supplied via the bias resistor and external bias supply circuit to the opposite side. The amount of charge flowing will depend on the charge discharged by capacitor, i.e. amount of light incident on the target. One can obtain the output signal by taking the voltage across bias resistor R_b.

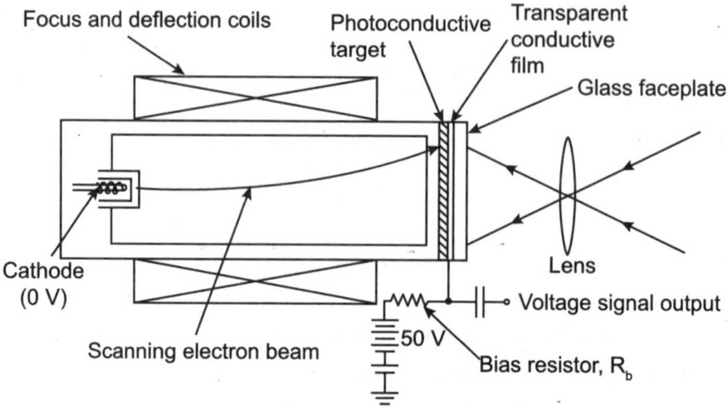

Figure 18.60 Vidicon tube

One of the major problem with vidicon is its relatively high dark current, i.e. it gives rise to poor signal to noise ratio at low light irradiance. A device that exhibits very low dark currents is called **plumbicon**. It is essentially identical with vidicon except for the nature of the photosensitive layer. The plumbicon consists of a thin film pin structure formed from lead oxide (Figure 18.61(a)). The transparent SnO_2 layer acts as the N type contact while the other surface has excess of oxygen, which causes it to be P type. The region in between (typically 15 μm thick) is effectively an intrinsic semiconductor. With no illumination, potentials across the device cause it to be in reverse-bias (thus very little dark current will flow). Any illumination of the film generates electron-hole pairs; these will then flow to the opposite sides of the structure and reduce the amount of stored charge. Plumbicon is widely used in colour TV studio cameras.

The energy gap of PbO is about 2 eV and therefore the red sensitivity of the device is poor; this can be improved by adding a thin layer of PbS (E_g = 0.4 eV) to the electron beam side of the target. Any red light not absorbed by PbO is then absorbed by PbS.

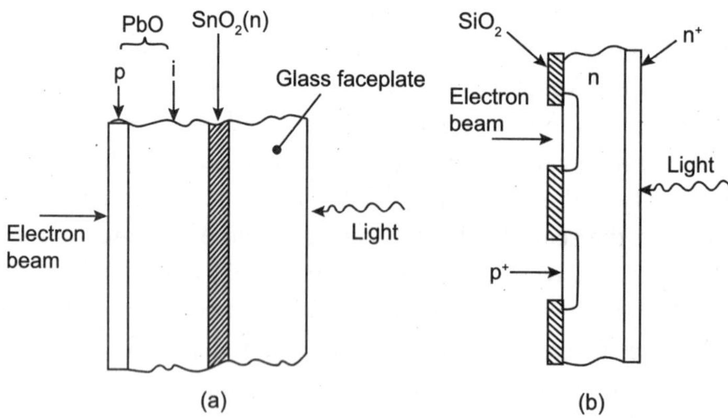

Figure 18.61 Structures for vidicon-type imaging tubes, using (a) lead oxide based pin structure (plumbicon) and (b) a discrete array of silicon PN junction diodes

There is a replacement of lead oxide target layer by an array of silicon diodes (Figure 18.61(b)). Dark currents become very small, and the devices exhibit very uniform sensitivities over the range 0.45 to 0.85 μm with good tolerance to high light levels.

18.9 Applications

18.9.1 Solar Cells

A solar cell is a photodiode operated in its photovoltaic mode. In its most common applications, it is used to convert light energy to electrical energy and thereby serves as a DC voltage source.

In solar power application, solar cells are connected in series/parallel arrays. Parallel connections increase the total current supplied by the array, whereas series connections increase the voltage.

Figure 18.62 show a solar panel. If each solar cell is capable of delivering about 0.8 A of current at 0.5 V of voltage, the entire 5 × 4 array would be capable of supplying a total of 4 A of current at a voltage of 2.0 V. Current generated in solar panels can also be used to charge batteries, so that the stored power can be consumed when it is dark. Thus, Solar cells can be used to store energy during the day to light street lamps at night.

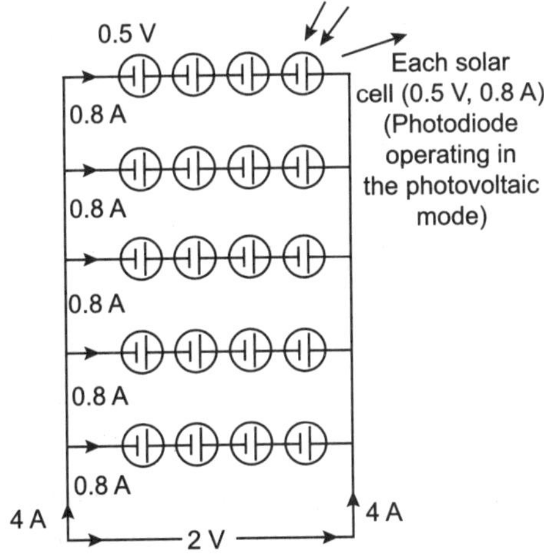

Figure 18.62 Solar panel

18.9.2 PN Junction Solar Cell

A PN junction solar cell is shown in Figure 18.63. It consists of a shallow PN junction formed on the surface, a front ohmic contact stripe and fingers, a back ohmic contact that covers the entire back surface and an antireflection coating on the front surface. When the cell is exposed to light, then only those photons whose energy is greater than forbidden gap contributes energy E_g to the cell output. Energy greater than E_g is wasted as heat.

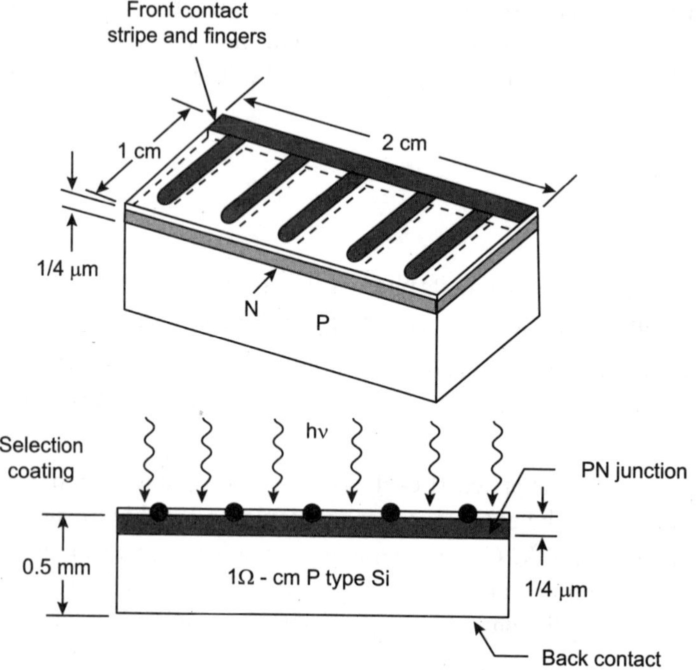

Figure 18.63 Schematic of PN junction solar cell

The ideal I-V characteristics of a solar cell device is given by

$$I = \text{Net PN junction current} = I_s\left[e^{qV/RT} - 1\right] - I_L \tag{15}$$

where I_L is current due to photogenerated carriers

and
$$J_s = \frac{I_s}{A} = qN_cN_V\left[\frac{1}{N_a}\frac{D_n}{\tau_N} + \frac{1}{N_d}\sqrt{\frac{D_P}{\tau_P}}\right]e^{-E_g/\text{kT}} \tag{16}$$

where A is the cross-sectional area. A plot of Eqn. 16 is shown in Figure 18.64. As seen from Figure 18.64, the curve passes through the fourth quadrant and, therefore, power can be extracted from the device.

Now, let $I = 0$ in Eqn. (15), the voltage at which current becomes zero is called open circuit voltage. It is represented as V_{OC}, i.e.

$$I_s\left[e^{qV_{OC}/kT} - 1\right] = I_L$$

$$V_{OC} = \frac{kT}{q}\ln\left(\frac{I_L}{I_s} + 1\right) \tag{17}$$

Hence, for a given I_L, V_{OC} increases logarithmically with decreasing I_s. The output power is

$$P = IV$$

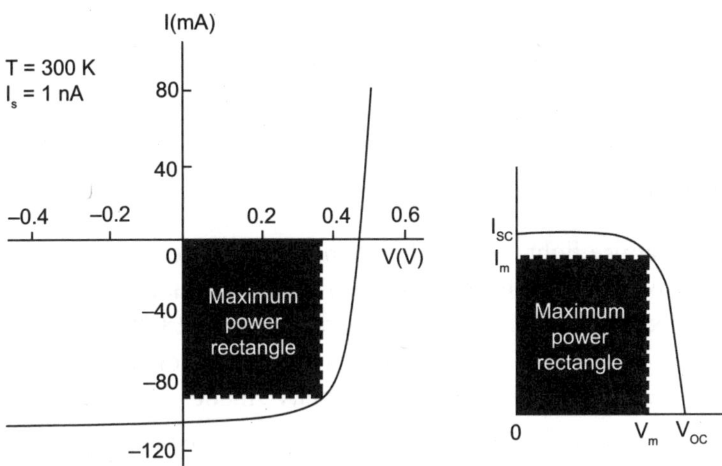

Figure 18.64 I-V characteristic of PN junction solar cell

Using Eqn. (15), we have

$$P = I_s V[e^{qV/kT} - 1] = I_L V \tag{18}$$

Maximum power can be obtained by differentiating Eqn. (18) w.r.t. V and setting it equal to zero

$$\frac{dP}{dV} = 0$$

Using Eqn. (18), one obtains

$$I_s[e^{qV/kT} - 1] = I_s V \frac{q}{kT}(e^{qV/kT} - 1) - I_L = 0$$

Putting $V \to V_m$, we have

$$V_m = \frac{kT}{q} \ln\left[\frac{1 + (I_L / I_s)}{1 + (qV_m / kT)}\right]$$

$$= V_{OC} - \frac{kT}{q} \ln\left[1 + \frac{qV_m}{kT}\right] \tag{19}$$

Maximum current is obtained from Eqn. (17) and is given as:

$$I_m = I_s \left(\frac{qV_m}{kT}\right) e^{qV_m/kT}$$

$$= I_L \left[1 - \frac{1}{qV_m / kT} \right] \tag{20}$$

The maximum power is obtained by multiplying Eqns. (19) and (20), i.e.

$$P_m = I_m V_m = I_L \left[V_{OC} - \frac{kT}{q} \ln\left(1 + \frac{qV_m}{kT}\right) - \frac{kT}{q} \right] \tag{21}$$

18.10 Optocouplers

These are devices that use light to optically couple a signal between two electrically isolated points. These are also called **optoisolators**. These devices include a light-emitting device and a light-receiving device, both available in one package. Figure 18.70 shows an optocoupler that uses an LED and a photodiode. The two devices are placed together in an opaque case. The photodiode operates in reverse-biased photoconductive mode.

A control circuit forward-biases the LED. When the LED lights up, light strikes the photodiode, thereby increasing its conductivity. This results in current flowing through load resistor R_L. When the LED is turned off, conductivity of the photodiode is low and there is no current through the load. Load current is controlled by the control circuit even though there is no direct electrical connection. Such a system is particularly useful for isolating control circuit noise from a sensitive electronic load circuit.

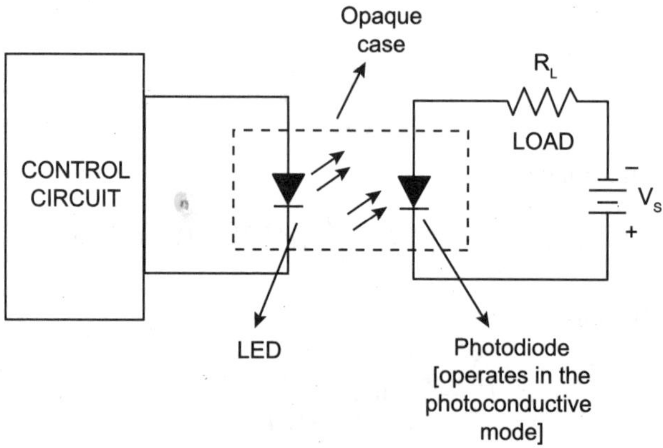

Figure 18.65 An optocoupler used to control load current

18.10.1 Photomultipliers

Photoelectrons in a photomultiplier are accelerated towards a series of electrodes (called dynodes) which are maintained at successively higher potentials with respect to the cathode.

On striking a dynode surface, each electron causes the emission of several secondary electrons which in turn are accelerated towards the next dynode and continue the multiplication process. Thus, if on average δ secondary electrons are emitted at each dynode surface for each incident electron, and if there are N dynodes overall, then the total current amplification factor between the cathode and anode is given by

$$G = \delta^N \tag{1}$$

Considerable amplification is possible. If we take, for example, $\delta = 5$ and $N = 9$, we obtain a gain of 2×10^6. The variation of δ with interdynode voltage for a typical dynode material is shown in Figure 18.66.

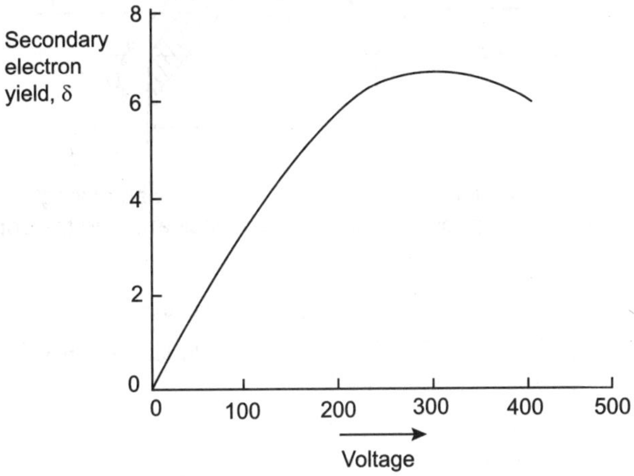

Figure 18.66 Variation of secondary electron yield δ with interdynode voltage for Be-Cu dynodes

18.10.2 Photoconductive Detectors

An electron may be raised from the valence band to the conduction band in a semiconductor where energy gap is E_g by the absorption of a photon of frequency ν provided that

$$h\nu \geq E_g$$

or in terms of wavelength,

$$\lambda \leq \frac{hc}{E_g} \tag{2}$$

We define bandgap wavelength,

$$\lambda_g = \frac{hc}{E_g} \tag{3}$$

As long as the electron remains in the conduction band, it will cause an increase in conductivity of the semiconductor. This is the phenomenon of photoconductivity, which is the basic mechanism operative in photoconductive detectors. For convenience, we suppose the semiconductor material to be in the form of a slab of width W, length L and thickness D with electrodes on opposite ends (Figure 18.67). An external potential across the electrodes is usually provided by the simple circuit shown in Figure 18.68.

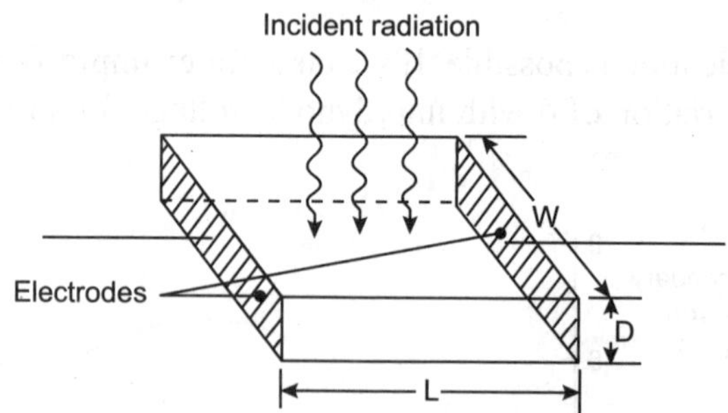

Figure 18.67 Geometry of slab of photoconductive material. The slab of length L, width W and thickness D has electrodes on opposite faces; radiation incident onto the upper surface.

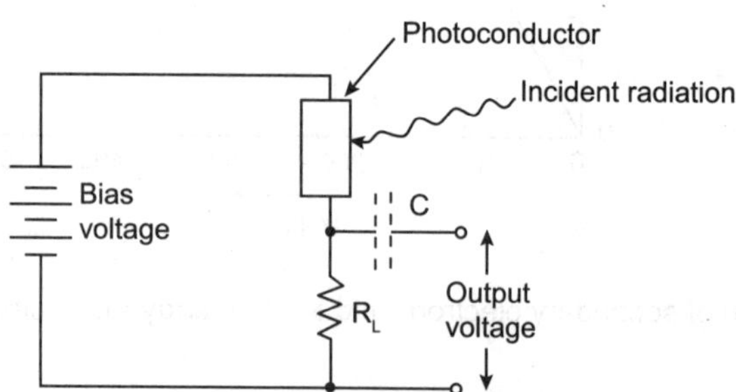

Figure 18.68 Photoconductor bias circuit. The photoconductor is placed in a series circuit comprising a voltage source, a load resistor R_L, and the photoconductor. Changes in the resistance of photoconductor cause changes in voltage appearing across R_L. If only the AC component of voltage is required then a blocking capacitor C may be placed.

18.10.3 Characteristics of Particular Photoconductive Materials

Cadmium sulfide (CdS) and cadmium selenide (CdSe) are both used for low cost, visible radiation sensors, e.g., light meters for cameras. These devices usually have high photoconductive gains (10^3 to 10^4) but poor response times (about 50 ms). The response time is in fact strongly dependent on the illumination level, being much reduced at high levels, a behaviour indicative of the presence of traps. Figure 18.69 shows the construction of CdS photoconductive cell. A film of the material in polycrystalline form is deposited on an insulating substrate and the electrodes are formed by evaporating a suitable metal, such as gold, through a mask to give the comb-like pattern. This geometry, which results in a relatively large area of sensitive surface and a small interelectrode spacing, helps to give a high sensitivity to the device.

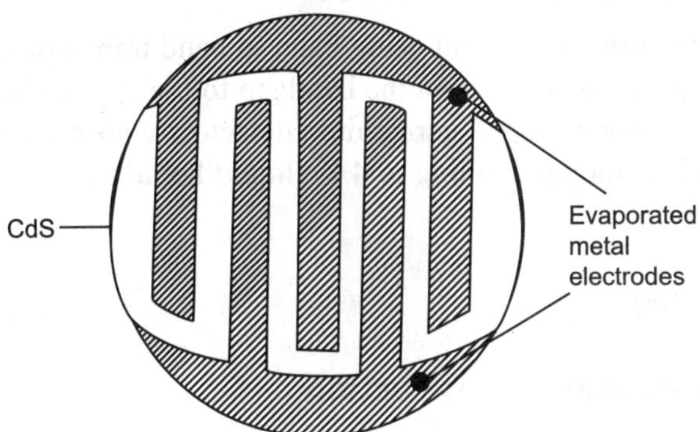

Figure 18.69 Top view of typical electrode geometry of CdS photoconductive cell

PbS is a well known near-infrared detector material having a useful wavelength response from 1 to 3.4 μm. By varying the growth conditions, the detector characteristics can be varied widely with expected trade off between gain and frequency response. A typical response time would be about 200 μs. In the region of 2 μm, it is one of the most sensitive photodetectors. The wavelength response may be extended to 4 μm by cooling to – 30°C (unlike most other photoconductive materials, the energy gap of PbS decreases with decreasing temperature). However, cooling also has an effect of reducing both the overall gain and frequency response. The internal dark resistance of these detectors is usually quite high (= 1 MΩ). This is an advantage when using the bias circuit of Figure 18.68, since then large values for the bias resistor R_L can be used, resulting in relatively high output voltage signals.

Indium antimonide (InSb) These detectors are usually formed from single crystals and tend to have low impedances (= 50 Ω). Consequently, their output voltages tend to be small. They have a wavelength response extending out to 7 μm and exhibit response times of around 50 ns. Operation at room temperature is possible. However, much improved noise performance results on cooling to liquid nitrogen temperature (77 K), although the peak wavelength response then shifts to 5 μm.

Mercury cadmium telluride ($HgCd_{1-x}Te$) This may be thought of as an alloy composed of the semi-metal HgTe and the semiconductor CdTe. Semi-metals have overlapping valence and conduction bands and may be regarded as having a negative bandgap (for HgTe, the bandgap is –0.3 eV). Consequently, depending on the composition of the alloy, a semiconductor can be formed with a bandgap varying between zero and 1.6 eV (bandgap of pure CdTe). Detectors are available whose peak sensitivities lie in the range 5 to 14 μm. This region is of particular importance, since it covers the peak emission wavelengths of bodies at or somewhat above ambient temperature and also corresponds to a region of good atmospheric transmission. Usually cooling to 77 K or below is necessary for satisfactory noise performance. However, detectors operating at the lower end of this wavelength range are sometimes used in conjunction with a thermoelectric cooler. Photoconductive gains of up to 500 are possible, while at low temperatures and under low illumination, the effects of traps can increase the gain further. As compared to indium antimonide, device resistances tend to be low.

In doped semiconductors, rather than using band to band transitions, it is also possible to use transitions from impurity levels within the bandgap to the appropriate band edge. Typical among these are zinc and boron doped germanium detectors whose wavelength response can extend from 20 to 100 μm, though cooling to 4 K (liquid helium temperatures) is essential to reduce background noise.

18.11 Photoresistor

18.11.1 Basic Operation and Construction

A photoresistor is a light detecting device and is also called a photoconductive cell or light-dependent resistor. It is a passive device composed of a semiconductor material that changes resistance when its surface is exposed to light. The symbol for a photoresistor is shown in Figure 18.70.

In conventional semiconductor devices, electron-hole pairs are created by heat energy. In photoresistors, the semiconductor material is light sensitive and free electrons are created by light energy. With the creation of free electrons, the resistance of the material drops. The greater the intensity of light, the greater will be the number of free electrons and the smaller will be its resistance.

Photoresistors are the simplest and least expensive in the class of optoelectronic devices. A typical construction is shown in Figure 18.70. The light-sensitive semiconductor material is arranged in a zigzag strip whose ends are attached to external terminals. The semiconductor material is either CdS or CdSe. A glass or transparent cover is attached for light to pass through to the device. The resistance of a photoresistor is largest when it is not conducting and is called **dark resistance**.

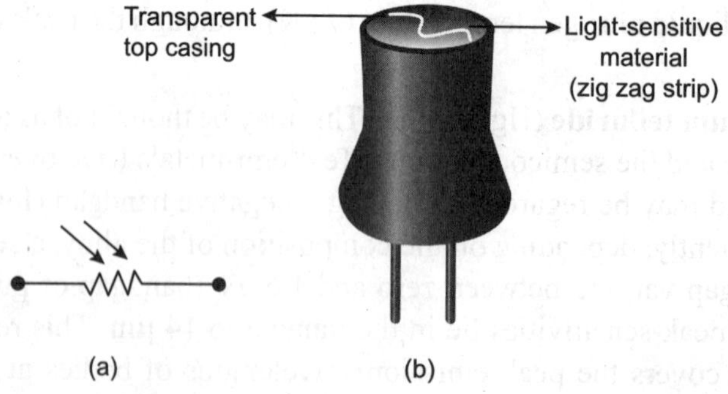

Figure 18.70 (a) Symbol of a photoresistor, (b) structure of photoconductive cell

18.11.2 Applications

Photoresistors are used in lighting controls, automatic door openers, alarm-activating circuits, and other such applications. An example of a photoresistor in streetlight control circuit is shown in Figure 18.71. The opening and closing of relay contact of Figure 18.71 is controlled by a photoresistor. A minimum amount of energizing current flowing through the relay coil is required to activate the switching.

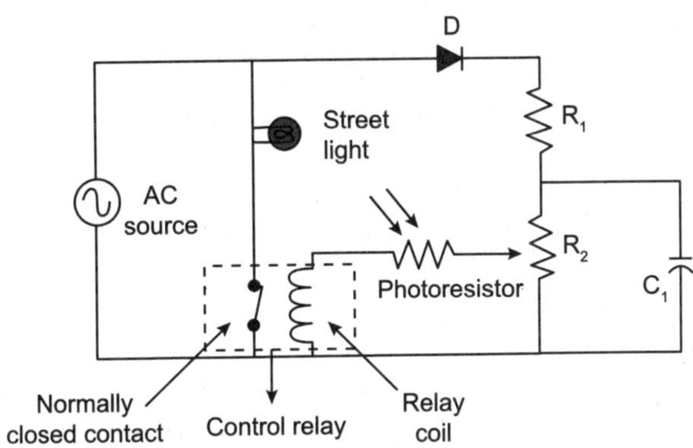

Figure 18.71 Photoresistor used to control a streetlight

The relay contact is normally closed. At night, when there is no light incident on the photoresistor, its resistance is large, making current through the photoresistor and coil very small. The relay is, therefore, not activated, and the contact remains in its normally closed position. This causes the streetlight to be connected to the AC source.

However, during daylight (when there is sufficient light striking the photoresistor), its resistance drops, thereby increasing current through the coil. This current is sufficient to energize the relay coil, and therefore the relay contact changes its position from the normally closed position to the open position. This results in the streetlight being cut off from the source. Note that diode D, resistors R_1 and R_2 and capacitor C_1 form a half wave rectifier circuit with the filter that supplies a DC voltage to the photoresistor and relay circuit.

Example 1 Calculate the ratio of rates of spontaneous and stimulated emission for a tungsten filament lamp operating at a temperature of 2000 K. Take average frequency = 5×10^{14} Hz.

Solution

$$R = \exp\left(\frac{h\upsilon}{kT}\right) - 1$$

or

$$R = \exp\left[\frac{6.6\times10^{-34}\times5\times10^{14}}{1.38\times10^{-23}\times2000}\right] + 1$$

$$= e^{12} + 1 = 1.5 \times 10^5$$

Example 2 Estimate the relative populations of two energy levels such that a transition from higher to lower will give a visible radiation.

Solution

Average wavelength of visible radiation = 550 nm

$$\therefore\ E_2 - E_1 = \frac{hc}{\lambda} = \frac{6.6\times10^{-34}\times3\times10^{8}}{550\times10^{-9}}$$
$$= 3.6\times10^{-19}\,J = 2.25\text{ eV}$$

Assuming room temperature = 300 K and $g_1 = g_2$, we have

$$\frac{N_2}{N_1} = \exp\left[\frac{-3.6\times10^{-19}}{1.38\times10^{-23}\times300}\right]$$

$$\approx e^{-87} = 10^{-37}$$

Example 3 A single crystal of silicon sample having width of 0.3 μm is illuminated with a monochromatic light of energy 3 eV. The incident power is 10 MW. Calculate (i) total energy absorbed by material per second, (ii) the rate of excess thermal energy dissipated to the lattice, and (iii) the number of photons per second given off from the recombination of intrinsic transition. Given $\alpha = 4\times10^4$/cm. [B. Tech]

Solution

(i) Energy absorbed/sec is given by

$$\phi_o[1-\exp(4\times10^4\times0.3\times10^{-4})]$$
$$=10^{-12}[1-\exp(4\times10^4\times0.3\times10^{-4})]$$
$$=10^{-2}[1-e^{1.2}] = 6.3\text{ mW}$$

(ii) The portion of each photon energy that is converted into heat is obtained as

$$\frac{h\upsilon - E_g}{h\upsilon} = \frac{3-1.12}{3} = 62\%$$

Obviously, the amount of energy dissipated/sec to lattice is 62% × 6.3 = 3.9 mW

(iii) Number of photons/sec from recombination is

$$\frac{2.4}{1.6\times10^{-19}\times1.12} = 1.3\times10^{16}\text{ photon/sec}$$

$\therefore$ Recombination radiation = 6.3 – 3.9

= 2.4 mW.

Example 4 Calculate the time taken for electrons to diffuse through a 5 μm thick layer of P type silicon. Given minority carrier diffusion coefficient $D_C = 3.4 \times 10^{-3} m^2 s^{-1}$.

Solution

$$\tau_{\text{Diff}} = \frac{d^2}{2D_c}$$

$$= \frac{(5\times10^{-6})^2}{2\times3.4\times10^{-3}} = 3.7\times10^{-9}\text{s}$$

Example 5 Estimate frequency bandwidth limitations arising from diode capacitance. Assume the typical values: $A = (1 \text{ mm}^2)$, $\varepsilon_r = 11.7$, $N_d = 10^{21}$ m^{-3} and $V = 10$ V.

Solution

We have

$$C_J = \frac{A}{2}(2e\varepsilon_r\varepsilon_o N_d)^{1/2}V^{-1/2}$$

$$= \frac{10^{-6}}{2}\left[2(1.6\times10^{-19})(11.7)(8.85\times10^{-12})10^{21}\right]^{1/2}(10)^{-1/2} \approx 30 \text{ pF}$$

Example 6 The reflection coefficients for a semiconductor laser are equal and have the value 0.5. The length of cavity is 1 μm. Assume that there are no additional loss terms. Calculate the gain of the laser. [B. Tech]

Solution No internal loss means $d_i = 0$; we have

$$g = \frac{1}{2L}\ln\left(\frac{1}{r_1r_2}\right)$$

$$= \frac{1}{2\times1}\ln\left(\frac{1}{1.5\times0.5}\right)$$

$$= 6.93\times10^3 \text{ /cm}$$

Example 7 Calculate the gain of a photoconductor device of length 100 μm and cross-sectional area of 10^{-7} cm^2. Assume the bar is of N type with minority carrier lifetime 10^{-6} sec. Take $\mu_n = 1350$ cm^2/V sec and applied voltage is 10 V. [B. Tech]

Solution

$$t_n = \frac{L}{\mu_n E} = \frac{L^2}{\mu_n V} = 7.41\times10^{-9} \text{ sec}$$

$$\text{Gain} = \frac{\tau_p}{t_n}\left(1+\frac{\mu_p}{\mu_n}\right) = 1.83\times10^2$$

Summary

1. Optoelectronics is a branch of electronics which deals with light sensitive devices. When light energy is incident upon these devices, electric current is generated. These devices are called photonic or optoelectronic devices.
2. There are two types of optoelectronic devices: (i) light (emitted) source type, e.g. LEDs, and laser diodes, (ii) light detector type, e.g. photodiodes, phototransistors, photoresistors, photovoltaic cell and LCDs.
3. The phenomenon of emission of electrons from a metal surface under the effect of light radiations is termed a photoelectric effect. The minimum frequency of light that can cause photoelectric effect is called threshold frequency.
4. A photoconductive cell is a two-terminal semiconductor device, whose resistance decreases with the intensity of light falling on it, and as a result free electrons are available. It is also called light (photo) dependent resistance (LDR).
5. An LED is a specially fabricated semiconductor PN junction diode that emits monochromatic light when forward-biased. LED has voltage drop-rating of about 2 V and current rating of about 20 mA.
6. The internal quantum efficiency of an LED is the fraction of diode current that will produce luminescence.
7. The external quantum efficiency of an LED is defined as the ratio of the number of photons emitted from the device per incident injected electron.
8. The brightness of light emitted by an LED depends upon current flowing through it.
9. LEDs have replaced small incandescent light bulbs as display devices. The long life expectancy, ruggedness, and much lower power consumption have contributed to the choice of LEDs over incandescent light bulbs. Some of the applications of LEDs are seven-segment display, polarity tester, and continuity tester.
10. Laser is an acronym of light amplification by stimulated emission of radiation.
11. Laser diodes are light-emitting devices capable of emitting laser beams. One of the commonly used laser diodes is the injection laser diode (ILD).
12. Laser light sources are monochromatic. All the light waves emitted by a laser diode are in phase with each other. Due to resonance, it emits a narrow beam of coherent, intense focused and monochromatic light.
13. A photodiode is a PN junction which generates significant current when light is incident upon it.
14. A phototransistor is like a photodiode except that photocurrent produced in a transistor is more than that produced in the photodiode, because it is multiplied in transistor by a factor β.

15. The sensitivity of photodiodes and phototransistors is expressed as the ratio of photocurrent output (in amperes) to radiant energy (in watts).
16. A photoresistor is a semiconductor light-detector device without any PN junction diode.
17. A solar cell (also called photovoltaic cell) converts light energy into electric energy.
18. A photomultiplier tube uses secondary emission to give multiplied output.
19. An optocoupler or optoisolator is an arrangement of coupling (or isolating) a light emitting source from a light dependent device (light detector) through a transparent insulating medium.
20. An LCD consists of a thin film (about 15 μm) of a conductive crystal in liquid form. Normally the crystal is transparent. However, when energized by an AC source in the presence of light, it changes to provide black-tone or gray readout.

Review Questions

1. What is optical light spectrum?
   (a) Frequency and wavelength
   (b) Visible versus invisible
   (c) Infrared and ultraviolet
2. The infrared band is given as 30 to 400 THz. What is this in wavelength? Answer the same question for visible light and ultraviolet light.
3. Define the difference between coherent and incoherent light.
4. Discuss the operation of LED.
   (a) How does it emit light? (b) When does it emit light?
5. A certain piece of equipment shows a green light when it is working properly and red light when it is not working properly. The light appears to come from the same device. How might this work?
6. A particular fiber-optic application will use a very long length of fiber-optic cable. Would this application use an LED or a laser diode? Why?
7. Discuss the two modes of operation of a photodiode.
8. A certain control circuit is being used to control the operation of a high-power motor starter. The motor starter generates enormous amounts of noise that might interfere with the control system. Discuss how the circuit might be designed to isolate noise from the controls.
9. Redraw the circuit of Figure 18.72 so that it works with a normally open contact instead of normally closed.

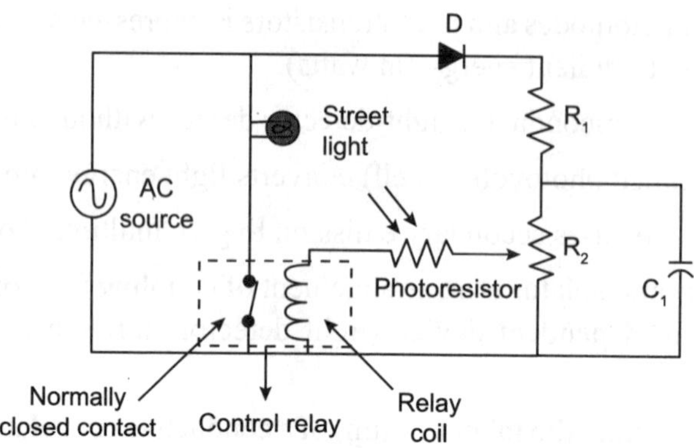

Figure 18.72 Photoresistor used to control a streetlight

10. You wish to operate a door opener with a light source. Draw a block diagram to show how this might be done.
11. What are optoelectronic devices? Illustrate your answer with suitable examples.
12. What is photoelectric effect? Write names of its types and explain each.
13. Describe the working of a photoconductive cell and mention its important applications.
14. What is an LCD? Write its applications, make a comparison of LCD and LED.
15. Explain the principle of solar cell and describe its working.
16. Describe the working of an optocoupler. Mention its uses.
17. Explain the working of a photomultiplier tube and mention its uses.
18. What is the minimum frequency to cause photoconduction?
19. Mention where the liquid crystal diode is practically used?
20. Show that when the simple bias circuit (Figure 18.73) is used with a photoconductive detector to detect small signal levels, then maximum voltage output signals across R_L are obtained when R_L is equal to detector resistance R_D.

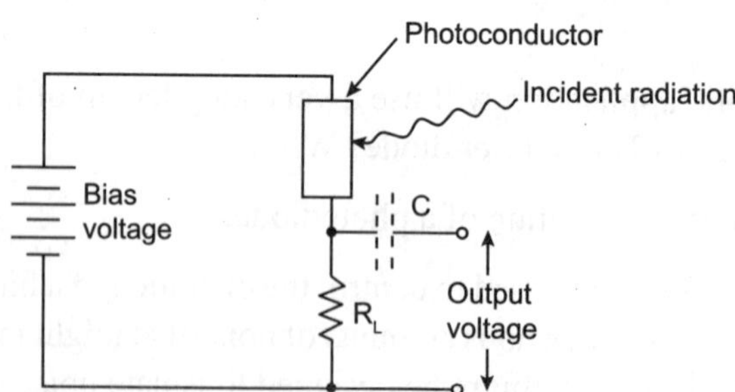

Figure 18.73 Photoconductor bias circuit. The photoconductor is placed in a series circuit comprising a voltage source, a load resistor R_L, and the photoconductor. Changes in the resistance of photoconductor cause changes in voltage appearing across R_L. If only the AC component of this voltage is required, then a blocking capacitor C may be placed as shown

21. Show that gain G of a photoconductive detector can be written as $Ge = \tau_{c_c} / \tau_d$ where τ_c is minority carrier lifetime, and τ_d is given by

$$\frac{1}{\tau_d} = \frac{1}{\tau_n} + \frac{1}{\tau_p}$$

τ_n and τ_p being the times taken for electrons and holes to drift across the photodetector.

22. Show that responsivity of a photodiode detector can be written as

$$R(f) = \frac{R(o)}{(1 + 4\pi^2 f^2 C^2 R_L^2)}$$

where C is diode capacitance and R_L load resistance.

Short Answer Questions

1. Do all optoelectronic devices produce light, either visible or invisible?

   Ans: No

2. Is an LED a specially fabricated PN junction diode that works in forward-bias?

   Ans: Yes

3. Does laser diodes find application in compact disk players and laser printers?

   Ans: Yes

4. Can a photovoltaic cell be used to measure the intensity of illumination?

   Ans: Yes

5. What is a photovoltaic cell?

   Ans: This is a device that converts light energy directly into electric energy.

6. What is a autocoupler?

   Ans: This is an assembly in a single package that contains a light emitting diode and a photosensitive device.

7. What do you understand by photodiode sensitivity?

   Ans: It is the ratio of photocurrent output (in amperes) to radiant energy (in watts) and expressed in amps/watt.

8. What is a laser diode?

   Ans: A laser diode is a heavily doped PN junction diode.

9. What is photoemissive or photoelectric effect?

   Ans: When radiation with a wavelength less than a critical value is incident upon a particular metal (sodium, potassium, etc.) surface, electrons are found to be emitted, this is called photoemissive or photoelectric effect.

10. What is a solar cell?

    Ans: The solar cell is basically a PN junction detector operated under conditions such that it can deliver power into an external load. It is used to convert light energy to electrical energy and thereby serves as a DC voltage source.

11. What is a phototransistor?

    Ans: A phototransistor is a light detecting device, which is also called a photo-sensor.

12. Why there has been interest in the development of white LEDs for general illumination?

    Ans: LEDs are three times as efficient as incandescent lamps and can last ten times longer. A white LED requires LEDs of red, green and blue colours.

13. What is the advantage of using heterojunction LEDs?

    Ans: Since a photon emitted to a radiative recombination process of direct bandgap material in one part of the device and reabsorbed prior it reaches the surface, one obvious choice is to put the PN junction close to the surface to avoid reabsorption. However, the presence of surface states, producing non-radiative recombination channels, can result in a reduction in radiative efficiency of the device. To solve this problem, heterojunction is used instead of avoiding reabsorption. A commonly used heterojunction is that of GaAs and AlGaAs. A heterojunction LED is designed in such a way that recombination occurs on the narrow gap semiconductor. Light is subsequently emitted through the AlGaAs window into the external environment. The bandgap of AlGaAs is greater than of GaAs and therefore most of the light will transmit through the AlGaAs without being absorbed.

14. Explain the importance of infrared LEDs.

    Ans: Under infrared LEDs fall GaAs LEDs which can emit light of wavelength near 0.9 μm and many III-IV compounds such as quaternary $Ga_x\ In_{1-x}\ Asy\ P_{1-y}$ LEDs, which emit light from 1.1 to 1.6 μm. Infrared LEDs have an important application in optoisolators where an input or control signal is decoupled from the output. Infrared LEDs also find application for transmission of optical signal through an optical fiber.

Objective Questions

1. Light is a form of ______________. **[Ans:** electromagnetic radiation]
2. Optoelectronic devices can be divided into two categories (i) ______________ (ii) ______________. [**Ans:** (i) light generating, (ii) light detecting devices]
3. A light-emitting diode generates ______________ when ______________ biased. [**Ans:** light, forward]
4. When the PN junction is reverse-biased, the LED does not conduct and there is no release of ______________. [**Ans:** light]
5. LEDs are made of semiconductor materials that are ______________ to light. [**Ans:** semitransparent]

6. LEDs have replaced small incandescent bulbs as ______________ devices.
   [Ans: display]
7. An individual LED is commonly used as a power level ______________ in a circuit.
   [Ans: indicator]
8. Series resistor limits current to the LED and often referred to as ______________.
   [Ans: current limiter]
9. Multicolour LEDs are usually two LEDs connected in ______________.
   [Ans: antiparallel]
10. Diode made of gallium arsenide releasing energy by way of heat and infrared light is called ______________.
    [Ans: infrared emitting diode]
11. Laser diodes are ______________ diodes. **[Ans:** light emitting]
12. A photodiode is a ______________ device that contains a semiconductor.
    [Ans: light-receiving]
13. When operating in the photovoltaic mode, the photodiode generates a ______________ in response to light. **[Ans:** voltage]
14. In photoconductive mode, conductance of the diode changes when ______________ is applied. **[Ans:** light]
15. A solar cell is a photodiode operated in its ______________ mode.
    [Ans: photovoltaic]
16. Optocouplers are devices that use light to optically couple a signal between two electrically ______________ points. **[Ans:** isolated]
17. An LED emits light when
    (a) unbiased (b) reversed-biased
    (c) forward-biased (d) placed in dark **[Ans:** (c)]
18. A laser diode is a P N junction which is
    (a) heavily doped and heavily forward-biased
    (b) lightly doped and lightly forward-biased
    (c) heavily doped and light forward-biased
    (d) lightly doped and heavily forward-biased **[Ans:** (a)]
19. An optocoupler uses
    (a) Four LEDs (b) two LEDs
    (c) three LEDs (d) one LED **[Ans:** (d)]
20. The infrared LEDs find application in
    (a) burglar alarm systems (b) display boards
    (c) calculators (d) display of time **[Ans:** (a)]

Chapter 19

Digital Electronics

19.1 Introduction

Under ideal conditions, a semiconductor diode when forward-biased acts like a closed switch, since maximum (infinite) current flows through it, and acts like an open switch when the diode is reverse-biased since minimum (zero) current flows through it. This unique characteristic of a semiconductor diode is employed in the design of basic logic circuits known as gates.

Logic gates are at the heart of modern branch of electronics known as digital electronics. Logic gates are either open or closed, corresponding to the switching OFF or ON condition of the electrical device known as switch.

Using binary number system, number 0 is designated as when gate is closed or switch is OFF and number 1 is designated as when gate is open or (device) switch is ON.

In digital electronics, a gate is a logic circuit with one output and one or more inputs. Although modern systems perform highly complex or sophisticated jobs, they consist of arrangements/ assembly of simple logic circuits. When computer decisions are of the yes or no type, it is known as two-state logic. So logic circuits can be in one of two states such as OFF or ON, Low or High, 0 or 1. Here we use the binary number system. Since it has only two digits (0 and 1) which correspond to Low and High states, respectively. In this respect, a toggle switch is also a simple example of two-state device.

19.2 Boolean Algebra

Augustaus De Morgan came close to finding the between logic and mathematics. But it was George Boole (1854) who put it all together. He devised a new kind of algebra that replaced Aristotle's verbal method. In 1938, Shannon applied the new algebra to telephone switching circuits. It was because of Shanmon's work that later on engineers made use of Boolean algebra to analyze and design computers.

19.3 Logic Circuits

We are quite familiar with analogue circuits. In analogue circuits, adjacent points on the load line in an amplifier may be used so that output voltage is continuous and because of this output voltage can have many values. But logic circuits (digital circuits) are quite different. Almost all digital circuits are designed for two-state operation. The two-state situation uses binary number system (0, 1). The circuit comprises of gates. These gates are sometimes called logic circuits.

We shall first discuss the number system and then boolean algebra related to binary numbers.

19.4 Number Systems

19.4.1 Decimal Number System

The decimal number system comprises 10 digits: 0, 1, 2, 3, 4, 5, 6, 7, 8, 9.

A collection of these digits makes a number like (a) 23, 54, 123, 3460, etc. and (b) 21.5, 246.1 etc., which in general has two parts – (i) integer and (ii) fraction set apart by radix point (.) Every number system has rules defined for operations like addition subtraction, multiplication. In example (a), the numbers depicted are integers or whole numbers only and in (b) the numbers depicted have two parts, i.e. both integer and (decimal) fractional.

In general, we can express any number as

$$(N)_b = d_{n-1}d_{n-2}\ldots d_i\ldots d_1d_0 \cdot d_{-1}\, d_{-2}\, d_{-3}\ldots\, d_{-f}\ldots\, d_m$$

Integer portion ↑ Fractional portion

Radix point

where

N – a number, b is radix or base of the number

n – number of digits in integer portion

m – number of digits in fractional portion

d_{n-1} – most significant digit (MSD)

d_m – least signification digit (LSD)

Accordingly in the example,

3460 – 3 is MSD and 0 is LSD

246.1 – 2 is MSD and 1 is LSD

Here digits are placed side by side and each position of digit in the number is assigned a "weight or index" of importance by same predesigned recycle.

(i) The number 3460 can be expressed as

$$3 \times 10^3 + 4 \times 10^4 + 6 \times 10^1 + 0 \times 10^0$$

and so we say 0 is in units place, 6 is in tens place, 4 is in hundred place and 3 is in thousand place.

(ii) The number 246.1 can be expressed as

$$2 \times 10^2 + 4 \times 10^1 + 6 \times 10^0 + 1 \times 10^{-1}$$

Here, 2 is in hundred place, 4 in tens place, 6 in units place.

As we know, the decimal number system comprises 10 digits, i.e. 0.1 to 9. After 9, we repeat the series 0, 1, 2,...9 followed by 1 as shown below.

10, 11, 12, 13 ... 19

Here again the series is repeated followed by 2 as shown below.

20, 21, 22, ... 29

Continuing this way we go up to 99, where again we repeat the series followed by 10 and we have

100, 101, 102, ... 110, 111, ... 999

and then 1000, 1001, 1002,... 999

19.4.2 Binary Number System

In this system, there are only two digits namely 0 and 1. This system has base or radix 2. The digits 0 and 1 are called bits. Here the sequence of numbers is 0 and 1 and we follow the rule as in decimal number system and we have

10, 11

So to generate more numbers we repeat and get

100, 101, 110, 111

1000, 1001 ..., 1111

In this way, we can generate next higher numbers. Table 19.1 represents decimal numbers and their equivalent binary number.

Table 19.1

Decimal number	Binary number
0	0000
1	0001
2	0010
3	0011
4	0100
5	0101
6	0110
7	0111
8	1000

Table 19.1 (Continued)

Decimal number	Binary number
9	1001
10	1010
11	1011
12	1100
13	1101
14	1110
15	1111

The binary number 111 can be expressed

$$111 = 100 + 10 + 1$$

Similarly, decimal number 7 can be expressed as

$$7 = 4 + 2 + 1$$
$$= 1 \times 2^2 + 1 \times 2^1 + 1 \times 2^0$$

19.4.3 Binary to Decimal Conversion

For a binary number, its equivalent decimal number can be found by the following steps.

Step 1 – Write the binary number.

Step 2 – Write the respective weights (1, 2, 4, 8, etc.) of the digits (bits) directly under their respective positions, working from right to left.

Step 3 – If a zero appears in a digit, then cross out the respective decimal weight for that position.

Step 4 – Add the remaining weights to get the final result.

The steps are further made clear by the following example:

For integers:

Example 1 Convert 101 to its decimal equivalent.

Step 1 – 1 0 1

Step 2 – 4 2 1 or $1 \times 2^2 + 1 \times 2^1 + 1 \times 2^0$

Step 3 – 4 ~~2~~ 1 or $1 \times 2^2 + \cancel{1 \times 2^1} + 1 \times 2^0$

Step 4 – $4 + 0 + 1 = 5$

Thus 5 is the decimal equivalent of 101.

Example 2 Convert 11101 to its decimal equivalent.

Step 1 – 1 1 1 0 1

Step 2 – 16 8 4 2 1

Step 3 – 16 8 4 $\cancel{2}$ 1

Step 4 – $16 + 8 + 4 + 0 + 1 = 1 \times 2^4 + 1 \times 2^3 + 1 \times 2^2 + 1 \times 2^0 = 29$

Thus 29 is the decimal equivalent of 11101.

For fractions:

Here the step 2 for writing weights of the corresponding digits are $\frac{1}{2}, \frac{1}{4}, \frac{1}{8}, \frac{1}{16}$ etc. So in powers of 2, the weights are 2^{-1}, 2^{-2}, 2^{-3}, 2^{-4}... from left to right, which in decimal form are 0.5, 0.25, 0.125, 0.0625, respectively.

Example 3 Convert 0.101 to its decimal equivalent.

Step 1 – 0.101

Step 2 – $0.5 + 0.25 + 0.125$

Step 3 – $0.5 + \cancel{0.25} + 0.125$

Step 4 – $0.5 + 0 + 0.125 = 0.625$

Thus, 0.625 is the decimal equivalent of 0.101.

Example 4 Convert 0.1101 to its decimal equivalent.

Step 1 – 0.1101

Step 2 – $0.5 + 0.25 + 0.125 + 0.0625$

Step 3 – $0.5 + 0.25 + \cancel{0.125} + 0.0625$

Step 4 – $0.5 + 0.25 + 0 + 0.0625$

Thus, 0.8125 is the decimal equivalent of 0.1101.

For mixed numbers (i.e. number comprising integer as well as fraction) the integer part is dealt separately and fractional part is dealt separately.

Example 5 Consider 10101.101.

Step 1 – 10101.101

Step 2 – $1 \times 2^4 + 1 \times 2^3 + 1 \times 2^2 + 1 \times 2^0 \cdot 2^{-1} + 2^{-3} + 2^{-3}$

Step 3 – $1 \times 2^4 + \cancel{1 \times 2^3} + 1 \times 2^2 + \cancel{1 \times 2^1} + 1 \times 2^0 + 2^{-1} + \cancel{2^{-2}} + 2^{-3}$

Step 4 – $[16 + 0 + 4 + 0 + 1] \cdot [0.5 + 0 + 0.125]$

$= 21.625$

Thus, 21.625 is the decimal equivalent of 10101.101.

19.4.4 Decimal to Binary Conversion

One way to convert a decimal number into its binary equivalent is to reverse the steps mentioned previously. For this, express the given number as a sum of powers of 2 and then write 1s and 0s in the appropriate position.

Example 6

$$
\begin{aligned}
9 &= 8 + 0 + 0 + 1 \\
&\Rightarrow 1 \times 2^3 + 1 \times 2^2 + 1 \times 2^1 + 1 \times 2^0 \\
&\Rightarrow 1 \times 2^3 + \cancel{1 \times 2^2} + \cancel{1 \times 2^1} + 1 \times 2^0 \\
&\Rightarrow 1\,0\,0\,1
\end{aligned}
$$

Example 7

$$
\begin{aligned}
31 &= 16 + 8 + 4 + 2 + 1 \\
&\Rightarrow 1 \times 2^4 + 1 \times 2^3 + 1 \times 2^2 + 1 \times 2^1 + 1 \times 2^0 \\
&\Rightarrow 1\,1\,1\,1\,1
\end{aligned}
$$

19.5 Double Dabble Method

Here we progressively divide the decimal number by 2, writing down the remainder after each division. Then the remainders are taken in reverse order. These form the binary number.

For integers:

Example 8 Convert 13 to its equivalent binary.

Step 1 Divide number by 2

$$
\begin{array}{r|l}
 & \ \ 6 \\ \hline
2 & 13 \\
 & \underline{12} \qquad 1 \text{ first remainder}
\end{array}
$$

Step 2 Divide 6 by 2

$$
\begin{array}{r|l}
 & \ 3 \\ \hline
2 & 6 \\
 & \underline{6} \qquad 0 \text{ second remainder}
\end{array}
$$

Step 3 Divide number by 2

$$
\begin{array}{r|l}
 & \ 1 \\ \hline
2 & 3 \\
 & \underline{2} \qquad 1 \text{ third remainder}
\end{array}
$$

Step 4 Divide 1 by 2

This is not possible since in quotient we get 0 and thus remainder is 1. Hence reading we have 1101.

Example 9 Convert 31.65 to its binary equivalent. Here integer part is 31 and fractional part is 0.65.

For integer: 31

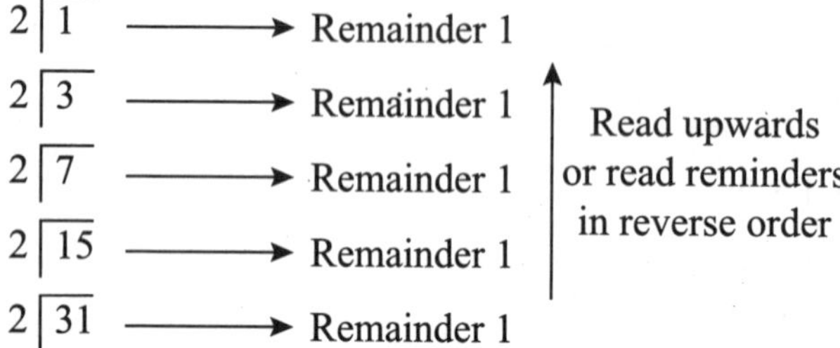

For fraction: We have to multiply by 2 instead of dividing as in case of integers and take respective carry and read in forward direction.

$0.65 \times 2 = 1.30 \rightarrow 0.3$ with carry 0
$0.3 \times 2 = 0.6 \rightarrow 0.6$ with carry 1
$0.6 \times 2 = 1.2 \rightarrow 0.2$ with carry 0
$0.2 \times 2 = 0.4 \rightarrow 0.4$ with carry 0
$0.4 \times 2 = 0.8 \rightarrow 0.8$ with carry 0
$0.8 \times 2 = 1.6 \Rightarrow 0.6$ with carry 1

Read forward

Thus, the binary equivalent of 31.65 is 11111 · 101001

Useful equivalents Table 19.2 represents some decimal-binary equivalents.

Table 19.2

Decimal	Binary
1	1
3	11
7	111
15	1111
31	11111
63	111111
127	1111111
255	11111111

From the chart it is evident that whenever a binary number is represented by all 1s (consists of only 1s), its decimal equivalent is given by the following formula.

$$\text{Decimal} = 2^n - 1$$

where n is the number of bits present in binary.

Example 10

1. Binary number → 1111
   Here we have 4 bits
   So equivalent decimal number $= 2^4 - 1$
   $= 16 - 1$
   $= 15$

2. Binary number → 1111111
   Here we have 7 bits
   So equivalent decimal number $= 2^7 - 1$
   $= 128 - 1$
   $= 127$

19.6 Octal Numbers

The octal number system has a base of 8 since it has 8 digits. The normal digits followed are 0, 1, 2, 3, 4, 5, 6, 7. These digits 0 through 7 have exactly the same significance as in decimal system. Here we follow the rule as in decimal and repeat digits to get numbers of higher order.

Octal	Decimal
10	8
11	9
12	10
13	11
14	12
15	13
16	14
17	15
20	16
21	17
22	18
23	19
24	20
25	21
26	22
27	23
-	-

19.6.1 Octal to Decimal Conversion

In the octal number system, each digit corresponds to a power of 8. Accordingly,

$$8^3 \quad 8^2 \quad 8^1 \quad 8^0 \;.\; 8^{-1} \quad 8^{-2} \quad 8^{-3} \ldots$$

↑
Octal point

Example 11 Convert octal 23 to equivalent decimal number.

$$2\times 8^1 + 3\times 8^0 = 16+3 = 19$$

Example 12 Convert octal 257 to equivalent decimal number.

$$2\times 8^2 + 5\times 8^1 + 7\times 8^0 = 128+40+7 = 175$$

19.6.2 Decimal to Octal Conversion (Octal Dabble Method)

It is very similar to double dabble method. Here instead of dividing by 2[(the base of binary number) we divide the decimal number by 8 (the base of octal numbers] writing down the remainders after each division. The remainders in reverse order form the octal number.

Example 13 Convert 175 to octal.

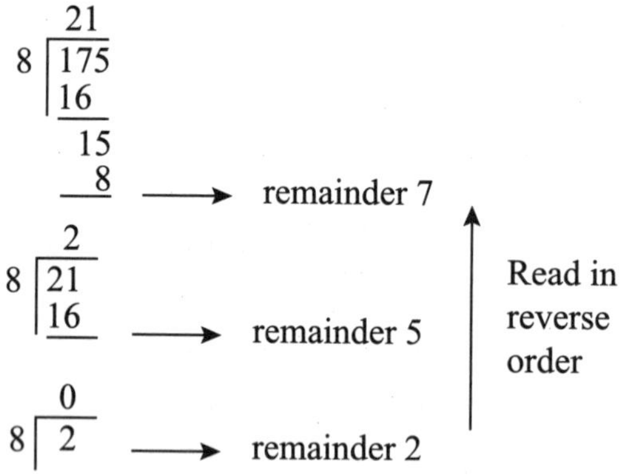

```
     21
 8 | 175
   | 16
     15
      8  ——→  remainder 7     ↑
                              |  Read in
      2                       |  reverse
 8 | 21                       |  order
   | 16  ——→  remainder 5     |

      0
 8 |  2  ——→  remainder 2
```

Thus, we get 257 which is the required octal number.

Split the given numbers into integer part and fractional part and apply double dabble method for each part separately. So the equivalent octal number is 257.

Fraction Here we multiply by 8 instead of 2 as in binary. Write the carry into integer position and read in forward direction.

Example 14 Convert 0.75.

$0.85 \times 8 = 6.80 \Rightarrow 0.8$ with a carry 6
$0.80 \times 8 = 6.40 \Rightarrow 0.4$ with a carry 6
$0.40 \times 8 = 3.20 \Rightarrow 0.2$ with a carry 3
$0.20 \times 8 = 1.6 \Rightarrow 0.6$ with a carry 1
$0.60 \times 8 = 4.8 \Rightarrow 0.8$ with a carry 4

↓ Read in forward direction

Thus, we get .66314, which is the required octal number.

19.6.3 Octal to Binary Conversion

The base of octal numbers is 8 and it is third power of 2 (base of binary numbers). So we can convert an octal to binary by changing each octal digit to its binary equivalent.

Example 15 Change octal 34 to its binary equivalent.

3 4
↓ ↓

Binary equivalent of each digit → 011 100

Here each octal digit is converted to its binary form; so the binary equivalent octal 34 is 011100.

Example 16 Change 357 to its binary equivalent.

3 5 7
↓ ↓ ↓

Binary equivalent of each digit → 011 101 111

So the binary equivalent of octal 357 is 0 1 1 1 0 1 1 1 1

For mixed numbers We follow the same procedure gives i.e. we convert each octal digit to its equivalent binary number. For example, 53.672 gives 101011·110111010.

19.6.4 Binary to Octal Conversion

To convert binary to octal, we follow the above steps in reverse order. Starting at the binary point, convert each group of three digits to its equivalent octal (0s are added at each end if necessary).

Example 17 Convert 1011·01101 to its octal equivalent. We rearrange the given number in group of three (starting from right to left in case of integer and left to right in case of fraction).

$\underline{001}\ \underline{001} \cdot \underline{011}\ \underline{010}$
↓ ↓ ↓ ↓

Octal equivalent of each digit → 1 3 · 3 2

Thus, the octal equivalent of the given number is 13.32.

19.7 Hexadecimal Numbers

Hexadecimal numbers are extensively used in microprocessors work just like how binary numbers work in computers. They are much shorter than binary numbers, and so it is easy to remember them. Here, the sequence of numbers does not stop at 9 as in the case of decimal numbers but continue to A, B, C, D, E, F. Thus, the total number of digits is 16 right from 0, 1, 2, … 9, A B C D E F. It has a base of 16.

Table 19.3 Decimal, binary and hexadecimal numbers

Decimal	Binary	Hexadecimal
0	0000	0
1	0001	1
2	0010	3
3	0011	4
4	0100	5
5	0101	6
6	0110	7
7	0111	8
8	1000	9
9	1001	9
10	1010	A
11	1011	B
12	1100	C
13	1101	D
14	1110	E
15	1111	F

19.7.1 Hexadecimal to Binary Conversion

To convert a hexadecimal number to binary equivalent, convert each hexadecimal digit to its 4-bit equivalent using the code given in Table 19.3.

Example 18 Convert 9AF to its primary equivalent.

$$\begin{array}{ccc} \underline{9} & \underline{A} & \underline{F} \\ \downarrow & \downarrow & \\ 1001 & 1010 & 1111 \end{array}$$

Thus, the equivalent binary number is 1001 1010 1111.

Example 19 Convert hexadecimal number C5E2 to its binary equivalent.

$$\begin{array}{cccc} \underline{C} & \underline{5} & \underline{E} & \underline{2} \\ \downarrow & \downarrow & \downarrow & \downarrow \\ 1100 & 0101 & 1110 & 0010 \end{array}$$

Thus, equivalent binary number is 1100010111100010.

19.7.2 Binary to Hexadecimal Conversion

Here also we use the code given in Table 19.3.

Example 20 Convert binary 1000 1100 to its equivalent hexadecimal.

$$\begin{array}{cc} \underline{1000} & \underline{1100} \\ \downarrow & \downarrow \\ 8 & C \end{array}$$

Thus, the equivalent hexadecimal number is 8C.

19.7.3 Hexadecimal to Decimal Conversion

In the hexadecimal number system, each digit position corresponds to a power of 16. The weights of digit position in the system are as follows.

$$16^3\ 16^2\ 16^1\ 16^0 \cdot 16^{-1}\ 16^{-2}\ 16^{-3} \ldots$$

$$\uparrow$$

Hexadecimal point

So to convert from hexadecimal to decimal, multiply each hexadecimal digit by its weight and add the decimal of the digits A, B, C, D, E, F.

Example 21 Convert hexadecimal F8E6.39 to its equivalent decimal number.

$$\begin{aligned} \text{F8E6.39} &= F(16^3) + 8(16^2) + E(16^1) + 6(16^0) + 3(16^{-1}) + 9(16^{-2}) \\ &= 15(16^3) + 8(16^2) + 14(16^1) + 6(16^0) + 3(16^{-1}) + 9(16^{-2}) \end{aligned}$$

$$= 61440 + 2048 + 224 + 6 + 0.1875 + 0.0352$$
$$= 63718.2227$$

19.7.4 Decimal to Hexadecimal Conversion

Similar to the double dabble method, we divide by 16 instead of 2 and take the remainders, then read in reverse order keeping in mind that remainders greater than 9 have to be changed to hexadecimal digits (here 10 becomes A and 15 becomes F).

Example 22 Convert 2479 to its hexadecimal equivalent.

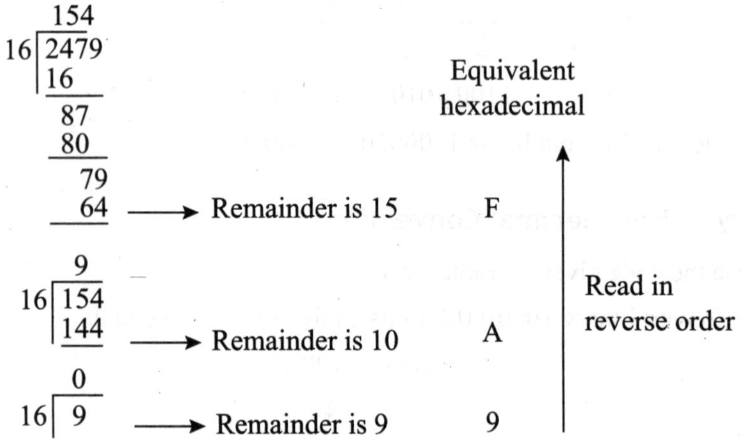

```
      154
  16 | 2479                                  Equivalent
     | 16                                    hexadecimal
       ----                                       ↑
        87                                        |
        80                                        |
       ----                                       |
         79                                       |
         64  ──→ Remainder is 15          F       |
       ----                                       |
          9                                       |  Read in
  16 | 154                                        |  reverse order
     | 144   ──→ Remainder is 10          A       |
       ----                                       |
          0                                       |
  16 |   9   ──→ Remainder is 9           9       |
```

Equivalent hexadecimal number is 9AF.

Example 23 Convert 65535 to its equivalent hexadecimal and binary number.

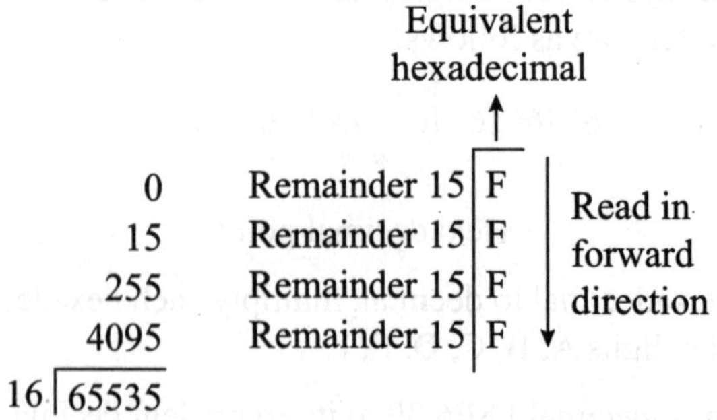

```
                               Equivalent
                               hexadecimal
                                    ↑
         0     Remainder 15 | F  |  Read in
        15     Remainder 15 | F  |  forward
       255     Remainder 15 | F  |  direction
      4095     Remainder 15 | F  ↓
16 | 65535
```

Thus the equivalent hexadecimal number is FFFF.

For binary conversion, write the equivalent binary number of each digit obtained in decimal system.

$$\begin{array}{cccc} 15 & 15 & 15 & 15 \\ \downarrow & \downarrow & \downarrow & \downarrow \\ 1111 & 1111 & 1111 & 1111 \end{array}$$

Thus, the equivalent binary number is 1111 1111 1111 1111

19.8 ASCII

To get information into and out of the computer, we need to use some kind of alphanumeric code (for letters, numbers and other symbols). For this, an international code has been formulated and followed throughout globally. This international code is known as **American Standard Code for Information Interchange (ASCII)**. It comprises 7-bits – $X_6\ X_5\ X_4\ X_3\ X_2\ X_1\ X_0$, where each X is 0 or 1 (Table 19.4).

Table 19.4 ASCII

				$X_6X_5X_4$					
X_3	X_2	X_1	X_0	010	011	100	101	110	111
0	0	0	0	SP	0	@	P	.	P
0	0	0	1	!	1	A	Q	a	q
0	0	1	0	"	2	B	R	b	r
0	0	1	1	#	3	C	S	c	s
0	1	0	0	$	4	D	T	d	t
0	1	0	1	%	5	E	U	e	w
0	1	1	0	&	6	F	V	f	v
0	1	1	1	,	7	G	W	g	w
1	0	0	0	(	8	H	X	h	x
1	0	0	1	)	9	I	Y	i	y
1	0	1	0	*	:	J	Z	j	z
1	0	1	1	+	;	K		k	
1	1	0	0	,	<	L		1	
1	1	0	1	–	=	M		m	
1	1	1	0	*	>	N		n	
1	1	1	1	/	?	O		o	

19.8.1 Parity Bit

The ASCII code is used for sending digital data over telephone lines. It is known that 1-bit error may occur in the transmitted data. To catch these errors, a parity bit is added and transmitted along with others (original bit). Then a parity checker at the receiving end can test for even or odd parity, whichever parity has been pre-arranged between the sender and the receiver. Since ASCII uses 7 bits, the addition of a parity bit to the transmitted data produces an 8-bit number in this format:

$$X_7X_6X_5X_4X_3X_2X_1X_0$$

↑

Parity bit

This is an ideal length because most digital equipment are set to handle bytes of data.

We have more similar codes, for example, Excess-3 code and Gray code, to name a few.

A description of these codes is beyond the scope of this chapter.

19.8.2 Binary Addition

We are familiar with addition in case of decimal numbers, e.g. $3 + 5 = 8$

We know that computer circuits process binary numbers only. For this, the following four rules are to be kept in mind.

$$0 + 0 = 0 \quad \text{(i)}$$

$$0 + 1 = 1 \quad \text{(ii)}$$

$$1 + 0 = 1 \quad \text{(iii)}$$

$$1 + 1 = 10 \quad \text{(iv)}$$

Equations (i) to (iii) are identical with decimal addition. But equation (iv) is different. Here if we follow decimal addition, then

$$1 + 1 = 2$$

i.e. ◍ + ◍ = ◍ ◍

If you have one pebble and bring another, it results in two pebbles so two pebbles; in binary represent 10.

19.8.3 Subscripts

We have four systems of numbers, namely decimal, binary, octal and hexadecimal, and each system uses a different way to code physical quantity. But how do we know which code is being used? In other words, how can we tell if 10 is a decimal or binary or octal or hexadecimal number.

To identify the system, we use the following subscripts:

2 → binary

8 → octal

10 → decimal

16 → hexadecimal

Accordingly

11_2 → 11 in binary system

23_8 → 23 in octal system

Following this we have

$1_2 + 1_2 = 10_2$ (10 in binary system)

$7_8 + 1_8 = 10_8$ (10 in octal system)

$9_{10} + 1_{10} = 10_{10}$ (10 in decimal system)

$F_{16} + 1_{16} = 10_{16}$ (10 in hexadecimal system)

19.8.4 Larger Binary System

Column-by-column addition applies when the number comprises of more than two digits.

Example 24 Add 11100 to 11010.

Step 1 - 1st (digit) LSD → 0

Step 2 - 2nd digit (column) → 1

Step 3 - 3rd digit (column) → 1

Step 4 - 4th digit (column) → 0 with a carry 1

Step 5 - 5th digit column we have 1 + 1 + 1; it gives 10 with carry of 1

So, the result is 1 1 0 1 1 0.

In the above steps, we have used the basic four rules (Eqn. i to iv).

Now consider the addition of 8-bit numbers.

$$\begin{array}{r} 0101\ \ 0111 \\ +\,0011\ \ 0101 \end{array}$$

Following the basic rules, the answer is 1000 1100

Let us consider hexadecimal addition using the above example. Numbers are 57 and 35; so the answer is

$$\begin{array}{r} 57 \\ \underline{+35} \\ 8C \end{array}$$

Generally, letter H is used as suffix to show hexadecimal number system; accordingly we have 8CH.

19.8.5 Binary Subtraction

Similar to addition, here too we have four basic rules:

$$0 - 0 = 0 \quad \text{(v)}$$
$$1 - 0 = 1 \quad \text{(vi)}$$
$$1 - 1 = 0 \quad \text{(vii)}$$
$$10 - 1 = 1 \quad \text{(viii)}$$

For larger numbers, we subtract column by column similar as in decimal numbers.

Example 25 Subtract 1010 from 1101.

$$\begin{array}{r} 1101 \\ -1010 \\ \hline \end{array}$$

Step 1 – 1st column 1 minus zero → 1

Step 2 – 2nd column borrow 1 from third column.
So 10 minus 1 → 1

Step 3 – 3rd column 0 minus 0 → 1

Step 4 – 4th column, 1 minus 1 → 0

So the answer is 0011.

Example 26 Subtract 125_{10} from 200_{10}. Find their equivalent in hexadecimal, binary and then subtract in sequence.

Solution

For decimal system,
$$\begin{array}{r} 200 \\ -125 \\ \hline 75 \end{array}$$

The equivalent of 200 and 125 in hexadecimal is C8 and 7D, respectively.

Their equivalent in binary is 1100 1000 and 0111 1101, respectively.

For binary

So we have
$$\begin{array}{r} 1100\ 1000 \\ -\ 0111\ 1101 \\ \hline \end{array}$$

Subtracting column by column, we have 0100 1011

For hexadecimal

We have

$$\begin{array}{r} C\ 8 \\ -7\ D \\ \hline 4\ B \end{array}$$

1's compliment The 1's compliment of a binary number is the number that results when we compliment the given binary (i.e. invert the given number).

Example 27 Let $X_3\,X_2\,X_1\,X_0 = 1000$

Its compliment is $\overline{X_3}\,\overline{X_2 X_1 X_0} = 0111$

It also applies to larger binary numbers comprising 8 or 16 bits or larger.

2's compliment It results when 1 is added to 1's compliment of given number; so

$$2\text{'s compliment} = 1\text{'s compliment} + 1$$

Example 28 Let the given number be 1011,

1's compliment of 1011 $\rightarrow$ 0100

Hence 2's compliment is 0100 + 1 $\rightarrow$ 0101

19.9 Gates

19.9.1 AND Gate

Figure 19.1 shows a diode circuit with switch input and load R connected to supply voltage 5 V (say).

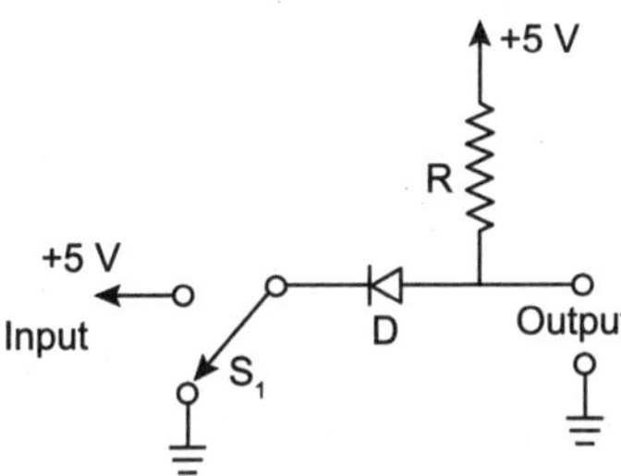

Figure 19.1

When switch S_1 is in ground position, then diode D is forward-biased and output is very low (0.7 V). Thus, we have a low output voltage when input is zero (low). On the other hand, when switch S_1 is at (+)5 V, net voltage across the resistor is zero since no current flows through it, i.e. diode is non-conducting, and output voltage is (+)5 V (high). Thus, we have a high output voltage when input voltage is high. So, we can say in the above arrangement that output follows input.

19.9.2 AND Gate (Two Input)

Figure 19.2 shows a similar arrangement having two diodes D_1 and D_2 with their respective switches S_1 and S_2. Here we have two input and one output.

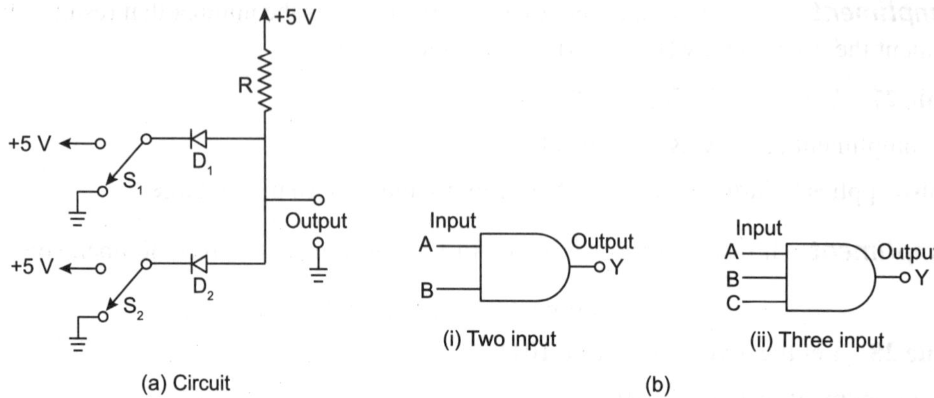

Figure 19.2 AND gate

Function

(i) When both switches S_1 and S_2 are in ground state, both diodes D_1 and D_2 are conducting; so output is low.

(ii) When switch S_1 is set to (+)5 V and switch S_2 is still in ground state, the output is still low since diode D_2 is conducting.

(iii) When switch S_2 is set to (+)5 V and switch S_1 is in ground state, the output is again low since diode D_1 is conducting.

(iv) When both switches S_1 and S_2 are set to (+)5 V, respectively, none of the diodes conducts and output is high.

Such an arrangement shown in Figure 19.2, represents a two-input AND gate. Its symbol (logic circuits) is shown in Figure 19.2(b). By adding more diodes with their own switches, we can have three (Figure 19.2b(ii)) or more input AND gates.

Regardless of the number of inputs an AND gate may have, the operation of the circuit is same, i.e. to get a High input all inputs must be High.

If any input is Low then output is always Low. We have discussed the function of AND gate using a diode logic circuit (Figure 19.2) where diode was used as a switch. However, the same function can be achieved using a diode-transistor logic circuit better known as D.T.L. circuit.

19.9.3 D.T.L. Circuit

Here the number of diodes used is equal to the number of inputs. As shown in Figure 19.2(c) where the number of inputs is three (A, B, C), so the diodes used are D_1, D_2 and D_3. Each diode acts as polarity controlled switch and transistor functions as emitter follower.

If any input V_1, V_2, or V_3 is at low state, say $V_2(0)$, then that diode is in conducting state and so voltage at the base of transistor is zero; and it will be in the cut off state so output V_0 is zero.

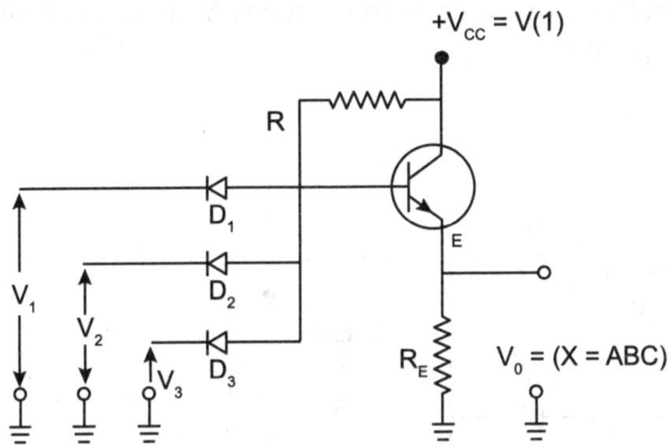

Figure 19.2(c)

When all the inputs V_1, V_2, V_3 are in high state, the diodes are reverse-biased and voltage at base in High, so the transistor is in saturation state and output voltage V_0 is high, i.e. $V_0 = 1$.

The action or function of a logic circuit is usually summarized in the form of Truth Table. These are tables which show the output state for all possible combinations of input signal states. Table 19.5 shows the truth label for two-input AND gate which summarizes all the actions described earlier.

Table 19.5 Truth table for AND gate

Input		Output
A	*B*	*Y*
0 (Low)	0 (Low)	0 (Low)
0 (Low)	1 (High)	0 (Low)
1 (High)	0 (Low)	0 (Low)
1 (High)	1 (High)	1 (High)

The above truth table can be represented by a Boolean equation:

$$Y = A \cdot B$$

Here (·) dot sign represents the action of AND gate. It reads as *Y* equals *A* AND *B*.

The function of (two input) AND gate is equivalent to two toggle switches S_1 and S_2 connected in series with a bulb and a battery (Figure 19.2(d)).

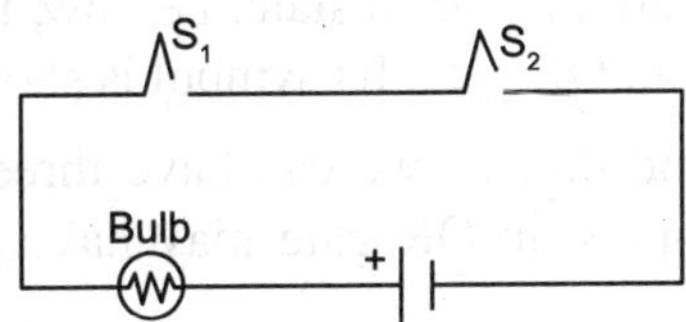

Figure 19.2(d)

The bulb will glow (ON) only when both switches S_1 and S_2 are ON. The bulb will remain OFF when either switch is OFF.

19.9.4 OR-Gate

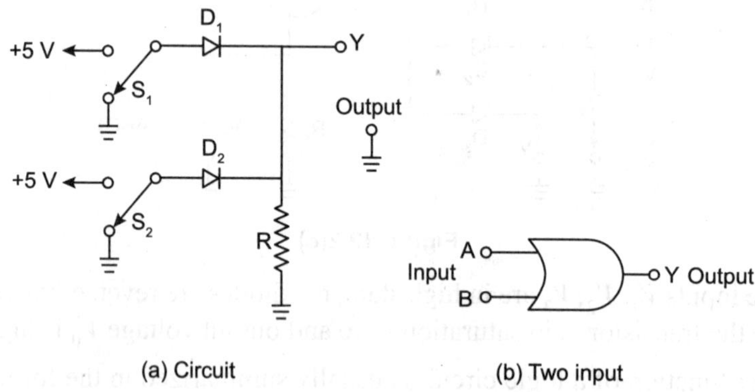

Figure 19.3 OR gate

Figure 19.3 represents the circuit of OR gate. Here also we have two diodes D_1 and D_2 connected to two switches S_1 and S_2, respectively. We can connect these diodes to two separate high inputs (+5 V) or to ground through these switches.

Function

(i) When both switches S_1 and S_2 are in ground state (low input), the diodes are non-conducting and output *Y* is low.

(ii) When switch S_1 is set to (+5 V) high state and S_2 is still connected to ground state, diode D_1 conducts and current flows through resistor *R*, thereby there is voltage across it, i.e. output *Y* is high.

(iii) When switch S_2 is set to (+5 V) high state and S_2 is connected to ground state diode D_2 conducts and current flows through resistor R_1 thereby there is voltage drop across it, i.e. output *Y* is high.

(iv) When both switches S_1 and S_2 are set to (+5 V) or in high state, both the diodes conduct and current flows through resistor *R*, thereby there is voltage drop across it, i.e. output *Y* is high.

We may conclude that when either input is high or both the inputs are high, the output is also high, and when both inputs are in ground state, i.e. low, then output is also low. Such an arrangement is known as two-input OR gate. Its symbol is shown in Figure 19.7(b).

By adding more switches and diodes, we can have three, four or more input OR gate. Regardless of the number of inputs an OR gate may have, the operation of the circuit is similar.

Truth Table Table 19.6 shows the truth table for two input OR gate.

Table 19.6 Truth table for OR gate

Input		Output
A	*B*	*Y*
0 (Low)	0 (Low)	0 (Low)
1 (High)	0 (Low)	1 (High)
0 (Low)	1 (High)	1 (High)
1 (High)	1 (High)	1 (High)

The above truth table is represented by Boolean equation as:

$$Y = A + B$$

Here (+) sign represents the action of OR gate. It reads as Y equals A or B. The function of two-input OR gate is equivalent to two toggle switches S_1 and S_2 connected in parallel with a bulb and a battery (Figure 19.4).

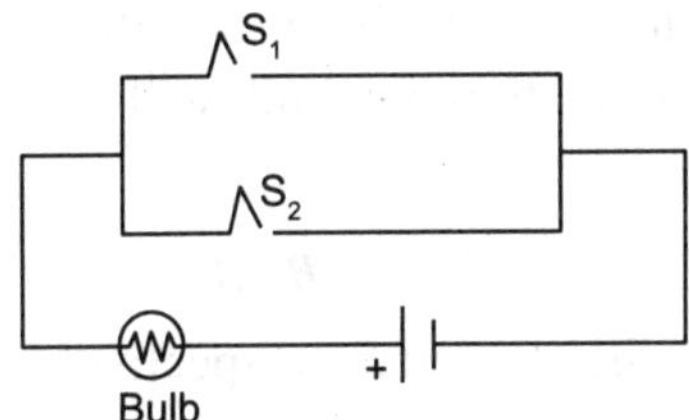

Figure 19.4

As is clear from the circuits the bulb glows when either or both switches S_1 and S_2 are ON. The bulb remains (OFF) when both switches are OFF.

The function of OR gate has been discussed using a diode logic circuit.

1. The following three rules of ordinary algebra apply in Boolean algebra.

   (i) Commutative law If the inputs in the OR gate or AND gate are interchanged, then also output remains same, i.e.

$$A + B = B + A \quad (1)$$

$$\text{and } A \cdot B = B \cdot A \quad (2)$$

The above law holds good for (+) addition or (·) multiplication operation.

   (ii) Associative law According to this law, whatever group we select the variables from, the output remains the same,

   (a) For OR gate,

$$A + B + C = A + (B + C) = (A + B) + C \quad (3)$$

(b) For AND gate,

$$ABC = A(BC) = (AB)C \tag{4}$$

(iii) Distributive law Here, the Boolean equation remains unchanged as for example

$$A(B + C) = AB + AC \tag{5}$$

and

$$A + (B \cdot C) = (A + B)(A + C) \tag{6}$$

The communicative law as expressed by Eqn. (1 and 2) can be further illustrated by the following circuit (Figure 19.4(a)).

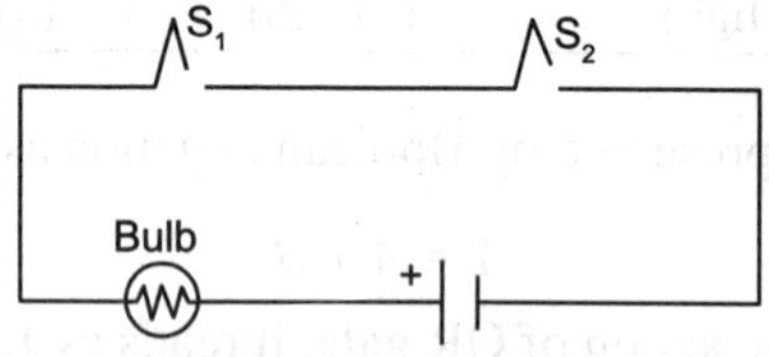

Figure 19.4(a)

Here switches S_1 and S_2 are communicated in series and it is not important which may they are joined from the circuit's operation point of view. In other words, we could interchange their positions and still have the same operation, i.e. bulb will glow when both of them are ON. So this idea can also be expressed as

$$A \cdot B = B \cdot A$$

where A and B represent two switches S_1 and S_2, respectively.

Similarly, the other expression Eqn. (1) can be further explained by the following circuit (Figure 19.4(b)).

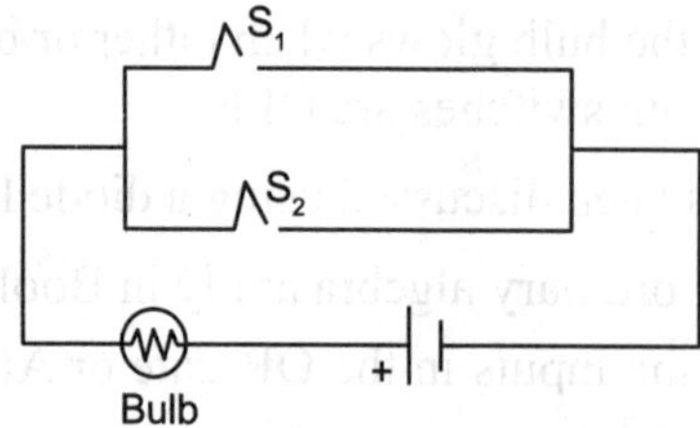

Figure 19.4(b)

Here the two switches S_1 and S_2 are connected in parallel, and from the circuit operation point of view they can be interchanged and yet have the same operation, i.e. bulb will glow when either of them is ON. So this operation is expressed as

$$\mathrm{A + B = B + A} \tag{i}$$

where A and B represent the two switches

and

$$A \cdot B = B \cdot A \tag{ii}$$

The law as expressed by Eqn. (5) can be further illustrated by the following circuit (Figure 19.4(c)).

(i) (ii)

Figure 19.4(c)

From the circuit, we find switches S_2 and S_3 are in parallel, i.e. they represent OR operation whereas switch S_1 is in series either with S_2 or S_3. So this either combination represents AND operation. Hence it be easily seen that the circuit (Figure 19.4(c)) represents the L.H.S. of Eqn. (6) and R.H.S. of Eqn. (5) is represented by the circuit shown in Figure (19.3(b)). Hence the two are equivalent.

Similarly, from the circuit (Figure 19.4(d)),

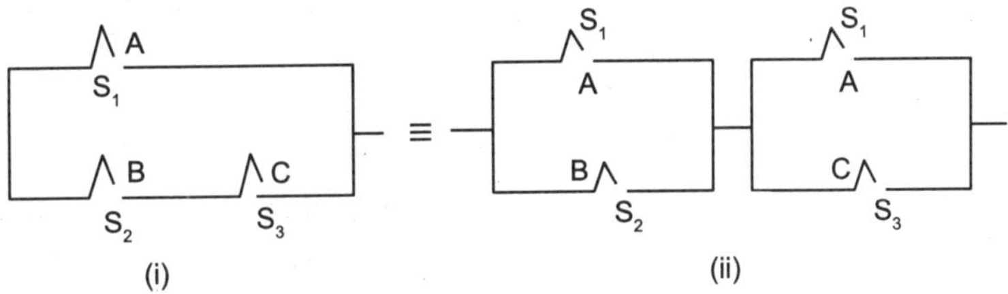

Figure 19.4(d)

we can easily identify Figure (19.4(d-ii)) with L.H.S. of Eqn. (6) and Figure (19.4(d-ii))with R.H.S. of Eqn. (6).

19.9.5 NOT Gate (Inverter)

A NOT gate circuit is simply an inverter circuit as shown in Figure 19.5. The circuit represents an amplifier circuit biased to cut off whose output is always out of phase with input.

In the absence of input signal, the transistor is cutoff and output is $+V_{CC}$, i.e. high.

When a positive pulse is applied to the base of transistor, it conducts and voltage at the collector drops. We say output is low. Thus, in the above circuit, we find that output is high when input is low and output is low when input in high. In a way it is an inverter circuit. Here we have one input and one output.

A mathematical statement of the characteristics of inverter (NOT) circuit is

If input $V_{in} = A$, then $V_{out} = \overline{A}$, which we spell as A bar.

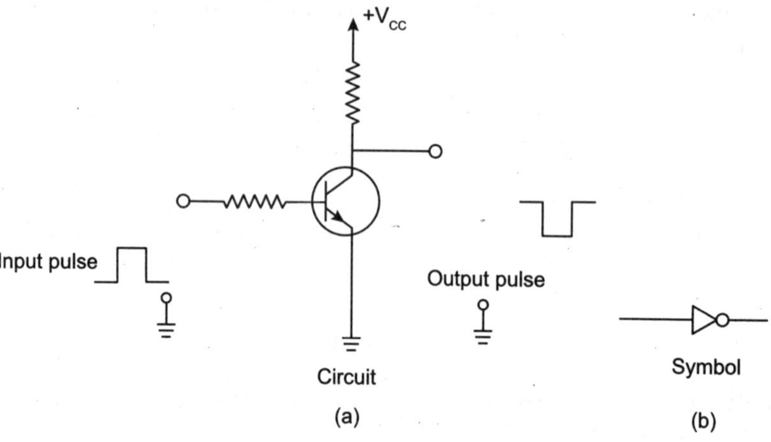

Figure 19.5

Truth table for the above circuit is given in Table 19.7.

Table 19.7 Truth table for NOT gate

Input	Output
0 (Low)	1 (High)
1 (High)	0 (Low)

Note: It is to be noted in the above circuits (truth table) that we follow positive logic where binary 0 represents Low state and binary 1 represents High state.

19.10 Mixed Gates

19.10.1 NOR Gate

The logic circuit in Figure 19.6 is called a NOT-OR gate because the output is given by

$$Y = \overline{A + B}$$

It is read as *Y* equals NOT *A* OR *B*. It may also be read as *Y* equals compliment of *A* OR *B*. The bubble on the output (Figure 19.6(b)) is a reminder of inversion that takes place after ORing.

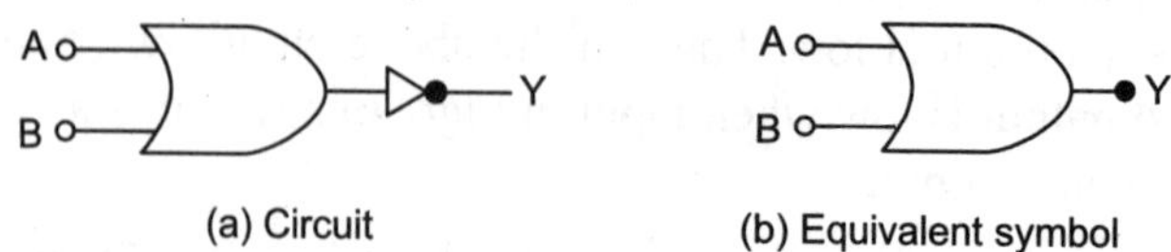

Figure 19.6 NOR Gate ≡ NOT + OR ≡ OR + NOT

Truth table for NOR gate is shown in Table 19.8.

Table 19.8 Truth table for NOR gate

Input		$A + B$	Output
A	B		$Y = \overline{A+B}$
0	0	0	1 (High)
0	1	1	0 (Low)
1	0	1	0 (Low)
1	1	1	0 (Low)

From the truth table, it is clear that output is High only when both inputs are Low. If any input is High, output is always Low.

19.10.2 Bubbled AND Gate

Figure 19.7 shows inverters (NOT gate) connected in each input to the AND gate. Its equivalent symbol is shown in Figure 19.7(b).

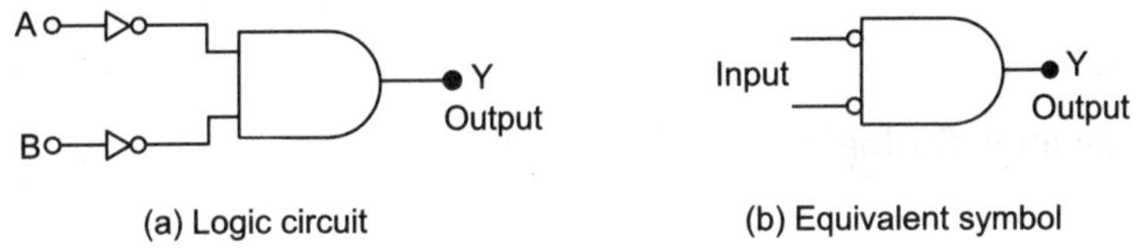

(a) Logic circuit (b) Equivalent symbol

Figure 19.7 Bubble AND gate

Bubbles on the input side are a reminder of inversion that takes place before AND operation.

The logic circuit (Figure 19.7(a)) can be analyzed as under:

(i) When input $A = 0$ (Low) and $B = 0$ (Low). The inverters produce high inputs at the AND gate.

So output $Y = 1$ (High)

(ii) When input $A = 0$ (Low) and $B = 1$ (High).

The upper inverter produces a high input and lower inverter produces a low input to the AND gate. So output $Y = 0$ (Low).

(iii) When input $A = 1$ (High) and $B = 0$ (Low). The upper inverter produces a low input and lower inverter produces a high input to the AND gate. So output $Y = 0$ (Low).

(iv) When input $A = 1$ (High) and $B = 1$ (High). The two inverters produce low inputs to the AND gate. So output $Y = 0$ (Low).

The above can be summarized in the Table 19.9.

Table 19.9 Truth table for bubbled AND gate

Input		Output
A	*B*	*Y*
0	0	1 (High)
0	1	0 (Low)
1	0	0 (Low)
1	1	0 (Low)

Table 19.9 matches Table 19.8. So we conclude that NOR gate and bubbled AND gate give the same output for common inputs. In this respect, the two circuits are interchangeable. Hence we can interchange any logic circuit with NOR gate.

19.11 De Morgan's First Theorem

The Boolean equation for logic circuit (Figure 19.6) is

$$Y = \overline{A + B} \tag{1}$$

The Boolean equation for logic circuit Figure 19.7 is

$$Y = \overline{A} \cdot \overline{B} \tag{2}$$

The first equation describes a NOR gate and second equation describes a bubbled AND gate. From Tables 19.9 and 19.8 for the two gates, we find that output is same for common inputs. We can equate the R.H.S. of the above equations and have

$$\overline{A + B} = \overline{A} \cdot \overline{B} \tag{3}$$

The above identity is known as De Morgan's first theorem.

It says, the compliment of a sum equals the product of compliments. Figure 19.8(a) is the graphical representation of De Morgan's first theorem.

A
B
Y ≡
A
B
Y

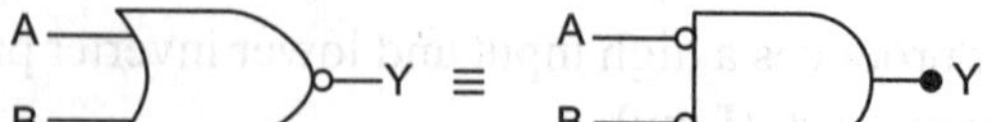

Figure 19.8 (a) Logic circuit representation of De Morgan's first theorem (NOR gate is equivalent to bubbled AND gate(b))

The function of NOR gate can be understood using diode-transistor logic circuit (Figure 19.8b).

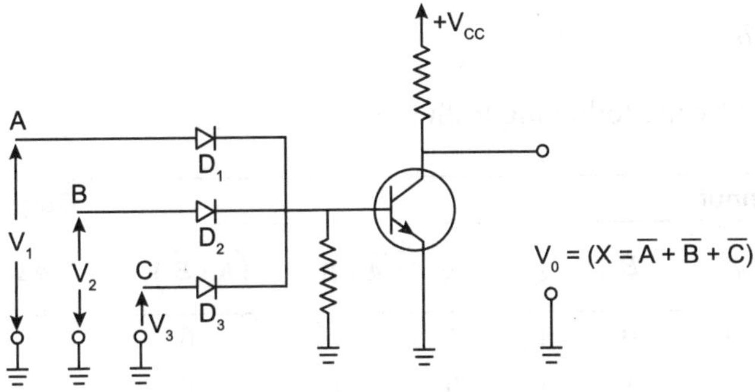

Figure 19.8 (b) NOR gate circuit

When any input *A*, *B* or *C* is in high (1) state, the diode conducts and output is in low (0) state. Since the base of the transistor is in positive state, transistor is in saturation state and output (collector) voltage is zero. So we conclude that when any input *A*, *B* or *C* is low (0), state the transistor is in cutoff state and so output (collector) voltage is high (1).

19.11.1 Function of NOR Gate

By combining NOR gates in a certain sequence, we can have the same results as those obtained from OR, AND and NOT gates individually. To show the equivalence, we make use of the following rules.

(a) $A + A = A$

which is verified by the following truth table.

Input		Output
A	A	$A + A$
0	0	0
1	1	1

(b) $\overline{\overline{A + B}} = A + B$

which is verified by the following truth table.

Input		**A + B**	$\overline{A+B}$	**Output**
A	**B**			$\overline{\overline{A+B}}$
0	0	0	1	0
0	1	1	0	1
1	0	1	0	1
1	1	1	0	1

(c) $\overline{\left(\overline{A}+\overline{B}\right)} = AB$

which is verified by the following truth table.

Input						Output
A	B	$\overline{A}$	$\overline{B}$	$\overline{A}+\overline{B}$	$\overline{\left(\overline{A}+\overline{B}\right)}$	AB
0	0	1	1	1	0	0
0	1	1	0	1	0	0
1	0	0	1	1	0	0
1	1	0	0	0	1	1

(i) Function of NOR gate as NOT gate

This function can be accomplished by keeping the two inputs of NOR gate same, i.e. either A or B. If A, then output will be $\overline{A}$. The result resembles that of NOT gate.

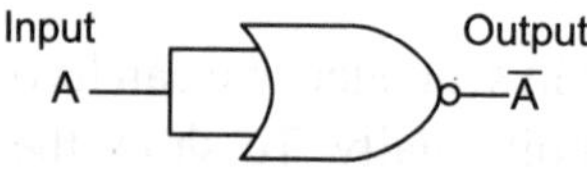

Figure 19.9(a)

(ii) Function of NOR gate as AND gate

Here NOR gates are employed as shown in Figure 19.9(b). The process is accomplished in two steps to get the result.

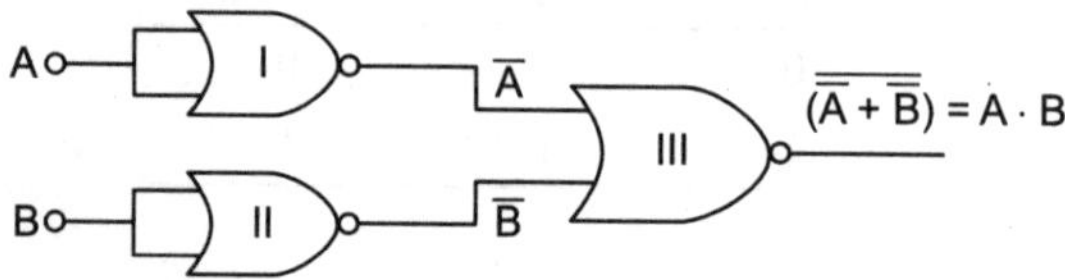

Figure 19.9(b)

The first two NOR gates I and II are used as NOT gates. Their respective output are $\overline{A}$ and $\overline{B}$· These are fed as input to the third NOR gate. The output of this (iii) gate is $\overline{\left(\overline{A}+\overline{B}\right)}$ which is equivalent to AB as per rule (c). Thus, by using three NOR gates in the above sequence, we can get the same result as that of a single AND gate.

(iii) Function of NOR gate as OR gate

Here two NOR gates are employed as shown in Figure 19.9(c) and the process is accomplished in two steps to get the result.

Figure 19.9(c)

The output of first NOR gate is $\overline{A+B}$. This output is employed as single input to second NOR gate which acts like a NOT gate. So the output is $A + B$ as per rule (b).

19.12 NAND Gate

The logic circuit (Figure 19.10) is called a NOT-AND gate because the output is given by $Y = \overline{A \cdot B}$

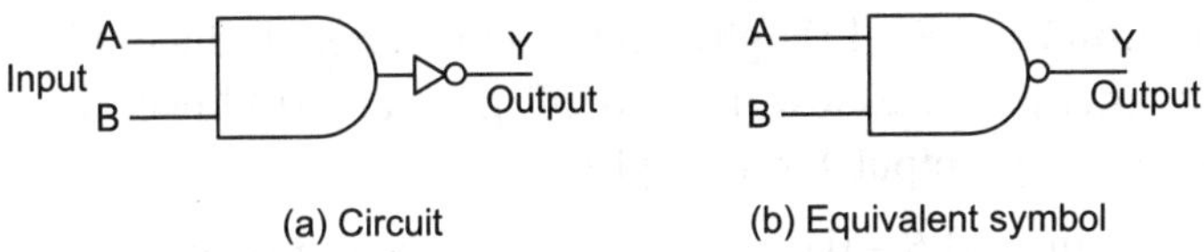

(a) Circuit (b) Equivalent symbol

Figure 19.10 NAND gate = NOT + AND ≡ AND + NOT

The above equation is read as NOT *A* AND *B*. It may also be read as *Y* equals compliment of *A* AND *B*. In a way, it is equivalent of an AND gate followed by NOT gate.

The truth table for NAND gate is shown in Table 19.10.

Table 19.10

Input		***AB***	**Output** $Y = \overline{A \cdot B}$
A	B		
0	0	0	1
0	1	0	1
1	0	0	1
1	1	1	0

19.13 Bubbled OR Gate

The logic circuit (Figure 19.11) shows inverters (NOT gate) connected in the input to OR gate. Its equivalent symbol is shown in Figure 19.11(b).

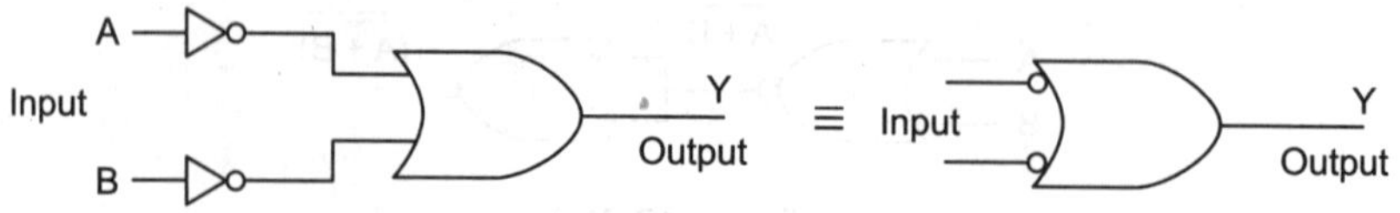

(a) Logic circuit (b) Equivalent circuit

Figure 19.11 Bubbled OR gate

Bubbles on the input side are a reminder of inversion that takes place before OR operation.

The logic circuit (Figure 19.11(a)) can be analyzed as under:

(i) When $A = 0$ (low) and $B = 0$ (low). The respective inverters produce high inputs at the OR gate. So output $Y = 1$ (High).

(ii) When $A = 0$ (Low) and $B = 1$ (High). (a) Upper inverter produces high input to OR gate; (b) lower inverter produces low input to OR gate; and (b) lower inverter produces low input to OR gate. So output $Y = 1$ (High).

(iii) When $A = 1$ (High) and $B = 0$ (Low). The upper inverter produces low input to OR gate and lower inverter produces high input to OR gate. So output $Y = 1$ (High).

(iv) When both inputs A and B are high. The respective inverters produce low inputs at each line. So output $Y = 0$ (Low).

The above function can be summarized in Table 19.11.

Table 19.11

Input		Output
A	B	Y
0	0	1
0	1	1
1	0	1
1	1	0

Table 19.10 matches Table 19.11. So we conclude that NAND gate and bubbled OR gate gives the same output for common inputs. In this respect, the two circuits are interchangeable. Hence we can interchange any logic circuit with NAND gate by a bubbled OR gate. Conversely, if a logic circuit has bubbled OR gate, it can be replaced by a NAND gate.

19.14 De Morgan's Second Theorem

The Boolean equation for logic circuit (Figure 19.12) is

$$Y = \overline{AB} \tag{4}$$

The Boolean equation for logic circuit (Figure 19.11(a)) is

$$Y = \overline{A} + \overline{B} \tag{5}$$

The first equation describes a NAND gate and second equation describes a bubbled OR gate. From truth tables 19.10 and 19.11 for the two gates, we find that the output is same for common inputs. We can equate the R.H.S. of Eqns. (4) and (5) and have

$$\overline{AB} = \overline{A} + \overline{B} \tag{6}$$

The above identity is known as *De Morgan's second theorem.*

It says, the complement of a product equals the sum of complements. Figure 19.12 is logic circuit representation of De Morgan's second theorem.

Figure 19.12 Graphical representation of De Morgan's second theorem (NAND gate is equivalent to Bubbled OR gate)

19.14.1 Functions of NAND Gate

By combining NAND gates in a certain sequence, we can have the same result as those obtained from OR, AND and NOT gates individually. To show the equivalence, we make use of the following rules:

(a) $A \cdot A = A$

which is verified by the following truth table.

Input		Output
A	A	Y
0	0	0
1	1	1

(b) $\overline{\overline{A+B}} = AB$

which is verified by the following truth table.

Input					Output
A	B	$A+B$	$\overline{A+B}$	$\overline{\overline{A+B}}$	AB
0	0	0	1	0	0
0	1	1	0	1	0
1	0	1	0	1	0
1	1	1	0	1	1

(c) $\overline{\overline{A} \cdot \overline{B}} = A + B$

which is verified by the following truth table.

Input		$\overline{A}$	$\overline{B}$	$\overline{A} \cdot \overline{B}$	$\overline{\overline{A} \cdot \overline{B}}$	Output $A + B$
A	*B*					
0	0	1	1	1	0	0
0	1	1	0	0	1	1
1	0	1	0	0	1	1
1	1	0	0	0	1	1

(i) Function of NAND gate as NOT gate

The function can be accomplished by keeping the two inputs same i.e. *A* or *B*. If A, then output will be $\overline{A}$.

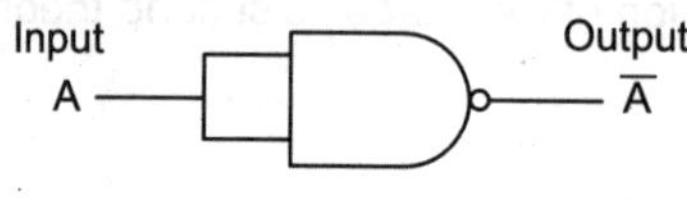

Figure 19.13(a)

(ii) Function of NAND Gate as AND Gate

This function can be accomplished by joining two NAND gates as shown in Figure 19.13(b).

Figure 19.13(b)

Here the output of NAND gate I is $\overline{A \cdot B}$ which is input for NAND gate II and output is AB, which is the output of a simple AND gate having *A* and *B* as inputs.

(iii) Function of NAND gate as OR gate

This function case be accomplished by connecting three NAND gates as shown in Figure 19.14(a).

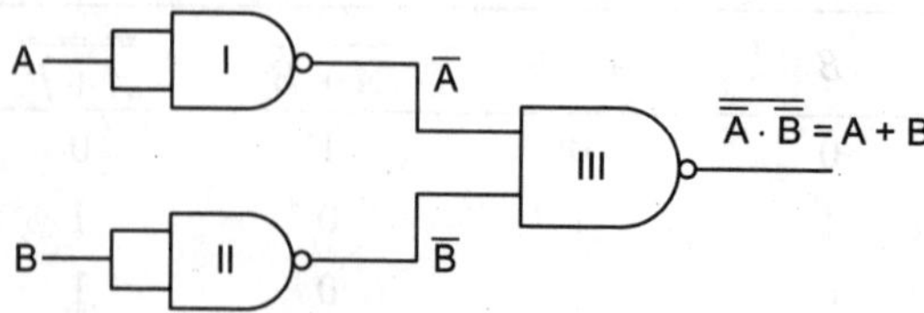

Figure 19.14(a)

It is based on De Morgan's second theorem.

$$\overline{\overline{A}\cdot\overline{B}} = A + B$$

It is accomplished in two steps. In the first step, the NAND gates I and II are used as NOT gates for each input. So their respective outputs are $\overline{A}$ and $\overline{B}$. These two are inputs for NAND gate III and its output is $A + B$. According to the De Morgan's theorem the function of NAND gate can be understood using the diode-transistor logic circuit (Figure 19.14(b)).

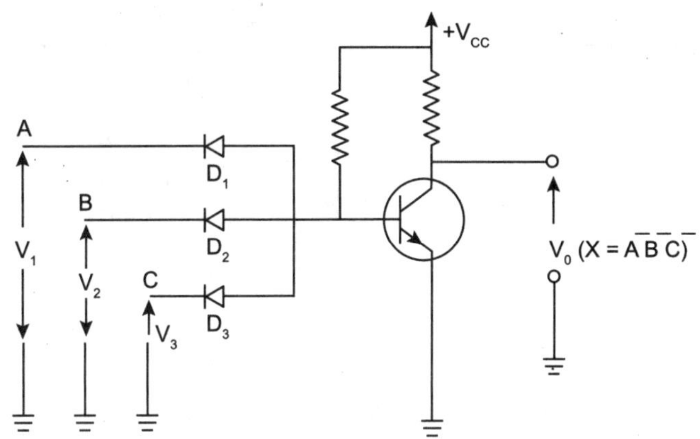

Figure 19.14(b) NAND gate circuit

When any of the inputs A, B or C is in Low (0) state, the output of AND gate, i.e. voltage at the base of transistor, is zero (Low state) and output *X* at the collector of transistor is 1 (High state) since the base of transistor is in zero state. When all the inputs *A*, *B* and *C* are in high (1) state, output *X* at the collector is zero (Low state). Since the base of transistor is in High state and transistor is in saturation state, collector voltage is zero.

Presently, most of the logic circuits use complimentary MOSFET (CMOS) instead of DTL or DL circuits. It is very popular in MSI and LSI area.

A CMOS is obtained by connecting a P-channel and an M-channel MOSFET in series, with their drains tied together. The input is supplied to common gate and output is taken at common drain.

19.15 CMOS Inverter (NOT gate)

The basic CMOS logic circuit is an inverter as shown in Figure 19.14(c). For this circuit, the logic levels are 0 V (Low) and V_{CC} (High 1). When input $V_i = V_{CC}$ (or high state 1), T_1 turns ON and T_2 turns OFF. Therefore, output $V_0 = 0$ (Low state). Since the two transistors are connected in series, current I_D is very small. On the other hand, when input $V_i = 0$ (Low state), transistor T_1 turns OFF and T_2 is ON giving an output voltage $V_0 = V_{CC}$ (High state 1) and drain current I_D is again very small. In either logic state, T_1 or T_2 is OFF and quiescent power dissipation, which is the product of OFF leakage current and V_{CC}, is very small.

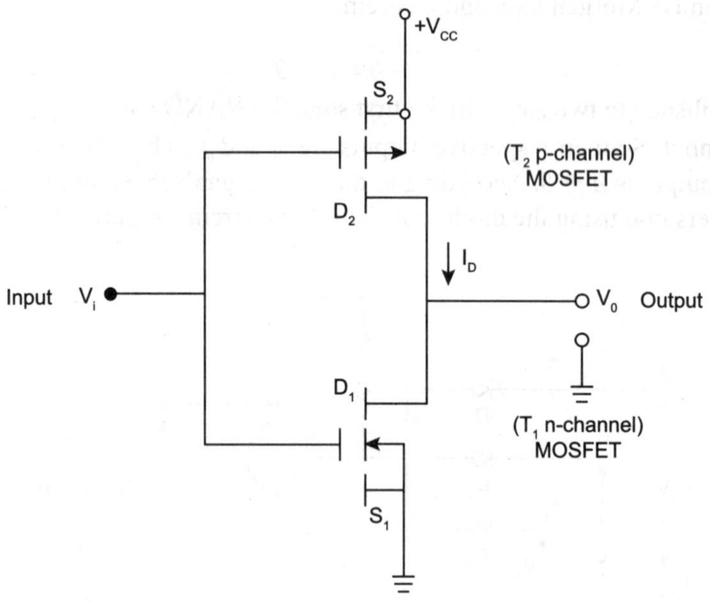

Figure 19.14(c) CMOS inverter (NOT gate)

19.15.1 NAND Gate (using CMOS)

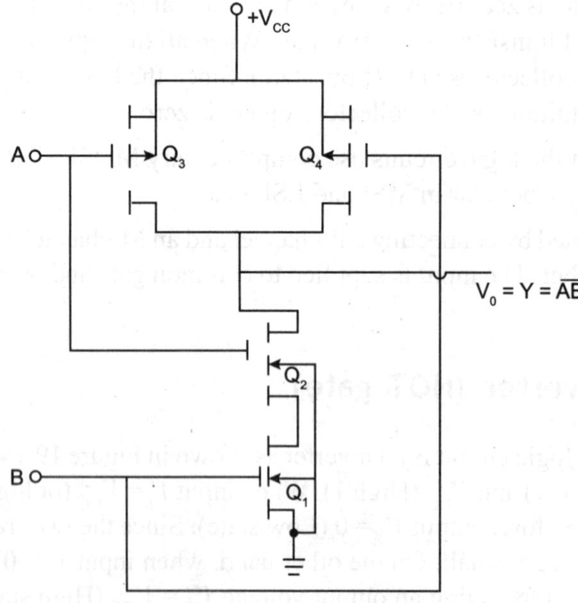

Figure 19.14(d) 2-input CMOS NAND gate

19.15.2 NOR Gate (using CMOS)

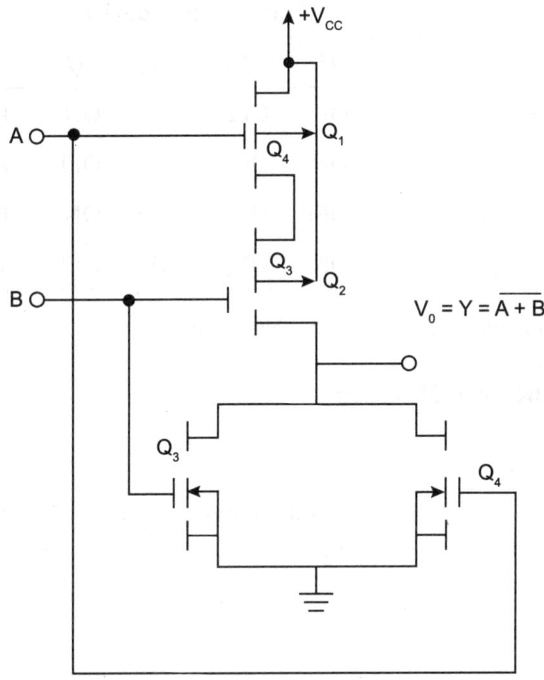

Figure 19.14(e) 2-input CMOS NOR gate

A CMOS NAND gate is shown in Figure 19.14(d). A CMOS NOR gate is shown in Figure 19.14(e).

The operation of NAND gate can be understood from Table 19.12.

Table 19.12 Operation of CMOS NAND gate

Input		State of MOS device				Output
A	*B*	Q_1	Q_2	Q_3	Q_4	
0 (Low)	0 (Low)	OFF	OFF	ON	ON	V_{CC} (High)
0 (Low)	V_{CC} (High)	ON	OFF	ON	OFF	V_{CC} (High)
V_{CC} (High)	0 (Low)	OFF	ON	OFF	ON	V_{CC} (High)
V_{CC} (High)	V_{CC} (High)	ON	ON	OFF	OFF	0 (Low)

The above table resembles Table 19.10 for NAND gate. The operation of NOR gate can be verified in a similar way by Table 19.13.

Table 19.13 Operation of CMOS NOR gate

Input		State of MOS device				Output
A	B	Q_1	Q_2	Q_3	Q_4	Y
0 (Low)	0 (Low)	OFF	OFF	ON	ON	V_{CC} (High)
0 (Low)	V_{CC} (High)	ON	OFF	ON	OFF	V_{CC} (High)
V_{CC} (High)	0 (Low)	OFF	ON	OFF	ON	V_{CC} (High)
V_{CC} (High)	V_{CC} (High)	ON	ON	OFF	OFF	0 (Low)

The above table resembles Table 19.8 for NOR gate.

19.15.3 Exclusive OR or XOR Gate

Statement I

An exclusive OR gate operates on the principle that output is high (1) if one and only one input assumes High (1) state. Obviously the output of the two input exclusive OR gates is high only when either $A = 1$ and $B = 0$ or $A = 0$ and $B = 1$.

It can be expressed as

$$Y = A\overline{B} + \overline{A}B \tag{7}$$

Its logic circuit is shown in Figure 19.15(a) and equivalent symbol is shown in Figure 19.15(b).

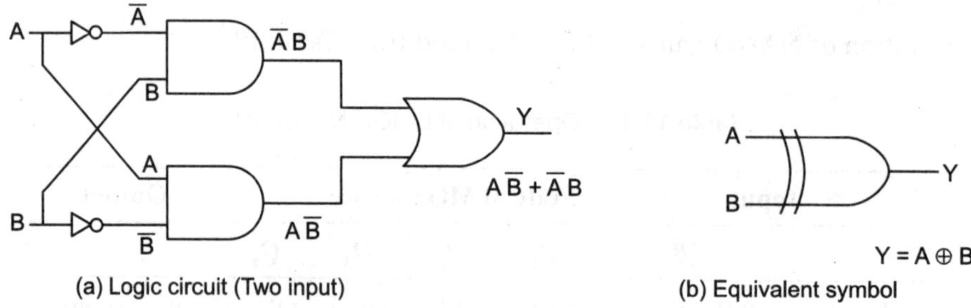

(a) Logic circuit (Two input) (b) Equivalent symbol

Figure 19.15 Exclusive OR gate

Eqn. (7) is generally expressed as

$$Y = A \oplus B$$

Here it is important to note the symbol plus encircled for XOR gate operation. Using 2-input exclusive OR gate as building blocks, we can have an exclusive OR gate with any number of inputs. Figure 19.16 represents such as arrangement for an exclusive OR gate having three inputs and one output. Figure 19.16(b) shows an arrangement having 6 inputs and output.

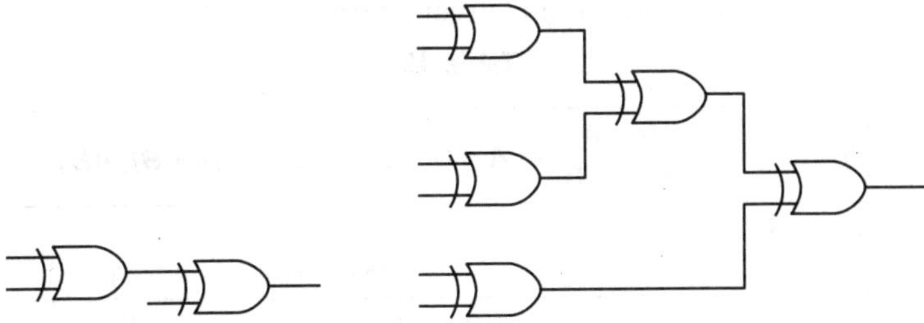

(a) Exclusive OR gate (three input) (b) Exclusive OR gate (six input)

Figure 19.16 Exclusive OR gate with more than two inputs

Truth table for XOR gate having two inputs is shown in Table 19.14.

Table 19.14

Input		Output $Y = A \oplus B$	$A\bar{B}$	$B\bar{A}$	$A\bar{B} + B\bar{A}$
A	B				
(1)	(2)	(3)	(4)	(5)	(6)
0	0	0	0	0	0
0	1	1	0	1	1
1	0	1	1	0	1
1	1	0	0	0	0

From the above truth table, we find that the resulting column (6) is same as that of column (3).

Statement II

An exclusive OR gate operates on the principle that output is high (1) only if either $A = 1$ (High) or $B = 1$ (High) but both inputs together are not high (1). It may be expressed as

$$Y = (A + B)\,(\overline{AB})$$

Its logic circuit is shown in Figure 19.17.

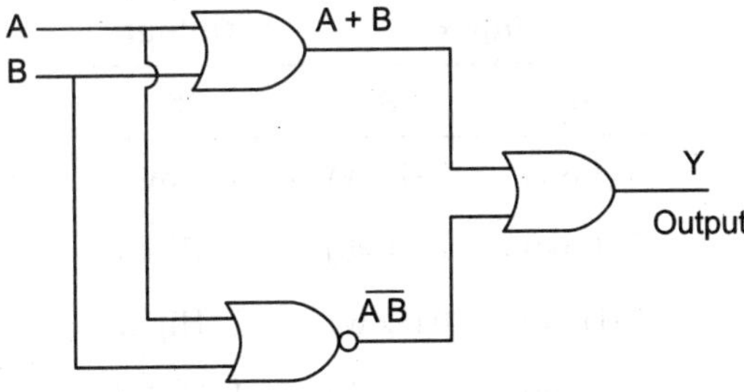

Figure 19.17 Exclusive OR gate logic circuit (Another version)

Truth table for the above expression is given in Table 19.15.

Table 19.15

Input		Output $Y = A \oplus B$	$A+B$	AB	$\overline{AB}$	$(A+B)(\overline{AB})$
A (1)	B (2)	(3)	(4)	(5)	(6)	(7)
0	0	0	0	0	1	0
0	1	1	1	0	1	1
1	0	1	1	0	1	1
1	1	0	1	1	0	0

From the truth table, we find that the result of column (7) is same as obtained in column (3).

It is to be noted that logic circuits for a gate are not unique and they depend on how the process is executed as explained in Figure 19.18 and Figure 19.19.

19.16 Positive and Negative Logic

All the gates described so far are based on positive logic, where binary 0 (zero) is used for low voltage (state) and binary 1 (one) is used for high voltage (state).

There is another code known as negative logic where binary 0 stands for high voltage (state) and binary 1 stands for low voltage (state). Even though this code appears unnatural, negative logic has may uses. The following discussion introduces some terminology and concepts for both types of logic.

19.16.1 Positive and Negative Gates

An OR gate in positive logic system becomes an AND gate in negative logic system as can be easily seen from its truth table (Table 19.16).

Table 19.16

Input		Output
A	B	Y
0 (Low)	0 (Low)	0 (Low)
0 (Low)	1 (High)	1 (High)
1 (High)	0 (Low)	1 (High)
1 (High)	1 (High)	1 (High)

In the above truth table, we find that output is high when either input or both inputs are high. As defined in positive logic system, binary 0 stands for low and binary 1 stands for high, and so these binary numbers have been put along with the respective state.

However, in negative logic system, binary 1 stands for low voltage (state) and binary 0 stands for high voltage. So the above table can be represented as.

Table 19.17 Truth table for negative AND gate

Input		Output
A	*B*	*Y*
1 (Low)	1 (Low)	1 (Low)
1 (Low)	0 (High)	0 (High)
0 (High)	1 (Low)	0 (High)
0 (Low)	0 (High)	0 (High)

In the above table, output *Y* is 1 only when both *A* and *B* are 1. This statement sounds like AND gate (in positive logic). Thus, in negative logic system, we call it negative AND.

So we conclude that gates are defined by the way they process binary 0s and 1s. If we use binary 1 for low (voltage) state and binary 0 for high (voltage) state, then the above truth table represents AND gate in negative logic system.

In a similar way, we can represent Table 19.5 for AND gate in positive logic as Table 19.18.

Table 19.18 Truth table for negative OR gate

Input		Output
A	*B*	*Y*
0 (High)	0 (High)	0 (High)
1 (Low)	0 (High)	1 (Low)
0 (High)	1 (Low)	1 (Low)
1 (Low)	1 (Low)	1 (Low)

In the above table, output *Y* is 1 only when either input *A* or *B* is 1 or both inputs are 1. This statement sounds like OR gate in positive logic. Thus, in negative logic system, we call it negative OR gate.

By analyzing the inputs and outputs in terms of 0s and 1s, we find the following equivalence.

Positive OR ↔ Negative AND
Positive AND ↔ Negative OR
Positive NOR ↔ Negative NAND
Positive NAND ↔ Negative NOR

Table 19.19 summarizes these gates and their definitions in terms of voltage (state) level. These definitions are always valid.

Table 19.19 Voltage definitions of basic gate

Gate	Definition
Positive OR/Negative AND	Output is High if any input is high
Positive AND/Negative OR	Output is High when all inputs are high
Positive NOR/Negative NAND	Output is Low if only input is high
Positive NAND/Negative NOR	Output is Low when all inputs are high

Figure 19.18 shows the equivalent symbols of gates in positive and negative logic system.

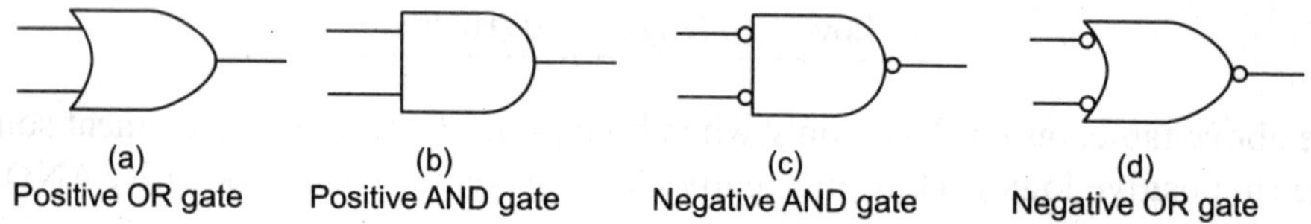

Figure 19.18 Equivalent of symbols of gates in positive and negative logic system

Figure 19.19 shows the logic equivalence of various gates described previously.

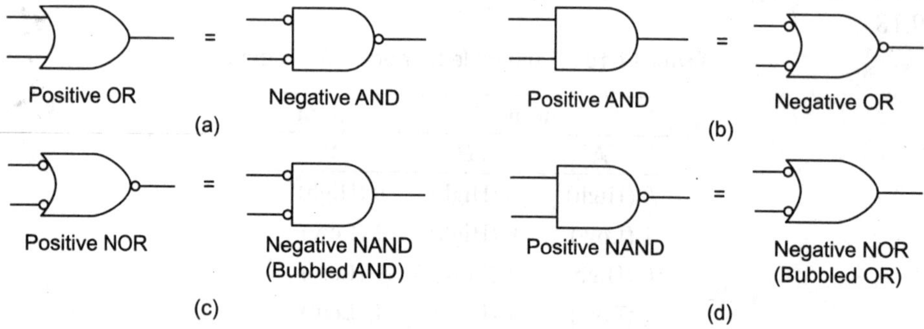

Figure 19.19 Logic equivalence

Basic rules for single variable

(i) Rules for OR operation

$$A + 0 = A \tag{1}$$

$$A + 1 = 1 \tag{2}$$

$$A + A = A \tag{3}$$

$$A + \overline{A} = 1 \tag{4}$$

(ii) Rules for AND operation

$$A \cdot 0 = 0 \tag{5}$$

$$A \cdot 1 = A \tag{6}$$

$$A \cdot A = A \tag{7}$$

$$A \cdot \overline{A} = 0 \tag{8}$$

It is to be noted that each rule valid for OR operation is the dual for each rule valid for AND operation and vice versa. In a way, they follow duality theorem.

19.16.2 Duality Theorem

Starting with a Boolean relation, one can derive another relation by

(i) Changing each OR sign to an AND sign

(ii) Changing each AND sign to an OR sign

(iii) Complementing any 0 or 1 appearing in the expression

For instance, for $A + 0 = A$, the dual relation is $A \cdot 1 = A$

The above dual property is obtained by (i) changing OR sign to AND sign and (ii) by complementing 0 to get 1.

Table 19.20 Boolean algebraic theorems

S. No.	Single variable theorem	
1.	$A + 0 = A$	Follows OR gate
2.	$A \cdot 1 = A$	Föllows AND gate
3.	$A + 1 = 1$	Follows OR gate
4.	$A \cdot 0 = 0$	Follows AND GATE
5.	$A + A = A$	Follow OR gate
6.	$A \cdot A = A$	Follow AND gate
7.	$A + \overline{A} = 1$	Follow OR gate
8.	$A \cdot \overline{A} = 0$	Follow AND gate
9.	$A \cdot (B + C) = AB + AC$	
10.	$A + BC = (A + B) \cdot (A + C)$	
11.	$A + AB = A$	
12.	$A \cdot (A + B) = A$	
13.	$A + \overline{A}B = A + B$	
14.	$A \cdot (\overline{A} + B) = AB$	

15. $AB + A\overline{B} = A$

16. $(A + B) \cdot (A + \overline{B}) = A$

17. $AB + \overline{A}C = (A + C)\,(\overline{A} + B)$

18. $(A + B) \cdot (\overline{A} + C) = AC + \overline{A}B$

19. $AB + \overline{A}C + BC = AB + \overline{A}C$

20. $(A + B) \cdot (\overline{A} + C) \cdot (B + C) = (A + B) \cdot (\overline{A} + C)$

De Morgan's theorem

21. $\overline{A \cdot B \cdot C \cdot \ldots} = \overline{A} + \overline{B} + \overline{C} + \ldots$

22. $\overline{A + B + C + \ldots} = \overline{A} \cdot \overline{B} \cdot \overline{C} \cdot \ldots$

Further, it is to noted Eqn. (4) can be verified. Assume $A = 1$ then $\overline{A} = 0$.

So,

$$A + \overline{A} = 1 + 0$$
$$= 1$$

Similarly, if $A = 0$ then $\overline{A} = 1$

So,

$$A + \overline{A} = 0 + 1$$
$$= 1$$

Eqn. (8) can also be verified.

Assume $A = 1$ then $\overline{A} = 0$

So,

$$A \cdot \overline{A} = 1 \cdot 0$$
$$= 0$$

Similarly, if $A = 0$ then $\overline{A} = 1$.

So,

$$A \cdot \overline{A} = 0 \cdot 1$$
$$= 0$$

In addition to the above theorems, we have

(i) Complimentary rule: $\overline{A} = A$

(ii) Absorption rule:

(a) Theorem (11) can be proved as

$$
\begin{aligned}
A + AB &= A[1 + B] && \text{by distributive law}\\
&= A \cdot 1 && \text{by theorem (3)}\\
&= A && \text{by theorem (2)}\\
\therefore\ A + AB &= A
\end{aligned}
$$

(b) Theorem (13) can be proved as

$$
\begin{aligned}
A + \overline{A}B &= A(B + 1) + \overline{A}B && \text{by theorem 3}\\
&= AB + A + \overline{A}B\\
&= B(A + \overline{A}) + A && \text{by distributive law}\\
&= B + A && \text{by theorem (7)}\\
\therefore\ A + A\overline{B} &= A + B
\end{aligned}
$$

(c) Theorem (15) can be proved as

$$
\begin{aligned}
A + A\overline{B} &= A(B + 1) + A\overline{B} && \text{by applying theorem (3)}\\
&= AB + A + A\overline{B}\\
&= AB + A\overline{B} + A\\
&= A(B + \overline{B}) + A && \text{by distributive law}\\
&= A + A && \text{by theorem (7)}\\
&= A && \text{by theorem (5)}
\end{aligned}
$$

By applying the above theorems, we can simplify many expressions.

Example 1 Simplify $(A + B)(A + C)$.

Solution

$$
\begin{aligned}
(A + B)(A + C) &= AA + AC + AB + BC && \text{by distributive law}\\
&= A + AC + AB + BC && \text{by theorem (6)}\\
&= A(1 + C) + AB + BC && \text{by distributive law}\\
&= A \cdot 1 + AB + BC && \text{by theorem (3)}\\
&= A(B + 1) + BC && \text{by distributive law}\\
&= A + BC\\
\therefore\ (A + B)(A + C) &= A + BC
\end{aligned}
$$

Example 2 Simplify, $AB + A(B + C) + B(B + C)$.

Solution

$$\begin{aligned} AB + A(B+C) + B(B+C) &= AB + AB + AC + BB + BC \\ &= AB + AC + B + BC \\ &= AB + AC + B(1+C) \\ &= AB + AC + B \\ &= B(A+1) + AC \\ &= B \cdot 1 + AC \\ &= B + AC \end{aligned}$$

$$\therefore \; AB + A(B+C) + B(B+C) = B + AC$$

Example 3 Simplify, $[A\overline{B}(C + BD) + \overline{A}B]C$.

Solution

$$\begin{aligned} [A\overline{B}(C+BD) + \overline{AB}]C &= [A\overline{B}C + A\overline{B}BD + \overline{AB}]C \\ &= [A\overline{B}C + O + \overline{AB}]C \\ &= [A\overline{B}C + \overline{AB}]C \\ &= A\overline{B}CC + \overline{AB}C \\ &= A\overline{B}C + \overline{AB}C \\ &= \overline{B}C[A + \overline{A}] \\ &= \overline{B}C \end{aligned}$$

Example 4 Prove that:

$$(\overline{A} + B + C)(\overline{A} + \overline{B} + C) = \overline{A} + C$$

Solution

$$\begin{aligned} \text{L.H.S. } (A+B+C)(A+B+C) &= AA + AB + AC + BA + BB + BC + CA + C\overline{B} + CC \\ &= A + AB + AC + BA + O + BC + CA + CB + C \\ &= A[B+B]A + AC + BC + CB + C \\ &= \overline{A} + \overline{A} + \overline{A}C(C) + BC + C(B+1) \\ &= \overline{A} + \overline{A}(1+C) + BC + C \\ &= \overline{A} + \overline{A} + C(B+1) \\ &= \overline{A} + C \end{aligned}$$

Example 5 Prove that:

$$(A+C)(A'+B) = AB + A'C$$

Solution

$$\begin{aligned}\text{L.H.S. } (A+C)(A'+B) &= AA' + AB + CA' + BC \\ &= 0 + AB + CA' + BC \\ &= AB + (A+A')BC + CA' \\ &= AB + ABC + A'BC + CA' \\ &= AB + ABC + AC'(B+1) \\ &= AB + ABC + A'C \\ &= AB(C+1) + A'C \\ &= AB + A'C\end{aligned}$$

19.17 Fundamental Products

Figure 19.20 shows four possible ways to AND two input signals that are in complimented and uncomplimented form. The output for each such combination is called fundamental products.

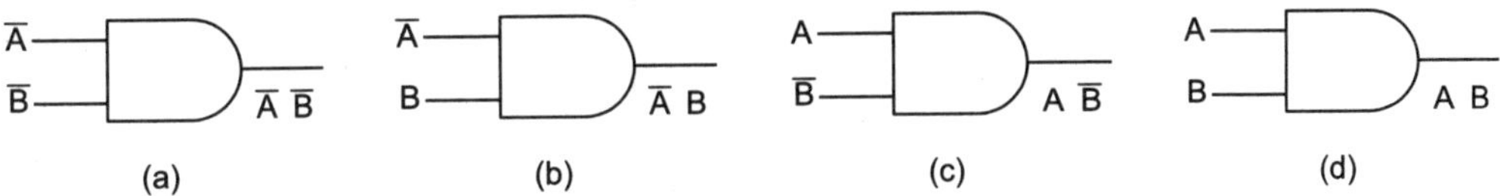

Figure 19.20 ANDing two variables and their compliments

Table 19.12 lists each fundamental product next to the input conditions producing a high output. For instance, $\overline{A}\overline{B}$ is high when A and B both are low; similarly, $\overline{A}B$ is high when A is low and B is high.

Table 19.21 Fundamental products for two inputs

A	*B*	Fundamental product
0	0	$\overline{A}\overline{B}$
0	1	$\overline{A}B$
1	0	$A\overline{B}$
1	1	AB

The idea of fundamental products applies to three or more input variables. Assuming three variables A, B, C and their compliments, we have eight ways to AND the three variables and their compliments. The resulting fundamental products are $\overline{A}\,\overline{B}\,\overline{C}$, $\overline{A}\,\overline{B}C$, $\overline{A}B\overline{C}$, $\overline{A}BC$, $A\overline{B}\,\overline{C}$, $A\overline{B}C$, $AB\overline{C}$, ABC. As an illustration, we have the following representation for first three fundamental products.

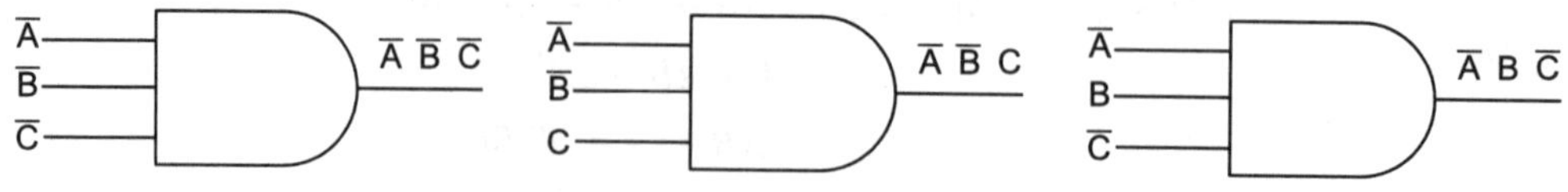

Figure 19.21(a)

Table 19.13 summarizes the fundamental products listing each one next to the input condition that results in a high output. For instance, when $A = 1$, $B = 0$, $C = 0$, the fundamental product results in an output $Y = A\overline{B}\,\overline{C} = 1\cdot\overline{0}\cdot\overline{0} = 1$.

Table 19.22 Fundamental products for three inputs

A	B	C	Fundamental product
0	0	0	$\overline{A}\,\overline{B}\,\overline{C}$
0	0	1	$\overline{A}\,\overline{B}C$
0	1	0	$\overline{A}B\overline{C}$
0	1	1	$\overline{A}BC$
1	0	0	$A\overline{B}\,\overline{C}$
1	0	1	$A\overline{B}C$
1	1	0	$AB\overline{C}$
1	1	1	ABC

19.17.1 Sum of Product Equation

For a given truth table, the sum of products equation can be written. For this, one has to locate each output 1 in truth table and write down the fundamental product. For instance, in case of three variables, the first output 1 appears for the input of $A = 0$, $B = 1$, $C = 1$. The corresponding fundamental product is $\overline{A}BC$. Continuing this way, one can identify all the fundamental products as shown in Table 19.22. To get the sum of products equation, all that one has to do is OR the fundamental products of Table 19.22 and so we have

$$Y = \overline{A}BC + A\overline{B}C + AB\overline{C} + ABC \tag{8}$$

19.17.2 Logic Circuit

After one obtains the sum of products equation, one can derive the logic circuit by drawing the AND –OR network. In Eqn. (8), each product is the output of a 3-input AND gate. Furthermore, the logical sum *Y* is output of four input OR gates (Figure 19.21(b)).

This AND-OR circuit is one solution to the design problem. The truth table for the circuit is shown in Table 19.23.

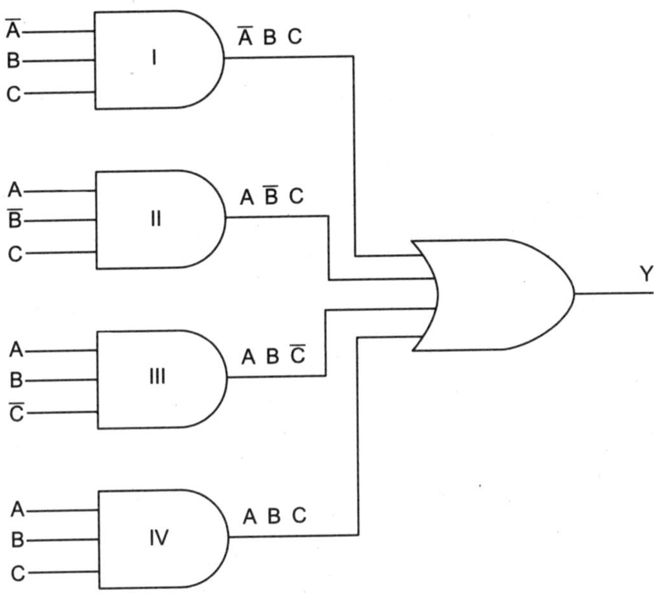

Figure 19.21(b) AND-OR solution

Table 19.23 Truth table for circuit (AND-OR)

A	*B*	*C*	*Y*	
0	0	0	0	
0	0	1	0	
0	1	0	0	
0	1	1	1	$\rightarrow \overline{A}BC$
1	0	1	1	$\rightarrow A\overline{B}C$
1	1	0	1	$\rightarrow AB\overline{C}$
1	1	1	1	$\rightarrow ABC$

19.17.3 Karnaugh Map

A Karnaugh map is a visual display of fundamental products needed for a sum of products solution. We can convert a turth table into its Karnaugh map as described below.

Let the truth table be

Table 19.24

A	*B*	*Y*
0	0	0
0	1	0
1	0	1
1	1	1

Now begin by drawing the following in sequence (Figure 19.22):

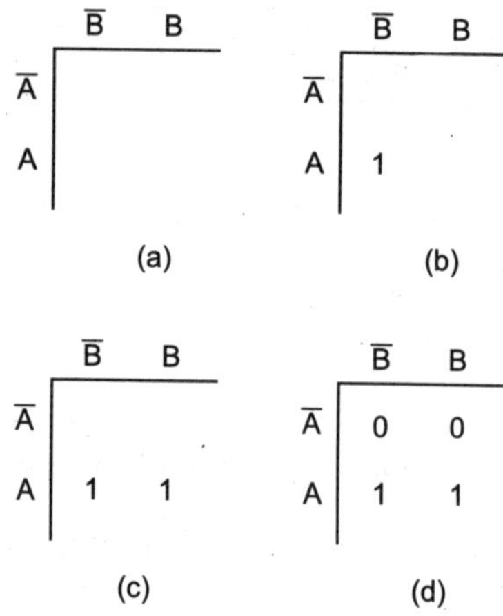

Figure 19.22 Construction of Karnaugh map

Note the variables and compliments. The vertical column has $\overline{A}$ followed by $A1$ and the horizontal row has $\overline{B}$ followed by B. Note the output 1s in truth table (Table 19.15). The first output 1 appears for A = 1 and B = 0. The fundamental product for this is $A\overline{B}$. Enter this fundamental product on Karnaugh map as shown in Figure 19.22(b). This 1 represents the product $A\overline{B}$ because 1 is in row A and column $\overline{B}$.

Similarly, Table 19.24 has an output appearing 1 for inputs of $A = 1, B = 1$. The fundamental product is AB which can be entered on Karnaugh map as shown in Figure 19.22(c). The final step in drawing the Karnaugh map is to enter 0s in the remaining space (Figure 19.22(b)).

Just like a two-variable Karnaugh map, we can also have a Karnaugh map for three variables (Figure 19.23).

	$\overline{C}$	C
$\overline{A}\,\overline{B}$		
$\overline{A}\,B$		
$A\,B$		
$A\,\overline{B}$		

(a)

	$\overline{C}$	C
$\overline{A}\,\overline{B}$		
$\overline{A}\,B$	1	
$A\,B$	1	1
$A\,\overline{B}$		

(b)

	$\overline{C}$	C
$\overline{A}\,\overline{B}$	0	0
$\overline{A}\,B$	1	0
$A\,B$	1	1
$A\,\overline{B}$	0	0

(c)

Figure 19.23 Construction of three-variable Karnaugh map

First draw a blank map for Table 19.25. Then draw the Karnaugh map (Figure 19.23).

Table 19.25

A	*B*	*C*	*Y*
0	0	0	0
0	0	1	0
0	1	0	1
0	1	1	0
1	0	0	0
1	0	1	0
1	1	0	1
1	1	1	1

The vertical column is labelled $\overline{A}\,\overline{B}, \overline{A}B$, AB and $A\overline{B}$. With this order, only one variable changes from complimented to uncomplemented form (or vice versa) as one moves downward.

Now look for output 1s in Table 19.16. Output 1s appear for *ABC* inputs of 010, 110 and 111. The fundamental products for these inputs are $\overline{A}B\overline{C}$, $AB\overline{C}$ and *ABC*, respectively. So enter 1s for these products (Figure 19.23(b)). The final step is to enter 0s in the remaining spaces (Figure 19.23(c)).

19.18 Four-Variable Maps

Many digital computers and systems process 4-bit numbers. For instance, some digital chips will work with nibbles like 0000, 0001, 0010, etc. For this region, logic circuits are often designed to handle 4-input variables (or their compliments). We shall now study the formation of four-variable Karnaugh map.

Suppose the given truth table is

A	*B*	*C*	*D*	*Y*
0	0	0	0	0
0	0	0	1	1
0	0	1	0	0
0	0	1	1	0
0	1	0	0	0
0	1	0	1	0
0	1	1	0	1
0	1	1	1	1
1	0	0	0	0
1	0	0	1	0
1	0	1	0	0
1	0	1	1	0
1	1	0	0	0
1	1	0	1	0
1	1	1	0	1
1	1	1	1	0

For this, we start by drawing a blank map (Figure 19.24(a)) where the vertical column comprises $\overline{A}\,\overline{B}, \overline{A}B, AB$ and $A\overline{B}$ and the horizontal column comprises $\overline{C}\,\overline{D}, \overline{C}D, CD$ and $C\overline{D}$. Looking at the table, we have output 1s appearing for *ABCD* inputs of 0001, 0110 and 1110. The fundamental products of these input conditions are $\overline{A}\,\overline{B}\,\overline{C}D, \overline{A}BC\overline{D}$, $\overline{A}BCD$ and $ABC\overline{D}$. So we enter 1s on the Karnaugh map (Figure 19.24 (b)). The final step is to enter 0s in the complete map (Figure 19.24 (c)).

	$\overline{C}\,\overline{D}$	$\overline{C}D$	CD	$C\overline{D}$
$\overline{A}\,\overline{B}$				
$\overline{A}B$				
AB				
$A\overline{B}$				

(a)

	$\overline{C}\,\overline{D}$	$\overline{C}D$	CD	$C\overline{D}$
$\overline{A}\,\overline{B}$		1		
$\overline{A}B$			1	1
AB				1
$A\overline{B}$				

(b)

	$\overline{C}\,\overline{D}$	$\overline{C}D$	CD	$C\overline{D}$
$\overline{A}\,\overline{B}$	0	1	0	0
$\overline{A}B$	0	0	1	1
AB	0	0	0	1
$A\overline{B}$	0	0	0	0

(c)

Figure 19.24

19.18.1 Karnaugh Map Simplification

Before we go to the method for simplification of a Karnaugh map, one must be familiar with few definitions. The following terms are usually used in simplifying Karnaugh maps. The terms are pair, quad and octals.

Pair Where two 1s occur, which are (i) horizontally adjacent (one after the other) in the row they form a pair or (ii) vertically one after the other in the column, they form a pair.

A few examples are:

	$\overline{C}\,\overline{D}$	$\overline{C}\,D$	$C\,D$	$C\,\overline{D}$
$\overline{A}\,\overline{B}$	0	0	0	0
$\overline{A}\,B$	0	0	0	0
$A\,B$	0	0	(1	1)
$A\,\overline{B}$	0	0	0	0

The two 1s occur horizontally adjacent to each other in the above map for which the sum of fundamental products is:

$$Y = ABCD + ABC\overline{D}$$

It may be simplified as

$$\begin{aligned} Y &= ABC(D + \overline{D}) \\ &= ABC \end{aligned}$$

since $D + \overline{D} = 1$

We can conclude that a pair of 1s like that shown above means the sum of products equation will have a variable and its compliment, which will drop out like $(D + \overline{D})$.

We can also visualize a vertical pair as shown in the following Karnaugh map:

	$\overline{C}\,\overline{D}$	$\overline{C}\,D$	$C\,D$	$C\,\overline{D}$
$\overline{A}\,\overline{B}$	0	0	0	0
$\overline{A}\,B$	0	0	0	0
$A\,B$	0	0	0	(1
$A\,\overline{B}$	0	0	0	1)

Here the two 1s occur in a vertical column. The sum of fundamental products is

$$\begin{aligned} Y &= AB\,C\overline{D} + A\overline{B}\,C\overline{D} \\ &= AC\overline{D}(B + \overline{B}) \\ &= AC\overline{D} \end{aligned}$$

since $B + \overline{B} = 1$

Here also we find a variable and its compliments. So variable B and its compliment $\overline{B}$ drop out and the result is

$$Y = AC\overline{D}$$

Quad A quad is a group of four 1s, that are horizontally or vertically together. The 1s may be end to end or in the form of a square as shown in the following Karnaugh map.

	$\overline{C}\,\overline{D}$	$\overline{C}D$	CD	$C\overline{D}$
$\overline{A}\,\overline{B}$	0	0	0	0
$\overline{A}B$	0	0	0	0
AB	1	1	1	1
$A\overline{B}$	0	0	0	0

(a)

	$\overline{C}\,\overline{D}$	$\overline{C}D$	CD	$C\overline{D}$
$\overline{A}\,\overline{B}$	0	0	0	0
$\overline{A}B$	0	0	0	0
AB	0	0	1	1
$A\overline{B}$	0	0	1	1

(b)

A quad eliminates two variables and their compliments; so the sum of products is simplified as explained below for map (a).

$$\begin{aligned} Y &= AB\overline{C}\,\overline{D} + AB\overline{C}D + ABCD + ABC\overline{D} \\ &= AB\overline{C}[\overline{D} + D] + ABC[D + \overline{D}] \\ &= AB\overline{C} + ABC \\ &= AB[\overline{C} + C] \\ &= AB \end{aligned}$$

Similarly, for map (b), when we move horizontally, D is the variable which changes from uncomplimented to complimented form. When we move vertically, B is the variable which changes the form. Here, variables A, C remain unchanged. Thus, the simplified equation for sum of products is

$$Y = AC$$

Octet This is a group of eight 1s occurring adjacent to each other as shown below:

	$\overline{C}\,\overline{D}$	$\overline{C}D$	CD	$C\overline{D}$
$\overline{A}\,\overline{B}$	0	0	0	0
$\overline{A}B$	0	0	0	0
AB	1	1	1	1
$A\overline{B}$	1	1	1	1

Here three variables and their compliments are eliminated. On simplifying, we can visualize an octet as two quals. For quads, we have

$$Y = AC + AC$$
$$= A[C + C]$$
$$= A \qquad \text{since } C + C = 1$$

Hence, in the three variables and their compliments drop out of the corresponding product.

19.18.2 Karnaugh Simplification

If a pair exists, it eliminates one variable and its compliment; a quad eliminates two variables and their compliments; an octet eliminates three variables and their compliments. So whenever we have a Karnaugh map, the following procedure is followed for its simplification:

1. Visualize an octet if any, encircle it
2. Visualize a quad if any, encircle it
3. Visualize a pair if any, encircle it

Write the Boolean equation and simplify it further if possible.

Example 1

	$\overline{C}\,\overline{D}$	$\overline{C}D$	CD	$C\overline{D}$
$\overline{A}\,\overline{B}$	0	1	1	1
$\overline{A}B$	0	0	0	1
AB	1	1	0	1
$A\overline{B}$	1	1	0	1

(a)

	$\overline{C}\,\overline{D}$	$\overline{C}D$	CD	$C\overline{D}$
$\overline{A}\,\overline{B}$	0	1	1	1
$\overline{A}B$	0	0	0	1
AB	1	1	0	1
$A\overline{B}$	1	1	0	1

(b)

We can easily visualize that there are

(a) Two quads

(b) One pair as shown in map (b)

(i) The pair exists in first row; it represents the simplified form $\overline{A}\,\overline{B}$ D

(ii) The lower quad (first two columns and last two rows) represents the simplified form $A\overline{C}$.

(iii) The right quad (last column) represents the simplified form $C\overline{D}$

By ORing these results, we have

$$Y = \overline{A}\,\overline{B}D + A\overline{C} + C\overline{D}$$

19.18.3 Overlapping Groups

Here we use the same 1 more than once. The following Karnaugh map illustrates this:

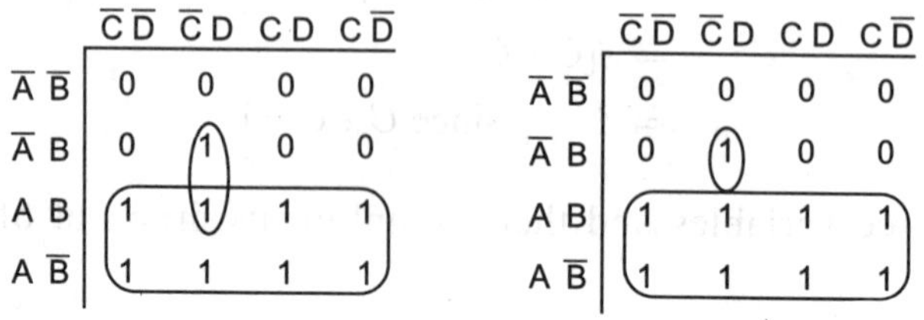

	$\bar{C}\bar{D}$	$\bar{C}D$	CD	$C\bar{D}$
$\bar{A}\bar{B}$	0	0	0	0
$\bar{A}B$	0	1	0	0
AB	1	1	1	1
$A\bar{B}$	1	1	1	1

(a)

	$\bar{C}\bar{D}$	$\bar{C}D$	CD	$C\bar{D}$
$\bar{A}\bar{B}$	0	0	0	0
$\bar{A}B$	0	1	0	0
AB	1	1	1	1
$A\bar{B}$	1	1	1	1

(b)

In map (a), the 1s representing the fundamental product $AB\bar{C}D$ is part of the octet. The simplified equation is

$$Y = A + B\bar{C}D$$

If we do not follow the above procedure, a more complicated equation for map (b) results

$$Y = \bar{A} + \bar{A}B\bar{C}D$$

So always overlap groups if possible. That is, use 1s more than once to get the largest group.

19.18.4 Rolling the Map

Consider the following map:

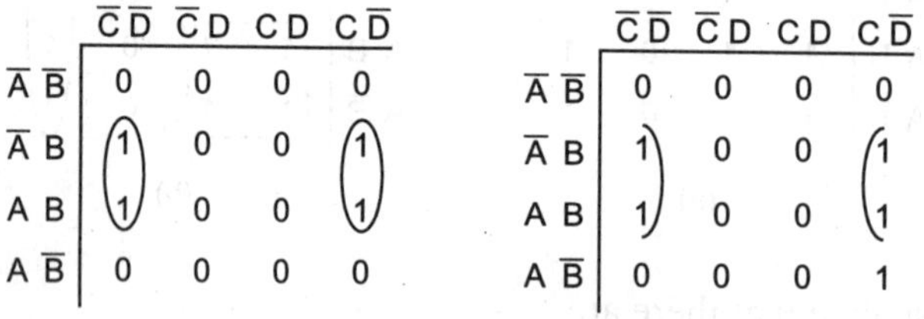

	$\bar{C}\bar{D}$	$\bar{C}D$	CD	$C\bar{D}$
$\bar{A}\bar{B}$	0	0	0	0
$\bar{A}B$	1	0	0	1
AB	1	0	0	1
$A\bar{B}$	0	0	0	0

(a)

	$\bar{C}\bar{D}$	$\bar{C}D$	CD	$C\bar{D}$
$\bar{A}\bar{B}$	0	0	0	0
$\bar{A}B$	1	0	0	1
AB	1	0	0	1
$A\bar{B}$	0	0	0	1

(b)

The pairs result into equation

$$Y = B\bar{C}\bar{D} + BC\bar{D} \tag{i}$$

If we visualize such that by rolling the map the left side touches the right side, the two pairs form a quad. To indicate this, draw half circles (map (b)) around each pair. Thus, a quad results into the equation

$$Y = B\bar{D}$$

We can justify the above procedure because Eqn. (i) can be further simplified as below:

$$Y = B\overline{C}\,\overline{D} + BC\overline{D}$$
$$= B\overline{D}[\overline{C} + C]$$
$$= B\overline{D}$$

19.18.5 Eliminating Redundant Groups

If done with encircling groups, eliminate redundant group if any. A redundant group consists of a group whose 1s are already used by other groups. We can understand it by the following example.

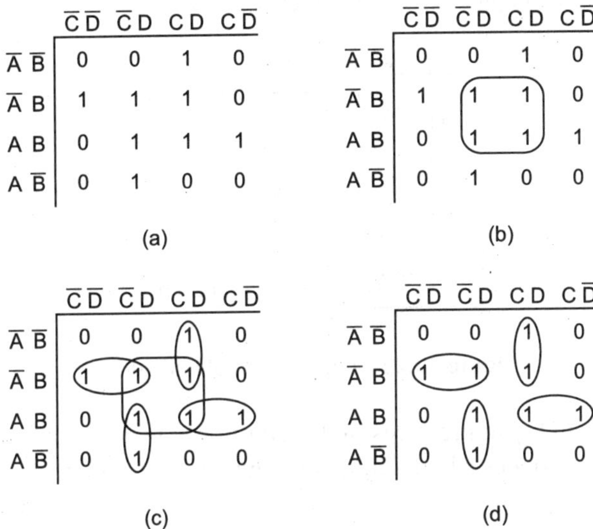

Encircle the quad to get map (b). Next, group the remaining 1s into pairs by overlapping map (c). Here all the 1s of the quad are used by pairs. Because of this, the quad is reduced and can be eliminated to get map (d). Here all the 1s are covered by the pairs.

In summary, Karnaugh map method of simplifying Boolean equations can be listed as:

1. Enter 1 on the Karnaugh map for each fundamental product that produces 1 output in the truth table. Enter 0s elsewhere.
2. Encircle octets, quads, pairs; remember to roll and overlap to get the largest groups possible.
3. If any isolated 1s remain, encircle each.
4. Eliminate any redundant group.
5. Write the Boolean equation by ORing the products corresponding to the encircled groups.

19.19 Arithmetic Building Blocks

We may develop a logic circuit that performs 8-bit arithmetic on both positive and negative members. But first we need to cover three basic circuits that are used as building blocks. These building blocks are (i) half adder, (ii) full adder and (iii) controller inverter.

19.19.1 Half Adder

When we add two binary numbers, we start with the least significant column (bit). It means that we have to add two bits, with the possibility a carry. The circuit used for this is called half adder.

Figure 19.25 shows the logic circuit of half adder. The output of exclusive OR gate is called SUM and the output of AND gate is called CARRY.

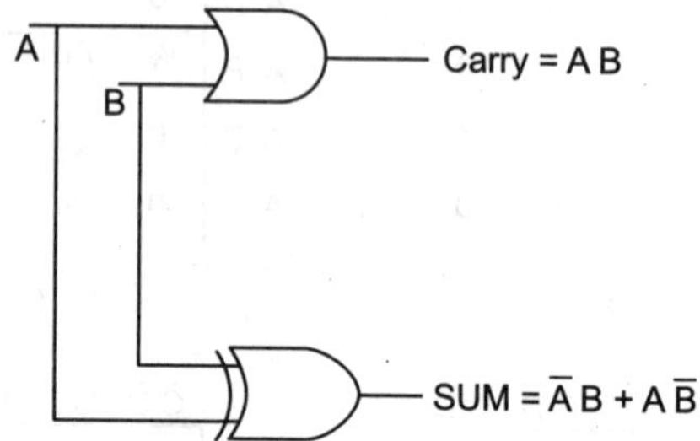

Figure 19.25 Logic circuit of half adder

As we know, the AND gate produces high output only when both input *A* and *B* are high. The *exclusive OR* gate produces a high output only when either input is high but not when both inputs are high.

Table 19.26 shows the truth table for half adder. On examining each entry in the truth table, we find that a half adder performs binary addition of two digits. It does electronically, what we do mentally when we add two bits. Its action is as under:

Table 19.26 Truth table for half adder

A	*B*	Carry (*C*)	SUM (*S*)
0	0	0	0
0	1	0	1
1	0	0	1
1	1	1	0

First entry

Input $A = 0, B = 0$

We know 0 plus 0 is 0 with a carry of 0

Half adder gives SUM = 0 and carry = 0

Second entry

Input $A = 0, B = 1$

We know 0 plus 1 is 1 with carry 0

Half adder gives SUM = 1 and carry = 0

Third entry

Input $A = 1, B = 0$

We know 1 plus 0 is 1 with carry 0

Half adder gives SUM = 1 and carry = 0

Fourth entry

Input $A = 1, B = 1$

We know 1 plus 1 is 0 with carry 1

Half adder gives SUM = 0 and carry = 1

Thus, we find that the result for each entry of mental operation and logic circuit of half adder matches. The only difference is about the speed of operation. The operation of a logic circuit is million times faster than normal mental function.

19.19.2 Full Adder

Half adder is used for two binary numbers; for higher order columns, we use full adder. It comprises a logic circuit which can add 3 bits at a time. Here the third bit is the carry from lower column. It implies that we need a logic circuit with three inputs and two outputs. The simplest form of such a logic circuit is shown in Figure 19.26.

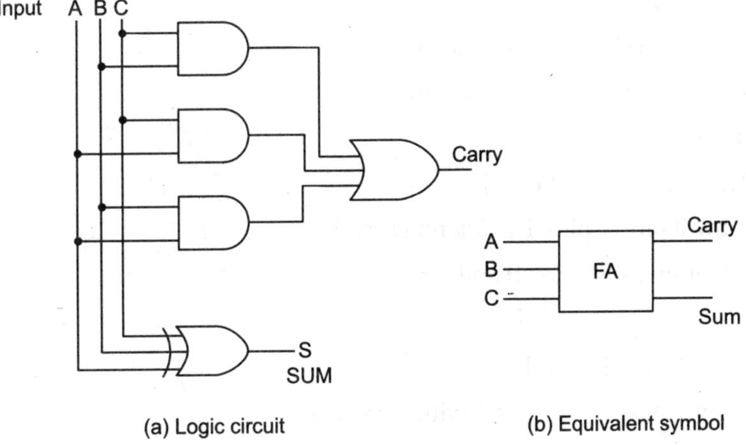

(a) Logic circuit

(b) Equivalent symbol

Figure 19.26 Full adder

Table 19.27 shows the truth table for full adder. We can easily check this table for validity. For instance, carry is high in Figure 19.26(a) when two or more of the inputs *ABC* is high; this agrees with the "carry" column of Table 19.27. Also when an odd number of high *ABC* input drives the exclusive OR gate, it produces a high output; this verifies the SUM column of truth table.

Table 19.27 Truth table for full adder

A_n	B_m	C_{n-1}	Carry (C_n)	SUM (S_n)
0	0	0	0	0
0	0	1	0	1
0	1	0	0	1
0	1	1	1	0
1	0	0	0	1
1	0	1	1	0
1	1	0	1	0
1	1	1	1	1

On examining each entry in Table 19.27, we find that a full adder performs binary addition of 3 bits; it does electronically what we do mentally for 3 bits. Its action for some selected entries can be verified.

First entry

Input $A = 0, B = 0, C = 0$

We know 0 plus 0 plus 0 is 0 with a carry 0

Full adder gives SUM = 0 and carry = 0

Second entry

Input $A = 0, B = 0, C = 1$

We know 0 plus 0 plus 1 is 1 with carry 0

Full adder gives SUM = 1 and carry = 0

Fourth entry

Input $A = 0, B = 1$ and $C = 1$

We know 0 plus 1 plus 1 is 0 with carry 1

Full adder gives SUM = 0 and carry = 1

Last entry

Input $A = 1, B = 1, C = 1$

We know 1 plus 1 plus 1 is 1 with a carry of 1

Full adder gives SUM = 1 and carry = 1

Thus, we find that result of each entry, i.e. mental operation and logic circuit, of full adder matches. The only difference is speed, which is million times faster in case of a logic circuit.

19.19.3 Controller Inverter

The logic circuit for controlled inverter is shown in Figure 19.27.

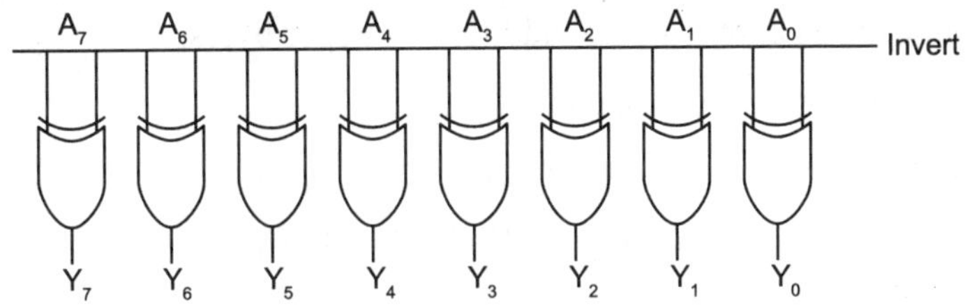

Figure 19.27 Logic circuit of controlled inverter

Function

When invert is low, it transmits the 8 bit input to output; when invert is high, it transmits 1's compliment. For instance, if the input number is

$$A_7 A_6 \ldots A_0 = 01101110$$

a low invert produces

$$Y_7 Y_6 \ldots Y_0 = 0110\ 1110$$

but a high invert produces

$$Y_7 Y_6 \ldots Y_0 = 1001\ 0001$$

The controlled inverter is important because it is a step in the right direction. During a subtraction process, we first need to take 2's compliment of the subtrahend, then we can add the complimented subtrahend to get the answer.

19.19.4 Adder-Subtractor

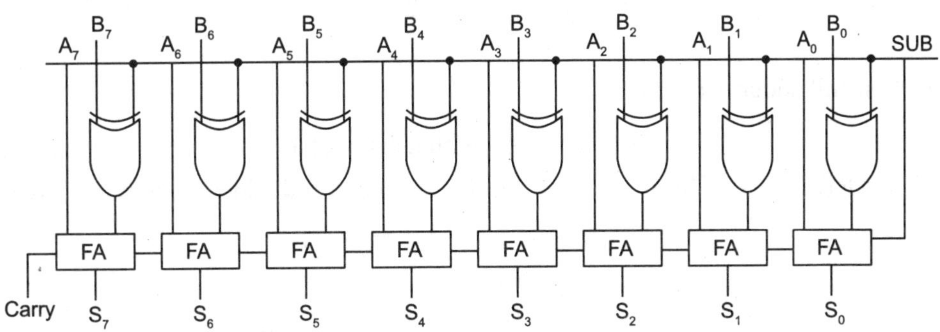

Figure 19.28 Binary adder-substractor logic circuit

We can connect full adders as shown in Figure 19.28 to add or subtract binary numbers. The circuits is laid out from right to left, similar as how we add binary numbers. Therefore, the least significant column is on the right and most significant column is on the left. The boxes labelled *FA* are full adders with least significant column.

The carry out of each full adder is the CARRY IN to next higher full adder. The numbers being processed are $A_7, A_6 \ldots A_1, A_0$ and $B_7, B_6 \ldots B_1, B_0$ while the answer is $S_7, S_6 \ldots S_1, S_0$ 8-bit arithmetic, the final carry is carry into the addition of upper bytes.

Addition Here is how addition appears.

$$\begin{array}{r} A_7\ A_6\ A_5\ A_4\ A_3\ A_2\ A_1\ A_0 \\ +\ B_7\ B_6\ B_5\ B_4\ B_3\ B_2\ B_1\ B_0 \\ \hline S_7\ S_6\ S_5\ S_4\ S_3\ S_2\ S_1\ S_0 \end{array}$$

During addition, the SUB signal is deliberately kept in the low state. Therefore, the binary number $B_7, B_6 \ldots B_0$ passes through the controller inverter with no change. The full adders then produce the correct output sum. They do this by adding the bits in each column, passing carry's to next higher column, and so on. For instance, starting at the LSB, the full adder adds A_0, B_0 and SUB. This produces the sum S_0 and a carry out to next high full adder. The next full adder then adds A_1, B_1 and the CARRY IN to produce S_h and a CARRY OUT. A similar action occurs for each of the remaining full adders, and the connect sum appears at the output lines.

For instance, suppose we have to add + 125 and –67. Then $A_7, A_6 \ldots A_0 = 0111\ 1101$ and $B_7, B_6 \ldots B_0 = 1011\ 1101$. So the problem is

$$\begin{array}{r} 0111\ 1101 \\ +\ 1011\ 1101 \\ \hline ? \end{array}$$

Since SUB = 0 during an addition, the CARRY IN to the least significant column is 0, so

$$\begin{array}{r} 0 \leftarrow \text{SUB} \\ 0111\ 1101 \\ +\ 1011\ 1101 \\ \hline \end{array}$$

Steps

The first full adder performs

$$0 + 1 + 1 = 0 \text{ with a carry of } 1$$

This CARRY OUT of becomes CARRY IN for second full adder; so now we have

$$\begin{array}{r} 1 \leftarrow \text{carry} \\ 0111\ 1101 \\ 1011\ 1101 \\ \hline 10 \end{array}$$

So in the second column, we have

$$1 + 0 + 0 = 1 \text{ with carry } 0$$

This CARRY OUT becomes CARRY IN for third full adder and the process goes on. The addition process end with final carry

$$\begin{array}{r} 0111\ 1101 \\ +\,1011\ 1101 \\ \hline 1\ 1011\,1010 \end{array}$$

During bit arithmetic, the last carry is ignored. Hence the final answer is

$$S_7, S_6 \ldots S_0 = 0011\ 1010$$

It is equivalent to +58 which is the algebraic sum of decimal numbers +125 and –67.

Subtraction During subtraction, the signal SUB is deliberately put into high state. Therefore, the controller inverter produces the 1's compliment of B_7, B_6... B_0. Furthermore, because SUB is the CARRY IN for first full adder, the circuit processes data like

$$\begin{array}{cccccccc} & & & & & & & 1 \leftarrow \text{SUB} \\ A_7 & A_6 & A_5 & A_4 & A_3 & A_2 & A_1 & A_0 \\ \overline{B_7} & \overline{B_6} & \overline{B_5} & \overline{B_4} & \overline{B_3} & \overline{B_2} & \overline{B_1} & \overline{B_0} \\ \hline S_7 & S_6 & S_5 & S_4 & S_3 & S_2 & S_1 & S_0 \end{array}$$

For instance, the numbers are + 82 and +17. Then $A_7\, A_6\, A_5\, A_4\, A_3\, A_2\, A_1\, A_0 = 0101\ 0010$ and $B_7\, B_6\, B_5\, B_4\, B_3\, B_2\, B_1\, B_0 = 0001\ 0001$. The controller inverter produces 1's compliment of B, which is 1110 1110 since SUB = 1. During subtraction, the circuit performs the following addition.

$$\begin{array}{r} 0101\ 0010 \\ +\,1110\ 1110 \\ \hline 0\ 0100\ 0001 \end{array}$$

For 8-bit arithmetic, the final carry is ignored, and the final answer is

$$S_7\, S_6 \ldots S_0 = 0100\ 0001$$

This answer is equivalent to +65, which is the algebraic difference between +82 and +17.

Binary multiplication and division Binary multiplication is done with addition instructions and division with subtraction instruction. Therefore, an adder-subtractor, is all that is required for addition, subtraction, multiplication and division.

For example, multiplication is equivalent to repeated addition. Consider

$$8 \times 4 = ?$$

Multiplying 8 by 4 is adding 8 four times.

$$8 + 8 + 8 + 8 = ?$$

One way to multiply 8 by 4 is to programme a computer to add 8 until a total of four 8's have been added, and this approach is known as programmed multiplication by repeated addition.

19.20 Logic Circuits

Digital circuits are classified into two broad categories:

(i) Combinational circuits

(ii) Sequential circuits

19.20.1 Combination Circuits

In combinational circuits, the outputs at any instant of time depends upon the inputs present at that instant of time. It means there is no memory in these circuits.

19.20.2 Sequential Circuits

In sequential circuits, the output at any instant of time depends upon the present input as well as past inputs/outputs. It means there are elements to store past information, and such elements are known as memory.

A sequential circuit may have a combinational logic subsystem. Combinational circuits have already been discussed. In this chapter, we shall briefly discuss sequential circuits.

19.20.3 Sequential Logic Circuits

These circuits comprise memory. Memory in a computer is one of the vital elements. These elements can store binary words. Second-generation computers employ magnetic core memory. With the advancement of integrated circuit technology, IC memories built on semiconductor chips have almost replaced core memory. The most widely used memory element is *flip-flop* or *latch*.

Digital elements like flip-flop remain in one of two stable states denoted by 1 (High) or 0 (Low). Flip-flops are capable of storing or retaining information in memory binary digits. Transition from one stable state to another may occur, however, on application of clock pulses (CLK). These pulses are a train of rectangular pulses, the pulse time (t_p) being small compared with time period (T) of flip-flop circuit shown in Figure 19.29.

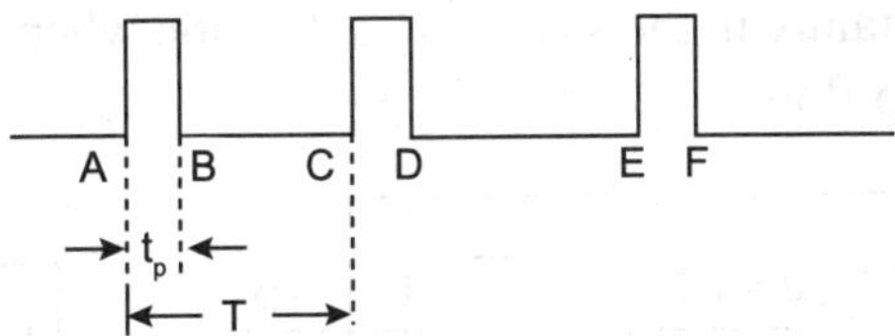

Figure 19.29

R-S Flip-Flop It comprises a bistable multivibrator. It may be constructed using two NOR gates interconnected as shown in Figure 19.30.

S (Set) and *R* (Reset) are controlling signal pulses that trigger transition from one state to another.

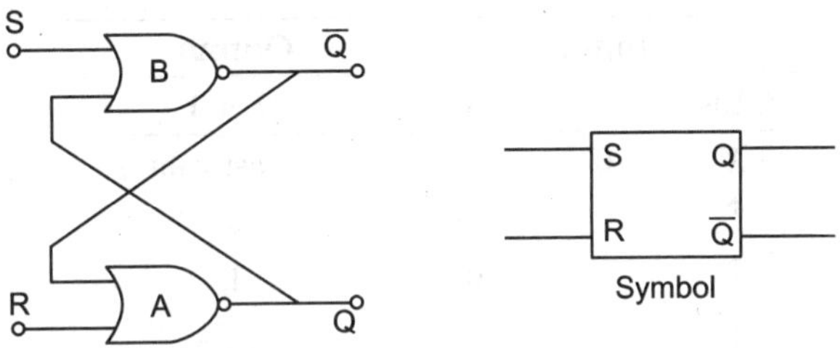

Figure 19.30 *R-S* flip-flop circuit (using NOR gate)

Input		Output
S	***R***	
0	0	Q_n (Last value)
1	0	1
0	1	0
1	1	?

As is evident from truth table, when both inputs are low, *Q* output is stable in whatever state it was previously. Pulsing the *S* input high sets the *Q* output low and so $\overline{Q}$ output is high; pulsing the *R* input has just the opposite effect. However, when both inputs go from high to low, together their output state is unpredictable.

A simple *R-S* flip-flop has following demerits:

(1) Its operation for (1, 1) state is unpredictable.

(2) It is transparent, i.e. when enabled, the output changes when ever the input changes.

J-K Flip-Flop The uncertainty in the state of *R-S* flip-flop when $S = R = 1$ can be eliminated by converting it into *J-K* flip-flop.

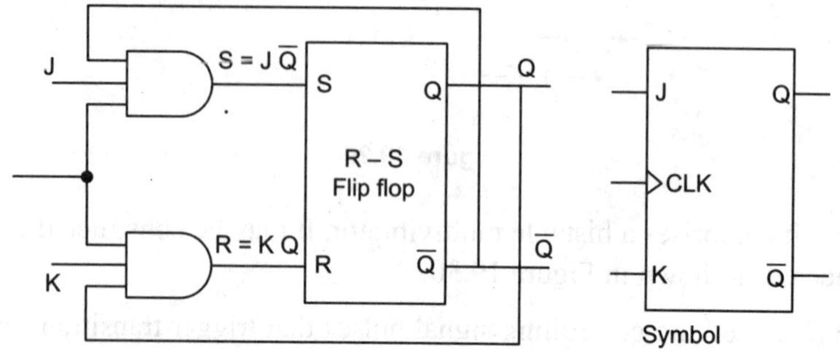

Figure 19.31 *J-K* flip-flop circuit

Input			Output
CLK	J	K	$Q_n + 1$
X	0	0	Q_n (Last value)
↑	0	1	0
↑	1	0	1
↑	1	1	$\overline{Q_n}$

Figure 19.31 shows the circuit of *J-K* flip-flop. Its symbol and truth table is also shown in Figure 19.31.

As is evident from the truth table, the 1, 1 state rather than being unpredictable, causes the output to $\overline{Q}$ (toggle) such that on receipt of the clock input the output changes to opposite (inverted) state.

These flip-flops are available with preset and clear inputs whose symbols are as shown in Figure. 19.32.

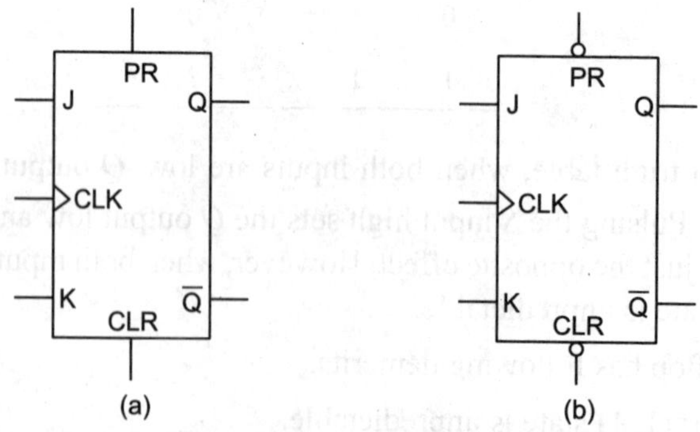

Figure 19.32 (a) Positive preset and clear, (b) inverted preset and clear

It provides a means of setting the flip-flop into a known condition independent of clock and *J-K* inputs. The clock input may be arranged to control transitions using the following two ways.

Edge-Triggered Flip-Flops In edge-triggered flip-flops, one edge is used to define the time at which ouput changes. Depending on the device, either the rising or falling edge may be used. The *J-K* inputs must be present for a minimum time (set up time) before the clock edge occurs, and remain for a minimum time (hold time) after clock edge. The output will be delayed for a period of time after clock edge (propagation delay).

Pulse-triggered flip-flop or master-slave J-K flip-flop Figure 19.33 shows one way to build master-slave *J-K* flip flop.

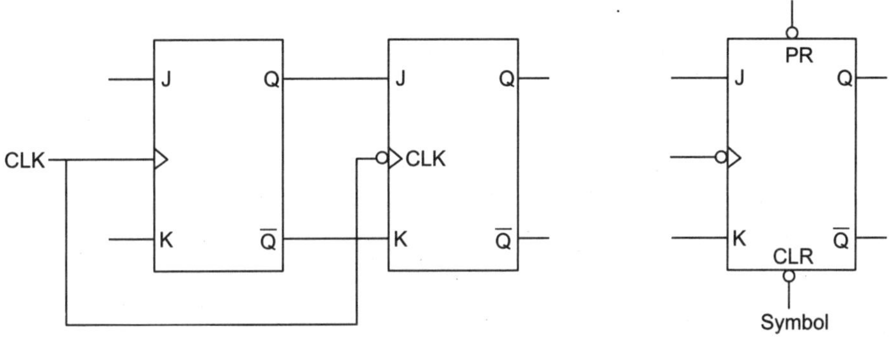

Figure 19.33 Master slave *J-K* flip-flop

The above configuration provides another way to avoid racing. The master is positive edge-triggered and the slave is, negative edge-triggered. Therefore, the master responds to its *J* and *K* inputs before the slave. If $J = 1$ and $K = 0$, the master sets on the positive clock edge. The high Q output of master drives the *J* input of slave, so when negative edge of clock hits, the slave sets copying the action of master.

If $J = 0$ and $K = 1$, the master resets on the leading edge of clock. The high $\overline{Q}$ output of master goes to *K* input of slave. Therefore, the arrival of the clock's trailing edge forces the slave to reset. Again the slave has copied the master.

If the master's *J* and *K* inputs are both high, it toggles on the positive clock edge and the slave then toggles on the negative clock edge. Regardless of what the master does, the slave copies the master: if the master sets, the slave sets, and if the master resets, the slave resets.

The *R-S* flip-flop has two data inputs *R* and *S*. To store a high bit, we need a high *S*, to store a high bit we need a high *S*; to store a low bit we need a high *R*. The generation of two signals to drive a flip-flop is a disadvantage in many applications. Furthermore, the forbidden condition of both *R* and *S* high may occur inadvertently. This has led to the *D* flip-flop, a circuit which needs only a single data input.

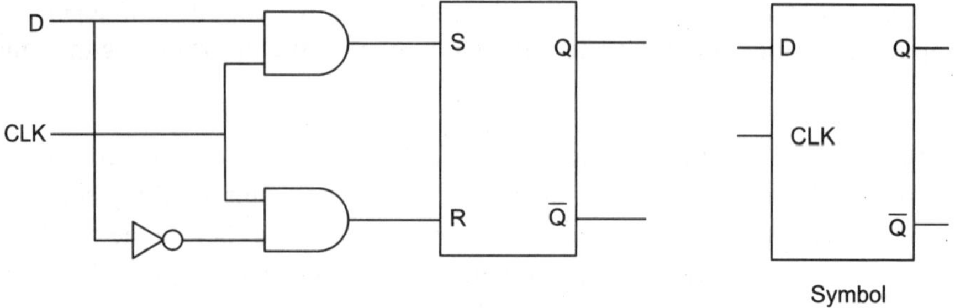

Figure 19.34 *D* flip-flop

Figure 19.34 shows a simple way to build a *D* flip-flop. This kind of flip-flop prevents the value of *D* from reaching the *Q* output until a clock pulse occurs. The action of the circuit is as follows:

When clock is low, both AND gates are disabled; therefore, *D* can change value without affecting the value of *Q*. On the other hand, when clock is high, both AND gates are enabled. In this case, *Q* is forced to equal the value of *D*. When clock again goes low, *Q* retains or stores the last value of *D*.

In general, a *D* flip-flop is a bistable circuit whose *D* input is transferred to the output after a clock pulse is received.

19.20.4 Register

Computers employ a number of registers. A register is composed of a group of memory elements (flip-flop), their number depending upon the size of binary word it can hold. One can regard a register as one word, short time memory.

I.C flip-flops are currently being used to make registers. Besides flip-flops, a register also contains additional gates to execute data processing operations.

19.20.5 Shift Register

A shift register is a very important digital building block. Registers are often used momentarily to store binary information appearing at the output of an encoding matrix. This is a serially connected arrangement of flip-flops that store data. Data entered into the serial input moves through the shift register under the control of clock pulse. Output data is available either serially at several output points, delayed by *n* clock pulses, or in parallel form using the output of each flip-flop. They are also available with parallel load facility (PISO) and either forward or reverse shifting. It can be used as counter. Depending upon the mode of data shifting, a shift register is classified into four types:

(i) Series in series out

(ii) Series in parallel out

(iii) Parallel in parallel out

(iv) Parallel in series out

To explain its function, we consider a four-bit register of serial-in-parallel-out type. Each flip-flop is *R-S* (or *J-K*) master-slave type. The stage which stores the most significance bit is transformed into a *D*-type latch by connecting an inverter between *S* and *R*. The binary serial data, say 1011 (LSB), is to be entered into the register via *S* terminal.

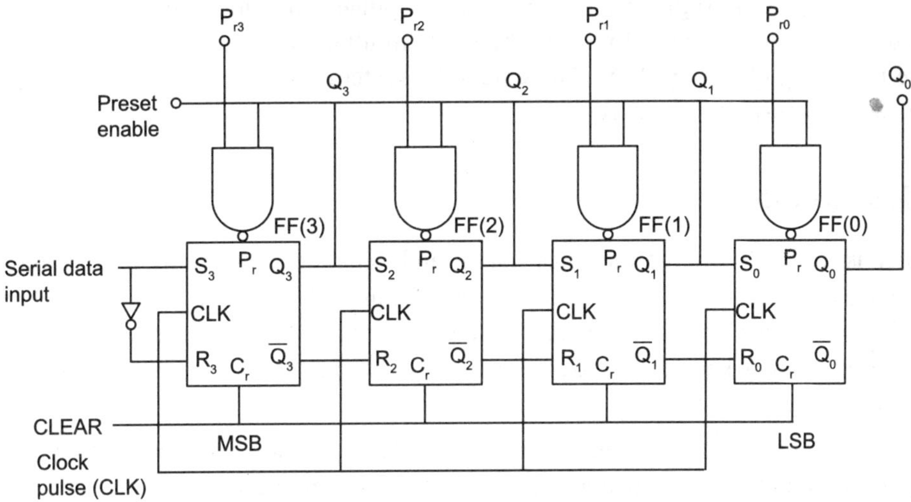

Figure 19.35 A four-bit shift register

To start with, the flip-flops are cleared by applying a clear (C) input so that every output Q_0, Q_1, Q_2, Q_3 is zero. Then set $Cr = 1$ and $Pr = 1$ (by maintaining the preset enable at (0)). The serial data train and clock pulse are now switched ON. The LSB, which is 1 here, is entered into flip-flop (FF)3 when the clock pulse (CP) assumes a (1) from a (0) by the action of *D* type FF. Thus, after the clock pulse, $Q_3 = 1$ and $Q_2 = Q_1 = Q_0 = 0$.

Table 19.28

Clock pulse	Serial data input	Q_3	Q_2	Q_1	Q_0
		0	0	0	0
1	1 →	1	0	0	0
			1	0	0
2	1 →	1		1	
			1		1
3	0 →	0		1	
			0		
4	1 →	1			

When the second clock pulse appears, the state of Q_3 is shifted to the master latch of FF(2) by the operation of an *S-R* flip-flop. At the end of second clock pulse, the bit in each master shifts to its slave and results in to $Q_3 = Q_1$, $Q_2 = 1$ and $Q_1 = Q_0 = 0$. The state of the register after each pulse is shown Table 19.28. It is evident from the table that after the third clock pulse, Q_2 has shifted to Q_1, Q_3 to Q_2 and the third input bit 0 has entered FF(3) so that $Q_3 = 0$. Similarly, after the fourth clock pulse, Q_3 has shifted to Q_2, Q_2 to Q_1 and Q_1 to Q_0 and the fourth input bit 1 has entered FF(3) resulting in $Q_3 = 1$; clearly the binary input word 1011 is installed in the register after fourth clock pulse. The clock pulses are stopped after the word is registered. The output may be read simultaneously since they are available on separate lines (Q_3, Q_2, Q_1, Q_0). As data enters the system serially, but come out in parallel, it is called a serial-in-parallel-out register.

19.20.6 Counters

A counter is probably one of the most useful and versatile subsystems in a digital system. A counter driven by a clock can be used to count the number of clock cycles. Since clock pulses occur at known intervals, the counter can be used as an instrument for measuring time and thereby period or frequency is known. There are basically two types of counters – (i) synchronous and (ii) asynchronous.

The ripple counter is simple and straightforward in operation and construction and usually requires minimum hardware. It does, however, have a limitation of speed. Each flip-flop is triggered by the previous flip-flop, and thus the counter has a cumulative settling time. Such counters are called series or asynchronous counters.

An increase in speed of operation can be achieved by the use of parallel or synchronous counters. Here every flip-flop is triggered by the clock (in synchronism) and thus settling time is simply equal to the delay time of a single flip-flop. The increase in speed is at the cost of hardware which is not that simple as in the case of asynchronous counters.

Summary

1. Almost all digital circuits are designed for two-state operation which means the signal voltage is either at low (0) or high (1) level. These circuits comprise gates and better known as logic circuits.
2. (i) An OR gate has two or more inputs; produces only one output. The output is high when one of the input signals is high and the output is low when all input signals are low. It represents addition in algebra. A plus sign is the symbol for OR gate.

   (ii) An AND gate has two or more inputs. It produces only one output. The output is high when all inputs are high and the output is low when input is low. It represents multiplication in algebra. A dot sign is the symbol for AND gate.

(iii) An inverse is one type of logic; it produces output which is complimentary of input. It is represented by NOT gate. Generally, an overbar is the algebraic symbol for NOT operation.

(iv) An AND-OR circuit always produces a sum of products equation while OR-AND circuit produces a product of sums equation.

(v) NOR gate is equivalent to OR gate followed by inverter (NOT gate).

(vi) NAND gate is equivalent to AND gate followed by inverter (NOT gate).

3. De Morgan's theorem:

   1st theorem

$$\overline{A+B} = \overline{A}\,\overline{B}$$

   2nd theorem

$$\overline{AB} = \overline{A} + \overline{B}$$

4. (i) Because of De Morgan's first theorem, a NOR-NOR circuit is equivalent to an OR-AND circuit.

   (ii) De Morgan's 2nd theorem says the NAND gate is equivalent to bubbled OR gate.

   (iii) The NAND gate is the universal gate since any logic circuit can be built with it.

5. Duality theorem: Starting with a Boolean relation, we can derive another Boolean relation by

   (i) Changing each OR sign to an AND sign

   (ii) Changing each AND sign to an OR sign

   (iii) Complementing 0 or 1 appearing in the relation.

6. Each gate is represented by a truth table, e.g.

OR Gate			**AND gate**			**NOT gate**	
Input		**Output**	**Input**		**Output**	**Input**	**Output**
A	*B*	*Y*	*A*	*B*	*Y*		
0	0	0	0	0	0	0	1
0	1	1	0	1	0	1	0
1	0	1	1	0	0		
1	1	1	1	1	1		

7. By gating the truth table, one can identify the fundamental products that produce output 1's. By ORing these products, we can get a sum of products equation for the truth table. A sum of products equation always results in an AND-OR circuit or its equivalent NAND-NAND circuit.

8. (i) The Karnaugh method of simplification starts by converting a truth table into Karnaugh map. Here we encircle all the octets, quads and pairs. This allows the user to write a simplified Boolean equation and draw a simplified logic circuit.
   (ii) One way to get the product of sums circuit is to compliment the Karnaugh map and write the simplified Boolean equation for $\overline{Y}$.
9. In writing a truth table for a gate, we follow the positive logic, i.e. 0 for low state and 1 for high state. But we can also have the reverse conversion, and accordingly we have the negative logic. Accordingly, we have the following equivalent.

   Positive OR ↔ Negative NAND

   Positive AND ↔ Negative OR

   Positive NOR ↔ Negative NAND

   Positive NAND ↔ Negative NOR
10. For performing mathematical operations like addition, subtraction, etc., we use the logic circuits. For this, we three have building blocks:
    (a) Half adder
    (b) Full adder
    (c) Controller inverter
    (a) A half adder is used where we have two inputs/two binary numbers. It adds two bits at a time.
    (b) A full adder is used where we have more than two binary numbers, i.e. higher order columns. It has three inputs and two outputs. It adds three bits at a time.
    (c) We also have a binary adder-substractor logic circuit by connecting a controller inverter and a full adders in a certain sequence; thus we have adder-subtractor circuits.
11. A flip-flop is an electronic circuit that has two stable states. It is said to be bistable. A basic *R-S* flip or latch can be constructed by connecting two NAND gates or two NOR gates with feedback connection. A signal at set input (*S*) of an *R-S* flip-flop will force the *Q* output to become 1 while a signal at reset (*R*) input will force *Q* to become *O*, i.e. $\overline{Q}$.

Review Questions

1. Explain the process of conversion of decimal number to binary number and vice versa by suitable examples (integral, fractional number, and mixed number).
2. Describe the method of addition and subtraction of binary numbers by suitable examples (integral, fractional numbers and mixed numbers).
3. Can we convert an octal number into binary number? If yes, explain.
4. What do you mean by DD logic and TT logic? How these logics are used to function as an OR gate? Explain.

5. What is an OR gate? Give its truth table. Draw a suitable circuit using diodes which represents the function of OR gate.
6. What is an AND gate? Give its truth table. Draw a suitable circuit using diodes, which represents the function of AND gate.
7. What do you mean by (i) diode logic, (ii) diode-transistor logic. Explain the two citing suitable examples.
8. State De-Morgan's theorem. Represent each theorem by suitable logic circuit.
9. What are universal gates? Why they are called so? Explain by giving suitable example for each case.
10. State duality theorem. Mention its use by giving two examples.
11. For the logic expression $Y = A\overline{B} + \overline{A}B$
    (a) Obtain the truth table
    (b) Name the operation performed
    (c) Realize the operation using AND, OR and NOT gate
    (d) Realize the operation using NAND gate only.
12. Prove the following:
    (a) A positive logic AND gate operation is equivalent to negative logic OR gate operation.
    (b) A positive logic NAND gate operation is equivalent to negative logic NOR gate operation.
13. Prove the following Boolean algebraic theorems:
    (a) $A + \overline{A}\cdot B + A\cdot\overline{B} = A + B$
    (b) $A\cdot B + \overline{A}\cdot B + \overline{A}\cdot\overline{B} = \overline{A} + B$
    (c) $\overline{A}BC + A\overline{B}C + AB\overline{C} + ABC = AB + BC + CA$
14. Prove the above using truth table operation.
15. Prove the following using De-Morgan's theorems.
    (a) $AB + CD = \overline{\overline{AB}\cdot\overline{CD}}$
    (b) $(A+B)\cdot(C+D) = \overline{\overline{(A+B)} + \overline{(C+D)}}$

    And hence prove the following statements:

    (i) An AND-OR configuration is equivalent to NAND-NAND configuration.
    (ii) An OR-AND configuration is equivalent to NOR-NOR configuration.
16. Verify if the following operations are commutative and associative:
    (a) AND (b) OR (c) EXOR
17. Verify if the following operations are commutative but not associative.
    (a) NAND (b) NOR

18. Define the terms pairs, quads and octets as used in a Karnaugh map. Illustrate your answer with suitable example.
19. What do you mean by 'fundamentals products'. Illustrate your answer for (i) two inputs, (ii) three inputs.
20. Construct a Karnaugh map for the following truth table.

Input		Output
A	*B*	*Y*
0	0	0
0	1	1
1	0	1
1	1	

Input			Output
A	*B*	*C*	*Y*
0	0	0	0
0	0	1	0
0	1	0	1
0	1	1	0
1	0	0	0
1	0	1	0
1	1	0	1
1	1	1	1

21. Can we use NAND gate to perform the function of OR gate? If yes, how it can be achieved?
22. What do you mean by the term truth table of a logic gate? Illustrate your answer by making use of two suitable examples.
23. Prove that the following two logic circuits are equivalent.

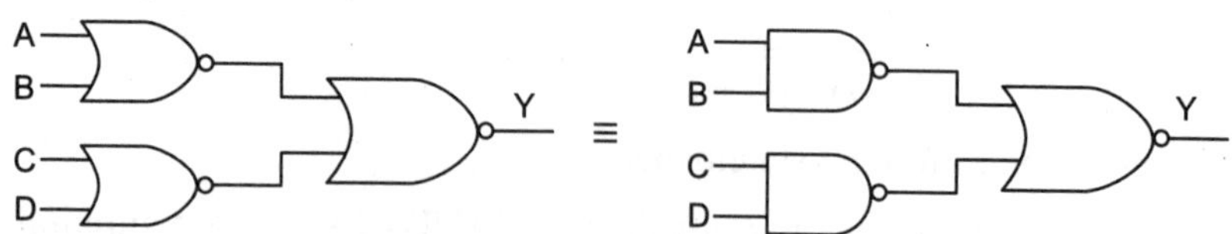

24. Draw the logic circuit which represents the Boolean equation

$$Y = A\overline{B}C + AB\overline{C}$$

25. Prove that output of a two-input exclusive OR gate is:

$$Y = \overline{A}B + A\overline{B}$$

Show the logic circuit of above Boolean equation.

26. What are combinational and sequential circuits? Illustrate your answer with suitable example for each.
27. What is the basic element of a sequential circuit? Describe its function and purpose. Give its truth table.
28. What do you mean by the term flip-flop. How many types of it you are familiar with? Describe the function of *JK* flip-flop.
29. Draw the logic circuit of master-slave *JK* flip-flop. Discuss its function.

Numerical

1. Determine the equivalent decimal number for each of the following binary numbers:
   (a) 110101 (b) 101101 (c) 1111 1111
   (d) 1001·0101 (e) 0·10101

   [**Ans:** 53, 45, 255, 9·3125, 0.6562]
2. Determine the equivalent binary number for each of the following decimal numbers:
   (a) 37 (b) 255 (c) 15 (d) 25.5 (e) 0.6875

   [**Ans:** 00100101, 11111111, 1111, 11001.10, 0.10101]
3. Add the following groups of binary numbers:
   (a) 1011 + 1101 (b) 1010.1101 + 101.0100
4. Perform subtraction using 2's compliment method:
   (a) 01000 – 01001 (b) 0011.1001 + 0001.1110
5. How many nibbles are there in each of the following:
   (a) 1001 (b) 1100 1111 (c) 1010 1010 1111 [**Ans:** 1, 2, 3]
6. Convert the following binary numbers to octal numbers.
   (a) 1010 1111 (b) 1101·0110111 [**Ans:** 257, 15·331]
7. Convert the following hexadecimal numbers to binary numbers:
   (a) E5 (b) B4D (c) 7AF4

   [**Ans:** 1110 0101, 1011 0100, 0111 1010 0100]
8. Give the sum of the following:
   (a) $3_8 + 7_8 =$
   (b) $4_{16} + C_{16} =$
   (c) $5_8 + 6_8$
   (d) $8_{16} + F_{16} =$ [**Ans:** (a) 12_8, (b) 10_{16} (c) 13_8 (d) 17_{16}]

9. Show the 8-bit addition of the following decimal numbers in 2's compliment representation:

   (a) +45, +56 (b) +89, –34 (c) +67, –98

   [**Ans:** (a) 00101101 | sum
   – 00111000 | 01100101

   (b) 01011001 | sum
   – 11011110 | 00110111

   (c) 01000011 | sum
   – 10011110 | 11100011

10. What is 2's compliment of the following:

    (a) 0000 1111 (b) 0101 1010 (c) 1011 1110

    [**Ans:** 1111 0001, 1010 0110, 0100 0010]

11. Give the ASC11 for each of the following:

    (a) 7 (b) W (c) f (d) y

    [**Ans:** 0110111, 1010111, 1100110, 1111001]

Short Answer Questions

1. Name the different types of number systems.

   Ans: We are familiar with four types of number systems. They are decimal number system, binary number system, hexa decimal number system, octal number system.

2. Which number system is generally used in computers?

   Ans: Binary number system.

3. What is the characteristic of digital signal? How it differs from analogue signal?

   Ans: A digital signal is either low or high as specified, whereas an analogue signal is continuous and can have any value as specified range.

4. What is the radix of decimal number and binary?

   Ans: The radix or weight of decimal number is 10 and that of binary number is 2.

5. What is a BCD?

   Ans: BCD is a natural binary number code in which decimal digits 0 through 9 are represented by their binary equivalents using four bits.

6. What do you mean by logic circuit?

   Ans: Just like the electric circuit, a logic circuit comprises gates connected in as sequence as per the logic equation.

7. State De Morgan's theorem.

   Ans: De Morgan's 1st theorem states that the compliment of a sum equals the product of compliments.

$$\overline{A+B+C+\ldots}=\overline{A}\cdot\overline{B}\cdot\overline{C}\ldots$$

De Morgan's 2nd theorem states that the compliment of a product equals the sum of compliments.

$$\overline{A\cdot B\cdot C}=\overline{A}+\overline{B}+\overline{C}$$

8. What do you mean by positive logic and negative logic?

   Ans: In positive logic, 0 is considered a low state and 1 is considered as high state of a signal, whereas in negative logic 0 is considered as high state and 1 is considered as low state of a signal.

9. Mention the equivalent binary number of decimal number 15.

   Ans: The equivalent binary number of 15 is 1 1 1 1.

10. Which is equivalent gate of AND gate in negative logic?

    Ans: The equivalent gate of AND gate in negative logic is negative OR gate.

11. Which mathematical operation is represented by OR gate and AND gate?

    Ans: OR gate represents addition and AND gate represents multiplication.

12. Two toggle switches are connected in series in a electric circuit. Which gate represents their function?

    Ans: The function of two toggle switches in series in an electric circuit represents OR gate.

13. State duality theorem.

    Ans: It states that starting a Boolean relation,

    (i) change each OR sign to an AND sign, (ii) change each AND sign to an OR sign, (iii) compliment any 0 or 1 appearing in the relation.

14. What is a Karnaugh map?

    Ans: A Karnaugh map is a visual display of the fundamental products needed for the sum of products solution.

Objective Questions

1. 101·101 is a number in binary system. Its equivalent in decimal system is

   (a) 5·625 (b) 5·125

   (c) 6·525 (d) 6·125 [**Ans:** (a)]

2. In decimal system, a given number is 43. Its equivalent in binary system is

   (a) 100011 (b) 101011

   (c) 110011 (d) 111011 [**Ans:** (b)]

3. Universal gates are:
   (a) OR and AND gate
   (b) OR and NOR gate
   (c) NOR and NAND gate
   (d) AND and NAND gate [**Ans:** (c)]

4. The radix in binary number system and hexadecimal number system is
   (a) 2 and 16 respectively
   (b) 2 and 10 respectively
   (c) 8 and 10 respectively
   (d) 10 and 16 respectively [**Ans:** (a)]

5. The equivalent gate of AND gate in negative logic is
   (a) positive NOR gate
   (b) positive NAND gate
   (c) negative OR gate
   (d) negative NAND gate [**Ans:** (c)]

6. In a NAND gate, all the inputs are High (1). The output is
   (a) 1 (High) (b) zero (Low)
   (c) infinity (High) (d) cannot be predicted [**Ans:** (b)]

7. The function of positive OR gate can also be accomplished by
   (a) AND gate (b) NOT gate
   (c) NAND gate (d) none of the above [**Ans:** (c)]

8. By using Boole an algebra, the simplified form of $(A + B)(B + C)$ is
   (a) $A + BC$ (b) $B + AC$
   (c) $C + AB$ (d) $A + B$ [**Ans:** (b)]

9. 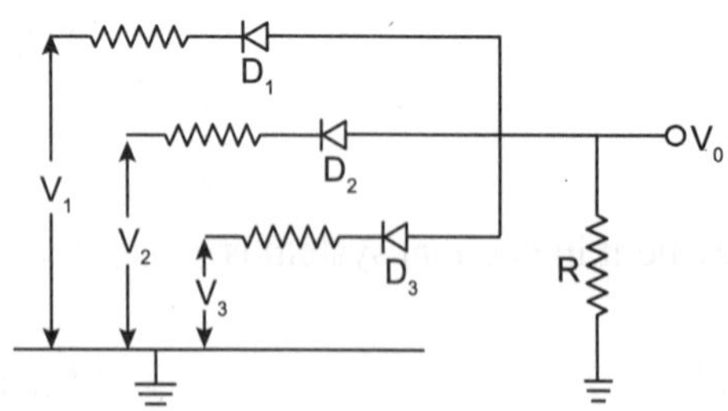

   The above circuit represents the function of
   (a) NOR gate (b) AND gate
   (c) NAND gate (d) OR gate [**Ans:** (d)]

10.

	$\bar{C}\bar{D}$	$\bar{C}D$	CD	$C\bar{D}$
$\bar{A}\bar{B}$	0	1	1	1
$\bar{A}B$	0	0	0	1
AB	1	1	0	1
$A\bar{B}$	1	1	0	1

The above Karnaugh Map comprises
(a) one pair and two quads (b) five pairs
(c) three pairs and one quad (d) two pairs and one quad **[Ans:** (a)]

11. In a two-input gate, if both inputs are low or high, the output is always Low. The above statement holds for
(a) OR gate (b) NOR gate
(c) NAR gate (d) NAND gate **[Ans:** (b)]

12. A NOR gate is equivalent to
(a) OR gate (b) bubbled OR gate
(c) bubbled AND gate (d) AND gate **[Ans:** (c)]

13. The truth table for a gate is

Input		Output
0	0	1
0	1	1
1	0	1
1	1	0

The above truth table holds for
(a) OR gate (b) bubbled OR gate
(c) XOR gate (d) bubbled AND gate **[Ans:** (c)]

14. The function of NOR gate can be accomplished by using
(a) diode logic circuit only (b) transistor logic circuit only
(c) diode-transistor logic circuits (d) transistor-transistor logic circuit **[Ans:** (c)]

15. For a Quad four 1^s appear in a Karnaugh Map
(a) in a row (b) in column
(c) side by side in row and column (d) all of the above **[Ans:** (d)]

Mark the Following Statements True/False.

1. Binary number system is the most popular system used in computers. [**Ans:** True]
2. Boolean algebra follows the rules of ordinary algebra. [**Ans:** True]
3. AND gate demonstrates the multiplication process. [**Ans:** True]
4. Two toggle switches connected in series with a bulb in an electric circuit represent the function of OR circuit. [**Ans:** False]
5. NOR gates when connected in a certain sequence can function as any other specific gate. [**Ans:** True]
6. Truth table for XOR gate and OR gate is similar. [**Ans:** False]
7. Positive logic OR gate is equivalent to negative logic AND gate. [**Ans:** True]
8. Two CMOS can be connected so that the combination functions as NOT gate. [**Ans:** True]
9. NOR gate is equivalent to bubbled AND gate. [**Ans:** True]
10. A Karnaugh map consists of quads only. [**Ans:** False]

Fill in the Blanks.

1. _______________ gate is universal gate. [**Ans:** NAND]
2. AND gate represents _______________ of inputs mathematically.

   [**Ans:** multiplication]
3. _______________ gate represents addition of inputs mathematically.

   [**Ans:** OR]
4. In AND gate, the output is _______________ if _______________ inputs are high.

   [**Ans:** High, all]
5. In _______________ gate, the output is High when _______________ input is high.

   [**Ans:** XOR, one]
6. All functions of _______________ gates can be accomplished by _______________ gate.

   [**Ans:** All, NAND]
7. Mathematical functions like _______________ can be performed by _______________.

   [**Ans:** addition, OR]
8. In writing a truth table for _______________, we use _______________ logic.

   [**Ans:** gate, positive]

Chapter 20

Microprocessors

20.1 Introduction

The past two decades have witnessed a remarkable growth in integrated circuit technology. The development of large scale integrated circuits (LISICs) and very large scale integrated circuits (VLSICs) coincided with the design and manufacture of highly sophisticated devices called *microprocessors*. A microprocessor is a digital IC which can perform arithmetic, logic and control operations. It acts as a CPU in a digital microcomputer. It is sometimes referred to as brain of the personal computer. All the circuitry of a microprocessor is contained within a single IC chip.

20.2 Microcomputer

Microprocessors are used in microcomputers. They make up the central processing unit of a computer. In addition to the microprocessor, a computer contains other components and circuits which permit it to carry on its operations. Thus, microcomputers contain memory units to store information and instructions (programme), input/output circuits to drive peripheral devices such as keyboards or mechanical switches for entering information and printers or cathode-ray tubes for reading out information. They also contain internal clocks for timing and synchronization of all the circuits.

Some of the more recent microprocessors are actually self-contained microcomputers (not including the peripheral devices) and are used as computers by designers and hobbyists and in many modern appliances such as microwave ovens, washing machines and automobiles.

20.3 Microprocessors vs Hard Wired Logic Devices

A microprocessor is a logic device but different from other logic circuits such as AND, NAND, OR and NOR gates. The action of these simple devices cannot be changed, i.e data is presented to their input and the output is the result of "logic" operations and cannot be changed. As such, the logic of these devices is called "hard wired" logic. On the other hand, the logic operations of microprocessors can be changed by instructions (programme) presented to them. Thus, a microprocessor can be made to perform a multiplicity of logical operations by simply changing its programme.

The design of a microprocessor, to make certain that each of its circuits is operating properly alone and in conjunction with other circuits, may take a team of engineers and technicians many months to accomplish.

20.4 Principle of Working

Microprocessors and microcomputers operate by making use of binary numbers. A single binary digit is called "bit" e.g. 0 or 1. When 4 bits are combined in a binary number, e.g. 1110, the resulting number is called a "nibble". Similarly, an eight-bit binary number is called a "byte", while a 16-bit binary number is called a "word". A "long word" comprises 32 bits.

Microprocessors are designed to work on a specific length of binary numbers. The more advanced units are 16-bit processors, i.e. they work on "words". Some work on long words, others on bytes and some on nibbles. The common 6800 microprocessor is an 8-bit processor operating on bytes.

20.4.1 Microprocessor and Computer Terminology

Address It is a code that identifies the location or destination of data.

ALU (Arithmetic-logic unit) It is one the basic logic units in a microprocessor. It adds, subtracts, and performs digital logic. (e.g. AND operation).

Assembly language It consists of a set of symbols describing computer operations with address and memory locations.

Buffer It refers to a circuit placed between two circuit components to prevent interaction between them. A buffer is also a term used for temporary storage where data is assembled during data transfer.

CPU (Central processing unit) It is an important subsystem of a computer which controls the interpretation and execution of instructions (programme) fed to the computer.

Compiler It is a programme which changes high-level computer code into machine language code. The microprocessor and computer respond only to machine language.

Data It comprises information that can be processed or produced by a computer.

Decode It means to interpret a coded instruction.

DMA (Direct memory access) It is a means of gauging direct entry to a memory location in order to store or retrieve data.

EPROM (Erasable programmable read only memory) It is a programmable read only memory (PROM) which can be erased and used again. Most EPROMs can be erased by ultraviolet light.

Input/output (I/O) It refers to peripheral equipment used to communicate with a computer. It also refers to data used in communication.

LSD (Least significant digit) It is the extreme right-hand digit in a number. The digit carriers the least numerical weight. In decimal system, it is known as unit digit.

Machine language It is a system of codes, in binary form to represent instructions and data in a particular computer or microprocessor.

Memory It is an essential component in a digital system where information which can be recalled is stored.

MSD (Most signification digit) It is the extreme left-hand digit in a number. This digit carries the maximum numerical weight.

OP code It refers to the machine language instruction for operation which is to be performed. Thus (+) and (–) signs in a computer instruction are OP code.

Programme It constitutes a set of instructions, arranged in sequence, which orders microprocessor or computer to perform a certain job (operation). The programme details every step to be taken and the order in which it is to be carried out.

RAM (Random access memory) It is storage device which can hold and produce on demand any data entered into it. A RAM can read from but not written to.

ROM (Read only memory) It is a device where data is permanently stored. It is the basic programme memory and can only read from but not written to.

Register It is a temporary storage device which holds one word or one byte of data, depending on the microcomputer.

Software It refers to programmes, languages and procedures for a computer system. Hardware refers to the physical components of the computer.

Vocabulary It is a list of operating codes or instructions which a programmer may use in writing a programme for a particular computer.

20.5 Architecture of 8085 Microprocessor

Figure 20.1 represents the block diagram of 8085 mircroprocessor only major circuit groupings are shown. These blocks constitute the architecture, i.e. internal construction, of the processor. It consists of the following major circuits.

1. ALU (arithmetic logic unit)
2. Accumulators

3. Various registers, including data and address registers
4. Data bus, address bus
5. Programme counter
6. Instruction encoder and control
7. Stack pointer
8. Condition code register

It is to be noted that this processor by itself does not constitute a microcomputer; external programme memory and external clock and I/O must be added to make it operate as a microcomputer.

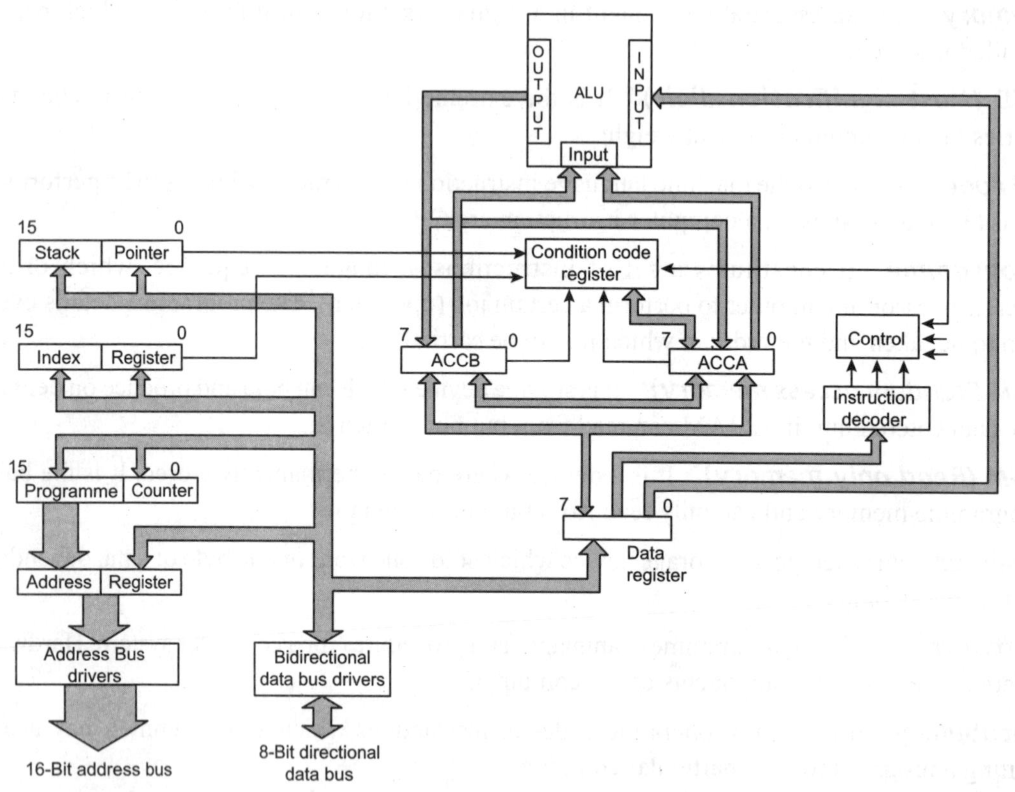

Figure 20.1 Block diagram of 8085 microprocessor

The functions of the various blocks are briefly discussed as under:

ALU This unit performs arithmatic and logic operations on sets of 8-digit binary numbers that are presented to it by addressed data memory (RAM) and by internal accumulators. The ALU can be made to add the input binary numbers in various ways. The control and timing signals are generated within the microprocessor according to the programmed instructions based on pulses supplied by an external clock.

Registers These are temporary memory devices where information in the form of bits can be entered and stored as long as the unit is not shut off. When desired, information in these registors can be read and then used in calculations performed in the ALU. These RAM devices are also known as scratch pad memory. The 8085 consists of six general-purpose registers to store 8-bit data during the execution of programmes. They are marked B, C, D, E, H and L.

Accumulator This is a special register, to speed up the performance of operations and to simplify the internal construction of microprocessor in order that many instructions can be performed on the data in a particular register. This register is called accumulator because it will hold the SUM or accumulation of bits after addition. The 8085 contains two accumulators. It is an 8-bit register used to store 8-bit data. In accordance with results stored in the accumulator, five flip-flops called flags contained in the ALU are set or reset. The flags are affected by arithmetic and logic operations performed in the ALU. The flags represent data conditions in the accumulator. The five flags are called zero (Z), carry (CY), sign (S), parity (P) and auxiliary carry (AC). The zero (Z) and carry (CY) flags are very common.

Data bus, address bus A bus consists of a group of conductors on which digital information is transferred. In the 6800 there are eight conductor data bus and 16 conductor address bus. The data bus is bidirectional, that is information can be sent along its lines from the 8085 into an external memory device or the information can be sent into the 8085 from an external memory. The address bus that transfers only address information into the external memory is unidirectional.

Programme counter The programme counter keeps track of the address at which the next instruction is found. Usually the next consecutive address calls for the next instruction. However, under the programme control, the order of instructions can be changed in two ways. In the first, some instructions branch to other address locations depending on the outcome of some test. The second possible exception is a special address into the programme counter. In this way, the programme can be changed when required. It is a 16-bit register containing memory addresses. Thus, it is a memory pointer.

Instruction decoder/control The machine code sent to the microprocessor by the ROM must be interpreted, and the microprocessor must take proper action. This function is entirely internal to the microprocessor. After interpreting the instruction the instruction decoder sends out necessary signals to assure that instructions are carried out as ordered.

Stock pointer It is a special register used to identify addresses in the stack where information is to be stored. It is a 16-bit register containing 16-bit memory addresses; so in a way it is a memory pointer. The stack is a spiral set of locations in the internal 8085 memory which are set aside as operating registers. The stack pointer contains the address of next available empty location in the stack. Data is normally entered into or taken from the stack one location after another. The stack pointer makes possible the use of subroutines in a programme. Subroutines are consecutive programme steps which achieve a specific result. A log programme may require the use of a specific subroutine many times. The stack pointer and the stack make this possible.

Condition code register This special register keeps track of special conditions in the coded form. There are 6 bits in the condition code register.

Bit 1 is called overflow, abbreviated V

Bit 2 is called zero, abbreviated Z

Bit 3 is called negative, abbreviated N

Bit 4 is called interrupt, abbreviated I

Bit 5 is called half carry, abbreviated H

Bit 6 is called carry, abbreviated C

In the various steps of addition or subtraction, these condition indicators are important. These indicators are often called flags, since they are used to flag or warn the operator of possible problems. The condition code is important because it permits the operator to change the programme flow depending on the zero operation. For example, we know that divisible by zero is not permissible. Thus, before calling the subroutine divide, we check the zero (Z) flag. The instruction BEQ checks the Z flag and branches to the address which is the start of the instruction if the flag shows that the dividend is zero. Otherwise, the programme will continue to divide.

20.6 Timing the 8085 (Clock)

The programme is carried out in the microprocessor in an orderly fashion. This is assured by a clock which is external to 6800. The clock pulses are used to synchronize the entire system. Thus, a rise of clock pulses may signal that a new instruction is to be obtained. The fall of pulse may signal that the data obtained is now valid. The next rising pulse may start the execution of an instruction, and so on. Most instructions require a number of clock cycles to be completed. One measure of the efficiency of a microprocessor is the speed of clock. Another measure is how few clock cycles are required to complete most instructions.

20.7 I/O Ports

Microprocessors normally communicate with the user by way of input/output ports. Specific pins on the microprocessor are arranged in groups so that they may be accessed together. A group of pins used for output is called output port. In the 6800 system, the microprocessor does not contain I/O ports. Therefore, communication with the user is accomplished by way of peripheral interface adaptor (PIA).

20.7.1 Additional Features

Interfacing Additional logic circuits called "interfacing devices" are required to interconnect peripherals with the microprocessor. The devices include buffers, decoders, encoders and latches.

For this bus-oriented system, where several components share the same bus lines, tri-state logic circuits are required.

A tri-state logic (circuit) device has three states: (i) two logic status (logic 0, logic 1) and (ii) a high impendence state as if disconnected from the system. When the enable is high, the tri-state device acts like an inverter. The microprocessor communicates with one peripheral at a time by enabling the tri-state line of the interfacing device.

I/O devices or peripherals can be interfaced with the 8085 microprocessor in two ways: (i) peripheral I/O and (ii) memory mapped I/O. In the peripheral I/O, an 8-bit address is employed by the MPU to identify an I/O. In and out instructions are used for data transfers. On the other hand, in the memory mapped I/O, a 16-bit address is used by the MPU to address an I/O and associated instructions for data transfer.

20.7.2 Interrupts

The 8085 microprocessor can respond to external device or signals for four operations: (i) RESET (ii) INTERRUPT (iii) READY and (iv) HOLD. With the activation of reset, all the internal operations are suspended and the programme counter is cleared. Programme execution can begin at zero memory address. When interrupt is activated, the processor is interrupted from executing the current programme and directed to execute another brief programme and then to resume the current programme.

20.8 The ET 3400 Trainer

ET 3400 (Figure 20.2) is a versatile device that uses the 8055 microprocessor in conjunction with external memory ICs, a clock, a display unit (six 7-segment LEDs), a 17 keyboard, a series of buffers, test sockets, etc.

A keyboard is used for entering data and programme. Each of the 15 keys (1 through F) has a dual function. Note that above each key designation (digit or letter) is printed a group of small type letters, such as ACCA above the 1, ACCB above 2 and so on. With power ON, when the operator presses key 1, either 1 will appear some where in the display or letters ACCA will appear. Which one appears depends on what key the operator has pressed just before pressing the key 1. For example, if the operator has just pressed the RESET key and then presses key 1, the trainer performs the ACCA function and ACCA will appear in the display. The trainer performs the function specified on the pressed key. The functions of concern to us are the following:

(i) ACCA shows the contents in accumulator A; the display will show ACCA followed by two hexadecimal digits.

(ii) ACCB shows the contents of accumulator B in the same manner as above. Note, however, that B appears as a lower case letter 'b'. The aim is to prevent confusing an uppercase B in the display with number 8. Thus, in reading Hex numbers on the trainer, one should not confuse 6 with B.

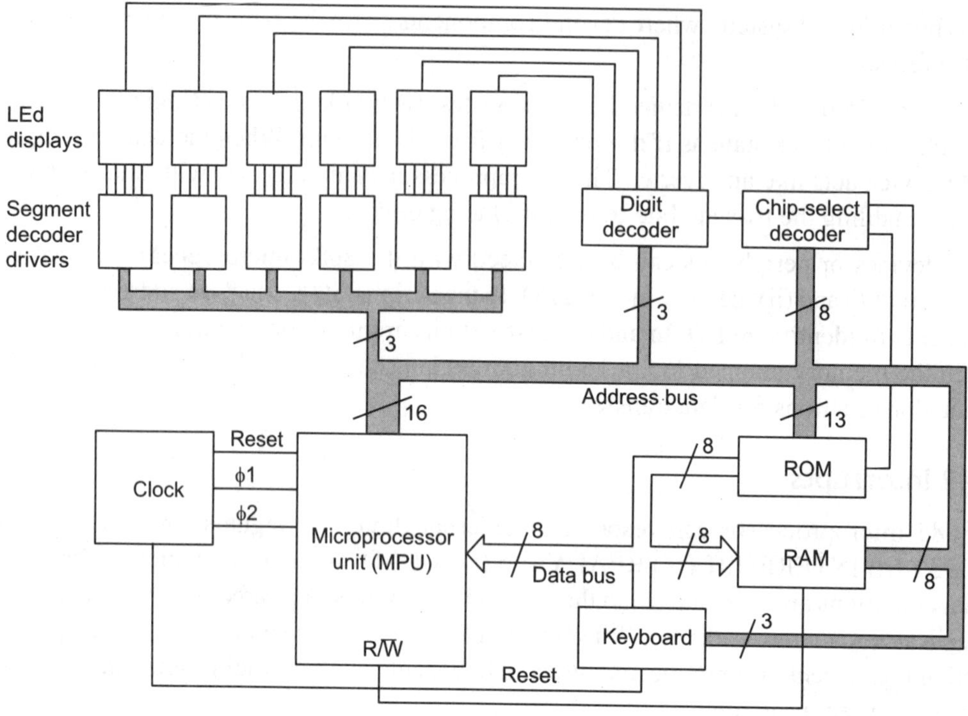

Figure 20.2 Block diagram of trainer

(iii) PC causes the system to display the contents in the programme counter. The display will be PC followed by four digits stored in the programme counter. The trainer uses the hexadecimal numbering system. The programme center can go up to the hex number FFFF, which in decimal system is represented by 65,535. This is the highest number which this particular trainer can address.

(iv) CC is the display of condition code. It uses the six LED displays to indicate the value of each of the condition code bits. The letter next to each display LED (lower left corner) shows the letter (HINZVC) to which each bit corresponds.

(v) SP is the stack pointer. Its display is SP followed by the four-digit number in stack pointer.

(vi) RTI stands for instruction RETURN from interrupt. It is a command to the microprocessor to resume operation where last stopped.

(vii) SS is also a command to the microprocessor. It stands for SINGLE STEP. It causes the programme to advance a step at a time following the instructions to the branches or jumps, or just the next instruction, depending on the programme.

(viii) BR is the acronym for BREAK POINT and forces the microprocessor to stop at any address we choose. The display is - - - - - br when we first press the key. We can now enter any address by pressing the digits of that address. Remember that the address is not in hexadecimal. When the programme is running, it will stop at the address we set as break point. This permits us to find problems that we may have in our programme.

(ix) AUTO is used to enter programme information. The display which appears on the trainer when this key is first pressed is - - - - - Ad. We can again enter any memory address we choose. This will be the starting address of the programme we wish to enter. Suppose that we wish to start at 0000 - - . We then enter the digits 0000 and our display changes to 0000 ---. If we now enter the two digits corresponding to the machine code for the first instruction, the address will automatically change to 0001 -- and the trainer will wait for us to enter the next instruction or operand as required by the programme. This arrangement is a great help in entering the programme quickly. The only way to stop is to press the RESET key.

(x) EXAM stands for examine memory. Pressing this key again asks us to enter an address by displaying - - - - Ad. As soon as we enter an address of four digits, the last two digits change to show the contents of the memory at the address. Now we have a number of options. If we press E again, we can enter new address. Pressing FWD will advance the address to the next location (increment the address) and display the memory contents of the preceding address. Pressing CHAN will change the display by replacing the contents with - - - . We can now enter any value we wish into the displayed memory location. This arrangement makes it possible to check our programme and correct any errors we may have made when it was first entered into the AUTO mode.

(xi) CHAN can also be used to change any of the registers we displayed before. Suppose we are displaying the ACCA register and our display reads ACCA 32. If we wish the contents of accumulator A to 23 (hexadecimal), we press CHAN getting ACCA - - -. Then we press in turn keys 2 and then 3 and we have entered 23 into accumulator A and then display will show ACCA 23.

(xii) INDEX will display the contents of the index register IN xxxx (x stands for digit) in a similar way to that of programme counter display.

(xiii) DO is an instruction to perform the programme in memory, starting at the address we enter. For example, suppose we wish the microprocessor to start carrying out the instructions starting at the address 0001. We press DO and the display that appears is - - - - do. We now enter the address 0001 and the trainer will start executing the programme beginning at the address shown.

20.9 Microprocessor Unit (MPU)

In order to write a programme efficiently, we should know the steps that a particular microprocessor takes in executing an instruction. Figure 20.3 shows the internal architecture of a microprocessor unit, that is the features are involved in executing a programme step.

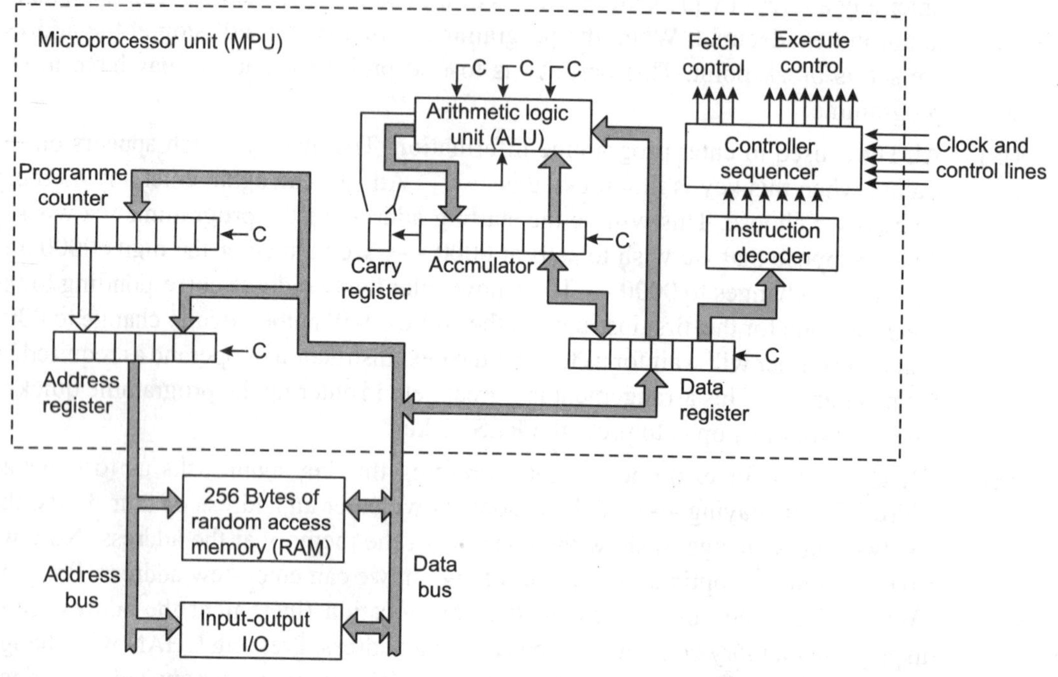

Figure 20.3 Internal architecture of an elementary microprocessor unit

It should be noted that the architecture of different microprocessors differs widely. For example the length of the word may differ from 4 to 32 bits. The 8085 microprocessor uses 8-bit byte. The number of internal registers can also vary widely. Some microprocessors contain both RAM and ROM inside the MPU, and as such these devices are called microcomputers.

After a programme has been entered, each MPU acts to carry out the details of the programme. To that end, the various MPUs are similarly arranged. Each has its own programme counter (PC) as shown in Figure 20.3, which keeps a track of the address for next instruction. At a given time in the clock cycle, the microprocessor is ready to fetch the next instruction. At that time the address in PC is transferred to the address register and placed on address bus. The ROM is signaled to accept, and it places the instructions stored at the address on the data bus. Next, the MPU transfers this instruction to the instruction decoder. The instruction decoder decides what action the MPU must perform to execute the instruction.

Execution of instructions is the next phase in the microprocessor operation. The nature of execution depends on the instructions. Consider the fact that some instructions are single byte while some others are multibyte. For example, the single byte instruction NOP, meaning no operation, just tells the MPU to do noting except increment the programme counter during execution. After NOP, the next instruction will be fetched as soon as the clock is at the proper point in its cycle. On the other hand, the JSR (jump to subroutine) instruction, to which the microprocessor can respond, requires 3 bytes including 2 bytes of address for the operand following the instruction, so that the microprocessor will know the address to which it must jump. The JSR instruction requires many steps from the microprocessor, as listed below:

1. The present address must be placed on the stack. To accomplish this, (a) the address in the stack pointer must be placed on the address bus; (b) R/W must be pulled low to tell the RAM that MPU is writing; (c) the present address must be placed on the data bus; (d) the stack pointer must be decremented.
2. The next byte in ROM must be fetched and held in data register because it is part of the new address to be placed in the programme counter.
3. The programme counter must be incremented again.
4. The address is placed in the address register.
5. The data register information is transferred to the most significant byte (MSB) of programme counter.
6. The least significant byte (LSB) of the new address is obtained from the next ROM location and transferred through the data register to the lower byte of programme counter. Now, the MPU is ready to fetch the next instruction from new programme location.

20.9.1 MPU Instruction

Every instruction must be interpreted by the MPU and executed. The instruction set is a means of communication between the programmer and MPU. A particular microprocessor might have a large number of instructions and most of them can be executed with different forms of addressing.

In order to programme the microprocessor efficiently, each of the instructions must be first understood by the programmer. The instruction set can be considered a language. The difference between our spoken language and a microprocessor or computer language is primarily the degree of exactness. A computer cannot work with vague statements. A computer statement/instruction will be exactly the same way every time it is presented to it. The biggest difficulty is that a beginner has to think concisely and to formulate programme accurately. The second step is to draw a flowchart of steps required to accomplish the desired goal. Further, this flowchart is translated step by step into the assembly language.

It is to be remembered that a computer language is a set of binary numbers, so an assembly language is used to help the programmer. This first step is to associate a specific instruction with a mnemonic. A mnemonic consists of two letters, usually the first letter(s) or word(s) of the required operation.

The relationship of letters to the words of instruction makes mnemonic easier to remember than the binary code of operation. For example, the mnemonic for "no operation" is "NOP". The machine code in binary for that instruction is 0000,0001; in Hex it is 01. By and large we have five addressing modes. These are (i) immediate, (ii) inherent, (iii) direct, (iv) indexed and (v) extended.

20.9.2 MPU and Clock

The MPU performs all the logic and arithmetic functions as well as processes and keeps track of data and programme instructions. To operate the MPU, an external clock is required. The clock generates two timing signals equal in amplitude and frequency but 180° out of phase. The microprocessor initiates and stops different actions depending on the transitions, i.e., polarity change of two signals.

20.9.3 Address and Data Buses

The microprocessor connects to the rest of the circuitry through two buses. The width of the bus is shown by the number of connecting lines in each bus. Thus, /8 on the data bus indicates that there are 8 lines of data going from microprocessor to RAM, ROM and keyboard. On the other hand, the address bus coming out of the microprocessor is 16 lines wide. In addition, 3 lines go to the segment decoder/drivers, 3 other lines go to the digit decoder, 8 lines go to the chip select decoding circuit, 13 address lines go to the ROM, and 8 address lines go to the RAM.

The number of address lines determines the number of different addresses that can be written to a device.

20.9.4 RAM and ROM

The arrows on the data bus indicate that data can go from ROM to microprocessor, from key board to microprocessor, and both from RAM to microprocessor and from microprocessor to RAM; that is, the data bus is bidirectional between RAM and microprocessor. We know that a RAM can be both written into and read from, whereas a ROM can only be read from and it cannot be written into.

Decoders Figure 20.2 shows three different decoders; chip-select decoder, digit decoder and segment decoder. The address bus connects to each of them. Each has a unique function.

The results of logical and arithmetical actions of the microprocessor are shown on LED displays.

The digit decoder determines which LED display is to be lit. The segment decoder decides which segment of an LED is to be lit. It is to be noted that LEDs are not connected directly to the address bus. They do directly go to the segment and digit decoders which receive their input from address bus.

20.9.5 Programming the Microprocessor

To understand microprocessor programming, we must first look at the hardware of the trainer. Figure 20.2 shows the block diagram of the trainer.

It includes MPU (microprocessor unit), which is the heart of the system, a clock, a keyboard, a RAM, a ROM, six seven-segment LED display units, a digit decoder, a chip select decoder, an address bus and a data bus.

20.9.6 Introduction to Programming

In writing a programme, the programmer must follow these steps.

1. Define the requirements for the system; that is, decide exactly what the system is to accomplish.
2. Draw a flowchart showing what the input to the system must be, the steps the system must take to accomplish the objective, and finally what the system must do after the objective is achieved.
3. Write the mnemonic for each step on the flowchart.
4. Translate the mnemonic into machine code which the system recognizes.
5. Enter the machine code into memory.
6. Check the programme.
7. Run the programme.

20.9.7 Checking the Programme

After a programme has been entered, the next step is to check it for typographical errors, logical errors, and hardware errors which may some times occur. Errors in a programme are called bugs, and so the process of correcting errors in a programme is called debugging.

Many trainers (ET3460) have two functions for debugging; one is SS (single step functions). In this operating mode, we can advance the programme one instruction at a time, so that we can examine the contents of memory accumulated, or other registers at each step. Since we know what the programme is supposed to do at each step, we can locate the bug because that is the point at which the programme either does not perform the expected operation or performs it improperly.

An obvious problem of SS through a programme is the time required. It is specially true when a section of programme is repeated many times. If we know that a portion of the programme works correctly, we would like to run it at normal speed and stop at a point that has not been checked yet. For this purpose, a second function on the trainer called BR is used for debugging. BR stands for *break point*. We can set an address as a break point, and the programme will stop at that point when it is started with the 'DO' command. Then, we can single-step from that point 'ON'.

After the programme has been checked and debugged, it can be sucessfully run to give the desired result.

20.9.8 A Sample Programme

Consider adding two hexadecimal numbers. Let us write the first statement "Add two hex numbers". This statement is incomplete since it does not define the numbers to be added, where they are, or where they are to be placed in the memory of the trainer and must also know where the result is it to be stored.

Let the two numbers be 21_{16} and 35_{16}. Let 21 be stored in memory address say 0090, and 35 in memory location say 0092. With these details we can draw a flowchart for the programme. In this chart, two terms, "AUGEND" and "ADDEND" must be defined. Augend is the number to which another number addend is to be added.

The flow chart instructs the programmer first to enter number 21 into address 0090 and then 35 into address 0091. These steps are not really a part of the programme. They simply supply data to the MPU. The programme starts with NOP instruction (No operation) and then directs the programmer to write the instruction for loading the augend into accumulator (A_{CC} A), then adding the addend to the number in accumulator B, and storing the result in memory address 0092.

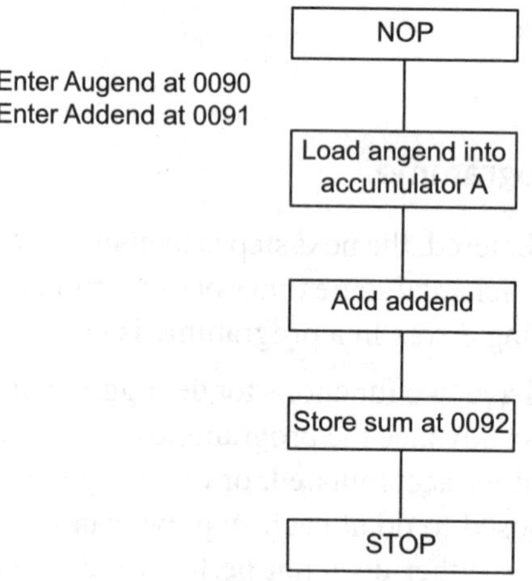

Figure 20.4 Flow chart for adding two numbers

We are now ready to write mnemonics for the instructions in the programme. Now let us convert the mnemonics into machine code starting with the address 0000.

Programme A

Address	OP code	Mnemonics and addressing mode
0000	01	NOP (Inherent)
0001	96	LDDA (Direct)
0002	90	Operand
0003	9B	ADDA (Direct)
0004	91	Operand
0005	97	STTA (Direct)
0006	92	Operand
0007	3E	WAI (Inherent)

Enter data and address as follows:

Address	Data
0090	21
0091	35

After entering the programme and its data, the programme should be checked for errors. The programme can then be executed by using SS or DO functions. If the DO function is used to execute the programme, the display will appear blank after execution. By resetting the trainer and checking the contents of address 0092, the sum 56 will be displayed.

Summary

1. A microprocessor is a digital IC which can perform arithmetic, logic and control operations. It acts as the CPU in a digital microcomputer.
2. In addition to the CPU, a computer contains the following major circuits: I/O circuit and devices, and a clock. The most sophisticated microprocessors are in a way microcomputers.
3. A microprocessor is different from basic logic gates whose output cannot be changed since it depends on the input state, but in a microprocessor the output can be changed by instructions presented to it. It works when instructions are presented in binary code.
4. The architecture of a microprocessor contains among other circuits the following: ALU, two accumulators, data and address registers, data and address buses, a programme counter, an instruction decoder control, a stack pointer, and a condition code register.
   (a) ALU performs arithmetic and logic circuits.
   (b) Registers are temporary storage devices.
   (c) Accumulator is a special storage register which holds data on which arithmetic operations are performed. It also holds the final result.
   (d) Bus is a group of conductors over which digital information is transferred.
5. In the execution of a programme, the programme counter keeps a record of the address at which next instruction is found.
6. The machine code instructions which a micro processor receives are interpreted and decoded by the decoder/control. Timing synchronization for the circuits of a microprocessor is provided by an external clock.
7. The trainer usually uses the hexadecimal code.

Review Questions

1. Show the architecture of 8085 microprocessor in a block diagram. Explain briefly each of the major circuits.
2. What is the difference between a microprocecssor and a microcomputer?

3. What is a bus? Name the different buses used by a microprocessor and explain their functions.
4. Why are data bus lines bidirectional? How many data lines are required in a (i) 8-bit microprocessor, (ii) 16-bit microprocessor.
5. Which of the following more closely approximates a microcomputer – the 8085 microprocessor or ET 3400 Trainer? Explain.
6. Explain the terms:
   (a) Mnemonic
   (b) Assembly language
   (c) Machine language
   (d) Bus
7. Mention the methods to execute a programme using a ET 3400 trainer. What advantage, if any, is there in executing a programme by the SS method over the DO method? Mention its demerits.
8. What are the basic segments of a microprocessor? Briefly mention their functions. What type of 8-bit microprocessor is widely used? Do we have a 16-bit microprocessor?
9. How would you change the last two digits in the final display with the use of a trainer? Write the instructions to achieve it.
10. What steps you must perform on the ET 3400 trainer to display the contents of accumulator B?

Short Answer Questions

1. What is a microprocessor?

   Ans: A microprocessor is a logic device whose logic operation can be changed by a set of instructions. It acts as the CPU in a digital computer.

2. What do you mean by the term ALU?

   Ans: The term ALU means arithmetic and logic unit.

3. What are registers?

   Ans: Registers are temporary storage devices.

4. What is an accumulator?

   Ans: An accumulator is a special register which holds the data on which arithmetic operations are to be performed. It also holds the final result of these operations.

5. What is a bus?

   Ans: A bus is a group of conductors over which digital information is transferred in a microprocessor.

6. In what code programming instruction are generally made? Why?

   Ans: Programming codes are generally made in Hex code for hexadecimal machines, because hex code is easier to remember.

7. What is the difference between microprocessor and a microcomputer?

   Ans: A microprocessor unit (MPU) is a digital IC component which contains all the circuits which make up a digital microcomputer. The microprocessor makes up the central processing unit of a computer. The computer in addition consists of circuits which permit it to carry on its operations; for this they contain memory units to store information and instruction and I/O circuits to drive peripheral devices such as key boards or mechanical switches.

Objective Questions

1. A microprocessor is a
   (a) logic device like other gates
   (b) logic device but unlike other gates
   (c) microcomputer
   (d) logic device using universal gates **[Ans: (b)]**
2. A microprocessor trainer uses
   (a) binary number code
   (b) hexadecimal number code
   (c) decimal number code
   (d) octal number code **[Ans: (b)]**
3. Both RAM and ROM
   (a) are different from each other
   (b) are same
   (c) are slightly different from each other
   (d) are different from each other but lead to the same result **[Ans: (a)]**
4. An accumulator is
   (a) a special type of register in microprocessor
   (b) a register like other registers in microprocessor
   (c) a special type of register in a trainer
   (d) a register like other registers in a trainer **[Ans: (a)]**
5. A word consists of
   (a) 4 binary digits
   (b) 2 binary digits
   (c) 8 binary digits
   (d) 16 binary digits **[Ans: (d)]**

Mark the Following Statements True/False.

1. The 8055 microprocessor is a self-contained an microcomputer. **[Ans:** False]
2. A microprocessor is a hard-wired logic device. **[Ans:** False]
3. The 8055 microprocessor does not contain internal clock. **[Ans:** True]
4. The instruction and address decoders make it possible for the microprocessor to keep the operations on buses. **[Ans:** True]
5. The keyboard on a trainer generates hexadecimal numbers. **[Ans:** True]
6. The utility programmes are permanently entered in ROM. **[Ans:** True]
7. A computer language is a set of binary numbers. **[Ans:** True]

Fill in the Blanks.

1. The data bus between the microprocessor and RAM is ____________. **[Ans:** bidirectional]
2. The keyboard of ET3400 ____________ generates ____________ digits. **[Ans:** trainer, hexadecimal]
3. A computer language is a set of ____________ numbers. **[Ans:** binary]
4. The instruction set of 8055 microprocessor lists ____________ for each instruction, the modes and the ____________ for each mode. **[Ans:** mnemonics, OP codes]
5. A ____________ binary number is called ____________. **[Ans:** 16-bit, word]
6. A section of the computer that controls the interpretation and execution of a programme is called ____________. **[Ans:** CPU]
7. Programme instructions for the ____________ trainer are written in ____________ code. **[Ans:** 3400, Hex]
8. A set of instructions that order a ____________ to do certain job is called ____________. **[Ans:** microcomputer, programme]

Chapter

21 Integrated Circuits

21.1 Introduction

The methods employed in manufacturing semiconductor devices determine their electrical characteristics and thus the applications for which they are to be used. For example, low current, fast switching transistors must be in a way designed different than high power transistors.

An integrated circuit consists of many components like resistors, diodes and transistors making up one or more complete circuits in one small package. The major fabrication techniques for ICs are (i) monolithic, (ii) thin film, (iii) thick film and (iv) hybrid construction.

21.2 Processing of Semiconductor Materials

The first step in manufacturing a semiconductor device like transistor is the preparation of the base material, which is silicon or germanium. Silicon is one of the commonest elements on earth. It occurs in the form of silicon dioxide and as silicates or mixture of silicon and other materials. Similarly, germanium is derived from zinc or copper ores. When converted to bulk metal, both contain a large number of impurities in varying amounts. So both metals must be carefully refined before they can be used in the manufacture of desired devices.

A semiconductor material is normally polycrystalline after it has been refined. It means it is made up of many individual formations of atoms with no overall fixed pattern or relationship between them. For use in transistor, the material must be converted into a single crystal. It must be made to follow single atomic pattern throughout.

21.2.1 PN Junction Formation

After a single crystal with electronic grade purity has been grown, the fabrication of the semiconductor device primarily consists of making PN junctions. Many techniques have been developed to grow junctions. Junction growing being a metallurgical process requires a precise control of time and temperature.

There are three popular methods of preparing junction, namely (1) grown junction, (2) alloyed junction and (3) diffusion junction.

Diffusion and epitaxial growth Among the other methods of fabrication, the technique of diffusion has become very popular because precise control can be exercised over various parameters during the manufacturing process. In this technique, a slab of pure semiconductor (silicon) doped with a particular type of impurity is heated to a high temperature in the presence of an impurity of opposite type in gaseous form.

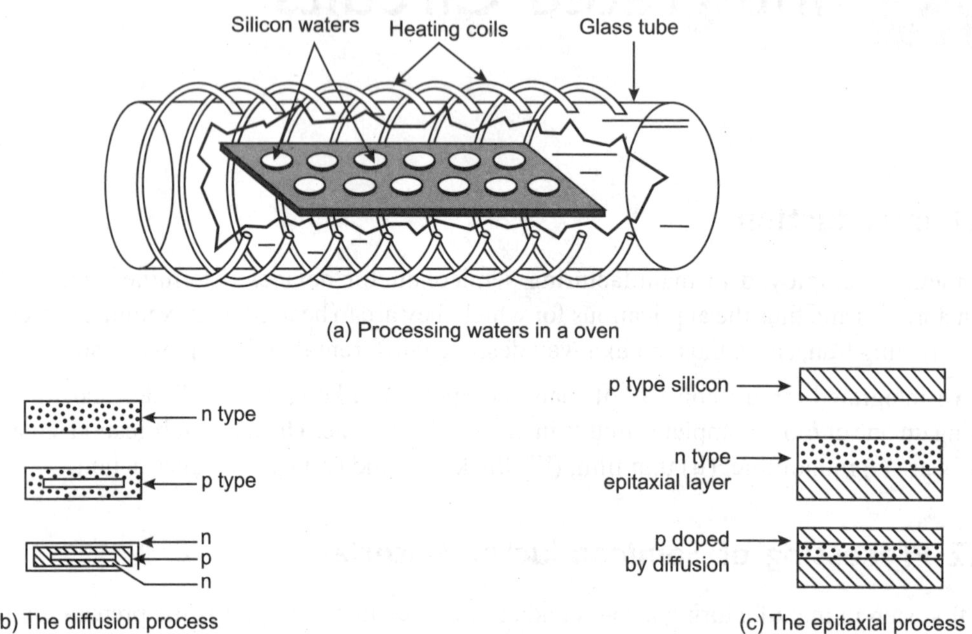

Figure 21.1 Diffusion and epitaxial growth techniques

Some of impurities are absorbed or diffused into each wafer. The outer layer of N type material is then converted into P type. The process can be continued by further heating the material in an atmosphere of N type impurities. Thus, the wafer can have an outer N type layer with P type layer just below it and an N type center. This is the diffused process illustrated in Figure 21.1(a,b). Since the process is very slow (about 2.5 μm/hour), very narrow diffused regions can be accurately produced by careful timing. The epitaxial process (Figure 21.1(c)) is very similar to diffusion process except that atoms of germanium (or silicon) are contained in an atmosphere of gas surrounding the semiconductor wafer. The semiconductor atoms in gas grow on the wafer in the form of a very thin layer. This layer is single-crystal material and may be P type or N type according to the impurity content in gas. The epitaxial layer may then be doped by diffusion process.

21.2.2 Transistor Fabrication

A number of techniques are available for the fabrication of transistor. The most common methods are (i) alloy junction, (ii) grown junction and (iii) diffused junction.

Alloy junction transistor A single crystal of N type wafers is cut into many small sections or dices whose thickness is of the order of 10 to 100 micron (1 micron = 10^{-4} cm), which forms the substrate of one transistor. A small pellet of P type material (say indium) is melted on one side of the substrate until it partially penetrates and forms an alloy with the substrate. It creates a P-N junction (Figure 21.2). This process is repeated on the other side of substrate to constitute finally a P-N-P transistor. The area covered by emitter section is also quite small, about 300 microns, and that of collector is 500 microns.

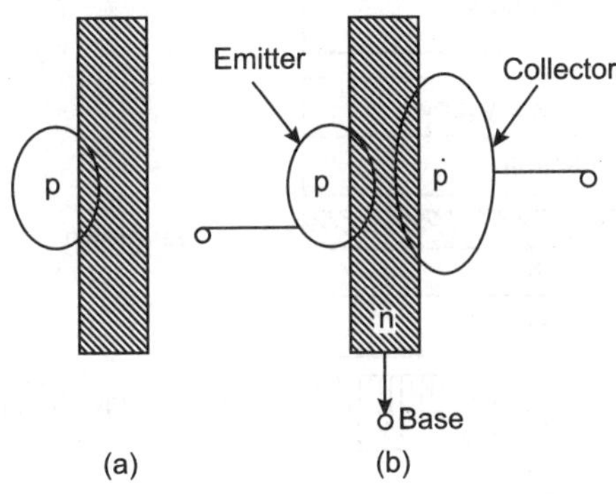

Figure 21.2 (a) Growth of P-N-P alloy transistor

Diffusion junction Diffusion technique is becoming very popular because it offers a precise control on geometry and results into better transistor characteristics and improved performance. Different junctions are grown by the diffusion of impurities in to the semiconductor wafer. A schematic block of a transistor is shown in Figure 21.2(b). This technique is known as **Silicon Planar Technology**.

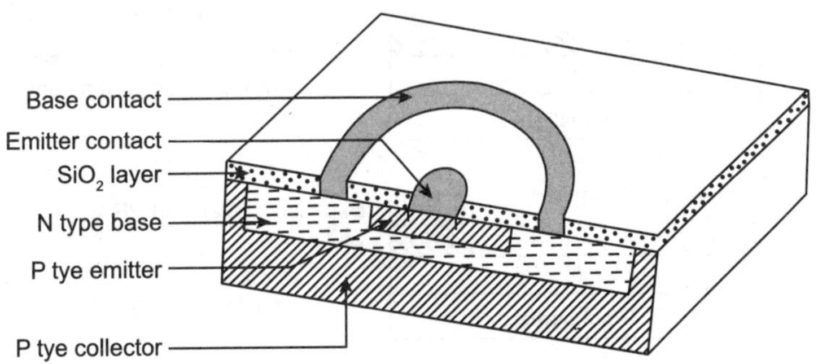

Figure 21.2 (b) Diffused planar transistor (sectional view)

21.3 Silicon Planar Technology

Silicon has become a very popular semiconductor because (i) its temperature range extends up to 200°C, and due to its large band gap no electron-hole pair is generated at such high ambient temperature; (ii) it is available in abundance in nature: (iii) it occurs in the form of SiO_2 which is a good insulator and can be used as mask in the preparation of semiconductor devices. SiO_2 can be simply deposited on silicon by heating. The use of mask prevents the diffusion of impurities where they are not required. However, wherever diffusion of impurities is required, the selected region on silicon chip is left unexposed. This provides a lot of flexibility and control in device fabrication.

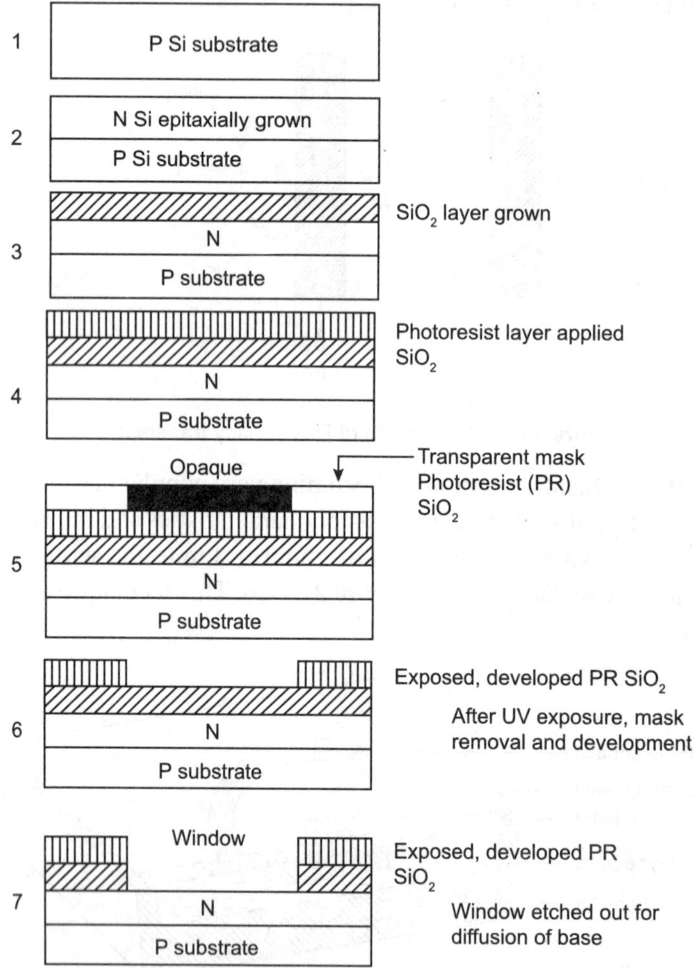

Figure 21.3 (a) Steps in silicon planar technology for preparing NPN transistor

A P type silicon wafer is taken to act as substrate as shown in Figure 21.3(a). A thin layer of N type silicon of required resistivity is epitaxially grown on P type silicon substrate. An epitaxial layer is grown by thermal decomposition of silicon-bearing compounds such as silane, silicon tetrachloride and trichlorosilane. It is possible to provide single crystal layers of uniform resistivity over the semiconductor substrate. The resistivity of epitaxial layer can be chosen independently of the substrate.

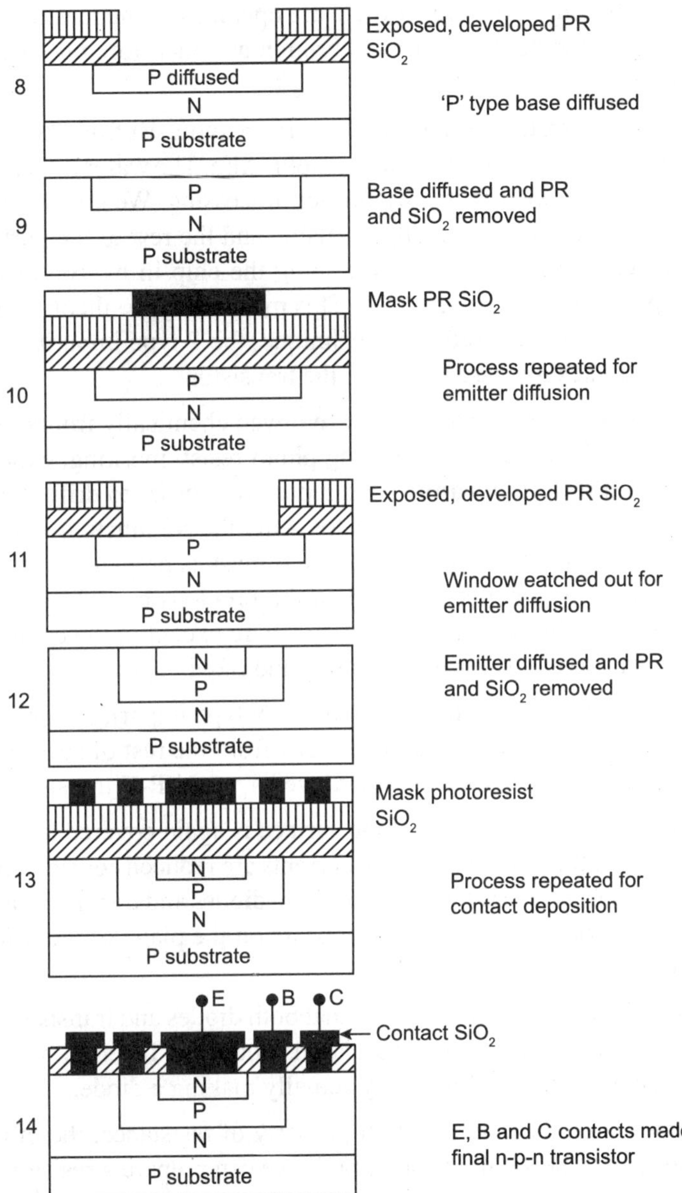

Figure 21.3 (b) Steps in silicon planar technology for making NPN transistor

A film of photoresist is applied over SiO_2 layer formed over the epitaxial layer. The property of photoresist is that it gets polymerized when exposed to ultraviolet light. Thus, a mask is prepared in which portions which are not required to be doped are kept transparent and portions where doping is required are made non-transparent to the ultraviolet light. Usually, the pattern is drawn on large scale and then reduced by photographic techniques. The mask is then placed over the photoresist and ultraviolet light is allowed to fall over it. The ultraviolet light polymerizes the photoresist wherever there are openings in the mask. The nontransparent portions do not allow ultraviolet light to pass through and therefore photoresists below these portions are unexposed.

After ultraviolet exposure, the mask is removed and photoresist from the unexposed portions is dissolved by developing the wafer like a photographic film. The wafer is then fixed in a solution so that exposed portions are not affected by further processing. We now have a wafer in some portions of which the oxide layer is directly available and the rest of the portion is covered by polymerized, developed and fixed resist. By etching the chip in hydrofluoric acid, the oxide layer from the unexposed portions is removed. This makes N type silicon surface available for doping with a P type impurity. By diffusion technique, the P type impurity is now diffused in selected regions. This makes the base region of the transistor.

The developed and fixed photoresist is now removed chemically from other portion. Thus, we have a NP junction. The process of applying photo resist, masking, exposing, developing and fixing is known as **photolithography**. It is repeated in order to diffuse a N type impurity into P region and thus giving rise to NPN transistor. After the NP junction is ready, SiO_2 is again deposited over the wafer and photoresist is applied. A mask is prepared with a non-transparent region in the center which is smaller in size than the previous closed region. The ultraviolet exposure now polymerizes the photoresistor every where except in the centre. The unexposed photoresist is dissolved and the wafer in developed and fixed.

The SiO_2 is etched out to expose the P region. An N type impurity in desired concentration is now diffused. The developed and fixed photoresist from the rest of surface is removed. The collector, base and emitter contacts are made and we have a N-P-N transistor.

Except for metallic contacts, the rest of the surface is again covered by SiO_2 layer. In fact, due to the oxide layer over the junction, surface currents are reduced considerably. These surface leakage currents add to the reverse saturation current in diodes and transistors and thus adversely affect the device characteristics. The junction appears on the plane of the wafer and hence the name planar technology.

From the above working process, it is clear that both diodes and transistors can be prepared easily. Further we also know that a reverse-biased diode can function as capacitor. Thus, a capacitor can be made using this technique by actually making a diode.

A semiconductor material has an inherent property of resistance, therefore by choosing a proper width and length of the semiconductor path, we can prepare a resistor of desired value.

We now understand that MOS devices are easier to prepare than BJT devices using the planar technique, because they require only a deposition of SiO_2 layer and a metallic gate.

21.4 Integrated Circuits (ICs)

An integrated circuit is one in which resistors, capacitors, diodes, transistors and other passive and active components are integrated in a small semiconductor (silicon) chip, such that they become automatically a part of the chip. The size of the chip may be smaller than $2 \times 2 \times 0.1$ mm. ICs may be entirely self contained, requiring only input and output connections and a supply voltage to function. Alternatively, a few external components may have to be connected to make the circuit in package operational.

So far, we have studied "**discrete**" circuits, in which the various components of the circuit are 'joined' together by wires or 'plated' on a printed circuit board (PCB), where each component of the circuit is distinct or separate from other components. We may note that ICs are a leap towards miniaturization of devices. Interestingly, an IC chip of dimension $2 \times 2 \times 0.1$ mm may contain thousands of components. Special techniques are used for the fabrication of a number of components in a small chip.

The disadvantages of discrete circuits are:

(i) They occupy a large space.
(ii) Due to soldering joints their reliability is poor.
(iii) When inductors and transformers are used, they become bulky.

In comparison to discrete circuits, ICs have the following advantages:

(i) Due to the absence of solder joints, they have superior reliability. The components are connected through special techniques.
(ii) They need lesser space due to very small size.
(iii) Their weight is less.
(iv) IC can be operated at comparatively high temperature.
(v) The power consumption of ICs is very low.
(vi) When ICs are processed in a large number at a time, their cost of production is low.
(vii) Temperature difference between different parts of an IC is small.
(viii) ICs have close matching of components and their temperature coefficients. The disadvantages of integrated circuits are as follows:
    (i) When any component of IC does not work, then it is not possible to replace the component. The entire IC chip has to be thrown out. In the case of discrete circuits, one can replace a particular component by a new one.
    (ii) One cannot fabricate heavy components like inductors, transformers, etc. within the chip. These are to be connected externally within the circuit. However, in case of discrete circuits, one can solder heavy components into the board itself.
    (iii) The maximum power rating for which ICs can be produced is low, whereas no such limit exists for discrete circuits.
    (iv) Parasitic capacitance is formed due to the isolation of components from each other, and due to this high-frequency response is restricted.
    (v) It is difficult to achieve low noise and high voltage operations with ICs.
    (vi) In case of ICs, the values of resistors and capacitors are voltage dependent.
    (vii) High-grade PNP assembly is not easy.

The differences between discrete circuits and ICs can be summarized as follows:

(i) Discrete circuits use individual resistors, diodes, transistors, capacitors, and other devices to complete the circuit function. These individual parts are usually mounted on some sort of a circuit board and then interconnected using wires or solder traces.

(ii) To accomplish the same function, discrete circuit assemblies are usually more expensive than an equivalent IC.

(iii) Although IC assemblies do not eliminate the need for circuit boards, assembly soldering, and testing, they allow the same circuit to be produced for lower cost and a more complex circuit to be produced for the same cost.

(iv) With ICs, the number of discrete parts can be reduced.

(v) ICs are smaller and use less power, in addition to being less expensive to manufacture.

(vi) Because the electronics involved in IC assemblies often require fewer alignment steps, ICs can be set up and calibrated for lower cost.

(vii) Because ICs allow for fewer discrete parts in a piece of equipment, IC systems tend to be more reliable.

Integrated circuits may be classified in terms of their function as **analogue** or **digital**. Analogue ICs (also termed linear ICs) may be amplifiers, voltage regulators, etc., Digital ICs contain transistors which are normally either in a switched OFF or switched ON state. Logic gates, counting circuits and the like are available in IC form. The ICs are broadly classified into following two classes:

(i) **Linear ICs** These deals with linear (analogue) signals. A linear signal is that which varies in direct proportion to its input.

(ii) **Non-linear/Digital ICs** These deals with non-linear/digital signals. A digital signal is that in which two discrete voltage levels (0 and 1) are used. The logical circuits are known as Gate which are made of both "**passive**" and **active** components.

Gate It is used to perform logic operations like addition, multiplication, etc. in computers and Gate is a piece of hardware or an electronic circuit which has more than one input but only one output.

One may sub-classify digital ICs as:

(i) **VLSI (Very larg Scale Integration)** This contains 1000 or more logic gates or components.

(ii) **LSI (Large Scale Integration)** This contains 100–1000 logic gates or components.

(iii) **MSI (Medium Scale Integration)** This contains less than 12 gates or components.

We may note that very very large scale integration (VVLSI) and extra large scale integration are also available. These are summarized as follows:

IC class	App No. of components per chip	Component nomenclature
SSI	12	Resistors, diodes, transistors
MSI	12 to 100	Resistors, diodes, transistors
LSI	More than 100	Mostly MOSFET
VLSI	More than 1000	MOSFET
ULSI	Nearly 10^6	Components INTEL 486 and Pentium Microprocessor

Schematics The internal features for ICs are seldom shown in schematics. However, it is more important to know what the IC is supposed to do and how it works as part of the complete circuit. Figure 21.4 shows two typical ways of illustrating an IC. The schematic and a few voltage specifications are all technician usually needs to verify proper operation of the IC.

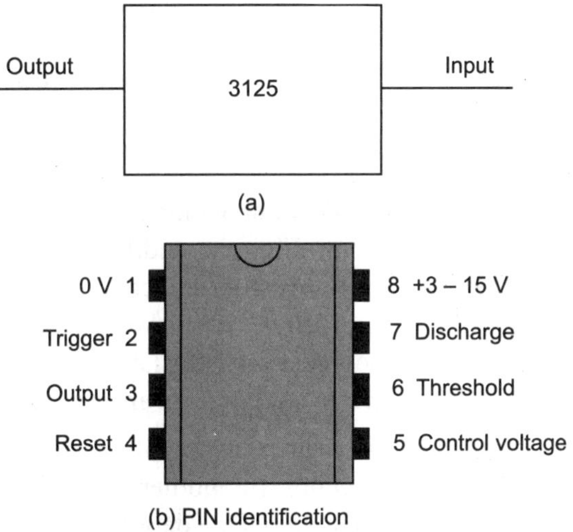

Figure 21.4 Usual way of showing an IC on a schematic

Fabrication The manufacturing of ICs starts in a radio frequency furnace. The P-type silicon wafers are processed by the technique of photolithography. This process creates the desired P type and N type zones in the substrate (wafer). Later in the process, the numerous circuit functions in the IC are electrically insulated from each other by **isolation diffusion** (forming an area to block current flow).

21.4.1 Monolithic Integrated Circuits

In a monolithic integrated circuit, all the components are fabricated by diffusion process on a single chip of silicon such that they become part of a single P type thin wafer. Inter connections among the components are provided on the surface of the structure and external connecting wires are taken out to the terminals. Although a monolithic circuit has distinct disadvantages, yet the vast majority of integrated circuits use this type of construction because it is most economical for mass production. As an example, two transistors are shown in Figure 21.4 in the same substrate. The active and passive components of the integrated circuit are connected within the chip.

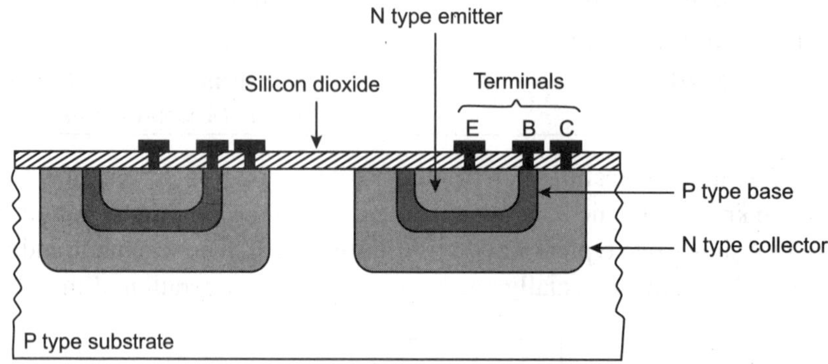

Figure 21.5 Two transistors on the same chip

21.4.2 Thin Film Integrated Circuits

Thin film integrated circuits are constructed by depositing films of a conducting material on the surface of a glass or ceramic base. By controlling the width and thickness of the films and by using different materials for their resistivity, resistors and conductors are fabricated. Capacitors are produced by sandwiching a film of insulating oxide between two conducting films. Inductors are made by depositing a spiral formation of film. Transistors and diodes can be produced by thin film technique, but usually tiny discrete components are connected into the circuit.

One method to produce thin films is vacuum evaporation, in which vaporized material is deposited on a substrate contained in vacuum. In another method, called cathode sputtering, atoms from a cathode made of desired film material are deposited on a substrate located between the cathode and anode.

21.4.3 Thick Film Integrated Circuits

These are also known as printed thin film circuits. In this process, silk screen printing techniques are employed to create the desired circuit pattern on a ceramic substrate. The screens are actually made of fine stainless steel wire mesh, and the inks have conductive, resistive, or dielectric properties. After printing, the circuits are high temperature-fired in a furnace to fuse the films to the substrate. Thick film passive components (resistors, etc.) are fabricated in the same way as those in thin film circuits. As with thin film circuits, active components are added as separate devices. A portion of a thick film circuit is shown in Figure 21.5.

Integrated circuits produced by thin or thick film techniques usually have better component tolerance and give higher frequency performance than monolithic integrated circuits.

21.4.4 Hybrid or Multichip Integrated Circuits

This structure is shown in Figure 21.5. The circuit is constructed by interconnecting a number of individual chips. The active components are diffused transistors or diodes. The passive components may be a group of diffused resistors or capacitors on a single chip or they may be thin film components. Wiring or metallized pattern provides connections between the chips.

Like thin and thick film ICs, multichip circuits usually have better performance than monolithic chips. Although the process is too expensive for mass production, multichip techniques are quite economical for small quantities and are frequently used. We may note that the base or platform on which the components are formed is called "substrate" or 'wafer'.

Some important points about ICs are:

(i) On the surface of the chip, no component of an IC is visible.

(ii) A high-power microscope is needed to examine connections in a chip. The branch of engineering dealing with ICs and similar other microdevices is termed as **microelectronics**.

(iii) The size of an IC is thousand times smaller than a that of discrete circuit. A complete microprocessor can be built in a single chip of size 9/32″ × 1/4″, which contains more than 257×10^3 components.

(iv) Whenever a component fails, the entire IC should be replaced in a more economical approach.

(v) In the manufacturing of ICs, automation is becoming extremely important, since automation reduces the amount of handling and obviously the possibility of contamination.

(vi) The number of components that can be formed per unit length of the chip is termed as IC density. Nowadays, manufacturers use microns (micron = 1 micrometer = 1 mm = 10^{-6} m) to specify the size of an IC.

(vii) In the production of IC, cleanliness is the most important factor. Even the smallest dust particles can spoil the quality of the product. Therefore, the workers prior to entering the room must wear special apron.

(viii) There are ICs that contain only linear circuits or only digital circuit. However, there are hybrid ICs that contain linear as well as digital circuits. Among linear/digital ICs are **comparator circuits**, digital/analogue converters, timer circuits, etc. A comparator IC is that which compares two linear input voltages. The output may be digital. ICs that convert digital signals into analogue or vice versa find application in aerospace equipment, compact disc players, etc. A **timer IC** is a linear or digital circuit that which can be used in car alarms, ON/OFF of lights in houses and other timing applications. Amongst the most popular timers IC is '555 TIMER.

21.4.5 Physical and Electrical Characteristics of IC

It is to be noted that various components in an IC are not discrete components wired together in a miniature circuit. Rather an IC is a complete circuit formed by an intricate microphotolithographic process on a silicon chip no larger than the head of a pin. The IC "roadmap" of circuits can only be seen under high magnification.

21.4.6 Construction

One of the techniques described earlier is followed to construct all the circuit elements, namely transistors, diodes, resistors and capacitors. The various elements are connected by depositing vaporized aluminum in the desired pattern to form a circuit. A typical layout of a chip is shown in Figure 21.5.

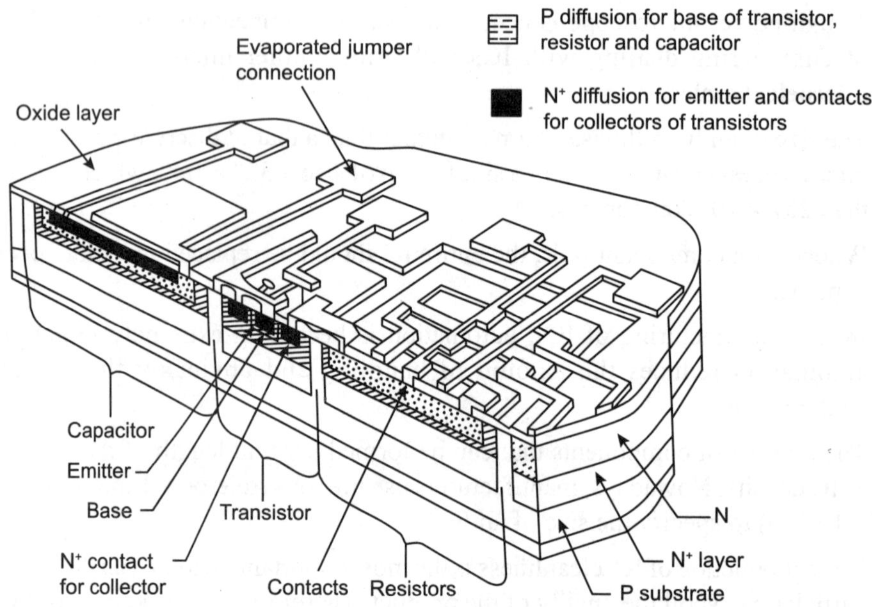

Figure 21.6 IC chip cross-section

A completed disk may have as many as 200 ICs on it. The chips are separated, tested and mounted on ceramic or metal bases. Aluminium wires, about one-third the thickness of human hair, are wound between IC contacts and header leads. Then the package is sealed.

An important consideration in IC economics is flexibility. Chips are designed to be used in more than one circuit configuration. Their arrangements involve various discrete external components. For example, the same chip in different external circuit arrangements may be used as (i) high frequency wide band amplifier, (ii) low frequency amplifier or (iii) high frequency amplifier. This flexibility is facilitated by the fact that circuit leads on the body of IC make it possible to break in on the circuit arrangement at critical points within the IC configuration.

As an example, we can briefly describe CA 3020 as a monolithic integrated circuit which can be used as a linear audio amplifier. Its functional and schematic diagram is shown in Figure 21.7.

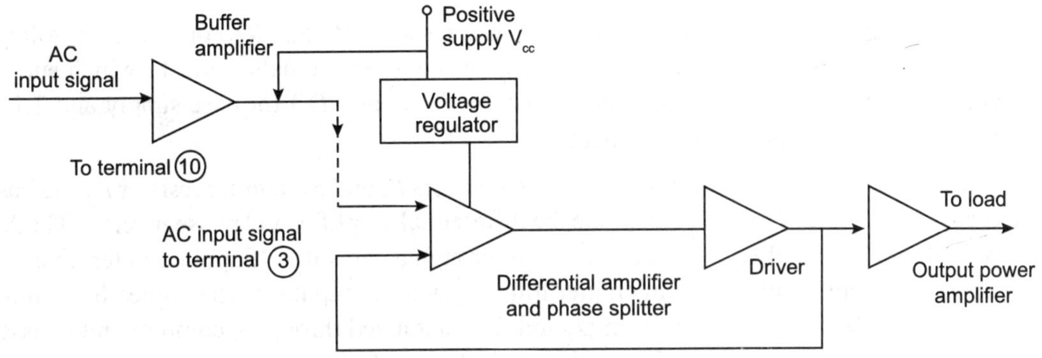

Figure 21.7 (a) Functional block diagram of CA 3020

It is a multipurpose, multifunctional power amplifier designed for use in portable and fixed audio communication equipment and servo control systems. The schematic is shown in Figure 21.7(a) and its functional block diagram is shown in Figure 21.7(b).

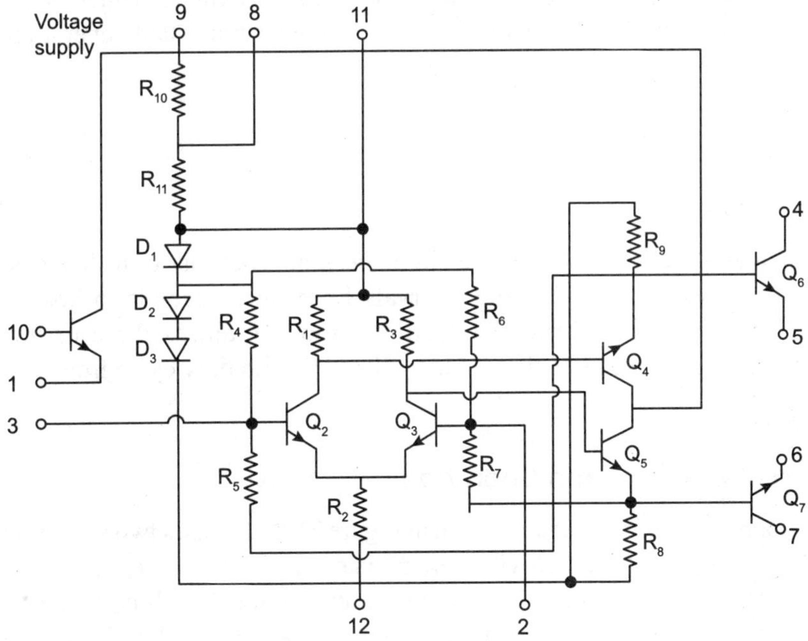

Figure 21.7 (b) Schematic diagram of CA 3020 audio amplifier

This direct coupled amplifier performs as preamplifier, phase inverter, driver and power output functions without a transformer. It can operate from +3 V to +9 V power server. The power output is determined by supply voltage. A temperature-compensating voltage regulator permits operation over a range of –55°C to +125°C.

Series-connected diodes D_1, D_2, D_3 together with resistors R_{11} and R_{10} constitute the voltage regulator. The external power supply voltage is connected to terminals 9 and 12 which supply relatively constant voltage of 1.4 V (function of Diode D_1 and D_2) for base supply and 2.1 V (function of D_1 and resistor R_{11}) for collectors.

Transistors Q_2 and Q_3 together with collector resistors R_1 and R_3, emitter resistor R_2 and base biasing resistors R_4, R_5, R_6 and R_7 constitute the differential amplifier and phase inverter. The AC input signal may be supplied to terminal 3 or 10 through a capacitor. If applied to terminal 10, Q_1 is operated as buffer amplifier, an emitter follower whose output is then coupled to terminal 3; the input of Q_2 and Q_3 coupling from Q_2 and Q_3 is achieved through a common unbypassed emitter resistor R_2. Two equal amplitude signals 180° out of phase are developed at collectors of Q_2 and Q_3. These signals are direct coupled to the bases of emitter followers of amplifiers Q_4 and Q_5. Negative feedback to Q_2 is provided by R_5 and R_7 for DC and AC stability of Q_2 and Q_3. The signal is direct coupled from emitters of Q_4 and Q_5 to the bases of Q_6 and Q_7; power amplifiers Q_6 and Q_7 operate in a class B push pull mode to deliver power to the load, which may be a high-impedance centretapped loudspeaker fed directly.

Another illustration of IC is NE555 timer introduced by Signetics which is a 8-pin IC that can be connected to external components for either stable or monostable operation.

21.5 NE555 IC Timer

21.5.1 Introduction

The NE555 IC timer is very popular among circuit designers because of its low cost, versatility and stable time delays. It allows timing intervals from microseconds to hours. Depending on the preferred output waveform, the oscillator mode requires three or more external components. Frequencies from less than 1 Hz to 500 kHz with duty cycle from 1% to 99% can be achieved.

21.5.2 Internal Circuitry and Operation

The major components of 555 timer are shown in Figure 21.8. It houses two voltage comparators, a flip-flop (also called a bistable multivibrator), resistor divider network, a discharge transistor, and an output amplifier with up to 200 mA current capability. The three divider resistors are of equal value. This network sets the trigger comparator at one-third of V_{cc} and the threshold comparator trip point at two-thirds of V_{cc}. V_{cc} may range from 4.5 to 16 V.

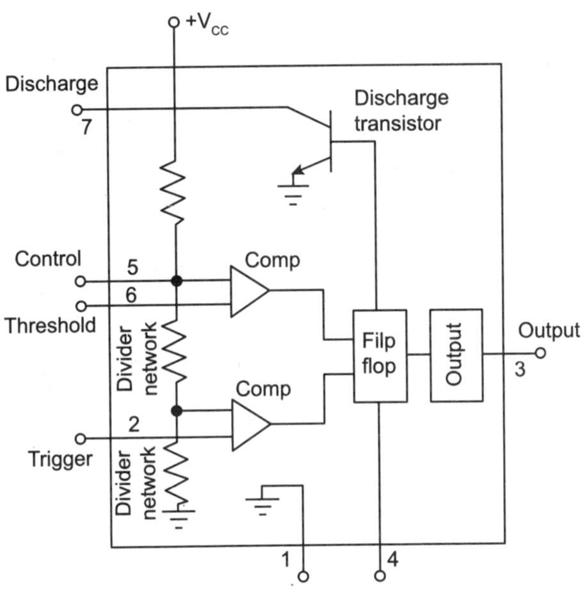

Figure 21.8 Diagram of NE555 IC timer

In Figure 21.8, assume that V_{cc} = 12 V. Given this, the trigger point will be 4 V (1/3 × 12 V) and the threshold point will be 8 V (2/3 × 12). When pin 2 falls below 4 V, the trigger comparator output changes states and sets the flip-flop to the high state, and output pin 3 goes high.

If pin 2 returns to a value higher than 4 V, the output stays high because the flip-flop knows that it was set. On the other hand, if pin 6 climbs above 8 V, the threshold comparator changes states and resets the flip-flop to its low state. This causes the output (pin 3) to go low and the discharge transistor to be turned on.

Note that the output of 555 timer is digital, that is either high or low. When it is low, it is near ground potential, and when it is high it is close to V_{cc}. In Figure 21.8, pin 6 is usually connected to a capacitor that is part of an external RC timing network. If capacitor voltage exceeds 2/3 V_{cc}, the threshold comparator will reset the flip-flop to a low state. This turns on the discharge transistor, which can be used to discharge the external capacitor in preparation for the next timing cycle.

The reset, pin 4, allows for direct access to the flip-flop. This pin overrides the various timer functions and pins. The reset can be used to stop a timing cycle. The reset pin is a digital input, and when it is taken low, it resets the flip-flop, turns on the discharge transistor, and drives the output pin 3 low. The reset function is not usually needed, and thus pin 4 is normally tied to V_{cc}.

Monostable mode The IC timer is connected in monostable (one-shot) mode in Figure 21.9. This mode produces an RC-controlled output pulse that goes high when the device is triggered. The timer is considered negative-edge triggered, which means the timing cycle begins at t_1 when the trigger input falls below 1/3 V_{cc}.

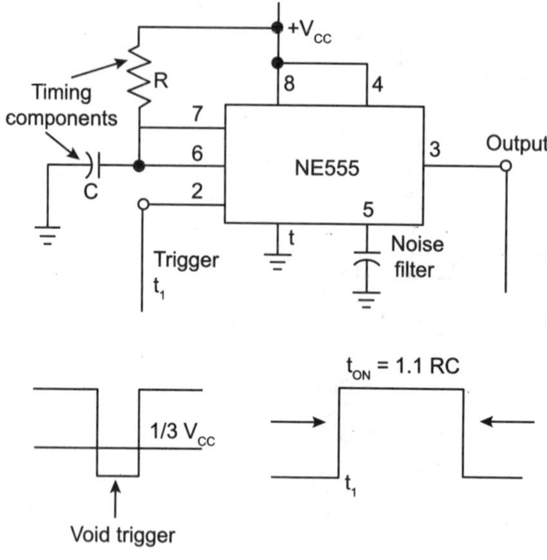

Figure 21.9 Monostable mode

Once the trigger input is greater than 1/3 V_{cc}, the time-out period begins, which means the trigger pulse cannot be wider than the output pulse. When the trigger pulse is wider, the trigger input will need to be AC coupled. Using Figure 21.10, you can see that the 0.1 μF coupling capacitor and the 4 kΩ resistor differentiate the input trigger pulse. The effective width of trigger pulse is decreased by this pulse differentiation (AC coupling).

The width of output pulse is RC controlled in the monostable (one-shot) circuit, and the timing capacitor begins charging through the timing resistor when the timer is triggered. Recall that as capacitor voltage reaches 2/2 V_{cc}, the flip-flop is reset. This turns on the discharge transistor, and the capacitor is emptied in preparation for the next cycle. This results in output pulse width equaling 1.1 time constants. The formula for finding output pulse is $t_{ON} = 1.1$ RC

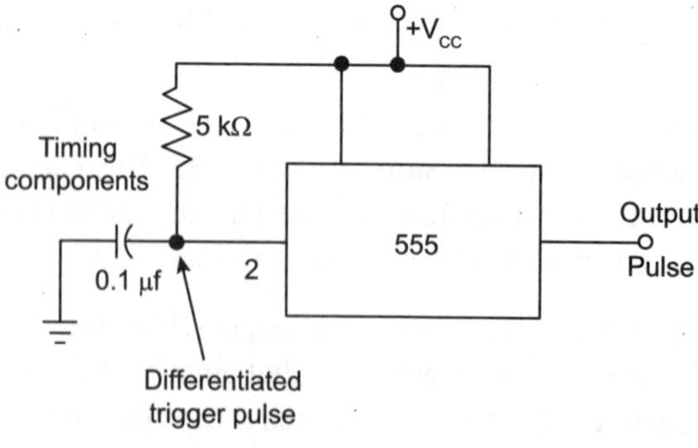

Figure 21.10 AC-coupled trigger pulse

Example 1 Find the output width for Figure 21.10, $R = 5\ \text{k}\Omega$ and $C = 0.1\ \mu\text{F}$. The pulse width is equal to 1.1 time constants.

Solution

$$t_{\text{ON}} = 1.1 \times R \times C = 1.1 \times 5\ \text{k}\Omega \times 0.1\ \mu\text{F} = 0.55\ \text{m sec}$$

Figure 21.11 shows a typical application for 555 connected as a monostable (one shot) timer. This circuit may be used in applications such as automatic shutoff timer for photographic exposure lamp. When the momentary NO switch is pressed and released, the 555 timer output at pin 3 goes high. The 555 output drives an inexpensive solid-state relay block (a triac switch may be built from discrete components, or a suitable relay may be used), which turns on the exposure lamp.

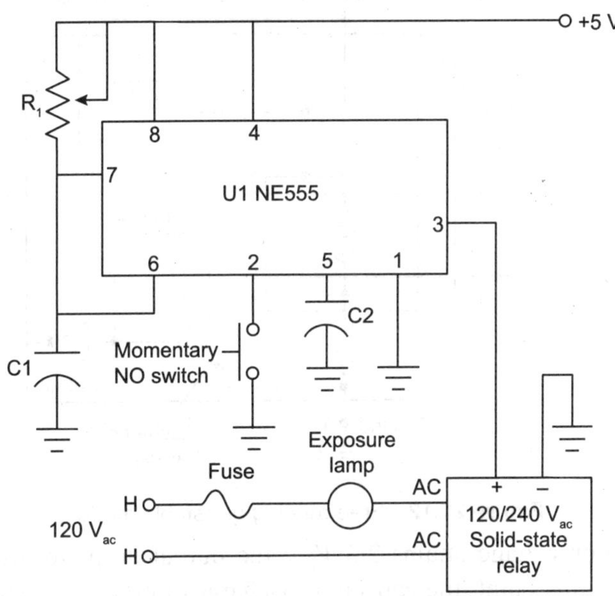

Figure 21.11 Practical exposure lamp timer circuit

When the 555 one-shot output drops low (time out), the exposure lamp turns off. The potentiometer at 555 pin 7 may be calibrated, and it may be labelled in minutes and seconds so that the operator may select the lamp on time, press the switch, and walk away. Long time constants often require electrolytic capacitors. The accuracy of electrolytic capacitor values is notoriously poor. Therefore, calibration is required. In the circuit of Figure 21.11, time constant values must be high because exposure times may need to be quite high. If we select a maximum time of 15 minutes and capacitance of 1,000 μF, the potentiometer value would be calculated as follows:

$$T = 1.1 \times RC \Rightarrow R = \frac{t}{1.1 \times C}$$

$$R = \frac{15 \text{ minutes}}{1.1 \times 1{,}000 \ \mu\text{F}} = \frac{15 \text{ min.} \times 60 \text{ sec/min}}{1.1 \times 1{,}000 \ \mu\text{F}} = 818 \text{ k}\Omega$$

The potentiometer, R_1, may be rounded up to 1 mega ohm. C_2 used for stability may be any convenient value from 0.001 μF to 0.1 μF.

Astable mode An example of time configured in astable (free running) mode is shown in Figure 21.12 with the trigger (pin 2) being tied to the threshold (pin 6). As the circuit is turned on and timing capacitor is discharged, the timer begins charging through series combination (R_1 and R_2).

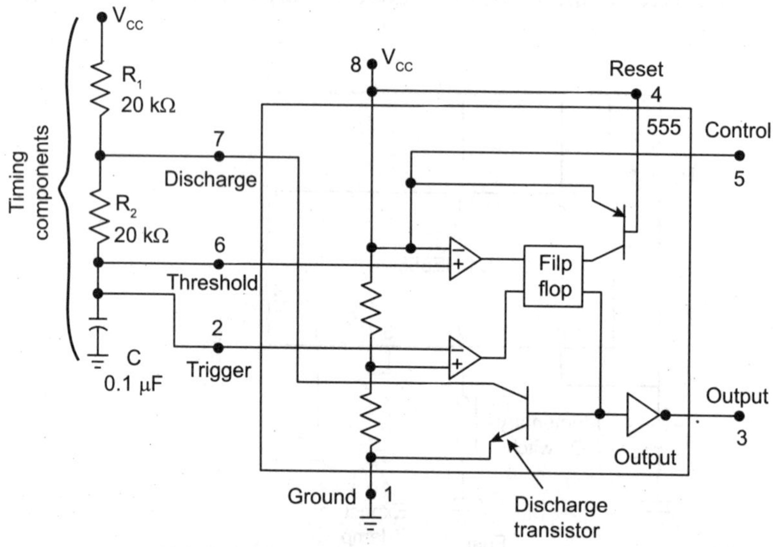

Figure 21.12 Free running or astable mode

When the capacitor voltage attains 2/3 V_{CC}, the output drops to low and the discharge transistor turns on. At this point, the capacitor discharges through R_2. The output changes to high and the discharge transistor turns off as the capacitor reaches 1/3 V_{CC}. Now the capacitor starts charging through R_1 and R_2. This cycle will continuously repeat as the capacitor charges and discharges.

The following formula calculates time the output is held high in an astable circuit through two resistors.

$$t_{\text{high}} = 0.69\,(R_1 + R_2)\,C$$

Example 2 Suppose both timing resistors in Figure 21.12 are 20 kΩ and the timing capacitor is 0.1 μF. Find the time the output will remain high.

Solution

$$t_{\text{high}} = 0.69\,(R_1 + R_2)\,C = 0.69\,(20 \text{ k}\Omega + 20 \text{ k}\Omega)0.1 \ \mu\text{F} = 2.76 \text{ m sec}$$

Because the discharge path is passing through only one resistor (R_2), the time the output is held low is found by using a similar formula.

$$t_{high} = 0.69\ (R_2)\ C$$

Example 3 Using the same data from Example 2, with 20 kΩ resistor and 0.1 μF timing capacitor, find the time for which output will remain low.

Solution

$$t_{low} = 0.69\ (R_2)\ C = 0.69 \times 20\ \text{k}\Omega \times 0.1\ \mu\text{F} = 1.38\ \text{m sec}$$

Because the resistors are equal, the time held low is half the time held high. Because the times held high and low are different, the output waveform is **nonsymmetrical**. When you add t_{high} and t_{low}, the result is the total period (which is needed to calculate the output frequency). The output frequency is equal to the reciprocal of total period, which can be found by using the following formula.

$$f_o = \frac{1.45}{(R_1 + 2R_2)C}$$

Example 4 Using the data from Example 3, where resistors are 20 kΩ and timing capacitors are 0.1 μF, find the output frequency.

$$f_o = \frac{1.45}{(R_1 + 2R_2)C} = \frac{1.45}{(20\ \text{k}\Omega + 40\ \text{k}\Omega)0.1\ \mu\text{F}} = 241.67\ \text{Hz}$$

The duty cycle is the portion (percentage) of time the output is high. This is found by taking the time the output is high and dividing it by the total period of waveform. The formula is:

$$D \frac{R_1 + R_2}{R_1 + 2R_2} \times 100$$

Example 5 Using the information from Figure 21.12 and two 20 kΩ timing resistors, calculate the duty cycle of this rectangular waveform.

Solution

$$D = \frac{20\ \text{k}\Omega + 20\ \text{k}\Omega}{20\ \text{k}\Omega + 2(20\ \text{k}\Omega)} \times 100 = 66.7\%$$

Study Figure 21.12 because the timing capacitor charges through two resistors and discharges only through R_1; this circuit can produce only a rectangular wave. A square wave, which is a rectangular wave with 50% duty cycle, cannot be produced with this circuit. As R_1 gets smaller relative to R_2, the duty cycle will get closer to 50% but will not reach 50%. Making R_1 equal to 0 Ω in the formula will result in a 50% duty cycle. However, this would damage the IC because there would be no current limiting for the internal discharge transistor.

An alternative circuit that allows duty cycles of 50% or less is shown in Figure 21.13. A diode, which bypasses R_2 in the charging circuit, has been added in parallel with R_1. The previous formulas are modified slightly for the altered circuit.

$$t_{high} = 0.69\,(R_1)C$$

$$t_{low} = 0.69\,(R_2)C$$

$$f_o = \frac{1.45}{(R_1 + R_2)C}$$

$$D\% = \frac{R_1}{R_1 + R_2} \times 100$$

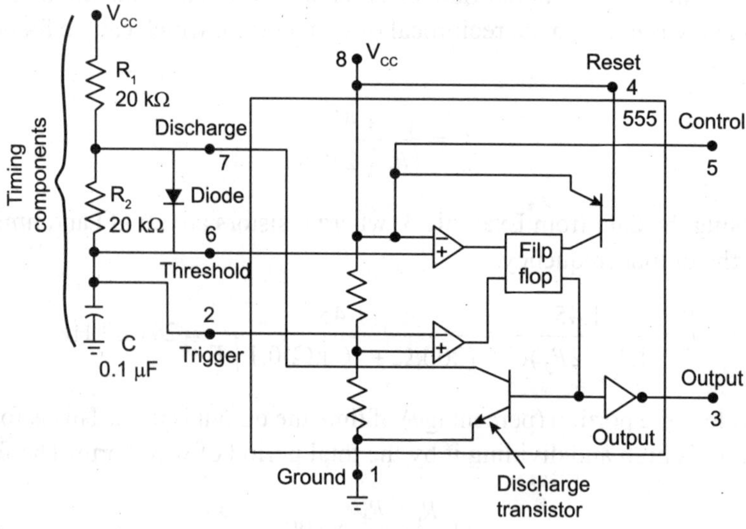

Figure 21.13 Bypass diode around R_2

It is now possible to find resistor values that will produce square waves.

Example 6 Find the resistor values that will produce a 2 kHz square wave given a timing capacitor of 0.01 μF.

Solution

Beginning with the formula for output frequency,

$$f_o = \frac{1.45}{(R_1 + R_2)C}$$

We have $R_1 + R_2 = \dfrac{1.45}{f_o \times C} = \dfrac{1.45}{2\text{ kHz} \times 0.01\ \mu\text{F}} = 72.5\text{ k}\Omega$

Because a square wave has 50% duty cycle, each resistor must be the same. Therefore, each resistor must be half of 72.5 kΩ.

$$R_1 = R_2 = \frac{72.5\ \text{k}\Omega}{2} = 36.25\ \text{k}\Omega$$

Figure 21.14 illustrates a common application of 555 timer connected as an estable oscillator. In this circuit, the 555 output from pin 3 drives an infrared light emitting diode (IR LED). An IR LED, phototransistor or photodiode located across an opening is directed toward the IR LED. If an object blocks the infrared path, the IR LED detector circuit will trigger and halt a motor, snap a photo, sound an alarm, increment a counter, and so on.

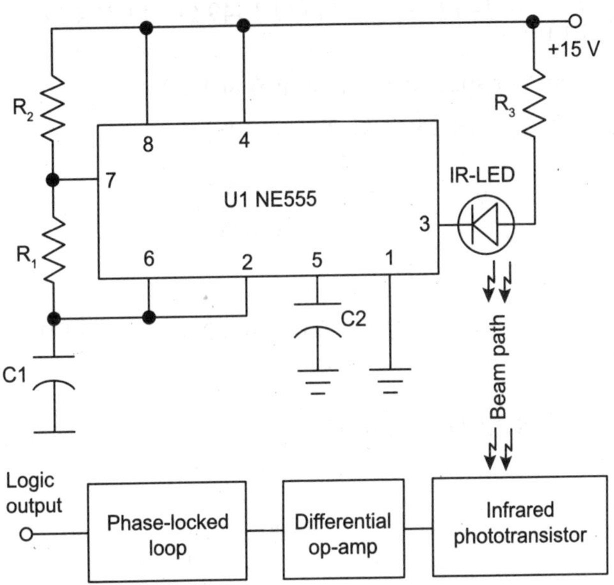

Figure 21.14 Practical IR LED transmitter circuit

The IR LED is pulsed on and off by an astable 555 so that other sources of IR (the sun, for example) will not cause accidental triggering of the detector circuit. An additional reason for pulsing, especially at a low duty cycle, is that the power rating of IR LED (an average measurement) may be extended, thereby expanding the range and improving the reliability of the system. Solving for component value is accomplished as follows:

Select the pulse frequency: 1 kHz

Find the period: $t = \dfrac{1}{f} \Rightarrow t = 1{,}000$ microseconds

Select output low time: $t_{(\text{low})} = 100$ microseconds

Select "output high time: $t_{(\text{high})} = 900$ microseconds

Select C_1:0.10 μF

$$\text{Find } R_1 : t_{(\text{low})} = 0.69(R_1)C_1 \Rightarrow R_1 = \frac{100\ \mu\text{S}}{0.69 \times C_1} = 1449\ \Omega$$

$$\text{Find } R_2 : t_{(\text{high})} = 0.69(R_1 + R_2) \Rightarrow R_1 + R_2 = \frac{t_{(\text{high})}}{0.69 \times C_1}$$

$$R_2 = \frac{t_{(\text{high})}}{0.69 \times C_1} - R_1 \Rightarrow R_2 = \frac{t_{(\text{high})}}{0.69 \times C_1} - R_1 \Rightarrow$$

$$R_2 = \frac{900\ \mu\text{S}}{0.69 \times 0.1\ \mu\text{F}} - 1449\ \Omega = 13{,}043\ \Omega\ 1449\ \Omega = 11.594\ \Omega$$

Select IR LED current per specification: 50 mA at 1.2 V.

$$\text{Find } R_3 : V_{R3} = V_{CC} - V_{\text{OR-LED}} \Rightarrow 15\ \text{V} - 1.2\ \text{V} = 13.8\ \text{V}$$

$$R_3 = \frac{V_{R3}}{I_{R3}} \Rightarrow R_3 = \frac{13.8\ \text{V}}{50\ \text{mA}} = 276\ \Omega$$

Round off R_3 to 270 Ω.

$$P_{R3} = I^2 \times R_3 \Rightarrow P_{R3} = 50^2\ \text{mA} \times 270\ \Omega = 0.69\ \text{W}$$

Round off P_{R3} to 1 W.

We may note that the IR LED in Figure 21.14 is driven only when 555 pin 3 is low. The duty cycle is then

$$\text{Duty cycle\%} = \frac{t_{(\text{low})}}{t_{(\text{low})} + t_{(\text{high})}} \times 100 \Rightarrow \frac{100\ \mu\text{s}}{1000\ \mu\text{s}} \times 100 = 10\%$$

If the frequency is critical, a potentiometer could be substituted for R_2. When the IR light beam crosses to the receiver, it is detected by the phototransistor, amplified and decoded to produce a digital low at the output of phase-locked loop (PLL). The PLL is shown as part of the detector block.

Time-delay mode The versatility of 555 applications is shown in Figure 21.15. This configuration creates a stable timing output with the addition of a transistor and two diodes to the RC timing network. The frequency can be varied over a wide range while maintaining a constant 50% duty cycle.

When the output is high, the transistor is biased into saturation by R_1 so that the charging current passes through the transistor and R_2 to C_2. When the output goes low, the 555 internal discharge transistor (pin 7) cuts off transistor Q_1 and discharges the capacitor through R_2 and D_2. The high and low periods are equal. The values of capacitor (C_1) and resistor (R_1) are dependent on the type or length of timing desired.

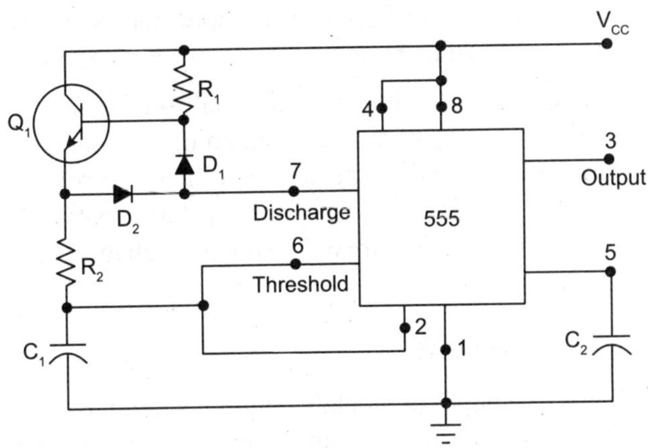

Figure 21.15 Time delay mode

Example 7 If $R = 260\ \text{k}\Omega$ and $C = 25\ \mu\text{F}$, calculate the time delay.

Solution

Time delay is calculated using the following formula:

$$t_{\text{delay}} = 1.1 \times R \times C = 1.1 \times 260\ \text{k}\Omega \times 25\ \mu\text{F} = 7.15\ \text{sec}$$

In this mode, if the trigger signal goes high before the IC times out, the output will not go low. Security alarms take advantage of this feature by allowing time to clear an area before the alarm is armed. Another use of 555 timer is as voltage-controlled oscillator (Figure 21.16)

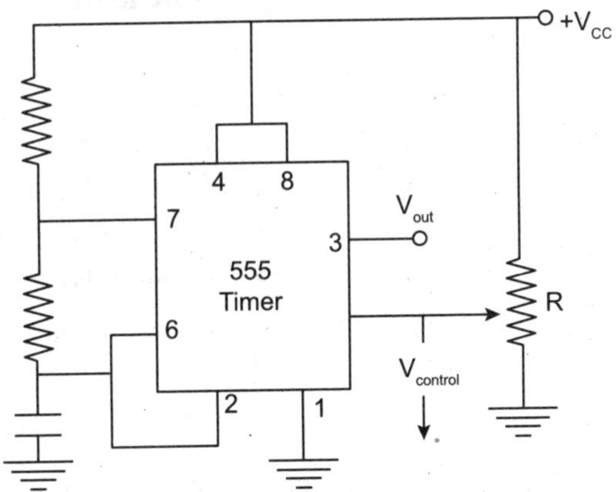

Figure 21.16 Voltage-controlled oscillator

21.5.3 Other Integrated Circuits

In the earlier chapters, we discussed about some of the most widely used ICs such as differential amplifiers and operational amplifiers. ICs can also replace transistor stages such as an intermediate frequency (IF) amplifier. The use of an IC (with multiple transistors) will normally provide improved performance, such as more gain, better selectively, greater noise rejection and so on.

Liner ICs that provide more than one function are usually called subsystem ICs. An example of a subsystem IC is a television sound system containing a limiter, an FM detector, an IF amplifier, a regulated power supply, and an electronic volume control. This type of IC reduces the number of parts needed in system applications. With fewer parts, cost, circuit board real estate, power dissipation and maintenance are reduced while reliability and operating speed are improved.

21.5.4 Digital Integrated Circuits

Digital ICs are logic circuits, the building blocks of digital computers and calculators. In digital electronics, a gate is a logic circuit with one output and one or more inputs; an output signal occurs for a certain combination of input signals. In a way it is a unidirectional device. We are familiar with the unidirectional property of solid-state semiconductor diodes, where we studied that when a diode is forward-biased, it acts like a closed switch, i.e. it permits current to flow in the circuit and when reverse-biased it acts like an open switch and permits no or very little current to flow in the circuit. Thus, this unique property of a semiconductor diode is employed in the design of basic logic circuits. The most common logic circuits or gates are OR, AND, NOT, etc. but the universal gates are NOR and NAND gates.

Modern computers perform many jobs from scientific computation to business accounting. Computers are also used in automatic control of manufacturing process. Although they are very complex, they consist of arrangements of single logic circuits. Since computer decisions are of the Yes/No type, in a way they are also known as **two-state logic**.

21.6 Digital Signal Processing

One application of digital signal processing (DSP) involves changing AM analogue signals (such as music) into a stream of numbers. The circuit then performs various arithmetic operations on those numbers and changes the resulting numbers back into an improved or enhanced signal. With DSP, many different things can be achieved. Some examples are:

(i) Eliminate unwanted signal characteristics or components (such as echoes, noise or interference)

(ii) Compress or expand a signal

(iii) Enhance a signal to provide special effects (such as surround sound)

(iv) Balance the different frequency components

(v) Demodulate an encoded signal

Analogue circuits can also accomplish these effects. However, DSP is preferred because the technology is small and inexpensive compared to analogue equipment. Other benefits of DSP include the following:

1. Changes can be made through software adjustment rather than component changes. This reduces the cost and time spent on making modifications.
2. Digital designs are more reliable, with maintenance and calibration being easier to perform and less costly.
3. They are not sensitive to environmental changes (temperature and humidity) and are not affected by aging as are analogue circuits.

A DSP system includes the following stages (Figure 21.17).

(i) Amplifier
(ii) Low-pass input filter, also called anti-aliasing filter.
(iii) Sample and hold circuit
(iv) Analogue-to-digital converter
(v) Memory
(vi) Microprocessor
(vii) Digital-to-analogue converter
(viii) Low-pass output filter

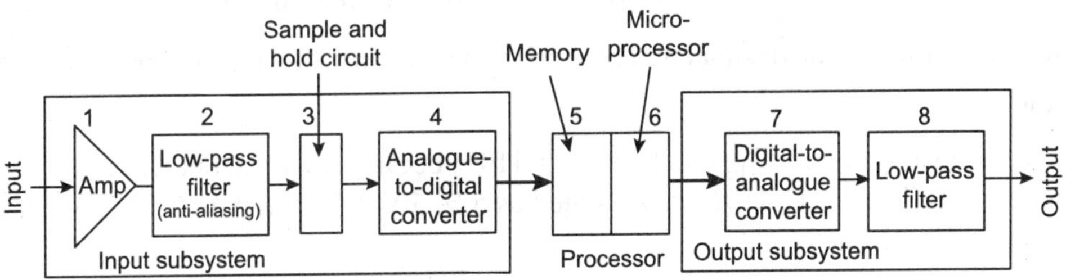

Figure 21.17 A digital signal processing system

Today, DSP has many applications in addition to audio signal processing. For example, a DSP system may be configured to convert one video format to another. Simple digital instructions may programme the generation of multiple signal waveforms simultaneously, or make voltage and frequency measurements at multiple test points. DSP ICs can be configured by software to perform a seemingly endless variety of functions.

In a rare event of DSP failure, replacement is required and an identically programmed DSP system must be made available. These components, with all of their advantages, are usually very expensive. Thus, the technician needs to be confident that his analysis of a defective DSP system is correct prior to proceeding with replacement.

21.6.1 Analogue-to-Digital Conversion

Analogue-to-digital (A/D) and digital-to-analogue (D/A) conversion is required in many IC applications. A digital signal processor has both in its circuitry. Analogue signals are continuous;

that is, their voltage changes smoothly over time. On the other hand, a discrete signal can change in voltage value at any point in time.

To make digital processing possible, the continuous signal must be changed into a digital signal. Changing the continuous signal into a discrete signal uses a process called sampling. Basically, "snapshots" (sample) of the signal are taken at different points in time. The higher the sampling rate, the better the discrete signal will represent the continuous signal. Conversely, the discrete signal will be worse as the sampling rate becomes lower. This will result in poor quality of sound or picture.

Before the signal is received for sample-and-hold process, it first passes through an anti-aliasing filter, which helps remove interference and high-frequency noise. After sampling, the signal is sent to the A/D converter. An A/D converter converts each signal sample into a string of bits.

Bits are binary values. The value 0 is represented by a voltage. The value 1 is usually represented by a more positive voltage than the value 0. The digital logic IC family dictates the voltage limits for 1s and 0s. It is common today to find a variety of logic levels on a single circuit card assembly. The old standard TTL family of ICs defines 0s as –0.5 V through +0.8 V and 1s as +2.4 V through +5.5 V. This is generally called "5 volt logic". Rather than discussing voltage levels, we equate term 1 with term HI and term 0 with term LO.

Example 8 How can the decimal number 253 be defined in TTL voltage terms?

Solution

From the number system, $253_{10} - FD_{16} = 1111\ 1101_2$ because 1s are about 5 V in TTL and 0s are about 0 V in TTL, 1111 1101 can be stated as 5 V, 5 V, 5 V, 5 V, 5 V, 05 V, 5 V. How can 253 be defined as HI_s and LO_s?

$$253_{10} = 1111\ 1101_2 = \text{HI, HI, HI, HI, HI, HI LO, HI}$$

The smallest increment in a binary number is a single bit. The next larger grouping of values used in the binary system is a grouping of four bits, called a **nibble**. The combination of two nibbles is called a **byte**. Bytes can be grouped into **words**. Words are defined by the system in which they are used.

A word can be 1 byte or more. It is common to find words of 2 bytes (16 bits), 4 bytes (32 bits), and even more. It is also possible to encounter words with an odd number of bits, such as 10. A small A/D (or ADC) or D/A (or DAC) or memory may use 10 or 12 bits to a word.

Some example of bytes are:

(i) $0000\ 0000_2$ (this byte represents the decimal 0)

(ii) $0000\ 0100_2$ (this represents the decimal 4)

(iii) $1000\ 0111_2$ (this byte represents the decimal 135)

(iv) $1111\ 1111_2$ (this byte represents the decimal 255)

After the A/D has converted the signals to bits, bytes and words, they are sent to the memory and then (when called upon) to the microprocessor, which performs calculations specified by the software programme. From there, it is sent to the DAC and then on to a low pass filter. A good example of this application is the telephone answering machine.

21.7 Phase Locked Loop: ICPLL

The PLL discrete circuitry known for many years began to find massive application in the early 1980s with the popularity of single-chip ICs such as the Signetics NE565, 567 and others. A PLL is an assemblage of four basic electronic blocks consisting of a phase detector, filter, DC amplifier, and voltage-controlled oscillator (VCO) (Figure 21.18).

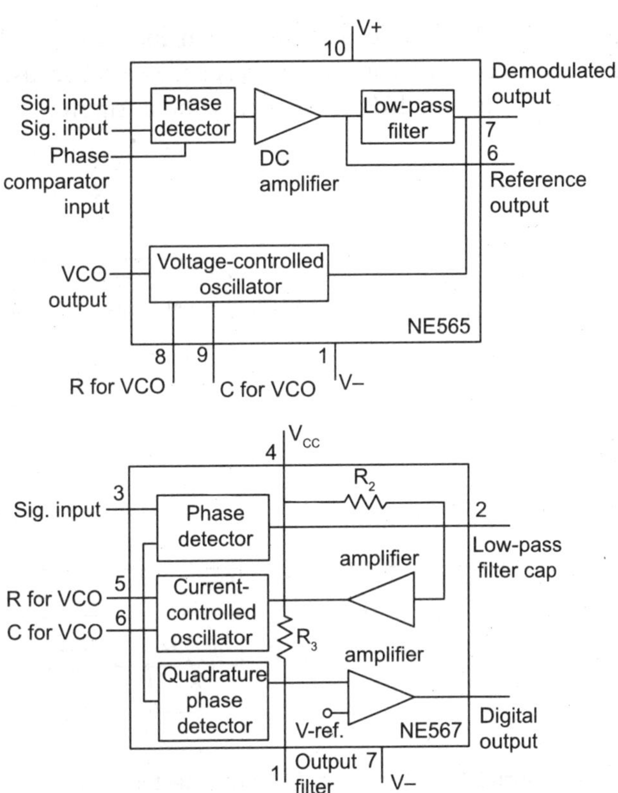

Figure 21.18 Block diagram of NE665 and 567 PLLs

Dramatic reductions in circuit board real estate, component count and cost are benefits of single-IC PLL.

PLLs can perform a wide variety of functions, including frequency synthesis, FM (phase) detection, frequency multiplication and division, tone decoding and signal regeneration. The manner in which a minimum of external components is applied in conjunction with the PLL determines the function and operating characteristics of PLL circuits.

21.7.1 FM Detection

Figure 21.19 shows a PLL. FM detector pin 2 is the capacitively coupled single-ended input to the PLL detector. Pin 3, the differential input pin, is grounded through a matching 560 Ω resistor. Audio output is taken from pins 6 and 7. Pins 10 and 1 are differential power points. The free running VCO frequency is determined by the capacitor at pin 9, and resistance at pin 8 with the following formula.

Select $R = 5.5$ kΩ and IF = 455 kHz. Find C.

$$f_o = \frac{1.2}{4\ RC} \Rightarrow C = \frac{1.2}{4 \times 5.5\ \text{k}\Omega \times 455\ \text{kHz}} = \frac{1.2}{1 \times 10^{10}} = 120\ \text{pF}$$

The series-connected 1 kΩ variable resistor shown in Figure 21.19 allows first-time finetuning of VCO to 455 kHz center frequency. This is necessary because the components are not sufficiently accurate to ensure 455 kHz resonance without trimming.

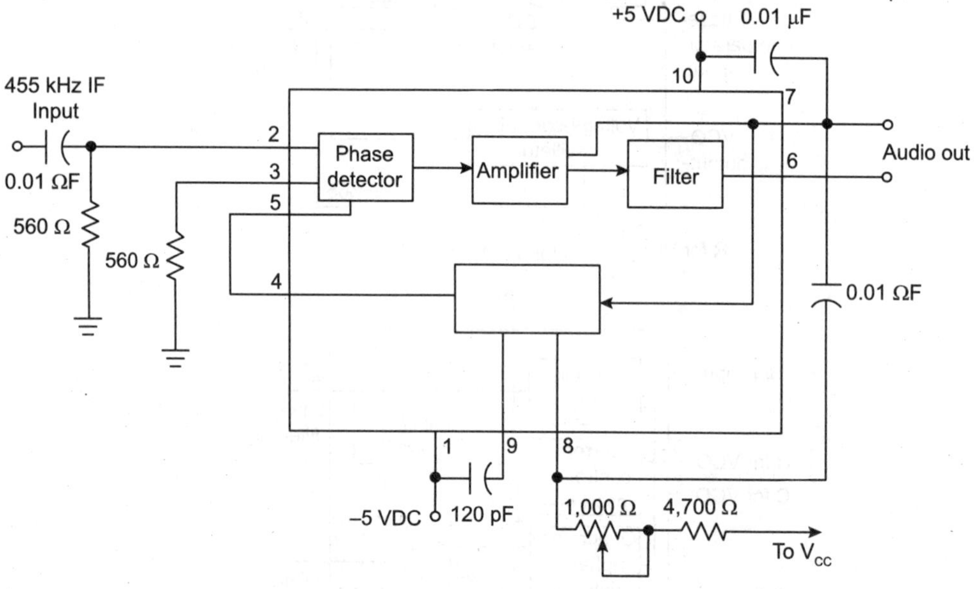

Figure 21.19 Practical PLL FM detector circuit

21.7.2 Frequency Synthesis

The block diagram shown in Figure 21.20 is a PLL application of tremendous importance. It shows a PLL and auxiliary devices that form a frequency synthesizer. Synthesizers are highly accurate signal frequency sources employing as few as one crystal oscillator, but capable of producing multiple stable output frequencies.

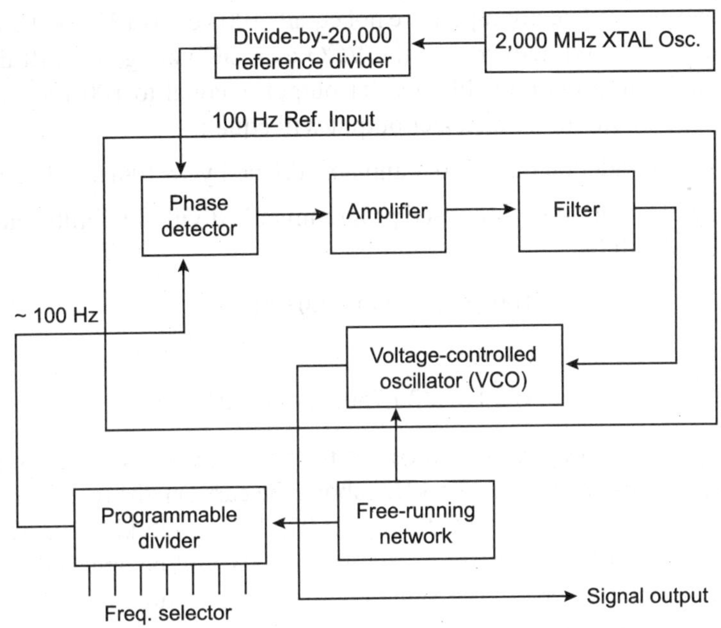

Figure 21.20 PLL frequency synthesizer

In Figure 21.20, a crystal oscillator of 2,000 MHz drives a frequency divider. The frequency divider produces

$$\frac{2{,}000\ \text{MHz}}{20{,}000} = 100\ \text{Hz}$$

This constant 100 Hz signal is one of the two signals introduced into the PLL phase detector. The VCO output is routed to two points: the synthesizer output and a programmable divider circuit. The programmable divider can be externally controlled to divide by a variety of combinations. Assume the divider in Figure 21.20 can furnish divided-bys of 9,950 through 20,750 in incremental steps of 100.

Further, assume that the programmable divider is currently set to divide by 12,250. Programmable divider output is delivered to the remaining phase detector input. The free-running network is simply the *R* and *C* components that set the VCO free-running frequency to approximately the middle of the range of desired synthesizer output frequencies. If the VCO is free running at approximately 1,535 kHz, then

$$\text{Programmable divider output} = \frac{f_{VCO}}{\text{Programmable divide-by}}$$

$$\frac{1{,}535\ \text{kHz}}{12{,}250} = 125.306\ \text{Hz}$$

Because the two-phase detector inputs are not equal (100 versus 125.306 Hz), the difference will be amplified, filtered and delivered to the VCO as a DC voltage to shift the VCO output frequency down until the programmable divider output is equal to 100 Hz. This occurs very quickly. From that point on, the synthesizer output will equal:

f_{VCO} = Programmable divider output × Programmable divide-by ⇒ 100Hz × 12,250 = 1,225 kHz

Using the same calculation for proposed programmable divide-by limits, the band of signal frequencies produced will be:

$$100 \text{ Hz} \times 9{,}950 = 995 \text{ kHz}$$

through

$$100 \text{ Hz} \times 20{,}750 = 2{,}075 \text{ kHz}$$

If the IF of a superheterodyne radio receiver is 455 kHz and the local oscillator operates between 995 (above the carrier) and 2,075 kHz (above the carrier), the received frequency band would be 995 kHz – 455 kHz = 540 kHz through 2,075 kHz – 455 kHz = 1,620 kHz. In other words, the hypothetical PLL synthesizer of Figure 21.20 would be appropriate for a digitally tuned AM broadcast band receiver. The same configuration is suitable for use as the master oscillator of a radio transmitter.

21.7.3 Tone Decoding

The 567 tone decoder of Figure 21.21 finds application where recognition of one or more tones is desired. Pin 8 of 567 drives from high to ground when PLL VCO lock occurs with an incoming signal. Typical applications of 567 tone decoder PLL include touch-tone decoding, wireless intercoms and ultrasonic control.

Input to the PLL of Figure 21.21 can easily be connected to the audio circuit of a broadcast band receiver. It is preferable that input signal level be greater than 200 mV. Only when the Emergency Broadcast System (EBS) activates the dual-tone alert signal (853 and 960 Hz) will the circuit of Figure 21.21 produce a LOGIC HIGH output from pin 3 of 4001 NOR logic gate. Component values for EBS alert tone decoder circuit are calculated as follows:

Assign 853 Hz to PLL U1

$$f_{o1} = \frac{1.1}{R_1 C_1} \Rightarrow R_1 \approx \frac{1.1}{f_{o1} \times C_1}$$

Given $f_{o1} = 853$ Hz

Assume $C_1 = 0.22\ \mu\text{F}$

$$R_1 = \frac{1.1}{853 \text{ Hz} \times 0.22\ \mu\text{F}} = 5{,}861 \text{ k}\Omega$$

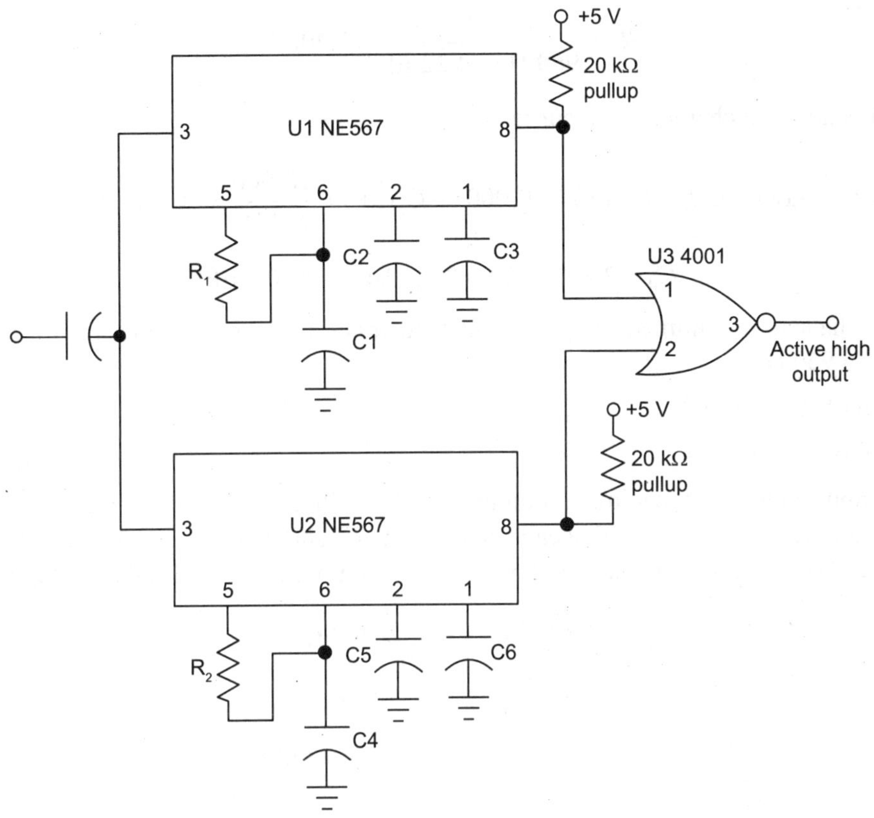

Figure 21.21 Practical PLL EBS alert tone decoder circuit

From bandwidth characteristic curve,

For 5% bandwidth: $f_o \times C_2(\mu\text{F}) = 10{,}000 \Rightarrow C_2(\mu\text{F}) = \dfrac{10{,}000}{853\ \text{Hz}} = 11.7\ \mu\text{F}$

R_1 will have to be trimmed, using a potentiometer and frequency counter on pin 6 to center the VCO on 853 Hz.

Round off C_2 to 12 μF.

Round off C_3 to 22 μF.

Assign 960 Hz to PLL U_2

$$f_{o2} = \frac{1.1}{R_2 C_4} \Rightarrow R_2 \approx \frac{1.1}{f_{o2} \times C_4}$$

Given f_{o2} = 960 Hz

Assume C_4 = 0.22 μF

$$R_2 \approx \frac{1.1}{960\ \text{Hz} \times 0.22\ \mu\text{F}} \approx 5.208\ \text{k}\Omega$$

From bandwidth characteristic curve,

For 5% bandwidth: $f_o \times C_5(\mu\text{F}) = 10{,}000 \Rightarrow C_2(\mu\text{F}) = \dfrac{10{,}000}{960\ \text{Hz}} = 10.4\ \mu\text{F}$

$$C_6 = 2 \times C_5\,(\mu\text{F}) = 2 \times 10.4\ \mu\text{F} = 20.8\ \mu\text{F}$$

R_1 will have to be trimmed using a potentiometer and frequency counter on pin 6 to center the VCO on 960 Hz.

Round off C_2 to 10 μF.

Round off C_3 to 22 μF.

The input coupling capacitor is common to both PLLs. We use the rule of thumb for coupling and bypass capacitance that capacitive reactance should be no greater than 10% of load impedance. The specification sheet on 567 states that input resistance is 20 kΩ. Therefore,

$$X_{C(\text{input coupling cap})} = \frac{1}{10} \times R_{\text{input}} = \frac{1}{10} \times 20\ \text{k}\Omega = 2\ \text{k}\Omega$$

$$C_{(\text{input coupling cap})} = \frac{1}{2 \times \pi \times f \times X_c} = \frac{1}{2 \times \pi \times 853\ \text{Hz} \times 2\ \text{k}\Omega}$$

$$C_{(\text{input coupling cap})} = 0.093\ \mu\text{F}$$

Round off the input coupling capacitor to 0.1 μF.

21.8 IC Operational Amplifiers

An Op-Amp is a high-gain, direct-coupled differential amplifier whose response characteristics are externally controlled by negative feedback from the output to the input. They are widely used in computers and can perform various mathematical operations like summing, integration and differentiation. A common Op-Amp is 741C which is generally used. A schematic circuit diagram and its pin configuration is shown in Figure 21.22 (a, b). Its detailed operation is beyond the scope of this chapter.

Logic circuits can be in one of two states (i.e. Yes/No): ON/OFF or High/Low or Magnetized/Unmagnetized, etc. Electrically a toggle switch is a simple example of two-state device. Presently, most logic circuits are available as ICs. Earlier digital ICs used only resistors and transistors known as resistor-transistor logic (RTL). This type of IC is now obsolete. Next came diode-transistor logic (DTL). They consist of diodes and transistors in various designs for OR gates, AND gates and other logic circuits. However, these are rapidly heading towards obsolescence.

Since 1964, transistor-transistor logic (TTL) has become commercially viable and as such they are very popular these days.

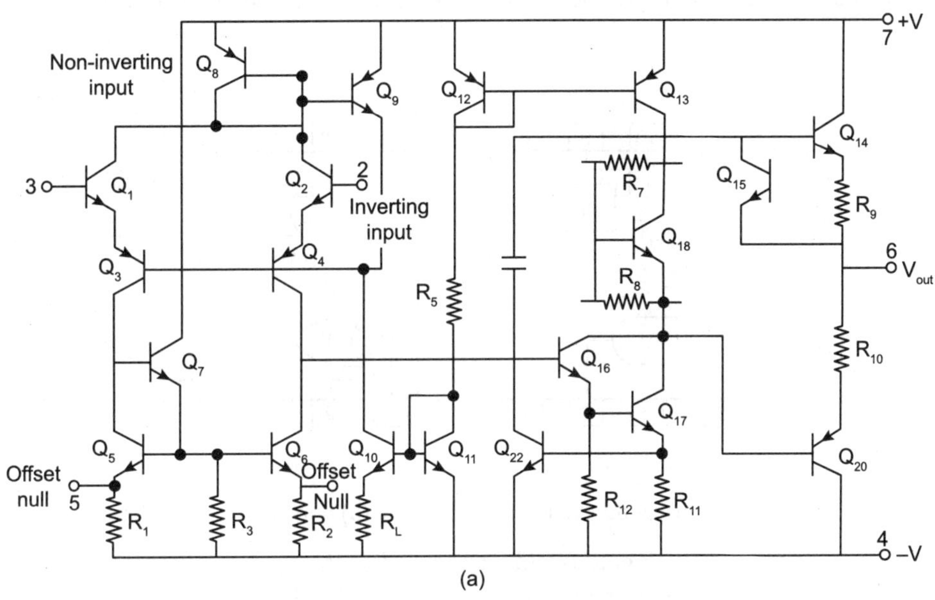

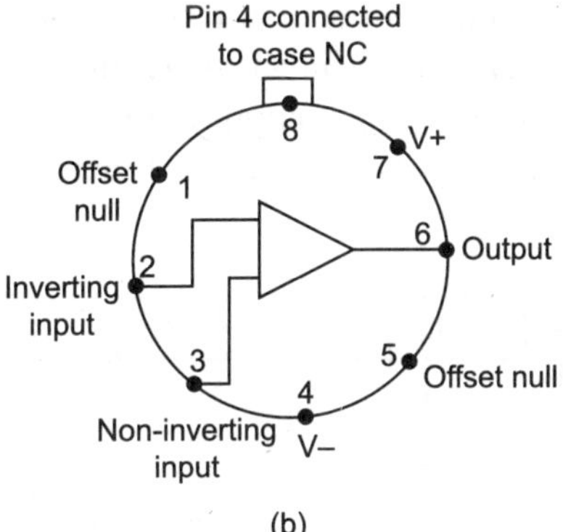

Figure 21.22 Circuit diagram of 741C IC

Figure 21.23 shows symbolically a 7408, one of the many available ICs in TTL family. It is dual-in-line package and contains four AND gates each gate having two inputs and one output. On account of this, it is called quad two input AND gate. Before using such ICs, one must be familiar with which serial numbers are for input side and which are connected to output. In addition, one pin is related to power supply and another pin is ground. In Figure 21.23, top view of quad two input AND gates and quad two input OR gates are shown symbolically.

The operation of such ICs depends upon the operation of AND gate or OR gate of which it is made up of; so their use is related to the function of individual gate. As such one must be familiar with the truth table of gate. For such TTL devices to work properly, the supply voltage must be between +4.75 V and +5.25 V. This is why +5 V is the nominal supply voltage specified for TTL devices. The common ground of the chip is pin no. 7. The other pins are for input and output.

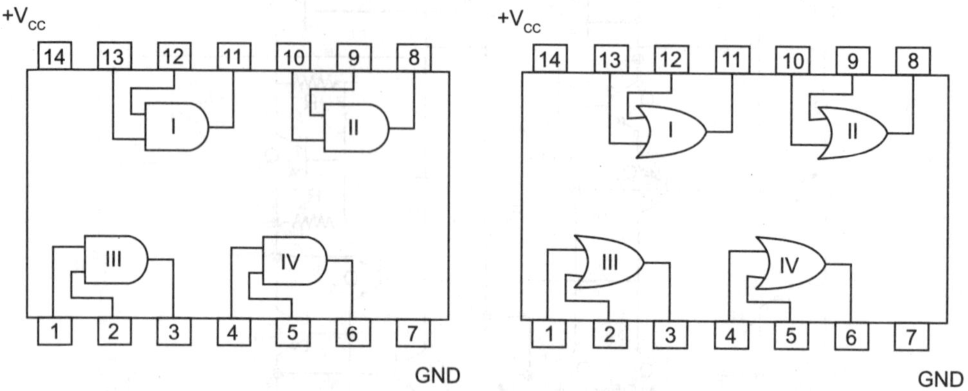

Figure 21.23 (a) Quad two input AND gates, (b) quad two input OR gates

It should be noted that each gate is independent of each other. In other words, they can be connected to each other as per requirement of other TTL device. Before using any such device, one must be familiar with boolean algebra. As per requirement, one can also have two TTL devices, one containing AND gate and other comprising OR gate. In Figure 21.23, we have shown gates having two inputs but still they have a single output as in the case of 7427 IC which comprises three NOR gates each having three inputs as shown in Figure 21.24.

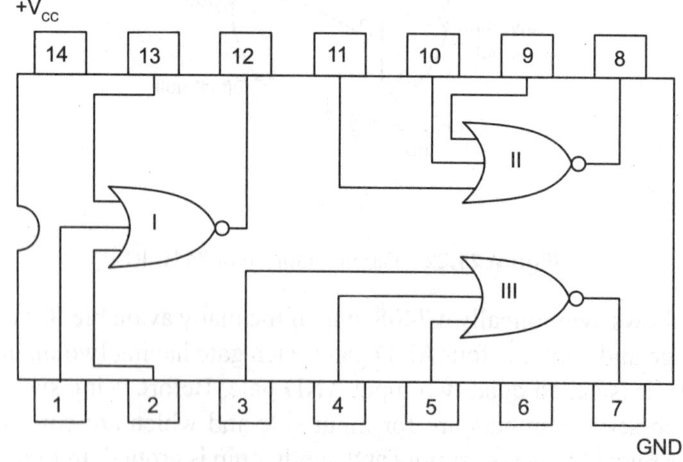

Figure 21.24 Top view of 7427 IC (comprising three NOR gates) each having three inputs

A list of few TTL devices is given in Table 21.1. It also represents their specific functions.

Table 21.1

Number	Detail/Function
7400	Quad 2 input NAND gates
7402	Quad 2 input NOR gates (flip-flop)
7404	Hex inverters (flip-flop)
7408	Quad 2 input AND gates (Full adder)
7410	Triple 3 - input NAND gates
7411	Triple 3 - input AND gates
7420	Dual - 4 input NAND gates
7426	Quad- 2 input TTL - MOS interface NAND gates
7430	8 input NAND gates
7432	Quad 2 input OR gate (Full adder)
7441	BCD to decimal decoder-nixie driver
7448	BCD to seven segment decoder divers
74153	Dual 4/1 multiplexer
7472	J K master slave flip-flop
7476	Dual J K master slave flip-flop
7489	64- bit random access read write memory
7494	4 bit shift register
7496	5 bit parallel in parallel out shift register
74148	Priority encoder
74152	8 channel data selector-multiplexer
74107	Dual J K master slave flip-flop
74136	Quad 2 input exclusive OR gate

21.8.1 Transistor and IC Packing

Low-power transistors may simply be encapsulated in resin and connecting leads left protruding (Figure 21.25(a)). This has the advantage of economy, but offers a limited range of operating temperature. In another method of low-power transistor packing, the device is hermetically sealed in a metal can (Figure 21.25 TO-5 or TO-18 package). The transistor is first mounted with its collector in contact with a heat conducting metal plate. Wires that are insulted from the plate pass through for emitter and base connections. The collector connecting wire is then welded directly to the heat conducting plate, and the covering metal can is finally welded to the plate.

High-power transistors are encapsulated in a sealed can (TO-3). In this case, however, the heat conducting plate is much larger and is designed for mounting directly out heat sink (Figure 21.26). Connecting pins are provided for the base and the emitter and collector connection is made by means of the metal plate. This is not the only form of power transistor package. For higher power dissipation, there are stud mounted devices (Figure 21.26(b)).

Like all semiconductor devices, ICs are also packaged to provide mechanical protection and terminals for electrical connection. Several standard packages in general use are illustrated in Figure 21.27. The plastic dual inline package is much cheaper and widely used for general industrial applications where high temperature performance is not required.

Microminiaturization in electronics is progressing every day. A monolithic IC has about 3000 components on 1 cm^2 silicon wafer. We now have large scale integration and very large scale integration. A VLSI chip has more than 100000 component per cm^2.

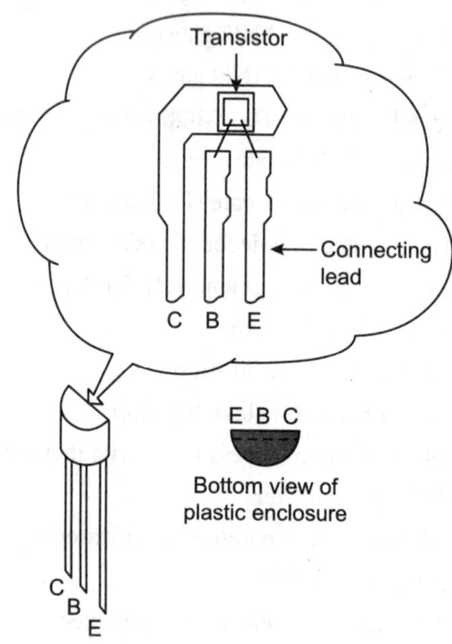

Figure 21.25 (a) Plastic encapsulated transistor (TO-92 package)

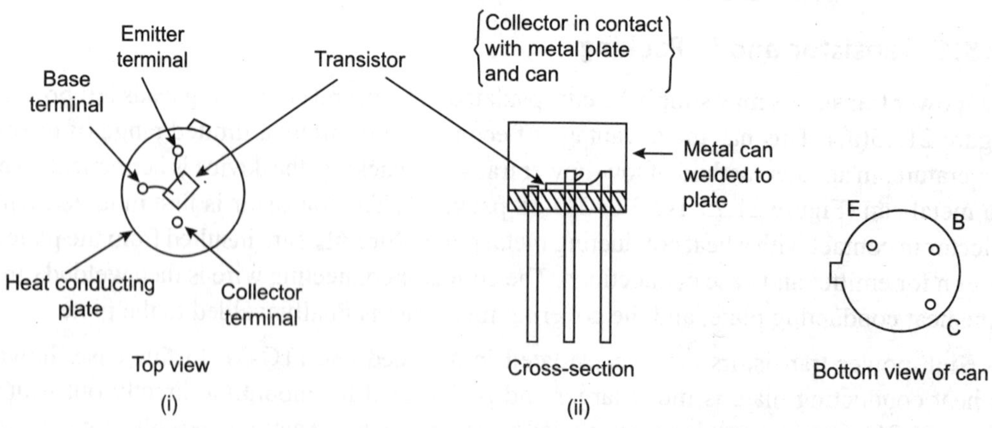

Figure 21.25 (b) Low-power transistors may be (i) plastic encapsulated (ii) contained in a metal can

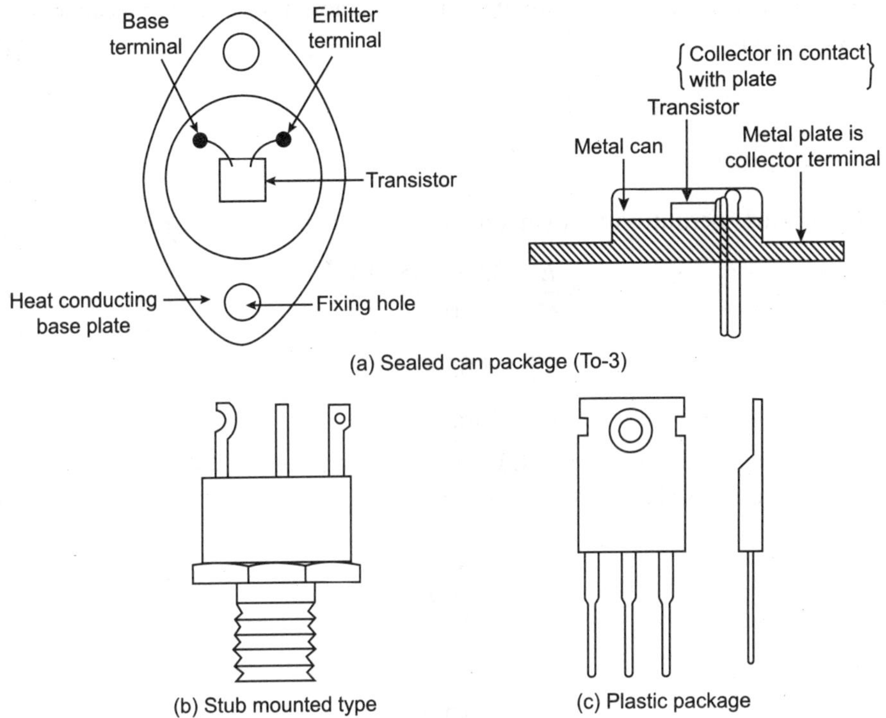

(a) Sealed can package (To-3)

(b) Stub mounted type

(c) Plastic package

Figure 21.26 High-power transistors are contained in a relatively large package that can be bolted to a beat sink for power dissipation.

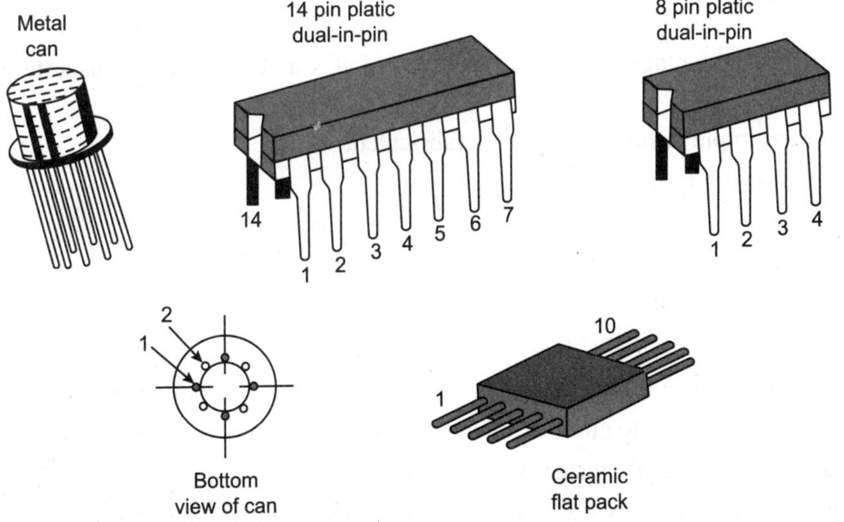

Figure 21.27 Integrated circuits are usually packaged in top-hat-style metal cans, plastic dual-in-line containers or ceramic flat packs

Example 9 Show that the required silicon oxide thickness for designing a capacitor of value $C = 10\ \mu f$ is 0.345 nm. [B. Tech]

Solution

We have
$$C = \frac{d\varepsilon_{ox}}{d}$$

where ε_{ox} is dielectric permittivity of silicon dioxide

or
$$d = \frac{E}{C} = \frac{3.9 \times 8.85 \times 10^{-18}}{10 \times 10^{-6}}$$
$$= 3.9 \times 8.85 \times 10^{-9}$$
$$= 0.345 \text{ nm}$$

Example 10 Taking sheet resistance 1 kΩ, show that the length of semiconductor bar to design a resistance of value 12 kΩ having width 10 μm is 120 μm. [B. Tech]

Solution

We have
$$R = R\left(\frac{L}{W}\right)$$

or
$$\frac{12}{1} = \frac{L}{10\ \mu\text{m}}$$
$$L = 120\ \mu\text{m}$$

Review Questions

1. What are the advantages and disadvantages of ICs as compared to discrete component circuits?
2. Describe in simple terms how an IC is fabricated.
3. Use the following terms in sentences:
   a. Bit
   b. Word
   c. Nibble
   d. Byte
4. Explain why active components are preferred on chip than passive components.
5. In comparison to discrete circuits, integrated circuits are more desirable. Why it is so?
6. What do you understand by tranny of numbers?
7. What do you understand by wafer size and feature size? What is the basic difference between wafers and chips?
8. Explain the major advantages of monolithic circuit over hybrid circuit.

9. What is the main purpose of clean room in the fabrication of microelectronic circuits?
10. Explain the purpose of using photoresist material in the fabrication of a device. What is the importance of mask in fabrication?
11. What is the purpose of allowing ultraviolet rays to fall on photoresist material?
12. Explain the purpose of performing etching during fabrication process.
13. Using illustrations, explain the fabrication process for monolithic integrated circuits. Discuss their merits and demerits.
14. Briefly explain how diodes, resistors and capacitors are fabricated in monolithic ICs.
15. Briefly explain thin film and thick film methods of IC manufacture. Give their merits and demerits.
16. Which of the three operating modes should be used (NE555) for the following?
    (a) Steady square-wave output
    (b) A switch that can trigger a given action after fixed time delay
    (c) A circuit that can be used to determine how long a particular operation requires
17. What are the advantages of digital signal processing as compared to analogue processing?
18. Using Figure 21.28, describe the basic purpose and operation of each of the blocks in a digital signal processor.

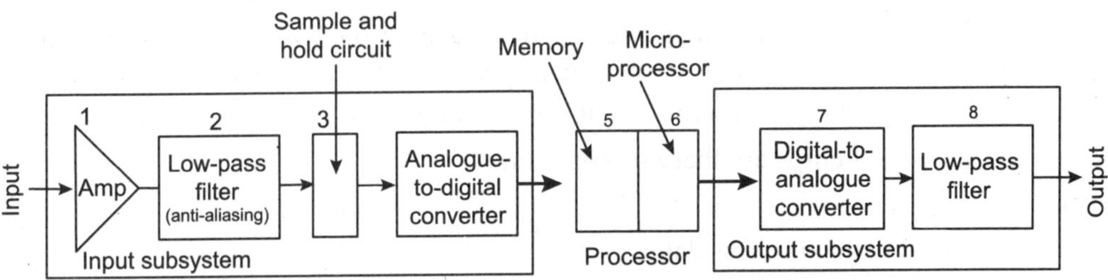

Figure 21.28 Diagram of a digital signal processing system

Short Answer Questions

1. What do you understand by tyranny of numbers?

   Ans: Advanced circuits contain so many components and connecting them was virtually impossible. This problem is known as tyranny of numbers.

2. What do you understand by microelectronics?

   Ans: This is the branch of electronic technology devoted to the design and development of extremely small electronic devices that consume very little electric power. Although the term microelectronics is sometimes used to describe discrete electronic components assembled in an extremely small and compact form, it is often taken as a synonym for integrated circuit (IC) technology. Discrete components that can be made using microelectronic techniques include resistors, capacitors and transistors.

3. What are hybrid microcircuits?

   Ans: These contain a mixture of discrete and integrated circuits on a single substrate.

4. What do you understand by semiconductor device fabrication?

   Ans: Semiconductor device fabrication is the process used to create chips, the integrated circuits that are present in electrical and electronic devices. It is a multiple-step sequence of photographic and chemical processing during which electronic circuits are gradually created on a wafer made of pure semiconducting material.

5. What do you understand by wafers?

   Ans: A wafer is made out of extremely purified silicon that is grown into monocrystalline cylindrical ingots using Czochralski process. These ingots are then sliced into 0.75 mm thick wafers and polished to obtain a very regular and flat surface.

6. Mention the necessary steps to produce a desired semiconductor IC from wafers.

   Ans: In general, the steps can be grouped into four areas:

   1. Frontend processing
   2. Test
   3. Back-end processing
   4. Packaging

7. What is chip?

   Ans: Several hundred identical ICs are made at a time on a thin wafer several hundred centimeters wide, and the wafer is subsequently sliced into individual ICs called chips. The terms chip, silicon chip, microchip and IC are synonymous.

8. How many types of chips are there by function?

   Ans:

   (i) Logic chips which process data.

   (ii) Memory chips which store data and instructions either temporarily or permanently. RAM chips are the computer's main memory.

   (iii) Microcontroller is a single chip that contains all the components of a computer, including the processor, non-volatiles memory (ROM), volatile memory (RAM), I/O control unit and timing clock.

   (iv) Bio chip uses tiny strands of DNA to latch onto and quickly recognize thousands of genes at a time, intended for use in biological environment.

   (v) Analogue/digital and signal processing chips: A/D converters and D/A converters are chips that convert signals. A related chip is a digital signal process or (DSP) that performs fast instruction sequences commonly used in such applications.

9. What is a clean room?

   Ans: A clean room is a specially designed room where extremely effective air filters and air circulation systems change air completely up to 10 times a minute. Such rooms are used for chip production.

10. How many types of integrated circuits are there?

Ans:

(i) **Monolithic** have single solid structure. Monolithic ICs are constructed within a single wafer of semiconductor material. Monolithic circuits contain all the components in a single rigid structure which can be batch fabricated, i.e. hundreds of identical circuits can be built simultaneously on a silicon wafer.

(ii) **Hybrid ICs** are one of the less well-known aspects of electronics. Hybrid combines active and passive components on a substrate to form a single packaged module.

Hybrids come in two ways, like thin films and thick films. These terms referto the "film" of materials that are deposited on the substrate to form both interconnects and passive components. Presently, the distinction between two types is blurring.

Hybrid may be analogue or digital and often used in specialized or custom applications.

11. What is the advantage of thin film technology?

Ans: This technology allows for greater precision and miniaturization and generally preferred when space is an important consideration. The film interconnection patterns and resistors can be vacuum deposited on a glass substrate.

12. What is a photomask?

Ans: A photomask is the actual size of a chip, replicated many types to fit on a round silicon wafer up to 12 inch in diameter.

13. What is photolithography?

Ans: In photolithography, a highly energetic ultraviolet light is shown through a mask (mask describes parts of the chip) and the UV light will only hit areas that are not covered by the mask. When the film is developed, the areas hit by light are removed. Now, the chip has unprotected and protected area forming a pattern, which is the first step to the final components of the chip.

14. Write the expression for resistance of bar shaped resistor.

Ans: $R = l\dfrac{L}{A} = R\dfrac{L}{W}$

where R is called sheet resistance. Sheet resistance (R) is determined by implantation or diffusion process and the ratio (L/W) is determined by pattern dimensions.

15. Write the expression for IC capacitance per unit area.

Ans: $C = \dfrac{\varepsilon_{ox}}{d}\ \text{F/cm}^2$

where ε_{ox} is dielectric permittivity of silicon dioxide and d is thickness of the oxide.

16. Why a single crystal silicon is used for ICs than polycrystalline silicon?

Ans: A single crystal silicon substrate possess a high level of purity and perfection. Moreover, it has no defects associated with grain boundaries. It is found in polysilicon. Such defects limit the lifetime of minority carriers. Single silicon substrate may also have a high degree of chemical purity, a high degree of crystalline perfection and high structure uniformity.

17. How one can obtain a high-grade starting single crystal silicon?

    Ans: This involves the following two major steps:

    (i) Refinement of raw material (e.g. quartzite) into electronic grade polycrystalline silicon (EGS) using a complex multistage process.

    (ii) Growing a single-crystal silicon from this EGS either by Czchralski (CZ process) or float zone process.

18. How many types of photoresist are there?

    Ans: There are actually two types of photoresist: (i) The photoresist which is initially insoluble and become soluble after exposure to UV light is called *positive photoresist.* (ii) The photoresist which is initially soluble and becomes insoluble (hardened) after exposure to UV light is called *negative photoresist*. Negative photoresists are more sensitive to light, but their photolithographic resolution is not as high as that of positive photoresists. This is why negative photoresists are less commonly used in the manufacturing of high-density ICs.

Fill in the Blanks.

1. The process in which chemicals are used to remove the photoresist hit areas by UV light is called ______________. [**Ans:** etching]
2. By using ______________ one can realize high frequency, high voltage and low saturation voltage transistors. [**Ans:** epitaxial layer]
3. Single crystal silicon is used for integrated circuits than ______________ silicon. [**Ans:** polycrystalline]
4. For an enhancement mode N-channel device, ______________ ions are implanted in the channel to increase the threshold voltage. [**Ans:** boron]
5. A photoresist which is initially insoluble and becomes soluble after exposure to UV light is called ______________ photoresist. [**Ans:** positive]
6. For accurate generation of high-density patterns required in submicron devices, ______________ lithography is used instead of optical lithography. [**Ans:** electron beam]

Appendix

I Physical Constants

Charge of an electron	e	1.60×10^{-19} coulombs
Mass of an electron	m	9.09×10^{-31} Kg
e/m ratio of an electron	e/m	1.759×10^{11} C/Kg
Plank's constant	h	6.626×10^{-34} J-sec
Boltzman's constant	$\bar{\mathrm{k}}$	1.381×10^{-23} J/°K
	k	8.62×10^{-5} ev/°K
Avogadro's number	N_A	6.023×10^{23} molecules/mole
Velocity of light	c	3×10^{8} m/sec
Permeability of free space	μ_o	1.257×10^{-6} H/m
Permittivity of free space	ε_o	8.85×10^{-12} F/m
Intrinsic concentration in silicon at 300 °K	n_i	$1.5 \times 10^{10}/\text{cm}^3$
Intrinsic resistivity in silicon at 300 °K	r_i	230,000 W–cm
Mobility of electronics in silicon	m_n	1300 cm^2/V–sec
Mobility of holes in silicon	m_p	500 cm^2/V–sec
Energy gap at in silicon at 300 °K		1.1 eV

Appendix

II Resistor and Capacitor Values

Typical standard resistor values (± 10% Tolerance)

Ω	Ω	Ω	kΩ	kΩ	kΩ	MΩ	MΩ
–	10	100	1	10	100	1	10
–	12	120	1.2	12	120	1.2	–
–	15	150	1.5	15	150	1.5	15
–	18	180	1.8	18	180	1.8	–
–	22	220	2.2	22	220	2.2	22
2.7	27	270	2.7	27	270	2.7	–
3.3	33	330	3.3	33	330	3.3	–
3.9	39	390	3.9	39	390	3.9	–
4.7	47	470	4.7	47	470	4.7	–
5.6	56	560	5.6	56	560	5.6	–
6.8	68	680	6.8	68	680	6.8	–
–	82	820	8.2	82	820	–	–

Typical standard capacitor values (± 10% tolerance)

pF	pF	pF	pF	μF	μF	μF	μF	μF	μF	μF
5	50	500	5000		0.05	0.5	5	50	50	5000
–	51	510	5100		–	–	–	–	–	–
–	56	560	5600		0.056	0.56	5.6	56	–	5600
–	–	–	6000		0.06	–	6	–	–	6000
–	62	620	6200		–	–	–	–	–	–
–	68	680	6800		0.068	0.68	6.8	–	–	–
–	75	750	7500		–	–	–	75	–	–
–	–	–	8000		–	–	8	80	–	–
–	82	820	8200		0.082	0.82	8.2	82	–	–
–	91	910	9100		–	–	–	–	–	–
10	100	1000		0.01	0.1	1	10	100	1000	10,000
–	110	1100		–	–	–	–	–	–	–
12	120	1200		0.012	0.12	1.2	–	–	–	
–	130	1300		–	–	–	–	–	–	–
15	150	1500		0.015	0.15	1.5	15	150	1500	
–	160	1600		–	–	–	–	–	–	–
18	180	1800		0.018	0.18	1.8	18	180	–	
20	200	2000		0.02	0.2	2	20	200	2000	
24	240	2400		–	–	–	–	240	–	
–	250	2500		–	0.25	–	25	250	2500	
27	270	2700		0.027	0.27	2.7	27	270	–	
30	300	3000		0.03	0.3	3	30	300	3000	
33	330	3300		0.033	0.33	3.3	33	330	3300	
36	360	3600		–	–	–	–	–	–	
39	390	3900		0.039	0.39	3.9	39	–	–	
–	–	4000		0.04	–	4	–	400	–	
43	430	4300		–	–	–	–	–	–	
47	470	4700		0.047	0.47	4.7	47	–	–	

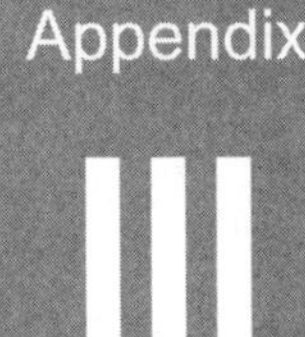

Colour Codes for Electronic Components

Resistor Colour Code

First three bands				Fourth band	
Black	0	Blue	6	Gold	± 5%
Brown	1	Violet	7	Silver	± 10%
Red	2	Grey	8	None	± 20%
Orange	3	White	9		
Yellow	4	Silver	0.01		
Green	5	Gold	0.1		

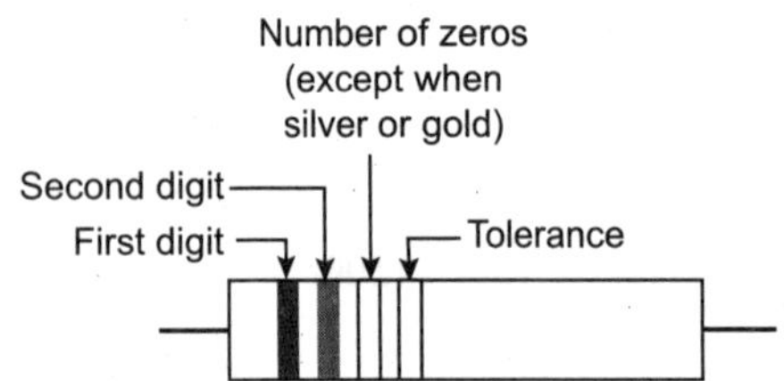

Capacitor Colour Code

Colour	Figure significant	Tolerance (%)
Black	0	20
Brown	1	1
Red	2	2
Orange	3	3
Yellow	4	4
Green	5	5
Blue	6	6
Violet	7	7
Grey	8	8
White	9	9
Silver	0.01	10
Gold	0.1	5

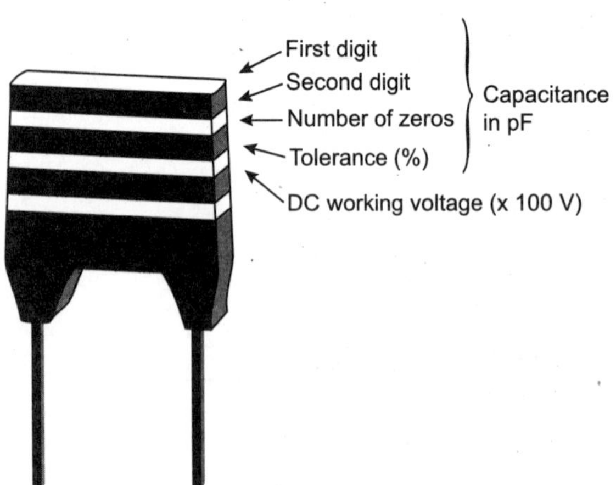

Inductor Colour Code

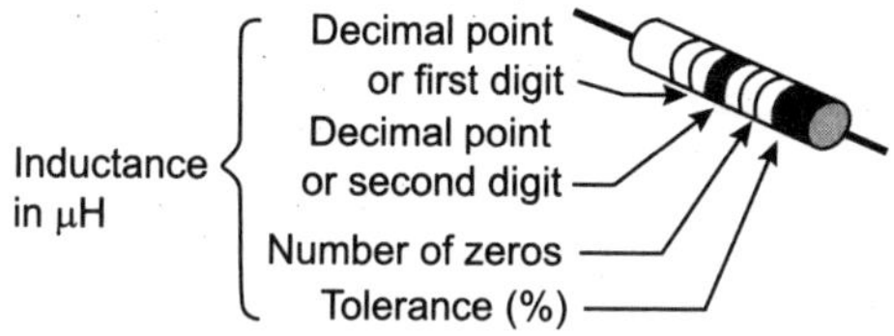

Inductance in μH: Decimal point or first digit; Decimal point or second digit; Number of zeros; Tolerance (%)

Colour	Significant figure	Tolerance (%)
Black	0	
Brown	1	
Red	2	
Orange	3	
Yellow	4	
Green	5	
Blue	6	
Violet	7	
Grey	8	
White	9	
Silver		10
Gold	Decimal point	5
No Band		20

Colour Code Memory Aid

Memory aid	Colour	Number
Black	Black	0
Bruins	Brown	1
Relish	Red	2
Ornery	Orange	3
Young	Yellow	4
Greenhorns	Green	5
Blue	Blue	6
Violets	Violet	7
Growing	Grey	8
Wild	White	9
Smell	Silver	0.01 10%
Good	Gold	0.1 5%

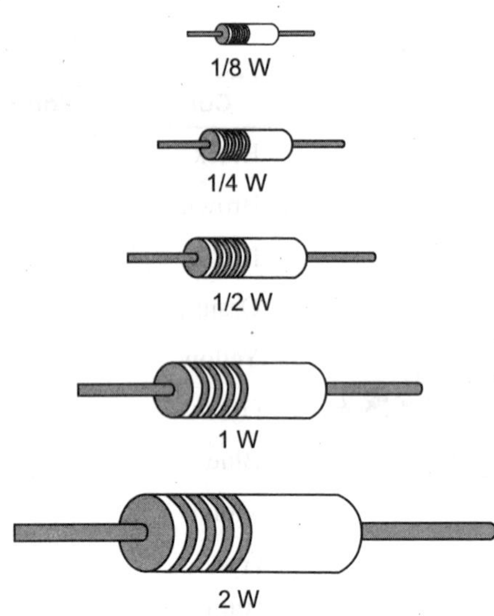

Figure A-1.1 Relative size of carbon composition in resistors with various power ratings

Specifications of Power Transistors

Device No.	Type	P_D(W)	I_C(A)	V_{CEO}(V)	V_{CBO}(V)	h_{FE}MIN	Max	f_TM.HZs
2N 6688	NPN	200	20	200	300	20	80	20
2N 3442	NPN	117	10	140	160	20	70	0.08
BUX 39	NPN	120	30	90	120	15	45	8
ECP 149	PNP	30	4	40	50	30	–	2.5

Darlington Pair

2N 6052	PNP	150	12	100	100	750	–	4
2N 6059	NPN	150	12	100	100	750	–	4

Suggested Readings

1.	David A. Bell	Electronics Devices and Circuits, (3rd ed.), Prentice Hall (PHI)
2.	J. Millman and C.C Halia	Electronics Devices and Circuits, Tata McGraw Hill (THM)
3.	S.L. Kakani and K.C. Bhandari	A Text Book of Electronics (3rd Ed.), New Age Int. Pub. (2009)
4.	S.L. Kakani and K.C. Bhandari	Electronics: Theory and Applications (4th ed.), New Age Pub. (2010)
5.	Boylstal and Nehlasky	Electronic Devices and Circuits
6.	B.G. Streetman and Sanjay Banerjee	Solid State Electronic Devices (4th Ed.), PHI (1995)
7.	R.S Sedha	Applied Electronics, S. Chand
8.	Allen Mottershed	Electronic Devices and Circuits (Prentice Hall)
9.	N.N. Bhargava, D.C Kulshresta and D.C Gupta	Basic Electronics and Linear Circuits, TMH
10.	J. Wilson and J.F.B. Hawkes	Optoelectronics (2nd Ed.), PHI (1992)
11.	A.P. Malviono	Electronic Principles, TMH (6th ed.)
12.	John Powers	An Introduction to Fibre Optic Systems, Irwin Publications (1997)
13.	Jeff Hecht	Understand Fiber Optics, BPB Pub. (1997)
14.	NIIT	Basic of Electronic Devices, PHI (2004)
15.	S. Klein	Introduction to Semiconductors, CENGAGE Learning (2008)

(Cont.)

16.	A.K. Singh	Electronics Devices and Integrated Circuits, PHI (2008)
17.	P.J. Paul	Electronics Devices and Circuits, New Age (2003)
18.	S.P. Bali	Solid State Devices and Circuits, New Age (1995)
19.	T.L. Floyd	Electronic Devices, Pearson Education (2002)
20.	S. Salivahanan, et al.	Electronic Devices and Circuits, Tata McGraw Hill (1998)
21.	S.M. Size	Physics of Semiconductor Devices (2nd Ed.) (1999)
22.	NIIT	Communication Systems, PHI (2004)
23.	NIIT	Basics of Communication Systems, PHI (2004)
24.	J. Das, S.K. Mullick and P.K, Chatterjee	Principles of Digital Communications, New Age (1986)
25.	John and Ryder	Electronics Fundamentals and Applications (Prentice Hall)
26.	Milman and Halkias	Integrated Electronics (McGraw Hill)
27.	John M. Senior	Optical Fibre Communications, Pearson (2009)

Subject Index

A

B

C

D

E

M

N

O

P

Q

R

S

T

U

V

W

Z